HUMAN ANATOMY

ROBERT CAROLA

JOHN P. HARLEY

Professor of Physiology
Eastern Kentucky University

CHARLES R. NOBACK

Professor of Anatomy and Cell Biology
College of Physicians and Surgeons
Columbia University

INTERNATIONAL EDITION

McGRAW-HILL, INC.

New York • St. Louis • San Francisco
Auckland • Bogotá • Caracas • Lisbon • London
Madrid • Mexico • Milan • Montreal • New Delhi
Paris • San Juan • Singapore • Sydney • Tokyo • Toronto

INTERNATIONAL EDITION

Copyright © 1992.
Exclusive rights by McGraw-Hill Inc. for manufacture and export.
This book cannot be re-exported from the country to which it is
consigned by McGraw-Hill.

When ordering this title use

ISBN 0-07-112562-0

Parts of this book were taken from
Human Anatomy and Physiology.

HUMAN ANATOMY

    2 3 4 5 6 7 8 9 0   VNH   VNH   9 0 9 8 7 6 5 4 3 2

ISBN 0-07-010527-8

This book was set in Janson by York Graphic Services, Inc.
The editors were Kathi M. Prancan and Holly Gordon;
the development editor was Deena Cloud;
the designer was Gayle Jaeger;
the cover illustrator was Robert J. Demarest;
the production supervisor was Janelle S. Travers.
The photo editor was Kathy Bendo.
Von Hoffmann Press, Inc., was printer and binder.

Photo credits begin on page C.1.

Library of Congress Cataloging-in-Publication Data

Carola, Robert.
   Human anatomy / Robert Carola, John P. Harley, Charles R. Noback.
      p.   cm.
   Includes bibliographical references and index.
   ISBN 0-07-010527-8
   1. Human anatomy.   I. Harley, John P.   II. Noback, Charles
Robert, (date).   III. Title.
[DNLM: 1. Anatomy.   QS 4 C292ha]
QM23.2.C35   1992
611—dc20
DNLM/DLC
for Library of Congress   91-19649

About the Cover

After deciding on an active pose for the "anatomical person"
cover, I began the task of ensuring anatomical accuracy. This
process entails much more than simply filling in the muscles and
bones. For example, when the left arm is raised straight on the side
with the thumb forward, as shown in the illustration, the arm
cannot be lifted any higher without outward rotation. If you try this
yourself, you will find that as you raise your arm, there is a point
at which the upper arm bone turns. The rotation is necessary so
that the greater tuberosity of the humerus does not impinge on the
acromion process of the scapula.

   A common problem in anatomical illustrations arises when the
artist uses a skeleton as a model. Skeletons are generally wired
together and hung from a support. In this case, the pelvis hangs
too low and rotates forward, abnormally increasing the space
between the lowest rib and the top of the ilium. Again, you can
check the relative positions of the ribs and hipbones on your own
body.

   An understanding of both form and function is essential to
accurate depiction. This is what makes medical illustration
interesting and dynamic.

Robert J. Demarest

Chapter-Opening Photographs

1 chromosomes in dividing cell, **2** desmosomes (red) in human
skin tissue, **3** human cheek cells, **4** human elastic cartilage, **5** basal
cells of epidermis, **6–9** bone tissue, **10–11** skeletal muscle fibers,
12–15 neurons in the cerebral cortex, **16** rods and cones in human
retina, **17** human thyroid tissue, **18–20** red blood cells,
21 lymphocyte, **22** capillary networks around alveoli of lung,
23 stomach tissue, **24** urea crystals, **25** ciliated epithelium of
uterine tube, **26** chromosomes in dividing cell

WEIGHT

Metric	=	English (USA)
milligram	=	0.02 grain (0.000035 oz)
gram	=	0.04 oz
kilogram	=	35.27 oz, 2.20 lb
metric ton (1,000 kg)	=	1.10 tons

English (USA)	=	Metric
grain	=	64.80 mg
ounce	=	28.35 g
pound	=	453.60 g, 0.45 kg
ton (short) 2,000 lb	=	0.91 metric ton (907 kg)

VOLUME

Metric	=	English (USA)
cubic centimeter	=	.06 cu in
cubic meter	=	35.3 cu ft, 1.3 cu yd
milliliter*	=	0.03 oz
liter	=	2.12 pt, 1.06 qt, 0.27 gal

English (USA)	=	Metric
cubic inch	=	16.39 cc
cubic foot	=	0.03 m^3
cubic yard	=	0.765 m^3
ounce	=	0.03 liter, 3 ml*
pint	=	0.47 liter
quart	=	0.95 liter
gallon	=	3.79 liters

*1 liter ÷ 1,000 = milliliter or cubic centimeters (10^{-3} liter); i.e., 1 ml = 1 cc

PRESSURE

Metric	=	English (USA)
kilograms per square centimeter × 14.2231	=	pounds per square inch

English (USA)	=	Metric
pounds per square inch × .0703	=	kilograms per square centimeter

TEMPERATURE

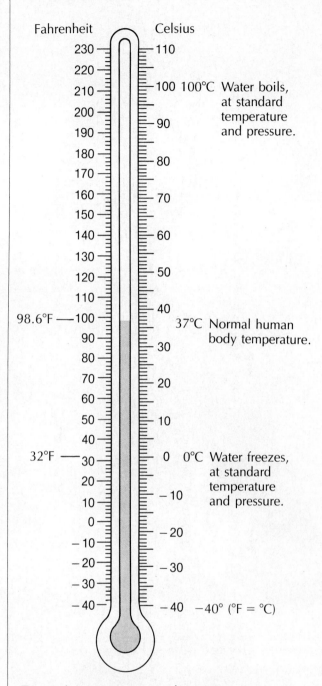

To convert temperature scales:

Fahrenheit to Celsius: $°C = \dfrac{5}{9}(°F - 32)$

Celsius to Fahrenheit: $°F = \dfrac{9}{5}°C + 32$

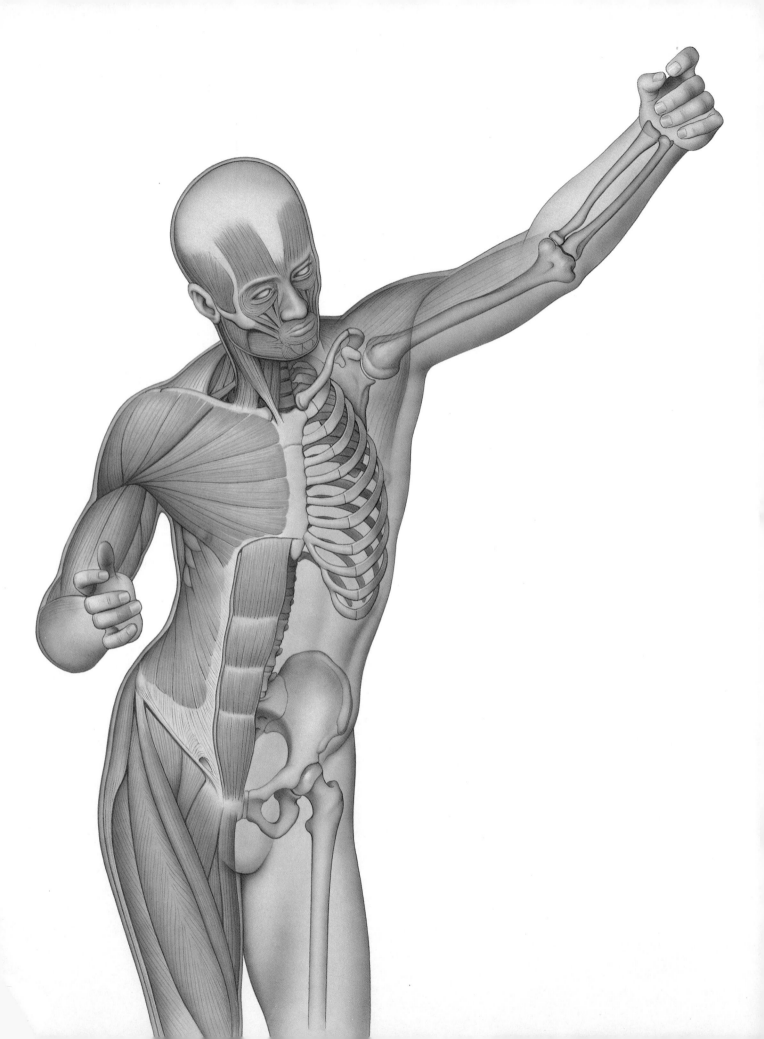

ABOUT THE AUTHORS

Robert Carola is a science writer who has written six textbooks. He also writes for the Smithsonian Institution, Fisher-Price, Exxon Corporation, Mobil Corporation, Michigan Bell Telephone, IBM, and many other corporate clients. He is a member of the National Association of Science Writers.

John P. Harley is a professor of biological sciences at Eastern Kentucky University. He holds a B.A. degree from Youngstown State University, and M.A. and Ph.D. degrees from Kent State University. He has received two distinguished teaching awards at Eastern Kentucky University where he has taught human anatomy and physiology for over 20 years.
Dr. Harley has coauthored three other college textbooks: *Human Anatomy and Physiology* (with Robert Carola and Charles R. Noback) for McGraw-Hill, *Microbiology* (with Lansing Prescott and Donald Klein), and *Zoology* (with Steve Miller) for Wm C. Brown Publishers.

Charles R. Noback has been Professor of Anatomy and Cell Biology at the College of Physicians and Surgeons, Columbia University for 40 years. Dr. Noback is the coauthor of *The Human Nervous System,* Third Edition; *The Nervous System: Introduction and Review,* Third Edition (both published by McGraw-Hill); and *The Human Nervous System: Introduction and Review,* Fourth Edition (with N. L. Strominger and Robert J. Demarest), for Lea & Febiger. He has contributed a section on the Human Nervous System to the *Encyclopaedia Britannica.*

This book is dedicated to our wives
Leslie Carola, Jane Harley, and Eleanor Noback

CONTENTS IN BRIEF

CONTENTS

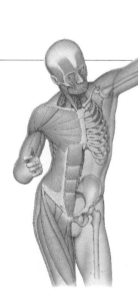
1   INTRODUCTION TO ANATOMY   2

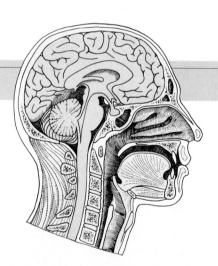

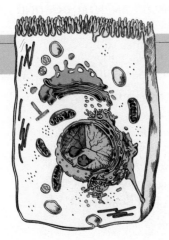

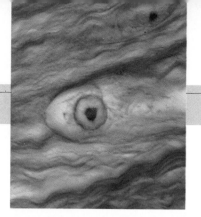

4   TISSUES OF THE BODY   84

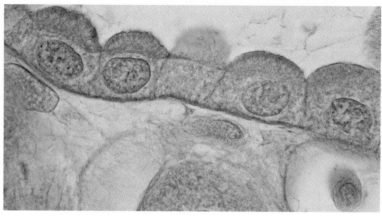

PART II

PROTECTION, SUPPORT, AND MOVEMENT   109

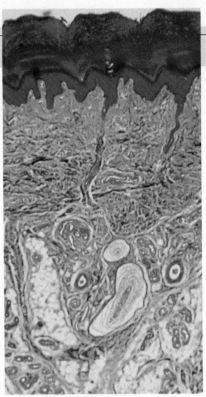

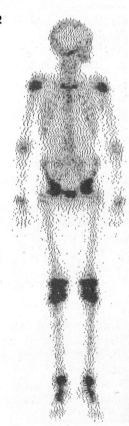

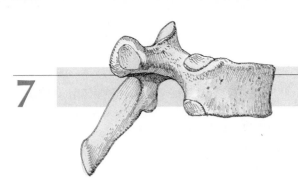

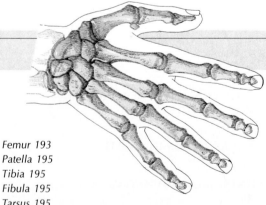

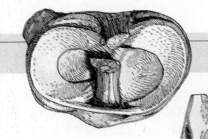

9 ARTICULATIONS 202

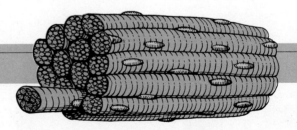

10 MUSCLE TISSUE 228

11 THE MUSCULAR SYSTEM 243

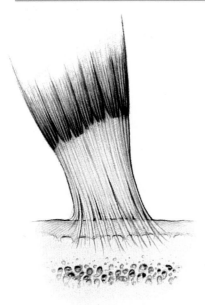

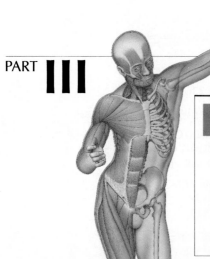

PART III

CONTROL, COMMUNICATION, AND COORDINATION 291

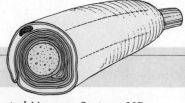

12   NERVOUS TISSUE   292

13   THE SPINAL CORD AND SPINAL NERVES   317

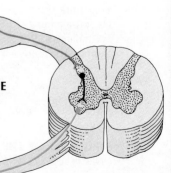

14 THE BRAIN AND CRANIAL NERVES 347

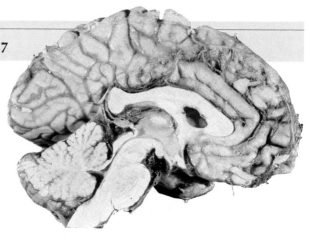

15 THE AUTONOMIC NERVOUS SYSTEM 393

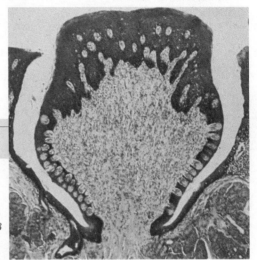

16 THE SENSES 407

17   THE ENDOCRINE SYSTEM   445

PART IV

TRANSPORT, MAINTENANCE, AND REPRODUCTION   463

18   THE CARDIOVASCULAR SYSTEM: BLOOD   464

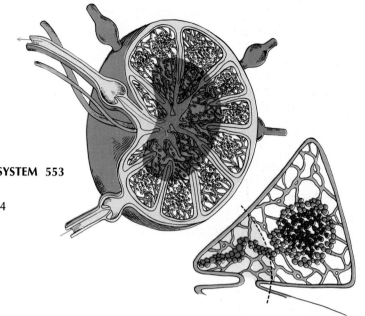
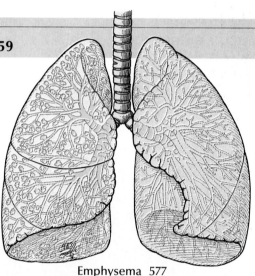

23

THE DIGESTIVE SYSTEM 582

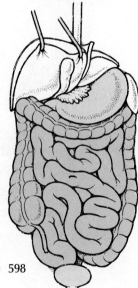

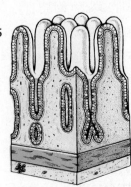

24

25

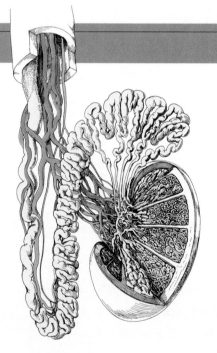

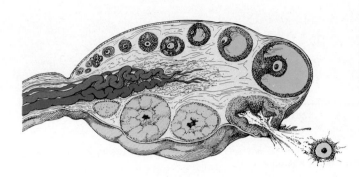

26 DEVELOPMENTAL ANATOMY: A LIFE SPAN APPROACH 670

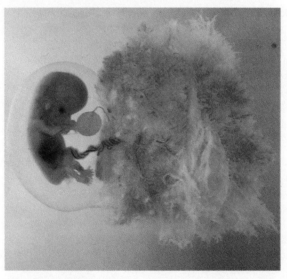

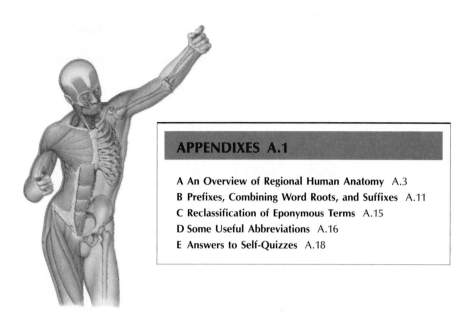

AN INVITATION TO HUMAN ANATOMY

It may seem that the story of anatomy needs to be told only once. After all, how much do bones and muscles change? But, actually, the study of human anatomy *has* changed since Leonardo da Vinci and Andreas Vesalius produced their exquisite drawings. Even the venerable *Gray's Anatomy* may no longer be the last word. Undoubtedly, the authors of every textbook try to make their book accurate and up-to-date. But any book, including *Human Anatomy*, can be made more inviting to student and teacher alike. That was our major goal as we wrote this text. We focused our efforts toward four major categories: content and organization, pedagogy, the illustration program, and the supplements package. Following are some examples of how we tried to accomplish our overall goal.

CONTENT AND ORGANIZATION

The title for each unit represents one of the four concepts that is used to organize the information in this text.

Part I: **How the Body Is Organized** (Chapters 1–4). This unit introduces the student to an historical overview of the science of anatomy, as well as to the basic terminology used in discussing anatomical structures, surfaces, and locations. In addition, the basic organization of the body at the level of cells and tissues is presented.

Part II: **Protection, Support, and Movement** (Chapters 5–11). The interactions of the integumentary, skeletal, and muscular systems form the basis of this unit. Carefully designed illustrations and tables have been coordinated with the textual material to provide a visually effective summary of major bone and muscle regions.

Part III: **Control, Communication, and Coordination** (Chapters 12–17). This unit focuses on the two regulatory systems of the body, the nervous system and the endocrine system. We start by describing the structure of nervous tissue and progress to topics that describe how the electrical and chemical interactions of these systems control the functions of these and other body systems.

Part IV: **Transport, Maintenance, and Reproduction** (Chapters 18–26). The remaining systems of the human body are involved in the transport of nutrients, water, wastes, and in reproduction and development. Three chapters on the circulatory system focus on the blood (Chapter 18), the heart (Chapter 19), and the blood vessels (Chapter 20). The lymphatic, respiratory, digestive, urinary, and reproductive systems chapters focus on how each system is involved in maintaining homeostasis, and the final chapter describes a life span approach to developmental anatomy.

CHAPTER FEATURES

The **Chapter Introductions** present an interesting and informative framework that draws the student into each chapter and introduces the basic ideas that will be explored in the chapter.

Boxed Essays make anatomy come alive and help the student to learn anatomical information in a meaningful way. See "Knee Injuries—The 'Achilles Heel' of Athletes" (page 207) or "Left Brain, Right Brain" (pages 370–371).

Over 100 in-text **Clinical Applications** are marked with a special caduceus icon (✝) in the text margins for easy identification.

Up-to-Date Technology, such as fetal surgery and new ways of exploring the body, is presented concisely and clearly.

Current Anatomical Terminology is used, with the eponymous term in parentheses the first time it is used; for example: *perforating canals* (Volkmann's canals). Appendix C lists eponymous terms and their current terminology, according to the fifth edition of *Nomina Anatomica.*

The entire running text contains **Information That Can Prove Useful to Students** as well as helping them understand aspects of human anatomy. For example, on page 223 you will find a description of how to return a jaw to its proper location after it is dislocated by a vigorous yawn, and you will learn on page 438 about the prevention and relief of motion sickness without medication.

Information about **Exciting and Current Areas of Human Interest** such as space travel are related to the usual structure and function of the body.

Only Enough Physiology Is Included to Support Discussions of Anatomy. The authors have tried to make this an *anatomy* book for beginning students. The presentation is concise, yet comprehensive enough to use for future reference.

The **Developmental Anatomy** related to each body system is presented at the end of each chapter for those instructors who choose to integrate developmental anatomy into their course. Also, at the end of selected chapters is a discussion of the **Effects of Aging** on the system.

"When Things Go Wrong" is a section at the end of each chapter that provides the student with enough pathology to reinforce the textual material. This section also supplies useful information to students majoring in health professions.

PEDAGOGY

Itemized Lists reinforce the understanding of complex structures and processes, and make it easy for the student to isolate and remember the most important concepts of the process. Processes are also often described in accompanying photos and drawings (see FIGURE 25.17 on page 660). See also the step-by-step discussion of Intramembranous Ossification on pages 136 and 138, with illustrations on page 137.

Question-and-Answer Boxes provide useful and easy-to-remember anatomical information. For example, "Why is a tattoo so difficult to remove?" is answered in a way that helps the student to understand an important characteristic of the layers of the skin. Or, "Why can a sneeze be stopped by pressing between the upper lip and the base of the nose?" is answered in a way that relates this action to a nerve response.

Footnotes are meant to reinforce learning by presenting interesting material about the subject at hand. Some are informative, providing Greek or Latin derivatives of anatomical terms. Others are fun: "every cell in the human body contains genetic information that, if translated into words, would fill 3000 books of 1000 pages each, with 1000 words on each page."

Many **Tables Are Illustrated,** providing a dynamic learning tool for the student through the association of text with visual images, such as TABLE 9.1 Classification of Joints (pages 204–205).

Word Origins are used selectively to help the student remember important concepts. Macrophage: *macro* means "large" and *phagein* is Greek for "to eat," so a macrophage is a "big eater."

Word Pronunciations are also used selectively to help the student lose any fear of mispronouncing a word during class discussions, when the climate should be as relaxed as possible.

Chapter Summaries are presented in an itemized form, reviewing one main section at a time. The summaries are placed at the end of each chapter, immediately preceding "Study Aids for Review." Page numbers are included for each main section for easy reference and text location.

"Study Aids for Review" at the end of each chapter is a self-contained Study Guide for the student (see pages 82–83). Each review section contains Medical Terminology, Understanding the Facts, Understanding the Concepts, and a Self-Quiz, often including "A Second Look." Answers to the Self-Quizzes can be found in Appendix E.

"A Second Look" is a special feature in the questions section at the end of the chapter. It is a labeling exercise designed to reinforce the student's understanding of anatomical photographs and drawings within the chapter (see page 227).

Appendixes and Endpapers provide the student with useful, easy-to-access information such as metric conversion tables and a list of in-text clinical applications.

ILLUSTRATION PROGRAM

Illustrated Summaries create inviting study aids by presenting all the relevant material about a specific subject (text, table, and illustrations) in self-contained packets of information; see the discussion of epithelial tissues on pages 88–93, mitosis on pages 72–74, and muscles of the shoulder on pages 266–267. Also, see Chapter 2, Surface Anatomy.

Color-Coded Illustrations are consistent throughout the book to make identification easy for the student. For example, the parietal bone is always light blue, and the temporal bone is always a darker blue. Also, all lateral views of anatomical drawings have been drawn showing the *right* side for consistent orientation.

Visual Roadmaps guide the student through intricate material. Many of the book's illustrations move from the macro to the micro (see FIGURE 10.1 on page 230) or complement drawings with a matching photograph (see FIGURE 5.6 on page 119) to help the student visualize, understand, and retain textual material.

Especially Commissioned Photographs of selected body parts (see pages 338 and 485) present a dramatic view of the human body.

A special portfolio of **Origins and Insertions** of muscles gives the student an easy-to-use reference guide when studying bones and joints (see FIGURE 11.25 on pages 284–286).

ACKNOWLEDGMENTS

Human Anatomy by Carola, Harley, and Noback has been judged to be the most up-to-date anatomy book available. Much of the credit for that aspect of our book goes to the many expert reviewers of the manuscript and art, including:

Paul Biersuck
Nassau Community College

Richard Bowen
Colorado State University

Marion C. Diamond
University of California — Berkeley

Judy A. Donaldson
Northern Seattle Community College

William Farrar
Eastern Kentucky University

Jackson Jeffrey
Virginia Commonwealth University

David A. Kaufmann
University of Florida

Sherwin Mizell
Indiana University — Bloomington

Thomas M. O'Connor
Washburn University

Michael Postula
Parkland College

Richard L. Potter
California State University — Northridge

Lynn Romrell
University of Florida College of Medicine

P. George Simone
Eastern Michigan University

Barrie D. Smith
University of Missouri

Ronald E. Smith
Modesto Junior College

Kenneth T. Wilkins
Baylor University

We feel that the art in *Human Anatomy* is extraordinary, and we would like to thank the following artists for achieving such a high level of excellence:

Robert J. Demarest
Cover/part-opening illustration

Marsha J. Dohrmann
Nervous system

Carol Donner
Reproductive systems

John V. Hagen
Muscular system

Neil O. Hardy
Digestive system

Steven T. Harrison
Cardiovascular and urinary systems

Jane Hurd
Respiratory system

Joel Ito
Integumentary, endocrine, and lymphatic systems; cells and embryology

George Schwenk
Skeletal system

The authors are indebted to the many people at McGraw-Hill who have supported us in every possible way and have consistently striven to make this a special book, including Seibert Adams, Deena Cloud, Holly Gordon, Gayle Jaeger, Denise Schanck, and especially our sponsoring editor, Kathi Prancan. There are no words to thank Kathi adequately.

Robert Carola
John P. Harley
Charles R. Noback

SUPPLEMENTS PACKAGE

Supplementary materials include the most complete selection of teaching and learning aids ever presented in a Human Anatomy text. The following materials are available:

FOR THE STUDENT

A *Student Study Guide*, prepared by John P. Harley, which includes a number of learning activities, has been designed to help the student master the many facets of human anatomy. For each chapter there is a list of prefixes and suffixes, an outline of the chapter for the student to complete based on major heads and subheads of the text and including page references to the text, a labeling exercise, a listing of major terms, informational essays called "Your Career in Health," and a post-test with an answer key.

Two-hundred *Flash Cards*, prepared by Barbara Cocanour, are packaged in a separate box to self-test the understanding of the muscles, skeletal parts, nerves, and blood vessels.

Radiographic Anatomy: A Working Atlas, prepared by Harry W. Fisher, M.D., is a complete atlas of human anatomy as seen through today's imaging technologies. Approximately 200 beautifully reproduced radiographs, CAT scans, and high resolution NMR images are accompanied by clearly labeled line drawings that highlight the anatomical structure.

The *Laboratory Manual* was prepared by Joseph P. Farina, Ted Namm, Alease S. Bruce, and Barbara Cocanour. This lab manual presents student-tested laboratory exercises designed to accommodate laboratory sessions of various lengths and focuses. The 26 chapters are organized into five major categories: levels of organization; protection, movement, and support; control and integration; homeostatic systems; and continuity of life. Each chapter begins with a complete list of objectives followed by a general introduction. The material in each chapter is subdivided into exercises that may be adapted to specific laboratory sessions. Each chapter ends with a selection of comprehensive Study Questions.

The *Activities Manual* was prepared by Barbara Cocanour and William Farrar. This self-study tool offers a coloring book, crossword puzzles, quotation puzzles, and anatomical flash cards. These are designed to help achieve mastery of anatomical and physiological information. The coloring book includes modified illustrations from the text to be colored in and labeled. Brief definitions and descriptions as well as a self-test are included for each illustration. The crossword puzzles appear within each section and provide an entertaining way to recall important definitions. A quotation puzzle appears at the end of each unit and combines clues from all the crossword puzzles in that unit. The flash cards are designed to review the muscles, skeletal parts, nerves, and blood vessels.

FOR THE INSTRUCTOR

Instructor's Resource Manual, prepared by John P. Harley, contains a variety of supplementary teaching aids, such as instructional enrichment sections, listings of pertinent anatomy films and videocassettes, listing of software programs in anatomy, topics for discussion and library research, and answers to the text sections Understanding the Facts and Understanding the Concepts.

Test Bank Manual, prepared by John P. Harley, contains 75 test items per chapter, for each of the 26 chapters. Each question has been class tested and checked for level, clarity and validity by Valerie Nelson, Testing Specialist for American College Testing Corporation. Answers are given at the end of the manual. The test questions are also available on diskette for IBM PC, Apple II, and Macintosh computers.

Overhead Transparencies. Over 100 color transparencies of important illustrations, photographs, and electron micrographs from the text are available free to adopters. Lettered callouts are consistently large and bold so they can be viewed easily, even from the back of a large lecture room.

Lecture Outlines, prepared by John P. Harley, contain complete outlines for each chapter in large type. These can be used as classroom handouts or made into overhead transparencies for projection. Each outline covers the major headings and subheadings as found in the textbook. Enough space has been allowed for the addition of ancillary notes or lecture material.

VisiQuizzes, prepared by John P. Harley, contain approximately 200 visual labeling quizzes taken from figures in the text. These can be copied for class handout or made into overhead transparencies.

The *Human Anatomy and Physiology Image Library* is available on Macintosh CD-ROM and on Macintosh 3.5-inch floppy disk. A database of over 600 images will allow instructors to create their own HyperCard tutorials, transparencies, tests, and classroom handouts.

Atlas of Human Anatomy, prepared by Frank H. Netter, is an illustrated atlas containing 514 full-color line drawings of the major structures of the human body.

1

HOW THE BODY IS ORGANIZED

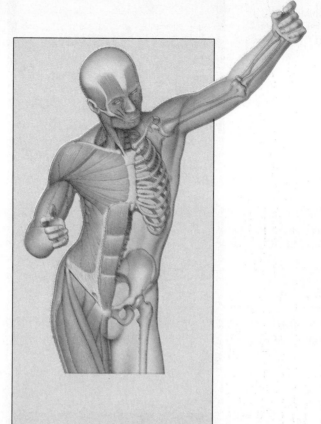

1

Introduction to Anatomy

2

Surface Anatomy

3

Cells: The Basic Units of Life

4

Tissues of the Body

1

Introduction to Anatomy

LEARNING OBJECTIVES

1 Define anatomy and describe some of its subspecialties.

2 Identify the organizational levels of the body, and explain how they are related.

3 Briefly describe the four tissue groups that make up the body.

4 Identify the systems of the body, and explain a major function of each system.

5 Describe the anatomical position, and discuss its relationship to the directional terms of the body.

6 Distinguish between the relative directional terms, regions, planes, and cavities of the body.

7 Describe the major noninvasive techniques of exploring the body.

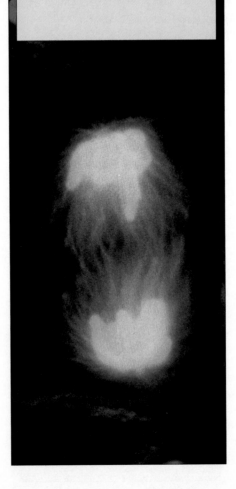

Royal Academy lecture in the eighteenth century.

Nothing fascinates the human mind more than the human body. Human beings have been curious about the structure and function of their bodies for thousands of years, wondering how fingers work, or exactly where the heart is, or why the arms bulge when something heavy is lifted. Today, the study of the human body is more sophisticated than ever before, and answers to these questions and many more are known. In fact, we have probably learned more about ourselves in this century than in all the previous centuries put together.

WHAT IS ANATOMY?

Human anatomy (from the Greek word meaning *to cut up*) is the study of the many *structures* that make up the body, and how those structures relate to each other. In contrast, **physiology** is the study of the *functions* of those structures.

The study of anatomy includes many subspecialties, from gross anatomy and dissection to the most recent forms of microscopic and ultramicroscopic anatomy.

Regional anatomy is the study of specific regions of the body (such as the head or neck), while *systemic anatomy* is the study of different systems, such as the reproductive or digestive systems. Regional and systemic anatomy are both examples of **gross anatomy,** which includes any branch of anatomy that can be studied without a microscope. In contrast, **microscopic** and **ultramicroscopic anatomy** require the use of microscopes. Some other subdivisions of anatomy are *embryological anatomy*, or *embryology*, the study of prenatal development; *developmental anatomy*, the study of human growth and development from fertilized egg to mature adult; and *radiographic anatomy*, or *radiology*, the study of the structures of the body using x-rays.

As you study anatomy, you will see that form and function go together. For example, some parts of the body have specific shapes that make them suitable to perform their specific functions. Think of teeth. Three different kinds of teeth do three very different jobs. *Flat* molars grind food, *sharp* canines tear, and *pointed* incisors cut. As you progress through this book, try to keep in mind the crucial connection between anatomy and physiology: form and function.

A Short History of Human Anatomy

As far as we know, anatomy is the oldest medical science. Cave paintings of the early Stone Age, about 30,000 years ago,* show a simple knowledge of the anatomy of animals, and it is assumed that these cave dwellers applied some of their anatomical knowledge to their own bodies. The civilizations of the Babylonians, Assyrians, Egyptians, Chinese, and Hindus made no serious attempt to learn anatomy because they were interested in the supernatural world, not the natural one, and their cultures placed strong religious restrictions against debasing the body. Any anatomical dissections that were performed on animals were made to "study" organs in an effort to predict the future and to tell fortunes.

Anatomy in Ancient Greece

The systematic study of anatomy may have begun in the fifth century B.C., with the work of two Greek scientists, **Alcmaeon** (ca. 500 B.C.) in Italy and **Empedocles** (ca. 490–430 B.C.) in Sicily, where Greek culture and science flourished. Alcmaeon was probably the first person to dissect the human body for research purposes, and he is also given credit for proposing that the brain is the center of intelligence. Empedocles, who believed that the heart distributed life-giving heat to the body, initiated the idea that an ethereal substance called *pneuma,* which was both life and soul, flowed through the blood vessels. Although such early anatomists were often incorrect, their work was essential to the development of later scientists.

Anatomical inferences without dissection continued in Greece with **Hippocrates** (ca. 460–377 B.C.), who is known as the Father of Medicine. (Many medical students still take the Hippocratic Oath upon graduation from medical school.) He might also be called the Father of Holistic Medicine, since he advocated the importance of the relationship between patient, physician, and disease in the diagnosis and treatment of illness. This philosophy was rejected at a time when diseases were still thought to be punishments from the gods.

*The Stone Age, the earliest known period of human culture, is characterized by the use of small stone tools.

Hippocrates' knowledge of internal anatomy was severely limited by the lack of dissections, and not until **Aristotle** (384–322 B.C.) did physicians begin to dissect animals carefully enough to deduce even the barest essentials of human anatomy. Aristotle corrected many of the anatomical errors of his predecessors, but because he was primarily a philosopher rather than a physician, he depended more on logical deduction than on observation and experimentation. His scanty knowledge of the inner workings of the human body led to many gross inaccuracies; for example, he believed that the brain cooled the heart by secreting "phlegm," and that the arteries contained only air. Nevertheless, he had an enormous influence on scientists for hundreds of years.

The Beginnings of Modern Anatomy

With the decline of Greek influence on the mainland, Alexandria became the transplanted center of Greek culture. It was there that the Greeks **Herophilus** (ca. 335–280 B.C.) and **Erasistratus** (ca. 310–250 B.C.) conducted the first systematic dissections of the human body. Herophilus established the brain as the center of intelligence, distinguished between veins and arteries, and made many other accurate observations about the structure of the human body, especially the nervous system. He conducted the first public dissection and is supposed to have taught the first female medical student. Erasistratus, an intense rival of Herophilus, was more interested in physiology than anatomy and studied the process of circulation in the body. He believed that pneuma, or vital air, was carried by the arteries. The written works of both Greeks were lost when the library at Alexandria was destroyed in A.D. 272, but their ideas were found in the writings of the Roman Celsus (30 B.C.–A.D. 30) and the physician Claudius Galenus, popularly known as Galen.

Galen (ca. A.D. 129–199), considered to be the greatest ancient physician after Hippocrates, was born in Pergamon in Asia Minor (now Pergama in Turkey). His early knowledge of anatomy derived from his studies in Asia

Minor, Greece, and Alexandria, and after his return to Pergamon, his job as chief physician to the gladiators. He later gained prominence in Rome as physician to the court of Emperor Marcus Aurelius, where he wrote prolifically about human anatomy. Galen's dissections of African monkeys (human dissections were still forbidden) provided him with enough related information about humans so that he described correctly many brain structures, the structural differences between veins and arteries, and many other structures of the human body, including heart valves. He also observed that muscles contract in response to a stimulus from nerves, and demonstrated *experimentally* that the arteries carry blood, not air.

Despite Galen's improvements on earlier anatomical studies and his other achievements, he is often remembered for the fact that the Catholic church did not allow his ideas to be criticized; thus many of his erroneous ideas were perpetuated and major progress in the field of anatomy was halted until the sixteenth century.

The twelfth and thirteenth centuries saw a gradual reawakening of valid scientific investigation after the barren years of the Dark Ages. The first true university was founded in Bologna in the twelfth century, and a medical faculty was established there by 1156. By the end of the thirteenth century, the demand for accurate information was so great that the medical dissection of human corpses began in earnest. Anatomists at this time were still conditioned to revere the outdated notions of Aristotle and Galen, and if an autopsy revealed a deviation from prior teachings, the anatomists concluded that the body was abnormal.

The fourteenth century brought a more scientific attitude to the study of the human body. To some extent, artists, rather than scientists, set the pace in revealing new aspects of human anatomy. **Leonardo da Vinci** (1452–1519) was undoubtedly the most industrious artist, producing hundreds of anatomical drawings made from dissections; unfortunately he had little influence on the anatomists of his time.

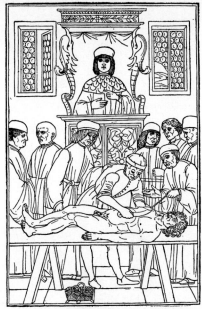

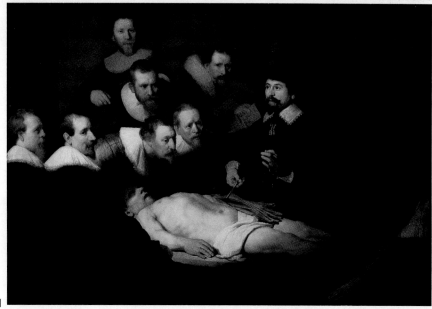

Two views of dissection. **[A]** A fifteenth-century woodcut shows an anatomy lesson at the University of Padua. The professor is seated high above the proceedings as two assistants perform the actual dissection according to his directions. **[B]** By the seventeenth century, anatomy professors were performing their own dissections, as shown in this 1632 painting by Rembrandt *(The Anatomy Les-* *son)* of Professor Nicolaes Tulps performing a "public anatomy" for a small group of surgeons in Amsterdam. (Although Rembrandt's painting is an acknowledged masterpiece, it contains a serious anatomical error. The group of muscles shown actually has its origin on the medial aspect of the elbow, not on the lateral side as illustrated.)

Five years before the death of Leonardo, the true "Father of Anatomy" was born. This was **Andreas Vesalius** (1514–1564), who at the age of 29 published his seven-volume *De humani corporis fabrica (On the Structure of the Human Body)*, in which he carefully integrated text and drawings made from dissections, setting anatomy on a new course toward the scientific method. (The drawings were made by Jan Calcar, a student of Titian.) Another significant scientific event occurred in the same year, 1543, when the Polish astronomer Nicolaus Copernicus (1473–1543) published his view that the earth revolved around a stationary sun.

The publication of the *Fabrica* was a major scientific event because it was instrumental in overcoming the authority of the Catholic church. For the first time, anatomy was placed on an objective level, and Galen's inaccuracies were exposed. Unfortunately, Vesalius's ideas were originally rebuked by anatomists because they challenged Galen and others.

The Contributions of William Harvey
The English physician and anatomist **William Harvey** (1578–1657) studied at the University of Padua (the newly established center of medical research) several years after Vesalius taught there. In 1628, Harvey published *An Anatomical Treatise on the Motion of the Heart and Blood in Animals,* in which he described for the first time how blood is pumped by the contractions of the heart, circulates throughout the body, and returns to the heart. Both the accurate plan of the circulation and the idea that the heart is a pump were enormous breakthroughs that helped overcome the primitive ideas of Aristotle and Galen once and for all. Although Harvey's discovery was attacked by Galen's steadfast followers, it was difficult to argue against Harvey's methods of first-hand observation and experimentation. Harvey had not only made a most important anatomical discovery, he had also demonstrated a logical and scientific approach that set the standard for future anatomical research. From then on, physicians and anatomists considered structure *and* function when investigating the human body. Such research was aided by microscopes, beginning with those produced by the Dutch microscopist, Antonie van Leewenhoek (1632–1723), which enabled scientists to examine the cells, tissues, and fluids of the body.

Modern Anatomy
To many, gross human anatomy is associated with *Gray's Anatomy*, originally published by the English surgeon Sir Henry Gray in 1858. Since then the book has had several authors and has evolved into the current thirty-seventh edition in Great Britain and the thirtieth edition in the United States, each with its own character.

Radiological advances in the twentieth century have allowed scientists to make remarkable connections between anatomy and physiology, and researchers are integrating the study of anatomy with other disciplines, including biochemistry, genetics, and biophysics. Physicians now have access to advanced technology such as CAT and PET scanners, and magnetic resonance imaging (MRI), all of which go far beyond microscopy and x-rays. These techniques permit physicians to look inside the body without performing surgery, yet another major breakthrough in the history of anatomy.

FIGURE 1.1 STRUCTURAL LEVELS OF THE BODY

Specialized cells and tissues of the respiratory system help the lungs and other parts of the system to perform their functions of breathing and gas exchange at an optimum level of efficiency. The same levels of organization are found in all body systems.

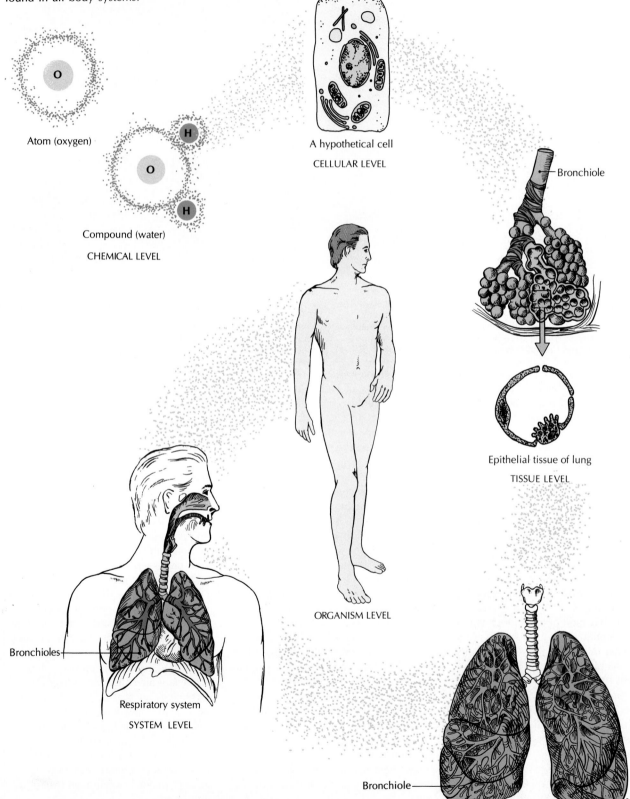

Atom (oxygen)

Compound (water)

CHEMICAL LEVEL

A hypothetical cell

CELLULAR LEVEL

Bronchiole

Epithelial tissue of lung

TISSUE LEVEL

ORGANISM LEVEL

Bronchioles

Respiratory system

SYSTEM LEVEL

Bronchiole

Lungs and bronchial trees

ORGAN LEVEL

FROM ATOM TO ORGANISM: STRUCTURAL LEVELS OF THE BODY

The human body has different structural levels of organization, starting with atoms, molecules, and compounds, and increasing in size and complexity to the cells, tissues, organs, and systems that make up the complete organism [FIGURE 1.1].

Atoms, Molecules, and Compounds

At its simplest level, the body is composed of **atoms** (from *atomos*, the Greek word for indivisible), the basic structural units of all matter. Scientists have identified about 100 different kinds of atoms. Each kind is an *element*. The most common elements in living things are carbon, hydrogen, oxygen, nitrogen, and phosphorus.

When two or more atoms combine, they form a **molecule**. For example, when two oxygen atoms combine, they form an oxygen molecule, O_2. A molecule containing atoms of more than one element is called a **compound**. Water (H_2O) and carbon dioxide (CO_2) are compounds, as are the carbohydrates, proteins, and lipids (fats) that are so important to our bodies.

Cells

Cells are the smallest independent units of life, and all life as we know it depends on the many chemical activities of cells. Some of the basic functions of cells are growth, metabolism, irritability, and reproduction. Cells can vary in size from a sperm, which is about 5 micrometers (5-millionths of a meter) long, to a nerve cell with thin fibers that may be more than a meter long. Human cell structure and function are discussed in detail in Chapter 3.

Tissues

Tissues are made up of many similar cells that perform a specific function. They are divided into four groups: epithelial, connective, muscle, and nervous [FIGURES 1.2 and 1.3].

Epithelial tissue is found in the outer layer of the skin and in the linings of organs, blood and lymph vessels, and body cavities. It is well-suited to its protective function because the cells are closely packed and arranged in sheets [FIGURE 1.3A], and because it can add new cells when the old ones become worn or damaged. Some epithelial tissues are specialized for secretion, and release specific chemical substances.

FIGURE 1.2 SOME BASIC TISSUES AND STRUCTURES

This multilevel "step dissection" through the right lower leg (mid-calf) shows typical tissue layers, bones, and a bundle of nerves and blood vessels.

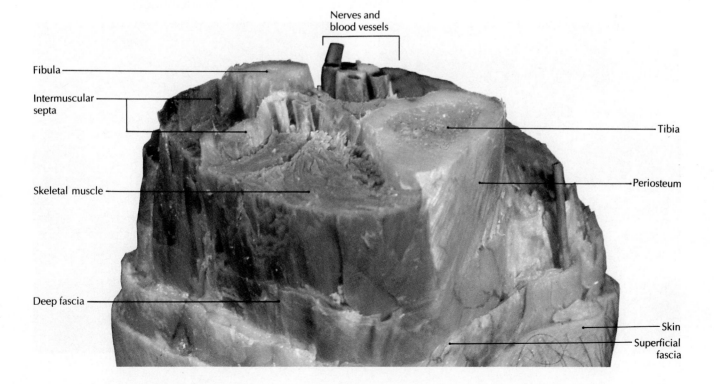

FIGURE 1.3 REPRESENTATIVE TYPES OF BODY TISSUES

[A] A highly magnified photograph of epithelial tissue. The arrangement of flat, overlapping sheets is well suited for the shingles of a roof or the skin of your body, both of which protect the inside from the outside environment. One hair is visible. ×150. [B] Meshed fibers of connective tissue are seen in this photograph of interwoven collagen fibers. Connective tissue is abundant in the body and can resist stress in several directions. ×3100. [C] Cardiac muscle tissue is found only in the heart. It consists of separate, but interconnected, cells, each with a centrally located nucleus. ×200. [D] Nervous tissue consists of branched nerve cells with slender processes that span relatively long distances in the body. ×200.

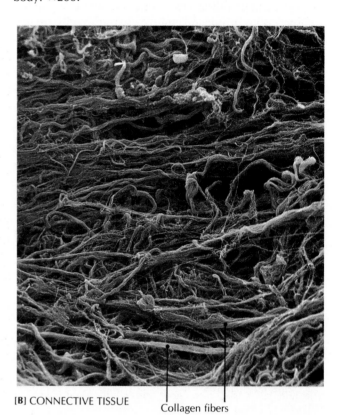

[A] EPITHELIAL TISSUE

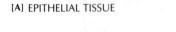
Hair

[B] CONNECTIVE TISSUE   Collagen fibers

Branched nerve cells

[C] CARDIAC MUSCLE TISSUE

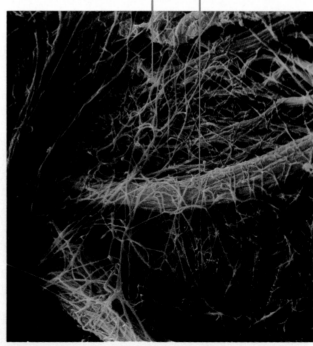

[D] NERVOUS TISSUE

Connective tissue, in addition to other functions, connects and supports most parts of the body. It is found in skin, and also comprises the major portions of bones and tendons. Some connective tissues contain fibers that form a strong mesh [FIGURE 1.3B]. Connective tissue is the most widely distributed of all body tissues.

Muscle tissue produces movement through its unique ability to contract, or shorten. *Skeletal muscle tissue* is found in the limbs, trunk, and face; *smooth muscle tissue* is found in the digestive tract, eyes, blood vessels, and ducts; and *cardiac muscle tissue* is found only in the heart [FIGURE 1.3C].

Nervous tissue is found in the brain, spinal cord, and nerves. It responds to various types of stimuli and transmits nerve impulses (messages) from one area of the body to another. Its long nerve fibers are well-adapted to these functions [FIGURE 1.3D].

Organs

The next level of complexity above the tissue level is the organ. An **organ** is an integrated collection of two or more kinds of tissue that work together to perform a specific function. The stomach is a good example: Epithelial tissue lines the stomach and helps protect it; smooth muscle tissue churns food, breaking it down into smaller pieces and mixing it with digestive chemicals; nervous tissue transmits nerve impulses that initiate and coordinate the muscle contractions; and connective tissue helps hold all the other tissues together.

Systems

A **system** is a group of organs that work together to perform a major body function. The respiratory system, for example, contains several organs that provide a mechanism for the exchange of oxygen and carbon dioxide between the air outside the body and the blood inside. All body systems are specialized, and their functions are coordinated* to produce a dynamic and efficient **organism** — your body.

*Although the external environment changes constantly, the internal environment of a healthy body is relatively stable, a state known as **homeostasis** (ho-mee-oh-STAY-sihss; Gr. "staying the same"). Homeostasis requires a chemical balance between the inside and outside of the body cells. To achieve this balance, the composition of the **extracellular fluid** — the fluid that surrounds and bathes the cells — must remain fairly constant.

ASK YOURSELF

1 What are the four types of body tissues?

2 What are the main differences between organs and tissues?

3 How does a system depend on organs?

BODY SYSTEMS

The structures of each system are closely related to their functions. The body systems are illustrated and described briefly in FIGURES 1.4 through 1.14 and in the table below.

Body system	Major functions

INTEGUMENTARY (skin, hair, nails, sweat and oil glands)

Covers and protects internal organs; helps regulate body temperature; contains sensory receptors.

SKELETAL (bones, cartilage)

Supports body, protects organs; provides lever mechanism for movement; manufactures red blood cells.

MUSCULAR (skeletal, smooth, cardiac muscle)

Allows for body movement; moves materials through body parts; produces body heat.

NERVOUS (brain, spinal cord; peripheral nerves; sensory organs)

Regulates most body activities; receives and interprets information from sensory organs; initiates actions by muscles.

ENDOCRINE (ductless glands)

Endocrine glands secrete hormones, which regulate many chemical actions within cells.

CARDIOVASCULAR (heart, blood, blood vessels)

Heart pumps blood through vessels; blood carries materials to tissues; transports tissue wastes for excretion.

LYMPHATIC (glands, lymph nodes, lymph, lymphatic vessels)

Returns excess fluid to blood; part of immune system.

RESPIRATORY (airways, lungs)

Provides mechanism for breathing, exchange of gases between air and blood.

DIGESTIVE (stomach, intestines, other digestive structures)

Breaks down large food molecules into small molecules that can be absorbed into blood; removes solid wastes.

URINARY (kidneys, ureters, bladder, urethra)

Eliminates metabolic wastes; helps regulate blood pressure, acid-base and water-salt balance.

REPRODUCTIVE (ovaries, testes, reproductive cells, accessory glands, ducts)

Reproduction, heredity.

BODY SYSTEMS

FIGURE 1.4 INTEGUMENTARY SYSTEM

The integumentary system consists of the skin and all the structures derived from it, including hair, nails, sweat glands, and oil glands. The skin envelops the body, providing a protective barrier between the internal organs and the outside environment; it is also involved in regulation of body temperature, elimination of wastes, and synthesis of vitamin D. The skin contains receptors for touch, pain, pressure, heat, and cold.

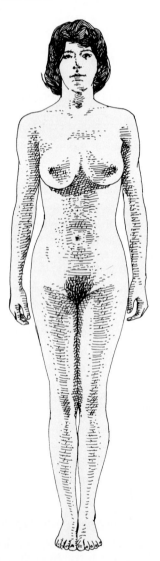

FIGURE 1.5 SKELETAL SYSTEM

The skeletal system consists of bones and cartilage. This illustration shows a typical bone, cartilage, and joint (a place where two bones meet). The skeletal system supports the body and protects the organs, provides a system of levers and points of attachment for muscles, and manufactures blood cells in the bone marrow. Bone tissue also contains the body's main reserve supplies of calcium and phosphorus.

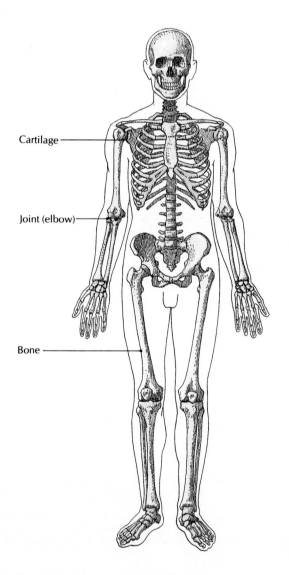

FIGURE 1.6 MUSCULAR SYSTEM

The muscular system consists of three different types of muscles: skeletal (some of the major ones are shown in the drawing), smooth, and cardiac (heart). It also includes tendons (the fibrous cords of connective tissue that attach muscles to bones). Muscles allow movement; maintain posture; move blood, food, urine, and other materials through various parts of the body; and produce body heat.

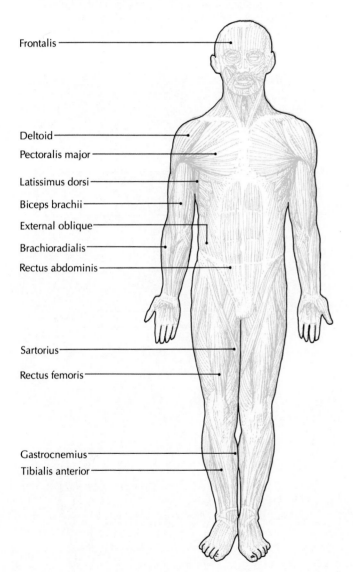

Frontalis

Deltoid
Pectoralis major

Latissimus dorsi
Biceps brachii
External oblique
Brachioradialis
Rectus abdominis

Sartorius

Rectus femoris

Gastrocnemius
Tibialis anterior

FIGURE 1.7 NERVOUS SYSTEM

The nervous system consists of the *central nervous system* (the brain and spinal cord) and the *peripheral nervous system* (nerves that connect the brain and spinal cord with the rest of the body). It also includes the sensory organs. The nervous system receives and interprets information from the sensory organs and initiates instructions for actions that are carried out by the muscles and glands; it regulates most body activities.

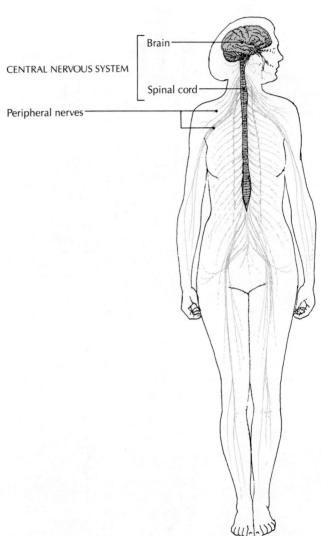

CENTRAL NERVOUS SYSTEM

Brain

Spinal cord

Peripheral nerves

(Body Systems continue on following pages)

BODY SYSTEMS *(Continued)*

FIGURE 1.8 ENDOCRINE SYSTEM

The endocrine system is the second major regulating system in the body, working closely with the nervous system. It is composed of ductless glands, each with a distinctive function. The locations of the major endocrine glands are shown in the drawing. (The parathyroids are not shown, because they are located on the posterior side of the thyroid gland.) Endocrine glands release their secretions, called *hormones,* directly into the bloodstream. Hormones regulate chemical reactions within cells, growth and development, stress and injury responses, reproduction, and many other critical functions.

FIGURE 1.9 CARDIOVASCULAR SYSTEM

The cardiovascular system consists of the heart, blood, and blood vessels. The illustration shows the heart, aorta, an artery (red), and a vein (blue); capillaries, which connect arteries and veins, are not shown. The heart pumps blood through a complex system of blood vessels. The blood within the vessels transports oxygen, nutrients, hormones, and other substances to all the body tissues, and it transports waste products from the body tissues to the kidneys and lungs for excretion. Blood is involved in fighting infections and also helps regulate body temperature and water balance.

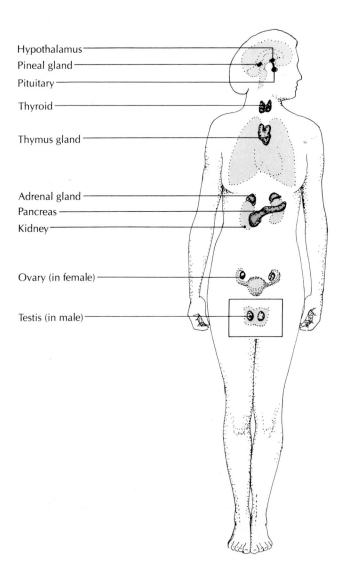

Hypothalamus
Pineal gland
Pituitary
Thyroid
Thymus gland
Adrenal gland
Pancreas
Kidney
Ovary (in female)
Testis (in male)

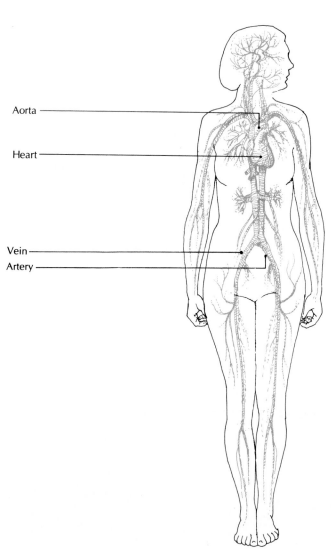

Aorta
Heart
Vein
Artery

FIGURE 1.10 LYMPHATIC SYSTEM

The lymphatic system is made up of glands (including the spleen, tonsils, and thymus gland), lymph nodes, and a network of thin-walled vessels that carry a clear fluid called *lymph*. The lymphatic system returns excess fluid and proteins to the blood and is an important part of the immune system. It helps to defend the body against harmful microorganisms and tumor cells and produces white blood cells that fight disease.

FIGURE 1.11 RESPIRATORY SYSTEM

The respiratory system is composed of the nose; a system of airways that includes the pharynx, larynx, and trachea; the lungs; and the muscles that help move air into and out of the body, the most important of which is the diaphragm. The respiratory system is concerned with the mechanics of breathing, and also provides a mechanism for the exchange of gases between blood and air.

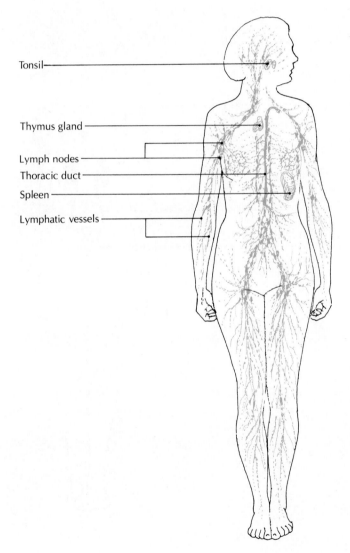

Tonsil

Thymus gland

Lymph nodes

Thoracic duct

Spleen

Lymphatic vessels

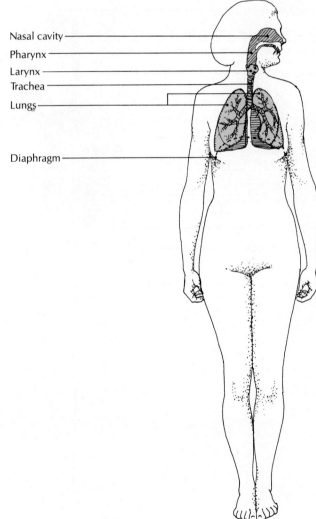

Nasal cavity

Pharynx

Larynx

Trachea

Lungs

Diaphragm

(Body Systems continue on following pages)

BODY SYSTEMS *(Continued)*

FIGURE 1.12 DIGESTIVE SYSTEM

The digestive system includes the teeth, tongue, salivary glands, esophagus (a tube leading from the mouth to the stomach), stomach, small and large intestines, liver, gallbladder, and pancreas. It is compartmentalized, with each part adapted to a specific function. The overall function of the system is to break down large food molecules physically and chemically until they are small enough to be absorbed into the bloodstream from the small intestine. Solid, undigested wastes are removed from the body through the anus.

FIGURE 1.13 URINARY SYSTEM

The urinary system consists of the kidneys, which produce urine; ureters, which carry urine to the urinary bladder where it is stored; and the urethra, which conveys urine to the outside. The urinary system rids the body of the wastes of cell metabolism; it also helps to regulate blood pressure and the composition and volume of blood and to maintain the body's acid-base and water-salt balance. The female urinary system is shown here; the male urethra passes through the penis.

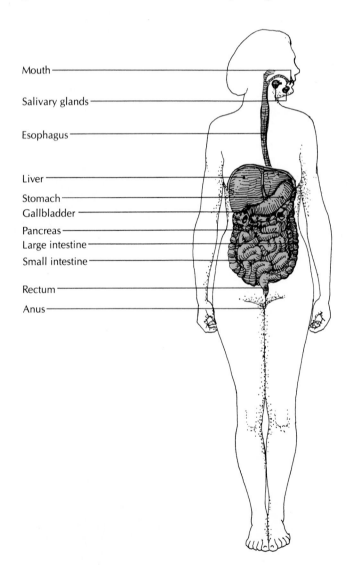

Mouth
Salivary glands
Esophagus
Liver
Stomach
Gallbladder
Pancreas
Large intestine
Small intestine
Rectum
Anus

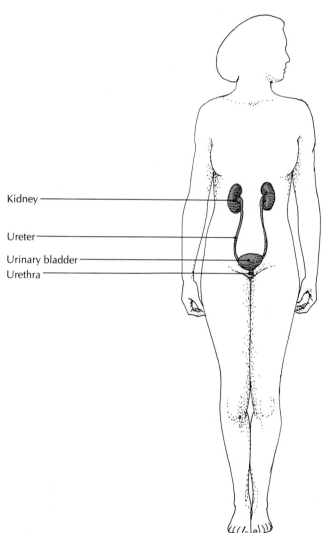

Kidney
Ureter
Urinary bladder
Urethra

FIGURE 1.14 REPRODUCTIVE SYSTEMS

[A] The female and **[B]** the male reproductive systems. The systems are enlarged in the drawings for clarity. Each of the sexes has reproductive organs (ovaries or testes) that secrete sex hormones and produce reproductive cells (eggs or sperm) and a set of ducts and accessory glands and organs, such as the uterus, vagina, prostate gland, and penis. The reproductive systems maintain the human species through reproduction and heredity.

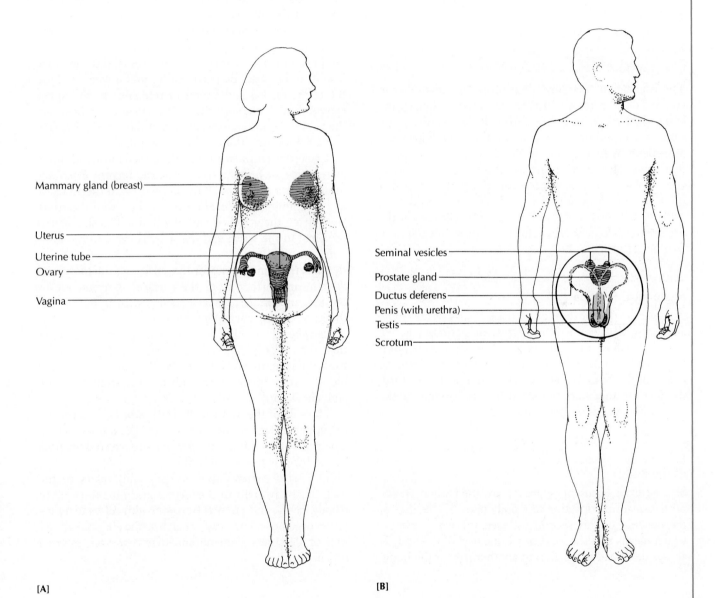

[A]   [B]

ANATOMICAL TERMINOLOGY

The language of anatomy will probably be unfamiliar to you at first. But once you have an understanding of the basic word roots, combining word forms, prefixes, and suffixes, you will find that anatomical terminology is not as difficult as first imagined. For instance, if you know that *cardio* refers to the heart and that *myo* means muscle, you can figure out that *myocardium* refers to the muscle of the heart. See Appendix B for a detailed list of prefixes, suffixes, and combining word forms. The Glossary at the back of this book can also help you learn the meanings of anatomical terms.

Anatomical Position

To describe the location of particular parts of the body, anatomists have defined the ***anatomical position*** [FIGURE 1.15A], which is universally accepted as the starting point for positional references of the body. In the anatomical position, the body is standing erect and facing forward, the feet are together, and the arms are hanging at the sides with the palms facing forward.

Relative Directional Terms of the Body

Standardized terms of reference are used when anatomists explain the location of a body part. Notice that in the heading of this section we used the word *relative*, which means that the location of one part of the body is always described in relation to another part of the body.

For instance, when you use standard anatomical terminology to locate the head, you say, "The head is *superior* to the neck," instead of saying, "The head is *above* the neck." When using directional terms, it is always assumed that the body is in the anatomical position.

Like so much else in anatomy, directional terms are found in pairs. If there is a term that means "above," there is also a term that means "below." If the thigh is ***superior*** to the knee, the knee is ***inferior*** to the thigh. The term ***anterior***, or ***ventral***, means toward the front of the body, and ***posterior***, or ***dorsal***, means toward the back of the body [FIGURE 1.15B]. The toes are anterior to the heel, and the heel is posterior to the toes.

Medial means nearer to the imaginary midline of the body or a body part, and ***lateral*** means farther from the midline. The nose is medial to the eyes, and the eyes are lateral to the nose. Be particularly careful when you use the terms medial and lateral in reference to the upper extremities. Remember that in the anatomical position, the palms are turned forward so that the little-finger side is medial and the thumb side is lateral.

The terms *proximal* and *distal* are used mostly for body extremities, such as the arms, legs, and fingers. ***Proximal*** means nearer the trunk of the body (toward the attached end of a limb), and ***distal*** means farther from the trunk (away from the attached end of a limb). The shoulder is proximal to the wrist, and the wrist is distal to the forearm.

Superficial means nearer the surface of the body, and ***deep*** means farther from the surface. ***External*** means outside, and ***internal*** means inside; these terms are not the same as superficial and deep.

Peripheral is used at times to describe structures other than internal organs that are located or directed away from the central axis of the body. Peripheral nerves and blood vessels, for instance, radiate away from the brain and spinal cord.

The sole of the foot is called the ***plantar*** surface, and the upper surface of the foot is called the ***dorsal*** surface. The palm of the hand is the ***palmar*** surface, and the back of the hand is the dorsal surface.

The term ***parietal*** (puh-RYE-uh-tuhl) refers to the walls of a body cavity or the membrane lining the walls of a body cavity, and ***visceral*** (VIHSS-er-uhl) refers to an internal organ or a body cavity (such as the abdominal cavity), or describes a membrane that covers an internal organ.

FIGURE 1.15 RELATIVE DIRECTIONAL TERMS OF THE BODY

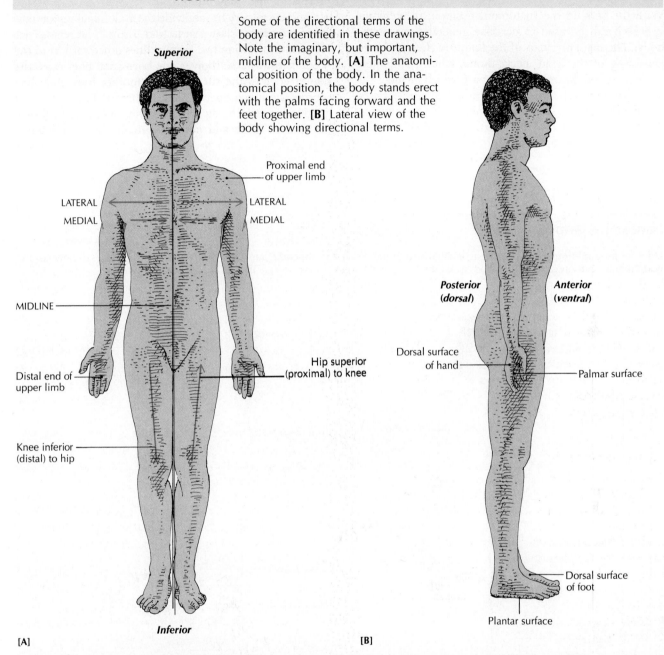

Some of the directional terms of the body are identified in these drawings. Note the imaginary, but important, midline of the body. **[A]** The anatomical position of the body. In the anatomical position, the body stands erect with the palms facing forward and the feet together. **[B]** Lateral view of the body showing directional terms.

Superior

Proximal end of upper limb

LATERAL

LATERAL

MEDIAL

MEDIAL

MIDLINE

Distal end of upper limb

Hip superior (proximal) to knee

Knee inferior (distal) to hip

Inferior

[A]

Posterior (dorsal)

Anterior (ventral)

Dorsal surface of hand

Palmar surface

Dorsal surface of foot

Plantar surface

[B]

Term	Definition and example
Superior (cranial)	Toward the head. The leg is superior to the foot.
Inferior (caudal)	Toward the feet or tail region. The foot is inferior to the leg.
Anterior (ventral)	Toward the front of the body. The nose is anterior to the ears.
Posterior (dorsal)	Toward the back of the body. The ears are posterior to the nose.
Medial	Toward the midline of the body. The nose is medial to the eyes.
Lateral	Away from the midline of the body. The eyes are lateral to the nose.
Proximal	Toward (nearer) the trunk of the body or the attached end of a limb. The shoulder is proximal to the wrist.
Distal	Away (farther) from the trunk of the body or the attached end of a limb. The wrist is distal to the forearm.
Superficial	Nearer the surface of the body. The ribs are more superficial than the heart.
Deep	Farther from the surface of the body. The heart is deeper than the ribs.
Peripheral	Away from the central axis of the body. Peripheral nerves radiate away from the brain and spinal cord.

Body Regions

With the body in the anatomical position, the regional approach can be used to describe general areas of the body. The main divisions of the body are the ***axial*** part, consisting of the head, neck, thorax (chest), abdomen, and pelvis; and the ***appendicular*** part, which includes the *upper extremities* (shoulders, upper arms, forearms, wrists, and hands) and the *lower extremities* (thighs, legs, ankles, and feet) [FIGURE 1.16A]. (The extremities are also called *limbs.*) FIGURE 1.16B and C presents additional technical terms for the body regions.

It is customary to subdivide the abdominal region into nine regions, as shown in FIGURE 1.17A. The regions are located by drawing two vertical lines downward from the centers of the collarbones, one horizontal line across the lower edge of the rib cage, and another horizontal line across the upper edges of the hipbones. The abdominal region may also be divided into four quadrants, as shown in FIGURE 1.17B and described in the list that follows.

FIGURE 1.16 BODY REGIONS

[A] The basic portions of the body, as located within the axial and appendicular parts. **[B]** Ventral and **[C]** dorsal views of the body regions present a more detailed list of the many terms used to locate specific parts of the body.

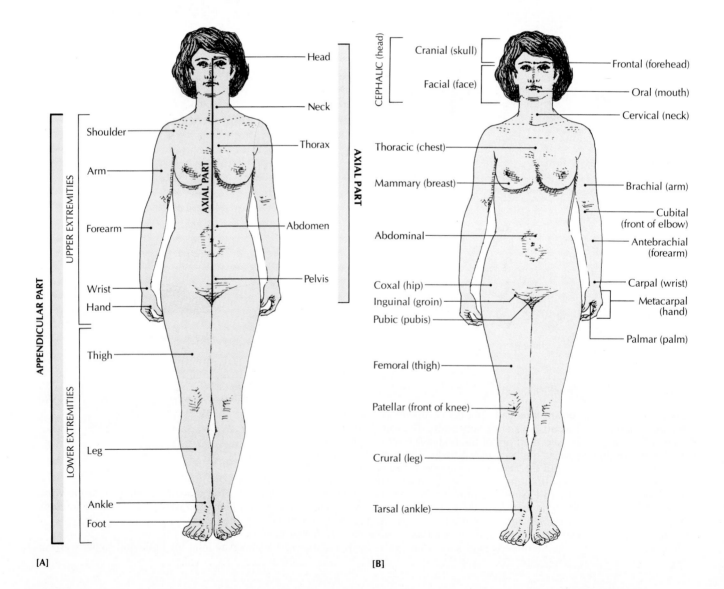

[A]

[B]

1 *Upper abdomen:* right hypochondriac region, epigastric region, left hypochondriac region; roughly the upper third of the abdomen.

2 *Middle abdomen:* right lumbar (lateral) region, umbilical region, left lumbar (lateral) region; roughly the middle third of the abdomen.

3 *Lower abdomen:* right iliac (inguinal) region, hypogastric (pubic) region, left iliac (inguinal) region; roughly the lower third of the abdomen.

FIGURE 1.17 ABDOMINAL SUBDIVISIONS

[A] The nine subdivisions of the abdominal region. **[B]** The four quadrants of the abdominal region.

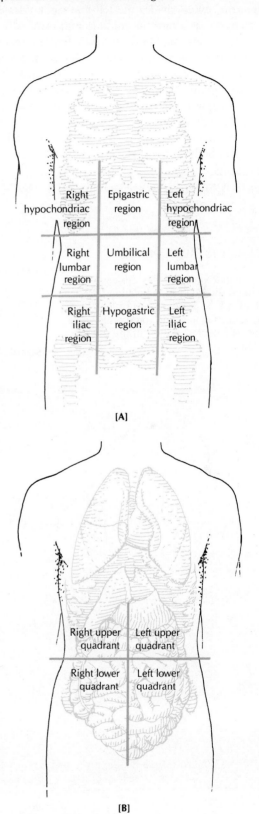

Right hypochondriac region | Epigastric region | Left hypochondriac region
Right lumbar region | Umbilical region | Left lumbar region
Right iliac region | Hypogastric region | Left iliac region

[A]

Right upper quadrant | Left upper quadrant
Right lower quadrant | Left lower quadrant

[B]

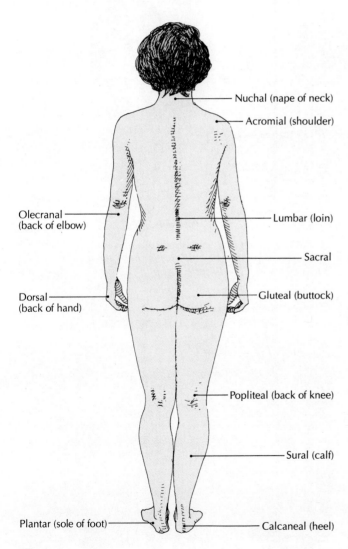

Nuchal (nape of neck)

Acromial (shoulder)

Olecranal (back of elbow)

Lumbar (loin)

Sacral

Dorsal (back of hand)

Gluteal (buttock)

Popliteal (back of knee)

Sural (calf)

Plantar (sole of foot)

Calcaneal (heel)

[C]

Body Planes

For further identification of specific areas, the body can be divided by imaginary flat surfaces, or **planes.** The **midsagittal plane** divides the left and right sides of the body lengthwise along the midline into externally symmetrical sections [FIGURE 1.18A]. If a longitudinal plane is placed off-center and separates the body into asymmetrical left and right sections, it is called a **sagittal plane.** A plane passing through the body lengthwise at right angles to the midsagittal plane is a **frontal,** or **coronal, plane.** A frontal plane divides the body into asymmetrical anterior and posterior sections [FIGURE 1.18B]. A **transverse plane** divides the body horizontally into upper (superior) and lower (inferior) sections. A transverse (or *horizontal*) plane is at right angles to the midsagittal, sagittal, and frontal planes. Transverse planes do not produce symmetrical halves.

The system of planes is also used with *parts* of the body, including internal parts. FIGURE 1.19A shows how cross sections, oblique sections, and longitudinal sections of internal parts are made. If your laboratory manual or any other book refers to a drawing of a sagittal *section,* a frontal *section,* or a transverse *section,* you should be aware of what is actually being shown and how it relates to its corresponding plane. FIGURE 1.19B shows a cut along the midsagittal plane of the head. Such a cut produces an exposed surface of the head called a **midsagittal section.** A cut along a frontal plane produces a **frontal section;** FIGURE 1.19C shows a frontal section of the brain. A cut along the transverse plane — in the case of FIGURE 1.19D, across the abdomen — produces a **transverse section.**

FIGURE 1.18 BODY PLANES

The imaginary body planes are an additional source of identification and location. **[A]** Representations of the midsagittal, sagittal, and transverse planes. **[B]** Representation of the frontal plane.

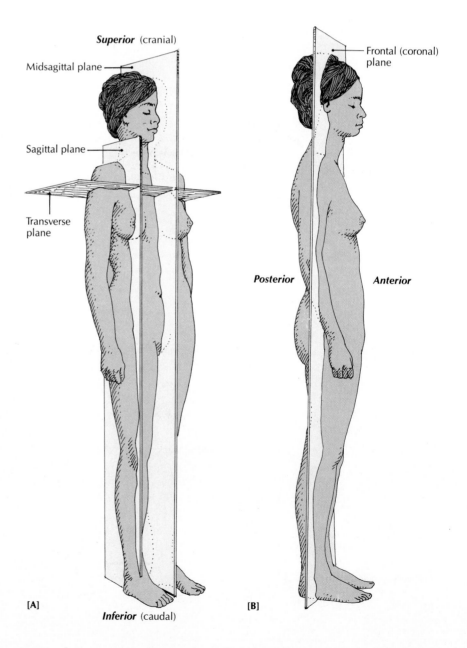

Superior (cranial)

Midsagittal plane

Sagittal plane

Transverse plane

Frontal (coronal) plane

Posterior

Anterior

[A]

[B]

Inferior (caudal)

FIGURE 1.19 BODY SECTIONS

[A] Three different ways to cut sections for microscopic examination. [B] A cut along the midsagittal plane of the head (smaller drawing) produces a midsagittal section of the head. [C] A cut along the frontal plane of the brain (smaller drawing) produces a frontal section. [D] A cut along the transverse plane of the abdomen (smaller drawing) results in a transverse section, in this case looking down into the lower portion of the cut body.

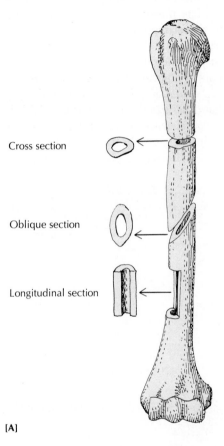

Cross section

Oblique section

Longitudinal section

[A]

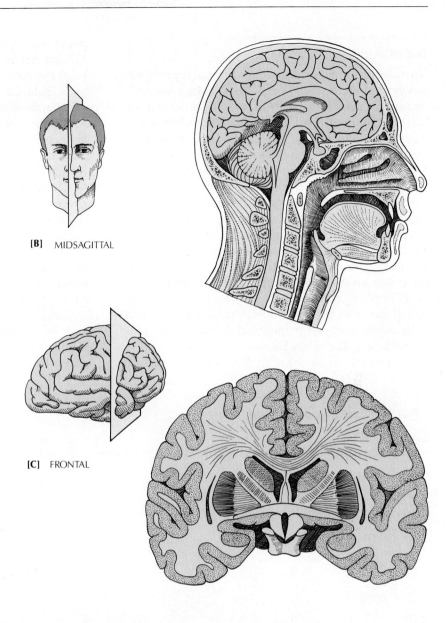

[B] MIDSAGITTAL

[C] FRONTAL

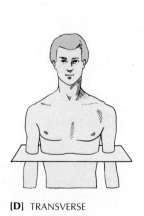

[D] TRANSVERSE

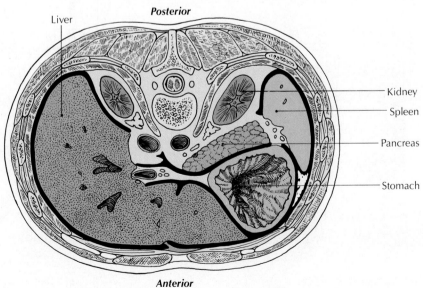

Liver

Posterior

Kidney

Spleen

Pancreas

Stomach

Anterior

Body Cavities

The cavities of the body house the internal organs, commonly referred to as the *viscera*. The two main body cavities are the **ventral (anterior) cavity** and the smaller **dorsal (posterior) cavity** [FIGURE 1.20A, B]. Each of these main cavities has subdivisions. For example, the dorsal cavity contains the *cranial cavity*, which contains the brain, and the *spinal (vertebral) cavity*, which contains the spinal cord.

The ventral cavity is separated by the dome-shaped diaphragm into the superior *thoracic cavity* and the inferior *abdominopelvic cavity*. These two cavities contain many organs, and also the pericardial, pleural, and peritoneal cavities. The *pericardial cavity* contains the heart; the *pleural cavities* contain the lungs; and the *peritoneal cavity* contains the stomach, liver, uterus, intestines, and other organs. Each of these cavities contains some lubricating fluid that facilitates movements, such as the muscular contractions of the intestines during digestion and of the uterus during childbirth.

An analogy can be made to illustrate the concept of the pleural, pericardial, and peritoneal cavities. Imagine a balloon that has been blown up so that it is still soft. Press your fist into the balloon. No matter how much you press, your fist cannot enter the cavity of the balloon as long as the surface of the balloon is intact. The space within the balloon is equivalent to the cavities, and your fist is equivalent to the organs. The portion of the balloon in contact with your fist would be the *visceral* ("organ") *peritoneum*, and the rest of the balloon would be the *parietal* ("wall") *peritoneum*.

Thoracic cavity The **thoracic cavity** is protected by the rib cage and its associated musculature, the sternum (breastbone) anteriorly, and the 12 thoracic vertebrae posteriorly. It consists of the right and left *pleural cavities*, each of which contains a lung, and the *mediastinum* (mee-dee-as-TIE-nuhm; L. *medius*, middle), the partition of tissues and organs that separates the lungs [FIGURE 1.20B]. The mediastinum contains the heart and its attached blood vessels, as well as the esophagus, trachea, thymus gland, lymph nodes, thoracic duct, phrenic and vagus

---

FIGURE 1.20 MAIN BODY CAVITIES

[A] The dorsal and ventral cavities and the smaller cavities within them; midsagittal section. [B] Frontal section showing the thoracic and abdominopelvic cavities and the cavities and mediastinum within them.

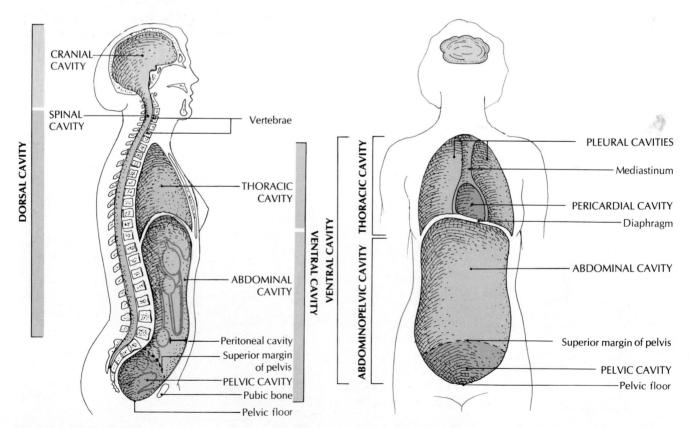

[A] MIDSAGITTAL SECTION

[B] FRONTAL SECTION

nerves — all the contents of the thoracic cavity except the lungs and pleural cavities.

The heart and pericardial cavity are enclosed by a double-walled sac called the *pericardium*, and each lung and pleural cavity are enclosed by a double-walled sac called the *pleura*. An inflammation of the the pleura is called *pleurisy*.

Abdominopelvic cavity The *abdominopelvic cavity* extends from the diaphragm inferiorly to the floor of the pelvis. It is divided by an imaginary line at the superior margin of the pelvis into the superior *abdominal cavity* and the inferior *pelvic cavity*. The abdominal cavity contains the stomach, intestines, liver, spleen, and gallbladder, as well as other structures, including the superior portion of the peritoneal cavity. The pelvic cavity contains the urinary bladder, rectum, internal portions of the male and female reproductive systems, and the inferior portion of the peritoneal cavity.

Body membranes *Membranes* are layers of epithelial and connective tissue that line the body cavities and cover or separate certain regions, structures, and organs. The three main types of membranes are *mucous*, *serous*, and *synovial*. They are described in Chapter 4.

ASK YOURSELF

1 What is the anatomical position of the body?

2 What is the difference between superior and anterior?

3 How does a sagittal plane differ from a midsagittal plane?

4 What organs are within the two main body cavities?

5 How is the thoracic cavity separated from the abdominopelvic cavity?

NEW WAYS OF EXPLORING THE BODY

Until just a few years ago, a physician who wanted to look *inside* the body to make an accurate diagnosis was limited by relatively old-fashioned equipment. X-rays can be used to assess the damage to broken bones and help in the diagnosis of diseases, such as lung cancer, that would be difficult or impossible to detect without surgery. But because x-rays can only show a flat picture, without indicating depth or isolating different layers of an organ, they do not always distinguish between healthy and diseased tissue.

Instruments for peering into the body, such as the endoscope and laparoscope, are basically periscopelike tubes with viewing devices at both ends. The latest versions of such scopes project the image of the internal organ being viewed onto a television screen. Because such instruments can be inserted into body openings or tiny surgical incisions, the need for major exploratory surgery is reduced drastically.

CAT (Computer-Assisted Tomography) and PET (Positron-Emission Tomography) Scanning

In 1973 the CAT (computer-assisted tomography*) scanner revolutionized the diagnosis of upper-body and brain disorders. This instrument projects a cross section of the internal body part being examined onto a television screen. Barely 10 years later, a second revolution came with the PET (positron-emission tomography) scanner, which not only shows physical disorders as a CAT scan does, but can also be useful for the diagnosis of some metabolic imbalances.

A *CAT scan* combines x-rays with computer technology to show cross-sectional views of internal body structures. Before the actual scanning procedure begins, the patient is usually injected with a low-level radioactive tracer through an intravenous catheter in the arm. The patient lies on a table, and is passed slowly through a circular opening in the scanning device [FIGURE 1.21]. A low-intensity x-ray beam then rotates 180 degrees across the width of the body. The degree to which the x-rays are absorbed by various body tissues (which have different

*The word *tomography* means "a technique for making a picture of a section or slice of an object," from the Greek *tomos*, a cut or section + *graphein*, to write or draw.

FIGURE 1.21 PATIENT UNDERGOING A CAT SCAN

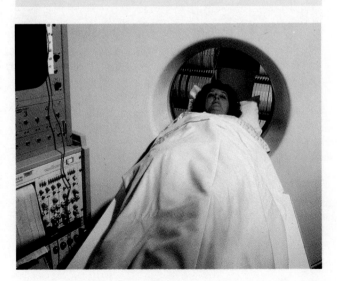

absorption rates) is recorded by detectors placed opposite the x-ray beam. The detectors convert the modified beams into electronic signals, which are fed into a computer. The computer analyzes the changes in the x-ray beams and shows them in high-resolution images on a television screen as a *tomogram,* a picture of a predetermined plane section of an organ or part of the body [FIGURE 1.22]. The video images can be preserved on film and examined one "slice" at a time to create a three-dimensional picture in the physician's mind. Although computers that produce three-dimensional images are still in their developmental stages, some dramatic three-dimensional images can be obtained [FIGURE 1.23].

A CAT scan helps physicians to detect blood clots, tumors, and other physical damage, but it does not reveal how an organ is functioning. A ***PET scan,*** in contrast, reveals the metabolic state (level of chemical activity) of the organ by indicating the rate at which its tissues consume injected biochemicals such as glucose (also called blood sugar). PET scans have been used to detect cancers (since some malignant tumors consume glucose at a faster rate than healthy tissue), and to study the metabolic patterns of the cardiovascular system. But their most dramatic application has been the examination of the brain.

FIGURE 1.23 THREE-DIMENSIONAL CAT SCAN IMAGE

The image shows a twisted and compressed spinal bone (vertebra) (center) following a motorcycle accident. Note the ruptured disks between the vertebrae.

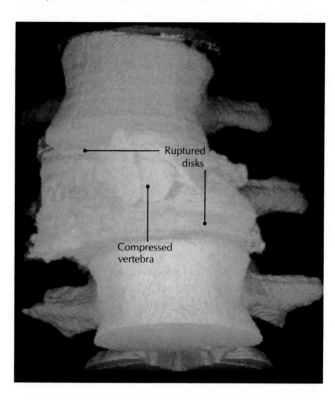

Ruptured disks

Compressed vertebra

FIGURE 1.22 CAT SCAN SHOWING AN OPTIC NERVE LESION

Optic nerve lesion

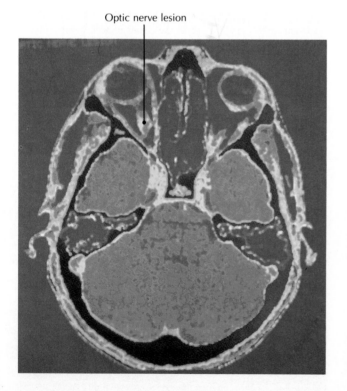

FIGURE 1.24 PATIENT UNDERGOING A PET SCAN

The PET scanner is fringed with gamma-ray detectors.

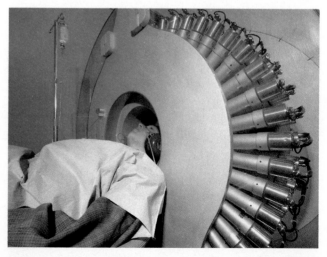

The glucose that is administered to the patient contains a radioactive tracer. The "tagged" glucose travels to the brain, and is consumed by brain cells. Once inside the brain, the radioactive tracer begins to decay, emitting positively charged subatomic particles called *positrons*. When the positrons meet negatively charged subatomic particles called electrons in the brain cells, the positrons emit energy in the form of gamma rays. These rays are detected by scanning devices that surround the patient's head [FIGURE 1.24]. The detectors relay the gamma-ray information to a computer, which reconstructs color-coded, cross-sectional images that indicate the rate of metabolic activity [FIGURE 1.25]. The processing of each image takes about 5 minutes.

Normal brain activity can be seen in reaction to light or movement; the extent of damage due to a stroke or epileptic seizure can be determined; and mental disorders, such as schizophrenia, manic-depression, and senile dementia, can be diagnosed. Early studies of depression indicate a decreased brain metabolism, especially on the left side of the brain, where language and analytic functions are processed.

FIGURE 1.26 DYNAMIC SPATIAL RECONSTRUCTOR

A dynamic spatial reconstructor (DSR) produced these oblique views of branching arteries within the lung.

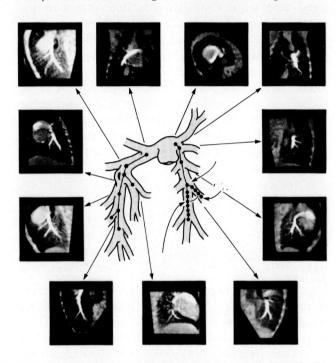

FIGURE 1.25 BRAIN PET SCAN

[A] A PET scan of a patient with senile dementia (Alzheimer's disease) shows a decreased rate of metabolism (blue and green on the color scale). [B] A PET scan of normal individual.

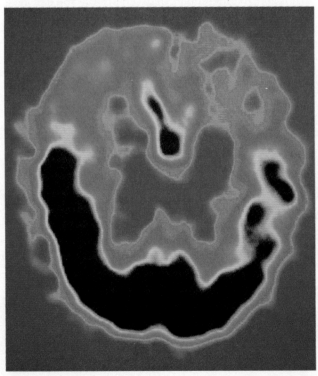

[A]

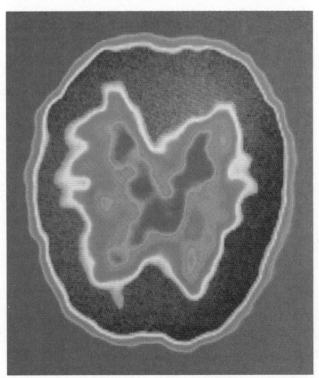

[B]

Dynamic Spatial Reconstructor (DSR)

The *dynamic spatial reconstructor,* or **DSR,** produces three-dimensional computer-generated images of the active brain. The DSR has 28 revolving x-ray machines, each firing 60 times a second. What sets DSR apart from other new exploratory techniques is that it can look under, over, around, and *inside* an organ [FIGURE 1.26]. With a DSR, it is possible to view the flow of blood through an organ, and then replay it in slow motion or stop action. By watching the flow of blood through the brain, a physician may be able to predict an impending stroke, providing a giant stride toward the *prevention* of disease instead of its treatment.

Magnetic Resonance Imaging (MRI)

Another tool for diagnosing disorders is *magnetic resonance imaging,* or **MRI.** (This technique is sometimes called *nuclear magnetic resonance,* or *NMR.*) Like the CAT scanner, MRI produces basically anatomical images, but the images are exceptionally clear, and show the difference between healthy and diseased tissue.

To produce an image with MRI, the patient lies on a nonmagnetic stretcher and is slipped inside the magnetic scanner, where the body is enveloped in a magnetic field. The strong magnetic field causes positively charged subatomic particles called protons in the body's many hydrogen atoms to line up in rows parallel to the magnetic charge. Then radio waves of a specific frequency are introduced at right angles to the magnetic field, causing the protons to "wobble" out of line for a fraction of a second. When the radio signal is turned off, the protons return to their upright position. The lag between the two upright positions is called the *relaxation time.* It is different for each type of tissue, so it is possible to tell which part of the body the signals come from. (Cancerous tissue, for example, has a longer relaxation time than normal tissue.) Each relaxation time is read and interpreted by a computer, which then assigns to each part of the object a different shade of gray or a different color in the computer-generated image [FIGURE 1.27].

Magnetic resonance imaging is very expensive, but it has several advantages over CAT scans and other x-ray techniques:

1 It does not usually use potentially dangerous x-rays, dyes, or radioactive tracers.

2 It does not register bone, which typically obscures the soft tissue in x-rays.

FIGURE 1.27 MAGNETIC RESONANCE IMAGING SCAN

A color-enhanced MRI scan of a 7-month-old child shows a malignant tumor between the right kidney and spinal column pressing on the spinal cord.

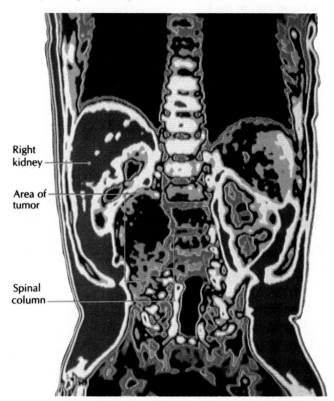

Right kidney

Area of tumor

Spinal column

3 It is able to differentiate between healthy and damaged myelin sheaths (which surround nerves), making possible an early diagnosis of multiple sclerosis, a disease that exhibits damaged myelin sheaths.

4 It can reveal atherosclerosis early by showing the build-up of fatty deposits (plaque) within blood vessels.

5 It can show some brain tumors that are invisible on CAT scans.

Digital Subtraction Angiography (DSA)

Digital subtraction angiography (Gr. *angeion,* vessel), or **DSA,** is a technique that produces three-dimensional pictures of blood vessels. It is especially effective in revealing blocked vessels [FIGURE 1.28]. After an image of the heart is produced by a digital x-ray scanner, blood vessels leading to the large arteries of the heart are injected with a substance that shows up in x-rays. A second x-ray picture of the heart shows the opaque substance flowing through the arteries. A computer then subtracts the first image from the second, producing a final image that indicates any blocked vessels. Photographs are made

FIGURE 1.28 DIGITAL SUBTRACTION ANGIOGRAPHY

A picture of the heart and its left coronary artery made by digital subtraction angiography reveals a constriction (arrow). The constriction has blocked about 60 percent of the blood flow to the lower heart. The appearance of the heart was enhanced by an opaque dye.

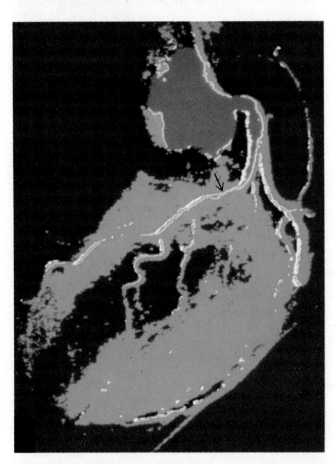

from several angles, and the computer converts the images into a digital code and analyzes the extent of blockage in a vessel by comparing the various images. The computer also measures the rate of blood flow to the heart. With this information, physicians are often able to predict an imminent heart attack, and take appropriate action to prevent it.

Ultrasound (Sonography)

The noninvasive exploratory technique known as *ultrasound* (also called *sonography*) sends pulses of ultrahigh-frequency sound waves into designated body cavities. From the echoes that result, pictures can be constructed of the tissue or organ under investigation. The latest ultrasound systems include a computer that presents moving pictures on a television screen. Recently improved ultrasound images, such as the one of a fetus shown in FIGURE 1.29, not only allow physicians to have a clear

FIGURE 1.29 ULTRASOUND

An ultrasound image of the head and neck of a 19-week-old fetus.

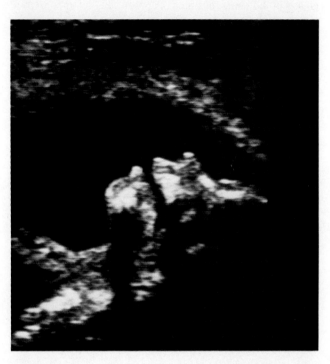

look at a developing fetus, they also permit a differentiation of tissues for the accurate diagnosis of heart disease and cancer.

Ultrasound offers an apparently harmless alternative to amniocentesis,* which presents some known risk to the fetus. Ultrasound can also be used in connection with amniocentesis, however, to help the physician guide the needle used in amniocentesis away from the placenta and fetus. Some practical applications of ultrasound include the detection of ectopic pregnancy (the implantation of a fertilized egg away from its normal location in the uterus) and multiple pregnancy, the assessment of fetal growth and development, the disclosure of some birth defects and probable miscarriages, and the assistance of physicians during fetal surgery.

Thermography

Thermography ("heat writing") is a technique that reveals chemical reactions that are taking place within the body based on heat changes in the skin. The thermographic camera can detect these normally invisible infra-

---

*Diagnostic ultrasound may be capable of transforming normal cells into a precancerous state. Investigations on this subject are continuing. Amniocentesis is a technique that obtains representative cells of a developing fetus by sampling the fluid in which the fetus floats. It is used to test for genetic defects.

FIGURE 1.30 THERMOGRAM

This thermogram reveals cancer in the left breast of a woman (red areas) by indicating a marked temperature increase in the cancerous area.

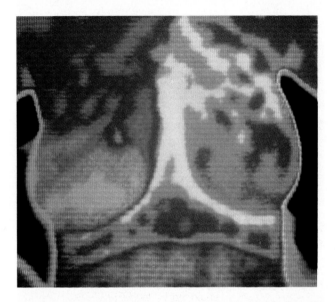

red radiations, convert them into electrical impulses, and record them as color-coded thermograms [FIGURE 1.30]. Some malignant tumors give off more heat than normal tissues do, and can be detected with thermography. In fact, thermography was first used in 1965 to detect breast cancer. Arthritis, blood circulation problems, and other disorders can also be detected with thermography.

ASK YOURSELF

1 What is an important diagnostic limitation of x-rays?

2 How does a PET scan provide different information than a CAT scan about the state of an organ?

3 How can the dynamic spatial reconstructor help to prevent disorders?

4 What are some of the advantages of magnetic resonance imaging over CAT scans?

5 What is the significance of the word *subtraction* in the term digital subtraction angiography?

6 How is ultrasound helpful in detecting cancer?

CHAPTER SUMMARY

What Is Anatomy? (p. 3)

1 The study of **anatomy** deals with the *structure* of the body; **physiology** explains the *functions* of the parts of the body.

2 Some subdivisions of anatomy are *regional anatomy, systemic anatomy, embryological anatomy, developmental anatomy,* and *radiographic anatomy.* Any branch of anatomy that can be studied without a microscope is called *gross anatomy; microscopic anatomy* requires the use of a microscope.

From Atom to Organism: Structural Levels of the Body (p. 7)

1 At its simplest level, the body is composed of **atoms,** the basic units of all matter. When two or more atoms combine, they form a **molecule.** If a molecule is composed of more than one element, it is a **compound.**

2 *Cells* are the smallest independent units of life. Some of the basic functions of cells are metabolism, irritability, growth, and reproduction.

3 *Tissues* are composed of many similar cells that perform a specific function. Tissues are classified into four types: *epithelial, connective, muscle,* and *nervous.*

4 An **organ** is an integrated collection of two or more kinds of tissues that combine to perform a specific function.

5 A **system** is a group of organs that work together to perform a major body function. All the body systems are specialized within themselves and coordinated with each other to produce an **organism.**

Body Systems (p. 9)

1 The **integumentary system** consists of the skin and all the structures derived from it. The main function of the skin is to protect the internal organs from the external environment.

2 The **skeletal system** consists of bones, certain cartilages and membranes, and joints. It supports the body, protects the organs, enables the body to move, manufactures blood cells in the marrow

within the bone, and stores calcium and phosphorus.

3 The **muscular system** consists of muscles and tendons. It allows for movement and generates a large amount of body heat.

4 The **nervous system** consists of the *central nervous system* and the *peripheral nervous system;* it also includes special sensory organs. The nervous system is the body's main control and regulatory system.

5 The **endocrine system** comprises a group of ductless glands that secrete *hormones.* Hormones regulate chemical reactions within cells *(metabolism),* growth and development, stress and injury responses, reproduction, and many other critical functions.

6 The **cardiovascular system** consists of the heart, blood, and blood vessels. An important function of the cardiovascular system is to transport oxygen and other necessary substances throughout the body and to transport wastes to the lungs and kidneys for removal.

7 The *lymphatic system* returns excess fluid and proteins to the blood, and helps protect the body from disease.

8 The *respiratory system* accomplishes the process of breathing and also provides a mechanism for the exchange of gases between blood and air.

9 The *digestive system* breaks down food physically and chemically into molecules small enough to be absorbed from the small intestine into the bloodstream. Solid, undigested wastes are removed through the anus.

10 The *urinary system* eliminates wastes produced by cells and regulates fluid balance in the body.

11 The *reproductive systems* have organs that produce specialized reproductive cells that make it possible to maintain the human species.

Anatomical Terminology (p. 16)

1 When the body is in the *anatomical position,* it is erect and facing forward, with the feet together, arms hanging at the sides, and palms facing forward.

2 Some basic pairs of anatomical terms used to describe the relative position of parts of the body include *superior/inferior, anterior/posterior, medial/lateral, proximal/distal, external/internal,* and *superficial/deep.*

3 The main *regions* of the body are the *axial* part, consisting of the head, neck, thorax, abdomen, and pelvis; and the *appendicular* part, which includes the upper and lower extremities. The abdominal region is divided into nine subregions.

4 The body, and parts of the body, may be divided by imaginary *planes,* including the *midsagittal, sagittal, frontal,* and *transverse.*

5 The *body cavities* house the internal organs. The two main body cavities are the *ventral* and the *dorsal.* The dorsal cavity contains the *cranial and spinal cavities,* and the ventral body cavity is separated by the diaphragm into the superior *thoracic* and inferior *abdominopelvic cavities.*

6 *Membranes* are layers of epithelial and connective tissue that line the body cavities and cover or separate certain regions, structures, and organs. The three main types of membranes are *mucous, serous,* and *synovial.*

New Ways of Exploring the Body (p. 23)

1 Several noninvasive techniques are replacing x-rays and major exploratory surgery as effective diagnostic tools.

2 *Computer-assisted tomography,* or *CAT scanning,* combines x-rays with computer technology to produce cross-sectional views of internal body structures.

3 *Positron-emission tomography,* or *PET scanning,* can reveal the metabolic state of an organ by measuring the rate at which tissues consume chemical substances such as glucose.

4 The *dynamic spatial reconstructor (DSR)* produces three-dimensional computer-generated images to reveal the flow of blood through the brain.

5 *Magnetic resonance imaging (MRI)* envelops the patient in a strong magnetic field, and measures differences between healthy and unhealthy tissue.

6 *Digital subtraction angiography (DSA)* uses a digital computer to produce three-dimensional pictures that effectively show blockages of blood vessels. The computer also measures the rate of blood flow to the heart.

7 *Ultrasound* is an apparently harmless exploratory technique that employs pulses of ultrahigh-frequency sound waves to produce images of the contents of body cavities.

8 *Thermography* reveals differences in metabolic activity within the body based on heat changes in the skin. It is useful in detecting cancer, arthritis, and circulatory problems.

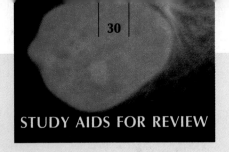

STUDY AIDS FOR REVIEW

UNDERSTANDING THE FACTS

1 What are anatomy and physiology, and how are they related?

2 What are some anatomical disciplines?

3 What does a body system do?

4 What is the anatomical position?

5 What are the meanings of superior, inferior, anterior, and posterior?

6 Define parietal and visceral.

7 What are the two main regions of the body?

8 What is a transverse plane? What is a frontal plane?

9 Name the two main subdivisions of the ventral cavity.

10 What is one major advantage of PET scanning over CAT scanning?

11 How does dynamic spatial reconstruction differ from CAT scanning?

UNDERSTANDING THE CONCEPTS

1 What would be a reason for the size variations of different types of body cells?

2 How are the different structural levels of the body related?

3 If a bullet entered the right side of the body just below the armpit and exited at the similar point on the left side, which cavities would be pierced (in order) and which organs probably damaged?

4 Explain why some noninvasive exploratory techniques can be used to predict heart attacks and other disorders.

SELF-QUIZ

Multiple-Choice Questions

1.1 The dorsal cavity contains the
a heart   c spinal cord   e b and c
b brain   d a and b

1.2 The abdominal cavity contains the
a heart and lungs
b reproductive organs and urinary bladder
c liver, spleen, and stomach
d urinary bladder and lungs
e testes and ovaries

1.3 Which of the following correctly describes the anatomical position?
a palms facing forward   d feet together
b body standing   e all of the above
c body facing forward

1.4 In anatomical terminology, "below" is referred to as
a bottom   c posterior to   e medial
b inferior to   d lateral

1.5 A plane that divides the body into anterior and posterior parts is a/an
a medial plane   d transverse plane
b coronal, or frontal, plane   e oblique plane
c sagittal plane

True–False Questions

1.6 Ultrasound is also known as sonography.

1.7 There are five general categories of tissues in the body.

1.8 Epithelial tissue is the most widely distributed of all body tissues.

1.9 The term *inferior* means "toward the back."

1.10 The liver is located in the right hypochondriac region of the abdomen.

Completion Exercises

1.11 _____ tissue may also be called supportive tissue.

1.12 Hair, nails, and sudoriferous (sweat) glands are part of the _____ system.

1.13 Nerves that go to and from the brain and spinal cord are part of the _____ nervous system.

1.14 The lymphatic system is made up of a network of thin-walled vessels that carry a clear fluid called _____.

1.15 The sole of the foot is called the _____ surface.

1.16 The plane that would be at right angles to both a sagittal and a frontal plane is a _____ plane.

1.17 The shoulder is _____ to the hand.

1.18 The kneecaps can be seen on the anterior, or _____, surface of the body.

1.19 The pleural cavities contain the _____.

1.20 The pericardial cavity contains the _____.

Matching

1.21 _____ proximal   a the soles of the feet
1.22 _____ lateral   b nearer the outside of the body
1.23 _____ superior
1.24 _____ ventral   c toward the head
1.25 _____ medial   d toward the back
1.26 _____ distal   e toward the side of the body
1.27 _____ dorsal   f toward the midline of the body
1.28 _____ inferior
1.29 _____ plantar   g nearer the trunk of the body
1.30 _____ superficial   h farther from the point of origin
   i toward the front of the body
   j toward the feet

2

Surface Anatomy

LEARNING OBJECTIVES

1 Describe some ways in which a knowledge of surface anatomy can be useful.

2 Describe diagnostic situations appropriate to auscultation, inspection, palpation, and percussion.

3 Name the major surface features of the head.

4 Distinguish between the cranial and facial skulls.

5 Identify the major surface features of the neck.

6 Identify the major surface features of the trunk.

7 Describe the major surface features of the upper and lower extremities.

The branch of gross anatomy that describes the structures that are visible or touchable on the body surface is called *surface anatomy*. Using a knowledge of surface anatomy, the positions of many organs and other internal structures can be described in relation to surface structures. The kidneys, spleen, and pancreas, for example, can all be located by using surface structures. Surface anatomy can also be used to diagnose some internal conditions. For example, it is possible to learn something about the condition of the lungs by *percussing* (tapping) the chest or to detect an infection by *palpating* (touching) the lymph nodes [TABLE 2.1].

Study of surface anatomy reveals body contours and anatomical structure, the degree of thinness or obesity, and the position of structures associated with bone fractures. Observations can be made of body movements and the gait during walking. The following sections show how such anatomical information is obtained from the head and neck; trunk (thorax, back, abdomen, pelvis); upper extremity (shoulder girdle, arm, forearm, wrist, hand); and lower extremity (pelvic girdle, thigh, leg, ankle, foot).

Every surface area and region of the body is identified by a specific anatomical term that describes precisely the location of anatomical structures. Some regions are named after underlying or adjacent bones (for example, sternal, frontal, temporal), while other regions are named for underlying muscles (sternocleidomastoid, deltoid, pectoral) or specialized structures (umbilical, oral, nasal).

The surface anatomy of the various regions of the human body is illustrated and described in FIGURES 2.1 through 2.9. The structures of the head are detailed in FIGURE 2.1 and those of the neck in FIGURE 2.2. The trunk and thorax, the back, and the abdomen and pelvis are shown in FIGURES 2.3 through 2.5, respectively. The surface anatomy of the upper extremity is shown in FIGURES 2.6 and 2.7 and that of the lower extremity in FIGURES 2.8 and 2.9.

TABLE 2.1 USE OF SURFACE ANATOMY IN DIAGNOSIS

Method	Description
Auscultation (L. *auscultare*, to listen to)	The diagnostic interpretation of the physical condition of various internal organs (especially the heart and lungs) by using a stethoscope to listen to the sounds arising from those organs. For example, the heartbeat, opening and closing of heart valves, and airflow into and out of the lungs can be heard with a stethoscope.
Inspection (L. *inspectare*, to look into)	The visual examination of the external body, noting such conditions as the shape of the chest, any asymmetry of the body, and the rate and depth of breathing.
Palpation (L. *palpare*, to touch)	The examination of the external body by touch for the purpose of diagnosing an internal condition. For example, palpation can reveal the texture of the skin and amount of underlying fat, help evaluate the sensitivity of a bruise or injury, and determine the pulse rate from the radial artery in the wrist.
Percussion (L. *percutere*, to strike hard)	A diagnostic technique for examining the external body (usually the thorax or abdomen) by tapping it sharply with the fingers and interpreting the resonance of the sound produced. For example, if a lung is filled with liquid it produces a dull note when tapped; an air-filled, normal lung produces a hollow sound.

FIGURE 2.1 HEAD

The **head,** or **cephalic region,** can be divided anatomically into (1) the *cranial skull,* which supports and protects the brain and its membranes, and (2) the *facial skull,* which forms the lower jaw (mandible, jawbone) and the framework for the cavities of the eyes, nose, mouth, and throat [FIGURE 2.1A]. The **skull** is usually defined as the skeleton of the head.

The **cranial skull** is divided into the *frontal region,* which includes the forehead; the *parietal region,* which is the crown or vertex of the skull; the *temporal region,* or side of the skull; and the *occipital region,* or base of the skull.

The **facial skull** is composed of four main regions: (1) the *auricular region* includes the external ears [FIGURE 2.1B]; (2) the *ocular region* includes the eyes and associated structures that protect the eyes [FIGURE 2.1C]; (3) the nasal region includes all the structures associated with the nose and nasal cavities [FIGURE 2.1D]; and (4) the *oral region* includes the mouth and associated structures, such as the tongue and salivary glands. Other surface structures of the head are shown in FIGURE 2.1D and E.

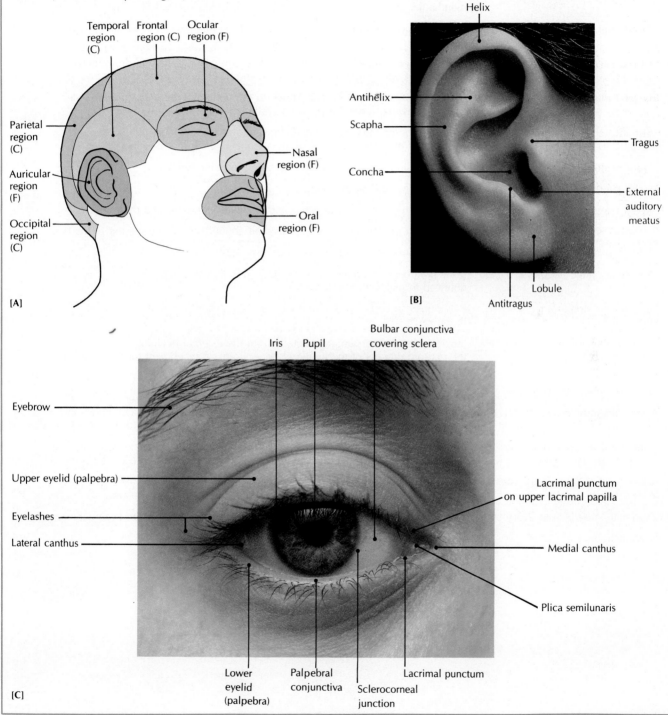

(Figure 2.1 continues on following page)

FIGURE 2.1 HEAD (Continued)

SOME SURFACE STRUCTURES OF THE HEAD

Structure	Description
External auditory meatus [FIGURE 2.1B, E]	Passage or opening of the external ear.
External occipital protuberance [FIGURE 2.1E]	Palpable protuberance at the posterior midline junction of the head and neck, just superior to the easily felt soft nuchal groove.
Facial artery	Pulsations can be felt on the jawbone (mandible), about 2.5 cm anterior to the mandibular angle; pressure on this artery helps reduce facial bleeding.
Frontal eminence [FIGURE 2.1D, E]	Slight rounded prominence on either side of the midline of the upper forehead, extending from the hairline to eyebrow.
Mandible [FIGURE 2.1E]	Jawbone; horseshoe-shaped skeleton of the lower jaw, extending from the mandibular fossa of one temporal bone to the mandibular fossa of the other.
Masseter muscle [FIGURE 2.1D, E]	Muscle extending from the cheekbone to the body and angle of the mandible; used in chewing; it can be palpated easily when the teeth are clenched.
Mastoid process of temporal bone [FIGURE 2.1E]	Projection of the temporal bone downward and forward from behind the ear; develops after birth from the pull of the sternocleidomastoid muscle.
Nasion [FIGURE 2.1D, E]	Depression in the midline where the nose meets the forehead.
Nucha [FIGURE 2.1E]	Nape (scruff, back) of the neck.
Orbital margin	Formed by the frontal, zygomatic, and maxillary bones around the eyes.
Parietal eminence [FIGURE 2.1E]	Structure on the lateral surface of the skull about 5 cm above the top of the ear.
Philtrum [FIGURE 2.1D]	Vertical groove extending between the nose and upper lip.
Superciliary ridge [FIGURE 2.1D, E]	Bony protuberance above each eye orbit, usually covered by eyebrows; more pronounced in males than in females.
Superficial temporal artery [FIGURE 2.1E]	Pulsations can be felt where artery crosses the zygomatic arch immediately in front of the ear.
Superior nuchal line [FIGURE 2.1E]	Slightly curved ridge running laterally from the external occipital protuberance to the mastoid process of the temporal bone; point of attachment for the trapezius and sternocleidomastoid muscles.
Symphysis menti [FIGURE 2.1D]	Midline union of the two halves of the mandible; forms the point of the chin.
Temporalis muscle [FIGURE 2.1E]	When the teeth are clenched, this muscle bulges on the lateral skull above the zygomatic arch.
Temporomandibular joint (TMJ) [FIGURE 2.1D, E]	Joint in front of the ear, where the head of the mandible articulates with the mandibular fossa of the temporal bone; the head of the mandible can be felt distinctly when the mouth is opened wide.
Vertex [FIGURE 2.1E]	Highest and most easily palpable portion of skull in the sagittal plane.
Zygomatic arch [FIGURE 2.1D, E]	Bony arch extending forward in front of the ear and ending in the zygomatic bone.
Zygomatic bone	Cheekbone; it forms the lateral third of the orbital margin.

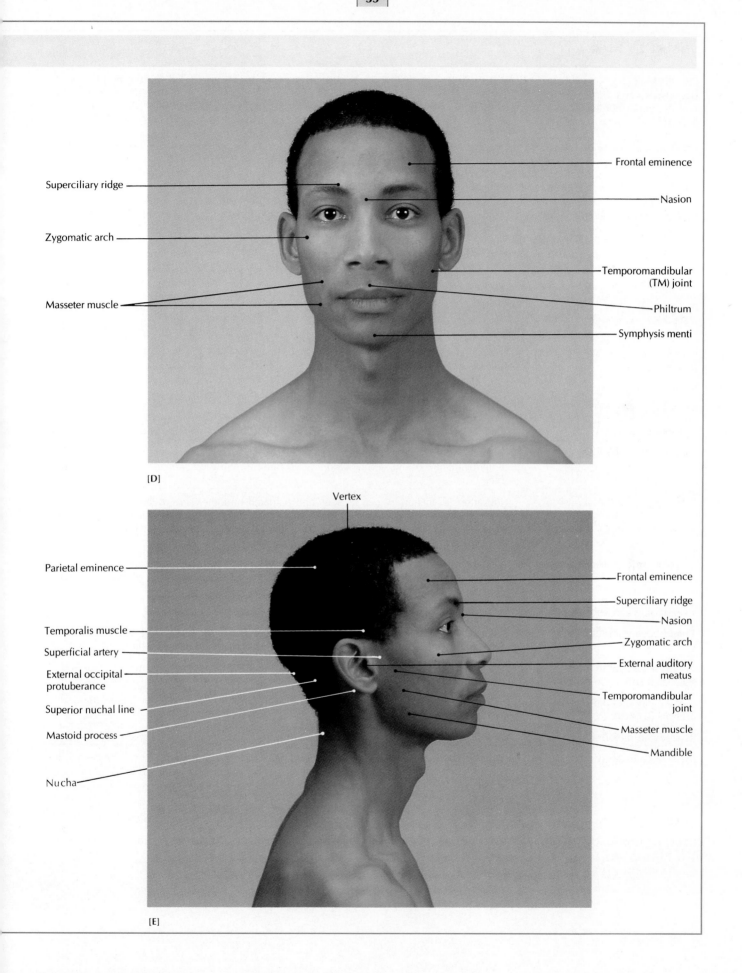

Superciliary ridge

Zygomatic arch

Masseter muscle

Frontal eminence

Nasion

Temporomandibular
(TM) joint

Philtrum

Symphysis menti

[D]

Vertex

Parietal eminence

Temporalis muscle

Superficial artery

External occipital
protuberance

Superior nuchal line

Mastoid process

Nucha

Frontal eminence

Superciliary ridge

Nasion

Zygomatic arch

External auditory
meatus

Temporomandibular
joint

Masseter muscle

Mandible

[E]

FIGURE 2.2 NECK

The **neck** (*nucha*, NOO-kuh, or *collum*) is the region of the body between the lower margin of the mandible (jawbone) and the upper border of the clavicle (collarbone) [FIGURE 2.2A, B].

Anterior and posterior triangles
The neck region is divided into the anterior and posterior triangles by the sternocleidomastoid muscle, which is palpable over its entire length as it runs obliquely from the mastoid process under the ear to the front of the clavicle.

The **anterior triangle** is bounded by the body of the mandible, the sternocleidomastoid muscle, and the midline of the body. It is subdivided into the digastric (submandibular), carotid, muscular, and submental (suprahyoid) triangles.

The **posterior triangle** is bounded in front by the sternocleidomastoid muscle, behind by the anterior border of the trapezius muscle, and inferiorly by the clavicle.

Lymph nodes
The lymph nodes in the neck are usually not palpable, but they enlarge during an infection and become palpable beyond the anterior border of the sternocleidomastoid muscle. Also, infections in the head, neck, upper extremity, or chest can cause enlargement of the lymph nodes above the clavicle. These enlarged nodes can be palpated on the posterior border of the sternocleidomastoid muscle near its attachment to the clavicle.

SOME SURFACE STRUCTURES OF THE NECK

Structure	Description
Anterior triangle [FIGURE 2.2A, B]	Region bounded by the median line of the neck from the chin to the manubrium, the anterior margin of the sternocleidomastoid muscle, and the lower margin of the mandible.
Common carotid artery	Pulsations of this artery can be seen and palpated on either side of the larynx (voice box), slightly superior to it. The artery descends deep into the sternocleidomastoid muscle; it divides into external and internal carotid arteries at the superior border of the thyroid cartilage.
Cricoid cartilage [FIGURE 2.2A]	Lies inferior to the thyroid cartilage at the junction of the larynx with the trachea; at the level of vertebra C6. Its anterior arch can be palpated at the midline. Marks the upper limit of the trachea.
External jugular vein [FIGURE 2.2A]	Extends caudally from behind the angle of the mandible to the middle of the clavicle.
Hyoid bone [FIGURE 2.2A]	Lies between the inferior border of the mandible and the upper border of the thyroid cartilage; can be felt as the first bony structure below the midline of the chin.
Posterior (lateral) triangle [FIGURE 2.2A]	Formed by the trapezius muscle posteriorly, the sternocleidomastoid muscle anteriorly, and the collarbone at its base.
Sternocleidomastoid muscle [FIGURE 2.2A, B]	Can be felt throughout its length as it passes from the sternum and clavicle to the mastoid process.
Thyrohyoid membrane	Membrane extending between hyoid bone and thyroid cartilage (Adam's apple).
Thyroid cartilage [FIGURE 2.2A]	Adam's apple; triangular cartilage in the center of the neck region, 1 or 2 cm below the hyoid bone; it can be felt to rise and then lower during swallowing. The "Adam's apple" is actually only part of the thyroid cartilage, and is technically called the *laryngeal prominence*.
Trachea [FIGURE 2.2A]	Windpipe; cartilaginous tube extending from the base of the larynx to the lungs.
Trapezius muscle [FIGURE 2.2A, B]	Occupies the area between the base of the neck and the shoulder; can be felt by shrugging the shoulder.

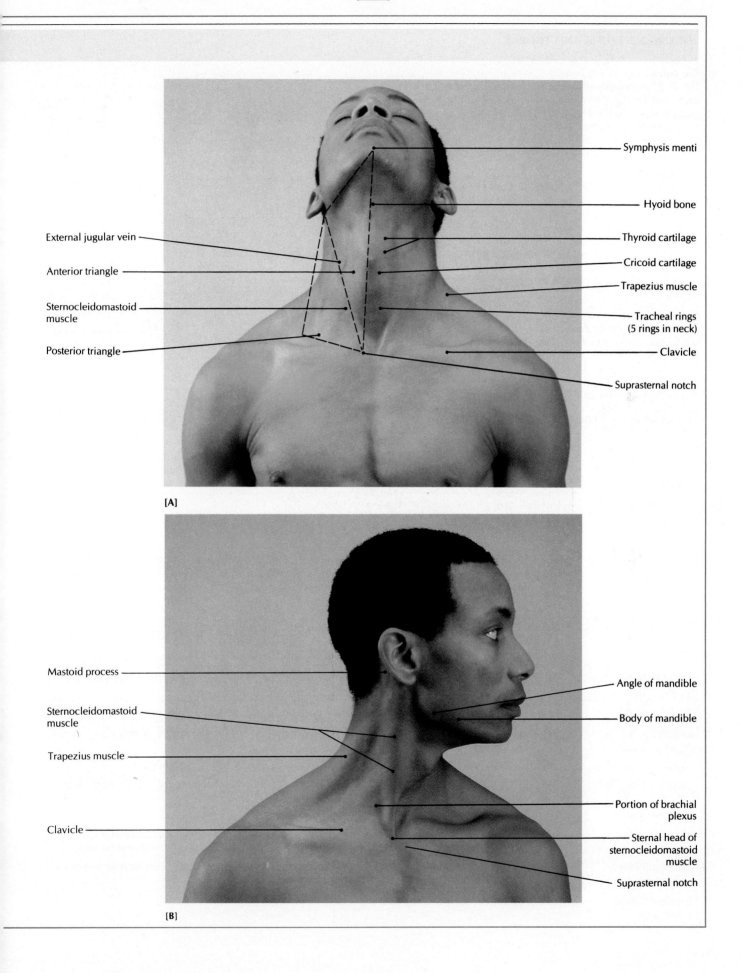

Symphysis menti

Hyoid bone

Thyroid cartilage

Cricoid cartilage

Trapezius muscle

Tracheal rings
(5 rings in neck)

Clavicle

Suprasternal notch

External jugular vein

Anterior triangle

Sternocleidomastoid
muscle

Posterior triangle

[A]

Mastoid process

Sternocleidomastoid
muscle

Trapezius muscle

Clavicle

Angle of mandible

Body of mandible

Portion of brachial
plexus

Sternal head of
sternocleidomastoid
muscle

Suprasternal notch

[B]

FIGURE 2.3 TRUNK AND THORAX

The trunk

The *trunk* is composed of the thorax (chest), back, abdomen, and pelvis. The surface anatomy of these body regions is described in FIGURES 2.3 through 2.5.

Several lines of orientation are used in describing the positions of the surface structures of the thorax or of other body parts. These lines are shown in FIGURE 2.3A.

The thorax

The bony *thorax* is composed of 12 thoracic vertebrae, 12 pairs of ribs, and the sternum (breastbone) [FIGURE 2.3B, C]. Except for rib 1, which lies deep to the clavicle (collarbone), the ribs can be easily palpated on the side of the thorax inferior to the axilla (armpit). Rib 12 is sometimes too short to be palpable. In palpating the rib cage, note that a typical rib slopes downward, while its attached cartilage turns upward. A rib and its cartilage make up a *costa* (L. rib).

Apex heartbeat

The heartbeat can be heard best in the lower left region of the thorax, called the *apex*. The apex beat is usually heard in the fifth intercostal space (the space between ribs 5 and 6), which is about 9 cm from the midline (just below the nipple in males). If the apex is well to the left of the midclavicular line, it usually indicates an enlarged heart.

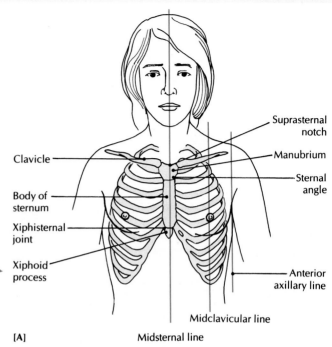

[A]

SOME SURFACE STRUCTURES OF THE THORAX

Structure	Description
Anterior axillary fold [FIGURE 2.3B]	Formed by the lower border of the pectoralis major muscle; anterior boundary of axilla (armpit).
Areola [FIGURE 2.3B]	The dark ring surrounding the nipple of a breast.
Clavicle [FIGURE 2.3B, C]	Collarbone; part of the upper extremity.
Costal margin [FIGURE 2.3B]	Lower boundary of the thorax; formed by cartilages of ribs 7 to 10 and ends of cartilages 11 and 12, with lowest part formed by rib 10 at the level of vertebra L3. The liver is located under the right costal margin of the rib cage, and the stomach is under the left costal margin.
Deltopectoral triangle [FIGURE 2.3B]	Triangle formed by the deltoid and pectoralis muscles.
Nipples [FIGURE 2.3B]	In the male, nipples usually lie in the fourth intercostal space (the space between ribs), about 10 cm from the midline; the position in the female is variable, depending on breast size and shape.
Pectoralis major muscle [FIGURE 2.3B]	Major muscle of thorax, extending across entire upper chest; flexes, extends arm, rotates and adducts arm medially.
Ribs [FIGURE 2.3A, C]	Bony cage encircling thorax; all ribs articulate posteriorly with vertebral column.
Sternum [FIGURE 2.3A, C]	Breastbone; palpable, flat bone forms anterior aspect of the rib cage in midline of thorax.
Body	Middle portion of sternum, articulating with the costal cartilages of ribs 2 through 10.
Suprasternal notch	Notch in superior margin of sternum (manubrium); easily palpable between the medial ends of the clavicle at the midline.
Manubrium	Superior portion of sternum, articulating with clavicle and rib 1.
Sternal angle	Angle formed by the joint between the manubrium and body of sternum (manubriosternal joint); site of attachment for rib 2, making it a reliable landmark for counting and identifying the ribs.
Xiphoid process	Inferior portion of sternum that attaches only to the body of the sternum; lies just inferior to the seventh costal cartilage, usually aligning with vertebrae T10 and 11.
Xiphisternal joint [FIGURE 2.3A]	Joint between the body and xiphoid process of sternum.

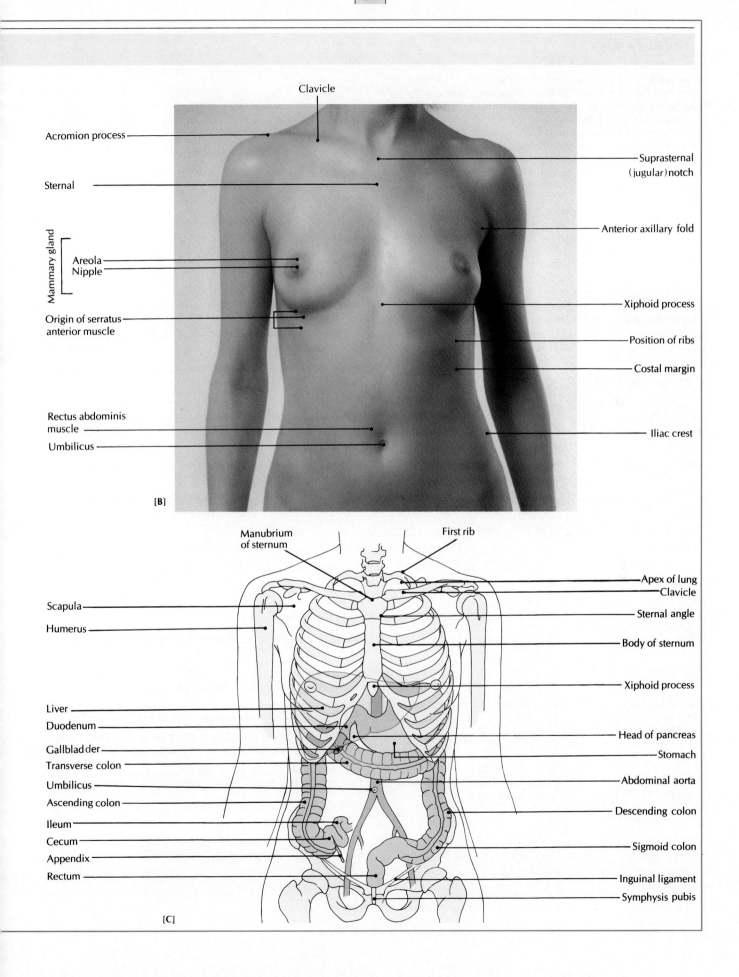

Clavicle

Acromion process

Sternal

Mammary gland
Areola
Nipple

Origin of serratus anterior muscle

Rectus abdominis muscle

Umbilicus

[B]

Suprasternal (jugular) notch

Anterior axillary fold

Xiphoid process

Position of ribs

Costal margin

Iliac crest

Manubrium of sternum

First rib

Scapula

Humerus

Liver

Duodenum

Gallbladder

Transverse colon

Umbilicus

Ascending colon

Ileum

Cecum

Appendix

Rectum

[C]

Apex of lung
Clavicle

Sternal angle

Body of sternum

Xiphoid process

Head of pancreas

Stomach

Abdominal aorta

Descending colon

Sigmoid colon

Inguinal ligament

Symphysis pubis

FIGURE 2.4 BACK

The **back** is the posterior surface of the trunk. It extends from the top of the skull to the inferior tip of the coccyx (tailbone) and includes the posterior aspect of the neck.

When palpating the back with the flat of the hand, normal muscles feel firm to the touch. A spastic muscle feels harder than normal. Because it is contracted, it pulls on the vertebral column, producing a concavity near the spine. Observation and palpation of the back can reveal abnormal curves in the spine and any asymmetry between the two sides of the body.

The *triangle of auscultation* is a site where the superficial back muscles are absent. As a result, physicians use this space to monitor heart and breathing sounds with a stethoscope because the sounds are not muffled by overlying muscles.

To avoid injury to the spinal cord, spinal taps to obtain cerebrospinal fluid surrounding the spinal cord are usually made between the third and fourth lumbar vertebrae, well below the ending of the spinal cord.

SOME SURFACE STRUCTURES OF THE BACK

Structure	Description
Erector spinae muscles	Flank the furrow down the middle of the back over the spinal column; move the spinal column, keep the spine erect when sitting or standing.
Infraspinatus muscle	Extends from the upper arm (humerus) to the shoulder blade (scapula); stabilizes shoulder joint, rotates arm laterally.
Latissimus dorsi muscle	Widest back muscle, extending from the spines of lower thoracic, lumbar, and sacral vertebrae, iliac crest, and ribs 7 to 12 to the proximal end of the humerus close to the shoulder joint; helps move arm, scapula.
Posterior axillary fold	Formed by the junction of the latissimus dorsi and teres major muscles along the side of the trunk; posterior boundary of axilla.
Scapula	Shoulder blade; part of the upper extremity.
Teres major muscle	Located inferior to infraspinatus muscle, extending from the inferior angle of the scapula to the proximal end of the humerus; stabilizes and moves upper arm.
Trapezius muscle	Triangular muscle extending from the cervical and thoracic vertebrae to the lateral third of the clavicle and the spine of the scapula.
Vertebrae	Bones that make up the spine; have bony protuberances called spinous processes that extend posteriorly along the length of the spinal column; most of these can be palpated.
Cervical	First palpable spinous process is that of C7 (vertebra prominens) at the base of the neck; other cervical processes (C1 to 6) cannot be palpated because they are embedded in muscle and tendons.
Thoracic	All 12 thoracic vertebrae join with ribs; the spinous process of T3 is at the level of the spine of the scapula, that of T7 is at the level of the inferior angle of the scapula. T1 has the most prominent spinous process.
Lumbar	Spinous process of L4 is at the highest point of the iliac crest; lumbar vertebrae are situated in the "small of the back."
Vertebra prominens	Palpable spinous process of the C7 vertebra.

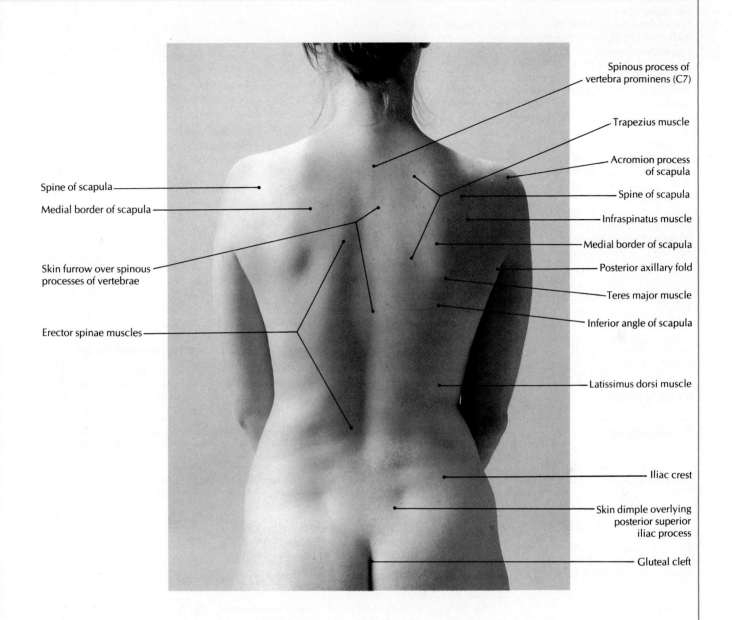

Spinous process of
vertebra prominens (C7)

Trapezius muscle

Acromion process
of scapula

Spine of scapula

Spine of scapula

Infraspinatus muscle

Medial border of scapula

Medial border of scapula

Posterior axillary fold

Skin furrow over spinous
processes of vertebrae

Teres major muscle

Inferior angle of scapula

Erector spinae muscles

Latissimus dorsi muscle

Iliac crest

Skin dimple overlying
posterior superior
iliac process

Gluteal cleft

FIGURE 2.5 ABDOMEN AND PELVIS

The abdomen

The **abdomen** (AB-doh-mehn; L. belly) is the part of the body that lies below the diaphragm in the anterior thorax and above the superior margin of the pelvis [FIGURE 2.5A]. The shape and contours of the abdominal wall result largely from the musculature, but the wall can be modified by accumulations of subcutaneous fat. The subdivisions and quadrants of the abdominal region are illustrated in FIGURE 1.17.

The pelvis

The **pelvis** is the region of the trunk inferior to the abdomen [FIGURE 2.5B]. The **pelvic girdle** is formed by the fused hipbones (ilium, ischium, pubis, sacrum, and coccyx). The pelvic girdle transmits the weight of the head, trunk, and upper limbs to the lower limbs, and it supports and protects the internal organs. Because the hipbones are surrounded by massive muscles, only a few bony sites are palpable.

SOME SURFACE STRUCTURES OF THE ABDOMEN AND PELVIS

Structure	Description
Anterior superior iliac spine [FIGURE 2.5A]	Located at the anterior margin of the iliac crest, at the upper lateral end of the fold of the groin.
External oblique muscle [FIGURE 2.5A]	Lateral to the rectus abdominis muscle, inferior to the serratus anterior muscle, and overlapped posteriorly by the latissimus dorsi muscle; helps hold in abdominal wall, flexes the trunk, and rotates the vertebral column laterally.
Iliac crest [FIGURE 2.5A, B]	Superior margin of ilium of hipbone; readily palpable at the level of L4 vertebral spinous process. A contusion of the iliac crest or the region immediately above it results in a *hip pointer*.
Linea alba [FIGURE 2.5A]	Tendinous ridge of the midline groove between the rectus abdominis muscles, extending from the xiphoid process to the symphysis pubis; interrupted by the navel, becoming thinner thereafter.
McBurney's point [FIGURE 2.5A]	Usual site for surgical access to the appendix during an appendectomy; located about 5 cm along a line from the anterior superior iliac spine to the navel. Pressure on this point evokes tenderness in acute appendicitis.
Mons pubis [FIGURE 2.5B]	Rounded fleshy prominence over the symphysis pubis.
Rectus abdominis muscle [FIGURE 2.5A]	Powerful muscle on either side of the midline (midsternal line), extending from the lower rib cage to the pubis; flexes lumbar vertebral column, depresses rib cage, helps stabilize pelvis during walking and running.
Symphysis pubis [FIGURE 2.5B]	Joint between the bodies of the anterior pubic bones; palpated as a bony protuberance at the midline, superior to the external genitals. When the urinary bladder is full, its superior wall can be palpated through the anterior abdominal wall above the symphysis pubis.
Tendinous inscription [FIGURE 2.5A]	Three fibrous bands running across the rectus abdominis muscles, forming transverse furrows in a rippling effect; the topmost is located just below the xiphoid process, the lowermost at the umbilical level, and the middle one in between.
Umbilicus [FIGURE 2.5A, B]	Navel; located at the midline, usually at the level of the intervertebral disk between vertebrae L3 and L4; site of attachment of the umbilical cord in the fetus.

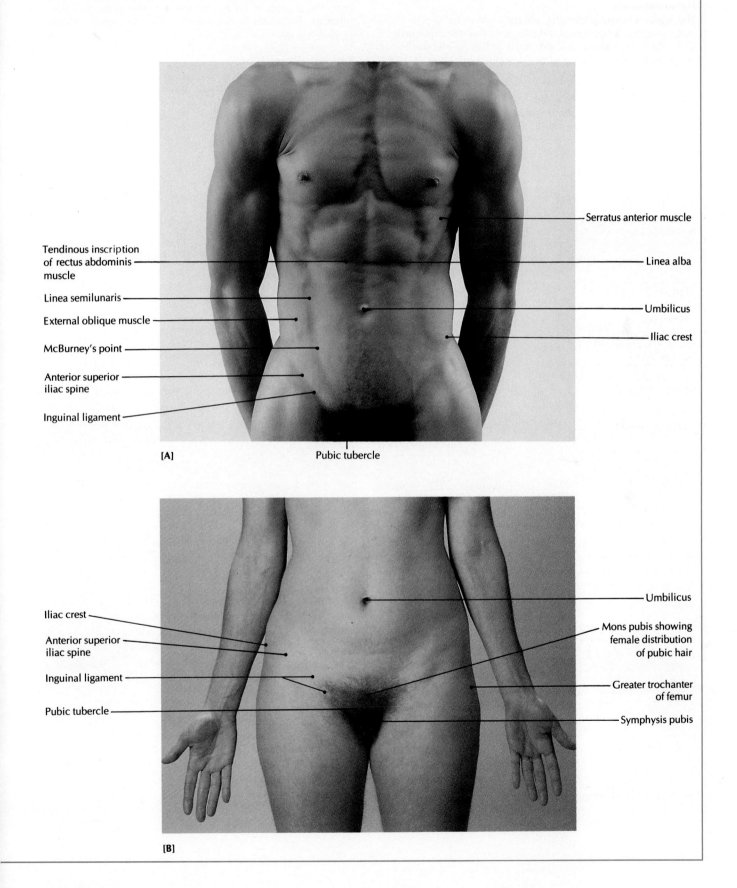

Tendinous inscription of rectus abdominis muscle

Linea semilunaris

External oblique muscle

McBurney's point

Anterior superior iliac spine

Inguinal ligament

Serratus anterior muscle

Linea alba

Umbilicus

Iliac crest

[A]

Pubic tubercle

Iliac crest

Anterior superior iliac spine

Inguinal ligament

Pubic tubercle

Umbilicus

Mons pubis showing female distribution of pubic hair

Greater trochanter of femur

Symphysis pubis

[B]

FIGURE 2.6 PECTORAL GIRDLE, AXILLA, ARM, AND ELBOW

Upper extremity

The *upper extremity* includes the pectoral girdle, axilla, arm, forearm, wrist, hands, and fingers. The pectoral girdle, axilla, arm, and elbow are described in this figure, while the forearm, wrist, and hand are described in FIGURE 2.7.

Pectoral girdle and axilla

The *pectoral girdle* is commonly called the *shoulder girdle,* while the *axilla* is the *armpit* [FIGURE 2.6A]. The bony framework of the shoulder girdle consists of the clavicle in front, the scapula in back, and the proximal portion of the humerus. The clavicle and scapula attach the upper extremity to the trunk.

Arm and elbow

The *arm,* commonly called the upper arm, extends from the shoulder to the elbow [FIGURE 2.6A, B]. The long bone of the arm, the *humerus,* extends from the shoulder joint to the *elbow,* which is the hinge-type joint between the humerus and forearm.

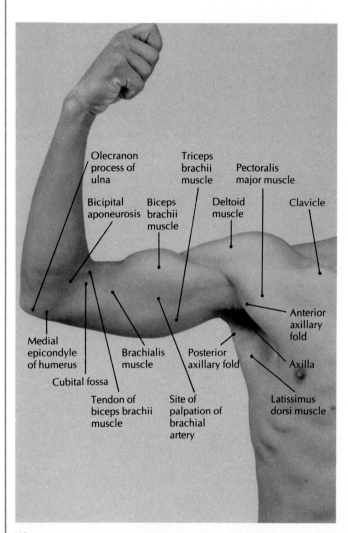

[A]

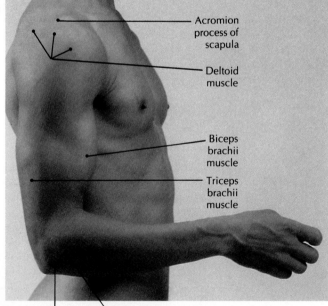

[B] Olecranon process of ulna | Lateral epicondyle of humerus

SOME SURFACE STRUCTURES OF THE PECTORAL GIRDLE, AXILLA, ARM, AND ELBOW

Structure	Description
Anterior axillary fold [FIGURE 2.6A]	Formed by the lower margin of the pectoralis major muscle.
Axilla [FIGURE 2.6A]	Armpit; pyramid-shaped space between the upper part of the arm and the side of the chest; contains many blood vessels, nerves, lymph nodes. The head of the humerus can be palpated through the floor of the axilla.
Axillary artery	Pulsations of this blood vessel can be felt high in the axilla; cords of the brachial plexus can be palpated around the artery.
Biceps brachii muscle [FIGURE 2.6A, B]	Located on the anterior surface of upper arm; helps flex forearm.
Bicipital aponeurosis [FIGURE 2.6A]	Tendinous band that can be palpated as it leaves the tendon of the biceps brachii muscle and extends to the medial side of the forearm; felt most easily if the elbow joint is flexed.
Brachial artery [FIGURE 2.6A]	A continuation of the axillary artery that begins at lower border of the teres major muscle and terminates by dividing into the ulnar and radial arteries opposite the neck of the radius; provides main arterial supply to the arm and elbow. Pulsations of this artery in anterior elbow are used for measuring blood pressure.
Clavicle [FIGURE 2.6A]	Collarbone; palpable S-shaped strut extending horizontally from the sternum (sterno-clavicular joint) medially to the acromion of the scapula (acromioclavicular joint) laterally; holds the arm away from the trunk. The most frequently fractured bone in the body; a fractured clavicle causes the shoulder to fall medially, forward, and downward.
Deltoid muscle [FIGURE 2.6A, B]	Triangular muscle that makes up rounded prominence covering the shoulder joint; involved in several arm movements, especially abduction.
Elbow	Joint between the upper arm and forearm.
Humerus	The arm bone located between the shoulder and elbow.
Lateral epicondyle	Prominent enlargement on the lateral side of the distal end of the humerus.
Medial epicondyle	Prominent enlargement on the medial side of the distal end of the humerus.
Median cubital vein	Can be seen and palpated across the cubital fossa in the anterior aspect of the elbow, connecting the lateral cephalic vein and the medial basilic vein; a common site for venous blood sampling and transfusions.
Scapula	Shoulder blade; flat, triangular bone lying on the posterior thoracic wall between ribs 2 and 7.
Acromion [FIGURE 2.6B]	Lateral end of the scapular spine, articulating with the clavicle; forms the point of the shoulder and is the attachment site for chest and arm muscles.
Coracoid process	Directed anteriorly, it projects over the glenoid fossa and can be palpated about 2.5 cm below and slightly medial to the outer end of the clavicle; serves as attachment for coracoclavicular ligament and several muscles.
Glenoid fossa (cavity)	Shallow cavity below the acromion; acts as a socket for the head of the humerus at the glenohumeral (shoulder) joint (not shown).
Spine	Horizontally oriented process, ending with the acromion.
Tendon of biceps brachii muscle [FIGURE 2.6A]	Can be palpated as it extends downward into the cubital fossa; felt most easily if the elbow joint is flexed.
Triceps brachii muscle [FIGURE 2.6A, B]	Located on posterior surface of upper arm; helps extend forearm.
Ulnar nerve	Can be palpated as a rounded cord behind the medial condyle of the humerus; commonly called the "funny bone" or "crazy bone" because a tingling feeling may be felt in the little finger if the nerve is stricken or palpated (not shown).

FIGURE 2.7 FOREARM, WRIST, AND HAND

The lower part of the upper extremity includes the forearm, wrist, and hand. The **forearm** is the region between the elbow and wrist [FIGURE 2.7A]. It contains two long bones: the *ulna* is the longer of the two, located on the side of the little finger; the *radius* is located lateral to the ulna, on the thumb side. The **wrist,** or **carpus,** is composed of eight short bones connected by ligaments [FIGURE 2.7A–C]. The **hand** is made up of the *metacarpal bones,* while the bones of the fingers and thumb are called *phalanges* (fuh-LAN-jeez) [FIGURE 2.7B–C].

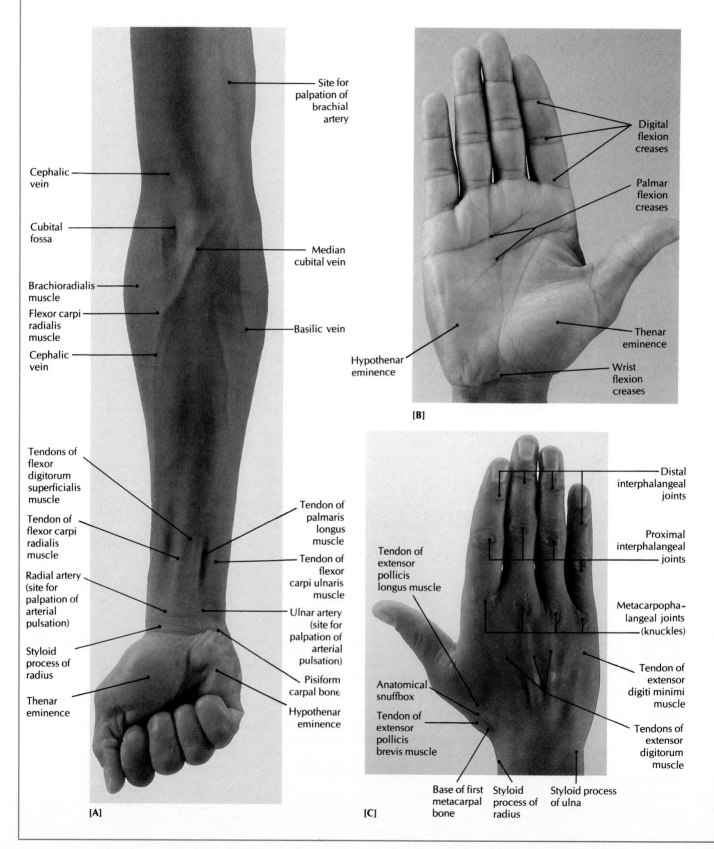

SOME SURFACE STRUCTURES OF THE FOREARM, WRIST, AND HAND

Structure	Description
"Anatomical snuffbox" [FIGURE 2.7C]	Depression at the base of the thumb between the tendons of the extensor pollicis longus and brevis muscles, so called because snuff was placed in this space before sniffing; the pulse of the radial artery can be palpated in the depression. Tenderness in the snuffbox may indicate a fracture of the scaphoid bone, which underlies the depression.
Brachioradialis muscle [FIGURE 2.7A]	Located at lateral aspect of forearm; helps flex forearm.
Cubital fossa [FIGURE 2.7A]	Triangular depression in anterior aspect of elbow; bound laterally by the brachioradialis muscle and medially by the pronator teres muscle; the tendon of the biceps brachii muscle can be felt as it passes downward into the fossa, and the bicipital aponeurosis can be felt as it leaves the tendon medially.
Digital flexion creases [FIGURE 2.7B]	Horizontal skin creases between joints of fingers on anterior (palmar) surface.
Dorsal venous arch	Branches of cephalic vein on the dorsal surface of hand.
Flexor carpi radialis muscle [FIGURE 2.7A]	Extends from the elbow to the pisiform and hamate metacarpal bones in the palm, in the center of the anterior surface of forearm; flexes and abducts wrist; contains pisiform bone.
Flexor carpi ulnaris muscle [FIGURE 2.7A]	Extends from the elbow to a metacarpal bone in the palm, on the medial aspect of the forearm; flexes and adducts wrist.
Hypothenar eminence [FIGURE 2.7A, B]	Muscular pad on the palm of the hand at the base of the little finger; formed by the muscles that move the little finger.
"Knuckles" [FIGURE 2.7C]	Bony knobs on the dorsal aspect of any joint of any finger when the fist is closed, especially the distal ends of metacarpal bones; the metacarpophalangeal joints.
Metacarpal bones	Bones of the hand (excluding the thumb and fingers; see Phalanges).
Palmar flexion creases [FIGURE 2.7B]	Skin creases of the palm.
Phalanges	Bones of the fingers and thumb.
Pisiform bone [FIGURE 2.7A]	Medial bone of the wrist; can be palpated distal to the styloid process of the ulna.
Radial artery [FIGURE 2.7A]	Pulsations can be felt just medial to the styloid process of the radius proximal to the wrist in the distal forearm; common location for taking the pulse.
Radius	One of the two bones of the forearm, located lateral to the ulna, on the thumb side.
Styloid process [FIGURE 2.7A]	Sharp, bony extension on distal end of radius.
Tendon of extensor digiti minimi muscle [FIGURE 2.7C]	Tendon on dorsal aspect of forearm and hand aligning with little finger; extensor digiti minimi muscle extends the finger.
Tendons of extensor digitorum muscle [FIGURE 2.7C]	Tendons on dorsal aspect of forearm in line with the three middle fingers (II, III, IV); extensor digitorum muscle extends the fingers and wrist.
Tendon of extensor pollicis brevis muscle [FIGURE 2.7C]	Tendon close to styloid process of radius on dorsal lateral surface of wrist at wrist line; extensor pollicis brevis muscle extends thumb.
Tendon of extensor pollicis longus muscle [FIGURE 2.7C]	Tendon close to styloid process of ulna on dorsal surface of wrist aligning with thumb; extensor pollicis longus muscle extends thumb.
Tendon of flexor carpi radialis muscle [FIGURE 2.7A]	Tendon on anterior surface of wrist and forearm, lateral to the tendon of palmaris longus muscle; flexor carpi radialis muscle flexes and abducts wrist.
Tendon of palmaris longus muscle [FIGURE 2.7A]	Tendon on anterior surface of wrist parallel to tendon of flexor carpi radialis muscle, but closer to ulna; palmaris longus muscle helps flex wrist. Superficial to median nerve; used to locate median nerve.
Thenar eminence [FIGURE 2.7A, B]	Muscular pad on the palmar surface at the base of the thumb; formed by the muscles that move the thumb.
Ulna	Longer of the two bones of the forearm, located on the side of the little finger.
Olecranon process	Prominent bony expansion of the ulna at the elbow.
Styloid process [FIGURE 2.7C]	Sharp, bony extension on distal end of ulna.
Wrist flexion creases [FIGURE 2.7B]	The proximal and distal transverse creases on the anterior surface of the wrist.

FIGURE 2.8 BUTTOCKS, THIGH, AND KNEE

Lower extremity
The **lower extremity** is composed of the pelvic girdle, buttocks, thigh, leg, ankle, foot, and toes. The buttocks, thigh, and knee are described in this figure, while the leg, ankle, and foot are described in FIGURE 2.9. The pelvic girdle is described in FIGURE 2.5.

Buttocks
The **buttocks** (BUTT-uhks), or gluteal region, are bounded superiorly by the iliac crest and inferiorly by the fold of the buttock at the top of the thigh [FIGURE 2.8A]. The buttocks are composed largely of the gluteal muscles and fat.

Thigh and knee
The **thigh,** or *upper leg,* extends between the hip and knee and contains the **femur,** the longest bone in the body [FIGURE 2.8B]. The **knee** is the joint between the thigh and the lower leg [FIGURE 2.8C], the most complex joint in the body.

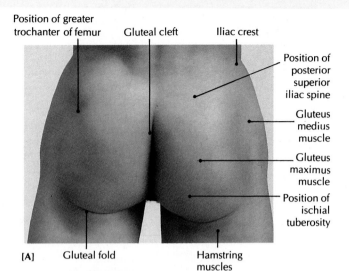

[A]

Position of greater trochanter of femur — Gluteal cleft — Iliac crest — Position of posterior superior iliac spine — Gluteus medius muscle — Gluteus maximus muscle — Position of ischial tuberosity — Gluteal fold — Hamstring muscles

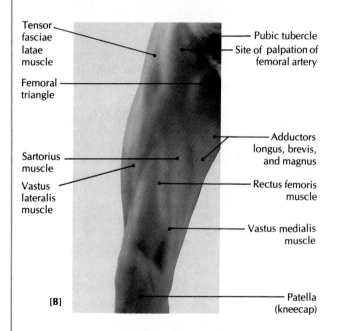

[B]

Tensor fasciae latae muscle — Femoral triangle — Sartorius muscle — Vastus lateralis muscle — Pubic tubercle — Site of palpation of femoral artery — Adductors longus, brevis, and magnus — Rectus femoris muscle — Vastus medialis muscle — Patella (kneecap)

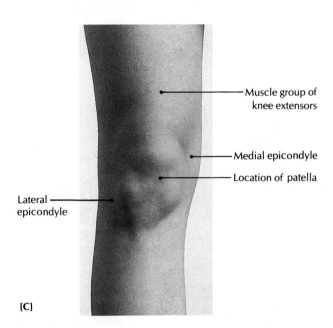

[C]

Muscle group of knee extensors — Medial epicondyle — Location of patella — Lateral epicondyle

SOME SURFACE STRUCTURES OF THE BUTTOCKS, THIGH, AND KNEE

Structure	Description
Adductor magnus muscle [FIGURE 2.8B]	Located on medial aspect of thigh below buttock; adducts and helps flex (its anterior fibers) and extend (its posterior fibers) thigh.
Coccyx	Tailbone; its inferior surface and lower tip can be felt about 2.5 cm above the anus.
Common peroneal nerve	Only palpable nerve of lower extremity; can be palpated where it passes over the lateral aspect of the neck of the fibula. A blow on the lateral side of the leg or a poorly fitted plaster cast can traumatize this nerve, resulting in impaired control of the foot.
Femoral artery [FIGURE 2.8B]	Pulsations can be felt in upper thigh within femoral triangle, just below the inguinal ligament midway between anterior superior iliac spine and pubic tubercle.
Femoral triangle [FIGURE 2.8B]	Depression below the fold of the groin; the base is formed by the inguinal ligament, the lateral border by the sartorius muscle, and the medial border by the adductor longus muscle.
Femur	Thighbone, located between the hip and knee; strongest, heaviest, longest bone in the body.
Gluteal cleft [FIGURE 2.8A]	The deep midline depression that separates the buttocks.

Structure	Description
Gluteal fold [FIGURE 2.8A]	The inferior margin of the buttock in the standing position; *does not* correspond to the inferior border of the gluteus maximus muscle.
Gluteus maximus muscle [FIGURE 2.8A]	Forms most of the buttock; extends and laterally rotates hip joint.
Gluteus medius muscle [FIGURE 2.8A]	Superior and lateral to the gluteus maximus, extending from the ilium to the greater trochanter of the femur; abducts and medially rotates hip joint and steadies the pelvis; frequent site of intramuscular injection.
Greater trochanter of femur [FIGURE 2.8A]	Projection of proximal end of femur; can be palpated on the superior lateral aspect of the thigh, preferably when the subject is lying down; site for attachment of gluteus medius and gluteus minimis muscles.
Hamstring muscles [FIGURE 2.8A]	Semitendinosus, semimembranosus, and biceps femoris; located on posterior thigh; extend hip and flex knee during walking.
Ischial tuberosity [FIGURE 2.8A]	Inferior protuberance of the lower hipbone (ischium); can be palpated in lower portion of the buttock. The medium-sized rounded protuberance is covered by the gluteus maximus muscle when standing, but extends below the inferior border of the muscle when sitting. We sit on the ischial tuberosity, not the gluteus maximus.
Knee [FIGURE 2.8C]	Joint between the thigh and leg. For structures of the joint to be palpable, knee muscles must be relaxed; excess fluid in the joint may be palpated.
Lateral condyle of femur [FIGURE 2.8C]	Lateral projection of distal end of femur; readily palpable.
Lateral condyle of tibia	Lateral projection of proximal end of tibia; readily palpable.
Medial condyle of femur [FIGURE 2.8C]	Medial projection of distal end of femur; readily palpable.
Medial condyle of tibia [FIGURE 2.8C]	Medial projection of proximal end of tibia; readily palpable.
Patella [FIGURE 2.8C]	Kneecap; triangular, flattened bone within the tendon of the quadriceps femoris muscle; increases leverage of knee, protects knee joint. When the quadriceps femoris muscle is relaxed and the leg is extended, the patella can be shifted from side to side.
Patellar tendon [FIGURE 2.8C]	Continuation of quadriceps femoris tendon distal to patella and inferior to knee.
Popliteal artery	Pulsations can be felt in the popliteal fossa with the leg flexed at the knee joint.
Popliteal fossa	Diamond-shaped space behind the knee; bounded above by the medial and lateral hamstrings and below by the two heads of the gastrocnemius muscle; located within the fossa are the popliteal artery and sciatic nerve; pulsations of the artery can be felt.
Posterior superior iliac spine [FIGURE 2.8A]	Located at the posterior margin of the iliac crest, level with the S2 vertebra beneath a surface dimple formed by the attachment of skin and underlying tissue to bone.
Pubic tubercle [FIGURE 2.8B]	Palpable in the male on the upper border of the body of the pubis; site of medial attachment of the inguinal ligament.
Quadriceps femoris muscles	"Quads"; composed of rectus femoris and vastus lateralis, intermedius, and medialis muscles in the anterior thigh; a muscle with four heads.
Rectus femoris muscle [FIGURE 2.8B]	Extends from anterior inferior iliac spine through the patellar ligament to the tibial tuberosity; extends leg at knee joint, flexes thigh at hip joint.
Saphenous vein	By compressing this vein in the upper thigh, it can be seen along its entire subcutaneous course on the medial aspect of the leg and thigh.
Sartorius muscle [FIGURE 2.8B]	Extends from the anterior superior iliac spine to upper medial surface of tibia; flexes leg at knee joint and thigh at hip joint. Popularly called the "tailor" muscle because tailors used to sit in a cross-legged position, stretching the muscle across the thigh.
Vastus lateralis muscle [FIGURE 2.8B]	Extends from greater trochanter of femur through the patellar tendon to tibial tuberosity; extends leg at knee joint.
Vastus medialis muscle [FIGURE 2.8B]	Extends from the upper two-thirds of the body of femur through the patellar tendon to tibial tuberosity; extends leg at knee joint.

FIGURE 2.9 LEG, ANKLE, AND FOOT

The *leg,* which extends between the knee and the ankle, contains two long bones: the *tibia* (shinbone) is located on the anterior and medial side of the leg; the *fibula* lies parallel and lateral to the tibia [FIGURE 2.9A, B]. The **ankle** is the joint between the leg and foot. The ankle is formed by 7 *tarsal bones,* which join the five *metatarsal bones* of the **foot;** the metatarsals connect to the 14 *phalanges,* or toe bones [FIGURE 2.9C, D].

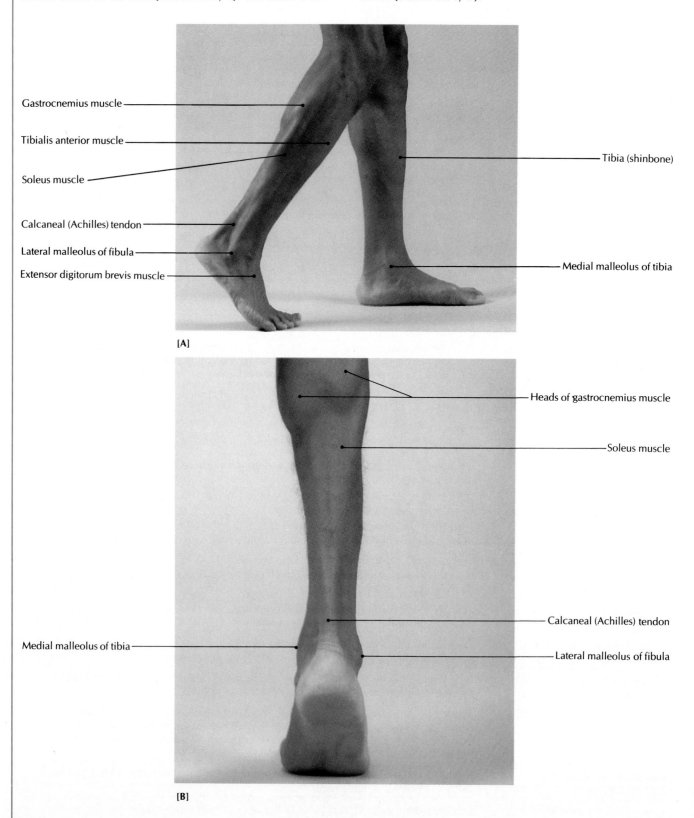

Gastrocnemius muscle

Tibialis anterior muscle

Soleus muscle

Calcaneal (Achilles) tendon

Lateral malleolus of fibula

Extensor digitorum brevis muscle

Tibia (shinbone)

Medial malleolus of tibia

[A]

Heads of gastrocnemius muscle

Soleus muscle

Calcaneal (Achilles) tendon

Medial malleolus of tibia

Lateral malleolus of fibula

[B]

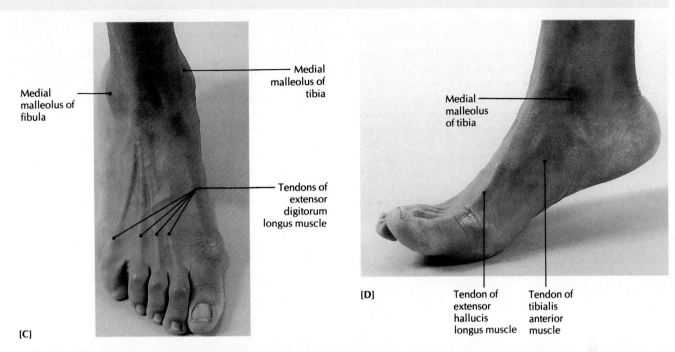

Medial malleolus of fibula

Medial malleolus of tibia

Tendons of extensor digitorum longus muscle

Medial malleolus of tibia

Tendon of extensor hallucis longus muscle

Tendon of tibialis anterior muscle

[D]

[C]

SOME SURFACE STRUCTURES OF THE LEG, ANKLE, AND FOOT

Structure	Description
Calcaneal (Achilles) tendon [FIGURE 2.9A, B]	Prominent common tendon above the posterior aspect of the ankle, between the tapering bellies of the gastrocnemius, soleus, and plantaris muscles. It inserts on the calcaneus (heelbone). If the calcaneal tendon is severed, it is impossible to walk.
Calcaneus	Heelbone at the posterior border of the calcaneal tendon; most posterior tarsal (ankle) bone.
Dorsal venous arch	Superficial veins on the lateral dorsal surface of the foot; they join to form the saphenous veins in the leg.
Extensor digitorum brevis muscle [FIGURE 2.9A]	Indicated by the soft area on the dorsal surface of the foot.
Fibula	Lower leg bone between knee and ankle; parallel and lateral to tibia.
Gastrocnemius muscle [FIGURE 2.9A, B]	"Calf muscle" that forms the contour on the posterior aspect of the leg below the knee joint; plantar flexes ankle joint, flexes knee joint.
Lateral malleolus of fibula [FIGURE 2.9A, B]	Projection of distal end of the fibula; forms bony projection of the lateral ankle.
Medial malleolus of tibia [FIGURE 2.9A, B]	Projection of the distal end of the tibia; forms bony projection of the medial ankle.
Soleus muscle [FIGURE 2.9A, B]	Extends from below the knee to join the calcaneal tendon; deep to gastrocnemius; plantar flexes ankle joint.
Tendon of extensor hallucis longus muscle [FIGURE 2.9D]	Tendon aligned with the great toe; the muscle dorsiflexes ankle joint, extends great toe.
Tendons of extensor digitorum longus muscle [FIGURE 2.9C]	Tendons aligned with the four toes; the muscle dorsiflexes ankle joint, extends toes at metatarsophalangeal and interphalangeal joints.
Tibia [FIGURE 2.9A]	Lower leg bone between knee and ankle; parallel and medial to fibula; its anterior aspect forms the "shin."
Tibialis anterior muscle [FIGURE 2.9A]	Extends from the lateral condyle and surface of tibia to the first metatarsal foot bone; dorsiflexes ankle joint, inverts foot.
Tibial tuberosity	Prominent protuberance on the proximal anterior surface of the tibia, where the patellar tendon is attached.

Anatomical Variations

We take it for granted when we recognize people instantly. But why is it so easy to tell Aunt Rose from Aunt Marie? Because the shapes of heads, noses, ears, mouths, nostrils, eyes, arms, legs, and torsos are always different enough (except perhaps with identical twins) for us to be able to distinguish one person from another. Human structure is not rigidly consistent. Instead, some anatomical variations are normal and even expected.

There are many such variations besides the typical facial differences. For example, one person in 20 has an extra pair of ribs. We say that the body contains 206 bones, but that is only the most common number, not the constant one. The same muscle may have a different nerve or blood supply in two different people, or a small muscle may be lacking altogether. The shape of a particular bone may vary from one person to another, and connections of muscles to bones may be different. The size and placement of blood vessels varies relatively often, while nerves are remarkably constant. Variations are found most often from one side of a person's body to the other. But as long as the body functions properly, such variations are to be expected, and are not even noted as being variations. In contrast, major anatomical variations, such as extra fingers or toes, are considered abnormalities, rather than variations.

CHAPTER SUMMARY

Surface anatomy is the study of body structures that can be seen or felt on the body surface. These structures can provide diagnostic landmarks. Diagnostic methods that require knowledge of surface anatomy include *auscultation* (listening), *inspection* (looking), *palpation* (touching), and *percussion* (tapping).

Head and Neck (pp. 33, 36)
1 The head and neck can be divided anatomically into several regions. The *head (cephalic) region* contains the cranial and facial skulls.

2 The *cranial skull* supports and protects the brain and its meninges. It is divided into the *frontal, parietal, temporal,* and *occipital regions.*

3 The *facial skull* includes the mandible and forms the framework for the cavities of the eyes, nose, mouth, and throat. It is composed of the *ocular, auricular, nasal,* and *oral regions.*

4 The region of the head that is normally covered with hair, including the thin layer of skin and underlying tissue covering the skull, is the *scalp.*

5 The *neck* is the region between the lower margin of the mandible and the upper border of the clavicle. It is divided into the *anterior* and *posterior triangles.*

Trunk (pp. 38, 40, 42)
1 The *trunk* is composed of the thorax, back, abdomen, and pelvis.

2 The bony *thorax* is composed of 12 thoracic vertebrae, 12 pairs of ribs, and the sternum. The *back* is the posterior surface of the trunk. The *abdomen* lies below the diaphragm of the anterior thorax and above the superior margin of the pelvis below. The *pelvis* is the region of the trunk inferior to the abdomen.

Upper Extremity (pp. 44, 46)
1 The *upper extremity* is formed by the shoulder girdle, axilla, arm, forearm, wrist, and hand.

Lower Extremity (pp. 48, 50)
1 The *lower extremity* is formed by the pelvic girdle, buttocks, thigh, leg, ankle, and foot.

STUDY AIDS FOR REVIEW

UNDERSTANDING THE FACTS

1 What is the difference between palpation and percussion?
2 What regions of the head comprise the cranial skull?
3 Distinguish between the anterior and posterior triangles of the neck.
4 How would you locate the apex beat of the heart?
5 How are the abdomen and pelvis defined? The upper and lower extremities?
6 How can the tibia be distinguished from the fibula by utilizing surface anatomy?

UNDERSTANDING THE CONCEPTS

1 What is one way a projection from a major bone, such as the mastoid process, can form?
2 What is the difference, if any, between the thyroid cartilage and the "Adam's apple"?
3 What is the significance of the costal margins in locating the liver and stomach?
4 If you are administering cardiopulmonary resuscitation, why should you avoid hitting the xiphoid process of the sternum?
5 Why is it easy to see and palpate muscles and bones on some people and not others?
6 What would happen to the shoulder if the clavicle were fractured?
7 Is a wristwatch actually worn on the wrist? Explain.

SELF-QUIZ
Multiple-Choice Questions
2.1 The region that is *not* part of the facial skull is the
  a ocular region
  b nasal region
  c nuchal region
  d auricular region
2.2 The muscle that divides the neck into the anterior and posterior triangles is the
  a tibialis anterior
  b external oblique
  c triceps brachii
  d sternocleidomastoid

2.3 Which of the following structures is not palpable?
  a clavicle
  b iliac crest
  c lateral condyle of tibia
  d C4 vertebra
2.4 The body region that lies between the diaphragm and the superior margin of the pelvis is the
  a nucha
  b abdomen
  c thorax
  d symphysis pubis
2.5 The structure that is *not* associated with the sternum is the
  a serratus anterior muscle
  b xiphoid process
  c manubrium
  d suprasternal notch

A SECOND LOOK

In the following photograph, label the iliac crest, gluteal fold, ischial tuberosity, and a frequent site of intramuscular injections:

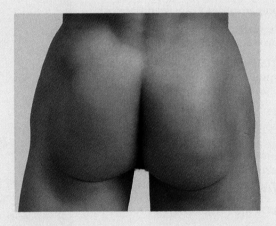

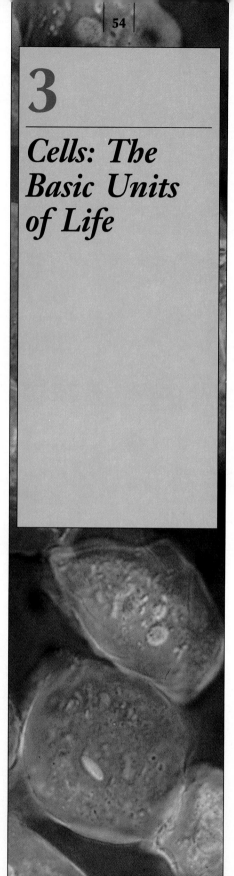

3
Cells: The Basic Units of Life

LEARNING OBJECTIVES

1 Describe, locate, and list the general functions of the principal structures of a cell.

2 Diagram the fluid-mosaic model of membrane structure.

3 Describe a chromosome.

4 Describe DNA replication.

5 Compare the events associated with each of the stages of mitosis.

6 Discuss the major hypotheses about aging.

7 Distinguish between benign and malignant neoplasms.

8 Explain how cancer spreads from its original site in the body.

9 Briefly explain the main hypotheses about how cancer develops.

We all begin life when a single sperm unites with a single egg—a fusion of two highly specialized cells. The resulting fertilized egg is the forerunner of all future cells in the body. The adult human body consists of more than 50 trillion (50 million million) cells. Most of these cells are specialized in structure and function. No matter what their specialized functions, most cells carry on such life-supporting activities as breaking down food molecules, excreting wastes, reproducing, generating the energy-rich compound ATP, engulfing foreign materials, and creating new cell structures [FIGURE 3.1, TABLE 3.1].

WHAT ARE CELLS?

As you saw in Chapter 1, cells are the smallest independent units of life. Your body cells have different sizes, shapes, and colors, but cells such as those that make up the liver are probably fairly typical. Several hundred liver cells would fit into a cube smaller than the small letters in this sentence. Early *cytologists* (scientists who study cells) thought that the interior of cells consisted of a homogeneous fluid, which they called *protoplasm*. Today, the word "protoplasm" is used only in a very general way; instead, scientists divide cells into four basic parts [FIGURE 3.1]:

1 The *plasma membrane* is the outer boundary of the cell.

2 *Cytoplasm* is the portion of the cell outside the nucleus and within the plasma membrane. Metabolic reactions take place here with the aid of specialized structures called *organelles*. The fluid portion of the cytoplasm is called *cytosol*.

3 The *nucleus* is the control center of a cell. It is a clearly defined body that is separated from the surrounding cytoplasm by a *nuclear envelope*. Within the nucleus are the chromosomes, which contain the genes that direct all cell activities.

4 *Nucleoplasm* is the material within the nucleus.

ASK YOURSELF

1 What are the four basic parts of a cell?

2 How do nucleoplasm and cytoplasm differ?

3 What is cytosol?

FIGURE 3.1 A GENERALIZED CELL

A highly simplified version of a cell is cut open to show the various structures and organelles (several of which are also cut) within the plasma membrane. Organelles include the endoplasmic reticulum, ribosomes, mitochondria, and nucleus, among other structures. The nucleus is sliced open to show the nucleoli inside.

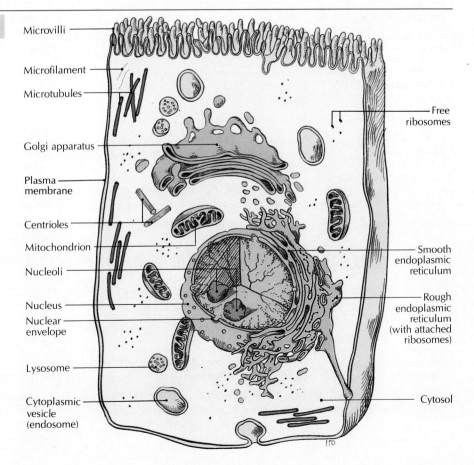

TABLE 3.1 CELL STRUCTURES AND THEIR FUNCTIONS

Structure	Description	Main functions
Centrioles [FIGURE 3.11]	Located within centrosome; contain nine triple microtubules.	Assist in cell reproduction; form basal body of cilia and flagella.
Cilia, flagella [FIGURE 3.12]	Cilia are short and threadlike; flagella are much longer.	Cilia move fluids or particles past fixed cells. Flagella provide means of movement for sperm.
Cytoplasm [FIGURE 3.1]	Semifluid enclosed within plasma membrane; consists of fluid cytosol and intracellular structures such as organelles.	Dissolves soluble proteins and substances necessary for cell's metabolic activities; houses organelles, vesicles, inclusions, and lipid droplets.
Cytoplasmic inclusions	Substances temporarily in cytoplasm.	Food materials or stored products of cell metabolism, including pigments, fats and carbohydrates, and secretions; also foreign substances.
Cytoplasmic vesicles (endosomes) [FIGURE 3.1]	Membrane-bound sacs.	Store and transport cellular materials.
Cytoskeleton [FIGURE 3.9]	Flexible cellular framework of interconnecting microfilaments, intermediate filaments, microtubules, other organelles.	Provides support; assists in cell movement; site for binding of specific enzymes.
Cytosol	Fluid portion of cytoplasm; enclosed within plasma membrane; surrounds nucleus.	Houses organelles; serves as transporting medium for secretions; site of many metabolic activities.
Deoxyribonucleic acid (DNA)	Nucleic acid that makes up the chromosomes.	Controls heredity and cellular activities.
Endoplasmic reticulum (ER) [FIGURE 3.5]	Membrane system extending throughout cytoplasm.	Internal transport and storage; rough ER serves as point of attachment for ribosomes; smooth ER produces steroids.
Golgi apparatus [FIGURE 3.7]	Flattened stacks of disklike membranes.	Packages proteins for secretion.
Intermediate filaments [FIGURE 3.9]	Elongated, fibrillar structures composed of protein subunits.	Key elements of cytoskeleton in most cells.
Lysosomes [FIGURE 3.1]	Small, membrane-bound spheres.	Digest materials; decompose harmful substances; play a role in cell death.
Microfilaments [FIGURE 3.9]	Solid, rodlike structures containing the protein actin.	Provide structural support and assist cell movement.
Microtubules [FIGURES 3.10 and 3.11]	Hollow, slender, cylindrical structures.	Support; assist movement of cilia and flagella and chromosomes; transport system.
Mitochondria [FIGURE 3.8]	Sacs with inner folded, double membranes.	Produce most of the energy (ATP) required for cell; site of cellular respiration.
Nucleolus [FIGURE 3.13]	Rounded mass within nucleus; contains RNA and protein.	Preassembly site for ribosomes.
Nucleus [FIGURE 3.13]	Large spherical structure surrounded by a double membrane; contains nucleolus and DNA.	Contains DNA that makes up the cell's genetic program and controls cellular activities.
Peroxisomes	Membranous sacs containing oxidative enzymes.	Carry out metabolic reactions and destroy hydrogen peroxide, which is toxic to cell.
Plasma membrane [FIGURE 3.2]	Outer bilayered boundary of cell; composed of lipids and proteins.	Regulates passage of substances into and out of cell; cell-to-cell recognition.
Ribonucleic acid (RNA) [FIGURES 3.6, 3.9]	Three types of nucleic acid involved in transcription and translation of genetic code.	Messenger RNA (mRNA) carries genetic information from DNA; transfer RNA (tRNA) is involved in amino-acid activation during protein synthesis; ribosomal RNA (rRNA) is involved in ribosome structure.
Ribosomes [FIGURE 3.6]	Small structures containing RNA and protein; some attached to rough ER.	Sites of protein synthesis.

CELL MEMBRANES

The thin membrane that forms the outermost layer of a cell is called the *plasma membrane.* It maintains the boundary and integrity of the cell by keeping the cell and its contents separate and distinct from the surrounding environment. The plasma membrane is a complex, selective structure that allows only certain substances to enter and leave the cell. Specific receptor molecules on the surface enable it to interact with other cells and with certain substances in its external environment.

Membranes that enclose or actually make up some organelles *inside* the cell are similar to the plasma membrane. They also help regulate cellular activities.

Structure of Cell Membranes

In 1972, the *fluid-mosaic model* of membrane structure was formulated. According to this model, the membrane is a fluid double layer, or bilayer, composed mainly of proteins and phospholipids [FIGURE 3.2].

The phospholipid bilayer forms a fluid "sea," similar in consistency to vegetable oil, in which specific proteins float like icebergs. Being fluid, the membrane is in a constant state of flux, shifting and changing, yet at the same time retaining its basic structure and properties. The word "mosaic" refers to the many different proteins in the phospholipid bilayer.

The following are the most important points about the fluid-mosaic model:

1 *Phospholipids,* shown like balloons, each with two strings, are arranged in a double layer consisting of sheets two molecules thick. This arrangement is called a *bimolecular layer,* or *bilayer.*

2 Phospholipid molecules have one electrically charged end and one uncharged end. The charged ends can extend into the watery cytosol inside the cell and into the similarly watery external environment (extracellular fluid), while the uncharged ends face each other in the middle of the double layer.

3 The "tails" (the *lipid,* or fatty, portion) of the phospholipid molecules are attracted to each other and are repelled by water — they are *hydrophobic,* or "water-fearing." As a result, the polar, or charged, spherical "heads" (the *phosphate* portion) of the phospholipid molecules line up over the outer and inner cell surfaces. The heads are *hydrophilic,* or "water-attracting." The lipid portions compose the center of the membrane between the two phosphate layers.

4 FIGURE 3.2A shows how the *protein* portions are embedded in the cell membranes like tiles in a mosaic, or like icebergs floating in the fluid phospholipids. Some proteins protrude only into the phospholipid, some extend all the way into the cytoplasm, some are partially sunken in the outer layer, and some are partially sunken in the inner layer.

5 Protruding from some of the proteins are antennalike *surface carbohydrates,* most of which are small, complex, nonrepeating sequences of simple sugars. Compounds consisting of surface carbohydrates linked to surface proteins are called *glycoproteins,* while surface carbohydrates linked to surface lipids form *glycolipids.* Overall, these carbohydrates and the portions of the proteins on the surface of the plasma membrane make up the *glycocalyx,* or *cell coat,* the carbohydrate-rich peripheral zone of the cell surface. The glycoproteins and glycolipids enable one cell to recognize another cell. Apparently, glycoproteins of similar cells "fit together" chemically, whereas a foreign cell is not "recognized" and is repelled or even destroyed.

6 Finally, the plasma membrane of human cells contains *cholesterol.* The flexibility (fluidity) of the membrane is determined by the ratio of cholesterol to phospholipids. The plasma membranes of human cells contain relatively large amounts of cholesterol (as much as one molecule for every phospholipid molecule), which enhances the mechanical stability of the bilayer.

Functions of Cell Membranes

In general, cell membranes serve several important functions: (1) they separate the inside of a cell from the outside environment, (2) they separate cells from one another, (3) they separate the various parts within cells, (4) they provide an abundant surface on which chemical reactions occur, and (5) they regulate the passage of materials into and out of cells and from one part of a cell to another.

Cells let some things in and keep others out, a quality called *selective permeability* (L. *permeare: per,* through + *meare,* pass). Some substances, such as water and lipids, pass through plasma membranes readily. Others, such as potassium or calcium, pass only with the expenditure of energy by the cell, while still others cannot pass at all.

In some cells, the surface area of the plasma membrane is greatly increased by the presence of many *microvilli* [FIGURE 3.1]. These slender extensions of the plasma membrane function in either the absorption or secretion of molecules. The increase in surface area allows many more molecules to be transported into or out of the cell.

A S K   Y O U R S E L F

1 To what do the terms fluid and mosaic refer in the fluid-mosaic model of cell membranes?

2 What is a selectively permeable membrane?

FIGURE 3.2 THE FLUID-MOSAIC MODEL

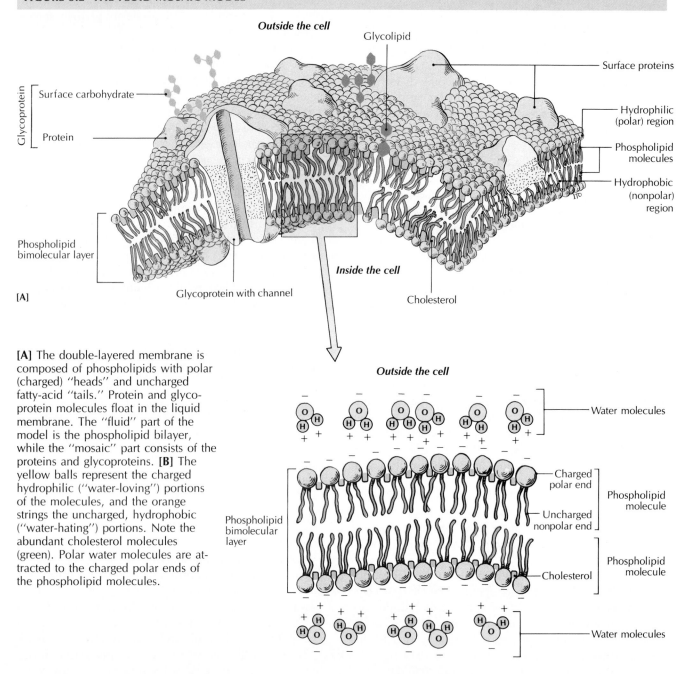

[A] The double-layered membrane is composed of phospholipids with polar (charged) "heads" and uncharged fatty-acid "tails." Protein and glycoprotein molecules float in the liquid membrane. The "fluid" part of the model is the phospholipid bilayer, while the "mosaic" part consists of the proteins and glycoproteins. **[B]** The yellow balls represent the charged hydrophilic ("water-loving") portions of the molecules, and the orange strings the uncharged, hydrophobic ("water-hating") portions. Note the abundant cholesterol molecules (green). Polar water molecules are attracted to the charged polar ends of the phospholipid molecules.

MOVEMENT ACROSS MEMBRANES

When molecules pass through a cell membrane *without the expenditure of energy by the cell*, the process is called *passive transport.* Passive transport processes include simple diffusion, facilitated diffusion, osmosis, and filtration, as described in TABLE 3.2.

In passive transport processes, certain small molecules (solutes) can move across plasma membranes in response to a difference in their concentrations on either side of the membrane (that is, inside and outside the cell). This difference in concentration is called a ***concentration gradient.*** In passive transport, molecules always move from an area where their concentration is high to an area where their concentration is lower. They are said to

TABLE 3.2 COMPARISON OF PASSIVE AND ACTIVE MOVEMENT ACROSS CELL MEMBRANES

Type of movement	Description	Example in body
PASSIVE MOVEMENT		
	Molecules move "down" the concentration gradient. No cell energy required.	
Simple diffusion	Molecules spread out randomly from areas of high concentration to areas of lower concentration until they are distributed evenly. Movement is directly through membrane or integral channel protein; however, a membrane need not be present.	Inhaled oxygen is transported into lungs and diffuses through lung cells into bloodstream.
Facilitated diffusion	Carrier proteins in plasma membrane temporarily bind with molecules, allowing them to pass through membrane, via protein channels, from areas of high concentration to areas of lower concentration.	Specific amino acids combine with carrier proteins to pass through plasma membrane into cell.
Osmosis	Water molecules move through selectively permeable membrane from areas of high water concentration to areas of lower water concentration.	Water moves into a red blood cell when concentration of water molecules outside cell is greater than it is inside cell.
Filtration	Hydrostatic pressure forces small molecules through selectively permeable membrane from areas of high pressure to areas of lower pressure.	Blood pressure inside capillaries is higher than in surrounding tissue fluid. This pressure forces water and dissolved small particles through capillary walls and into surrounding tissues. Blood pressure forces water and dissolved wastes into kidney tubules during formation of urine.
ACTIVE MOVEMENT		
	Cell energy allows substances to move through selectively permeable membrane from areas of low concentration to areas of higher concentration. Requires a living cell and energy in the form of ATP.	
Active transport	Carrier proteins in plasma membrane bind with ions or molecules to assist them across membrane "against" the concentration gradient.	Sodium ions are pumped out of resting nerve cells, although their concentration is much higher outside the cells.
Endocytosis [FIGURES 3.3 and 3.4]	Membrane-bound vesicles enclose large molecules, draw them into cytoplasm, and release them.	
Pinocytosis	Plasma membrane encloses small amounts of fluid droplets and takes them into cell.	Kidney cells take in tissue fluids in order to maintain fluid balance.
Receptor-mediated endocytosis	Extracellular molecules bind with specific receptor on plasma membrane, causing the membrane to invaginate and draw molecules into cell.	Intestinal epithelial cells take up large molecules.
Phagocytosis	Plasma membrane forms a pocket around a solid particle or cell and draws it into cell.	White blood cells engulf and digest harmful bacteria.
Exocytosis [FIGURE 3.4]	Vesicle (with undigested particles) fuses with plasma membrane and expels particles from cell through plasma membrane.	Nerve cells release chemical messengers.

FIGURE 3.3 TWO TYPES OF ENDOCYTOSIS

Endocytosis may involve **[A]** pinocytosis, where extracellular molecules are taken into a cell through an invaginated vesicle, or **[B]** receptor-mediated endocytosis, where external molecules (called *ligands*) are bound to specific receptors on the plasma membrane before they are taken into the cell.

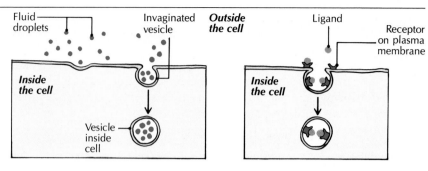

[A] PINOCYTOSIS

[B] RECEPTOR-MEDIATED ENDOCYTOSIS

FIGURE 3.4 PHAGOCYTOSIS AND EXOCYTOSIS

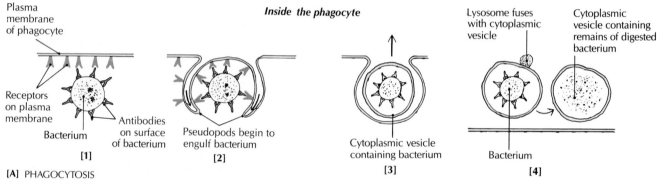

[A] PHAGOCYTOSIS

[A] Phagocytosis. A possible "zipper interaction" mechanism for phagocytosis showing (1) how antibodies on the surface of a bacterium bind to receptors on the plasma membrane of a phagocyte. (2) The bacterium is engulfed by pseudopods of the phagocyte. If only part of the surface of the bacterial cell is coated with antibodies, the pseudopods will not completely surround the cell, and the cell will not be engulfed. (3) The bacterium is enclosed within a cytoplasmic vesicle. (4) A lysosome fuses with the cytoplasmic vesicle and releases digestive enzymes that break down the bacterium. **[B]** Exocytosis. Unwanted particles engulfed and stored within a vesicle are transported to the plasma membrane and expelled. (1) The cytoplasmic vesicle with its particles approaches the plasma membrane. (2) The vesicle attaches to the plasma membrane. (3) The membranes of the cell and vesicle fuse, and the vesicle membrane begins to "unzip." (4) The point of contact between the two membranes widens slightly, allowing the hydrophobic regions of both membranes to be in contact. (5) The fused membranes open. (6) The fusion of membranes is complete, and the unwanted particles are expelled from the cell.

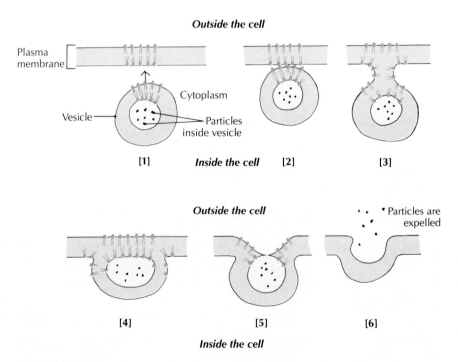

[B] EXOCYTOSIS

move "down" the concentration gradient, which is in keeping with their natural tendency to "spread out" until their concentration is equal throughout the area.

Active processes, on the other hand, move substances through selectively permeable membranes *from areas of low concentration* on one side of the membrane *to areas of higher concentration* on the other side. Such movement is "against" the concentration gradient and against the natural tendency of molecules to "spread out." Because of this, active movement requires energy. Among the active processes are active transport, endocytosis, and exocytosis, as described in TABLE 3.2 and shown in FIGURES 3.3 and 3.4.

ASK YOURSELF

1 How does passive transport work?

2 How do simple diffusion and facilitated diffusion differ?

3 Where do carrier proteins function?

4 What is osmosis?

5 How do active and passive processes of movement across plasma membranes differ?

6 What is active transport?

7 What is the difference between endocytosis, phagocytosis, and exocytosis?

CYTOPLASM

The *cytoplasm* is the portion of a cell outside the nucleus and within the plasma membrane. It is composed of two distinct parts, or phases: (1) The *particulate phase* consists of well-defined structures, including membrane-bound vesicles and organelles, lipid droplets, and *inclusions* (solid particles such as glycogen granules). (2) The *aqueous phase* consists of the fluid cytosol, in which the organelles and vesicles are suspended. The cytosol also contains various dissolved substances required for the cell's metabolic activities. Like most other body materials, the cytosol is mostly water (70 to 85 percent), but its other typical components are proteins (10 to 20 percent), lipids (about 2 percent), complex carbohydrates, nucleic acids, amino acids, vitamins, and electrolytes (substances such as potassium that dissolve in water and provide certain chemicals for cellular reactions).

In addition to the many organelles that were known before the advent of modern electron microscopes, we now know that cytoplasm also contains a *cytoskeleton* — a flexible lattice arrangement that supports the organelles and provides the machinery for the movement of cells and their organelles.

ORGANELLES

Suspended within the cytosol and nucleoplasm are various membrane-bound structures called *organelles.* Each organelle has a specific structure and function, as described in the following sections.

Endoplasmic Reticulum

The *endoplasmic reticulum* (en-doh-PLAZ-mik reh-TICK-yoo-luhm), or *ER*, is a labyrinth of membrane-bound flattened sheets, sacs, and tubules that branch and spread throughout the cytoplasm. In some places, the membranes of the ER are connected with the nuclear envelope. FIGURE 3.5A shows the position of the endoplasmic reticulum in a simplified cell.

The ER functions as a series of channels that carry various materials throughout the cytoplasm. It also serves as a storage site for enzymes and other proteins and as a point of attachment for ribosomes (structures that play an important part in forming proteins) [FIGURE 3.5B, C].

The space within the channels and enclosed by the ER membranes is called the *ER lumen.* It is thought to be a single, continuous space. When endoplasmic reticulum does not have attached ribosomes, it is called *smooth ER.* Smooth ER serves as a site for the production of steroids, detoxifies a wide variety of organic molecules, and stores calcium in skeletal muscle cells. *Rough ER* refers to endoplasmic reticulum that does have attached ribosomes on its outer face.

Ribosomes

Ribosomes (RYE-boh-sohmz) function in *protein synthesis* — the manufacture of proteins from amino acids. Ribosomes contain almost equal amounts of protein and a special kind of ribonucleic acid, *ribosomal RNA (rRNA).* Ribosomes are actually composed of two subunits with variable shapes [FIGURE 3.6A], but are usually pictured as slightly flattened spheres with "tucked-in" sections around the middle [FIGURE 3.6B].

Not all ribosomes are attached to the ER. Some float freely within the cytoplasm. Whether ribosomes are free or attached, they are usually grouped in clusters called *polyribosomes* ("many ribosomes") [FIGURE 3.6C]. These clusters are connected by a strand of a single molecule of specialized nucleic acid called *messenger ribonucleic acid (messenger RNA* or *mRNA).*

FIGURE 3.5 ENDOPLASMIC RETICULUM

[A] Note that the endoplasmic reticulum (blue) actually touches the outer nuclear envelope, where a transfer of materials takes place. Rough ER and smooth ER (without attached ribosomes) are shown. Ribosomes can be seen in the drawing [B] and in the electron micrograph [C]. ×25,000.

Endoplasmic
reticulum

[A]

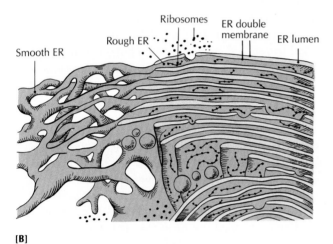

[B]

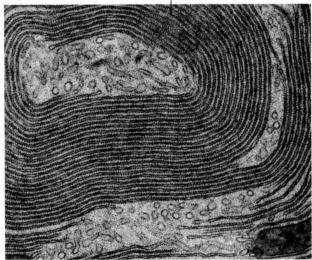

Rough
endoplasmic
reticulum

[C]

FIGURE 3.6 RIBOSOMES

[A] A current model of a ribosome with its two subunits. Apparently, the mRNA molecule is held in a channel between the large and small subunits. This representation is based on electron micrographs. [B] Simplified version of a ribosome, showing sites on the large subunit (A-site and P-site) for the binding of tRNA molecules, and the binding site for mRNA on the small subunit. [C] Electron micrograph of a string of ribosomes (a polyribosome) connected by a strand of mRNA. ×125,000.

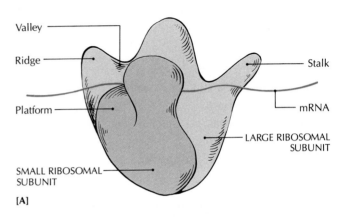

Valley

Ridge

Platform

SMALL RIBOSOMAL
SUBUNIT

Stalk

mRNA

LARGE RIBOSOMAL
SUBUNIT

[A]

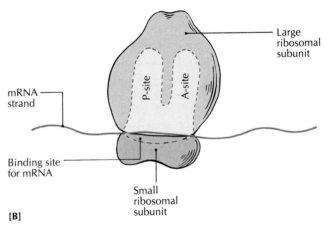

mRNA
strand

Binding site
for mRNA

P-site

A-site

Large
ribosomal
subunit

Small
ribosomal
subunit

[B]

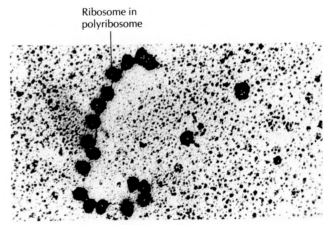

Ribosome in
polyribosome

[C]

Golgi Apparatus

The *Golgi apparatus* (GOAL-jee), or *Golgi complex* (named for Camillo Golgi, who discovered it in 1898), is a collection of membranes associated physically and functionally with the ER. Golgi apparatuses are usually located near the nucleus [FIGURE 3.7]. They are composed of flattened stacks of membrane-bound *cisternae* (siss-TUR-nee; Gr. *kiste*, basket), closed spaces serving as fluid reservoirs [FIGURE 3.7B]. The Golgi apparatus functions in the "packaging" and secretion of glycoproteins. Many of these glycoproteins are probably components of the glycocalyx.

As proteins are synthesized by ribosomes attached to the rough ER, some of the proteins are sealed off in little packets called *transfer vesicles.* These vesicles bud off, and fuse with the forming face of the Golgi apparatus close to the nucleus. Within the Golgi apparatus, the proteins can be concentrated, modified, or compacted. Eventually, the proteins are packaged into relatively large *secretory vesicles*, which are released from the cisterna of the mature face closest to the plasma membrane. When some of the vesicles reach the plasma membrane, they fuse with it; the contents of some of the vesicles are then released as secretory proteins to the outside of the cell by exocytosis. Other vesicles contain membrane proteins of the cell and proteins that remain within the cell.

Golgi apparatuses are most apparent and abundant in cells that are active in secretion, such as the pancreatic cells that secrete digestive enzymes, and nerve cells, which secrete transmitter substances.

Lysosomes and Microbodies

Lysosomes (Gr. "dissolving bodies") are small, membrane-bound organelles that contain digestive enzymes called *acid hydrolases.* These enzymes can digest proteins, nucleic acids, lipids, and carbohydrates under acidic conditions. The component molecules of the destroyed substances are discharged into the cytoplasm, where they are used in the synthesis of new materials. The acid hydrolases are synthesized in the ER, transported to the Golgi apparatus for processing, and then transported to the lysosomes in secretory vesicles.

Lysosomes are found in almost all cells, especially those in tissues that experience frequent cell divisions, such as the lungs, liver, and spleen, and in white blood cells. Lysosomes are sometimes considered to be the digestive system of the cell because they can take in and break down complex molecules, microorganisms, and damaged cell parts. If only a single acid hydrolase is absent from the lysosomes, serious genetic diseases can occur because the foreign material that enters a cell cannot be digested.

Lysosomes are often confused with *microbodies* be-

FIGURE 3.7 GOLGI APPARATUS

[A] A simplified diagram of a cell, cut to show the position of a Golgi apparatus. **[B]** A three-dimensional drawing of a Golgi apparatus, showing how the transfer vesicles from the ER merge with the Golgi apparatus, and how secretory vesicles bud off from the opposite side of the Golgi apparatus on their way to delivering secretions. **[C]** An electron micrograph of a Golgi apparatus showing secretory vesicles. ×50,000.

[A]

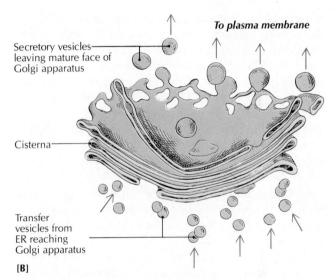

[B]

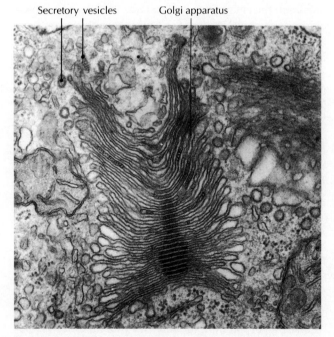

[C]

cause they both appear as dense, granular structures. There are two types of microbodies, but only one type, **peroxisomes,** is found in animal cells. Peroxisomes are membrane-bound organelles that contain enzymes that use oxygen to carry out their metabolic reactions. Peroxisomes are so named because they can both form hydrogen peroxide as they oxidize various substances and then destroy the peroxide with the enzymes they contain. The destruction of peroxides is crucial to the cell because they are toxic products of many metabolic processes, especially fatty-acid oxidation. Peroxisomes are numerous in liver and kidney cells, where they are important in detoxifying certain compounds, such as alcohol.

Mitochondria

Mitochondria (my-toe-KAHN-dree-uh; Gr. *mitos,* a thread; sing. *mitochondrion*) are organelles with double membranes found throughout the cytoplasm [FIGURE 3.8A]. Their main function is the conversion of energy stored in carbon-containing molecules into the high-energy bonds of ATP. (ATP is the only usable energy source for many cellular activities.)

The internal structure of a mitochondrion is compatible with its function. A complex inner membrane folds and doubles in upon itself to form incomplete partitions called *cristae* (KRIS-tee; L. crests) [FIGURE 3.8B, C]. The space between the cristae is called the *matrix.* The cristae greatly increase the surface area available for chemical reactions. Enzymes for these reactions are on the cristae and in the matrix.

The number of mitochondria per cell varies from as few as a hundred to as many as several thousand, depending upon the energy needs of the cell. Muscle cells, which convert chemical energy into mechanical energy, have more mitochondria than relatively inactive cells. Mitochondria contain their own DNA and ribosomes and are able to reproduce. They usually multiply when a cell needs increased amounts of ATP.

Microtubules, Intermediate Filaments, and Microfilaments

In most cells, a flexible latticed framework called the **cytoskeleton** is formed by microfilaments, intermediate filaments, and microtubules [FIGURE 3.9]. This framework extends throughout the cytoplasm, interconnecting its components with organelles, holding them more or less in place. The cytoskeleton also plays a role in cell movement and is a site for the binding of specific enzymes.

The most conspicuous elements of the cytoskeleton, **microtubules,** are hollow, slender, cylindrical structures found in most cells [FIGURE 3.10A, B]. Each microtubule is made up of spiraling subunits of a protein called *tubulin*

FIGURE 3.8 MITOCHONDRIA

[A] A drawing of mitochondria (cut) in a simplified cell. **[B]** A three-dimensional cutaway drawing of a mitochondrion. **[C]** An electron micrograph of a sectioned mitochondrion showing the outer membrane and the folds, or cristae, of the inner membrane. ×95,000.

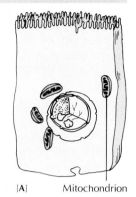
[A] Mitochondrion

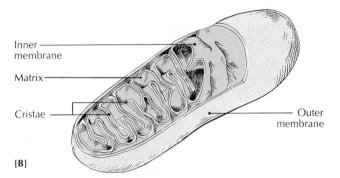
[B]

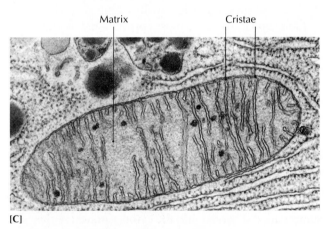
[C]

[FIGURE 3.10C]. Microtubules have several functions within a cell:

1 They are associated with intracellular movement, especially cilia and flagella (whiplike appendages) and with chromosome movement when the cell is dividing.

2 They are part of a transport system within the cell—for example, they help to move materials along the outer surface of each tubule through the cytoplasm.

3 They form part of the cytoskeleton.

4 They are involved in the overall shape changes that cells go through during periods of cell specialization.

FIGURE 3.9 CYTOSKELETON

This diagram of the cell cytoskeleton shows how fine microtrabecular strands and other structural units form a three-dimensional network that supports the principal structures of the cytoskeleton and other cellular structures.

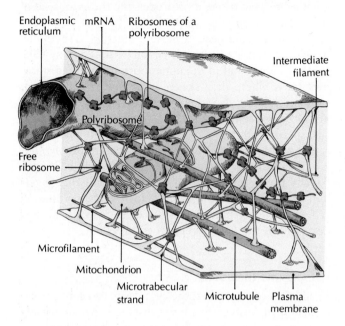

Intermediate filaments help to maintain the shape of cells and the spatial organization of organelles. They are a chemically heterogeneous group of structures whose subunits are made up of five different kinds of proteins. Intermediate filaments received their name because they are "intermediate" in diameter between the thin and thick filaments in muscle cells, where they were first described.

Solid cytoplasmic *microfilaments* are also found in most cells. They are composed of strings of protein molecules, including the protein actin. In nonmuscle cells, actin microfilaments provide mechanical support for various intracellular structures and are part of contractile systems responsible for many intracellular movements.

Centrioles

Near the nucleus of a cell lies a specialized region of cytoplasm called the *centrosome*. The centrosome contains two small structures called *centrioles* that lie at right angles to each other [FIGURE 3.11]. Centrioles are bundles of cylinders made up of microtubules, which resemble drinking straws. Each centriole is composed of nine triplets of microtubules that radiate from the center like the spokes of a wheel. Centrioles are duplicated when the cell undergoes cell division and are involved with the movement of chromosomes during cell division.

Cilia and Flagella

Cilia (SIHL-ee-uh; L. originally "eyelids," later "eyelashes"; sing. *cilium*) and *flagella* (fluh-JELL-uh; L. "small whips"; sing. *flagellum*) are threadlike appendages of

FIGURE 3.10 MICROTUBULES

[A] Microtubules in a simplified cell. [B] An electron micrograph showing microtubules from the brain. ×100,000. [C] A drawing of a portion of a microtubule showing the arrangement of its spiraling protein.

[A]

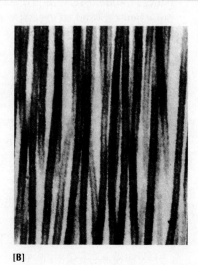

[B]

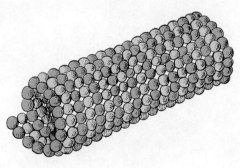

[C]

FIGURE 3.11 CENTRIOLES

[A] The position of centrioles near the cell nucleus. [B] A drawing of a pair of centrioles. Each has nine triplets of microtubules and lies at a right angle to the other. [C] An electron micrograph of a cross section of a centriole. ×400,000.

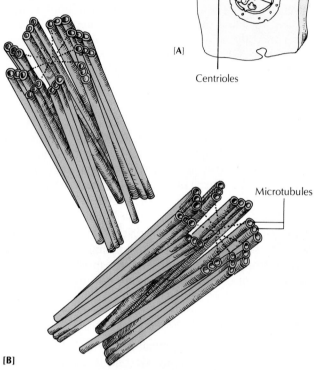

[A]

Centrioles

Microtubules

[B]

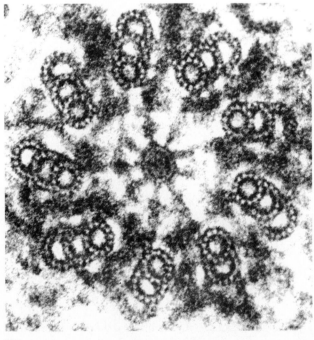

[C]

some cells. The shafts of cilia and flagella are composed of nine pairs of outer microtubules and two single microtubules in the center [FIGURE 3.12A]. Both types of shafts are anchored by a *basal body*, a specialized structure that acts as the template for the nine-plus-two arrangement of microtubules in the shaft. The basal body is structurally identical to a centriole; it protrudes into the cytoplasm of the cell to which it is attached.

Groups of cilia beat in unison, creating a rhythmic, wavelike movement in one direction [FIGURE 3.12B]. Ciliated cells are found in the upper portions of the respiratory tract, where they sweep along mucus and foreign substances. They also appear in portions of the female reproductive system [FIGURE 3.12C], where they move mucus, sperm, and eggs along the uterine (Fallopian) tube.

Flagella have the same internal structure as cilia, but are much longer. In human beings, the only cells with flagella are sperm, which normally have only one flagellum each. The tail of a sperm helps the free-swimming cell to move relatively great distances to reach an egg deep within the female reproductive system.

Cytoplasmic Inclusions

In addition to organelles, the cytoplasm also contains *cytoplasmic inclusions,* chemical substances that are usually either food material or the stored products of cellular metabolism. Unlike organelles, inclusions are not permanent components of a cell and are constantly being destroyed and replaced. Carbohydrates, for example, are stored mainly in liver and skeletal muscle cells in the form of *glycogen inclusions.* When the body needs energy, the glycogen is converted into glucose, the body's main source of energy. Lipids are also stored in cells as a reserve source of energy.

Also present in cytoplasm are several *pigments* (substances that produce color), probably the best known being *melanin.* Melanin protects the skin by causing it to tan when exposed to excessive ultraviolet radiation from the sun; melanin is also found in the hair and eyes.

ASK YOURSELF

1 What is the difference between smooth and rough ER?

2 How does a Golgi apparatus receive protein from the ER?

3 What do lysosomes do?

4 How is the internal structure of a mitochondrion compatible with its function?

5 Describe the internal structure of cilia and flagella.

6 What is a cytoplasmic inclusion?

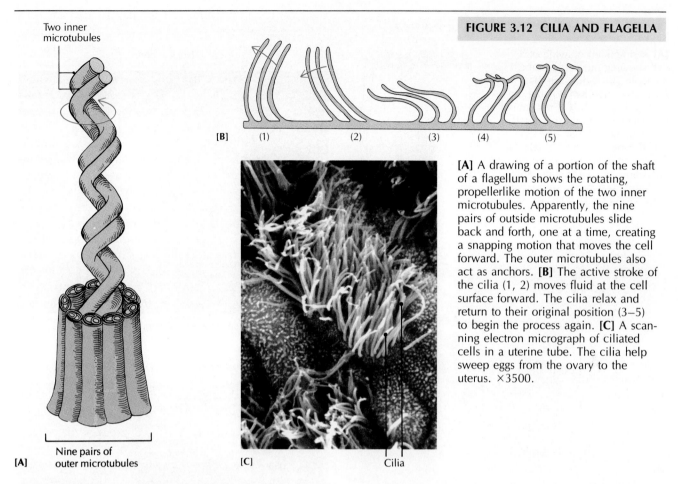

FIGURE 3.12 CILIA AND FLAGELLA

Two inner microtubules

Nine pairs of outer microtubules

[A] [B] (1) (2) (3) (4) (5) [C] Cilia

[A] A drawing of a portion of the shaft of a flagellum shows the rotating, propellerlike motion of the two inner microtubules. Apparently, the nine pairs of outside microtubules slide back and forth, one at a time, creating a snapping motion that moves the cell forward. The outer microtubules also act as anchors. **[B]** The active stroke of the cilia (1, 2) moves fluid at the cell surface forward. The cilia relax and return to their original position (3–5) to begin the process again. **[C]** A scanning electron micrograph of ciliated cells in a uterine tube. The cilia help sweep eggs from the ovary to the uterus. ×3500.

THE NUCLEUS

The *nucleus* (NOO-klee-uhss; L. nut, kernel; pl. *nuclei*, NOO-klee-eye) contains DNA and is the control center of the cell. It has two important roles:

1 It directs the chemical reactions in the cell; genetic information in the DNA is used in the synthesis of enzymes, which determine the cell's metabolic activities.

2 It contains hereditary information, which is transferred during cell division from one generation of cells to the next, and eventually from one generation of organisms to the next.

Almost all cells have nuclei. Mature red blood cells, which do not have nuclei, are incapable of producing mRNA. Thus they cannot synthesize new proteins, duplicate themselves, or perform some typical cellular activities. Red blood cells have a relatively short life span.

Depending on cell type, nuclei vary in shape, location, and number. For example, in some tall cells the nucleus is elongated and found at the base; in adipose (fat) cells, the nucleus is flattened against the cell membrane. But nuclei are generally somewhat spherical and located near the center of a cell [FIGURE 3.13A].

Nuclear Envelope

The boundary of the nucleus appears somewhat like the plasma membrane, but with some important differences. It is not one bilayered membrane, but two distinct membranes with a discernible "space" (compartment) between them, leading cytologists to call it a *nuclear envelope*. It contains many openings called *nuclear pores* [FIGURE 3.13B–D], giving the nucleus something of the appearance of a whiffle ball. Materials enter and leave the nucleus through these pores. The outer membrane of the nuclear envelope may be continuous with the membranes of the endoplasmic reticulum in places.

Nucleolus

Cell nuclei usually contain at least one dark, somewhat spherical mass called a *nucleolus* (noo-KLEE-oh-luhss; "little nucleus"; pl. *nucleoli*, noo-KLEE-oh-lye) [FIGURE 3.13]. The nucleolus contains protein, as well as DNA and RNA, and it is not bound by a membrane. Ribosomes may be partially synthesized in the nucleolus

FIGURE 3.13 NUCLEUS AND NUCLEAR ENVELOPE

[A] A simplified drawing of a cell showing the nucleus and nucleoli. [B] A drawing that shows two nucleoli floating in the nucleoplasm. [C] A scanning electron micrograph of a nucleus showing the nuclear envelope and its pores. ×15,000. [D] An electron micrograph of the nucleus and its constituents. Some of the nuclear pores are marked with arrows. ×21,000.

Nucleoli

[A] Nucleus

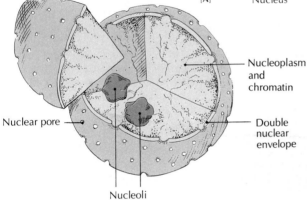

Nuclear pore

Nucleoli

[B]

Nucleoplasm and chromatin

Double nuclear envelope

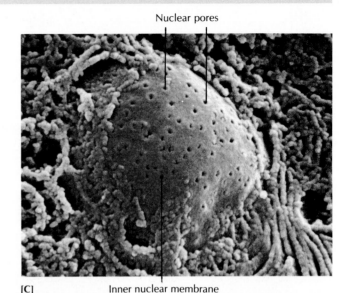

Nuclear pores

[C] Inner nuclear membrane

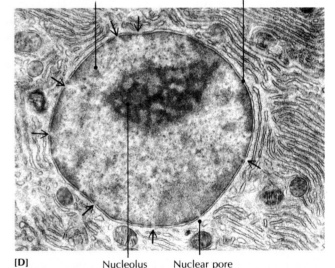

Nucleoplasm and chromatin   Double nuclear envelope

[D] Nucleolus Nuclear pore

before they pass into the cytoplasm through the pores of the nuclear envelope.

Chromosomes

The interior mass of the nucleus is the *nucleoplasm*, which contains, in a cell that is not dividing, genetic material called *chromatin* [FIGURE 3.13B, D]. Chromatin consists of a combination of DNA and protein. During a type of cell division called *mitosis* (described later in this chapter), strands of protein-rich chromatin become arranged in coiled threads called **chromosomes** (Gr. "colored bodies"); the chromosomes contain the hereditary material in segments of DNA called *genes*.* Genes are located along the chromosomes in a specific sequence and position. In fact, the sequence and position of the chemical subunits (nucleotides) that make up the genes create a *genetic code* that determines heredity and is responsible for protein structure.

*According to researchers at the California Institute of Technology, every cell in the body contains genetic information that, if translated into words, would fill 3000 books of 1000 pages each, with 1000 words on each page.

ASK YOURSELF

1 What is the main difference between a plasma membrane and nuclear envelope?

2 Why are the pores in the nuclear envelope important?

3 Distinguish between chromatin, nucleoplasm, and chromosomes.

4 Describe the structure and function of a nucleolus.

5 What constitutes a nucleotide?

The Structure of Nucleic Acids

To understand how DNA functions, you must be familiar with the structure of *nucleic* (noo-KLAY-ihk) *acids*. There are two types of nucleic acids: deoxyribonucleic acid (DNA) and ribonucleic acid (RNA). DNA makes up the chromosomes and is found only in the nucleus and mitochondria. RNA is present both in the nucleus and in the cytoplasm. DNA determines which proteins a cell will synthesize. The proteins (all enzymes are proteins) determine the structure of the cell, as well as its functions. Thus, DNA controls a cell's activities by determining which proteins are synthesized. The various kinds of RNA actually carry out the synthesis of proteins directed by the DNA.

Nucleic acids are the largest organic molecules in the body. Their size results from the bonding together of small units called **nucleotides** (NOO-klay-uh-tides). Each nucleotide has three components: a phosphate group, a five-carbon sugar, and a **nitrogenous base** [FIGURE 3.14]. The nitrogenous base is bonded to the sugar. DNA and RNA have different sugars. RNA contains ribose, a sugar that has five atoms of oxygen. Because the sugar in DNA contains one fewer oxygen atoms, it is called *deoxy*ribose.

DNA contains four kinds of bases: the double-ring *purines* (adenine and guanine) and the single-ring *pyrimidines* (thymine and cytosine). Adenine, guanine, thymine, and cytosine are usually referred to as A, G, T, and C. RNA contains the nitrogenous base uracil (U) instead of thymine. In both double-stranded DNA and single-stranded RNA, the nucleotides are connected by weak hydrogen bonds between the nitrogenous bases.

The components of DNA have been known for almost 100 years, but the structure of the DNA molecule was not discovered until 1953, when it was proposed by James D. Watson and Francis H. C. Crick. DNA has the shape of a twisted ladder, or *double helix* [FIGURE 3.14]. The "uprights" of the ladder are made of chains of alternating phosphate groups and deoxyribose sugars. The "rungs" of the ladder consist of pairs of nitrogenous bases connected to each other by hydrogen bonds and bonded to the sugars on the uprights by covalent bonds.

The bonding between the nitrogenous bases is very specific: adenine and thymine bond only with each other, as do guanine and cytosine. In RNA, uracil always bonds with adenine. Because of the specific pairing of the bases in the two strands of a DNA molecule, the strands are said to be *complementary*. For example, a thymine in one strand is always paired with an adenine in the other, and a guanine is paired with a cytosine. A purine always binds to a pyrimidine. Thus, the "rungs" are always of equal length. The sequence of the base pairs in DNA molecules determines the genetic make-up of each person.

Genes, chromosomes, and nucleic acids A *gene* is a segment of DNA that controls a specific cellular function, either by directing the synthesis of a particular protein or by regulating the activity of other genes. Genes are the hereditary units that carry hereditary traits. A **chromosome** is a nucleoprotein (nucleic acid + protein) structure in the nucleus of a cell that contains the genes.

Before a cell divides, the DNA in its nucleus must produce a perfect copy of itself. This process is called **replication** because the double strand of DNA makes a *replica*, or duplicate, of itself. The accurate replication of DNA is essential to ensure that each daughter cell receives the same genetic information as the parent cell.

THE CELL CYCLE

In the next minute, about three billion cells in your body will die. If all is well, three billion new cells will be created during that same minute. In this way, homeostasis is maintained. The correct number and kind of new cells are produced when existing, healthy cells divide. Each original (parent) cell becomes two daughter* cells, both genetically identical to the parent.

This reproduction through duplication and division is the key to development and growth. The single fertilized egg that marks the beginning of your life doubles, then each of its daughter cells doubles, and so on until a predetermined number is reached. During the earlier stages of human growth, the doubling process is accompanied by cell specialization. The process in which cells become specialized in structure and function is called *differentiation*. Once cells specialize, their functions seldom change. For example, when a cell differentiates into an epithelial cell, it will never be any other kind of cell, and neither will its daughter cells.

The process of *cell division* marks the beginning and end of the **cell cycle,** or life span, of a single cell [FIGURE 3.15]. The cell cycle is that period from the beginning of one cell division to the beginning of the next cell division. Before cell division can begin, the parent cell must double its mass and contents. For cell division to be complete, both the nucleus and the cytoplasm must divide in such a way that each daughter cell gets one complete set of chromosomes and all the necessary cytoplasmic constituents and organelles.

Thus cell division involves three major processes: (1) *interphase* — a period of cell growth during which the

---

*The use of the word "daughter" has nothing to do with gender. A "daughter" cell is not necessarily a *female* cell, but merely an *offspring*, a new cell formed from a parent cell or cells.

FIGURE 3.14 THE REPLICATION OF DNA

The replication of a twisted, double-stranded DNA molecule begins with the untwisting and separation of its two complementary strands. The "unzipping" of the double strands takes place when enzymes break the hydrogen bonds connecting the bases of the two strands. Once the bases on each strand are exposed, the enzyme *DNA polymerase* attaches to each of the separated strands and moves along it one base at a time. The enzyme selects free nucleotides from the nucleoplasm, and forms new hydrogen bonds with the exposed complementary bases on the unzipped strands. It is this principle of *complementary bonding* that ensures exact replication.

In this way, each separated strand of the original twisted DNA molecule acts as a mold, or *template,* to make an exact copy, not of itself, but of the *other* strand of the unzipped DNA molecule. As soon as the new bonds are formed, the original strand and the newly formed complementary strand begin to twist together. So the complicated activities of untwisting, unzipping, rezipping, and retwisting all occur simultaneously along different portions of the DNA molecule. In this drawing, the untwisted ends of the original double DNA strand are serving as templates for the formation of new strands, which form from free nucleotides in the cytoplasm. When replication is finished, each of the two newly formed DNA molecules will contain one strand from the original molecule and one new strand.

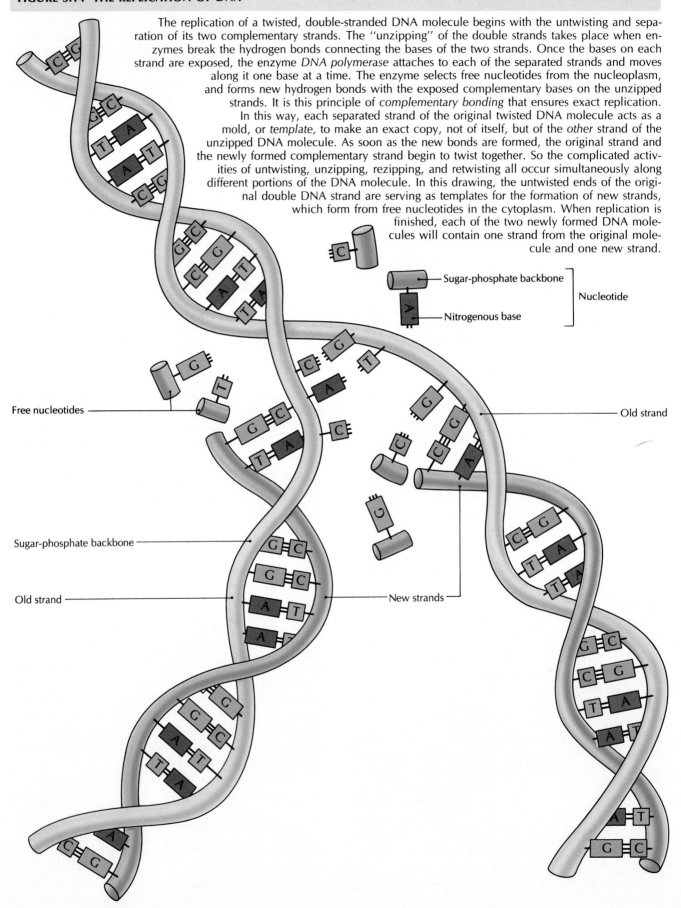

FIGURE 3.15 THE CELL CYCLE

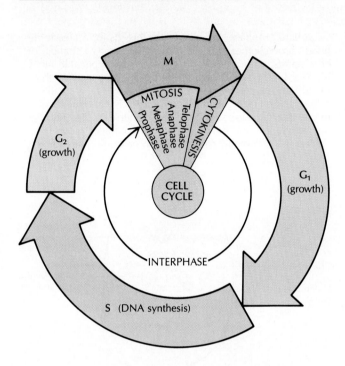

DNA in the nucleus replicates; (2) *mitosis* — the period during which the nucleus, with its genetic material, divides; and (3) *cytokinesis* — the division of the cytoplasm into two distinct but genetically identical cells. All three processes are described in the following sections.

Interphase

Often erroneously described as a resting period between cell divisions, **interphase** [FIGURE 3.16] is actually a period of great metabolic activity, occupying about 90 percent of the total duration of the cell cycle. It is the period during which the normal activities of the cell take place: growth, cellular respiration, and RNA and protein synthesis. It also sets the stage for cell division since *DNA replication occurs during interphase*.

Visually, interphase is distinguished by the presence in the nucleus of one or more nucleoli and thin chromatin threads; chromatin is the material of the chromosomes in dispersed form. There are three distinct stages of interphase:

1 G_1: Immediately after a new daughter cell is produced, it enters a period of growth also known as the "first gap," because it represents the "gap" between cell division and DNA replication. Following cell division, one centriole pair separates and begins to replicate, as do other organelles.

2 *S*: After G_1, the cell enters the S phase (for *synthesis*,

FIGURE 3.16 INTERPHASE

During interphase, the material of the chromosomes is in the form of chromatin threads.

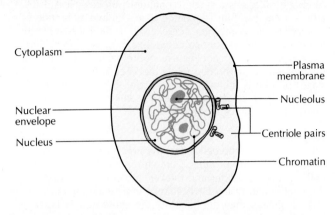

because DNA is synthesized then), and the DNA of the chromosomes replicate. Two pairs of centrioles appear during this phase.

3 G_2: With the S phase past and the chromosomes doubled, the cell goes into a "second gap" phase. The centriole pairs start to move apart as a prelude to the next cell division, and structures directly associated with the next step, mitosis, are assembled.

Mitosis and Cytokinesis

Once the DNA of the parent cell has been duplicated, the actual division of the cell can begin. This division has two parts; the first is nuclear division, or *mitosis*, and the second is *cytokinesis*, the division of the cytoplasm [FIGURE 3.17]. Mitosis and cytokinesis together are referred to as the *M phase* [FIGURE 3.15]. **Mitosis** accomplishes two things: (1) the arrangement of all cellular material for equal distribution between the daughter cells, and (2) the actual *division and equal distribution of DNA to each new cell*. Mitosis is triggered by the replication of DNA, but it is not known what initiates the DNA replication.

A S K   Y O U R S E L F

1 What is cellular differentiation?

2 Why is it necessary for DNA to replicate before mitosis begins?

3 How does complementary bonding occur in DNA?

4 What occurs during interphase?

5 What are the stages of mitosis?

6 What does mitosis accomplish?

7 What is cytokinesis?

FIGURE 3.17 MITOSIS AND CYTOKINESIS

MITOSIS

Although mitosis is a single, continuous process, it is usually described in four sequential stages known as *prophase, metaphase, anaphase,* and *telophase.* For the sake of clarity, the drawings that accompany the text show only four chromosomes and no organelles except the centrioles. The light micrographs show human white blood cells in various stages of mitosis. Human cells contain 46 chromosomes, but the principles of mitosis are the same no matter how many chromosomes are involved.

Prophase

After interphase, the cell enters the first stage of mitosis, **prophase.** In *early prophase,* the chromatin threads begin to coil, becoming shorter and thicker. (Before coiling begins, the total length of the DNA strands in a single human nucleus is greater than the length of the entire human body, but the strands are too thin to be seen with an ordinary light microscope.) The nucleoli and nuclear envelope begin to break up, and bursts of microtubules, called *asters* (L. stars), begin to radiate from the centrioles.

By *late prophase,* the chromatin threads form clearly defined chromosomes. Each chromosome has two strands, or **chromatids,** with a full complement of the replicated DNA formed during the S stage of interphase. Each pair of chromatids is joined somewhere along its length by a small spherical structure called a *centromere.* The fragments of the nuclear envelope and nucleoli disperse in the endoplasmic reticulum, and newly formed microtubules move in among the chromatid pairs. The two centriole pairs move apart.

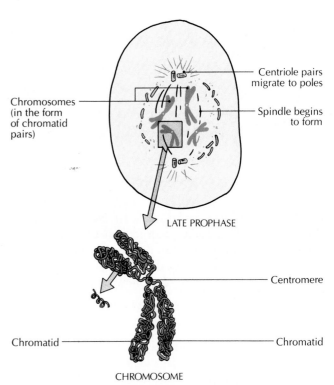

Centriole pairs migrate to poles

Spindle begins to form

Chromosomes (in the form of chromatid pairs)

LATE PROPHASE

Centromere

Chromatid

Chromatid

CHROMOSOME

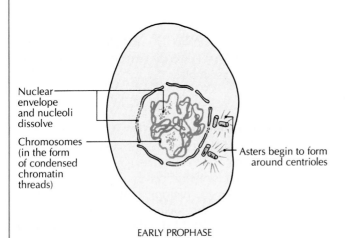

Nuclear envelope and nucleoli dissolve

Chromosomes (in the form of condensed chromatin threads)

Asters begin to form around centrioles

EARLY PROPHASE

By the end of prophase, the centriole pairs have moved to opposite ends, or *poles,* of the nucleoplasm. The position of the centrioles at the poles determines the direction in which the cell divides. Between the centrioles, the microtubules form a *spindle* that extends from pole to pole. The chromatid pairs move to the center of the spindle. The asters, spindle, centrioles, and microtubules are called the *mitotic apparatus.*

THE MITOTIC APPARATUS

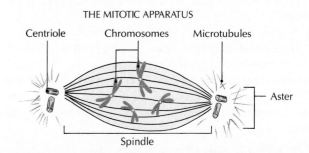

Centriole   Chromosomes   Microtubules

Aster

Spindle

Metaphase

During *metaphase,* the centromere of each chromatid pair attaches to one of the microtubules of the spindle. The centrioles are pushed apart as the spindle lengthens, and the double-armed chromosomes are pulled to the center of the nucleoplasm, lining up across the center of the spindle. Toward the end of metaphase, the centromeres double, so that each chromatid has its own centromere. At this point, each chromatid may be considered a complete chromosome, each with its double-stranded DNA molecule.

Anaphase

The shortest stage of mitosis is *anaphase.* In *early anaphase,* the chromosome pairs separate, and the members of each pair begin to move toward opposite poles of the cell. The microtubules are instrumental in moving the chromosomes toward the poles. By *late anaphase,* the poles themselves have moved farther apart. At this stage in human mitosis, 46 chromosomes are near one pole, and 46 are near the opposite pole.

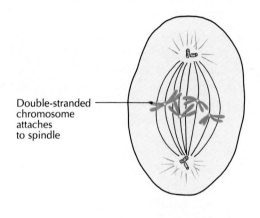

Double-stranded chromosome attaches to spindle

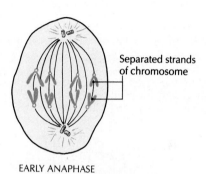

Separated strands of chromosome

EARLY ANAPHASE

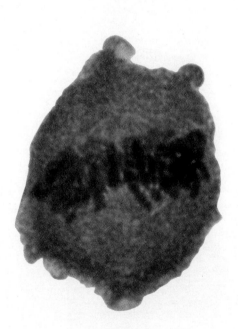

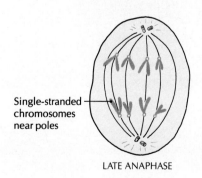

Single-stranded chromosomes near poles

LATE ANAPHASE

(Figure 3.17 continues on following page)

FIGURE 3.17 MITOSIS AND CYTOKINESIS (Continued)

MITOSIS (Continued)	CYTOKINESIS

Telophase

In **telophase,** the chromosomes arrive at the poles. In *early telophase,* the chromosomes lose their distinctive rodlike form and begin to uncoil. They appear as they did during interphase. The spindle and asters dissolve. By *late telophase,* fragments from the endoplasmic reticulum spread out around each set of chromosomes, forming a new nuclear envelope. The cell begins to pinch in at the middle. New nucleoli form from the nucleolar regions of the chromosomes. *Mitosis* is over, but *cell division* is not.

The third major phase of the cell cycle, and the final stage of cell division, is **cytokinesis** (Gr. *cyto,* cell + *kinesis,* movement), the separation of the cytoplasm into two parts. Before cytokinesis, the two newly formed nuclei still share the same cytoplasmic compartment. The separation is accomplished by *cleavage,* a pinching of the plasma membrane. The *cleavage furrow,* where the pinching occurs, looks as though someone tied a cord around the middle of the cell and pulled it tight. Cytokinesis may begin during telophase or anaphase, and the doubling of the centrioles may begin before cytokinesis is complete. The two new, genetically identical daughter cells formed are each about half the size of the original parent cell.

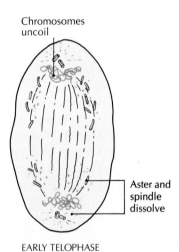

Chromosomes uncoil

Aster and spindle dissolve

EARLY TELOPHASE

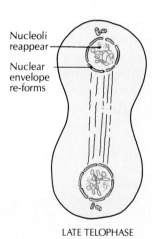

Nucleoli reappear

Nuclear envelope re-forms

LATE TELOPHASE

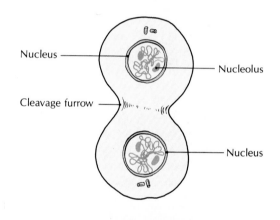

Nucleus

Nucleolus

Cleavage furrow

Nucleus

MEIOSIS

When we lose cells we replace them through mitosis, and when we grow from a single cell to an adult body with over 50 trillion cells, we also do it through mitosis. But when we reproduce, we use another process of cell division called **meiosis** (mye-OH-sihss; Gr. *meioun,* to diminish).

As you just learned, *mitosis* guarantees that new daughter cells receive exactly the same genetic information as the parent cell. If the parent cell has 23 *pairs* of chromosomes, or 46 chromosomes, each daughter cell will also have 23 *pairs* of chromosomes.

However, sex cells, or *gametes,* are not formed by mitosis. If a sperm contained 46 chromosomes and an ovum contained 46 chromosomes, their combination during fertilization would produce a cell with 92 chromosomes. Such a cell would not be a normal human cell. The problem is solved when *meiosis reduces the number of chromosomes in gametes to half*—each sperm has 23 chromosomes (not 23 *pairs*), and each ovum has 23 chromosomes (not 23 *pairs*). When the gametes (ovum and sperm) unite, they produce a cell with 46 chromosomes. This cell is called a *zygote.*

Comparison of Mitosis and Meiosis

Meiosis is divided into the same phases as mitosis: prophase, metaphase, anaphase, and telophase. However, in mitosis, the chromosomes replicate once and the nucleus divides once, while in meiosis, a single replication of the chromosomes is followed by *two* nuclear divisions [FIGURE 3.18].

The daughter cells produced by mitosis have the same number of chromosomes as the parent cell, or the **diploid** (*2n*) number. The diploid number is 46 in human beings. Body cells, that is, all cells except sex cells, are diploid. The daughter cells produced by meiosis have only half the parental number, or the **haploid** (*1n*) number. Gametes, with only 23 chromosomes, are haploid.

A critical part of meiosis occurs early in the sequence, when each double-stranded chromosome from the male parent lines up with a double-stranded chromosome from the female parent. Each male-female set of four chromatids is called a **homologous pair,** or a **tetrad** (Gr. *tetra,* four). At this point, portions of a male chromatid may be exchanged with portions of its female homologue, a process called **crossing-over.** Crossing-over results in a rearrangement of genes. The great variety among human beings is due to the unlimited possibilities of such genetic variation through meiosis and sexual reproduction.

FIGURE 3.18 COMPARISON OF MITOSIS AND MEIOSIS

In the simplified example shown here, there are four chromosomes: two long ones (one from the male parent and one from the female parent) and two short ones (one from the male parent and one from the female parent). Mitosis, with one division, results in two cells identical to the original one, with four chromosomes in each cell. Meiosis, with two divisions, results in four cells with two chromosomes in each cell.

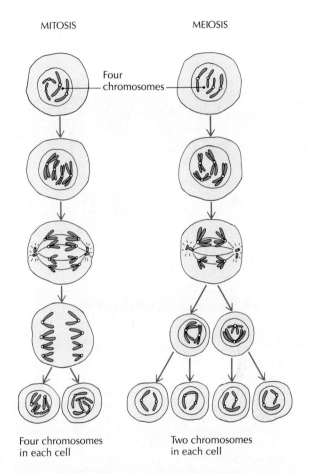

MITOSIS   MEIOSIS

Four chromosomes

Four chromosomes in each cell   Two chromosomes in each cell

DEVELOPMENT OF PRIMARY GERM LAYERS

The one-celled zygote produced by fertilization undergoes repeated cell divisions, creating a cluster of hundreds of identical daughter cells. Once the mass of cells becomes implanted in the uterus, cell division stops temporarily. When cell division starts again, *all the cells produced are not alike.* **Differentiation,** the process by which cells develop into specialized tissues and organs, begins. Soon the *inner cell mass* which is now called an *embryo,* is rearranged into three distinct **primary germ layers:** endoderm, ectoderm, and mesoderm [FIGURE 3.19 and TABLE 3.3].

FIGURE 3.19 FORMATION OF PRIMARY GERM LAYERS

The primary germ layers are the ectoderm, endoderm, and mesoderm. These layers give rise to all body tissues.

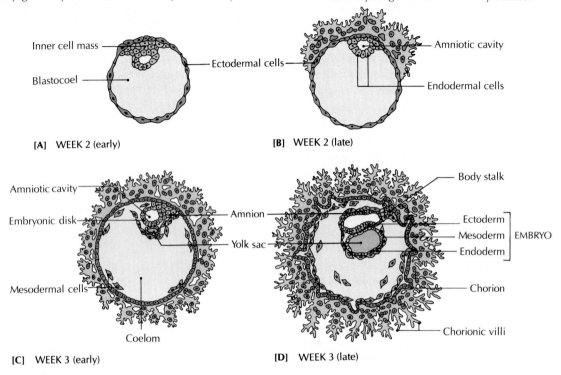

TABLE 3.3 MAJOR STRUCTURES DERIVED FROM PRIMARY GERM LAYERS

Endoderm	Mesoderm	Ectoderm

Epithelium of:
  Pharynx, larynx, trachea, lungs
  Tonsils, adenoids
  Thyroid, thymus, parathyroids
  Esophagus, stomach, intestines,
    liver, pancreas, gallbladder;
    glands of alimentary canal
    (except salivary)
  Urinary bladder, urethra (except terminal male portion),
    prostate, bulbourethral glands
  Vagina (partial)
  Inner ear, auditory tubes
	All muscle tissue (cardiac, smooth, skeletal), except in iris and sweat glands
	All connective tissue (fibrous, adipose, cartilage, bone, bone marrow)
	Synovial and serous membranes
	Lymphoid tissue: tonsils, lymph nodes
	Spleen
	Blood cells
	Reticuloendothelial system
	Dermis of skin
	Teeth (except enamel)
	Endothelium of heart, blood vessels, lymphatics
	Epithelium of: gonads, reproductive ducts; adrenal cortex; kidneys, ureters; coelom, joint cavities
		All nervous tissue
		Epidermis of skin; hair follicles, nails
		Epithelium and myoepithelial cells of sweat glands, oil glands, mammary glands
		Lens of the eye
		Receptor cells of sense organs
		Enamel of teeth
		Adrenal medulla
		Anterior pituitary
		Epithelium of: salivary glands, lips, cheeks, gums, hard palate, nasal cavity, sinuses
		Lower third of anal canal
		Terminal portion of male urethra
		Vestibule of vagina and vestibular glands
		Muscle of iris

The exact sequence of the formation of germ layers is difficult to ascertain, especially since the layers of ectoderm and endoderm form almost simultaneously [FIGURE 3.19A, B]. During the second week of development, the outer embryonic layer is formed. This is the *ectoderm* ("outside skin"), which will give rise to the outer layers of skin, hair, fingernails, tooth enamel, the nervous system, and other epithelial structures. Also during the second week, a flattened layer of cells appears on the inner cell mass, facing the inner cavity, or *blastocoel*. This layer of cells is the *endoderm* ("inside skin"). From this layer will develop the lining of the digestive tract and the glands and organs associated with it, the lining of the respiratory tract, and the lining of the urinary bladder, urethra, and vagina.

The third week is a period of rapid development for the embryo, and it is also the time when the prospective mother will miss her menstrual period and suspect that she is pregnant. About the fifteenth day, the inner cell mass develops into the three-layered **embryonic disk** [FIGURE 3.19C]. The cells of the embryonic disk fill the area between the endoderm and ectoderm and eventually form the bulk of the embryo. This final layer is the **mesoderm** ("middle skin") [FIGURE 3.19D], the forerunner of connective tissue in lower skin layers, bone, muscle, blood, and the epithelium of some internal organs.

CELLS IN TRANSITION: AGING

Much evidence points to the idea that aging is a normal, genetically programmed process involving a series of interrelated cellular events that accumulate until a change becomes noticeable and permanent.

Scientists have speculated that if the three main causes of death in old age—heart disease, cancer, and stroke—were eliminated, the average human life expectancy would be extended only 5 to 10 years beyond the present 72. If this is true, there must be a cause of aging and death besides disease. As more and more people reach old age, the study of aging, **gerontology,** becomes more and more important.

Part of the aging process includes the death of cells, or **necrobiosis** (Gr. *nekros,* corpse, death + *biosis,* way of life). Necrobiosis is a natural death, as opposed to **necrosis,** which is the death of a cell or tissue resulting from irreversible damage caused by disease or accident.

Hypotheses of Aging

There are several hypotheses that attempt to explain the basic mechanisms of aging. At least some of these mechanisms are probably at work at the same time to produce what we call aging. In other words, there is probably no single hypothesis that can explain aging.

One of the most important signs of aging is the **free-radical damage** that occurs within cells. Metabolic activity in mitochondria produces superoxide radicals. These negatively charged oxygen molecules are normally converted into hydrogen peroxide by a specific enzyme, and the potentially dangerous hydrogen peroxide is converted into oxygen and water by another enzyme. As cells age, however, the necessary enzymes are not produced fast enough, and superoxide radicals and hydrogen peroxide accumulate. These highly reactive substances can produce harmful molecules that impair the normal functions

of phospholipids, proteins, nucleic acids, and enzymes.

Another hypothesis of aging concerns **limits on cell division.** Normal cells of the body are not programmed to grow and reproduce forever, and certain cells are capable of only a predetermined number of divisions. Connective tissue cells from human embryos divide about 50 times before they slow down and stop. Similar cells from a 40-year-old person stop after about 40 divisions, while cells from an 80-year-old stop after about 30 doublings. The permanent cessation of normal cell division is called **cell senescence.**

Occasional **mutations** (changes in DNA structure) occur as the genetic message in DNA is passed along to RNA. Many of these mutations are corrected during RNA processing, but some are expressed in the production of a defective enzyme or other protein. Defective enzymes cause a cell to operate inefficiently, and eventually to die. The more an error is copied, the more severe it becomes. DNA has a built-in repair mechanism that corrects most of these defects before they become serious enough to be noticed. However, the "repair DNA" itself does not remain totally effective for an indefinite time. Eventually, the repair mechanism breaks down, but the mutations persist. Cells die, and "aging" occurs.

The *amount* of DNA may also play a role in aging. **Redundant** or "extra," **DNA** exists in the nucleus of a cell. If a gene is damaged, it can be repaired immediately by an identical sequence of redundant DNA, and no permanent damage is done. After a period of time, however, the redundant DNA is used up, and harmful or damaged genes are free to function. Eventually, changes associated with aging appear in the body.

Other factors, such as the strain on cells, the accumulation of cellular wastes, hormone imbalances, and cell starvation probably contribute to the aging process. Also, the body's immune system may begin to destroy normal cells instead of foreign ones, producing certain cardiovascular disorders, diabetes, rheumatoid arthritis, or some cancers.

The **thickening of collagen** seems to be a considerable factor in aging. The connective tissue *collagen* (KAHL-uh-juhn; Gr. *kolla,* glue + *genes,* born, produced) is the most common protein in the body. As the body ages, the amount of the protein *elastin* in elastic fibers decreases and the collagen present thickens, resulting in hardening of the arteries, stiffening of the joints, sagging muscles, wrinkling of the skin, and other changes that eventually slow down and kill the body.

Glycosylation is the chemical bonding of glucose to proteins without the aid of enzymes. It has been suggested that glycosylation adds glucose randomly to growing polypeptide chains, resulting in the formation of cross-links between protein molecules. Glycosylation could explain the cross-linking of proteins that contributes to the stiffening and nonresilience of aging tissues.

A Final Word about Growing Old

Many people in our society have been conditioned to regard old age as a time when many functions, such as sexual desire, memory, and learning ability, are inevitably decreased or even lost altogether. However, such drastic changes in mental and physical prowess are usually the result of serious disease or psychological problems and are not necessarily a part of the normal aging process. Proper diet, physical activity, and mental attitude will overcome many of the routine problems of old age. Sometimes we give up before our bodies do.

ASK YOURSELF

1 What is the difference between necrobiosis and necrosis?

2 What are some of the current hypotheses of aging?

CELLS OUT OF CONTROL: CANCER

Normal human body cells usually divide at exactly the rate required for them to replace the dying ones. Cancer cells are different. They lack a controlling mechanism that "tells" them when to stop reproducing. *Cancer* (L. crab, because to early anatomists the swollen veins surrounding the affected body part looked like the limbs of a crab) occurs when cells grow abnormally and then spread beyond the original site.

Neoplasms

When cells do not stop reproducing and die after the typical number of cell divisions, they form an abnormal growth of new tissue called a tumor, or *neoplasm* (Gr. *neos*, new + L. *plasma*, form). Neoplasms are either benign or malignant.

The cells of a *benign neoplasm* (L. *bene*, well) do not differ greatly in structure from those in normal tissue. Although the cells grow abnormally, their nuclei divide almost like normal cells, and there are usually few chromosomal changes. A benign neoplasm remains enclosed within a capsule of thick connective tissue and does not spread beyond its original site [FIGURE 3.20A]. A benign neoplasm can generally be located rather easily and either removed surgically or treated with another appropriate procedure without further problems. But if they are not removed in time, benign neoplasms occasionally become malignant.

The most important characteristic of a typical *malignant neoplasm* (L. *malus*, bad) is that its cells tend to

break out of their connective tissue capsule and invade neighboring tissue [FIGURE 3.20B, C]. Malignant neoplasms, or cancers, grow rapidly in an uncontrollable and disorderly pattern that can be recognized by obvious changes in chromosome, cell, and tissue structure [FIGURE 3.21].

Malignant neoplasms are usually classified according to the type of body tissue in which they originate:

1 A *carcinoma* (L. cancerous ulcer; *oma*, tumor) originates in *epithelial tissues*. Carcinomas are the most common type of cancer, occurring most often on the skin and in the lungs, breasts, intestines, stomach, mouth, and uterus. They are usually spread by way of the lymphatic system.

2 A *sarcoma* (Gr. *sarkoun*, to make fleshy) originates in *muscle cells* or in *connective tissue*. Sarcomas are usually spread through the bloodstream and may be found anywhere in the body.

3 A *mixed-tissue neoplasm* derives from tissue that is capable of differentiating into either epithelial or connective tissue and, as a result, is composed of several types of cells.

4 Cancers that do not fit into the categories above include the various *leukemias* derived from tissues that produce certain abnormal blood cells, and cancers that arise from cells of the nervous system.

How Cancer Spreads

The spread of a malignant neoplasm is called *metastasis* (muh-TASS-tuh-sihss; Gr. *meta*, involving change + *stasis*, state of standing). Metastasis begins when cancer cells break away from the original site, or *primary neoplasm*, and extend into normal tissues. Once cancer cells move from the original site, they may enter a body cavity, such as the thoracic or abdominal cavity, or they may enter the bloodstream or the lymphatic system and be transported away from the primary neoplasm. They can then establish new malignant *secondary neoplasms* elsewhere in the body.

As cancer cells spread, or *metastasize*, they encounter defensive cells of the body's immune system and are sometimes killed. Cancer is lethal when the body can no longer fight off the ever-increasing number of malignant cells, which disrupt and starve healthy cells.

Cancers usually progress slowly. In epithelial tissues, for example, the first stage is usually *dysplasia*, a condition in which dividing epithelial cells are not restricted to the basal layers as usual and show some abnormal differentiation. Dysplasia in cells of the uterine cervix can usually be detected in a "Pap smear." *Carcinoma in situ*

FIGURE 3.20 DIFFERENCE BETWEEN BENIGN AND MALIGNANT NEOPLASM

[A] A benign neoplasm is made up of cells that grow abnormally, but remain enclosed within a sturdy capsule of connective tissue. **[B]** The cells of a malignant neoplasm burst out of the capsule and can spread to other areas of the body. **[C]** Cells that break away from a malignant neoplasm may enter the lymphatic system and eventually the bloodstream through small openings in lymphatic and blood vessels. Once in the bloodstream, cancer cells may establish secondary neoplasms far from the primary neoplasm.

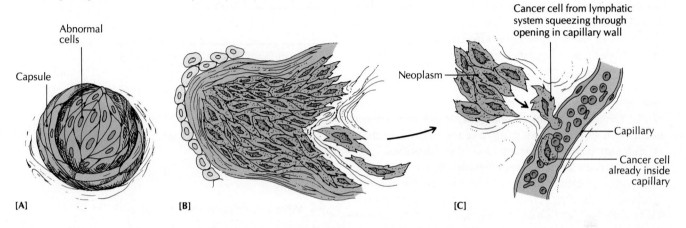

FIGURE 3.21 THE USUAL PROGRESSION OF EPITHELIAL CANCER CELLS

Cancer of the uterine cervix, showing the progression from dysplasia to a full-blown malignant carcinoma, where cancerous cells invade the underlying connective tissue and begin to spread.

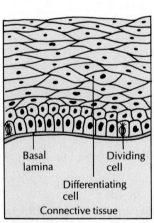 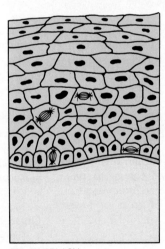  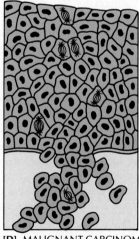

[A] NORMAL [B] DYSPLASIA [C] CARCINOMA *IN SITU* [D] MALIGNANT CARCINOMA

(L. *"cancer in place"*) is an abnormal condition in which apparently undifferentiated dysplastic epithelial cells proliferate in all tissue layers. The final stage in the progression of epithelial cancer reveals a ***malignant carcinoma,*** where cancerous epithelial cells break through the basal lamina to invade the underlying connective tissue.

Causes of Cancer

In some cases, the *tendency* to develop cancer may be inherited, but cancer researchers believe that most cancers are caused by repeated exposures to cancer-causing agents, or ***carcinogens*** (kar-SIHN-uh-jehnz). The main categories of carcinogens are *chemicals, radiation,* and *vi-*

ruses. Carcinogens disrupt the homeostasis of normal cells, and may eventually cause them to become cancerous. Most cancers are probably the result of prolonged exposure to a *combination* of several carcinogens.

Several hypotheses for the development of cancer have been proposed. One hypothesis suggests that carcinogens cause genetic changes, or mutations, in DNA. Current thinking is that certain genes help prevent cancer until they are made ineffective by a mutation, or that an extensive rearrangement of genetic material caused by chemicals or viruses produces effective mutations. Another hypothesis suggests that the forerunner of a cancer-causing gene—a *proto-oncogene* (Gr. *onkos*, tumor)—is a gene that helps regulate normal cell division. But when a proto-oncogene undergoes mutation it can be transformed into an **oncogene,** which can enter a normal cell and transform it into a cancer cell. More than 50 different proto-oncogenes have been identified.

Considerable evidence supports the idea that at least some types of cancer are caused by viruses. Viruses are involved in Burkitt's lymphoma (Epstein-Barr virus), warts (benign papillomas of the skin), cervical cancer (herpes type II and certain papilloma viruses), and hepatocellular (liver) cancer (hepatitis B virus).

Treatment of Cancer

The main treatments for cancer are surgery, radiation therapy, and drug therapy (chemotherapy); *combination therapy* uses more than one type of therapy. Surgery and radiation therapy have been used together for some time, and physicians are now including follow-up chemotherapy, especially for cancer of the breast, bones, large intestine, lung, and stomach.

ASK YOURSELF

1 What is the difference between a benign and malignant neoplasm?

2 What is a carcinoma? A sarcoma?

3 What is the difference between a primary and secondary neoplasm?

4 What are carcinogens, and what are their main categories?

5 What is the "oncogene" hypothesis?

6 What are the main types of cancer treatment?

CHAPTER SUMMARY

What Are Cells? (p. 55)

1 Cells can range widely in size, color, shape, and function.

2 All cells have four basic parts: the **nucleus, nucleoplasm, cytoplasm,** and the **plasma membrane.**

3 The **nucleus,** the control center of the cell, contains the chromosomes that direct the reproduction of, and contain the genetic blueprint for, new cells. The nucleus is surrounded by a nuclear envelope and contains **nucleoplasm.**

4 The portion of the cell surrounding the nucleus is **cytoplasm.** The fluid portion of the cytoplasm is called *cytosol;* it contains various subcellular structures called *organelles* and *inclusions.*

5 The **plasma membrane** forms the outer boundary of the cell.

Cell Membranes (p. 57)

1 Cell membranes are selective screens that allow certain substances to get into and out of cells. That quality is called **selective permeability.**

2 Cell membranes, which are composed mainly of phospholipids and proteins, are described by a dynamic **fluid-mosaic model.**

Movement across Membranes (p. 58)

1 **Passive transport** occurs when molecules use their own energy to pass through a cell membrane from areas of high concentration to areas of lower concentration.

2 Examples of passive processes are **simple diffusion, facilitated diffusion, osmosis,** and **filtration.**

3 **Active processes** of movement across membranes require energy from the cell, and can move substances through a selectively permeable membrane from areas of low concentration to areas of higher concentration.

4 Active processes include **active transport, endocytosis, phagocytosis,** and **exocytosis.**

Cytoplasm (p. 61)

1 The **cytoplasm** of a cell is composed of two parts, or phases. The *particulate phase* consists of membrane-bound vesicles and organelles, lipid droplets, and inclusions. The *aqueous phase* consists of fluid cytosol. The typical components of cytosol are water, nucleic acids, proteins, carbohydrates, lipids, and inorganic substances.

Organelles (p. 61)

1 The **endoplasmic reticulum (ER)** creates a series of channels for transport, stores enzymes and other proteins, and provides a point of attachment for ribosomes. Smooth ER does not contain ribosomes.

2 **Ribosomes** are the sites of protein synthesis.

3 **The Golgi apparatus** aids in the synthesis of glycoproteins. It also aids in the packaging of products to be secreted by the cell.

4 *Lysosomes* digest nutrients, and clean away dead or damaged cell parts.

5 *Mitochondria* produce most of the energy (in the form of ATP) required by a cell for metabolic activities.

6 *Microtubules, intermediate filaments,* and *microfilaments* (the *cytoskeleton*) provide a transport system, a supportive framework, and assist with organelle and chromosome movement.

7 *Centrioles* assist in cell reproduction and are involved with the movement of chromosomes during cell division.

8 *Cilia* are appendages that help move solids and liquids past fixed cells, and **flagella** help free-swimming cells move.

9 *Cytoplasmic inclusions* are chemical substances that are usually either basic food material or the stored products of the cell's metabolic activities.

The Nucleus (p. 67)

1 The *nucleus* of a cell contains DNA, which controls the cell's genetic program, including heredity, protein structure, and other metabolic activities.

2 The *nuclear envelope* contains many *nuclear pores* that allow material to enter and leave the nucleus.

3 Within the nucleus, *chromosomes* store the segments of DNA called *genes*, which contain the hereditary information for protein synthesis.

4 The *nucleolus* is a preassembly point for ribosomes.

The Cell Cycle (p. 69)

1 The process in which cells become specialized in structure and function is called **differentiation.**

2 When DNA *replicates,* a copy of itself is made to ensure that each new cell receives the same genetic information as the parent cell.

3 *Mitosis* distributes all cellular material from the parent cell to the daughter cells equally; it includes the actual division of the DNA. It is divided into four sequential stages: *prophase, metaphase, anaphase,* and *telophase.*

4 *Cytokinesis* is the process of cytoplasmic division that produces two daughter cells after the nucleus divides during mitosis.

Meiosis (p. 75)

Meiosis reduces the number of chromosomes in sex cells to half, with each sperm having 23 chromosomes and each ovum having 23. When ovum and sperm unite, they produce a cell with 46 chromosomes. A cell with the full number of chromosomes is **diploid,** and a cell with only half is **haploid.**

Development of Primary Germ Layers (p. 75)

1 The fusion of ovum and sperm nuclei produces the diploid *zygote.* The daughter cells, produced by a process of cell division called **cleavage,** are *blastomeres.* During the third day, about the time the zygote is entering the uterus, a solid ball of about 16 blastomeres is formed, which is called the *morula.*

2 After two or three days, the morula develops into a fluid-filled hollow sphere called a *blastocyst.* The blastocyst is composed of about 100 cells. A grouping of cells at one pole is the *inner cell mass* from which the embryo will grow.

3 *Implantation* of the embryo in the uterine lining occurs about a week after fertilization.

4 During week 2, the blastocyst develops into the *bilaminar embryonic disk,* composed of the endoderm and ectoderm. The zygote is now an *embryo.*

5 The *differentiation* of cells into specialized tissues and organs begins shortly after the start of implantation. The inner cell mass of the blastocyst is rearranged into the *primary germ layers:* endoderm, ectoderm, and mesoderm.

6 *Endoderm* develops mainly into the innermost organs, *ectoderm* into the outer layers of skin and the nervous system, and *mesoderm* into connective tissue, bone, muscle, blood, and some inner epithelium.

Cells in Transition: Aging (p. 77)

1 Part of the aging process includes the natural death of cells, or **necrobiosis.**

2 Aging is not a single process, but a series of interrelated events that accumulate until a change becomes noticeable and permanent.

3 No single theory can explain aging. Among the mechanisms that produce aging are free-radical damage, genetically determined limits on cell division, DNA mutations, loss of redundant DNA, thickening of collagen, and glycosylation.

Cells Out of Control: Cancer (p. 78)

1 When cells reproduce faster than the normal rate, and when they do not die after the typical number of cell divisions, they form an abnormal growth of tissue called a *neoplasm.*

2 The cells of a *benign neoplasm* grow abnormally, but do not differ significantly in structure from those in normal tissue and do not spread beyond their original site. The cells in a *malignant neoplasm,* or *cancer,* have the ability to invade surrounding tissues.

3 Malignant neoplasms are usually classified according to the type of body tissue in which they originate. *Carcinomas* originate in epithelial tissue, *sarcomas* in connective tissue, and *mixed-tissue neoplasms* in tissue that is capable of differentiating into either epithelial or connective tissue.

4 The spread of cancer cells beyond the primary neoplasm and the establishment of secondary neoplasms in other parts of the body is called *metastasis.*

5 The main categories of *carcinogens* (cancer-causing agents) are *chemicals, radiation,* and *viruses.* It is thought that most cancers are caused by repeated and prolonged exposures to a *combination* of several carcinogens.

6 Hypotheses to explain the basic cause of cancer involve DNA mutations, oncogenes (cancer genes), and viruses.

7 The main treatments of cancer are *surgery, radiation therapy,* and *chemotherapy. Combination therapy* uses more than one type of therapy to treat cancer.

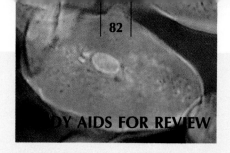

DY AIDS FOR REVIEW

MEDICAL TERMINOLOGY

ADENOMA (Gr. *aden*, gland + *-oma*, tumor) An epithelial tumor of glandular origin, usually benign or of low-grade malignancy.

ANAPLASIA (Gr. *ana*, reversion + *plasia*, growth) Reversion of cells to a less differentiated form.

BIOPSY (Gr. *bios*, life + *-opsy*, examination) The microscopic examination of tissue removed from the body.

GERIATRICS (Gr. *geras*, old age + *iatrikos*, physician, pertaining to a specific kind of medical treatment) The medical study of the physiology and ailments of old age.

HYPERPLASIA (Gr. *hyper*, above + *plasia*, growth) Enlargement of an organ or tissue due to an increase in the number of cells.

HYPERTROPHY (Gr. *hyper*, above + *trophe*, nourishment) Enlargement of an organ or tissue due to an increase in the size of cells.

MELANOMA (Gr. *melas*, black + *-oma*, tumor) A neoplasm made up of cells containing a dark pigment, usually melanin.

METAPLASIA (Gr. *meta*, involving change + *plasia*, growth) The change of cells from a normal state to an abnormal one.

ONCOLOGY (Gr. *onkos*, mass, tumor + *logy*, the study of) The study of neoplasms; the study of cancer.

PROGERIA (L. *pro*, before + Gr. *geras*, old age) Premature old age due to the early cessation of cell division.

UNDERSTANDING THE FACTS

1 What is the most abundant substance in cytosol?

2 What is the relationship between the quantity of protein in a plasma membrane and its metabolic activity?

3 Distinguish between the centrosome and centrioles.

4 What is an important difference between DNA and RNA as far as their locations within a cell are concerned?

5 Define *mitosis*.

6 Does DNA replication occur during mitosis? Explain.

7 During the process of cell division, when are the chromosomes first visible?

8 Which structures of a dividing cell are composed of microtubules?

9 What are the three main causes of death in old age?

10 What is the free-radical theory of aging?

11 What is probably the most fundamental difference between cancer cells and normal cells?

12 Malignant neoplasms are usually classified according to the tissues in which they originate. List these categories.

13 Name some of the more important carcinogens.

UNDERSTANDING THE CONCEPTS

1 What would happen if your cells lacked peroxisomes? Why?

2 What clues can you derive as to the function of a cell from the number of mitochondria it contains? What about the lysosomes?

3 Why is a "cell" that lacks a nucleus destined to have a relatively short life? Give one example of such a cell.

4 Discuss the following hypotheses about aging: (a) DNA mutations, (b) redundant DNA, (c) limit on cell divisions, (d) collagen thickening, (e) glycosylation.

5 Some cancers take many years to develop after exposure to a carcinogen or group of carcinogens. How does this fact relate to the mutation hypothesis?

6 Explain how cancer spreads from the primary site to secondary sites in the body.

7 Discuss the following possible causes of cancer: (a) mutation hypothesis, (b) "cancer-gene" hypothesis, (c) virus hypothesis.

SELF-QUIZ

Multiple-Choice Questions

3.1 Which of the following provides a barrier to the movement of materials across the plasma membrane?
   a proteins
   b surface carbohydrates
   c lipids
   d internal carbohydrates
   e all of the above

3.2 The endoplasmic reticulum
   a is the site of ATP synthesis
   b may have ribosomes attached to its surface
   c may not have ribosomes attached to its surface
   d contains acid hydrolases
   e b and c

3.3 Which of the following statements about centrioles is/are true?
   a They are located near the nucleus.
   b They play a role in cell division.
   c They play a role in the distribution of genetic material.
   d They contain protein.
   e All of the above.

3.4 Cell division in human body cells may be conveniently thought of as occurring in two stages:
a the G_1 phase and the G_2 phase
b anaphase and telophase
c the S and G_1 phases
d nuclear division and chromosome division
e mitosis and cytokinesis

3.5 The major components of the plasma membrane are
a lipids and carbohydrates
b proteins and carbohydrates
c lipids, proteins, and cholesterol
d carbohydrates and cholesterol
e carbohydrates and glycoproteins

3.6 The nucleolus
a is the site of ribosomal RNA synthesis
b has a surrounding membrane
c contains ATP used for chromosomal duplication
d is smaller in secretory cells than in nonsecretory cells
e specifies the chemical structure of enzymes

True-False Statements

3.7 In a cell that is not dividing, the genetic material is in the form of chromosomes.

3.8 Nucleoli contain DNA, RNA, and protein.

3.9 Division of the cell nucleus is called cytokinesis.

3.10 Asters are microtubules that radiate from the chromosomes during prophase.

3.11 Cytoplasmic inclusions are chemical substances that are usually basic food materials or the stored products of metabolism.

3.12 One hypothesis for the development of cancer suggests that carcinogens cause mutations in DNA.

3.13 Aging is a genetically programmed process.

Completion Exercises

3.14 The carbohydrate-rich zone of a plasma membrane is called the _____.

3.15 The characteristic of a membrane that allows some substances to enter the cell and that blocks others is called _____.

3.16 The _____ is a complex labyrinth of flattened sheets and double membranes that spreads through the cytoplasm.

3.17 _____ are small membrane-bound organelles that contain acid hydrolases.

3.18 The spread of a neoplasm is termed _____.

3.19 Any agent that causes cancer is called a/an _____.

3.20 A _____ is a change in the genetic material.

Matching

3.21 ___ Golgi apparatus	a	nuclear division
3.22 ___ ATP	b	energy for cellular activity
3.23 ___ DNA		
3.24 ___ ribose	c	asters
3.25 ___ mitochondrion	d	a secretory organelle
3.26 ___ microtubules	e	nucleic acid subunits
3.27 ___ lysosomes	f	makes up a gene
3.28 ___ microbodies	g	peroxisomes
3.29 ___ nucleotides	h	site of ATP production
3.30 ___ mitosis	i	digestive enzymes
3.31 ___ cytokinesis	j	an RNA sugar
3.32 ___ chromosome replication	k	cytoplasmic division
	l	interphase

A SECOND LOOK
Label the following structures in the drawing:

plasma membrane
nucleolus
mitochondrion
endoplasmic reticulum
Golgi apparatus
nuclear membrane
ribosome
microvilli

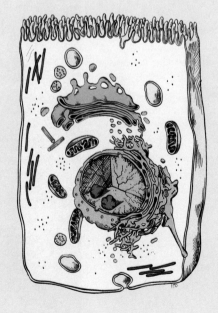

4

Tissues of the Body

LEARNING OBJECTIVES

1 Define a tissue.

2 Give an overview of the major characteristics and functions of epithelial tissues.

3 Describe the general classification of epithelial tissues.

4 Describe the fibers, ground substance, and cells of connective tissues.

5 Discuss the main types of connective tissues, including cartilage, and give their functions.

6 Discuss the main types of muscle and nervous tissue and describe their functions.

7 Describe the three types of membranes found within the body.

8 Explain the role of collagen in several disorders of connective tissues.

9 Describe the various kinds of biopsies and tissue transplantations.

In a multicellular organism, such as the human body, individual cells differentiate during development to perform special functions. These specialized cells carry out their functions as multicellular masses called **tissues.** The study of tissues, *histology* (Gr. *histos,* web), provides many excellent examples of how structures in the body have cellular arrangements (structures) that are closely related to their functions. In this chapter we look primarily at epithelial and connective tissues. Osseous (bone) tissue is described in detail in Chapter 6, muscle tissue (including heart muscle) in Chapter 10, and nervous tissue in Chapter 12.*

EPITHELIAL TISSUES

Epithelial tissue [epp-uh-THEE-lee-uhl; L. *epi-*, on, over, upon + Gr. *thele,* nipple), or *epithelia,* exists in many structural forms, but in general it either *covers* or *lines* something. The skin covers the external surface of the body, and epithelial tissue lines the gastrointestinal tract, ducts of glands, the mouth, and other body cavities. Another common function of epithelial tissue is the *secretion* of substances that lubricate parts of the body or take part in vital chemical reactions within the body. To suit these functions, epithelial tissues are typically made up of renewable flat sheets of cells that have surface specializations adapted for their specific roles. Modified secretory epithelia are classified as *glandular epithelia.*

Functions of Epithelial Tissues

The typical functions of epithelial tissues are *absorption* (by the lining of the small intestine, for example), *secretion* (by glands), *transport* (by kidney tubules), *excretion* (by sweat glands), and *protection* (by the skin). The size, shape, and arrangement of epithelial cells are directly related to these specific functions. Throughout this chapter you will see how these varied functions are enhanced by the different structures of epithelial tissues.

General Characteristics of Epithelial Tissues

A covering or lining epithelial tissue may be single-layered or multilayered, and its cells may have different shapes. However, all kinds of epithelial tissues, except glandular tissues, share certain characteristics.

*Several other kinds of tissues are described with their relevant systems in separate chapters rather than being placed in one chapter. The authors have chosen this organization because they believe that it gives the reader a better understanding of how anatomy and physiology work together and how the parts of the body make up a cohesive whole.

Cell shapes and junctions The cells that make up epithelial tissues are relatively regular in shape and are closely packed in continuous sheets. There is little or none of the extracellular material known as *matrix* between epithelial cells. The framework of the matrix that does exist for epithelial tissue is composed of *ground substance,* consisting of glycoproteins. Instead of depending on a matrix for support, the tight-fitting epithelial cells are held in place by strong adhesions formed between the plasma membranes of adjacent cells. The specialized parts that hold the cells together are known as **junctional complexes;** they enable groups of cells to function as a unit. There are three main kinds of junctional complexes: tight junctions, spot desmosomes, and gap junctions.

A *tight junction* is the site of close connection between two plasma membranes, with little or no extracellular space between the cells [FIGURE 4.1A]. This "nonleaky" junction creates a permeability barrier that stretches across a continuous layer of epithelial cells and keeps material either in or out. Tight junctions are found in the epithelial tissues of the urinary bladder, where they hold urine within the bladder.

A *spot desmosome* (Gr. *desmos,* binding) is a junction with no direct contact between adjacent plasma membranes [FIGURE 4.1B]. It is often referred to as a "spot weld." The plasma membranes are joined by a criss-crossed network of *intercellular filaments.* Thicker *tonofilaments* run parallel to the cell's surface, strengthening the internal framework of the cell. Spot desmosomes are common in skin, where great stress is constantly being applied to the junctions. Several other types of desmosomes exist.

A *gap junction* is formed from several links of channel protein connecting two plasma membranes [FIGURE 4.1C]. Gap junctions are found in many locations, including intestinal epithelial cells, where they allow the flow of ions and small molecules between adjacent cells.

Basement membranes Epithelial cells are anchored to underlying connective tissue by a basement membrane. The extracellular **basement membrane** consists of (1) a *basal lamina* (a homogeneous layer lacking fibers) and (2) a deeper layer containing reticular fibers [FIGURE 4.2; see also FIGURES 4.1 and 4.3A]. The terms *basement membrane* and *basal lamina* are often used synonymously, but that is incorrect. By firmly attaching the basal surface of the epithelial cells to the underlying connective tissue, the basement membrane assures that the epithelium is held in position and reduces the possibility of tearing. Basement membranes vary in thickness throughout the body. They are quite thin in the skin and intestines and rather thick in the trachea. The principal roles of the basement membrane are to provide elastic support and to act as a partial barrier for diffusion and

FIGURE 4.1 JUNCTIONAL COMPLEXES OF EPITHELIAL CELLS

A unique feature of epithelial cells is the presence of various functionally and structurally distinct intercellular junctions between their plasma membranes. The three main kinds of junctions are shown here in simplified form. **[A]** A tight junction. **[B]** A spot desmosome. **[C]** A gap junction.

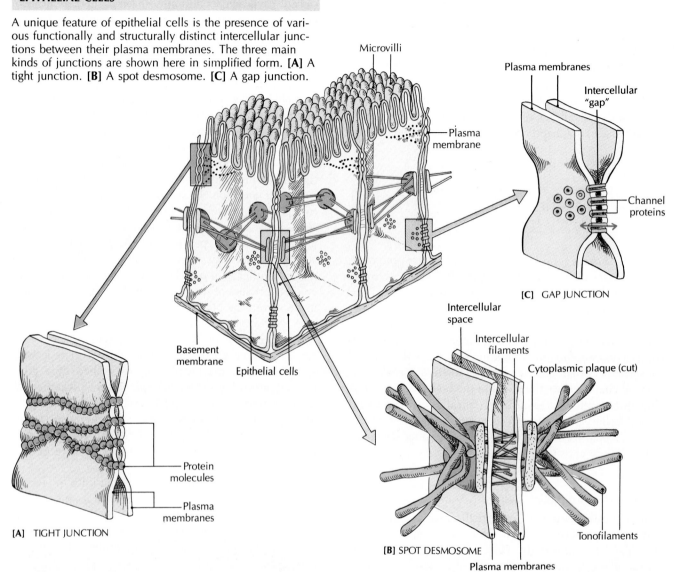

filtration. The basal lamina also provides a surface along which epithelial cells may migrate during cell renewal and wound healing.

Lack of blood vessels Epithelial tissues do not contain blood vessels. Oxygen and other nutrients diffuse through the selectively permeable basement membranes from capillaries in the underlying connective tissue. Wastes from the epithelial tissues diffuse into the connective tissue capillaries. Although epithelia have no blood vessels, they may contain nerves.

Surface specializations Epithelial tissues usually have several types of surface specializations. Some epithelial cells, like the ones that line the ventral body cavities and

blood vessels, have smooth surfaces, but most epithelial cells have irregular surfaces that result from the complex folding of their outer membranes. This extended folding produces **microvilli,** fingerlike projections that greatly increase the absorptive area of the cell [FIGURE 4.3A, B]. Microvilli are usually found on cells that are involved in absorption or secretion. The increased surface area of microvilli, coupled with the presence of surface enzymes that can break down large molecules, aids the absorptive process. The shape of microvilli on the free surface of epithelial cells gives rise to the term **brush border.** As seen in FIGURE 4.3A, the elaborate folding of the plasma membrane may occur on lateral and basal surfaces of cells as well as on free surfaces. Some epithelial cells, such as those of the respiratory system, have **cilia** projecting

FIGURE 4.2 COMPONENTS OF THE BASEMENT MEMBRANE

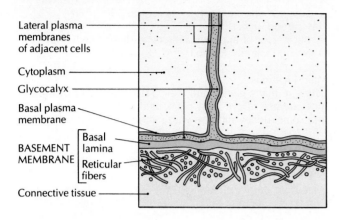

Lateral plasma membranes of adjacent cells

Cytoplasm

Glycocalyx

Basal plasma membrane

BASEMENT MEMBRANE { Basal lamina / Reticular fibers

Connective tissue

FIGURE 4.3 MICROVILLI

[A] Elaborate folding of the plasma membrane is evident not only in the microvilli (the outer surface) but also on the lateral and basal surfaces. The clusters of mitochondria at the base indicate an area of cellular activity. Notice the tight junction between cells. [B] A scanning electron micrograph of the small intestine where the extra surface area created by the microvilli aids the absorption of digested food into the bloodstream. The microvilli resemble tufts of a shag rug, an appropriate image, since villi is Latin for "shaggy hairs." ×30,000.

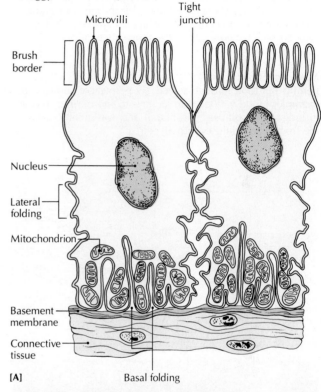

Microvilli   Tight junction

Brush border

Nucleus

Lateral folding

Mitochondrion

Basement membrane

Connective tissue

[A]   Basal folding

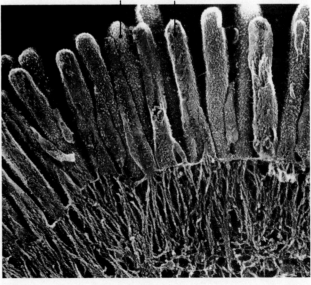

Microvilli forming brush border

[B]

from their free surfaces. The rapid lashing of the cilia propels fluids over the surface of the epithelium.

Regeneration Because epithelial tissue is located on surfaces, it is subject to constant injury. Fortunately, it is also capable of constant regeneration through cell division. When epithelial cells are damaged, they are shed and replaced by new cells. These cells are specialized to divide and move into the damaged area, where they can take on the function of the cells they replace. The rate of renewal depends on the type of epithelium. For example, the outer layer of the skin and the epithelial lining of the intestinal villi, where frequent friction is inevitable, are entirely replaced every few days. In contrast, the cells in glands and the lining of the respiratory tract are usually replaced every 5 to 6 weeks.

General Classification of Epithelial Tissues

Epithelial tissues are generally classified according to the *arrangement of cells* and the *number of cell layers,* and subdivided further by the *shape of the cells* in the superficial (top or outer) layer. In this section we discuss the two main groupings of epithelial tissues: *simple* and *stratified.*

If an epithelial tissue is made up of a single layer of cells, it is called *simple;* if its cells are arranged in two or more layers, it is *stratified.* The basic shapes of the superficial cells are *squamous* (SKWAY-muhss) (flat), *cuboidal* (like a cube), and *columnar* (elongated). Epithelial tissues are classified according to their cell arrangement and shape, using combinations of these terms.

These general types of epithelia, as well as some exceptions to this classification, are summarized and illustrated in FIGURES 4.4 through 4.7.

FIGURE 4.4 SIMPLE EPITHELIUM

SIMPLE SQUAMOUS EPITHELIUM (L. *squama,* scale)

Description
A single layer of flat cells that are wider than they are thick; nuclei are flattened and are parallel to the surface.

Main locations
Lining of lymph vessels, blood vessels, and heart (lining called *endothelium*); serous membranes lining peritoneal, pleural, pericardial, and scrotal body cavities (this epithelium, called *mesothelium,* is derived from mesoderm and has microvilli); glomerular (Bowman's) capsule in kidneys; lung air sacs (alveoli); small excretory ducts of many glands; membranes of inner ear.

Major functions
Permits diffusion through selectively permeable surface; for example, blood is filtered in kidneys to form urine; oxygen in lung alveoli diffuses into blood, and waste carbon dioxide from blood diffuses into alveoli.

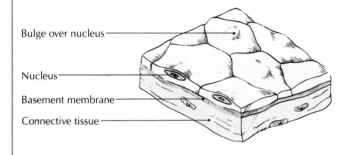

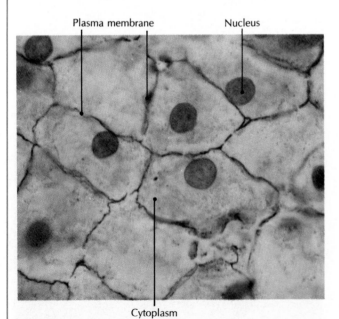

ABDOMINAL CAVITY; light micrograph, ×1000

SIMPLE CUBOIDAL EPITHELIUM

Description
A single layer of approximately cube-shaped cells; nucleus large and centrally located.

Main locations
Lining of many glands and their ducts; surface of ovaries; inner surface of eye lens; pigmented epithelium of eye retina; some kidney tubules (with microvilli on cell surface).

Major functions
Secretion of mucus, sweat, enzymes, other substances; absorption of fluids, other substances.

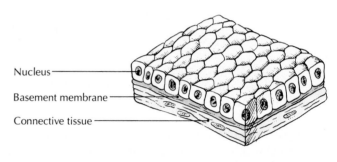

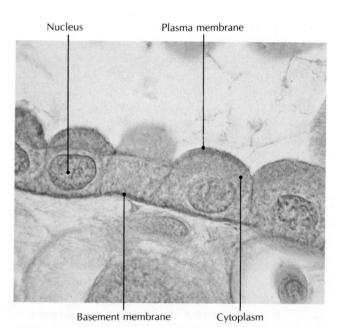

KIDNEY; light micrograph, ×900

SIMPLE COLUMNAR EPITHELIUM

Description
Single layer of cells that are taller than they are wide; looks like simple cuboidal epithelium with elongated cells; large, oval-shaped nucleus usually located at base of cell; may secrete mucus (goblet cells) may have cilia or microvilli on free surface of cell.

Main locations
Stomach, intestines (with microvilli), digestive glands, and gallbladder.

Major functions
Secretion, absorption, protection, lubrication; mucus and cilia combine to trap and sweep away foreign substances; cilia may also help to move objects through a duct, such as an egg in the uterine tube.

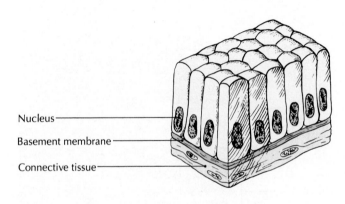

Nucleus

Basement membrane

Connective tissue

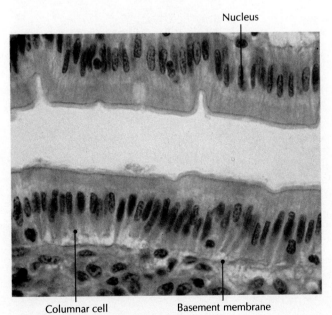

Nucleus

Columnar cell   Basement membrane

BILE DUCT; light micrograph, ×400

GOBLET CELLS (in Simple Columnar Epithelium)

Description
Mucus produced by goblet cells accumulates near the top of cell, causing the cell to bulge in shape of a goblet; goblet cells are interspersed in layer of columnar cells.

Main locations
Digestive tract, upper respiratory tract, uterine tube, uterus.

Major functions
Modified to secrete mucus from spaces at top of cell; mucus protects and lubricates walls of digestive tract.

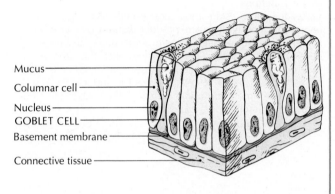

Mucus

Columnar cell

Nucleus

GOBLET CELL

Basement membrane

Connective tissue

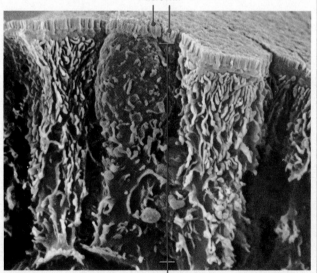

Microvilli

Goblet cell

EPITHELIAL LINING OF RAT SMALL INTESTINE; scanning electron micrograph, ×3000

FIGURE 4.5 STRATIFIED EPITHELIUM

STRATIFIED SQUAMOUS EPITHELIUM

Description
Several layers of cells; only superficial layer composed of flat squamous cells; underlying basal cells are modified cuboidal or columnar cells that are pushed upward from near basement membrane to replace dying superficial cells; shedding process in outer layer is *desquamation*.

Main locations
Areas where friction or possibility of cellular injury or drying occurs, such as epidermis, vagina, mouth and esophagus, anal canal, distal end of urethra.

Major functions
Protection. In the skin, replacement cells rising to the surface from underlying layers produce **keratin** (KER-uh-tihn; Gr. *keras,* horn), a tough protein; their nuclei and organelles disappear, and by the time they reach the surface, they consist mainly of keratin. These *keratinized* cells flake off as the next batch of cells reaches the surface. Cells in a moist environment are *nonkeratinized;* they retain their nuclei and organelles.

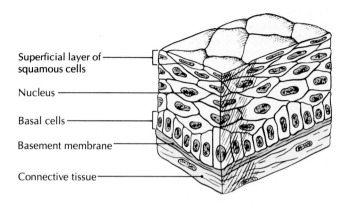

Superficial layer of squamous cells

Nucleus

Basal cells

Basement membrane

Connective tissue

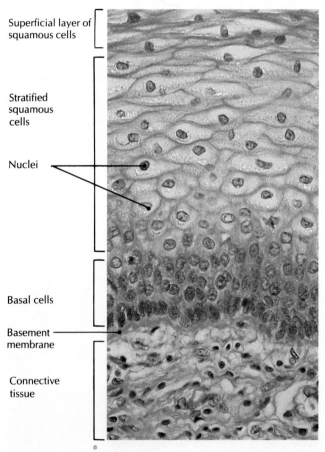

Superficial layer of squamous cells

Stratified squamous cells

Nuclei

Basal cells

Basement membrane

Connective tissue

VAGINA; light micrograph, ×400

STRATIFIED CUBOIDAL EPITHELIUM

Description
Multilayered arrangement with superficial layer composed of cuboidal cells; superficial cells are larger than basal cells; distinct basement membrane.

Main locations
Ducts of sudoriferous (sweat) glands, sebaceous (oil) glands, developing epithelium in ovaries and testes.

Major function
Secretion.

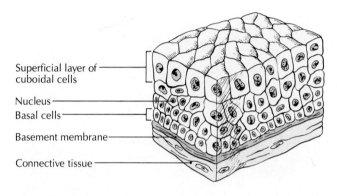

Superficial layer of cuboidal cells
Nucleus
Basal cells
Basement membrane
Connective tissue

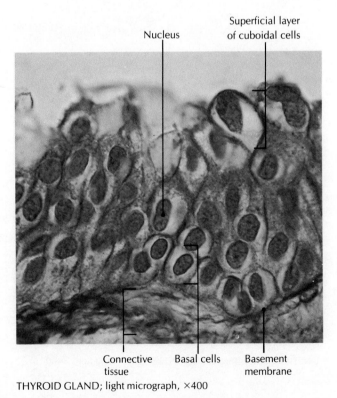

Nucleus
Superficial layer of cuboidal cells
Connective tissue
Basal cells
Basement membrane

THYROID GLAND; light micrograph, ×400

STRATIFIED COLUMNAR EPITHELIUM

Description
Multilayered arrangement with superficial layer composed of tall, thin columnar cells; underlying cuboidal-type cells in contact with basement membrane; sometimes ciliated, as in larynx and nasal surface of soft palate.

Main locations
Moist surfaces, such as larynx, nasal surface of soft palate, parts of pharynx, urethra, and excretory ducts of salivary glands and mammary glands.

Major functions
Secretion and movement of materials over cell surface.

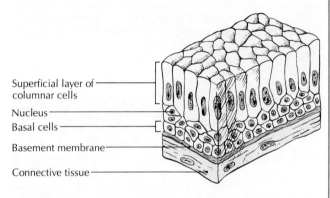

Superficial layer of columnar cells
Nucleus
Basal cells
Basement membrane
Connective tissue

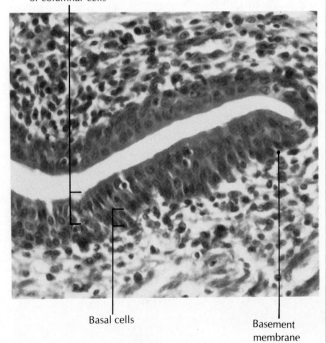

Superficial layer of columnar cells
Basal cells
Basement membrane

MALE URETHRA; light micrograph, ×400

FIGURE 4.6 ATYPICAL EPITHELIUM

PSEUDOSTRATIFIED COLUMNAR EPITHELIUM (*pseudo* = false)

Description
Single layer of cells of varying height and shape; nuclei at different levels give false impression of multilayered structure; all cells in contact with basement membrane, but not all reach superficial layer; when ciliated, called *pseudostratified ciliated columnar epithelium.*

Main locations
Large excretory ducts, most of male reproductive tract, nasal cavity and other respiratory passages, and part of ear cavity.

Major functions
Protection, secretion, and movement of substances across surfaces; mucus from cells in respiratory tract traps foreign substances; mucus moved to throat by sweeping action of cilia and either coughed out or swallowed and later eliminated in feces.

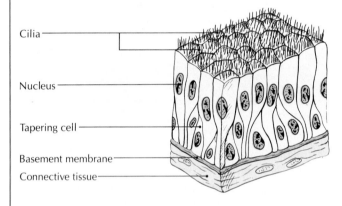

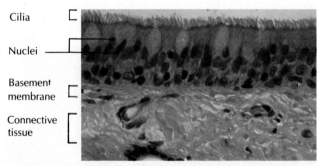

RESPIRATORY SYSTEM; light micrograph, ×600

TRANSITIONAL EPITHELIUM

Description
Stratified epithelium; shape of surface cells changes; they appear round when unstretched, flat when stretched; relaxed basal cells are cuboidal or slightly columnar, with layer above basal cells made up of irregularly shaped elongated cells; accordion-pleated membranes allow stretching and contracting.

Main locations
Urinary tract, where it lines bladder, ureters, urethra, and parts of kidney.

Major function
Allows for changes in shape.

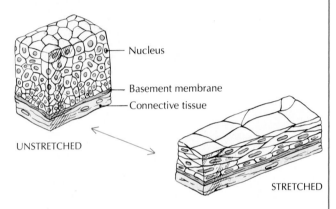

UNSTRETCHED

STRETCHED

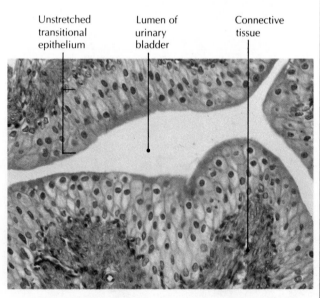

UNSTRETCHED URINARY BLADDER; light micrograph, ×400

FIGURE 4.7 GLANDULAR EPITHELIUM

Description
Epithelial cells modified to perform secretion. Main types are exocrine and endocrine; exocrine glands have ducts, endocrine glands secrete hormones directly into bloodstream where they travel to target cells. All glandular tissue is derived embryologically from epithelium.

Main locations
Various sites throughout body, including skin glands such as sweat and mammary glands; digestive glands such as salivary glands; endocrine glands such as thyroid.

Major functions
Synthesis, storage, and secretion of chemical products.

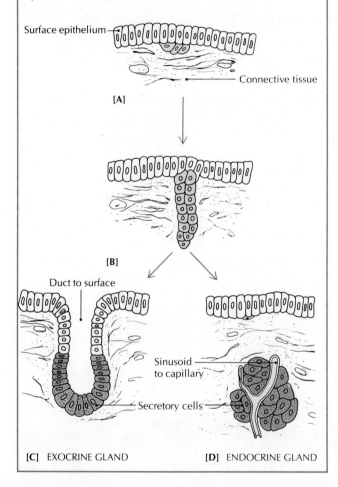

Surface epithelium

Connective tissue

[A]

[B]

Duct to surface

Sinusoid to capillary

Secretory cells

[C] EXOCRINE GLAND

[D] ENDOCRINE GLAND

Exocrine Glands

Exocrine glands have ducts that carry their secretions to openings on the body surfaces. The most typical ways of classifying exocrine glands are described below.

Unicellular and multicellular glands The only example of a *unicellular* (one-celled) *gland* in the human body

is the *goblet cell* [FIGURE 4.4, page 89], found in the lining of the intestines, other parts of the digestive system, the respiratory tract, and the conjunctiva of the eye. A goblet cell produces a carbohydrate-rich glycoprotein called *mucin* (MYOO-sihn), which later is secreted in the form of *mucus*. A *multicellular gland* contains many cells.

Simple and compound glands A gland with only one *unbranched* duct is a *simple gland.* A gland with a *branched* duct system, resembling a tree trunk and its branches (although upside down), is a *compound gland.*

Tubular, alveolar, and tubuloalveolar glands These three categories are concerned with the shapes of the secretory portion of the glands. If the secretory portion is *tubular*, the gland is a *tubular gland.* If the secretory portion is *rounded*, the gland is an *alveolar* (L. *alveolus*, hollow cavity), or *acinar* (ASS-ih-nur; L. *acinus*, grape, berry) *gland.* Glands with both tubular and alveolar secretory portions are *tubuloalveolar glands.* The secretory portion of a *simple coiled tubular gland* is coiled and tubular, and the secretory portion of a *simple branched tubular gland* is branched and tubular. TABLE 4.1 lists the types of glands developed from combinations of the forms above and shows their general shapes.

Mucous, serous, and mixed glands Glands may be classified according to their secretions, as well as their structures. *Mucous glands* secrete thick mucus; *serous glands* secrete a thinner, watery substance. The secretions of serous glands generally contain enzymes. *Mixed glands* contain mucous and serous cells and produce mucus and serous secretions.

Merocrine and holocrine glands Glands may also be classified according to how they release their secretions. *Merocrine glands* release their secretions without breaking the plasma membrane, using the process of *exocytosis.* The pancreas is a merocrine gland that secretes digestive enzymes. When *holocrine glands* release their secretions, whole cells become detached, die, and actually become the secretion. The sebaceous (oil) glands in the skin are the only holocrine glands in the body.

ASK YOURSELF

1 What are some of the general characteristics of epithelial tissues?

2 What are the three basic shapes of epithelial cells?

3 What are three exceptions to the basic system of classifying epithelial tissues?

4 How do endocrine and exocrine glands differ?

TABLE 4.1 CLASSIFICATION OF MULTICELLULAR EXOCRINE GLANDS

Type of gland	General shape	Shape of secretory portion	Location
MULTICELLULAR SIMPLE GLANDS			
Simple tubular		Straight and tubular	Intestinal glands (crypts of Lieberkühn)
Simple coiled tubular		Coiled	Sudoriferous (sweat) glands
Simple branched tubular		Branched and tubular	Mouth, tongue, and esophagus
Simple alveolar (acinar)		Rounded	Seminal vesicle glands (in male reproductive system)
Simple branched alveolar		Branched and rounded	Sebaceous (oil) glands
MULTICELLULAR COMPOUND GLANDS			
Compound tubular		Tubular	Mucous glands of mouth, bulbourethral glands (in male reproductive system), kidney tubules, and testes
Compound alveolar (acinar)		Rounded	Mammary glands
Compound tubuloalveolar		Tubular and rounded	Salivary glands, glands of respiratory passages, and pancreas

CONNECTIVE TISSUES

Connective tissues are aptly named because they connect other tissues together. Unlike epithelial tissues, connective tissues have a great deal of extracellular, fibrous material that helps to support the cells of the other tissues. In fact, when cartilage and bone are included in the overall classification of connective tissues, the whole group is often referred to as *supporting* tissues. Besides their supportive function, connective tissues also form protective sheaths around hollow organs and are involved in storage, transport, and repair.

Connective tissues vary greatly in their structure and function. For this reason, we have grouped together the more generalized types—such as loose, dense, and elastic—under the section Connective Tissue Proper. The more specialized forms—cartilage, bone (osseous), and blood—are treated separately. In this chapter we discuss cartilage in some detail but give only a brief overview of osseous tissue and blood. Bones and bone tissue are described in Chapter 6, the skeletal system in Chapters 7 and 8, and blood in Chapter 18.

All connective tissues consist of fibers, ground substance, cells, and some extracellular fluid [FIGURE 4.8]. The extracellular fibers and ground substance are known as the **matrix** (MAY-trihks). The arrangement, function, and composition of elements in the matrix vary in different kinds of connective tissues, as you will see in this and the following sections.

Fibers of Connective Tissues

Connective tissues are usually classified according to the arrangement and density of their extracellular fibers and the nature and consistency of the ground substance in which the fibers are embedded. Fibers of connective tissue may be collagenous, reticular, or elastic.

Collagenous fibers *Collagenous fibers* are whitish fibers composed mainly of the protein **collagen** (G. *kolla*, glue + *genes*, born, produced). Sturdy, flexible, and practically unstretchable, collagenous fibers are the most common type of fiber and are found in all kinds of connective tissue [FIGURE 4.10, page 99]. The arrangement of collagenous fibers varies from loose and pliable, as in the loosely woven connective tissue that supports most of the organs, to tightly packed and resistant, as in tendons. Generally, these fibers look wavy when they are not under pressure. No matter how collagenous fibers are arranged, they are extremely well suited for support. A collagenous fiber is actually a bundle of parallel *fibrils*, slender fibers composed of even smaller microfibrils. The number of fibrils in a bundle varies.*

---

*A high number of collagenous fibers can make meat tough. But when collagen is boiled in water, it turns into a soft gelatin. This is why tough meat simmered in a soup or stew becomes more tender. In contrast to this softening process, leather is toughened, or *tanned*, by the addition of tannic acid, which converts collagen into a firm, insoluble material.

FIGURE 4.8 CONNECTIVE TISSUE

A diagrammatic rendering of connective tissue, showing a representative sample of cells and fibers bathed in tissue fluid.

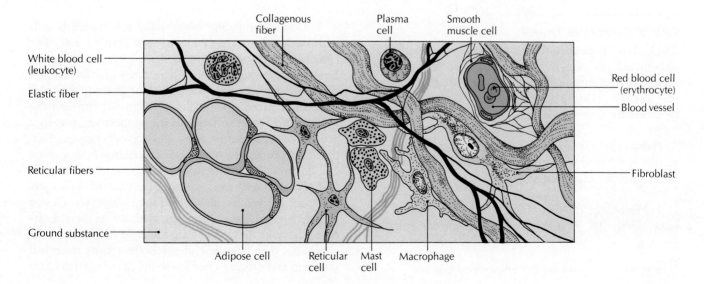

Reticular fibers *Reticular fibers* form delicately branched networks (*rete* is Latin for net), as compared to the thicker bundles of collagenous fibers [FIGURE 4.10, page 100]. Reticular fibers have the same molecular structure as collagenous fibers and practically the same chemical composition; they also contain fibrils and microfibrils. Reticular fibers help support other cells.

Elastic fibers Yellowish *elastic fibers* appear singly, never in bundles, though they do branch and form networks [FIGURE 4.10, page 99]. They contain microfibrils, but do not have the coarseness of collagenous fibers. Like rubber bands, elastic fibers stretch easily when pulled and return to their original shape when released. The protein *elastin* gives elastic fibers the resilience that organs like the skin need to move, stretch, and contract.

Ground Substance of Connective Tissues

Ground substance of connective tissue is a homogeneous, extracellular material that ranges in consistency from a semifluid to a thick gel. Whereas fibers give connective tissue its strength and elasticity, ground substance provides a suitable medium for the passage of nutrients and wastes between the cells and bloodstream. The main ingredients of ground substance are glycoproteins and proteoglycans,* including hyaluronic acid, chondroitin sulfate, and other specific sulfates. These compounds not only bind water and other tissue fluids that are needed for the exchange of nutrients and wastes, but also act as lubricants and shock absorbers. Hyaluronic acid forms a tough, protective mesh that helps prevent invasion by microorganisms and other foreign substances. Interestingly, some bacteria can produce an enzyme, called *hyaluronidase*, that breaks down the mesh of hyaluronic acid and allows the bacteria to enter the connective tissue.

Cells of Connective Tissues

The cells of connective tissues can usually be classified as either *fixed cells*, such as fibroblasts and adipose cells, or *wandering cells*, such as plasma cells, mast cells, and granular leukocytes (white blood cells). Macrophage cells may be either fixed or wandering.

Fixed cells *Fixed cells* have a permanent site and are usually concerned with long-term functions such as synthesis, maintenance, and storage:

1 *Fibroblasts* (L. *fibra*, fiber; Gr. *blastos*, growth) are large, long, flat, branching cells that appear spindle-shaped [FIGURES 4.8 and 4.10]. Their nuclei are light-colored and larger than the nuclei of other connective tissue cells. Fibroblasts are the most common cells in connective tissue and the only cells found in tendons. They synthesize and secrete the *matrix* materials (fibers and ground substance) and are considered to be secretory cells. In this way, fibroblasts assist wound healing.

2 *Adipose (fat) cells* synthesize and store lipids. A mature adipose cell accumulates so much fat that the nucleus and cytoplasm are flattened against the sides of the cell, forming a thin film around the rim. Adipose cells may appear singly or in clusters. Single cells are so bloated with one large droplet of stored fat that they look like spherical drops of oil [FIGURE 4.10, page 100].

3 *Macrophage cells (macrophages)* are active phagocytes ("cell eaters"). The name *macrophage* is descriptive: *macro* means large, and *phage*, as you have seen, comes from the Greek word *phagein*, to eat; so a macrophage is literally a "big eater." Macrophages have an irregular shape and short cytoplasmic extensions (processes) [FIGURE 4.9A]. They are normally fixed along the bundles of collagenous fibers and are then referred to as *fixed macrophages*. But in cases of inflammation, they become detached from the fibers, change their shape to resemble amoebas, and begin to move about actively as *free macrophages*. Free macrophages are scavengers that engulf and destroy foreign material in the bloodstream and tissues.

4 *Reticular cells* are flat and star-shaped with long processes. They come in contact with each other through these processes. The cells form the cellular framework in such netlike structures as bone marrow, lymph nodes, the spleen, and other lymphoid tissues. Reticular cells play a role in the immune response.

Wandering cells *Wandering cells* in connective tissues are usually involved with short-term activities such as protection and repair:

1 *Leukocytes (white blood cells)* are roundish cells with nuclei that have various shapes [FIGURE 4.8]. The cytoplasm may or may not contain secretory granules. Leukocytes multiply drastically in times of infection. Although leukocytes circulate in the bloodstream, they perform their protective function by moving through the walls of tiny blood vessels into the connective tissue by a process called *diapedesis*. Most leukocytes are phagocytes; they help protect the body against infection by engulfing and destroying harmful microorganisms.

2 *Plasma cells* are one specific type of leukocyte. They are derived from activated B-lymphocytes. Plasma cells are oval-shaped, with large, dark nuclei located off-center [FIGURE 4.9B]. They are the main producers of the antibodies that help defend the body against microbial infection and cancer. They are found in connective tissue

---

*Proteoglycans have a protein core with sulfate side chains.

FIGURE 4.9 MACROPHAGE

[A] A scanning electron micrograph of a macrophage ingesting red blood cells. The macrophage is covered with microvilli. ×2500. [B] Plasma cells.

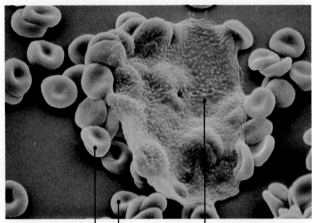

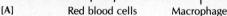

[A]   Red blood cells   Macrophage

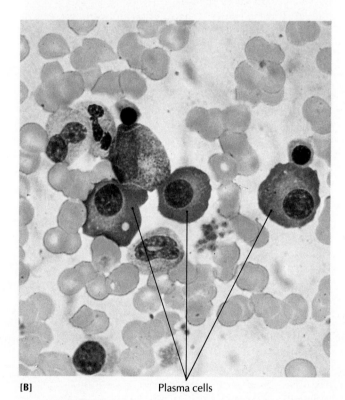

[B]   Plasma cells

under the moist epithelial linings of the respiratory and intestinal tracts, where microorganisms may enter through breaks in the epithelial membrane. Plasma cells are also present in serous membranes, lymphoid tissue, and in areas of chronic infection.

3 *Mast cells* are relatively large cells with irregular shapes and small, pale nuclei [FIGURE 4.8]. They are often found near blood vessels. Their cytoplasm is crowded with secretory granules, which are bound by membranes. These granules contain *heparin*, a polysaccharide that prevents blood from clotting as it circulates throughout the body, and *histamine*, a protein that increases vascular permeability and contracts smooth muscle. Mast cells are also effective in fighting long-term infections and inflammations.

4 *Macrophages* become mobile when stimulated by inflammation; they are listed here as a reminder of their frequent wanderings as phagocytes. Free macrophages are part of the extensive **reticuloendothelial system,** which is made up of an army of specialized cells concerned with producing antibodies and removing dead cells, tissue debris, microorganisms, and foreign particles from the fluids and matrix of body tissues.

ASK YOURSELF

1 What are the major components of connective tissues?

2 What are the three kinds of connective tissue fibers?

3 What types of cells are found in connective tissue?

Connective Tissue Proper

Connective tissues are usually classified according to the arrangement and density of their extracellular fibers [FIGURE 4.10]. *Loose (areolar) connective tissue* has irregularly arranged fibers and more tissue fluid and cells than fibers. *Dense (collagenous) connective tissue* has more fibers, and their arrangement is more regular. The following types of connective tissue are usually grouped under the heading connective tissue proper: embryonic, loose, dense, elastic, adipose, and reticular. Except for *embryonic connective tissue*, all the connective tissues we consider here are known as *adult connective tissues*.

ASK YOURSELF

1 What are the main kinds of connective tissues?

2 What is the most common kind of connective tissue?

3 How do regular and irregular dense connective tissues differ?

4 Why is elastic connective tissue well suited to the stomach?

FIGURE 4.10 TYPES OF CONNECTIVE TISSUE: CONNECTIVE TISSUE PROPER

EMBRYONIC (Mesenchymal, Mucous Connective)

Description
Unspecialized packing material; cells vary in shape from stars to spindles; very little mesenchymal tissue remains after cellular differentiation in embryo; called *mesenchyme* during first two months of prenatal development, *mucous connective tissue* from two months to birth.

Main locations
Mesenchyme: around developing bones and under skin of embryo; mucous connective tissue: in umbilical cord of fetus.

Major functions
Mesenchyme differentiates into supporting tissues, tissues of blood vessels, blood, and smooth muscle. Mucous connective tissue forms padding for blood vessels in umbilical cord (where it is sometimes called *Wharton's jelly*) and prevents cord from snarling when fetus turns within uterus.

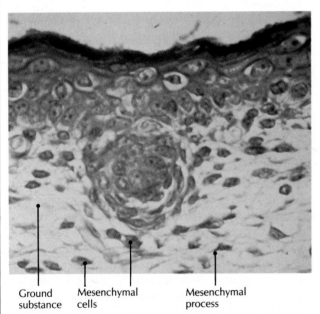

Ground   Mesenchymal   Mesenchymal
substance   cells   process

MESENCHYMAL TISSUE IN EMBRYO; light micrograph, ×400

LOOSE (Areolar) (L. *areola*, "open place")

Description
Irregular, loosely woven fibers in a semifluid base; contains extracellular spaces filled with ground substance and tissue fluid and many types of cells, especially fibroblasts and macrophages; most common connective tissue.

Main locations
Most parts of body, especially around and between organs; wraps around nerves and blood vessels.

Major functions
Supports tissues, organs, blood vessels, nerves; forms subcutaneous layer that connects muscles to skin; allows for movement; acts as protective packing material between organs. Tissue fluid is medium for exchange of substances between capillaries and tissue cells; ground substance acts as barrier against harmful microorganisms.

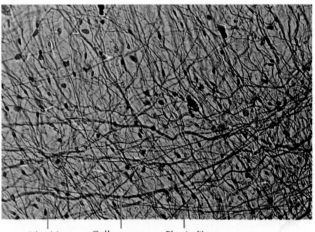

Fibroblast   Collagenous   Elastic fiber
              fiber

LOOSE AREOLAR TISSUE; light micrograph, ×64

DENSE (Collagenous)

Description
Tightly packed with coarse, collagenous fibers; distinctly white; classified as regular or irregular, depending on arrangement of fibers. Dense regular tissue has collagenous fibers arranged in parallel rows of thick, strong bundles; irregular tissue has thick bundle fibers arranged randomly in tough, resilient meshwork that can be stretched in more than one direction, as in the middle layer (dermis) of skin.

Main locations
Regular: tendons, ligaments, aponeuroses.
Irregular: dermis of skin, capsules of many organs; covering sheaths of nerves, tendons, brain, spinal cord; deep fibrous coverings of muscles.

Major functions
Provides support and protection; connects muscles to bones (tendons) and bones to bones (ligaments).

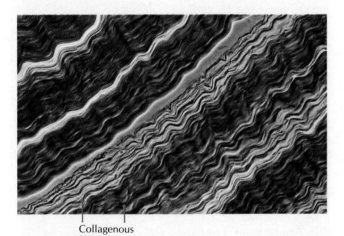

Collagenous
fibers

DENSE REGULAR CONNECTIVE TISSUE; light micrograph, ×100

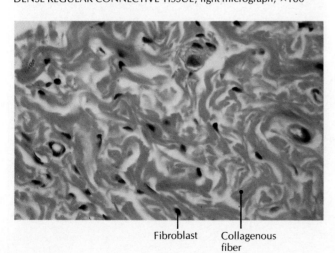

Fibroblast   Collagenous
fiber

DENSE IRREGULAR CONNECTIVE TISSUE; light micrograph, ×310

ELASTIC

Description
Yellow elastic fibers branch freely, resembling a fishing net; spaces between are filled by fibroblasts and some collagenous fibers.

Main locations
Walls of hollow organs such as stomach; walls of largest arteries; some parts of heart; trachea, bronchi; ligaments between neural arches of spinal vertebrae; vocal cords; other parts of body where stretching, support, and suspension are required.

Major functions
Allows stretching, and provides support and suspension.

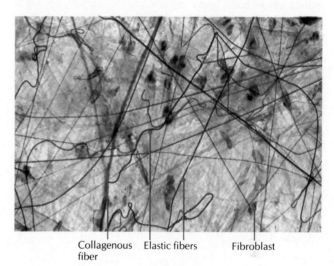

Collagenous   Elastic fibers   Fibroblast
fiber

ELASTIC CONNECTIVE TISSUE; light micrograph, ×200

(Figure 4.10 continues on following page)

FIGURE 4.10 TYPES OF CONNECTIVE TISSUE: CONNECTIVE TISSUE PROPER *(Continued)*

ADIPOSE (Fat)

Description
Consists of clustered adipocytes (cells specialized for fat storage); contains little or no extracellular substance; clusters of spherical, fat-bloated cells supported by collagenous and reticular fibers. Makes up about 10 percent of adult body weight; distributed differently in males and females.

Main locations
Beneath epidermis (outer layer of skin), around organs, many other sites.

Major functions
Provides reserve food supply by synthesizing and storing fat; cushions and protects organs; provides insulation against loss of body heat.

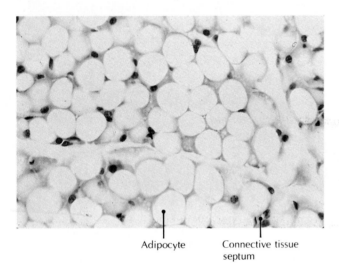

Adipocyte   Connective tissue
                 septum

ADIPOSE TISSUE; light micrograph, ×64

RETICULAR

Description
Lattice of fine, interwoven threads that branch freely, forming connecting and supporting framework; spaces between fibers are filled by reticular cells.

Main locations
Interior of liver, spleen, lymph nodes, tonsils, stomach, intestines, trachea, bronchi; supporting adipose cells.

Major functions
Forms connecting and supporting framework of reticular fibers for lymph nodes, bone marrow, spleen, thymus gland.

Reticular cells   Reticular fibers   Collagen
in fibers

RETICULAR CONNECTIVE TISSUE; light micrograph

Cartilage

Cartilage is a specialized connective tissue that provides support and aids movement at joints. Like other connective tissues, it consists of cells, fibers, and ground substance. Cartilage cells are called **chondrocytes** (KON-droh-sites; Gr. *khondros*, cartilage), and they are embedded within the matrix in small cavities called **lacunae** (luh-KYOO-nee; L. cavities). Cartilage does not contain blood vessels. Instead, oxygen, nutrients, and cellular wastes diffuse through the matrix. The three types of cartilage—hyaline, fibrocartilage, and elastic—are described and illustrated in FIGURE 4.11.

Bone as Connective Tissue

Bone is also a type of connective tissue. It is very similar to cartilage in that it consists mostly of matrix material that contains lacunae and specific cell types. However, unlike cartilage, bone is a highly vascular tissue. Bone histology is treated in detail in Chapter 6.

Blood as Connective Tissue

Blood cells and blood-forming tissues are mentioned in this section on connective tissue not because they connect anything—they do not—but because they have the same embryonic origin (mesenchyme) as the more typical connective tissues and because blood has the three components of any connective tissue (cells, fibers, and ground substance). Blood is treated extensively in Chapter 18.

A S K   Y O U R S E L F

1 What is the function of cartilage?

2 What are the three types of cartilage?

3 Why is blood classified as connective tissue?

MUSCLE TISSUE: AN INTRODUCTION

Muscle tissue helps the body, and parts of the body, to move. *Skeletal muscle* is attached to bones and makes body movement possible; the rhythmic contractions of *smooth muscle* create a churning action (as in the stomach), help to propel material through a tubular structure (as in the intestines and ureters), and control the size changes in hollow organs, such as the urinary bladder and uterus; and the contractions of *cardiac muscle* result in the heart beating. Muscle tissue is discussed more fully in Chapter 10.

NERVOUS TISSUE: AN INTRODUCTION

The *nervous tissue* of the nervous system is composed of several different types of cells, including impulse-conducting cells called *neurons* (the fundamental units of the nervous system), nonconducting *neuroglia* (protective, supportive, and nourishing cells), and *peripheral glial cells* (neurilemma cells), which form various types of sheaths and help to protect, nourish, and maintain cells of the peripheral nervous system. Nervous tissue is described more fully in Chapter 12.

MEMBRANES

Membranes are thin, pliable layers of epithelial and/or connective tissue. They line body cavities, cover surfaces, or separate or connect certain regions, structures, or organs of the body. The three kinds of membranes are mucous, serous, and synovial.

Mucous Membranes

Mucous membranes line passageways that open to the outside of the body, such as the nasal and oral cavities, and tubes of the respiratory, digestive, urinary, and reproductive systems. These membranes are made up of a layer of loose connective tissue covered by varying kinds of epithelial tissue. For example, the small intestine is lined with simple columnar epithelium, and the oral cavity is lined with stratified squamous epithelium. Glands in the mucous membranes secrete protective, lubricating mucus, which consists of water, salts, and the sticky protein mucin. Mucus traps invading microorganisms.

Serous Membranes

Serous membranes are double membranes of loose connective tissue covered by a layer of simple squamous epithelium (known as *mesothelium*). They line some of the walls of the thoracic and abdominopelvic cavities and cover the organs that lie within these cavities. Serous membranes include the *peritoneum*, *pericardium*, and *pleura:*

1 The *peritoneum* lines the peritoneal cavity and covers the organs inside the cavity. It also lines the abdominal and pelvic walls.

2 The *pericardium* lines the pericardial cavity with parietal pericardium, and covers the heart with a sac of visceral pericardium.

FIGURE 4.11 TYPES OF CONNECTIVE TISSUE: CARTILAGE

HYALINE CARTILAGE (Gr. *hyalos,* "glassy")

Description
Network of collagenous fibers with many spaces filled with ground substance; translucent, pearly blue-white appearance; strong, able to support weight; most common type of cartilage. Usually enclosed within fibrous ***perichondrium,*** which is composed of outer connective tissue layer and inner layer of cells capable of differentiating into chondrocytes.

Main locations
Trachea, larynx, bronchi, nose, costal (rib) cartilages, ends of long bones.

Major functions
Forms major part of embryonic skeleton; reinforces respiratory passages; aids free movement of joints; assists growth of long bones; allows rib cage to move during breathing.

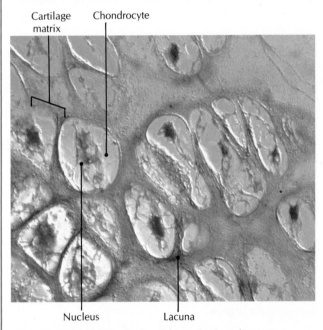

HYALINE CARTILAGE; scanning electron micrograph, ×1200

FIBROCARTILAGE

Description
Consists of bundles of resilient, pliable collagenous fibers that leave little room for ground substance; usually merges with hyaline cartilage or fibrous connective tissue; no sheath; resembles dense connective tissue.

Main locations
Intervertebral disks, fleshy pad between pubic bones (symphysis pubis); lining of tendon grooves, rims of some sockets (for example, hip and shoulder joints), some areas where tendons and ligaments insert into ends of long bones; fibrocartilage disks located at temporomandibular (jaw) joint, knee joint (where they are called *menisci*), wrist joint, and sternoclavicular joint between clavicle and sternum; type of cartilage formed after injury.

Major functions
Provides support and protection; collagenous fibers provide durability and strength to resist tension. Where fibrous tissue is subjected to great pressure it is replaced by fibrocartilage.

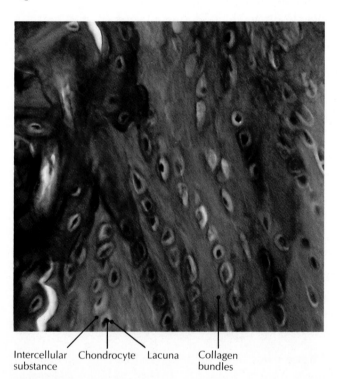

FIBROCARTILAGE; light micrograph, ×450

ELASTIC CARTILAGE

Description
Appears dark when stained because of dense elastic fibers, which are scattered in ground substance; cells can produce elastic and collagenous fibers; enclosed within perichondrial sheath; more flexible and elastic than hyaline cartilage.

Main locations
Areas that require lightweight support and flexibility, such as external ear, epiglottis in throat, auditory tube connecting middle ear to upper throat, larynx, nasopharynx.

Major functions
Provides flexibility and lightweight support.

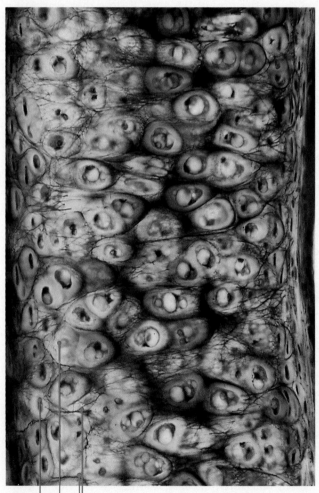

Chondrocytes   Elastic fibers

ELASTIC CARTILAGE; light micrograph, ×400

3 The *pleura* lines the pleural cavity, covers the lungs, and lines the wall of the thorax. (*Pleura* means "side" or "rib," but is normally used in connection with the lungs.)

The peritoneum, pericardium, and pleura are all double-layered membranes with a thin space between. These spaces are the peritoneal, pericardial, and pleural cavities, respectively. This space between serous membranes receives secretions of *serous fluids* that act as protective lubricants around organs and help to remove harmful substances through the lymphatic system. *Pleurisy* is a disease that occurs when the pleural membranes become inflamed and stick together, making breathing painful.

Organs in the abdominopelvic cavity are suspended from the cavity wall by fused layers of visceral peritoneum. These fused tissues are called *mesenteries,* or visceral ligaments. Besides providing a point of attachment for organs, mesenteries also permit vessels and nerves to connect with their organs through the otherwise impenetrable lining of the abdominopelvic cavity. Many abdominal and pelvic organs, such as the stomach, spleen, jejunum and ileum portions of the small intestine, uterus, and ovaries, are connected to the abdominopelvic wall by a mesentery. (In the case of the stomach, this mesentery is called an *omentum;* in the uterus it is known as the *broad ligament.*) But the kidneys, pancreas, and some other structures are attached to the posterior wall of the abdominopelvic cavity by a mesentery *behind* the peritoneal cavity. For this reason, the kidneys and pancreas are considered to be *retroperitoneal organs,* or located behind the peritoneum (*retro* means "behind").

Synovial Membranes
Synovial membranes line the cavities of movable joints, such as the knee or elbow, and other similar areas where friction needs to be reduced. They are composed of loose connective and adipose tissues covered by fibrous connective tissue. No epithelial tissue is present. The inner surface of synovial membranes is generally smooth and shiny, and the synovial fluid is a thick liquid with the consistency of egg white. It helps to reduce friction at the movable joints, and in fact, the synovial fluid together with the covering of the bones produces a mechanism that is practically frictionless.

ASK YOURSELF
1 Where are mucous membranes found?

2 What are the three kinds of serous membranes?

3 What are mesenteries?

4 How do synovial membranes aid movement?

WHEN THINGS GO WRONG

The prevention and treatment of certain diseases require an understanding of the structure and function of **collagen,** the important protein that forms the fibrous network in almost all tissues of the body.

The *pathogenesis* (development of disease conditions) of certain inherited and acquired diseases of connective tissues is now understood with respect to collagen synthesis and metabolic dysfunction caused by a lack of vitamin C (ascorbic acid) in the diet. In **scurvy,** collagen fibers are not formed. As a result, bone growth is abnormal, capillaries rupture easily, and wounds and fractures do not heal. In **rheumatoid arthritis,** collagen in articular cartilage (cartilage within a joint) is destroyed by specific enzymes from the inflamed synovial membranes. In **arteriosclerosis,** some of the damage to blood vessels is caused by the secretion of large amounts of collagen.

Biopsy

A **biopsy** (Gr. *bios,* life + *-opsy,* examination) is the microscopic examination of tissue removed from the body, for the purpose of diagnosing a disease, especially cancer. Several types of biopsies can be performed, depending on the specific diagnostic requirement. In a *skin* or *muscle biopsy,* a small piece of the tissue is removed, under a local anesthetic, and the small wound is then stitched closed. In a *needle biopsy,* a biopsy needle is inserted through a small incision into the skin or organ to be examined, and a tissue sample is removed. In an *aspiration biopsy,* cells are sucked from a tumorous tissue through a special needle. Needle biopsies usually require only a local anesthetic and can be guided by ultrasound or CAT scanning if necessary. During an *endoscopic biopsy,* an endoscope (see Chapter 1) is passed into an accessible hollow organ, such as the stomach or urinary bladder. A forceps attached to the penetrating end of the endoscope snips a tissue sample from the lining of the organ. Endoscopy usually requires sedation.

In an *open biopsy,* the surgeon actually opens a body cavity to obtain a sample of diseased tissue. This is necessary when other types of biopsies are impractical, and symptoms indicate that the biopsy will be positive and the diseased organ will have to be removed. An *excisional biopsy* is used to remove a lump in a tissue or organ. After removal, the tissue is examined microscopically according to standard procedures.

Marfan's Syndrome

Marfan's syndrome, or arachnodactyly (Gr. "spider fingers"), is a rare genetic disease of connective tissue that causes skeletal and cardiovascular defects and eye problems, including nearsightedness (caused by an abnormal position of the eye lens), detached retina, and glaucoma. Affected people grow very tall and thin; characteristic skeletal deformities include abnormally long arms (arm span usually exceeds height), legs, fingers, and toes; concave or convex chest and spine due to the excessive growth of ribs; and weak tendons, ligaments, and joint capsules, resulting in easily dislocated joints ("double-jointedness"). In almost all cases, the heart valves or aorta (the major blood vessel leading from the heart) are abnormal. Marfan's syndrome usually becomes apparent at about age 10, and patients usually die from cardiovascular complications, especially those related to the expansion or constriction of the aorta, before they reach 50. The exact cause of Marfan's syndrome is unknown, but it probably results from abnormalities in elastin (the essential component of elastic connective tissue) and collagen. Symptomatic defects may be treated surgically. Abraham Lincoln is supposed to have had Marfan's syndrome, but there is little confirming evidence.

Systemic Lupus Erythematosus

Systemic lupus erythematosus (SLE) is an autoimmune disease that affects collagen in the lining of joints and other connective tissues. Women are affected about nine times more often than men. The first signs of the disease, which can be fatal in extreme cases, are a general malaise, fever, a so-called butterfly rash on the face, appetite loss, sensitivity to sunlight, and pains in the joints. SLE causes cells in the immune system (B cells) to produce excess antibodies, called autoantibodies, which attack healthy cells. The basal layer of skin begins to deteriorate, and any organ can be affected. Because the body is unable to remove the autoantibodies, they may settle in such vital organs as the brain, heart, kidneys, and lungs, ultimately causing serious tissue damage.

Tissue Transplantations

The process of surgically *transplanting* healthy tissue to replace diseased or defective tissue is also called *grafting.* There are four basic types of tissue transplantations:

1 An **allograft** (Gr. *allos,* another) is a tissue transplant from one person (often recently deceased) to another person *not genetically identical.* The recipient usually begins to reject an allograft after 15 to 25 days when the transplantation site becomes infiltrated with *graft-rejection cells* that recognize and destroy the foreign cells of the newly grafted tissue.

2 A *heterograft* (Gr. *heteros*, other) is a tissue transplant from an animal into a human being. Although rejection may occur after a week or so with skin allografts and heterografts, they provide protection until the patient's own skin is ready to be transplanted.

3 An *autograft* (Gr. *autos*, self) is a tissue transplant from one site on a person to a different site on the *same person*. When performing a skin autograft operation, the surgeon does not have to connect the blood vessels of the graft to those of the host skin because the graft skin is thin enough to be nourished by blood seeping into it from the host skin. If the graft is not rejected, blood vessels from the host grow into the grafted skin.

4 A *syngeneic graft* is a tissue transplant from one person to another, *genetically identical*, person. (In other words, from one identical twin to another.)

Part of the body's rejection response is the proliferation of activated "killer" T cells, which are specialized white blood cells (lymphocytes) of the immune system that help destroy foreign microorganisms that enter body tissues. The most successful tissue transplant involves the cornea of the eye precisely because the cornea, having no blood supply to provide entry to T cells, is not affected by the rejection response of the immune system. *Immunosuppressive drugs*, such as cyclosporine, can be used to suppress tissue rejection by reducing the body's supply of killer T cells, but most immunosuppressive drugs also reduce the patient's ability to fight infection.

A process called *tissue typing* is used to match donor and recipient tissues as closely as possible in an attempt to minimize the rejection response.

CHAPTER SUMMARY

Epithelial Tissues (p. 85)

1 *Epithelial tissues* may be in the form of sheets that *cover surfaces* or *line body cavities or ducts,* or they may be modified to become *glandular epithelia* that are organized into glands.

2 The typical functions of epithelial tissues are *absorption, secretion, transport, excretion,* and *protection*.

3 All types of epithelial tissues except glandular tissues share certain characteristics: (1) the cells are *regular in shape* and *packed in continuous sheets,* (2) the tissue is usually anchored to the underlying connective tissue by a *basement membrane,* (3) the tissue *has no blood vessels,* (4) there may be several types of *surface specializations* (microvilli and cilia), and (5) the cells are capable of *regeneration*.

4 Epithelial tissues are classified according to the *arrangement of cells,* the *number of cell layers,* and the *shape of the cells* in the superficial layer.

5 Epithelial tissues one-cell-layer thick are **simple;** tissues with two or more cell layers are **stratified.** The basic shapes of the superficial cells are **squamous, cuboidal,** and **columnar.**

6 *Simple squamous epithelium* consists of flat cells in a single layer and is suited for diffusion or filtration.

7 *Simple cuboidal epithelium* has a single layer of approximately cube-shaped cells. It functions in secretion and absorption.

8 *Simple columnar epithelium* is a single layer of elongated cells, with some cells modified to secrete mucus and others modified for absorption, protection, and movement of materials over cell surface.

9 *Stratified squamous epithelium* usually has at least three layers of cells; only the superficial layer contains flat squamous cells. Underlying *basal cells* can replace damaged superficial squamous cells. Its function is primarily protective.

10 *Stratified cuboidal epithelium* is a multilayered arrangement of somewhat cube-shaped secretory cells.

11 *Stratified columnar epithelium* has a superficial layer with a regular arrangement of columnar cells; the smaller cuboidal cells underneath are in contact with the basement membrane. Its functions are secretion and movement.

12 Exceptions to the basic system of classification are **pseudostratified columnar epithelium, transitional epithelium,** and **glandular epithelium.**

13 *Pseudostratified columnar epithelium* has all of its cells in contact with the basement membrane, but not all reach the surface. Functions include protection and secretion.

14 *Transitional epithelium* is composed of cells capable of stretching as needed, thus allowing for changes in shape.

15 *Glandular epithelium* has cells specialized for the synthesis, storage, and secretion of their product. The two main types of glands are *exocrine* and *endocrine.*

16 *Endocrine glands* do not have ducts. They release their secretions into the bloodstream. *Exocrine glands* release their secretions into ducts.

17 Exocrine glands are classified as **unicellular** or **multicellular; simple** or **compound; tubular, alveolar,** or **tubuloalveolar; mucous, serous,** or **mixed;** and **merocrine** and **holocrine.**

Connective Tissues (p. 95)

1 Besides their supportive and protective functions, connective tissues also form sheaths around hollow organs; involved in storage, transport, repair.

2 All connective tissues consist of fibers, ground substance, and cells. Fibers and ground substance make up the **matrix.**

3 Types of connective tissue fibers are **collagenous, reticular,** and **elastic.**

4 *Ground substance* is a homogeneous extracellular material providing a medium for the passage of nutrients and wastes between cells and the blood.

5 The cells of connective tissues are usually classified as either *fixed cells (fibroblasts, adipose cells,* and *reticular cells)* or *wandering cells (plasma cells, mast cells,* and *leukocytes). Macrophage cells* may be either fixed or wandering. Fixed cells are usually concerned with *long-term functions.* Wandering cells are usually involved with *short-term activities.*

6 The *reticuloendothelial system* contains specialized cells that are active in defending the body against microorganisms and are involved with other cells in producing antibodies. These cells remove undesirable matter from the fluids and matrix of body tissues.

7 Connective tissues are usually classified according to the arrangement and density of their fibers. The typical connective tissues are *loose* (areolar), *dense* (collagenous), *elastic, adipose,* and *reticular.*

8 *Embryonic connective tissue* is called *mesenchyme* in the embryo and *mucous connective tissue* in the fetus.

9 *Cartilage* is a specialized connective tissue that provides support and facilitates movement at the joints.

10 *Chondrocytes* are cartilage cells, which are housed in *lacunae,* cavities within the matrix.

11 The three types of cartilage are *hyaline cartilage, fibrocartilage,* and *elastic cartilage.*

12 Hyaline cartilage and elastic cartilage are usually enclosed within a fibrous sheath called a *perichondrium.*

Membranes (p. 101)

1 *Membranes* are thin, pliable layers of epithelial or connective tissue that line body cavities, cover surfaces, or separate or connect certain regions, structures, or organs of the body.

2 *Mucous membranes* line body passageways that open to the outside.

3 *Serous membranes* line the pericardial, pleural, and peritoneal cavities within the thoracic and abdominopelvic cavities, and cover the organs that lie within these cavities. Serous membranes include the *peritoneum, pericardium,* and *pleura.*

4 *Synovial membranes* line the cavities of joints and other similar areas.

STUDY AIDS FOR REVIEW

MEDICAL TERMINOLOGY

ATROPHY (AT-roe-fee; Gr. *atrophos*, ill-nourished; *a-*, without + *trophe*, nourishment) A wasting of body tissue, usually caused by a lack of use, such as when a broken arm is healing within an immobilizing cast, or when faulty blood circulation produces insufficient cell nutrition.

UNDERSTANDING THE FACTS

1 Is pseudostratified epithelium correctly named? Why?

2 Describe the relationship of structure and function of transitional epithelium.

3 How is connective tissue classified?

4 What is the main ingredient of ground substance, and what are its functions?

5 What is the reticuloendothelial system? How important is it in the body?

6 Compare the three types of cartilage.

7 In what parts of the body are serous membranes found?

UNDERSTANDING THE CONCEPTS

1 Where in the body would you expect to find epithelial tissue?

2 We speak of cube-shaped cells, but as you observe these cells, what do you note? How many right angles can you discover in or on your body?

3 Is there a correlation between location and function of epithelial tissues and their ability to be renewed? Explain.

4 The anterior surface of the lens of the eye is covered with simple squamous epithelium. Is this a proper tissue type for this location?

5 What would happen if the body were deficient in the wandering cells of connective tissue? Could this situation be life-threatening?

SELF-QUIZ

Multiple-Choice Questions

4.1 Connective tissues are involved in
  a support   d repair
  b storage   e all of the above
  c transport

4.2 Which of the following is *not* a function of epithelial tissues?
  a absorption   d excretion
  b transport   e sensory reception
  c contraction

4.3 Which of the following is *not* a general characteristic of epithelial tissues?

  a regular shaped cells
  b little matrix present
  c cells tightly joined to each other
  d ground substance absent
  e junctional complexes present

4.4 The principal role(s) of the basement membrane is/are
  a to provide elastic support
  b to act as a barrier for diffusion and filtration
  c to separate epithelium from underlying connective tissue
  d a and b
  e a, b, and c

4.5 Which of the following structures may be found on the apical surface of simple epithelium?
  a cilia   d spot desmosomes
  b microvilli   e all but d
  c brush border

4.6 Epithelial tissues are classified according to the
  a arrangement of cells   d a and b
  b number of cell layers   e a, b, and c
  c shape of the cells

4.7 Transitional epithelium is
  a found in the gallbladder
  b stratified with surface cells that are larger and more rounded than those of underlying layers
  c ciliated
  d a and b
  e a, b, and c

4.8 Simple squamous epithelium is found in
  a the alveoli of the lungs
  b the pericardium
  c the pleura
  d the glomerular (Bowman's) capsule
  e all of the above

4.9 Epithelium whose function is chiefly protective is
  a stratified   d a and b
  b easily regenerated   e a, b, and c
  c often keratinized

4.10 Which of the following lines the abdominopelvic cavity and covers the organs inside the cavity?
  a peritoneum   d synovial membrane
  b pericardium   e all but d
  c pleura

True-False Statements

4.11 Epithelial tissues contain blood vessels.

4.12 Stratified cuboidal epithelium is involved in secretion.

4.13 Simple columnar epithelium may contain cilia.

4.14 Epithelial tissue may be classified as either covering, lining, or glandular.

4.15 The only example of a unicellular gland in the human body is a goblet cell.

4.16 If the secretory portion of a gland is tubular, the gland is an alveolar gland.

4.17 Loose connective tissue is also called areolar tissue because it contains tiny extracellular spaces.

4.18 Dense irregular connective tissue has thick fiber bundles that are arranged randomly in a tough, resilient meshwork.

4.19 Elastic connective tissue is found throughout the body in many different organs.

4.20 Mucus connective tissue is found only in the umbilical cord of the fetus, where it is called Wharton's jelly.

Completion Exercises

4.21 Epithelium that makes up the serous membranes of various body cavities is called _____ .

4.22 Overall, simple cuboidal cells are active in absorption and _____ .

4.23 _____ epithelia have nuclei situated at different levels, giving a multilayered impression.

4.24 Fibers of connective tissue may be collagenous, reticular, or _____ .

4.25 The most common cells in connective tissue, and the only cells found in tendons, are _____ .

4.26 Connective tissues are usually classified according to the arrangement and density of their _____ .

4.27 When embryonic connective tissue is found in the embryo it is called _____ .

4.28 Cartilage cells are called _____ .

4.29 _____ cartilage is the most prevalent type of cartilage in the human body.

4.30 The important protein that forms the fibrous network in almost all tissues of the body is _____ .

Matching

4.31 _____ ground substance

4.32 _____ elastic fibers

4.33 _____ reticular fibers

4.34 _____ collagenous fibers

4.35 _____ fibroblasts

4.36 _____ macrophages

4.37 _____ reticular cells

4.38 _____ mast cells

a thick bundles of whitish fibrils composed of collagen

b main ingredients are glycoproteins and proteoglycans

c synthesize and store lipids

d allows hollow organs and other structures to stretch

e delicately formed networks of inelastic fibrils

f produce heparin and histamine

g synthesize matrix materials

h engulf and destroy foreign bodies

i help form macrophages

j main producers of antibodies

Problem-Solving/Critical Thinking Questions

4.39 A surgeon performing abdominal surgery first cuts through the skin, then loose _____ tissue, then _____ , to reach the _____ lining of the inside wall of the abdomen.

a connective, muscle, parietal peritoneum

b adipose, connective, visceral peritoneum

c muscle, fat, visceral pleura

d areolar, smooth muscle, parietal pericardium

e connective, epithelium, elastic cartilage

A SECOND LOOK

Identify the type of tissue illustrated in each of the drawings.

 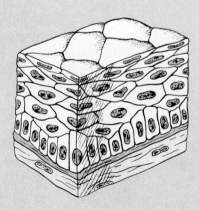 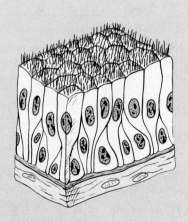

II

PROTECTION, SUPPORT, AND MOVEMENT

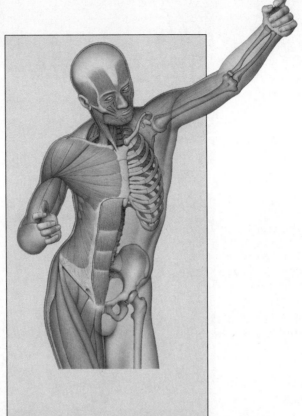

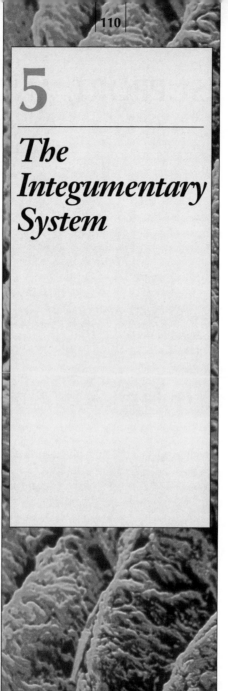

5

The Integumentary System

LEARNING OBJECTIVES

1 Name the five layers of the epidermis.

2 Describe the structures that are found in the dermis.

3 Explain why skin has different colors.

4 Describe the structure and function of sudoriferous glands and sebaceous glands.

5 Describe the structure of hair.

6 Explain how hair develops and grows.

7 Describe the structure and function of nails.

8 Contrast first-, second-, and third-degree burns.

9 Describe several common skin disorders and one or more types of skin cancer.

The ***integumentary system*** (L. *integumentum*, cover) consists of the skin and its derivatives, which include several types of glands, hair, and nails. The system functions in protection, and in the regulation of body temperature, excretion of waste materials, synthesis of vitamin D_3, and reception of various stimuli perceived as pain, pressure, and temperature.

SKIN

The ***skin*** is the largest organ in the body, occupying almost 2 sq m (21.5 sq ft) of surface area. It varies in thickness on different parts of the body, from less than 0.5 mm on the eyelids to more than 5 mm on the middle of the upper back. A typical thickness is 1 to 2 mm.

Skin has two main parts: the *epidermis* is the outermost layer of epithelial tissue; the *dermis* is a thicker layer of connective tissue beneath the epidermis [FIGURE 5.1A].

Epidermis

The outer layer of the skin, called the ***epidermis*** (Gr. *epi*, over + *derma*, skin), is stratified squamous epithelium. The epidermis contains no blood vessels, but most of it is so thin, however, that even minor cuts reach the dermis and draw blood. In the thick epidermis of the palms of the hands and the soles of the feet, there are five typical layers (strata). Starting with the innermost layer, they are stratum basale, stratum spinosum, stratum granulosum, stratum lucidum, and stratum corneum [FIGURE 5.1B]. The stratum spinosum and stratum basale together are known as the *stratum germinativum* (jer-mih-nuh-TEE-vuhm) because they generate new cells. In parts of the body other than the palms and soles, only the stratum corneum and stratum germinativum are regularly present.

Stratum basale The ***stratum basale*** (L. *basis*, base) rests on the basement membrane next to the dermis. It consists of a single layer of columnar or cuboidal cells. Like the stratum spinosum, it undergoes cell division, producing new cells to replace those being shed in the exposed superficial layer.

Stratum spinosum The ***stratum spinosum*** (L. *spinosus*, spiny) is composed of several layers of polyhedral (many-sided) cells that have delicate "spines" protruding from their surface. The interlocking spinelike projections help support this binding layer. Active protein synthesis takes place in the cells of the stratum spinosum, indicating that cell division and growth are oc-

curring. Some new cells are formed here and are pushed to the surface to replace the keratinized cells of the stratum corneum.

Stratum granulosum The ***stratum granulosum*** (L. *granum*, grain) lies just below the stratum lucidum. It is usually two to four cells thick. The cells contain granules of keratohyaline. The process of keratinization, associated with the dying of cells, begins in this layer.

Stratum lucidum The ***stratum lucidum*** (L. *lucidus*, bright, clear) consists of flat, translucent layers of dead cells. The stratum lucidum appears only in the palms of the hands and soles of the feet, acting as a protective shield against the ultraviolet (UV) rays of the sun and preventing sunburn of the palms and soles.

Stratum corneum The ***stratum corneum*** (L. *corneus*, cornified tissue) is a flat, relatively thick layer of dead cells arranged in parallel rows. This keratinized layer of the epidermis consists of soft keratin (as compared with the hard keratin in fingernails and toenails), which helps keep the skin elastic. The soft keratin also protects living cells underneath from being exposed to the air and drying out. The cells in the stratum corneum are constantly being shed through normal abrasion. They are replaced by new cells that are formed by cell division and pushed up from the germinative layers below. For example, cells on the soles of the feet are completely replaced every month or so.

Dermis

Most of the skin is composed of ***dermis*** ("true skin"), a strong, flexible connective tissue meshwork of collagenous, reticular, and elastic fibers. *Collagenous fibers*, which are formed from the protein collagen, are very thick and give the skin much of its toughness. (The thickest dermis is located on the back, thighs, and abdomen, in that order.) Although *reticular fibers* are thinner, they provide a supporting network. *Elastic fibers* give the skin flexibility. The cells of the dermis are mostly *fibroblasts*, *fat cells*, and *macrophages*, which digest foreign substances. Blood vessels, lymphatic vessels, nerve endings, hair follicles, and glands are also present. The dermis is composed of two layers that are not clearly separated. The thin *papillary layer* is directly beneath the epidermis; the deeper, thicker layer is called the *reticular layer*.

Q: *What is a blister?*

A: When the skin is burned or irritated severely, tiny blood vessels in the dermis widen and some clear plasma leaks out. The plasma accumulates between the dermis and epidermis in a fluid-filled pocket, or blister.

FIGURE 5.1 THE SKIN

[A] A "textbook" drawing of the skin's structure, showing its many components in an ideal arrangement. The stratum lucidum would not appear on hairy skin. An enlarged portion of the epidermis is shown in the small drawing (top).
[B] A vertical section of the thick epidermis of the fingertip. ×100.

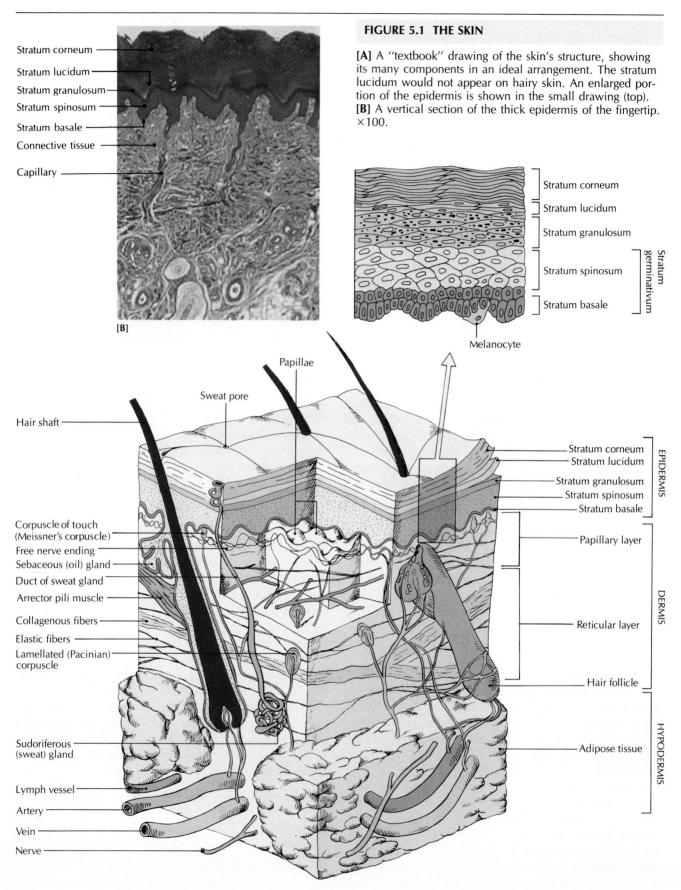

[B]

[A]

We All Get Wrinkles

Everybody's skin gets wrinkled. The length and depth of a wrinkle may differ from person to person, but no one is exempt. In fact, wrinkling is so predictable that we can expect to wrinkle in the same place, and at the same time, as the generation ahead of us. It is relatively easy to estimate a person's age merely by studying the wrinkles on the face and neck.

Wrinkling usually starts in the mid-twenties in the areas of the greatest facial expression: around the eyes, mouth, and brow. The familiar "crow's feet," lines radiating from the corner of each eye,

usually appear by the age of 30. During the thirties and forties, new lines appear between the eyebrows and in front of the ears, followed in the fifties by wrinkles on the chin and bridge of the nose. New wrinkles appear on the upper lip in the sixties and seventies, and all of the established wrinkles become more pronounced. In the eighties, the ears become elongated, and we can see the "long ears" of old age.

Wrinkling happens when the protein *elastin* in the elastic tissue of the dermis loses its resiliency and degenerates into *elacin*. This causes the dermis to become

more closely bound to the underlying tissue. Also, the layer of fat beneath the skin diminishes, causing the skin to sag into wrinkles. The wrinkling process is reinforced because an adult's aging epidermis does not produce new cells as readily as it did during childhood and adolescence. The epidermal cells are simply not as vital as they used to be.

Exposure to the ultraviolet rays in sunlight speeds up the wrinkling process, causing the collagen and elastin in the dermis to degenerate. As a result, the skin becomes slack and prematurely aged.

Papillary layer The *papillary* (L. dim. *papula*, pimple) *layer* of the dermis consists of loose connective tissue with thin bundles of collagenous fibers. The papillary layer is so named because of the papillae (tiny, fingerlike projections) that join it to the ridges of the epidermis [FIGURE 5.2]. Most of the papillae contain loops of capillaries that nourish the epidermis; others have special nerve endings, called corpuscles of touch (Meissner's corpuscles), that serve as sensitive touch receptors. A double row of papillae in the palms and soles produce ridges that help keep the skin from tearing, improve the gripping surfaces, and produce distinctive fingerprint patterns in the finger pads. (Fingerprint patterns in the epidermis follow the corrugated contours of the dermis underneath. No two people, not even identical twins, have exactly the same fingerprints.)

Reticular layer The *reticular* ("netlike") *layer* of the dermis is made up of dense connective tissue, with coarse collagenous fibers and fiber bundles that crisscross to form a strong and elastic network. You can understand how tough the reticular layer is when you realize that it is the part of an animal's skin that is processed commercially to make leather.

Although the collagenous fibers of the dermis appear to be arranged randomly, there is actually a dominant pattern. In fact, different directional patterns are found

in each area of the body [FIGURE 5.3]. The resulting lines of tension over the skin are known as *cleavage (Langer's) lines.* They are very important during surgery because an incision made *across* the lines causes a gaping wound that heals more slowly and leaves a deeper scar than an incision made *parallel* to the directional lines. An incision made parallel to the cleavage lines heals faster because it disrupts the collagenous fibers only slightly, and the wound needs only a small amount of scar tissue to heal.

Embedded in the reticular layer are many blood and lymphatic vessels, nerves, free nerve endings, fat cells, oil glands, and hair roots. Receptors for deep pressure (Pacinian corpuscles) are distributed throughout the der-

Q: *Why does a healing wound itch?*

A: When the skin is injured there is usually some damage to superficial nerves. After about a week or so, the nerves begin to fire impulses as new nerve endings begin to grow. The firings feel like an itch on the skin, which usually subsides after a few days.

FIGURE 5.2 DERMAL PAPILLAE

The boundary between the epidermis and dermis is marked by dermal papillae, which extend into the epidermis.

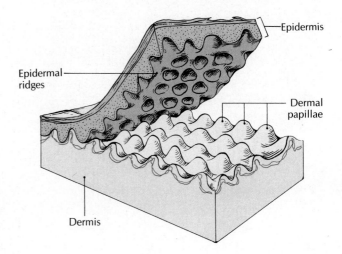

mis and subcutaneous layer. The deepest region of the reticular layer contains smooth muscle fibers, especially in the genital and nipple areas and at the base of hair follicles.

Although the dermis is highly flexible and resilient, it can be stretched beyond its limits. During pregnancy, for example, collagenous and elastic fibers may be torn. Characteristic abdominal "stretch marks" result from the repairing scar tissue.

Hypodermis

Beneath the dermis lies the **hypodermis** (Gr. *hypo*, under), or **subcutaneous layer** (L. *sub*, under + *cutis*, skin). It is composed of loose, fibrous connective tissue. The hypodermis is generally much thicker than the dermis and richly supplied with lymphatic and blood vessels and nerves. Also within the hypodermis are the coiled ducts of sweat glands and the bases of hair follicles. The boundary between the epidermis and dermis is distinct; that between the dermis and hypodermis is not.

Where the skin is freely movable, hypodermis tissue fibers are scarce. Where it is attached to underlying bone or muscle, the hypodermis contains tightly woven fibers. In some areas of the hypodermis where extra padding is desirable, as in the breasts or heels of the feet, thick sheets of fat cells are present. The distribution of fat in this layer is largely responsible for the characteristic body contours of the female.

FIGURE 5.3 CLEAVAGE (LANGER'S) LINES

Anterior [A] and posterior [B] views.

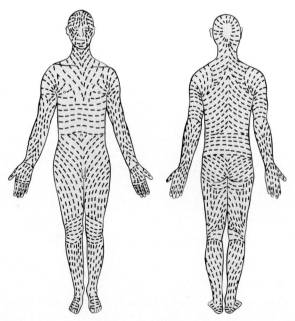

[A]   [B]

Functions of Skin

An obvious function of the skin is to cover and protect the inner organs. But this is only one of its many functions.

Protection The skin acts as a stretchable protective shield that prevents harmful microorganisms and foreign material from entering the body, and prevents the loss of body fluids. These functions are made possible by two features: (1) layered sheets of flat epithelial tissue that act like shingles on a roof, and (2) a nearly waterproof layer of keratin in the outer layer of the skin. Skin also plays an active role in defending the body against disease.

Temperature regulation Although the skin is almost waterproof and solid enough to keep out the water when you take a bath, it is also porous enough to allow some chemicals in* and to allow sweat to escape from sudoriferous glands through ducts ending at the skin surface as tiny pores.

When sweat is excreted through the pores and then evaporates from the surface of the skin, a cooling effect occurs. On very humid days you are not cooled easily because the outside air is already almost full of water, and sweat merely builds up on your skin instead of evaporating. On cold days your skin acts as a sheet of insulation that helps retain body heat and keep your body warm. In addition, the dermis contains a dense bed of blood vessels. On warm days the vessels dilate (enlarge), and heat radiation from the blood is increased; in this way, body heat is lost.

Excretion Through perspiration, small amounts of waste materials, such as urea, are excreted through the skin. Up to 1 g of waste nitrogen may be eliminated through the skin every hour.

Synthesis The skin helps to screen out harmful excessive ultraviolet rays from the sun, but it also lets in some necessary ultraviolet rays, which convert a chemical in

*Unfortunately, several harmful substances can be absorbed through the skin and introduced into the bloodstream. Among the most common intruders are metallic nickel, mercury, many pesticides and herbicides, and a skin irritant called urushiol, found in poison ivy, poison oak, and poison sumac. Even the smoke from burning these plants may be enough to produce skin eruptions.

Q: *Why is a tattoo so difficult to remove?*

A: Tattoo dyes are injected below the stratum germinativum, into the dermis, where the dyed cells do not move outward toward the surface as the skin is shed and replaced. In the same way, scars and stretch marks do not disappear easily because they involve the dermis.

No Fun in the Sun

Why is sunbathing potentially dangerous? The process of tanning is activated by the ultraviolet rays in sunlight (called *ultraviolet B*, or *UVB*). These tanning rays can kill some skin cells, damage others so that normal secretions are stopped temporarily, increase the risk of skin cancer and mutations by affecting the DNA in the nuclei of cells, and cause the immune system to falter. Ultraviolet rays can also damage enzymes and cell membranes and interfere with cellular metabolism. If tissue destruction is extensive, toxic waste products and other cellular debris enter the bloodstream and produce the fever associated with sun poisoning.

Ultraviolet rays cause tiny blood vessels below the epidermis to widen, allowing an increased flow of blood. The abundance of blood colors the skin red. Ultraviolet rays also stimulate melanocytes to produce melanin. The cells begin to divide more rapidly than usual, creating new cells that travel toward the surface in an attempt to repair the damaged skin. Ordinarily, these new cells take about four weeks to reach the surface, where they die and are shed unnoticed in the course of a normal day's activities. New cells produced in response to skin damage, however, are so numerous that they reach the surface in four or five days. At this point, sheets of old,

sunburned skin begin to peel off, and the pigment-rich melanocytes, which have traveled to just under the surface of the skin, produce the first signs of a tan. The tanned skin helps to prevent further ultraviolet damage by absorbing and scattering the harmful rays. Unfortunately, many people think that you have to burn before you can tan, but dermatologists (physicians who specialize in treating skin problems) disagree. Instead, they suggest short periods of sunbathing (about 10 min) until the skin has a protective tan.

---

the skin called 7-dehydrocholesterol into vitamin D_3 *(cholecalciferol)*.* Vitamin D in the diet is vital to the normal growth of bones and teeth. A lack of ultraviolet light and vitamin D impairs the absorption of calcium from the intestine into the bloodstream. When children are deprived of sunshine, they generally become deficient in vitamin D. Unless they receive cholecalciferol from another source, they develop rickets, a disease that may deform the bones permanently.

Sensory reception The skin is an important sensory organ, containing sensory receptors that respond to heat, cold, touch, pressure, and pain. Skin helps to protect us through its many nerve endings, which keep us responsive to things in the environment that might harm us: a hot stove, a sharp blade, a heavy weight. These nerve endings also help us to sense the outside world, so that adjustments can be made to maintain homeostasis.

Color of Skin

Skin color is determined by three factors: the presence of **melanin** (Gr. *melas*, black), a dark pigment produced by

*The general term "vitamin D" actually refers to a group of steroid vitamins, including vitamin D_3 *(cholecalciferol)* and vitamin D_2 *(calciferol, or ergocalciferol)*.

Q: *What causes the skin of Orientals to appear yellow?*

A: In addition to melanin, skin contains a yellow pigment called carotene. It is usually overshadowed by melanin, but since Orientals have little melanin, their skin has a yellow tint.

specialized cells called *melanocytes;* the accumulation of the yellow pigment *carotene;* and the color of the blood reflected through the epidermis. Melanocytes are usually located in the deepest part of the stratum basale, where their long processes extend under and around the neighboring cells. Melanin is found in all areas of the skin, but the skin is darker in the external genitals, the nipples and the dark area around them, the anal region, and the armpits. In contrast, there is hardly any pigment in the palms and soles. Melanin is present not only in skin but also in hair and in the iris and retina of the eyes.

The main function of melanin is to screen out excessive ultraviolet rays, especially protecting the cell nuclei and their genetic material. Extra protection is provided when melanin is darkened by the sun and transferred to the outer skin layers, producing a "suntan" that is less sensitive to the sun's rays than previously unexposed skin. In a sense, "freckles" can be thought of as a permanent tan produced by sun-darkened spots of melanin.

All races have some melanin in their skins. Although the darker races have slightly more melanocytes than light-skinned people do, the main reason for their darkness is the wider distribution of melanin in the skin beyond the deepest portion of the stratum basale into higher levels of the epidermis. Also, the melanocytes of a dark-skinned person probably produce more melanin than the melanocytes of a light-skinned person. A person who is genetically unable to produce *any* melanin is an *albino* (L. *albus*, white).

How a Wound Heals

When the skin is cut, the healing process begins immediately. This process, which can take from a week to a month (depending on the severity of the cut), proceeds step by step:

1 Blood vessels are severed along with the dermis and epidermis. Blood cells leak from the severed vessels into the wound. Blood cells called *platelets* (thrombocytes) and a blood-clotting protein called *fibrinogen* help to start a blood clot. A network of fibers containing trapped cells forms, and the edges of the wound begin to join together again. Tissue-forming cells called *fibroblasts* begin to approach the wound [FIGURE 5.4A].

2 Less than 24 hours later, the clotted area becomes dehydrated, and a scab forms. Phagocytes called *neutrophils* are attracted from blood vessels into the wound area and ingest cellular debris, microorganisms, and other foreign material. Epidermal cells at the edge of the wound begin to divide and start to build a bridge across the wound. White blood cells called *monocytes* migrate toward the wound from surrounding tissue [FIGURE 5.4B].

3 Two or three days after the wounding, monocytes enter the wound and ingest the remaining foreign material. Epidermal cells complete the bridge of new skin under the scab [FIGURE 5.4C]. When a totally new epidermal surface has been formed, the protective scab is sloughed off. Fibroblasts build scar tissue with collagen.

4 About 10 days after the wounding, the epidermis has been restored, and the scab is gone [FIGURE 5.4D]. Some monocytes and neutrophils remain in the area, but their work is usually completed by this time. Tough scar tissue continues to form, and bundles of collagen build up along the stress lines of the original cut.

Developmental Anatomy of Skin

The epidermis and dermis of the skin are derived from different embryological tissues: the outer epidermis is derived from surface ectoderm (epithelium), and the inner dermis is derived from mesenchyme (connective tissue) beneath the surface ectoderm [FIGURE 5.5A]. At about seven weeks after conception, the surface ectodermal cells form a superficial layer of simple squamous epithelium called the *periderm* [FIGURE 5.5B]. As the cells of the periderm undergo keratinization and begin to be shed, they are replaced by cells from the basal layer, which later forms the stratum germinativum.

At about 11 weeks, the mesenchymal cells begin to produce collagenous and elastic fibers in the dermis [FIGURE 5.5C]. As these fibers migrate upward they form the *dermal papillae* that project into the epidermis [see FIGURE 5.2]. The five distinct layers of the epidermis are recognizable by month 4.

Melanocytes are formed from melanoblasts that migrate upward toward the junction of the dermis and epidermis. Melanocytes begin producing melanin shortly before birth. The replacement of peridermal cells continues until about week 21, after which the periderm disappears. At birth [FIGURE 5.5D], all mature skin layers are present.

FIGURE 5.4 HOW A WOUND HEALS

[A] At the site of a wound, the epidermis, dermis, and blood vessels are severed. Fibroblasts move toward the wound area. **[B]** A scab forms. Epithelial cells at the edge of the wound begin to divide. Neutrophils begin to ingest cellular debris; monocytes move toward the wound area.

[C] New skin is formed under the scab. Monocytes ingest the remaining foreign material in the wound area. Fibroblasts begin to build scar tissue. **[D]** The scab is gone. Scar tissue continues to form.

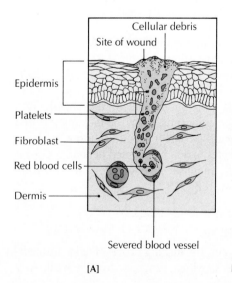
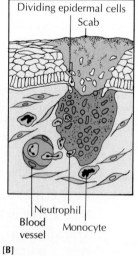
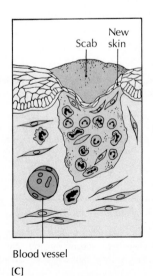
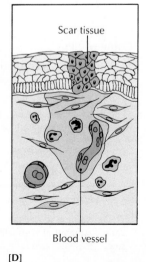

FIGURE 5.5 DEVELOPMENTAL ANATOMY OF SKIN

[A] Four weeks, **[B]** seven weeks, **[C]** 11 weeks, **[D]** newborn.

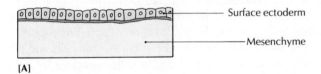

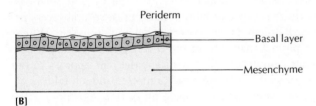

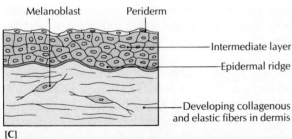

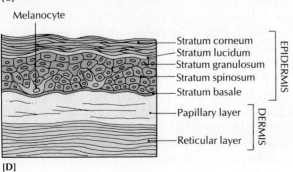

GLANDS OF THE SKIN

The glands of the skin are the sudoriferous and sebaceous glands.

Sudoriferous (Sweat) Glands

Sudoriferous glands (L. *sudor*, sweat) are also known as *sweat glands.* Two types of sudoriferous glands exist, eccrine (EKK-rihn) and apocrine (APP-uh-krihn). *Eccrine glands* (Gr. *ekkrinein*, to exude, secrete) are small sudoriferous glands. They are distributed over nearly the entire body surface. (There are no sudoriferous glands on the nail beds, margins of the lips, eardrums, inner lips of the vulva, or the tip of the penis.) These glands are generally of the simple, coiled tubular type [see TABLE 4.1]; the secretory portion of the gland is embedded in the hypodermis, and a hollow, corkscrewlike duct leads up through the dermis to the surface of the epidermis [FIGURE 5.1A]. The duct generally straightens out somewhat as it reaches the surface. All of these ducts combined would be about 9.5 km (6 mi) long.

Apocrine, or *odiferous, glands* are found in the armpits, the dark region around the nipples, the outer lips of the vulva, and the anal and genital regions. These are larger and more deeply situated than the eccrine glands. The female breasts are apocrine glands that have become adapted to secrete and release milk instead of sweat. Apocrine glands become active at puberty and enlarge just before menstruation. Eccrine glands respond to heat, but apocrine glands in the skin respond to stress (including sexual activity) by secreting sweat of a characteristic odor.

The *ceruminous glands* in the outer ear canals are also apocrine skin glands. They secrete the watery component of *cerumen* (ear wax). Cerumen helps to trap foreign substances before they can enter the deeper portions of the ear.

Q: *Why do your fingertips and toes pucker when you take a bath?*

A: The thick skin on the underside of your fingers and toes is capable of absorbing a great deal of water, and because your palms and soles are not coated with sebum, water soaks in easily through the epidermis. The dermis does not expand, however, so the epidermis becomes corrugated temporarily. Also, when you get out of the tub the excess water begins to evaporate. When it does, some of the water that was there before your bath is also removed, and your skin becomes dried out and wrinkled until the body fluid is replaced. (The same thing happens to a grape when it is dried out to make a raisin.)

Sebaceous (Oil) Glands

Sebaceous, or *oil, glands* (L. *sebum,* tallow, fat) are simple, branched alveolar glands found in the dermis. Their main functions are lubrication and protection. They are connected to hair follicles [FIGURE 5.1A]. The secretory portion is made up of a cluster of cells, polyhedral in the center of the cluster and cuboidal on the edges. Secretions are produced by the breaking down of the interior cells, which become the oily secretion called *sebum,* found at the base of the hair follicle.

Sebum functions as a permeability barrier, an emollient (skin-softening agent), and a protective agent against bacteria and fungi.

When sebaceous glands become inflamed and accumulate sebum, the gland opening becomes plugged, and a *blackhead* results. If the plugging is not relieved, a pimple or even a sebaceous cyst may develop. The "blackness" of blackheads is caused by air-exposed sebum, not dirt. Blackheads and pimples often accompany hormonal changes during puberty, when an oversecretion of sebum may enlarge the gland and plug the pore. The resulting skin disease is called *acne vulgaris.* (See When Things Go Wrong.)

Openings from sebaceous glands are found all over the surface of the skin except in the palms and soles, where they would be a nuisance.

A S K   Y O U R S E L F

1 What are the two types of skin glands?

2 What are the structure and functions of sudoriferous glands?

3 What are the structure and functions of sebaceous glands?

HAIR

Hair is composed of cornified threads of cells, a specialization that develops from the epidermis. Because hair arises from the skin, it is considered to be an appendage of the skin. It covers the entire body, except for the palms, soles, lips, tip of the penis, inner lips of the vulva, and nipples.

Functions of Hair

Obviously, hair does not function as an insulative covering in humans as it does in many other mammals, but human hair does serve some protective functions. Scalp hair provides some insulation against cold air and the heat of the sun. Scalp hair also protects us from bumps. Eyebrows act as cushions in protecting the eyes and also help reduce glare and prevent sweat from running into the eyes. Eyelashes act as screens against foreign particles. Tiny hairs in the nostrils trap dust particles in inhaled air. Other openings in the body, such as the ears, anus, and vagina, are also protected by hair.

Hair is often used for diagnostic testing. For example, dry, brittle hair may indicate an underactive thyroid gland, since a healthy thyroid secretes enough hormones to produce healthy hair. Hair samples can also be used to detect metallic poisons such as arsenic, and drugs such as heroin and cocaine.

Structure of Hair

The two parts of a hair are the *shaft,* which is the portion that protrudes from the skin, and the *root,* the portion embedded in the skin. Hair consists of epithelial cells arranged in three layers. From the inside out of a hair shaft (the portion that protrudes from the skin), these are the *medulla, cortex,* and *cuticle* [FIGURE 5.6A, B]. The medulla is composed of soft keratin, and the cuticle and cortex are composed of hard keratin. A strand of hair is stronger than an equally thick strand of nylon or copper.

The *medulla* forms the central core of the hair, and it usually contains loosely arranged cells separated by air "cells" or liquid in its extracellular spaces. The *cortex* is the thickest layer of hair, consisting of several layers of cells. Pigment in the cortex gives hair its color. When hair pigment fades from the cortex and the medulla becomes completely filled with air, the hair appears to be gray to white, with whiteness representing a total loss of pigment. The *cuticle* is made up of thin squamous cells that overlap to create a scalelike appearance [FIGURE 5.6B, C]. The cuticle can be softened or even dissolved by chlorine in pool water. The greatest damage occurs over sustained periods of time, as the effects of the chlorine accumulate.

The lower portion of the root, located in the hypodermis, enlarges to form the *bulb.* The bulb is composed of a *matrix* of epithelial cells. The bulb pushes inward along its bottom surface to form a papilla of blood-rich connective tissue. The entire bulb is enclosed within a tubular *follicle* (L. *folliculus,* little bag). The hair follicle consists of three sheaths: (1) an inner epithelial root sheath. (2) an outer epithelial root sheath, and (3) a connective tissue sheath [FIGURE 5.6A].

Just before a hair is to be shed, the matrix cells gradually become inactive and eventually die. The root of the hair becomes completely keratinized and detaches from the matrix. This cornified root bulb begins to move alone the follicle until it stops near the level of the sebaceous gland. The papilla atrophies, and the outer root sheath collapses.

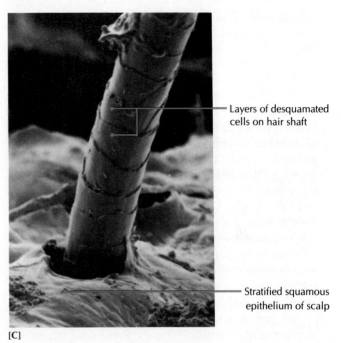

FIGURE 5.6 HAIR

[A] Longitudinal section of human hair, showing the hair root, follicle, and bulb. [B] The microscopic structure of hair. Notice how the scales of the cuticle all face upward. [C] A scanning electron micrograph of a hair shaft on the scalp. ×4375.

Cortex
Medulla
Cuticle

[B]

Hair shaft

EPIDERMIS

DERMIS

HYPODERMIS

Sebaceous (oil) gland
Arrector pili muscle

Hair root
Inner epithelial root sheath
Outer epithelial root sheath
Connective tissue sheath

HAIR FOLLICLE

Hair matrix
Hair papilla

HAIR BULB

[A]

Layers of desquamated cells on hair shaft

Stratified squamous epithelium of scalp

[C]

A new hair begins to grow when new cells from the outer root sheath start to develop near the old papilla. A new matrix develops, and a new hair starts to grow up the follicle, pushing the old hair out of the way if it hasn't been shed already.

Since the hair follicle arises from the epidermis, projecting at an angle, the hair shaft usually points away from the bulb. For this reason, hair covers the scalp more efficiently than if it grew at right angles to the scalp. The shape of the follicle openings determines whether your hair is straight, wavy, or curly. A round opening (in cross section) produces straight hair; an oval opening produces wavy hair; and a spiral-shaped opening produces curly hair.

An interesting part of the hair follicle is the bundle of smooth muscle attached about halfway down the follicle. As you saw in FIGURES 5.1A and 5.6C, a sebaceous gland nestles between the hair follicle and the muscle like a child sitting in the crook of a tree. This muscle is the *arrector pili* muscle. When it contracts, it pulls the follicle and its hair to an erect position, elevates the skin

above, and produces a "goose bump" on the skin. The contracting muscle also forces sebum from the sebaceous gland.

Growth, Loss, and Replacement of Hair

We are constantly shedding and (we hope) replacing our hair. A hair is shed when its growth is complete. Just before a hair is to be shed, the matrix cells gradually become inactive and eventually die. The root of the hair becomes completely keratinized and detaches from the matrix. This cornified root bulb begins to move along the follicle until it stops near the level of the sebaceous gland. The papilla atrophies, and the outer root sheath collapses.

After a short period of rest, a new hair begins to grow when new hair cells from the outer root sheath start to develop near the old papilla. A new matrix develops, and a new hair starts to grow up the follicle, pushing the old hair out of the way if it hasn't been shed already.

Each type of hair has its own life cycle. Even the same type of hair (hair on the scalp, for example) grows at staggered rates, so no area of the body sheds all of its old hair at the same time. Hair grows faster at night than during the day, and faster in warm weather than in cold. The rate of hair growth also varies with age, usually slowing down as we get older. The fastest growth rate occurs in women between ages 16 and 24. Hair textures and locations also affect growth rates. Coarse, black hair grows faster than fine, blond hair. Scalp hairs last much longer (3 to 5 years) than eyebrow and eyelash hairs (about 10 weeks).

Human hair follicles alternate between growing and resting phases. Scalp hairs, which grow about 0.4 mm a day, usually last 3 to 5 years before they are shed and the follicles go into a resting phase of 3 to 4 months. Plenty of scalp hair is visible at any given time, however, because 80 to 90 percent of the follicles in the scalp are in the growing phase at the same time. (The average scalp contains about 125,000 hairs.) Because healthy, active follicles keep producing new hair, it is not unusual to lose 50 to 100 hairs a day without becoming bald. A person might lose and replace more than 1.5 million hairs in a lifetime. Some hairs have a longer resting phase than a growing phase. Eyebrow hairs, for instance, grow 1 to 2 months before the follicles rest for the next 3 to 4 months. Eyelashes have an even shorter growing phase.

Although it seems that men have more body hair than women do, they actually have about the same number of hair follicles, about two million. Male hair is coarser, and therefore more obvious. Baldness in men seems to occur when there is a genetic predisposition toward baldness. If the father is bald, the son will probably be bald too, with a type of baldness called *pattern baldness.* Baldness also results from an abnormally large amount of testosterone (the male sex hormone) and an abnormally small amount of estrogen (a female sex hormone). Baldness can also be caused by disease, stress, malnutrition, and many other external factors.

Developmental Anatomy of Hair

The first hairs begin to develop during the third fetal month, when the epidermis begins to project downward into the dermis. Eventually these downgrowths form hair follicles that produce hairs. By the fifth month, the fetus

Q: *How long would scalp hair grow if left uncut?*

A: If scalp hair were left uncut, it would grow to a length of about 1 m (3.25 ft) in a lifetime. (Many exceptions of longer hair have been recorded.) If it did not grow in cycles that include inactivity and shedding, scalp hair would reach about 7.5 m (24.5 ft). Contrary to what some people believe, hair does not grow after a person dies. Instead, the skin shrinks, making the hair look longer.

is covered with fine, downy hair called *lanugo* (luh-NOO-go; L. *lana,* fine wool). Later in the fifth month, eyebrows and head hairs become apparent. Lanugo hair is shed before birth, except on the eyebrows and scalp, where it becomes thicker. After the baby is five or six months old, this hair is replaced by coarser hair, and the rest of the body becomes covered with a film of delicate hair called *vellus* (L. "fleece" or "coarse wool").

Coarse hair appears at puberty in the genital area and armpits, and young men also develop chest and facial hair. Body hair usually thickens during puberty. The early coarse hair of the eyebrows and scalp, as well as the hair that appears at puberty, is called *terminal hair.* Although the vellus and terminal hairs seem to be "new," no new hair follicles are developed after birth.

A S K   Y O U R S E L F

1 What are some of the functions of hair?

2 What are the three layers of a hair called?

3 What is a hair follicle?

NAILS

Nails, like hair, are modifications of the epidermis. Also, like the cuticle and cortex of a hair, they are made of hard *keratin.* Nails are composed of flat, cornified plates on the dorsal surface of the distal segment of the fingers and toes [FIGURE 5.7A]. They appear pink because the nail is translucent, allowing the red color of the vascular tissue underneath to show through. White nails may be a sign of liver disease. The proximal part of the nail, the *lunula* (L. "little moon"; commonly called "half-moon"), is white because of the "red" capillaries in the underlying dermis do not show through.* The lunula usually does not appear on the little finger.

The body of the nail consists of keratinized dead cells. It is the part that shows; the root is the part hidden under the skin folds of the nail groove [FIGURE 5.7B]. The nail ends with a *free edge* that overhangs the tip of the finger or toe. The nail rests on an epithelial layer of skin called the *nail bed;* the thicker layer of skin beneath the nail root is the *matrix,* the area where new cells are generated for nail growth and repair. If the nail is injured, a new one will grow as long as the living matrix is intact. After birth, fingernails grow faster than toenails, with an aver-

*Some anatomists contend that the lunula is white because it is the portion of the nail formed first. Relatively young nail substance like the lunula is generally whiter, thicker, and more opaque than the more mature nail that extends over the nail bed.

FIGURE 5.7 FINGERNAIL STRUCTURE

[A] The basic components of a fingernail.
[B] A drawing of a sagittal section of a fingernail.

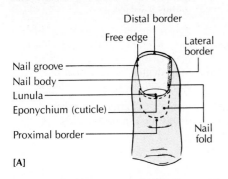

[A]

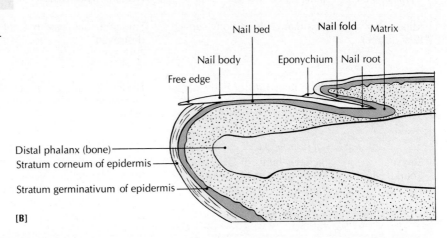

[B]

age growth rate of 0.5 mm a week. Both fingernails and toenails grow faster in warm weather than in cold.

The developing nail is originally covered by thin layers of epidermis called the ***eponychium*** (epp-oh-NICK-ee-uhm; Gr. *epi*, upon + *onyx*, nail). The eponychium, or *cuticle*, remains at the base of the mature nail.

EFFECTS OF AGING ON THE INTEGUMENTARY SYSTEM

With old age, the skin usually wrinkles as the elastic tissue becomes less resilient, and the fatty layer and supportive tissue beneath the skin decrease in thickness. As sebaceous and sudoriferous glands decrease their activity, the skin becomes dehydrated. Dry, brittle skin leads to more frequent bacterial skin infections. Also, sensitivity to changes in temperature increases as the skin becomes less able to maintain a constant body temperature. Blotching of skin color often occurs because of the irregular growth of pigment cells, probably due to exposure to sunlight and other ultraviolet radiation. Smooth, flat, brown areas called *lentigines* (lehn-TIHJ-ih-neez; sing. *lentigo*, L. lentil) (commonly, and incorrectly, called "liver spots") may form on the face or back of hands; damaged capillaries may give rise to small, bright red bumps, known as cherry angiomas, and also to larger, purple blotches. Raised brown or black dime-sized growths called *seborrheic keratoses* may appear on the epidermal layer of the skin.

In addition to the typical inherited baldness in some men, there is a thinning of the hair in both sexes because of the atrophy of hair follicles. Hair usually becomes gray and brittle, and more hair appears in the nostrils, ears, and eyebrows. Postmenopausal women may grow longer hair on the upper lip and chin due to an increase in androgens and a decrease in estrogens.

WHEN THINGS GO WRONG

Burns
Burns occur when skin tissues are damaged by heat, electricity, radioactivity, or chemicals. The seriousness of burns can be classified according to (1) *extent* (how large an area of the body is involved), and (2) *depth* (how many layers of tissue are injured) [TABLE 5.1].

A *first-degree burn* (such as a sunburn) may be red and painful, but it is not serious. Generally, it damages only the epidermis but does not destroy it. Such a burn responds to simple first-aid treatment, including cold water and sterile bandages. Butter and other greasy substances should never be used on *any* burn, because they may actually help to "cook" the damaged skin even further. Instead, the burn should be flushed or immersed in cold water (not ice water), or cold compresses (not ice) should be applied. Cold water helps to reduce pain, swelling, fluid loss, and infection, and also limits the extent of the damage.

A *second-degree burn* destroys the epidermis, and also causes some cell destruction in the dermis. Oozing blisters form, and scarring usually results. After a second-degree burn, the body may be able to regenerate new skin. Second-degree burns require prompt medical attention. If left untreated, a second-degree burn can progress to a third-degree burn.

A *third-degree burn* involves the epidermis, dermis, and underlying tissue. Because the skin cannot be regenerated, this kind of burn must be treated with surgery and

TABLE 5.1 CLASSIFICATION OF BURNS

Type of burn	Surface area affected	Depth of tissue damage	Major effects
Minor (first-degree)	Less than 10% of body surface.	Epidermis damaged but not destroyed.	Mild swelling, reddening, pain; injured cells peel off and skin heals without scarring, usually within 2 weeks.
Serious (second-degree)	More than 15% of body surface for an adult, 10% for a child.	Epidermis and part of dermis destroyed. New skin may regenerate.	Red or mottled appearance, blisters, swelling, wet surface due to plasma loss. Greater pain than third-degree burn (which destroys sensitive nerve endings).
Severe (third-degree)	Includes burns of face, eyes, hands, feet, genitals, and more than 20% of body surface. Prompt medical attention required.	All skin layers destroyed; deep tissue destruction. Nerve endings in skin destroyed. Skin cannot be regenerated. Surgery and skin grafts necessary.	White or charred appearance; severe loss of body fluids.

Estimating the extent of skin burns.
The *Lund-Browder method* for estimating the extent of skin burns for **[A]** a 5-year-old child, and **[B]** an adult. Except for the perineum and buttocks, percentages shown include both anterior and posterior portions of the body. **[C]** The "rule of nines" is also used for estimating the extent of skin burns. Except for the perineum, percentages shown include both the anterior and posterior portions of the body; the buttocks are included in the posterior trunk.

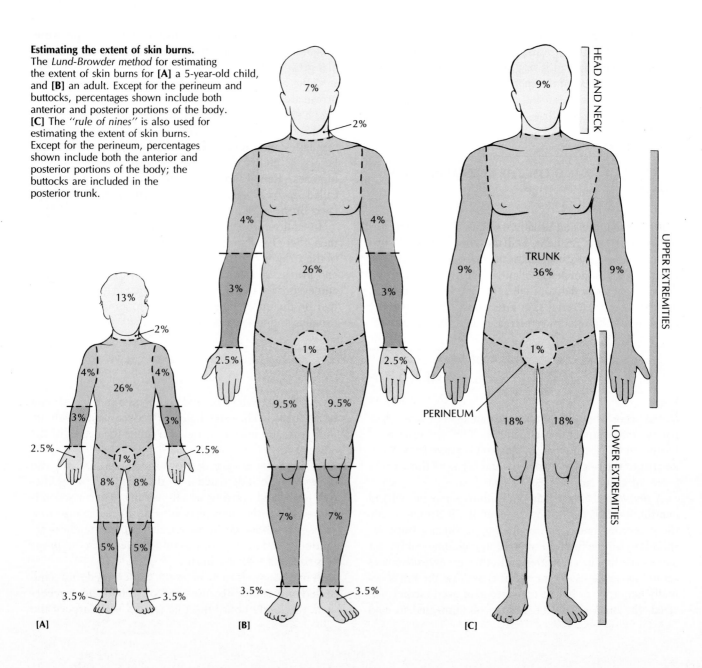

skin grafting. Ordinarily, the victim is in shock but feels no pain because nerve endings in the burned area have been destroyed. The damaged area is charred or pearly white, and fluid loss is severe.

One method of estimating the extent of burns on the skin is the **Lund-Browder method** (see drawings). This method assigns surface-area percentages to certain parts of the body. For example, if both the anterior and posterior areas of one foot are burned, the burn is said to extend over 3.5 percent of the body. Another method is known as the *"rule of nines"* (see drawing). It is generally less accurate than the Lund-Browder method. Physicians and therapists use these methods in conjunction with the specific area burned, the depth of the burn, and the age of the patient, especially in trauma and emergency centers.

A burn is the most traumatic injury the body can receive. Besides causing obvious tissue damage, serious burns expose the body to microorganisms, hamper blood circulation and urine production, and create a severe loss of body water, plasma, and plasma proteins that can produce shock. In fact, a major burn causes homeostatic imbalances in every system of the body. A severe burn leaves the skin more vulnerable to microbial infection than other types of wounds do. This happens because neutrophils, the skin's specialized infection-fighting cells, are practically immobilized by the burn. Instead of rushing to the infection site and releasing disease-fighting chemicals, the traumatized neutrophils release their chemicals prematurely. Because the chemicals interfere with the chemical signal from the infection site, the neutrophils do not know where to go, and only a few make it to the infected area.

Some Common Skin Disorders

Acne *Acne vulgaris* ("common acne") is most common among adolescents, when increased hormonal activity causes the sebaceous glands to overproduce sebum. When the flow of sebum is increased, dead keratin cells may become clogged in a follicle. These plugs at the skin opening are called either *blackheads* (open *comedones* — sing. *comedo* — which protrude from the follicle and are not covered by the epidermis) or *whiteheads* (closed comedones, which do not protrude from the follicle and are covered by the epidermis). The blocked follicle may become infected with bacteria, which secrete enzymes that convert the clogged sebum into free fatty acids. These acids irritate the lining of the follicle and eventually cause the follicle to burst. When the acid and sebum seep into the dermis, they cause an inflammation that soon appears on the surface of the skin as a pus-filled papule called a "pimple" [TABLE 5.2]. Picking and scratching merely spread the infection and may produce scarring.

Acne appears mostly on the face, chest, upper back, and shoulders. The problem generally affects young men more severely than young women, probably because the causative hormones are *androgens*, male hormones found in much greater abundance in males than in females.

The most advanced form of acne is **cystic acne**, which produces deep skin lesions called *cysts*. It is produced when sebaceous glands secrete excessive amounts of oil that nourish the infectious bacteria that cause acne in the first place.

Bedsores *Bedsores* (*decubitus ulcers*) are produced in areas where the skin is close to a bone and undergoes constant pressure, usually from the weight of the body itself. The pressure causes blood vessels to be compressed, depriving the affected tissue of oxygen and nutrition, and often leading to cell death. The most typical problem areas are the hips, elbows, tailbone, knees, heels, ankles (see photo), and shoulder blades. The first signs of bedsores are warm, reddened spots on the skin. Later, the spot may become purplish, indicating that blood vessels are being blocked and circulation is impaired. Actual breaks in the skin may follow, and bacterial infection is common if the lesions are left untreated. Cleanliness and dryness are important in preventing bedsores, as is changing the position of the patient frequently.

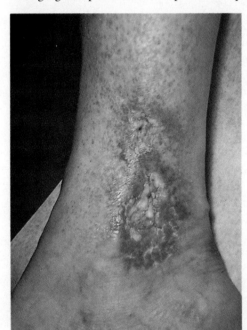

Bedsores. A decubitus ulcer on the right medial ankle.

Birthmarks and moles *Birthmarks* and *moles* are common skin lesions. The technical name for a birthmark is a **vascular nevus** (L. "birthmark"; plural *nevi*). A **nevus flammeus,** or *port-wine stain*, is a pink to bluish-red lesion that usually appears on the back of the neck. Its cause is not known. A **hemangioma**, or *strawberry mark*, affects only superficial blood vessels. Strawberry marks are usually present at birth, but may appear at any time. The most common sites are the face, shoulders, scalp, and neck.

TABLE 5.2 TYPES OF SKIN LESIONS (TISSUE ALTERATIONS)

Name	Example	Description
Bleb, bulla	Second-degree burn	A fluid-filled elevation of the skin
Cyst	Epidermal cyst	A mass of fluid-filled tissue, extending to dermis or hypodermis
Macule	Freckle; flat, pigmented mole	A discolored spot, not elevated or depressed
Papule	Acne, measles	Raised, red area resembling small pimples
Pustule	Acne vulgaris	Raised, pus-filled pimple
Tumor	Dermatofibroma	Fibrous benign tumor of the dermis
Vesicle	Blister, chicken pox, herpes simplex	A small sac filled with serous fluid
Wheal	Mosquito bite, hives	Local swelling, itching

The **common mole,** or **nevus,** is a benign lesion (in most cases) that usually appears before the age of 5 or 6, but it may appear any time up to about 30 years of age. Moles that darken, enlarge, bleed, or appear after a person is 30 should be checked by a physician, since an occasional mole may be transformed into a cancerous growth. Moles start out as flat brown or black spots, and typically enlarge and become raised later, especially during adolescence and pregnancy. The tendency to have moles is thought to be an inherited characteristic. It is not unusual for a mole to contain hair.

Psoriasis *Psoriasis* (suh-RYE-uh-sihss; Gr. itch) is a condition marked by lesions that are red, dry, and elevated and are covered with silvery, scaly patches. The cause of psoriasis is unknown, but there is general agreement that heredity plays a role. Attacks can be brought on by trauma, cold weather, pregnancy, hormonal changes, and emotional stress. The disease occurs when skin cells move from the basal layer to the stratum corneum in only 4 days instead of the usual 28. As a result, the cells do not mature, and the stratum corneum becomes flaky. The most usual sites of psoriasis are the elbows and knees, scalp, face, and lower back. It is most common in adults, but may occur at any age.

Allergic responses *Poison ivy, poison oak,* and *poison sumac* all cause skin irritations when contact is made with those plants, which contain *urushiol,* a powerful skin irritant. (It is interesting that these plants have no effect the first time a person is exposed to them.) Exposed parts of the body usually begin to redden several hours (or even several days) after exposure. Redness, itching, and swelling generally progress to vesicles (raised, red sacs), blisters, and finally a dry crust after serous fluid oozes from the blisters. Some people are so allergic to urushiol that they become affected by the smoke of the burning plants.

Warts *Warts (verrucae)* are benign epithelial tumors caused by viruses. A wart appears as a raised area of the skin with a pitted surface. It is usually no darker than the skin color, except on the soles of the feet (plantar warts), where it is often yellowish. (Warts may appear darker than the skin because dirt becomes lodged in the surface crevices.) Although they may appear anywhere on the body and on people of all ages, they are most common on the hands of children. This is probably so because the skin of the hands is likely to be irritated often, and a child's immune system is not yet effective against the virus. Warts are transmitted by direct contact, and they may be spread to other parts of the body by scratching and picking. They usually disappear after a year or so, but may be removed by a physician.

Skin Cancer

The two most common forms of skin cancer are *basal cell epithelioma* (*-oma* means tumor), also called basal cell carcinoma (see photo), and *squamous cell carcinoma.* The most serious type of skin cancer is *malignant melanoma.* All three forms can be prevented to a great degree by avoiding overexposure to the ultraviolet rays in sunlight. Other causes include arsenic poisoning, radiation, and burns. It is believed that people who have moles may have an increased risk of developing melanomas.

Basal cell epithelioma *Basal cell epithelioma* generally appears on the face, where sudoriferous glands, sebaceous glands, and hair follicles are abundant. It occurs

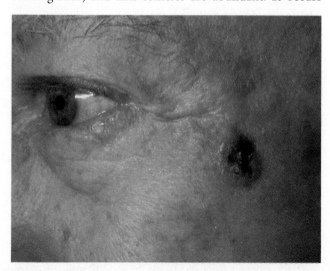

Basal cell epithelioma. The most common form of skin cancer.

most frequently in fair-skinned males over 40. Three types of lesions are typical: (1) *noduloulcerative lesions* are small and pinkish during the early stage; eventually they enlarge and become ulcerated and scaly. (2) *Superficial basal cell epitheliomas* frequently erupt on the back and chest. These lightly pigmented areas are sharply defined and slightly elevated. They are associated with exposure to substances that contain arsenic. (3) *Sclerosing basal cell epitheliomas* are waxy, yellowish-white patches that appear on the head and neck.

Squamous cell carcinoma *Squamous cell carcinoma* usually appears as premalignant lesions, typically in the keratinizing epidermal cells of the lips, mouth, face, and ears. Unlike basal cell epithelioma, squamous cell carcinoma may metastasize actively, especially when the lesions occur on the ears and lower lip. Squamous cell carcinoma is most common in fair-skinned males over 60.

Malignant melanoma *Malignant melanoma* involves the pigment-producing melanocytes. It usually starts as small, dark growths resembling moles that gradually become larger, change color, become ulcerated, and bleed easily. As with basal cell epithelioma and squamous cell carcinoma, the incidence of malignant melanoma is highest among fair-skinned persons, and it is slightly more common among women than among men. Besides the usual causes, malignant melanomas seem to be stimulated by hormonal changes during pregnancy. Most melanomas can be cured if they are treated early.

Bruises
Bruises (Middle Eng. *brusen*, to crush) appear "black-and-blue" because a hard blow to the surface of the skin breaks capillaries and releases blood into the dermis. Although the blood is red, it creates a black-and-blue mark on the surface because the skin filters out all but the blue light that reflects off the bruise and makes it appear dark blue or purplish. Ordinarily, the darker the bruise, the deeper the blood has seeped. Bruises sometimes turn yellow or green after several days. This is usually an indication that the spilled red blood cells have begun to break down into their components. Iron in the blood often gives the bruise a greenish color, as the decaying red pigment hemoglobin is transformed into a yellowish substance called *hemosiderin*, indicating that the bruise is in its final stages. Phagocytic white blood cells move into the affected area and ingest the hemosiderin and other debris, and the tissue returns to its normal color.

CHAPTER SUMMARY

Skin (p. 111)

1 Skin is the major part of the *integumentary system,* which also includes the hair, nails, and glands of the skin.

2 The *epidermis* of the skin is the outermost layer of epithelial tissue; the *dermis* is the thicker layer of connective tissue beneath the epidermis.

3 A typical arrangement of strata in the epidermis, from the outermost to the deepest, is: *stratum corneum, stratum lucidum, stratum granulosum, stratum spinosum,* and *stratum basale.* The stratum spinosum and stratum basale are collectively known as the *stratum germinativum.*

4 The *dermis* is composed of the thin *papillary* layer and the deeper, thicker *reticular* layer.

5 The *hypodermis,* or *subcutaneous layer,* lies beneath the dermis.

6 Skin serves as a stretchable protective shield, and is also involved in the regulation of body temperature and excretion of waste materials, and synthesis of vitamin D_3; it screens out harmful ultraviolet rays, and contains sensory receptors.

7 Skin color results from the presence of the pigments *melanin* and *carotene,* and the color of the underlying blood.

8 The wound-healing process involves the formation of blood clots by platelets and fibrinogen, phagocytosis by neutrophils and monocytes, and the growth of new epidermal and collagenous tissue.

Glands of the Skin (p. 117)

1 The glands of the skin are the *sudoriferous (sweat) glands* and the *sebaceous (oil) glands.* Sudoriferous glands are either *eccrine* glands, small sudoriferous glands that secrete sweat, or *apocrine* glands, which become active at puberty and produce thicker secretions.

2 The sudoriferous glands function mainly in *temperature regulation;* the function of sebaceous glands is *lubrication.*

Hair (p. 118)

1 *Hair* is composed of keratinized threads of cells that develop from the epidermis. Because of its origin, hair is considered to be an appendage of the skin.

2 Hair consists of epithelial cells arranged in three layers. From the inside out of a hair shaft, these are the *medulla, cortex,* and *cuticle.*

3 The hair *shaft* protrudes from the skin, the *root* is embedded beneath the skin and is enclosed within a *follicle.*

4 The first hairs begin to develop during the third fetal month. Terminal hairs appear at puberty in the genital area and armpits, and young men also develop chest and facial hair.

5 Human hair is constantly being shed and replaced; hair follicles alternate between growing and resting phases.

Nails (p. 120)

1 *Nails,* like hair, are modifications of the epidermis.

2 Composed of hard keratin, nails are flat, cornified plates on the dorsal surface of the distal segment of the fingers and toes.

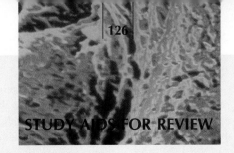

MEDICAL TERMINOLOGY

ALBINISM An inherited condition present at birth, and characterized by a lack of color in skin, hair, and eyes due to a lack of melanin.

ALOPECIA (L. "mange of a fox," "baldness") Loss of hair resulting in baldness.

ATHLETE'S FOOT (tinea pedis) A fungal infection that produces lesions of the foot, often accompanied by itching and pain.

CAFÉ AU LAIT SPOTS Light brown, flat spots on the skin; also known as von Recklinghausen's disease.

CALLUS (L. "hard") An area of the skin that has become hardened by repeated external pressure or friction.

CELLULITIS Inflammation of the skin and subcutaneous tissue, with or without the formation of pus.

CORN (L. *cornu*, horn) An area of the skin, usually on the feet, where the horny cells of the epidermis have become hardened by external pressure or friction, from ill-fitting shoes, for instance.

DANDRUFF A scaly collection of dried sebum and flakes of dead skin on the scalp.

DERMABRASION A surgical technique to scrape away areas of the epidermis, usually to remove acne scars. Facial skin is anesthetized and frozen at −30°C (−86°F) for 15 to 30 min, and the skin is smoothed with a high-speed diamond wheel. A typical smoothing procedure takes about 20 min.

DERMATOLOGIST A physician who specializes in the treatment of skin disorders.

DERMATOLOGY The medical study of the physiology and pathology of the skin.

DERMATOME (Gr. *derma*, skin + *tomos*, a cutting) An instrument used in cutting thin slices of the skin, as in skin grafting; a skin segment supplied by sensory nerve fibers used to diagnose the extent of spinal injuries.

ECZEMA (Gr. eruption) An inflammatory condition of the skin, producing red, papular, and vesicular lesions, crusts, and scales; characterized by itching and scaling.

ELECTROLYSIS Destruction of hair follicles by inserting a needle conducting an electric current into the follicle.

HIRSUTISM (HUR-soot-ism; L. *hirsutus*, bristly) An excessive growth of hair.

HIVES A skin eruption of wheals, usually associated with intense itching.

IMPETIGO (*impetigo contagiosa*) (ihm-puh-TIE-go; L. "an attack") A contagious infection of the skin, caused by staphylococci or streptococci bacteria and characterized by small red macules that become pus-filled.

KELOID (Gr. clawlike) An overgrowth of scar tissue, usually occurring after surgery or injury.

KERATOSIS (Gr. *keras*, horn) A thickening and overactivity of the cornified cells of the epidermis.

LIVER SPOTS Dark, flat patches on the skin, surrounded by lighter areas. The patches signal the skin's attempt to protect itself from ultraviolet damage. The condition occurs most often in people over 30. (Also called *age spots*; the correct term is *lentigines*.)

MELASMA (Gr. *melas*, black) A condition characterized by dark patches on the skin, usually brought on by stress, sensitivity to chemicals or sunlight, physical injury, and hormonal changes (such as in the "mask of pregnancy").

RINGWORM (tinea) A fungal infection, also known as *dermatophytosis*, characterized by red papules that spread in a circular pattern; affects tissues of the skin, nails, hair, and scalp.

SCABIES A contagious skin disease caused by the mite *Sarcoptes scabiei*, a parasite that lays eggs under the skin, causing itching and the eruption of vesicles.

VESICATION The formation of blisters.

VITILIGO (viht-ihl-EYE-go) An inherited condition characterized by splotchy white patches on the skin due to the destruction of melanocytes.

UNDERSTANDING THE FACTS

1 What layers make up the epidermis?
2 What structures are found in the dermis?
3 What three factors determine skin color?
4 What are some of the functions of hair?
5 What determines whether your hair is straight, wavy, or curly?
6 What is the relationship of blackheads and acne vulgaris to sebaceous glands? Why is acne vulgaris most common during adolescence?
7 What is the most serious form of skin cancer?

UNDERSTANDING THE CONCEPTS

1 How does the structure of the skin relate to its functions of protection, temperature control, waste removal, radiation protection, vitamin production, and environmental responsiveness?
2 When you receive a hypodermic injection, the needle passes through several layers of skin. List the basic layers and all subdivisions, in sequence from the outside in, that would be penetrated.
3 What is the function of the stratum germinativum?

4 Compare the structures of sudoriferous and sebaceous glands.

5 Explain how hair develops and grows.

6 Assume that you are a student nurse assigned to help with a severe burn victim. Because of the knowledge you have gained from your anatomy class, you know that this patient faces several life-threatening problems. Explain these problems.

SELF-QUIZ
Multiple-Choice Questions

5.1 Which of the following skin layers undergoes the most active rate of cell division?
  a stratum basale   **d** a and b
  b stratum spinosum   **e** a, b, and c
  c stratum granulosum

5.2 Which of the following fibers gives skin its flexibility?
  a collagenous   **d** a and b
  b reticular   **e** a, b, and c
  c elastic

5.3 The cells of the dermis are
  a fibroblasts   **d** histocytes
  b fat cells   **e** all of the above
  c macrophages

5.4 Which of the following is/are found in the reticular layer of the skin?
  a blood and lymph vessels   **d** sebaceous
  b nerves   glands
  c sensory nerve endings   **e** all of the above

5.5 Skin gets its color from
  a carotene   **d** a and b
  b underlying blood vessels   **e** a, b, and c
  c melanin

5.6 A fluid-filled elevation of the skin is called a/an
  a macule   **d** papule
  b wheal   **e** eruption
  c bleb

5.7 A burn in which the epidermis is damaged but not destroyed is a
  a first-degree burn   **c** third-degree burn
  b second-degree burn

True-False Statements

5.8 The skin is the largest organ in the human body.

5.9 Most of the skin consists of the epidermal layer.

5.10 In the skin, the papillary layer of the dermis is also known as the subepithelial layer.

5.11 Ultraviolet rays (UVB) stimulate melanocytes in the epidermis to produce melanin.

5.12 The most common form of skin cancer is a malignant melanoma.

5.13 The dermis consists only of a thin reticular layer.

Completion Exercises

5.14 The outer layer of the skin is the _____.

5.15 In the skin, the stratum spinosum and stratum basale together are known as the _____.

5.16 The outermost layer of the epidermis is the _____.

5.17 The papillary layer of the skin is so-named because of its tiny fingerlike projections called _____.

5.18 The lines of tension over the skin are known as _____.

5.19 Another name for the subcutaneous layer of the skin is the _____.

5.20 Melanin is produced by specialized skin cells called _____.

Problem-Solving/Critical Thinking Questions

5.21 *Pathology report:* A triangular segment of skin presenting an irregular pigmented lesion with irregular margins. Histological sections show a nest of bizarre and pleomorphic (variations in shape) malignant cells in the epidermis. Malignant cells are invading the papillary dermis but not into the reticular dermis. Final diagnosis:
  a mole   **d** malignant
  b hemangioma   melanoma
  c basal cell carcinoma   **e** squamous cell carcinoma

A SECOND LOOK

In the following drawing, identify the arrector pili muscle, dermis, sudoriferous gland, and the stratum spinosum.

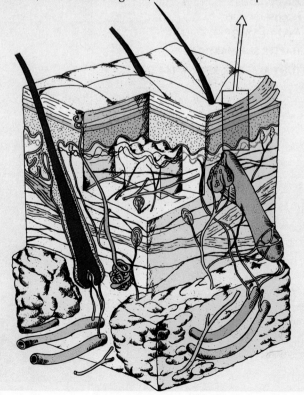

6

Bones and Bone Tissue

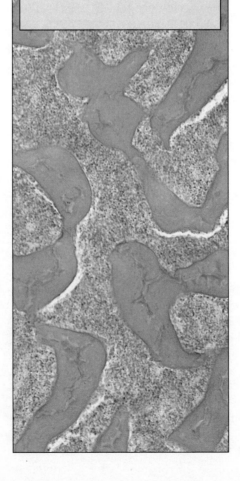

LEARNING OBJECTIVES

1 Explain how bones are classified by shape, and relate their forms to their functions.

2 Describe the gross anatomy of a typical bone.

3 Describe the composition of bone matrix.

4 Describe the composition of compact and spongy (cancellous) bone tissue.

5 Relate the structure of the osteon to its function.

6 State the functions of osteogenic cells, osteoblasts, osteocytes, osteoclasts, and bone-lining cells.

7 Compare intramembranous and endochondral ossification.

8 Explain how bones grow in length and diameter.

From a structural point of view, the human skeletal system consists of two main types of supportive connective tissue: cartilage and bone. The histology of cartilage was described in Chapter 4. This chapter deals specifically with the histology, gross anatomy, and mechanical functions of bone.

Bone (osseous) tissue is specialized connective tissue that has the strength of cast iron and the lightness of pine wood. We usually think of bone tissue as comprising the skeleton that supports and protects our internal organs, but its functions go far beyond that. Bone tissue contains the body's main reserve of calcium and phosphate, and bone marrow serves as the site for the manufacture of red blood cells and some white blood cells. Bones themselves aid movement by providing points of attachment for muscles, and transmitting the force of muscular contraction from one part of the body to another during movement. Living bone is not dry or brittle. It is a moist, changing, productive tissue that is continually resorbed (dissolved and assimilated), re-formed, and remodeled (replaced or renewed).

TYPES OF BONES AND THEIR MECHANICAL FUNCTIONS

Bones are usually classified by shape as flat, irregular, long, sesamoid, or short [FIGURE 6.1A–E]. Accessory bones are included here as a separate group. The shapes of bones are generally related to their mechanical functions, those involving the support or movement of body parts. For example, a long bone acts as a lever, a short bone is usually a connecting bridge between other bones, and a flat bone is a protective shell.

Long Bones

A bone is classified as a ***long bone*** when its length is greater than its width. The most obvious long bones are in the arms and legs (the longest is the femur), but some are relatively short, as in the fingers and toes. Long bones act as levers that are pulled by contracting muscles. It is this lever action that makes it possible for the body to move.

Short Bones

Short bones have about equal dimensions in length, width, and thickness, but they are shaped irregularly. They occur only in the wrists (carpal bones) and ankles (tarsal bones), where only limited movement is required. Short bones are almost completely covered with articular surfaces, where one bone moves against another in a

joint. However, there are some nonarticular areas on short bones where nutrient blood vessels enter, where tendons attach bones to muscles, and where ligaments connect bones.

Flat Bones

Flat bones are actually thin or curved more often than they are flat. Such bones include the ribs, scapulae (shoulder blades), sternum (breastbone), and bones of the cranium (skull). (See FIGURE 7.1 for illustrations of the ribs, scapulae, and sternum.) The shape of flat bones usually facilitates muscle attachment, and the gently curved bones of the skull form a protective enclosure for the brain.

Irregular Bones

Irregular bones do not fit neatly into any other category. Examples are the vertebrae, many facial bones, and the hipbones. The vertebrae have extensions that protrude from their main bony elements and serve as sites for muscle attachment. These bones support the spinal cord and protect it against compression forces. *Pneumatic bones*, such as the maxillary bone of the skull, have air-filled cavities.

Sesamoid Bones

Sesamoid bones are small bones embedded within certain tendons, the fibrous cords that connect muscles to bones. Sesamoid bones usually occur where tendons pass over the joint of a long bone, as in the wrist or knee. Typical sesamoid bones include the patella, or kneecap, which is within the tendon of one of the thigh muscles (quadriceps femoris), and the pisiform carpal bone within the tendon of a wrist muscle (flexor carpi ulnaris). Besides helping to protect the tendon, sesamoid bones help the tendon overcome compression forces, thus increasing the mechanical efficiency of joints. Sesamoid bones are so called because they sometimes resemble sesame seeds. Their number varies from person to person.

Accessory Bones

Accessory bones are most commonly found in the feet. They usually occur when developing bones do not fuse completely. Unfused accessory bones may look like extra bones or broken bones in x-rays, and it is important not to confuse them with actual fractures. *Sutural (Wormian*) bones* are accessory bones that occur as small bone clusters between the joints of the flat bones of the skull [see

---

*Wormian bones are named after Olaus Worm, the seventeenth-century Danish anatomist.

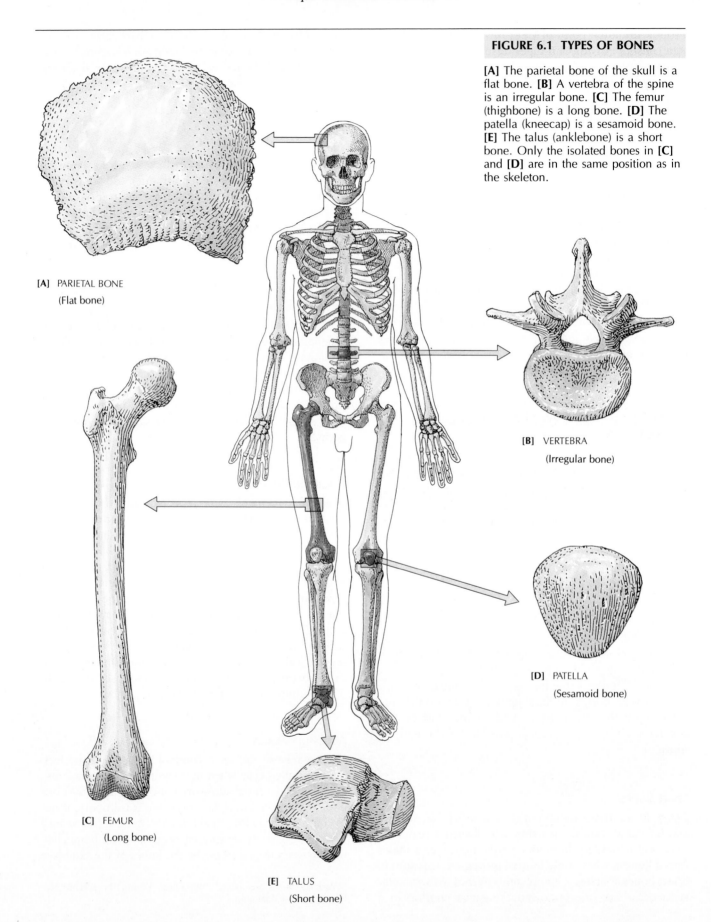

FIGURE 6.1 TYPES OF BONES

[A] The parietal bone of the skull is a flat bone. [B] A vertebra of the spine is an irregular bone. [C] The femur (thighbone) is a long bone. [D] The patella (kneecap) is a sesamoid bone. [E] The talus (anklebone) is a short bone. Only the isolated bones in [C] and [D] are in the same position as in the skeleton.

[A] PARIETAL BONE
(Flat bone)

[B] VERTEBRA
(Irregular bone)

[C] FEMUR
(Long bone)

[D] PATELLA
(Sesamoid bone)

[E] TALUS
(Short bone)

FIGURE 7.4B]. The number of sutural bones in an individual varies. In general, accessory bones add some slight support and protection to the area of the skeleton where they are found.

GROSS ANATOMY OF A TYPICAL LONG BONE

The long bones of the body (for example, the humerus, tibia, and radius) provide an excellent descriptive model for the gross anatomy of a typical bone. Most adult long bones have a tubular shaft, called the *diaphysis* (dye-AHF-uh-siss; Gr. "to grow between"). The diaphysis is a hollow cylinder with walls of compact bone tissue. The center of the cylinder, the *medullary cavity,* is filled with marrow. At each end of the bone is a roughly spherical *epiphysis* (ih-PIHF-uh-siss; Gr. "to grow upon") of spongy bone tissue [FIGURE 6.2]. The epiphysis is usually wider than the shaft. The flat bones and irregular bones of the trunk and limbs have many epiphyses. The long bones of the fingers and toes have only one epiphysis.

Separating these two main sections at either end of the bone is the *metaphysis.* It is made up of the *epiphyseal (growth) plate* and the adjacent bony trabeculae of spongy bone tissue on the diaphyseal side of the long bone. The growth plate is a thick plate of hyaline cartilage that provides the framework for the synthesis of the spongy bone tissue within the metaphysis. The epiphyseal plates are the only places where long bones continue to grow in length after birth.

The medullary cavity running through the length of the diaphysis contains *yellow marrow,* which is mostly fat. The porous latticework of the spongy epiphyses is filled with *red marrow.* The red bone marrow, also known as *myeloid tissue* (Gr. *myelos,* marrow), manufactures red blood cells, which give the marrow its color. Lining the medullary cavity of compact bone tissue and covering the trabeculae of spongy bone tissue is the *endosteum,* the membrane that lines the internal cavities of bones.

Covering the outer surface of the bone (except in the joint) is the *periosteum,* a fibrous membrane that has the potential to form bone during growth periods and in

FIGURE 6.2 GROSS ANATOMY OF A TYPICAL LONG BONE

A typical long bone showing key anatomical features; the interior is partially exposed.

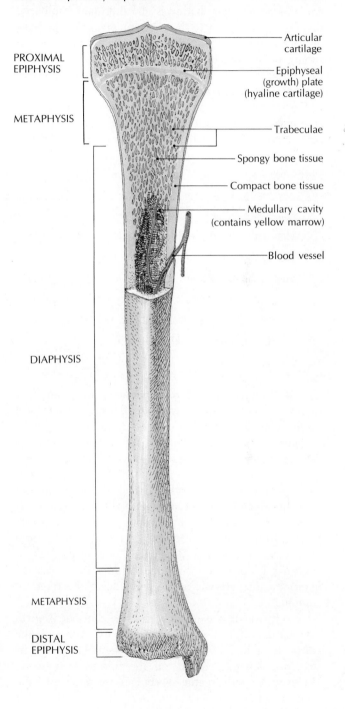

PROXIMAL EPIPHYSIS

METAPHYSIS

DIAPHYSIS

METAPHYSIS

DISTAL EPIPHYSIS

Articular cartilage

Epiphyseal (growth) plate (hyaline cartilage)

Trabeculae

Spongy bone tissue

Compact bone tissue

Medullary cavity (contains yellow marrow)

Blood vessel

fracture healing. The periosteum is often attached to the underlying bone by collagenous fibers called *periosteal perforating (Sharpey's) fibers* [see FIGURE 6.4A]. The fibers penetrate the inner layer of the periosteum and become embedded in the matrix of the bone. The periosteum contains nerves, lymphatic vessels, and many capillaries that provide nutrients to the bone and give living bone its distinctive pink color. Nutrients reach the marrow and spongy bone tissue by means of an artery that penetrates the compact bone tissue through a small opening called the **nutrient foramen.**

One of the few places where the periosteum is absent is at the joint surface of the epiphyses. There it is replaced by articular cartilage, which provides a slick surface that reduces friction and allows the joint to work smoothly.

A S K   Y O U R S E L F

1 What is the diaphysis? The epiphysis? The metaphysis?

2 What is the specific function of red bone marrow?

3 What is the endosteum? The periosteum?

BONE (OSSEOUS) TISSUE

The human body is about 62 percent water, but bone, or osseous, tissue contains only about 20 percent water. As a result, bones are stronger and more durable than skin or eyeballs, for instance. Like other types of connective tissue, osseous tissue is composed of cells embedded in a *matrix* of ground substance and fibers. However, osseous tissue is more rigid than other tissues because its homogeneous organic ground substance also contains inorganic salts, mainly calcium phosphate and calcium carbonate. In the bones, these compounds plus others form hydroxyapatite crystals. When the body needs the calcium or phosphate that is stored within the bones, the hydroxyapatite crystals ionize and release the required amounts.

A network of collagenous fibers in the matrix gives osseous tissue its strength and flexibility. Although the hardness of bone comes from the inorganic salts, its structure depends equally on the fibrous framework. When water and organic substances are removed from bone, it can crumble into a powdery chalk. On the other hand, if the inorganic salts are removed by a process called *decalcification*, the bone becomes so flexible that it can be tied into a knot.

The older we get, the less organic matter and the more inorganic salts we have in our bones. Because of this shifting proportion of matrix to salts, the bones of

older people are less flexible and more brittle than the bones of children.

Most bones have an outer shell of *compact bone tissue* enclosing an interior of *spongy bone tissue*, except where the spongy tissue is replaced by a marrow cavity or by air spaces called *sinuses*. For example, the leg bones contain marrow cavities, and some of the irregular skull bones contain sinuses that make the bones light. Irregular bones vary in the amount of spongy and compact tissue present. Short bones consist of spongy bone and marrow cavities.

The flat bones of the cranium (the part of the skull enclosing the brain) consist of two thin plates, called *tables*, of compact bone tissue with a layer of spongy bone tissue sandwiched between them [FIGURE 6.3]. The layer of spongy tissue contains marrow and veins (the diploic veins) and is called the **diploë** (DIHP-low-ee; Gr. "double"). Because of this protective arrangement, the outer table can be fractured without harming the inner table and brain.

Compact Bone Tissue

The compact bone tissue that forms the outer shell of a bone is very hard and dense (like ivory), and appears to the naked eye to be solid, but it is not. It contains cylinders of calcified bone known as **osteons** (Gr. *osteon*, bone), or **Haversian systems.** These cylinders are made up of concentric layers, or **lamellae** (luh-MELL-ee; sing. *lamella*, luh-MELL-uh), of bone [FIGURE 6.4]. The term *lamellae* is derived from the Latin word for "thin plates." These lamellae are arranged like wider and wider drinking straws, each one nestled inside the next wider one. The structure of osteons provides the great strength needed to resist typical, everyday compressive forces on long bones.

In the center of the osteons are **central canals** (*Haversian canals**), longitudinal channels that contain blood vessels, nerves, and lymphatic vessels. Central canals usually have branches called **perforating canals** (*Volkmann's canals*) that run at right angles to the central canals and extend the system of nerves and vessels outward to the *periosteum* (outer covering) and inward to the *endosteum* (inner lining) of the bony marrow cavity. Unlike central canals, perforating canals are not enclosed by concentric lamellae.

Lamellae contain *lacunae* (luh-KYOO-nee; sing. *lacuna*, luh-KYOO-nuh; L. cavities, pods), or little spaces, which house the *osteocytes*, or bone cells. Radiating like spokes from each lacuna are tiny *canaliculi* (KAN-uh-lick-yuh-lie; L. dim. *canalis*, channel) that contain the slender exten-

---

*Note that a *Haversian system* and an osteon are the same structure; the *central (Haversian) canal* is just the longitudinal channel.

FIGURE 6.3 FLAT BONES OF THE CRANIUM

[A] In the flat bones of the cranium, two tables (plates) of compact bone tissue surround a center of spongy bone tissue and marrow, the diploë. **[B]** An enlarged view showing more detail.

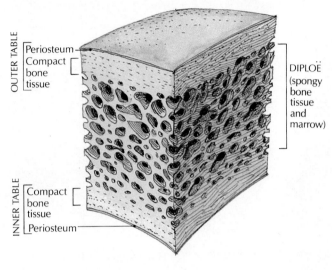

[A]

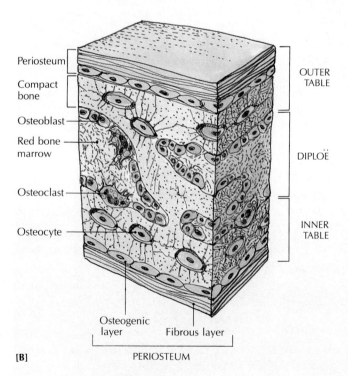

[B]

sions of the osteocytes. Nutrients and waste products can pass to and from the blood vessels in the central canals (1) by normal processes of intracellular transport within each osteocyte, (2) over gap junctions from one osteocyte to another, or (3) possibly via tissue fluid in the tiny spaces between the osteocytes and their surrounding lacunae.

Spongy (Cancellous) Bone Tissue

Spongy, or *cancellous* (latticelike), *bone tissue* is in the form of an open, lacy pattern that withstands maximum stress and supports shifts in weight distribution [FIGURE 6.5]. Prominent in the interior structure of spongy bone tissue are *trabeculae* (truh-BECK-yuh-lee; L. dim. *trabs*, beam), tiny spikes of bone tissue surrounded by bone matrix that has *calcified*, or become hardened by the deposition of calcium salts. The trabeculae form along the lines of greatest pressure or stress. Such an arrangement provides the greatest strength with the least weight. Spongy bone tissue is found inside most bones.

ASK YOURSELF

1 What does the bone matrix contain?

2 What is the structure of compact bone tissue? Of spongy bone tissue?

3 What is the difference between an osteon and a central canal?

BONE CELLS

Bones contain five types of cells that are capable of changing their roles as the needs of the body change in the growing and adult skeletons.

1 *Osteogenic (osteoprogenitor) cells* are found mostly in the deep layers of the periosteum and in bone marrow. These cells have a high mitotic potential and can be transformed into bone-forming cells *(osteoblasts)* or bone-destroying cells *(osteoclasts)* during healing.

2 *Osteoblasts* (Gr. *osteon*, bone + *blastos*, bud or growth) synthesize and secrete unmineralized ground substance called *osteoid* (Gr. *osteon*, bone + *eidos*, form). When calcium salts are deposited in the fibrous osteoid, it calcifies into bone matrix. Osteoblasts act as pump cells to move calcium and phosphate into and out of bone tissue, thereby calcifying or decalcifying it. Osteoblasts are usually found in the growing portions of bones, including the periosteum.

FIGURE 6.4 COMPACT BONE TISSUE

[A] An enlarged longitudinal section of compact bone tissue showing blood vessels, canals, and other internal structures. [B] An enlargement of a single osteon with lacunae, canaliculi, and a central (Haversian) canal visible. [C] An en- larged osteocyte (bone cell) inside a lacuna. [D] A scanning electron micrograph of compact bone tissue showing an osteon with its concentric lamellae housing lacunae and canaliculi. ×1000.

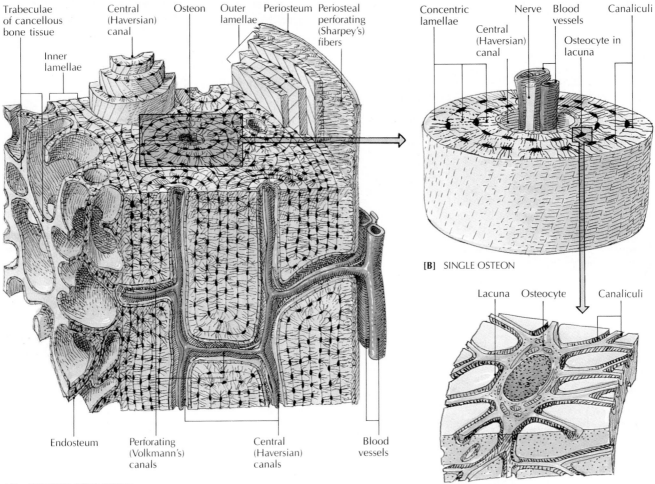

[A] COMPACT BONE TISSUE

[B] SINGLE OSTEON

[C] SINGLE BONE CELL

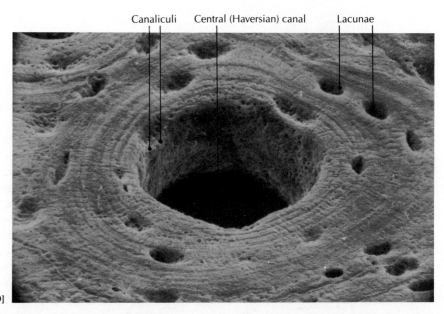

[D]

FIGURE 6.5 A SECTION THROUGH A HUMAN HIP JOINT

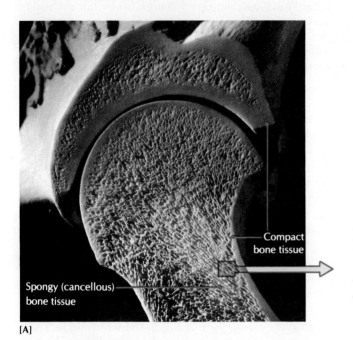

[A]

Compact bone tissue

Spongy (cancellous) bone tissue

[A] The porous, spongy (cancellous) portion of the bone has a streaked appearance, which indicates how the bone is built up in the direction of the greatest stress. **[B]** A scanning electron micrograph of trabeculae in cancellous bone. ×4.

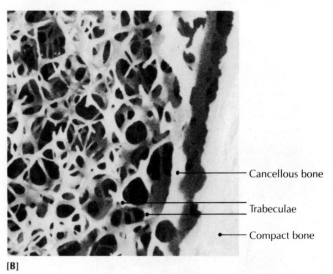

[B]

Cancellous bone

Trabeculae

Compact bone

3 *Osteocytes* are the main cells of fully developed bones. Each osteocyte has a cell body that occupies a lacuna within the bone matrix and long cytoplasmic processes that extend through the matrix via canaliculi. These processes are interconnected by gap junctions between neighboring osteocytes. Osteocytes are derived from the osteoblasts that have secreted bone tissue around themselves. They, along with osteoclasts, play an active role in homeostasis by helping to release calcium from bone tissue into the blood, thereby regulating the concentration of calcium in the body fluids. Osteocytes also seem to keep the matrix in a stable and healthy state by secreting enzymes and maintaining its mineral content.

4 *Osteoclasts* (Gr. *klastes*, breaker) are multinuclear giant cells that move about on bone surfaces, resorbing (dissolving and assimilating) bone matrix from sites where it is either deteriorating or not needed. They are usually found where bone is resorbed during its normal growth.*

5 *Bone-lining cells* are found on the surface of most bones in the adult skeleton. These cells are believed to be derived from osteoblasts that cease their physiological activity and flatten out on the bone surface. These cells may have several functions. They may serve as osteogenic cells that can divide and differentiate into osteoblasts. Most probably, they serve as an ion barrier around

*To remember the difference between osteoblast and osteoclast, remember the *b* in osteoblast as standing for "building."

bone tissue. This barrier helps regulate the movement of calcium and phosphate into and out of the bone matrix, which in turn helps control the deposition of hydroxyapatite in the bone tissue.

> ### ASK YOURSELF
> **1** What are the five kinds of bone cells?
>
> **2** What are the functions of osteoblasts? Osteocytes? Osteoclasts?

DEVELOPMENTAL ANATOMY AND GROWTH OF BONES

Bones develop through a process known as *ossification* (osteogenesis). Since the primitive "skeleton" of the human embryo is composed of either hyaline cartilage or fibrous membrane, bones can develop in the embryo in two ways: *intramembranous ossification* or *endochondral ossification*. However, in both cases, bones are formed from a pre-existing "connective tissue skeleton." Bone is the same no matter how it develops. Only the bone-making sequence is different.

FIGURE 6.6 shows the sites of intramembranous and endochondral ossification. Some bones, such as the skull, develop by intramembranous ossification, and other

FIGURE 6.6 OSSIFICATION

[A] The skull of a 3½-month-old fetus, showing the spongy bone radiating from the centers of intramembranous ossification; right lateral view. **[B]** The sites of intramembranous ossification (red) and endochondral ossification (blue) in a 10-week-old fetus.

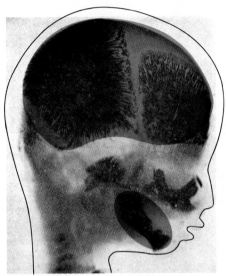

[A]

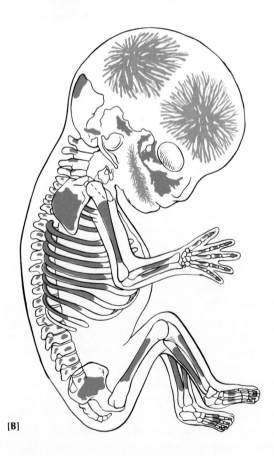

[B]

bones, such as those in the arms and legs, develop by endochondral ossification. FIGURE 6.7 compares fetal and adult skeletons. The ages when particular bones develop are shown on the skeletons.

Intramembranous Ossification

If bone tissue (spongy or compact) develops directly from mesenchymal (embryonic connective) tissue, the process is called ***intramembranous ossification.**** The vault (arched part) of the skull, flat bones of the face (including those lining the oral and nasal cavities), and part of the clavicle (collarbone) are formed this way. The skull is formed relatively early in the embryo to protect the developing brain. Some flexible tissue still exists between the flat skull bones at birth, so when the baby is born its skull is flexible enough to pass unharmed through the mother's birth canal.

From the time of the initial bone development, intramembranous ossification spreads rapidly from its center until large areas of the skull are covered with protecting and supporting bone. The first rapid phase begins when an *ossification center* (see item 3 in the following list) first appears from the eighth through the twelfth fetal week, and lasts through the end of the fifteenth fetal week, when the area is entirely covered [FIGURE 6.7A].

During the subsequent growth of the skull, the bones develop at a slower rate. The second phase continues into adolescence. Because the brain reaches full size by about the tenth year, the development of the protective skull vault is completed early in adolescence. In contrast, the bones of the face do not reach their adult size until the end of the adolescent growth spurt, anywhere from the age of 14 to 17.

Because intramembranous ossification is studied most easily in the skull, we describe the process there.

1 A layer of embryonic connective tissue called *mesenchyme* forms between the developing scalp and brain.
2 A plentiful supply of blood vessels arises in the mesenchyme, where some mesenchymal cells are already connected to neighboring cells by long, thin fibers, or processes [FIGURE 6.8A].
3 Bone tissue development begins when thin strands that will eventually become branching trabeculae appear in the matrix. At about the same time, the mesenchymal cells become larger and more numerous, and their processes thicken and connect with other embryonic connective tissue cells, forming a ring of cells around a blood vessel [FIGURE 6.8B]. The site of this ring formation is a

---

*Intramembranous means "within the membrane," and ossification means "bone formation." The term intramembranous ossification was originally used because this layer of mesenchyme was thought to be a sheet of membrane.

FIGURE 6.7 SOME DIFFERENCES BETWEEN THE FETAL AND ADULT SKELETONS

[A] Epiphyses begin to appear in the fetal skeleton about the fifth fetal week and continue to form until the sixteenth fetal week. [B] In the adult skeleton, zones of growth disappear, and epiphyses fuse with diaphyses between the ages of about 13 and 25.

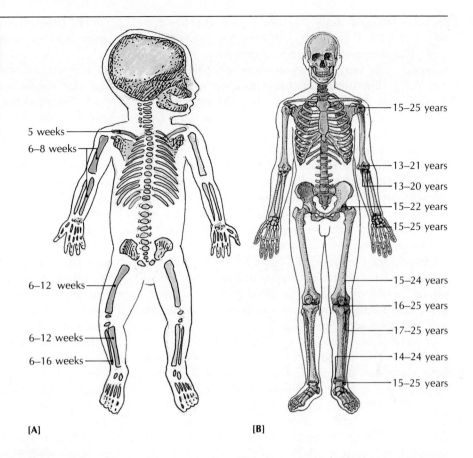

5 weeks
6–8 weeks
6–12 weeks
6–12 weeks
6–16 weeks

[A]

15–25 years
13–21 years
13–20 years
15–22 years
15–25 years
15–24 years
16–25 years
17–25 years
14–24 years
15–25 years

[B]

FIGURE 6.8 INTRAMEMBRANOUS OSSIFICATION

[A] The mesenchyme forms.
[B] The mesenchymal cells enlarge and form a ring around a blood vessel. [C] The mesenchymal cells differentiate into osteoblasts that secrete osteoid. [D] As the osteoid calcifies into bone matrix, it entraps osteocytes in lacunae. [E] The osteoclasts remove small areas of bone (calcium) from the walls of the lacunae.

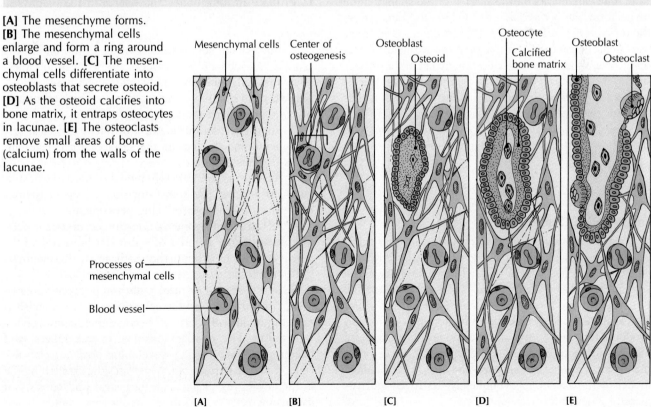

Mesenchymal cells
Center of osteogenesis
Osteoblast
Osteoid
Osteocyte
Calcified bone matrix
Osteoblast
Osteoclast

Processes of mesenchymal cells

Blood vessel

[A]   [B]   [C]   [D]   [E]

center of osteogenesis (also known as an **ossification center**). It begins about the second month of prenatal life.

4 The mesenchymal cells differentiate from osteogenic cells into osteoid-secreting osteoblasts [FIGURE 6.8C]. The osteoblasts begin to cause calcium salt deposits that form the spongy, latticelike bone matrix. (The trabeculae of cancellous bone tissue develop later from the spongy matrix.)

5 In this process of calcification, some osteoblasts become trapped within lacunae in the developing matrix. The entrapped osteoblasts, now called *osteocytes*, preserve the integrity of the matrix, and also release calcium ions as they are needed [FIGURE 6.8D]. These osteocytes remain in contact with the osteoblasts through *canaliculi*, tiny channels formed when the bone matrix is deposited around the long cell processes of the osteocytes. As bone-forming osteoblasts change into osteocytes, they are replaced by new osteoblasts that develop from the osteogenic cells in the surrounding connective tissue. In this way, the bone continues to grow.

6 While the osteoblasts are synthesizing and mineralizing the matrix, osteoclasts are playing a role in bone resorption by removing small areas of bone from the walls surrounding the lacunae [FIGURE 6.8E].

7 The trabeculae continue to thicken into the dense network that is typical of cancellous bone tissue. The collagenous fibrils deposited on the trabeculae crowd nearby blood vessels, which eventually condense into blood-forming bone marrow.

8 The osteoblasts on the surface of the spongy bone tissue form the *periosteum*—the membrane that covers the outer surface of the bone. It is made up of an inner osteogenic layer (with osteoblasts) and a thick, fibrous outer layer. The inner layer eventually creates a protective layer of compact bone tissue over the interior cancellous tissue. Once intramembranous ossification has stopped, the osteogenic layer becomes inactive, at least temporarily. It becomes active again when necessary—to repair a bone fracture, for example.

Endochondral Ossification

When bone tissue develops by replacing hyaline cartilage, the process is known as **endochondral ossification.** The term *endochondral* (Gr. *endo*, within + *khondros*, cartilage) means "inside the cartilage" and is used to describe this type of bone formation because it takes place where the cartilage model is eroded. Endochondral ossification produces long bones and all other bones not formed by intramembranous ossification. Remember, however, that the *cartilage itself is not converted into bone.* The cartilage model of the skeleton is completely destroyed and replaced by newly formed bone.

The stages of endochondral bone formation at the *cellular level* are described below.

1 The ossification center is a cartilaginous matrix that includes *chondrocytes* (cartilage cells) in lacunae.

2 The chondrocytes secrete alkaline phosphatase, an enzyme that triggers a chemical reaction in the matrix that causes mineralization. As the ossification process begins, the chondrocytes and lacunae enlarge, while the cartilage between the chondrocytes becomes mineralized (calcified) into cartilaginous spicules. These spicules act as scaffolds, upon which the osteoblasts lay down the osteoid matrix prior to ossification.

3 The calcified matrix blocks the diffusion of nutrients to the chondrocytes, which begin to die. Eventually, they are resorbed, leaving irregular cavities in the ossifying matrix. Once the bone is formed, these cavities contain the bone marrow.

4 *Pluripotent cells* (which have the potential to divide into several distinct cell types) lining the cavities begin to differentiate into osteoblasts and osteocytes. The osteoblasts deposit osteoid on the mineralized cartilage cores, forming a thin layer of spongy bone tissue. The osteoclasts are involved in the process of bone resorption during this stage.

Endochondral ossification is slower than intramembranous ossification. FIGURE 6.9 depicts endochondral ossification at the *tissue level.* The most typical sequence takes place in long bones and begins in the diaphyseal area.

1 Bone tissue begins to develop in a limb bud in the embryo. About six to eight weeks after conception, mesenchymal cells multiply rapidly and bunch together in a dense central core of precartilage tissue, which eventually forms the cartilage model [FIGURE 6.9A].

2 Soon after, an outline of a primitive *perichondrium* (the membrane that covers the surface of cartilage) appears [FIGURE 6.9B].

3 As it grows, the cartilage model is invaded by capillaries, triggering the transformation of the perichondrium into the bone-producing periosteum.

4 As intramembranous ossification occurs in the periosteum, a hollow cylinder of trabecular bone, called the *bone collar*, forms around the cartilage of the diaphysis [FIGURE 6.9C].

5 By the second or third month of prenatal development, the **primary center of ossification** is established near the middle of what will become the diaphysis [FIGURE 6.9D]. The cartilaginous matrix mineralizes, and begins to disintegrate. A developing *periosteal bud*, derived from the periosteum, penetrates the area left by the disintegrating cartilage. The periosteal bud consists of osteogenic cells that form the ossification center, blood

FIGURE 6.9 ENDOCHONDRAL OSSIFICATION AND LONGITUDINAL GROWTH IN A LONG BONE

Cartilage is shown in light blue, mineralized cartilage in darker blue. **[A]** Cartilage model. **[B]** The perichondrium forms. **[C]** The perichondrium is replaced by the periosteum. The bone collar forms. **[D]** Blood vessels enter the matrix. The primary center of ossification is established. **[E]** Blood vessels spread through the matrix. **[F]** The marrow cavity forms. **[G]** After birth, blood vessels enter the proximal epiphyseal cartilage, creating a secondary center of

ossification. **[H]** Another secondary center of ossification forms in the distal epiphyseal cartilage. **[I]** Some cartilage remains on the joint edge of each epiphysis to form articular cartilage. **[J]** As growth in length of the bone slows, the distal epiphyseal plate disappears, and the epiphyseal line forms on the interior of the epiphysis. **[K]** The proximal plate disappears somewhat later.

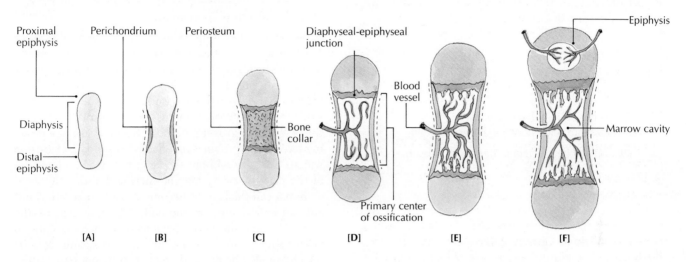

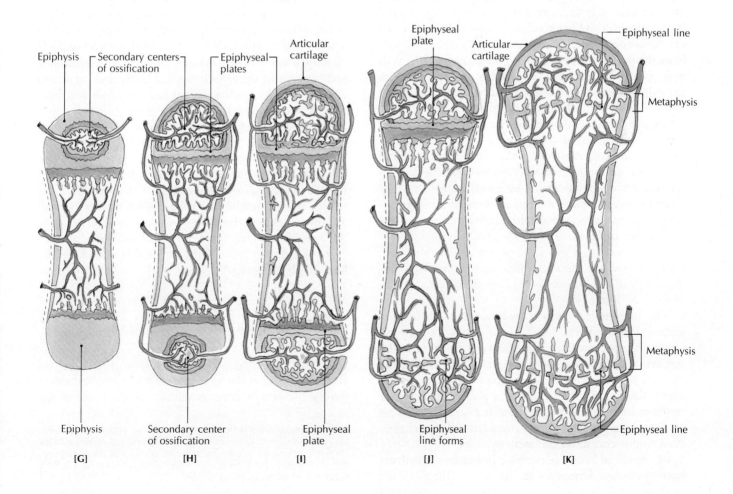

vessels that form the vascular network in the future marrow, and pluripotent cells that form the progenitors of red blood cells, certain white blood cells, and other cells. At the same time, blood vessels develop in the periosteum and branch into the diaphysis [FIGURE 6.9D, E]. The diaphysis now has a well-developed bone collar under the periosteum and an increasing amount of marrow in the center of the shaft [FIGURE 6.9F]. The epiphyses, however, still consist of cartilage.

> ### A S K   Y O U R S E L F
>
> **1** What is ossification?
>
> **2** What is the basic difference between intramembranous and endochondral ossification?
>
> **3** Can cartilage be converted into bone? Explain.

Longitudinal Bone Growth After Birth

After birth, chondrocytes in the epiphyses begin to mature and enlarge in the same way earlier chondrocytes did in the diaphysis. Here, too, the matrix mineralizes and the chondrocytes die. Blood vessels and osteogenic cells from the periosteum enter the epiphyses, where the cells develop into osteoblasts. The centers of this activity in the epiphyses are called *secondary,* or *epiphyseal, centers of ossification* [FIGURE 6.9G, H]. Although a few secondary centers may be present before birth, most do not appear until childhood or even adolescence.

Some cartilage remains on the outer (joint) edge of each epiphysis to form the *articular cartilage* necessary for the smooth operation of the joints [FIGURE 6.9I]. The *epiphyseal (growth) plate,* the thicker plate of cartilage between the epiphysis and diaphysis on either end of the bone, also remains throughout the growth period. It provides the framework for development of the cancellous bone tissue inside the *metaphysis,* the region of the epiphyseal plate and bony trabeculae on the diaphyseal side of the long bone.

The longitudinal growth of bone after birth takes place in small spaces in the cartilaginous epiphyseal plates. During growth, cartilage cells are generated in the epiphyseal plates, thickening the plates and moving the superior portion of the epiphysis upward, thus producing lengthwise growth. The epiphyseal plates generate new cartilage cells until the age of about 17. Depending on the bone, the growing period stops altogether any time from adolescence until the early twenties.

As growth in length of the long bone slows, the distal epiphyseal plate disappears. At this point — the *fusion,* or *closure,* of the epiphysis — all that remains of the epiphyseal plate is a thin epiphyseal line [FIGURE 6.9J]. The proximal epiphyseal plate disappears somewhat later [FIGURE 6.9K]. By the time longitudinal growth ends, the epiphyseal cartilage is completely replaced by osseous tissue.

How Bones Grow in Diameter

Although no new longitudinal bone growth takes place after the age of 25, the *diameter* of bones may continue to increase throughout most of our lives. Through intramembranous ossification, the osteogenic cells in the periosteum deposit new bone tissue beneath the periosteum, while most of the old bone tissue erodes away. (Imagine the periosteum growing outward like the bark of a growing tree.) The combination of cell deposition and erosion widens the bone without thickening its walls. The width of the marrow cavity may be increased also.

Bones can thicken, or become denser, to keep up with any physical changes in the body that may increase the stress, or load, the bones have to bear. For example, in people who greatly increase the size of their muscles, the bones can also be strengthened at the same time. If the bones are not made stronger they may fracture because they are unable to cope with the increased pull of the stronger muscles. In the same way, the bones in people who are overweight (and have increased the load on their skeletons) can thicken enough to offset the additional stress caused by the extra pounds. This thickening occurs as a result of the mechanical tension that stimulates osteoblastic activity and, in turn, more bone (calcium) deposition. Such thickening is often seen in the ribs, femurs, tibiae, and radii.

In an opposite way, a person who has one leg in a cast but continues to put weight on the other leg finds that the inactive leg becomes thin in only a month or so. That is because the muscles in the injured leg have atrophied, and the bones have decalcified by about 30 percent, while the active leg has remained normal. Bone loss occurs because of inactivity, immobility, or anything that takes the load off the skeleton.*

BONE MODELING AND REMODELING

Bone *modeling* is the alteration of bone size and shape during the bone's developmental growth. It alters the

---

*When astronauts are subjected to prolonged weightlessness in space, their bones begin to degenerate and lose calcium and other minerals. Dietary supplements of calcium, and even attempts at strenuous exercise, do not offset the degenerative effects of a lack of normal, everyday loads (stress) on bones.

amount and distribution of bone tissue in the skeleton, and determines the form of bones. Modeling occurs on different bone surfaces and at different rates during growth. Although bone is dense and hard, it is able to replace or renew itself through a process called *remodeling.* Remodeling occurs throughout adult life as a response to a variety of stresses.

Modeling

A dramatic example of bone modeling is the shaping of the skull bones. Bone yields to such soft structures as blood vessels and nerves by forming grooves, holes, and notches, not only in the fetus, when bone forms around blood vessels and nerves, but even later on, especially when vessels or nerves are rerouted because of injuries or other traumatic changes.

As the body grows to adulthood, the brain and cranium continue to grow. As the cranium enlarges, the curvature of its four major bones must decrease. The changing of the curvature is accomplished by the growth of successive layers of new bone on the outer surface of the skull bones and, at the same time, the resorption of old bone at the inner surface of the skull bones. When the cranium begins to shrink in old age (the brain mass also diminishes), bone is resorbed without being replaced. Most growth patterns are under hormonal control.

Remodeling

Just as for many other tissues of the body, the structural units of bone tissue must either be replaced or renewed to maintain homeostasis. This remodeling occurs through the selective resorption of old bone and the simultaneous production of new bone [FIGURE 6.10]. For example, new bone may grow, and its internal patterns may change to accommodate added body weight or other stresses. In contrast, bone that is not used tends to lighten by losing bone cells. Remodeling occurs at specific locations on the bone called *foci* (FOH-sye; sing. *focus*). Within each focus are specialized cells that make up a bone-remodeling unit that is responsible for resorption and for the refilling of the bone focus with new bone. Approximately 5 to 10 percent of the skeleton is remodeled each year.

An example of remodeling can be seen when braces are placed on teeth or when a tooth is extracted. When the teeth are repositioned during orthodontic treatment, force is placed on the bony tooth sockets. The slow shift of the teeth during their realignment takes place when some bone is resorbed on one side of the socket and added onto the other side. When a tooth is extracted, pressure on the socket is reduced, and some bone in the socket is resorbed. This reduction of pressure and bone mass allows the neighboring teeth to shift slightly.

FIGURE 6.10 BONE REMODELING

The process of bone deposition and resorption during the remodeling of a typical long bone.

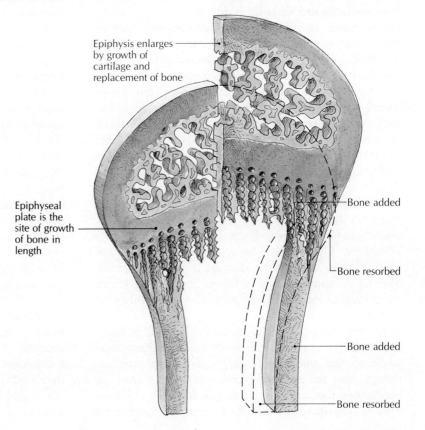

Epiphysis enlarges by growth of cartilage and replacement of bone

Epiphyseal plate is the site of growth of bone in length

Bone added

Bone resorbed

Bone added

Bone resorbed

THE EFFECTS OF AGING ON BONES

As people age, bones may undergo a loss of calcium, a decrease in calcium utilization, and a decreased ability to produce materials for the bone matrix. These metabolic changes make bones brittle so that they break more easily than the bones of a young person. The most common fractures in older people occur in the clavicle, femur, wrist bones, and humerus. (Fractures in the wrist and arm often occur when the person attempts to break a fall.) Bone marrow decreases, and *osteoporosis*, a metabolic disorder, leads to a loss of skeletal mass and density, especially in postmenopausal women (see When Things Go Wrong). Adverse changes in bones usually begin earlier in females (usually in the early forties) than in males (late fifties or early sixties) because of estrogen deficiency.

WHEN THINGS GO WRONG

Osteogenesis Imperfecta

Osteogenesis imperfecta is an inherited condition in which the bones are abnormally brittle and subject to fractures. The basic cause of this disorder is a decrease in the activity of the osteoblasts during bone formation. In some cases, the fractures occur during prenatal life, and the child is born with the deformities. In other cases, the fractures occur as the child begins to walk. However, the tendency to fracture bones is reduced after puberty.

Osteomalacia and Rickets

Osteomalacia (ahss-teh-oh-muh-LAY-shee-uh) (Gr. *osteon*, bone + *malakia*, soft) and *rickets* (variant of Gr. *rhakhitis*, disease of the spine) are skeletal defects caused by a deficiency of vitamin D, which leads to a widening of the epiphyseal growth plates, an increased number of cartilage cells, wide osteoid seams, and a decrease in linear growth. A deficiency of vitamin D may result from an inadequate diet, an inability to absorb vitamin D, or from too little exposure to sunlight.

Rickets is a childhood disease. It occurs less frequently than it used to, primarily because of improved dietary habits. It is most common in black children, not necessarily because of inadequate diets, but because highly pigmented skin absorbs fewer of the ultraviolet rays in sunlight. These rays are needed to convert 7-dehydrocholesterol in the skin to vitamin D (cholecalciferol). Skeletal deformities such as bowed legs, knock-knees, and a bulging forehead are typical in a young child with rickets.

Osteomalacia is the adult form of rickets and is sometimes referred to as "adult rickets." It leads to *demineralization*, an excessive loss of calcium and phosphorus. Although the skeletal deformities of rickets may be permanent, the similar skeletal abnormalities of osteomalacia may disappear with large doses of vitamin D.

Osteomyelitis

Osteomyelitis (Gr. *osteon*, bone + *myelos*, marrow) is an inflammation of bone and/or an infection of the bone marrow that can be either chronic or acute. It is frequently caused by *Staphylococcus aureus* and other bacteria, which can invade the bones or elsewhere in the body. Bacteria may reach the bone through the bloodstream or through a break in the skin from an injury. Although the disease often remains localized, it can spread to the marrow, cancellous tissue, and periosteum. *Acute osteomyelitis* is usually a blood-carried disease that most often affects rapidly growing children. *Chronic osteomyelitis*, more prevalent in adults, is characterized by draining sinuses and spreading lesions. Prompt use of antibiotics, such as vancomycin, is effective in treating the disease.

Osteoporosis

Osteoporosis (Gr. *osteon*, bone + *poros*, passage) is a loss of bone mass that can make the bones so porous that they crumble under the ordinary stress of moving about. (People with osteoporosis don't "fall and break a hip." They usually break their hip—or, more often, the neck of the femur—while walking, and then fall when the hip support is gone.) Most victims of osteoporosis are postmenopausal women, usually over age 60. After menopause, the ovaries produce little, if any, estrogen, one of the female hormones. Without estrogen, old bone is destroyed faster than new bone can be remodeled, so the bone becomes porous and brittle. Besides a lack of estrogen, osteoporosis can be hastened by smoking, a diet lacking in calcium or vitamin D, and lack of exercise. Postmenopausal estrogen therapy has proved to be relatively successful. Osteoporosis can cause vertebrae to crumble, producing a "dowager's hump" in the upper back. Most people over 70 are markedly shorter than they used to be. This height loss occurs when aging vertebrae and other small bones lose mass, pressing closer together as intervertebral disks become thinner.

Osteosarcomas

Osteosarcomas (Gr. *sark*, flesh + *oma*, tumor), or *osteogenic sarcomas,* are forms of bone cancer. Such malignant bone tumors are rare. Because the incidence of osteosarcomas is higher in growing adolescents than in children or adults, and because the adolescents affected are often taller than average, there is some speculation that areas of rapid growth are most vulnerable. No definite cause is known. Localized pain and tumors are common signs of malignancy.

The most common form of bone cancer is a *myeloma,* in which malignant tumors in the bone marrow interfere with the normal production of red blood cells. Anemia, osteoporosis, and fractures may occur. Myeloma occurs more frequently in women than in men.

Paget's Disease

Paget's disease (osteitis deformans) is a progressive bone disease in which a pattern of excessive bone destruction followed by bone formation contributes to thickening of bones. This deformity usually involves the skull, pelvis, and lower extremities. The cause is unknown. Paget's disease occurs after the age of 40 and most typically in the sixties.

CHAPTER SUMMARY

Types of Bones and Their Mechanical Functions (p. 129)

1 Bones may be classified according to their shape as *long, short, flat, irregular,* or *sesamoid. Accessory bones* are a minor category.

2 The shapes of bones are related to their functions. The mechanical functions of bones include support, protection, and movement.

Gross Anatomy of a Typical Long Bone (p. 131)

1 Most long bones consist of a tubular shaft called a *diaphysis,* with an *epiphysis* at either end of the bone.

2 Separating the diaphysis and epiphysis at each end of the bone is the *metaphysis.* It is made up of the *epiphyseal (growth) plate* and adjacent bony trabeculae of spongy bone tissue.

3 The epiphyseal plates and metaphyses are the only places where long bones continue to grow in length after birth.

Bone (Osseous) Tissue (p. 132)

1 *Bone,* or *osseous, tissue* is composed of cells embedded in a matrix of ground substance, inorganic salts, and collagenous fibers. Inorganic salts give bone its hardness, and organic fibers and ground substance give it strength and flexibility.

2 Most bones have an outer shell of compact bone tissue surrounding spongy bone tissue.

3 *Compact bone tissue* is made up of *osteons,* concentric cylinders of calcified bone. Within the osteons are **central (Haversian) canals** that contain nerves, lymphatic vessels, and blood vessels. These canals are connected with the outer surfaces of bones through **perforating (Volkmann's) canals.**

4 *Spongy,* or *cancellous, bone tissue* forms a lacy pattern designed to withstand stress and support shifts in weight. Tiny spikes of bone tissue called *trabeculae,* surrounded by calcified matrix, give spongy bone its latticelike appearance.

5 The *periosteum* is a fibrous membrane that covers the outer surfaces of bones, except at joints. It contains bone-forming cells, nerves, and vessels.

Bone Cells (p. 133)

1 Bones contain five types of cells: *osteogenic cells,* which can be transformed into osteoblasts or osteoclasts; *osteoblasts,* which synthesize and secrete new bone matrix as needed; *osteocytes,* which help maintain homeostasis; *osteoclasts,* giant cells responsible for the resorption of bone; and *bone-lining cells,* which have several diverse functions.

2 Bone cells change their roles as the needs of the body change in the growing and adult skeletons.

Developmental Anatomy and Growth of Bones (p. 135)

1 Bones develop through *ossification* (osteogenesis). If bone is formed from mesenchymal tissue (embryonic connective tissue), the process is called *intramembranous ossification.*

2 If bone develops by replacing a cartilage model, the process is called *endochondral ossification.*

3 The centers of growth activity after birth are the *secondary (epiphyseal) centers of ossification.*

4 Bones grow in diameter through intramembranous ossification, as osteogenic cells deposit new bone tissue beneath the periosteum and old bone tissue erodes.

5 The growth of the diameter of bones continues through most of life.

Bone Modeling and Remodeling (p. 140)

1 The alteration of bone size and shape during the bone's developmental growth is called **modeling.**

2 Bone replaces itself through a process called **remodeling.**

The Effects of Aging on Bones (p. 142)

1 Aging bones become brittle from loss of calcium, decreased calcium utilization, and the decreased ability to produce matrix material.

2 Skeletal problems affect women more than men.

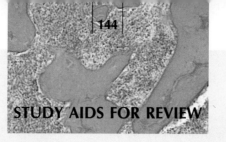

STUDY AIDS FOR REVIEW

UNDERSTANDING THE FACTS

1 What are some shapes of bones and their related functions?

2 What is the function of inorganic salts in bone tissue?

3 Compare the structure and function of central canals and perforating canals.

4 Distinguish between a diaphysis and an epiphysis.

5 What is the epiphyseal plate, and what are its functions?

6 What main roles do osteocytes play?

7 How do the functions of osteoblasts and osteoclasts differ?

8 Describe the bone disorders of osteomalacia, osteomyelitis, and osteoporosis.

UNDERSTANDING THE CONCEPTS

1 Compare the structure of compact and spongy bone tissue.

2 Relate the structure of an osteon to its function.

3 Why does the skull vault reach its final size and development before the face does?

4 What is the relationship between physical activity and bone development? Between body weight and bone development?

SELF-QUIZ

True-False Statements

6.1 Spongy, or cancellous, bone tissue consists of an open, lacy pattern of bony tissue.

6.2 At the center of the osteon is a lacuna.

6.3 The epiphysis contains periosteal tissue.

6.4 The endosteum is a membrane that lines the internal cavities of bones.

6.5 In endochondral ossification, cartilage is converted into bone.

6.6 Secondary centers of ossification are present in bone at birth.

6.7 The epiphyseal-metaphysical complex is another name for the growth plate.

6.8 Both bone and cartilage are derived from embryonic mesenchyme.

6.9 Both bone and cartilage contain blood vessels.

6.10 Bone diameter continues to increase throughout most of our lives.

Completion Exercises

6.11 The two types of connective tissue found in the human skeleton are _____ and _____ .

6.12 Another name for spongy bone is _____ .

6.13 The functional unit of compact bone is the _____ .

6.14 The small bone clusters that occur between the joints of the flat bones of the skull are called _____ bones.

6.15 The thin plates found in flat bones are known as _____ .

6.16 Red bone marrow manufactures _____ .

6.17 Yellow bone marrow is composed mostly of _____ .

6.18 Bone periosteum is often attached to underlying bone by collagenous fibers called _____ fibers.

6.19 In bone, the longitudinal canal that contains nerves, lymphatic vessels, and blood vessels is the _____ canal.

6.20 Osteocytes are found in _____ .

6.21 Giant cells found where bone is reabsorbed are called _____ .

6.22 Bones develop through a process known as _____ .

6.23 If bone develops directly from mesenchymal tissue, the process is called _____ .

Matching

6.24 _____ trabeculae

6.25 _____ diaphysis

6.26 _____ epiphysis

6.27 _____ metaphysis

6.28 _____ long bones

6.29 _____ flat bones

6.30 _____ short bones

6.31 _____ accessory bones

6.32 _____ lamellae

6.33 _____ perforating canal

6.34 _____ hydroxyapatite

a contains yellow bone marrow

b contains the epiphyseal plate

c tiny spikes of bone surrounded by matrix

d a tubular shaft

e the end of a bone

f found in the feet

g length greater than width

h found in the wrists

i contains a diploë

j found within tendons

k a branch off of a central canal

l young cells that resemble bone

m center of osteogenesis

n inorganic salt of bone

o concentric layers

7

The Axial Skeleton

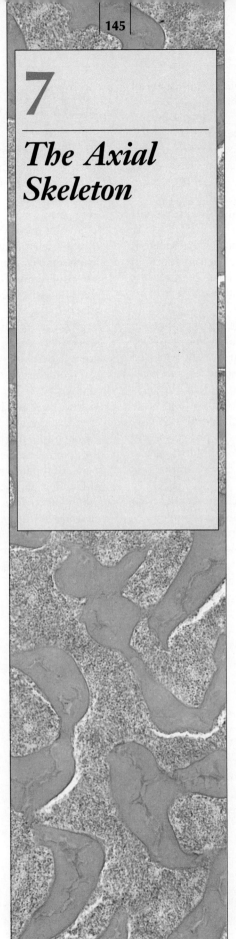

LEARNING OBJECTIVES

1 Describe the structural/mechanical functions of the skeleton.

2 Identify the general features and surface markings of bones, and relate them to function.

3 Distinguish between the axial and appendicular divisions of the skeleton, and identify their components.

4 Define the terms *suture*, *fontanel*, and *foramen*, and give examples of each.

5 Identify the bones of the skull and describe their locations and main features.

6 Describe the main parts of a typical vertebra.

7 Identify the five sections of the vertebral column, and describe their locations and main features.

8 Identify the bones of the thorax, and describe their locations and main features.

9 Describe the parts of a typical rib.

10 Describe some types of fractures of the human skeleton.

We are born with as many as 300 bones, but many of them fuse during childhood to form the adult skeleton of 206 bones. There are exceptions. For example, 1 person in 20 has an extra rib, and the number of small sutural (Wormian) bones in the skull varies from person to person. But for the most part, the bone count of 206 is the accepted one.

Other than the bones that come in pairs, no two are alike. Bones differ in size, shape, weight, and even composition. This diversity of form is directly related to the many structural or mechanical functions of the skeleton.

The most obvious function of the skeleton is to hold up the body. Without a bony skeleton to support our bag of skin and inner organs, we would collapse into a formless heap.

Besides supporting the inner organs, the skeleton surrounds and protects many of them within body cavities. For instance, the heart and lungs are safely enclosed within a roomy rib cage that offers protection and freedom of movement at the same time, and the brain is cushioned within the cranium, a shock-absorbing bone case that is designed to protect the brain from the many bumps of everyday life. The skeleton also protects many passageways. For instance, the bony scaffolding of the nasal region supports and protects the airway for the passage of air during breathing.

The bones of the skeleton also act as a system of levers for the pulley action of muscles. This lever-pulley arrangement provides attachment sites on bones for muscles, tendons, and ligaments, and allows us to move our entire bodies, or just one finger or toe.

This chapter and the next present an introduction to the skeletal system itself, concentrating on its two main parts: the central "anchor" of the axial skeleton, and the peripheral limbs of the appendicular skeleton. Throughout the chapters, each specific bone bears the same color in each illustration. For example, the temporal bone is always dark blue.

GENERAL FEATURES AND SURFACE MARKINGS OF BONES

The surface features of a bone, as well as its overall shape, often give clues about the bone's function. TABLE 7.1 describes some of the more important features of bones and their functions. For each feature there is an illustrated example.

Some outgrowths, or *processes* (such as tuberosities) may be attachment sites for the tendons of muscles or the ligaments that connect bones. A long, narrow ridge (linea aspera of femur) or a more prominent ridge (iliac crest of pelvis) may be the site where broad sheets of muscles are attached. Large, rounded ends (condyle of femur) or large depressions (glenoid fossa of scapula) indicate adaptations of joints between bones. Grooves, foramina (holes), or notches are usually formed on bones to accommodate such structures as blood vessels, nerves, and tendons.

Processes are a response to the continuous force exerted by muscles on a bone. For example, a muscle is attached to a trochanter, tuberosity, tubercle, or a ridge. If the muscle is removed, the ridge will eventually disappear. All the foramina in the skull, through which nerves or blood vessels pass, were present before the bone was in the vicinity. The bone responded by not invading the territory of the nerve or blood vessel (for example, the foramen ovale for the mandibular nerve). The radial groove on the humerus is present because the humerus yields as the radial nerve presses on it slightly.

ASK YOURSELF

1 What is a process on a bone? Give examples of processes, openings, and depressions.

2 Give an example of how a particular feature or marking on a bone is related to its function.

TABLE 7.1 SOME GENERAL FEATURES OR MARKINGS OF BONES

Feature or marking	Description and example
OPENINGS (HOLES) TO BONES	
Canal or meatus (L. channel) [FIGURE 7.4F]	Relatively narrow tubular channel, opening to a passageway. Carotid canal, external auditory meatus.
Fissure [FIGURE 7.4A]	Groove or cleft. Superior orbital fissure.
Foramen (L. opening) [FIGURE 7.4F]	Natural opening into or through a bone. Foramen magnum in occipital bone.

Feature or marking	Description and example
DEPRESSIONS ON BONES	
Fossa (L. trench) [FIGURE 7.4F]	Shallow depressed area. Mandibular fossa.
Groove or sulcus (L. groove) [FIGURE 8.4]	Deep furrow on the surface of a bone or other structure. Intertubercular and radial grooves of humerus.
Notch [FIGURE 8.5A]	Deep indentation, especially on the border of a bone. Radial notch of ulna.
Paranasal sinus (L. hollow) [FIGURE 7.7]	Air cavity within a bone in direct communication with nasal cavity. Maxillary sinus.
PROCESSES WHERE A BONE FORMS A JOINT WITH AN ADJACENT BONE	
Condyle (L. knuckle) [FIGURE 8.10]	Rounded, knuckle-shaped projection; concave or convex. Condyles of femur.
Facet (Fr. little face) [FIGURE 7.17B]	Small, flat surface. Head and tubercle of ribs.
Head (caput) (L. head) [FIGURES 8.4, 8.5, 8.12]	Expanded, rounded surface at proximal end of a bone; often joined to shaft by a narrowed neck, and bearing the ball of a ball-and-socket joint. Head of femur, humerus, radius.
Trochlea (L. pulley) [FIGURE 8.4]	Grooved surface serving as a pulley. Trochlea of humerus.
PROCESSES OF CONSIDERABLE SIZE ATTACHED TO CONNECTIVE TISSUE	
Cornu (L. horn) [FIGURE 7.9]	Curved, hornlike protuberance. Cornu of hyoid bone.
Crest or lip ridge [FIGURE 8.8A]	Wide, prominent ridge, often on the long border of a bone. Iliac crest.
Line or linea [FIGURE 8.10]	Narrow, low ridge. Linea aspera of femur.
Eminence (L. to stand out) [FIGURE 8.12]	Projecting part of bone, especially a projection from the surface of a bone. Intercondylar eminence of tibia.
Epicondyle [FIGURE 8.4]	Eminence upon a bone above its condyle. Epicondyles of humerus.
Malleolus (L. hammer) [FIGURE 8.12]	Hammer-shaped, rounded process. Malleoli of tibia and fibula.
Spine (spinous process) [FIGURES 7.11, 8.3C]	Sharp, elongated process. Spine of a vertebra and of scapula.
Sustentaculum [FIGURE 8.13C]	Process that supports. Sustentaculum tali of calcaneus bone of the foot.
Trochanter [FIGURE 8.10]	Either of the two large, roughly rounded processes found near the neck of the femur. Trochanters of femur.
Tubercle (L. small lump) [FIGURE 8.4]	Small, roughly rounded process. Tubercles of humerus.
Tuberosity (L. lump) [FIGURE 8.9]	Medium-sized, roughly rounded, elevated process. Ischial tuberosity.

Some anatomical markings of a typical long bone.

FIGURE 7.1 AXIAL AND APPENDICULAR SKELETONS

[A] Anterior view of the skeleton. The axial skeleton is shown in orange. **[B]** Posterior view of the skeleton. The appendicular skeleton is shown in green.

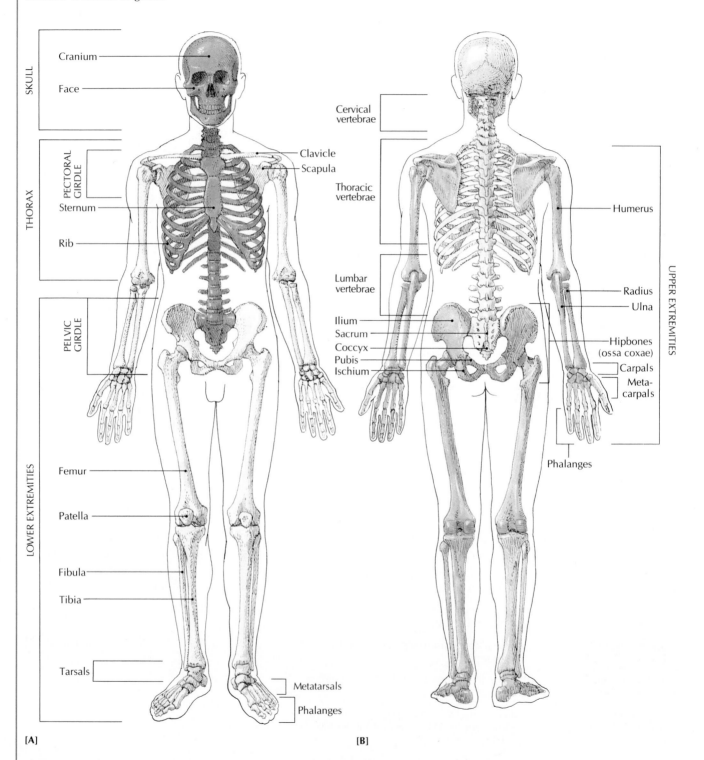

SKULL
— Cranium
— Face

THORAX
PECTORAL GIRDLE
— Clavicle
— Scapula
— Sternum
— Rib

PELVIC GIRDLE

LOWER EXTREMITIES
— Femur
— Patella
— Fibula
— Tibia
— Tarsals
— Metatarsals
— Phalanges

Cervical vertebrae
Thoracic vertebrae
Lumbar vertebrae
Ilium
Sacrum
Coccyx
Pubis
Ischium

UPPER EXTREMITIES
— Humerus
— Radius
— Ulna
— Hipbones (ossa coxae)
— Carpals
— Metacarpals
— Phalanges

[A]   [B]

DIVISIONS OF THE ADULT SKELETON (206 BONES)

AXIAL SKELETON (80 BONES)

Skull (29 bones)*
Cranium	8
Parietal (2)	
Temporal (2)	
Frontal (1)	
Ethmoid (1)	
Sphenoid (1)	
Occipital (1)	
Face	14
Maxillary (2)	
Zygomatic (malar) (2)	
Lacrimal (2)	
Nasal (2)	
Inferior nasal concha (2)	
Palatine (2)	
Mandible (1)	
Vomer (1)	
Ossicles of ear	6
Malleus (hammer) (2)	
Incus (anvil) (2)	
Stapes (stirrup) (2)	
Hyoid	1

Vertebral column (26 bones)
Cervical vertebrae	7
Thoracic vertebrae	12
Lumbar vertebrae	5
Sacrum (5 fused bones)	1
Coccyx (3 to 5 fused bones)	1

Thorax (25 bones)†
Ribs	24
Sternum	1
	80

APPENDICULAR SKELETON (126 BONES)

Upper extremities (64 bones)
Pectoral (shoulder) girdle	4
Clavicle (2)	
Scapula (2)	
Arm‡	2
Humerus (2)	
Forearm	4
Ulna (2)	
Radius (2)	
Wrist	16
Carpals (16)	
Hand and fingers	38
Metacarpals (10)	
Phalanges (28)	

Lower extremities (62 bones)
Pelvic girdle	2
Fused ilium, ischium, pubis	
Thigh	4
Femur (2)	
Patella (2)	
Leg‡	4
Tibia (2)	
Fibula (2)	
Ankle	14
Tarsals (14)	
Foot and toes	38
Metatarsals (10)	
Phalanges (28)	
	126

Total (Axial and Appendicular)   206

*The number of skull bones is sometimes listed as 22, when the ossicles of the ears (6 bones) and the single hyoid bone are counted separately. Technically, the hyoid bone is not part of the skull.
†The thoracic vertebrae are sometimes included in this category.

‡Technically, the term *arm* refers to the upper extremity between the shoulder and elbow; the *forearm* is between the elbow and wrist. The upper part of the lower extremity, between the pelvis and knee, is the *thigh*; the *leg* is between the knee and ankle.

DIVISIONS OF THE SKELETON

The skeleton has two major divisions: the axial skeleton and the appendicular skeleton [FIGURE 7.1]. The *axial skeleton* is so named because it forms the longitudinal *axis* of the body. It is made up of the skull, vertebral column, sternum, and ribs. The *appendicular skeleton* is composed of the upper and lower extremities, which include the pectoral (shoulder) and pelvic girdles that at-tach the upper and lower appendages to the axial skeleton (see Chapter 8).

ASK YOURSELF

1 What are the two main divisions of the skeleton, and where are they joined together?

2 What bones make up the axial skeleton?

3 How many bones make up the axial skeleton?

THE SKULL

At the top of the axial skeleton is the *skull,* usually defined as the skeleton of the head, with or without the mandible (lower jaw) [TABLE 7.2]. Other bones closely associated with the skull are typically pharyngeal (throat) bones; besides the mandible, they include the auditory bones (ear ossicles) and the hyoid bone. (In this book we will treat the mandible, ear ossicles, and hyoid as associated bones of the skull.)

The skull can be divided into (1) the *cranial skull,* which supports and protects the brain, its surrounding membranes (meninges), and the cerebrospinal fluid; and (2) the *facial skull,* which forms the framework for the nasal cavities and the oral (mouth) cavity.

The bones of the skull form a supporting framework that combines a minimum of bony substance and weight with a maximum of strength and support. Bony capsules surround and protect the brain, eyes, inner ear, and nasal passages [FIGURE 7.2]. The skull is lightened by small cavities called *paranasal sinuses.*

Bony buttresses (like the delicate but strong buttresses that support the walls of many Gothic cathedrals) within the skull can sustain the enormous pressures exerted by the teeth during biting and chewing. To prevent such pressures from crushing the facial skeleton, three but-tresses extend from the teeth upward through the facial skull.

Besides containing buttresses, the skull has numerous *foramina* (L. openings; sing. *foramen,* fuh-RAY-muhn), openings of various sizes through which nerves and blood vessels pass. The main foramina of the skull are shown in FIGURE 7.4A, B, F, and G.

Sutures and Fontanels

The skull contains 29 bones, 11 of which are paired. Except for the mandible, ear ossicles, and hyoid, they are joined by *sutures* (L. *sutura,* seam), wriggly, seamlike joints that make the skull bones of an adult immovable [FIGURE 7.3]. Four major cranial skull sutures are:

1 The *coronal,* between the frontal and parietal bones.

2 The *lambdoidal,* between the parietal and occipital bones.

3 The *sagittal,* between the right and left parietal bones.

4 The *squamous,* between the temporal and parietal bones.

Because the skull bones are joined by pliable membranes rather than tight-fitting sutures during fetal life and early childhood, it is relatively easy for the skull bones to move and overlap as the infant passes through the mother's narrow birth canal. Some of the larger membranous areas between such incompletely ossified bones are called *fontanels* [FIGURE 7.3]. They allow the skull to expand as the child's brain completes its growth and development during the first few years of postnatal life. The membrane-filled fontanels (often called "soft spots") are:

1 The *frontal (anterior) fontanel,* located between the angles of the two parietal bones and the two sections of the frontal bones. This diamond-shaped fontanel is the largest, and usually does not close until 18 to 24 months after the child is born. The pulsating blood flow in the arteries of the brain can be felt at the frontal fontanel. If childbirth is difficult, a pulse monitor may be placed at this site to monitor the baby's heartbeat.

2 The *occipital (posterior) fontanel,* located between the occipital bone and the two parietal bones. This smaller, diamond-shaped fontanel closes about two months after birth.

3 The two *sphenoidal (anterolateral) fontanels,* situated at the junction of the frontal, parietal, temporal, and sphenoid bones. They usually close about three months after birth.

FIGURE 7.2 SKULL CAVITIES

Coronal (frontal) section through the skull, illustrating the bony capsules surrounding the brain (cranial cavity), eye (orbit), maxillary sinus, and nose (nasal skeleton surrounding nasal cavity).

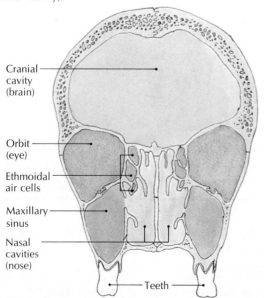

Cranial cavity (brain)

Orbit (eye)

Ethmoidal air cells

Maxillary sinus

Nasal cavities (nose)

Teeth

FIGURE 7.3 SUTURES AND FONTANELS

[A] Skull of a newborn infant illustrating the location of the fontanels; superior and right lateral views. [B] Sutures in an adult skull; superior and right lateral views. Note the absence of fontanels, which have closed.

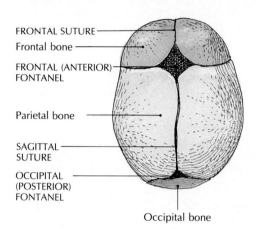
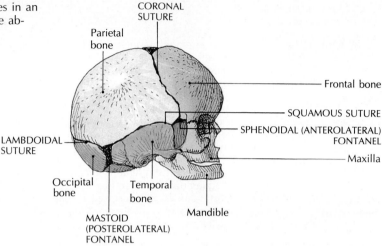

[A]

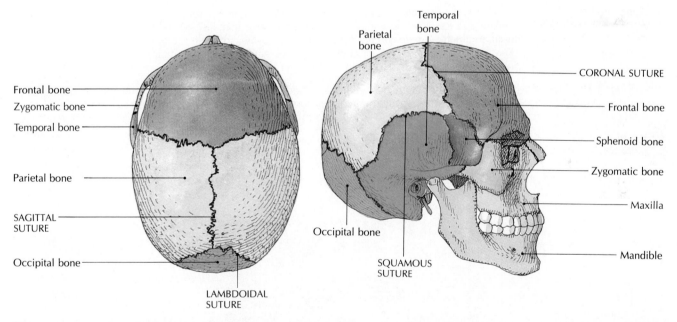

[B]

4 The two *mastoid (posterolateral) fontanels*, situated at the junction of the parietal, occipital, and temporal bones. They begin to close about two months after birth, but do not close completely until about 12 months after birth.

Bones of the Cranium

The eight bones of the cranium are the frontal bone, the two parietal bones, the occipital bone, the two temporal bones, the sphenoid bone, and the ethmoid bone. The sutural bones are also considered as part of the cranium.

The cranium consists of a roof and a base. The roof, called the **calvaria** (not calvarium), is made up of the squamous (L. *squama*, scale, which indicates that the bone is thin and relatively flat) part of a frontal bone (frontal squama), the parietal bones, and the occipital bone above the occipital protuberance [FIGURE 7.4A to E]. The **base** of the cranium is made up of portions of the ethmoid, sphenoid, and occipital bones. As viewed internally from above [FIGURE 7.4G], the base has three depressions, or *fossae:* the anterior, middle, and posterior fossae.

TABLE 7.2 BONES OF THE SKULL (29 BONES)

Bone	Description and function

CRANIAL SKULL (8 BONES)

Ethmoid (1) [FIGURES 7.4D,E, 7.6]
Base of cranium, anterior to body of sphenoid. Made up of horizontal cribriform plate, median perpendicular plate, paired lateral masses; contains ethmoidal sinuses, crista galli, superior and middle conchae. Forms roof of nasal cavity and septum, part of cranium floor; site of attachment for membranes covering brain.

Frontal (1) [FIGURE 7.4A,C,D,E,G]
Anterior and superior parts of cranium, forehead, brow areas. Shaped like large scoop; frontal squama forms forehead; orbital plate forms roof of orbit; supraorbital ridge forms brow ridge; contains frontal sinuses, supraorbital foramen. Protects front of brain; contains passageway for nerves, blood vessels.

Occipital (1) [FIGURE 7.4B–G]
Posterior part of cranium, including base. Slightly curved plate, with turned-up edges; made up of squamous, base, and two lateral parts; contains foramen magnum, occipital condyles, hypoglossal canals, atlantooccipital joint, external occipital crest and protuberance. Protects posterior part of brain; forms foramina for spinal cord and nerves; site of attachment for muscles, ligaments.

Parietal (2) [FIGURE 7.4A–G]
Superior sides and roof of cranium, between frontal and occipital bones. Broad, slightly convex plates; smooth exteriors and internal depressions. Protect top, sides of brain; passageway for blood vessels.

Sphenoid (1) [FIGURES 7.4A–G, 7.5]
Base of cranium, anterior to occipital and temporal bones. Wedge-shaped; made up of body, greater and lesser lateral wings, pterygoid processes; contains sphenoidal sinuses, sella turcica, optic foramen, superior orbital fissure, foramen ovale, foramen rotundum, foramen spinosum. Forms anterior part of base of cranium; houses pituitary gland; contains foramina for cranial nerves, meningeal artery to brain.

Temporal (2) [FIGURE 7.4A–E,G]
Sides and base of cranium at temples. Made up of squamous, petrous, tympanic, mastoid areas; contain zygomatic process, mandibular fossa, ear ossicles, mastoid sinuses. Form temples, part of cheekbones; articulate with lower jaw; protect ear ossicles; site of attachment for neck muscles.

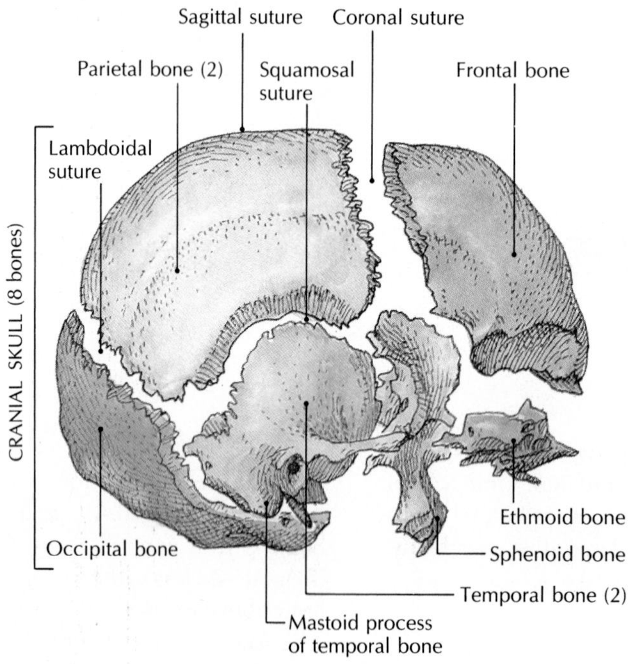

[A] **Exploded view of cranial skull.** Eight bones; right lateral view.

Bone	Description and function

FACIAL SKULL (14 BONES)

Inferior nasal conchae (2) [FIGURES 7.4A,E, 7.6A]	Lateral walls of nasal cavities, below superior and middle conchae of ethmoid bone. Thin, cancellous, shaped like curved leaves.
Lacrimal (2) [FIGURE 7.4A,D,F]	Medial wall of orbit, behind frontal process of maxilla. Small, thin, rectangular; contains depression for lacrimal sacs, nasolacrimal tear duct.
Mandible (1) [FIGURES 7.4A–E, 7.8]	Lower jaw, extending from chin to mandibular fossa of temporal bone. Largest, strongest facial bone; horseshoe-shaped horizontal body with two perpendicular rami; contains tooth sockets, coronoid, condylar, alveolar processes, mental foramina. Forms lower jaw, part of temporomandibular joint; site of attachment for muscles.
Maxillae (2) [FIGURE 7.4A–F]	Upper jaw and anterior part of hard palate. Made up of zygomatic, frontal, palatine, alveolar processes; contain infraorbital foramina, maxillary sinuses, tooth sockets. Form upper jaw, front of hard palate, part of eye sockets.
Nasal (2). [FIGURE 7.4A,C,D,E]	Upper bridge of nose between frontal processes of maxillae. Small, oblong; attached to a nasal cartilage. Form supports for bridge of upper nose.
Palatine (2) [FIGURE 7.4B,E]	Posterior part of hard palate, floor of nasal cavity and orbit; posterior to maxillae. L-shaped, with horizontal and vertical plates; contain greater and lesser palatine foramina. Horizontal plate forms posterior part of hard palate; vertical plate forms part of wall of nasal cavity, floor of orbit.
Vomer (1) [FIGURE 7.4A,B,E,F]	Posterior and inferior part of nasal septum. Thin, shaped like plowshare. Forms posterior and inferior nasal septum dividing nasal cavities.
Zygomatic (malar) (2) [FIGURE 7.4A,C,D,F]	Cheekbones below and lateral to orbit. Curved lateral part of cheekbones; made up of temporal process, zygomatic arch; contain zygomaticofacial and zygomaticotemporal foramina. Form cheekbones, outer part of eye sockets.

OTHER SKULL BONES (7)

Hyoid (1) [FIGURE 7.9]	Below root of tongue, above larynx. U-shaped, suspended from styloid process of temporal bone; site of attachment for some muscles used in speaking, swallowing.
Ossicles of ear Incus (2) Malleus (2) Stapes (2)	Inside cavity of petrous portion of temporal bone. Tiny bones shaped like anvil, hammer, stirrup, articulating with one another and attached to tympanic membrane. Convey sound vibrations from eardrum to oval window (see Chapter 16).

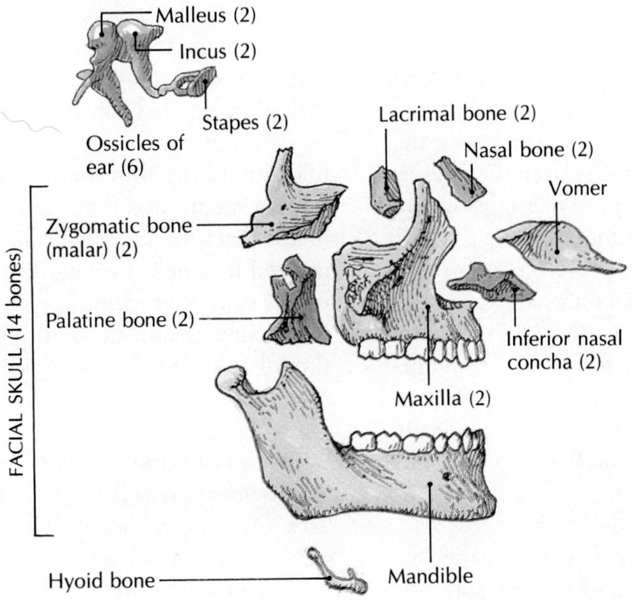

[B] Exploded view of facial skull, ear ossicles, and hyoid bone. The facial skull has 14 bones; the 6 ossicles are enlarged for clarity. Right lateral view.

FIGURE 7.4A SKULL, ANTERIOR VIEW

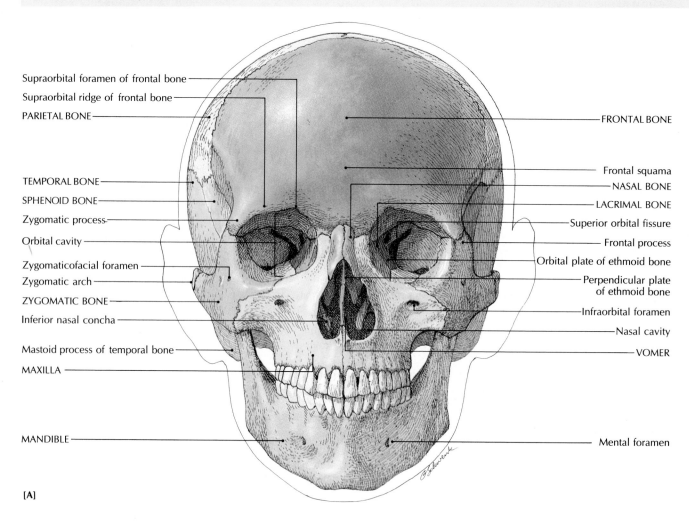

Supraorbital foramen of frontal bone
Supraorbital ridge of frontal bone
PARIETAL BONE

TEMPORAL BONE
SPHENOID BONE
Zygomatic process
Orbital cavity
Zygomaticofacial foramen
Zygomatic arch
ZYGOMATIC BONE
Inferior nasal concha
Mastoid process of temporal bone
MAXILLA

MANDIBLE

FRONTAL BONE

Frontal squama
NASAL BONE
LACRIMAL BONE
Superior orbital fissure
Frontal process
Orbital plate of ethmoid bone
Perpendicular plate of ethmoid bone
Infraorbital foramen
Nasal cavity
VOMER

Mental foramen

[A]

Frontal bone As shown in FIGURE 7.4A, C, and D, the large, scoop-shaped *frontal bone* forms the forehead and the upper part of the orbits (eye sockets). It forms the shell of the forehead as the *frontal squama*, the roof of the orbits as the *orbital plate*, and protrudes over the eyes as the *supraorbital ridges* to form the eyebrow ridges where the squamous and orbital portions meet.

The supraorbital nerve and blood vessels pass through the small supraorbital foramen (often only a notch) in the supraorbital ridge. Lining the inner surface of the frontal bone are many depressions, which follow the convolutions of the brain. At birth, the frontal bone has two parts, separated by a *frontal suture*. This suture generally disappears by the age of six, when the frontal bones become united [FIGURE 7.3A].

Parietal bones FIGURE 7.4A through G shows the two *parietal bones* (L. *paries*, wall), which form the superior

sides, roof, and part of the back of the skull. These broad, slightly convex bones have smooth exteriors, but like the frontal bone, they have internal depressions that accommodate the convolutions of the brain and blood vessels that supply the thin meninges that cover the brain. It is seldom mentioned that the parietal bones also form part of the back of the calvaria. In each parietal bone is a parietal foramen. Passing through the foramen is an emissary vein that connects the superior sagittal venous dural sinus inside the skull with scalp veins outside the skull. Emissary veins act as safety valves, by which blood may flow in either direction.

Occipital bone FIGURE 7.4B–G shows the location of the *occipital bone* (L. *occiput*, in back of the head, *ob*, in back of + *caput*, head). This bone forms the posterior part of the cranial skull.

The occipital bone is a slightly curved plate that con-

FIGURE 7.4B SKULL, POSTERIOR VIEW

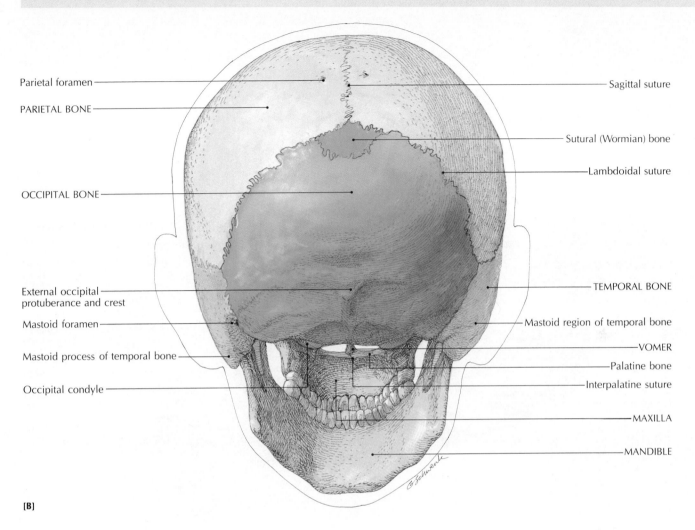

Parietal foramen

PARIETAL BONE

OCCIPITAL BONE

External occipital protuberance and crest

Mastoid foramen

Mastoid process of temporal bone

Occipital condyle

Sagittal suture

Sutural (Wormian) bone

Lambdoidal suture

TEMPORAL BONE

Mastoid region of temporal bone

VOMER

Palatine bone

Interpalatine suture

MAXILLA

MANDIBLE

[B]

sists of a squamous part, a base part, and two lateral parts. At its base is a large oval opening called the *foramen magnum*, where the spinal cord passes through to join the medulla oblongata of the brain. On either side of the foramen magnum are the *occipital condyles* and the paired *hypoglossal canals*, through which the hypoglossal nerves pass. These nerves stimulate the muscles of the tongue. The *atlantooccipital joint* between the occipital condyles and the first cervical vertebra (the atlas) permits the head to nod up and down in the "yes" movement.

The *external occipital crest* and the *external occipital protuberance* can be felt easily at the base of the occipital bone. Both landmarks are sites of attachment for muscles and ligaments.

Temporal bones FIGURE 7.4A–E, G shows the paired *temporal bones* (pertaining to the temples at the sides of the head). Together with portions of the sphenoid

bone, they form part of the sides and base of the cranium.

Each temporal bone has four parts: the squamous, petrous, and tympanic portions, and the mastoid region. The largest, and the superior part, is the *squamous portion*. The *zygomatic process* of the temporal bone joins the temporal process of the zygomatic bone to form the slender *zygomatic arch* in the lateral part of the cheekbone [FIGURE 7.4A, C, D, F]. The *mandibular fossa* is located posterior to the zygomatic arch. The fossa forms the socket that, together with the condyle of the mandible, forms the temporomandibular (jaw, TM) joint.

The *petrous portion* of the temporal bone is in the floor of the *middle cranial fossa*, wedged between the occipital and sphenoid bones [FIGURE 7.4G]. Projecting downward from the inferior surface of the petrous portion is the *styloid process* [FIGURE 7.4D]. The elongated styloid process is the site for the attachment of muscles

FIGURE 7.4C PHOTO OF SKULL, RIGHT LATERAL VIEW

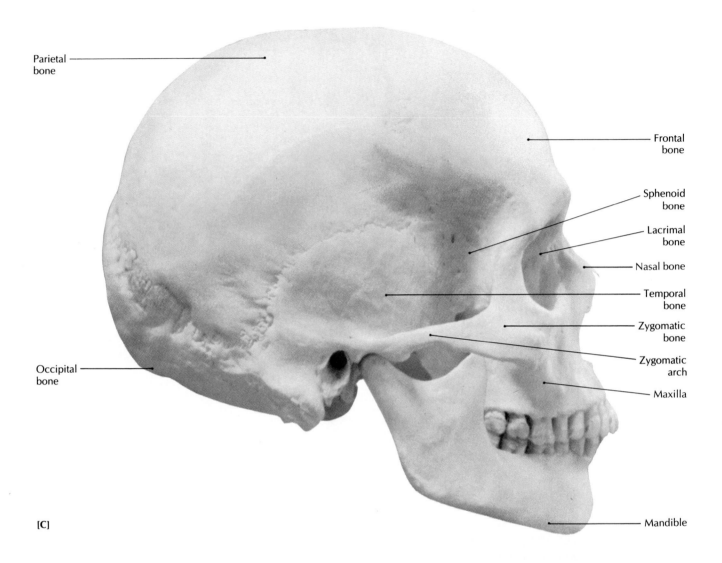

Parietal bone

Frontal bone

Sphenoid bone

Lacrimal bone

Nasal bone

Temporal bone

Zygomatic bone

Zygomatic arch

Occipital bone

Maxilla

Mandible

[C]

FIGURE 7.4D SKULL, RIGHT LATERAL VIEW

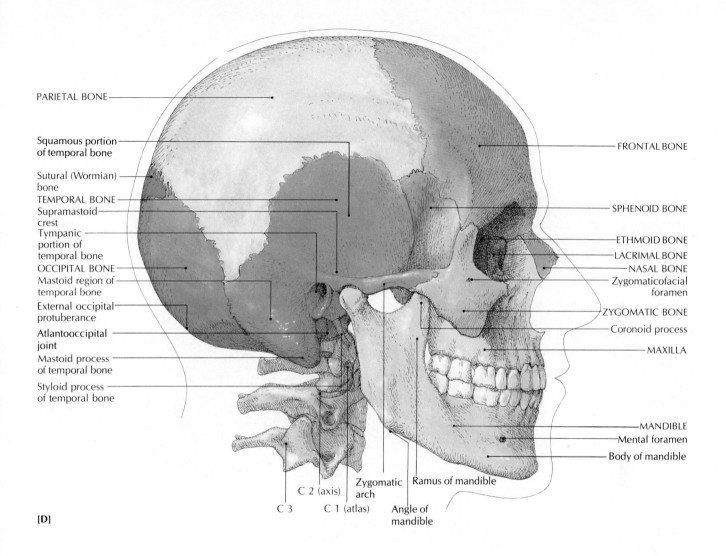

PARIETAL BONE

Squamous portion of temporal bone

Sutural (Wormian) bone

TEMPORAL BONE

Supramastoid crest

Tympanic portion of temporal bone

OCCIPITAL BONE

Mastoid region of temporal bone

External occipital protuberance

Atlantooccipital joint

Mastoid process of temporal bone

Styloid process of temporal bone

FRONTAL BONE

SPHENOID BONE

ETHMOID BONE

LACRIMAL BONE

NASAL BONE

Zygomaticofacial foramen

ZYGOMATIC BONE

Coronoid process

MAXILLA

MANDIBLE

Mental foramen

Body of mandible

C 2 (axis)

C 3

C 1 (atlas)

Zygomatic arch

Ramus of mandible

Angle of mandible

[D]

and ligaments involved with some movements of the lower jaw, tongue, and pharynx during speaking, swallowing, and chewing. Encased within the temporal bone are the three tiny auditory bones of the middle ear, the *malleus, incus,* and *stapes,* and the sensory receptors for hearing (cochlea) and balance (semicircular ducts, utricle, saccule).

The ***tympanic portion*** of the temporal bone [FIGURE 7.4D] forms part of the wall of the ***external auditory canal,*** or ***meatus*** (the external opening of the ear), and part of the wall of the tympanic cavity, both of which are involved with the transmission and reso-

nance of sound waves. The tympanic portion is the site for the attachment of the tympanic membrane (eardrum).

The ***mastoid region*** is actually the posterior portion of the temporal bone. Each prominent ***mastoid process*** in the posterior part of the temporal bone is located behind the ear [FIGURE 7.4B, D]; it provides the point of attachment for a neck muscle. These processes contain air spaces called *mastoid air cells* or *sinuses,* which connect with the middle ear. Because of this connection, infections within the middle ear can spread to the sinuses and cause an inflammation called *mastoiditis.*

FIGURE 7.4E SKULL, RIGHT MEDIAL VIEW

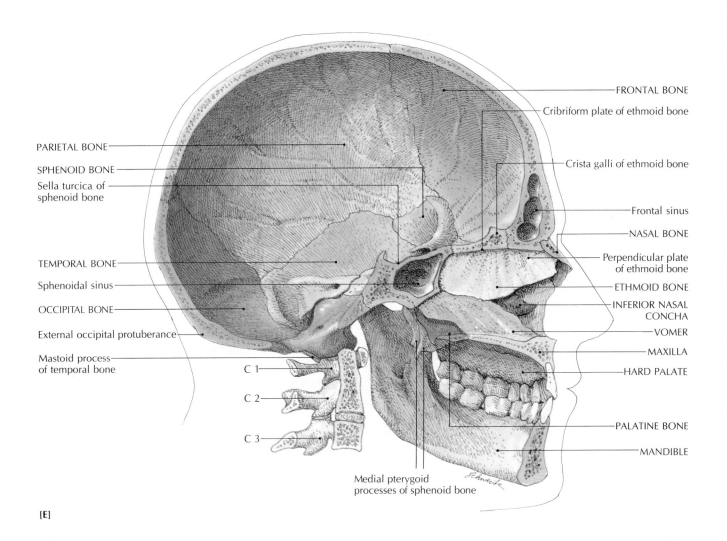

PARIETAL BONE

SPHENOID BONE

Sella turcica of
sphenoid bone

TEMPORAL BONE

Sphenoidal sinus

OCCIPITAL BONE

External occipital protuberance

Mastoid process
of temporal bone

C 1

C 2

C 3

Medial pterygoid
processes of sphenoid bone

FRONTAL BONE

Cribriform plate of ethmoid bone

Crista galli of ethmoid bone

Frontal sinus

NASAL BONE

Perpendicular plate
of ethmoid bone

ETHMOID BONE

INFERIOR NASAL
CONCHA

VOMER

MAXILLA

HARD PALATE

PALATINE BONE

MANDIBLE

[E]

FIGURE 7.4F SKULL, INFERIOR VIEW

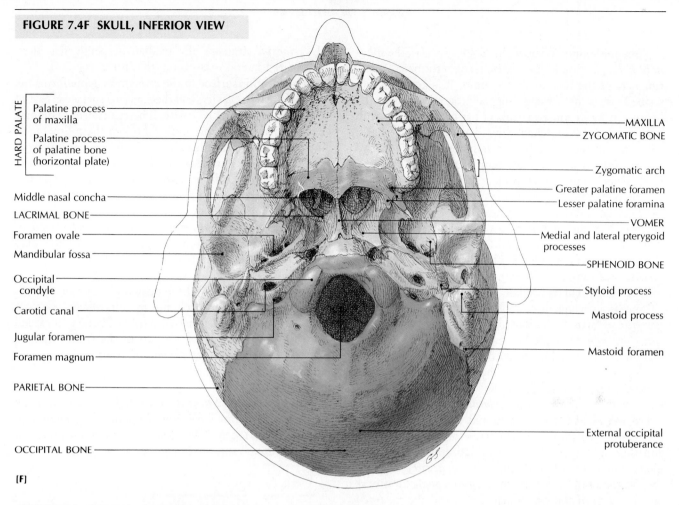

HARD PALATE

Palatine process of maxilla

Palatine process of palatine bone (horizontal plate)

Middle nasal concha

LACRIMAL BONE

Foramen ovale

Mandibular fossa

Occipital condyle

Carotid canal

Jugular foramen

Foramen magnum

PARIETAL BONE

OCCIPITAL BONE

MAXILLA

ZYGOMATIC BONE

Zygomatic arch

Greater palatine foramen

Lesser palatine foramina

VOMER

Medial and lateral pterygoid processes

SPHENOID BONE

Styloid process

Mastoid process

Mastoid foramen

External occipital protuberance

[F]

FIGURE 7.4G SKULL, SUPERIOR VIEW

The floor of the cranium and the anterior, middle, and posterior fossae are viewed from above. Dashed lines indicate boundaries of fossae.

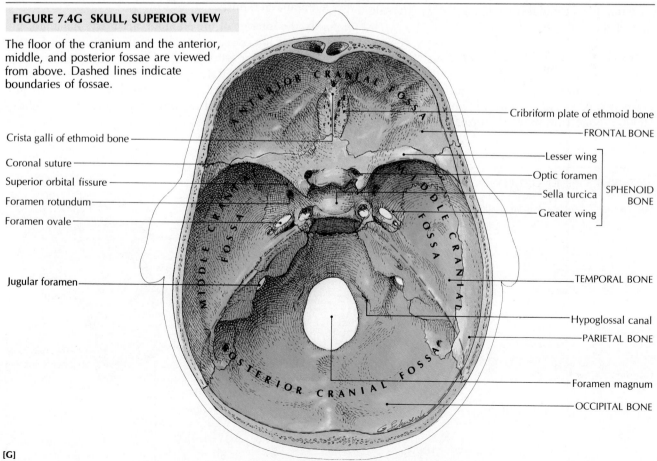

Crista galli of ethmoid bone

Coronal suture

Superior orbital fissure

Foramen rotundum

Foramen ovale

Jugular foramen

Cribriform plate of ethmoid bone

FRONTAL BONE

Lesser wing

Optic foramen

Sella turcica

Greater wing

SPHENOID BONE

TEMPORAL BONE

Hypoglossal canal

PARIETAL BONE

Foramen magnum

OCCIPITAL BONE

ANTERIOR CRANIAL FOSSA

MIDDLE CRANIAL FOSSA

POSTERIOR CRANIAL FOSSA

[G]

Sphenoid bone As shown in FIGURE 7.4A–G, the *sphenoid bone* (SFEE-noid; Gr. *sphen*, wedge) forms the anterior part of the base of the cranium. It is generally described as being "wedge-shaped" or "wing-shaped," because it looks somewhat like a bat or a butterfly with two pairs of outstretched wings [FIGURE 7.5]. It has a cube-shaped central *body*, two *lesser lateral wings*, two *greater lateral wings*, and a pair of *pterygoid processes* (Gr. *pteron*, wing), which project downward.

The pterygoid processes form part of the walls of the nasal cavity, and the undersurface of the body forms part of the roof of the nasal cavity. A key feature of the sphenoid is a deep depression within its body that houses and protects the pituitary gland. This depression is the *sella turcica* (SEH-luh TUR-sihk-uh), so called because it is said to resemble a Turkish saddle (L. *sella*, saddle).

The lesser wings form a part of the floor of the *anterior cranial fossa*. At the base of each lesser wing is the *optic foramen*, through which the optic nerve and ophthalmic artery pass into the orbit.

The greater wings form a major part of the *middle cranial fossa* [FIGURE 7.4G]. Between the greater and lesser wings is the *superior orbital fissure*, located between the middle cranial fossa and the orbit. Through the fissure pass three cranial nerves for the eye muscles and the sensory nerve to the orbital and forehead regions. Three important foramina in the greater wing are the *foramen rotundum* and the *foramen ovale* for different divisions of the trigeminal nerve, and the *foramen spinosum* for the middle meningeal artery to the side and roof of the skull [FIGURE 7.5].

Ethmoid bone The *ethmoid bone* (Gr. *ethmos*, strainer) forms the roof of the nasal cavity and the medial part of the floor of the anterior cranial fossa between the orbits [FIGURES 7.4D, E, and 7.6]. The light and delicate ethmoid bone consists of a horizontal cribriform plate (L. *cribrum*, strainer), the median perpendicular plate, and paired lateral masses, or labyrinths.

The inferior surface of the *cribriform plate* forms the roof of the nasal cavity, and the superior surface is part of the floor of the cranial cavity. The cribriform plate may have as many as 20 foramina. Through these tiny openings the filaments of the olfactory (smell) nerve pass from the mucous membrane of the nose to the brain.

The *median perpendicular plate* forms a large part of the vertical nasal septum between the two nasal cavities. The *crista galli*, or cockscomb, is a vertical protuberance in the center of the cribriform plate. It provides a site of attachment for a membrane (falx cerebri of the dura mater) that covers the brain.

Each *lateral mass* has a plate that forms the lateral wall of a nasal cavity and the orbital plate of an orbit. It also contains the superior and middle *conchae* (KONG-kee; L. "conch shell"), curved scrolls of bone that extend into the nasal cavity and are part of the lateral wall. In each lateral mass are *ethmoidal air cells* that open and drain into the nasal cavity.

Sutural bones *Sutural bones* (also called *Wormian* bones) are separate small bones found in the sutures of the calvaria [FIGURE 7.4B]. They are found most often in the lambdoidal suture. Their number may vary.

Paranasal Sinuses

In the interior of the ethmoid, maxillary, sphenoid, and frontal bones are four pairs of air cavities called *paranasal sinuses* [FIGURE 7.7]. They are named after the bones where they are located. The sinuses have two main functions. They act as resonating chambers for the voice and, being air-filled cavities, they lighten the weight of the skull bones.

FIGURE 7.5 SPHENOID BONE

See FIGURE 7.4G for position in skull.

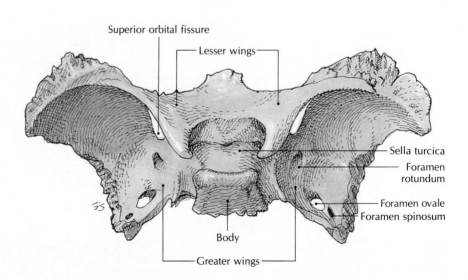

Superior orbital fissure
Lesser wings
Sella turcica
Foramen rotundum
Foramen ovale
Foramen spinosum
Body
Greater wings

FIGURE 7.6 ETHMOID BONE

[A] In the cranium, right medial view.
[B] Right lateral view. [C] Superior view.

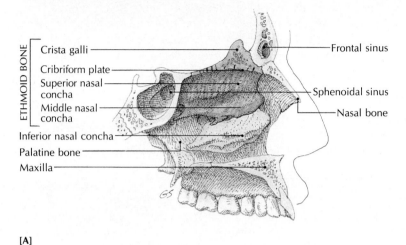

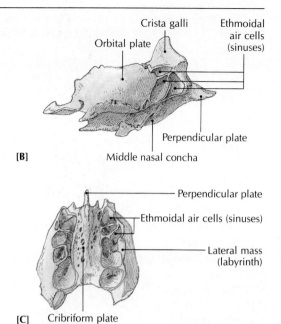

The paranasal sinuses are formed after birth as outgrowths from the nasal cavity when the spongy part of the bone is resorbed, leaving air-filled cavities. The *maxillary* and *sphenoidal sinuses* are not fully formed until adolescence.

The walls of the paranasal sinuses are lined with mucous membranes, which communicate with the mucous membranes of the nasal cavity through small openings. An inflammation of the mucosa of the nasal cavity is called rhinitis, or a "head cold." When the inflammation occurs in the mucous membranes of the sinuses, or when the sinuses react adversely to foreign substances (allergy), the result is *sinusitis*. The *ethmoidal air cells* (one of the paranasal sinuses) are made up of many small cavities in the ethmoid bone, between the orbit and the nasal cavity.

The *maxillary sinuses* are the largest of the paranasal sinuses. Mucus within the sinus drains into the nasal cavity through a single opening, which is high enough so that drainage is difficult when the nasal mucosa is swollen. Thus, fluid may accumulate and cause maxillary sinusitis.

The *sphenoidal sinuses* are contained within the body of the sphenoid bone. The size of the sphenoidal sinuses varies greatly, and they may extend into the occipital bone.

The *frontal sinuses* are separated by a bony septum that is often bent to one side. Each sinus extends above the medial end of the eyebrow and anteriorly into the medial portion of the roof of the orbit [FIGURE 7.7].

Bones of the Face

The facial skull is anterior and inferior to most of the cranial skull, to which it is attached. Its main functions are to support and protect the structures associated with the nasal, oral, orbital, and pharyngeal cavities. It also provides a pair of joints between the condyles of the mandible and temporal bones (the TM joints), which permit the lower jaw to open and close. Except for the mandible, all facial bones are united by sutures and are immovable.

The facial skull is composed of 14 irregularly shaped bones [FIGURE 7.4A, C–F]. They include the two inferior nasal conchae, vomer, two palatine bones, two maxillae, two zygomatic bones, two nasal bones, two lacrimal bones, and mandible. The single hyoid bone and the six ossicles of the ear are considered separately from cranial or facial bones.

Inferior nasal conchae FIGURES 7.4A, E and 7.6A show the location of the *inferior nasal conchae* in the lateral wall of each nasal cavity, below the superior and middle conchae of the ethmoid bone. These conchae are thin, bony plates, shaped somewhat like the curved leaves of a scroll. They increase the surface area of the nasal mucosa, a ciliated membrane with many blood vessels. Some of the fluid portion of the blood leaks out of these vessels, continuously moistening (humidifying) the nasal cavity. Cilia in the nasal cavities cleanse the air, and the blood vessels warm it.

Vomer The *vomer* (L. plowshare), shown in FIGURE 7.4A, B, E, and F, forms part of the nasal septum. The single vomer, as its name suggests, is shaped like a plowshare. It is very thin, and it is often bent to one side.

FIGURE 7.7 PARANASAL SINUSES

The paranasal sinuses include the frontal, ethmoidal, sphenoidal, and maxillary sinuses. The ethmoidal air cells, referred to as sinuses, look like honeycombs. **[A]** Anterior view. **[B]** Right medial view. **[C]** Photo of cleared specimen, anterior view.

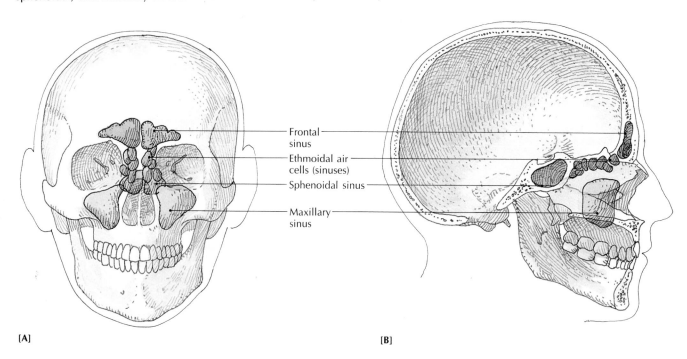

Frontal sinus

Ethmoidal air cells (sinuses)

Sphenoidal sinus

Maxillary sinus

[A]

[B]

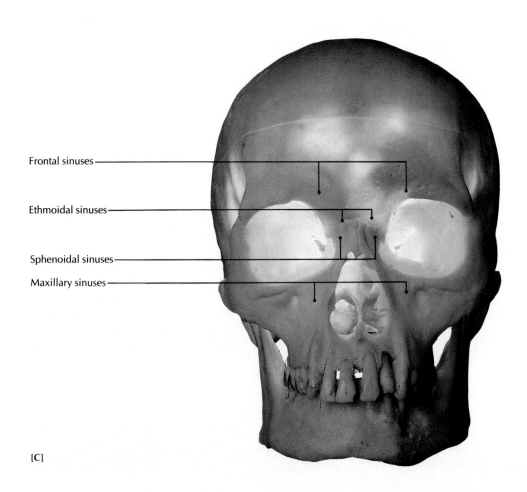

Frontal sinuses

Ethmoidal sinuses

Sphenoidal sinuses

Maxillary sinuses

[C]

If the septum is pushed too far to one side and one nasal cavity is much smaller than the other, the condition is called a *deviated septum*. With such a condition, an allergy attack or a severe head cold may swell the nasal mucosa enough to close the smaller cavity or even both nasal cavities.

Palatine bones FIGURE 7.4B and E show the paired *palatine bones* (referring to the *palate*, or roof of the mouth). These bones lie behind the maxillae (upper jaws), and form the posterior part of the hard palate, parts of the floor and walls of the nasal cavity, and the floor of the orbit. The hard palate is composed of the palatine processes of the palatine and maxillary bones.

Each palatine bone is L-shaped, and has a horizontal plate and a vertical plate. The horizontal plate, the palatine process, forms the posterior part of the hard palate. The vertical plate forms the posterior lateral wall of the nasal cavity and part of the floor of the orbit. Nerves that serve the palate pass through the *greater* and *lesser palatine foramina* in the posterior lateral corner of the horizontal plate. In some dental procedures, dentists infiltrate these foramina with anesthetics.

Maxillae FIGURE 7.4A through F show the two *maxillae* (L. upper jaw), which join to form the upper jaw. Each maxilla has a hollow body containing a large maxillary sinus and four processes: (1) The *zygomatic process* extends along the lateral orbital border, and articulates with the zygomatic bone to form the *zygomatic arch*. (2) The *frontal process* extends along the medial wall of the orbit, where it articulates with the frontal bone. (3) The *palatine process* extends horizontally to meet the palatine process of the other maxilla and form the anterior part of the hard palate. (4) The *alveolar process* joins the alveolar process of the other maxilla to form the *alveolar arch*, which contains the bony sockets for the upper teeth. Beneath the orbital margin is the *infraorbital foramen*, through which pass the sensory infraorbital nerve and blood vessels. The lengthening of the face just before adolescence is caused by the growth of the maxillae.

Zygomatic bones The *zygomatic bones* are the cheekbones [FIGURE 7.4A, C, D, F]. They lie below and lateral to the orbit. Each of the two zygomatic bones acts as a tie by connecting the maxilla with the frontal bone above and the temporal bone behind. Through the *zygomaticofacial* and *zygomaticotemporal foramina* in this bone pass the nerves bearing the same names.

Nasal bones As shown in FIGURE 7.4A and C through E, the two *nasal bones* lie side by side between the frontal processes of the maxillae. These two small, oblong bones unite to form the supportive bridge of the upper nose. In addition, they articulate with the frontal, ethmoid (per-

pendicular plate), and maxillary bones (frontal process), and are attached to the lower nasal cartilage.

Lacrimal bones Each *lacrimal bone* (L. *lacrima*, tear) is a thin bone located in the medial wall of the orbit behind the frontal process of the maxilla [FIGURE 7.4A, D, F]. The rectangular lacrimal bones are the smallest facial bones. In a depression of each bone is a *lacrimal sac*, which collects excess tears from the surface of the eye. Tears from the lacrimal sacs drain through the nasolacrimal ducts and foramen into the nasal cavity, sometimes causing a runny nose.

Mandible As shown in FIGURES 7.4A–E and 7.8, the *mandible* (L. *mandere*, to chew) is the bone of the lower jaw. It is a single bone that extends from the mandibular fossa of one temporal bone to the mandibular fossa of the other, forming the chin. It is the largest and strongest facial bone. The right and left halves of the mandible fuse together in the center at the *symphysis menti* during the first or second year of life.

Except for the ossicles of the ear and the hyoid bone, the mandible is the only movable bone in the skull. It can be raised, lowered, drawn back, pushed forward, and moved from side to side. Try it the next time you are chewing food.*

The mandible consists of a horseshoe-shaped horizontal *body* joined to two perpendicular upright portions called *rami* (RAY-mye; L. branches; singular, *ramus*). The site where the body joins each ramus is known as the *angle* of the mandible. At the superior end of each ramus are two processes separated by a deep depression called the *mandibular notch*: (1) the *coronoid process* is the attachment site for the temporalis muscle, and (2) the head of the *condylar process* articulates with the mandibular fossa of the temporal bone to form the temporomandibular (TM) joint. The inferior alveolar nerve enters the bone through the *mandibular foramen*, passes through the mandible, and a branch emerges as the mental nerve through the *mental foramen*, below the first molar tooth. Branches of the inferior alveolar nerve supply the teeth of the lower jaw. Blood vessels also pass through the mental foramen. The superior edge of the body of the mandible is the *alveolar process*, which contains the sockets for the lower teeth.

Ossicles of the Ear

Within the petrous portion of the temporal bone is the middle ear cavity, containing three pairs of tiny auditory

*Because the joint between the mandible and the temporal bone is slightly loose, it can be dislocated forward by a severe blow to the jaw. The head of the mandible may even slip forward during a yawn, locking the jaw in an open position. The condylar process can be realigned with the mandibular fossa of the temporal bone by depressing the jaw.

FIGURE 7.8 MANDIBLE

[A] Right lateral view. [B] The mandible in the infant at birth (top), in the adult (center), and in the aged (bottom). Note how the alveolar process disappears in old age after the teeth have been lost.

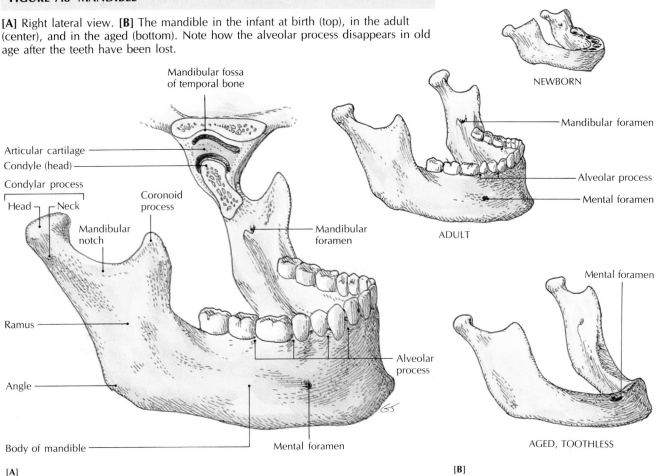

bones, or **ossicles.** [see FIGURE 16.5]. The ossicles, which are connected, transmit sound waves from the tympanic membrane (eardrum) to the inner ear. These bones are named according to their shapes. The outermost and largest bone is the **malleus** (L. hammer), or hammer. It is attached to the tympanic membrane. The middle bone is the **incus** (L. anvil), or anvil. The innermost bone, the **stapes** (L. stirrup), is shaped like a stirrup. The stapes is the smallest bone in the body. It fits into a tiny membranous oval window, which separates the middle and inner ears.

Hyoid Bone

FIGURE 7.9 shows the location of the U-shaped **hyoid bone** located inferior to the root of the tongue and superior to the larynx. When the chin is held up, the bone can be felt above the "Adam's apple," or thyroid cartilage. The hyoid bone does not articulate directly with any other bone. Instead, it is held in position by muscles and the stylohyoid ligaments, which extend from the styloid process of each temporal bone to the hyoid.

In the center of the hyoid bone is a **body,** and projecting backward and upward, a pair of **lesser** and **greater horns** or **cornua** (sing. *cornu*). The hyoid supports the tongue and provides attachment sites for muscles used in speaking and swallowing. In fact, the bone moves in a rotary motion (forward, up, back, and down) during the swallowing sequence. Because it is somewhat freely suspended, the hyoid bone can be held between the index finger and thumb and moved gently from side to side.

ASK YOURSELF

1 How is the skull usually defined? What are the two typical divisions of the skull?

2 What are sutures? Name the four major skull sutures.

3 What are fontanels? Give some examples.

4 What are the eight bones of the cranium?

FIGURE 7.9 HYOID BONE

[A] Anterior view, in position in the neck, superior to the larynx. Note how the hyoid is suspended from the stylohyoid ligaments. **[B]** Anterior view. **[C]** Right lateral view.

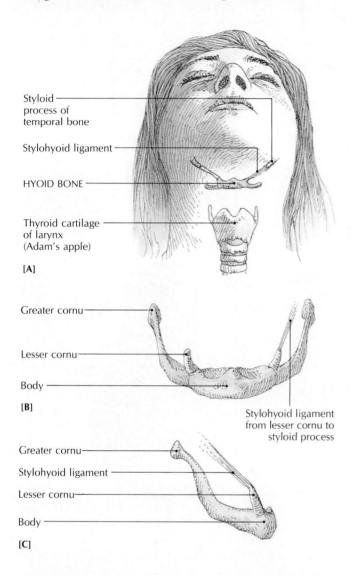

Styloid process of temporal bone

Stylohyoid ligament

HYOID BONE

Thyroid cartilage of larynx (Adam's apple)

[A]

Greater cornu

Lesser cornu

Body

[B]

Stylohyoid ligament from lesser cornu to styloid process

Greater cornu

Stylohyoid ligament

Lesser cornu

Body

[C]

THE VERTEBRAL COLUMN

The skeleton of the trunk of the body consists of the vertebral column (the spine or spinal column), the ribs, and the sternum (breastbone). These bones, along with the skull, make up the axial skeleton.

The *vertebral column* is actually more like an S-shaped spring than a column [FIGURE 7.10]. It extends from the base of the skull through the entire length of the trunk. The spine is composed of 26 separate bones called *vertebrae* (VER-tuh-bree; L. something to turn on, from *vertere*, to turn), which are united by a sequence of

fibrocartilaginous *intervertebral disks* to form a strong but flexible support for the neck and trunk. The vertebral column is stabilized by ligaments and muscles that permit twisting and bending movements, but limit some other movements that might be harmful to either the spinal column or the spinal cord.

In addition to protecting the spinal cord and spinal nerve roots and providing a support for the weight of the body, the spinal column helps the body keep an erect posture. The resilient intervertebral disks act as shock absorbers when the load upon the spinal column is increased, and they also allow the vertebrae to move without damaging each other. The disks account for about one-quarter of the length of the vertebral column and are thickest in the cervical and lumbar areas, where movement is greatest. The vertebral column is also the point of attachment for the muscles of the back.

Curvatures of the Vertebral Column

When viewed from the side, the adult vertebral column exhibits four curves: (1) a forward cervical curve, (2) a backward thoracic curve, (3) a forward lumbar curve, and (4) a backward sacral curve [FIGURE 7.10B]. The thoracic and sacral curves are known as *primary curves* because they are present in the fetus. The cervical and lumbar curves are *secondary*, because they do not appear until after birth. The cervical curve appears about 3 months after birth when the infant begins to hold up its head and continues to increase until about 9 months, when the child can sit upright; the lumbar curve appears at 12 to 18 months, when the child begins to walk.

The curves of the vertebral column provide the spring and resiliency necessary to cushion such ordinary actions as walking, and they are critical for maintaining a balanced center of gravity in the body.

The lumbar curve tends to be more pronounced in the female than in the male. During the later stages of pregnancy some women tend to increase the degree of the lumbar curve in an effort to maintain a balanced center of gravity. Backaches sometimes occur because of this exaggerated curve and the increased pressure on the posterior lumbar region.

A Typical Vertebra

Vertebrae are irregularly shaped bones, but the bones of each region are similar in shape. Except for the first and second cervical vertebrae, the sacrum, and the coccyx, they all have a similar structure. A "typical" vertebra consists of a *body*, a *vertebral (neural) arch*, and several *processes*, or projections [FIGURE 7.11]. The body supports weight, the arch protects the spinal cord, and the processes allow movement of the vertebral column.

FIGURE 7.10 THE VERTEBRAL COLUMN

[A] Anterior view. Note that the vertebrae are numbered in sequence within each region, starting from the head. [B] Right lateral view. Note the four spinal curves.

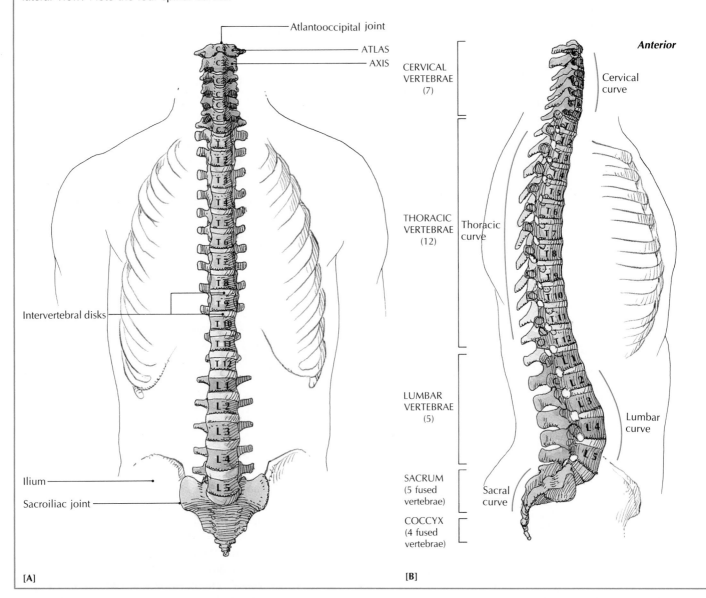

[A] [B]

Vertebral body The disk-shaped *vertebral body* is located anteriorly in each vertebra. The bodies of vertebrae from the third cervical to the first sacral become progressively larger in response to bearing increasing weight. The upper and lower ends of each body are slightly larger than the middle. These roughened ends articulate with the fibrocartilaginous intervertebral disk located between the bodies of the vertebrae above the sacrum. The anterior edge of the body contains tiny holes through which blood vessels pass.

Vertebral (neural) arch The *vertebral arch* is posterior to the vertebral body. Each arch has two thick *pedicles* (PED-uh-kuhlz; L. little feet), which form the lateral walls, and two *laminae* (LAMM-uh-nee; sing. *lamina*, LAMM-uh-nuh; L. thin plates), which form the posterior walls of the arch [FIGURE 7.11A]. The arch and the body together create the *vertebral foramen.* The sequence of all the vertebral foramina forms the *vertebral* (spinal) *canal,* which encloses the spinal cord and its surrounding meninges, nerve roots, and blood vessels. The vertebral

167

The Vertebral Column

BONES OF THE VERTEBRAL COLUMN (26 BONES)

Bones	Number of bones*	Description and function
Cervical vertebrae C1–C7 [FIGURE 7.12]	7	First (atlas), second (axis), and seventh vertebrae are modified; third through sixth are typical; all contain transverse foramina. Atlas supports head, permits "yes" motion of head at joint between skull and atlas; axis permits "no" motion at joint between axis and atlas.
Thoracic vertebrae T1–T12 [FIGURE 7.13]	12	Bodies and transverse processes have facets that articulate with ribs; laminae are short, thick, and broad. Articulate with ribs; allow some movement of spine in thoracic area.
Lumbar vertebrae L1–L5 [FIGURE 7.14]	5	Largest, strongest vertebrae; adapted for attachment of back muscles. Support back muscles; allow forward and backward bending of spine.
Sacrum (5 fused bones) [FIGURE 7.15]	1	Wedge-shaped, made up of five fused bodies united by four intervertebral disks. Support vertebral column; give strength and stability to pelvis.
Coccyx (3 to 5 fused bones) [FIGURE 7.15]	1	Triangular tailbone, united with sacrum by intervertebral disk. Vestige of an embryonic tail.
	26	

*In a child there are 33 separate vertebrae, the 9 in the sacrum and coccyx not yet being fused.

arch protects the spinal cord in the same way that the cranium protects the brain.

Each pedicle contains one vertebral notch on its inferior border and one on its superior border [FIGURE 7.11C]. These notches are arranged so that together they form an *intervertebral foramen*, through which the spinal nerves and their accompanying blood vessels pass.

Vertebral processes Seven *vertebral processes* extend from the lamina of a vertebra. Three of these processes, the *spinous process* and the paired *transverse processes*, are the attachment sites for vertebral muscles. These processes act like levers, helping the attached muscles and ligaments move the vertebrae.

The other four processes are two superior and two inferior *articular processes.* All four processes join directly to other bones. The superior articular processes of one vertebra articulate with the inferior articular process of the vertebra above [FIGURE 7.11C]. The articular processes prevent vertebrae from slipping forward and restrict movement between two vertebrae. Movement between two vertebrae occurs at the intervertebral disk and at the paired joints between the articular processes.

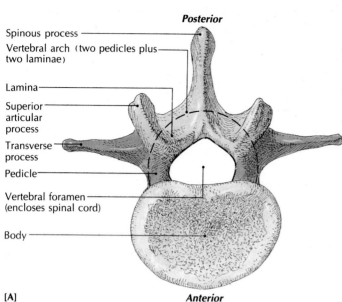

[A]

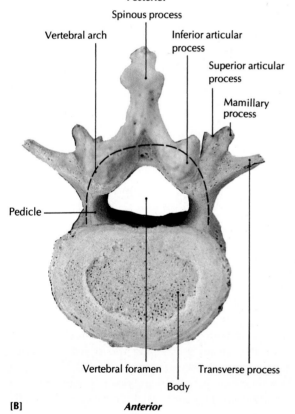

[B]

FIGURE 7.11 TYPICAL VERTEBRAE

[A] Drawing of a typical vertebra, illustrating its parts. The vertebral arch (dashed line) is made up of two pedicles and two laminae. Superior view. [B] Photo of a typical vertebra (L5), superior view; actual size. [C] Right lateral view. The sequence of vertebral foramina forms the vertebral canal, through which the spinal cord passes. The cauda equina is located caudal to the spinal cord. Both drawings are shown approximately actual size.

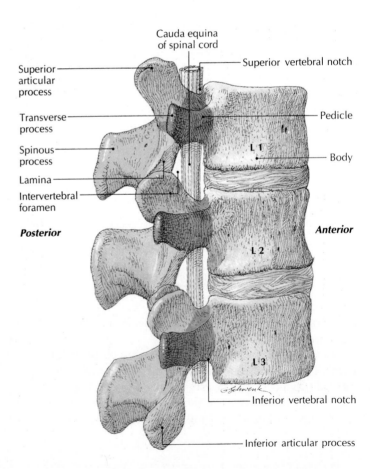

[C]

Cervical Vertebrae

The ***cervical vertebrae*** (L. *cervix*, neck) are the seven (C1 to C7) small neck vertebrae between the skull and the thoracic vertebrae [FIGURES 7.10A, B and 7.12]. They support the head and enable it to move up, down, and sideways. Each of these vertebrae has a pair of openings called the ***transverse foramina***, which are found only in these vertebrae. In all but the seventh, each foramen is large enough for the passage of a vertebral artery. The third to sixth cervical vertebrae are similar in shape, with a small, broad body, and short, forked (bifid) spines. Their articular facets are always positioned in the same way. However, the first, second, and seventh cervical vertebrae are not typical.

First cervical vertebra (atlas) The first cervical vertebra is called the ***atlas*** because it supports the head, as the Greek god Atlas supported the heavens on his shoulders. It is ringlike, with a short anterior and a long posterior arch, with large lateral masses on each side [FIGURE 7.12A]. Although the atlas is the widest of the

FIGURE 7.12 CERVICAL VERTEBRAE

[A] The atlas, or first cervical vertebra; superior view. [B] The axis, or second cervical vertebra; anterior view. [C] A typical cervical vertebra; superior view. Note that the vertebral artery passes through each transverse foramen, and the spinal cord through each vertebral foramen. [D] The articulation of the dens of the axis with the atlas, and their relationship to the other cervical vertebrae; posterior view.

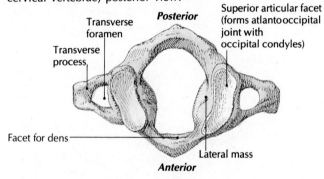

[A] ATLAS (FIRST CERVICAL VERTEBRA)

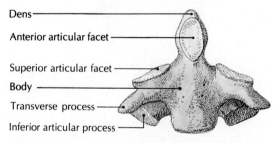

[B] AXIS (SECOND CERVICAL VERTEBRA)

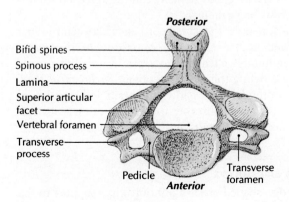

[C] TYPICAL CERVICAL VERTEBRA

[D] CERVICAL VERTEBRAE

cervical vertebrae, it has no spine and no body. It articulates above with the occipital condyles and below with the second cervical vertebra (axis). This *atlantooccipital joint* allows a nodding "yes" motion with the head.

Second cervical vertebra (axis) The second cervical vertebra is known as the *axis* because it forms a pivot point for the atlas to move the skull in a twisting "no" motion. The pivot is formed by a peglike protrusion from the body called the *dens* (Gr. tooth), which extends through the opening in the atlas [FIGURES 7.12B, D]. The dens, or *odontoid process*, is held in position against the anterior arch and away from the spinal cord by a strong transverse ligament of the atlas.

Quick death from hanging occurs when the transverse ligament snaps and the dens crushes the lower medulla oblongata and the adjacent spinal cord. A small tear of this ligament may produce *whiplash* symptoms, when the head is snapped backward during a violent automobile accident.

Seventh cervical vertebra The seventh cervical vertebra has an exceptionally long, unforked spinous process with a tubercle at its tip that can be felt, and usually seen, through the skin [FIGURE 7.12D]. This vertebra is known as the *vertebra prominens.*

Thoracic Vertebrae

All 12 *thoracic vertebrae* (Gr. *thorax*, breastplate) increase in size as they progress downward. The first four (T1 to T4) are similar to the cervical vertebrae, and the last four (T9 to T12) have certain features of lumbar vertebrae. Only the middle four (T5 to T8) are considered typical. Their bodies, viewed from above, are heart-shaped, and they have circular vertebral foramina [FIGURE 7.13]. Each thoracic vertebra has three articular (*costal*) facets on each side that provide attachments for the ribs [see FIGURE 7.17B]. Because there are 12 thoracic vertebrae and 12 corresponding flexible intervertebral disks, the total mobility of this region is considerable, even with the presence of the ribs and sternum.

Lumbar Vertebrae

The five *lumbar vertebrae* (L1 to L5) (L. *lumbus*, loin) are the largest and strongest vertebrae. They are situated in the "small of the back," between the thorax and the pelvis [FIGURES 7.1 and 7.10]. Their large kidney-shaped bodies have short, blunt, four-sided spinous processes, which are adapted for the attachment of the lower back muscles [FIGURE 7.14]. The arrangement of the facets on the articular processes of each vertebra maximizes forward and backward bending, but lateral bending is limited, and rotation is practically eliminated. The lumbar

FIGURE 7.13 THORACIC VERTEBRAE

[A] Right lateral view. **[B]** Superior view. The intervertebral notches between adjacent vertebrae form the intervertebral foramen, through which a spinal nerve passes.

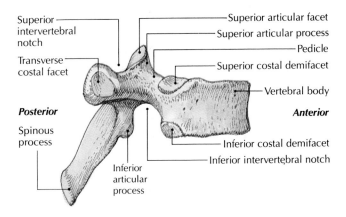

[A]

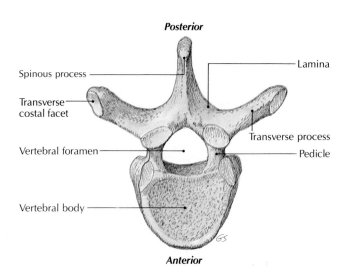

[B]

vertebrae have no transverse foramina. The transverse processes are long and slender, and the vertebral foramina are oval or triangular.

In most people, the spinal cord ends between the first and second lumbar vertebrae. Therefore, a *lumbar puncture* (commonly called a *spinal tap*) can usually be made safely just below this point to obtain *cerebrospinal fluid*. This clear fluid, which bathes the brain and spinal cord and acts as a shock absorber for the central nervous system, is useful in diagnosing certain diseases.

Sacrum and Coccyx

The wedge-shaped *sacrum* not only gives support to the vertebral column, it also provides strength and stability

FIGURE 7.14 LUMBAR VERTEBRAE

[A] Right lateral view. [B] Superior view. [C] X-ray, anterior view.

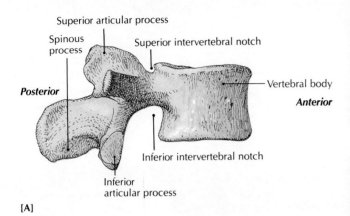

[A]

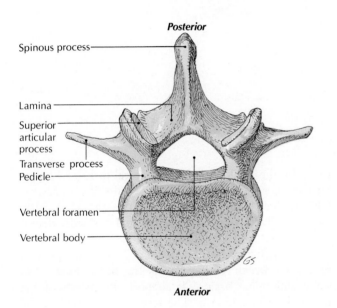

[B]

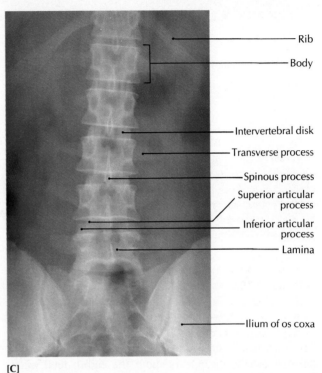

[C]

to the pelvis. It is composed of five vertebral bodies fused in an adult into one bone by four ossified intervertebral disks [FIGURE 7.15]. The sacrum is curved, forming a concave surface anteriorly and a convex surface posteriorly. The four *transverse lines* on the otherwise smooth concave anterior surface indicate the fusion sites of the originally separate vertebrae. At the lateral ends of these lines are two parallel rows of four ventral *sacral foramina.* On the dorsal convex surface of the sacrum are four pairs of *sacral foramina.* The areas lateral to these foramina are known as the *lateral masses.*

The sacrum articulates above with the last lumbar vertebra (L5), below with the coccyx, and laterally from the *auricular surfaces* with the two iliac bones of the hip to form the *sacroiliac joints.*

The fused laminae and short spines form the roof of the *sacral canal.* Because the laminae of the lower sacral vertebrae (S4 and S5) are often absent or do not meet in the midline, a gap called the *sacral hiatus* is present at these levels [FIGURE 7.15B]. Caudal anesthesia, which spreads upward and acts directly on the spinal nerves, can be injected into the sacral hiatus.

The projecting anterior edge of the first sacral vertebra is called the *sacral promontory.* It is used as a landmark for making pelvic measurements. Just above the promontory is the *lumbosacral joint* between the fifth lumbar vertebra and the sacrum. The sacrum is tilted at this point, so that the articulation forms the *lumbosacral angle.*

The *coccyx,* or tailbone, consists of three to five (usu-

FIGURE 7.15 SACRUM AND COCCYX

[A] Ventral view. [B] Dorsal view.

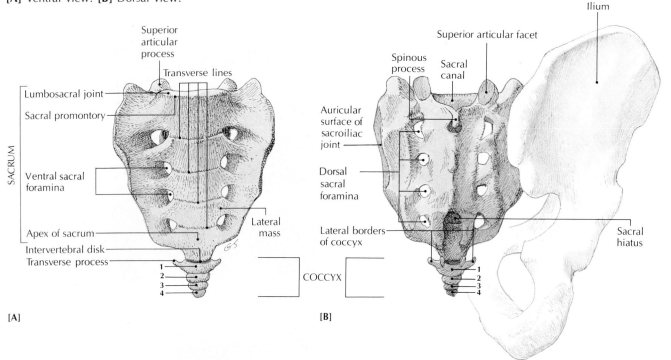

ally four) fused vertebrae, the vestiges of an embryonic tail that usually disappears about the eighth fetal week. Triangular in shape, it has an *apex, base, pelvic* and *dorsal surfaces*, and two *lateral borders* [FIGURE 7.15]. Its base articulates with the lower end of the sacrum by means of a fibrocartilaginous intervertebral disk.

1 How many vertebrae are there in the vertebral column?

2 What are the major functions of the vertebral column? What functions are served by the intervertebral disks?

3 What is the function of the four natural curves of the vertebral column?

4 From the neck down, what are the five vertebral regions called?

5 What are the parts of a typical vertebra?

THE THORAX

The *thorax* (Gr. breastplate) is the chest, which is part of the axial skeleton. The thoracic skeleton is formed poste-

riorly by the bodies and intervertebral disks of 12 thoracic vertebrae, and anteriorly by 12 pairs of ribs, 12 costal cartilages,* and the sternum. The thorax is fairly narrow at the top and broad below, and it is wider than it is deep [FIGURE 7.16].

The cagelike thoracic skeleton is a good example of a functional structure. It protects the heart, lungs, and some abdominal organs. It supports the bones of the pectoral girdle and arm. The lever arms formed by the ribs and costal cartilages provide a flexible mechanism for breathing. Also, the sternum provides a point of attachment for the ribs.

Sternum

The *sternum* (Gr. *sternon*, breast), or breastbone, is the midline bony structure of the anterior chest wall [FIGURES 7.16A, 7.17A]. It resembles a dagger in the adult, and consists of a manubrium, body, and xiphoid process.

The *manubrium* (L. handle) has a pair of *clavicular notches*, which articulate with the clavicle (collarbone). This *sternoclavicular joint* is the only site of direct at-

*Although ribs 11 and 12 are not attached to the sternum, they are tipped anteriorly with costal cartilage.

FIGURE 7.16 SKELETON OF THE THORAX

[A] Anterior view. [B] Posterior view. Note the sternum. Ribs and vertebrae are numbered.

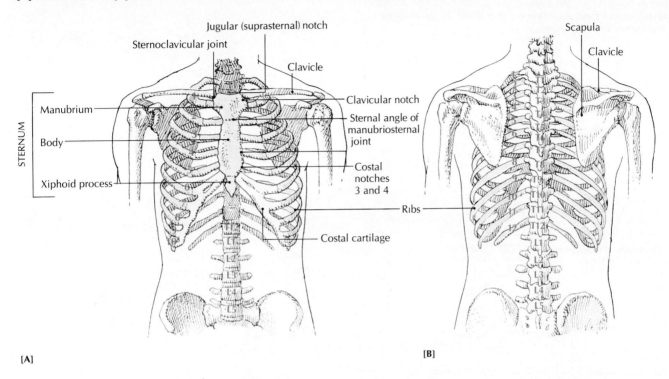

[A]

[B]

tachment of the pectoral girdle to the axial skeleton. At the lateral *costal notches* the sternum articulates with the costal cartilages of the first ribs and part of the second ribs. On its upper border is the *suprasternal notch.* The manubrium is united with the body of the sternum at the movable ***manubriosternal joint,*** which acts as a hinge to allow the sternum to move forward during inhalation. The joint may become ossified in the elderly, so that movements at the joint become restricted. The manubriosternal joint forms a slight angle, called the ***sternal angle,*** which can be felt through the skin. Because this angle is opposite the second rib, it is a reliable starting point for counting the ribs.

The ***body*** of the sternum is about twice as long as the manubrium. It articulates with the second through tenth pairs of ribs.

The ***xiphoid process*** (ZIFF-oid; Gr. sword-shaped) is the smallest, thinnest, and most variable part of the sternum. It is cartilaginous during infancy, but usually is almost completely ossified and joined to the body of the sternum by the fortieth year. The xiphoid process does not articulate with any ribs or costal cartilages, but several ligaments and muscles (including abdominal muscles) are attached to it. The joint between the body of the sternum and the xiphoid process is the *xiphisternal joint.* This joint is fragile, and if struck during cardiopulmonary resuscitation, the xiphoid process may break off.

Because the sternum is relatively accessible, a physician may insert a needle into its marrow cavity to obtain a specimen of red blood cells that are developing there. Such a procedure is called a *sternal puncture.*

Ribs

The ***ribs*** are curved, slightly twisted strips of bone that form the widest and major part of the thoracic cage [FIGURE 7.17]. There are usually 12 pairs of ribs, all of which articulate posteriorly with the vertebral column. The ribs increase in length from rib 1 to rib 7 and decrease from 8 to 12. The space between the ribs is the *intercostal space.* It contains the intercostal muscles.

The upper seven pairs (ribs 1 to 7) connect directly with the sternum by their attached strips of *costal cartilage,* which are made of hyaline cartilage. These ribs are called ***true ribs,*** because they attach to both the vertebrae and the sternum. The lower three pairs of ribs (8 to 10) are known as ***false ribs*** because they are attached to the sternum only indirectly. Of the false ribs, 8 to 10 have costal cartilages that connect with each other and also with the cartilage of rib 7. Ribs 11 and 12 are called ***floating ribs*** because they attach only to the vertebral column [FIGURE 7.17A].

Ribs 3 to 9 are *typical ribs* [FIGURE 7.17B]. Each typical rib has a wedge-shaped *head* on the end next to the spine.

FIGURE 7.17 RIBS

[A] Anterior view. [B] Typical rib, costal cartilage, thoracic vertebra and sternum, showing the articulations of those structures; superior view.

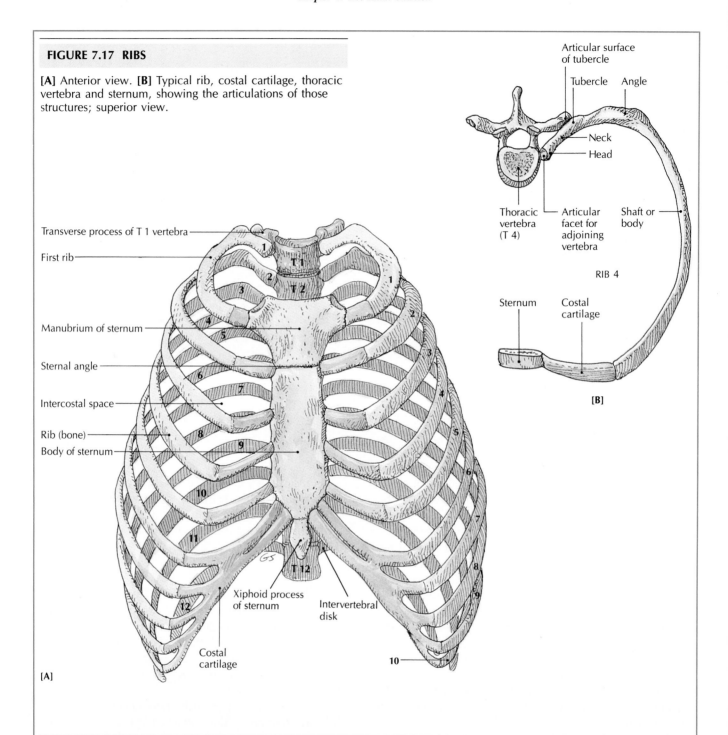

Bone	Description and Functions
Ribs (12 pairs)	Long, curved, varying in length and width; ribs 1–7 (true vertebrosternal ribs) attach directly to sternum; 8–12 (false ribs) do not attach directly to sternum; 8–10 (vertebrochondral ribs) attach to rib 7 cartilage; 11–12 attach only to vertebrae. With sternum and thoracic vertebrae, ribs form strong but lightweight cage to protect heart, lungs, other organs. With costal cartilages, ribs provide flexible mechanism for breathing and support bones of upper extremities.
Sternum	Dagger-shaped, about 15 cm long; made up of manubrium, body, xiphoid process; articulates with ribs 1–7 directly, 8–10 indirectly. Provides anterior attachment site for ribs; with ribs, forms protective cage for heart, lungs, other organs; supports bones of upper extremities.

The head articulates with an intervertebral disk and body of an adjacent vertebra. The short, flattened *neck* is located between the head and the tubercle. The *tubercle* is located between the neck and the body of a typical rib. It forms an articular surface with a transverse process on a vertebra. The *shaft*, or *body*, is the main part of the rib. It curves sharply forward after its junction with the tubercle to form a distinct *angle*, and then arches downward until it joins the costal cartilage. On the lower border of the rib is a *costal groove*, forming a protective passageway for intercostal blood vessels and nerves.

ASK YOURSELF

1 What are the components of the thorax?

2 What are the functions of the thoracic cage?

3 What are the main parts of the sternum? Where are the axial skeleton and appendicular skeleton connected to each other?

4 How many pairs of ribs are there? What is a true rib? A false rib? A floating rib?

5 What are the main parts of a typical rib?

WHEN THINGS GO WRONG

Fractures

A *fracture* is a broken bone. Children have fractures more often than adults because children have slender bones and are more active. Fortunately, the supple, healthy bones of children mend faster and better than the more brittle bones of older people. (A femur broken at birth is fully united within three weeks, but a similar break in a person over 20 may take four or five months to heal completely.) Usually, broken bones that are reset soon after injury have an excellent chance of healing perfectly because the living tissue and adequate blood supply at the fracture actually stimulate a natural repositioning.

In elderly people, bones contain relatively more calcified bone and less organic material. Consequently, old bones lose their elasticity and they break more easily. A fall that a child hardly notices can be serious in an elderly person. "To fall and break a hip," a common disaster among the elderly, could frequently be better stated, "to break a hip and fall," because the fragile old bones may break merely under the strain of walking, making the legs give way. The hip (actually the neck of the femur, the most fragile part in elderly people) may be broken before the body hits the ground.

A fractured bone goes through several stages of heal-

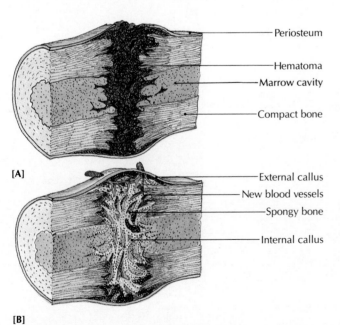

[A]

Periosteum

Hematoma

Marrow cavity

Compact bone

External callus

New blood vessels

Spongy bone

Internal callus

[B]

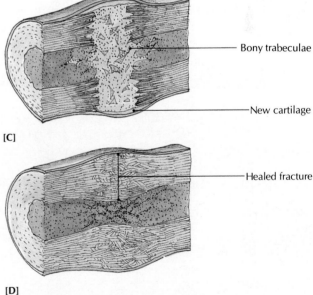

[C]

Bony trabeculae

New cartilage

Healed fracture

[D]

How a fracture heals. **[A]** Hemorrhaging occurs when tissue in the periosteum, osteon system, and marrow cavity are damaged. A hematoma (blood clot) forms several hours later. **[B]** Fibroblasts enter the damaged area, and a hard callus forms a few days later. **[C]** Osteogenic cells differentiate into new bony trabeculae, knitting the new fragments together. New cartilage forms on the outer collar of the fracture. **[D]** The cartilage is gradually replaced by spongy bone, and the fracture is eventually repaired with new compact bone.

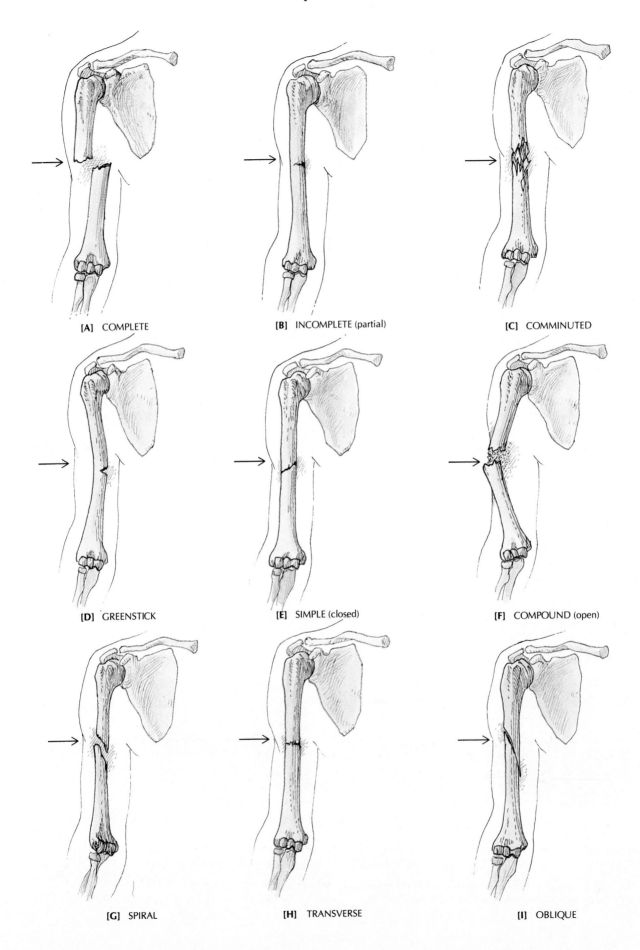

[A] COMPLETE

[B] INCOMPLETE (partial)

[C] COMMINUTED

[D] GREENSTICK

[E] SIMPLE (closed)

[F] COMPOUND (open)

[G] SPIRAL

[H] TRANSVERSE

[I] OBLIQUE

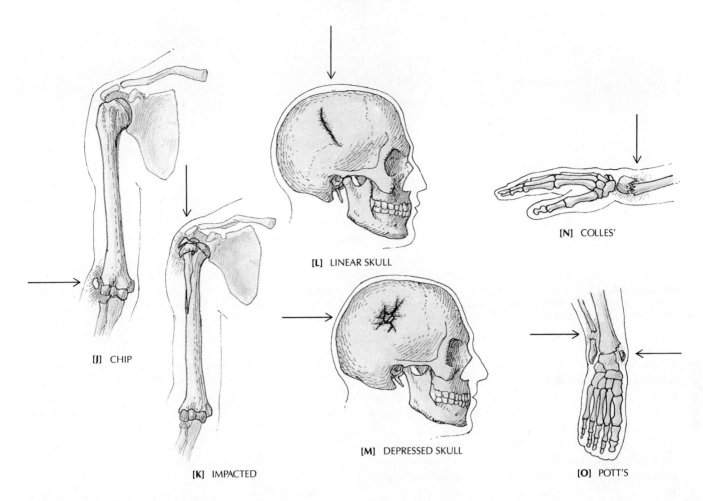

[J] CHIP

[K] IMPACTED

[L] LINEAR SKULL

[M] DEPRESSED SKULL

[N] COLLES'

[O] POTT'S

Kinds of fractures. Fractures can be classified according to the type and complexity of the break, the location of the break, and certain other special features. The following commonly used types and classifications are shown in the respective illustrations.
[A] *Complete.* The bone breaks completely into two pieces.
[B] *Incomplete (partial).* The bone does not break completely into two or more pieces.
[C] *Comminuted.* The bone is splintered or crushed into small pieces.
[D] *Greenstick.* The bone is broken on one side and bent on the other; common in children.
[E] *Simple (closed).* The bone is broken but does not break through the skin.
[F] *Compound (open).* The bone is broken and cuts through the skin.

[G] *Spiral.* The bone is broken by twisting.
[H] *Transverse.* The bone is broken directly across, at a right angle to the bone's long axis.
[I] *Oblique.* The bone is broken on a slant, at approximately a 45-degree angle to the bone's long axis.
[J] *Chip.* The bone is chipped where a protrusion is exposed.
[K] *Impacted.* The bone is broken when one part is forcefully driven into another, as at a shoulder or hip.
[L] *Linear skull.* The skull is broken in a line, lengthwise on the bone.
[M] *Depressed skull.* The skull is broken by a puncture, causing a depression below the surface.
[N] *Colles'.* The distal end of the radius is displaced posteriorly.
[O] *Pott's.* The distal part of the fibula and medial malleolus are broken.

ing. But even before healing can begin properly, the fragments of the broken bone must be manipulated, or *reduced*, back into their original positions by a physician. Usually the bone is immobilized by a cast, splint, or traction, and in severe cases, surgery and a continuing program of physical therapy may be necessary.

Fractures of the Vertebral Column

Many fractures of the vertebral column may be serious in themselves, but the real danger lies in injury to the spinal cord, which can result in paralysis or death. (Spinal cord injuries will be discussed in Chapter 13.)

The most common type of fracture is a *compression fracture,* which crushes the body of one or more vertebrae. Compressions often occur where there is the greatest spinal mobility: the middle or lower regions of the vertebral column, and near the point where the lumbar and thoracic regions meet. Although a compression fracture crushes the body of a vertebra, the vertebral arches and the ligaments of the spine remain intact. As a result, the spinal cord is not injured.

In contrast to compression fractures, *extension fractures* and dislocations involve a pulling force, usually affecting the posterior portions of the vertebral column.

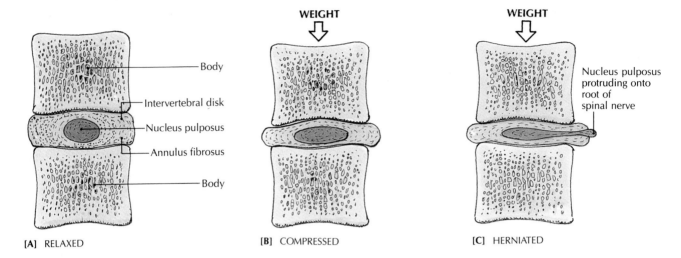

WEIGHT

WEIGHT

Body

Intervertebral disk

Nucleus pulposus

Annulus fibrosus

Body

Nucleus pulposus protruding onto root of spinal nerve

[A] RELAXED

[B] COMPRESSED

[C] HERNIATED

Herniated disk. **[A]** A relaxed intervertebral disk. **[B]** Heavy weight on the vertebral column compresses the pulpy nucleus pulposus in the center of a disk. **[C]** Excessive weight may cause the outer ring (annulus fibrosus) of the disk to rupture, allowing the nucleus pulposus to protrude onto the root of a spinal nerve.

When the neck is severely hyperextended (bent backward), as in a "whiplash" injury, the atlas may break at several points, and further extension may break off the arch of the axis. An even greater force may rupture one of the ligaments and *annulus fibrosus* (outer ring) of the C2/C3 intervertebral disk. Such a great force separates the skull, atlas, and axis from the rest of the vertebral column, and the spinal cord is usually severed in the process.

The *interspinous ligaments* connect the spinous processes, and restrict the range of movement of vertebrae. Some "whiplash" injuries may result when rapid stretching of these ligaments produces small tears in them. Hyperextension injuries usually do not occur in the thoracic region because of the support of the ribs and the relative immobility of the thoracic vertebrae.

Fracture usually accompanies dislocations of vertebrae because the thoracic and lumbar articular processes interlock. The spinal cord is not necessarily severed when cervical vertebrae are dislocated because the vertebral canal in the cervical region is wide enough to allow some displacement without damaging the cord.

The primary treatment for injuries of the vertebral column is usually immobilization, which allows the bone to heal and prevents damage to the spinal cord. Surgery may be necessary to relieve pressure or repair severely damaged vertebrae or tissues. Exercises to strengthen back muscles are ordinarily prescribed after the fracture is healed.

The anterior longitudinal ligament interconnects the vertebral bodies and intervertebral disks along the anterior surface of the vertebral column. This broad ligament helps prevent herniated intervertebral disks from being directed anteriorly. The ligament can actually be used to *splint* a fractured vertebra when the trunk is cast in extension. When a vertebral fracture is suspected, the vertebral column should be kept in extension. Emergency paramedics keep the back in extension when removing a crash victim from a car, since flexion of the vertebral column can cause further injury to the spinal cord.

Herniated Disk

A **herniated disk** (also called ruptured or slipped disk) occurs when the soft, pulpy center (*nucleus pulposus*) of an intervertebral disk protrudes through a weakened or torn surrounding outer ring (*annulus fibrosus*) on the posteriolateral side of the disk. The nucleus pulposus pushes against a spinal nerve, or occasionally, on the spinal cord itself. This produces a continuous pressure on the spinal cord, which may cause permanent injury. Actually, nothing "slips"; the nucleus pulposus pushes out. Herniated disks occur most often in adult males. They may be caused by a straining injury or by degeneration of the intervertebral joint. The lumbar region is usually affected, but herniation may occur anywhere along the spine. Sharp pain usually accompanies a herniated disk, and because roots of spinal nerves can be involved, the pain may radiate beyond the primary low back area to the buttocks, legs, and feet.

Because the posterior longitudinal ligament, located on the posterior aspect of the vertebral bodies and intervertebral disks, is relatively narrow, herniated disks pass on only one side of the ligament. As a result, a herniated disk puts pressure on nerve roots on one side only, producing sciatic nerve pain on only one side of the body, usually in a lower limb.

Treatment usually consists of bed rest, sometimes including traction. Heat applications, a regulated exercise program, and muscle-relaxing or pain-killing drugs may be prescribed. If such traditional treatment is ineffective, surgery is done to remove the protruding nucleus pulposus. To gain access to the hernia, a portion of the vertebral arch (lamina) is removed. Such an operation is called a *laminectomy*.

If a spinal fusion is required, bone chips from the ilium are placed over the laminae. The bone that develops from the chips fuses to form a splint.

Cleft Palate

Cleft palate is the common name for a defect that occurs when the structures that form the palate do not fuse before birth. As a result, there is a gap in the midline of the roof of the mouth. Such a gap creates a continuity between the oral and nasal cavities. When a gap is forward, on the upper lip, the lip is separated, and a "hare lip" ("cleft lip") results. A hare lip never appears at the midline, and may be paired. An infant with a cleft palate may have difficulty suckling. In most cases, the defect can be repaired, at least partially, by surgery.

Microcephalus and Hydrocephalus

Sometimes the calvarial bones, along with the fontanels, close earlier or later than expected. If they fuse too early, brain growth may be retarded by the excessive pressure. This condition is called *microcephalus* (Gr. *mikros*, small + *kephale*, head). This condition can be alleviated by removing the bone and widening the sutures. A contrasting problem is *hydrocephalus* (Gr. *hudor*, water + head), commonly called "water on the brain." It is usually a congenital condition in which an abnormal amount of cerebrospinal fluid accumulates around the brain and within the brain ventricles (cavities), causing an enlargement of the skull and pressure on the brain. Fusion of the skull sutures is delayed by the increased volume of the cranial cavity.

Physicians attend to the problem by inserting replaceable plastic tubes with pressure valves in the brain, which drain away excess fluid into a vein or body cavity. The valve prevents excessive rises in pressure and allows sufficient fluid to be retained. Even with drainage procedures,

mental retardation and vision loss may occur. Without drainage, the condition is usually fatal.

Spina Bifida

Spina bifida (SPY-nuh BIFF-ih-duh; L. *bifidus*, split into two parts), or cleft spine, affects one of every 1000 children. In its severe form, it is the most common crippler of newborns. Spina bifida is a condition in which the neural arches of one or more vertebrae do not close completely during fetal development. In serious cases, the spinal cord and nerves in the area of the defective vertebrae form a fluid-filled sac, called a myelomeningocele, which protrudes through the skin. Because the myelomeningocele is covered by only a thin membrane, the protruding spinal nerves are easily damaged or infected.

Spinal Curvatures

Three abnormal spinal curvatures are kyphosis, lordosis, and scoliosis. *Kyphosis* (Gr. hunchbacked) is a condition where the spine curves backward abnormally, usually at the thoracic level. A characteristic "hunchbacked" or "roundbacked" appearance results. Adolescent kyphosis is the most common form. It generally results from infection or other disturbances of the vertebral epiphysis during the active growth period. Adult kyphosis (hunchback) is generally caused by a degeneration of the intervertebral disks, resulting in collapse of the vertebrae, but many other causes, such as poor posture and tuberculosis of the spine, may be responsible.

Lordosis (Gr. bent backward), also known as "swayback," is an exaggerated forward curvature of the spine in the lumbar area. Among the causes of lordosis are the great muscular strain of advanced pregnancy, an extreme "potbelly" or general obesity that places abnormal strain on the vertebral column, tuberculosis of the spine, rickets, and poor posture.

The most common spinal curvature is *scoliosis* (Gr. crookedness), an abnormal lateral curvature of the spine in the thoracic, lumbar, or thoracolumbar portion of the vertebral column. It is interesting to note that curves in the thoracic area are usually convex to the right, and lumbar deformities are usually convex to the left.

CHAPTER SUMMARY

The skeleton supports the body. It also protects the inner organs and passageways, and it acts as a system of levers that allows us to move. The bones supply reserve calcium and phosphate, and blood cells are produced within the bone marrow.

General Features and Surface Markings of Bones (p. 146)

The markings on the surface of a bone are related to the bone's function. Some important features of bones are **processes,** or outgrowths on bones, such as a condyle, crest, or trochanter; **openings** to bones, such as a foramen or meatus; and **depressions** on bones, such as a fossa, groove, or notch.

Divisions of the Skeleton (p. 149)

1 The skeleton (206 bones) is divided into two major portions: the axial skeleton (80 bones) and the appendicular skeleton (126 bones). They are joined together at the pectoral girdle and pelvic girdle to form the overall skeleton.

2 The **axial skeleton** forms the longitudinal axis of the body. It is made up of the skull, vertebral column, sternum, and ribs.

3 The **appendicular skeleton** is composed of the upper and lower extremities, which include the pectoral and pelvic girdles.

4 Each **upper extremity** consists of the pectoral (shoulder) girdle, upper arm bone, two forearm bones, and the wrist and hand bones. Each **lower extremity** consists of the pelvic (hip) girdle, upper leg (thigh) bone, two lower leg bones, and the ankle and foot bones.

The Skull (p. 150)

1 The **skull** is usually defined as the skeleton of the head, with or without the mandible. The skull can be divided into the **cranial skull** and the **facial skull.** The skull *protects* many structures, including the brain and eyes; *provides points of attachment* for muscles involved in eye movements, chewing, swallowing, and other movements; and *supports* various structures, such as the mouth, pharynx, and larynx.

2 Except for the mandible (and the ear ossicles and hyoid), the skull bones are joined together by **sutures,** seamlike joints that make the bones of an adult skull immovable. The 4 major skull sutures are the *coronal, lambdoidal, sagittal,* and *squamous.*

3 The membrane-covered spaces between incompletely ossified sutures are the four **fontanels;** *anterior, posterior, anterolateral, posteriolateral.*

4 The 8 **cranial bones** are the *frontal,* 2 *parietal, occipital,* 2 *temporal, sphenoid,* and *ethmoid.* The sutural bones are also considered part of the cranium.

5 In the interior of the ethmoid, maxillary, sphenoid, and frontal bones are four pairs of air cavities called **paranasal sinuses.**

6 The 14 **facial bones** are 2 *inferior nasal conchae, vomer,* 2 *palatines,* 2 *maxillae,* 2 *zygomatic,* 2 *nasal,* 2 *lacrimal,* and *mandible.*

The Vertebral Column (p. 165)

1 The **vertebral column,** or spine, is the skeleton of the back. The spine is composed of 26 separate bones called *vertebrae.*

2 The vertebral column protects the spinal cord and nerves, supports the weight of the body, and keeps the body erect. **Intervertebral disks** act as shock absorbers and protect the vertebrae.

3 The adult vertebral column has four *curves,* which provide spring and resiliency.

4 A "typical" vertebra consists of a body, a vertebral (neural) arch, and several processes. The arch and the body meet to form an opening called the **vertebral foramen.** The sequence of foramina forms the **vertebral canal,** which encloses the spinal cord. The processes are attachment sites for muscles and ligaments.

5 The **cervical vertebrae** are the 7 between the skull and the thorax. The **atlas** supports the head and permits the "yes" motion; the **axis** permits the "no" motion. The 12 **thoracic vertebrae** articulate with the ribs. The 5 **lumbar vertebrae** are the largest and strongest vertebrae and provide attachments for lower back muscles. The adult **sacrum,** composed of 5 fused vertebral bodies, supports both the spinal column and the pelvis. The **coccyx** consists of 3 to 5 fused vertebrae.

The Thorax (p. 172)

1 The **thorax,** or chest, is formed by the bodies and intervertebral disks of 12 thoracic vertebrae posteriorly, 12 pairs of ribs, 12 costal cartilages, and the sternum anteriorly.

2 The thoracic cage protects inner organs, provides a point of attachment for some bones and muscles of the upper extremities, and provides a flexible breathing mechanism.

3 The **sternum,** or breastbone, consists of a manubrium, body, and xiphoid process. The articulation of the manubrium with the clavicle is the upper attachment of the axial skeleton to the appendicular skeleton.

4 There are usually 12 pairs of **ribs,** all of which articulate posteriorly with the vertebral column. The *true ribs* (1 to 7) attach to the vertebrae and sternum, but the *false ribs* (8 to 10) attach directly only to the vertebral column. Ribs 11 and 12 are called *floating ribs* because they are not even indirectly attached to the sternum or ribs above.

5 A *typical rib* is composed of a head, neck, and shaft.

STUDY AIDS FOR REVIEW

MEDICAL TERMINOLOGY

ABLATION (L. *ablatus*, removed) The surgical removal of part of a structure, such as part of a bone.

BONE BIOPSY (Gr. *bios*, life + *opsia*, examination) The surgical removal of a small piece of bone for microscopic examination.

BONE-MARROW TEST The withdrawal of bone marrow from the medullary cavity of a bone for microscopic examination.

CERVICAL RIB An overdevelopment of the costal projection of the seventh cervical vertebra. It resembles a rib, and can be a separate bone.

CRANIOTOMY The surgical cutting or removal of part of the cranium.

CREPITATION (L. *crepitare*, to crackle) The grating sound caused by the movement of fractured bones or by other bones rubbing together.

KINESIOLOGY (Gr. *kinema*, motion) The study of movement and the active and passive structures involved.

OSTECTOMY The surgical excision of a bone.

OSTEOCLASIS The surgical refracture of an improperly healed broken bone.

OSTEOPLASTY The surgical reconstruction or repair of a bone.

PELVIMETRY The measurement of the pelvic cavity and birth canal by a physician prior to the birth of a child. The procedure determines if the opening of the mother's lesser pelvis is large enough to allow the passage of the child's head and shoulders.

POTT'S DISEASE Tuberculosis of the spine, which may result in a partial destruction of vertebrae and a spinal curvature.

REDUCTION The nonsurgical manipulation of fractured bones to return (reduce) them to their normal positions.

REPLANTATION The reattachment of a severed limb.

SPINAL FUSION The fusion of two or more vertebrae.

SPONDYLITIS (Gr. *spondulos*, vertebra) Inflammation of one or more vertebrae.

UNDERSTANDING THE FACTS

1 What are the unpaired bones of the skull?
2 What is included in an extremity?
3 What are the two main divisions of the skull?
4 Which fontanel persists longest after birth?
5 Which are the movable bones of the skull?
6 What are the main functions of the spinal column?
7 What is the difference between the pelvic girdle and the pelvis?
8 What is spina bifida?

UNDERSTANDING THE CONCEPTS

1 What are some of the differences in the skeleton of the child and adult? The adult male and female?
2 How is timing important in the development of the skull?
3 In what ways do cervical, thoracic, and lumbar vertebrae differ?
4 What problems would you have if your spinal column lacked its normal curvature?

SELF-QUIZ
Multiple-Choice Questions

7.1 Unlike other vertebrae, thoracic vertebrae have
   a facets for rib   c laminae
   attachment   d transverse processes
   b pedicles   e a and c

7.2 A prominent ridge or border on the surface of a bone is a
   a condyle   d line
   b crest   e trochanter
   c head

7.3 A large, roughly rounded process found only on the femur is the
   a crest   d tuberosity
   b trochanter   e ridge
   c spine

7.4 The hard palate is made up of the
   a lacrimal bone
   b ethmoid bone
   c zygomatic bones
   d maxillae and palatine bones
   e sphenoid bones

7.5 The greater sciatic notch is on the
   a ilium   d hyoid bone
   b pubis   e coccyx
   c ischium

7.6 The major bone at the posterior part of the base of the skull is the
   a sphenoid   d lacrimal
   b occipital   e zygomatic
   c temporal

7.7 Which of the following is *not* a component of the axial skeleton?
   a sacrum   d vertebra
   b patella   e mandible
   c sternum

7.8 All of the following are skull bones except the
   a frontal bone   d temporal bone
   b zygomatic bone   e hyoid bone
   c parietal bone

7.9 Which of the following bones is least involved in protecting the brain?
a frontal   d parietal
b temporal   e occipital
c mandible

7.10 Which of the following skull bones is not paired?
a parietal   d zygomatic
b nasal   e temporal
c frontal

True-False Statements

7.11 An example of a bone process is a tuberosity.

7.12 A condyle is a small flat surface.

7.13 A natural opening into or through a bone is a fissure.

7.14 The sagittal suture is between the parietal and the occipital bones.

7.15 The frontal bone of the skull forms the forehead and the upper part of the eye orbits.

7.16 The occipital bone forms the posterior part of the cranial skull.

Completion Exercises

7.17 Wormian bones are also called _____ bones.

7.18 In the interior of the ethmoid, maxillary, sphenoid, and frontal bones are four pairs of air cavities called _____.

7.19 The bone that is shaped like a plowshare and that forms part of the nasal septum is the _____.

7.20 The upper jaw consists of two _____.

7.21 The skeleton of the lower jaw is the _____.

7.22 The three ear ossicles are the _____, _____, and _____.

7.23 The first cervical vertebra is the _____.

7.24 The _____ consists of three to five fused bones at the end of the vertebral column.

7.25 A human has _____ pairs of ribs.

Problem-Solving/Critical Thinking

7.26 Fill in the blanks: Mr. Ramey and Mr. Deaton were involved in a serious car accident on their way home from a fishing trip. Mr. Ramey had pronounced swelling of the upper right side of his head. X-rays showed a fracture of the _____ bone. X-rays also showed Mr. Deaton fractured his "tail-bone." Correctly, this bone is called the _____.

7.27 The hyoid bone is _____ to the thyroid cartilage.
a distal   d posterior
b superior   e inferior
c anterior

7.28 The most prominent bony landmark on the posterior aspect of the neck is the
a spinous process of the seventh cervical vertebra
b transverse process of the sixth cervical vertebra
c the hyoid bone
d the transverse process of the atlas
e the spinous process of the axis

A SECOND LOOK

1 Identify the bone in the following drawing:

2 Identify the types of fractures in these two drawings:

8

The Appendicular Skeleton

LEARNING OBJECTIVES

1 Identify the bones of the upper extremities, and describe their locations, features, and functions.

2 Identify the bones of the lower extremities, and describe their locations, features, and functions.

3 Explain the major differences between the male and female pelvis.

The ***appendicular skeleton,*** although its name is derived from the Latin word *appendere,* to hang from, should not be thought of as consisting of only the hanging parts — the arms and legs.* In fact, the appendicular skeleton is composed of the *upper extremities,* which include the scapula (shoulder blade) and clavicle (collarbone) of the upper limb or pectoral girdle, in addition to the arms, forearms, and hands, and the *lower extremities,* which include the hipbone of the lower limb or pelvic girdle, as well as the thighs, legs, and feet.

THE UPPER EXTREMITIES (LIMBS)

The skeleton of the ***upper extremity,*** or ***limb,*** consists of 64 bones [FIGURE 8.1]. These include the scapula and clavicle of the shoulder girdle, the humerus of the arm, the radius and ulna of the forearm, the carpals of the wrist, the metacarpals of the palm, and the phalanges of the fingers. The upper extremity is connected to and supported by the axial skeleton by only one joint and many muscles. The joint is the sternoclavicular joint between the manubrium of the sternum and the clavicle. The muscles form a complex of suspension bands from the vertebral column, ribs, and sternum to the pectoral girdle.

Pectoral Girdle

The upper limb girdle is known as the ***pectoral girdle,*** or ***shoulder girdle*** [FIGURE 8.1]. It consists of the clavicle and scapula. The clavicle is a long bone that extends from the sternum to the scapula in front of the thorax. The triangular scapula is located behind the thoracic cage.

The pectoral girdle is designed more for mobility than for stability. It is held in place, and surrounded by, muscles and ligaments. The stability of the shoulder is provided by these muscles and ligaments, rather than by the shape of the bones and joints.

Clavicle The paired ***clavicles*** (L. key), or collarbones, are located at the root of the neck. The clavicle is a horizontal double-curved, long bone, with a rounded medial end and a flattened lateral end [FIGURE 8.2]. The entire length of the clavicle can be felt through the skin. The medial part of the clavicle is curved anteriorly, and the lateral part is curved posteriorly. The medial, or *sternal,* end of the clavicle articulates with the manubrium just above the first rib at the ***sternoclavicular joint,*** connect-

ing the axial and appendicular skeletons. The lateral, or *acromial,* end articulates with the ***acromion process*** of the scapula. This ***acromioclavicular joint*** involves the clavicle in all movements of the scapula.

The clavicle is held in place by strong ligaments at both ends. At the medial end is the ***costal tuberosity,*** where the costoclavicular ligament is attached. At the lateral end is the ***conoid tubercle,*** where the coracoclavicular ligament is attached. The clavicle is also a point of attachment for muscles of the pectoral girdle and neck. Large blood vessels and some nerves for the upper limb pass below the anterior curvature.

The main function of the clavicle is to act as a strut to hold the shoulder joint and arm away from the thoracic cage, allowing the upper limb much freedom of movement. Because of its vulnerable position and relative thinness, the clavicle is broken more often than any other bone in the body, and when it is, the whole shoulder is likely to collapse.

Scapula The ***scapula*** (L. shoulder), or shoulder blade (because of its resemblance to the blade of a shovel), is located on the posterior thoracic wall between the second and seventh ribs.

The flat, triangular *body* of the scapula has an obliquely oriented *spine* that can be felt on the posterior surface [FIGURE 8.3]. The prominent ridge of the spine separates the supraspinous fossa from the infraspinous fossa. The spine and fossae provide attachment sites for muscles that move the arm. The spine ends in the large, flat ***acromion,*** which forms the point of the shoulder. It articulates with the clavicle and is an attachment site for chest and arm muscles. Below the acromion is the shallow ***glenoid fossa,*** or ***cavity,*** which acts as a socket for the head of the humerus. This articulation forms the ***glenohumeral*** (shoulder) ***joint.*** Just medial to the upper end of the glenoid fossa is the anteriorly directed ***coracoid process,*** which serves as an attachment for the coracoclavicular ligament and several muscles of the arm and chest (origins of pectoralis minor, short head of biceps brachii, and coracobrachialis muscles).

Figure 8.3 shows the location of the three borders and two angles of the scapula. The ***medial border*** runs parallel to the vertebral column. The ***superior*** and ***inferior angles*** are located at the ends of the medial border. Extending upward from the inferior angle to the glenoid fossa is the ***lateral border.*** Running horizontally from the glenoid fossa to the superior angle is the ***superior border.***

Bones of the Arm, Forearm, and Hand

The bones of the arm, forearm, and hand are the humerus, ulna, radius, carpals (wrist), metacarpals, and phalanges (fingers).

---

*In terms of the appendicular skeleton, an *extremity* and a *limb* mean the same thing; a limb is not an arm or leg. Remember that the upper limb, or extremity, includes the scapula and clavicle *as well as* the arm, forearm, wrist, and hand.

FIGURE 8.1 THE UPPER EXTREMITY (64 BONES)

The upper extremity includes the shoulder (pectoral) girdle.
[A] Anterior view. **[B]** Posterior view.

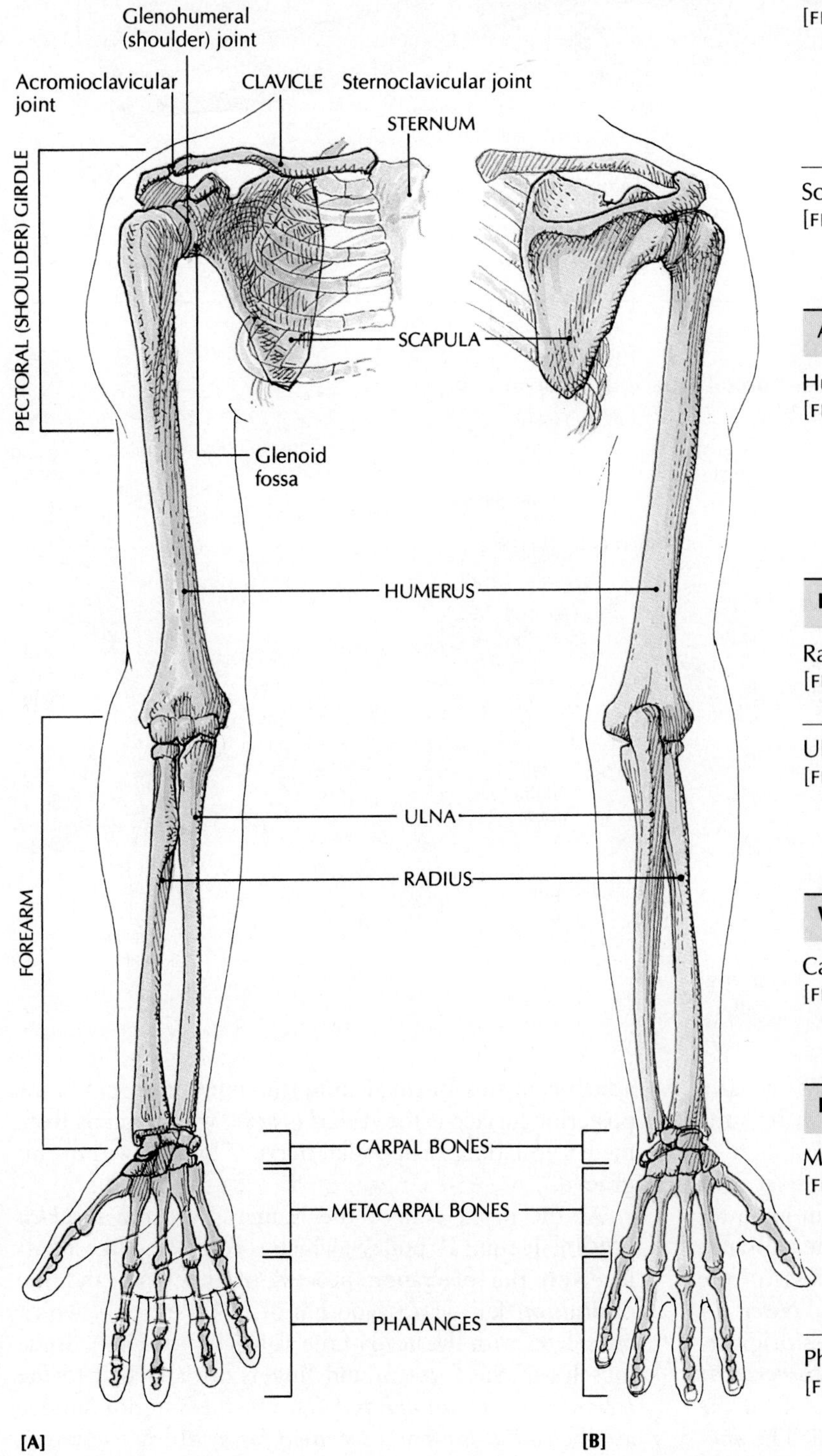

Bones	Description and functions
SHOULDER (PECTORAL) GIRDLE	
Clavicle (2) [FIGURE 8.2]	Collarbone; double-curved, long bone with rounded medial end and flattened lateral end; held in place by ligaments. Holds shoulder joint and arm away from thorax so upper limb can swing freely.
Scapula (2) [FIGURE 8.3]	Shoulder blade; flat, triangular bone with horizontal spine separating fossae. Site of attachment for muscles of arm and chest.
ARM	
Humerus (2) [FIGURE 8.4]	Longest, largest bone of upper limb; forms ball of ball-and-socket joint with glenoid fossa of scapula. Site of attachment for muscles of shoulder and arm, permitting arm to flex and extend at elbow.
FOREARM	
Radius (2) [FIGURE 8.5]	Shorter of two bones in forearm. Allows forearm to rotate in radial motion.
Ulna (2) [FIGURE 8.5]	Larger of two bones in forearm; large proximal end consists of olecranon process (prominence of elbow). Forms hinge joint at elbow.
WRIST	
Carpals (16) [FIGURE 8.6]	Small short bones; in each wrist, 8 carpals in 2 transverse rows of 4. With attached ligaments, allow slight gliding movement.
HANDS AND FINGERS	
Metacarpals (10) [FIGURE 8.6]	Five miniature long bones in each hand in fanlike arrangement; articulate with fingers at metacarpophalangeal joint (the knuckle). Aid opposition movement of thumb; enable cupping of hand.
Phalanges (28) [FIGURE 8.6]	Miniature long bones, 2 in each thumb, 3 in each finger; articulate with each other at interphalangeal joint. Allow fingers to participate in stable grips.

FIGURE 8.2 RIGHT CLAVICLE

Anterior view.

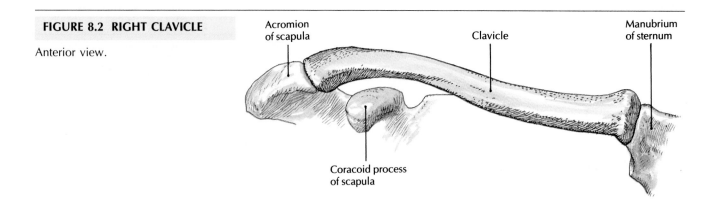

FIGURE 8.3 RIGHT SCAPULA

[A] Anterior view, costal surface. **[B]** Posterior view, dorsal surface. **[C]** Lateral view.

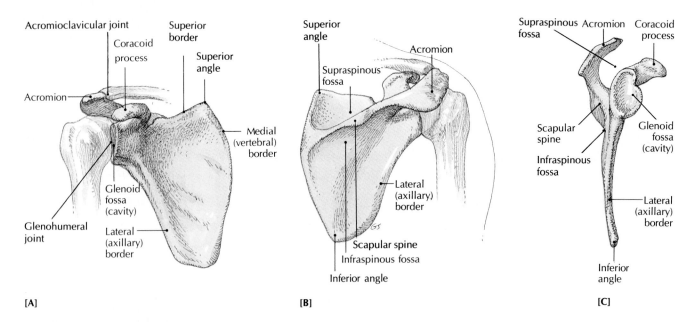

[A]   [B]   [C]

Humerus The *humerus* (L. upper arm) is the arm bone located between the shoulder and elbow. It is the longest and largest bone of the upper limb [FIGURE 8.4].

The *shaft*, or body, of the humerus is cylindrical in its upper half and flattened from front to back in its lower half. The *head* of the humerus is the ball of the ball-and-socket glenohumeral joint with the glenoid fossa of the scapula. Close to the head are the *greater* and *lesser tubercles,* which are attachment sites for muscles originating on the scapula. Passing within the *intertubercular groove* between the two tubercles is the tendon of the long head of the biceps muscle for the forearm. The *anatomical neck* is located between the head and the tubercles. The *surgical neck,* so named because it is the site of frequent fractures of the upper end of the humerus, is located just below the tubercles. The *deltoid tuberosity,* which is halfway down the lateral side of the shaft, is the

attachment site for the deltoid (shoulder) muscle. On the posterior surface is the *radial groove,* which spirals from medial to lateral. The radial nerve is located within this groove.

At the distal end of the humerus are the *trochlea* (TROHK-lee-uh; L. pulley), which is connected like a pulley with the olecranon process of the ulna, and the *capitulum* (kuh-PITCH-yoo-luhm; L. little head), which articulates with the head of the radius [FIGURE 8.4]. Some muscles of the forearm and fingers are attached to the *lateral* and *medial epicondyles.* On the anterior surface are the *radial fossa* and *coronoid fossa,* which accommodate the head of the radius and the coronoid process of the ulna, respectively, when the arm is bent at the elbow. On the posterior surface is the *olecranon fossa,* which accommodates the olecranon process of the ulna when the arm is straightened out.

FIGURE 8.4 RIGHT HUMERUS

[A] Anterior view. [B] Posterior view. [C] Photograph of upper and lower ends of right humerus, anterior view.

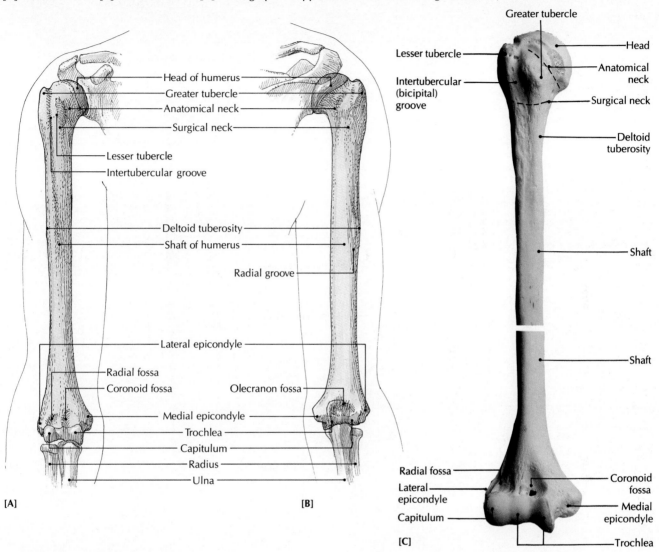

Ulna The *ulna* (L. elbow) is the longer of the two long bones of the forearm, between the elbow and the wrist. It is medially located (on the side of the little finger).

The large proximal end of the ulna consists of the *olecranon process,* which curves upward and forward to form the *semilunar,* or *trochlear, notch* [FIGURE 8.5A]. This half-moon-shaped depression articulates with the trochlea of the humerus to form the hinged elbow joint. The *coronoid process* on the anterior surface of the ulna forms the lower border of the trochlear notch. The *radial notch* on the lateral surface of the coronoid process is the articulation site for the head of the radius. The ulna is narrow at its distal end, with a small head and a short, blunt peg called the *styloid process,* which is attached to a fibrocartilaginous disk that separates the ulna from the carpus.

Radius The *radius* (L. spoke of a wheel) is the long bone that is located lateral to the ulna (on the thumb side).

At the proximal end of the radius, its disk-shaped *head* articulates with the capitulum of the humerus [FIGURE 8.4A] and the radial notch of the ulna [FIGURE 8.5A]. The *radial tuberosity* on the medial side is the attachment site for the biceps brachii muscle. An interosseous membrane connects the shafts of the ulna and radius. The *shaft* becomes broader toward its large distal end, which articulates with two of the lunate carpal bones in the hand [FIGURE 8.6A]. The wrist joint is called the *radiocarpal* joint because it joins the forearm (radius) and wrist (carpus). At the lower end of the radius are the prominent *styloid process* (which may be felt on the outside of the wrist where it joins the hand) and the U-shaped *ulnar notch* into which fits the head of the ulna.

FIGURE 8.5 RIGHT RADIUS AND ULNA

[A] Anterior view. [B] Posterior view. [C] Right lateral view of articulations with the humerus at the elbow.

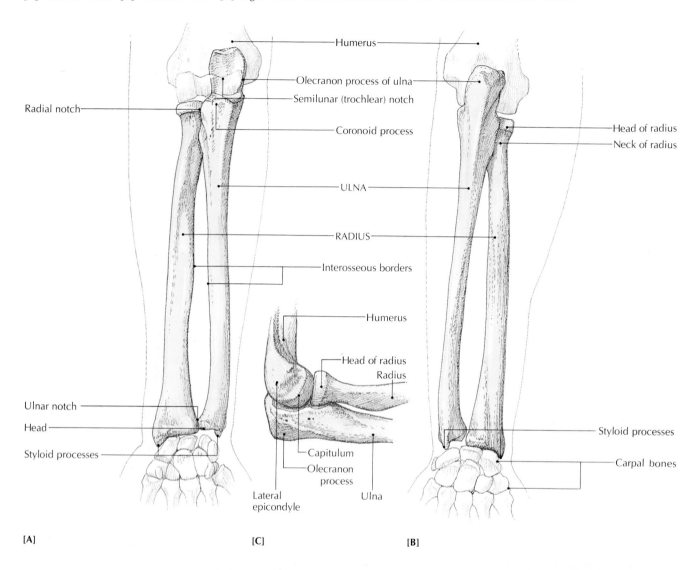

[A]   [C]   [B]

Carpus (wrist) The wrist, or *carpus* (Gr. *karpos,* wrist), is composed of eight short bones, connected to each other by ligaments that restrict their mobility primarily to a gliding movement.* As shown in FIGURE 8.6, the carpals are arranged in two transverse rows of four bones each. In the proximal row, from lateral to medial position, are the *scaphoid, lunate, triquetrum,* and *pisiform.* In the distal row are the *trapezium, trapezoid, capitate,* and *hamate.*

The carpal bones as a unit are shaped so that the back of the carpus is convex, and the palmar side is concave. A

*The wrist is the region between the forearm and the hand (i.e., distal to the forearm). A "wrist" watch is not usually worn around the wrist; instead it encircles the distal end of the forearm, just proximal to the head of the ulna.

connective tissue bridge called the *flexor retinaculum* stretches between the hamate and pisiform and the trapezium and scaphoid bones. This bridge converts the palmar concavity into a tunnel—called the *carpal tunnel.* Nine long tendons and the median nerve pass through the carpal tunnel from the forearm to the hand. (See the discussion of *carpal tunnel syndrome* on page 342.)

The easily felt small pisiform is a clinically useful landmark. It is actually a sesamoid bone within the tendon of the flexor carpi ulnaris muscle.

Metacarpus and phalanges The 5 *metacarpal* bones make up the skeleton of the palm of the hand, or *metacarpus,* and the 14 *phalanges* are the finger bones [FIGURE 8.6].

FIGURE 8.6 BONES OF THE RIGHT HAND

[A] Palmar (ventral, anterior) aspect. [B] Dorsal (posterior) aspect. [C] Photograph of right hand, dorsal aspect.

Key: **S** = Scaphoid   **L** = Lunate   **TRI** = Triquetral   **P** = Pisiform   **TRU** = Trapezium   **TRO** = Trapezoid   **C** = Capitate   **H** = Hamate

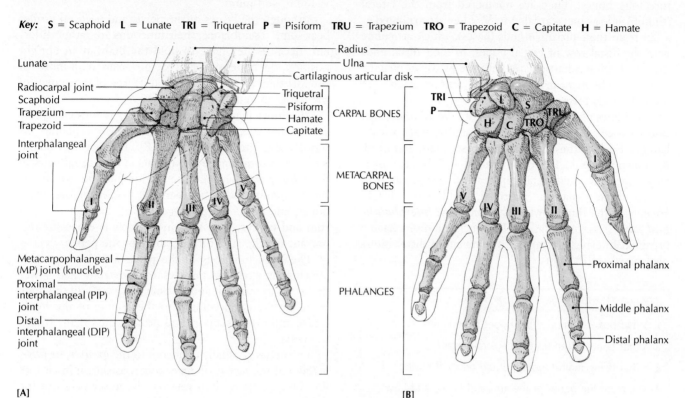

[A]   [B]

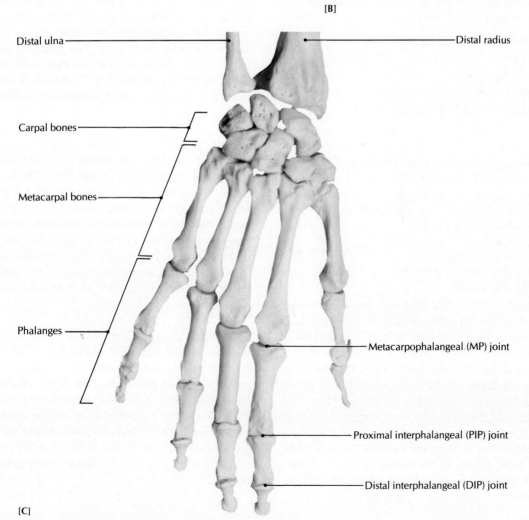

[C]

The *metacarpal* (L. behind the wrist) *bones* are miniature long bones. They are numbered from the lateral (thumb) side as metacarpals I to V. They are arranged as a fan from their proximal ends (bases), which articulate with the distal row of carpal bones, to their distal ends (heads). Each head articulates with the proximal phalanx of a digit. The *metacarpophalangeal joint* (referred to as the MP joint) forms a "knuckle."

The bones of the digits (fingers) are the 14 *phalanges* (fuh-LAN-jeez; Gr. line of soldiers; singular, *phalanx*, FAY-langks). Each of these bones has a base, shaft, and head. As shown in FIGURE 8.6, the thumb (digit I) has two phalanges (proximal and distal), and each finger (digits II to V) has three phalanges (proximal, middle, and distal). Except for the thumb, which has only one *interphalangeal joint* (referred to as the IP joint), the digits have a proximal interphalangeal (PIP) joint and a distal interphalangeal (DIP) joint.

A S K   Y O U R S E L F

1 What bones comprise the upper extremities?

2 What bones make up the pectoral girdle?

3 What is the main function of the pectoral girdle?

4 What are the bones of the arm and hand, and how many of each type of bone are there?

5 What are the functions of the ulna and radius?

THE LOWER EXTREMITIES (LIMBS)

As part of the appendicular system, the lower extremities or limbs consist of 62 bones [FIGURE 8.7]. These include the hipbones of the pelvic girdle, the femur of the thigh, the tibia and fibula of the leg, the tarsal bones of the ankle, the metatarsals of the foot, and the phalanges of the toes.

Pelvic Girdle and Pelvis

The lower limb girdle, called the *pelvic girdle,* is formed by the right and left hipbones, which are also known as the *ossa coxae* (L. hipbones; sing. *os coxa*). The hipbones, sacrum, and coccyx form the pelvis. The paired hipbones are the broadest bones in the body. They are formed in the adult by the fusion of the ilium, ischium, and pubis. These three bones are separate in infants, children, and young adolescents, but generally fuse between the ages of 15 and 17. On the lateral surface of each hipbone is a deep cup, called the *acetabulum* (ass-eh-TAB-yoo-luhm; L. vinegar cup), which is the socket of the ball-and-socket joint with the head (ball) of the femur [FIGURE

8.8B]. The acetabulum is formed by parts of the ilium, ischium, and pubis.

Although the bones of the pelvic and pectoral girdles bear some resemblance, their functions are rather different. Because they have to hold the body in an upright position, the bones of the pelvic girdle are built more for support than for the exceptional degree of movement of the upper extremities.

The bowl-shaped *pelvis* (L. basin) is formed by the sacrum and coccyx posteriorly and the two hipbones anteriorly and laterally [FIGURE 8.8]. The pelvis is bound into a structural unit by ligaments at the lateral pairs of *sacroiliac joints* between the sacrum and the two ilia, at the *symphysis pubis* between the bodies of the pubic bones, and at the *sacrococcygeal joints* between the sacrum and coccyx. The symphysis pubis is especially important for the structural security of the pelvis.

The basic functions of the pelvis are (1) to provide attachment sites for muscles of the trunk and lower limbs, (2) to transmit and transfer the weight of the body from the vertebral column to the femurs of the lower limbs, and (3) to support and protect the organs within the pelvis.

The pelvis is usually divided into the *greater,* or *false, pelvis* and the *lesser,* or *true, pelvis,* as shown in FIGURE 8.8. Obstetricians often refer to the lesser pelvis in the female as the "obstetric pelvis" because it is the critical region during childbirth, providing the opening through which the baby must pass. The junction between two pelves is the *pelvic brim,* which surrounds the superior pelvic aperture (pelvic inlet). This brim is the bony ring extending from the *sacral promontory* to the top of the symphysis pubis.

Ilium The *ilium* (L. flank) is the largest of the three fused hipbones [FIGURES 8.8 and 8.9]. It forms the easily felt lateral prominence of the hip. (Think of the *l* in ilium; it will help you to remember that it occupies the *l*ateral position in the ossa coxae.) On its superior border is the *iliac crest,* which ends anteriorly as the *anterior superior iliac spine.* On the medial posterior part of the ilium is the *auricular surface* that articulates with the sacrum. Below this auricular surface is the *greater sciatic notch.* The internal surface of the ilium is the concave *iliac fossa.*

Ischium The *ischium* (IHSS-kee-uhm; L. hip joint) is the lowest and strongest bone of the ossa coxae. It is formed by the lower lateral portion of the acetabulum and the *ischial tuberosity,* the bony prominence that bears the weight of the body when we are seated [FIGURE 8.9]. Extending from the body of the ischium are the slender *superior* and *inferior rami.* Above the body is the *ischial spine,* which is located between the greater sciatic notch of the ilium and the *lesser sciatic notch* of the ischium.

FIGURE 8.7 THE LOWER EXTREMITY (62 BONES)

[A] Anterior view. [B] Posterior view. Note that the pelvic (hip) girdle is part of the lower extremity.

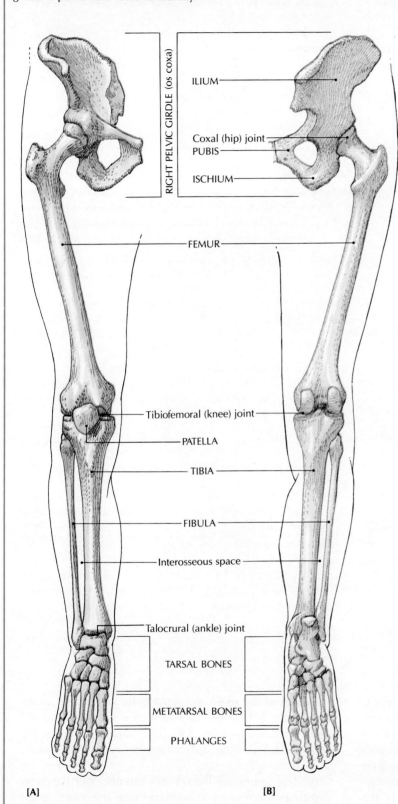

RIGHT PELVIC GIRDLE (os coxa)

ILIUM

Coxal (hip) joint
PUBIS
ISCHIUM

FEMUR

Tibiofemoral (knee) joint

PATELLA

TIBIA

FIBULA

Interosseous space

Talocrural (ankle) joint

TARSAL BONES

METATARSAL BONES

PHALANGES

[A] [B]

Bone	Description and functions
PELVIC GIRDLE	
Hipbone (os coxa) (2) [FIGURE 8.9]	Irregular bone formed by fusion of ilium, ischium, pubis; with sacrum and coccyx forms pelvis; forms socket of ball-and-socket joint with femur. Site of attachment for trunk and lower limb muscles; transmits body weight to femur.
THIGH	
Femur (2) [FIGURE 8.10]	Thighbone; typical long bone; longest, strongest, heaviest bone; forms ball of ball-and-socket joint with pelvic bones; provides articular surface for knee. Supports body.
Patella (2) [FIGURE 8.11]	Kneecap; sesamoid bone within quadriceps femoris tendon. Increases leverage for quadriceps muscle by keeping tendon away from axis of rotation.
LEG	
Fibula (2) [FIGURE 8.12]	Smaller long bone of lower leg; articulates proximally with tibia and distally with talus. Bears little body weight, but gives strength to ankle joint.
Tibia (2) [FIGURE 8.12]	Larger long bone of lower leg; articulates with femur, fibula, talus. Supports body weight, transmitting it from femur to talus.
ANKLE	
Tarsals (14) [FIGURE 8.13]	Ankle, heelbones; short bones; 7 in each ankle including talus, calcaneus, cuboid, navicular, 3 cuneiforms; with metatarsals, form arches of foot. Bear body weight; raise body and transmit thrust during running and walking.
FOOT AND TOES	
Metatarsals (10) [FIGURE 8.13]	Miniature long bones; 5 in each foot; form sole; with tarsals, form arches of feet. Improve stability while standing; absorb shocks; bear weight; aid in locomotion.
Phalanges (28) [FIGURE 8.13]	Toes; miniature long bones; 2 in each big toe, 3 in each other toe; arranged as in hand. Provide stability during locomotion.

FIGURE 8.8 MALE AND FEMALE PELVES

[A] Male pelvis; anterior view. **[B]** Female pelvis; anterior view. Note the wider pelvic aperture (inlet of true pelvis) in the female.

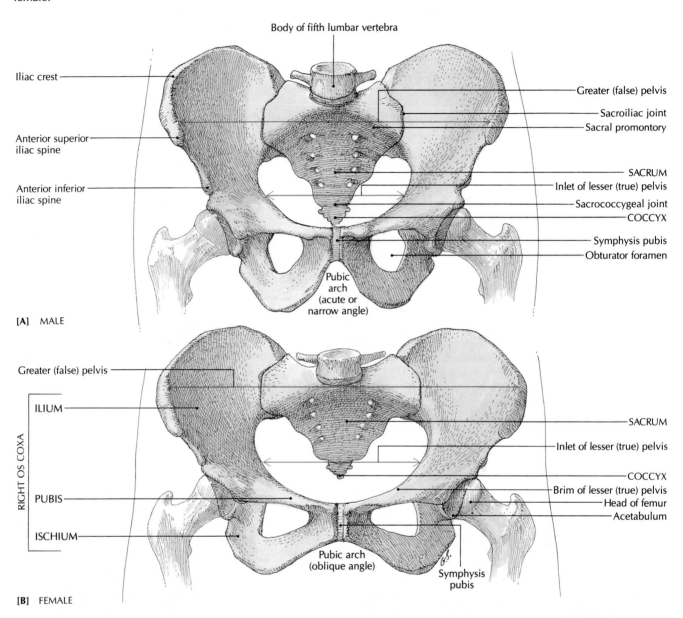

Body of fifth lumbar vertebra

Iliac crest

Anterior superior iliac spine

Anterior inferior iliac spine

Greater (false) pelvis
Sacroiliac joint
Sacral promontory

SACRUM
Inlet of lesser (true) pelvis
Sacrococcygeal joint
COCCYX

Symphysis pubis
Obturator foramen

Pubic arch (acute or narrow angle)

[A] MALE

Greater (false) pelvis

RIGHT OS COXA

ILIUM

PUBIS

ISCHIUM

SACRUM
Inlet of lesser (true) pelvis
COCCYX
Brim of lesser (true) pelvis
Head of femur
Acetabulum

Pubic arch (oblique angle)

Symphysis pubis

[B] FEMALE

Pubis The bilateral body of the *pubis* (L. *pubes*, adult) is joined together in front to form the *symphysis pubis* [FIGURES 8.8 and 8.9]. Extending from the body of the pubis are the *superior* and *inferior rami,* which join with the ilium and ischium. The *pubic tubercle* is located on the body of the pubis. A large opening called the *obturator foramen* is bounded by the rami and bodies of the pubis and ischium. Nerves and blood vessels pass through this foramen into the thigh.

Pelves of the male and female Because of the structure of the female pelvis, a woman is able to carry and deliver a child [FIGURE 8.8]. The female pelvis usually shows the following differences from the male pelvis:

1 The bones are lighter and thinner, and the bony markings are less prominent because the muscles are smaller.

2 The sacrum is less curved and is set more horizon-

FIGURE 8.9 RIGHT OS COXA (HIPBONE)

[A] Right medial view. The borders of the different colored areas illustrate the lines of fusion of sutures at the junctions of the pubis, ischium, and ilium. [B] Right lateral view.

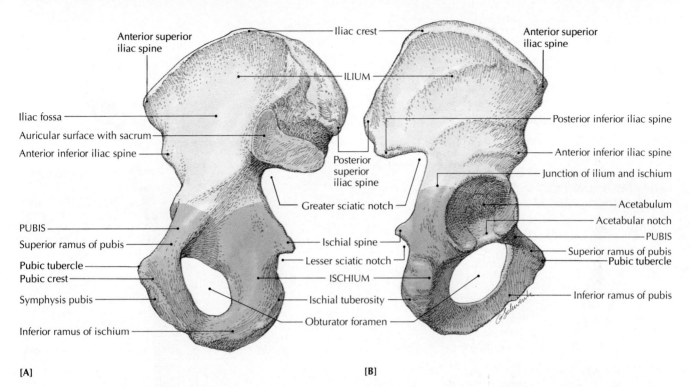

[A]

[B]

tally, which increases the distance between the coccyx and the symphysis pubis, and makes the sacrum broader. The pubic rami are longer, and the ischial tuberosities are set further apart and turned outward. As a result, the *pubic arch* has a wider angle. (This angle is the easiest criterion for distinguishing male from female skeletons.) The combination of all these features creates wider and shallower hips.

3 The pelvis has larger openings. The true pelvis and these openings surround and define the size of the birth canal.

Bones of the Thigh, Leg, and Foot

Completing our study of the lower extremities are the bones of the thigh, leg, and foot, including the femur, patella, tibia, fibula, tarsal bones, metatarsal bones, and phalanges.

Femur The *femur* (FEE-mur; L. thigh), or thighbone, is located between the hip and the knee. It is the strongest, heaviest, and longest bone in the body. (A person's height is usually about four times the length of the femur.) This strong bone plays an important part in supporting the body, and provides mobility via the hip and

knee joints. It can withstand a pressure of 3500 kg/sq cm (1200 lb/sq in.), more than enough to cope with the pressures involved in normal walking, running, or jumping.

The proximal end of the femur consists of a head, neck, and greater and lesser trochanters [FIGURE 8.10]. The *head*, which forms slightly more than half a sphere, articulates with the acetabulum (socket) of the hipbone. In the center of the head is a small depression called the *fovea capitis*, where the ligament and a blood vessel of the head of the femur are attached. If the blood vessel is ruptured due to trauma, the head of the femur may deteriorate. The thin *neck*, connecting the head to the shaft, is a common site for fractures of the femur, especially in elderly people. At the junction between the head and neck, laterally the *greater trochanter* (Gr. "to run") and medially the *lesser trochanter* are the sites of attachment for some large thigh and buttock muscles.

The *shaft*, or body, of the femur is slightly bowed anteriorly. It is fairly smooth, except for a longitudinal posterior ridge called the *linea aspera* (L. rough line), which provides attachment sites for muscles. At the end of the femur toward the knee are *medial* and *lateral condyles*, which articulate with the tibia. Above the condyles are *medial* and *lateral epicondyles*. Epicondyles are attachment sites for muscles, and condyles function in the

FIGURE 8.10 RIGHT FEMUR

[A] Anterior view. The fovea capitis is the site for the attachment of the ligament of the head of the femur [see FIGURE 9.5].
[B] Photograph of upper and lower ends of right femur, posterior view.

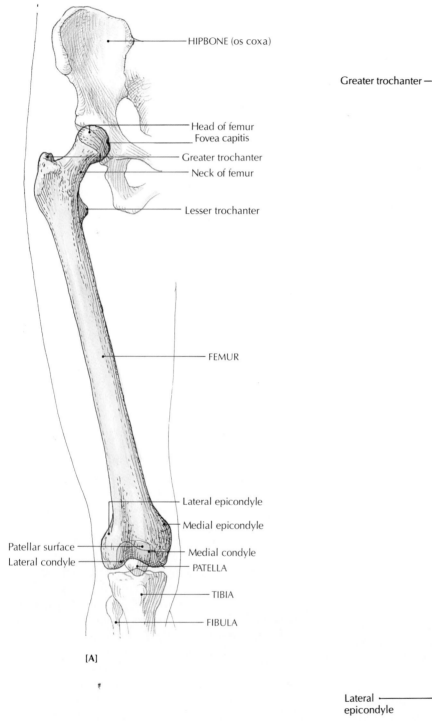

HIPBONE (os coxa)

Head of femur
Fovea capitis
Greater trochanter
Neck of femur

Lesser trochanter

FEMUR

Lateral epicondyle
Medial epicondyle
Patellar surface
Medial condyle
Lateral condyle
PATELLA
TIBIA
FIBULA

[A]

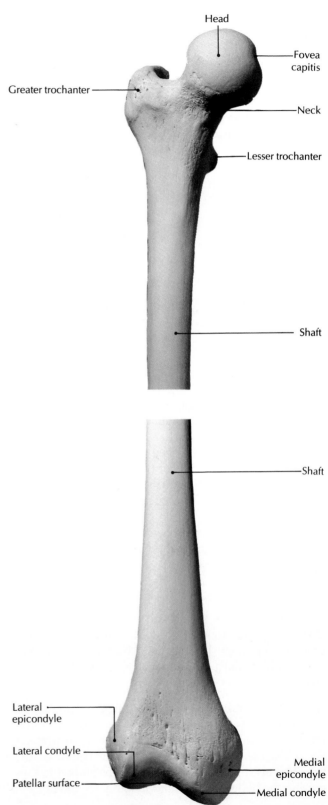

Head
Fovea capitis
Greater trochanter
Neck
Lesser trochanter

Shaft

Shaft

Lateral epicondyle
Lateral condyle
Patellar surface
Medial epicondyle
Medial condyle

[B]

movement of joints. The medial condyle is larger than the lateral condyle, so that when the knee is planted during walking, the femur rotates medially to "lock" the knee. Between the condyles on the anterior surface is a slight groove that separates the articular surface from the *patellar surface.* When the leg is bent or extended, the patella (kneecap) slides along this groove (patellar surface).

Patella The *patella* (L. little plate), or kneecap, is located within the quadriceps femoris tendon. This is the tendon of the muscle that extends the leg from the knee. Facets on the deep surface of the patella fit into the groove between the condyles of the femur.

The patella is the largest sesamoid bone in the body. It protects the knee, but more importantly, it increases the leverage for the action of the quadriceps femoris muscle by keeping its tendon further away from the axis of rotation of the knee. The slightly pointed *apex* of the patella lies at the inferior end, and the rounded *base* is at the superior border [FIGURE 8.11].

Tibia The *tibia* (L. pipe), or shinbone, is located on the anterior and medial side of the leg, between the knee and the ankle. It is the second longest and heaviest bone in the body (the femur is first). This bone supports the body weight, transmitting it from the femur to the talus bone at the ankle joint.

At the proximal end of the tibia the *medial* and *lateral condyles* articulate with the condyles of the femur at the knee joint [FIGURE 8.12]. On the proximal anterior surface is the prominent *tibial tuberosity,* where the patellar tendon is attached.

At the distal end of the tibia is the *medial malleolus* (muh-LEE-oh-luhss), which articulates medially with the head of the talus. The junction of the talus and medial malleolus forms the easily felt prominence on the medial side of the ankle.

The tibia and fibula are attached throughout their lengths by an *interosseous membrane.* They articulate at both the proximal and the distal tibiofibular joints. The site of the distal articulation is the *fibular notch* on the tibia.

Fibula The *fibula* (L. pin or brooch) is a long, slender bone parallel and lateral to the tibia [FIGURE 8.12]. It probably is so named because together with the tibia it somewhat resembles the clasp of a pin. The head at its proximal end articulates with the lateral condyle of the tibia, but not with the femur. It articulates distally with the talus.

The slender shaft of the fibula bears little, if any, body weight, and it is not involved in the knee joint. However, the security of the ankle joint depends largely upon the seemingly delicate fibula.

The medial malleolus of the tibia and the prominent *lateral malleolus* of the fibula both articulate tightly with the head of the talus. However, the fibula and talus are even more firmly bound together by ligaments to form a *mortise* (socket), which strengthens the ankle joint *(talocrural joint)* but limits the movement there to bending the foot up or down. Because the head of the talus is slightly wider anteriorly, and the mortise is at the widest part of the talus, the ankle is most stable when the foot is dorsiflexed (as when a skier is crouching down). When the foot is extended, as when standing on tiptoe, the ankle joint is less stable because the mortise is at a narrower part of the talus.

Tarsus The *tarsus* is composed of the seven proximally located tarsal bones of the foot. They are classified as short bones.

The foot, unlike the hand, has relatively little free movement between its bones. The fingers, with their manipulative and gripping roles, are the functionally dominant structures in the hand. However, it is not the toes, but the tarsal and metatarsal bones, with their weight-bearing and locomotive roles, that are the functionally significant structures of the foot. The bones work together as a lever, helping to raise the body and to transmit the thrust during walking and running.

The tarsal bones are the talus, calcaneus, cuboid, navicular, and three cuneiforms [FIGURE 8.13].

1 The *talus* (TAY-luhss; L. ankle), or anklebone, is the central and highest foot bone. It articulates with the tibia and fibula to form the ankle joint. Together with the calcaneus it receives the weight of the body.

2 The *calcaneus* (kal-KAY-nee-us; L. heel) is commonly called the heelbone. It is the largest tarsal bone and is suited to supporting weight and adjusting to irregularities of the ground. It acts as a lever, providing a site of

---

FIGURE 8.11 RIGHT PATELLA

[A] Anterior view. [B] Posterior view.

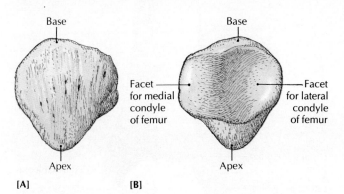

Base

Base

Facet for medial condyle of femur

Facet for lateral condyle of femur

Apex

Apex

[A]

[B]

FIGURE 8.12 RIGHT TIBIA AND FIBULA

[A] Anterior view. [B] Photograph of upper and lower ends of right tibia and fibula, posterior view.

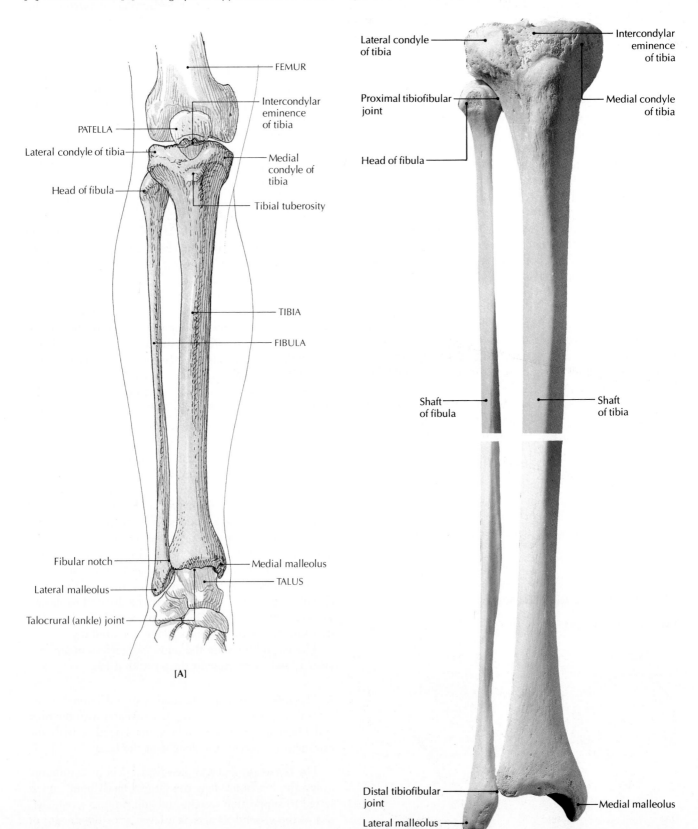

[A]

FEMUR
Intercondylar eminence of tibia
PATELLA
Lateral condyle of tibia
Medial condyle of tibia
Head of fibula
Tibial tuberosity
TIBIA
FIBULA
Fibular notch
Medial malleolus
Lateral malleolus
TALUS
Talocrural (ankle) joint

[B]

Lateral condyle of tibia
Intercondylar eminence of tibia
Proximal tibiofibular joint
Medial condyle of tibia
Head of fibula
Shaft of fibula
Shaft of tibia
Distal tibiofibular joint
Lateral malleolus
Medial malleolus

FIGURE 8.13 BONES OF THE RIGHT FOOT

[A] Superior view. [B] Photograph of bones of right foot, superior view. [C] Right medial view. [D] Lateral view.

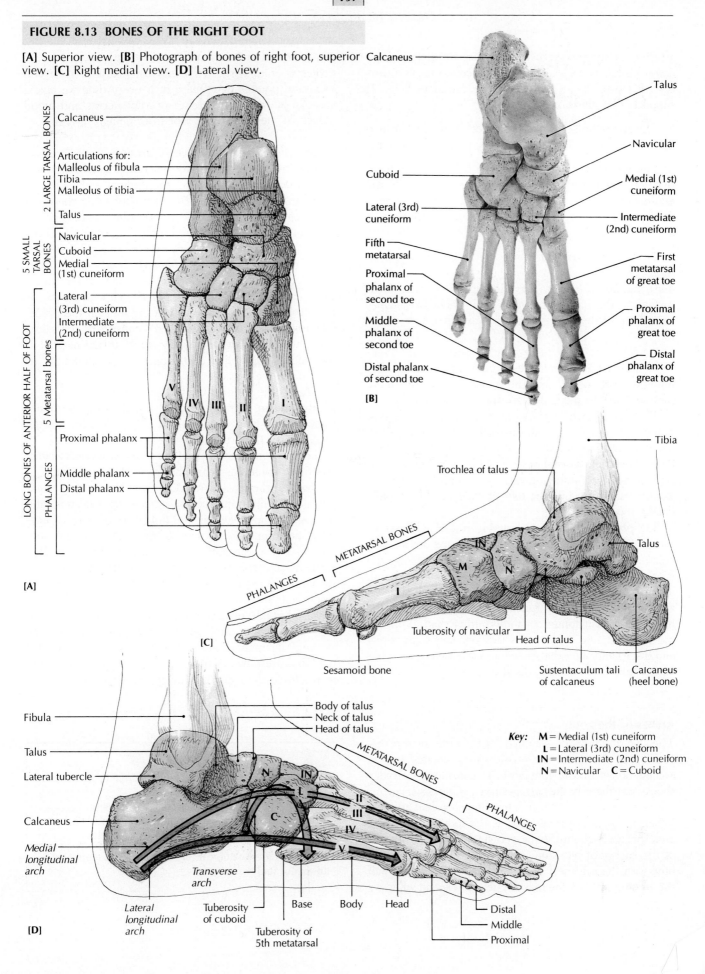

[A]

2 LARGE TARSAL BONES
- Calcaneus
- Articulations for:
  - Malleolus of fibula
  - Tibia
  - Malleolus of tibia
- Talus

5 SMALL TARSAL BONES
- Navicular
- Cuboid
- Medial (1st) cuneiform
- Lateral (3rd) cuneiform
- Intermediate (2nd) cuneiform

LONG BONES OF ANTERIOR HALF OF FOOT
- 5 Metatarsal bones
  - V IV III II I
- PHALANGES
  - Proximal phalanx
  - Middle phalanx
  - Distal phalanx

[B]

- Calcaneus
- Talus
- Cuboid
- Navicular
- Lateral (3rd) cuneiform
- Medial (1st) cuneiform
- Intermediate (2nd) cuneiform
- Fifth metatarsal
- First metatarsal of great toe
- Proximal phalanx of second toe
- Proximal phalanx of great toe
- Middle phalanx of second toe
- Distal phalanx of great toe
- Distal phalanx of second toe

[C]

- Trochlea of talus
- Tibia
- METATARSAL BONES
- Talus
- PHALANGES
- I
- IN
- M
- N
- Sesamoid bone
- Tuberosity of navicular
- Head of talus
- Sustentaculum tali of calcaneus
- Calcaneus (heel bone)

[D]

- Fibula
- Talus
- Lateral tubercle
- Calcaneus
- Medial longitudinal arch
- Body of talus
- Neck of talus
- Head of talus
- N
- IN
- L
- METATARSAL BONES
- I II III IV V
- PHALANGES
- Lateral longitudinal arch
- Transverse arch
- Tuberosity of cuboid
- Tuberosity of 5th metatarsal
- Base
- Body
- Head
- Distal
- Middle
- Proximal

Key: M = Medial (1st) cuneiform
L = Lateral (3rd) cuneiform
IN = Intermediate (2nd) cuneiform
N = Navicular C = Cuboid

attachment for the large calf muscles through the *calcaneal* (Achilles) *tendon.* The sustentaculum tali is a process that projects medially from the calcaneus bone. It supports the talus bone.

3 The *cuboid* bone is usually, but not necessarily, shaped like a cube. It is the most laterally placed tarsal bone.

4 The *navicular* (L. little ship) is a flattened oval bone that some people think looks like a boat because of the depression on its posterior surface that houses the head of the talus. The three cuneiform bones line up on its anterior surface.

5 The three *cuneiform* bones (KYOO-nee-uh-form; L. wedge) are referred to as *medial* (first), *intermediate* (second), and *lateral* (third). They line up along the anterior surface of the navicular bone. Their wedgelike shapes contribute to the structure of the transverse arch of the foot.

Metatarsus and phalanges The 5 metatarsal (L. behind the ankle) bones form the skeleton of the sole of the foot, and the 14 phalanges are the toe bones.

The *metatarsus* consists of five metatarsal bones, numbered I to V from the medial side [FIGURE 8.13]. They are miniature long bones consisting of a proximal base, a shaft, and a distal head. The tarsus and metatarsus are arranged as arches, primarily to improve stability during standing, while the toes provide a stable support during locomotion.

The *phalanges* (fuh-LAN-jeez; Gr. line of soldiers), or toes, like the metatarsals, are miniature long bones, with a base, shaft, and head. There are 14 phalanges in each foot, arranged similarly to the phalanges of the hand. The phalanges of the foot, however, are much shorter, and are quite different functionally from those in the hand. They contribute to stability rather than to precise movements. The big toe has two phalanges (proximal and distal) and the other four toes each have three phalanges (proximal, middle, distal).

Arches of the Foot

The sole of your foot is arched for the same reason that your spine is curved: the elastic spring created by the arched bones of the foot absorbs enormous everyday shocks just the way the spring-curved spine does. A running step may flatten the arch by as much as half an inch. When the pressure is released as the foot is raised, the arch springs back into shape, returning about 17 percent of the energy of one step to the next. The arches combine with the calcaneal tendons to reduce by about half the amount of work the muscles need to expend when we

run. Without arches, our feet would be unable to move properly or to cushion the normal pressure of several thousand pounds per square inch every time we take a step. In addition, the arches prevent nerves and blood vessels in the sole of the foot from being crushed.

Longitudinal and transverse arches There are two longitudinal arches of the foot: (1) the medial longitudinal arch, and (2) the lateral longitudinal arch. The so-called *transverse arch* is located roughly along the distal tarsal bone and the tarsometatarsal joints. It runs across the foot between the heel and the ball of the foot. Technically, it is not an arch because only one side contacts the ground. The high *medial longitudinal arch* consists, in order, of the calcaneus, talus, navicular, three cuneiforms, and metatarsals I, II, and III [FIGURE 8.13D]. It runs from the heel to the ball of the foot on the inside. The low *lateral longitudinal arch* consists of the calcaneus, cuboid, and metatarsals IV and V. It runs parallel to the medial longitudinal arch from the heel to the ball of the foot on the outside.

The arches are held in place primarily by strong ligaments, and in part by attached muscles and the bones themselves. The ligaments are largely responsible for the resiliency of the arches.

The *plantar* (underside of the foot) *calcaneonavicular ligament,* extending from the front of the calcaneus to the back of the navicular, is known as the "spring" ligament. It is important because it keeps the calcaneus and navicular bones together and holds the talus in its keystone position. The security of the high arch and, in a way, of the whole foot, depends on this ligament. If it weakens, and the calcaneus and navicular spread apart, the talus will sag and occupy the space between them; the result is a "flat foot" ("fallen arch"). The tendon of the powerful tibialis posterior muscle comes from the medial side of the foot and passes immediately below the "spring" ligament on its way to its lateral insertion. This tendon is presumed to act as an additional spring, helping to reinforce the ligament during excessive strain.

ASK YOURSELF

1 What bones compose the lower extremities?

2 What bones compose the pelvic girdle?

3 Why are male and female pelves different? What are the differences between male and female pelves?

4 What is the function of the arches of the foot? Name the major arches.

WHEN THINGS GO WRONG

Bunion

A **bunion** (Old Fr., *buigne*, bump on the head) is a lateral deviation of the big toe toward the second toe, accompanied by the formation of a swollen and inflamed bursa and callus at the bony prominence of the first metatarsal. It may be caused by poorly fitted shoes that compress the toes. Bunions are most common among women.

Shin Splints

Some joggers and long-distance runners suffer from **shin splints,** a painful condition of the anterior part of the tibia. The pain is caused when the sheaths of swollen leg muscles block the normal blood circulation of the muscles, which in turn causes more swelling and pain. This condition typically afflicts poorly conditioned people who overuse their muscles and trained athletes who do not warm up properly or go too far beyond their usual limits. In some individuals, shin splints may be caused by tendinitis of the tibialis posterior muscle. In these cases, the resulting imbalance may be corrected with different running shoes or corrective shoe supports (orthotics).

CHAPTER SUMMARY

The Upper Extremities (Limbs) (p. 184)

1 Each **upper extremity,** or **limb,** of the appendicular skeleton includes the scapula and clavicle of the pectoral girdle, humerus of the arm, radius and ulna of the forearm, carpal bones of the wrist, metacarpals of the palm, and phalanges of the fingers. Some functions of the upper extremity include *balancing* while the body is moving, *grasping* of objects, and the *manipulation* of objects.

2 The **pectoral girdle,** or **shoulder girdle,** consists of the clavicle and scapula.

3 The (upper) arm bone is the humerus, and the forearm bones are the ulna and radius. In each wrist are 8 carpals, in each palm 5 metacarpals, in each thumb 2 phalanges, and in each finger 3 phalanges.

The Lower Extremities (Limbs) (p. 190)

1 The skeleton of the **lower extremity** consists of the hipbones of the pelvic girdle, the femur of the thigh, the tibia and fibula of the leg, the tarsal bones of the ankle, the metatarsals of the foot, and the phalanges of the toes. Among the functions of the lower extremity are *movement,* such as in walking, and *balancing,* such as in standing.

2 The **pelvic girdle** is formed by the paired hipbones (ossa coxae), which help hold the body in an upright position. The *pelvis* is formed by the sacrum, coccyx, and hipbones.

3 The **hipbones** are formed in the adult by the fusion of the ilium, ischium, and pubis.

4 The female pelvis is lighter and wider than the male pelvis to enable a woman to carry and deliver a child.

5 The bones of the legs and feet are the femur (2), patella (2), tibia (2), fibula (2), tarsus (14), metatarsus (10), and phalanges (28).

6 The **arches** of the foot provide strength and resiliency. The two true arches are the *medial longitudinal arch* and the *lateral longitudinal arch.* The so-called *transverse arch* contacts the ground on only one side.

MEDICAL TERMINOLOGY

BUNION An inflamed protrusion of the medial side of the joint of the big toe.

COLLES' FRACTURE A specialized type of fracture that occurs at the distal end of the radius. It often occurs when a person extends the hands to break a fall.

GENU VALGUM (L. *genu*, knee + *valgus*, bowlegged) Bowleggedness; a deformity typical of rickets.

GENU VARUM (L. *varus*, bent inward) Knock-kneed.

POTT'S FRACTURE A fracture and dislocation of the distal fibula. It is usually caused by the forceful turning outward of the foot, which destabilizes the ankle joint.

SYNDACTYLISM (Gr. *syn*, together + *daktulos*, finger) The whole or partial fusion of two or more fingers or toes.

UNDERSTANDING THE FACTS

1 Name the parts of the appendicular skeleton.

2 Name the bones of the upper extremity. Of the lower extremity.

3 Describe the structure and functions of the pectoral girdle.

4 How do the phalanges of the hand differ from the phalanges of the feet in structure? In function?

5 Describe the structure and functions of the pelvic girdle.

6 Identify each of the following:
 a pectoral girdle
 b sternoclavicular joint
 c scapula
 d glenohumeral joint
 e olecranon process
 f radial tuberosity
 g flexor retinaculum
 h ossa coxae
 i acetabulum
 j symphysis pubis

7 The bone that holds the shoulder and arm away from the thorax, enabling the arm to swing freely, is the
 a carpal
 b metacarpal
 c humerus
 d clavicle

8 The bone that enables the forearm to rotate is the
 a humerus
 b radius
 c ulna
 d clavicle
 e scapula

UNDERSTANDING THE CONCEPTS

1 How are the extremities attached to the axial portion of the body?

2 Describe how the weight of the upper portion of the body is transmitted to the feet when the body is in a standing position.

3 How can an anthropologist determine whether a skeleton is that of a male or female?

SELF-QUIZ
Multiple-Choice Questions

8.1 Which of the following is *not* part of the appendicular skeleton?
 a scapula
 b clavicle
 c radius
 d ribs
 e tibia

8.2 The deltoid tuberosity, radial groove, and intertubercular groove are structural features of the
 a ulna
 b humerus
 c femur
 d tibia
 e scapula

8.3 The nonweight-bearing bone of the lower extremity is the
 a tibia
 b fibula
 c talus
 d navicular
 e calcaneus

8.4 The cuneiforms are
 a metatarsal bones
 b phalanges
 c pelvic bones
 d carpal bones
 e tarsal bones

8.5 The metacarpophalangeal joint forms the
 a shoulder
 b knuckle
 c ankle
 d wrist
 e symphysis pubis

8.6 Which of the following is *not* a tarsal bone?
 a navicular
 b calcaneus
 c cuboid
 d talus
 e coccyx

8.7 The fibula is lateral to the
   a tibia   **c** femur   **e** talus
   b ulna   **d** radius

8.8 The heelbone is the
   a fibula   **c** navicular   **e** lunate
   b talus   **d** calcaneus

True-False Statements

8.9 Anatomically, the "arm" extends between the shoulder and the elbow, and the leg extends between the knee and ankle.

8.10 The carpal bones form part of the ankle.

8.11 The phalanges are the bones of the fingers.

8.12 The ossa coxae are the hipbones.

8.13 The symphysis pubis is the joint between the sacrum and the ilium.

8.14 The head of the femur forms a ball-and-socket joint with the acetabulum of the hipbone.

8.15 The patella is located within the tendon of the quadriceps femoris muscle.

Completion Exercises

8.16 The bone extending between the shoulder and the elbow is the _____.

8.17 The shinbone is the _____.

8.18 The medial malleolus is part of the _____.

8.19 The bones of the wrist are the _____ bones.

8.20 The largest sesamoid bone in the body is the _____.

8.21 The tibia and fibula are attached along their lengths by a(n) _____.

8.22 The opening in the pelvis through which a baby passes during birth is called the _____.

8.23 The hipbones are formed by the fusion of the _____, _____, and _____.

8.24 The pelvis is formed by the _____, _____, and _____.

8.25 Muscles of the thigh and buttocks attach to the trochanters of the _____.

A SECOND LOOK

In the following drawings, identify the radius, ulna, tibia, and fibula.

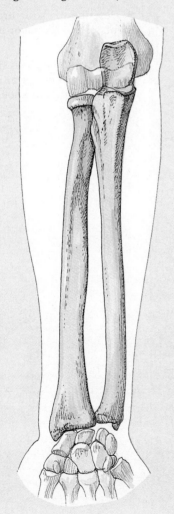

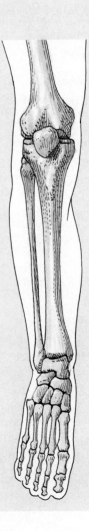

9
Articulations

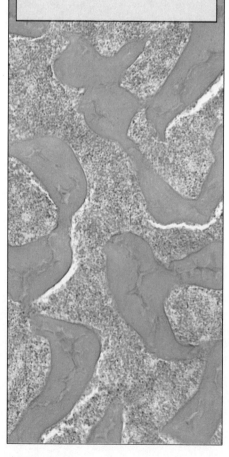

LEARNING OBJECTIVES

1 Define articulation, and explain the basis for the two principal methods of classifying articulations.

2 Describe the three types of fibrous joints, and give an example of each.

3 Describe the two types of cartilaginous joints, and give an example of each.

4 Describe the four basic structural features of a synovial joint and their functions.

5 Give the functions of bursae and tendon sheaths.

6 Explain those factors that inhibit movement at a synovial joint.

7 Define the basic terms used in describing movements that take place at synovial joints.

8 Differentiate among uniaxial, biaxial, and multiaxial joints.

9 Describe the structure and movements of the six major types of synovial joints.

10 Describe the basic structure and possible movements of the jaw, shoulder, hip, and knee joints.

Bones give the body its structural framework, and muscles give it its power, but movable joints provide the mechanism that allows the body to move. A joint, or *articulation* (L. *articulus*, small joint), is the place where two adjacent bones or cartilages meet, or where adjacent bones and cartilages are joined. Although most joints are movable, some are not. The effectiveness of the articular system involves the exquisite coordination of the nervous, muscular, and skeletal systems. To fully appreciate the usefulness of movable joints, try to walk or sit without bending your knees, or eat dinner without bending your elbow or wrist.

CLASSIFICATION OF JOINTS

Joints are classified by two methods. One way to classify joints is by the extent of their *function*, that is, their *degree of movement*. According to this system, a joint may be immovable, slightly movable, or freely movable. An immovable joint, called a *synarthrosis* (Gr. *syn.*, together + *arthrosis*, articulation), is an articulation in which the bones are rigidly joined together. A slightly movable joint, called an *amphiarthrosis* (Gr. *amphi*, on both sides), allows limited motion. A freely movable joint is a *diarthrosis* (Gr. *dia*, between).

Another way to classify joints is by their *structure*. This classification is based on the presence or absence of a joint cavity and the kind of supporting tissue that binds the bones together. Based on structure, three types of joints are recognized: *fibrous, cartilaginous,* and *synovial*.

TABLE 9.1 sums up the classification of joints based on structure. As you can see, fibrous joints are generally synarthroses, cartilaginous joints are generally amphiarthroses, and synovial joints are generally diarthroses. However, the structural and functional categories are *not always* equivalent.

FIBROUS JOINTS

Fibrous joints lack a joint cavity, and fibrous connective tissue unites the bones. Because they are joined together tightly, fibrous joints are mainly immovable in the adult, although some of them do allow slight movement. Three types of fibrous joints are generally recognized: sutures, syndesmoses, and gomphoses.

Sutures

A *suture* is usually such a tight union in an adult that movement rarely occurs between the two bones. Because sutures are found only in the skull, they are sometimes called "skull type" joints [see FIGURE 7.3B]. Movement at sutures can occur in fetuses and young children because the joints have not yet grown together. In fact, the flexibility of the skull is necessary for a newborn baby to be able to pass through its mother's narrow birth canal. The flexibility of cranial sutures in fetuses and children is also important to allow for the growth of the brain. In adults, the fibers of connective tissue between bones are replaced by bone, and the bones become permanently fused. This fusion provides complete protection for the brain from external factors. Such a sealed joint is called a *synostosis* (Gr. *syn*, together + *osteon*, bone).

Syndesmoses

When bones are close together but not touching, and are held together by collagenous fibers or interosseous ligaments, the joint is called a *syndesmosis* (sihn-dehz-MOH-sihss; Gr. to bond together: *syn* + *desmos*, bond). The amount of movement, if any, in a syndesmosis depends on the distance between the bones and the amount of flexibility of the fibrous connecting tissue. Ligaments (composed of collagenous connective tissue) make a firm articulation at the inferior tibiofibular joint so that very little movement occurs there. This limited movement adds strength to the joint. In contrast, the interosseous ligaments between the shafts of the radius and the ulna allow much more movement, including the twisting (pronation, supination) of the forearm [TABLE 9.1].

Gomphoses

A *gomphosis* (gahm-FOH-sihss; Gr. *gomphos*, bolt) is a fibrous joint made up of a peg and a socket. The root of each tooth (the peg) is anchored into its socket by the fibrous periodontal ("around the tooth") ligament, which extends from the root of the tooth to the alveolar processes of the maxillae and mandible.

CARTILAGINOUS JOINTS

In *cartilaginous joints,* the bones are united by a plate of hyaline cartilage or a fibrocartilaginous disk [TABLE 9.1]. These joints lack a joint cavity and permit little or no movement. The two types of cartilaginous joints are the synchondrosis and the symphysis.

TABLE 9.1 CLASSIFICATION OF JOINTS

Structural and functional classification*	Type of structure and movement†	Examples
FIBROUS		
Fibrous connective tissue unites articulating bones; no joint cavity. *Mostly immovable, some slightly movable; usually synarthroses.* 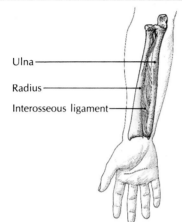 Ulna / Radius / Interosseous ligament	**Suture:** found only in skull; fibrous tissue between articulating bones in children, but permanently fused in adults. *Some movement in fetuses and young children; immovable in adults.*	Cranial sutures, such as coronal suture between frontal and parietal bones.
	Syndesmosis: articulating bones held together (but not touching) by fibrous or interosseous ligaments. *Slight movement: twisting of forearm (pronation, supination).*	Inferior tibiofibular joint; interosseous ligament between shafts of radius and ulna.
	Gomphosis: a peg fitting into a socket. *Mostly immovable; very slight movement of teeth in their sockets.*	Roots of teeth in alveolar processes of mandible and maxillae.
CARTILAGINOUS		
Articulating bones united by plate of hyaline cartilage or fibrocartilaginous disk. *Mostly slightly movable, some immovable; usually amphiarthroses.* 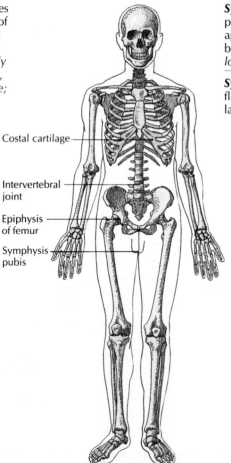 Costal cartilage / Intervertebral joint / Epiphysis of femur / Symphysis pubis	**Synchondrosis:** temporary joint composed of hyaline cartilage joining diaphysis and epiphysis of growing long bones. *Immovable; permits growth of long bones.*	Epiphyseal plate of femur; union of manubrium and body of sternum.
	Symphysis: bony surfaces bridged by flattened plates or disks of fibrocartilage. *Slight movement.*	Symphysis pubis, manubriosternal joint, intervertebral joints between bodies of vertebrae.

*Note that synarthrosis and fibrous, amphiarthrosis and cartilaginous, and diarthrosis and synovial are *not always* synonymous.
†See TABLE 9.2 for an explanation of the terms for different types of movement.

Structural and functional classification*	Type of structure and movement†	Examples

SYNOVIAL

Articulating bones moving freely along smooth, lubricated articular cartilage; enclosed within flexible articular capsule. *Freely movable; usually diarthroses.*

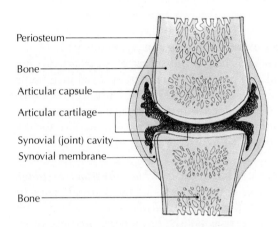

Periosteum
Bone
Articular capsule
Articular cartilage
Synovial (joint) cavity
Synovial membrane
Bone

SYNOVIAL JOINT WITHOUT ARTICULAR DISK

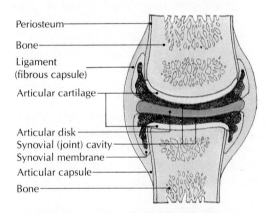

Periosteum
Bone
Ligament (fibrous capsule)
Articular cartilage
Articular disk
Synovial (joint) cavity
Synovial membrane
Articular capsule
Bone

SYNOVIAL JOINT WITH ARTICULAR DISK

UNIAXIAL: movement of bone about one axis of rotation.

Hinge: convex surface of one bone fitted into concave surface of other. *Flexion, extension.*

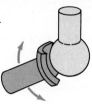 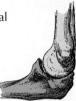

Elbow, interphalangeal joints, knee, ankle.

Pivot: central bony pivot surrounded by collar of bone and ligament. *Supination, pronation, rotation.*

Proximal radioulnar joint, atlantoaxial joint.

BIAXIAL: movement of bone about two axes of rotation.

Condyloid (ellipsoidal): modified ball-and-socket. *Flexion, extension, abduction, adduction, circumduction.*

  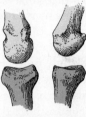

Metacarpophalangeal (knuckle) joints, except thumb.

MULTIAXIAL: movement of bone about three axes.

Gliding: essentially flat articular surfaces. *Simple gliding movement within narrow limits.*

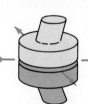

Between articular processes of vertebrae, acromioclavicular joint, some carpal and tarsal bones.

Saddle: opposing articular surfaces with both concave and convex surfaces that fit into one another. *Abduction, adduction, opposition, reposition.*

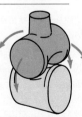

Carpometacarpal joint of thumb.

Ball-and-socket: globelike head of one bone fitted into cuplike concavity of another bone. *Flexion, extension, internal rotation, lateral rotation, abduction, adduction, circumduction.*

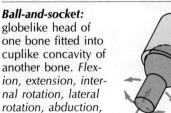 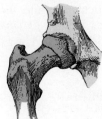

Shoulder joint, hip joint.

Synchondroses

A *synchondrosis* (sihn-kahn-DROH-sihss; Gr. *syn* + *chondros*, cartilage) is also called a *primary cartilaginous joint*. The chief function of a synchondrosis is to permit growth, not movement. It is a temporary joint composed of an epiphyseal plate of hyaline cartilage that joins the diaphysis and epiphysis of a growing long bone. A synchondrosis is eventually replaced by bone when the long bone stops growing, and it then becomes a synostosis. However, a few synchondroses are not replaced by synostoses and are still present in the adult. One such articulation is the sternoclavicular joint, where the clavicle, the first costal cartilage, and the manubrium of the sternum are joined [see FIGURES 7.17 and 7.16].

Symphyses

A *symphysis* (SIHM-fuh-sihss; Gr. growing together) is sometimes called a *secondary synchondrosis*. In such an articulation the two bony surfaces are covered by thin layers of hyaline cartilage. Between them are disks of fibrocartilage that serve as shock absorbers. Fibrocartilage is a dense mass of collagenous fibers filled with cartilage cells and a scant cartilage matrix, so it has the firmness of cartilage and the toughness of a tendon. One of these slightly movable joints is the symphysis pubis, the midline joint between the bodies of the pubic portions of the paired hipbones. During late pregnancy, the symphysis pubis is relaxed somewhat to allow for the necessary displacement of the mother's hipbones during childbirth. Two other examples are the slightly movable manubriosternal joint at the sternal angle, and the intervertebral disks between the bodies of the vertebrae.

SYNOVIAL JOINTS

Most of the permanent joints in the body are synovial. Of all the types of joints, *synovial joints* allow the greatest range of movement [TABLE 9.1]. Such free movement is possible because the ends of the bones are covered with a smooth hyaline *articular cartilage*, the joint is lubricated by a thick fluid called *synovial fluid* (Gr. *syn* + *ovum*, egg),* and the joint is enclosed by a flexible *articular capsule*. A synovial joint has a joint cavity.

Articular capsules that are reinforced with collagenous fibers are called *fibrous capsules*. The portion of a fibrous capsule reinforced by a thick layer of collagenous fibers is called a *ligament*.

---

*Synovial fluid gets its name because its thick consistency resembles the white of an egg.

Typical Structure of Synovial Joints

Because synovial joints allow more free movement than any other type of joint, they are also more complicated in structure. In general, *the more flexible a joint is, the less stable it is.* You will see how this rule applies when we look at some examples of major joints later in this chapter. The four essential structures of a synovial joint are the synovial cavity, the articular cartilage, the articular capsule, and ligaments.

Synovial cavity A *synovial cavity,* or joint cavity, is the space between the two articulating bones, but does not include the articular cartilage. It contains folds of *synovial membrane* that sometimes contain pads of fat. These fatty pads help to fill spaces between articulating bones, and also reduce friction. The synovial membrane secretes the thick synovial fluid that lubricates the synovial cavity. Combined with the articular cartilage, synovial fluid provides an almost friction-free surface for the easy movement of joints.

Articular cartilage An *articular (hyaline) cartilage* caps the surface of the bones facing the synovial cavity. In a living body, the articular cartilage has a silvery-blue luster and appears as polished as a pearl. Because of its thickness and elasticity, the articular cartilage acts like a shock absorber. If it is worn away (which occurs in some joint diseases), movement becomes restricted and painful. The cartilage itself is insensitive to feeling, since it has no nerve supply, but the other portions of the joint are supplied with pain-receptor nerve fibers.

Within several synovial joints there are *articular disks,* or fibrocartilaginous disks (see illustration in TABLE 9.1). Their roles vary depending upon the joint. These disks may act as shock absorbers to reduce the effect of shearing (twisting) upon a joint and to prevent jarring between bones. They also adjust the unequal articulating surfaces of the bones so that the surfaces fit together more evenly.

In some joints, the fibrocartilaginous disk forms a complete partition, dividing the joint into two cavities. In the knee joint, the fibrocartilages, called the **medial** and **lateral menisci** (muh-NISS-eye; sing., *meniscus*; Gr. *meniskos*, crescent), are crescent-shaped wedges that form incomplete partitions [FIGURE 9.1B]. The menisci serve to cushion as well as guide the articulating bones. Many athletes, especially football players, tear these menisci, commonly referred to as torn cartilages. (See "Knee Injuries—The 'Achilles Heel' of Athletes.")

Articular capsule An *articular capsule* lines the synovial cavity in the noncartilaginous parts of the joint. This fibrous capsule is lax and pliable, permitting considerable

Knee Injuries—The "Achilles Heel" of Athletes

One of the most frequent sports injuries is commonly called torn cartilage (usually the medial meniscus) of the knee. When the cartilage is torn, it may become wedged between the articular surfaces of the femur and tibia, causing the joint to "lock." Until recently, such an injury might sideline an athlete for months, or even end his or her career altogether. Fifteen or twenty years ago, a football player recovering from surgery to repair knee cartilage that was torn on the opening day of the season would remain on crutches for about two months and would be unable to play for the remainder of the season. Now, a diagnostic and surgical technique called **arthroscopy** ("looking into a joint"), or arthroscopic surgery, makes it possible for an athlete with torn cartilage to return to the football field in about half the usual time.

Designed in Japan in 1970, arthroscopy permits a surgeon to place a lighted scope about the size of a pencil into the joint capsule to view the structural damage. If the damage is local and there do not seem to be complications, the surgeon makes another quarter-inch incision and inserts microsurgical instruments to clear away damaged cartilage. Most incisions are closed with a single stitch, or even just a Band-Aid. The patient is usually encouraged to walk lightly on the day of the operation.

Such rapid recovery does not give leg muscles time to atrophy. So besides having minor surgery instead of major tissue damage during conventional open surgery, the patient does not have to worry about rehabilitating weakened muscles.

Another common knee injury is a tear of the anterior and/or posterior cruciate ligaments. When the cruciate ligaments are torn, the knee joint becomes nonfunctional. To repair the cruciate ligaments, holes are drilled through the femur. Sutures are stitched to the damaged ligaments, and the sutures are passed out through the holes. The ligaments are secured to the bone by knotting the sutures outside the femur. If the collateral ligament is separated from the bone, it is reattached by being sutured to the bone. Six to eight weeks of healing is usually followed by about a year of rehabilitative therapy.

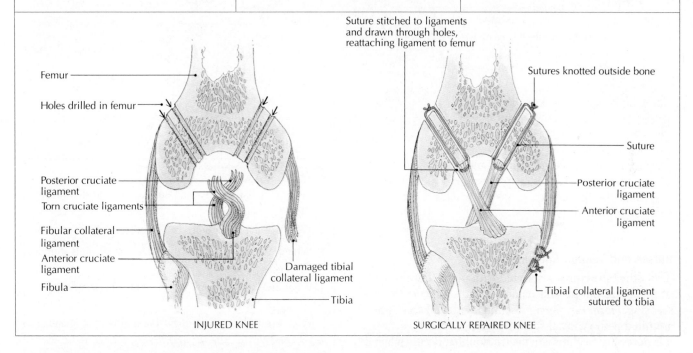

INJURED KNEE — Femur; Holes drilled in femur; Posterior cruciate ligament; Torn cruciate ligaments; Fibular collateral ligament; Anterior cruciate ligament; Fibula; Damaged tibial collateral ligament; Tibia

SURGICALLY REPAIRED KNEE — Suture stitched to ligaments and drawn through holes, reattaching ligament to femur; Sutures knotted outside bone; Suture; Posterior cruciate ligament; Anterior cruciate ligament; Tibial collateral ligament sutured to tibia

movement. The inner lining of the capsule is the synovial membrane, which extends from the margins of the articular cartilages. The outer layer of the capsule is a fibrous membrane that extends from bone to bone across a joint and reinforces the capsule. So-called "double-jointed" people have loose articular capsules.

Ligaments *Ligaments* are fibrous thickenings of the articular capsule that join one bone to its articulating mate. They vary in shape, and even in strength, depending upon their specific roles. Most ligaments are considered inelastic, and yet they are pliable enough to permit movement at the joints. However, ligaments tear rather than stretch under excessive stress, as when an ankle is sprained. For example, a sprained ankle results from excessive inversion of the foot, which causes a partial tearing of the anterior talofibular ligament and the calcaneofibular ligament. Torn ligaments are extremely painful, and are accompanied by immediate local swelling. In general, however, ligaments are strong enough to

FIGURE 9.1 BURSAE AND TENDON SHEATHS

[A] Bursa in shoulder, positioned to reduce friction between articulating surfaces. **[B]** Bursae in knee, positioned to facilitate movement of tendons, muscles, and skin. **[C]** Diagram of a tendon sheath.

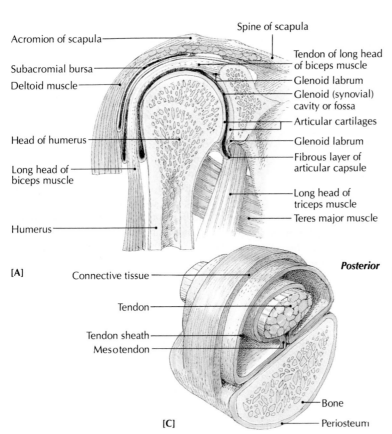

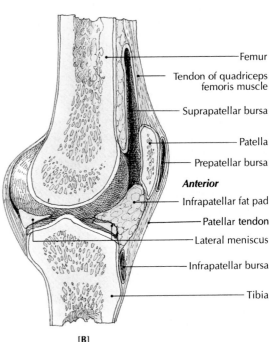

prevent any excessive movement and strain, and a rich supply of sensory nerves helps prevent a person from stretching the ligaments excessively when the joints are being overworked.

Bursae and Tendon Sheaths

Two other structures associated with joints, but not part of them, are bursae and tendon sheaths. ***Bursae*** (BURR-see; sing. *bursa*; Gr. purse) resemble flattened sacs. They are filled with synovial fluid. Bursae are found wherever it is necessary to eliminate the friction that occurs when a muscle or tendon rubs against another muscle, tendon, or bone [FIGURE 9.1A, B]. They also function to cushion certain muscles and to facilitate the movement of muscles over bony surfaces. Bursitis results when bursae become inflamed.

A modification of a bursa is the ***tendon*** (synovial) ***sheath*** surrounding long tendons that are subjected to constant friction. Such sheaths surround the tendons of the wrist, palm, and finger muscles. Tendon sheaths are long, cylindrical sacs filled with synovial fluid. Like bursae, tendon sheaths reduce friction and permit tendons to slide easily.

ASK YOURSELF

1 How do synarthroses, amphiarthroses, and diarthroses compare?

2 What are the three types of fibrous joints?

3 Where are the two kinds of cartilaginous joints found in the body?

4 What are the four main components of a synovial joint?

5 What are the functions of bursae and tendon sheaths?

MOVEMENT AT SYNOVIAL JOINTS

Any movement produced at a synovial joint is limited in some way. Among the limiting factors are the following:

1 *Interference by other structures.* For example, lowering the shoulder is limited by the presence of the thorax.

2 *Tension exerted by ligaments of the articular capsule.* For example, the thigh cannot be hyperextended at the hip joint because the iliofemoral ligament becomes taut as it passes in front of the hip joint.

3 *Muscle tension.* For example, when the knee is straight, it is more difficult to raise the thigh than when the knee is bent, because the stretched hamstring muscles exert tension on the back of the thigh.

Terms of Movement

Muscles and bones work together to allow different types of movement at different joints. Representative movements at synovial joints and the terms used to describe those movements are shown in TABLE 9.2.

A S K   Y O U R S E L F

1 What three factors restrict movement at synovial joints?

2 How does circumduction form a "cone of movement"?

3 How does rotation differ from supination and pronation?

4 What movement is necessary to open the mouth?

Uniaxial, Biaxial, and Multiaxial Joints

In any movement about a joint, one member of a pair of articulating bones moves in relation to the other, with one bone maintaining a fixed position, and the other bone moving about an axis (or axes), called an *axis of rotation*. For example, at the shoulder joint (the glenohumeral joint between the scapula and the humerus), the scapula is fixed in relation to the movement of the humerus.

When the movement of a bone at a joint is limited to rotation about one axis, as is the elbow joint, the joint is said to be *uniaxial* [FIGURE 9.2A]. In other words, the forearm may be flexed or extended from the elbow, moving in only one plane. When two movements can take place about two axes of rotation, as in the radiocarpal (wrist) joint, the joint is *biaxial* [FIGURE 9.2B]. The hand can be flexed or extended in one plane and moved from side to side (abduction and adduction) at the wrist in a second plane. When three independent rotations occur about three axes of rotation, as in the shoulder joint, the joint is *multiaxial,* or *triaxial* [FIGURE 9.2C]. The arm movements at the shoulder joint occur in three directions: flexion and extension, abduction and adduction, and medial and lateral rotation.

TYPES OF SYNOVIAL JOINTS

Synovial joints are freely movable joints that are classified according to the shape of their articulating surfaces and the types of joint movements those shapes permit. (Once again, structure influences function.) Six types of synovial joints are recognized: hinge, pivot, condyloid (ellipsoidal), gliding, saddle, and ball-and-socket. (See TABLE 9.1 for a complete classification of joints.)

Hinge Joints

Hinge joints roughly resemble the hinges on the lid of a box. The convex surface of one bone fits into the concave surface of another bone, so that only a uniaxial, back-and-forth movement occurs around a single (transverse) axis [FIGURE 9.2A]. These uniaxial joints have strong collateral ligaments in the capsule around the joint. The rest of the capsule is thin and lax, permitting flexion and extension, the only movements possible in these joints. Hinge joints are found in the elbow, finger, knee, and ankle.

Pivot Joints

Another type of uniaxial joint is the *pivot joint,* which is only able to rotate around a central axis. Pivot joints are composed of a central bony pivot surrounded by a collar made partly of a bone and partly of a ligament. Pivot joints have a rotational movement around a long axis through the center of the pivot (like the hinges of a gate).

The atlantoaxial joint between the atlas and the axis is a pivot joint in which the collar (composed of the anterior bony arch of the atlas and the transverse ligament) rotates around the pivot, which is the dens of the axis. The "no" movement of the head occurs at the atlantoaxial pivot joint.

Condyloid Joints

Condyloid (Gr. *condylus,* knuckle) *joints* are modifications of the multiaxial ball-and-socket joint [TABLE 9.1]. However, because the ligaments and muscles around the joint limit the rotation to two axes of movement, the joint is classified as *biaxial.* Examples of condyloid joints are the metacarpophalangeal joints (knuckles) of the fingers (except the thumb). In condyloid joints the axes are at right angles to each other, permitting the usual flexion and extension of a hinge joint, as well as abduction, adduction, and circumduction. However, rotational movement is not permitted.

TABLE 9.2 TYPES OF MOVEMENTS AT SYNOVIAL JOINTS

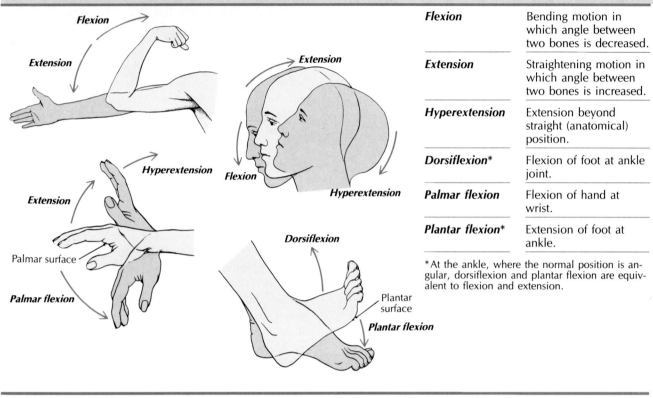

Flexion	Bending motion in which angle between two bones is decreased.
Extension	Straightening motion in which angle between two bones is increased.
Hyperextension	Extension beyond straight (anatomical) position.
Dorsiflexion*	Flexion of foot at ankle joint.
Palmar flexion	Flexion of hand at wrist.
Plantar flexion*	Extension of foot at ankle.

*At the ankle, where the normal position is angular, dorsiflexion and plantar flexion are equivalent to flexion and extension.

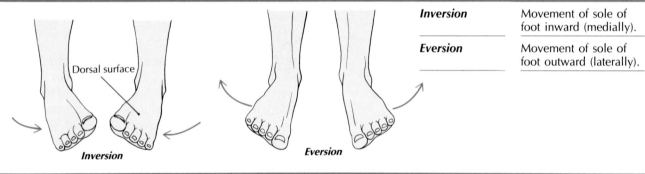

Inversion	Movement of sole of foot inward (medially).
Eversion	Movement of sole of foot outward (laterally).

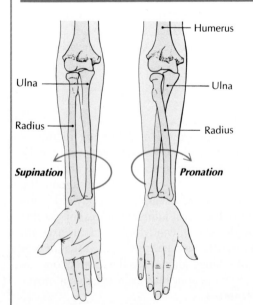

Supination	Pivoting movement of forearm in which radius is "rotated" to become parallel to ulna.
Pronation	Pivoting movement of forearm in which radius is "rotated" diagonally across ulna.

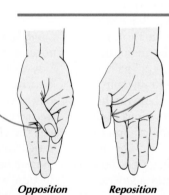

Opposition	Angular movement in which thumb pad is brought to touch and to oppose a finger pad of the extended fingers; occurs only at carpometacarpal joint of thumb.
Reposition	Movement that returns thumb to anatomical position; opposite of opposition.

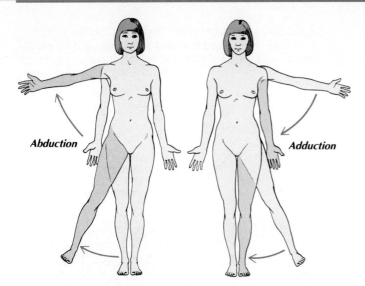

Abduction   Movement of limb away from midline of body; movement of fingers or toes away from longitudinal axis of hand or foot.

Adduction   Movement of limb toward or beyond midline of body; movement of fingers or toes toward longitudinal axis of hand or foot.

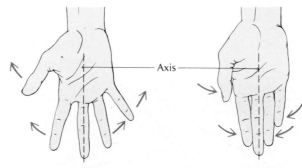

Abduction   Adduction

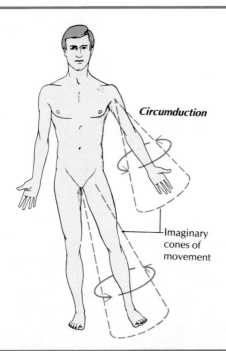

Circumduction

Imaginary cones of movement

Circumduction   Movement in which distal end of bone moves in circular motion while proximal end remains stable; accomplished by successive flexion, abduction, extension, and adduction.

Rotation   Movement of body part (usually entire extremity) around its own axis without any displacement of its axis.

Medial (internal)   Movement in which ventral surface of extremity rotates toward midline of body.

Lateral (external)   Movement in which ventral surface rotates outward away from midline.

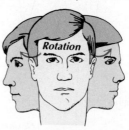

Rotation

Protraction
Forward movement.

Retraction
Backward movement.

Depression
Lowering a body part.

Elevation
Raising a body part.

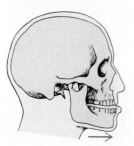

Protraction

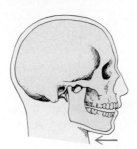

Retraction

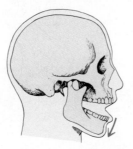

Depression

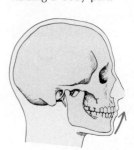

Elevation

FIGURE 9.2 PLANES OF MOVEMENT AT SYNOVIAL JOINTS

[A] Uniaxial movement at elbow joint, illustrating axis for flexion and extension. [B] Biaxial movement at radiocarpal (wrist) joint, illustrating axes for flexion and extension and for abduction and adduction. [C] Multiaxial movement at glenohumeral (shoulder) joint, illustrating axes for flexion and extension, for abduction and adduction, and for medial rotation and lateral rotation.

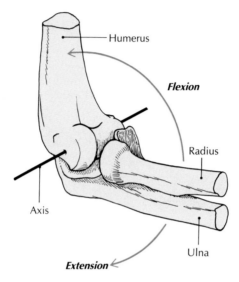

[A] UNIAXIAL (elbow)

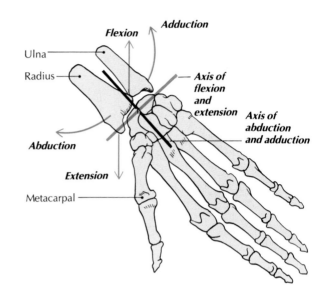

[B] BIAXIAL (wrist)

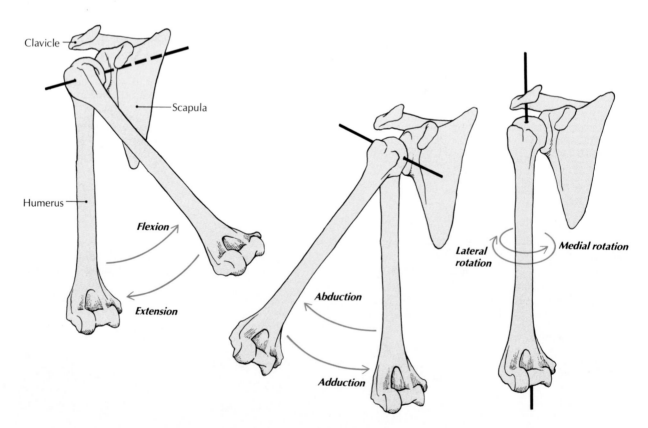

[C] MULTIAXIAL (shoulder)

Gliding Joints

Gliding joints are almost always small and are formed by essentially flat articular surfaces, so that one bone *slides* on another bone with a minimal axis of rotation, if any. Examples include the joints between the articular processes of adjacent vertebrae and the joints between some carpal and tarsal bones.

Saddle Joints

The *saddle joint* is so named because the opposing articular surfaces of both bones are shaped like a saddle, that is, they have both concave and convex areas at right angles to each other. Movement is permitted in several directions, and the joint is considered *multiaxial*. Movements are abduction, adduction, opposition, and reposition. The carpometacarpal joint of the thumb is the best example of a saddle joint in the body.

Ball-and-Socket Joints

Ball-and-socket joints are composed of a globelike head of one bone that fits into a cuplike concavity of another bone. This is the most freely movable of all joints, permitting movement along three axes of rotation. Actually, an almost infinite number of axes are available, and the joint is classified as *multiaxial*. The movements of a ball-and-socket joint are flexion, extension, medial (internal) rotation, lateral (external) rotation, abduction, adduction,

and circumduction. The shoulder and hip joints are ball-and-socket joints.

ASK YOURSELF

1 Compare uniaxial, biaxial, and multiaxial joints.

2 What is the difference in the direction of hinge motion in hinge joints and pivot joints?

3 What is an example of a condyloid joint?

4 Which type of joint is the most movable?

DESCRIPTION OF SOME MAJOR JOINTS

The jaw, shoulder, hip, and knee are major joints of the body, each with its own distinctive anatomical characteristics. The following sections describe and illustrate these four joints in detail. TABLE 9.3 summarizes major articulations of the body.

Temporomandibular (Jaw) Joint

The *temporomandibular* or *TM, joint* (jaw) is the only movable joint of the head. It is a synovial joint with a combination of hinge and gliding surfaces [FIGURE 9.3].

FIGURE 9.3 THE TEMPOROMANDIBULAR JOINT

The temporomandibular (jaw or TM) joint is actually two joints. The upper joint is for gliding movements (protraction and retraction), and the lower joint is for hinge movements (elevation and depression). Note how the relation of the head of the mandible and articular disk to the articular tubercle changes when the mouth closes [A] and opens [B]. Right lateral views.

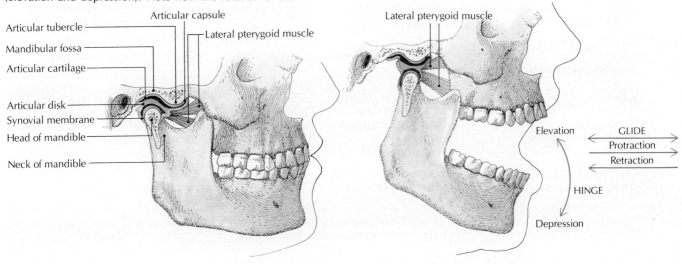

[A]   [B]

TABLE 9.3 SOME MAJOR ARTICULATIONS OF THE HUMAN BODY

Joint and classification	Articulation	Type of structure and movement
JOINTS ASSOCIATED WITH THE SKULL [FIGURES 7.4C, 7.4D, 9.3]		
Temporomandibular (jaw, TM) joint Diarthrosis (synovial)	Between head of mandible and mandibular fossa of temporal bones; divided into two compartments by articular disk.	Hinge (lower compartment); gliding (upper compartment). Simultaneous hinge and gliding movements open and close jaw. Protraction, retraction, grinding.
Atlantooccipital joint Diarthrosis (synovial)	Between atlas and occipital bone of skull.	Hinge. "Yes" movement of head.
JOINTS OF THE VERTEBRAL COLUMN [FIGURE 7.10]		
Atlantoaxial joints (paired) Medial: Diarthrosis (synovial)	Between dens of axis and anterior arch of atlas and its transverse ligament.	Pivot. "No" movement of head.
Lateral: Diarthrosis (synovial)	Between articular processes of atlas and axis.	Gliding. "No" movement of head.
Intervertebral joints Diarthrosis (synovial)	Paired joints between articular processes of adjacent vertebrae.	Gliding. *Neck:* considerable variety and range of movement.
Amphiarthrosis (fibrocartilaginous)	Unpaired joints (intervertebral disks) between adjacent bodies of vertebrae.	Synchondrosis. *Thorax:* considerable variety, limited range. *Lumbar:* essentially flexion, extension, lateral bending.
JOINTS ASSOCIATED WITH THE RIBS AND STERNUM [FIGURE 7.17]		
Costovertebral joints Diarthrosis (synovial)	Between articular facets on heads and tubercles of ribs and costal facets on transverse processes of thoracic vertebrae.	Gliding. Rotation
Sternocostal joints Synarthrosis (fibrous)	Between rib 1 costal cartilage and sternum.	Syndesmosis (rib 1). No movement.
Diarthrosis (synovial)	Between ends of costal cartilages of ribs 2–7 and concavities on sides of sternum.	Gliding (true ribs). Rotation of true ribs (2–7).
Interchondral joints Diarthrosis (synovial)	Between successive costal cartilages of ribs 5–9.	Gliding. Adjustment during respiration.
Manubriosternal joint Amphiarthrosis (cartilaginous)	Between manubrium and body of sternum.	Symphysis. Slight movement at fibrocartilaginous disk between manubrium and sternal angle.
Xiphisternal joint Amphiarthrosis (cartilaginous)	Between xiphoid process and body of sternum.	Symphysis. Slight movement.
JOINTS OF THE CLAVICLE AND PECTORAL (SHOULDER) GIRDLE [FIGURES 8.2, 8.3, 9.4]		
Sternoclavicular Diarthrosis (synovial)	Between medial end of clavicle, manubrium, and first costal cartilage; divided into two compartments by articular disk.	Gliding. Elevation, depression, protraction, retraction.
Acromioclavicular Diarthrosis (synovial)	Between lateral end of clavicle and medial surface of acromion.	Gliding. Essentially an accommodation between movement of clavicle and scapula.
Coracoclavicular Synarthrosis (fibrous)	Between clavicle and coracoid process of scapula; connected by coracoclavicular ligament.	Syndesmosis. Prevents separation of clavicle from scapula.
JOINTS OF THE ARM AND FOREARM [FIGURES 8.1, 8.5, 9.4]		
Glenohumeral (shoulder) joint Diarthrosis (synovial)	Between head (ball) of humerus and glenoid cavity (socket) of scapula.	Ball-and-socket. Flexion, extension, abduction, adduction, medial and lateral rotation, circumduction.

Joint and classification	Articulation	Type of structure and movement
Elbow Diarthrosis (synovial)	Between trochlea and capitulum of humerus, trochlear notch of ulna, and head of radius.	Hinge. Flexion, extension.
Radioulnar articulation Proximal: Diarthrosis (synovial)	Between head of radius and radial notch of ulna.	Pivot. Pronation, supination.
Distal: Diarthrosis (synovial)	Between head of ulna and ulnar notch of radius.	Pivot. Pronation, supination.
Radiocarpal (wrist) joint Diarthrosis (synovial)	Between distal end of radius (and articular disk) and proximal row of carpal bones (scaphoid, lunate, triquetrum).	Ellipsoidal. Radial abduction, ulnar adduction, flexion, extension, hyperextension, circumduction.

JOINTS OF THE HAND [FIGURE 8.6]

Midcarpal joints Diarthrosis (synovial)	Between proximal and distal rows.	Hinge. Slight flexion, extension.
Carpometacarpal joint of thumb Diarthrosis (synovial)	Between trapezium and proximal end of first metacarpal bone.	Saddle. Abduction, adduction, opposition, reposition.
Metacarpophalangeal (knuckle) joints Diarthrosis (synovial)	Between heads of metacarpal bones and bases of proximal phalanges.	Condyloid. Flexion, extension, abduction, adduction, circumduction.
Interphalangeal joints Diarthrosis (synovial)	Between heads of phalanges and concave base of adjacent phalanges.	Hinge. Flexion, extension.

JOINTS OF THE PELVIS [FIGURES 8.8, 9.5]

Sacroiliac joint Anterior: Diarthrosis (synovial)	Between sacrum and ilium on anterior side.	Gliding. Slight gliding and rotary movement; gives resilience to joint.
Posterior: Synarthrosis (fibrous)	Between sacrum and ilium on posterior side.	Symphysis. Slight movement; gives security to joint.
Symphysis pubis Amphiarthrosis (fibrocartilaginous)	Between bodies of pubic bones.	Symphysis. Almost immovable, but accommodates during childbirth.
Coxal (hip) joint Diarthrosis (synovial)	Between head (ball) of femur and acetabulum (socket) of os coxa (hipbone).	Ball-and-socket. Flexion, extension, abduction, adduction, medial and lateral rotation, circumduction.

JOINTS OF THE LEG AND ANKLE [FIGURES 8.7, 8.12, 9.6]

Tibiofemoral (knee) joint Diarthrosis (synovial)	Between medial and lateral condyles of distal femur and medial and lateral condyles of proximal femur.	Modified hinge. Flexion, extension, some rotation ("screw-home" action at end of extension).
Talocrural (ankle) joint Diarthrosis (synovial)	Between socket for talus (distal tibia flanked by medial and lateral malleolus of tibia) and upper surface of talus.	Hinge. Dorsiflexion, plantar flexion (hyperextension).

JOINTS OF THE FOOT [FIGURE 8.13]

Subtalar Diarthrosis (synovial)	Posterior joint between talus and calcaneus.	Three joints articulate as a unit; axis of rotation forms a line called the subtalar axis. Eversion, inversion.
Talocalcaneonavicular Diarthrosis (synovial)	Combined anterior joint between talus and calcaneus and joint between talus and navicular.	
Transverse tarsal Diarthrosis (synovial)	Combined joint between calcaneus and cuboid and joint between talus and navicular.	
Tarsometatarsal Diarthrosis (synovial)	Between four anterior tarsal bones and bases of metatarsal bones.	Gliding. Slight movement.
Metatarsophalangeal Diarthrosis (synovial)	Between heads of metatarsal bones and bases of proximal phalanges.	Condyloid. Flexion, extension, abduction, adduction, circumduction.
Interphalangeal Diarthrosis (synovial)	Between heads of phalanges and concave bases of adjacent phalanges.	Hinge. Flexion, extension.

The structure of the jaw is well fitted to the movements involved in chewing, biting, speaking, and so on. The articular surface of the temporal bone has a concave back part, called the *mandibular fossa,* and a convex part, called the *articular tubercle.* The joint cavity between the mandible and the temporal bone is divided into two compartments by a fibrocartilaginous *articular disk,* which is fused to the *articular capsule.* The capsule is attached above to the temporal articular surface and below to the neck of the mandible.

When the mouth is closed, the convex head of the mandible fits into the mandibular fossa. Two movements occur at the TM joint while the mandible is being depressed and the mouth is opening. The first movement is the forward gliding that takes place in the upper compartment of the joint cavity as the disk and the head of the mandible move onto the articular tubercle. The second movement is a hingelike rotation that takes place in the lower compartment as the head of the mandible rotates on the articular head (condyle).

In the acts of protraction (protrusion) and retraction of the mandible, the head of the mandible and the articular disk slide forward and then backward on the articular surface of the temporal bone. The grinding action of the teeth is produced when protrusion and retraction of the mandible alternate from one side to the other.

Glenohumeral (Shoulder) Joint

The *glenohumeral,* or *shoulder, joint* is a multiaxial, ball-and-socket, synovial joint (diarthrosis). The head of the humerus (the "ball") articulates with the shallow, concave *glenoid fossa* (the "socket") of the scapula [FIGURE 9.4, TABLE 9.3]. As is typical of a ball-and-socket joint, the movements of flexion, extension, abduction, adduction, medial rotation, lateral rotation, and circumduction occur at the shoulder joint.

The shoulder joint has a greater range of movement than any other joint in the body, but what it gains in mobility, it loses in stability and security. As a result, the shoulder joint is dislocated more than any other joint. (See When Things Go Wrong at the end of this chapter.)

Q: *What causes the "popping" sound when you crack your knuckles?*

A: There are at least three likely possibilities: (1) When a joint is contracted, small ligaments or muscles may pull tight and snap across the bony protuberances of the joint. (2) When the joint is pulled apart, air can pop out from between the bones, creating a vacuum that produces a popping sound. (3) When the fluid pressure of the synovial fluid is reduced by the slow articulation of a joint, tiny gas bubbles in the fluid may burst, producing the sound.

The extensive range of movement results mainly from the laxity of the articular capsule and the shallowness of the glenoid fossa.

The shallow glenoid fossa is deepened slightly by a fibrocartilaginous rim called the *glenoid labrum.* The *articular capsule,* looser than in any other important joint, completely envelops the articulation. It is attached to the glenoid labrum of the scapula, to the articular margin of the head, and to part of the anatomical neck of the humerus. The articular capsule is under tension during adduction and lateral rotation, but the *coracohumeral ligament* between the scapula and humerus becomes taut during those movements, strengthening the joint considerably.

The strength and stability of the shoulder joint depend also on four muscles, which nearly encircle the joint and hold the head of the humerus in the glenoid fossa. These joint-reinforcing muscles — the *supraspinatus, infraspinatus, teres minor,* and *subscapularis* (sometimes referred to as SITS muscles, for the first letter of each) — are collectively called the *rotator cuff muscles.* The tendons of the SITS muscles and an overall connective tissue sheath (the joint capsule) comprise the so-called *rotator cuff* itself. The rotator cuff acts as a dynamic "ligament" keeping the head of the humerus pressed into the glenoid fossa of the scapula, thus providing major stability in the shoulder joint. However, the rotator cuff offers little support to the joint inferiorly, and as a result, dislocations generally occur inferiorly. Rotator cuff injuries are common in baseball pitchers because pitching motions strain the rotator cuff muscles and their tendons.

Of the several bursa in the shoulder region, the large subacromial bursa is important because it is associated with subacromial bursitis. (See When Things Go Wrong at the end of this chapter.)

Coxal (Hip) Joint

The *coxal,* or *hip, joint* is a multiaxial, ball-and-socket, synovial joint (diarthrosis). The head of the femur articulates with the acetabulum of the hipbone (os coxa) [FIGURE 9.5].

The bones of the hip are arranged in such a way that the joint is one of the most secure, strong, and stable articulations in the body. The large, globular *head* (ball) of the femur fits snugly into the deep hemispherical *acetabulum* (socket). The acetabulum is made deeper by the fibrocartilaginous *acetabular labrum,* which forms a complete circle around the socket. The deep socket holds the head of the femur securely, but because of such stability the movement at the hip joint is not as free as it is at the ball-and-socket shoulder joint. A notch on the inferior aspect of the acetabulum is bridged by the *transverse acetabular ligament.* The acetabulum faces down-

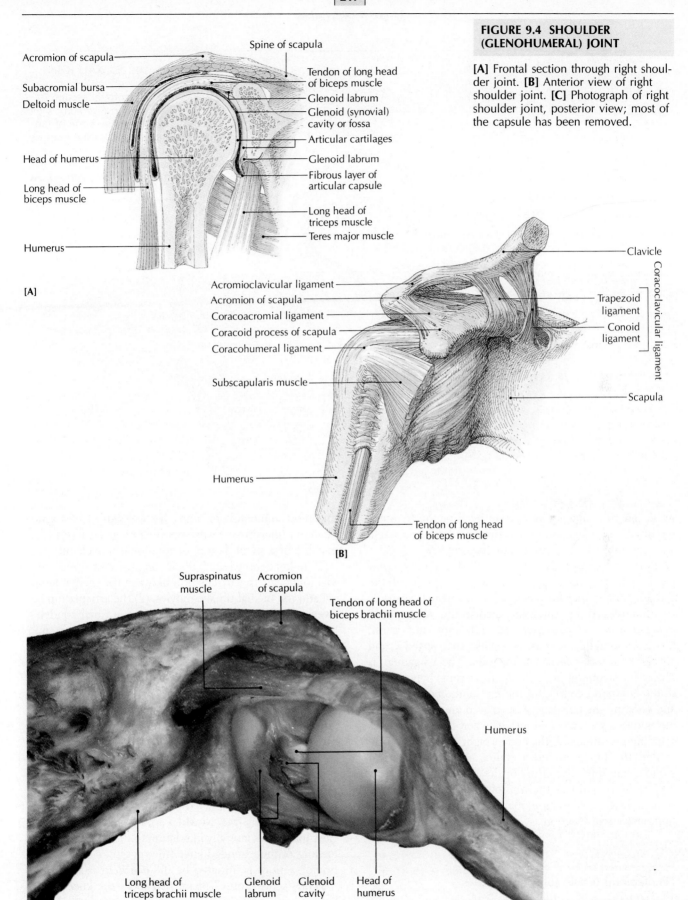

Spine of scapula

Acromion of scapula

Subacromial bursa

Deltoid muscle

Head of humerus

Long head of biceps muscle

Humerus

Tendon of long head of biceps muscle

Glenoid labrum

Glenoid (synovial) cavity or fossa

Articular cartilages

Glenoid labrum

Fibrous layer of articular capsule

Long head of triceps muscle

Teres major muscle

[A]

FIGURE 9.4 SHOULDER (GLENOHUMERAL) JOINT

[A] Frontal section through right shoulder joint. [B] Anterior view of right shoulder joint. [C] Photograph of right shoulder joint, posterior view; most of the capsule has been removed.

Clavicle

Acromioclavicular ligament

Acromion of scapula

Coracoacromial ligament

Coracoid process of scapula

Coracohumeral ligament

Subscapularis muscle

Humerus

Trapezoid ligament

Conoid ligament

Coracoclavicular ligament

Scapula

Tendon of long head of biceps muscle

[B]

Supraspinatus muscle

Acromion of scapula

Tendon of long head of biceps brachii muscle

Humerus

Long head of triceps brachii muscle

Glenoid labrum

Glenoid cavity

Head of humerus

[C]

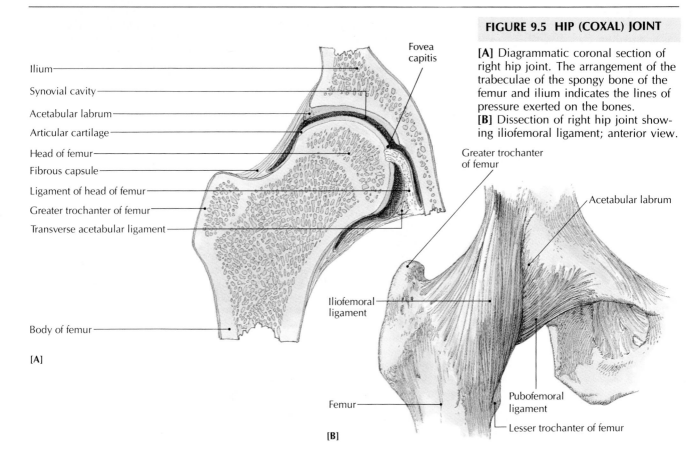

Ilium

Synovial cavity

Acetabular labrum

Articular cartilage

Head of femur

Fibrous capsule

Ligament of head of femur

Greater trochanter of femur

Transverse acetabular ligament

Body of femur

[A]

Fovea capitis

Greater trochanter of femur

Acetabular labrum

Iliofemoral ligament

Femur

Pubofemoral ligament

Lesser trochanter of femur

[B]

FIGURE 9.5 HIP (COXAL) JOINT

[A] Diagrammatic coronal section of right hip joint. The arrangement of the trabeculae of the spongy bone of the femur and ilium indicates the lines of pressure exerted on the bones.
[B] Dissection of right hip joint showing iliofemoral ligament; anterior view.

ward, laterally, and forward. The head of the femur is mounted on the neck so that it forms a 125-degree angle with the long shaft of the femur [FIGURE 9.5].

The fibrous capsule in the hip is thick and tense compared with the relatively thin and lax capsule of the shoulder. The outer fibrous layer of the articular capsule is strengthened by three ligaments: the *iliofemoral, pubofemoral,* and *ischiofemoral.* The most important, the iliofemoral ligament, is shaped like an inverted Y, and covers the anterior surface of the joint. The Y ligament is one of the strongest in the body. It prevents hyperextension (backward bending) of the hip joint, becoming taut and resisting the tensile stresses placed upon it when the hip is fully extended. As a result, when the body is in a standing position, with the body weight centered slightly behind the hip joint, erect stance can be maintained without any muscular effort in the hip region.

The movements at the hip joints are flexion, extension, abduction, adduction, medial rotation, lateral rotation, and circumduction.

Tibiofemoral (Knee) Joint

The *tibiofemoral,* or *knee, joint* is the largest and most complex joint in the body, as well as one of the weakest

and most vulnerable to injury [FIGURE 9.6]. It is a synovial joint (diarthrosis) with modified hinge structure. It is capable of a small degree of rotational movement.

The knee joint is actually a composite of three synovial joints: (1) the articulation between the medial femoral and the medial tibial condyles, (2) the articulation between the lateral femoral and the lateral tibial condyles, and (3) the articulation between the patella and the femur (medial and lateral tibiofemoral joints and patellofemoral joint).

Flexion and extension are the primary movements at the knee joint, with a slight amount of medial and lateral rotation. With the knee flexed, the leg can be rotated laterally and medially. A rotational "screw-home" movement occurs just as the knee assumes full extension. If this "screw-home" action takes place when the foot is not free, as in the act of standing up, the femur is the bone that rotates medially in relation to the tibia until the knee is locked. If the "screw-home" action takes place when the foot is free, as when a punter kicks a football, the tibia rotates laterally in relation to the femur. The unlocking of the extended knee (the reverse of the "screw-home" action) is initiated by the *popliteus muscle.*

The "screw-home" phase that locks the knee occurs because the articular surface of the medial condyle is longer than that of the lateral condyle. As a result, the

FIGURE 9.6 KNEE (TIBIOFEMORAL) JOINT

[A] Anterior view of right knee joint. The patella has been removed, and the patellar tendon has been cut. [B] Photograph of right knee joint, anterior view; the patellar tendon has been removed. [C] Sagittal section showing bursae of knee. [D] Cruciate ligaments of the knee joint in flexed knee, lateral view. The anterior cruciate ligament is slack during flexion.

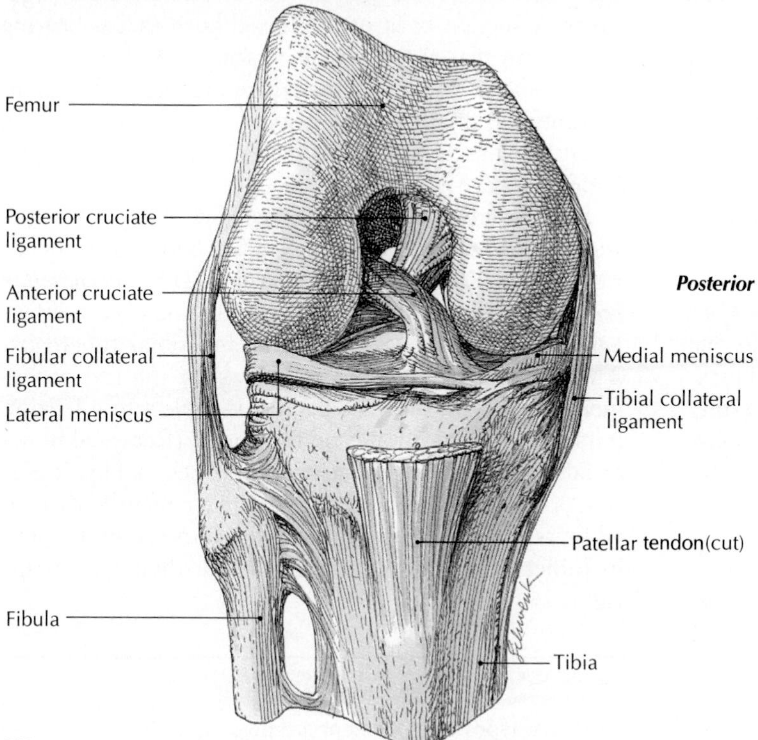

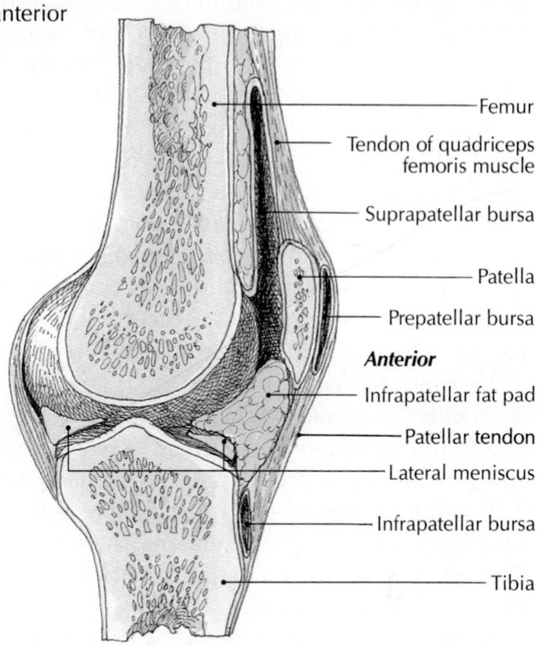

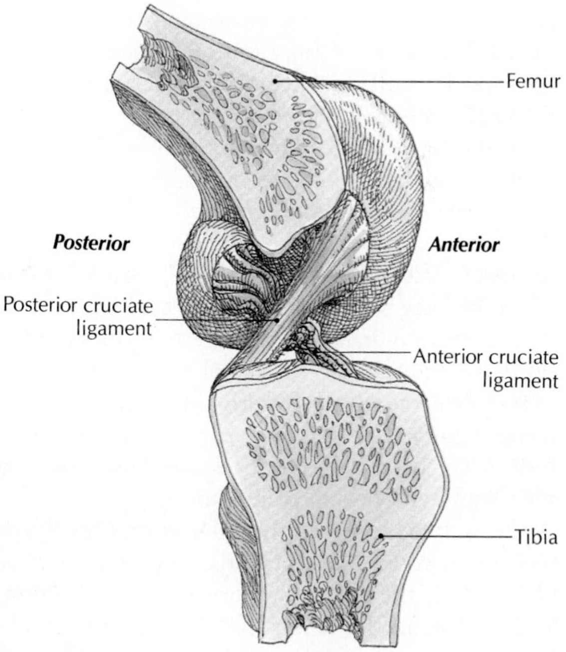

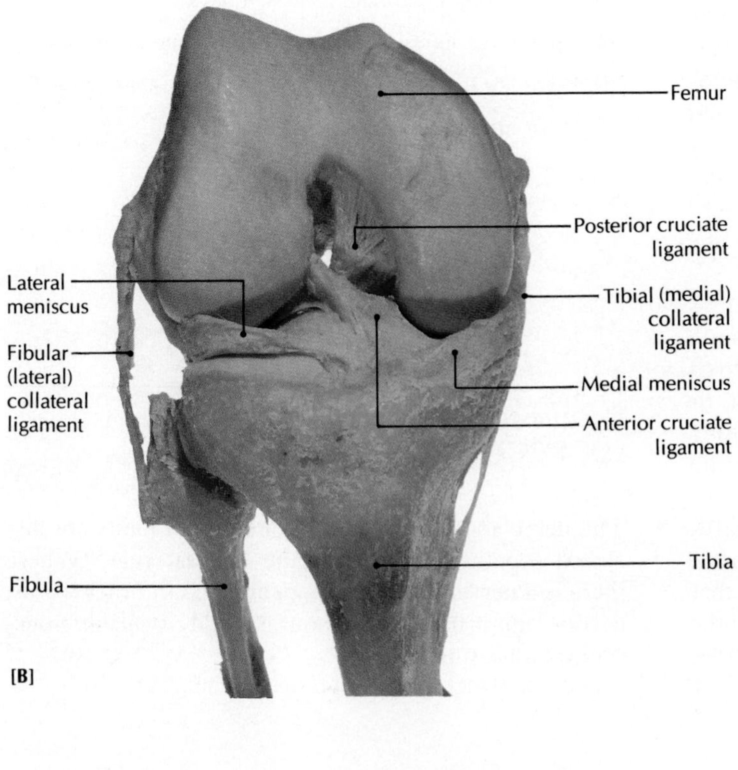

lateral condyle uses up its articular surface just before full extension is realized. The completion of extension occurs as the medial condyle continues to rotate on its longer articular surface, accompanied by the "screw-home" action and the locking of the knee. During this final phase, the lateral condyle acts as a pivot.

Several anatomical features are basic to knee movements. The curved condyles of the femur articulate with the flattened condyles of the tibia, allowing some rotation along with flexion and extension. Because the pelvis is wide, the shaft of the femur is medially directed from its proximal (hip) end to its distal (knee) end to assume an oblique set at the knee joint. Except in bowlegged people, the tibias of both legs are usually parallel.

In front, the synovial membrane is lax when the knee is extended. The membrane is extensive enough so that no strain is exerted on it when the knee is flexed. Thickenings within the articular capsule include the *tibial* (medial) *collateral ligament* and the *fibular* (lateral) *collateral ligament* extending from the sides of the femur. Both are strong ligaments that are slack during flexion and taut during extension.

A common knee injury to football players is caused by a blow on the lateral side of the leg, which produces excessive stress on the medial side of the knee that may cause a strained (stretched) or torn tibial collateral ligament. Such an injury may be accompanied by damage to the medial meniscus. Injuries to the knee usually involve the three Cs: collateral ligaments, cruciate ligaments, and cartilages (menisci).

In the center of the knee joint, between the condyles of the femur and tibia, are two ligaments: the *anterior* and *posterior cruciate ligaments.* They are strong cords that cross each other like an X. The anterior cruciate ligament is taut when the knee is fully extended and slack when the knee is flexed. It prevents the backward dislocation of the femur, forward dislocation of the tibia, and excessive extensor or rotational movements, especially hyperextension. The posterior cruciate ligament becomes tighter as flexion proceeds. It prevents the forward dislocation of the femur, backward dislocation of the tibia, and hyperflexion of the knee.

Of the two ligaments, the anterior cruciate ligament is torn much more often. The damage may come from a blow driving the tibia forward relative to the femur, from a blow driving the femur back relative to the tibia, or from excessive hyperextension of the knee. The resulting instability of the knee is so serious that use of the limb may be impaired severely.

Within the knee joint are two crescent-shaped fibrocartilaginous plates called the *lateral* and *medial menisci.* They act as a cushion between the ends of bones that meet in a joint. The menisci at the knee joint deepen the articular surfaces between the femoral and tibial condyles and increase the security of the joint by adjusting the nonmatching surfaces of the tibial and femoral condyles.

Injury to the medial meniscus occurs about 20 times more often than to the lateral meniscus because the medial meniscus is attached to the medial collateral ligament. A sudden twist of the flexed knee that is bearing weight can tear the medial meniscus.

In addition to the ligaments, the muscles and tendons surrounding the knee joint also play a role in stabilizing it. Of the muscles, the most important is the quadriceps femoris.

Several bursae cushion the knee joint. They are (1) the *suprapatellar bursa* above the patella, between the quadriceps femoris muscle and the femur, (2) the *prepatellar bursa* between the skin and the patella, and (3) the *infrapatellar bursa* between the skin and tibial tuberosity. The prepatellar bursa, which allows for the free movement of the skin over the patella, is subject to a friction bursitis, known as "water on the knee." (Repeated blows or falls may also cause "water on the knee.") This condition occurs in miners and others who frequently work on all fours. Bursitis of the infrapatellar bursa is often found in roofers and others who kneel with their trunks upright.

ASK YOURSELF

1 What types of movement are possible at the temporomandibular joint?

2 What major structures make up the ball-and-socket joint of the shoulder?

3 How does the arrangement of the articulating bones of the hip joint contribute to its stability?

4 What three joints make up the composite knee joint?

5 What are the primary movements at the knee joint?

6 What are the functions of the menisci?

NERVE SUPPLY AND NUTRITION OF SYNOVIAL JOINTS

The nerve and blood supplies of synovial joints are discussed together because of the general rule "Where there is a nerve, there is also an artery." In other words, nerves cannot function without a readily available nourishing blood supply.

Nerve Supply of Joints

The same nerves supplying the muscles that move a joint also send out branches that supply the skin over the muscle attachments and the joint itself. One joint may be supplied by the branches of several nerves. Many *sensory nerve fibers* of these nerves terminate as nerve endings in the fibrous capsules, ligaments, and synovial membranes of the joints. Sensory nerve fibers relay information about the joint activity to the spinal cord and brain. This information is processed in the spinal cord and brain, which initiate "messages" (impulses) that are conveyed by *motor fibers* of these nerves to the muscles controlling the movements of the joints.

Blood and Lymph Supply of Joints

All the tissues of synovial joints receive nutrients either directly or indirectly from blood vessels, except the articulating portions of the articular cartilages, disks, and menisci. (No articulating cartilage has a direct blood supply, for example.) The articulating areas are nourished indirectly by the synovial fluid that is distributed over the surface of the articular cartilage. The flow of synovial fluid is stimulated by movement at the joints, and this is why physical activity is essential for the maintenance of healthy joints. Under normal conditions, only enough synovial fluid is secreted to produce a thin film over the joint surfaces. If a joint becomes inflamed, however, the secretion of synovial fluid may become overstimulated, causing the fluid to accumulate and producing swelling and discomfort at the joint.

Arteries near a synovial joint send out branches that join together freely on the outer surface of the joint. These branches penetrate the fibrous capsule and ligaments and form a branching network of capillaries that spreads throughout the synovial membrane. These capillaries are so numerous and so close to the surface of the synovial membrane that it is relatively easy for a hemorrhage to occur into the articular cavity, even as the result of only a minor injury. The arteries also extend into the fatty pads and the nonarticulating portions of the articular cartilage, disks, and menisci.

ASK YOURSELF

1 How is the nonarticulating portion of a joint nourished?

2 Why is a hemorrhage at a joint likely after an injury?

THE EFFECTS OF AGING ON JOINTS

With age, there is usually a decrease in the synovial fluid in joints, and the cartilage in joints also becomes thinner. Arthritis, bursitis, and other diseases of the joints are prevalent, with some of them being attributed to autoimmune conditions. Ligaments become shorter and less flexible, bending the spinal column into a hunched-over position.

WHEN THINGS GO WRONG

Arthritis

The word **arthritis** means "inflammation (*-itis*) of a joint (*arthro-*)." Arthritis may affect any joint in the body, but the most common sites are the shoulders, knees, neck, hands, low back, and hips. About 30 million Americans have arthritis seriously enough to require medical treatment. Arthritis is most common in men over 40, postmenopausal women, blacks, and people in low-income groups. Elderly people in general are most seriously affected. There are as many as 25 different specific forms of arthritis. The most common types are described in the following paragraphs.

Gouty arthritis *Gouty arthritis,* or simply *gout,** is a metabolic disease resulting from chemical processes in the body. One product of the metabolism of purine is uric acid, which is normally excreted in urine. Sometimes, however, the body produces too much uric acid or the kidneys do not remove enough of it. The excess uric acid combines with sodium to form needle-sharp crystals of sodium urate salt that settle in soft body tissues and cause inflammation and pain. Any joint can be affected by

*The word *gout* comes from the Latin word *gutta,* meaning "drop," because it was once thought that gout was caused by drops of unhealthy humors, or poisons, dripping into the joints.

gout, but the most common sites are the big toe and, oddly, since it is not a joint at all, the cartilage of the rim of the ear, just above the earlobe. The pain is sudden and intense, usually starting in the metatarsophalangeal (first) joint of the big toe. The pain may progress to the instep, ankle, heel, knee, or wrist joints. Crystal deposits in the kidneys may lead to kidney stones or even kidney failure.

Osteoarthritis *Osteoarthritis,* or *degenerative arthritis,* is the most common form of arthritis. It is a chronic, progressive degenerative disease, usually caused by a breakdown of chondrocytes during the normal "wear and tear" of everyday living. Osteoarthritis often affects weight-bearing joints, such as the hip, knee, and lumbar region of the vertebral column. It occurs equally in men and women. Osteoarthritis usually begins as a normal part of aging, but it may also develop as a result of damage to joints, including infection or metabolic disorders. Typical symptoms are pain in the joints, morning stiffness, grating of the joints, and restricted movement. Osteoarthritis of interphalangeal joints produces irreversible bony growths in the distal joints (Heberden's nodes) and in the proximal joints (Bouchard's nodes).

Rheumatoid arthritis *Rheumatoid arthritis* is the most debilitating form of chronic arthritis. It is an auto-

immune disease, in which certain T cells attack the joint cartilage. It usually involves matching joints on opposite sides of the body and their surrounding bursae, tendons, and tendon sheaths. Most movable joints can be affected, but the fingers, wrists, and knees are most susceptible. Rheumatoid arthritis affects three times as many women as men and is most prevalent in women between 35 and 45. The disease usually starts with general symptoms, such as fatigue, low-grade fever, and anemia, before it begins to affect the joints. Eventually the tissue in the joint capsule becomes thickened (the tissue is then called a *pannus*) as a result of inflammation and the accumulation of synovial fluid. Soon the disease invades the inte-

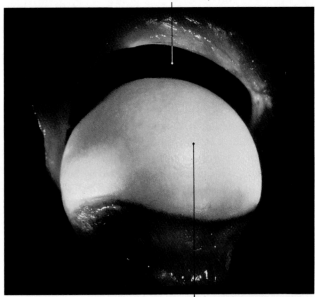
Glenoid fossa of scapula

[B] Head of humerus

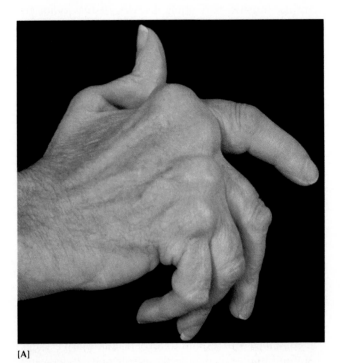
[A]

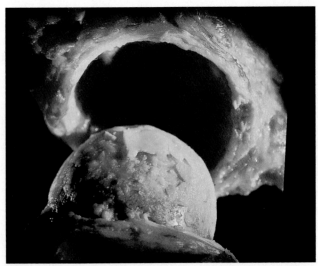
[C]

[A] Rheumatoid arthritis destroys articular cartilage and produces crippling deformities. Photographs of [B] the head of a healthy humerus, and [C] the diseased joint of a person with rheumatoid arthritis. The articular cartilage has been removed.

rior of the joint and interferes with the nourishment of the articular cartilage. Slowly, the cartilage is destroyed, and crippling deformities are produced (see photos). Fibrous tissue then develops on the articulating surfaces of the joint and makes movement difficult (a condition called *fibrous ankylosis;* Gr. *ankulos,* bent). In the final stage, the fibrous tissue becomes calcified and forms a solid fusion of bone, so that the joint is completely non-functional *(bony ankylosis).*

Bursitis

Bursitis is an inflammation of one or more *bursae.* Such an inflammation produces pain and swelling and restricts movement. It occurs most frequently in the subacromial bursa near the shoulder joint, the prepatellar bursa near the knee joint, and the olecranon bursa of the elbow. The pain that is associated with subacromial bursitis severely limits the mobility of the shoulder, especially during the initial stages of abduction or flexion of the shoulder joint.

Bursitis may be caused by an infection or by physical stress from repeated friction related to a person's activities (miner's knee, tennis elbow) or from repeated blows or falls.

Dislocation

A *dislocation,* or *luxation* (L. *luxare,* to put out of joint), is a displacement movement of bones in a joint that causes two articulating surfaces to become separated. A partial dislocation is called a *subluxation.* A dislocation usually is the result of physical injury, but it can also be congenital (as in hip dysplasia) or the side effect of a disease such as Paget's disease (see Chapter 6).

The *shoulder joint* is dislocated more than any other joint, partly because of the shallowness of the glenoid cavity, which holds the head of the humerus, and the loose capsule. The *hip joint,* in contrast to the shoulder, is seldom dislocated because of the stability produced by the secure fit of the head of the femur inside the deep acetabulum, the strong articular capsule, and the strong ligaments and muscles that surround the joint.

The *TM(jaw) joint* may also become dislocated. Following a vigorous laugh or yawn, the head of the mandible may move so far forward that it glides in front of the articular tubercle, with an accompanying spasm of the temporalis muscle that jams the mandible under the zygomatic bone and locks the jaw. As a result of such a dislocation, the mouth cannot be closed, except by a knowledgeable second party who understands how to relocate the joint.

To return the jaw to its proper location, it must be pushed downward to overcome the pull of the temporalis, masseter, and medial pterygoid muscles. As the condyle is pushed back and over the tubercle to its proper location, the jaw snaps back into place. When performing such a maneuver it is important *not* to force the jaw down by exerting pressure with the thumbs on the patient's teeth, because the sudden release of the jaw is accompanied by an opposition of the molars that is powerful enough to damage the thumbs seriously. In the correct procedure, downward force is generally exerted on the mandibular ridge just below the teeth.

Immediate reduction (relocation) of the displaced joint prevents edema, muscle spasm, and further damage to tissues, nerves, and blood vessels. After reduction, the injured joint may need to be immobilized for two to eight weeks. (However, see New Therapy for Damaged Joints below.)

Temporomandibular Joint (TMJ) Syndrome

The pain/dysfunction syndrome that places the chewing muscles and the TMJ muscles under painful stress is called *temporomandibular joint (TMJ) syndrome.* It may be caused by the clenching of teeth (bruxism), a misaligned bite (malocclusion), arthritis, or, more typically, psychologically related stress. Stress management, and heat and muscle-relaxation treatments may help the condition. Dental braces or other attempts to improve the bite should be temporary and reversible.

Sprain

A *sprain* is a tearing of ligaments that follows the sudden wrench of a joint. It is usually followed by pain, loss of mobility (which may not occur until several hours after the injury), and a "black-and-blue" discoloration of the skin, caused by hemorrhaging into the tissue surrounding the joint. In a slight sprain, the ligaments heal within a few days, but severe sprains (usually called "torn ligaments") may take weeks or even months to heal. In some cases, the ligament must be repaired surgically. A sprained ankle is the most common joint injury.

New Therapy for Damaged Joints

Ordinarily, damaged joints are immobilized in a cast for several weeks after surgery. According to recent reports, however, joints heal faster after surgery when they are kept in constant motion, rather than being immobilized. When joints move, they stimulate the flow of synovial fluid, which prevents the synovial membrane from adhering to cartilage in the joint. Constant motion at the joint can be accomplished by a new technique called *continuous passive motion* (CPM), in which a motorized apparatus moves the joint backward and forward gently. Usually, the apparatus does not disturb sleep after a short period of adjustment, and the muscles are not fatigued because the machine does all the work.

Generally, CPM patients are ready for physical therapy about 7 to 10 days after surgery. Immobilized patients usually begin therapy about six weeks after surgery, when their plaster cast is removed.

CHAPTER
SUMMARY

Bones give the body its structural framework, and muscles give it its power, but joints (articulations) provide the mechanism that allows the body to move.

Classification of Joints (p. 203)

1 Joints may be classified by function and degree of movement: an immovable joint is a **synarthrosis;** a slightly movable joint is an **amphiarthrosis;** a freely movable joint is a **diarthrosis.**

2 Based on structure or the type of tissue that connects the bones, joints may be classified as *fibrous, cartilaginous,* or *synovial.*

Fibrous Joints (p. 203)

1 Fibrous connective tissue unites the bones in **fibrous joints.**

2 Three types of fibrous joints are **sutures, syndesmoses,** and **gomphoses.**

Cartilaginous Joints (p. 203)

1 In **cartilaginous joints,** bones are united by a plate of hyaline cartilage or a softer fibrocartilaginous disk.

2 The two types of cartilaginous joints are **synchondroses** and **symphyses.**

Synovial Joints (p. 206)

1 *Synovial joints* are the articulations where the bones move easily upon each other. The ends of the bones are plated with a smooth **articular cartilage,** lubricated by **synovial fluid,** and bound together by an **articular capsule.**

2 The synovial joint is composed of the **synovial cavity, articular cartilage, articular capsule,** and **ligaments.**

3 *Bursae* are sacs filled with synovial fluid that helps eliminate friction when a muscle or tendon rubs against another muscle, tendon, or bone. A **tendon sheath** is a modification of a bursa that helps reduce friction around tendons.

Movement at Synovial Joints (p. 208)

1 Movement at synovial joints may be restricted by interference from other structures, tension exerted by ligaments of the articular capsule and muscles.

2 In any movement about a joint, one member of a pair of articulating bones maintains a fixed position, and the other bone moves in relation to it about an axis (or axes).

3 When movement is restricted to rotation about one axis, the joint is said to be **uniaxial.** When two movements can take place about two axes of rotation, the joint is **biaxial.** A **multiaxial** (triaxial) joint has three independent rotations about three axes.

Types of Synovial Joints (p. 209)

1 *Synovial joints* are classified on the basis of the shape of their articulating surfaces and the types of movements those shapes permit.

2 *Hinge joints* are uniaxial. The convex surface of one bone fits into the concave surface of another.

3 *Pivot joints* are uniaxial and have a rotational movement around a long axis through the center of the pivot.

4 *Condyloid joints* are modifications of ball-and-socket joints, with ligaments and muscles limiting movement to only two axes of rotation (biaxial).

5 *Gliding* (plane) *joints* are small biaxial joints formed by essentially flat surfaces.

6 *Saddle joints* are multiaxial joints in which both articulating bones have saddle-shaped concave and convex areas.

7 *Ball-and-socket joints* are multiaxial joints of the most movable type. The globelike head (ball) of one bone fits into the cuplike concavity (socket) of another bone.

Description of Some Major Joints (p. 213)

1 The *temporomandibular (jaw) joint,* the only movable joint of the head, is a multiaxial synovial joint. It includes the articular head of the mandible and the mandibular fossa and articular tubercle of the temporal bone.

2 The *glenohumeral (shoulder) joint* is a multiaxial ball-and-socket joint. The head of the humerus articulates with the shallow glenoid cavity of the scapula.

3 The *coxal (hip) joint* is a multiaxial, ball-and-socket, synovial joint. The head of the femur articulates with the acetabulum of the hipbone (os coxa).

4 The *tibiofemoral (knee) joint* is the largest, most complex, and one of the weakest joints in the body. It is a synovial, modified hinge joint. The knee joint is a composite of the articulation between (a) the medial femoral and medial tibial condyles, (b) the lateral femoral and lateral tibial condyles, and (c) the patella and the femur.

Nerve Supply and Nutrition of Synovial Joints (p. 220)

1 Joints cannot function without the effective stimulation of nerves, which in turn need a nourishing blood supply.

2 Some nerves of a joint are *sensory* nerves that convey information to the brain and spinal cord. Others are *motor* nerves that convey instructions from the brain and spinal cord to the muscles controlling joint movements.

3 All the tissues of synovial joints receive nutrients directly from blood vessels except the articulating cartilages, disks, and menisci, which are nourished by the synovial fluid.

The Effects of Aging on Joints (p. 221)

1 With age, synovial fluid usually decreases, articular cartilage becomes thinner, and ligaments become shorter and less flexible.

2 Arthritis and other joint diseases become prevalent.

STUDY AIDS FOR REVIEW

MEDICAL TERMINOLOGY

ANKYLOSIS (Gr. *ankulos*, bent + *osis*, condition of) Stiffness or crookedness in joints.

ARTHRODESIS (Gr. *arthro*, joint + *desis*, binding) The surgical fusion of the bones of a joint.

ARTHROGRAM (joint + Gr. *grammos*, picture) An x-ray picture of a joint taken after the injection into the joint of a dye opaque to x-rays.

ARTHROPLASTY (joint + Gr. *plastos*, molded) The surgical repair of a joint or the replacement of a deteriorated part of a joint with an artificial joint.

BURSECTOMY (bursa + *-ectomy*, removal of) The surgical removal of a bursa.

CAPSULORRHAPHY (capsule + Gr. *raphe*, suture) The surgical repair of a joint capsule to prevent recurrent dislocations.

CHRONDRITIS (Gr. *khondros*, cartilage + *-itis*, inflammation) Inflammation of a cartilage.

MENISCECTOMY (meniscus + *-ectomy*, removal of) The surgical removal of the menisci of the knee joint.

OSTEOPLASTY (Gr. *osteon*, bone + *plastos*, molded) The scraping away of deteriorated bone.

OSTEOTOMY (bone + *-tomy*, cutting of) The surgical cutting of bone, for example, the realignment of bone to relieve stress.

RHEUMATISM (Gr. *rheumatismos*, to suffer from a flux or stream) Any of several diseases of the muscles, tendons, joints, bones, or nerves. Generally replaced by the term *arthritis*.

RHEUMATOLOGY The study of joint diseases, especially arthritis.

SHOULDER SEPARATION A separation of the acromioclavicular joint, not the shoulder. It is usually a serious injury only when the accompanying ligaments are torn.

SUBLUXATION (L. *sub*, less than + *luxus*, dislocated) An incomplete dislocation.

SYNOVECTOMY (L. *synovia*, lubricating liquid + *-ectomy*, removal of) The surgical removal of the synovial membrane of a joint.

SYNOVITIS (*synovia* + *-itis*, inflammation) An inflammation of the synovial membrane.

UNDERSTANDING THE FACTS

1 What is an articulation?
2 In terms of functional classification, which of the three major joint types is the most movable? The least movable?
3 What are the three major joint types by structure?
4 What is a syndesmosis? A gomphosis? A synchondrosis?
5 What type, by structure, are most permanent joints?
6 Give three functions of articular disks.
7 What are the functions of synovial fluid?
8 Perform the following movements, and name the type of movement that is occurring:
  a move thumb pad to touch a finger pad
  b move arm so that it describes the surface of a cone
  c bring wrist close to shoulder
  d rotate hand and forearm, turning palm forward
  e move head indicating "no"
  f spread fingers
  g open mouth
9 Give an example of a uniaxial, biaxial, and multiaxial joint.
10 Why is the shoulder joint also called the glenohumeral joint?
11 What movements are possible at the shoulder joint?
12 How is the iliofemoral ligament important to the hip joint?
13 A torn cartilage of the knee, a common athletic injury, usually involves which specific structure?
14 Which parts of synovial joints have no nerve supply?
15 Why is physical activity so essential to proper nourishment of joints?

UNDERSTANDING THE CONCEPTS

1 What is the difference between a suture and a synostosis?
2 The articulating surfaces of joints are often almost friction-free. What is responsible for this smoothness?
3 What relationship does the depth of the socket of a joint have to its range of movement?
4 How does the structure of a pivot joint relate to its movement?
5 Why is it important for the hip joint to be very secure, strong, and stable?
6 Why do injuries to joint ligaments and nonarticulating parts of a joint often produce intense pain, while injuries to the articulating portions do not?
7 Why is arthritis most common in the elderly?
8 The statement "Structures that encourage movement do not promote stability" appears in one form or another throughout this chapter. Now that you have studied this material on articulations, what does this statement mean to you?

SELF-QUIZ
Multiple-Choice Questions

9.1 Based on structure, which of the following types of joints contain a joint cavity?
   a fibrous
   b cartilaginous
   c synovial
   d suture
   e gomphoses

9.2 Which of the following is *not* a type of fibrous joint?
   a suture
   b syndesmoses
   c gomphoses
   d synchondrosis
   e a and b

9.3 A cartilaginous joint
   a contains a plate of hyaline cartilage
   b lacks a joint cavity
   c is slightly movable
   d a and b
   e a, b, and c

9.4 A synovial joint has
   a articular cartilage
   b synovial fluid
   c an articular capsule
   d a joint cavity
   e all of the above

9.5 Which of the following limits movement at synovial joints?
   a interference by other structures
   b tension exerted by ligaments of the articular capsule
   c muscle tension
   d a and b
   e a, b, and c

9.6 When the movement at a joint is limited to rotation about one axis, the joint is said to be
   a uniaxial
   b biaxial
   c multiaxial
   d triaxial
   e flexible

9.7 Which of the following is *not* a type of synovial joint?
   a hinge
   b pivot
   c condyloid
   d suture
   e saddle

9.8 Which of the following is a multiaxial joint?
   a ball-and-socket
   b saddle
   c gliding
   d a and b
   e a, b, and c

9.9 The temporomandibular joint is a
   a fibrous joint
   b cartilaginous joint
   c gliding joint
   d hinge joint
   e c and d

9.10 The shoulder joint is
   a multiaxial
   b a ball-and-socket joint
   c a synovial joint
   d the glenohumeral joint
   e all of the above

9.11 Which of the following is part of the knee joint?
   a the articulation between the medial femoral and the medial tibial condyles
   b the articulation between the lateral femoral and the lateral tibial condyles
   c the articulation between the patella and the femur
   d a and b
   e a, b, and c

9.12 The hip (coxal) joint is
   a multiaxial
   b a ball-and-socket joint
   c a synovial joint
   d a and b
   e a, b, and c

True-False Statements

9.13 A slightly movable joint is called an amphiarthrosis.

9.14 A freely movable joint is called a synarthrosis.

9.15 The amount of movement in a syndesmosis depends upon the distance between the bones and the flexibility of connecting tissue.

9.16 The chief function of a synchondrosis is to permit growth.

9.17 A symphysis is sometimes called a secondary synchondrosis.

9.18 In general, the more flexible a joint is, the more stable it is.

9.19 Bursitis is caused by inflamed tendon sheaths.

9.20 Hinge joints are uniaxial.

9.21 Condyloid joints are biaxial.

9.22 The TM joint is the only movable joint of the head.

9.23 The shoulder joint has more freedom of movement than any other joint in the body.

9.24 The knee (tibiofemoral) joint is the largest and most complex joint in the body.

Matching

9.25 ____ extension
9.26 ____ flexion
9.27 ____ abduction
9.28 ____ adduction
9.29 ____ depression
9.30 ____ retraction
9.31 ____ eversion
9.32 ____ pronation

a increasing the angle between two bones
b movement of big toe downward
c lowering a body part
d movement toward the midline
e bone moves in circular motion

f movement away from the midline
g backward pushing movement
h movement around an axis
i decreasing the angle between two bones
j pivoting movement of forearm

A SECOND LOOK

1 Identify the joints shown in the following drawings.

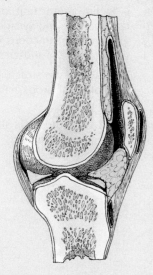
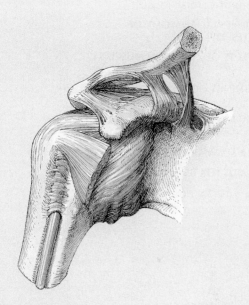

2 In the following drawing of the knee joint, label the anterior and posterior cruciate ligaments, lateral meniscus, and fibular collateral ligament.

3 Identify the movements shown in the following drawings.

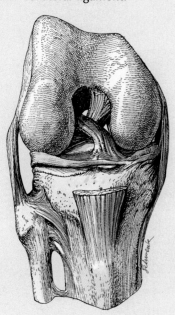
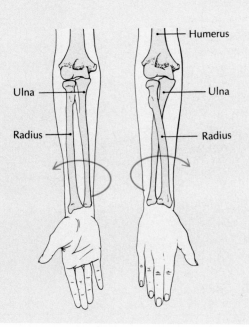

10

Muscle Tissue

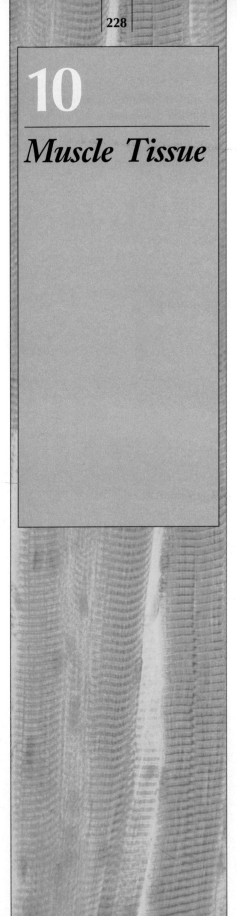

LEARNING OBJECTIVES

1 Identify the four important physiological properties and three general functions of muscle tissue.

2 List the three types of muscle tissue, and relate each to its special functions and characteristics.

3 Describe the cell structure and organization of skeletal muscle.

4 Identify three forms of fascia and five other connective tissue elements in skeletal muscle.

5 Describe the blood and nerve supply to skeletal muscles.

6 Define motor unit, neuromuscular junction, and synaptic cleft, and explain their functions in the nervous control of muscles.

7 Describe the structure and basic physiological properties of smooth muscle.

8 Describe the basic structure of cardiac muscle tissue.

Joints make a skeleton potentially movable, and bones provide a basic system of levers, but bones and joints cannot move by themselves. The driving force, the power behind movement, is muscle tissue. The three types of muscle tissue are smooth, cardiac, and skeletal.

The basic physiological property of muscle tissue is **contractility,** the ability to *contract,* or shorten. In addition, muscle tissue has three other important physiological properties. *Excitability* (or irritability) is the capacity to receive and respond to a *stimulus,* **extensibility** is the ability to be stretched, and **elasticity** is the ability to return to its original shape after being stretched or contracted. These four properties of muscle tissue are related, and all involve movement.

In the process of contracting, important work is done. For example, food is passed along the digestive tract by a series of rhythmic waves of **smooth muscle** contractions. The contractions of **cardiac muscle** pump blood with remarkable force and consistency from the heart to all parts of the body. As certain **skeletal muscles** in your body contract, your lower limbs move at the ankle, knee, and hip, and you walk or run. Muscular contractions also help to maintain body posture in a standing or sitting position, even when there is no obvious motion. Finally, the contractions of skeletal muscles produce much of the heat needed to maintain a homeostatic body temperature. To sum up, the general functions of muscle tissue are **movement, posture,** and **heat production.**

The muscular system presents no exception to the general rule that in the human body structure and function work together to accomplish the most advantageous results. In the following sections you will see how the anatomy of each muscle type allows it to perform its own special physiological function.

SKELETAL MUSCLE

Skeletal muscle tissue acquired its name because most of the muscles involved are attached to the skeleton and make it move.* It is also called *striated muscle* because its fibers or cells show alternating light and dark stripes, or striations, when viewed with a light microscope. Skeletal muscle has yet another descriptive name, *voluntary muscle,* because we can contract it when we want to. Although skeletal muscles can be contracted voluntarily, they are also capable of contraction *without* conscious control (involuntarily). Muscles are usually in a partially contracted state, which gives them *tonus,* or what we commonly refer

to as "muscle tone." Tonus is necessary to keep a muscle ready to react to the stimulus preceding a complete contraction, to hold parts of the body such as the head erect, and to aid in the return of blood to the heart.

Cell Structure and Organization

Skeletal muscle is composed of individual, specialized cells called **muscle fibers** [FIGURE 10.1]. These cells are known as fibers because they have a long, cylindrical shape and numerous nuclei. Thus, they look more like fibers than typical cells. Muscle fibers average 3.0 cm (1.2 in.) in length, but some may be more than 30 cm (12 in.) or as short as 0.1 cm (0.04 in.).[†] Diameters usually range from 0.01 cm (0.004 in.) to 0.001 cm (0.0004 in.)

Each skeletal muscle fiber is enclosed within a thin plasma membrane called the **sarcolemma** (Gr. *sarkos,* flesh + *lemma,* husk). The fiber contains several nuclei and a specialized type of cytoplasm called **sarcoplasm.** Within the sarcoplasm are many mitochondria and a large number of individual threadlike fibers known as **myofibrils** (*myo,* muscle), which run lengthwise and parallel to one another [FIGURE 10.2]. Wrapped around each myofibril, and running parallel to it, is the **sarcoplasmic reticulum.** This network of tubes and sacs contains calcium ions and is something like the smooth endoplasmic reticulum found in other types of cells. Crossing the sarcoplasmic reticulum at right angles are the **transverse tubules,** or *T tubules,* a series of tubular organelles that run across the fiber to the outside. The sarcolemma continues within the muscle fiber as the lining of the T tubule. Together, a T tubule and a *terminal cisterna* of the sarcoplasmic reticulum on either side of the tubule make up a **triad.**

The myofibrils are made up of many thick and thin threads called **myofilaments.** The thick myofilaments are composed of a fairly large protein called **myosin.** The thin myofilaments are composed mainly of a smaller protein, **actin,** and they also contain the proteins troponin and tropomyosin. The striped appearance of skeletal muscle tissue when viewed with a light microscope is caused by the arrangement of the actin and myosin strands. An overlapping of the thick myosin strands and the thin actin strands produces dark *A bands;* the thin actin strands alone appear as the light *I bands* [FIGURE 10.1D, E]. Cutting across each I band, like a dime in a stack of pennies, is a dark *Z line.* Within the A band is a somewhat lighter *H zone,* consisting of thick myosin

---

*The word *muscle* is based on the Latin *musculus,* which means "little mouse." Muscles were so named because the movement of muscles under the skin was thought to resemble a running mouse.

[†]The sartorius muscle of the anterior thigh is the longest muscle in the body. It may contain muscle fibers over 30 cm (12 in.) long, reaching from the hip to the knee. The shortest fibers are those of the stapedius muscle in the inner ear, which is shorter than 1 mm (0.04 in.); these fibers attach to the tiniest of bones, the stapes of the ear. The most powerful muscle in the body is probably the gluteus maximus of the buttocks.

[A] MUSCLE IN ARM

FIGURE 10.1 ANATOMY OF A SKELETAL MUSCLE

A progression from gross structure to molecular structure. **[A]** Muscle in arm. **[B]** Muscle fascicle (bundle of muscle fibers). **[C]** Muscle fiber (bundle of myofibrils). **[D]** Myofibril. Myofibrils, which are made up of myofilaments, show the banding that gives the designation *striated* to the muscle. **[E]** Myofilaments. The arrangement of myofilaments forms the sarcomere, the functional unit of muscle contraction, extending from Z line to Z line. **[F]** Light micrograph of skeletal muscle fibers. ×600. **[G]** Electron micrograph of skeletal muscle showing sarcomere. ×20,000.

Nucleus   Muscle fiber

[B] MUSCLE FASCICLE

[C] MUSCLE FIBER

A band   I band   A band   I band

Z line   M line   Z line

SARCOMERE

[D] MYOFIBRIL

I band   A band   I band

Actin
Myosin

Z line   M line   Z line

H zone

SARCOMERE

[E] MYOFILAMENTS

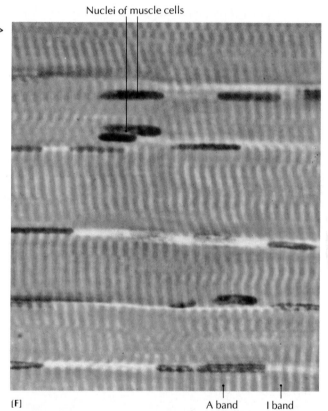

Nuclei of muscle cells

[F]

A band   I band

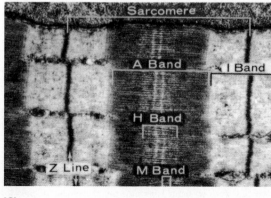

Sarcomere

A Band   I Band

H Band

Z Line   M Band

[G]

FIGURE 10.2 THREE-DIMENSIONAL DRAWING OF A SMALL PORTION OF A MUSCLE FIBER

The myofibril to the right has the sarcoplasmic reticulum peeled away to show the A and I bands, H zone, and M and Z lines.

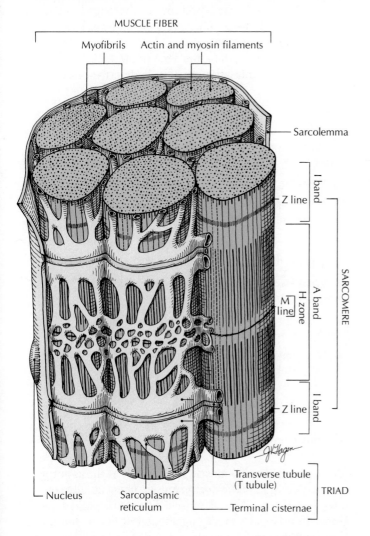

MUSCLE FIBER
Myofibrils   Actin and myosin filaments
Sarcolemma
I band
Z line
A band
H zone
M line
SARCOMERE
I band
Z line
Transverse tubule (T tubule)
TRIAD
Nucleus   Sarcoplasmic reticulum
Terminal cisternae

strands only. Extending across the H zone is a delicate *M line,* which connects adjacent thick filaments. The fundamental unit of muscle contraction is the *sarcomere* (Gr. *meros,* part), which is made up of a section of the muscle fiber that extends from one Z line to the next one.

Q: *Why is one muscle larger than another?*

A: Ordinarily, one muscle is larger than another because it contains more bundles of fibers. The largest muscles are found where large, forceful movements are common, such as in the back and legs. Delicate muscles, such as those in the eye, generally produce precise movements. When a muscle is enlarged by exercise, the fibers increase in diameter, but the *number* of fibers remains the same.

Fasciae

The fibrous connective tissue that covers the skeletal muscles and holds them together is part of a network called *fascia* (FASH-ee-uh; pl. *fasciae,* FASH-ee-ee; L. band). Fascia appears in two major forms: superficial and deep.

The *superficial fascia* is located deep to the dermis of the skin and is found especially in the scalp, palms, and soles. Varying in thickness, the superficial fascia is generally composed of loose connective tissue containing blood vessels, nerves, lymphatic vessels, many fat cells, and in the face, the muscles of facial expression. The superficial fascia provides a protective layer of insulation, and allows the skin to move freely over deeper structures.

Below the superficial fascia is the *deep fascia,* made up of several layers of dense connective tissue. Extensions of the deep fascia extend between muscles and groups of muscles. It contains blood vessels, nerves, lymphatic vessels, and small amounts of fat. The deep fascia is found in fibrous structures such as tendon sheaths.

Other Connective Tissue Associated with Muscles

The fascia that surrounds a muscle is a connective tissue sheath called the *epimysium* [FIGURE 10.3]. Extending inward from the epimysium is a layer of connective tissue, the *perimysium,* that encloses bundles of muscle fibers. These bundles of muscle fibers are called *fascicles.* Further extensions of the connective tissue, called *endomysium,* wrap around each muscle fiber. The sarcolemma is located deep to the endomysium. The three sheaths of connective tissue contain many blood vessels, lymphatic vessels, and nerves.

Besides serving as packing material around muscles, protecting and separating them, the connective tissue sheaths provide a point of attachment to bones and other muscles. Extending from the sheaths that cover muscles or muscle fibers are tendons and aponeuroses. A *tendon* is a strong cord of fibrous connective tissue that extends from the muscle to the periosteum of the bone, with many of its collagenous periosteal perforating fibers anchored into the bone. An *aponeurosis* is a broad, flat sheet of dense connective tissue that attaches to two or more muscles that work together or to the coverings of a bone.

Q: *How much of a person's body weight is taken up by skeletal muscles?*

A: Skeletal muscles comprise about 40 to 50 percent of an adult male's total body weight, and about 30 to 40 percent of an adult female's total body weight.

FIGURE 10.3 STRUCTURE OF A SKELETAL MUSCLE

Cross section of a skeletal muscle showing how each muscle fiber is surrounded by the endomysium, each fascicle by the perimysium, and each muscle (group of fascicles) by the epimysium.

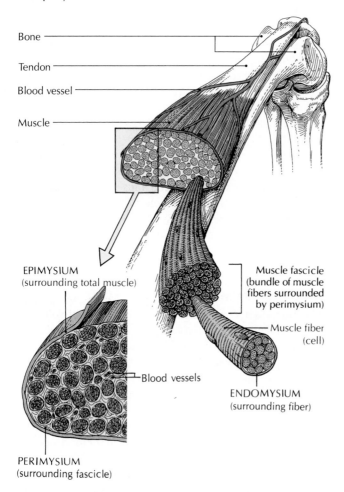

Bone

Tendon

Blood vessel

Muscle

EPIMYSIUM
(surrounding total muscle)

Muscle fascicle
(bundle of muscle
fibers surrounded
by perimysium)

Muscle fiber
(cell)

Blood vessels

ENDOMYSIUM
(surrounding fiber)

PERIMYSIUM
(surrounding fascicle)

Blood and Nerve Supply

Muscles are supplied with blood by arteries that penetrate the connective tissue coverings. These arteries branch out into tiny, thin-walled blood vessels called *capillaries*, which carry an abundant supply of oxygen-rich blood to the muscles. In fact, each muscle fiber is sup-

Q: *What causes a hand or foot to "fall asleep"?*

A: Pressure on a limb (as when you cross your legs, or fall asleep on your arm) may temporarily cut off its blood supply or the steady stream of nerve impulses that help to maintain muscle tone. Sensation may be lost as long as pressure remains on the nerve or blood vessel, but when the blockage is removed, the nerve impulses are effective once more, and the blood supply is returned. The combination of these renewed processes may cause an uncomfortable "pins and needles" feeling until normal functions resume completely.

plied with oxygen and glucose by several capillaries that surround individual muscle cells. As you will see later, without a steady and adequate oxygen supply, the muscles would be unable to contract properly, and if oxygen is cut off for too long, the muscles weaken and die. Accumulated wastes are removed from the muscles via capillaries and carried in the blood toward the heart by veins.

Each skeletal muscle fiber is contacted by at least one nerve ending. One motor nerve fiber can stimulate several muscle fibers at the same time [FIGURE 10.4A]. This important neuromuscular junction is described next.

ASK YOURSELF

1 What are the four important properties of muscle tissue?

2 What are the three types of muscle tissue?

3 Why is skeletal muscle also known as striated or voluntary muscle?

4 Where is a sarcomere located within a muscle?

5 What is the function of fasciae?

6 What is the difference between a fascicle and a fascia?

7 What are the main connective tissue sheaths associated with muscles?

8 How and why are muscles supplied with nerves and blood vessels?

Nervous Control of Muscle Contraction

Although some small muscle fibers (such as those in the eye) may contract individually, muscle fibers usually contract in groups. Skeletal muscle fibers are packed together into fascicles averaging about 150 fibers, and the fibers within each fascicle are controlled by a single *motor neuron* (a nerve cell that stimulates a muscle). In the powerful leg muscles, one motor neuron may stimulate over 1000 fibers. A motor neuron, together with the muscle fibers it innervates, is called a *motor unit* (Figure 10.4).

Motor end plates The site where a motor nerve ending contacts a muscle fiber is called a *neuromuscular* (nerve + muscle), or *myoneural* (muscle + nerve), *junction.* The end branches of the motor neuron, known as *axon terminals*, gain access to the muscle fiber through the endomysium. At the point of contact between the muscle fiber and the motor neuron, the muscle fiber membrane forms a *motor end plate.* The motor end plate is the specialized portion of the sarcolemma of a muscle fiber. It surrounds the terminal end of the axon. As shown in FIGURE 10.4D, mitochondria are particularly

FIGURE 10.4 NEUROMUSCULAR JUNCTION

[A] Scanning electron micrograph of a motor neuron fiber terminating on several muscle fibers. The axon terminals form flattened motor end plates on the surface of the muscle fibers. ×1000. [B] Schematic drawing of a neuromuscular junction. [C] Enlarged drawing showing the motor end plate in detail. [D] Enlarged drawing of the synaptic cleft between the sarcolemma of the muscle fiber and the axon terminal. [E] Electron micrograph of a neuromuscular junction. ×17,400.

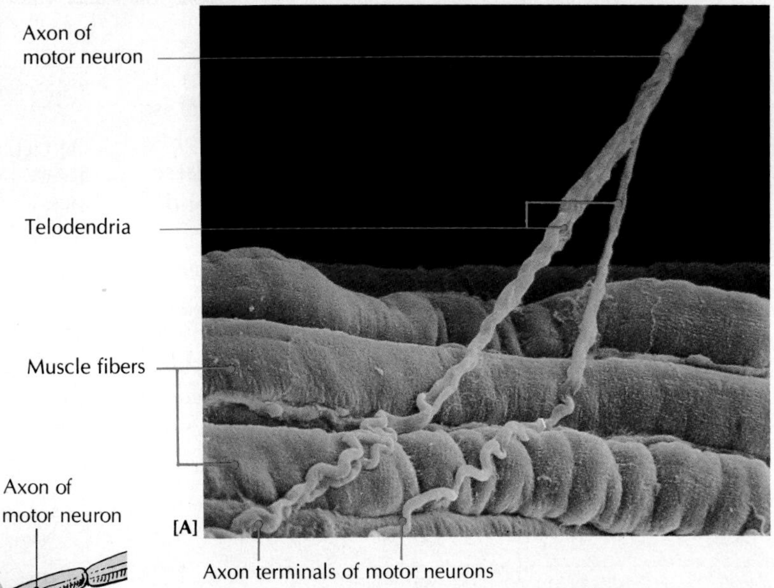

Axon of motor neuron

Telodendria

Muscle fibers

[A]

Axon terminals of motor neurons

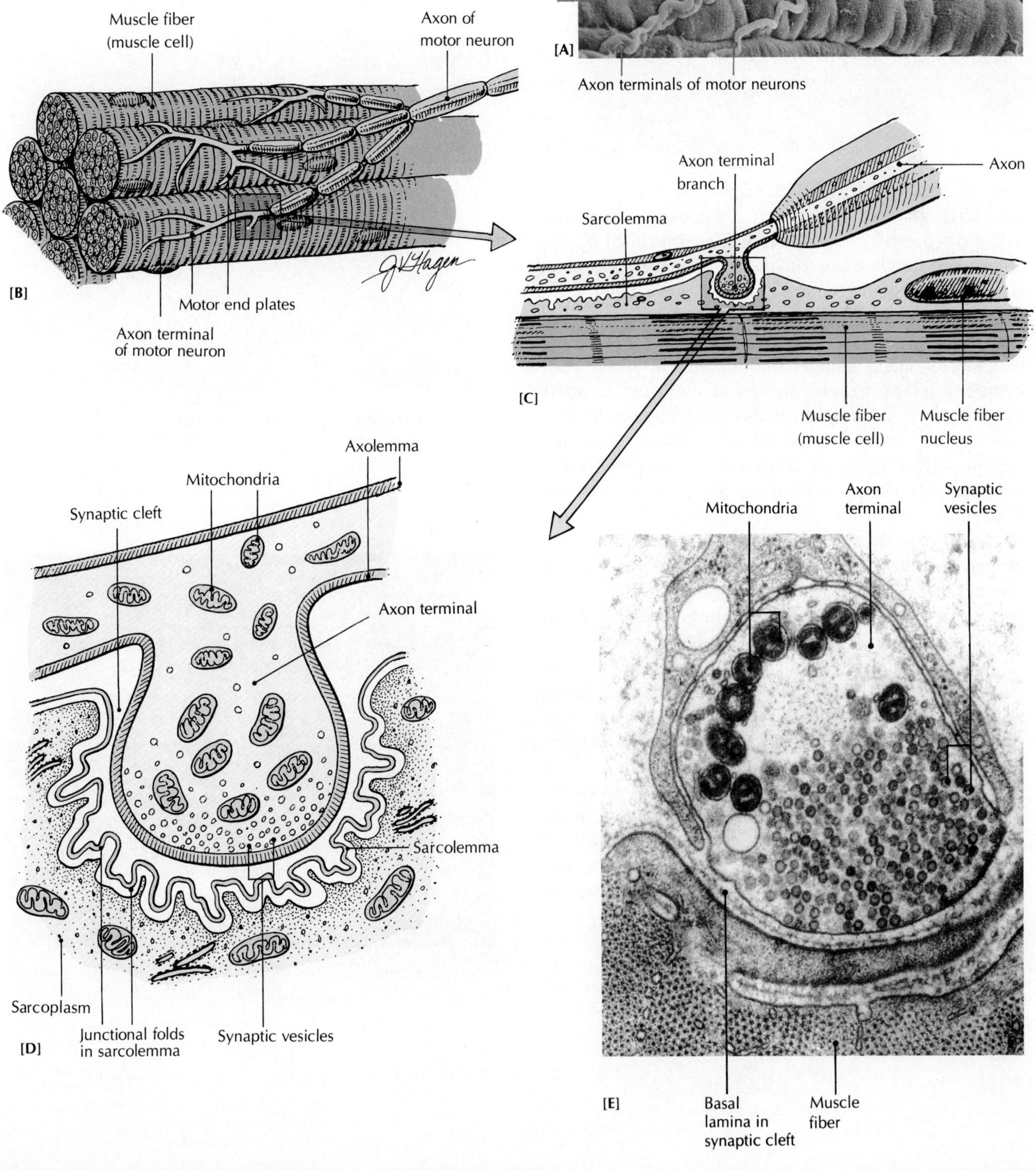

Muscle fiber (muscle cell)

Axon of motor neuron

[B]

Motor end plates

Axon terminal of motor neuron

Axon terminal branch

Axon

Sarcolemma

[C]

Muscle fiber (muscle cell)

Muscle fiber nucleus

Axolemma

Mitochondria

Synaptic cleft

Axon terminal

Sarcolemma

Sarcoplasm

Junctional folds in sarcolemma

Synaptic vesicles

[D]

Mitochondria

Axon terminal

Synaptic vesicles

[E]

Basal lamina in synaptic cleft

Muscle fiber

abundant near a motor end plate. The invaginated area of the sarcolemma under and around the axon terminal is called the *synaptic gutter,* and the clefts inside the folds along the sarcolemma are called *subneural clefts.* These clefts greatly increase the surface area of the synaptic gutter.

At the motor end plate, nerve endings are separated from the sarcolemma of the muscle fiber by a tiny gap called a *synaptic cleft* [FIGURE 10.4D, E]. When impulses (messages) passing along the nerve reach the axon terminal, they cause the release of a chemical transmitter substance. The chemical transmitter diffuses across the synaptic cleft and initiates events that cause the muscle fiber to contract.

SMOOTH MUSCLE

Smooth muscle tissue is so called because it does not appear striped when viewed with a microscope. It is also called *involuntary muscle* because it is controlled by the autonomic nervous system, the involuntary division of the nervous system.* Smooth muscle tissue is found most commonly in the circulatory, digestive, respiratory, and urogenital systems.

Unlike skeletal muscle, smooth muscle is not connected to bones. Instead, slender smooth muscle fibers generally form sheets in the walls of large, hollow organs, such as the stomach and urinary bladder. Although smooth muscle fibers are often arranged in parallel layers, the exact arrangement of the fibers varies from one location to another [FIGURE 10.5A]. In the walls of the intestines, for example, smooth muscle fibers are arranged at right angles to each other, one layer running longitudinally and the next wrapped around the circumference of the tubular intestine. These layers work together in a coordinated action to supply the constrictions that move the intestinal contents toward the anus prior to defecation. In the urinary bladder and uterus, the layers are poorly defined and are oriented in several different directions. Connective tissue outside the layers extends into the spaces between muscle fibers and binds them into bundles. In contrast to the walls of hollow organs, smooth muscle fibers are wrapped around some blood vessels like tape around a rubber hose [FIGURE 10.5B]. This arrangement is appropriate, since contraction and relaxation of the smooth muscle fibers in blood vessels change the diameter of the vessels.

*Although smooth muscle is usually controlled involuntarily, techniques such as biofeedback may actually allow a person to regulate some functional activities of smooth muscle. (See "Biofeedback and the Autonomic Nervous System" on page 402.)

[A] Light micrograph of a cross section of smooth muscle tissue. ×760. **[B]** Scanning electron micrograph of smooth muscle wrapped around a vein. Nerve axons can be seen as thin white lines. ×1000.

Nucleus of smooth muscle cell

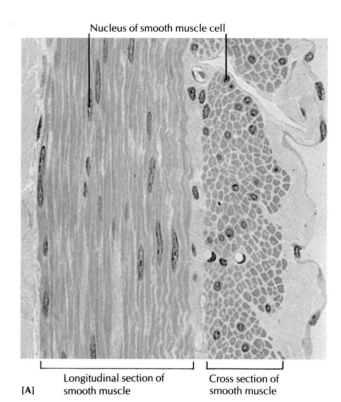

[A] Longitudinal section of smooth muscle | Cross section of smooth muscle

Smooth muscle surrounding vein | Neurolemmocyte

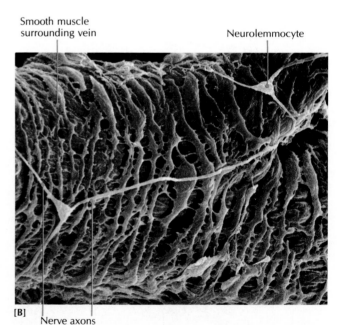

[B] Nerve axons

Properties of Smooth Muscle

The two main characteristics of smooth muscle are that (1) its contraction and relaxation periods are slower than those of any other type of muscle, and (2) its action is rhythmical. Its contractions may last for 30 seconds or more, but it does not tire easily. Such sustained contractions, plus its ability to be stretched far beyond its resting state, make smooth muscle well suited to the muscular control of the stomach and intestines, the urinary bladder, and the uterus, especially during pregnancy.

More smooth muscle is found in the digestive system than in any other place in the body. Smooth muscle cells that line the walls of the stomach and intestines contract and relax rhythmically to help move food along the digestive tract. After we swallow food, all muscular contractions in the digestive system are involuntary until we consciously initiate the process of defecation.

Structure of Smooth Muscle Fibers

Like the cells of skeletal muscle tissue, the cells of smooth muscle tissue are called *fibers*. Each fiber is long (but not nearly as long as skeletal muscle fibers), spindle-shaped, and slender. It contains only one nucleus, which is usually located near the center of the fiber, at its widest point. The shortest fibers (about 0.01 mm) are in the walls of blood vessels, and the longest (about 0.5 mm) are found in the uterus during the late stages of pregnancy. The fibers of the intestines have a more typical length of about 0.2 mm.

Although smooth muscle fibers are arranged differently from those in skeletal muscle, the basic contractile mechanism appears to be much the same. The actin and myosin myofilaments within the myofibrils are very thin, and are arranged more randomly than in skeletal muscle cells. As a result, smooth muscle cells lack striations. Also, the myosin in smooth muscle is chemically distinct from that in skeletal muscle. The sarcoplasmic reticulum of smooth muscle is poorly developed, and T tubules and Z lines are not present.

Based on their arrangements as separate bodies or in bundles, smooth muscle fibers are usually classified as either the single-unit type or the multiunit type.

Single-unit smooth muscle Most smooth muscle is the ***single-unit*** type, which is generally arranged in large sheets of fibers [FIGURE 10.6A]. It is also called *visceral* (L. *viscera,* body organs) *smooth muscle* because it surrounds the hollow organs of the body, such as the uterus, stomach, intestines, and urinary bladder, as well as some blood vessels. The fibers are in contact with each other along many gap junctions. This is critical, because when a muscle cell is stimulated, it contracts and spreads the stimula-

FIGURE 10.6 TYPES OF SMOOTH MUSCLE

[A] Single-unit, or visceral, type and [B] multiunit type.

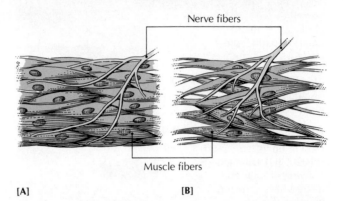

Nerve fibers

Muscle fibers

[A]   [B]

tion to adjacent cells. This method produces a steady wave of contractions, such as those that push food through the intestines. The smooth muscle fiber that receives the stimulus from a motor neuron initially and passes it on to adjacent fibers is known as the *pacemaker cell.*

Two types of contractions take place in single-unit smooth muscle: tonic and rhythmic. *Tonic contractions* cause the muscle to remain in a constant state of partial contraction, or *tonus.* Tonus is necessary in the stomach and intestine, for example, where food is moved along the digestive tract. It is also found in ring-shaped muscles called *sphincters,* which regulate the openings from one part of the digestive tract to the next. Tonus also prevents stretchable organs such as the stomach and urinary bladder from becoming stretched out of shape permanently.

Smooth muscle helps to retain the tension in the walls of expandable organs and tubes, such as the urinary bladder and blood vessels. When a smooth muscle fiber relaxes or rests, its tension is decreased or increased, respectively, for only a few minutes. Then the normal tension is restored.

Rhythmic contractions are a pattern of repeated contractions produced by the presence of self-exciting muscle fibers from which spontaneous impulses travel. The digestive system provides excellent examples of the two types of rhythmic contractions: (1) *mixing movements,* which resemble the kneading of dough, blend the swallowed food with digestive juices, and (2) *propulsive movements,* or *peristalsis,* propel the swallowed food from the throat to the anus. The peristaltic contractions of smooth muscle in the intestines and other tubular parts of the digestive tract form a tightening ring of muscle that moves along the tract, pushing the contents of the tract forward. FIGURE 10.7 shows how smooth muscle contracts like the folds of an accordion.

FIGURE 10.7 SMOOTH MUSCLE CONTRACTION

[A] Electron micrograph of partly contracted smooth muscle fibers. ×1000.
[B] Enlarged drawing of partly contracted fiber. **[C]** Electron micrograph of fully contracted smooth muscle fibers. ×1000. **[D]** Enlarged drawing of fully contracted fiber showing how bundles of myofilaments in the fiber contract. The dense bodies anchor the myofilaments and have characteristics similar to the Z lines in skeletal muscle.

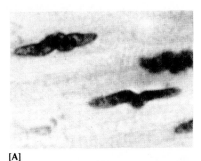

[A]

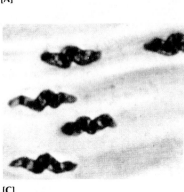

[C]

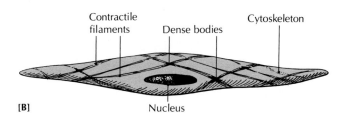

[B]

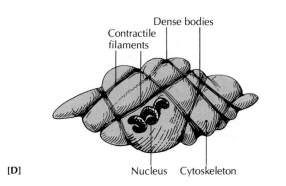

[D]

Multiunit smooth muscle *Multiunit smooth muscle* is so named because each of its individual fibers is stimulated by separate motor nerve endings. There are no gap-junction connections between the muscle cells, and each multiunit cell can function independently [FIGURE 10.6B]. Multiunit smooth muscle is found, for example, in the iris and ciliary muscles of the eye, where rapid muscular adjustments must be made for the eye to focus properly. The arrector pili muscles in the skin that cause the hair to stand on end and that produce "goose bumps" are also of the multiunit type, as are the muscles in the wall of the ductus deferens, the tube that carries sperm from the testes during ejaculation.

ASK YOURSELF

1 Why is smooth muscle also called involuntary muscle?

2 How does the sheetlike arrangement of smooth muscle cells help the muscle to perform its jobs?

3 Why is the slow and rhythmic action of smooth muscle important?

4 What is the difference between mixing movements and propulsive movements?

5 What is single-unit smooth muscle? Multiunit smooth muscle?

CARDIAC MUSCLE

Cardiac muscle tissue (Gr. *kardia*, heart) is found only in the heart. It contains the same type of myofibrils and protein components as skeletal muscle. Although the number of myofibrils varies in different parts of the heart, the contractile process is basically the same as that described for skeletal muscle. As expected in such hard-working tissue, cardiac muscle cells contain huge numbers of mitochondria.* The sarcoplasmic reticulum and T tubules are also evident.

*Mitochondria in cardiac muscle cells occupy about 35 percent of the cell, as compared with 2 percent in skeletal muscle cells. This reflects cardiac muscle's extreme dependence on aerobic metabolism (metabolism requiring oxygen).

Q: *Can cardiac muscle be regenerated?*

A: When cardiac muscle becomes injured, as in a heart attack, there is little chance that the muscle fibers will be able to regenerate. Any healing that does take place produces scar tissue, not muscle tissue.

FIGURE 10.8 CARDIAC MUSCLE

[A] Electron micrograph of cardiac muscle in longitudinal section, showing the intercalated disks between cardiac fibers. Notice the numerous mitochondria. ×12,000. [B] Detailed drawing of intercalated disks, showing two types of junctions involved.

Mitochondria

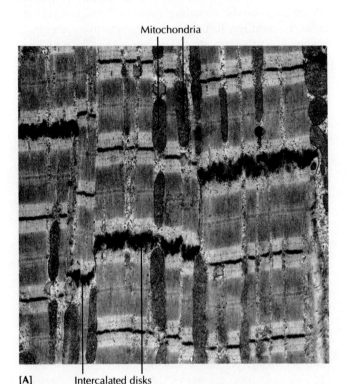

[A] Intercalated disks

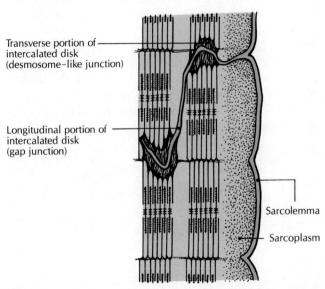

Transverse portion of intercalated disk (desmosome–like junction)

Longitudinal portion of intercalated disk (gap junction)

Sarcolemma

Sarcoplasm

[B]

Structure of Cardiac Muscle

When viewing with a light microscope you can see that cardiac muscle has striations similar to those of skeletal muscle. Although cardiac muscle cells are closely packed, they are *separate*, each with its own nucleus. The cells are joined end-to-end by specialized cell junctions called ***intercalated disks*** (L. *intercalatus*, to insert between) that attach one cell to another with desmosomes, connect the myofibril filaments of adjacent cells, and contain gap junctions that help synchronize the contractions of cardiac muscle cells by allowing electrical impulses to spread rapidly from one cell to the next. In FIGURE 10.8, intercalated disks appear as dark bands that are wider than the Z lines in skeletal muscle. Intercalated disks always occur at the location of the Z lines.

DEVELOPMENTAL ANATOMY OF MUSCLES

The differentiation (specialization) of cells begins very early during embryonic development, forming three distinct layers of germ cells: the inner *endoderm*, the middle *mesoderm*, and the outer *ectoderm*. Endodermal cells develop into the epithelium of the digestive tract (including the liver and pancreas) and respiratory tract. The ectoderm develops into nervous tissue and outer body regions such as the epidermis of the skin and its derivatives, including hair, sweat glands, nails, and sebaceous glands. Muscle (except that in the iris of the eye) and connective tissue (including cartilage and bone) are derived from ***mesodermal cells***. Primitive muscle cells are called *myoblasts*. Muscle development begins about the fifth week of embryonic development, and the muscles differentiate into their final shapes and relations throughout the body during the seventh and eighth weeks.

Many *skeletal muscles* develop from mesodermal tissue arranged in paired, segmented cell masses called ***somites*** [FIGURE 10.9A]. The somites are located on both sides of the central neural tube of the primitive nervous system. The inner and outer walls of the somites differentiate into distinct layers that form different parts of the embryo [FIGURE 10.9B]: (1) The outer layer of a somite is the ***dermatome*** (Gr. skin slice). It develops into connective tissue, including the dermis of the skin. (2) The inner layer, called the ***sclerotome*** (Gr. hard slice), breaks up into a mass of mesenchymal cells that migrate to the spaces surrounding the primitive spinal cord, where they eventually develop into connective tissue, cartilage, and bones associated with the vertebral column. (3) The middle layer of a somite is the ***myotome*** (Gr. muscle slice). The mesenchymal cells of the myotomes eventually dif-

FIGURE 10.9 DEVELOPMENTAL ANATOMY OF SKELETAL MUSCLES

[A] Paired somites surround the neural tube of a 23-day-old embryo; dorsal view. [B] Transverse section of a somite.

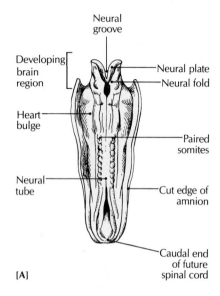

[A]

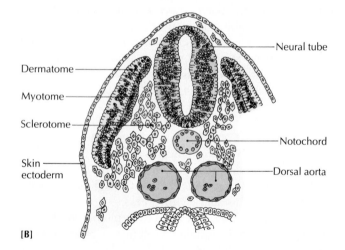

[B]

ferentiate into the striated skeletal muscle cells of some head muscles, and muscles of the neck, thorax, abdomen, and pelvis. The striated muscles of the extremities and most of the muscles of the head are derived from mesodermal cells other than those of myotomes.

Smooth muscle develops from primitive mesenchymal cells that migrate to the developing linings of hollow organs and vessels, including those in the digestive, vascular, respiratory, urinary, and reproductive systems. It also includes the arrector pili muscle associated with hair. *Cardiac muscle* develops from mesodermal cells that specialize early in embryonic life to eventually form the heart, migrating to the developing heart while it is still a simple tube.

THE EFFECTS OF AGING ON MUSCLES

As the body ages, muscles begin to weaken and become smaller and dehydrated. Fibrous tissue appears, and muscles become increasingly infiltrated with fat. Muscle reflexes slow down, and nocturnal cramps are common, especially among women. Cramps may be caused by a deficiency of sodium, improper blood flow, or other factors.

For many reasons, elderly people tend to move about less than they should. As a result of their general reduced mobility, some of the following problems may arise: (1) cramps, loss of muscle tone, and osteoporosis; (2) blood clots (embolisms) within blood vessels; (3) digestive disturbances, such as diarrhea, constipation, and general indigestion; some digestive problems may be related to decreased peristalsis; (4) accumulations of secretions within the lungs and the rest of the respiratory tract, disrupting the balance between carbon dioxide and oxygen; (5) kidney stones and infections of the urinary tract; and (6) bedsores in bedridden patients whose movements are restrained. Regular physical activity is one of the greatest aids to healthy muscles.

WHEN THINGS GO WRONG

Muscular Dystrophy

Muscular dystrophy is a general name for a group of inherited diseases that result in progressive weakness due to the degeneration of muscles. It is usually limited to the skeletal muscles, but cardiac muscle may also be affected. The characteristic symptoms include the degeneration and reduction in size of muscle fibers and an increase in connective tissue and fat deposits. Muscular dystrophy usually begins in childhood or early adolescence, but adults can also be afflicted.

Although the exact cause of muscular dystrophy is unknown, the defective gene that causes the *Duchenne* form of the disease has been identified.* Progressive deterioration cannot be stopped, at least not permanently. Treatment usually includes exercise, physiotherapy, braces, and, occasionally, surgery. It is known that patients show increased levels of creatine in the urine, and increased blood levels of the enzymes transaminase and creatine phosphokinase (CPK) are also detectable early.

Myasthenia Gravis

Myasthenia gravis ("grave muscle weakness") is a disease in which even the slightest muscular exertion causes extreme fatigue. Although myasthenia may become progressively worse, it is usually not fatal unless the respiratory muscles fail and breathing becomes impossible. There is evidence that myasthenia gravis is caused by the production of antibodies that block chemical receptors at the motor end plate.

Although it can occur at any age, myasthenia gravis

most commonly affects women between the ages of 20 and 40, and men over 40. It is about three times as common in women as in men.

Tetanus (Lockjaw)

Technically, *tetanus* (Gr. *tetanos*, stretched) is a disease that affects the nervous system, but it is discussed here because it produces spasms and painful convulsions of the skeletal muscles. Tetanus, commonly called *lockjaw* because of the characteristic tightening of the jaw muscles, is caused by the extracellular toxin produced by the bacterium *Clostridium tetani*. This bacterium produces an exotoxin 50 times stronger than poisonous cobra venom and 150 times stronger than strychnine. (An *exotoxin* is a toxin produced in the cell and released into the environment.) Most often, *C. tetani* is introduced into a puncture wound, cut, or burn by contaminated soil, especially soil that contains horse or cattle manure. These bacteria multiply only in anaerobic conditions, so deep puncture wounds are ideal for their growth. As the bacteria multiply, they produce the exotoxin that destroys surrounding tissue, entering the central nervous system by way of branching spinal nerves.

The early indication of tetanus is local pain and stiffening, but once the exotoxin begins to spread, painful spasms are felt in the muscles of the face and neck, chest, back, abdomen, arms, and legs. Prolonged convulsions may cause sudden death by asphyxiation.

Most children today receive a DPT vaccine, which permanently immunizes them against *d*iphtheria, *p*ertussis (whooping cough), and *t*etanus. If a child who has been immunized suffers an injury that might be conducive to a tetanus infection, a booster shot of tetanus toxoid is usually given.

*The Duchenne form of the disease affects only boys, generally of preschool age.

CHAPTER SUMMARY

1 The basic properties of muscle tissue are ***contractility, excitability, extensibility,*** and ***elasticity.***

2 The general functions of muscle tissue are movement, posture, and heat production. Muscles are specialized in their form to perform different functions. The three types of muscle tissue are ***skeletal, smooth,*** and ***cardiac.***

Skeletal Muscle (p. 229)

1 Most ***skeletal muscle tissue*** is attached to the skeleton. It is also called ***striated muscle*** because it appears to be striped, and ***voluntary muscle*** because it can be contracted voluntarily.

2 Skeletal muscle tissue is composed of individual cells called ***muscle fibers. Fascicles*** are groups of fibers, and ***muscles*** are groups of fascicles.

3 Skeletal muscle fibers contain several nuclei and a specialized type of cytoplasm called ***sarcoplasm,*** which contains many mitochondria and individual fibers called ***myofibrils.*** Each fiber is enclosed within a membrane, the ***sarcolemma.***

4 Each myofibril is composed of myofilaments containing the proteins ***myosin*** and ***actin.***

5 Myofibrils have alternating light and dark bands. The dark ***A bands*** contain myosin; the light ***I bands*** contain actin. Cutting across each I band is a ***Z line.*** Within the A band is a pale ***H zone,*** which contains the ***M line.***

6 A section of myofibril from one Z line to the next makes up a ***sarcomere,*** the fundamental unit of muscle contraction.

7 ***Fascia*** is a sheet of fibrous tissue enclosing muscles or groups of muscles. The major types of fasciae are *superficial* and *deep.*

8 A sheath of connective tissue called ***endomysium*** surrounds each muscle fiber, ***perimysium*** surrounds each fascicle, and ***epimysium*** encases muscles.

9 Muscles are supplied with blood by arteries. Each skeletal muscle fiber is contacted by at least one nerve ending.

10 A motor neuron, together with all the muscle fibers it stimulates, is a ***motor unit.*** The motor neuron endings make contact with muscle fibers at the ***motor end plate,*** and the actual impulse is transmitted chemically to the fibers across a small space called a ***synaptic cleft.*** A nerve ending contacts a muscle fiber at a ***neuromuscular junction.***

Smooth Muscle (p. 234)

1 ***Smooth muscle tissue*** is not striated. It is also known as ***involuntary muscle*** because it is controlled by the involuntary division of the nervous system.

2 The slow and rhythmic contractions of smooth muscle make it suitable for the contractile control of internal organs (but not the heart), especially the stomach, intestines, urinary bladder, and uterus. It also lines hollow vessels, including blood vessels.

3 ***Fibers of single-unit smooth muscle*** contract as a single unit in response to nervous stimulation. ***Multiunit smooth muscle*** consists of individual fibers that are stimulated by separate motor nerve endings.

Cardiac Muscle (p. 236)

1 ***Cardiac muscle tissue*** is found only in the heart. It is striated and involuntary.

2 Cardiac muscle cells are closely packed end to end, but remain separate, each with its own nucleus. ***Intercalated disks*** strengthen the junction between cells and facilitate the passing of an impulse from one cell to the next.

Developmental Anatomy of Muscles (p. 237)

1 Most muscle tissue is derived from embryonic ***mesodermal cells.***

2 Much skeletal muscle develops from mesodermal tissue arranged in paired cell masses called ***somites.***

3 The middle layer of a somite is the ***myotome,*** which develops into most of the skeletal muscle tissue in the body.

The Effects of Aging on Muscles (p. 238)

1 With aging, muscles become weaker, smaller, and dehydrated. Reflexes slow down, and cramps may occur.

2 Because elderly people are less physically active than they used to be, many seemingly unrelated problems may begin to occur, including cardiovascular, digestive, and respiratory problems.

STUDY AIDS FOR REVIEW

MEDICAL TERMINOLOGY

ASTHENIA Loss or lack of bodily strength; weakness.

CHARLEY HORSE A cramp or stiffness of various muscles in the body, especially in the arm or leg, caused by injury or excessive exertion.

CONVULSION An intense involuntary tetanic contraction (or series of contractions) of a whole group of muscles.

CRAMP A sudden, involuntary, complete tetanic muscular contraction causing severe pain and temporary paralysis, often occurring in the leg or shoulder as the result of strain or chill.

FIBRILLATION Uncoordinated twitching of individual muscle fibers with little or no movement of the muscle as a whole.

MYOCARDITIS Inflammation of cardiac muscle tissue.

MYOMA Benign tumor of muscle tissue.

MYOPATHY A general term for any disease of a muscle.

MYOSARCOMA Malignant tumor of muscle tissue.

MYOSITIS Inflammation of a muscle, usually a skeletal muscle.

PHYSIOTHERAPY Treatment of muscle weakness by physical methods, such as heat, massage, and exercise.

SPASM A sudden, involuntary contraction of a muscle or group of muscles.

TIC A habitual, spasmodic muscular contraction, usually in the face, hands, or feet, and often of neurological origin.

TREMOR A trembling or quivering of muscles.

UNDERSTANDING THE FACTS

1 What is the difference between extensibility and elasticity?

2 What produces the alternating dark and light bands in striated muscle?

3 What is the fundamental unit of contraction in skeletal muscle?

4 What is the difference between the epimysium and the perimysium?

5 What functions are served by blood vessels and nerves in skeletal muscles?

6 What are the basic properties of smooth muscle?

7 What is the most important characteristic of cardiac muscle?

8 Give a location of the following:

a voluntary muscle tissue	**e** sarcolemma
b fascia	**f** pacemaker cell
c endomysium	**g** intercalated disks
d transverse tubules	

UNDERSTANDING THE CONCEPTS

1 What is the function of the synaptic cleft in muscular contraction?

2 In what important ways do the structure and function of skeletal, smooth, and cardiac muscle differ?

SELF-QUIZ

True-False Statements

10.1 Smooth muscle contraction is controlled by the somatic nervous system.

10.2 Multiunit smooth muscle is so named because each of its individual fibers can be stimulated by separate nerve endings.

10.3 Each skeletal muscle fiber is contacted by at least one nerve ending.

10.4 In cardiac muscle tissue, intercalated disks occur in place of Z lines.

10.5 Skeletal muscles are covered by dense connective tissue that is an extension of the superficial fascia.

10.6 Fascicles are bundles of muscle fibers covered by perimysium.

10.7 Skeletal muscles are attached to bones by ligaments.

10.8 The sarcolemma of a skeletal muscle fiber is the endoplasmic reticulum.

10.9 Smooth muscle cells are multinucleate.

10.10 Skeletal muscle fibers contain myofilaments of actin and myosin.

Completion Exercises

10.11 A broad, flat sheet of dense connective tissue that attaches muscles together is a(n) _____.

10.12 A motor neuron, together with the muscle fiber it innervates, is called a _____.

10.13 The gap between a nerve ending and the sarcolemma of a muscle fiber is the _____.

10.14 Single-unit smooth muscle is also known as _____ smooth muscle.

10.15 Short, branching fibers are found in _____ muscle tissue.

10.16 Most muscle tissue develops from the embryonic _____ layer.

10.17 The type of muscle tissue found in a sphincter is _____ muscle tissue.

10.18 Tendons connect _____ to _____.

10.19 Skeletal muscle fibers are enclosed by a layer of connective tissue called the _____.

10.20 A motor end plate is found at the junction of a _____ and a muscle fiber.

A SECOND LOOK

1 In the following drawings, label the perimysium, epimysium, and endomysium.

2 What type of muscle tissue is shown in the following photograph?

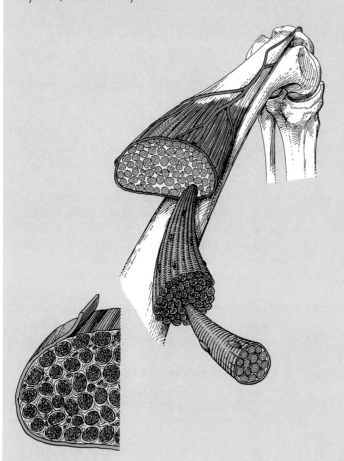

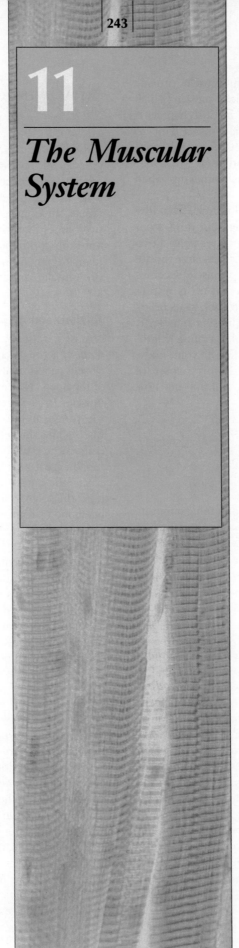

11

The Muscular System

LEARNING OBJECTIVES

1 Explain how muscles are named and give some examples.

2 Explain how muscles are attached to bones and how the attachment sites are related to the skeleton.

3. Describe three patterns of muscle fascicles, and explain how each pattern is related to the power and range of movement of a muscle with that pattern.

4 Describe the action of the following types of muscles: flexor, extensor, abductor, adductor, pronator, supinator, rotator, levator, depressor, protractor, retractor, tensor, sphincter, evertor, and invertor.

5 Define the roles of agonist, antagonist, synergist, and fixator, and explain how they are coordinated in a group of muscles.

6 Explain the principles of first-class, second-class, and third-class levers, and give an example of each type.

7 Identify the major skeletal muscles in each region of the body, and describe the origin, insertion, and action of each.

Bones operate as a system of levers, but muscles provide the power to make them move. A skeletal muscle (which is the only type of muscle described in this chapter) is attached to one bone at one end and, across a joint, to another bone at its other end. When the muscle contracts, one bone is pulled *toward* the other (as when your elbow joint bends), or *away* from each other (as when your elbow joint straightens).

Every movement caused by skeletal muscle, from lifting a heavy television set to scratching your nose, from running a mile to tapping your foot, involves at least two sets of muscles. When one muscle contracts, an opposite one is stretched. During contraction, a muscle becomes thicker and shortens by about 15 percent of its resting length. A contracted muscle is brought back to its original condition by contraction of its opposing muscle. It can be stretched to about 120 percent of its resting length.

There are about 600 *skeletal muscles* in your body, which make up the *muscular system.** In this chapter we examine the most important of them, and find out how they operate on a gross level.

HOW MUSCLES ARE NAMED

The name of a muscle generally tells something about its structure, location, or function. **Shape** is described in such muscles as the trapezius (shaped like a trapezoid), deltoid (triangular or delta-shaped), or gracilis (slender). **Size** is clearly indicated in the gluteus *maximus* (largest) and gluteus *minimus* (smallest) muscles. We know the **location** of the supraspinatus and infraspinatus muscles (above and below, *supra* and *infra*, the spine of the scapula) and the tibialis anterior (in front of the tibia).

Some muscles are named to indicate their **attachment sites.** For example, the sternohyoid muscle is attached to the sternum and hyoid bones. The number of **heads of origin** can be determined in muscles such as the biceps (*bi*, two) and triceps (*tri*, three). The **action** or function of a muscle is plain in the extensor digitorum, which extends the fingers (digits), and the levator scapulae, which raises (elevates) the scapula. The names of some muscles indicate the **direction of their fibers** with respect to the structures to which they are attached, for example, *transversus* (across) and *obliquus* (slanted or oblique).

Most muscles are named by a combination of the above methods. An example is the flexor digitorum profundus, which means the deep (profundus) flexor of the fingers (digitorum). The name of this muscle thus tells its depth, location, and action.

*We say "about" 600 skeletal muscles rather than giving a definite number because anatomists disagree about whether to count some muscles as separate or as pairs. It is generally agreed, however, that we have more than 600 skeletal muscles.

ATTACHMENT OF MUSCLES

Although the form and actions of skeletal muscles vary greatly, all have certain basic features in common. All contain a fleshy center, often the widest part, called the **belly** of the muscle and two ends that attach to other tissues. In Chapters 4 and 10, we described the connective tissues that form this attachment. It is important to note the continuity of the tissues that hold the muscle fibers together, bind the fibers into bundles, wrap an entire muscle, and attach muscle to bone.

Tendons and Aponeuroses

A muscle is usually attached to a bone (or a cartilage) by a **tendon,** a tough cord of connective tissue composed of closely packed collagen fibers.[†] A tendon is an extension of the deep fascia and/or the epimysium surrounding the muscle. It also extends into the periosteum that covers the bone, and into the bone as the periosteal perforating fibers. The thickest tendon in the body is the calcaneal (kal-KAY-nee-uhl) tendon, commonly called the Achilles tendon [FIGURE 11.1]. It attaches the calf muscles (gastrocnemius; gas-trahk-NEE-mee-uhss; Gr. "belly of the leg"; the soleus and plantaris are also calf muscles) to the heelbone (calcaneus). Tendons add length and thickness to muscles and are especially important in reducing strain on muscles.

In some parts of the body, such as the abdominal wall, the tendon spreads out in a broad, flat sheet called an **aponeurosis.** This sheetlike attachment is directly or indirectly connected with the various muscle sheaths, and in some cases with the periosteum covering a bone. Another example of an aponeurosis is the fibrous sheath beneath the skin of the palm, called the *palmar aponeurosis.*

Origin and Insertion of Muscles

When a muscle contracts, one of the bones attached to it remains stationary while the other bone moves along with the contraction. In reality, the mechanics are hardly ever that simple, but for convenience we say that the end of a muscle attached to the bone that does not move is the **origin,** and the point of attachment of the muscle on the bone that moves is the **insertion** [FIGURES 11.2 and 11.25]. Generally, the origin is the more proximal attachment (closer to the axial skeleton) and the insertion is the more distal attachment.

However, *origin* and *insertion* are only relative terms, and they can be reversed with the same muscle, depend-

[†] Although most muscles are attached to bones, not all are. Some, such as the muscles of facial expression, are attached to skin, and the lumbrical muscles of the hand are attached to tendons of other muscles.

FIGURE 11.1 THE MUSCLE-TENDON-BONE RELATIONSHIP

[A] The calcaneal (Achilles) tendon connecting the gastrocnemius (calf) muscle and the calcaneus bone. **[B]** Detail showing collagenous periosteal perforating (Sharpey's) fibers passing into the connective tissue, periosteum, and bone.

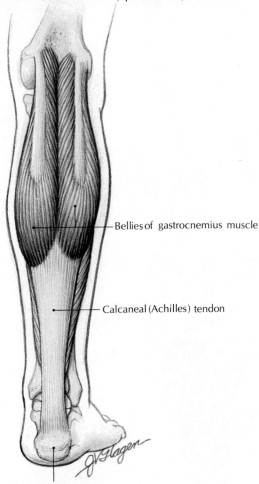

Bellies of gastrocnemius muscle

Calcaneal (Achilles) tendon

[A]   Calcaneus (heel bone)

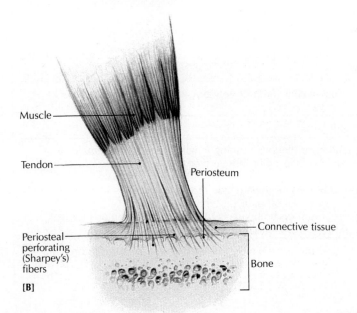

Muscle

Tendon

Periosteum

Periosteal perforating (Sharpey's) fibers

Connective tissue

Bone

[B]

ing upon the action. For example, muscles from the chest to the upper extremity generally move the extremity with the origin on the thoracic skeleton and the insertion on the extremity. But when a person climbs a rope, the same muscles pull the body up, so the extremity becomes the origin and the insertion is on the thoracic skeleton.

ASK YOURSELF

1 What is the fleshy center of a muscle called?

2 What is the difference between a tendon and an aponeurosis?

3 What is the function of a tendon?

4 What is the difference between the origin and insertion of a muscle?

ARCHITECTURE OF MUSCLES

The fibers of skeletal muscles are grouped in small bundles called fascicles, with the fibers within each running parallel to one another [see FIGURE 10.1]. The fascicles, however, are organized in various architectural patterns in different muscles. The specific pattern determines the *range of movement* and the *power* of the muscle. The greater the length of the belly, the greater the range of movement. Also, the greater the number of fibers, the

FIGURE 11.2 ORIGIN AND INSERTION OF ARM MUSCLES

The origins of the two heads of the biceps brachii muscle are on the scapula; the insertion is on the radius. Contraction of the biceps brachii flexes the elbow joint (arrow up).

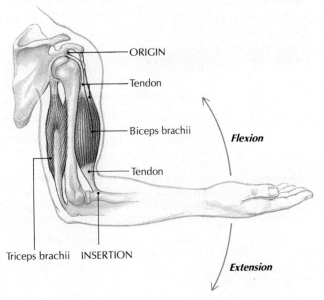

ORIGIN

Tendon

Biceps brachii

Tendon

Flexion

Triceps brachii   INSERTION

Extension

FIGURE 11.3 ARCHITECTURE OF MUSCLE

[A] A parallel strap muscle, sternohyoid. [B] A fusiform muscle, biceps brachii. [C] A unipennate muscle, flexor pollicis longus. [D] A bipennate muscle, rectus femoris. [E] A circular or sphincter muscle, orbicularis oris. [F] A multipennate muscle, deltoid.

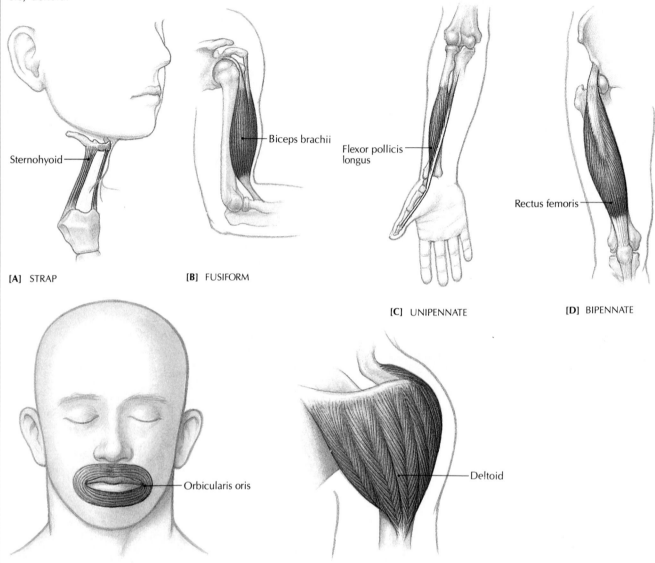

[A] STRAP

[B] FUSIFORM

[C] UNIPENNATE

[D] BIPENNATE

[E] CIRCULAR (SPHINCTER)

[F] MULTIPENNATE

Muscle type	Description and examples
Strap	Fascicles parallel to long axis; wide range of movement, but not very powerful. Sternohyoid (of neck), rectus abdominis (of abdominal wall).
Fusiform	Spindle-shaped with thick belly. Biceps brachii (arm).
Pennate	Short fascicles at angle (oblique) to a long tendon or tendons running length of muscle; resembles a feather. More fascicles attached directly to tendons than other muscle types; generally more powerful than other types.
Unipennate	Oblique fascicles on one side of tendon; direction of pull toward side of tendon with fascicles. Flexor pollicis longus (of thumb).
Bipennate	Oblique fascicles on both sides of tendon; equal pull on both sides of tendon. Rectus femoris (of leg).
Multipennate	Many oblique fascicles arranged along several tendons. Deltoid muscle (of shoulder).
Circular	Fascicles arranged in circular pattern around opening or structure. Orbicularis oris (of mouth), orbicularis oculi (of eye).

greater the total force generated by the muscle. The arrangement of fascicles and their tendons of attachment creates patterns such as strap muscles, fusiform muscles, pennate muscles, and circular muscles [FIGURE 11.3].

ASK YOURSELF

1 What are the four main patterns of fascicle arrangement?

2 What are the three types of pennate muscles?

3 Which arrangement of fascicles generally contributes to a greater range of movement?

INDIVIDUAL AND GROUP ACTIONS OF MUSCLES

Skeletal muscles can be classified according to the types of movement that they perform. TABLE 11.1 defines these actions and gives examples, and FIGURE 11.4 shows some of the major muscles.

Most movements, even those that seem simple, actually involve the complex interaction of several muscles or groups of muscles. Muscles produce or restrict movement by acting as agonists, antagonists, synergists, and fixators.

TABLE 11.1 CLASSIFICATION OF MUSCLES BASED ON ACTION

Muscle type	Definition of action	Example
Flexor	Bending so angle between two bones decreases.	Flexor pollicis longus [FIGURE 11.18A, D]
Extensor	Bending so angle between two bones increases.	Extensor carpi ulnaris [FIGURE 11.18B]
Dorsiflexor	Bending of foot dorsally (toward back of foot).	Extensor digitorum longus [FIGURE 11.23A]
Palmar flexor	Bending (flexing) wrist ventrally (toward palm).	Flexor carpi ulnaris [FIGURE 11.18A, B]
Plantar flexor	Bending (extending) foot at ankle toward sole of foot.	Gastrocnemius [FIGURE 11.23A, B]
Abductor	Movement away from midline of body or structure.	Abductor pollicis brevis [FIGURE 11.18A, D]
Adductor	Movement toward midline of body or structure.	Adductor pollicis [FIGURE 11.18D]
Pronator	Turning of forearm so palm faces downward.	Pronator teres [FIGURE 11.18A]
Supinator	Turning of forearm so palm faces upward.	Supinator [FIGURE 11.18A]
Rotator	Turning movement around a longitudinal axis.	Sternocleidomastoid [FIGURES 11.9, 11.11]
Medial (internal) rotator	Turning movement so anterior surface faces median plane.	Subscapularis [FIGURE 11.17C]
Lateral (external) rotator	Turning movement so anterior surface faces away from median plane.	Infraspinatus [FIGURES 11.16B, 11.17D]
Levator (elevator)	Movement in an upward direction.	Levator scapulae [FIGURE 11.16A, B]
Depressor	Movement in a downward direction.	Depressor labii inferioris [FIGURE 11.6A, B]
Protractor	Movement in a forward direction.	Lateral pterygoid [FIGURE 11.8B]
Retractor	Movement in a backward direction.	Temporalis (horizontal fibers) [FIGURE 11.8A]
Tensor	Makes a body part more tense.	Tensor fasciae latae [FIGURES 11.21A, 11.22A]
Sphincter	Reduces size of an opening (orifice).	Orbicularis oris [FIGURE 11.6]
Evertor	Turning movement of foot so sole faces outward.	Peroneus longus [FIGURE 11.23A]
Invertor	Turning movement of foot so sole faces inward.	Tibialis anterior [FIGURE 11.23B]

FIGURE 11.4 THE MUSCULAR SYSTEM

[A] Anterior view. **[B]** Posterior view.

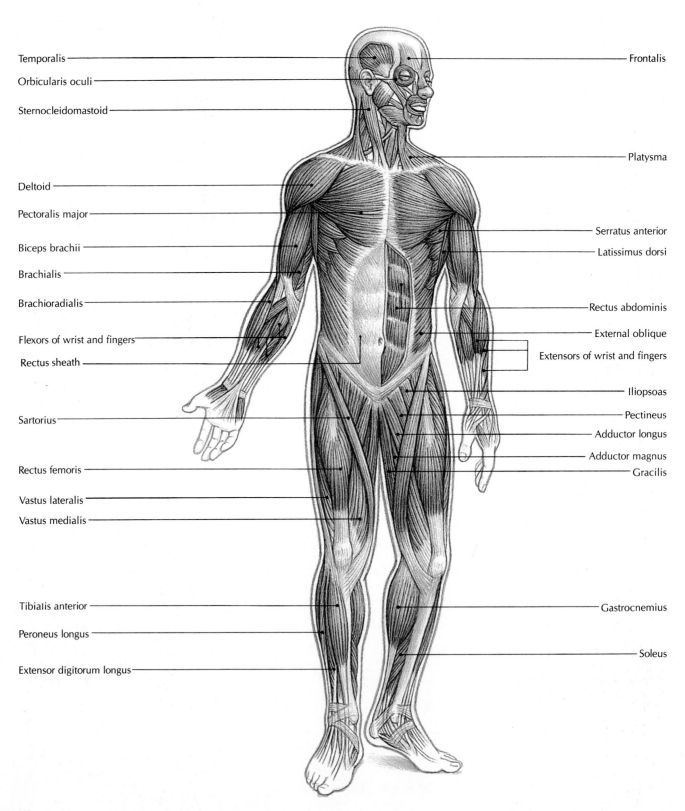

Temporalis

Orbicularis oculi

Sternocleidomastoid

Deltoid

Pectoralis major

Biceps brachii

Brachialis

Brachioradialis

Flexors of wrist and fingers

Rectus sheath

Sartorius

Rectus femoris

Vastus lateralis

Vastus medialis

Tibialis anterior

Peroneus longus

Extensor digitorum longus

Frontalis

Platysma

Serratus anterior

Latissimus dorsi

Rectus abdominis

External oblique

Extensors of wrist and fingers

Iliopsoas

Pectineus

Adductor longus

Adductor magnus

Gracilis

Gastrocnemius

Soleus

[A]

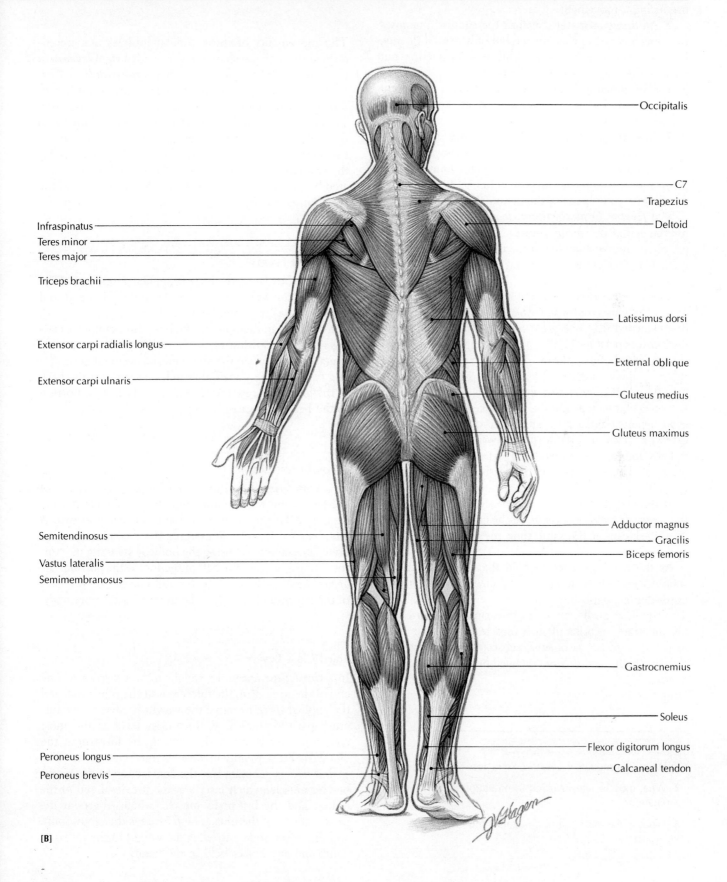

Occipitalis

C7

Trapezius

Deltoid

Infraspinatus

Teres minor

Teres major

Triceps brachii

Latissimus dorsi

Extensor carpi radialis longus

External oblique

Gluteus medius

Extensor carpi ulnaris

Gluteus maximus

Adductor magnus

Gracilis

Biceps femoris

Semitendinosus

Vastus lateralis

Semimembranosus

Gastrocnemius

Soleus

Flexor digitorum longus

Calcaneal tendon

Peroneus longus

Peroneus brevis

[B]

1 An *agonist* (Gr. *agonia*, contest, struggle), or *prime mover,* is the muscle that is primarily responsible for producing a movement.

2 An *antagonist* (Gr. "against the agonist") opposes the movement of a prime mover, but only in a subtle way. It does not oppose the agonist while it is contracting, but only at the end of a strong contraction to protect the joint. The antagonist helps produce a smooth movement by slowly relaxing as the agonist contracts, so it actually cooperates rather than "opposes."

3 A *synergist* (SIHN-uhr-jist; Gr. *syn,* together + *ergon,* work) works together with a prime mover by preventing movements at an "in-between" joint when a prime mover passes over more than one joint. In general, a synergist complements the action of a prime mover.

4 A *fixator,* or *postural muscle,* provides a stable base for the action of a prime mover. It usually steadies the proximal end (such as the arm), while the actual movement is taking place at the distal end (the hand).

The coordination of an action by a group of muscles can be demonstrated when a heavy object is held firmly in a clenched fist. The *prime movers* in this case are the long flexors [FIGURE 11.18A, D] of the fingers (flexor digitorum superficialis, flexor digitorum profundus, and flexor pollicis longus). When these muscles contract unopposed, the wrist also flexes. This undesired wrist flexion is eliminated, and the wrist is kept in a neutral or even hyperextended position by the contraction of the extensors [FIGURE 11.18B] of the wrist (extensor carpi radialis longus and brevis, and extensor carpi ulnaris muscles). These extensors of the wrist are the *synergists* in this action.

If the clenched fist is slowly flexed, the extensors of the wrist can act as *antagonists* in the control of the activity by relaxing at the same time the prime movers are contracting.

As the hand clenches to hold the heavy object, the shoulder and elbow joints are stabilized to a greater or lesser degree, depending upon the weight of the object. The shoulder girdle, arm, and forearm are stabilized by the integrated actions of such muscles as the pectoralis major, deltoid, supraspinatus, subscapularis, biceps brachii, brachialis, and triceps brachii. These muscles are the *fixators.*

ASK YOURSELF

1 What is a pronator? Levator? Retractor? Sphincter?

2 What muscles are involved in pronation and supination?

3 What is the main job of an agonist muscle? Of a synergist?

LEVER SYSTEMS AND MUSCLE ACTIONS

The movements of most skeletal muscles are accomplished through a system of levers, with a rigid *lever arm* pivoting around a fixed point called a *fulcrum* (L. *fulcire,* to support). Also, acting on every lever are two different *forces:* (1) the weight to be moved (resistance to be overcome) and (2) the pull or effort applied (applied force). In the body, the bone acts as the lever arm, and the joint as the fulcrum. The weight of the body part to be moved is the resistance to be overcome. The applied force generated by the contraction of the muscle, or muscles, at the insertion is usually enough to produce a movement.

First-Class Levers

In a *first-class lever,* the force is applied at one end of the lever arm, the weight to be moved is at the other end, and the fulcrum is at a point between the two [FIGURE 11.5A]. A crowbar is an example. In the body, an example is raising the facial portion of the head where the atlantooccipital joint between the atlas and the occipital bone acts as a *fulcrum.* The vertebral muscles inserting at the back of the head generate the *applied force.* The facial portion of the head is the *weight* to be moved.

Second-Class Levers

In a *second-class lever,* the applied force is at one end of the lever arm, the fulcrum is at the other end, and the weight to be moved is at a point between the two. A wheelbarrow is a classic example [FIGURE 11.5B]. In the body, an example is raising the body by standing on "tiptoe." In this action, the ball of the foot is the *fulcrum,* the *applied force* is generated by the calf muscles on the back of the leg, and the *weight* to be moved is the entire body.

Third-Class Levers

In a *third-class lever,* the weight to be moved is at one end of the lever arm, the fulcrum is at the other end, and the applied force to move the weight is close to the fulcrum [FIGURE 11.5C]. A third-class lever is the most common lever in the body. Flexing the forearm at the elbow to lift a weight involves a third-class lever. The *applied force* is generated by the contraction of the biceps brachii muscle, which inserts at the proximal end of the radius, and the brachialis muscle, which inserts at the proximal end of the ulna. The *fulcrum* is the elbow joint, and the *weight* to be moved is the weight of the forearm, hand, and any object held in the hand.

FIGURE 11.5 CLASSES OF LEVERS

[A] In a first-class lever, the fulcrum is between the force and the weight. [B] In a second-class lever, the weight is between the force and the fulcrum. [C] In a third-class lever, the force is between the weight and the fulcrum. The arrows indicate the applied force and the weight of the object to be moved.

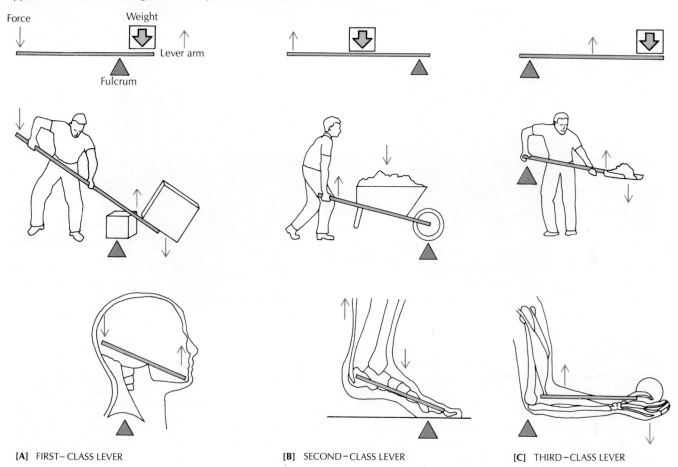

[A]   FIRST–CLASS LEVER   [B]   SECOND–CLASS LEVER   [C]   THIRD–CLASS LEVER

Leverage: Mechanical Advantage

If you have ever used a screwdriver to pry open a can of paint, you know that the screwdriver (lever) made the work easier. When muscles use a lever system, they also gain a mechanical advantage, which is called *leverage*.

The leverage of a muscle is improved as the distance from the insertion to the joint (fulcrum) is increased. In other words, a muscle with an insertion relatively far from a joint has *more power* than a muscle with an insertion closer to the joint because the longer lever arm between the joint and insertion produces more power. However, the longer lever arm must move a greater distance to produce this power, and the movement is slower. For example, the pectineus muscle, which is attached close to the axis of rotation at the hip joint, has less power than the adductor longus muscle, which is attached farther from the axis [FIGURE 11.22]. But the pectineus can produce a quicker movement than the adductor longus because it moves through a shorter distance.

ASK YOURSELF

1 What two forces are involved in a lever system?

2 What are the main differences in the three classes of levers? Give an example of each of the three types of levers in the body.

3 How does leverage operate in the human body?

SPECIFIC ACTIONS OF PRINCIPAL MUSCLES

In this section, which consists of FIGURES 11.6 through 11.25, the major muscles of the body are described and illustrated. The origins, insertions, innervation, and actions of these muscles are summarized in the accompanying tables.

FIGURE 11.6 MUSCLES OF FACIAL EXPRESSION

[A] Anterior view. Note that different layers of muscle are shown on each side of the face. [B] Lateral view.

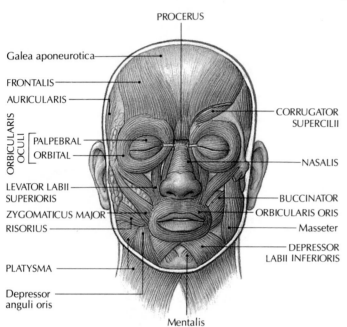

[A]

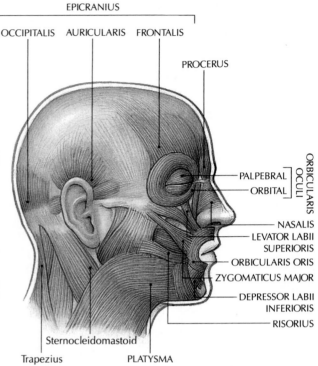

[B]

The **muscles of facial expression** are located not only in the face, but in the scalp and neck. Two features characterize these muscles—they are all innervated by the facial cranial nerve (VII), and they are used in the display of human emotions. They can move parts of the face because their principal insertions extend into the deep layers of the skin. Their origins may be located in a tendon, aponeurosis, bone, or the skin itself. Some of the facial muscles are arranged around the eyes, mouth, nose, and ears, and act as **sphincters** (ringlike muscles that normally constrict a bodily opening until relaxation is called for).

Facial muscles are used to express a wide variety of emotions. For example, when you smile, the muscles radiating from the angles of your mouth raise the corners of your upper lip. When you frown, certain muscles lower your eyebrows and depress the corners of your lower lip. (The cliché that it takes more effort to frown than to smile is true; you use 43 muscles to frown and only 17 to smile.)

Besides expressing emotion, the facial muscles have other uses. For example, the nasalis musculature dilates your nostrils as you inhale. The **buccinator muscle** (L. *bucca*, cheek) lining the cheek is essential for nudging food from between the cheek and teeth during chewing, and for blowing or sucking.* The scalp (epicranius) muscles originating from a broad, flat aponeurosis (galea aponeurotica) can wrinkle the forehead, producing a look of surprise or fright.

*When the buccinator muscles lose their tone and elasticity, the cheeks bulge uncontrollably when a person makes a blowing motion. This condition is known as "Gillespie's pouches," named for the jazz trumpeter Dizzy Gillespie, whose ineffective buccinator muscles cause his cheeks to bulge every time he blows into his trumpet. (However, the buccinator muscle was known as the trumpeter's muscle long before Dizzy Gillespie came along.)

Muscle	Origin	Insertion	Innervation	Action
MUSCLES OF SCALP				
Epicranius				
Frontalis	Galea aponeurotica	Fibers of orbicularis oculi	Temporal branches of facial nerve (VII)	Elevates eyebrows (surprised look); produces horizontal frown.
Occipitalis	Occipital bone	Galea aponeurotica	Posterior auricular branch of facial nerve (VII)	Draws scalp backward.
Auricularis	Galea aponeurotica	Vicinity of outer ear	Temporal branches of facial nerve (VII)	Wiggles ears in a few individuals.

Muscle	Origin	Insertion	Innervation	Action
MUSCLES OF EYELID AND ORBIT				
Orbicularis oculi Palpebral (eyelid)	Medially from medial palpebral ligament	Sphincter fibers oriented in concentric circles around orbital margin and in eyelid	Temporal and zygomatic branches of facial nerve (VII)	Closes eyelid (as in winking).
Orbital	Medial orbital margin (maxilla), medial palpebral ligament	Sphincter muscle	Temporal and zygomatic branches of facial nerve (VII)	Closes eyelid tightly (against bright light).
Corrugator supercilii	Brow ridge of frontal bone	Skin of eyebrow	Temporal branch of facial nerve (VII)	Pulls eyebrows together.
Levator palpebrae superioris	Central tendinous ring around optic foramen	Skin of upper eyelid	Oculomotor nerve (III)	Raises upper eyelid.
MUSCLES OF NOSE				
Nasalis	Maxilla medial and inferior to orbit	Lower region of cartilage of nose	Facial nerve (VII)	Widens nasal aperture (as in deep breathing).
Procerus	Lower part of nasal bone, lateral nasal cartilage	Skin between eyebrows	Buccal branches of facial nerve (VII)	Pulls eyebrow downward and inward.
MUSCLES OF MOUTH				
Orbicularis oris	Sphincter muscle within lips; encircles mouth and merges with other muscles of mouth.		Buccal and mandibular branches of facial nerve (VII)	Closes mouth; purses lips; significant role in speech.
Levator labii superioris	Above infraorbital foramen of maxilla	Skin of upper lip	Buccal branch of facial nerve (VII)	Raises lateral aspect of upper lip.
Zygomaticus major	Zygomatic bone	Skin at angle of mouth	Zygomatic or buccal branch of facial nerve (VII)	Draws angle of mouth upward (as in smiling).
Risorius	Fascia over masseter muscle	Skin at angle of mouth	Buccal branch of facial nerve (VII)	Draws angle of mouth laterally (as in grinning).
Depressor labii inferioris	Mandible	Skin of lower lip	Mandibular branch of facial nerve (VII)	Lowers lateral aspect of lower lip.
MUSCLE OF CHEEK				
Buccinator	Alveolar processes of mandible and maxilla, from pterygomandibular raphe (connective tissue) between sphenoid and mandible	Fibers of orbicularis oris	Buccal branch of facial nerve (VII)	Pulls cheek against teeth to move food during chewing and to make blowing or sucking motion.
MUSCLE OF NECK				
Platysma	Fascia of upper thorax over pectoralis major, deltoid muscles	Mandible, skin of lower face, fibers of orbicularis oris at angle of mouth	Cervical branch of facial nerve (VII)	Draws outer part of lower lip down and back (as in frowning); tenses skin of neck.

FIGURE 11.7 MUSCLES THAT MOVE THE EYEBALL

[A] Superior view, with horizontal and sagittal axes of the eyeball indicated. [B] Lateral view, with vertical axis of the eyeball indicated.

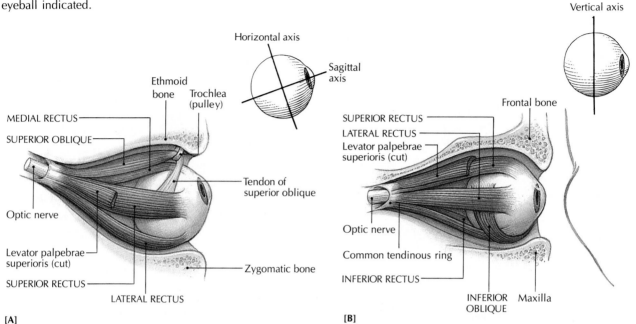

The movements of the eyeball are controlled by six rapidly responsive *extrinsic* (outside the eyeball) eye muscles. These muscles can rotate the eyeball around the horizontal, vertical, and sagittal axes passing through the center of the eyeball. The four *recti* ("upright") *muscles* insert in front of the horizontal axis, and the two *oblique muscles* insert behind the horizontal axis. The recti muscles pass slightly obliquely forward from their origin around the optic foramen. They form a cone as they spread out toward their insertions in the eyeball. The actions of the recti muscles can be understood by realizing that they pull on the eyeball in a plane slightly oblique to the visual (sagittal) axis. Squint (strabismus or cross-eye) is a disorder in which both eyes cannot be directed at the same point or object at the same time. It can result from an imbalance in the power and length of the recti muscles.

Muscle	Origin	Insertion	Innervation	Action
Superior rectus	Tendinous ring anchored to bony orbit around optic foramen	Superior and central part of eyeball	Oculomotor nerve (III)	Rolls eye upward; also adducts and rotates medially.
Inferior rectus	Tendinous ring anchored to bony orbit around optic foramen	Inferior and central part of eyeball	Oculomotor nerve (III)	Rolls eye downward; also adducts and rotates laterally.
Lateral rectus	Tendinous ring anchored to bony orbit around optic foramen	Lateral side of eyeball	Abducens nerve (VI)	Rolls (abducts) eye laterally (out).
Medial rectus	Tendinous ring anchored to bony orbit around optic foramen	Medial side of eyeball	Oculomotor nerve (III)	Rolls (adducts) eye medially (in).
Superior oblique	Sphenoid bone	Posterior and lateral to equator of eyeball under superior rectus*	Trochlear nerve (IV)	Depresses, abducts, and rotates eye medially.
Inferior oblique	Lacrimal bone	Posterior and lateral to equator of eyeball under lateral rectus	Oculomotor nerve (III)	Elevates, abducts, and rotates eye laterally.

*The tendon of the superior oblique changes its direction abruptly when it passes through the trochlea (a fibrocartilaginous pulley) located in the upper front of the bony orbit.

FIGURE 11.8 MUSCLES OF MASTICATION

[A] Temporal and masseter muscles; right lateral superficial view. **[B]** Medial and lateral pterygoid muscles; right lateral view deep to the mandible.

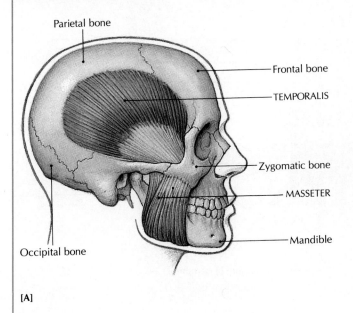

Parietal bone
Frontal bone
TEMPORALIS
Zygomatic bone
MASSETER
Mandible
Occipital bone

[A]

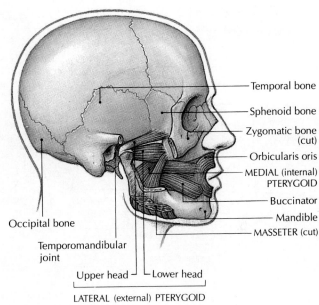

Temporal bone
Sphenoid bone
Zygomatic bone (cut)
Orbicularis oris
MEDIAL (internal) PTERYGOID
Buccinator
Mandible
MASSETER (cut)
Occipital bone
Temporomandibular joint
Upper head — Lower head
LATERAL (external) PTERYGOID

[B]

Four pairs of muscles produce biting and chewing movements: the **masseter, temporalis, lateral pterygoid** (Gr. "wing"; ter-uh-GOID), and **medial pterygoid.** The masseter (Gr. *maseter,* chewer) and the temporalis muscles lie close to the skin and can be felt easily when the teeth are clenched.

In chewing, the mandible may move from side to side in a grinding motion. This movement is controlled by the masseter and temporalis muscles on the same side as the direction of the movement, and by the lateral and medial pterygoid muscles on the opposite side.

Muscle	Origin	Insertion	Innervation	Action
Masseter	Zygomatic bone	Outer surface of angle and ramus of mandible	Mandibular division of trigeminal nerve (V)	Elevates mandible (closes mouth); slightly protrudes mandible.
Temporalis	Temporal fossa of temporal bone	Coronoid process of mandible	Mandibular division of trigeminal nerve (V)	Elevates mandible; retracts mandible (pulls jaw back).
Medial pterygoid	Medial surface of lateral pterygoid plate of sphenoid bone, maxilla	Inner surface of angle and ramus of mandible	Mandibular division of trigeminal nerve (V)	Elevates mandible; slightly protrudes mandible; draws jaw toward opposite side in grinding movements.
Lateral pterygoid	Lateral surface of lateral pterygoid plate of sphenoid bone	Condyle of mandible and fibrocartilage articular disk	Mandibular division of trigeminal nerve (V)	Protrudes mandible; depresses mandible (opens mouth); produces side-to-side movements during chewing and grinding.

FIGURE 11.9 MUSCLES THAT MOVE THE HYOID BONE

The sternocleidomastoid and sternohyoid muscles of the neck and the digastric muscle attached to the hyoid bone are shown on the right side of the diagram but not on the left; anterior view.

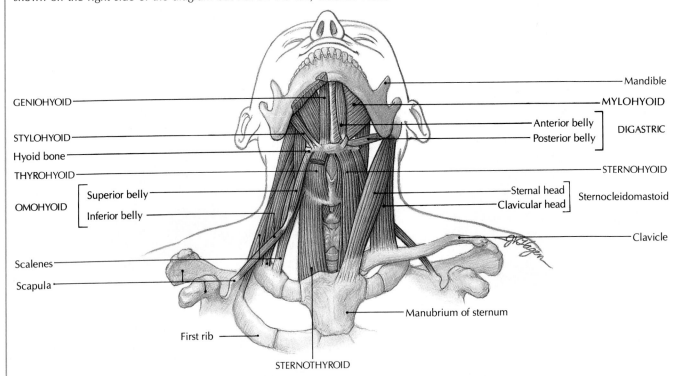

Many muscles associated with the mouth, throat (pharynx), and neck are attached to the hyoid bone. The **suprahyoid muscles** extend above the hyoid and attach it to the skull. The straplike **infrahyoid muscles** extend from the hyoid to skeletal structures below. The precise movements of the hyoid are controlled by coordinated muscular activity and are especially evident during swallowing, when the hyoid bone moves upward and forward and then downward and backward to its original position. The hyoid has attachments for some muscles of the tongue and larynx. When the mandible does not move, the hyoid is raised by the **mylohyoid, geniohyoid, digastric** (suprahyoid), and **stylohyoid muscles.** The infrahyoid muscles pull the hyoid downward. The diaphragmlike paired mylohyoid muscles form the floor of the mouth below the tongue. Both suprahyoid and infrahyoid muscles can aid in depressing the mandible.

FIGURE 11.10 MUSCLES THAT MOVE THE TONGUE

Right lateral view.

The muscles of the tongue are essential for normal speech and for the manipulation of food within the mouth. The *intrinsic muscles,* which are located within the tongue, are oriented in the horizontal, vertical, and longitudinal planes. They are able to fold, curve, and squeeze the tongue during speech and chewing. The action of the three *extrinsic muscles* of the tongue allow it to protrude **(genioglossus muscle)**, retract **(styloglossus muscle)**, and depress **(hyoglossus muscle).** All tongue muscles are innervated by the hypoglossal cranial nerve (XII).

Normally, the genioglossus muscle prevents the tongue from falling backward toward the throat, where it can cause suffocation. But when the muscle is paralyzed or totally relaxed during general anesthesia, it can block the respiratory passage. Anesthetists keep the tongue in a forward position by pushing the mandible forward, or by using a curved intubation tube. The tube should not be allowed to reach the larynx.

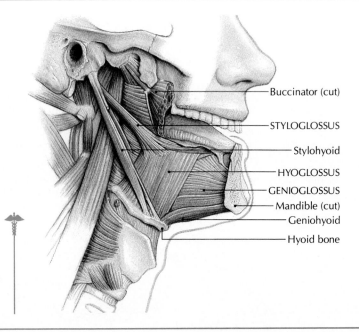

Muscle	Origin	Insertion	Innervation	Action
SUPRAHYOID MUSCLES (MUSCLES OF FLOOR OF MOUTH)				
Digastric				
Posterior belly	Mastoid process of temporal bone	Common tendon attached to body of hyoid	Facial nerve (VII)	Raises hyoid.
Anterior belly	Digastric fossa of mandible near symphysis		Mandibular division of trigeminal nerve (V)	Depresses mandible.
Stylohyoid	Styloid process of temporal bone	Body of hyoid	Facial nerve (VII)	Raises hyoid; pulls hyoid backward.
Mylohyoid	Mylohyoid line on internal surface of mandible	Hyoid bone, central raphe in floor of mouth	Mandibular division of trigeminal nerve (V)	Raises hyoid; forms and elevates floor of mouth.
Geniohyoid	Adjacent to symphysis of mandible	Body of hyoid	First cervical nerve	Pulls hyoid upward and forward; helps to open jaw. Depresses mandible when hyoid is fixed.
INFRAHYOID MUSCLES (STRAP MUSCLES)				
Sternohyoid	Manubrium of sternum	Body of hyoid	Cervical plexus*	Depresses hyoid.
Sternothyroid	Manubrium of sternum	Thyroid cartilage of larynx	Cervical plexus	Depresses thyroid and hyoid.
Thyrohyoid	Thyroid cartilage of larynx	Greater horn of hyoid	Cervical plexus	Depresses hyoid; raises thyroid cartilage.
Omohyoid				
Inferior belly	Superior border of scapula (near coracoid process)	Intermediate tendon	Cervical plexus	Depresses hyoid.
Superior belly	Intermediate tendon attached to medial end of clavicle	Body of hyoid	Cervical plexus	Depresses hyoid.

*The cervical plexus innervating these muscles is formed from nerves derived from the first three cervical nerves.

Muscle	Origin	Insertion	Innervation	Action
Genioglossus	Tubercle of anterior part of mandible beside midline	Fibers pass through tongue to insert in bottom of tongue	Hypoglossal nerve (XII)	Protrudes and depresses tongue.
Styloglossus	Tip of styloid process of temporal bone	Side of tongue		Retracts and elevates tongue.
Hyoglossus	Greater horn of hyoid bone	Side of tongue		Depresses tongue.
Intrinsic muscles	Arranged in three planes within tongue in horizontal, vertical, and longitudinal planes			Alter shape of tongue.

FIGURE 11.11 MUSCLES THAT MOVE THE HEAD

Posterior view.

In a sense, the head is balanced upon the vertebral column, with the atlas articulating with the occipital condyles. This *atlantooccipital* articulation is involved in flexing (bending forward) and extending (holding erect) the head, and is located about midway between the back of the head and the tips of the nose and chin. The head is normally bent slightly forward, with the neck muscles that insert in the occipital region (such as the splenius capitis and semispinalis capitis) partially contracted to prevent the head from falling forward. When the head flexes, the posterior neck muscles relax while the **sternocleidomastoid** and other anterior neck muscles contract. The *bilateral* (both sides) contraction of the **splenius capitis, longissimus capitis,** and **semispinalis capitis** extends the head. The *unilateral* (one side only) contraction of these muscles rotates the head, tilts the chin up, and turns the face toward the contracted side.

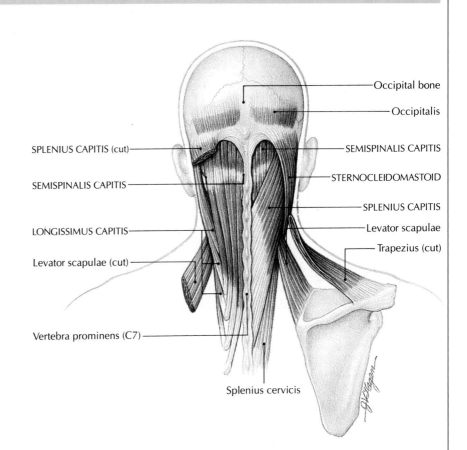

Muscle	Origin	Insertion	Innervation	Action
Sternocleido-mastoid	Sternum, clavicle	Mastoid process of temporal bone	Spinal accessory nerve (XI)	*Bilateral:* flex vertebral column, bringing head down. *Unilateral:* bends vertebral column to same side, drawing head toward shoulder and rotating head so chin points to opposite side.
Semispinalis capitis	Vertebral arches of C7 and T1 to T6 vertebrae	Occipital bone	Dorsal rami of cervical and upper thoracic nerves	*Bilateral:* extend head. *Unilateral:* extends head; turns face to opposite side.
Splenius capitis	Spines of C7 and T1 to T4 vertebrae, ligamentum nuchae (nape of neck)	Occipital bone, mastoid process of temporal bone	Dorsal rami of middle cervical nerves	*Bilateral:* extend head. *Unilateral:* bends head to same side; rotates face to same side.
Longissimus capitis	Transverse processes of T1 to T4 or T5 vertebrae	Mastoid process of temporal bone	Dorsal rami of middle and lower cervical nerves	*Bilateral:* extend head. *Unilateral:* bends head to same side; rotates face to same side.

FIGURE 11.12 INTRINSIC MUSCLES THAT MOVE THE VERTEBRAL COLUMN

Posterior view.

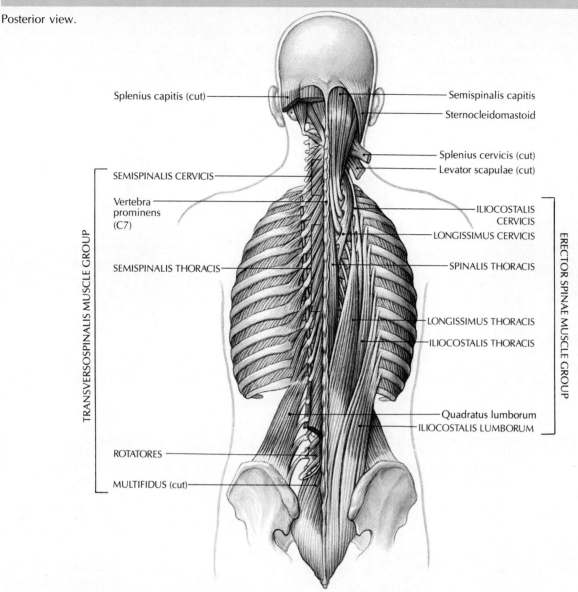

The vast numbers of small muscles and muscle bundles that make up the intrinsic back muscles are located in a pair of broad, longitudinally oriented gutters on either side of the spines of the vertebrae. They are usually organized into two major groups: the *superficial group,* called the **erector spinae,** or **sacrospinalis, muscles,** and the *deep group,* called the **transversospinalis muscles.**

The superficial group consists of overlapping muscle fascicles. They mount "up" the vertebral column, and each bundle spans from origin to insertion over five or so vertebrae. This muscle group is subdivided into three columns: (1) the lateral column, the **iliocostalis,** originates on the iliac crest and ribs, and then after spanning five levels inserts on a rib or transverse process of a cervical vertebra; (2) the middle column, the **longissimus,** originates and inserts on the transverse processes of the thoracic and cervical vertebrae and mastoid process of the skull; and (3) the medial column, the **spinalis,** extends from spinous process to spinous process. Muscles of the superficial group are named **lumborum, thoracis,** and **cervicis,** depending on the region they are associated with.

The deep group consists of small bundles of shorter muscle fascicles called the **transversospinalis muscles.** Each

bundle originates on a transverse process, extends upward and obliquely, and inserts on a spinous process. Some bundles **(rotatores)** run up one or two levels, others **(multifidus)** span two or three vertebrae, and still others **(semispinalis)** span four to five vertebrae. The semispinalis muscles are called **thoracis** or **cervicis,** depending on their location.

The intrinsic back muscles act to extend, bend laterally, and rotate the vertebral column and head. They also counteract the force of gravity, which tends to flex the vertebral column. The posture of the vertebral column is regulated by the vertebral muscles. Contrary to a common belief, these muscles are not active constantly. They are fairly relaxed during normal standing, and they are almost completely relaxed when the vertebral column is bent back (extension), because the ligaments of the vertebral column assume the load. The axiom to remember is: When ligaments suffice, the muscles will yield. In contrast, the vertebral muscles are most active when the vertebral column is bent forward. Extension and hyperextension of the column are used in forced inspiration (as in gasping for air). Gravity and the contraction of the rectus abdominus muscles flex the vertebral column.

FIGURE 11.13 MUSCLES USED IN QUIET BREATHING

[A] Note that the direction of the fibers of the internal and external intercostal muscles are at approximate right angles to each other. [B] Scalene muscles, which extend from the cervical vertebrae to the first two ribs. [C]–[E] "Bucket-handle" movement of ribs during breathing. Right lateral views.

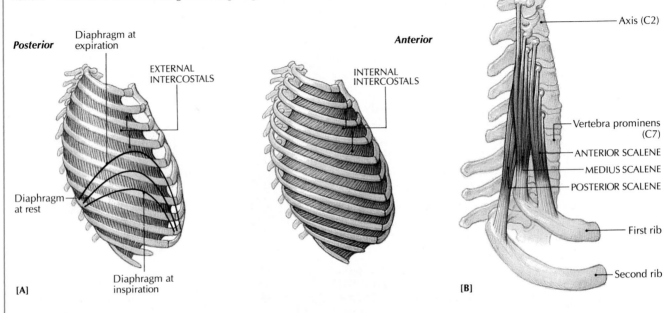

Muscle	Origin	Insertion	Innervation	Action
Diaphragm	Lower six costal cartilages, xiphoid cartilage	Central tendon of diaphragm	Phrenic nerve (from C3 to C5 nerves) innervates portion of diaphragm attached to lumbar vertebrae	Contraction pulls central tendon down to flatten dome and increase vertical length and volume of thorax, resulting in inspiration.
External intercostals	Each from lower border of each bony rib (except 12)	Upper border of next rib below	Intercostal nerves	Elevates ribs to increase all diameters of thorax, resulting in inspiration.
Internal intercostals	Each from upper border of each bony rib (except 1)	Lower border of each bony rib above	Intercostal nerves	Depress (lower) ribs for forced inspiration.
Internal interchondrals (portion of internal intercostals attached to costal cartilages)	Costal (cartilagenous) portion of ribs 2 to 11	Costal portion of ribs 1 to 10	Intercostal nerves	Elevates ribs during inspiration.
Scalenes Anterior	Fronts of transverse processes in cervical vertebrae	Rib 1	Branches of cervical nerves	Steady ribs 1 and 2 during breathing; may assist in elevating them.
Medius and posterior	Backs of transverse processes in cervical vertebrae	Medius to rib 1, posterior to rib 2	Branches of cervical nerves	Steady ribs 1 and 2 during breathing; may assist in elevating them.

Quiet, or normal breathing, involves the coordinated activity of several muscles or groups of muscles. Inspiration (breathing in) requires the enlargement of the rib cage, so that air pressure inside the thorax is less than the atmospheric pressure. Because of this imbalance, air is sucked into the lungs as the dome-shaped **diaphragm** contracts and moves down. This movement is accompanied by the relaxation of the muscles of the abdominal wall. The contraction of the **external intercostal muscles** raises the ribs, in what is called the "bucket-handle movement," which increases the transverse diameter of the rib cage.

In expiration (breathing out), the diaphragm relaxes and is raised to its higher resting position, arching up into the thorax, by the pressure generated in the abdominal cavity by the contractile tone of the muscles in the abdominal wall. The rib cage becomes smaller when the ribs are lowered, largely by elastic recoil, and expiration may also be aided by the contraction of the **internal intercostal muscles.** The **scalene muscles** also assist in respiration by steadying the first two ribs and possibly elevating them.

How can the rib cage change its shape and size to accommodate the shifting volume of the thorax? The answer lies in the "bucket-handle" and "pump-handle" movements of the ribs that result from the twisted shape of a typical rib and its cartilaginous connections to the sternum and vertebral column. Although the increase in size of the rib cage during inspiration requires muscular effort, the decrease during expiration is merely an elastic recoil, produced by the lungs and the costal cartilages.

In the bucket-handle movements of the ribs, there is a resultant increase in the transverse diameter of the thorax as the ribs move upward and outward. In the pump-handle movements there is an increase in the anteroposterior diameter of the thorax as the sternum moves forward and upward as the sternal angle (between the manubrium and the body of the sternum) decreases from 180 degrees (in expiration) to 160 degrees (in inspiration).

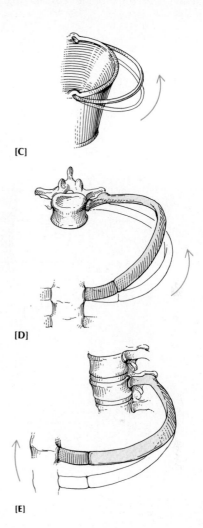

[C]

[D]

[E]

FIGURE 11.14 MUSCLES THAT SUPPORT THE ABDOMINAL WALL

In this anterior view, some of the muscles have been cut to show the different layers of muscles, from superficial to deep.

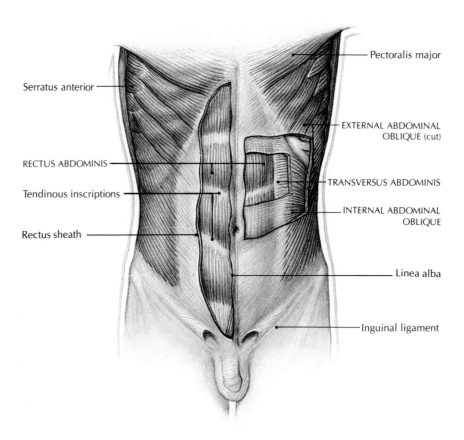

Pectoralis major

Serratus anterior

EXTERNAL ABDOMINAL
OBLIQUE (cut)

RECTUS ABDOMINIS

TRANSVERSUS ABDOMINIS

Tendinous inscriptions

INTERNAL ABDOMINAL
OBLIQUE

Rectus sheath

Linea alba

Inguinal ligament

The muscles of the abdominal wall act to support and protect the internal organs. Several layers of muscles running in different directions add strength to the abdominal wall:

1 As the outside layer of the anterior and lateral walls, the ***external abdominal oblique muscle*** extends forward and down from the ribs until it becomes an aponeurosis, which attaches to the linea alba at the midline. The ***linea alba*** (L. white line) is a tendon running from the xiphoid process of the sternum to the pubic symphysis.

2 The next layer inward is the ***internal abdominal oblique muscle,*** which runs forward and upward from the iliac crest and inguinal ligament, until it too becomes an aponeurosis that connects with the linea alba. The inguinal ligament extends from the anterior superior iliac spine to the pubis bone.

3 The innermost layer is the ***transversus abdominis muscle,*** which extends horizontally from the dorsolumbar fascia (the fascia of the lumbar region, attached to the lumbar vertebrae) to the inguinal ligament and ribs. After becoming an

aponeurosis (passing deep to the rectus abdominis), it continues to meet the linea alba.

4 Running lengthwise on either side of the linea alba is the ***rectus abdominis*** muscle, which extends from the rib cage and sternum to the pubic crest. It is enveloped by the rectus sheath, which is actually made up of the aponeuroses of the other three layers of muscles. The transverse bands crossing the rectus abdominis muscle are called *tendinous inscriptions.*

Together these four abdominal muscles act as a dynamic corset around the abdomen, supplying the necessary pressure for respiration, urination, defecation, and parturition (childbirth). In addition, they also play roles in flexion and lateral bending of the vertebral column.

On the posterior wall of the abdomen, the ***quadratus lumborum muscle*** extends from the twelfth rib to the posterior iliac crest. Besides allowing the trunk to bend toward one side, this muscle gives the vertebral column stability and plays a role in respiration.

Muscle	Origin	Insertion	Innervation	Action
MUSCLES OF ANTERIOR AND LATERAL WALLS				
Rectus abdominis	Pubic crest, symphysis pubis	Costal cartilages of ribs 5 to 7, xiphoid process of sternum	Branches of intercostal nerves (T7 to T12)	Powerful flexor of lumbar vertebral column; depresses rib cage, plays role in stabilizing pelvis against leverage of thigh muscles during walking and running.
External abdominal oblique	Ribs 5 to 12	Iliac crest, linea alba	Branches of intercostal (T8 to T12), iliohypogastric, and ilioinguinal nerves	Along with rectus abdominis, these muscles hold in and compress the abdominal contents, aiding in defecation, urination, and childbirth. Contract to protect abdominal contents against external blows; flex trunk, rotate vertebral column.
Internal abdominal oblique	Iliac crest, inguinal ligament, lumbodorsal fascia	Costal cartilages and ribs 8 to 12, linea alba, xiphoid process of sternum	Branches of intercostal (T8 to T12), iliohypogastric, and ilioinguinal nerves	
Transversus abdominis	Iliac crest, inguinal ligament, lumbar fascia, cartilages of ribs 7 to 12	Xiphoid process of sternum, linea alba, pubis	Branches of intercostal (T8 to T12), iliohypogastric, and ilioinguinal nerves	
MUSCLE OF POSTERIOR WALL				
Quadratus lumborum [FIGURE 11.12]	Iliac crest, iliolumbar ligament, transverse processes of lower lumbar vertebrae	Rib 12, transverse processes of upper lumbar vertebrae	Ventral rami of T12 to L4 nerves	*Unilateral:* flexes (bends) trunk toward same side. *Bilateral:* extends and stabilizes lumbar vertebral column; aids inspiration by increasing length of thorax.

FIGURE 11.15 MUSCLES THAT FORM THE PELVIC OUTLET

Inferior views: **[A]** male, **[B]** female.

The muscles of the pelvic outlet are often divided into two groups: those of the pelvic diaphragm and those of the perineum. The **pelvic diaphragm** (pelvic floor) is the funnel-shaped muscular floor of the pelvic cavity. Its role is to support the pelvic organs and thus prevent organs such as the urinary bladder, uterus, and rectum from falling down (undergoing prolapse) and moving through the diaphragm (as in a hernia). The **levator ani muscle** (consisting of the **pubococcygeus** and **iliococcygeus muscles**), the **coccygeus muscle,** and fascia comprise the pelvic diaphragm. The *urogenital (UG) diaphragm* is composed of fascia and the **deep transverse perinei muscles.** The UG diaphragm also functions as part of the **external sphincter of the urethra** and provides added reinforcement.

The urethra of both sexes and the vagina of the female pass through both the pelvic and UG diaphragms. The rectum passes through the pelvic diaphragm to become the anus. The tonus of the medial portion of the pubococcygeus muscle, called the *puborectalis,* acts as a major deterrent to fecal incontinence (the involuntary passage of feces). Certain muscles of these diaphragms participate in maintaining urinary continence (the ability to retain urine within the bladder until conditions are proper for urination).

The **perineum** (per-uh-NEE-uhm) is the diamond-shaped region extending from the symphysis pubis to the coccyx. The anterior half of the perineum, shaped like a triangle, is the urogenital diaphragm (or triangle). The posterior half, also triangular, is the anal region. The muscles of the urogenital triangle provide voluntary control of the urethra and the release of urine.

The perineum is of special importance to proctologists (physicians who treat disorders of the rectum and anus) and urologists (physicians who treat disorders of the urogenital organs). The female perineum is also of special importance to obstetricians and gynecologists, especially for the management of childbirth and disorders involving the ovaries, uterine tubes, uterus, and vagina.

Occasionally, labor and delivery during childbirth weaken the pelvic floor, and the pelvic diaphragm may be stretched or torn. Such a condition may be accompanied by a hernia or prolapse of the uterus or rectum. The damage can usually be repaired by surgery.

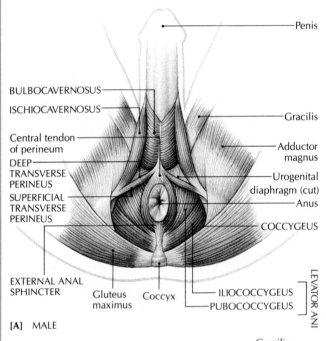

[A] MALE

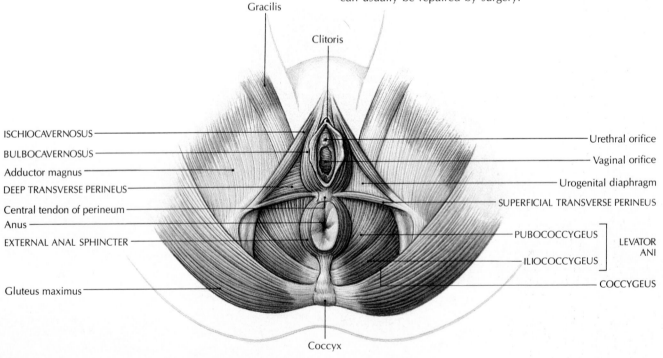

[B] FEMALE

Muscle	Origin	Insertion	Innervation	Action
MUSCLES OF PELVIC DIAPHRAGM (PELVIC FLOOR)				
Levator ani Pubococcygeus	Pubis	Coccyx, anal canal, central tendon of perineum and urethra	Branches of third and fourth sacral nerves and pudendal nerve (second, third, fourth sacral)	Supports pelvic viscera; helps maintain intra-abdominal pressure; acts as sphincter to constrict anus.
Iliococcygeus	Ischial spine	Coccyx	Branches of third and fourth sacral nerves and pudendal nerve (second, third, fourth sacral)	Same as pubococcygeus.
Puborectalis (pubic sling; most medial portion of levator ani; not illustrated)	Pubic arch	Pubic arch (loops around rectum to form sling)	Branches of fourth sacral nerve and pudendal nerve	Draws the rectum anteriorly (forward); acts as principal component for fecal continence.
Coccygeus	Ischial spine	Fifth sacral vertebra, coccyx	Branches of fourth and fifth sacral nerves and pudendal nerve (second, third, fourth sacral)	Same as pubococcygeus.
MUSCLES OF PERINEUM (UROGENITAL DIAPHRAGM)				
Superficial transverse perineus	Ischial tuberosity	Central tendon of perineum	Pudendal nerve (second, third, fourth sacral)	Probably acts to stabilize perineum.
External sphincter of urethra	Central tendon of perineum	Midline in male; vaginal wall in female (not illustrated)	Pudendal nerve (second, third, fourth sacral)	Under voluntary control, prevents urination when bladder wall contracts; relaxes during urination; can contract to cut off stream of urine.
Deep transverse perineus	Ischial rami	Central tendon	Pudendal nerve (second, third, fourth sacral)	Helps support perineum by steadying central perineal tendon.
Ischiocavernosus	Ischial tuberosity and rami of pubis and ischium	Corpus cavernosum of penis in males; clitoris in females	Pudendal nerve (second, third, fourth sacral)	Contributes to maintaining erection of penis and clitoris.
Bulbocavernosus (bulbospongiosus)	Central tendon of perineum	Fascia of urogenital triangle and penis in males; pubic arch, clitoris in females	Pudendal nerve (second, third, fourth sacral)	In male, contracts to expel last drops of urine, contracts rhythmically during ejaculation to propel semen along urethra; in female, acts as sphincter at vaginal orifice.
MUSCLES OF PERINEUM (ANAL REGION)				
External anal sphincter	Anococcygeal raphe	Central tendon of perineum	Pudendal nerve (second, third, fourth sacral)	Constant state of tonic contraction to keep anal orifice closed; closes anus during efforts not associated with defecation.

FIGURE 11.16 MUSCLES THAT MOVE THE SHOULDER GIRDLE

[A] Superficial muscles are shown on the right side of the body and deep muscles on the left; anterior view. [B] Superficial muscles are shown on the left side of body and deep on right; posterior view. [C] Superficial dissection of right shoulder and neck; anterior view.

Three joints contribute to the movements of the shoulder girdle. They are (1) the sternoclavicular joint between the clavicle and the manubrium of the sternum, (2) the acromioclavicular joint between the acromion of the scapula and the clavicle, and (3) the scapulothoracic joint between the subscapularis and serratus anterior muscles attached to the scapula and the thoracic wall. (The scapulothoracic joint is not a true anatomical joint.) The acromioclavicular joint is a sliding joint of accommodation. The movements of the two axes of the sternoclavicular joint are (1) elevation and depression, and (2) protraction and retraction.

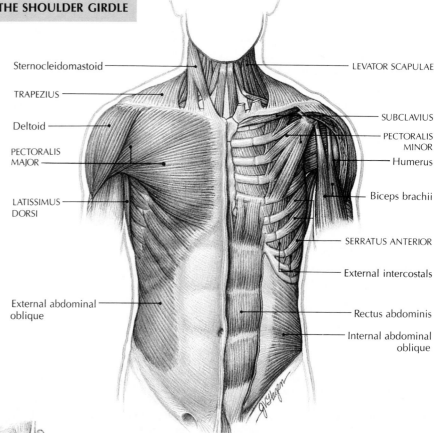

Left

[A]

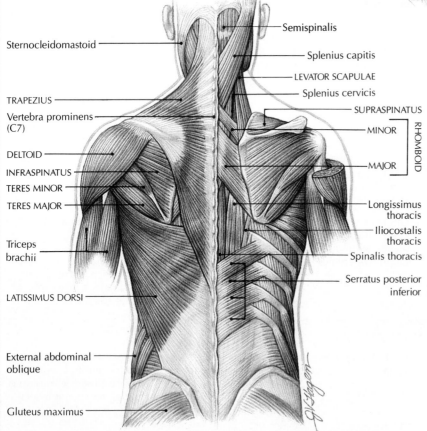

Left

[B]

The muscles of the shoulder girdle contract together to stabilize the girdle for certain movements, such as lifting a heavy object held in the hand. Muscles active during *elevation* include the upper fibers of the **trapezius, levator scapulae,** and **rhomboids.** Those active during *depression* include the **pectoralis major, pectoralis minor,** and **latissimus dorsi.** *Upward rotation* of the girdle (as in reaching upward) uses both the upper and lower fibers of the **trapezius** and lower fibers of the **serratus anterior.** *Downward rotation* uses the **levator scapulae, rhomboids major** and **minor, pectoralis major** and **minor,** and the **latissimus dorsi.** *Upward* and *downward rotary movements* of the scapula rotate around an axis located roughly in the center of the scapula just below the middle of the spine. The function of the **subclavius** is to steady the clavicle during shoulder movements. Because of its location, the subclavius can act as a protective cushion between a fractured clavicle and the subclavian vein and artery.

Muscle	Origin	Insertion	Innervation	Action
Trapezius	Occipital bone, ligamentum nuchae, spines of C7 to T12 vertebrae	Lateral third of clavicle; acromion and spine of scapula	Spinal accessory motor nerve, C3 to C4 sensory nerves	Steadies, elevates, retracts, and rotates* scapula.
Levator scapulae	Transverse processes of vertebrae C1 to C4	Upper vertebral border of scapula	Dorsal scapular nerve	Elevates and rotates* scapula downward.
Rhomboids: major and minor	Ligamentum nuchae, spines of C7 to T5 vertebrae	Medial border of scapula	Dorsal scapular nerve	Retract and elevate scapula.
Serratus anterior	Ribs 1 to 8, midway between angles and costal cartilages	Entire medial border of scapula	Long thoracic nerve	Steadies, rotates* upward, holds against chest wall, and protracts scapula.
Pectoralis minor	Ribs 3, 4, and 5	Coracoid processes of scapula	Medial pectoral nerve	Steadies, depresses, rotates* downward, and protracts scapula.
Pectoralis major	Medial half of clavicle (clavicular head) and sternum, ribs 1 to 6 (sternal head)	Lateral crest of intertubercular groove of humerus	Medial and lateral pectoral nerves	Adducts and rotates* arm medially. Clavicular head flexes shoulder joint (raises arm forward), sternal head extends shoulder joint (carries arm backward).
Latissimus dorsi	Spines of T7 to sacral vertebrae, iliac crest, ribs 7 to 12	Intertubercular groove of humerus	Thoracodorsal nerve (middle subscapular nerve)	Depresses, rotates* downward, and retracts scapula; also extends, adducts, and medially rotates humerus at shoulder (glenohumeral joint).
Subclavius	Rib 1 (median)	Clavicle	Nerve to subclavius	Draws clavicle toward sternoclavicular joints; also steadies clavicle.

*The axis of rotation of the scapula is located in the center of the scapula. When the glenoid fossa (at shoulder joint) is directed downward, the scapula is said to rotate downward. When the glenoid fossa is directed upward, the scapula is said to rotate upward.

FIGURE 11.17 MUSCLES THAT MOVE THE HUMERUS AT THE SHOULDER JOINT

[A] Superficial muscles; anterior view. [B] Superficial muscles; posterior view. [C] Deep muscles; anterior view. [D] Deep muscles; posterior view. All views show the right upper limb.

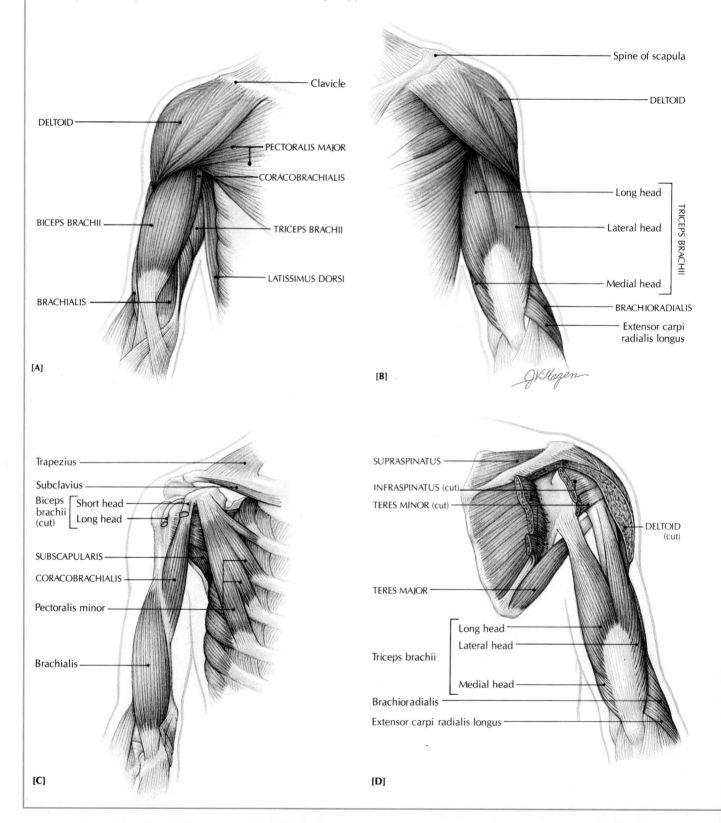

Many muscles contribute to the movements of the shoulder joint. The four SITS muscles—*supraspinatus, infraspinatus, teres minor,* and *subscapularis*—are the *rotator cuff muscles.* They are located adjacent to the articular capsule of the joint. The tendons of the SITS muscles (also called the *rotator cuff*) prevent instability and dislocations in an otherwise potentially unstable joint. The deltoid muscle overrides the SIT muscles (supraspinatus, infraspinatus, teres minor), and helps form the roundness of the shoulder. Other contributing muscles are noted below.

The shoulder joint has three axes of rotation, resulting in (1) flexion and extension, (2) abduction and adduction, and (3) medial and lateral rotation. A combination of all these movements in succession is *circumduction.*

The *flexors* of the shoulder joint include the **pectoralis major** (clavicular head), **deltoid** (clavicular part), **coracobrachialis,** and **biceps brachii.** The *extensors* include the **latissimus dorsi, teres major, pectoralis major** (sternal head), **deltoid** (spinous part), and **triceps brachii** (long head). The *abductors* are the **supraspinatus** and **deltoid** (acromial part), and the *adductors* are the **pectoralis major** (both heads), **latissimus dorsi, teres major, coracobrachialis,** and **triceps brachii** (long head). The *lateral rotators* include the **infraspinatus, teres minor,** and **deltoid** (spinous part). The *medial rotators* include the **pectoralis major** (both clavicular and sternal heads), **teres major, latissimus dorsi, subscapularis,** and **deltoid** (clavicular part).

Muscle	Origin	Insertion	Innervation	Action
ROTATOR CUFF MUSCLES				
Supraspinatus	Supraspinatus fossa of scapula	Greater tubercle of humerus	Suprascapular nerve	Stabilizes joint; abducts arm.
Infraspinatus	Infraspinatus fossa of scapula	Greater tubercle of humerus	Suprascapular nerve	Stabilizes joint; laterally rotates arm.
Teres minor	Lower border of lateral scapula	Greater tubercle of humerus	Axillary nerve	Stabilizes joint; laterally rotates arm.
Subscapularis	Subscapular fossa of scapula	Lesser tubercle of humerus	Upper and lower subscapular nerves	Stabilizes joint; medially rotates arm.
OTHER CONTRIBUTING MUSCLES				
Teres major	Inferior angle of scapula	Medial crest of intertubercular groove of humerus	Lower subscapular nerve	Stabilizes upper arm during adduction; adducts, extends, and medially rotates humerus.
Deltoid	Clavicle (lateral third), acromion process and spine of scapula	Deltoid tuberosity of humerus	Axillary nerve	Clavicular part: flexes and medially rotates arm; acromial part: abducts arm; spinous part: extends and laterally rotates arm.
Pectoralis major	Medial half of clavicle (clavicular head) and sternum, and ribs 1 to 6 (sternal head)	Lateral aspect of intertubercular groove of humerus	Medial and lateral pectoral nerves	Clavicular head: flexes, adducts, and medially rotates arm; sternal head: extends, adducts, and medially rotates arm.
Latissimus dorsi	Spines of lower thoracic, lumbar, and sacral vertebrae, iliac crest, and ribs 7 to 12	Intertubercular groove of humerus	Thoracodorsal (middle subscapular) nerve	Extends, adducts, and medially rotates arm; also depresses, retracts, and rotates scapula downward.
Coracobrachialis	Coracoid process of scapula	Middle third of humerus	Musculocutaneous nerve	Flexes and adducts arm.

FIGURE 11.18 MUSCLES THAT MOVE THE FOREARM AND WRIST

[A] Superficial muscles of right hand; ventral view. [B] Superficial muscles; dorsal view. (FIGURE 11.18 continues on page 272.)

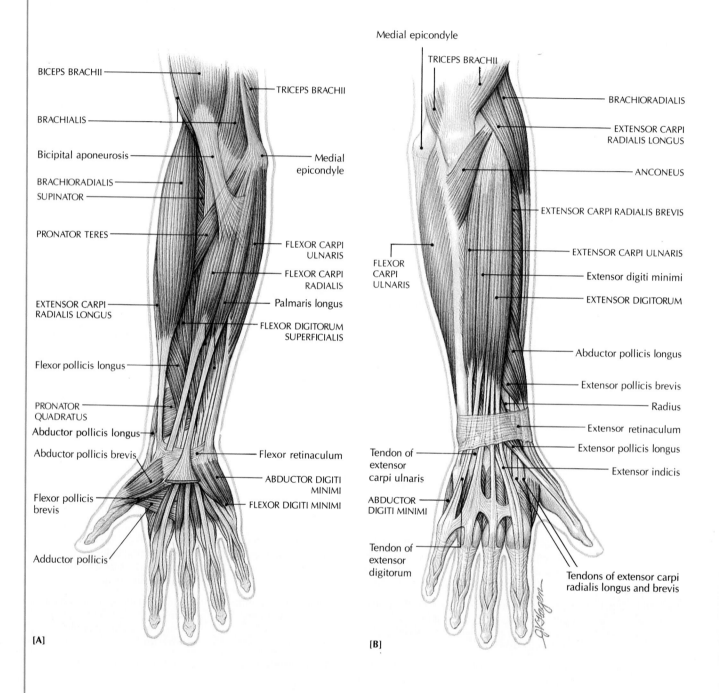

BICEPS BRACHII

BRACHIALIS

Bicipital aponeurosis

BRACHIORADIALIS
SUPINATOR

PRONATOR TERES

EXTENSOR CARPI
RADIALIS LONGUS

Flexor pollicis longus

PRONATOR
QUADRATUS
Abductor pollicis longus

Abductor pollicis brevis

Flexor pollicis
brevis

Adductor pollicis

TRICEPS BRACHII

Medial
epicondyle

FLEXOR CARPI
ULNARIS

FLEXOR CARPI
RADIALIS

Palmaris longus

FLEXOR DIGITORUM
SUPERFICIALIS

Flexor retinaculum

ABDUCTOR DIGITI
MINIMI

FLEXOR DIGITI MINIMI

[A]

Medial epicondyle

TRICEPS BRACHII

FLEXOR
CARPI
ULNARIS

Tendon of
extensor
carpi ulnaris

ABDUCTOR
DIGITI MINIMI

Tendon of
extensor
digitorum

BRACHIORADIALIS

EXTENSOR CARPI
RADIALIS LONGUS

ANCONEUS

EXTENSOR CARPI RADIALIS BREVIS

EXTENSOR CARPI ULNARIS

Extensor digiti minimi

EXTENSOR DIGITORUM

Abductor pollicis longus

Extensor pollicis brevis

Radius

Extensor retinaculum

Extensor pollicis longus

Extensor indicis

Tendons of extensor carpi
radialis longus and brevis

[B]

At the one-axis elbow joint, flexion and extension are the only movements. The *flexors* of the joint include the **biceps brachii, brachialis,** and **brachioradialis.** The *extensors* are the **triceps brachii** and the tiny **anconeus.**

At the proximal and distal radioulnar joints, pronation and supination are the only movements. The *pronators* include the **pronator teres** and the **pronator quadratus.** The *supinators* are the **supinator** and the **biceps brachii.** Supination is more powerful than pronation. That is why screws, caps on jars, and door handles are designed the way they are. (For left-handed people, the motion involved in moving these objects is *pronation.*)

The wrist joints include the (1) radiocarpal joint (between the radius and proximal row of carpal bones), where most of the movements occur, and (2) the midcarpal joint (between the proximal and distal rows of carpal bones), where a slight amount of flexion and extension occurs.

There is no movement between the distal carpal bones and the metacarpals. The radiocarpal joint is a two-axis condyloid joint where (1) flexion and extension (and hyperextension), and (2) abduction (radial deviation, toward the radius) and adduction (ulnar deviation, toward the ulna) occur.

The *flexor* muscles of the wrist joint include the **flexor carpi ulnaris, flexor carpi radialis,** and the **palmaris longus.** The *extensor* muscles are the **extensor carpi radialis longus, extensor carpi radialis brevis,** and the **extensor carpi ulnaris.** The *abductor* muscles include the **flexor carpi radialis, extensor carpus radialis longus** and **brevis,** and **abductor pollicis longus.** The *adductor* muscles are the **extensor carpi ulnaris** and the **flexor carpi ulnaris.** The contraction of any of these muscles in a sequential order can produce circumduction. Contracting in concert, these muscles act to stabilize the wrist for the effective use of the fingers.

Muscle	Origin	Insertion	Innervation	Action
Biceps brachii	*Long head:* supraglenoid tubercle of scapula *Short head:* coracoid process of scapula	Tuberosity of radius, bicipital aponeurosis	Musculocutaneous nerve	Flexes elbow; supinates forearm.
Brachialis	Anterior surface of humerus	Tuberosity and coronoid process of ulna	Musculocutaneous nerve	Flexes elbow.
Triceps brachii	*Long head:* infraglenoid tubercle of scapula *Lateral head:* posterior and lateral humerus above radial groove *Medial head:* posterior humerus below radial groove	Olecranon process of ulna	Radial nerve	Extends elbow.
Supinator	Lateral epicondyle of humerus, supinator crest of ulna	Lateral surface of radius (distal to head)	Radial nerve	Supinates forearm (rotates palm of hand anteriorly).
Pronator teres	Medial epicondyle of humerus, coronoid process of ulna	Midlateral surface of radius	Median nerve	Pronates forearm; flexes elbow.
Pronator quadratus	Anterior distal end of ulna	Anterior distal end of radius	Median nerve	Pronates forearm.
Brachioradialis	Lateral supracondylar ridge of humerus	Lateral surface of radius near base of styloid process	Radial nerve	Flexes elbow.
Anconeus	Posterior lateral epicondyle of humerus	Lateral surface of olecranon process of ulna	Radial nerve	Extends elbow.
Flexor carpi radialis	Medial epicondyle of humerus	Base of second and third metacarpals	Median nerve	Flexes and abducts wrist.
Flexor carpi ulnaris	Medial epicondyle of humerus, olecranon process of ulna	Pisiform, hamate, base of fifth metacarpal	Ulnar nerve	Flexes and adducts wrist.
Extensor carpi radialis longus	Lateral epicondyle of humerus	Base of second metacarpal	Radial nerve	Extends and abducts wrist.
Extensor carpi radialis brevis	Lateral epicondyle of humerus	Base of third metacarpal	Radial nerve	Extends and abducts wrist.
Extensor carpi ulnaris	Lateral epicondyle of humerus	Base of fifth metacarpal	Radial nerve	Extends and adducts wrist.

(Figure 11.18 continues on following page)

FIGURE 11.18 MUSCLES THAT MOVE THE FOREARM AND WRIST *(Continued)*

[C] Superficial dissection of right forearm and hand; dorsal view. [D] Deep muscles; ventral view.

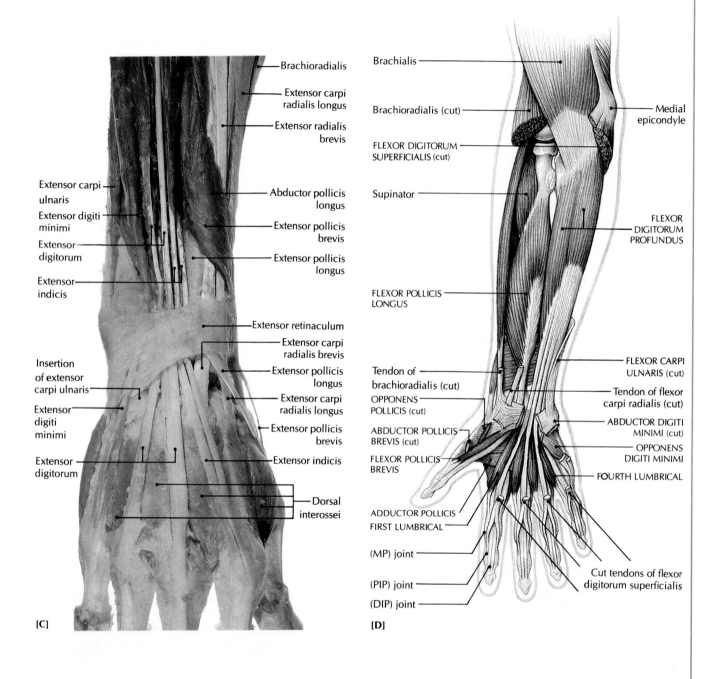

[C]

Brachioradialis

Extensor carpi radialis longus

Extensor radialis brevis

Extensor carpi ulnaris

Extensor digiti minimi

Extensor digitorum

Extensor indicis

Abductor pollicis longus

Extensor pollicis brevis

Extensor pollicis longus

Extensor retinaculum

Insertion of extensor carpi ulnaris

Extensor digiti minimi

Extensor digitorum

Extensor carpi radialis brevis

Extensor pollicis longus

Extensor carpi radialis longus

Extensor pollicis brevis

Extensor indicis

Dorsal interossei

[D]

Brachialis

Brachioradialis (cut)

FLEXOR DIGITORUM SUPERFICIALIS (cut)

Supinator

FLEXOR POLLICIS LONGUS

Tendon of brachioradialis (cut)

OPPONENS POLLICIS (cut)

ABDUCTOR POLLICIS BREVIS (cut)

FLEXOR POLLICIS BREVIS

ADDUCTOR POLLICIS

FIRST LUMBRICAL

(MP) joint

(PIP) joint

(DIP) joint

Medial epicondyle

FLEXOR DIGITORUM PROFUNDUS

FLEXOR CARPI ULNARIS (cut)

Tendon of flexor carpi radialis (cut)

ABDUCTOR DIGITI MINIMI (cut)

OPPONENS DIGITI MINIMI

FOURTH LUMBRICAL

Cut tendons of flexor digitorum superficialis

FIGURE 11.19 MUSCLES THAT MOVE THE THUMB

Muscles of the back of the thumb; medial side of right hand.

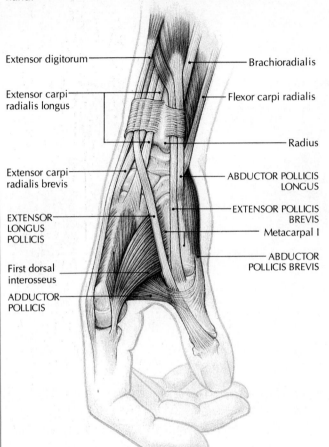

Extensor digitorum

Extensor carpi radialis longus

Extensor carpi radialis brevis

EXTENSOR LONGUS POLLICIS

First dorsal interosseus

ADDUCTOR POLLICIS

Brachioradialis

Flexor carpi radialis

Radius

ABDUCTOR POLLICIS LONGUS

EXTENSOR POLLICIS BREVIS

Metacarpal I

ABDUCTOR POLLICIS BREVIS

The movements of the thumb take place at (1) the carpometacarpal joint, which is a saddle joint with three axes of rotation, and (2) two hinge joints involving the phalanx. The movements at the saddle joint include (1) flexion and extension, (2) abduction and adduction, and (3) opposition and reposition. Crucial to the opposability of the thumb are opposition and reposition, and abduction and adduction. In opposition the metacarpal bone rolls toward the midline of the hand so that the ball of the thumb's distal phalanx can touch the balls of the distal phalanges of the other fingers. The muscular mound at the base of the thumb, the *thenar eminence*, is formed by the three thenar muscles: **abductor pollicis brevis, flexor pollicis brevis,** and **opponens pollicis.** At the carpometacarpal joint, the flexor is the flexor pollicis brevis. The *extensors* are the **extensor pollicis longus, extensor pollicis brevis,** and **abductor pollicis longus.** The *adductor* is the **adductor pollicis.** The *opposition* movement is by the **opponens pollicis** and the *reposition* movements by the **abductor pollicis longus** and **extensor pollicis longus** and **brevis.** At the phalangeal hinge joints, the *flexors* are the **flexor pollicis longus** and **flexor pollicis brevis,** and the *extensors* are the **extensor pollicis longus** and **brevis.**

Several features of the thumb contribute immensely to the versatility of the hand. The length of the thumb and the presence of the flexor pollicis longus and the opponens muscles permit the opposition movement and the great flexibility in using the thumb along with the other fingers.

The little finger is associated with the *hypothenar eminence*, formed by the flexor digiti minimi, abductor digiti minimi, and opponens muscles. They have roles similar to their counterparts of the thenar eminence.

Muscle	Origin	Insertion	Innervation	Action
Flexor pollicis longus	Anterior surface of radius; interosseous membrane	Base of distal phalanx of thumb	Median nerve	Flexes thumb.
Flexor pollicis brevis	Trapezium and adjacent region	Base of proximal phalanx of thumb	Median nerve	Flexes thumb; helps in opposition and reposition.
Opponens pollicis	Trapezium and adjacent region	First metacarpal of thumb	Median nerve	Rolls thumb toward midline of palm (opposition).
Abductor pollicis brevis	Scaphoid and trapezium bones and adjacent region	Base of proximal phalanx of thumb	Median nerve	Abducts thumb; helps in opposition.
Abductor pollicis	Second and third metacarpals and capitate bones	Base of proximal phalanx of thumb	Ulnar nerve	Adducts, flexes thumb.
Extensor pollicis longus	Dorsal surface of ulna; interosseous membrane	Base of distal phalanx of thumb	Radial nerve	Extends thumb; helps in reposition.
Extensor pollicis brevis	Dorsal surface of radius; interosseous membrane	Base of proximal phalanx of thumb	Radial nerve	Extends thumb; helps in reposition, adduction
Abductor pollicis longus	Dorsal surfaces of ulna and radius	Base of first metacarpal bone	Radial nerve	Abducts and extends thumb; helps in reposition.

FIGURE 11.20 MUSCLES THAT MOVE THE FINGERS (EXCEPT THE THUMB)

Deep intrinsic muscles of the right hand. **[A]** Palmar interossei; palmar view. **[B]** Dorsal interossei; dorsal view. **[C]** Lumbricals; palmar view.

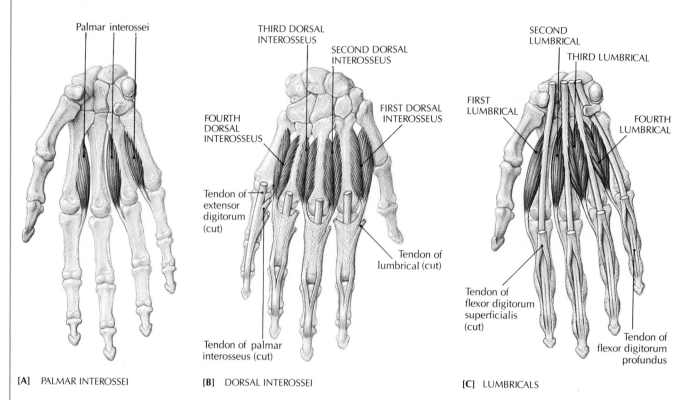

Palmar interossei

THIRD DORSAL INTEROSSEUS
SECOND DORSAL INTEROSSEUS
FOURTH DORSAL INTEROSSEUS
FIRST DORSAL INTEROSSEUS
Tendon of extensor digitorum (cut)
Tendon of lumbrical (cut)
Tendon of palmar interosseus (cut)

SECOND LUMBRICAL
THIRD LUMBRICAL
FIRST LUMBRICAL
FOURTH LUMBRICAL
Tendon of flexor digitorum superficialis (cut)
Tendon of flexor digitorum profundus

[A] PALMAR INTEROSSEI

[B] DORSAL INTEROSSEI

[C] LUMBRICALS

The movements of the fingers (except the thumb) are based on a remarkable interplay of several muscle groups, and especially on the role of the lumbrical muscles. To fully appreciate this interplay we must understand the movements at each joint and the anatomical relationships and actions of the tendons and muscles acting on those joints.

Each finger has three joints: (1) the *metacarpophalangeal joint* (MP or knuckle joint) between the metacarpal and proximal phalanges, (2) the *proximal interphalangeal joint* (PIP joint) between the proximal and middle phalanges, and (3) the *distal interphalangeal joint* (DIP joint) between the middle and distal phalanges. The metacarpophalangeal joint is a condyloid joint with two axes of rotation, permitting flexion and extension, abduction and adduction, and circumduction. The proximal and distal interphalangeal joints are hinge joints with one axis of rotation, permitting only flexion and extension.

The *flexors* of the metacarpophalangeal joints are seven **interosseous muscles** and four **lumbrical muscles,** and the *extensor* is the **extensor digitorum** (including the extensor digitorum indicus and digitorum minimus). The *abductors* and *adductors* are the interosseous muscles and the abductor digiti minimi. The four **dorsal interossei** abduct

(spread) the fingers (except the little finger*), and the three **palmar interossei** adduct the fingers (bring them back together). The movements of abduction and adduction of the fingers are defined in relation to the third (middle) finger, which acts as the axis of the hand. Thus, spreading of the fingers away from the third finger is abduction, and movement toward the middle finger is adduction. The **abductor digiti minimi** abducts (spreads) the little finger away from the fourth finger. Circumduction combines all of these actions in sequence. Because the **flexor digitorum profundus** and **flexor digitorum superficialis** pass on the palmar side of the MP joint, they contribute to *flexion* at the joint. The *extension* of the MP joint is carried out by the **extensor digitorum muscles,** whose tendons attach to the base of each proximal phalanx on the dorsal side of the axis. The extensors for the index finger and little finger are the **extensor digitorum indicus** and the **extensor digitorum minimus,** respectively.

The *flexor* of the proximal interphalangeal joint is the **flexor digitorum superficialis,** with assistance from the

*Each finger has its own anatomical name. The thumb is the *pollex* (L. thumb), next to it is the *index* finger (L. pointer), then the *medius* or middle finger, *annularis* (L. ring) or fourth finger, and *minimus* (L. least) or little finger.

flexor digitorum profundus. The *flexor* of the distal interphalangeal joint is the *flexor digitorum profundus.* The *extensors* of the PIP and DIP joints are the interossei and lumbricals, with minor assistance from the extensor digitorum.

The "paradox" of the lumbrical and interosseous muscles acting to flex the MP joint and to extend the PIP and DIP joints is explained by the relationship of the tendons of these muscles to the joints. The tendons of these muscles (1) pass on the palmar side of the MP joint, (flexing the joint), and (2) continue to join the extensor tendon on the dorsal side of each of the proximal phalanges II to IV (extending the PIP and DIP joints).

The roles of the lumbrical and interosseous muscles are critical in the extension of the PIP and DIP joints, and in the coordination of the finger movements. Although the tendons of the extensor digitorum muscles continue beyond the MP joint and attach to the base of the middle and distal phalanges on the dorsal sides, they contribute minimally to the extension of the PIP and DIP joints. This occurs because the force generated by the contraction of the attached muscles is almost completely expended at the MP joint. This complex arrangement between the flexors and extensors of the interphalangeal joints is all-important in producing delicate finger movements. The extensor activities of the lumbricals (largely) and interossei allow those muscles to act as effective antagonists in adjusting the activity of the flexor muscles during writing and other fine finger movements.

Muscle	Origin	Insertion	Innervation	Action
Flexor digitorum superficialis	Medial epicondyle of humerus, coronoid process of ulna, anterior border of radius	Both sides of middle phalanges of fingers 2 to 5	Median nerve	Flexes PIP and MP joints.*
Flexor digitorum profundus	Anterior and medial surface of ulna	Palmar surfaces of distal phalanges of fingers 2 to 5	Median nerve, ulnar nerve	Flexes PIP, DIP, and MP joints.*
Interossei (4 dorsal interossei and 3 palmar interossei)	Metacarpal bones	Bases of proximal phalanges and extensor expansions of fingers	Ulnar nerve	Dorsal interossei: abduct fingers. Palmar interossei: adduct fingers; extend PIP and DIP joints; flex MP joints.†
Lumbricals (4 muscles)	Tendons of flexor digitorum profundus in palm of hand	Extensor tendon (expansions) of digits distal to knuckles	Lateral two by median nerve; medial two by ulnar nerve	Flex MP joint (knuckles); extend PIP and DIP joints.†
Extensor digitorum	Lateral epicondyle of humerus	Dorsal surface of phalanges in fingers 2 to 5	Radial nerve	Extends MP joint (and PIP slightly).
Flexor digiti minimi	Hook of hamate bone and adjacent region	Medial side of proximal phalanx of little finger	Ulnar nerve	Flexes proximal phalanx of little finger (MP joint).
Opponens digiti minimi	Hook of hamate bone and adjacent region	Medial half of palmar surface of little finger	Ulnar nerve	Rotates little finger toward midline of hand (opposition).
Abductor digiti minimi	Pisiform bone and adjacent region	Medial side of proximal phalanx of little finger	Ulnar nerve	Abducts little finger away from fourth finger.

*Because of the pull they can generate, the long flexor muscles can help flex joints more proximal than that of their primary action.

†The extensor digitorum is not the primary extensor of the PIP and DIP joints because it expends most of its force at the MP joint. The lumbricals and interossei extend the PIP and DIP joints through their attachments to the extensor tendons.

FIGURE 11.21 MUSCLES THAT MOVE THE FEMUR AT THE HIP JOINT

[A] Right anterior view. [B] Right posterior view.

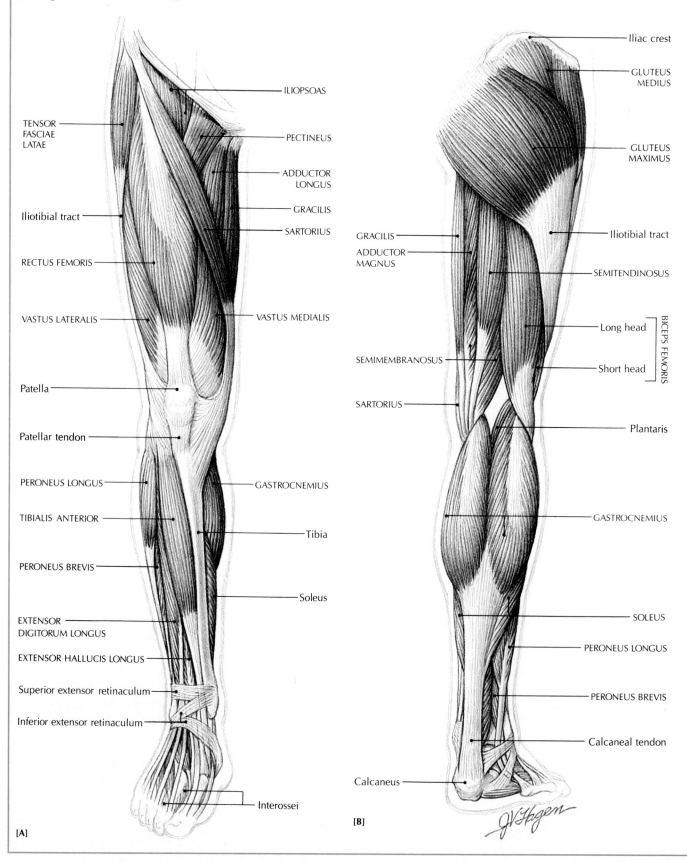

[A] Right anterior view labels: ILIOPSOAS, TENSOR FASCIAE LATAE, PECTINEUS, ADDUCTOR LONGUS, GRACILIS, Iliotibial tract, SARTORIUS, RECTUS FEMORIS, VASTUS LATERALIS, VASTUS MEDIALIS, Patella, Patellar tendon, PERONEUS LONGUS, GASTROCNEMIUS, TIBIALIS ANTERIOR, Tibia, PERONEUS BREVIS, Soleus, EXTENSOR DIGITORUM LONGUS, EXTENSOR HALLUCIS LONGUS, Superior extensor retinaculum, Inferior extensor retinaculum, Interossei.

[B] Right posterior view labels: Iliac crest, GLUTEUS MEDIUS, GLUTEUS MAXIMUS, Iliotibial tract, GRACILIS, ADDUCTOR MAGNUS, SEMITENDINOSUS, SEMIMEMBRANOSUS, BICEPS FEMORIS Long head, Short head, SARTORIUS, Plantaris, GASTROCNEMIUS, SOLEUS, PERONEUS LONGUS, PERONEUS BREVIS, Calcaneal tendon, Calcaneus.

The ball-and-socket hip joint, with its three axes of rotation, is capable of (1) flexion and extension, (2) abduction and adduction, (3) medial and lateral rotation, and (4) circumduction. The powerful muscles surrounding the hip joint are active during walking, running, and climbing. They also act to stabilize the joint.

The *hamstring muscles (semitendinosus, semimembranosus,* and *biceps femoris)* act as extensors of the hip and flexors of the knee during walking on the level; these muscles relax as the hip flexors contract. The *gluteus maximus* acts as an extensor of the hip only when power is required, as in moving against gravity. For example, it is used in rising from a seated position, in climbing upstairs or up a steep hill, and in running. The *psoas major* contributes to the stability of the hip and is a powerful flexor. (The psoas major and the *iliacus* are sometimes considered together as the *iliopsoas* muscle.) The *tensor fasciae latae* and gluteus maximus pull on the iliotibial tract (a modified fascial sheath located on the lateral thigh) extending beyond the knee joint. This muscle functions to brace the knee so the joint doesn't buckle while the other foot is off the ground during walking. The *adductor magnus* and *adductor longus* are powerful muscles used in kicking a soccer ball, for instance.

During locomotion the hip remains parallel to the ground through the action of the *gluteus medius* muscle. If this muscle is paralyzed on one side, a lurching gait results. If you place your hand on the gluteus muscle while walking, you will note that it contracts while the limb is grounded and relaxes when the limb is free and in motion. The contraction pulls down on the pelvis and prevents the opposite side from dropping, thereby preventing a lurch.

Muscle	Origin*	Insertion*	Innervation	Action
Iliopsoas†	*Iliacus:* iliac fossa of false pelvis *Psoas:* lumbar vertebrae	Lesser trochanter of femur	*Iliacus:* femoral nerve *Psoas:* directly by L2 and L3	Flexes hip joint; can flex and rotate vertebral column.
Pectineus	Superior ramus of pubis	Pectineal line below lesser trochanter on posterior femur	Femoral and obturator nerves	Flexes and adducts hip joint.
Adductor longus	Body of pubis	Middle third of linea aspera of femur	Obturator nerve	Adducts and flexes hip joint.
Adductor brevis	Body and inferior ramus of pubis	Proximal part of linea aspera of femur	Obturator nerve	Adducts and helps flex hip joint.
Adductor magnus	Inferior ramus of pubis to ischial tuberosity	Middle of linea aspera, adductor tubercle of femur	Obturator nerve and tibial portion of sciatic nerve	Adducts and helps flex and extend hip joint.
Gluteus maximus	Iliac crest, sacrum, coccyx	Iliotibial tract of tensor fasciae latae and gluteal tuberosity of femur	Inferior gluteal nerve	Extends and laterally rotates hip joint.
Gluteus medius	Ilium	Greater trochanter of femur	Superior gluteal nerve	Abducts and medially rotates hip joint; steadies pelvis.
Gluteus minimus (not illustrated)	Ilium	Greater trochanter of femur	Superior gluteal nerve	Abducts and medially rotates hip joint; steadies pelvis.
Tensor fasciae latae	Iliac crest, anterior superior iliac spine	Lateral condyle of tibia through iliotibial tract	Superior gluteal nerve	Abducts, flexes, and medially rotates hip joint; steadies trunk; extends knee joint.
Small lateral rotators (piriformis, obturator internus, quadratus femoris)	Sacrum, obturator foramen, ischial tuberosity	Greater trochanter of femur	Branch from sacral plexus for each muscle	Laterally rotates and adducts hip joint.

*The origin and insertion of a muscle acting on the hip and knee joints are reversible, depending on the movement. The insertion is on the femur when the thigh is moved and the pelvis is fixed in position, whereas the insertion is on the pelvis when the pelvis is moved and the thigh is fixed.

†The iliopsoas muscle is composed of an iliacus muscle and a psoas muscle.

FIGURE 11.22 MUSCLES THAT ACT AT THE KNEE JOINT

[A] Right anterior view. [B] Right posterior view.

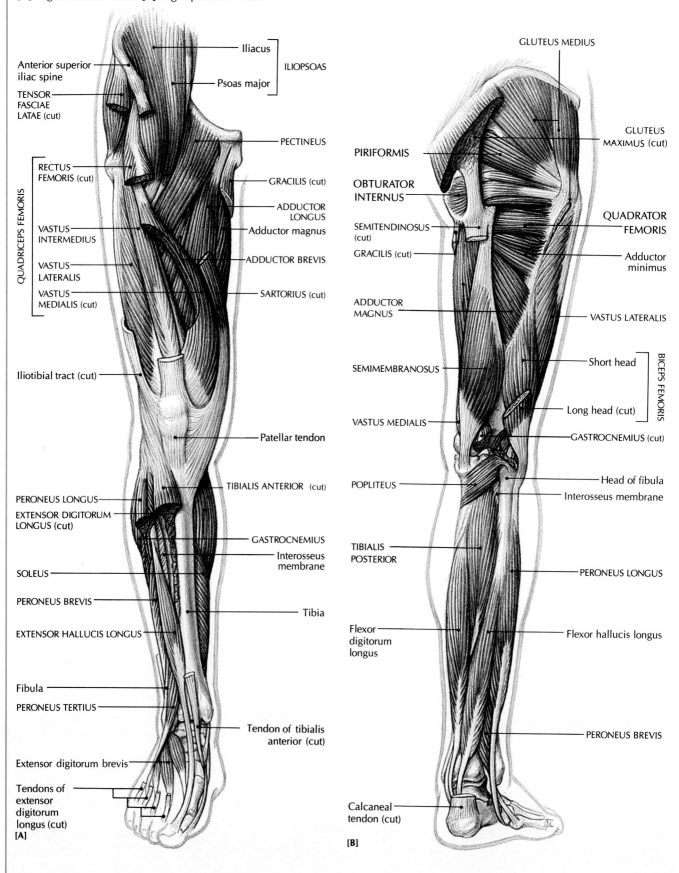

The knee is a hinge joint capable of flexion and extension, with a slight amount of medial and lateral rotation when not fully extended. During the final stage of full extension during walking, the femur rotates medially in relation to the tibia when the foot is planted on the ground. Try it. This is known as "locking" the knee to make it rigid. If the knee is extended while the limb is free, as when punting a football, the tibia rotates laterally in relation to the femur to lock the knee. Unlocking of the knee by the *popliteus muscle* is the first stage of flexion of the knee joint. The *hamstring muscles* are flexors *(biceps femoris, semimembranosus, semitendinosus),* lateral rotators *(biceps femoris),* and medial rotators *(semitendinosus* and *semimembranosus).* The *sartorius, gracilis,* and *gastrocnemius* contribute to flexion.

The *quadriceps femoris muscle* is made up of four parts. The patella, the bone to which the quadriceps femoris is attached, acts to place the attachment of the patella and quadriceps femoris further away from the flexion-extension axis of rotation. Such an adjustment increases the length of the "lever arm" of the joint by increasing the leverage and power of the strong quadriceps femoris.

Four bursae are located in the vicinity of the patella, and are important to knee movement. The *suprapatellar bursa* is located just above the patella, between the femur and tendon of the quadriceps femoris muscle. It permits free movement of the quadriceps tendon over the distal end of the femur and facilitates the full range of extension and flexion of the knee joint. The *infrapatellar bursa* is located just below the patella, between the tibia and the patellar tendon. The *subcutaneous prepatellar bursa* lies between the skin and the anterior surface of the patella. It allows free movement of the skin over the underlying patella during flexion of the knee joint, for example. The subcutaneous prepatellar bursa can become the victim of *friction bursitis,* as a result of rubbing between the skin and patella. When inflammation is chronic, the bursa can become distended with fluid in individuals who frequently work on all fours. The *subcutaneous infrapatellar bursa* lies between the skin and the fascia anterior to the tibial tuberosity. It allows the skin to glide over the tibial tuberosity and to withstand pressure when kneeling with the trunk upright (when praying, for example).

Muscle	Origin*	Insertion*	Innervation	Action
QUADRICEPS FEMORIS				
Rectus femoris	Anterior inferior iliac spine	Patella, through patellar tendon to tibial tuberosity	Femoral nerve	Extends leg at knee joint; flexes thigh at hip joint.
Vastus lateralis	Greater trochanter and linea aspera of femur	Patella, through patellar tendon to tibial tuberosity	Femoral nerve	Extends leg at knee joint.
Vastus intermedius	Anterior femur	Patella, through patellar tendon to tibial tuberosity	Femoral nerve	Extends leg at knee joint.
Vastus medialis	Linea aspera of femur	Patella, through patellar tendon to tibial tuberosity	Femoral nerve	Extends leg at knee joint.
HAMSTRING MUSCLES				
Biceps femoris	*Long head:* ischial tuberosity. *Short head:* linea aspera of femur	Head of fibula and lateral condyle of tibia	*Long head:* tibial nerve *Short head:* common peroneal nerve	Flexes and laterally rotates knee joint; extends thigh at hip joint.
Semitendinosus	Ischial tuberosity	Medial surface of proximal tibia	Tibial nerve	Flexes and medially rotates knee joint; extends thigh at hip joint.
Semimembranosus	Ischial tuberosity	Medial condyle of tibia	Tibial nerve	Flexes and medially rotates knee joint; extends thigh at hip joint.
OTHER MUSCLES				
Gracilis	Body and inferior ramus of pubis	Medial surface of proximal tibia	Obturator nerve	Flexes and medially rotates knee joint; adducts thigh at hip joint.
Sartorius	Anterior superior iliac spine	Medial surface of proximal tibia	Femoral nerve	Flexes leg at knee joint and thigh at hip joint.
Popliteus	Lateral condyle of femur	Posterior surface of tibia	Tibial nerve	Flexes, rotates leg at knee joint; unlocks knee joint.

*In the knee joint, the origin and insertion can be reversed depending upon whether the leg and foot are moved in relation to the thigh, as when one is seated and the leg freely moves, or whether the foot is planted and the thigh does much of the moving, as in squatting exercises or in rising from the seated position.

FIGURE 11.23 MUSCLES THAT MOVE THE FOOT

[A] Right lateral view. [B] Right medial view.

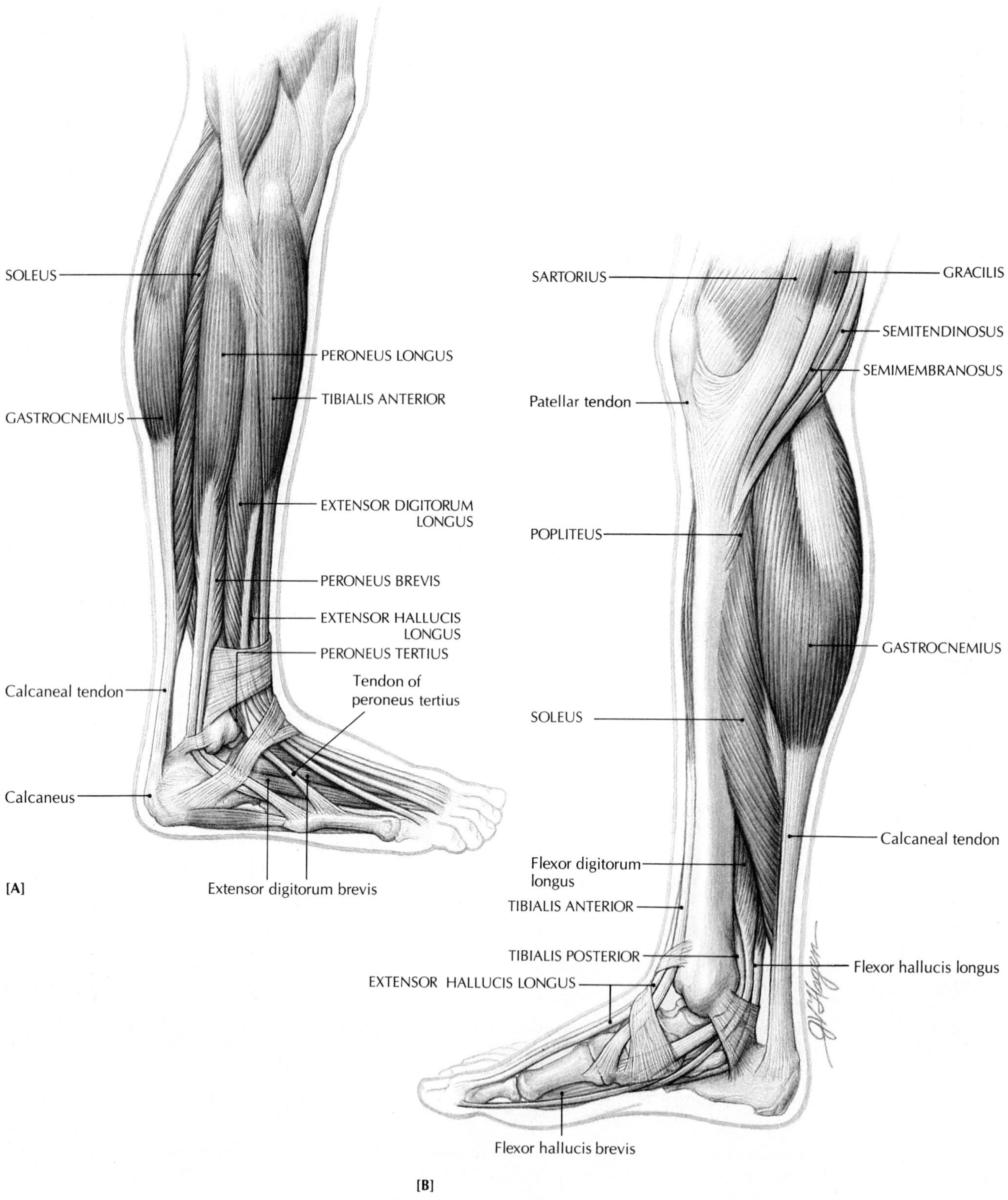

SOLEUS

PERONEUS LONGUS

TIBIALIS ANTERIOR

GASTROCNEMIUS

EXTENSOR DIGITORUM LONGUS

PERONEUS BREVIS

EXTENSOR HALLUCIS LONGUS

PERONEUS TERTIUS

Tendon of peroneus tertius

Calcaneal tendon

Calcaneus

Extensor digitorum brevis

[A]

SARTORIUS

GRACILIS

SEMITENDINOSUS

SEMIMEMBRANOSUS

Patellar tendon

POPLITEUS

GASTROCNEMIUS

SOLEUS

Calcaneal tendon

Flexor digitorum longus

TIBIALIS ANTERIOR

TIBIALIS POSTERIOR

Flexor hallucis longus

EXTENSOR HALLUCIS LONGUS

Flexor hallucis brevis

[B]

Specific Actions of Principal Muscles

The ankle (talocrural) joint is a hinge joint capable of plantar flexion (moves sole of foot down) and dorsiflexion. The *soleus* and *gastrocnemius* muscles, both powerful plantar flexors, are active during the take-off phase of the foot in walking and running. Because of the shortness of its muscle belly, the gastrocnemius cannot flex the knee and plantar flex the ankle joint at the same time, but it can do each independently. With the knee flexed, the soleus is the great plantar flexor of the ankle. In many individuals, these muscles are not active during ordinary standing.

The movements of eversion of the foot (turning so the sole faces outward) and inversion (turning so the sole faces inward) use the subtalar axis, or axis of Henke (the joint between the talus and calcaneus), when we walk over rough terrain. When the foot is planted on the ground, it assumes an awkward position somewhere between eversion and inversion. The muscles responsible for these movements are the invertors and evertors. Should these muscles be "caught off guard," a sprained ligament might result. The *tibialis anterior* and *posterior* muscles, whose tendons pass on the great-toe side of the subtalar axis, are invertors. The *peroneus longus, brevis,* and *tertius,* whose tendons pass on the little-toe side of the axis, are evertors.

Muscle	Origin	Insertion	Innervation	Action
Soleus	Head of fibula, medial border of tibia	Posterior surface of calcaneus	Tibial nerve	Plantar flexes ankle joint; steadies leg during standing.
Gastrocnemius	Posterior aspects of condyles of femur	Posterior surface of calcaneus	Tibial nerve	Plantar flexes ankle joint; flexes knee joint.
Peroneus longus	Head and lateral surface of fibula	Lateral side and plantar surface of first metatarsal and medial cuneiform	Superficial peroneal nerve	Plantar flexes ankle joint; everts foot.
Peroneus brevis	Lateral surface of fibula	Dorsal surface of fifth metatarsal	Superficial peroneal nerve	Plantar flexes ankle joint; everts foot.
Tibialis posterior	Interosseous membrane, fibula, and tibia	Bases of second, third, and fourth metatarsals, navicular, cuneiform, cuboid, calcaneus	Tibial nerve	Plantar flexes ankle joint; inverts foot.
Tibialis anterior	Lateral condyle and surface of tibia	First metatarsal, medial cuneiform	Deep peroneal nerve	Dorsiflexes ankle joint; inverts foot.
Extensor digitorum longus	Lateral condyle of tibia, anterior surface of fibula	Middle and distal phalanges of four outer toes	Deep peroneal nerve	Dorsiflexes ankle joint; extends toes at metatarsophalangeal and interphalangeal joints.
Extensor hallucis longus	Middle of fibula, interosseous membrane	Dorsal surface of distal phalanx of great toe	Deep peroneal nerve	Dorsiflexes ankle joint; extends great toe.
Peroneus tertius	Distal third of fibula, interosseous membrane	Dorsal surface of fifth metatarsal	Deep peroneal nerve	Dorsiflexes ankle joint; everts foot.

FIGURE 11.24 MUSCLES OF THE TOES

The drawings show superficial (first) **[A]**, second **[B]**, third **[C]**, and deepest **[D]** layers. Right plantar views.

The names and attachments of the small intrinsic muscles of the foot to the toes are somewhat similar to those in the hand. However, many of these muscles are ineffective in carrying out the presumed movements suggested by their names. Their primary functional role is to help maintain the stability of the resilient foot as it adjusts to the forces placed upon it.

The role of the short and long extensors of the toes is *dorsiflexion* (extension) at the metatarsophalangeal (MP) joints. The ***lumbricals*** are potential *extensors* of the interphalangeal (IP) joints. *Flexion* of the IP joints is accomplished by the ***flexor digitorum longus*** and the ***quadratus plantae.*** *Flexion* of the MP joints is the role of the ***interossei, lumbricals,*** and short flexors and abductors of the great and little toes. *Abduction* and *adduction* are accomplished by the ***interossei, abductor*** and ***adductor hallucis,*** and ***abductor digiti minimi.*** Apparently, the actions and functions of the lumbricals and interossei are not as important as their counterparts in the hand.

In walking, we plant the heel on the ground and then pass the weight forward on the lateral small-toe side of the foot, and then across the heads of the metatarsals to the metatarsal of the great toe and the pair of sesamoid bones of the two tendons of the flexor hallucis brevis for the take-off. Try it. During the take-off, the interossei muscles by simultaneously flexing the MP joints and extending the IP joints, permit transfer of some weight from the metatarsal heads to the toes. Note that the tendon of the flexor hallucis longus does not bear the weight at take-off, because it is protected by passing between the sesamoids of the flexor hallucis brevis.

Specifically, there are four basic muscle layers of the foot. From the outside (superficial) to the inside (deep), the first layer contains the ***abductor hallucis,*** which abducts the great toe (hallux); the ***flexor digitorum brevis,*** which flexes the second phalanges of the four small toes; and the ***abductor digiti minimi,*** which abducts the small toe. In the second layer are the ***quadratus plantae,*** which flex the terminal phalanges of the four small toes, and the ***lumbricals,*** which flex the proximal phalanges and also extend the distal phalanges of the four little toes. The third layer consists of the ***flexor hallucis brevis,*** which flexes the proximal phalanx of the great toe; the ***adductor hallucis,*** which abducts the great toe; and the ***flexor digiti minimi brevis,*** which flexes the proximal phalanx of the little toe. The fourth or innermost muscle layer contains (1) the four ***dorsal interossei,*** which abduct the second, third, and fourth toes, and flex the proximal and extend the distal phalanges, and (2) the three ***plantar interossei,*** which adduct the third, fourth, and fifth toes. Remember that the axis of abduction and adduction of the foot is along the second toe. (The second toe is abducted by two dorsal interossei.)

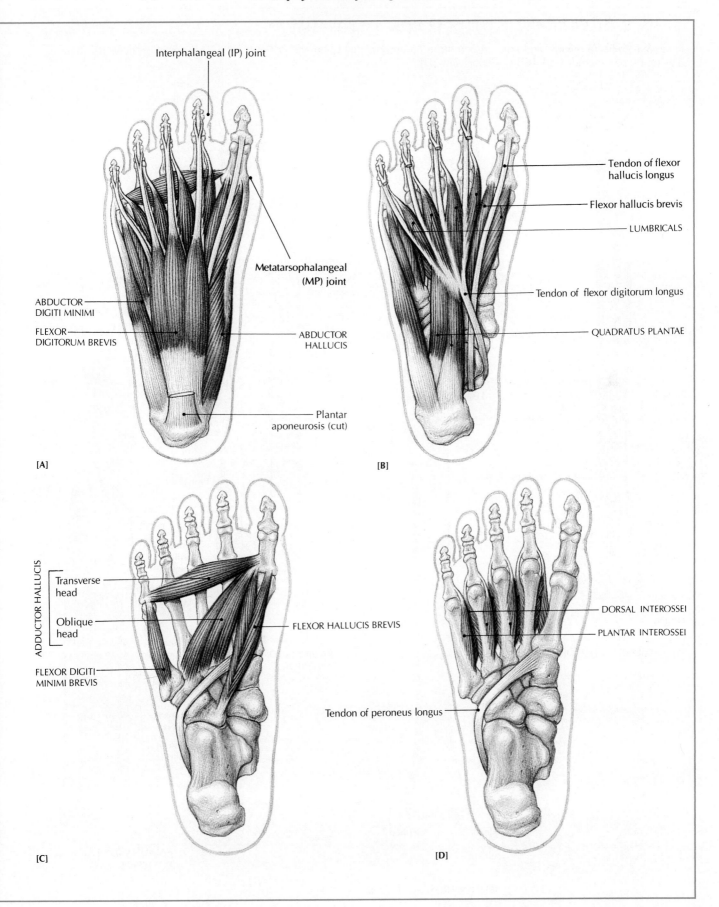

Interphalangeal (IP) joint

Metatarsophalangeal (MP) joint

ABDUCTOR
DIGITI MINIMI

FLEXOR
DIGITORUM BREVIS

ABDUCTOR
HALLUCIS

Plantar
aponeurosis (cut)

[A]

Tendon of flexor
hallucis longus

Flexor hallucis brevis

LUMBRICALS

Tendon of flexor digitorum longus

QUADRATUS PLANTAE

[B]

ADDUCTOR HALLUCIS

Transverse
head

Oblique
head

FLEXOR HALLUCIS BREVIS

FLEXOR DIGITI
MINIMI BREVIS

[C]

DORSAL INTEROSSEI

PLANTAR INTEROSSEI

Tendon of peroneus longus

[D]

FIGURE 11.25 ORIGINS AND INSERTIONS OF SOME MAJOR MUSCLES

Pink areas indicate origins, and blue areas indicate insertions. Right humerus, anterior view [A], posterior view [B]. Right radius and ulna, anterior view [C], posterior view [D].

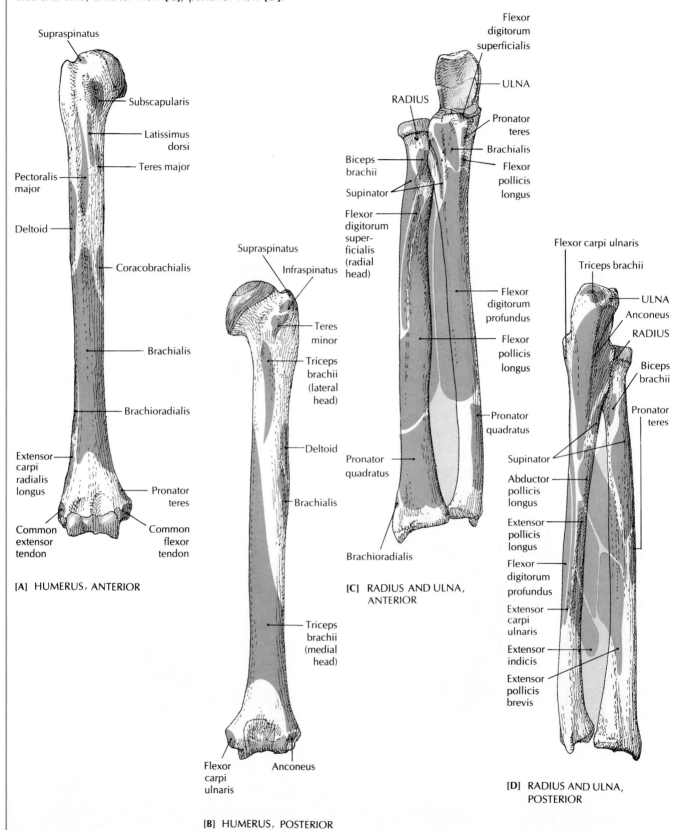

[A] HUMERUS, ANTERIOR

[B] HUMERUS, POSTERIOR

[C] RADIUS AND ULNA, ANTERIOR

[D] RADIUS AND ULNA, POSTERIOR

Right femur, anterior view [E], posterior view [F]. Right tibia and fibula, anterior view [G], posterior view [H].

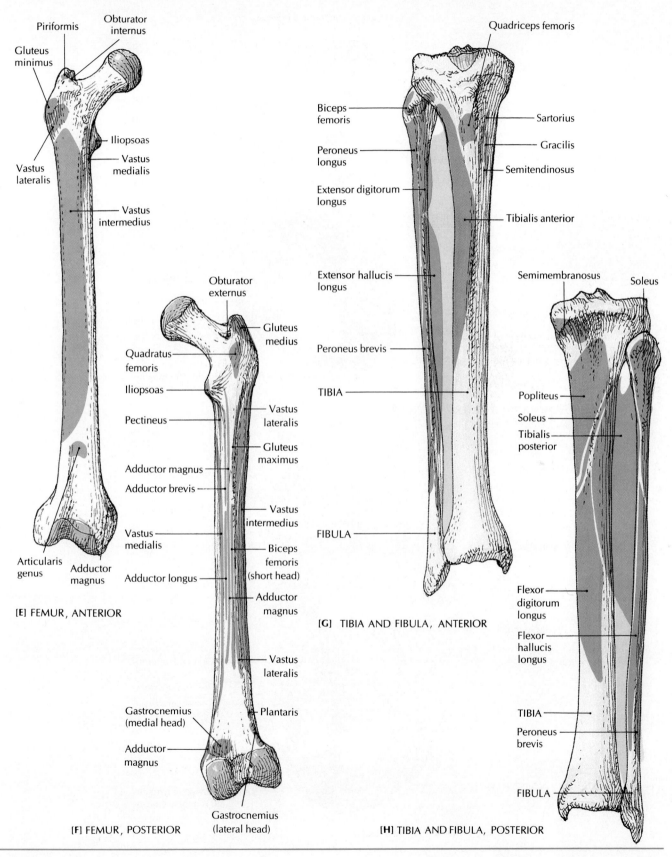

[E] FEMUR, ANTERIOR

[F] FEMUR, POSTERIOR

[G] TIBIA AND FIBULA, ANTERIOR

[H] TIBIA AND FIBULA, POSTERIOR

(Figure 11.25 continues on following page)

FIGURE 11.25 ORIGINS AND INSERTIONS OF SOME MAJOR MUSCLES *(Continued)*

Right hipbone, external surface **[I]**, internal surface **[J]**.

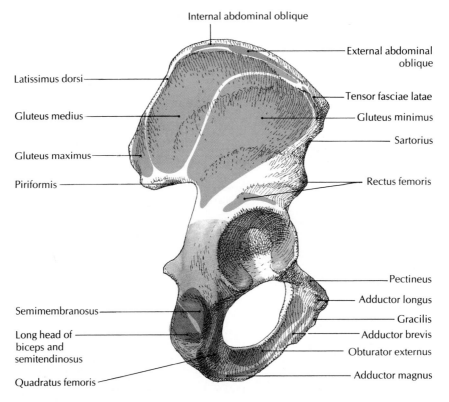

[I] HIPBONE, EXTERNAL SURFACE

- Internal abdominal oblique
- External abdominal oblique
- Latissimus dorsi
- Tensor fasciae latae
- Gluteus medius
- Gluteus minimus
- Sartorius
- Gluteus maximus
- Rectus femoris
- Piriformis
- Pectineus
- Adductor longus
- Semimembranosus
- Gracilis
- Long head of biceps and semitendinosus
- Adductor brevis
- Obturator externus
- Quadratus femoris
- Adductor magnus

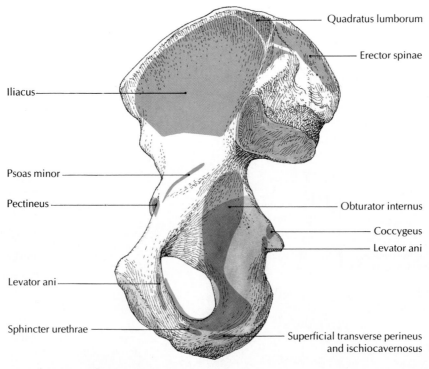

[J] HIPBONE, INTERNAL SURFACE

- Quadratus lumborum
- Erector spinae
- Iliacus
- Psoas minor
- Pectineus
- Obturator internus
- Coccygeus
- Levator ani
- Levator ani
- Sphincter urethrae
- Superficial transverse perineus and ischiocavernosus

WHEN THINGS GO WRONG

Intramuscular Injections

Because deep muscle tissue contains relatively few nerve endings for pain, irritating drugs are injected into a muscle instead of being taken orally or being injected under the skin. Also, when medication is injected intramuscularly, it is absorbed into the body rapidly because of the rich capillary beds in muscle tissue.

Intramuscular injections are usually given in the buttock (gluteus medius and gluteus maximus), the lateral part of the thigh, or the deltoid muscle of the arm. The most usual site for intramuscular injections is the gluteus maximus or gluteus medius muscle, about two or three inches below the ilac crest, in the upper outer quadrant of the buttock. When the gluteus maximus is used it is important to avoid injuring the nearby sciatic nerve. Also, drugs should not be injected directly into large blood vessels, no matter where they are located. A standing position tenses this muscle, and an injection into a tense muscle causes considerable pain. Therefore, the patient is usually placed in a prone position to relax the muscle before the injection.

Hernias

A *hernia* (L. protruded organ), or rupture, is the protrusion of any organ or body part through the muscular wall that usually contains it. *Inguinal* (IHNG-gwuh-null; L. groin) *hernias* are the most common type, occurring most often in males. Most inguinal hernias are caused by a weakness in the fascial wall that allows a loop of the intestines or fat to protrude in the area of the groin superior to the inguinal ligament and superior and medial to the pubic tubercle, to which the inguinal ligament is attached. *Hiatal* (hye-A-tuhl; L. *hiatus*, gap) *hernias* develop when a defective diaphragm allows a portion of the stomach to pass through the opening for the esophagus in the diaphragm into the thoracic cavity. *Femoral hernias* are most common in women. They occur just below the groin, where the femoral artery passes into the thigh. Usually a fatty deposit within the femoral canal creates a gap large enough for a loop of intestine or fat to bulge through. *Umbilical hernias* protrude at the navel. They are most common in newborns, obese women, and women who have had several pregnancies. *Incisional hernias* result from the weakening around a surgical wound that does not heal properly.

Tendinitis

Tendinitis (which you might think should be spelled tendonitis) is a painful inflammation of the tendon and tendon sheath, typically resulting from a sports injury or similar strain on the tendon. It may also be associated with musculoskeletal diseases such as rheumatism or with abnormal posture. Tendinitis occurs most often in the shoulder area, calcaneal tendon, or hamstring muscles.

Tennis Elbow

Tennis elbow is often thought of as a form of tendinitis, but it is not. Its medical name is *lateral epicondylitis*, the inflammation of either the forearm extensor and supinator tendon fibers where they attach to the lateral epicondyle of the humerus, or the lateral (ulnar) collateral ligament of the elbow joint. These conditions are caused by the premature degeneration of the common tendon (origin) of the extensor muscles, and produce tenderness over the anterior aspect of the lateral epicondyle. Tennis elbow afflicts people who rotate their forearms frequently or have chronically weak joints, and is aggravated by stretching the muscles during wrist flexion.

Tension Headache

Tension headaches, the most common type of headache, are often caused by emotional stress, fatigue, or other factors that produce painful muscular contractions in the scalp and back of the neck. Muscle spasms may constrict blood vessels, increasing general discomfort. Pain usually spreads from the back of the head to the area above the eyes, and may cause pain in the eyes themselves. The removal of toxic muscle wastes may be hampered by constricted blood vessels in the scalp, causing even further tenderness.

CHAPTER SUMMARY

How Muscles Are Named (p. 244)

The name of a muscle generally tells something about its location, action (or function), or structure, such as its size, shape, attachment sites, number of heads of origin, or direction of fibers.

Attachment of Muscles (p. 244)

1 A muscle is usually attached to a bone or cartilage by a *tendon.* Some tendons are expanded into a broad, flat sheet called an *aponeurosis.*

2 In addition to acting as attachments, tendons add useful length and thickness to muscles, reduce muscle strain, and add strength to muscle action.

3 The *origin* is the place on the bone that does not move when the muscle contracts, and the *insertion* is the place on the bone that does move when the muscle contracts. Generally, the origin is the attachment closer to the axial skeleton. Some muscles have more than one origin or insertion.

4 The origin and insertion of some muscles can be reversed, depending on which bone moves.

Architecture of Muscles (p. 245)

1 The fibers of skeletal muscles are grouped into small bundles called fascicles, with the fibers within each bundle running parallel to each other.

2 The specific architectural pattern of fascicles within a muscle determines the range of movement and power of the muscle.

3 Patterns of arrangement of fascicles and tendons of attachment include *strap, fusiform, pennate,* and *circular muscles.*

Individual and Group Actions of Muscles (p. 247)

1 Most movements involve the complex interactions of several muscles or even groups of muscles.

2 An *agonist,* or *prime mover,* is the muscle that is primarily responsible for producing a movement. An *antagonist* muscle helps produce a smooth movement by slowly relaxing as the agonist contracts. A *synergist* works with a prime mover by preventing movements at an "in-between" joint when the prime mover passes over more than one joint. A *fixator* provides a stable base for the action of the prime mover.

3 In different situations, the same muscle can act as a prime mover, antagonist, synergist, or fixator.

Lever Systems and Muscle Actions (p. 250)

1 Most skeletal muscle movements are accomplished through a system of levers. A lever system includes a rigid *lever arm* that pivots around a fixed point called a *fulcrum,* with the bone acting as lever arm and the joint as fulcrum. Every lever system includes two different *forces:* the resistance to be overcome and the effort applied.

2 There are three types of levers, referred to as first-, second-, and third-class levers.

3 The mechanical advantage that muscles gain by using a lever system is called *leverage.*

Specific Actions of Principal Muscles (p. 251)

See FIGURES 11.6 through 11.25.

STUDY AIDS FOR REVIEW

UNDERSTANDING THE FACTS

1 Are all muscles attached to bones? Explain.
2 May a muscle have more than one origin and insertion?
3 How are the fascicles arranged in a pennate muscle?
4 What role do fixator muscles play in group action?
5 Which class of lever is used most in the human body?
6 Distinguish between extrinsic and intrinsic muscles.
7 List the muscles that are involved in the movement of the eyeball.
8 List the muscles of mastication.
9 Name the muscles involved in closing the mouth. Raising the upper eyelid. Rolling the eye downward.
10 Which muscles are used in swallowing? Nodding "yes"?
11 Define linea alba.
12 What is the perineum?
13 Name the muscles that act on the shoulder girdle.
14 Name the four rotator cuff muscles. What do they do?
15 Finger movement is achieved mainly by the contraction of muscles in what area?
16 What muscles form the "hamstring" group?

UNDERSTANDING THE CONCEPTS

1 Give one or more examples of muscles named for their shape, size, action, heads of origin, and attachment site. Explain the meaning in each case.
2 Why may the tongue of a patient pose a problem for the anesthetist? Name the muscle involved and explain how the problem may be overcome.
3 Explain the structure of the muscles of the abdominal wall and how their structure is related to their functions.
4 When an incision is made in the abdominal wall for an appendectomy, three laminated muscles would be cut. List them in order.
5 In what ways are the muscles of the hands and feet alike? In what ways are they different?
6 What anatomical fact is responsible for a pianist's forearms becoming tired before the fingers?
7 Which characteristic of the human thumb makes it so important? Give examples of its importance in several daily activities.

SELF-QUIZ

Multiple Choice Questions

11.1 Which of the following is a muscle of the thorax?
a intercostal   d mentalis
b piriformis   e zygomaticus
c transversus

11.2 Which of the following is *not* a muscle of mastication?
a masseter   d lateral pterygoid
b temporalis   e mentalis
c medial pterygoid

11.3 All of the following are muscles of the eye *except* the
a superior rectus   d inferior oblique
b styloglossus   e medial rectus
c superior oblique

11.4 Which of the following muscles move the tongue?
a genioglossus   d a and b
b styloglossus   e a, b, and c
c buccinator

11.5 Movement of the shoulder girdle is accomplished by the
a levator scapulae   d trapezius
b rhomboid minor   e all of the above
c pectoralis minor

11.6 All of the following muscles move the humerus *except* the
a coracobrachialis   d deltoid
b brachialis   e supraspinatus
c teres major

11.7 Which of the following muscles move the wrist?
a flexor carpi radials d a and b
b flexor carpi ulnaris e a, b, and c
c extensor carpi ulnaris

11.8 Which of the following muscles move the forearm?
a supinator   d a and b
b pronator teres   e a, b, and c
c pronator quadratus

11.9 Which of the following is *not* a muscle of the abdominal wall?
a external oblique
b internal oblique
c transversus abdominis
d rectus abdominis
e quadratus lumborum

11.10 All of the following muscles move the femur *except* the
a iliopsoas   d abductor longus
b gluteus maximus   e adductor brevis
c gluteus medius

11.11 All of the following muscles are involved in moving the foot *except* the
a sartorius   d peroneus
b gastrocnemius   e tibialis
c soleus

11.12 The external oblique muscle
a originates from the anterior surface of the lower ribs
b inserts on the linea alba   d a and b
c flexes the spine   e a, b, and c

11.13 The adductor muscles of the thigh are the
a piriformis and obturator
b adductor brevis and obturator
c adductor longus, brevis, and magnus
d semitendinosis and deltoid
e sternohyoid and transversus

11.14 Muscle fascicles may be
a parallel   d a and b
b oblique   e a, b, and c
c circular

True-False Statements

11.15 The coracobrachialis muscle is located in the leg.

11.16 The biceps brachii muscle is located in the dorsal side of the arm.

11.17 The brachioradialis muscle attaches to the radius.

11.18 The scalene muscles elevate ribs 1 and 2.

11.19 The flexor digitorum profundus flexes the foot.

11.20 The rectus abdominis muscle flexes the vertebral column.

Completion Exercises

11.21 The widest part of a muscle is called its _____.

11.22 A muscle is usually attached to a bone (or a cartilage) by a _____.

11.23 Tendon sheaths that are broad and flat are called _____.

11.24 The end of a muscle attached to the bone that does *not* move is the _____.

11.25 The point of attachment of the muscle to the bone that moves is the _____.

11.26 The fibers of skeletal muscles are grouped into small bundles called _____.

11.27 A muscle action in which the angle between two bones is decreased is called _____.

11.28 The type of muscle that turns the forearm so that the palm faces downward is a/an _____.

11.29 The type of muscle that turns the foot inward is a/an _____.

11.30 In a _____-class lever, the force is applied at one end of the lever arm, the resistance at the other end, and the fulcrum is in between.

Matching

11.31 _____ strap muscles   a postural muscles
11.32 _____ fusiform muscle   b transverse fascicles
    c movement toward the midline
11.33 _____ pennate muscle   d opposes the movement of a prime mover
11.34 _____ circular muscle   e tapers at both ends
11.35 _____ agonist   f circular fascicles
11.36 _____ antagonist   g works with prime movers
11.37 _____ synergist
11.38 _____ fixator   h movement away from the midline
11.39 _____ abductor   i aneural muscles
11.40 _____ adductor   j parallel fascicles
    k prime mover
    l oblique fascicles

A SECOND LOOK

In the following drawing, label the deltoid, latissimus dorsi, pectoralis minor, and trapezius muscles.

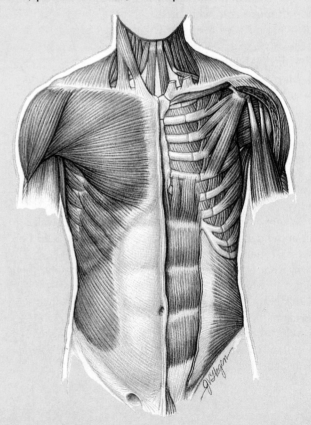

III

CONTROL, COMMUNICATION, AND COORDINATION

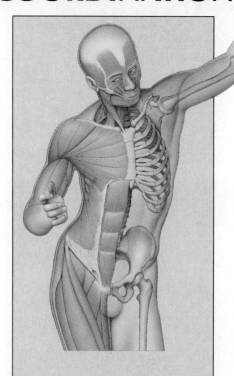

12

Nervous Tissue

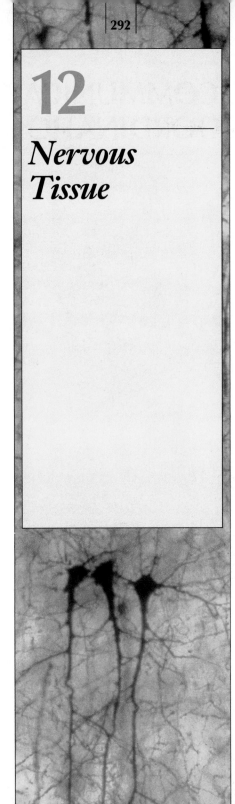

LEARNING OBJECTIVES

1 Describe the gross anatomical divisions of the nervous system, and list the main components of each division.

2 Describe the functional divisions and subdivisions of the peripheral nervous system.

3 Identify two basic properties of a neuron.

4 Describe the parts of a neuron and their functions.

5 Identify three factors that affect the speed of conduction of a nerve impulse.

6 Compare the three types of neurons based on function, and classify the functional components of neurons.

7 Identify the three types of neurons based on structure, and describe their four major segments.

8 Identify the associated cells of the central and peripheral nervous systems, and describe their major functions.

9 Explain the conditions necessary for regeneration of peripheral nerve fibers and how the regeneration process works.

10 Describe the initiation and conduction of impulses in a neuron.

11 Distinguish between electrical and chemical synapses.

12 Explain how a nerve impulse crosses a synapse.

13 Describe the four major groups of neurotransmitters.

14 Distinguish between diverging and converging neuronal circuits, and identify five types of neuronal circuits.

15 Describe one or more disorders involving neurons or the transmission of nerve impulses.

In order to maintain homeostasis, the body is constantly reacting and adjusting to changes in the outside environment as well as within the body itself (internal environment). Such environmental changes, or **stimuli,** initiate impulses in sensory receptors. These impulses (input) pass via nerves to the brain and spinal cord, where they are analyzed, combined, compared, and coordinated by a process called **integration.** After being sorted out, impulses are conveyed by nerves to the muscles and glands of the body. The nervous system expresses itself through muscles and glands, causing muscles to contract or relax, and glands to secrete or not secrete their products.

The nervous system and the endocrine system are the two major regulatory systems of the body, and both are specialized to make the proper responses to stimuli. The two systems work together continuously to maintain homeostasis, but the nervous system is the faster of the two. Stimuli received by the nervous system are processed rapidly through a combination of electrical impulses and chemical substances called *neurotransmitters* for communication between two nerve cells, between a nerve cell and a muscle cell, or between a nerve cell and glandular cells. The endocrine system (the subject of Chapter 17) must depend on slower chemical transmissions, using chemical substances called *hormones.* The endocrine system typically regulates such long-term processes as growth and reproductive ability, instead of the short, quick responses to stimuli controlled by the nervous system. In fact, the two systems are so closely interrelated that they can be considered a single regulatory agency.

Although the nervous system is a single, unified communications network, it is usually divided on a gross anatomical basis into the central nervous system and the peripheral nervous system. The **central nervous system** consists of the brain and spinal cord; the **peripheral nervous system** includes all the nerve cells and fibers that lie outside the brain and spinal cord. The peripheral nervous system connects the brain and spinal cord with the rest of the body.

NEURONS: FUNCTIONAL UNITS OF THE NERVOUS SYSTEM

The nervous system contains over 100 billion nerve cells, or **neurons** (Gr. nerve). Neurons are specialized to transmit impulses from short to relatively long distances, from one part of the body or central nervous system to another. They have two important properties: (1) *excitability,* or the ability to respond to stimuli, and (2) *conductivity,* or the ability to conduct a signal.

Parts of a Neuron

Neurons are among the most specialized types of cells. Although they vary greatly in shape and size, all neurons contain three principal parts, each associated with a specific function: the cell body, dendrites, and an axon [FIGURE 12.1].

Cell body The **cell body** of a neuron may also be called a *soma* (Gr. body) or a *perikaryon* (per-ih-KAR-ee-on; Gr. *peri,* near, around + *karyon,* kernel, nut). It may be star-shaped (stellate), roundish, oval, or even pyramid-shaped, but its distinguishing structural features are its complex, spreading processes (branches or fibers) that pass on to or receive impulses from other cells. Besides varying in shape, cell bodies may vary in size from about half the size of a red blood cell to almost 17 times the size of a red blood cell. A cell body has a large nucleus, which contains a prominent nucleolus, as well as several organelles that are responsible for metabolism, growth, and repair of the neuron. These organelles include chromatophilic substance, or Nissl bodies (made up of rough ER and free ribosomes), endoplasmic reticulum, lysosomes, mitochondria, neurofilaments (intermediate filaments), neurotubules (microtubules), and Golgi apparatus.

The neurotubules and neurofilaments of a cell body are threadlike protein structures that actually extend into and throughout the cell body and processes of the neuron, and run parallel to the long axis of each process. *Neurotubules* function in the intracellular transport of proteins and other substances from the cell body to the ends of the processes and also in the opposite direction. Vesicles may even be transported in both directions at the same time [FIGURE 12.2]. The proteins that are transported throughout the cell are needed to replace used-up protein in the neuron and for the regeneration of nerve fibers in the peripheral nervous system. *Neurofilaments* are semirigid structures that provide a skeletal framework for the axon. In Alzheimer's disease, many cell bodies in the cerebral cortex of the brain contain "tangles" of neurotubules and neurofilaments, forming *amyloid plaques.*

Dendrites and axons Neurons have two types of processes, dendrites and axons. Either process is sometimes referred to as a *nerve fiber.* As shown in FIGURE 12.1, the typical neuron has many short, threadlike branches called **dendrites** (Gr. *dendron,* tree), which are actually extensions of the cell body. As a result, the structures in the cell body are continuous with those in the dendrites. Dendrites conduct nerve impulses *toward* the cell body. A neuron may have as many as 200 dendrites.

FIGURE 12.1 A NEURON

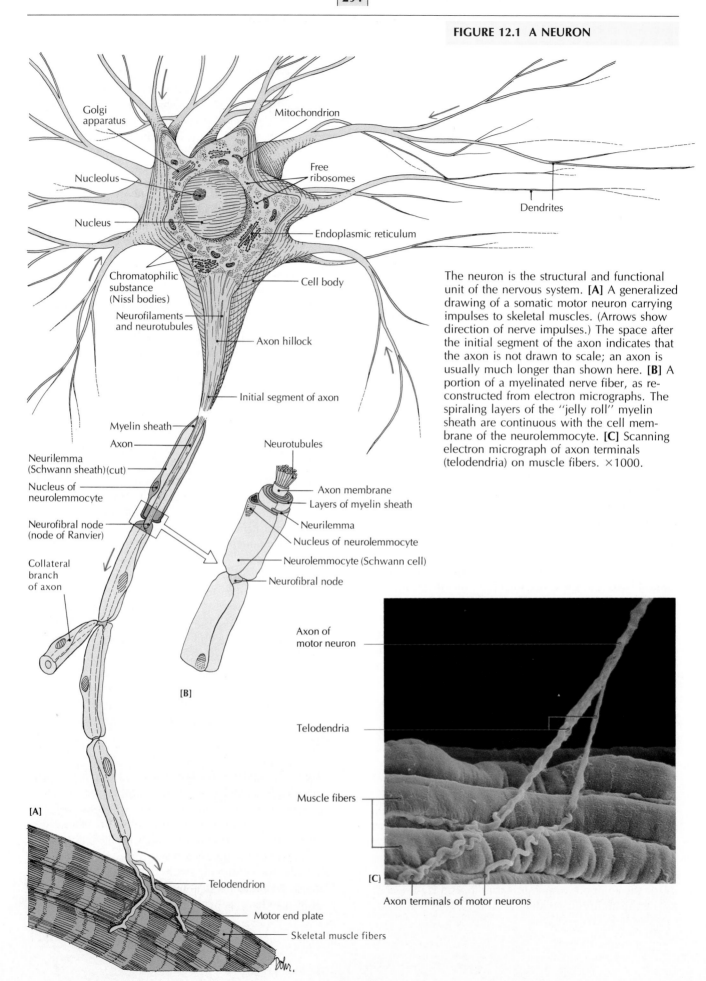

Golgi apparatus

Mitochondrion

Nucleolus

Free ribosomes

Nucleus

Dendrites

Endoplasmic reticulum

Chromatophilic substance (Nissl bodies)

Cell body

Neurofilaments and neurotubules

Axon hillock

Initial segment of axon

Myelin sheath

Axon

Neurotubules

Neurilemma (Schwann sheath)(cut)

Nucleus of neurolemmocyte

Axon membrane

Layers of myelin sheath

Neurofibral node (node of Ranvier)

Neurilemma

Nucleus of neurolemmocyte

Collateral branch of axon

Neurolemmocyte (Schwann cell)

Neurofibral node

[B]

[A]

Telodendrion

Motor end plate

Skeletal muscle fibers

The neuron is the structural and functional unit of the nervous system. [A] A generalized drawing of a somatic motor neuron carrying impulses to skeletal muscles. (Arrows show direction of nerve impulses.) The space after the initial segment of the axon indicates that the axon is not drawn to scale; an axon is usually much longer than shown here. [B] A portion of a myelinated nerve fiber, as reconstructed from electron micrographs. The spiraling layers of the "jelly roll" myelin sheath are continuous with the cell membrane of the neurolemmocyte. [C] Scanning electron micrograph of axon terminals (telodendria) on muscle fibers. ×1000.

Axon of motor neuron

Telodendria

Muscle fibers

[C]

Axon terminals of motor neurons

FIGURE 12.2 AXONAL TRANSPORT

Neurotransmitters and other substances to be transported along an axon between the cell body and synaptic terminal of a neuron are contained within vesicles that are moved the length of the axon along neurotubules that act like conveyor belts (arrows). Precursors of neurotransmitters and plasma membrane proteins are produced by the Golgi apparatus in the cell body. Surplus substances are packaged and transported back to the cell body, where they are degraded by lysosomes. Mitochondria are transported in either direction in response to the energy needs of the cell.

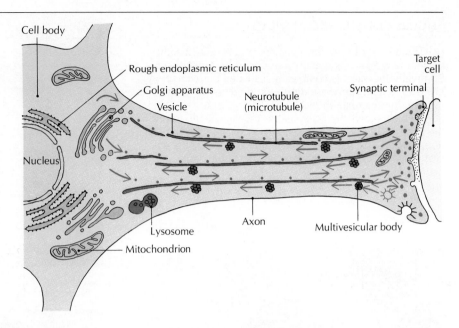

A neuron generally has just one **axon,** a slender process that extends from the cell body from less than a millimeter (as in the brain) to more than a meter (as in the axons of the sciatic nerve, which extend from the spinal cord to the foot). An axon carries nerve impulses *away* from the cell body, to the next nerve cell, muscle cell, or gland cell.

The cytoplasm of an axon is called *axoplasm*, and the plasma membrane of an axon is known as the *axolemma.* Unlike the cell body, the axon of a neuron does not contain chromatophilic substance, ribosomes, or Golgi apparatus.

In most neurons, the axon originates from a cone-shaped elevation of the cell body called the *axon hillock* [FIGURE 12.1A]. The thin part of the axon immediately after the axon hillock, called the *initial segment*, is important in generating a nerve impulse. The axon may have some side branches, called *collateral branches*, that leave the main axon. The axon and its collateral branches end in a spray of tiny branches called *telodendria*. The branches of the telodendria usually have tiny swellings called **end bulbs,** or **synaptic boutons** (Fr. button). A **synapse** (Gr. *synapsis*, a connection) is formed where an end bulb associates with the plasma membrane of another neuron.

An axon may be covered with a laminated lipid sheath of **myelin** (MY-ih-linn; Gr. *myelos*, marrow), which forms a thick pad of insulation called a **myelin sheath** [FIGURE 12.1A, B]. A nerve fiber that has a myelin sheath

is called a *myelinated fiber*. In the peripheral nervous system, the myelin sheath is made up of rolled layers of the plasma membrane of a type of peripheral nerve cell called a **neurolemmocyte,** or *Schwann cell*. The outer layer, or sheath, of these cells is known as the **neurilemma** (Gr. nerve + rind or husk) or *Schwann sheath* [FIGURE 12.3]. The myelin sheaths of axons in the central nervous system are formed by nonneural cells called *oligodendroglia*, each of which has many processes.

The myelin sheath is segmented, interrupted at regular intervals by gaps called **neurofibral nodes,** or *nodes of Ranvier*. The distance from one node to the next is an *internode*. The longer the internode, the thicker is the diameter of the sheath. The myelin sheath of each internode is formed by one neurolemmocyte. The myelin is absent at each neurofibral node. The maximum speed of a human nerve impulse in a myelinated fiber is about 120 m/sec (360 km/hr), with a minimum speed of about 3 m/sec.

A nerve fiber that does not have a myelin sheath is called an *unmyelinated fiber*. In a peripheral nerve, several (5 to 20) unmyelinated nerve fibers are enclosed singly within longitudinal invaginations in neurolemmocytes [FIGURE 12.3C]. Unmyelinated fibers are usually protected and nourished by the organ tissue in which they are located. These fibers conduct at relatively slow speeds, ranging from 0.7 to 2.3 m/sec.

As stated previously, a collection of cell bodies of neurons of peripheral nerves is called a *ganglion*. These cell bodies are surrounded by satellite cells, which are the equivalents of the neurolemmocytes of the nerves.

What is a nerve? In common usage, a **nerve** is simply a bundle of fibers enclosed in a connective tissue sheath, like many telephone wires in a cable. The anatomical term for bundles of fibers and their sheaths in the central

---

Q: *What are the relative sizes of the parts of a neuron?*

A: If the cell body of a motor neuron were enlarged to the size of a baseball, the axon would extend about a mile, and the dendrites would fill a large field house.

FIGURE 12.3 THE MYELIN SHEATH

[A] Three stages in the development of a myelin sheath, beginning with a single axon surrounded by a neurilemma. Eventually the axon is completely insulated by overlapping layers of myelin; arrows indicate the direction of growth. **[B]** Electron micrograph of a rat sciatic nerve in cross section, which shows the layers of the myelin sheath around the axon (×29,000); the boxed area is enlarged in the inset (top, ×129,000). **[C]** Nine unmyelinated nerve fibers enclosed within individual troughs (invaginations) of a neurolemmocyte.

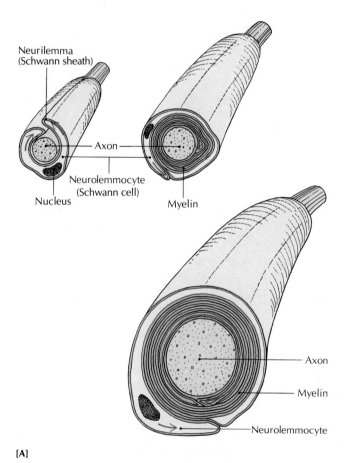

[A]

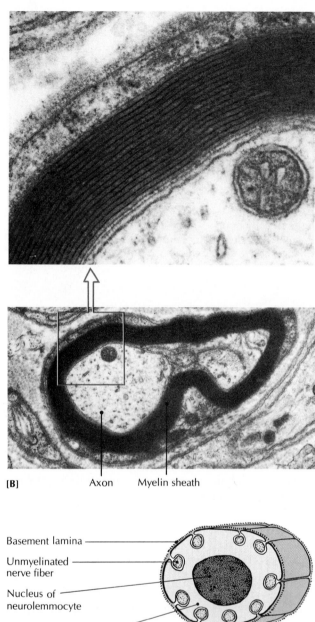

[B]   Axon   Myelin sheath

[C]

nervous system is *tracts*, and for those in the peripheral nervous system is *nerves*.

Types of Neurons: Based on Function

The neurons of the peripheral nervous system may be classified according to the direction in which they transmit nerve impulses. Neurons conveying information from sensory receptors in the body *to* the central nervous system are called **afferent**, or **sensory, neurons**. Afferent neurons, whose cell bodies are in clusters, or ganglia, located close to the central nervous system, also carry impulses from lower to higher centers in the central nervous system. The distal dendrites of these afferent fibers are the *sensory receptors* (or are connected to the sensory receptors), which are responsive to internal or external stimulation.

Efferent or **motor, neurons** or **fibers** convey nerve impulses from higher to lower centers of the central nervous system, or *away from* the central nervous system to the **effectors** (muscles and glands), where the response actually takes place. Most cranial and spinal nerves (nerves connected to the brain and spinal cord) are called *mixed nerves* because they are composed of hundreds, or even thousands, of afferent *and* efferent nerve fibers.

Interneurons lie mostly within the central nervous system, but a few are distributed between the central and peripheral nervous systems. Their functions are to carry impulses from sensory neurons to motor neurons and to process incoming neural information. Interneurons with long axons, called *relay neurons*, convey signals over long distances. Interneurons with short, usually branching, axons are called *local circuit neurons*. They convey signals locally, over short distances. Local circuit neurons play an important role in learning, emotions, and language. Complex functions such as learning and memory may depend on thousands of local circuit neurons. Less complicated activities, such as the simpler reflexes, may involve only a few or no interneurons, since sensory and motor neurons may be directly connected.

Types of Neurons: Based on Structure

Neurons may also be classified according to their structure, or more specifically, according to the number of

their processes. *Multipolar neurons* generally have many dendrites radiating from the cell body, but only one axon [FIGURE 12.4A]. Most of the neurons of the brain and spinal cord are multipolar. *Bipolar neurons* have only two processes [FIGURE 12.4B]. Essentially, all neurons in the adult develop from bipolar cells, which are the neurons of the embryonic nervous system. In the adult, bipolar neurons are located in only a few structures, including the retina of the eye, cochlear and vestibular nerves of the ear, and the olfactory nerve in the upper nasal cavity. *Unipolar neurons* are the most common sensory neurons in the peripheral nervous system [FIGURE 12.4C]. The single process of a unipolar neuron divides into two branches. One branch extends into the brain or spinal cord, conducting nerve impulses away from the cell body, and the other branch extends to a peripheral sensory receptor in some distal part of the body.

For all of their structural differences, most neurons are composed of four distinct functional segments, as shown in FIGURE 12.5:

1 The *receptive segment* in a multipolar neuron is composed of the cell body and its dendrites. It is the segment that receives a continuous bombardment of synaptic inputs from numerous other neurons on the receptor sites on its plasma membrane. These complex inputs are pro-

FIGURE 12.4 CLASSIFICATION OF NEURONS BASED ON STRUCTURE

[A] Multipolar neuron. [B] Bipolar neuron. [C] Unipolar neuron. Arrows indicate the direction of nerve impulses.

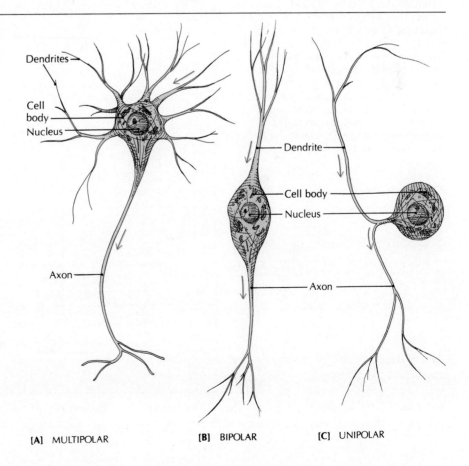

[A] MULTIPOLAR   [B] BIPOLAR   [C] UNIPOLAR

cessed in the receptive segment, and if sufficiently stimulated, the resolution of this processing is conveyed to the initial segment at the junction of the cell body and axon hillock.

2 The *initial segment* is the trigger zone of the neuron, where the processed neural information of the receptive segment is converted into a nerve impulse. As a result, this segment, with its low threshold, is critical for the generation of a nerve impulse.

3 The *conductive segment* (the axon) is specialized for the conduction of neural information as nerve impulses over relatively long distances in the body. It acts to convey results of the neural processing of the receptive segment via nerve impulses to the terminal transmissive segment.

4 The *transmissive* (effector) *segment* contains the axon terminals that convert the stimulation of the nerve impulse to release chemical neurotransmitters at its synapses. These chemicals exert influences upon the receptor sites on an effector cell (another neuron, a muscle cell, or a glandular cell).

A S K   Y O U R S E L F

1 What are the basic properties of neurons?

2 What are the main parts of a neuron?

3 What is myelin?

4 What is the difference between afferent and efferent neurons?

5 What is the main function of an interneuron?

6 What is the function of the transmissive segment of a neuron? The receptive segment?

PHYSIOLOGY OF NEURONS

Neurons are specialized to respond to stimulation. They do this by generating electrical impulses. These impulses are expressed as changes in the electrical potentials conducted along the plasma membranes of the dendrites, cell

---

FIGURE 12.5 FUNCTIONAL ORGANIZATION OF A TYPICAL NEURON

[A] Most neurons contain a receptive segment, an initial segment, a conductive segment, and a transmissive segment. The cell body may be located within the receptive segment, as in the lower motor neuron shown in **[B]**, or within the conductive segment, as in the sensory neuron shown in **[C]**.

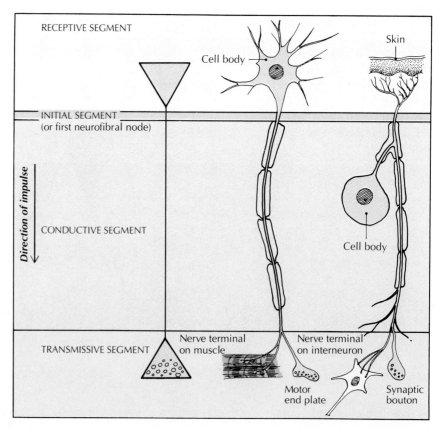

body, and axon of each neuron. The difference in potential across the plasma membrane of the neuron results from differences in the concentration of certain ions on either side of the membrane.

In the following sections we will explain how impulses are generated, conducted along a nerve fiber, and passed on to an adjacent neuron, muscle cell, or gland cell.

Characteristics of the Neuron Plasma Membrane

The plasma membrane of a neuron contains proteins that are specialized as *channels* and *pumps*. The *channels* act as pores through which ions can pass (for example, sodium ion channels). Associated with the channels are *receptor sites*, which open or close the channels in response to stimuli. Those channels that respond to chemicals (neurotransmitters such as acetylcholine) are known as *chemically-gated channels*, and those that respond to changes in electrical voltage are called *voltage-gated channels*. Other channels are associated with *pumps*, which use metabolic energy to move ions into and out of the cell (sodium ions that pass into the neuron through a sodium channel are driven by a sodium-potassium pump back through a channel to the outside of the neuron). Thus different regions of the plasma membrane are specialized to perform different roles.

Resting Membrane Potential

A "resting" neuron is one that is not conducting a nerve impulse, but *is* electrically charged. (It may be compared with a charged battery that does not generate an electrical current until it is switched on.) The plasma membrane of such a "resting" neuron is said to be *polarized,* meaning that there is a difference in electrical charge between the inside and the outside of the membrane. The intracellular fluid on the inner side of the membrane is negatively charged *in relation to* the positively charged extracellular fluid outside the membrane. The difference in the electrical charge across the membrane is called the *potential difference*, and in a "resting" neuron it is called the *resting membrane potential.*

The potential difference is measured in millivolts (mV).* Normally, the resting membrane potential of about -70 mV is due to the unequal distribution of various electrically charged ions (sodium, potassium, chloride, and protein molecules) and to the selective permeability of ion channels in the plasma membrane to the different ions.

---

*The resting membrane potential of -70 mV (less than one-tenth of a volt) means that the electrical charge on the inside of the plasma membrane measures $\frac{70}{1000}$ of a volt *less* than that on the outside. By convention, a minus sign denotes a negative charge inside the membrane.

The Na^+ and K^+ ions constantly diffuse through the ion channels of the plasma membrane, moving from regions of high concentration to regions of lower concentration. Yet their relative concentrations remain remarkably constant. The extracellular fluid usually contains about 10 Na^+ ions for every K^+ ion. Within the cell the ratio is reversed, with K^+ ions outnumbering Na^+ ions by at least 10 to 1.

Such ionic consistency is necessary if a neuron is to remain excitable and be able to respond to changes in its surroundings. How are these concentrations maintained? Part of the answer lies in a self-regulating transport system, located within the plasma membrane, known as the *sodium-potassium pump,* which is powered by ATP. The more sodium that leaks into the neuron through the plasma membrane, the more active the pump becomes, pumping actively to restore the ionic concentrations that maintain the resting membrane potential. The sodium-potassium pump transports three sodium ions out of the cell for every two potassium ions it brings into the cell. So even though both ions are positive, more positive ions are moving out through the membrane than are moving in. Partly for this reason, the inner side of the plasma membrane is more negative than the outer side.

The Mechanism of Nerve Impulses

The *change* in the electrical potential difference across a plasma membrane is the key factor in the initiation and subsequent conduction of a nerve impulse. The process of conduction differs only slightly between unmyelinated and myelinated fibers; the following steps describe conduction in unmyelinated fibers:

1 A stimulus that is strong enough to initiate an impulse in a neuron is called a *threshold stimulus.* When such a stimulus is applied to a polarized resting membrane of an axon, sodium ion channels into the cell open, and sodium ions rush in,† reversing the electrical charge at the point of stimulus [FIGURE 12.6A]. Thus, at the point of stimulus, the *inside* of the membrane becomes positively charged relative to the *outside*, a condition called *depolarization.* When a stimulus is strong enough to cause depolarization, the neuron is said to *fire.*

2 Once a small area on the axon is depolarized (change in voltage), it stimulates the adjacent area (which contains voltage-gated channels), and an *action potential,* or *nerve impulse,* is initiated and conducted along the plasma membrane.

3 Shortly after depolarization, the original balance of sodium and potassium ions is restored by the action of

---

†In the 1 millisecond that a channel is open, about 20,000 sodium ions flow through.

FIGURE 12.6 TRANSMISSION OF A NERVE IMPULSE ALONG AN AXON

[A] A threshold stimulus causes sodium ions to rush into the cell body of a resting neuron through the plasma membrane. This activity produces a reversal of charges along the axon membrane, resulting in *depolarization* and the initiation of an action potential (nerve impulse). [B] Almost immediately afterward, the original charges are restored, and the membrane is *repolarized*. A second impulse at the original point of impulse causes another depolarization as the first impulse travels along the axon.

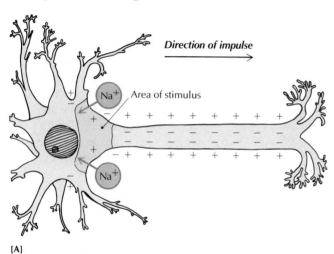

[A]

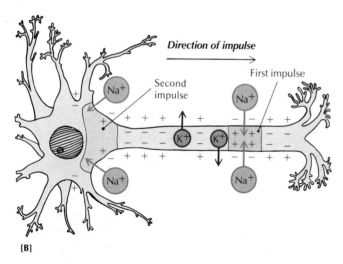

[B]

the membrane pumps; the relative electrical charges inside and outside the membrane are also restored. The membrane is then said to be **repolarized**. The transmission of a nerve impulse along the plasma membrane may be visualized as a *wave* of depolarization and repolarization [FIGURE 12.6B].

4 After each firing, there is an interval of from 0.5 to 1 msec before it is possible for an adequate stimulus to generate another action potential. This period is called the **refractory period.** Most nerve fibers are capable of generating about 300 impulses per second.

In a *myelinated* nerve, conduction is speeded up somewhat as the action potential appears to "jump" from one neurofibral node to the voltage-gated channels of the next node. For this reason, such conduction is known as **saltatory conduction** (L. *saltare*, to jump).

All-or-None Law

A stimulus of a minimum strength is necessary to initiate a nerve impulse, but an increase in stimulus strength above the minimum does not increase the strength of the impulse. The principle is the same as when a gun is fired: if the trigger is not pulled hard enough, nothing happens, but the minimum required pull on the trigger will fire the gun. Pulling the trigger even harder will not make the gun fire harder. The principle that states that a neuron will fire at full power or not at all is known as the **all-or-none law.** It applies primarily to action potentials in axons, but also applies to muscle-fiber contraction.

But what about the difference we can feel between a light touch and a strong one, or between a soft sound and a loud one? Don't those differences show that the all-or-none law is wrong? No, they do not. Such differences can be perceived when the *frequency* of the impulses, not their strength, is changed. Some nerve fibers can conduct at different frequencies per second. The more frequent the impulses, the higher the level of excitation. Other fibers respond to *different thresholds*, but all impulses carried on any given fiber are of the same strength. Also, the *number of neurons* involved can make a difference in how strong the stimulus is perceived to be. You feel the difference between a light push and a strong one, for example, because the strong push affects more of your neurons. But when it comes to a single axon in an experimental situation, the all-or-none law still applies.

Synapses

A nerve impulse travels along an axon and eventually reaches the branching axon terminals in the transmissive segment of the neuron. If the impulse is to be effective, it must be conveyed to another neuron, a muscle cell, or gland cell. In this discussion we will concentrate on the junction between the axon terminal of one neuron and the dendrite, cell body, or axon of the next neuron. This junction between neurons is called a **synapse** (SIN-aps; Gr. connection).

The neuron carrying the impulse *toward* a synapse is called the **presynaptic** ("before the synapse") **neuron.** It

---

Q: *How fast can people react to stimuli?*

A: Olympic male sprinters react to the starter's gun in about 12/100 of a second.

initiates a response in the receptive segment of a ***postsynaptic*** ("after the synapse") ***neuron*** leading *away from* the synapse [FIGURE 12.7]. The presynaptic cell is almost always a neuron, but the postsynaptic cell can be a neuron, muscle cell, or gland cell.

The transmission of a nerve impulse at a synapse can be either chemical or electrical, but chemical synapses are far more common than electrical ones. In the following sections we will discuss electrical synapses briefly, and then describe chemical synapses in greater detail.

Electrical synapses An *electrical synapse* is joined by a *gap junction* in which the two communicating cells are electrically coupled by tiny intercellular channels of adjacent plasma membranes called *connexons*. The free movement of ions occurs through the connexons between the presynaptic and postsynaptic cells. Because of the extremely low resistance associated with connexons, the nerve impulse virtually travels directly from the presynaptic to the postsynaptic cell. Also, since there are no chemical neurotransmitters that require time to react, the transmission of an electrical synapse is almost instantaneous.

In humans (and other mammals), electrical synapses (gap junctions) are found between cardiac cells, between smooth muscle cells of the intestinal tract, and between a few neurons—in the retina of the eye, for example.

Chemical synapses and neurotransmitters In a *chemical synapse,* two cells communicate by way of a chemical agent called a ***neurotransmitter,*** which is released by the transmissive segment of the presynaptic neuron [FIGURE 12.8]. A neurotransmitter is capable of changing the resting potential in the plasma membrane of the receptive segment of the postsynaptic cell.

Q: *How many synapses may a neuron have?*

A: Some neurons have only a few synapses, but a motor neuron may have as many as 2000, and a Purkinje cell (conducting myofiber) of the cerebellum may have as many as 200,000 on its dendrites alone.

When the nerve impulse of a presynaptic neuron reaches the end bulb of the axon, it depolarizes the presynaptic plasma membrane, causing voltage-gated calcium channels to open. As a result, calcium ions (Ca^{2+}) diffuse into the presynaptic terminal. As calcium ions rush through Ca^{2+} channels of the plasma membrane of the terminal ending, they cause the storage vesicles to fuse with the plasma membrane, releasing the neurotransmitter from the vesicles by exocytosis into the narrow *synaptic cleft** [FIGURE 12.8D]. One such neurotransmitter is the chemical *acetylcholine;* another is *norepinephrine.*

The structure of the synapse is such that impulses can travel in only one direction in the nervous system. (The precursor of the neurotransmitter is stored in the axons of the presynaptic neuron and the receptor sites for that transmitter are in the membrane of the postsynaptic neuron.) The transmitters can pass only from axon terminals across the synaptic cleft to the receptor cell, but not the other way. This directional control prevents neural effects from traveling in all directions at once. If they did, neural messages would be garbled and scrambled.

The short trip across the synaptic cleft takes less than a millisecond. That interval results in a *synaptic delay.* Some of the released neurotransmitter binds with *receptor protein sites* in the postsynaptic membrane, causing changes in the membrane's permeability to certain ions. As a result of the change in permeability in the postsynaptic membrane, sodium and potassium ions rush through their open ion channels and cause the same kind of depolarization that the presynaptic cell experienced. A wave of depolarization and repolarization passes along the postsynaptic neuron, and the nerve impulse continues along its path to an effector.

Once the neurotransmitter has bridged the synaptic cleft, it is quickly broken down. Acetylcholine, the most common neurotransmitter, is broken down into acetate and choline by the enzyme *acetylcholinesterase* and recy-

*This narrow gap is usually less than $\frac{1}{500}$ the width of a human hair (about 25 nanometers).

FIGURE 12.7 PRESYNAPTIC AND POSTSYNAPTIC NEURONS

Highly simplified representation of the relationship of presynaptic and postsynaptic neurons and the direction of nerve impulse transmission.

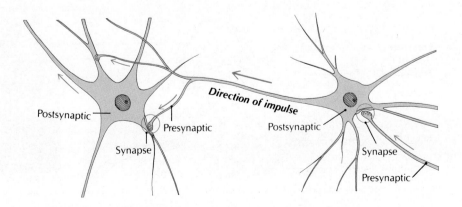

Postsynaptic

Presynaptic

Synapse

Direction of impulse

Postsynaptic

Synapse

Presynaptic

FIGURE 12.8 TRANSMISSION OF A NERVE IMPULSE ACROSS A SYNAPSE

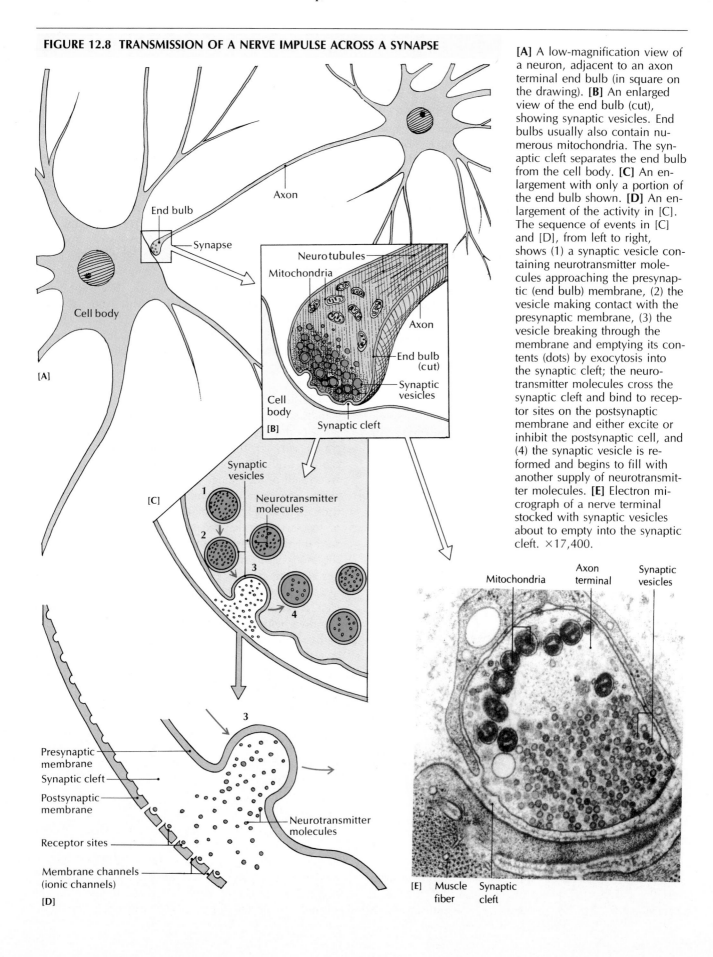

[A] A low-magnification view of a neuron, adjacent to an axon terminal end bulb (in square on the drawing). [B] An enlarged view of the end bulb (cut), showing synaptic vesicles. End bulbs usually also contain numerous mitochondria. The synaptic cleft separates the end bulb from the cell body. [C] An enlargement with only a portion of the end bulb shown. [D] An enlargement of the activity in [C]. The sequence of events in [C] and [D], from left to right, shows (1) a synaptic vesicle containing neurotransmitter molecules approaching the presynaptic (end bulb) membrane, (2) the vesicle making contact with the presynaptic membrane, (3) the vesicle breaking through the membrane and emptying its contents (dots) by exocytosis into the synaptic cleft; the neurotransmitter molecules cross the synaptic cleft and bind to receptor sites on the postsynaptic membrane and either excite or inhibit the postsynaptic cell, and (4) the synaptic vesicle is reformed and begins to fill with another supply of neurotransmitter molecules. [E] Electron micrograph of a nerve terminal stocked with synaptic vesicles about to empty into the synaptic cleft. ×17,400.

cled for future use. Some chemical substances, such as certain insecticides and "nerve gases," work by inhibiting the chemical action of acetylcholinesterase. Since acetycholine is not destroyed, nerve impulses take place continuously. Muscle cells remain in a state of contraction, and the breathing muscles are paralyzed. Death by asphyxiation results.

The effect of a neurotransmitter on the postsynaptic membrane may be to produce either an excitatory or inhibitory response. For example, the neurotransmitter acetylcholine evokes an excitatory response at the motor end plate in a skeletal muscle cell, but it evokes an inhibitory response at its junction with a cardiac muscle cell.

Many possible neurotransmitters are being investigated. At least 50 are present in the nervous system, and there may be over 100. In addition to acetylcholine, other neurotransmitters include various amino acids (gamma-aminobutyric acid, glycine, glutamic acid); monoamines (norepinephrine, dopamine, serotonin); and neuropeptides.

ASK YOURSELF

1 What is polarization?

2 What is a membrane potential?

3 What is a threshold stimulus?

4 What is an action potential?

5 What is the all-or-none law?

6 How do sodium-potassium pumps function?

7 What is a synapse?

8 How do presynaptic and postsynaptic neurons differ?

Degeneration and Regeneration of Nerve Fibers

The production of neurons begins before birth, and is not completed until infancy. Before this completion, some neurons *degenerate* and die, and they continue to die throughout our lives, never again equaling the number at infancy. (Such a loss of brain cells, for instance, is usually no problem in a healthy person, because we start out with many more neurons than we actually need.) But even though neurons are not replaced when they die, severed peripheral nerve *fibers* are sometimes able to regenerate if the cell body is undamaged, and if the neurilemma is intact. The only neurons that are replaced continuously are the olfactory (smell) neurons of the olfactory nerve.

When a peripheral nerve fiber is cut, a motor neuron has the capacity to regenerate its axon, and a sensory neuron has the capacity to regenerate its dendrites. First,

the part of the axon severed from its cell body begins to degenerate in the stump distal to the cut, and in a few weeks disappears altogether [FIGURE 12.9A].

As part of the process of **regeneration,** the cell body enlarges and chromatophilic substance increases its activity and produces extra protein, which is needed to support the growth of new branches, called *terminal sprouts*, from the proximal portion of the axon still connected to the cell body. At the same time, neurolemmocytes in the distal stump divide and arrange themselves into continuous cords of cells. These cords extend from the cut end of the stump to the sensory and motor nerve endings at the distal end of the nerve [FIGURE 12.9B].

Each regenerating fiber can form 50 or more terminal sprouts. The sprouts grow along the neurilemma cords, which act as guides the way a trestle guides the path of a growing vine. The actual guide is the basal lamina of each neurolemmocyte. The fiber proceeds to grow into the newly formed neurilemma cord (also called tube) [FIGURE 12.9B]. Most sprouts grow along cords and do not form viable functional contacts. These sprouts will degenerate eventually, as, for example, sprouts of motor fibers in a cord leading to sensory endings, or sprouts of sensory fibers in a cord leading to motor endings. A regenerating sprout that finally reaches a nerve ending compatible with its own functional role can make a physiological connection. A sensory fiber may connect with a similar sensory ending, or a fiber of a motor neuron may connect with a motor end plate or fibers of a skeletal muscle [FIGURE 12.9A]. Such "successful" sprouts become myelinated by surrounding neurolemmocytes.

Severed fibers may also receive support from adjacent undamaged fibers that send out *collateral sprouts* from neurofibral nodes [FIGURE 12.9B]. These collateral sprouts are attracted to the axonless neurilemma cords. The degenerating nerve fibers stimulate the healthy fibers (possibly by producing so-called *neurotrophic factors*) to form collateral sprouts, which are helpful in reinnervating degenerated muscle fibers.

Depending upon the severity of the damage to the nerve, and the distance of the injury from the nerve endings at the motor end plate, functional recovery of nerve fibers may take from about a month to more than a year.

Regeneration in the Central Nervous System

When axons of the central nervous system of mature human beings and other mammals are severed completely, they do not regenerate to a functional degree, except in a few exceptional cases. Damaged neurons make an attempt to regenerate axonal sprouts at the severed end, but they are unsuccessful in reaching their original terminals, probably for three reasons: (1) the sprouts cannot penetrate the glial scar tissue formed at the site of injury; (2) there is no basal lamina associated with the oligodendroglia in the central nervous system to guide the regeneration of fibers; and (3) oligodendroglia do not form continuous cords, and so cannot guide the fibers in an organized regeneration. This lack of functional regeneration can have serious effects, such as those that follow some strokes or spinal cord injuries.

FIGURE 12.9 DEGENERATION AND REGENERATION OF SEVERED PERIPHERAL MOTOR FIBERS

The regenerated portions of the fibers conduct nerve impulses more slowly than before because the distance between nodes (internodes) is shorter in regenerated nerves.

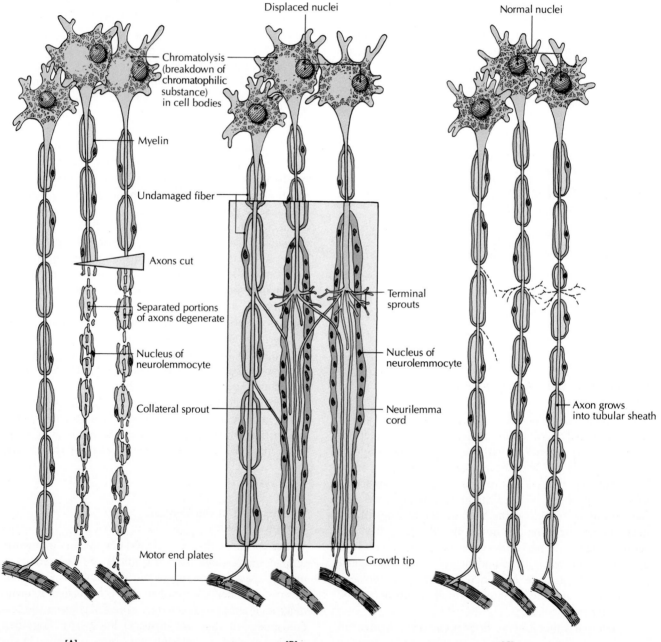

[A] [B] [C]

The Guiding "Glue" of the Nervous System

In 1982, a group of Rockefeller University researchers, led by Nobel laureate Gerald M. Edelman, discovered a cellular "glue" that may be involved in holding cells together. The glue in this "glue control model" is a glycoprotein of the plasma membrane of a neuron called a *cell adhesion molecule (CAM)*. The CAM found in neurons is called *N-CAM,* and is present in the plasma membranes of embryonic and adult neurons and their processes. N-CAMs are basically recognition molecules with the ability to recognize each other in adjacent neu-rons. During early development, neurons aggregate in prescribed groupings when N-CAM is turned on, and may migrate and have their processes grow when it is turned off. This enables the neurons to assemble into their final organization, and the processes make synaptic contact with other neurons and muscles when the N-CAM is turned on again. Developmental errors usually result when the mechanism that turns the N-CAM on and off is faulty.

Subsequent research has indicated that N-CAM may be the substance that guides the specific embryonic development and growth of neurons by providing binding sites between neurons and other structures such as muscles. Experiments with animals suggest that a pathway containing N-CAM guides the growth of axons from the eye to the brain. In contrast, a region of the brain where axons never grow (between the areas receiving input related to sight and smell) does not contain any N-CAM. N-CAM also seems to guide nerve growth at neuromuscular junctions.

ASSOCIATED CELLS OF THE NERVOUS SYSTEM

Associated cells are specialized cells in the central and peripheral nervous systems other than neurons. For the most part, they support the neurons in some way. Although the neuron is the functional unit of the nervous system, most of the cells in the system are not neurons, but neuroglial cells. Depending upon the region, neuroglial cells outnumber neurons by 10 to 50 times.

Neuroglia: Associated Cells of the Central Nervous System

In place of connective tissue, which is relatively sparse in the central nervous system, are the neuroglia. The **neuroglia** ("nerve glue"; Gr. *glia,* glue) are nonconducting cells that protect and nurture, as well as support, the nervous system.

Certain neuroglia (astrocytes and ependymal cells), along with the capillary bed of the central nervous system, form what is known as the *blood-brain barrier.* This barrier permits certain chemical substances to gain access to the neurons and slows down or even prevents other substances, such as penicillin, from reaching the neurons. In this way, the nonneural cells and blood vessels of the brain and spinal cord act to maintain the homeostasis of the environment immediately surrounding each neuron and its processes.

There are four types of neuroglia: astrocytes, oligodendrocytes, microglia, and ependyma [FIGURE 12.10].

Associated Cells of the Peripheral Nervous System

The peripheral nervous system consists mainly of nerves, nerve endings, organs of special sense, and *ganglia* (singular, *ganglion;* Gr. cystlike tumor, hence, nerve bundle). **Ganglia** are groups of cell bodies located *outside* the brain and spinal cord. They include (1) the *cranial* and *spinal (dorsal root) sensory ganglia,* which are located near the central nervous system, and (2) the *motor ganglia* of the autonomic nervous system, which are located either near the central nervous system or near or within the visceral organs.

The peripheral ganglia and nerves are composed of connective tissue, neurons, and associated cells. The *associated cells* are the **satellite cells** of the ganglia and the neurolemmocytes of the nerves. A sequence of these associated cells forms a single continuous layer that ensheaths each neuron. Satellite cells surround each cell body, and neurolemmocytes ensheath the axon and dendrites. Satellite cells and neurolemmocytes both nurture the neurons and are similar in structure and function. Neurolemmocytes form the myelin sheath of myelinated nerve fibers [see FIGURE 12.1]. They also ensheath the axons and dendrites of the unmyelinated fibers, but do not form myelin. The cell bodies are usually unmyelinated. The nerve fibers and ganglia are bound together in groups by connective tissue, which contains blood vessels that nourish the fibers and ganglia.

ASK YOURSELF

1 What are the main functions of neuroglia?

2 What are the four types of neuroglia?

3 What are the associated cells of the PNS?

FIGURE 12.10 NEUROGLIA

[A] Astrocyte with foot plates on a capillary. [B] Oligodendrocyte. [C] Microglial cell. [D] Ependyma.

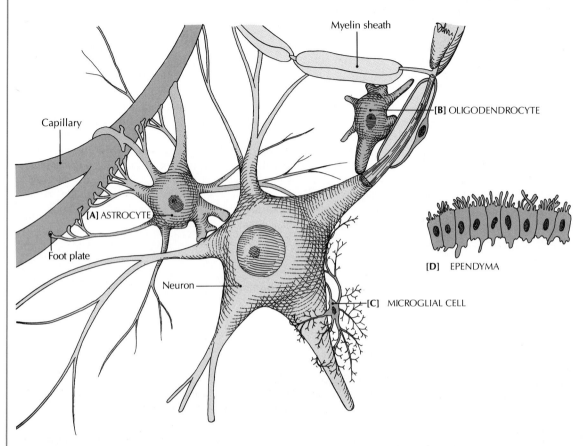

Type of cell	Description and functions
NEUROGLIA	Associated cells of CNS. Nonconducting. Protect, nourish, support cells of CNS.
Astrocyte	Largest, most numerous glial cell, with long, starlike processes. Sustains neurons nutritionally, helps maintain concentration of chemicals in extracellular fluid, provides packing material and structural support, helps regulate transfer of substances from capillaries to nervous tissue.
Oligodendrocyte	Relatively small, with several branching processes. Found in gray and white matter of CNS. Similar to neurolemmocyte. Produces and nurtures myelin sheath segments (internodes) of many nerve fibers (each process forming one intermodal segment). Provides supportive framework, supplies nutrition for neurons.
Microglial cell	Smallest glial cell. Usually found between neurons and along blood vessels. Macrophage, not a true glial cell. Removes disintegration products of damaged neurons.
Ependymal cell	Elongated cell. Arranged in single layer to line central canal of spinal cord and ventricles of brain. Helps to form part of inner membrane of neural tube during embryonic growth, secretes cerebrospinal fluid.
PERIPHERAL GLIAL CELLS	Associated cells of PNS. Provide various types of sheaths. Help to protect, nourish, maintain cells of PNS.
Satellite cell (capsule cell)	Essentially the same structure as neurolemmocyte. Forms capsule around cell bodies of neurons of peripheral ganglia; separates cell bodies from connective tissue framework of ganglion.
Neurolemmocyte (Schwann cell)	Flattened cell. Arranged in single layer along nerve fibers. Forms myelin sheaths of peripheral nerve fibers. Associated with both myelinated and unmyelinated fibers.

ORGANIZATION OF THE NERVOUS SYSTEM

The general organization of the human nervous system is shown in FIGURE 12.11. The brain and spinal cord make up the **central nervous system,** and all nerve cells outside the brain and spinal cord make up the **peripheral nervous system.** *Cranial nerves,* which are connected to the brain, and *spinal nerves,* which are connected to the spinal cord, are part of the peripheral nervous system.

The spinal cord and spinal nerves are described in detail in Chapter 13. The brain and cranial nerves are discussed in Chapter 14.

Central Nervous System

The brain and spinal cord of the central nervous system (CNS) are surrounded and protected by the skull and vertebral column, respectively. The central nervous system may be thought of as the body's central control system, receiving and interpreting or integrating all stimuli, and relaying nerve impulses to muscles and glands, where the designated actions actually take place.

Peripheral Nervous System

The nerve cells and their fibers that make up the peripheral nervous system (PNS) allow the brain and spinal cord to communicate with the rest of the body.

In terms of function, two types of nerve cells are present in the peripheral nervous system: (1) **afferent** (L. *ad,* toward + *ferre,* to bring), or **sensory, nerve cells** carry nerve impulses from sensory receptors in the body *to* the central nervous system, where the information is processed; (2) **efferent** (L. *ex,* away from), or **motor, nerve cells** convey information *away from* the central nervous system to the effectors (muscles and glands).

The peripheral nervous system may be further divided, on a purely functional basis, into the *somatic nervous system* and the *visceral nervous system* [FIGURE 12.11]. Each of these systems is composed of an afferent (sensory) division and an efferent (motor) division.

Somatic nervous system The **somatic nervous system** is composed of afferent and efferent divisions. The *afferent (sensory) division* consists of nerve cells that receive and process sensory input from the skin, skeletal muscles, tendons, joints, eyes, tongue, nose, and ears. This input is conveyed to the spinal cord and brain via the spinal and some cranial nerves, and is utilized by the nervous system at an unconscious level. On a conscious level, the sensory input is perceived as sensations such as touch, pain, heat, cold, balance, sight, taste, smell, and sound.

The *efferent (motor) division* of the somatic nervous system is composed of neuronal pathways that descend from the brain through the brainstem and spinal cord to influence the lower motor neurons of some cranial and spinal nerves. When these lower motor neurons are stimulated, they always excite (never inhibit) the skeletal muscles to contract. This system regulates the "voluntary" contraction of skeletal muscles. (As you saw in Chapter 10, not all such activity is actually voluntary or under our conscious control.)

Visceral nervous system The *visceral nervous system* is also composed of afferent and efferent divisions. The *afferent (sensory) division* includes the neural structures involved in conveying sensory information from sensory receptors in the visceral organs* of the cardiovascular, respiratory, digestive, urinary, and reproductive systems. Some of the input from these sensory receptors may be utilized on a conscious level and is perceived as sensations such as pain, intestinal discomfort, urinary bladder fullness, taste, and smell. The *efferent (motor) division,* more commonly known as the *autonomic nervous system,* includes the neural structures involved in the motor activities that influence the smooth muscle, cardiac muscle, and glands of the skin and viscera.

Autonomic nervous system The **autonomic nervous system** *(visceral efferent motor division)* is made up of nerve fibers from the brain and spinal cord that may either inhibit or excite smooth muscle, cardiac muscle, and glands. This system is the modulator, adjuster, and coordinator of "involuntary" visceral activities such as the heart rate and the secretions of glands. Many of these visceral activities can be carried out even if the organs are deprived of innervation by the autonomic nervous system. For example, the heart continues to contract (beat) without innervation. Although the heart can contract without innervation, the autonomic nervous system can either increase or decrease the strength of contraction and heart rate.

The autonomic nervous system is divided into two subsystems, the sympathetic and parasympathetic systems, which complement each other. The *sympathetic nervous system* stimulates activities that are mobilized during emergency and stress situations, the so-called "fight, fright, or flight responses." These responses include an acceleration of the heart rate and strength of contraction, an increase in the concentration of blood glucose, and an increase in blood pressure. In contrast, the *parasympa-*

---

*The word "visceral" pertains to a *viscus* (L. body organ), or one of several large body organs in any of the great body cavities; note that *viscera* is the plural of viscus. Do not confuse viscus with *viscous,* which means "to flow with relative difficulty."

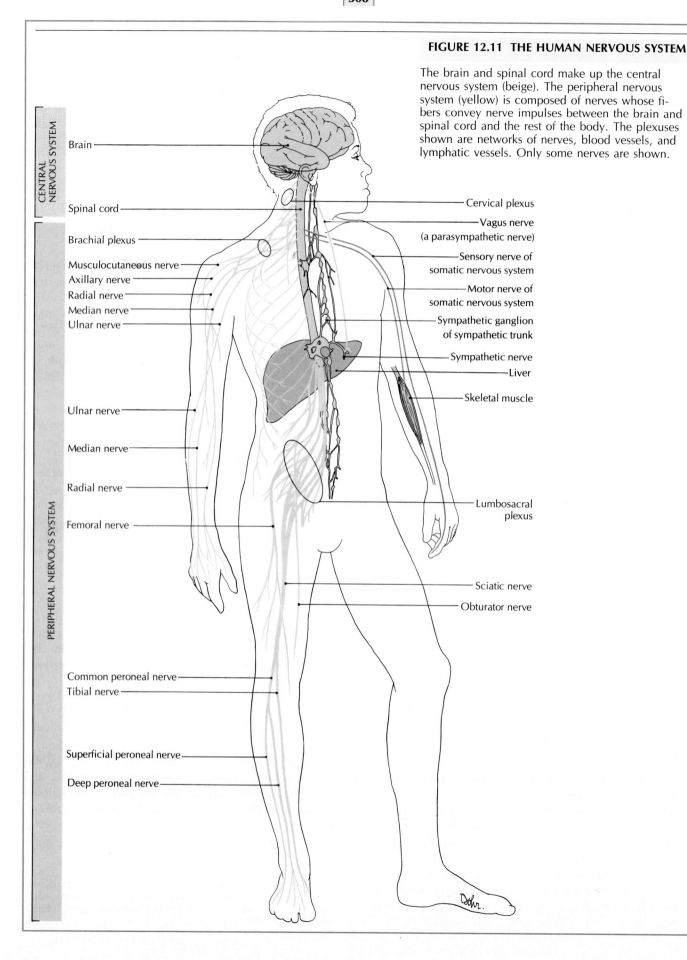

FIGURE 12.11 THE HUMAN NERVOUS SYSTEM

The brain and spinal cord make up the central nervous system (beige). The peripheral nervous system (yellow) is composed of nerves whose fibers convey nerve impulses between the brain and spinal cord and the rest of the body. The plexuses shown are networks of nerves, blood vessels, and lymphatic vessels. Only some nerves are shown.

CENTRAL NERVOUS SYSTEM

Brain

Spinal cord

PERIPHERAL NERVOUS SYSTEM

Brachial plexus

Musculocutaneous nerve
Axillary nerve
Radial nerve
Median nerve
Ulnar nerve

Ulnar nerve

Median nerve

Radial nerve

Femoral nerve

Common peroneal nerve
Tibial nerve

Superficial peroneal nerve

Deep peroneal nerve

Cervical plexus
Vagus nerve (a parasympathetic nerve)
Sensory nerve of somatic nervous system
Motor nerve of somatic nervous system
Sympathetic ganglion of sympathetic trunk
Sympathetic nerve
Liver
Skeletal muscle

Lumbosacral plexus

Sciatic nerve
Obturator nerve

Divisions	Functions

CENTRAL NERVOUS SYSTEM (CNS)

Brain and spinal cord.	Body's central control system. Receives impulses from sensory receptors, relays impulses for action to muscles and glands. Interpretive functions involved in thinking, learning, memory, etc.

PERIPHERAL NERVOUS SYSTEM (PNS)

Cranial and spinal nerves, with afferent (sensory) and efferent (motor) nerve cells.	Enables brain and spinal cord to communicate with entire body. **Afferent (sensory) cells:** Carry impulses from receptors to CNS. **Efferent (motor) cells:** Carry impulses from CNS to effectors (muscles and glands).
Somatic nervous system Characterized by axons (nerve fibers) of lower motor neurons that go directly from CNS to effector muscle without synapsing.	**Afferent (sensory) division:** Receives and processes sensory input from skin, skeletal muscles, tendons, joints, eyes, ears. **Efferent (motor) division:** Excites skeletal muscles.
Visceral nervous system Characterized by nerve fibers of motor neurons that go from CNS to interact with other nerve cells within a ganglion located outside CNS; nerve fibers of second nerve cells that go to effectors.	**Afferent (sensory) division:** Receives and processes input from internal organs of cardiovascular, respiratory, digestive, urinary, and reproductive systems.
Autonomic nervous system	**Efferent (motor) division** (*autonomic nervous system*): May inhibit or excite smooth muscle, cardiac muscle, glands.
Sympathetic nervous system	Relaxes intestinal wall muscles; increases sweating, heart rate, blood flow to skeletal muscles.
Parasympathetic nervous system	Contracts intestinal wall muscles; decreases sweating, heart rate, blood flow to skeletal muscles.

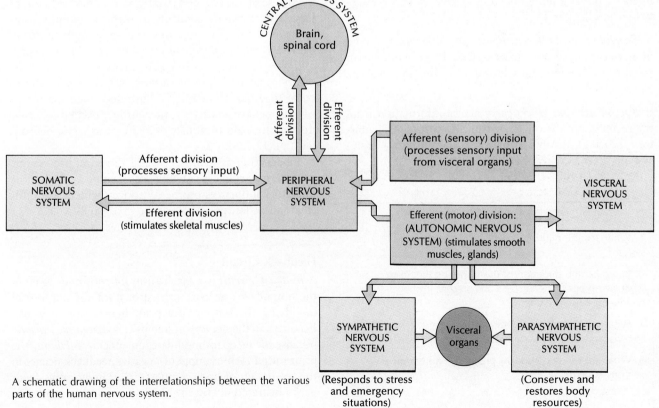

A schematic drawing of the interrelationships between the various parts of the human nervous system.

thetic nervous system directs activities associated with the conservation and restoration of body resources. These activities include a decrease in the heart rate and strength of contraction and the rise in gastrointestinal activities associated with increased digestion and absorption of food.

The autonomic nervous system is dealt with in greater detail in a chapter of its own, Chapter 15.

Classes of Peripheral Neurons

The neurons of the peripheral nervous system and their fibers may be classified as follows:

1 *General somatic afferent fibers* carry sensory information from the skin, skeletal muscles, joints, and connective tissues to the central nervous system.

2 *General visceral afferent fibers* carry information from the visceral organs to the central nervous system.

3 *General somatic efferent fibers* carry nerve impulses from the central nervous system to most of the skeletal muscles of the body. These impulses result in the contraction of skeletal muscles.

4 *General visceral efferent fibers* carry impulses from the central nervous system that modify the activities of the heart, smooth muscles, and glands. These are the fibers of the autonomic nervous system, which are discussed in Chapter 15.

5 *Special visceral efferent fibers* carry impulses from the brain to certain skeletal muscles, including the muscles of facial expression and muscles found in the jaw, pharynx (throat), and larynx.

6 *Special afferent fibers* carry neural information from the receptors of the olfactory (smell), optic (sight), auditory (hearing), vestibular (balance), and gustatory (taste) systems to the central nervous system.

ASK YOURSELF

1 What are the main components of the central nervous system?

2 Based on the direction of the impulses they carry, what two types of nerve cells are present in the peripheral nervous system?

3 What are the major differences between the somatic and visceral nervous systems?

4 What are the two subdivisions of the autonomic nervous system?

NEURONAL CIRCUITS

The nervous system is an exquisitely structured network of neurons arranged in synaptically connected sequences called **neuronal circuits** [FIGURE 12.12]. The circuits are formed by input, intrinsic, and relay neurons:

1 An *input neuron* is a nerve cell that conveys information from one group of neurons to another. Such a group is called a **nucleus** (an anatomical term), or a **neuron pool** (a physiological term).

2 An *intrinsic neuron* is an interneuron that is often located entirely within the nucleus or pool.

3 A *relay neuron* is an interneuron that projects from one nucleus to another. Within each nucleus there is synaptic organization, where processing occurs.

Several patterns and types of neuronal circuits exist. The circuits described in the following sections are the ones most commonly identified.

Divergence and Convergence

Some neurons in the central nervous system may synapse with as many as 25,000 other neurons. When the transmissive segment of a presynaptic neuron branches out to have many synaptic connections with the receptive segments of many other neurons, it is an example of *divergence* (L. *divergere*, to bend) [FIGURE 12.12A]. In diverging synapses, one neuron may excite or inhibit many others. The principle of *convergence* (L. *convergere*, to come together, to merge) is illustrated when the receptive segment of a postsynaptic neuron is excited or inhibited by the axon terminals of many presynaptic neurons [FIGURE 12.12B].

Fibers from as many as several thousand presynaptic neurons may converge on one postsynaptic neuron. If the excitatory influences on the receptive segment are sufficient, an action potential can be generated.

Feedback Circuit

A *feedback circuit* is a mechanism for returning some of the output of a neuron (or neurons) for the purpose of modifying the output of a prior neuron (or neurons). Feedback in the nervous system is called *negative feedback*, because the feedback modulates an effect by *inhibiting* the prior output. An example of negative feedback occurs in the regulation of body temperature. The negative feedback illustrated in FIGURE 12.12C occurs when the lower

FIGURE 12.12 NEURONAL CIRCUITS

Arrows indicate direction in which nerve impulses are propagated. **[A]** Principle of divergence. One presynaptic neuron branches and synapses with several postsynaptic neurons. **[B]** Principle of convergence. Several presynaptic neurons synapse with one postsynaptic neuron. **[C]** Simple feedback circuit in which the axon collateral branch of neuron A synapses with interneuron B, which, in turn, synapses with neuron A. **[D]** Parallel circuits in which neuron A synapses in neuron pools B-C and D-E, with each, in turn, projecting as parallel circuits. **[E]** Two-neuron sequence of afferent neuron synapsing with efferent neuron, found in a two-neuron reflex arc. **[F]** Three-neuron sequence of afferent neuron, interneuron, and efferent neuron, found in a three-neuron reflex arc.

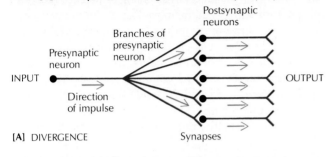

[A] DIVERGENCE

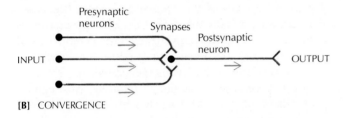

[B] CONVERGENCE

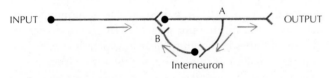

[C] FEEDBACK CIRCUIT

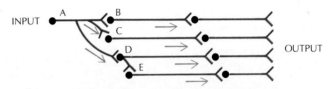

[D] PARALLEL CIRCUITS

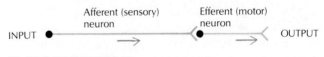

[E] TWO-NEURON CIRCUIT

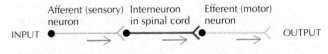

[F] THREE-NEURON CIRCUIT

motor neuron, which innervates a skeletal muscle, is excited. Its collateral axonal branch stimulates an interneuron through excitatory synapses. This interneuron then inhibits the lower motor neuron, readying it for restimulation by its presynaptic neurons.

Parallel Circuits

As a result of the neural processing of convergence and divergence, information is often conveyed by relay neurons to other neural levels through ***parallel circuits*** [FIGURE 12.12D]. Through these circuits, different forms of neural information can be relayed, and ultimately recombined at the same time, at other levels. For example, the sensory pathways for pain ascend to the brain from the spinal cord via two parallel systems: the spinothalamic pathway and the spinoreticulothalamic pathway. Much of the information conveyed via these pathways is recombined for processing at higher cerebral levels.

Two-Neuron Circuit

The simplest neuronal circuit is the ***two-neuron*** (mono-synaptic) ***circuit.*** It consists of a sequence of an afferent (sensory) neuron, one (*mono-*) synapse, and an efferent (motor) neuron [FIGURE 12.12E]. A familiar two-neuron sequence is the "knee-jerk" (patellar) extension reflex, where a tap on the patellar tendon of the flexed knee produces an extension of the knee.

Three-Neuron Circuit

A ***three-neuron circuit*** is a sequence of an afferent (sensory) neuron, an interneuron, and an efferent (motor) neuron [FIGURE 12.12F]. Such a chain is used in flexor reflexes, such as the flexion of the elbow when you touch a hot stove and pull your hand away. The circuit involved in the elbow flexion consists of (1) sensory neurons from the pain receptors in the hand to the spinal cord, (2) interneurons, located entirely within the spinal cord, which connect the sensory neurons with (3) the lower motor neurons to the biceps brachii muscle, which is stimulated to contract. There are synapses between sensory neurons and the interneurons and between the interneurons and the lower motor neurons. This type of circuit, where each neuron in the sequence is connected to many different neurons, is known as an *open circuit*.

ASK YOURSELF

1 How do diverging and converging circuits differ?

2 How do parallel circuits operate?

3 Besides the number of neurons involved, what is the difference between a two-neuron circuit and a three-neuron circuit?

THE EFFECTS OF AGING ON THE NERVOUS SYSTEM

During old age, neurons continue to die, but there is usually no shortage. Some neurons lose their processes, and other neurons are replaced by fibrous cells, but there is only a 10-percent decrease in the *velocity of nerve impulses*. However, reflexes do slow down. The many convolutions of the brain smooth out somewhat, and the blood supply to the brain is reduced. Although the number of brain cells is reduced, there is no appreciable loss of brain function unless the blood supply is cut off temporarily by a stroke or other disorder. A stroke may lead to the progressive loss of intellectual abilities, personality, and memory.

Short-term memory may be impaired somewhat in old age, especially if the blood supply to the brain is decreased by diseased arteries. There is usually little or no change in *learning ability*, although many elderly people have been conditioned to think that their mental faculties are seriously diminished by the aging process.

Several age-related diseases affect brain functions. In *Alzheimer's disease*, which often appears between 40 and 60 years of age; some spaces between parts of the brain are enlarged, degenerative changes occur (amyloid plaque formation), and memory loss is common. *Parkinson's disease*, a motor disability, is commonly called shaking palsy because it is characterized by tremors of the head and hands, slow movements, rigid joints, and sagging facial muscles. It is a common ailment that usually appears between 55 and 70, and occurs more often in men than in women.

Eyesight is usually impaired because of the degeneration of fibers in the optic nerve and an accumulation of injuries to the eyes. Common ailments are *presbyopia* (loss of the ability to focus on close objects), *cataract* (opaque density of the lens of the eye), and *glaucoma* (increased fluid pressure within the eyeball). Color perception is reduced as cones in the retina degenerate. People over 60 usually need twice as much light as a 40-year-old because some rods in the retina have degenerated.

Hearing, *smell*, and *taste* are all reduced. Most elderly people have adequate hearing, but high-pitched sounds become increasingly difficult to hear as the eardrum loses its elasticity and fibers in the auditory nerve degenerate. Hearing may also be impaired by the accumulation of hardened wax in the outer ear. *Otosclerosis* is a disease of the small bones in the middle ear in which the bones fuse together, making it impossible for them to vibrate properly and amplify sound waves that must reach the auditory nerve. The abilities to smell and taste begin to deteriorate at about 60, as the lining of the mucous membrane becomes thinner and less sensitive, and as the number of active taste buds is reduced.

Elderly people often sleep up to two hours less than they did when they were younger, and wake up more often during the night. The more-frequent awakenings may be due to breathing problems and, in men, the need to urinate because of an enlarged prostate gland.

WHEN THINGS GO WRONG

Multiple Sclerosis

Multiple sclerosis (MS) is a progressive demyelination of neurons that interferes with the conduction of nerve impulses and results in impaired sensory perceptions and motor coordination. Because almost any myelinated site in the brain and spinal cord may be involved, the symptoms of the disease may be diverse. With repeated attacks of inflammation at myelinated sites, scarring (sclerosis) takes place and some permanent loss of function occurs. The disease usually affects young adults between 18 and 40, and is five times more prevalent in whites than in blacks.

Although MS is very disabling, it progresses slowly, and most patients lead productive lives, especially during the recurring periods of remission. Among the typical symptoms are problems with vision, muscle weakness and spasms, urinary infections and bladder incontinence, and drastic mood changes. The specific cause of MS is not known, but two theories are currently held: (1) a slow-acting virus infects the CNS, and (2) the body's immune system attacks its own CNS (an autoimmune response).

Amyotrophic Lateral Sclerosis

Amyotrophic lateral sclerosis (ALS; commonly known as Lou Gehrig's disease; Gehrig was a professional baseball player with the New York Yankees who died of ALS) is a motor neuron disease. Both upper motor neurons and lower motor neurons are affected. ALS is confined to the voluntary motor system, with progressive degeneration of corticospinal tracts and alpha motor neurons. The disease causes skeletal muscles to atrophy.

In about 10 percent of the cases, ALS is inherited, affecting men and women almost equally. In other cases, ALS generally affects men four times more often than women, and is more common among whites than blacks.

Victims of ALS generally have weakened and atrophied muscles, especially in the hands and forearms. They may also exhibit impaired speech and difficulty in breathing, chewing, and swallowing, resulting in choking or excessive drooling. No effective treatment or cure exists, and the disease is invariably fatal. Its cause is unknown.

Peripheral Neuritis

Peripheral neuritis is a progressive degeneration of the axons and myelin sheaths of peripheral nerves, especially those that supply the distal ends of muscles of the limbs. It results in muscle atrophy and weakness, the loss of tendon reflexes, and some sensory loss. The disease is associated with infectious diseases such as syphilis and pneumonia, metabolic and inflammatory disorders such as gout and diabetes mellitus, nutritional deficiencies, and chronic intoxication, including lead and arsenic poisoning and chronic exposure to benzene, sulfonamides, and other chemicals. The prognosis is usually good if the underlying cause can be identified and removed.

Myasthenia Gravis

Myasthenia gravis (*myo*, muscle + Gr. *asthenia*, weak) is an autoimmune disease caused by antibodies (molecules that defend against a foreign substance) directed against acetylcholine receptors. The antibodies reduce the number of functional receptors or impede the interaction between acetylcholine and its receptors. As a result, skeletal muscles become chronically weak.

Parkinson's Disease

Parkinson's disease (Parkinsonism or "shaky palsy") is a motor disability characterized by symptoms such as stiff posture, tremors, and reduced spontaneity of facial expressions. It results from a deficiency of the neurotransmitter dopamine in certain brain neurons involved with motor activity. Parkinson's disease is discussed in greater detail in Chapter 14.

Huntington's Chorea

Huntington's chorea is a fatal hereditary brain disease, which has been associated with insufficient amounts of the neurotransmitter GABA. In most cases, the onset of the disease occurs in the fourth and fifth decades of life, usually after the victim has married and had children, and hence too late to avoid passing the affliction on to children.

CHAPTER SUMMARY

The body uses a combination of electrical impulses and chemical messengers to react and adjust to stimuli in order to maintain homeostasis.

Neurons: Functional Units of the Nervous System (p. 293)

1 *Neurons* are nerve cells specialized to transmit nerve impulses throughout the body. Their basic properties are **excitability** and **conductivity**.

2 A neuron is composed of a **cell body,** branching **dendrites,** and an **axon.** Dendrites conduct action potentials toward the cell body, and the axon usually carries impulses away from the cell body. Axons may be surrounded by a sheath of *myelin,* which enhances the speed of conduction.

3 *Afferent* (sensory) *neurons* of the peripheral nervous system carry information from sensory receptor cells to the central nervous system. *Efferent* (motor) *neurons* of the peripheral nervous system carry neural information away from the central nervous system to muscles and glands. *Interneurons* are connecting neurons that carry impulses from one neuron to another.

4 Neurons and their peripheral fibers may be classified as *somatic, visceral, general, special, afferent,* and *efferent.*

5 Neurons may be classified according to their *structure* as **multipolar, bipolar,** and **unipolar.**

6 Most neurons are composed of specific functional segments known as **re-**

ceptive, initial, conductive, and *transmissive.*

Physiology of Neurons (p. 298)

1 *Nerve impulses,* or *action potentials,* are conducted along a nerve fiber when specific changes occur in the electrical charges of the fiber membrane.

2 A resting nerve cell (one that is not conducting an impulse) is *polarized,* with the outside of its plasma membrane positively charged with respect to its negatively charged interior. In resting cells, the difference in electrical charge between the outside and inside of the plasma membrane is called the *resting membrane potential.*

3 The resting potential of a neuron is maintained by a *sodium-potassium pump* that regulates the concentration of sodium and potassium ions inside the plasma membrane.

4 A *threshold stimulus* is the minimal stimulus required to initiate a nerve impulse.

5 *Depolarization* is a decrease in the potential difference across the cell membrane that can make the outside of the cell negative relative to the inside. When a stimulus is strong enough to cause sufficient depolarization, the nerve cell fires. Depolarization can produce an action potential.

6 *Repolarization* is the restoration of the original electrical potential difference across the plasma membrane. The *refractory period* is the time during which a neuron will not respond to a second stimulus.

7 *Saltatory conduction* occurs when a nerve impulse jumps from node to node on a myelinated fiber.

8 The phenomenon of a neuron firing at full power or not at all is the *all-or-none law.*

9 A *synapse* is the junction between the axon terminal of one neuron and the dendrite, cell body, or specific parts of the axon of the next neuron. The cell carrying the impulse toward a synapse is a *presynaptic neuron;* it initiates a response in the receptive segment of a *postsynaptic neuron* leading away from the synapse.

10 Synapses can be electrical or chemical. In a chemical synapse, two cells communicate by way of a chemical agent called a *neurotransmitter,* such as acetylcholine or norepinephrine. In an electrical synapse, the electrical activity of one neuron spreads readily to the next neuron.

11 Neurotransmitters are classified in four groups: *acetylcholine, amino acids, monoamines,* and *neuropeptides.*

Associated Cells of the Nervous System (p. 305)

1 *Associated cells* are specialized cells in the central and peripheral nervous systems other than neurons. They nourish, protect, and support neurons.

2 *Neuroglia* are nonconducting cells that protect, nourish, and support neurons within the central nervous system. The types of neuroglia are astrocytes, oligodendrocytes, microglia, and ependymal cells.

3 Associated cells of the peripheral nervous system provide various types of sheaths. They are *satellite cells* (capsule cells) and *neurilemma cells* (Schwann sheaths).

4 *Regeneration* of severed peripheral nerves is usually possible if the cell body is undamaged and the neurilemma is intact.

Organization of the Nervous System (p. 307)

1 The nervous system may be divided on a gross anatomical basis into the *central nervous system* (the brain and spinal cord) and the *peripheral nervous system* (peripheral nerves and their ganglia located outside the central nervous system).

2 The peripheral nervous system contains *afferent* (sensory) nerve cells, *efferent* (motor) nerve cells, and sensory and autonomic ganglia.

3 The peripheral nervous system may be divided on a functional basis into the *somatic nervous system* and the *visceral nervous system,* each composed of afferent and efferent neurons, including their fibers. The *somatic afferent division* conveys sensory information from the skin, muscles, joints, ears, eyes, and associated structures; the *visceral afferent division* conveys sensory information from the viscera, nose (smell), and taste buds to the central nervous system. The *somatic efferent division* conveys impulses from the central nervous system for the regulation of the skeletal muscles. The *visceral efferent division,* also known as the *autonomic nervous system,* conveys impulses from the central nervous system involved in the regulation of cardiac muscle, smooth muscle, and glands. A special group of cranial nerves (V, VII, IX, and X) have motor nerve fibers that convey impulses from the brainstem to skeletal muscles of the jaw, pharynx, and larynx, as well as the muscles of facial expression. The autonomic nervous system is subdivided into the *sympathetic* and *parasympathetic* nervous systems, which complement each other.

Neuronal Circuits (p. 310)

1 Neurons are organized in networks arranged in synaptically connected sequences called *neuronal circuits.*

2 The principle of *divergence* is shown when the transmissive segment of a neuron branches to have many synaptic connections with the receptive segments of many other neurons. When the axon terminals from many presynaptic neurons synapse with the receptive segment of only one postsynaptic neuron, it is an example of *convergence.*

3 Neuronal circuits may be *divergent, convergent, feedback circuits, parallel circuits, two-neuron circuits,* or *three-neuron circuits.*

The Effects of Aging on the Nervous System (p. 312)

1 Although some neurons continue to die during old age, there is usually no shortage. Moreover, there is only about a 10-percent decrease in the velocity of nerve impulses.

2 Short-term memory may be impaired, especially if the blood supply to the brain is decreased.

3 Several age-related diseases (such as Alzheimer's disease and Parkinson's disease) affect brain functions adversely.

4 The senses of sight, hearing, smell, and taste are usually diminished. Presbyopia, cataract, and glaucoma are relatively common eye disorders of old age; otosclerosis is an ear disease in which the ear ossicles fuse.

5 Sleep is usually reduced and interrupted.

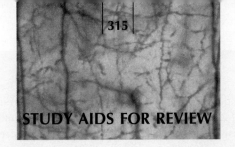

STUDY AIDS FOR REVIEW

UNDERSTANDING THE FACTS

1 Is it true or false that the somatic nervous system always excites skeletal muscles and never inhibits them?
2 Name the two types of effectors in the body.
3 What is another name for the sympathetic division of the autonomic system?
4 In what direction is a nerve impulse transmitted in the dendrite? The axon?
5 Distinguish between a nerve and a tract.
6 Give the location of the following (be specific):
  a peripheral nervous system
  b chromatophilic substance
  c axolemma
  d neurofibral nodes
  e satellite cells
  f neuroglia
7 Distinguish between the three types of neurons by structure.
8 What are the basic functions of astrocytes?
9 What is saltatory conduction?
10 What are the two kinds of synapses between neurons?

UNDERSTANDING THE CONCEPTS

1 Although impulses can travel in either direction along the fibers of a neuron, why do they travel in only one direction within the human nervous system?
2 How do the sympathetic and parasympathetic nervous systems complement each other?
3 In which direction would impulses be transmitted in a mixed nerve? Explain your answer.
4 How does the number of synapses in a neural circuit affect the total transmission time of a nerve impulse?
5 Explain why the synapse is so important in determining the direction of a nerve impulse.
6 In what ways is the nervous system involved in multiple sclerosis?

SELF-QUIZ
Multiple-Choice Questions
12.1 Which of the following is/are *not* part of the peripheral nervous system?
  a cranial nerves
  b spinal nerves
  c ganglia
  d spinal cord
  e c and d

12.2 The efferent division of the nervous system is commonly known as the
  a somatic efferent division
  b somatic afferent division
  c visceral efferent motor division
  d autonomic division
  e c and d
12.3 CNS axons cannot regenerate because
  a the sprouts cannot penetrate the glial scar tissue formed at the site of injury
  b there is no basal lamina in the CNS
  c oligodendroglia have no basis for reorganization
  d a and b
  e a, b, and c
12.4 Which of the following is *not* a major group of neurotransmitters?
  a acetylcholine
  b amino acids
  c monamines
  d neuropeptides
  e nucleic acids
12.5 A progressive degeneration of myelin sheaths in neurons of the brain and spinal cord is called
  a myasthenia gravis
  b multiple sclerosis
  c Parkinson's disease
  d Huntington's chorea
  e senility
12.6 The simplest neuronal circuit is the
  a feedback circuit
  b parallel circuit
  c two-neuron circuit
  d three-neuron circuit
  e divergence circuit

True-False Statements
12.7 The somatic nervous system is composed of afferent and efferent divisions.
12.8 The somatic afferent division of the nervous system regulates the voluntary contraction of skeletal muscles.
12.9 The axon of a nerve cell contains chromatophilic substance.
12.10 The myelin sheaths of axons in the CNS are formed by oligodendroglia.
12.11 Afferent, or sensory, nerve fibers have their cell bodies in ganglia located close to the CNS.
12.12 Glial cells are conducting cells of the nervous system.

Completion Exercises

12.13 At chemical synapses, impulses pass from one neuron to the next by means of _____.

12.14 The central nervous system consists of the _____ and _____.

12.15 The peripheral nervous system can be divided on a functional basis into the _____ and the _____ nervous systems.

12.16 The two subdivisions of the autonomic nervous system are the _____ and _____ systems.

12.17 The myelin sheath is interrupted at regular intervals by gaps called _____.

12.18 A collection of cell bodies of neurons of peripheral nerves is called a _____.

12.19 A stimulus that is strong enough to initiate an impulse in a nerve cell is called a _____.

12.20 The principle that states that a neuron will fire at full power or not at all is known as the _____.

12.21 In a myelinated nerve, the "jumping" of an action potential from one neurofibral node to the next is called _____.

12.22 The junction between neurons is called a _____.

Matching

12.23 _____ general somatic afferent fibers

12.24 _____ general visceral afferent fibers

12.25 _____ general somatic efferent fibers

12.26 _____ general visceral efferent fibers

12.27 _____ special afferent fibers

12.28 _____ astrocytes

12.29 _____ oligodendrocytes

12.30 _____ microglia

a form inner membrane of neural tube during embryonic growth
b carry information from visceral organs to CNS
c carry special senses information to CNS
d sustain neurons nutritionally
e same as satellite cells
f help remove disintegrating products of neurons
g modify activities of the heart, smooth muscles, glands
h similar to neurolemmocytes (Schwann cells)
i carry sensory information from the skin to CNS
j carry information to skeletal muscles

A SECOND LOOK

1 In the drawings below, label the bipolar neuron; in the multipolar neuron, label the axon, dendrites, and cell body.

2 In the micrograph, label the synaptic cleft, synaptic vesicles, axon terminal, and muscle fiber.

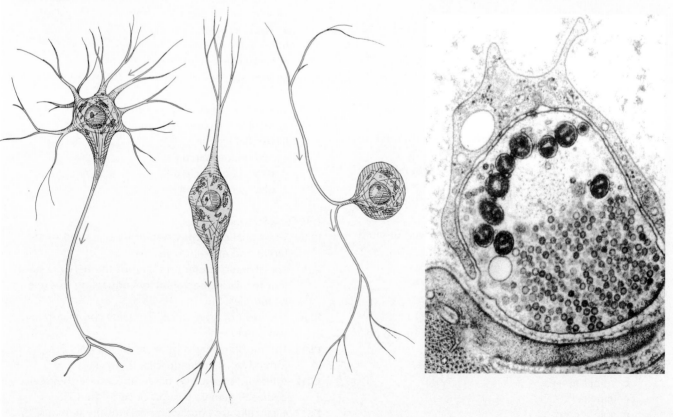

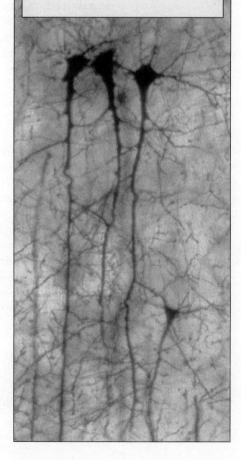

13

The Spinal Cord and Spinal Nerves

LEARNING OBJECTIVES

1 Describe the basic anatomy of the spinal cord and its protective coverings.

2 Explain the internal structure of the spinal cord, including the gray and white matter and nerve roots.

3 Describe the origin, termination, and functional roles of the major ascending and descending spinal tracts.

4 Define reflex, and list the basic components of a simple reflex arc and the stretch reflex.

5 Describe the function of a gamma motor neuron reflex arc.

6 Describe the components and function of withdrawal reflexes.

7 Explain how spinal nerves are named.

8 Describe the structure of the spinal nerves, their roots, sheaths, branches, and plexuses.

9 Identify the location and function of the major spinal nerves in the cervical, brachial, lumbar, and sacral plexuses.

10 Describe the role of intercostal nerves.

11 Describe the arrangement and function of dermatomes.

As you learned in Chapter 12, the nervous system is divided on a gross anatomical basis into the central and peripheral nervous systems. However, to give a clearer understanding of the relationships between the structure and function of the nervous system, we will focus here on the spinal cord and the spinal nerves, and in Chapter 14 on the brain and the cranial nerves.

The spinal cord, with its 31 pairs of spinal nerves, serves two important functions: (1) It is the connecting link between the brain and most of the body, and (2) it is involved in spinal reflex actions, both somatic and visceral. *Somatic spinal reflexes* involve a series of responses to receptors in the skin and skeletal muscles. These reflexes help us to move and maintain a correct posture. *Visceral spinal reflexes* occur in certain organs, for example, when the urinary bladder becomes distended and evokes the urge to urinate. Spinal reflexes also help to regulate blood pressure by affecting the smooth muscle in blood vessels, and they influence the action of glands. Thus, both voluntary and involuntary movement of the limbs, as well as certain visceral processes, depend on the spinal cord, a vital link between the brain and the rest of the body.

BASIC ANATOMY OF THE SPINAL CORD

The *spinal cord* is the part of the central nervous system that extends from the foramen magnum of the skull downward (caudally) for about 45 cm (18 in.) to the level of the first lumbar vertebra (L1) in adults. Its upper end is continuous with the lowermost part of the brain (the medulla oblongata). Its lower end tapers off as the cone-shaped *conus terminalis*, also called the *conus medullaris*, located in the vicinity of the first lumbar vertebra [FIGURE 13.1]. Extending caudally from the conus is a non-neural fiber called the *filum terminale*, which attaches to the coccyx. The filum consists mainly of fibrous connective tissue.

Surrounding and protecting the spinal cord is the bony vertebral column. The central canal of the vertebral column (the opening through which the spinal cord passes) has a diameter slightly larger than your index finger (about 1 cm). Inside the column, the cylindrical cord itself is a little thicker than a pencil. It is slightly flattened dorsally and ventrally, with two prominent enlargements known as the *cervical* and *lumbosacral* (or lumbar) *enlargements* [FIGURE 13.1]. Emerging from these enlargements are the spinal nerves that innervate the upper and lower limbs.

Of the 31 pairs of spinal nerves and roots, there are 8 cervical (C) nerve pairs, 12 thoracic (T), 5 lumbar (L), 5 sacral (S), and 1 to 3 coccygeal (Co). Each pair of spinal nerves typically passes through a pair of intervertebral foramina, located between two successive vertebrae, and then is distributed to a specific pair of segments of the body. Note that there is a pair of spinal nerves for each vertebra.

The roots of all nerves passing caudally below the conus terminalis (below L1 vertebral level) resemble flowing, coarse strands of hair. For this reason, the lumbar and sacral roots are collectively called the *cauda equina* (KAW-duh ee-KWY-nuh), which means "horse's tail" in Latin.

The spinal cord and the roots of its nerves are protected not only by the flexible vertebral column and its ligaments, but also by the spinal meninges and cerebrospinal fluid.

Spinal Meninges

The spinal cord is surrounded by three layers of protective membranes called *meninges* (muh-NIHN-jeez; Gr. plural of *meninx*, membrane) [FIGURE 13.2]. They are continuous with the same layers that cover the brain. The outer layer, called the *dura mater* (DURE-uh MAY-ter; L. hard mother), is a tough, fibrous membrane that merges with the filum terminale. The middle layer, the *arachnoid* (Gr. cobweblike), runs caudally to the S2 vertebral level, where it joins the filum terminale. The arachnoid is so named because it is delicate and transparent and has some connective strands running across the space that separates it from the innermost layer, the *pia mater* (PEE-uh MAY-ter; L. tender mother). The thin, highly vascular pia mater is tightly attached to the spinal cord and its roots. It contains blood vessels that nourish the spinal cord. A pair of fibrous bands of pia mater that extend from each side of the entire length of the spinal cord is the *denticulate ligament* [FIGURE 13.2]. This ligament is attached to the dura mater at regular intervals.

Between the dura mater and the periosteum of the vertebrae is the *epidural space* containing many blood vessels and some fat. Anesthetics can be injected into the epidural space below the S2 vertebral level to dull pain. This procedure, known as a *saddle block*, permits patients to be conscious during operations on the pelvic region, and also during painless childbirth.

Between the dura mater and the arachnoid is the *subdural space*, which is merely a slit, and contains no cerebrospinal fluid. The *subarachnoid space*, which separates the arachnoid and the pia mater, contains cerebrospinal fluid, blood vessels, and spinal roots.

Cerebrospinal Fluid

Cerebrospinal fluid (CSF) is a clear, watery ultrafiltrate solution formed primarily from blood in the choroid

FIGURE 13.1 SPINAL CORD, SPINAL NERVES, AND VERTEBRAL COLUMN

[A] Right lateral view of the spinal cord and vertebral column, showing the various spinal segments and the emergence of spinal nerves from the vertebral column. [B] Posterior view.

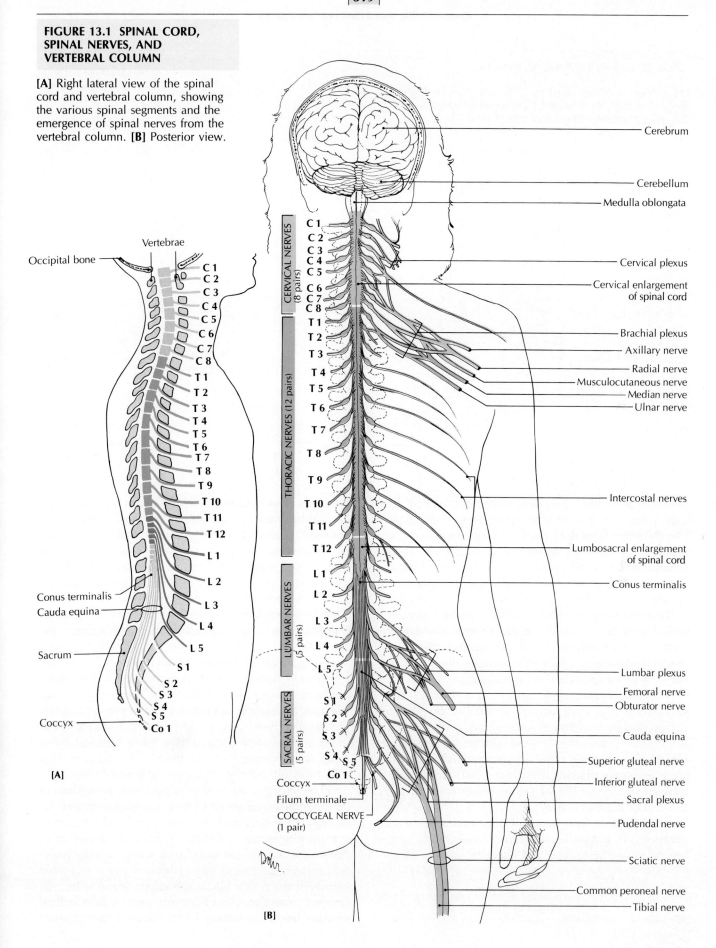

FIGURE 13.2 THE SPINAL MENINGES

[A] A drawing of the spinal meninges. [B] The top layers are peeled away to show the triple-layered meninges and the spaces between them; right lateral view.

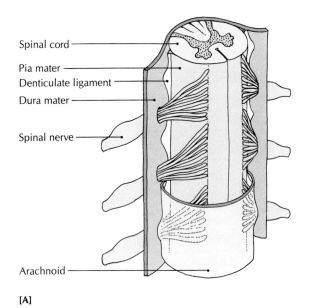

[A]

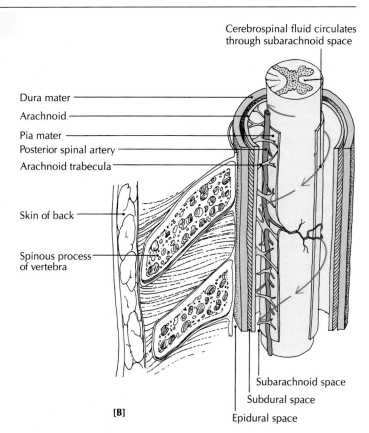

[B]

plexus of the brain.* The basic mechanism of CSF formation involves both an active transport system and passive diffusion. The fluid passes through small openings from the fourth ventricle into the subarachnoid spaces around the brain and spinal cord. Under normal conditions, it returns to the blood at about the same rate that it is formed (0.3 to 0.4 mL/min, or about 300 mL/day).

The cerebrospinal fluid provides a cushion that, together with the vertebral column, protects the delicate tissues of the spinal cord. It is also involved in the exchange of nutrients and wastes between the blood and the neurons of the brain and spinal cord.

The fluid is generally similar to blood plasma. It consists of water; a small amount of protein; oxygen and carbon dioxide in solution; sodium, potassium, calcium, magnesium, and chloride ions; glucose; a few white blood cells; and many other organic materials. The subarachnoid space contains about 75 mL (60 percent) of the total amount (about 125 mL) of the cerebrospinal fluid in the central nervous system.

Cerebrospinal fluid has many important clinical implications. In certain neurological disorders, alterations can occur in the cellular and chemical content of the

fluid, as well as its pressure. The fluid can be removed for analysis by inserting a needle into the subarachnoid space between L3 and L4, with the patient usually lying curled up on one side [FIGURE 13.3]. Such a *lumbar puncture*, or "spinal tap," is performed on this lower lumbar region to avoid injury to the spinal cord, which ends between L1 and L2. A *cisternal puncture* can be performed by inserting a needle between the occipital bone and atlas (C1), entering the cisterna cerebellomedullaris, and withdrawing spinal fluid.

A marked increase in white blood cells in the cerebrospinal fluid occurs in acute bacterial meningitis, and a moderate increase may indicate the presence of a viral infection or cerebral tissue damage. The protein (gamma globulin) content is increased in multiple sclerosis, and glucose levels are reduced during active bacterial infections. The presence of red blood cells indicates that blood has entered the subarachnoid space.

These changes can be determined from an analysis of cerebrospinal fluid obtained from a lumbar puncture. An increase in intracranial pressure may be the result of a brain tumor that obstructs the normal circulation of cerebrospinal fluid, causing it to back up. A lumbar puncture is never attempted when a brain tumor is suspected because if the tumor blocks the subarachnoid space, the removal of cerebrospinal fluid may cause a drop in fluid pressure below the tumor. The resulting higher pressure

*Because cerebrospinal fluid is formed in the brain and is a crucial part of the brain's physiology, it will also be discussed in the next chapter, which discusses the brain in detail.

above the block can force the medulla oblongata of the brain into the foramen magnum, compressing the medulla and killing the patient.

Internal Structure

A cross section of the spinal cord reveals a tiny central canal, which contains cerebrospinal fluid, and a dark portion of H-shaped or butterfly-shaped "gray matter" surrounded by a larger area of "white matter" [FIGURE 13.4]. The spinal cord is divided into more-or-less symmetrical left and right halves by a deep groove, called the *anterior median fissure*, and a median septum, called the *posterior median sulcus*. Extending out from the spinal cord are the ventral and dorsal roots of the spinal nerves. (The spinal cord and nerves cut vertically are shown in FIGURE 13.5; the roots are shown in FIGURE 13.6.)

Gray matter The *gray matter* of the spinal cord consists of nerve cell bodies and dendrites of association and efferent neurons, unmyelinated axons of spinal neurons, sensory and motor neurons, and axon terminals of neurons. The gray matter forms an H shape, and is composed of three pairs of columns of neurons running up and down the length of the spinal cord from the upper cervical level to the sacral level [FIGURE 13.4A]. The pairs of columns that form the two vertical bars of the H are called *horns*. The two that run dorsally are the **posterior horns,** which function in afferent input, and the two that run ventrally are the **anterior horns,** which function in efferent somatic output. In the thoracic and upper lumbar levels of the spinal cord (T1 to L2) there are also two small horns that extend laterally, the *lateral horns;* these horns contain the cell bodies of preganglionic efferent neurons of the sympathetic nervous system. The cross bar of the H is known as the **gray commissure** (L. "joining together"), and its nerve fibers function in cross reflexes. The gray matter is actually a pinkish-gray color because of a rich network of blood vessels.

White matter The *white matter* of the spinal cord gets its name because it is composed mainly of myelinated nerve fibers, and myelin has a whitish color. The white matter is divided into three pairs of columns, or *funiculi* (fyoo-NICK-yoo-lie; L. little ropes), of myelinated fibers that run the entire length of the cord. The funiculi consist of the anterior (ventral) column, the posterior (dorsal) column, the lateral column, and a commissure area [FIGURE 13.4B].

The bundles of fibers within each funiculus are subdivided into *tracts* called *fasiculi* (fah-SICK-yoo-lie; L. little bundles). *Ascending tracts* are made up of sensory fibers that carry impulses *up* the spinal cord to the brain [TABLE 13.1]; *descending tracts* of motor fibers transmit impulses from the brain *down* the spinal cord to the efferent neurons. The longer tracts carry nerve impulses up to the brain or down through the cord to neurons that innervate the muscles or glands. The shorter tracts convey impulses from one level of the cord to another. The major ascending and descending tracts are described in TABLES 13.1 (page 325) and 13.2 (page 326).

FIGURE 13.3 LUMBAR PUNCTURE, OR SPINAL TAP

The spine is in a flexed position **[A],** which separates the spinous processes of the vertebrae and allows the lumbar puncture needle to enter the subarachnoid space well below the termination of the spinal cord **[B].**

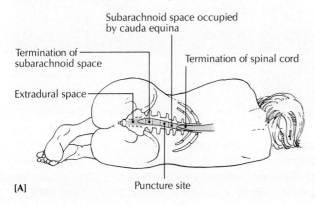

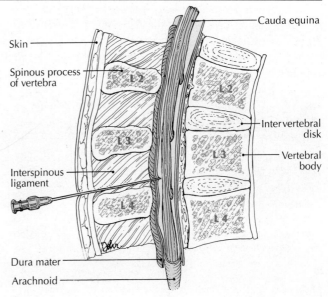

FIGURE 13.4 INTERNAL STRUCTURE OF THE SPINAL CORD

[A] Cross section of the spinal cord with the bones that protect it. Spinal nerves connect with the spinal cord in *pairs* at the ventral and dorsal roots. **[B]** Cross section of the spinal cord showing some prominent internal features, including columns, or funiculi, of myelinated nerve fibers. The insets show the composition of white matter (right) and gray matter (left).

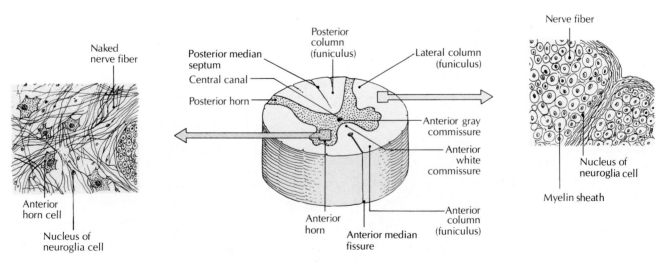

FIGURE 13.5 SPINAL CORD AND SPINAL NERVES

Photographs of [A] the spinal cord, and [B] spinal nerves sectioned vertically; posterior view.

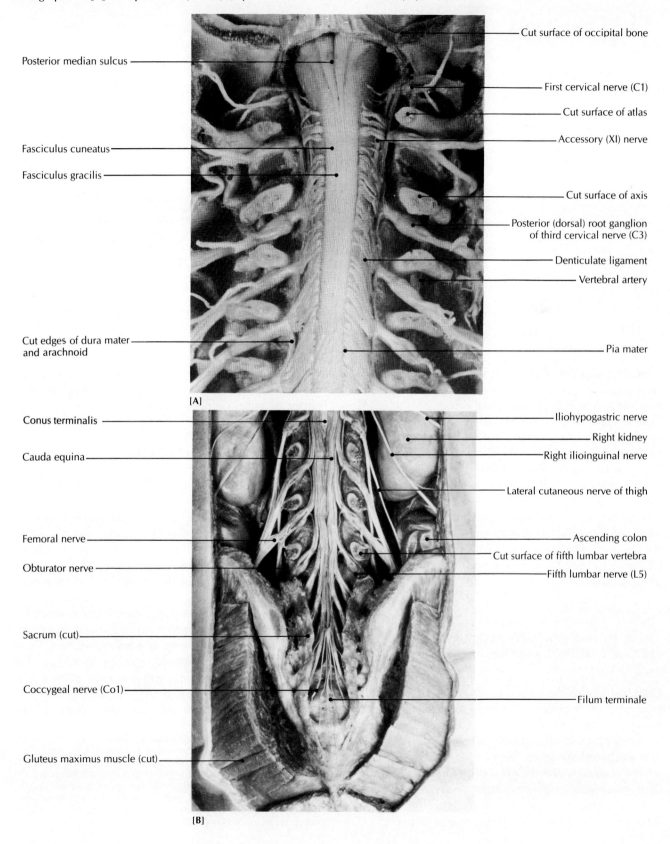

Posterior median sulcus

Fasciculus cuneatus

Fasciculus gracilis

Cut edges of dura mater and arachnoid

Cut surface of occipital bone

First cervical nerve (C1)

Cut surface of atlas

Accessory (XI) nerve

Cut surface of axis

Posterior (dorsal) root ganglion of third cervical nerve (C3)

Denticulate ligament

Vertebral artery

Pia mater

[A]

Conus terminalis

Cauda equina

Femoral nerve

Obturator nerve

Sacrum (cut)

Coccygeal nerve (Co1)

Gluteus maximus muscle (cut)

Iliohypogastric nerve

Right kidney

Right ilioinguinal nerve

Lateral cutaneous nerve of thigh

Ascending colon

Cut surface of fifth lumbar vertebra

Fifth lumbar nerve (L5)

Filum terminale

[B]

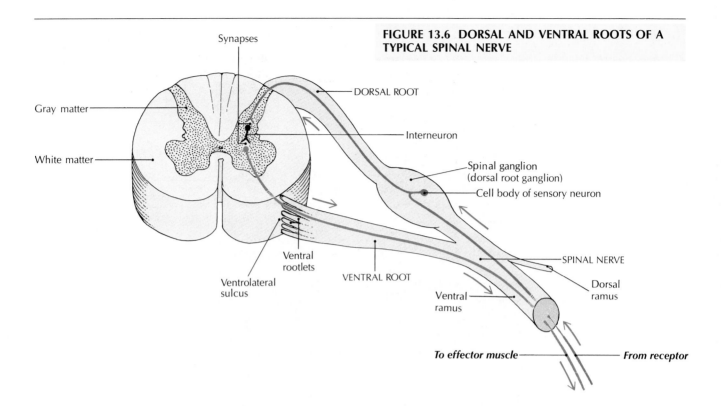

FIGURE 13.6 DORSAL AND VENTRAL ROOTS OF A TYPICAL SPINAL NERVE

Ventral and dorsal roots In the vicinity of the cord, each spinal nerve divides into a ventral (anterior, motor) root and a dorsal (posterior, sensory) root [FIGURE 13.6]. *Ventral roots* contain efferent nerve fibers* and convey motor information. *Dorsal roots* contain afferent nerve fibers, which enter the cord with sensory information. The ventral roots emerge from the spinal cord in a groove called the *ventrolateral sulcus,* and the dorsal roots enter the cord in the *dorsolateral sulcus* [FIGURE 13.4B]. The axons of anterior horn cells make up the ventral roots and lie *within* the gray matter of the cord. In contrast, groups of cell bodies whose axons make up the dorsal roots lie *outside* the cord in the *dorsal root ganglia,* or *spinal ganglia.*

FUNCTIONAL ROLES OF PATHWAYS OF THE CENTRAL NERVOUS SYSTEM

Each pathway of the central nervous system is basically composed of organized sequences of neurons. Upper motor neurons in the brain influence the activity of lower motor neurons in the cranial and spinal nerves. Some neurons have long axons (fibers) that terminate in *processing centers,* where the processing of neural information takes place. These processing centers may be con-

*Some fibers of ventral roots are now known to be afferent.

sidered as the "computers" of the pathways. They are called by one of several names: nucleus, ganglion, gray matter of spinal cord, or cortex.

General Somatic Efferent (Motor) System

The brain exerts active and subtle influences upon the activity of the skeletal muscles through descending motor pathways made up of the *upper motor neurons.*† Originating from cell bodies located in the cerebral cortex and brainstem, these upper motor neuron pathways act by regulating, modulating, and biasing the activity of the *lower motor neurons* of the cranial and spinal nerves.

Lower motor neurons The *lower motor neurons* include alpha and gamma motor neurons. *Alpha motor neurons* have cell bodies in the central nervous system. Their axons course through cranial and spinal nerves, and terminate to form the motor end plates innervating skeletal muscle fibers. These muscle fibers are also known as *extrafusal muscle fibers* because they are outside of the

†In the past, the motor pathways and associated neural circuits were called the pyramidal and extrapyramidal systems. The *pyramidal system* (because its fibers passed through the pyramids of the medulla oblongata) referred to the corticobulbar and corticospinal tracts; the *extrapyramidal system* consisted of the other upper motor neurons: the rubrospinal, reticulospinal, and vestibulospinal tracts. Because the pyramidal-extrapyramidal classification has little significance in modern neurology, its use is being abandoned.

TABLE 13.1 MAJOR ASCENDING (SENSORY) TRACTS FROM SPINAL CORD TO BRAIN

Tract and location in spinal cord	Origin	Description and course to thalamus of brain	Course from thalamus to termination in cerebral cortex	Sensations conveyed
Lateral spinothalamic, in anterior half of lateral column*	Posterior horn of spinal cord gray matter (location of cell bodies of second-order neurons)	Fibers of second-order neurons cross to the opposite side at each spinal cord level, ascend and terminate in thalamus.	Axonal fibers of third-order neurons with cell bodies in thalamus terminate in postcentral gyrus of cerebral cortex.	Pain and temperature
Spinoreticulothalamic, in lateral and anterior columns*	Posterior horn of spinal cord gray matter (location of cell bodies of second-order neurons)	Crosses to opposite side in spinal cord, ascends to and terminates in reticular formation of brainstem. After a sequence of several neurons, fibers terminate in thalamus.	Fibers of neurons of cell bodies in thalamus terminate in many areas of cerebral cortex.	Pain
Fasciculus gracilis and fasciculus cuneatus, in posterior column (posterior column-medial lemniscus pathway)	First-order neurons with cell bodies in dorsal root ganglia; ascend on same side of cord	Axons of first-order neurons terminate in nuclei gracilis and cuneatus in medulla of same side. Axonal fibers of second-order neurons with cell bodies in nuclei gracilis and cuneatus cross to opposite side to form *medial lemniscus,* which ascends to, and terminates in, thalamus.	Axonal fibers of third-order neurons with cell bodies in thalamus terminate in postcentral gyrus of cerebral cortex.	Touch-pressure, two-point discrimination, vibratory sense, position sense (proprioception);† also some light (coarse) touch
Anterior spinothalamic, in anterior column*	Posterior horn of spinal cord gray matter (location of cell bodies of second-order neurons)	Crosses to opposite side at each spinal cord level. Joins medial lemniscus of brainstem, which terminates in thalamus.	Axonal fibers of third-order neurons with cell bodies in thalamus terminate in postcentral gyrus of cerebral cortex.	Light (coarse) touch‡

Tract and location in cerebellum of brain	Origin	Description and course in cerebellum	Termination	Modality conveyed
Posterior (dorsal) spinocerebellar, in posterior half of lateral column	Posterior horn of spinal cord gray matter	Uncrossed tract. Enters cerebellum via inferior cerebellar peduncle.	Cerebellum	Unconscious proprioception§
Anterior (ventral) spinocerebellar, in anterior half of lateral column	Posterior horn of spinal cord gray matter	Some fibers cross to opposite side in spinal cord. Enter cerebellum via superior cerebellar peduncle.	Cerebellum	Unconscious proprioception

*These three tracts are collectively called the *anterolateral system.*

†*Touch-pressure* (deep touch) sensation is obtained by deforming the skin with pressure. The ability to discriminate between one or two points applied to the skin is called *two-point discrimination;* it is acute on the ball of a finger, but poor on the back of the body. *Vibratory sense* is the sensation of feeling vibrations when the stem of a vibrating tuning fork is placed on the bones of a joint. *Position sense* enables us to know where the various body parts are with our eyes closed.

‡*Light touch* is the sensation obtained by touching, but not deforming, the skin or by moving a hair.

§*Unconscious proprioception* (*proprio* = self) is the sensory information from the body, which does not result in any sensation. Such information about movement and limb position is continuously streaming in from the body and is utilized by the nervous system in the coordination of voluntary muscles during movements.

TABLE 13.2 MAJOR DESCENDING (MOTOR) TRACTS FROM BRAIN TO SPINAL CORD

Tract and location in spinal cord	Origin	Description	Termination	Motor impulse conveyed
Lateral (pyramidal) corticospinal, in lateral column	Cerebral cortex	Crosses to opposite side between medulla oblongata and spinal cord.	Gray matter, but primarily anterior horn	Manipulative movements of extremities, especially delicate finger movements.
Anterior (pyramidal) corticospinal	Cerebral cortex	Mainly uncrossed tracts; crosses near termination.	Anterior horn of spinal cord	Voluntary movements of axial musculature.
Rubrospinal, in lateral column	Nucleus ruber of midbrain	Crosses in midbrain to opposite side in midbrain.	Gray matter, but primarily anterior horn	Manipulative movements of extremities.
Medullary and pontine reticulospinal, in anterior column and anterior half of lateral column	Reticular formation of medulla oblongata and pons	Mainly uncrossed tracts.	Anterior horn	Maintenance of erect posture. Integrate movements of axial body and girdles of limbs, especially during postural movement.
Vestibulospinal, in anterior column	Vestibular nuclei of medulla oblongata	Uncrossed tract.	Anterior horn	Maintenance of erect posture. Integrates movements of body and limbs.

DESCENDING (MOTOR) TRACTS OF THE SPINAL CORD   ASCENDING (SENSORY) TRACTS OF THE SPINAL CORD

neuromuscular spindles [FIGURE 13.7]. Neuromuscular spindles are sensory receptors that monitor the extent and rate of muscle lengthening, and hence are called *stretch receptors*. *Gamma motor neurons* have their cell bodies within the central nervous system. Their axons pass through the cranial and spinal nerves to innervate the *intrafusal muscle fibers* inside the neuromuscular spindles. Intrafusal muscle fibers allow a stretch-reflex action to be made smoothly.

Because the lower motor neurons are the only neurons innervating the skeletal muscle fibers, they function as the *final common pathway*, the final link between the central nervous system and skeletal muscles. Lower motor neurons are located in both cranial and spinal nerves. Those in cranial nerves innervate the skeletal muscles associated with the movements of the eyes and tongue, and of chewing (mastication), facial expression, swallowing, and vocalizing.

Upper motor neurons Lower motor neurons are influenced by two sources, (1) sensory receptors in the body that are integrated into reflex pathways, and (2) **upper motor neurons** from the brain forming the "voluntary descending pathways." The *upper motor neuron pathways* comprise (1) the corticospinal (pyramidal) and corticobulbar tracts ("bulbar" refers to the pons and medulla oblongata) originating from neurons in the motor areas of the cerebral cortex, (2) the rubrospinal tract originating from neurons in the nucleus ruber in the midbrain, (3) the reticulospinal tracts originating from reticular nuclei of the lower brainstem (pons and medulla oblongata), and (4) the vestibulospinal tracts originating from vestibular nuclei in the lower brainstem. The nucleus ruber and reticular nuclei receive input from the motor areas of the cerebral cortex and other sources. The vestibular nuclei receive their primary inputs from the vestibular sensors and the cerebellum.

FIGURE 13.7 A NEUROMUSCULAR SPINDLE

The drawing shows the encapsulated intrafusal muscle fibers and the extrafusal muscle fibers outside the spindle.

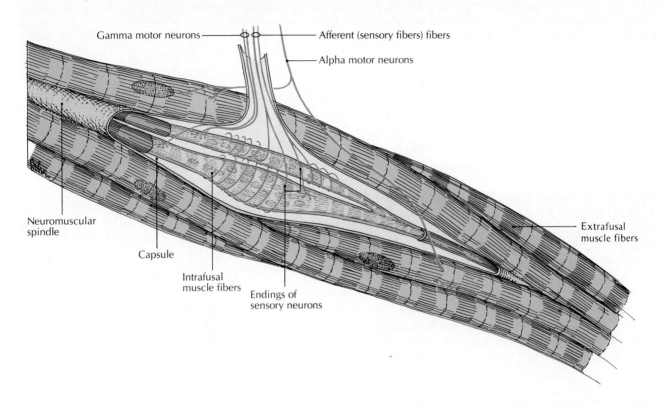

Gamma motor neurons

Afferent (sensory fibers) fibers

Alpha motor neurons

Neuromuscular spindle

Capsule

Intrafusal muscle fibers

Endings of sensory neurons

Extrafusal muscle fibers

In contrast to a lower motor neuron, which is located in both the central and peripheral nervous systems, an upper motor neuron is located wholly within the central nervous system. The upper motor neurons have significant roles in the maintenance of posture and equilibrium, control of muscle tone, and reflex activity.

Processing Centers

Each processing center consists of (1) the terminal branches of axons entering the center, (2) cell bodies and dendrites of neurons whose axons form the tract, and (3) intrinsic neurons whose dendrites, cell bodies, and axons are located wholly within the center. The interactions among neurons within the processing centers cause the input information to be altered (*processed*) in some way before it is conveyed to the next center. Many centers receive input from nerve cells originating in more than one location.

Some pathways contribute to the complex processing within the brain itself. These are neuronal circuits of interconnected sequences of a number of processing centers (some are called feedback circuits). They integrate and process information at the higher levels of brain functions, including behavioral activities, complex voluntary movements, and thinking.

Sensory Pathways

Some sensory pathways have sequences that are made up of three neurons. In some pathways, the neurons in the sequence are called first-, second-, and third-order neurons. A *first-order neuron* extends from the sensory receptor to the central nervous system; a *second-order neuron* extends from the spinal cord or brainstem to a nucleus in the thalamus; and a *third-order neuron* extends from the thalamus to a sensory area of the cerebral cortex.

A critical feature of many pathways is that the axons of a tract (for example, the second-order neuron of many sensory pathways) *cross over,* or *decussate,* from one side of the spinal cord or brainstem to the other side [FIGURE 13.8B and TABLES 13.1 and 13.2]. Because of this crossing over, one side of the body communicates with the opposite side of the brain. Knowing where a pathway crosses over helps a physician to locate the site of an injury (lesion) in the central nervous system. For example, the touch-pressure pathway decussates in the medulla oblongata (lower brainstem). A lesion in this pathway on one side of the spinal cord results in the loss of sensation on the *same* side of the body as the lesion, whereas a lesion above the decussation in the brain results in the loss of sensation on the *opposite* side of the body.

Many tracts are named after their centers of origin and termination. For example, the lateral spinothalamic tract (a tract in a pain pathway) originates in the gray matter of the spinal cord *(spino-)*, terminates in the thalamus of the cerebrum *(thalamic)*, and is located laterally in the spinal cord *(lateral)*.

Anterolateral System

The ***anterolateral system*** consists of the lateral spinothalamic tract, spinoreticulothalamic pathway, and anterior spinothalamic tract. This system involves general somatic sensory information that is conveyed from the body via the spinal nerves to the spinal cord.

Lateral spinothalamic tract The ***lateral spinothalamic tract*** [FIGURE 13.8A] is involved with the sensations of pain and temperature. It consists of the sequence of (1) first-order neurons whose axons terminate in the posterior horn of the spinal cord, (2) second-order neurons with cell bodies in the posterior horn and axons that decussate and ascend on the opposite side of the spinal cord as the lateral spinothalamic tract to the thalamus, and (3) third-order neurons with cell bodies in the thalamus and axons that ascend and terminate in the primary somatosensory area (postcentral gyrus) of the cerebral cortex.

Spinoreticulothalamic pathway The ***spinoreticulothalamic pathway*** [FIGURE 13.8A] is composed of (1) first-order neurons that terminate in the posterior horn and (2) second-order neurons with cell bodies in the posterior horn; some axons ascend on the same side (do not decussate), and other axons decussate and ascend on the opposite side to the reticular formation of the brainstem. After several sequences of reticular neurons, the pathway terminates in the thalamus. Neurons of the thalamus have axons that terminate widely in many areas of the cerebral cortex.

Anterior spinothalamic tract The ***anterior spinothalamic tract*** is a tract involved with the sensation of light-touch. It is similar to, and parallels, the course of the lateral spinothalamic tract (not illustrated).

Posterior Column-Medial Lemniscus Pathway

The ***posterior column-medial lemniscus pathway*** [FIGURE 13.8B] is involved with touch-pressure, vibratory sense, two-point discrimination, position sense (proprioception), and some light touch. Its first-order neurons have axons that ascend within the posterior columns (fasiculus gracilis and fasiculus cuneatus) without de-

FIGURE 13.8 SENSORY PATHWAYS

[A] The lateral spinothalamic tract and spinoreticulothalamic pathway of the anterolateral system. **[B]** The posterior column-medial lemniscus pathway.

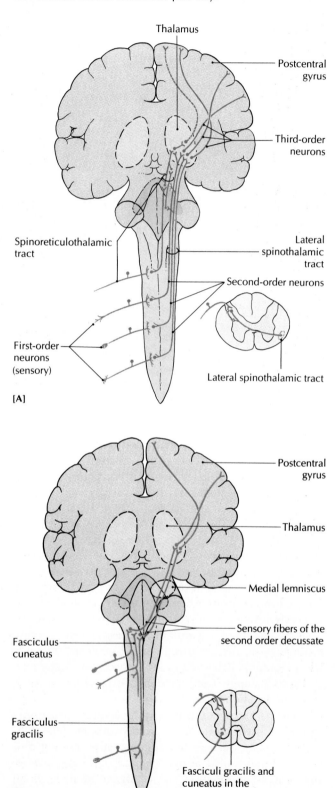

cussation and terminate in the nucleus gracilis and nucleus cuneatus of the lower medulla oblongata. (The fasciculus gracilis consists of fibers conveying impulses from the lower half of the body, including the lower limb, and the fasciculus gracilis consists of fibers from the upper half of the body, including the upper limb, of the same side.) The second-order neurons have cell bodies in the nuclei gracilis and cuneatus and axons that decussate in the lower medulla oblongata as the internal arcuate fibers, and then ascend as the medial lemnicus to the thalamus. The third-order neurons with cell bodies in the thalamus ascend and terminate in the primary somatosensory (postcentral gyrus) cortex.

SPINAL REFLEXES

In addition to linking the brain and most of the body, the spinal cord coordinates reflex action. A *reflex* (L. to bend back) is a predictable involuntary response to a stimulus, such as quickly pulling your hand away from a hot stove. A reflex involving the skeletal muscles is called a *somatic reflex*. A reflex involving responses of smooth muscle, cardiac muscle, or a gland is a *visceral* (autonomic) *reflex*. Visceral reflexes control the heart rate, respiratory rate, digestion, and many other body functions, as described in Chapter 15 (The Autonomic Nervous System). Both types of reflex action allow the body to respond quickly to internal and external changes in the environment in order to maintain homeostasis.

All *spinal reflexes*, that is, reflexes carried out by neurons in the spinal cord alone and not immediately involving the brain, are based on the sequence shown in FIGURE 13.9. Such a system is called a *reflex arc*. In a *monosynaptic* (one-synapse, two-neuron) *reflex arc*, the sensory and motor neurons synapse directly. More often, however, one or more *interneurons* synapse with the sensory and

motor neurons in a *polysynaptic reflex arc*. An example of a simple polysynaptic reflex is pulling your foot away when you step barefoot on a tack (the stimulus) [FIGURE 13.9].

Most reflexes in the human body are more complex than the one shown in FIGURE 13.9. The more usual reflexes are actually *chains* of reflexes, with the possibility of several skeletal muscles being activated almost simultaneously.

Reflex actions save time because the "message" being transmitted by the impulse does not have to travel from the receptor all the way to the brain. Instead, most reflex actions never travel any further than the spinal cord, though some extend to the brainstem.

Types of spinal reflexes include the following:

1 The *stretch (myotatic) reflex* is a two-neuron (monosynaptic) reflex arc. It acts to maintain our erect upright posture and stance by exciting the extensor muscles of the lower limbs, back, neck, and head. A well-known example is the *knee jerk*, or *patellar*, *reflex*, which is produced by tapping the patellar tendon of the relaxed quadriceps femoris muscle of the thigh [FIGURE 13.10]. Such a reflex is described as *ipsilateral* (ipsi, same + lateral, side) because the response occurs on the side of the body and spinal cord where the stimulus is received.

In the knee jerk, a tap on the patellar tendon suddenly stretches the quadriceps femoris tendon and its muscles and some of the neuromuscular spindles within the muscle. The neuromuscular spindles respond to changes in the length of a muscle. The stretched spindles excite sensitive nerve endings (stretch-gated receptors) of afferent fibers within the spindles, generating nerve impulses. The afferent fibers convey these nerve impulses to the L2 or L3 vertebral level of the spinal cord. Here the afferent neurons synapse with lower motor neurons called alpha motor neurons, which are among the largest of spinal neurons. The axons of the alpha neurons carry the impulse rapidly to the motor end plates of the quadriceps femoris muscle, stimulating it to contract, thus causing the leg to swing forward (extend).

2 The *gamma motor neuron reflex arc* acts to smooth out the movements of muscle contractions, or to sustain the contraction of a muscle such as in holding a heavy object, by regulating the state of contraction of the intrafusal fibers in neuromuscular spindles. The smoothing out is mediated through the activity of the sensory fibers innervating the intrafusal muscle fibers [FIGURE 13.7]. When a muscle is stretched, the spindle (a stretch receptor) stimulates the sensory neurons of the two-neuron stretch reflex [FIGURE 13.10]. After a muscle is stimulated by the alpha motor neuron of this reflex arc, the muscle spindles within the contracting muscle become shorter and, when shortened, reduce the firing rate of the sensory neurons. This reduction is followed by the relaxation of the muscle.

FIGURE 13.9 PATHWAY OF A SIMPLE THREE-NEURON REFLEX ARC

A reflex arc always starts with a sensory neuron and ends with a motor neuron. The arc pictured here begins with a tack pricking the skin surface of a big toe. The impulse (arrows) travels from the toe to the spinal cord and back to a muscle in the foot, which jerks away from the tack—a flexor, or withdrawal, reflex.

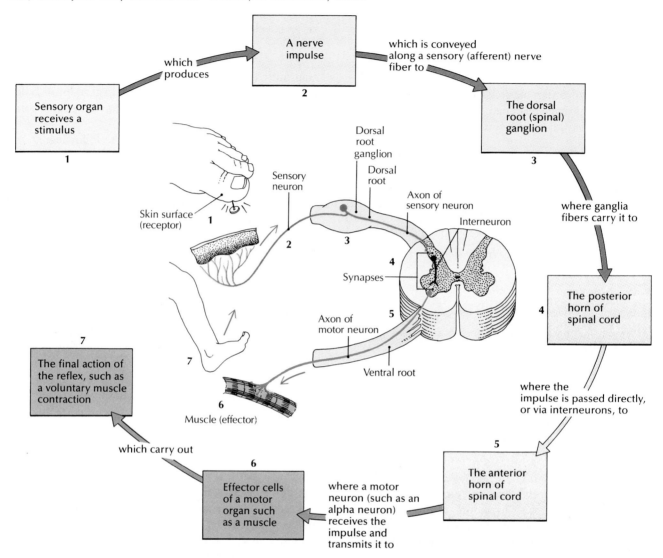

To smooth out the movement or sustain the contractions of the muscle by the two-neuron reflex, the nervous system utilizes gamma motor neurons. When the central nervous system stimulates alpha motor neurons, which stimulate a muscle to contract, it also stimulates a gamma motor neuron, which stimulates the intrafusal fibers of the spindle to contract. This contraction of the intrafusal fibers, which occurs as the spindle shortens, activates the sensory fibers of the neurons of the two-neuron stretch reflex to increase their firing rate, and thus maintain the contraction of the extrafusal fibers.

3 The ***withdrawal reflexes*** are also known as protec-tive or escape reflexes. They combine *ipsilateral reflexes* [FIGURE 13.11A] and *crossed extensor reflexes* [FIGURE 13.11B].

The crossed extensor reflex is important because it helps the lower limbs support the body. When the right leg moves up from the ground by flexion, the left leg is activated to extend so that it supports the body. Such a recip-rocal arrangement occurs when we walk or run. The spi-nal reflex circuits are also involved in the maintenance of body position and posture by stimulating motor nerves to regulate and sustain tonus of the body musculature.

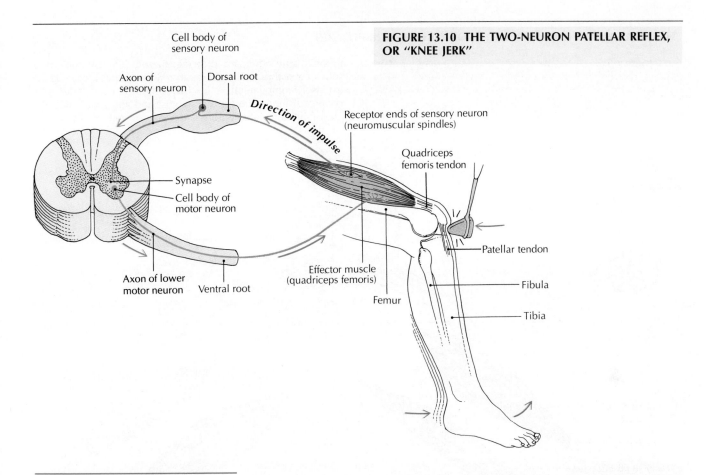

FIGURE 13.10 THE TWO-NEURON PATELLAR REFLEX, OR "KNEE JERK"

Functional and Clinical Aspects of Reflex Responses

In a resting skeletal muscle, some of the fibers are always partially contracted because of the continual stimulation to receptors of certain reflex arcs, especially the stretch reflex. This continuous state of contraction is known as *muscle tone* (tonus). Muscle tone can also be described as the minimal degree of contraction exhibited by a normal muscle at "rest." When an examiner passively manipulates a limb in a relaxed patient (flexion and extension), the muscle tone is expressed as the *amount of resistance* perceived by the examiner.

When some of the lower motor neurons or afferent fibers innervating neuromuscular spindles are injured so they cannot convey impulses, the muscle loses some of its tonus, a condition called *hypotonia*. During hypotonia a muscle has a decreased resistance to passive movement. Poliomyelitis (polio) is a viral infection of lower motor neurons. When this condition is followed by an absence of stimuli to the muscles, they lose much or all of their tonus. If a muscle is completely deprived of its motor innervation, it will lose all of its tonus, a condition called *atony*. *Hypertonia* is a condition in which a muscle has increased tonus. During hypertonia a muscle has an increased resistance to passive movement, as in the later stages following some strokes.

In addition to changes in tonus, lesions in the nervous system may result in changes in reflex activity (reflexia). *Hyporeflexia* is a condition in which a reflex is less responsive than normal, and *hyperreflexia* is one in which a reflex is more responsive than normal. Both conditions are associated with changes in the excitability of the stretch reflex. Hyporeflexia, like hypotonia, is caused by lesions of afferent fibers from the spindles, or of the lower motor neurons to a muscle. Hyporeflexia accompanies poliomyelitis.

The clinical symptoms and alteration in function just noted are the consequence of lesions to specific neural structures. A lesion of lower motor neurons results in a *lower motor paralysis*, and a lesion of upper motor neurons results in an *upper motor paralysis*. A lower motor neuron paralysis is characterized by (1) hypotonia or loss of tone (*atony*), (2) hyporeflexia or loss of reflex (*areflexia*) of the muscle or muscles innervated by the injured lower motor neurons, and (3) wasting of the muscles innervated (*atrophy*). Poliomyelitis and lesions of the anterior horn (site of the cell bodies of the lower motor neurons) result in a lower motor neuron paralysis. An upper motor neuron paralysis is characterized by hypertonia, hyperreflexia, and Babinski's reflex (big toe points up, other toes spread and curl), which is elicited by stroking the sole of the foot with a blunt probe.

FIGURE 13.11 WITHDRAWAL AND CROSSED EXTENSOR REFLEXES

In this highly schematic drawing, the pin-prick stimulus results in a withdrawal (ipsilateral) reflex **[A]** and a crossed extensor (contralateral) reflex **[B].** In both reflexes there is reciprocal innervation of the agonist and antagonist muscles. Inhibitory interneurons are shown in red.

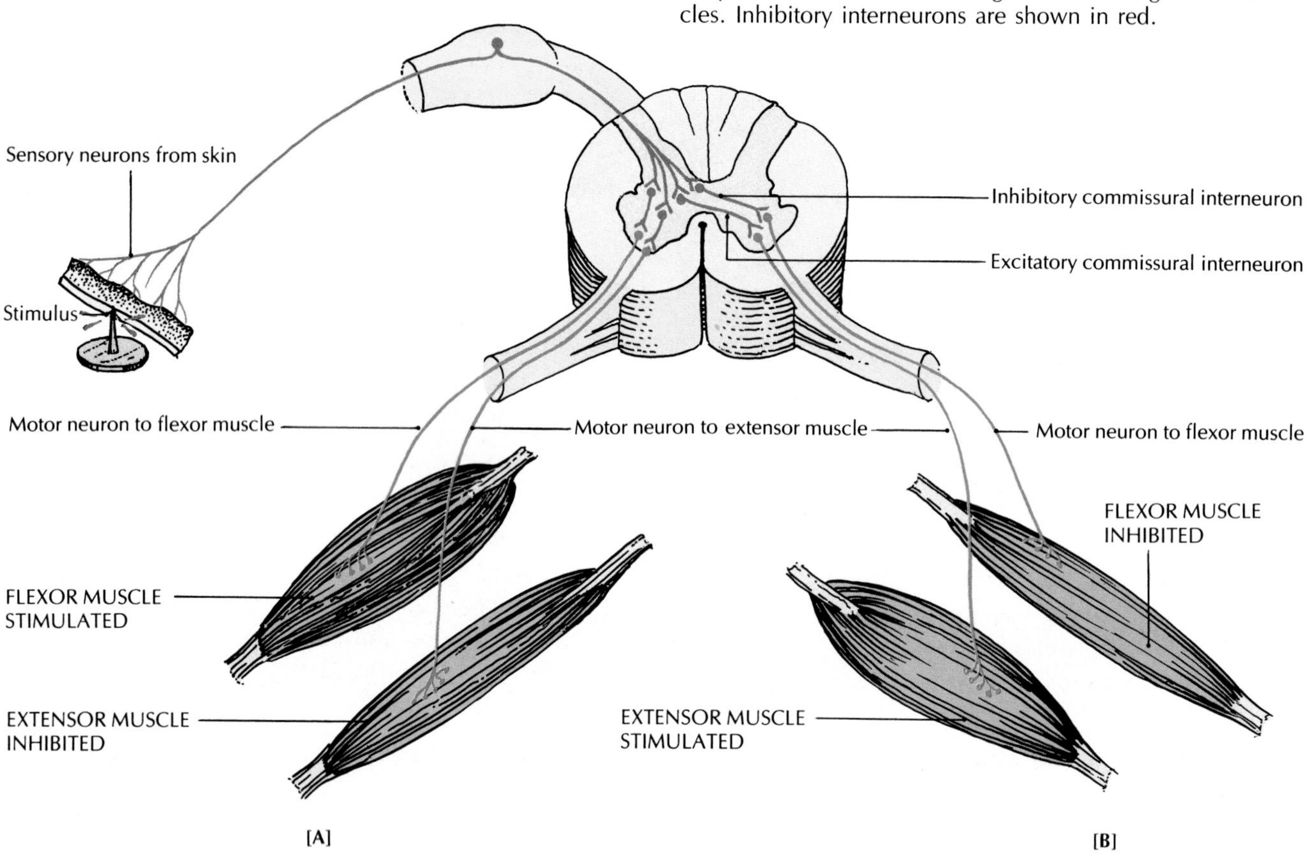

Sensory neurons from skin

Inhibitory commissural interneuron

Excitatory commissural interneuron

Stimulus

Motor neuron to flexor muscle — Motor neuron to extensor muscle — Motor neuron to flexor muscle

FLEXOR MUSCLE INHIBITED

FLEXOR MUSCLE STIMULATED

EXTENSOR MUSCLE INHIBITED

EXTENSOR MUSCLE STIMULATED

[A]

[B]

ASK YOURSELF

1 What is a reflex action?

2 What are the components of a stretch (myotatic) reflex?

3 How does a gamma motor neuron arc coordinate muscular contraction?

4 What is a withdrawal reflex arc?

5 How are reflex responses used for diagnostic purposes?

STRUCTURE AND DISTRIBUTION OF SPINAL NERVES

At each segment of the spinal cord, a pair of nerves branches and exits the H-shaped gray matter. One nerve

of the pair exits to the left, entering the left side of the body. The other nerve of the pair exits to the right, entering the right side of the body. Each nerve has a *ventral* (anterior) *root* and a *dorsal* (posterior) *root* [see FIGURE 13.6], which meet shortly after leaving the spinal cord to form a single *mixed nerve.* The spinal nerves are mixed, containing both sensory and motor fibers that, together with cranial nerves, form part of the peripheral nervous system.

How Spinal Nerves Are Named

The spinal nerves are named for their associated vertebra (cervical, thoracic, lumbar, sacral, or coccygeal) and numbered [see FIGURE 13.1A]. Most spinal nerves pass through an intervertebral foramen and then are distributed to a specific segment of the body. The first cervical nerve, however, passes between the occipital bone and the first cervical vertebra. The numbering of each cervical nerve corresponds to the vertebra *inferior* to its exit. For example, the third cervical (C3) nerve emerges above the third cervical vertebra. But because there are only seven cervical vertebrae, the eighth cervical (C8) nerve passes through the intervertebral foramen between the seventh cervical vertebra and the first thoracic vertebra. The numbering of each of the spinal nerves other than the cervical nerves corresponds to the vertebra *superior* to its exit from the vertebral column.

Structure of Spinal Nerves

Nerves of the peripheral nervous system are rougher and more cordlike than the tissue of the central nervous system because of three sheaths of connective tissue around the nerve fibers [FIGURE 13.12].

Each nerve is made up of nerve fibers enclosed in distinct bundles called *fascicles* [FIGURE 13.12A]. Surrounding the entire peripheral nerve to bind together the large number of fascicles is a tube of connective tissue called the *epineurium* ("upon the nerve"). Also within the epineurium are blood vessels and lymph vessels. Some small nerves do not have an epineurium. A thicker sheath of connective tissue, the *perineurium* ("around the nerve"), encases each fascicle of nerve fibers. Each of these fibers is also covered by the *endoneurium* ("within the nerve"), the interstitial connective tissue that separates individual nerve fibers.

Branches of Spinal Nerves

A short distance after the dorsal and ventral roots join together to form a spinal nerve proper, the nerve divides into several branches called *rami* (RAY-mye; singular *ramus*, RAY-muhss). These branches are the dorsal (posterior) ramus, ventral (anterior) ramus, meningeal ramus, and the rami communicantes [FIGURE 13.13].

The branches of the *dorsal ramus* (a mixed nerve) innervate the skin of the back, the skin on the back of the head, and the tissues and intrinsic (deep) muscles of the back. Branches of the *ventral ramus* (a mixed nerve) innervate the skin, tissues, and muscles of the neck, chest, abdominal wall, both pairs of limbs, and the pelvic area. The *meningeal ramus* innervates the vertebrae, spinal meninges, and spinal blood vessels. The *rami communicantes* (sing., *ramus communicans*) are composed of sensory fibers associated with the general visceral afferent system, and motor nerve fibers associated with the autonomic nervous system (general visceral efferent system) innervating the visceral structure.

Plexuses

The ventral rami of the spinal nerves (except T2 through T12) are arranged to form several complex networks of nerves called *plexuses* (sing. *plexus*; L. braided) [FIGURE 13.14]. In a plexus, the nerve fibers of the different spinal nerves are sorted and recombined, so that fibers associated with a particular peripheral nerve are composed of the fibers from several different rami. Plexuses include: (1) the *cervical plexus*, (2) the *brachial plexus*, (3) the *lumbar plexus*, (4) the *sacral plexus* (sometimes the lumbar and sacral plexuses are referred to collectively as the *lumbosacral plexus*), and (5) the *coccygeal plexus*. Note that the ventral rami of T2 through T12 do not form plexuses. Instead, each ramus innervates a segment of the thoracic and abdominal walls.

Cervical plexus The *cervical plexus* is composed of the ventral rami of spinal nerves C1 through C4. Its branches can be placed into four groups:

1 The *cutaneous sensory nerves* innervate the skin and scalp near the external ear, the skin of the sides and front of the neck, and the skin of the upper portions of the thorax and shoulder.

2 The branches of the *ansa* (looplike) *cervicalis* and its superior and inferior roots provide the motor innervation to the strap muscles attached to the hyoid bone in the neck.

3 The *phrenic nerves* from C3, C4, and C5 descend into the thorax to innervate the diaphragm.

4 Some nerves supply motor innervation to the prevertebral muscles of the neck. These muscles assist in flexion.

Lesions of both phrenic nerves cause paralysis of the diaphragm. The major symptoms are shortness of breath and difficulty in coughing or sneezing. A severe case of hiccups (spasmodic, sharp contractions of the diaphragm) can be relieved by surgically crushing the phrenic nerve in the lower neck, which temporarily paralyzes the diaphragm. (The nerve fibers regenerate in time.) A less drastic method is the injection of an anesthetic solution around the phrenic nerve. The anesthesia concentrates on the anterior surface of the middle third of the anterior scalene muscle, producing a temporary paralysis of half of the diaphragm.

Q: *What is a "broken neck"?*

A: A fatal "broken neck" occurs when the spinal cord is severed above the fourth cervical nerve. The lesion severs the motor tracts (upper motor neurons) from the brain. As a result, all muscles of respiration including the diaphragm (innervated by cervical nerves 4 and 5), and all thoracic and abdominal muscles (innervated by intercostal nerves) are paralyzed because they are deprived of upper motor neuron innervation. Death results because the victim cannot breathe.

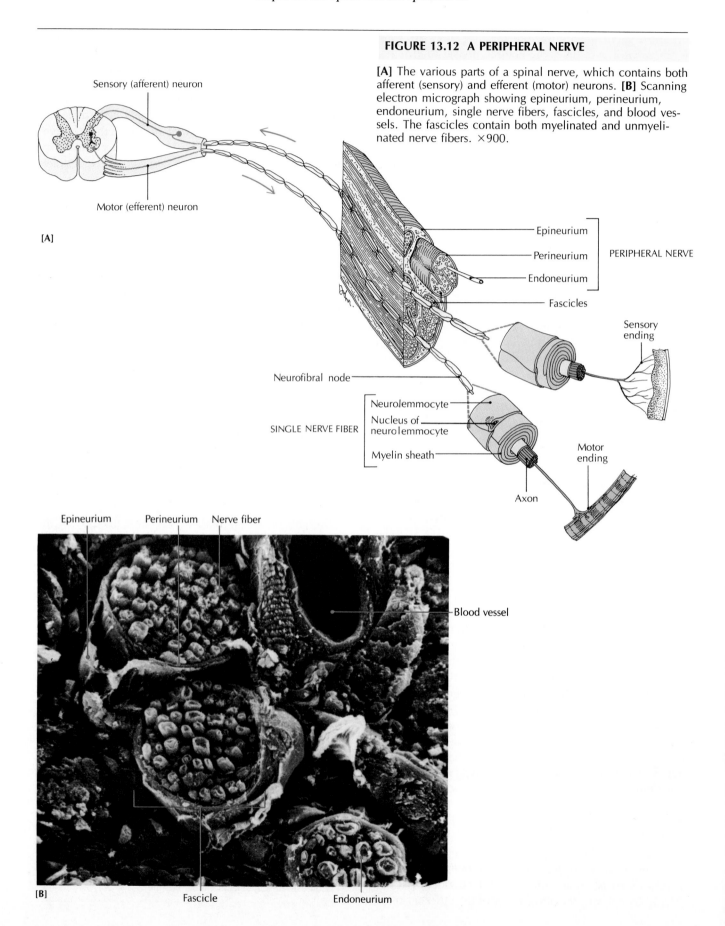

FIGURE 13.12 A PERIPHERAL NERVE

[A] The various parts of a spinal nerve, which contains both afferent (sensory) and efferent (motor) neurons. [B] Scanning electron micrograph showing epineurium, perineurium, endoneurium, single nerve fibers, fascicles, and blood vessels. The fascicles contain both myelinated and unmyelinated nerve fibers. ×900.

FIGURE 13.13 RAMI AND OTHER PARTS OF A SPINAL NERVE

Dorsal and ventral roots join to form a spinal nerve, in this case, a thoracic nerve. The nerve divides into a dorsal (posterior) ramus, a ventral (anterior) ramus, a meningeal ramus, and white and gray rami communicantes. Each sympathetic ganglion of the autonomic nervous system near the spinal cord is a paravertebral ganglion (adjacent to the spinal column), and each one near the abdominal viscera is a prevertebral ganglion.

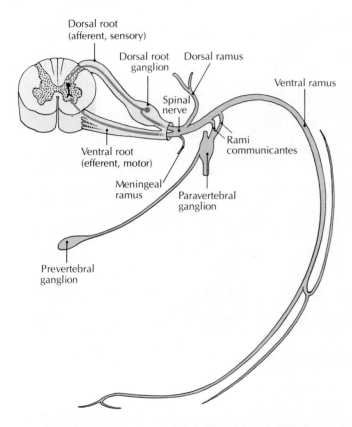

Brachial plexus The *brachial plexus* is made up primarily of the ventral rami of C5 to C8 and T1 spinal nerves [FIGURES 13.14 and 13.15]. The plexus extends downward and laterally, passing behind the clavicle and into the armpit. In its course downward, the brachial plexus consists of branches and recombinations that are called, in order, roots, trunks, divisions, cords, and main branches.

The brachial plexus gives rise to a number of nerves to the upper limb. Five major nerves constitute the terminal branches of the cords. These nerves—the axillary, ulnar, median, radial, and musculocutaneous—innervate the shoulder, arm, forearm, and hand.

Lumbar plexus The *lumbar plexus* is composed of fibers of the ventral rami of L1 to L4 nerves [FIGURE 13.14]. It supplies the anterior and lateral abdominal wall, external genitals, and the thigh. Two major nerves,

the femoral and obturator, and some lesser nerves (genitofemoral, ilioinguinal, iliohypogastric) are derived from the recombining branches.

The *femoral nerve* is composed of fibers from L2, L3, and L4. It innervates the muscles that flex the hip joint and extend the knee joint (knee jerk). Inflammation of the femoral nerve leads to lumbago (pain in the lumbar region). The *obturator nerve*, also composed of fibers from L2, L3, and L4, innervates the adductor muscles of the thigh, as well as the skin.

Sacral plexus The *sacral plexus* is composed of fibers of the ventral rami of spinal nerves L4, L5, and S1 through S3 [FIGURE 13.14]. It passes in front of the sacrum and into the regions of the buttocks. Several important nerves are derived from the plexus. The *sciatic nerve*, the thickest and longest nerve in the body, extends from the pelvic area to the foot, dividing above the knee into the *common peroneal* and *tibial nerves*. Together, those nerves innervate the thigh, leg, and foot muscles. Inflammation of the sciatic nerve, or *sciatica*, is described in When Things Go Wrong at the end of the chapter.

The *superior gluteal nerve* innervates the gluteus medius, gluteal minimus, and tensor fascia lata muscles, and the *inferior gluteal nerve* innervates the gluteus maximus muscle. The *pudendal nerve* innervates the voluntary muscles of the perineum, especially the sphincters of the urethra and anus. It is the pudendal nerve among others, that is blocked by local anesthesia in the "saddle block" procedure sometimes used during childbirth.

Coccygeal plexus The *coccygeal plexus* is formed by the coccygeal nerve (Co1) and sacral nerves S4 and S5. A few fine nerve filaments supply the skin in the coccyx region.

Intercostal Nerves

The *intercostal nerves* are the ventral rami of the second through twelfth thoracic nerves (T2 to T12). The intercostal nerves innervate muscles and skin in the thoracic and abdominal walls. After these nerves leave the intervertebral foramina, they take a course parallel to the ribs, but continue past them. Intercostal nerves T2 through

Q: *What is the "funny bone"?*

A: The "funny bone" is actually the exposed part of the ulnar nerve, which extends the entire length of the upper limb into the hand. The nerve is well protected by tissue everywhere but at the elbow, just behind the medial epicondyle of the humerus. When you hit your "funny bone," you are actually stimulating your ulnar nerve, so you feel a strange twinge. (Because the bone of the upper arm is the humerus, could the term "funny bone" have come from a play on words—humorous for humerus?)

FIGURE 13.14 SPINAL PLEXUSES AND SOME SPINAL NERVES

Anterior view.

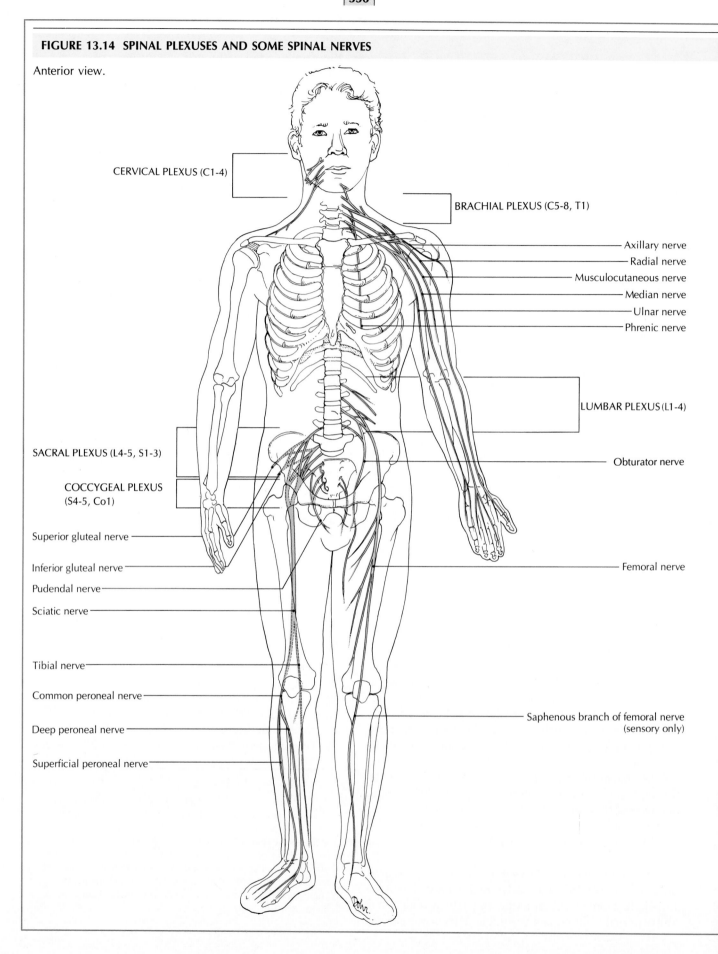

CERVICAL PLEXUS (C1-4)

BRACHIAL PLEXUS (C5-8, T1)

—— Axillary nerve
—— Radial nerve
—— Musculocutaneous nerve
—— Median nerve
—— Ulnar nerve
—— Phrenic nerve

LUMBAR PLEXUS (L1-4)

SACRAL PLEXUS (L4-5, S1-3)

COCCYGEAL PLEXUS
(S4-5, Co1)

—— Obturator nerve

Superior gluteal nerve ——

Inferior gluteal nerve ——

Pudendal nerve ——

Sciatic nerve ——

—— Femoral nerve

Tibial nerve ——

Common peroneal nerve ——

Deep peroneal nerve ——

—— Saphenous branch of femoral nerve
(sensory only)

Superficial peroneal nerve ——

Plexus	Components	Location	Major nerve branches	Regions innervated	Result of damage to specific nerves
Cervical	Ventral rami of C1 to C4 nerves	Neck region; origin covered by sterno-cleidomastoid	Cutaneous, muscular, communicating, phrenic, ansa cervicalis	Skin and some muscles of back of head, neck; diaphragm.	*Phrenic:* paralysis of diaphragm (respiratory muscles).
Brachial	Ventral rami of C5 to C8 and T1 nerves	Lower neck, axilla	Axillary, ulnar, median, radial, musculocutaneous	Muscles and skin of neck, shoulder, arm, forearm, wrist, hand.	*Axillary:* impaired ability to abduct and rotate arm. *Ulnar:* "clawhand" (fingers partially flexed). *Median:* "ape hand" (thumb adducted against index finger); impaired ability to oppose thumb; carpal tunnel syndrome. *Radial:* "wristdrop" (inability to extend hand at wrist).
Lumbar*	Ventral rami of L1 to L4 nerves	Interior of posterior abdominal wall	Femoral, obturator	Muscles, skin of abdominal wall.	*Femoral:* inability to extend leg; marked weakness in flexing hip; inflammation leads to lumbago (back pain).
Sacral*	Ventral rami of L4, L5, and S1 to S3 nerves	Posterior pelvic wall	Superior gluteal, inferior gluteal; sciatic nerve branches:		*Superior gluteal:* walk with lurching gait. *Inferior gluteal:* difficulty walking up stairs. *Sciatic:* severely impaired ability to extend hip, flex knee; impaired ability to plantar flex foot.
			Tibial	Buttocks, medial thigh, muscles and skin of posterior thigh, posterior leg, plantar (sole) of foot.	
			Common peroneal	Muscles and skin of lateral posterior thigh, anterior leg, dorsal foot.	*Common peroneal:* "footdrop" (inability to dorsiflex foot, toes).
			Pudendal	Voluntary sphincters of urethra, anus.	*Pudendal:* incontinence in urination, defecation.
Coccygeal	Co1 nerve, plus communications from S4 and S5 nerves	Coccyx region	A few fine filaments	Skin in coccyx region.	Possibly some loss of sensation in coccygeal region.

*Both are part of the lumbosacral plexus.

FIGURE 13.15 PHOTOGRAPH OF DISSECTED RIGHT BRACHIAL PLEXUS

All blood vessels have been removed, and some cords have been displaced to reveal the branches in this anterior view.

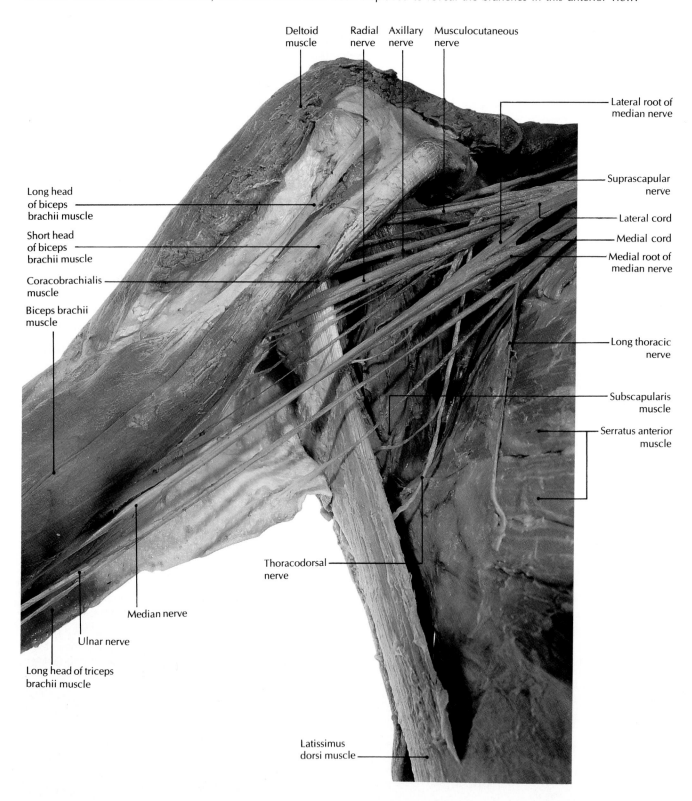

T6 innervate intercostal muscles, plus the skin on the lateral and anterior thoracic walls. Nerves T7 through T12 innervate the intercostal muscles, plus the abdominal wall and its overlying skin.

Dermatomes

The dorsal root of each spinal nerve is distributed to a specific region or segment of the body and supplies the sensory innervation to a segment of the skin known as a **dermatome** (Gr. *derma*, skin + *tomos*, cutting). There are 30 dermatomes, one for each spinal nerve, except C1, which does not innervate the skin [FIGURE 13.16]. The face and scalp in front of the ears are innervated by a cranial nerve called the trigeminal, whose three divisions (ophthalmic, maxillary, and mandibular nerves) each innervate a separate region.

When the function of even a single dorsal nerve root is interrupted, there is a faint but definite decrease of sensitivity in the dermatome. The area of diminished sensitivity is detected by using a light pin scratch to stimulate a pain sensation. Using this method, a map of the dermatomes can be used in locating the sites of injury to dorsal roots of the spinal cord.

The sensory innervations of adjacent dermatomes overlap in such a way that if a spinal nerve is cut, the loss of sensation in its dermatome is minimal. However, if one dorsal root is irritated, as in shingles (described in When Things Go Wrong), the resultant pain does spread to adjacent, overlapping dermatomes.

A S K   Y O U R S E L F

1 How are spinal nerves named?

2 What are the epineurium, perineurium, and endoneurium?

3 What is a ramus? A plexus?

4 How do the thoracic nerves differ from the other spinal nerves?

5 What nerves make up the cervical plexus? The brachial plexus? The lumbar and sacral plexuses?

6 What is a dermatome?

Q: *Why does your foot sometimes "fall asleep"?*

A: When you cross your legs or otherwise put pressure on one of your legs, you may press down on a nerve and temporarily cut off the feeling in your foot. When the local pressure is removed, you may feel a numbness we often call "pins and needles" as the nerve endings become reactivated. The same thing may happen when a blood vessel that supplies a peripheral nerve is temporarily forced shut. In that case, the "pins and needles" feeling occurs when the many tiny capillaries in the foot restore the nerve's blood supply.

DEVELOPMENTAL ANATOMY OF THE SPINAL CORD

The embryonic development of the nervous system begins with the formation of the *notochord*, a rod that defines the longitudinal axis of the embryo. As the notochord continues to develop, the embryonic ectoderm thickens to form the *neural plate*. On about the nineteenth day, the neural plate folds in on itself to form a *neural groove* with *neural folds* on each side [FIGURE 13.17A]. By day 21, the neural folds have fused, forming the *neural tube* [FIGURE 13.17B]. The cephalic end of the tube develops into the brain, and the caudal portion becomes the spinal cord. At the site of the closure forming the neural tube, there develops a paired segmented series of outgrowths called the **neural crest.** It contains the cells that form all the future neurons of the spinal (dorsal root) ganglia, satellite cells, neurolemmocytes, meninges, some cartilage cells, and multipolar (postganglionic) neurons of the autonomic ganglia [FIGURE 13.17D].

The spinal cord portion of the neural tube differentiates into three layers [FIGURE 13.17C]: (1) The inner *matrix (ependymal) layer*, which lines the central canal of the cord, contains the cells that develop into all neurons and macroglial (astroglia and oligodendroglia) cells of the spinal cord. All divisions of these cells occur within the matrix layer. Once a potential neuron or macroglial cell leaves the matrix layer it migrates to its final destination and does not divide. (2) The middle *mantle layer* develops into the butterfly-shaped gray matter, which contains the cell bodies of neurons, including their processes, and macroglia. (3) The outer *marginal layer* develops into the white matter of the spinal cord, which is composed of macroglial cells and myelinated and unmyelinated axons. The axons of the white matter ascend to other spinal levels and to the brain, and descend from the brain and to other levels of the spinal cord.

As shown in cross-sectional FIGURE 13.17C, the thickened inner side walls of the neural tube form the *lateral plates,* and the thin dorsal and ventral walls form the *roof plate* and *floor plate,* respectively. These plates are demarcated from each other by a groove, or sulcus, called the *sulcus limitans.* The sulcus divides the lateral plates into two portions: (1) a dorsal afferent (sensory) *alar plate,* whose neurons are involved with reflexes that convey signals to the brain, and (2) a ventral efferent (motor) *basal plate,* whose neurons are involved with reflexes that convey signals from the brain.

At the outset, the development of the spinal cord and vertebrae proceed at an even pace, and the spinal cord extends throughout the entire length of the vertebral column. But after the third fetal month the vertebral column continues to elongate faster than the spinal cord, and thus the spinal cord becomes relatively shorter than

FIGURE 13.16 DERMATOMES, WITH THEIR SPINAL NERVES

[A] Anterior view. [B] Posterior view.

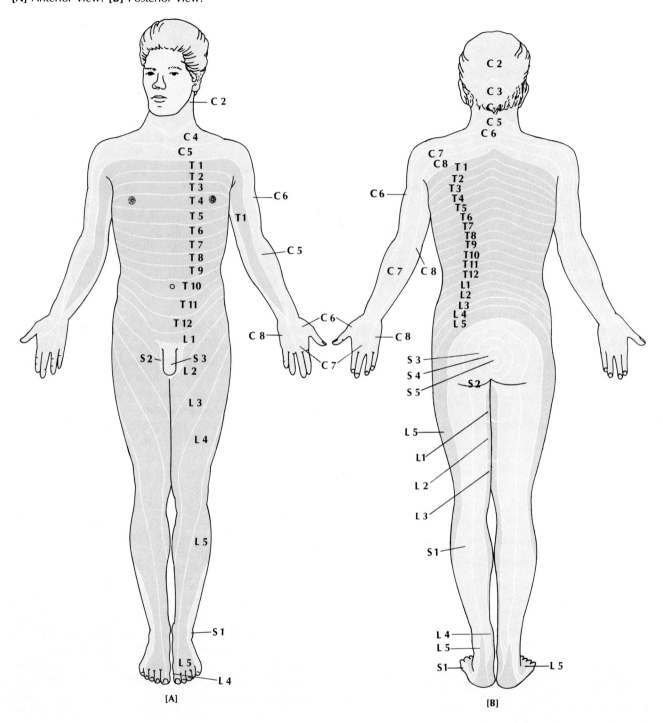

[A]   [B]

FIGURE 13.17 DEVELOPMENT OF THE SPINAL CORD

Transverse sections at [A] 19 days after fertilization, [B] 21 days, [C] 26 days, [D] 30 days.

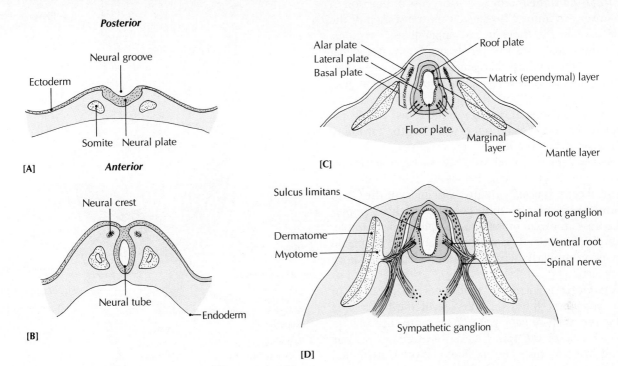

the vertebral column. The spinal cord of a newborn child ends at about the third lumbar vertebra, and extends to between the first and second lumbar vertebrae during adolescence and adulthood. As a result of the differing developmental rates of the spinal cord and vertebral column, the spinal nerves do not align with the intervertebral spaces they pass through, with several of the spinal nerves projecting downward.

WHEN THINGS GO WRONG

Injury and disease can severely impair the functioning of the spinal cord and spinal nerves. Only a few examples are given here.

Spinal Cord Injury

Spinal cord injury is any lesion of the spinal cord that bruises, cuts, or otherwise damages the neurons of the cord. Each year in the United States there are over 10,000 such injuries, most of them the result of motor vehicle accidents.

Transection, or severing, of the spinal cord may be complete or incomplete. Complete transections produce total paralysis of all skeletal muscles, bowel or bladder incontinence, loss of reflex activity (*areflexia*), and a total loss of sensation below the level of the injury. Injuries or diseases causing incomplete transections produce partial loss of voluntary movements and sensations below the level of the injury.

Immediately following a spinal cord injury, *spinal shock* occurs, lasting for several hours to several weeks. Spinal shock involves paralysis, areflexia, and loss of sensation above the level of the injury. In the case of partial transection, spinal shock is followed by a period of spasticity, exaggerated spinal reflexes, and a decreased sensitivity to pain and temperature. Babinski's reflex may also appear.

People with incomplete lesions may recover partially, but until recently there was no hope of any restoration of motor and sensory functions with complete transections. Researchers are presently experimenting with nerve grafts, nerve regeneration, enzymes, hormones, steroids, growth-associated proteins, and various drugs that offer some hope to an estimated 500,000 American victims of spinal cord injury.

Paraplegia is the motor or sensory loss of function in both lower extremities. It results from transection of the spinal cord in the thoracic and upper lumbar regions. When the spinal cord is severed completely at a spinal level below the cervical enlargement and above the upper lumbosacral enlargement, paralysis below the lesion generally occurs immediately afterward, impairing excretory and sexual functions. If damage to the spinal cord is incomplete, however, some sensory and motor capability will remain below the lesion.

Quadriplegia is a paralysis of all four extremities, as well as any part of the body below the level of injury to the spinal cord. It usually results from injury at the C8 to T1 level. Quadriplegia is more complicated than paraplegia because it affects other body systems. For example, the cardiovascular and respiratory systems may be unable to function properly because of insufficient respiratory muscle action.

Hemiplegia is an upper motor neuron paralysis of upper and lower limbs on one side of the body. It is usually the result of damage to upper motor neurons on only one side of the spinal cord above C5, or serious brain damage on the opposite side. It is often attributed to a lesion of the corticospinal tract.

Carpal Tunnel Syndrome

Carpal tunnel syndrome is the most common syndrome in which a peripheral nerve is entrapped. It results from the compression of the median nerve at the wrist, as the nerve passes through the *carpal tunnel* (see drawing). The carpal tunnel also provides passage from the forearm to the hand for the eight flexor tendons to the fingers (4 tendons of flexor digitorum superficialis, 4 of flexor digitorum profunda) and one tendon to the thumb (tendon of flexor pollicis longus). The carpal tunnel is formed by a concave arch on the palmar side of the carpal bones, which is covered by a ligament called the flexor retinaculum (transverse carpal ligament). Carpal tunnel syndrome is usually associated with arthritis of the wrist, dislocation of the lunate carpal bone, or inflammation of the tendon sheaths of the flexor tendons. Compression on the median nerve produces a tingling and loss of some sensation in the palmar surfaces of the thumb, and index and middle fingers, weakness, and pain that may worsen at night. Fine, coordinated movements of the hand may suffer, and the affected person is usually unable to make a fist. This condition is most common in people who use their fingers excessively in rigorous work, and is often found among operators of computer terminals. Treatment ranges from rest and corticosteroid injections to severing the flexor retinaculum at the wrist.

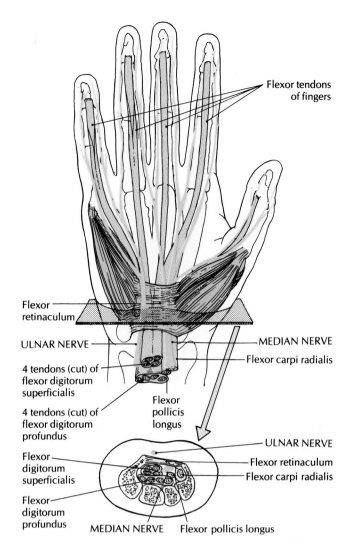

Carpal tunnel syndrome, as seen in this palmar view (bottom) and cross section (top) of the right hand.

Poliomyelitis

Poliomyelitis is a contagious viral infection that affects both the brain and spinal cord, and sometimes causes the destruction of neurons. The poliomyelitis virus shows a preference for infiltrating the lower motor neurons of the spinal cord and brainstem. The initial symptoms may be sore throat and fever, diarrhea, or painful back and limbs. In cases of *nonparalytic polio*, these symptoms disappear in less than a week. When motor neurons in the spinal cord are damaged, there is obvious lower motor paralysis of muscles within a few days. Paralysis may be limited to the limbs (especially the lower limbs), or it may also affect the muscles used for breathing and swallowing. The mortality rate for all types of poliomyelitis is 5 to 10 percent.

Treatment is supportive rather than curative. Medication is ineffective against the polio virus except as a preventive measure. The Salk vaccine, which became available in 1955, virtually eliminated the disease among those immunized, and the Sabin oral vaccine has been shown to be even more effective and easier to use.

Sciatica

Sciatica (sye-AT-ih-kuh) is a form of *neuritis* (nerve inflammation) characterized by sharp pains along the sciatic nerve and its branches. Pain usually extends from the buttocks to the hip, back, and posterior thigh, leg, ankle, and foot. One of the most common causes is pressure from a herniated intervertebral disk on a dorsal or ventral root of the sciatic nerve or one of its branching nerves, the tibial or common peroneal. Sciatic pain may also be caused by a tumor, by inflammation of the nerve or its sheath, or by disease in an adjacent area such as the sacroiliac joint.

The sciatic nerve may be injured by an intramuscular injection into the buttocks. The safe area for injection is the upper, outside quadrant (see drawing). An injection into the apparently safe area may pierce a nerve if the patient is standing up instead of lying face down.

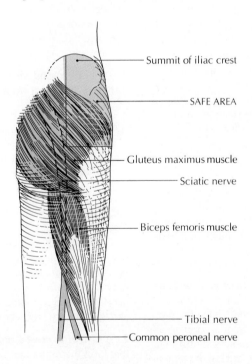

Summit of iliac crest

SAFE AREA

Gluteus maximus muscle

Sciatic nerve

Biceps femoris muscle

Tibial nerve

Common peroneal nerve

The safe area for intramuscular injection is indicated in green, and the area where the sciatic nerve is vulnerable is shown in red.

Shingles

Shingles, or *herpes zoster*, is an acute inflammation of the dorsal root ganglia. Shingles occurs when the virus that caused childhood chickenpox lies dormant in the ganglia of cranial nerves or the ganglia of posterior nerve roots, and then becomes reactivated and attacks the root ganglia.

Shingles begins with fever and a general feeling of illness, followed in two to four days by severe deep pain in the trunk, and occasionally in the arms and legs. A rash typically appears in a dermatome around one side of the chest, trunk, or abdomen, and the pain and rash always progress along the course of one or more spinal nerves (usually intercostal nerves) beneath the skin. Sometimes the ganglion of the trigeminal cranial nerve is affected, producing pain in the eyeball and a rash from the eyelid to the hairline. The rash usually disappears within two or three weeks, but some pain may persist for months. Shingles seldom recurs.

If the disease is diagnosed within 48 hours after its onset, it can usually be cured with the antiviral drug Zoverax.

Spinal Meningitis

Spinal meningitis is an inflammation of the spinal meninges, especially the arachnoid and pia mater, which increases the amount of cerebrospinal fluid and alters its composition. The cause may be either viral or bacterial. Meningitis may also follow a penetrating wound or an infection in another area. Meninges infected by bacteria produce large amounts of pus, which infiltrates the cerebrospinal fluid in the subarachnoid space between the arachnoid and pia mater. Pus is not usually formed when the infection is viral.

The first sign of meningitis is usually a headache, accompanied by fever, chills, and vomiting. Also evident may be rigidity in the neck region, abnormal reflexes, exaggerated deep tendon reflexes, and back spasms that cause the body to arch upward. Coma may develop. Diagnosis is usually made by analyzing cerebrospinal fluid withdrawn by a lumbar puncture. The fluid is examined for its content of protein, sugar, and bacteria. The chances of recovery are excellent if the disease is diagnosed early, and if the infecting microorganisms respond to antibiotics. The recovery rate is less encouraging with infants and the elderly, and most untreated victims of any age will die.

CHAPTER SUMMARY

The spinal cord is the connecting link between the brain and most of the body. It also controls many reflex actions. Thus, many voluntary and involuntary actions depend on it.

Basic Anatomy of the Spinal Cord (p. 318)

1 The **spinal cord** extends caudally from the brain for about 45 cm. Its upper end is continuous with the brain.

2 There are 31 pairs of spinal nerves and roots: 8 cervical, 12 thoracic, 5 lumbar, 5 sacral, and 1 coccygeal. The lumbar and sacral roots are called the **cauda equina.**

3 The spinal nerves emerging from the **cervical enlargement** of the cord innervate the upper limbs, and the nerves emerging from the **lumbosacral enlargement** innervate the lower limbs.

4 The spinal cord and spinal nerve roots are protected by the bony vertebral column and its ligaments, the triple-layered **spinal meninges** (inner pia mater, arachnoid, and outer dura mater), and the cerebrospinal fluid.

5 The **gray matter** of the spinal cord consists primarily of nerve cells. Three pairs of nerve cell columns in the gray matter are the **posterior horns, anterior horns,** and a median connecting column that forms the anterior and posterior **gray commissures.** In the T1 to L2 region, there are also two small **lateral horns.**

6 Each spinal nerve emerges from the cord as ventral rootlets that form a **ventral root,** and as dorsal rootlets that form a **dorsal root.** The cell bodies of neurons whose axons comprise the dorsal roots are located in **dorsal root ganglia.**

7 The **white matter** of the cord is composed mainly of bundles of myelinated nerve fibers. These bundles are **ascending** (sensory) **tracts** and **descending** (motor) **tracts.** The white matter is divided into longitudinal columns called **funiculi.**

Functional Roles of Pathways of the Central Nervous System (p. 324)

1 The brain influences the activity of skeletal muscles through descending **upper motor neurons** from the brain that regulate the activity of **lower motor neurons** of the peripheral nervous system.

2 Each pathway of the central nervous system can be viewed as consisting of sequences of **processing centers,** each center having terminal branches of axons entering the center, cell bodies and dendrites of neurons whose axons form the tract, and intrinsic neurons whose dendrites, cell bodies, and axons are located within the center.

3 Sensory pathways may contain **first-order, second-order,** and **third-order neurons** in their sequences.

4 The **anterolateral system** consists of the lateral spinothalamic tract, spinoreticulothalamic pathway, and anterior spinothalamic tract. The **posterior column-medial lemniscus pathway** is involved with touch-pressure, vibratory sense, two-point discrimination, position sense, and some light touch.

Spinal Reflexes (p. 329)

1 The spinal cord is involved with spinal reflex actions. A **reflex** is a predictable, involuntary response to a stimulus that enables the body to adapt quickly to environmental changes. The pathway consisting of a sensory cell, an effector cell, and usually one or more connecting nerve cells is a **reflex arc.**

2 The **stretch (myotatic) reflex** is a two-neuron, ipsilateral reflex. An example is the knee-jerk reflex. It consists of a neuromuscular spindle, an afferent neuron, an alpha motor neuron, and a voluntary muscle. It plays a role in maintaining body position and is integrated with other spinal reflexes in normal voluntary motor activities.

3 A **gamma motor neuron reflex arc** consists of gamma motor neurons, a neuromuscular spindle, afferent neurons, an alpha motor neuron, and a voluntary muscle. The gamma motor neuron arc acts to smooth out the movement of muscle contractions or to maintain the contraction when an object is being held or lifted.

4 The **withdrawal flexor reflexes** are protective, escape reflexes. The flexor reflex circuit is composed of sensory receptors, afferent neurons, spinal interneurons, alpha motor neurons, and voluntary muscles.

5 Reflexes such as the patellar reflex (knee jerk) and Babinski's reflex are used for diagnostic purposes.

Structure and Distribution of Spinal Nerves (p. 332)

1 At each segment of the spinal cord a pair of nerves is distributed to each side of the body. Each nerve has a **ventral** (anterior) **root** and a **dorsal** (posterior) **root.** Motor fibers emerge from the ventral root, and sensory fibers emerge as the dorsal root. Both roots meet to form a single **mixed nerve.**

2 Each nerve is made up of nerve fibers enclosed in bundles of connective tissue called **fascicles.** Several fascicles are held together by a sheath called the **epineurium,** each fascicle is encased by the **perineurium,** and each nerve fiber is covered by the **endoneurium.**

3 After a spinal nerve leaves the vertebral column, it divides into initial branches called **rami.** The ventral rami (except T2 to T12 nerves) are arranged to form networks of nerves called the cervical, brachial, lumbar and sacral (lumbosacral), and coccygeal **plexuses.**

4 The **intercostal nerves** are the T2 through T12 spinal nerves.

5 The dorsal root of each spinal nerve arises from a specific region or segment of the body and supplies sensory innervation to a segment of the skin called a **dermatome.**

Developmental Anatomy of the Spinal Cord (p. 339)

1 The embryonic development of the nervous system begins with the formation of the **notochord.** By day 21, the **neural tube** forms, with its cephalic and caudal ends developing into the brain and spinal cord, respectively.

2 The **neural crest** contains the cells that form the future neurons of all spinal ganglia, satellite and neurilemma cells of the peripheral nerves, and neurons of autonomic ganglia.

3 The spinal cord portion of the neural tube differentiates into the inner **matrix** (ependymal) **layer,** the middle **mantle layer,** and the outer **marginal layer.**

4 At about month 3, the vertebral column becomes longer than the spinal cord, causing the spinal nerves to be misaligned with the intervertebral spaces through which they pass.

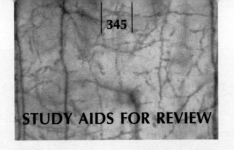
STUDY AIDS FOR REVIEW

MEDICAL TERMINOLOGY

ATAXIA (Gr. "not ordered") Loss of motor coordination due to disease of the nervous system or certain genetic disorders.

CORDOTOMY (Gr. *tomos*, a cut) Cutting into a nerve fiber tract (usually spinothalamic) of the spinal cord, usually to relieve pain.

CRYONEUROSURGERY (Gr. *kryos*, icy cold) The surgical destruction of nerve tissue by exposure to extreme cold.

EPIDURAL ANESTHESIA (caudal block) An injection of anesthesia into the epidural space outside the dura mater, which anesthetizes the nerves of the cauda equina.

FRIEDREICH'S ATAXIA Degenerative genetic disorder of the cerebellum and dorsal columns of the spinal cord.

HEMATOMYELIA Hemorrhaging within or upon the spinal cord.

MYELITIS (Gr. *myelo*, marrow, spinal cord) A general term for inflammation of the spinal cord. Various forms of myelitis range from poliomyelitis and leukomyelitis, which produce some motor and sensory dysfunction, respectively, to acute transverse myelitis, which affects the entire thickness of the cord and usually produces rapid degeneration.

MYELOGRAM X-ray of the spinal cord, which is obtained by injecting contrast fluid into the subarachnoid space surrounding the cord, a technique called *myelography*.

NERVE BLOCK An injection of anesthesia near a nerve that supplies the area to be treated.

NEURALGIA (*algia* = pain) Pain along the length of a nerve.

NEURECTOMY The removal of part of a nerve.

NEURITIS Inflammation of a nerve.

NEUROLYSIS (Gr. *lys*, loosening) Relief of tension upon nerve caused by adhesions.

NEUROTRIPSY (Gr. *trips*, friction) The surgical crushing of a nerve.

PALSY (L. paralysis, disabled) A condition marked by some paralysis, weakness, and loss of muscle coordination.

POSTLATERAL SCLEROSIS Degeneration of white matter in dorsal and lateral funiculi, caused by insufficient vitamin B_{12}.

REFLEX TESTING Stimulation of a nerve to determine if the appropriate reflex is operative.

SPINA BIFIDA (L. *bi*, two + *fidus*, split) Certain congenital defects producing the absence of a neural arch, with varying degrees of protrusion of a portion of the spinal cord or meninges.

SPINAL ANESTHESIA The injection of anesthesia into the epidural space to block the nerves below and numb the lower part of the body.

UNDERSTANDING THE FACTS

1 What are the major functions of the spinal cord?
2 Describe the basic anatomy of the spinal cord.
3 What is the filum terminale?
4 At what level does the adult spinal cord terminate?
5 What are the three layers of spinal meninges?
6 Each spinal nerve is composed of two roots. Name them and give their functions.
7 Name the major ascending and descending spinal tracts. Where are they located?
8 List in sequence the components of a simple reflex arc.
9 Define a mixed nerve.
10 Distinguish between epineurium, perineurium, and endoneurium.
11 Identify the four rami of a spinal nerve, and give their functions.
12 What two nerves are divisions of the sciatic nerve?
13 When a spinal nerve is cut, why may the sensation in its dermatome still be felt?
14 Give the location of the following (be specific):
   a conus terminalis
   b fasciculi
   c arachnoid
   d longest nerve in the body
   e "funny bone"

UNDERSTANDING THE CONCEPTS

1 What is the significance of the subarachnoid space?
2 How do the stretch (myotatic) reflex and the flexor reflexes help the body to maintain homeostasis?
3 Describe how a processing center in the spinal cord functions.
4 What general regions are innervated by the cervical, brachial, lumbar, and sacral plexuses?
5 How do the following disorders affect the nervous system?
   a poliomyelitis
   b sciatica
   c shingles
   d spinal meningitis

SELF-QUIZ
Multiple-Choice Questions

13.1 The major functions of the spinal cord include
   a linking the brain with most of the body
   b its involvement in spinal reflex actions
   c its autonomic motor functions
   d a and b
   e a, b, and c

13.2 The spinal cord and the roots of its nerves are protected by
  a the flexible vertebral column
  b the ligamentum flavum and the posterior longitudinal ligament
  c the spinal meninges
  d cerebrospinal fluid
  e all of the above

13.3 The gray matter of the spinal cord consists of
  a nerve cell bodies and dendrites
  b unmyelinated axons of spinal neurons
  c sensory and motor neurons
  d axon terminals of neurons
  e all of the above

13.4 The spinal cord funiculi consist of the
  a anterior (ventral) column **d** commissure area
  b posterior (dorsal) column **e** all of the above
  c lateral column

13.5 Processing centers of the central nervous system consist of
  a the terminal branches of axons entering the center
  b cell bodies and dendrites of neurons whose axons form the tract
  c intrinsic neurons whose dendrites, cell bodies, and axons are within the center
  d a and b
  e a, b, and c

13.6 The lateral spinothalamic tract
  a originates in the gray matter of the spinal cord
  b terminates in the thalamus
  c is located laterally in the spinal cord
  d is a tract in the pain pathway
  e all of the above

13.7 The reflex that helps to support the body against gravity is the
  a flexor reflex   **d** monosynaptic reflex
  b extensor reflex **e** crossed extensor reflex
  c bireflex

13.8 Which of the following is not a spinal cord plexus?
  a cervical   **d** lumbosacral
  b thoracic   **e** coccygeal
  c brachial

13.9 The spinothalamic tract conducts impulses
  a from the thalamus to the cerebral cortex
  b down the spinal cord from the thalamus
  c that stimulate movements in skeletal muscles
  d from the cerebrum
  e up the spinal cord to the thalamus

True-False Statements

13.10 The middle meninx is the pia mater.

13.11 The subdural space is between the dura mater and the periosteum of the vertebrae.

13.12 Anesthetics can be injected into the epidural space below the S2 vertebral level.

13.13 The fibrous band of pia mater that extends the entire length of both sides of the spinal cord is the denticulate ligament.

13.14 A first-order neuron extends from the thalamus to a sensory area of the cerebral cortex.

13.15 The patellar reflex is a good example of a stretch extensor reflex.

13.16 Alpha motor neurons are among the smallest of spinal neurons.

Completion Exercises

13.17 The neural terminal end of the spinal cord is called the _____.

13.18 The nonneural fibers at the end of the spinal cord are called the _____.

13.19 Groups of cell bodies whose axons make up the dorsal roots lie outside the spinal cord and are called _____.

13.20 In the spinal cord, the white matter is divided into three pairs of columns, or _____.

13.21 In the central nervous system, the crossing over of tracts from one side of the spinal cord or brainstem to the other side is called _____.

13.22 A _____ is a predictable involuntary response to a stimulus.

13.23 The numbering of each cervical nerve corresponds to the vertebra _____ its exit.

13.24 The ventral rami of the spinal nerves are arranged to form several networks of nerves called _____.

13.25 The dorsal root of each spinal nerve supplies sensory innervation to a segment of the skin known as a _____.

13.26 The _____ nerves are the T2 to T12 spinal nerves.

A SECOND LOOK

Label the following: dorsal root, ventral root, gray matter, white matter, spinal ganglion, and spinal cord.

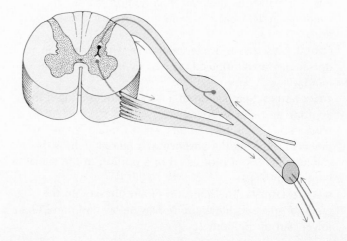

14

The Brain and Cranial Nerves

LEARNING OBJECTIVES

1 Describe the major divisions and subdivisions of the brain.

2 Compare the meninges of the brain with those of the spinal cord.

3 Explain the ventricular system and the circulation of cerebrospinal fluid in the brain.

4 Describe the nutritional needs of the brain and the blood-brain barrier.

5 Describe the three main anatomical divisions of the brainstem and their functions.

6 Explain the roles of the reticular formation and its neural pathways.

7 Describe the structure and functions of the cerebellum.

8 Describe the structure and functions of the cerebrum.

9 Describe the location and roles of the six cerebral lobes.

10 Explain the function of the limbic system.

11 Describe the major divisions of the diencephalon and their functions.

12 Name the 12 pairs of cranial nerves, and describe their locations and functions.

13 Define senility, Alzheimer's disease, cerebral palsy, cerebrosvascular accident, epilepsy, Parkinson's disease, dyslexia, encephalitis, and Bell's palsy and other disorders of cranial nerves.

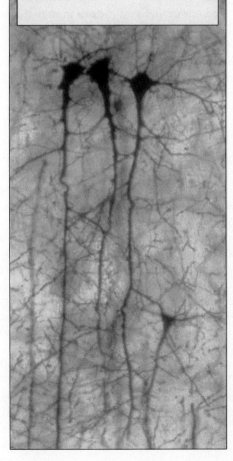

The human brain weighs only about 1400 g (3 lb), and yet it contains approximately 100 billion neurons, about the same number as stars in the Milky Way.* In addition, each neuron may have from 1000 to 10,000 synaptic connections with other nerve cells. There may be as many as 100 *trillion* synapses in the brain. It is tempting to compare the human brain with a computer. But there is really no comparison. Nothing we know can match the exquisite complexity of the brain.

Your brain does much more than help you to think and make decisions. It is your body's main key to regulating body processes, from cellular metabolism to the overall functioning of organs and systems. Your nerves and their specialized receptors may *receive* the stimuli of sound, touch, vision, smell, and taste, but you actually *experience* the sensation in your brain. Your brain is working even when you are sleeping to activate, coordinate, and regulate the body's many functions and their relationships to the outside world.

GENERAL STRUCTURE OF THE BRAIN

The brain is technically called the *encephalon* (en-SEFF-uh-lon; Gr. *en*, in + *kephale*, head). It has four major divisions: *brainstem, cerebellum, cerebrum,* and *diencephalon* [FIGURE 14.1]. The major divisions and structures of the brain are summarized in TABLE 14.1 (pages 354–355).

The **brainstem** is composed of the midbrain, pons, and medulla oblongata. The **cerebellum** is the coordinating center for skeletal-muscle movement. It is located posterior to the brainstem. The **diencephalon** is composed of the thalamus, hypothalamus, epithalamus, and ventral thalamus.

Probably the most obvious physical feature of the brain is the large pair of hemispheres that make up about 85 percent of the brain tissue. These are the two hemispheres of the **cerebrum.** Another obvious feature of the brain is the outer portion of the cerebrum, called the *cerebral cortex,* with its many folds or convolutions. Lying below the gray matter of the cerebral cortex are both white matter and deep, large masses of gray matter called the *basal ganglia.* Connecting the two cerebral hemispheres is a bundle of nerve fibers called the *corpus callosum,* which relays nerve impulses between the hemispheres. Emerging from the cerebrum are 2 of the 12 cranial nerves (I and II). The other 10 arise from the brainstem.

*To get some idea of the enormity of 100 billion, think of this: you would have to spend almost *$5.5 million a day* to spend $100 billion in 50 years. If you wanted to spend $100 billion in *one* year, you would have to spend $274 million *a day.* One hundred billion is a lot of neurons.

The brain is covered by the same three meninges that protect the spinal cord. Cerebrospinal fluid also flows through the brain from a series of cavities called *ventricles* into the subarachnoid spaces around the brain and spinal cord.

MENINGES, VENTRICLES, AND CEREBROSPINAL FLUID

The human brain is mostly water (about 75 percent in an adult). It has the consistency of gelatin, and if it were not supported, the brain would slump and sag. Fortunately, it has ample support. The brain is protected by the scalp, with its hair, skin, fat, and other tissues, and by the reinforced cranium, one of the strongest structures in the body. It also floats shockproof in cerebrospinal fluid, and is encased by three layers of cranial meninges.

Cranial Meninges

The dura mater, arachnoid, and pia mater are basically the same in the brain and spinal cord. However, there are some minor anatomical differences.

Dura mater The outermost cranial meninx is the **dura mater.** It consists of two fused layers: an inner dura mater that is continuous with the spinal dura mater, and an outer dura mater, which is actually the periosteal layer of the skull bones. The outer cranial dura mater is a tough, fibrous layer containing veins and arteries that nourish the bones. The inner dura mater extends into the fissure that divides the left and right hemispheres of the cerebrum (the *falx cerebri*) [FIGURE 14.2A] and reaches into the fissure between the cerebrum and cerebellum (*tentorium cerebelli).* By dividing the cranial cavity into three distinct compartments, the dura mater adds considerable support to the brain.

The fused inner and outer cranial dura mater is intimately attached to the bones of the cranial cavity. As a result, there is no epidural space between the membrane and the bones, as there is in the vertebral column. Between the inner dura mater and the arachnoid is a potential space called the **subdural space.** It does not contain cerebrospinal fluid. In certain locations, the two layers of dura mater separate to form channels [FIGURE 14.2A], which are lined with endothelium and contain venous blood. These spaces are the *dural sinuses of the dura mater,* which drain venous blood from the brain.

Arachnoid The middle layer of the meninges, between the dura mater and pia mater, is the **arachnoid,** a delicate connective tissue. Between the arachnoid and the pia mater is a network of trabeculae and the *subarachnoid*

FIGURE 14.1 MAJOR STRUCTURES OF THE BRAIN

[A] Right lateral view of the external surface of the brain, showing the cerebellum and four of the six cerebral lobes. **[B]** Right sagittal section. **[C]** Right sagittal section showing the major subdivisions of the brain and its connection to the spinal cord.

THE FOUR MAJOR DIVISIONS OF THE BRAIN	
BRAINSTEM	{ Midbrain / Pons / Medulla oblongata
CEREBELLUM	
CEREBRUM	
DIENCEPHALON	{ Thalamus / Hypothalamus / Epithalamus / Ventral thalamus

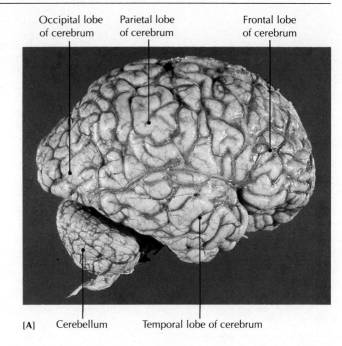

Occipital lobe of cerebrum Parietal lobe of cerebrum Frontal lobe of cerebrum

[A] Cerebellum Temporal lobe of cerebrum

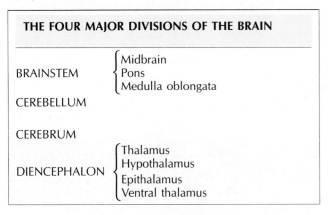

DIENCEPHALON
Thalamus Hypothalamus
CEREBRUM Corpus callosum

CEREBELLUM Medulla oblongata Pons Midbrain
BRAINSTEM

[B]

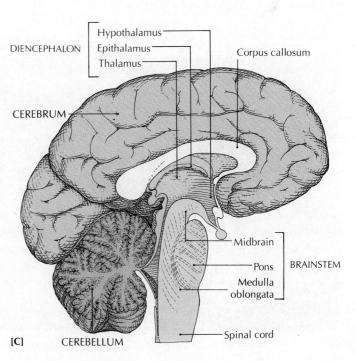

DIENCEPHALON { Hypothalamus / Epithalamus / Thalamus
Corpus callosum
CEREBRUM
Midbrain
Pons BRAINSTEM
Medulla oblongata
Spinal cord
[C] CEREBELLUM

space, which contains cerebrospinal fluid. The arachnoid contains no blood vessels of its own, but blood vessels are present in the subarachnoid space.

Pia mater The *pia mater* is the delicate innermost meningeal layer. It directly covers, and is attached to, the surface of the brain and dips down into the fissures between the raised ridges of the brain. Most of the blood to the brain is supplied by the large number of small blood vessels in the pia mater.

In head injuries, blood may flow from a severed blood vessel into the potential space between the skull and cranial dura mater (an extradural or epidural hemorrhage), into the potential subdural space (a subdural hemorrhage), into the subarachnoid space (a subarachnoid hemorrhage), or into the brain itself.

FIGURE 14.2 CRANIAL MENINGES

[A] Frontal section through the superior sagittal (dural) sinus. Note the supportive trabeculae in the subarachnoid space, the falx cerebri separating the cerebral hemispheres, and the triangular superior sagittal (dural) sinus. Most of the cerebrospinal fluid returns into the blood through the arachnoid villi. Note the detailed drawing. **[B]** Electron micrograph of cranial meninges. The space between the dura mater and arachnoid occurred during the processing of the specimen; it is normally only a potential space.

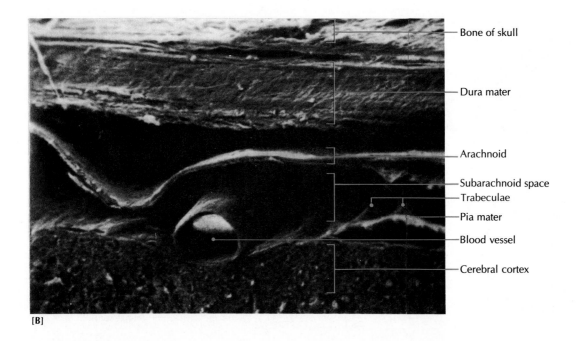

Ventricles of the Brain

Within the brain is a series of connected cavities called *ventricles* (L. little bellies). Each cranial ventricle is filled with cerebrospinal fluid and is lined by cuboidal or epithelial cells known as *ependyma*. A network of blood vessels called a *choroid plexus* is formed in several places where the ependyma contacts the pia mater. The four true ventricles are numbered from the top of the brain downward. They are the *left* and *right lateral ventricles* of the cerebral hemispheres, the *third ventricle* of the diencephalon, and the *fourth ventricle* of the pons and medulla oblongata [FIGURE 14.3].

Each lateral ventricle is connected to the third ventricle of the diencephalon through the small *interventricular foramen* (of Monro). The third ventricle is continuous with the fourth ventricle through a narrow channel called the *cerebral aqueduct* (of Sylvius) of the midbrain.

The Blood-Brain Barrier

Nowhere in the body is the need for homeostasis greater than in the brain, and nowhere else in the body is an organ so well protected against chemical imbalances. The mechanism for maintaining homeostasis in the brain is a special one, focusing on capillaries that supply blood to the brain. The structure of brain capillaries is different from capillaries in the rest of the body. The walls of other capillaries contain penetrable gaps that allow most substances to pass through. The walls of brain capillaries, however, are formed by endothelial cells that are joined by tight junctions (see drawing). The junctions actually merge the outer membrane layers of adjoining cells. The relatively solid walls of brain capillaries make up the ***blood-brain barrier.***

Because of the blood-brain barrier, all substances entering the brain from the blood must either diffuse through the endothelial cells or be conveyed via active transport through the plasma membranes rather than passing through spaces between cells. The blood-brain barrier helps to maintain the delicate homeostasis of neurons in the brain by restricting the entrance of potentially harmful substances from the blood, and by allowing essential nutrients to enter. The foot processes of astrocytes contact brain capillaries and neurons (see drawing). These astrocytes (1) store metabolites and transfer them from capillaries to neurons, (2) take up excess potassium ions from the extracellular fluid during intense neuronal activity, and (3) take up excess neurotransmitters.

Lipid-Soluble Molecules Pass

Because the plasma membranes of the endothelial cells are composed primarily of lipid molecules, lipid-soluble substances pass through the blood-brain barrier rather easily. Such substances include nicotine, caffeine, ethanol, and heroin. In contrast, water-soluble molecules such as sodium, potassium, and chloride ions are unable to cross the barrier without assistance. Molecules of essential water-soluble substances such as glucose (the brain's main energy source) and those amino acids that the brain cannot synthesize are "recognized" by carrier proteins and transported across the barrier. Some other substances, such as large protein molecules and most antibiotics cannot enter at all. For example, tetracycline crosses the barrier easily, but penicillin is admitted only in trace amounts.

Essential nutrients pass through the blood-brain barrier easily, assisted by carrier proteins that "recognize" the molecules of specific substances and bind to them before transporting them through the barrier. Transport systems not only carry essential substances through the barrier into the brain, they also remove surplus material. In this way, the transport systems contribute directly to the homeostatic regulation of the brain.

Some drugs pass through the blood-brain barrier easily. They include barbiturates; anesthetics such as sodium pentothal, ether, and nitrous oxide (laughing gas); carbon monoxide; cyanide; strychnine; hallucinogenic drugs such as LSD and mescaline; and alcohol in large quantities.

The Metabolic Component of the Blood-Brain Barrier

Recently, a "metabolic" component of the blood-brain barrier has been clarified. It complements the cellular blood-brain barrier by enzymatically altering substances that enter the endothelial cells of brain capillaries, converting them into a chemical form that cannot actually enter the brain. One such substance is L-dopa, a precursor of the important chemicals dopamine and norepinephrine.

Modifying the Barriers

How can scientists modify the barriers to allow antibiotics and other useful drugs to reach the brain in order to treat disorders such as Parkinson's disease and cancer? Most anticancer drugs fail to destroy brain tumors because insufficient amounts of the drugs are able to reach the tumors. One experimental approach aims to target specific brain tumors by linking anticancer drugs with lipid-soluble antibodies, immunologic molecules that can "recognize" and bond to cancer cells. When an antibody-drug complex attacks the tumor, the anticancer drug is also delivered at the precise tumor site. Another promising idea is to bond water-soluble drugs to lipid-soluble carrier molecules that can penetrate the blood-brain barrier. Once inside the brain tissue, the entire complex is modified enzymatically to become water-soluble, so that it is unable to escape from the brain tissue. At this point, the drug would be activated by enzymes in the brain by separating it from its carrier, allowing the drug to provide a sustained release.

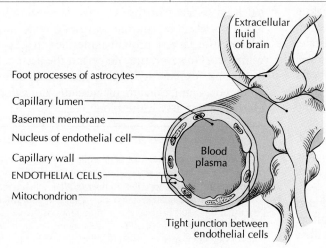

Foot processes of astrocytes

Capillary lumen

Basement membrane

Nucleus of endothelial cell

Capillary wall

ENDOTHELIAL CELLS

Mitochondrion

Extracellular fluid of brain

Blood plasma

Tight junction between endothelial cells

FIGURE 14.3 CIRCULATION OF CEREBROSPINAL FLUID

Right lateral view showing the ventricles of the brain and the flow of cerebrospinal fluid. The arrows indicate the direction of flow.

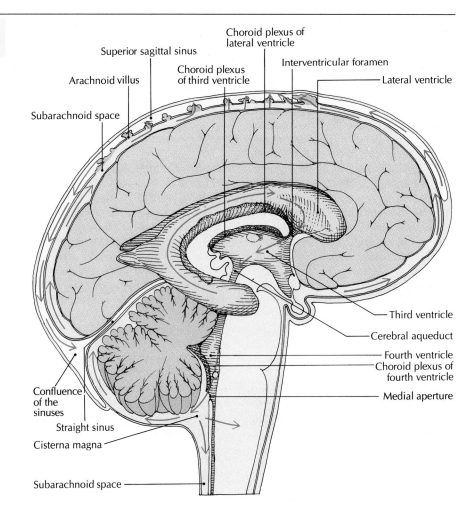

Superior sagittal sinus
Arachnoid villus
Subarachnoid space
Choroid plexus of third ventricle
Choroid plexus of lateral ventricle
Interventricular foramen
Lateral ventricle
Third ventricle
Cerebral aqueduct
Fourth ventricle
Choroid plexus of fourth ventricle
Medial aperture
Confluence of the sinuses
Straight sinus
Cisterna magna
Subarachnoid space

Cerebrospinal Fluid in the Brain

Cerebrospinal fluid (CSF) is a clear, colorless liquid, which is essentially an ultrafiltrate of blood. The fluid in the subarachnoid space provides a special environment in which the brain floats, cushioning it against hard blows and sudden movements. The weight of the brain floating in CSF is only about 14 percent of its actual weight.

Besides providing a protective buoyancy for the brain, the cerebrospinal fluid aids in maintaining the chemical environment of the central nervous system. It conveys excess components and unwanted substances away from the extracellular fluid and into the venous portion of the blood circulatory system.

An adult has about 125 mL of cerebrospinal fluid, with about 50 mL in the ventricles and 75 mL in the subarachnoid space around the brain and spinal cord.

Formation of cerebrospinal fluid The ependymal cells that line the ventricles and the pia mater, with its rich blood supply, form the *choroid plexuses,* with their intricate networks of capillaries. The choroid plexuses

are considered to be components of the blood-brain barrier [FIGURE 14.3]. Most of the cerebrospinal fluid is formed continuously at the choroid plexuses of the lateral, third, and fourth ventricles by a combination of diffusion and active transport.

The choroid plexuses are able to produce cerebrospinal fluid because of the selective permeability of their blood vessels. The choroid plexuses do not allow blood cells or the largest protein molecules to pass into the ventricles. However, they do permit the passage of traces of protein; oxygen and carbon dioxide in solution; sodium, potassium, calcium, magnesium, and chloride ions; glucose; and a few white blood cells.

Q: *Does the weight of the brain increase after childhood?*

A: Although the number of brain cells does not increase after infancy, they do grow in size and degree of myelination as the body grows. In addition, the number of glial cells increases after birth. These changes account, in large measure, for the fact that the adult brain is about three times as heavy as it was at birth. After the age of about 20, the brain begins to lose about one gram a year as neurons die and are not replaced.

Circulation of cerebrospinal fluid Ordinarily, cerebrospinal fluid moves from the ventricles inside the brain to the subarachnoid space outside the brain. The fluid flows slowly from the two lateral ventricles of the brain, where much of the fluid is formed, through the paired interventricular foramina to the third ventricle. From there, it passes through the cerebral aqueduct into the fourth ventricle. The fluid leaves the ventricular system by oozing through three openings, or apertures, in the roof of the fourth ventricle. From there, it passes into the *cisterna magna*, the subarachnoid space behind the medulla oblongata.

From the cisterna magna, some cerebrospinal fluid slowly makes its way down the spinal cord to the lumbar cistern. Most of the fluid, however, circulates slowly toward the top of the brain through the subarachnoid space. Here the spongelike *arachnoid villi*, which contain pressure-sensitive valves, permit a one-way bulk flow of cerebrospinal fluid into the superior sagittal sinus, which drains venous blood from the brain [FIGURES 14.2A and 14.3].

The total volume of cerebrospinal fluid is formed and renewed about three times a day. When the pressure of the cerebrospinal fluid in the subarachnoid space exceeds the venous pressure, the small channels open, and the fluid flows into the superior sagittal sinus, where it joins the venous blood. When the venous pressure exceeds the pressure in the subarachnoid space, the channels close.

A S K   Y O U R S E L F

1 How does the dura mater nourish the cranial bones?

2 What is contained in the subarachnoid space?

3 Which meningeal layer envelops the brain directly?

4 What are cranial ventricles?

5 What are the functions of cerebrospinal fluid?

6 Where is cerebrospinal fluid formed?

7 Describe the circulation of cerebrospinal fluid in the brain.

Q: *What is the difference between a concussion and a contusion?*

A: A *concussion* (L. shake violently) is an abrupt and momentary loss of consciousness after a violent blow to the head. It results from the sudden movement of the brain within the skull. The loss of consciousness may be due to the sudden pressure upon neurons essential to the conscious state, or to sudden changes in the polarization of certain neurons. A *contusion* (L. a bruising) is a cerebral injury that produces a bruising of the brain due to blood leakage. Unconsciousness follows a contusion, and may last minutes or hours. A concussion is rarely serious, but a contusion can be.

NUTRITION OF THE BRAIN

The nutrients needed by the brain can reach it only through the blood, and about one pint of blood is circulated to the brain every minute. Blood reaching the brain contains glucose as well as oxygen. A steady supply of glucose is necessary, not only because it is the body's chief source of usable energy, but also because the brain cannot store it. The brain also requires about 20 percent of all the oxygen used by the body, and the need for oxygen is high even when the brain is at rest.

Effects of Deprivation

A lack of either oxygen or glucose will damage brain tissue faster than any other tissue. Deprivation for even a few minutes may produce permanent brain damage. Lack of oxygen in body tissues (*hypoxia*) can occur even before birth, especially during the last four months of prenatal development.

Proper and continuous nourishment is so important to the brain that it has built-in regulating devices that make it almost impossible to constrict blood vessels that would reduce the incoming blood supply. It is likely that the brain's blood vessels are prevented from constricting by the products of cell metabolism, which are either formed in the brain itself or carried to the brain by the blood. An increase in carbon dioxide, or a decrease in oxygen, dilates the blood vessels leading to the brain. The opposite condition, *hyperoxia* (too much oxygen and too little carbon dioxide), is one of the few conditions that allows *constriction* of cerebral blood vessels. A hyperoxic person may suffer dizziness, mental confusion, convulsions, or even unconsciousness until the brain's regulatory system balances the supply of oxygen and carbon dioxide.

A S K   Y O U R S E L F

1 What two substances does the brain need continuously?

2 Why is it important that blood vessels to the brain not be constricted easily?

Q: *How long can the brain be deprived of oxygen before it becomes damaged?*

A: If the brain is deprived of oxygen for more than 5 seconds we lose consciousness. After 15 to 20 seconds, the muscles begin to twitch in convulsions, and after 9 minutes, brain cells are damaged permanently. (There is a big difference between holding your breath for 15 seconds, and depriving the brain of oxygen for the same amount of time, because the oxygen in the air within the lungs is available for exchange with the circulating blood.)

TABLE 14.1 MAJOR STRUCTURES OF THE BRAIN

Structure	Description	Major functions
Basal ganglia	Large masses of gray matter contained deep within each cerebral hemisphere.	Help to coordinate skeletal muscle movements by relaying information via thalamus to motor area of cerebral cortex to influence descending motor tracts.
Brainstem*	Stemlike portion of brain continuous with diencephalon above and spinal cord below. Composed of midbrain, pons, medulla oblongata.	Relays messages between spinal cord and brain, and from brainstem cranial nerves to cerebrum. Helps control heart rate, respiratory rate, blood pressure. Involved with hearing, taste, other senses.
Cerebellum	Second largest part of brain. Located behind pons, in posterior section of cranial cavity. Composed of cerebellar cortex, two lateral lobes, central flocculonodular lobes, medial vermis, some deep nuclei.	Processing center involved with coordination of muscular movements, balance, precision and timing of movements, body positions. Processes sensory information used by motor systems.
Cerebral cortex	Outer layer of cerebrum. Composed of gray matter and arranged in ridges (gyri), grooves (sulci), depressions (fissures).	Involved with most conscious activities of living. (See major functions of cerebral lobes.)
Cerebral lobes	Major divisions of cerebrum, consisting of frontal, parietal, temporal, occipital lobes (named for bones under which they lie), insula, and limbic lobe.	Frontal lobe involved with motor control of voluntary movements, control of emotional expression and moral behavior. Parietal lobe involved with general senses, taste. Temporal lobe involved with hearing, equilibrium, emotion, memory. Occipital lobe organized for vision and associated forms of expression. Insula may be involved with gastrointestinal and other visceral activities. Limbic lobe (along with the limbic system) is involved with emotions, behavioral expressions, recent memory, smell.
Cerebrospinal fluid	Fluid that circulates in ventricles and subarachnoid space.	Supports and cushions brain. Helps control chemical environment of central nervous system.
Cerebrum	Largest part of brain. Divided into left and right hemispheres by longitudinal fissure and divided into lobes. Also contains cerebral cortex (gray matter), white matter, basal ganglia.	Controls voluntary movements, coordinates mental activity. Center for all conscious living.
Corpus callosum	Bridge of nerve fibers that connects one cerebral hemisphere with the other.	Connects cerebral hemispheres, relaying sensory information between them. Allows left and right hemispheres to share information, helps to unify attention.
Cranial nerves	Twelve pairs of cranial nerves. Olfactory (I) and optic (II) arise from cerebrum; others (III through XII) arise from brainstem. Sensory, motor, or mixed.	Concerned with senses of smell, taste, vision, hearing, balance. Also involved with specialized motor activities, including eye movement, chewing, swallowing, breathing, speaking, facial expression.
Diencephalon	Deep portion of brain. Composed of thalamus, hypothalamus, epithalamus, ventral thalamus.	Connects midbrain with cerebral hemispheres. (See major functions of thalamus and hypothalamus.)

*In this book, the *brainstem* is made up of the midbrain, pons, and medulla oblongata. Some textbooks use variations of this definition.

Structure	Description	Major functions
Hypothalamus	Small mass below the thalamus; forms floor and part of lateral walls of third ventricle.	Highest integrating center for autonomic nervous system. Controls most of endocrine system through its relationship with the pituitary gland. Regulates body temperature, water balance, sleep-wake patterns, food intake, behavioral responses associated with emotion.
Medulla oblongata	Lowermost portion of brainstem. Connects pons and spinal cord. Site of decussation of descending corticospinal (motor) tract and an ascending sensory pathway from spinal cord to thalamus; emergence of cranial nerves VI through XII; movement of cerebrospinal fluid from ventricle to subarachnoid space.	Contains vital centers that regulate heart rate, respiratory rate, constriction and dilation of blood vessels, blood pressure, swallowing, vomiting, sneezing, coughing.
Meningeal spaces	Spaces associated with meninges. Potential epidural space between skull and dura mater; potential subdural space between dura mater and arachnoid; subarachnoid space between arachnoid and pia mater (contains cerebrospinal fluid).	Provide subarachnoid circulatory paths for cerebrospinal fluid, protective cushion.
Meninges	Three layers of membranes covering brain. Outer tough dura mater; arachnoid; inner delicate pia mater, which adheres closely to brain.	Dura mater adds support and protection. Arachnoid provides space between the arachnoid and pia mater for circulation of cerebrospinal fluid. Choroid plexuses, at places where pia mater and ependymal cells meet, are site of formation of much cerebrospinal fluid. Pia mater contains blood vessels that supply blood to brain.
Midbrain	Located at upper end of brainstem. Connects pons and cerebellum with cerebrum. Site of emergence of cranial nerves III, IV.	Involved with visual reflexes, movement of eyes, focusing of lens, dilation of pupils.
Pons	Short, bridgelike structure composed mainly of fibers that connect midbrain and medulla oblongata, cerebellar hemispheres, and cerebellum and cerebrum. Lies anterior to cerebellum and between midbrain and medulla oblongata. Site of emergence of cranial nerve V.	Controls certain respiratory functions. Serves as relay station from medulla oblongata to higher structures in brain.
Reticular formation	Complex network of nerve cells organized into ascending (sensory) and descending (motor) pathways. Located throughout core of entire brainstem.	Specific functions for different neurons, including involvement with respiratory and cardiovascular centers, regulation of individual's level of awareness.
Thalamus	Composed of two separate bilateral masses of gray matter. Located in center of cerebrum.	Intermediate relay structure and processing center for all sensory information (except smell) going to cerebrum.
Ventricles	Cavities within brain that are filled with cerebrospinal fluid. Left and right lateral ventricles in cerebral hemispheres, third ventricle in diencephalon, fourth ventricle in pons and medulla oblongata.	Provide circulatory paths for cerebrospinal fluid. Choroid plexuses associated with ventricles are site of formation of most cerebrospinal fluid.

BRAINSTEM

The **brainstem** is the stalk of the brain, and it relays messages between the spinal cord and the brain. Its three segments are the *midbrain, pons,* and *medulla oblongata* [FIGURE 14.4]. The brainstem is continuous with the diencephalon (the inferior region of the cerebrum) above and the spinal cord below. It narrows slightly as it leaves the skull, passing through the foramen magnum to merge with the spinal cord.

Other structures of the brainstem that are very important functionally include the *long ascending* and *descending tracts* between the spinal cord and parts of the cerebrum, which all pass through the brainstem. A network of nerve cell bodies and fibers called the *reticular formation* is also located throughout the core of the entire brainstem. It plays a vital role in maintaining life. Finally, all *cranial nerves* except I (olfactory) and II (optic) emerge from the brainstem. The sensory nuclei in which the sensory fibers of these cranial nerves terminate, and the motor nuclei from the motor fibers of these cranial nerves originate, are located in the brainstem.

Neural Pathways of Nuclei and Long Tracts

In the nervous system, the word **nucleus** (pl. *nuclei*) means a collection of nerve cell bodies *inside* the central nervous system. A similar collection of nerve cell bodies *outside* the central nervous system is a **ganglion** (pl. *ganglia*).

Adequate stimulation of the sensory receptors results in an action potential that is conveyed to the spinal cord and brain by afferent fibers. This afferent input is processed within the central nervous system in *nuclei* and *centers* (collections of cell bodies and dendrites). (The basal ganglia of the cerebrum are actually nuclei.) It is then conveyed through *tracts* (bundles of axons) to other nuclei and centers for further processing. These sequences of nuclei and tracts are called **pathways.**

Sensory (ascending) pathways extend upward from the spinal cord through the brainstem to the cerebrum and cerebellum. Such ascending pathways include those for pain, touch, and vision. *Motor (descending) pathways* extend from the cerebral cortex and cerebellum to the brainstem, and from the cerebral cortex and brainstem to the spinal cord. TABLE 13.1 on page 325 summarizes some of the major pathways.

Reticular Formation

Deep within the brainstem is a slender but complex network of nerve cells and fibers called the **reticular formation** (L. *reticulum,* netlike). It is organized into (1) *ascending (sensory) pathways* from ascending spinal cord tracts and from the cerebellum; (2) *descending (motor) pathways* from the cerebral cortex and hypothalamus; and (3) *cranial nerves* [FIGURE 14.5]. The reticular formation runs

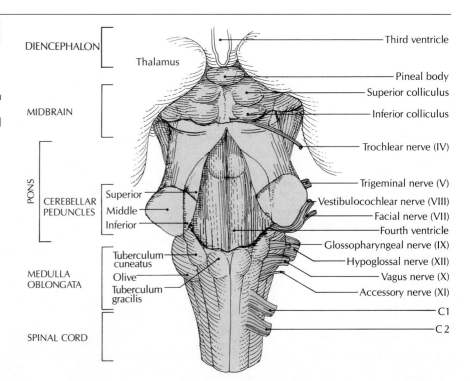

FIGURE 14.4 THE BRAINSTEM

All cranial nerves except the olfactory and optic nerves emerge from the brainstem. Only the cranial nerves on the right side of the body are shown in this dorsal view. The oculomotor and abducens nerves emerge on the ventral side.

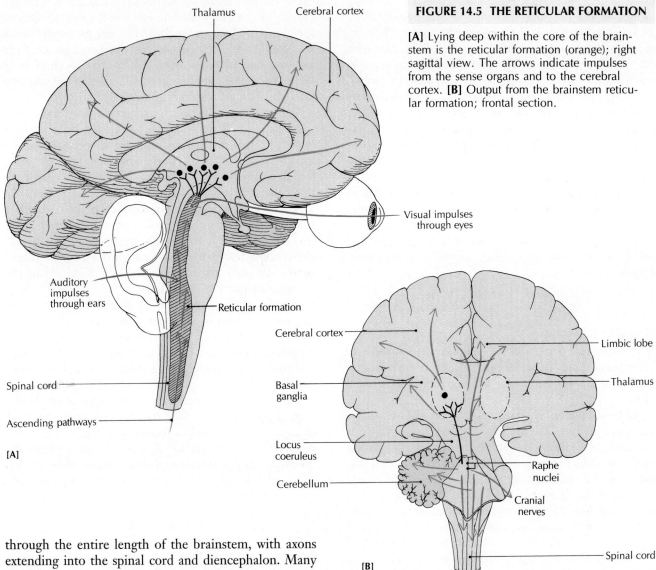

Thalamus

Cerebral cortex

FIGURE 14.5 THE RETICULAR FORMATION

[A] Lying deep within the core of the brainstem is the reticular formation (orange); right sagittal view. The arrows indicate impulses from the sense organs and to the cerebral cortex. [B] Output from the brainstem reticular formation; frontal section.

Visual impulses
through eyes

Auditory
impulses
through ears

Reticular formation

Spinal cord

Ascending pathways

[A]

Cerebral cortex

Limbic lobe

Basal
ganglia

Thalamus

Locus
coeruleus

Raphe
nuclei

Cerebellum

Cranial
nerves

[B]

Spinal cord

through the entire length of the brainstem, with axons extending into the spinal cord and diencephalon. Many neurons in the reticular formation have as many as 30,000 synaptic connections with other neurons in the central nervous system. A lesion in the reticular formation of the upper brainstem may result in loss of consciousness *(coma)*, which may last for months or even years.

Neurons within the reticular formation are organized into several groups, each having a specific and life-sustaining function. The reticular formation contains the respiratory and cardiovascular centers, which help to regulate such functions as breathing, heart rate, and the changing diameter of blood vessels. It also helps to regulate our level of awareness. When the effects of sensory stimuli pass through the brainstem on their way to the highest centers of the brain, they stimulate the reticular formation, which in turn results in the increased activity of the cerebral cortex.

The ascending pathways to the cerebrum are integrated in the ascending *reticular activating system*, which

influences the brain's state of arousal or wakefulness. The reticular formation also contains the spinoreticulothalamic pain pathway, which is associated with the dull and diffuse qualities of pain. Descending motor pathways include the reticulospinal pathways, which convey motor impulses to the spinal cord.

Reticular Activating System

The reticular formation of the brainstem receives a great variety of sensory information as input from our internal and external environments. Projections from the reticular formation are widespread. They influence most of the processing centers and cortical areas of the central nervous system. Two specialized nuclear groups in the brainstem are especially interesting: the *raphe nuclei*, located in the midsagittal region, and the *locus coeruleus*, located in the reticular formation of the pons [FIGURE 14.5B]. The neurons of the raphe nuclei contain the neurotransmitter serotonin, and those of the locus coeruleus contain norepinephrine. The neurons of the raphe nuclei and locus coeruleus are unusual because their axons project throughout the central nervous system and directly to the cerebral cortex and cerebellum.

The influences from the reticular formation are expressed functionally by the *reticular activating system* (RAS), which has roles associated with many behavioral activities. These roles include adjusting certain aspects of the sleep-wake cycle, awareness, alertness, levels of sensory perception, emotions, and motivation. The reticular activating system also helps the cerebellum to coordinate selected motor units to produce smooth, coordinated contractions of skeletal muscles, as well as maintaining muscle tonus.

Medulla Oblongata

The lowermost portion of the brainstem is the *medulla oblongata.* About the same length (3 cm) as the midbrain and pons, it is situated in the inferior part of the cranial cavity.

The medulla oblongata is continuous with the spinal cord, and extends from the foramen magnum to the pons [FIGURES 14.1 and 14.4]. Within the reticular formation of the medulla oblongata are the bundles of nuclei that make up the vital cardiac, vasomotor (constriction and dilation of blood vessels), and respiratory centers. The medulla oblongata monitors the level of carbon dioxide (as well as hydrogen ion concentration) in the body so closely that it will cause the respiration rate to double if the carbon-dioxide concentration in the blood rises by as little as .03 percent. The medulla oblongata also regulates vomiting, sneezing, coughing, and swallowing.

The medulla oblongata is connected to the pons by longitudinal bundles of nerve fibers. It is joined to the cerebellum by the paired bundles of fibers called the *inferior cerebellar peduncles* (peh-DUNG-kuhlz; L. little feet).

The ventral surface of the medulla oblongata contains bilateral elevated ridges called the *pyramids.* The pyramids are composed of the fibers of motor tracts from the motor cerebral cortex to the spinal cord. These *pyramidal*, or *corticospinal*, *tracts* cross over, or *decussate*, in the lower part of the medulla oblongata to the opposite side of the spinal cord, forming an X.* This crossing over of the motor nerve fibers in the medulla oblongata is

called the *pyramidal decussation* [FIGURE 14.6]. The pyramidal fibers that decussate become the lateral corticospinal tracts of the spinal cord. The few pyramidal fibers that do not decussate become the uncrossed fibers of the anterior corticospinal tract, which finally cross over within the spinal cord before terminating on the opposite side (see TABLE 13.2 on page 326).

Because almost all motor and sensory pathways cross over, *each side of the brain controls the opposite side of the body.* The left cerebral hemisphere affects the skeletal muscles on the right side of the body, and the right cerebral hemisphere is similarly linked to the left side of the body. (See "Left Brain, Right Brain" on pages 370–371.)

On the lateral anterior surface of each pyramidal tract is an olive-shaped swelling, appropriately called the *olive* (FIGURE 14.4). It is formed by the inferior olivary nucleus, whose fibers cross over and convey excitatory input signals through the inferior cerebellar peduncle to the opposite half of the cerebellum.

On the dorsal surface of the medulla oblongata are two pairs of bumps: the *tuberculum gracilis* (L. lump + slender) and the *tuberculum cuneatus* (L. wedge). These tubercula are formed by the relay nuclei (nuclei gracilis and cuneatus) in the posterior column-medial lemniscus pathway that convey touch and related sensations.

Several cranial nerves emerge from the medulla oblongata [FIGURE 14.7]. The rootlets of cranial nerve XII emerge from the groove located anterior to the olive. The rootlets of nerves IX, X, and XI emerge from the groove located posterior to the olive. Cranial nerves VI, VII, and VIII emerge from the junction between the medulla oblongata and pons.

Pons

Just superior to the medulla oblongata is the *pons* (L. bridge), so named because it forms a connecting bridge between the medulla oblongata and the midbrain, the uppermost portion of the brainstem [FIGURES 14.1 and 14.4]. The posterior portion of the pons is called the *dorsal pons*, and the anterior portion is the *ventral pons*.

Dorsal pons The *dorsal pons* consists of the reticular formation, some nuclei associated with cranial nerves, ascending pathways, and some fibers of descending pathways. Within the reticular formation are the *pneumotaxic* and *apneustic centers*, which help to regulate breathing. These respiratory centers are integrated with the respiratory centers of the medulla oblongata. The locations of the cranial nerve nuclei are shown in FIGURE 14.7.

The ascending pathways include the neurons of the ascending reticular system, and such sensory tracts as the medial lemniscus (touch-pressure, proprioception) and spinothalamic and trigeminothalamic tracts (pain, tem-

---

*Decussate comes from the Latin word decussare, which in turn is derived from dec, meaning "10." The Latin symbol for 10 is X, which represents the crossing over of the pyramidal tracts.

FIGURE 14.6 DECUSSATION OF NERVE FIBERS IN THE MEDULLA OBLONGATA

[A] Schematic drawing of ascending touch pathway and descending motor (corticospinal) tract; anterior view.
[B] Detail of decussation in the corticospinal tracts. The ascending touch pathway is composed of a sequence of three neurons. The second neuron in the sequence decussates in the medulla oblongata; anterior view. [C] The following pathways decussate: (1) corticopontocerebellar pathway from cerebral cortex to pontine nuclei to cerebellum; (2) cerebellothalamic pathway from cerebellum to thalamus

to motor cerebral cortex; (3) corticospinal (pyramidal) tract from motor cortex to spinal cord; anterior view. Ninety percent of the fibers cross over (pyramidal decussation) in the lower medulla oblongata, descend as the lateral corticospinal tract, and terminate in all levels of the spinal cord. Ten percent of the fibers descend as the uncrossed fibers of the anterior corticospinal tract, which cross over within the spinal cord before terminating in the cervical and upper thoracic levels of the spinal cord.

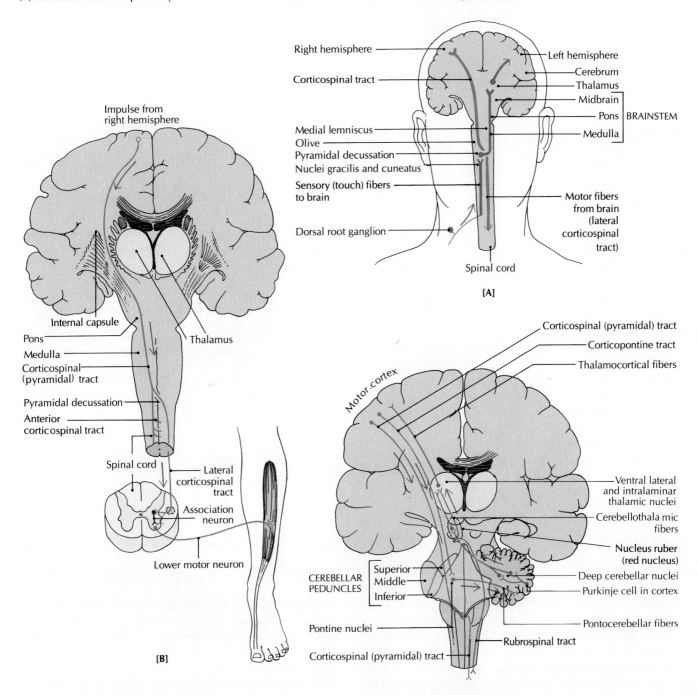

FIGURE 14.7 LOCATIONS OF CRANIAL NERVE NUCLEI IN THE BRAINSTEM

[A] Afferent cranial nerve nuclei. [B] Efferent cranial nerve nuclei.

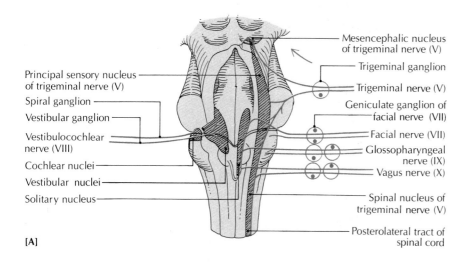

Mesencephalic nucleus of trigeminal nerve (V)
Trigeminal ganglion
Trigeminal nerve (V)
Geniculate ganglion of facial nerve (VII)
Facial nerve (VII)
Glossopharyngeal nerve (IX)
Vagus nerve (X)
Spinal nucleus of trigeminal nerve (V)
Posterolateral tract of spinal cord

Principal sensory nucleus of trigeminal nerve (V)
Spiral ganglion
Vestibular ganglion
Vestibulocochlear nerve (VIII)
Cochlear nuclei
Vestibular nuclei
Solitary nucleus

[A]

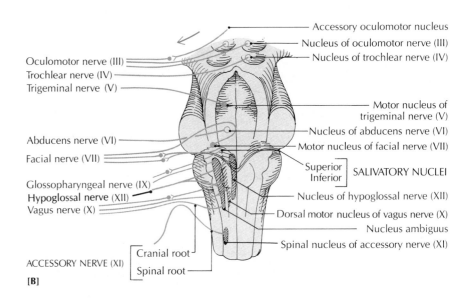

Accessory oculomotor nucleus
Nucleus of oculomotor nerve (III)
Nucleus of trochlear nerve (IV)
Oculomotor nerve (III)
Trochlear nerve (IV)
Trigeminal nerve (V)
Motor nucleus of trigeminal nerve (V)
Nucleus of abducens nerve (VI)
Motor nucleus of facial nerve (VII)
Abducens nerve (VI)
Facial nerve (VII)
Superior / Inferior SALIVATORY NUCLEI
Glossopharyngeal nerve (IX)
Hypoglossal nerve (XII)
Vagus nerve (X)
Nucleus of hypoglossal nerve (XII)
Dorsal motor nucleus of vagus nerve (X)
Nucleus ambiguus
Spinal nucleus of accessory nerve (XI)
ACCESSORY NERVE (XI)
Cranial root
Spinal root
[B]

perature, and light touch). The descending pathways are made up of the corticobulbar fibers, corticoreticular fibers, and rubrospinal tract. The corticobulbar fibers influence the motor nuclei of the cranial nerves. The corticoreticular fibers regulate the activity of the reticulospinal tracts that project to the spinal cord. The rubrospinal tract conveys motor impulses from one side of the midbrain to the other side of the body. Nerve V emerges from the ventrolateral aspect of the pons.

Ventral pons The *ventral pons* contains the pontine nuclei, which are the relay nuclei of the corticopontocerebellar pathway. The pontocerebellar fibers of this pathway cross over and convey excitatory influences to the cerebellum via the middle cerebellar peduncle [FIGURE 14.6C]. The corticopontocerebellar pathway is the means by which the cerebral cortex communicates with the cerebellum of the opposite side. The pyramidal tracts pass through the ventral pons.

Midbrain

The *midbrain*, or *mesencephalon*, is the segment of the brainstem located between the diencephalon and the pons [FIGURES 14.1 and 14.4]. It connects the pons and cerebellum with the cerebrum (forebrain). On the ventral surface of the midbrain is a pair of *cerebral peduncles*, made up of the pyramidal tracts (fibers to the motor nuclei of the spinal nerves within the spinal cord), corticobulbar fibers (motor fibers to the cranial nerve motor nuclei), and corticopontine fibers to the pons [FIGURE 14.10]. Emerging from the fossa between the peduncles on its ventral side is the pair of oculomotor nerves (cranial nerve III).

Passing through the midbrain is the cerebral aqueduct. The dorsal portion of the midbrain, situated above the aqueduct, is called the roof, or *tectum* (L. roof). The tectum contains four elevations called *colliculi* (L. little hills); the colliculi are known collectively as the *corpora quadrigemina* (L. bodies of four twins). The *superior* pair of colliculi are reflex centers that help to coordinate the movements of the eyeballs and head, regulate the focusing mechanism in the eyes, and adjust the size of the pupils in response to certain visual stimuli. The trochlear nerves (cranial nerve IV) emerge from the roof of the midbrain. Just posterior are the *inferior colliculi*, which are relay nuclei of the auditory pathways conveying influences to the thalamus and eventually to the auditory cortex.

The large *nucleus ruber (red nucleus)* is a major motor nucleus of the reticular formation [FIGURE 14.6C]. (It is so named because it has a reddish appearance in the fresh state.) This nucleus is the termination point for nerve fibers from the cerebral cortex and cerebellum. It also gives rise to the rubrospinal tract, which crosses over in the midbrain before descending to the spinal cord.

The corticorubrospinal pathway, along with the corticospinal tract, is involved in somatic motor activities. The heavily black-pigmented nucleus called the *substantia nigra* (L. *nigra*, black) is integrated into neural circuits with the basal ganglia of the cerebrum [FIGURE 14.11]. The substantia nigra is a basal ganglion. It has a role in Parkinson's disease.

ASK YOURSELF

1 What are the three major components of the brainstem?

2 What is the reticular formation?

3 What are cerebral peduncles?

4 How do ganglia and nuclei differ?

5 What is decussation?

CEREBELLUM

The main role of the *cerebellum* (L. little brain) is to regulate balance, timing and precision of body movements, and body positions. It processes input from sensory receptors in the head, body, and limbs. Through connections with the cerebral cortex, vestibular system, and reticular formation, the cerebellum refines and coordinates muscular movements. It does not initiate any movements, and is not involved in the conscious perception of sensations.

Anatomy of the Cerebellum

The cerebellum is located posterior to the pons in the posterior cranial fossa [FIGURE 14.8]. It is the second largest part of the brain, the cerebrum being the largest. The cerebellum is separated from the occipital lobes of the cerebrum by a fold of dura mater called the *tentorium cerebelli* and by the transverse cerebral fissure.

The cerebellum may be divided into three parts: (1) a midline portion, called the *vermis* (L. wormlike), (2) two small flocculonodular lobes (vestibular cerebellum), and (3) two large lateral lobes. The *flocculonodular lobes* (L. *flocculus*, little tuft of wool), together with the centrally placed worm-like *vermis*, play a role in maintaining skeletal muscle tone, equilibrium, and posture through their influence on the motor pathways that regulate the activity of the trunk muscles.

The much larger *lateral lobes*, or *hemispheres*, of the cerebellum help to smooth out muscle movement, and they synchronize the delicate and precise timing of the many skeletal muscles involved with any complex activity. Such synchronization is especially apparent in the movements of the upper and lower extremities.

The lobes of the cerebellum are covered by a surface layer of gray matter called the *cerebellar cortex* (L. bark or shell), which is composed of a network of billions of neurons. The cerebellar cortex is corrugated, with long, parallel ridges called *folia cerebelli*, which are more regular than the gyri of the cerebral cortex [FIGURE 14.8]. The folia are separated by *fissures* (deep folds). Under the cortex is a mass of white matter composed of nerve fibers. Lying deep within the white matter are the deep cerebellar nuclei, from which axons project out of the cerebellum to the cerebral cortex. A median section of the vermis reveals a branched arrangement of white matter called the *arbor vitae* (VYE-tee; L. tree of life).

The cerebellum is attached to the brainstem by three *cerebellar peduncles*. The *inferior cerebellar peduncle*, which connects the medulla oblongata to the cerebellum,

FIGURE 14.8 CEREBELLUM

[A] Right sagittal view. [B] Sagittal section, showing the arbor vitae. [C] Superior surface, showing the lobes and vermis. [D] Inferior surface showing the flocculonodular lobe.

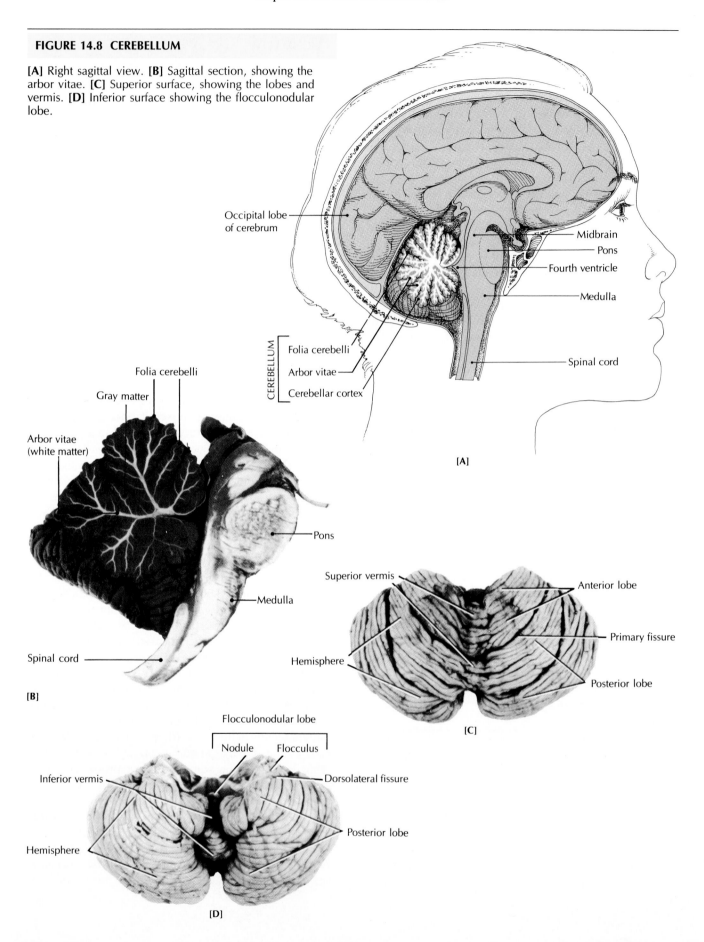

is a bundle of fibers originating in the spinal cord, medulla oblongata, and vestibular system, and terminating in the cerebellum. In addition, some fibers from the cerebellum pass through this peduncle to the vestibular nuclei. A major source of input to the cerebellum is via fibers from the inferior olivary nucleus (which forms the olive of the medulla oblongata). The fibers cross over and pass through the inferior cerebellar peduncle. The *middle cerebellar peduncle* is composed of pontocerebellar fibers passing from the pontine nuclei of the opposite side to the cerebellum [FIGURE 14.6C]. The *superior cerebellar peduncle*, which connects the cerebellum to the midbrain, is composed primarily of fibers from the deep nuclei of the cerebellum. These fibers cross over in the midbrain and terminate in the brainstem reticular formation, nucleus ruber, and thalamus.

Functions of the Cerebellum

The cerebellum integrates the contractions of skeletal muscles in relation to each other as they participate in a movement or series of movements. It is especially involved with coordinating agonists and antagonists in a cooperative way. The cerebellum smoothes out the action of each muscle group by regulating and grading muscle tension and tone in a precise and delicate way. Although the cerebellum *does not initiate any movements*, it participates in each movement through connections to and from the cerebral cortex.

The cerebellum continuously monitors sensory input from muscles, tendons, joints, and vestibular (balance) organs. These sensory *proprioceptive inputs* (the sense of the relative position of one body part to another) are derived on an unconscious level from neuromuscular spindles, tendon (Golgi) organs, the vestibular system, and other sensory endings.

One way to understand the function of the cerebellum is to observe the results of a cerebellar lesion, such as might be caused by a stroke or a tumor on one side of the cerebellum. In such a lesion, the reflexes are diminished, and the absence of perfect coordination is shown through tremors and jerky, puppetlike movements. The condition is called *ataxia* (Gr. lack of order). The patient has the symptoms on the same side of the body as the lesion because the fibers cross at two sites, making a "double cross" instead of the usual single crossing over [FIGURE 14.6A, C].

The double-cross expression of cerebellar activity is first conveyed (1) from the cerebellum to the cerebral cortex via a crossing in the midbrain; then (2) the control of the movement is conveyed by way of the pyramidal tract via a second crossing in the lower medulla oblongata to the opposite side of the body (the same side as the cerebellar lesion).

A S K   Y O U R S E L F
1 What are the parts of the cerebellum?
2 What is the cerebellar cortex?
3 What are the cerebellar peduncles?
4 What are the main functions of the cerebellum?

CEREBRUM

The largest and most complex structure of the nervous system is the *cerebrum* (suh-REE-bruhm; L. brain). It consists of two cerebral hemispheres [FIGURE 14.1]. Both hemispheres are composed of a cortex (gray matter), white matter, and basal ganglia. The cortex is further divided into six lobes: the frontal, parietal, temporal, occipital, limbic, and insula (central lobe) [FIGURE 14.9]. Each of the first four lobes contains special functional areas, including speech, hearing, vision, movement, and the appreciation of general sensations. The olfactory nerve and bulb are located beneath the frontal lobe [FIGURE 14.13].

All of our conscious living depends on the cerebrum. Popularly, it is considered the region where thinking is done, but no specific parts of the cerebrum have been identified as the exact areas of consciousness or intellectual learning.

Anatomy of the Cerebrum

The cerebrum has a surface mantle of gray matter called the *cerebral cortex*. The cortex is a thin (about 4.5 mm), convoluted covering containing over 50 billion neurons and 250 billion glial cells. The ridges of the cortex are called convolutions, or *gyri* (JYE-rye; sing., *gyrus*), which are separated by slitlike grooves called *sulci* (SUHL-kye; sing., *sulcus*) [FIGURE 14.9C]. The gyri and sulci increase the surface area of the cerebral cortex, resulting in a 3:1 proportion of cortical gray matter to the underlying white matter. Although each person has a specific pattern of gyri and sulci, the overall developmental design of the cerebrum produces a fairly consistent pattern of gyri and sulci.

Extremely deep grooves of depressions are called *fissures* (L. *fissus*, crack). The cerebral hemispheres are separated by the *longitudinal fissure*, and the cerebrum is separated from the cerebellum by the *transverse cerebral fissure*.

FIGURE 14.9 GYRI, SULCI, AND FISSURES OF THE CEREBRAL HEMISPHERES

[A] Superior view. **[B]** Right lateral view. **[C]** In this photo, the cranial meninges have been removed to show the convolutions of the cerebral cortex more clearly.

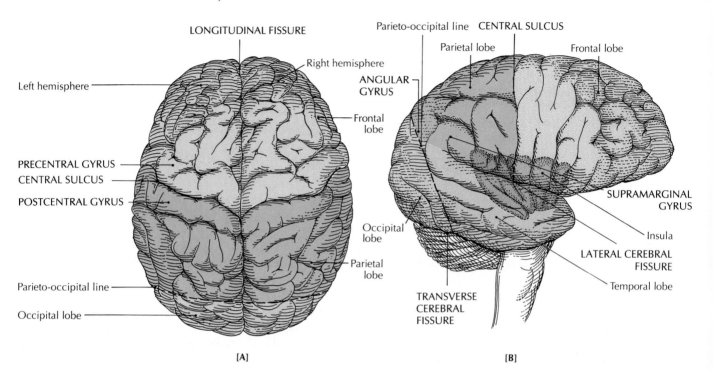

[A]

[B]

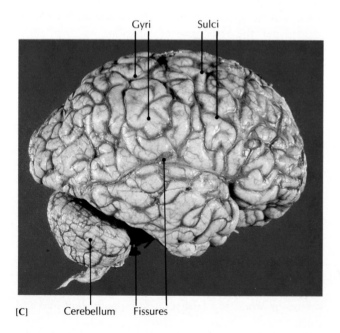

[C]

Beneath the cortex lies a thick layer of white matter. The white matter consists of interconnecting groups of axons projecting in two basic directions. One group projects *from the cortex* to other cortical areas of the same hemisphere and the opposite hemisphere, to the thalamus, to the basal ganglia, to the brainstem, or to the spinal cord. The other group projects *from the thalamus* to the cortex. Actually, the thalamus is functionally integrated with the cerebral cortex in the highest sensory and motor functions of the nervous system.

The white matter consists of three types of fibers (association fibers, commissural fibers, and projection fibers), all of which originate in cell bodies in the cerebral cortex [FIGURE 14.10]:

1 *Association fibers* link one area of the cortex to another area of the cortex of the *same hemisphere.*

2 *Commissural fibers* are the axons that project from a cortical area of one hemisphere to a corresponding cortical area of the *opposite hemisphere.* The two major cerebral commissures are the *anterior commissure* and the massive bundle of axons called the ***corpus callosum*** (L. hard body). Both connect the two cerebral hemi-

FIGURE 14.10 FIBER TRACTS IN THE WHITE MATTER OF THE CEREBRUM

[A] Right lateral view. [B] Anterior view.

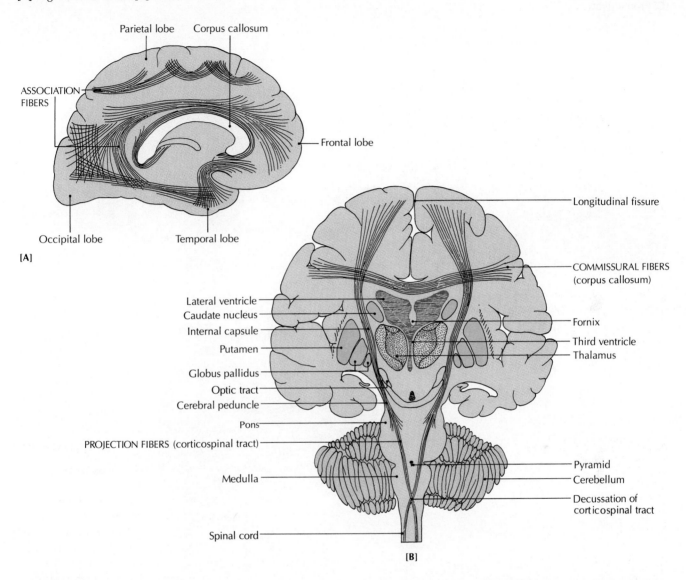

[A]

[B]

spheres and relay nerve impulses between them. The corpus callosum contains about 200 million nerve fibers.

3 *Projection fibers* include the axons that project from the cerebral cortex *to other structures* of the brain, such as the basal ganglia, thalamus, and brainstem, and to the spinal cord. The corticospinal tract from the motor cortex to the spinal cord is composed of projection fibers.

Deep within each cerebral hemisphere is a core of gray matter called the ***basal ganglia*** (which are actually nuclei). They include the *caudate* ("tail-shaped") *nucleus, putamen, globus pallidus* ("pale ball"), *subthalamic nucleus,* and *substantia nigra* [FIGURES 14.10, 14.11]. The ***caudate nucleus*** and the ***putamen*** (pyoo-TAY-muhn; L. prunings)

are called the ***striatum.*** The putamen and globus pallidus comprise the ***lentiform*** ("lens-shaped") ***nucleus.*** The basal ganglia are organized as an intricate network of neuronal circuits and processing centers. Along with the cerebellum, the basal ganglia act at the interface between the sensory systems and many motor responses, affecting motor, emotional, and cognitive behaviors. Clinically, the terms *movement disorders* and *malfunctioning* of the *basal ganglia* are essentially synonymous. Such disorders as Parkinson's disease and Huntington's chorea are expressions of basal ganglia disturbances.

The basic circuitry associated with the basal ganglia may be summarized as follows: The basal ganglia receive information from many areas of the cerebral cortex (cor-

FIGURE 14.11 BASAL GANGLIA

[A] Section through the brain showing the parts of the basal ganglia (orange). [B] Structures that make up the basal ganglia. [C] Frontal view.

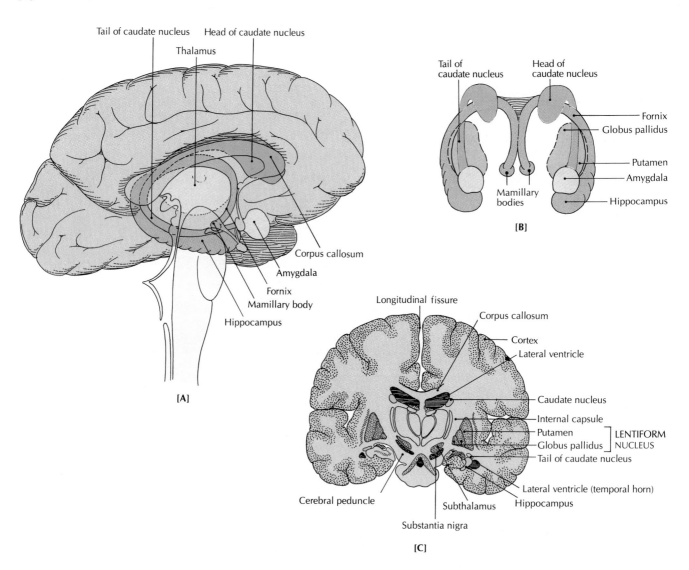

tex of somesthetic, motor, association and limbic areas), process and integrate these inputs, and then relay them to the thalamus which, in turn, projects its output to the motor, premotor, and limbic cortical areas. These areas then exert their influences on the upper motor neuron pathways and other systems that affect motor, emotional, and cognitive behaviors.

The abnormal involuntary movements associated with malfunctioning of the basal ganglia are called *dyskinesias*. These movements are considered to be expressions of *release phenomena*, in which certain inhibitory influences are reduced or lost. Consequently, certain neuronal centers are deprived of modulatory influences. An analogy is the brakes (inhibitors) that act to control a car. Without the use of brakes a moving car is deprived of their inhibi-

tory role. In Parkinson's disease there is a reduction in the neurotransmitter dopamine in the neurons with cell bodies in the substantia nigra and axons that terminate in the striatum (caudate nucleus and putamen). The resulting reduction in the "inhibitory" activity of dopamine elicits the symptoms of Parkinson's disease, which is characterized by rigidity, tremor, and akinesia. *Rigidity* is the increased tonus (hypertonus) in the skeletal muscles; *tremor* is the rhythmic oscillatory trembling movements, especially of the forearm and hand; and *akinesia* is the tendency to be immobile.

Huntington's chorea (L. dance) is a hereditary disorder characterized by jerky, irregular, brisk and graceful movements of the limbs, accompanied by involuntary grimacing and twitching of the face.

Functions of the Cerebrum

The two cerebral hemispheres have the same general appearance, but each has slightly different functions. One of the hemispheres, usually the left, is active in speech, writing, calculation, language comprehension, and analytic thought processes [FIGURE 14.12]. The other hemisphere, usually the right, is more specialized for the appreciation of spatial relationships, conceptual nonverbal ideas, simple language comprehension, and general thought processes. (See "Left Brain, Right Brain" on pages 370–371.) The left hemisphere sorts out the *parts* of things, while the right hemisphere concentrates on the *whole*. In a manner of speaking, the left hemisphere sees the trees, but not much of the forest, while the right hemisphere sees the forest, but not too many of the trees.

The specific functions of the cerebrum and their localized areas are discussed further in the following sections on the cerebral lobes.

Cerebral Lobes

Each cerebral hemisphere is subdivided into five lobes: the frontal, parietal, temporal, and occipital lobes and the insula (central lobe) [FIGURES 14.9 and 14.12]. The first four lobes are named for the skull bones covering them.

The frontal lobe is separated from the parietal lobe by the **central sulcus,** or *fissure of Rolando.* The **lateral cerebral sulcus** *(fissure of Sylvius)* divides the frontal and parietal lobes from the temporal lobe. The parietal and temporal lobes are separated from the occipital lobe by the arbitrary *parieto-occipital line* [FIGURE 14.9]. Buried deep in the central sulcus is the small central lobe, or *insula.*

Another subdivision of each hemisphere, the **limbic lobe,** is on the medial surface of the cerebral hemisphere [FIGURE 14.13]. It consists largely of the **cingulate gyrus** ("girdling convolution") and the **parahippocampal gyrus.** The **parietooccipital sulcus** on the medial surfaces separates the occipital lobe from the parietal lobe.

Frontal lobe The *frontal lobe* (also called the "motor lobe") is involved with two basic cerebral functions. One is the motor control of voluntary movements, including those associated with speech. The other is the control of a variety of emotional expressions and moral and ethical behavior. The motor activity is expressed through the *primary motor cortex* (Brodmann area 4), *supplementary* and *premotor motor cortex* (area 6), the *frontal eye field* (area 8), and *Broca's speech area** (areas 44, 45) [FIGURE 14.12].

*This area was named for Pierre Paul Broca, who first described it in the late nineteenth century.

FIGURE 14.12 CEREBRUM

A partial Brodmann's numbered map of the right lateral cerebral hemisphere. Brodmann's maps are useful mainly to show regions of different neural structure, but some areas also have a functional designation.

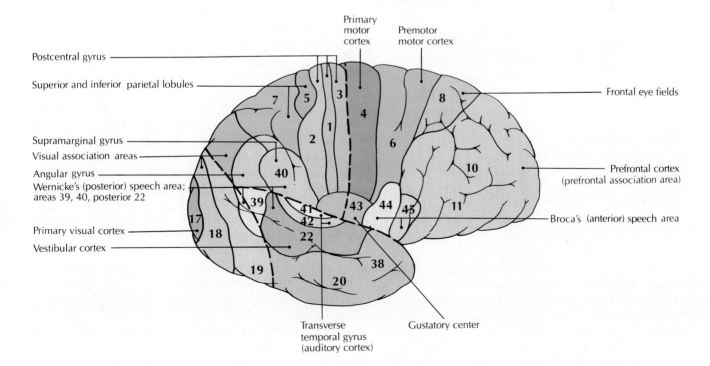

FIGURE 14.13 LIMBIC SYSTEM

[A] Simplified right lateral view. [B] Major components of the limbic system. [C] Median surface of the cerebral hemisphere; the limbic lobe is shaded.

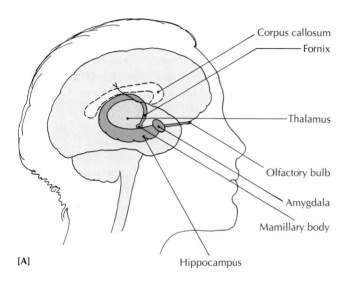

Corpus callosum
Fornix
Thalamus
Olfactory bulb
Amygdala
Mamillary body
Hippocampus

[A]

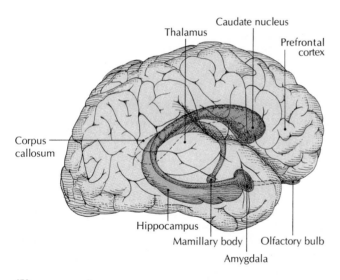

Caudate nucleus
Thalamus
Prefrontal cortex
Corpus callosum
Hippocampus
Mamillary body
Olfactory bulb
Amygdala

[B]

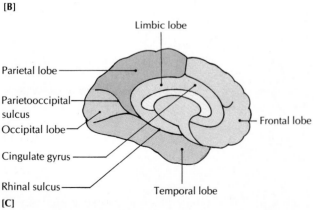

Limbic lobe
Parietal lobe
Parietooccipital sulcus
Occipital lobe
Cingulate gyrus
Rhinal sulcus
Frontal lobe
Temporal lobe

[C]

From the *primary,* *supplementary,* and *premotor motor cortex,* nerve impulses are conveyed through the motor pathways to the brain, brainstem, and spinal cord to stimulate the motor nerves of the skeletal muscles. FIGURE 14.14 shows the body drawn in proportion to the cortical space allotted to sensations and movement patterns of different parts of the body. Relatively large cortical areas are devoted to the face, larynx, tongue, lips, and fingers, especially the thumb. The thumb is so important for our dexterity that more of the brain's gray matter is devoted to manipulating the thumb than to controlling the thorax and abdomen. The disproportionate allotment of the cortical areas to these body parts reflects the delicacy with which facial expressions, vocalizations, and manual manipulations can be controlled.

The *frontal eye field* is a cortical area that regulates the scanning movements of the eyes, such as searching the sky to locate an airplane. *Broca's speech area (anterior speech area)* is critically involved with the formulation of words.

The cortex in front of the motor areas is called the *prefrontal cortex.* It has a role in the various forms of emotional expression. The functional role of the prefrontal cortex has been deduced largely from patients who have had the white matter of various areas of the prefrontal lobe surgically removed or cut, a procedure called a *prefrontal lobotomy.* Following the operation, the patients are usually less excitable and less creative than before, but vent their feelings frankly, without the typical restraint. Physical drive is lowered, but intelligence is not. An awareness of pain remains, but normal feelings associated with the bothersome aspects of pain are lost.

Parietal lobe The *parietal lobe* is concerned with the evaluation of the general senses, and of taste. It integrates and processes the general information that is necessary to create an awareness of the body and its relation to its external environment. The parietal lobe is composed of the *postcentral gyrus* (areas 3, 1, 2), which integrates general sensations, the *superior* and *inferior parietal lobules* (areas 5, 7), and the *supramarginal gyrus* (area 40) [FIGURE 14.12]. The postcentral gyrus is called the **primary somesthetic association area** (general sensory area) because it receives information about the general senses from receptors in the skin, joints, muscles, and body organs.

Q: *Do the most intelligent people have the largest brains?*

A: Not necessarily. Some of the most intelligent people in history had smaller-than-average brains, and one of the largest brains ever measured belonged to an idiot. Brain surface area may be a better indicator of intelligence, but it is by no means infallible.

FIGURE 14.14 CORTICAL SPACE ALLOTTED TO SENSATIONS

The human body is distorted in this model, so that the body parts are proportional to the space allotted in the brain to sensations from different parts of the body. If the eyes were drawn to scale, they would be larger than the entire body.

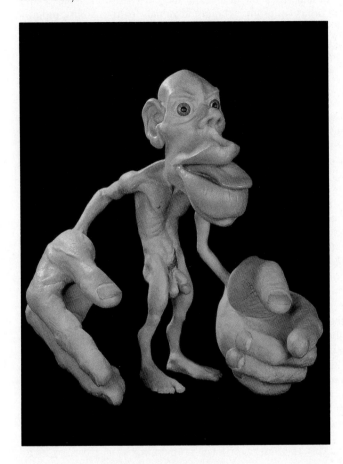

Temporal lobe The *temporal lobe* is the lobe located closest to the ears. It has critical functional roles in hearing, equilibrium, and to a certain degree, emotion and memory. The *auditory cortex* is located in the *transverse temporal gyrus* (of Heschl) (areas 41, 42), just anterior to the vestibular (equilibrium) area. Electrical stimulation of the auditory cortex elicits such elementary sounds as clicks and roaring, while the same stimulation of the vestibular area may produce feelings of dizziness and vertigo (a sense of rotation). The understanding of the spoken word involves the participation of the *superior temporal gyrus* (area 22).

Occipital lobe Although the *occipital lobe* is relatively small, it is important because it contains the visual cortex. It is made up of several areas organized for vision. The general region where the parietal, temporal, and occipital lobes mesh on the lateral surface is essential for the comprehension of the written and spoken word, and for an appreciation of the scheme and image of the body.

Visual images from the retina of the eye are transmitted and projected to the *primary visual cortex* (area 17). Each specific site in the retina is represented by a specific site in area 17. Visual information from this area is conveyed for further processing and elaboration to the visual association area (areas 18 and 19), and then to other areas such as the *angular gyrus* (area 39). Areas 18 and 19 are involved with assembling the features of a visual image and making it meaningful.

Wernicke's area (*posterior speech area*) (areas 39, 40, 22) is concerned with the understanding of the written word (area 39) and the spoken word (area 22), and with the ability to conceive the symbols of language. Lesions on the dominant side of Wernicke's area 39 (usually the left side) can produce *alexia*, a failure to recognize written words. (*Dyslexia* is associated with reading and writing disabilities; see page 388). At the same time, the primary visual perception remains intact, and the patient may accurately describe the shape, color, and size of an object.

Insula Little is known about the role of the *insula* (L. island), or *central lobe*, which appears to be an island in the midst of the rest of the brain. It seems to be associated with gastrointestinal and other visceral activities.

Limbic lobe and limbic system The *limbic lobe* (L. *limbus*, border) is the ring of the cortex, located on the medial surface of each cerebral hemisphere, and surrounding the central core of the cerebrum [FIGURE 14.13]. This lobe is composed of the cingulate gyrus, isthmus, parahippocampal gyrus, and uncus.

The *olfactory cortex*, which is involved with the perception of odors, is located in the region of the uncus. Irritation of this region may produce hallucinations of an odor (usually foul), along with fear and feelings of unreality. The cingulate sulcus separates the cingulate gyrus from the frontal and parietal lobes. The rhinal sulcus separates the temporal lobe from the parahippocampal gyrus and uncus.

The *limbic system* is defined in functional terms as an assemblage of cerebral (limbic lobe, portion of temporal lobe), diencephalic, and midbrain structures that are actively involved in memory and emotions and the visceral and behavioral responses associated with them. This physiologically defined system includes such structures as the limbic lobe, amygdala, hippocampus, and parts of other structures such as the thalamus and midbrain. Two of these structures are closely associated with the limbic lobe. One is the *hippocampus* (Gr. sea horse; so named because it is S-shaped), which is located medial to the

Left Brain, Right Brain

Because the corticospinal pathway crosses in the lower medulla oblongata at the base of the brain, each side of the brain controls actions on the *opposite* side of the body. This much has been known for at least 100 years. But recent experiments have begun to refine our knowledge of the inner workings of the brain.

Connecting the left and right cerebral hemispheres and relaying messages between them is the corpus callosum. In normal people, the corpus callosum makes it possible for the two sides of the brain to work together.

Split-Brain Experiments
In the early 1950s, Roger W. Sperry and his coworkers at the California Institute of Technology began experiments that strengthened the case for the "split-brain" theory. In 1981, Sperry received the Nobel Prize in Medicine for his work with split-brain patients, individuals whose corpus callosum had been severed in an effort (usually successful) to relieve epileptic seizures. (This procedure did not alter the patients' normal personality or behavior.)

In one experiment, when Sperry told split-brain patients to hold a spoon in their right hand (without looking at the spoon or their hand), the patients usually could describe the spoon without diffi-

culty. If the spoon was held in the *left* hand, however, the patients could *not* describe it. Why? Because the speech center is located in the *left* hemisphere, which also controls the actions of the *right* hand. Without a corpus callosum, objects held in the left hand could not be described because there was no way to transmit the information to the main speech center in the left hemisphere.

In a similar experiment, subjects faced a split picture screen, focusing on the midline. A picture of a spoon was projected onto the left side of the screen, and the image of a knife was shown on the right side for such a short time (a fraction of a second) before a reflex movement of the eyes could "see" the objects with the other half of the retina. When asked to describe what they saw, the subjects reported seeing a knife. (They actually saw both pictures, but the verbal side of the split brain—the left—could report verbally on only the picture on the right.) Then the subjects were asked to pick up the same object they saw on the screen—the knife. Instead of a knife, however, they picked up a spoon. When asked to identify the object, the split-brain subjects said they had picked up a knife. Although they *knew* it was a spoon, they were unable to *say* it. Without an intact corpus callosum to unify their attention and allow

the two hemispheres to *share* information, the subjects were hopelessly confused. Each half of the brain was competing for control, asserting its own sense of the outside world. (Most split-brain individuals learn to adjust, reducing their conflict enough so that they can begin to live a fairly normal life.)

Many experiments by Sperry and others have helped to prove that the left and right cerebral hemispheres are involved with different skills, including:

Left cerebral hemisphere

Right-hand control
Spoken language
Written language
Scientific skills
Numerical skills
Reasoning
Sorting out parts

Right cerebral hemisphere

Left-hand control
Music awareness
Recognition of faces and other three-dimensional shapes
Art awareness
Insight and imagination
Grasping the whole

parahippocampal gyrus in the floor of the lateral ventricle. The other is the **amygdala,** a complex of nuclei located deep within the uncus [FIGURE 14.13]. The amygdala and hippocampus are integrated with the functional expressions of the limbic system.

The hippocampus is involved with memory traces for recent events. Patients with Korsakoff's syndrome, in which lesions are present in the hippocampus, have a loss of short-term memory and lose their sense of time. They tend to become confused easily, forget questions that were just asked, and may reply to questions with irrelevant answers. All such expressions of the amygdala and hippocampus are associated with an interaction with other centers. For example, the impact of inputs of the sensory systems from the sensory cortical areas upon these structures can trigger a variety of emotional and behavioral responses.

DIENCEPHALON

The **diencephalon** (L. "between brain") is the deep part of the brain, connecting the midbrain with the cerebral hemispheres. It houses the third ventricle and is composed of the thalamus, hypothalamus, epithalamus, and ventral thalamus (subthalamus) [see FIGURE 14.1]. The bilateral thalami are separated by the third ventricle. The pituitary gland is connected to the hypothalamus.

Flanking the diencephalon laterally is the *internal capsule*, a massive structure of white matter made up of ascending fibers that convey information to the cerebral cortex, and descending fibers *from* the cerebral cortex.

Although the latest research supports the idea that each hemisphere controls the actions of the opposite side of the body, it disputes the notion that the brain is divided into neat compartments that totally control separate body functions. Apparently, such functions as speaking, for example, are not as localized as they were once thought to be. Radioactive tracers in the brains of experimental volunteers who were asked to speak out loud indicated that the flow of blood to the left hemisphere increased, as expected, but blood flow also increased in the right side. This is an important point. We must understand that the brain probably functions as a whole. It would be simplistic to believe that any given function exists in only one isolated part of the brain, without any relation to any other part.

Some Interesting Split-Brain Facts

Young children have speech potential in both hemispheres, and if the left hemisphere is badly damaged, they develop a speech center in the right hemisphere. If there are no unusual circumstances, speech dominance is fixed in the left hemisphere by the time a child is about 8 years old.

People usually turn their eyes and head to the right when they are thinking deeply.

When a person is playing the piano and humming the music at the same time, the right hand is more accurate than the left. When one melody is being played and another is being hummed, however, the *left* hand makes fewer mistakes than the right. This happens because if an activity calls for two parts of the brain to perform closely related tasks, a person functions best when both control centers are located in the same hemisphere. However, if the task demands less-related activities at the same time, a person performs best when the tasks are controlled by different hemispheres.

About 10 percent of the population is left-handed. A disproportionate number of great artists have been left-handed, showing a possible advantage in using the left hand for a creative action that is controlled by the opposite, right, hemisphere. Left-handed people also show less difference between their left and right hemispheres than right-handers do. Also, left-handers vary greatly among themselves in the way their brains process information.

Which face is happier? Most people choose the top one, even though the faces are mirror images. According to one theory, more people choose the top face because they are making the decision with their *right* brains, the side that controls emotion.

Many fibers of the internal capsule continue caudally into the ventral aspect of the brainstem [FIGURE 14.10].

Thalamus

The *thalamus* (Gr. inner chamber, so named because early anatomists thought it was hollow, resembling a room) is composed of two egg-shaped masses of gray matter covered by a thin layer of white matter. It is located in the center of the cranial cavity, directly beneath the cerebrum and above the hypothalamus. It forms the lateral walls of the third ventricle.

The thalamus is the intermediate relay point and processing center for all sensory impulses (except the sense of smell) ascending to the cerebral cortex from the spinal cord, brainstem, cerebellum, basal ganglia, and other sources. After processing the input, the thalamus relays its output to the cerebral cortex. The thalamus is involved with four major areas of activity:

1 *Sensory systems.* Fibers from the thalamic nuclei project into the sensory areas of the cerebral cortex, where the sensory input is "decoded" and translated into the appropriate sensation. For example, light is "seen," and sound is "heard." Crude sensations and some aspects of the general senses may also be brought to consciousness in the thalamus.

2 *Motor systems.* The thalamus has a critical role in influencing the motor cortex. Some thalamic nuclei receive neural input from the cerebellum and basal ganglia and then project into the motor cortex. The motor pathways that regulate the skeletal muscles innervated by the cranial and spinal nerves originate in the motor cortex.

3 *General neural background activity.* Background neurophysiological activities of the brain, such as the sleep-wake cycles and the electrical brain waves, are expressed in the cerebral cortex. These cortical rhythms are generated and monitored by thalamic nuclei, which receive much of their input from the ascending reticular systems in the brainstem.

4 *Expression of the cerebral cortex.* The thalamus, through its connections with the limbic system, helps to regulate many expressions of emotions and uniquely human behaviors. In fact, it is linked with the highest expressions of the nervous system, such as thought, creativity, interpretation and understanding of the written and spoken word, and the identification of objects sensed by touch. Such accomplishments are possible because of the two-way communication between the thalamus and the association area of the cortex.

Hypothalamus

The *hypothalamus* ("under the thalamus") lies directly under the thalamus [FIGURES 14.15 and 14.1]. It is a small region (about the size of a lump of sugar, and only 1/300 of the brain's total volume); it is located in the floor of the diencephalon and forms the floor and part of the lateral walls of the third ventricle. Extending from the hypothalamus is the *hypophysis (pituitary gland)*, which is neatly housed within the sella turcica in the body of the sphenoid bone.

The most important functions of the hypothalamus are the following:

1 *Integration with the autonomic nervous system.* The hypothalamus is considered to be the highest integrative center associated with the autonomic nervous system. Through the autonomic nervous system, it adjusts the activities of other regulatory centers, such as the cardiovascular centers in the brainstem. The hypothalamus modifies blood pressure, peristalsis (muscular movements that push food along the digestive tract) and glandular secretions in the digestive system, secretion of sweat glands and salivary glands, control of the urinary bladder, and rate and force of the heartbeat.

2 *Temperature regulation.* The hypothalamus plays a vital role in the regulation and maintenance of body temperature. Specialized nuclei within the hypothalamus monitor the temperature of the blood.

3 *Water and electrolyte balance.* The hypothalamus has a "thirst center" and a "thirst-satiety (satisfaction) center" that help to produce a balance of fluids and electrolytes in the body. These centers regulate the intake of water (through drinking) and its output (through kidneys and sweat glands).

FIGURE 14.15 HYPOTHALAMUS

Hypothalamic nuclei are shown, as are the relationships between the hypothalamus and hypophysis and between the hypothalamus and brainstem.

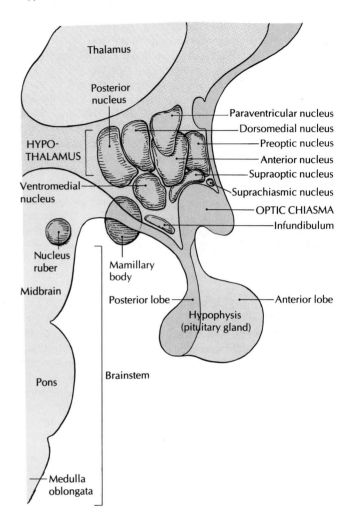

4 *Sleep-wake patterns.* The hypothalamus is integrated with the neural circuitry that regulates sleep-wake patterns and the state of awareness.

5 *Food intake.* The hypothalamus has a crucial role in food consumption. When neural centers collectively called the "appestat," or "hunger or feeding centers" are stimulated by low levels of glucose, fatty acids, and amino acids in the blood and possibly other factors, they activate appropriate body responses to satisfy the deficiency: you experience hunger. After food has been eaten, the "satiety (satisfaction) center" is stimulated to inhibit the feeding center.

6 *Behavioral responses associated with emotion.* The subjective feelings of emotion (pleasure, pain, anger, fear, love) are expressed as visible physiological and physical changes when the cerebral cortex activates the autonomic nervous system by way of the hypothalamus. The autonomic nervous system, in turn, is responsible for changes in the heart rate and blood pressure, blushing, dryness of the mouth, clammy hands, crying, gastrointestinal discomfort, fidgeting and many other emotional expressions.

7 Endocrine control. The hypothalamus produces the hormones oxytocin and antidiuretic hormone, which are released by the posterior lobe of the pituitary, as well as certain other hormones that control the release of hormones by the anterior lobe of the pituitary.

8 Sexual responses. The dorsal region of the hypothalamus contains specialized nuclei that respond to sexual stimulation of the genital organs. The sensation of an orgasm involves nerve activity within this center.

Epithalamus

The *epithalamus* is the dorsal portion of the diencephalon, located near the third ventricle. It contains the pineal body, a glandlike structure that extends outward from the posterior end of the epithalamus by way of a stalk. The *posterior commissure* consists of commissural fibers that connect with the superior colliculi of the midbrain.

Ventral Thalamus

The *ventral thalamus,* or *subthalamus,* located ventrally in the diencephalon, contains the subthalamic nucleus, which is one of the basal ganglia. According to recent evidence, the subthalamic nucleus is the driving force that regulates and modulates the output of the basal ganglia and its influences on motor activity.

ASK YOURSELF

1 What are the basal ganglia?

2 What is the cerebral cortex?

3 What is the benefit of cerebral gyri and sulci?

4 What is the function of the corpus callosum?

5 What are the six lobes of the cerebrum? Where is each located?

6 What are the roles of the limbic system and the limbic lobe?

7 What are the main parts of the diencephalon?

8 What are the chief functions of the thalamus? The hypothalamus? The epithalamus?

CRANIAL NERVES

The 12 pairs of *cranial nerves* are the peripheral nerves of the brain. Their names indicate some anatomical or functional feature of the nerve, and their numbers (in Roman numerals) indicate the sequential order in which they emerge from the brain.*

Cranial nerves I and II are nerves of the cerebrum, and nerves III through XII are nerves of the brainstem. (Part of nerve XI emerges from the cervical spinal cord.) Of the 10 brainstem nerves, one (VIII) is a purely sensory nerve, four are primarily motor nerves with some proprioceptive fibers (III, IV, VI, and XII), and five are mixed nerves containing both sensory and motor fibers (V, VII, IX, X, and XI) [FIGURE 14.16].

The *motor (efferent) fibers* of the cranial nerves emerge from the brainstem. They arise from bundles of neurons called *motor nuclei,* which are stimulated by nerve impulses from many outside sources, including the cortex of the cerebrum and the sense organs. Axons of motor neurons in cranial motor nuclei are in the form of cranial nerves. These axons have two roles. They either (1) stimulate skeletal muscles, or (2) synapse with ganglia of the autonomic nervous system (parasympathetic division). These ganglia, in turn, relay nerve impulses to cardiac muscle, smooth muscle, or glands.

The *sensory (afferent) fibers* of cranial nerves emerge from neurons with cell bodies outside the brain. These cell bodies are found either in *sensory ganglia,* which are groups of cell bodies situated on the trunks of cranial nerves, or are located in peripheral sense organs, such as the ear. Axons of sensory neurons enter the brain, synapse with assemblies of neurons called *sensory nuclei,* produce the appropriate sensation (hearing, for example, in the case of the ear), and play other sensory roles.

The cranial nerves are concerned with the *specialized (special) senses* of smell, taste, vision, hearing, and balance, and the *general senses.* They are also involved with the *specialized motor activities* of eye movement, chewing, swallowing, breathing, speaking, and facial expression. The vagus nerve is an exception, projecting fibers to organs in the abdomen and thorax. In the following sections you will see how individual nerves attend to these diverse functions.

*To remember the cranial nerves and their order, consider the following rhyme, in which each word begins with the same letter as the relevant cranial nerve: On (I) Old (II) Olympus (III) Towering (IV) Tops (V) A (VI) Finn (VII) and (VIII; Auditory) German (IX) Viewed (X) Some (XI; Spinal accessory) Hops (XII).

FIGURE 14.16 ORIGINS OF THE CRANIAL NERVES

Basal view of the brain (center), showing the distribution of cranial nerves to relevant body parts.

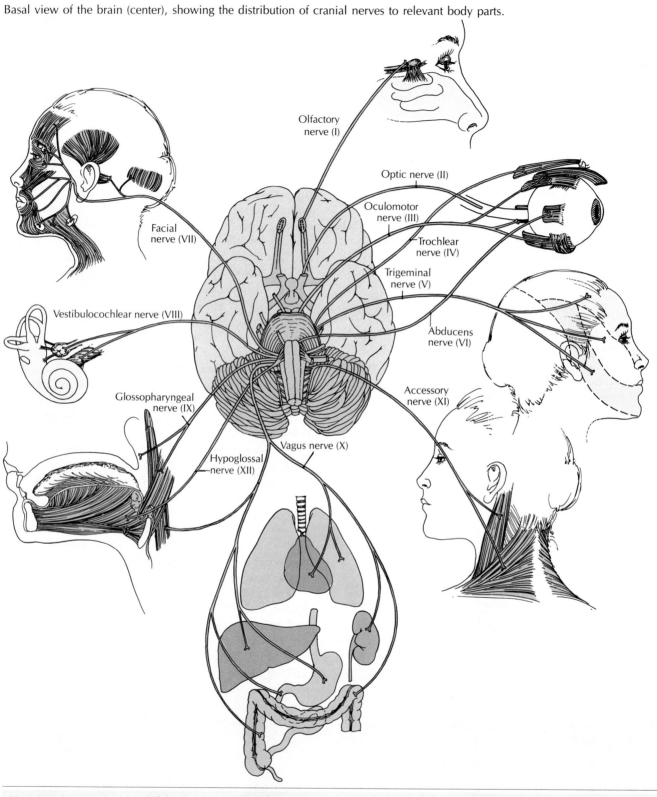

Nerve/type	Origin	Distribution	Function
I Olfactory (sensory) [FIGURE 14.17]	Nasal mucous membrane high in nasal cavities	Terminates in olfactory bulb of cerebrum.	Smell (olfaction).
II Optic (sensory) [FIGURE 14.18]	Retina of eye	Terminates in lateral geniculate body of thalamus and superior colliculus of midbrain.	Vision. Afferent limb of reflex of focusing by adjusting lens and constricting pupil.

III Oculomotor (motor) [FIGURE 14.18]	Midbrain	To all extrinsic muscles of eyeball except superior oblique and lateral rectus; also autonomic fibers to ciliary muscles of lens and constrictor muscles of iris.	Movements of eyeball, elevation of upper eyelid, constriction of pupil, focusing by lens (accommodation).
IV Trochlear (motor) [FIGURE 14.18]	Caudal midbrain	Innervates superior oblique muscle of eye.	Eye movements (down and out).
V Trigeminal [FIGURE 14.19] Ophthalmic nerve (V¹) (sensory)	General area of forehead, eyes	General area of forehead, eyes.	Conveys general senses from cornea of eyeball, upper nasal cavity, front of scalp, forehead, upper eyelid, conjunctiva, lacrimal (tear) gland.
Maxillary nerve (V²) (sensory)	General area of maxillary region	General area of maxillary region.	Conveys general senses from cheek, upper lip, upper teeth, mucosa of nasal cavity, palate, parts of pharynx.
Mandibular nerve (V³) (mixed)	*Sensory:* General area of mandibular region *Motor:* pons	*Sensory:* general area of mandibular region. *Motor:* innervates muscles of mastication.	*Sensory:* conveys general senses from tongue (not taste), lower teeth, skin of lower jaw. *Motor:* chewing.
VI Abducens (motor) [FIGURE 14.18]	Caudal pons	Innervates lateral rectus muscle of eye.	Abduction of eye (lateral movement).
VII Facial (mixed) [FIGURE 14.20]	Pons	*Sensory:* innervates taste buds of tongue. *Motor:* innervates muscles of facial expression; autonomic fibers to salivary glands, lacrimal glands.	*Sensory:* taste. *Motor:* salivation, lacrimation, movement of muscles of facial expression.
VIII Vestibulocochlear [FIGURE 14.21] Cochlear (sensory)	Cochlea of inner ear	Cochlea of inner ear.	Hearing.
Vestibular (sensory)	Semicircular ducts, utricle, and saccule of inner ear	Semicircular ducts, utricle and saccule of inner ear.	Equilibrium.
IX Glossopharyngeal (mixed) [FIGURE 14.22]	Medulla oblongata	*Sensory:* conveys taste from posterior third of tongue, general senses from upper pharynx. *Motor:* innervates stylopharyngeus muscle; autonomic fibers stimulate parotid gland.	Taste, other sensations of tongue. Secretion of saliva; swallowing.
X Vagus† (mixed) [FIGURE 14.23]	Medulla oblongata	Voluntary muscles of soft palate, cardiac muscle, smooth muscle in respiratory, cardiovascular, digestive systems.	Swallowing. Monitors oxygen and carbon dioxide concentrations in blood, senses blood pressure, other visceral activities of affected systems.
XI Accessory† (spinal accessory) (motor) [FIGURE 14.24]	Medulla oblongata, cervical spinal cord	Muscles of larynx, strenocleidomastoid, trapezius.	Voice production (larynx); muscle sense; movement of head, shoulders.
XII Hypoglossal* (motor) [FIGURE 14.25]	Medulla oblongata	Tongue muscles.	Movements of tongue during speech, swallowing; muscle sense.

*Contains some sensory proprioceptive fibers from extraocular muscles of eyes and tongue muscles, which leave nerve and join trigeminal nerve (see FIGURE 14.16).

†Some fibers of spinal accessory nerve innervating muscles of the larynx join the vagus nerve; hence, these muscles are usually said to be innervated by the vagus nerve.

FIGURE 14.17 DISTRIBUTION OF THE OLFACTORY NERVE (I)

Right lateral view.

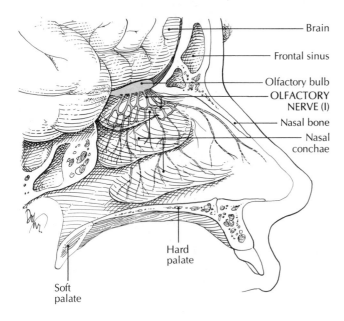

Brain

Frontal sinus

Olfactory bulb

OLFACTORY NERVE (I)

Nasal bone

Nasal conchae

Hard palate

Soft palate

Cranial Nerve I: Olfactory

The *olfactory nerve* (L. *olfacere,* to smell) is strictly a sensory nerve [FIGURE 14.17]. The 10 to 20 million bipolar neurons *(olfactory cells)* of the olfactory nerve are located high in the nasal cavities within nasal epithelium. They act both as chemoreceptors that sense odors and as conductors of impulses that ultimately result in the perception called smell. The unmyelinated axons of olfactory cells pass through the foramina of the cribriform plate of the ethmoid bone and synapse with neurons in the axons of the *olfactory bulb,* which is actually an "appendage" of the brain. The axons of the bipolar olfactory neurons synapse with neurons within the olfactory bulb. The axons of these olfactory bulb neurons form the *olfactory tract,* which conveys impulses to the primary olfactory cortex of the uncus of the cerebrum.

Cranial Nerve II: Optic

The *optic nerve* is a special sensory nerve [FIGURE 14.18]. It conveys impulses that result in vision and in reflexes associated with vision. Each optic nerve is actually a tract, composed of about a million axons that arise from the ganglion cells of the retina. In the retina, rods and cones (the photoreceptive cells), interneurons, and bipolar neurons form a circuitry that interacts with ganglion neurons.

Each optic nerve passes out of the orbit through the optic foramen into the cranial cavity [FIGURE 14.16]. While passing through the X-shaped *optic chiasma* (kye-AZ-muh; so named because its X shape reminded early anatomists of the Greek letter *chi* [KYE], which also has an X shape), the axons of the medial half of the retinas of both eyes cross the midline. The axons from the lateral half of the retinas of both eyes do not cross the midline. One-half of the fibers of each optic nerve of each eye then cross over to the other side to form the optic chiasma. After passing through the optic chiasma, the axons continue, forming the two *optic tracts.*

The axons of each optic tract terminate mainly in the lateral geniculate body of the thalamus. They synapse with neurons that project as the optic radiations to the primary visual cortex in the occipital lobe of the cerebrum. Other fibers terminate in the superior colliculus of the midbrain. The midbrain, through its connections with cranial nerves III, IV, and VI, is involved with subconscious visual reflexes (eye movements, pupillary responses).

The *pupillary response,* or *light reflex,* in which the pupil of the eye dilates or constricts in response to different amounts of light, is an unconscious reflex in which the stimulation comes directly from the retina to the superior colliculus and cranial nerve III (oculomotor). In contrast, *focusing,* or the *accommodation reflex,* is a conscious reflex in the sense that the observer can elect where to focus the eyes. Stimuli are derived from the visual cortex and conveyed via corticocollicular fibers to the superior colliculus and oculomotor nerve.

Cranial Nerve III: Oculomotor

Cranial nerves III (oculomotor), IV (trochlear), and VI (abducens) are classified as the *extraocular motor nerves* because they innervate the extraocular (extrinsic) muscles that move the eyeball. The *oculomotor nerve* (L. *oculus,* eye + motor, or "eye movement") innervates most of the muscles that move the eye. In addition, it innervates the levator palpebrae superioris muscle, which elevates the upper eyelid, and the smooth muscles within the eyeball that constrict the pupil (usually in response to bright light) and regulates the focusing of the lens.

The brainstem motor nucleus, including the parasympathetic fiber of the autonomic nervous system, of the oculomotor nerve is located in the midbrain [FIGURE 14.18]. The oculomotor nerve emerges from the ventral surface of the midbrain, passes through the superior orbital fissure, and enters the orbit. It innervates the superior rectus, inferior rectus, medial rectus, levator palpebrae superioris, and inferior oblique muscles. The sensory fibers from the proprioceptors (neuromuscular spindles) in the muscles innervated by cranial nerves III,

IV, and VI join the ophthalmic nerve, and have their cell bodies in the trigeminal ganglion [FIGURE 14.18].

The parasympathetic fibers of the oculomotor nerves innervate the constrictor muscles of the pupil (usually a response to bright light) and also the ciliary muscles, which regulate the tension on the lens for focusing (accommodation).

Cranial Nerve IV: Trochlear

The *trochlear nerve* (L. pulley) is so named because it innervates the superior oblique muscle, whose tendon passes through a cartilaginous pulleylike sling on its way to its insertion into the sclera (outer covering) of the eyeball. The nerve arises from its nucleus in the caudal midbrain, emerges from the dorsal surface of the midbrain, and passes forward through the superior orbital fissure into the orbit, where it innervates the superior oblique muscle. The trochlear nerve is long and slender, allowing it to follow a winding course [FIGURE 14.18].

When the trochlear nerve is injured, double vision may result, especially when looking downward. As a result, a person with an injured trochlear nerve might have trouble walking downstairs or stepping off a curb into the

FIGURE 14.18 DISTRIBUTION OF THE OPTIC, OCULOMOTOR, TROCHLEAR, AND ABDUCENS NERVES (II, III, IV, VI)

Right lateral view.

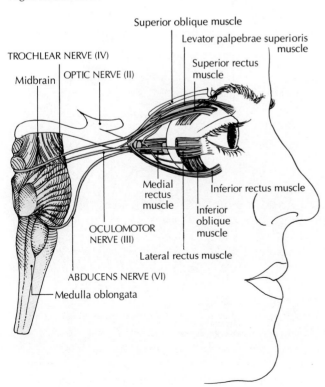

street. (See TABLE 14.2 on page 387 for an outline of some major disorders of the cranial nerves.)

Cranial Nerve V: Trigeminal

The *trigeminal nerve* (L. *tri*, three + *geminus*, twin, referring to its three major branches and two roots) is the largest, but not the longest, of the cranial nerves [FIGURE 14.19]. It is a mixed nerve, being the chief facial sensory nerve and the motor nerve of the muscles of mastication. The sensory nerves convey impulses of the general senses, similar to those found in the spinal nerves. All the sensory fibers terminate in trigeminal nuclei in the brainstem. The motor fibers originate from the trigeminal motor nucleus in the pons.

The sensory root has a sensory ganglion, located close to where the nerve emerges from the pons within the middle cranial fossa. The cell bodies of the sensory nerves are located in this *trigeminal ganglion* (the equivalent of the dorsal root ganglion of a spinal nerve).

After the trigeminal nerve emerges from the pons, it divides into three branches, the ophthalmic, maxillary, and mandibular nerves.

1 The *ophthalmic nerve* (V^1),* a sensory nerve, passes through the superior orbital fissure, through the orbit, and into the upper head. Its branches convey the general senses from the front of the scalp, forehead, upper eyelid, conjunctiva (inner membrane of the eyelid), cornea of the eyeball, and the upper nasal cavity.

2 The *maxillary nerve* (V^2), a sensory nerve, passes through the foramen rotundum. Its branches convey general sensations from the skin of the cheek, upper lip, upper teeth, mucosa of the nasal cavity, palate, and parts of the pharynx. It covers the general area of the maxillary bone.

3 The *mandibular nerve* (V^3), a mixed nerve, passes from the skull through the foramen ovale. Its branches convey general senses from the mucosa of the mouth (including the tongue, but not sensations of taste), lower teeth, and skin around the lower jaw. The motor fibers innervate muscles of mastication. The mandibular nerve is distributed to the general region of the mandible.

Cranial Nerve VI: Abducens

The *abducens nerve* (L. *abducere*, to lead away) is an extraocular motor nerve, together with cranial nerves III and IV. The brainstem nucleus is located in the caudal pons. The nerve emerges at the junction of the pons and medulla oblongata and passes forward through the supe-

*Branches of cranial nerves are designated by superscript numerals in the order of their branching from the main cranial nerve.

FIGURE 14.19 DISTRIBUTION OF THE TRIGEMINAL NERVE (V)

[A] Motor nerves. [B] Sensory nerves. Right lateral views.

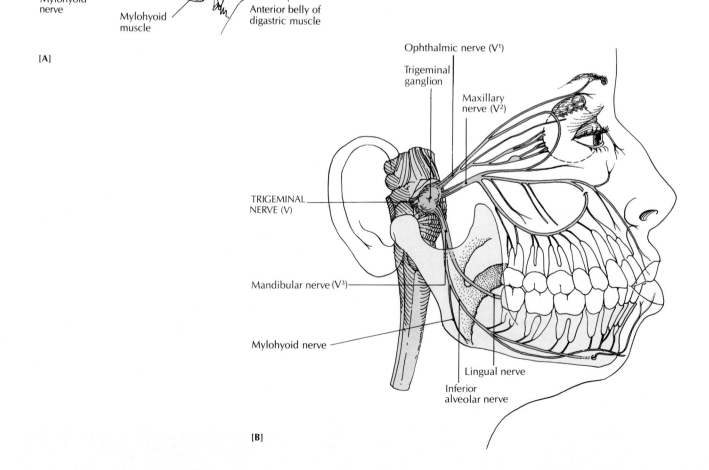

Motor branches to muscles of mastication (temporalis, masseter, external pterygoid, internal pterygoid, mylohyoid and anterior belly of digastric)

External pterygoid muscle

V¹
V²
V³

Masseter muscle

Mylohyoid nerve

Mylohyoid muscle

Anterior belly of digastric muscle

[A]

Ophthalmic nerve (V¹)

Trigeminal ganglion

Maxillary nerve (V²)

TRIGEMINAL NERVE (V)

Mandibular nerve (V³)

Mylohyoid nerve

Lingual nerve

Inferior alveolar nerve

[B]

FIGURE 14.20 DISTRIBUTION OF THE FACIAL NERVE (VII)

Right lateral view.

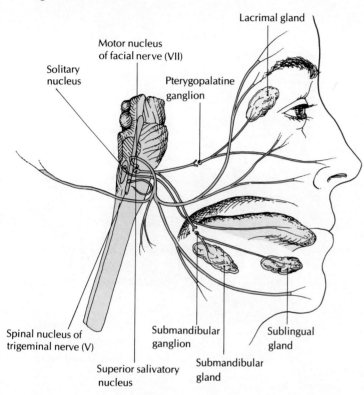

rior orbital fissure into the orbit, where it innervates the lateral rectus muscle to *abduct* (turn outward) the eye, hence the name *abducens* [FIGURE 14.18].

Cranial Nerve VII: Facial

The *facial nerve* is a mixed nerve. It emerges from the junction of the pons and medulla oblongata and passes through the internal auditory meatus and the facial canal of the temporal bone; its branches leave the skull from the stylomastoid foramen and other foramina [FIGURE 14.20]. The *geniculate ganglion* in the petrous portion of the temporal bone is the sensory ganglion of the facial nerve.

The sensory fibers of the chorda tympani nerve innervate the taste buds of the anterior two-thirds of the tongue. The fibers terminate in the solitary tract nucleus in the brainstem. Fibers project from this nucleus into the thalamus, and then to the gustatory (taste) area of the cerebral cortex.

The motor fibers originate from the motor nucleus of the facial nucleus in the pons and innervate all the muscles of facial expression. Parasympathetic (autonomic nervous system) efferent fibers originating in the pterygopalatine and submandibular ganglia innervate the lac-

rimal (tear) glands and the submandibular and sublingual salivary glands.

Cranial Nerve VIII: Vestibulocochlear

The *vestibulocochlear nerve* (L. entranceway + snail shell) is a sensory nerve composed of two nerves, the cochlear nerve and the vestibular nerve. The cochlear nerve conveys impulses concerned with hearing, and the vestibular nerve conveys information about equilibrium, or balance, and the position and movements of the head.

The fibers of the *cochlear nerve* arise at their synapses with the hair cells of the spiral organ in the snail-shaped cochlea of the inner ear [FIGURE 14.21]. The cell bodies of the cochlear nerve are located in the spinal ganglion near the cochlea. The axons of the cochlear neurons terminate in the cochlear nuclei of the upper medulla oblongata. Auditory pathways from the cochlear nuclei terminate in the medial geniculate body of the thalamus. From the thalamus, impulses are relayed via fibers of the auditory radiations to the auditory cortex in the temporal lobe of the cerebrum.

The *vestibular nerve* arises at its synapse with the hair cells in the semicircular canals, utricle, and saccule of the inner ear. The cell bodies of the vestibular nerve are located in the vestibular ganglion within the petrous part of the temporal bone. Its axons terminate in the vestibular nuclei in the upper medulla oblongata.

FIGURE 14.21 DISTRIBUTION OF THE VESTIBULOCOCHLEAR NERVE (VIII)

Right lateral view.

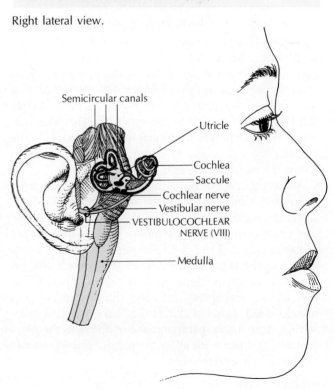

FIGURE 14.22 DISTRIBUTION OF THE GLOSSOPHARYNGEAL NERVE (IX)

Right lateral view.

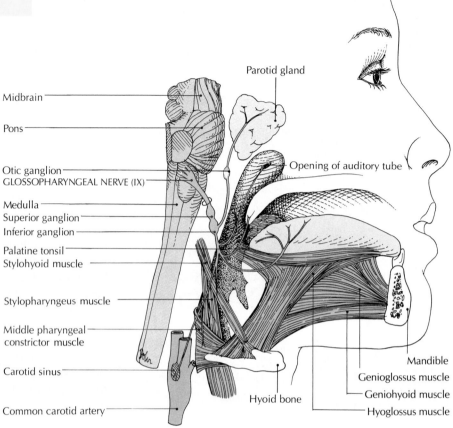

Cranial Nerve IX: Glossopharyngeal

The *glossopharyngeal nerve* (Gr. tongue + throat) is a small mixed nerve. After the nerve emerges from the upper medulla oblongata, it leaves the posterior cranial fossa through the jugular foramen [FIGURE 14.22]. Its branches are distributed to the region of the posterior third of the tongue and to the upper pharynx. The sensory fibers terminate in several sensory nuclei in the medulla oblongata, and its motor fibers originate from motor nuclei in the medulla oblongata.

Sensory fibers convey taste and general senses from the posterior third of the tongue and general senses from adjacent structures of the upper pharynx. When the sensory nerve in the back of the mouth is stimulated, the act of swallowing and the gag reflex are triggered. A small branch of the nerve terminating in the carotid sinus (in the neck) monitors blood pressure. Motor fibers (from the nucleus ambiguus) innervate the stylopharyngeus muscle, which helps to elevate the pharynx during swallowing. Autonomic (parasympathetic) fibers in the glossopharyngeal nerve stimulate the parotid gland (anterior to the ear) to secrete saliva.

Cranial Nerve X: Vagus

The *vagus nerve* (L. wandering) is a mixed nerve. It is so named because it has the most extensive distribution of any cranial nerve [FIGURE 14.23]. The vagus nerve is the longest cranial nerve, innervating structures in the head, neck, thorax, and abdomen. Two different types of motor fibers are apparent:

1 The fibers that innervate the skeletal muscles of the soft palate, pharynx, and larynx originate in a motor nucleus (nucleus ambiguus) in the medulla oblongata. These muscles are involved in swallowing and speaking.

2 Originating from another nucleus (dorsal vagal nucleus) in the medulla oblongata are fibers of the parasympathetic system, which convey impulses involved with the activity of cardiac muscle, smooth muscle, and exocrine glands (glands with ducts) of the cardiovascular, respiratory, and digestive systems.

The sensory fibers of the vagus nerve convey impulses from all the structures of the systems innervated by the motor fibers. The sensory fibers terminate in the solitary

FIGURE 14.23 DISTRIBUTION OF THE VAGUS NERVE (X)

Right lateral view.

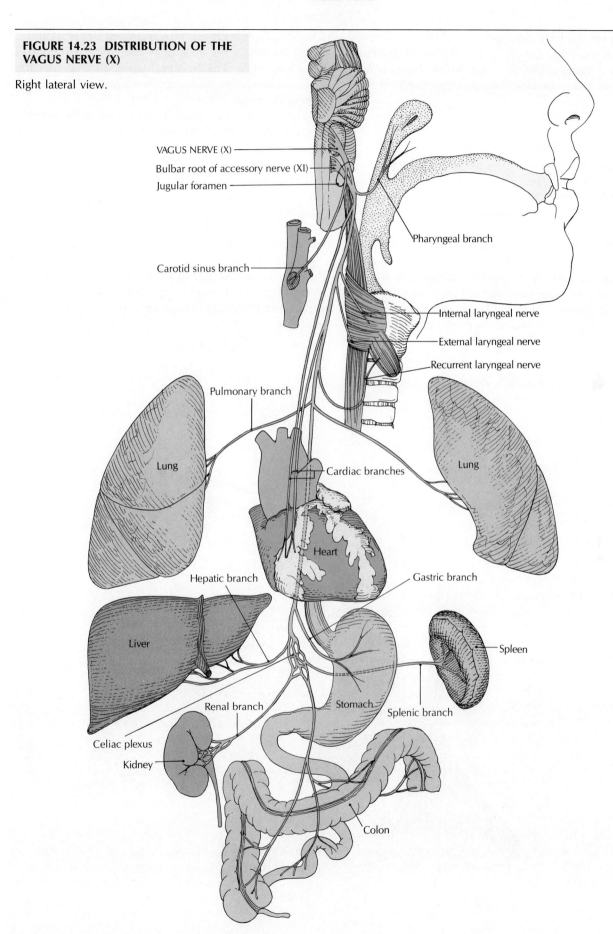

- VAGUS NERVE (X)
- Bulbar root of accessory nerve (XI)
- Jugular foramen
- Carotid sinus branch
- Pharyngeal branch
- Internal laryngeal nerve
- External laryngeal nerve
- Recurrent laryngeal nerve
- Pulmonary branch
- Lung
- Cardiac branches
- Lung
- Heart
- Hepatic branch
- Gastric branch
- Liver
- Spleen
- Renal branch
- Stomach
- Splenic branch
- Celiac plexus
- Kidney
- Colon

nucleus in the medulla oblongata. Some of the afferent fibers convey to higher brain centers the blood concentrations of oxygen and carbon dioxide, and blood pressure changes in the aorta, carotid sinus, and carotid body.

Cranial Nerve XI: Accessory

The *accessory* (or *spinal accessory*) *nerve* is a mixed nerve originating from both the medulla oblongata and the cervical spinal cord [FIGURE 14.24]. Both the bulbar (medullary) and spinal roots are composed primarily of motor fibers.

The *bulbar root* arises from a motor nucleus (nucleus ambiguus) in the medulla oblongata, joins the spinal root, and passes through the jugular foramen. The fibers of the bulbar root leave the fibers of the spinal root, and join the vagus nerve. Thus the larynx is actually innervated by the accessory nerve, which joins the vagus nerve.

The fibers eventually form the recurrent laryngeal nerve, which innervates the muscles of the larynx that control the vocal cords. The laryngeal nerve also supplies the sensory innervation of the upper trachea and the region around the vocal cords.

The *spinal root* originates from neurons in the anterior gray horn of the first five cervical spinal levels. The fibers of the spinal root emerge from the spinal cord, ascend within the vertebral canal, pass through the foramen magnum, join the bulbar root, and pass through the jugular foramen. The spinal fibers leave the bulbar root fibers and pass into the neck to innervate the sternocleidomastoid and trapezius muscles. Both muscles are involved with movements of the head.

The proprioceptive fibers from these muscles pass through the cervical nerves to the spinal cord. Other afferent fibers, for example those from the region of the vocal cords, terminate in trigeminal nuclei of the medulla oblongata.

FIGURE 14.24 DISTRIBUTION OF THE ACCESSORY NERVE (XI)

Right lateral view.

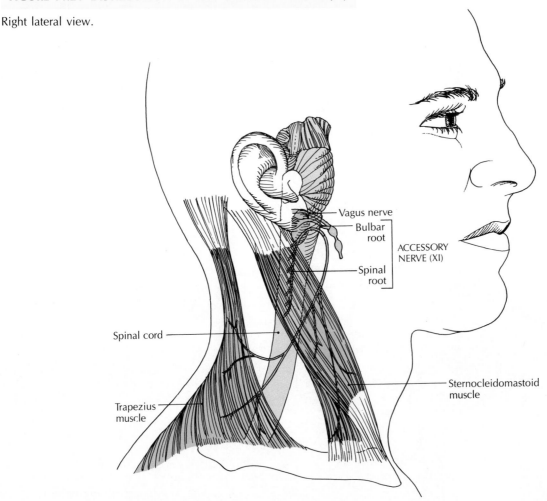

FIGURE 14.25 DISTRIBUTION OF THE HYPOGLOSSAL NERVE (XII)

Right lateral view.

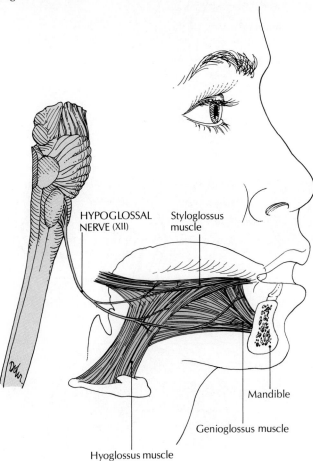

HYPOGLOSSAL NERVE (XII)

Styloglossus muscle

Mandible

Genioglossus muscle

Hyoglossus muscle

Cranial Nerve XII: Hypoglossal

The *hypoglossal nerve* ("under the tongue") is a motor nerve. It innervates the muscles of the tongue. It also contains some proprioceptive fibers of muscle sense. The motor fibers originate from the motor nucleus of the hypoglossal nerve in the medulla oblongata, emerge from the medulla oblongata, and pass through the hypoglossal canal into the region of the floor of the mouth [FIGURE 14.25].

The branches of the hypoglossal nerve innervate the intrinsic muscles of the tongue, and three extrinsic muscles: hypoglossus, genioglossus, and styloglossus. These muscles are important in manipulating food within the mouth, in speaking, and in swallowing. The proprioceptive fibers from the tongue join the mandibular nerve, and have their cell bodies in the trigeminal ganglion of cranial nerve V.

DEVELOPMENTAL ANATOMY OF THE BRAIN

The central nervous system develops from cells that first form a flat plate. This *neural plate* is soon transformed by the rolling of its edges into a structure called the *neural tube*, which will develop into the brain and spinal cord [FIGURE 14.26B]. The hollow part of the tube will become the ventricular system of the brain.

By about the fourth fetal week, the neural tube enlarges to form three cavities, or vesicles, called the *prosencephalon* (forebrain), *mesencephalon* (midbrain), and the *rhombencephalon* (hindbrain), which grow and differentiate according to their own patterns of development [FIGURE 14.26C]. By the fifth week, cranial nerves III through XII of the embryo have already begun to develop as offshoots from the brainstem. In the seven-week embryo, the prosencephalon has divided into two secondary vesicles, known as the *telencephalon* and the *diencephalon;* the rhombencephalon has also divided into two vesicles, called the *metencephalon* (cerebellum and pons) and the *myelencephalon* (medulla oblongata) [FIGURE 14.26E].

By the end of the third fetal month, the overall form of the brain is recognizable, but the surface is still smooth [FIGURE 14.26F]. *Sulci* begin to develop in the cerebrum during the fourth month, and the folds of the cerebellum are clearly formed in a six-month fetus [FIGURE 14.26G]. By the eighth month, the major sulci are present, and the occipital lobe dominates the underlying cerebellum [FIGURE 14.26H].

The brain weighs about 350 g at birth, and after a year it weighs about 1000 g. Brain growth slows down thereafter, until full growth is attained during puberty.

FIGURE 14.26 DEVELOPMENTAL ANATOMY OF THE BRAIN

[A, B] Dorsal views. [C–I] Right lateral views.

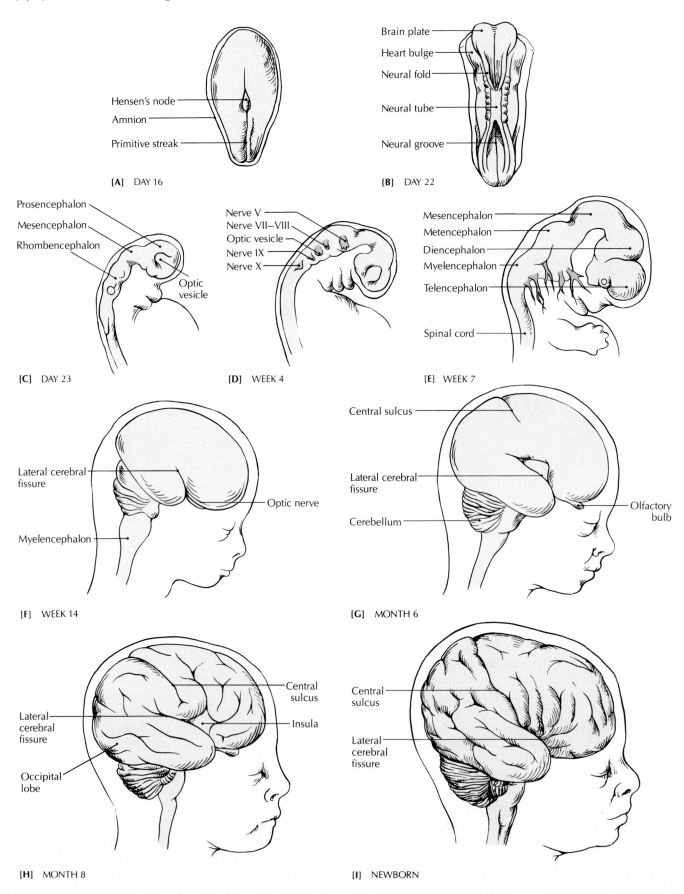

Hensen's node
Amnion
Primitive streak

[A]   DAY 16

Brain plate
Heart bulge
Neural fold
Neural tube
Neural groove

[B]   DAY 22

Prosencephalon
Mesencephalon
Rhombencephalon
Optic vesicle

[C]   DAY 23

Nerve V
Nerve VII–VIII
Optic vesicle
Nerve IX
Nerve X

[D]   WEEK 4

Mesencephalon
Metencephalon
Diencephalon
Myelencephalon
Telencephalon
Spinal cord

[E]   WEEK 7

Lateral cerebral fissure
Myelencephalon
Optic nerve

[F]   WEEK 14

Central sulcus
Lateral cerebral fissure
Cerebellum
Olfactory bulb

[G]   MONTH 6

Central sulcus
Lateral cerebral fissure
Occipital lobe
Insula

[H]   MONTH 8

Central sulcus
Lateral cerebral fissure

[I]   NEWBORN

WHEN THINGS GO WRONG

The nervous system expresses itself through the sensory system and associated sensations, or through the motor system and the activity of muscles and glands. Malfunctioning of any of these systems due to injury or disease can be noted through clinical signs and symptoms. The signs may be an increase, decrease, or distortion of a sensation, as, for example, greater pain, less or absent pain, or a distortion of pain, as in a phantom limb. (Following the amputation of a foot, for example, the patient can often feel excruciating pain where the foot used to be; so-called phantom pain.) A malfunction in the motor system may be indicated by greater, less, or distorted activity by the muscular and glandular systems. The enhancement of sensations and motor activities following a lesion can be explained in the following way: the injured structure may contain centers that regulate other centers by inhibiting them. Following the injury, the inhibition is decreased, and the ordinarily inhibited centers express themselves without the usual controls. The result may be a more intense feeling of sensations and greater or unwanted motor activity.

Because disturbances of the brain are so numerous, we will give only a brief description of some of the major ones here. We will also summarize some of the major disorders of cranial nerves.

Senility

Some brain neurons, which are irreplaceable, begin to deteriorate at a relatively early age. Because of a surplus of brain neurons, however, a shrinkage (atrophy) of the brain and a slowing down of mental processes usually do not occur until after the age of 60; sometimes they do not occur at all. Severe atrophy of the brain is commonly called "senility." The more technical name is ***senile dementia*** (L. *senex*, old man + *demens*, out of mind). It is characterized by progressive mental deterioration, including anxiety, irritability, difficulty with speech, and irrationality. In most cases, senility does not occur until after the age of 70. However, it should be emphasized that senility is a disease, and definitely *not* a normal condition of aging.

Alzheimer's Disease

In a few cases, brain atrophy may occur much earlier than usual during the normal course of aging, even as early as 30 or 40 years of age, with the symptoms being identical to senile dementia. Such a condition is known as ***Alzheimer's disease***, a neurodegenerative disease characterized by a progressive loss of memory, intellectual functions, and speech. Like senility, Alzheimer's disease is not a part of the normal aging process. Alzheimer's disease afflicts

more than 3 million Americans and kills about 120,000 of them a year, making it the fourth leading cause of death among the elderly (after cardiovascular disease, cancer, and stroke). Death usually occurs less than 10 years after the first symptoms appear.

The death of certain brain cells appears to be the basis for many of the symptoms associated with Alzheimer's disease. Autopsies of victims invariably reveal significant changes in the brain, including a reduced number of neurons, especially in areas related to thought processes, but also in neurons with cell bodies in a nucleus (basal nucleus of Meynert) located in the forebrain just anterior to the hypothalamus, and axons that release acetylcholine at their terminals in the cerebral cortex; an increased amount of a protein called *amyloid* ("starchlike") in and around blood vessels; and an accumulation of *tangled neurofilaments* and *neuritic plaques* (cellular debris and amyloid). Victims of Alzheimer's disease also have a reduced amount of the enzyme *protein kinase C*, which aids in the proper uptake of phosphate and the cascade of cellular events that follow. It is possible that increased amounts of calcium in the elderly brain help to inhibit protein kinase C. Finally, there is a marked reduction in the release of acetylcholine. This reduction occurs when certain cholinergic neurons with their cell bodies in the basal nucleus and terminating fibers in wide areas of the cerebral cortex are functionally impaired. Presumably, the loss of acetylcholine inhibits memory formation.

The cause of Alzheimer's disease is unknown. It may have a genetic basis, although the disease is not necessarily inherited, and most cases occur sporadically. There is no proven effective treatment at present.

Cerebral Palsy (CP)

Cerebral palsy (CP)* actually comprises a group of neuromuscular disorders that usually result from damage to a child either before it is born, during childbirth, or shortly after birth. Actually, the brain can be affected at any time from fertilization through infancy. The major types of cerebral palsy are *spastic*, *athetoid*, and *ataxic*. All three forms involve an impairment of skeletal motor activity to some degree, ranging from muscular weakness to complete paralysis. Related disorders such as mental retardation and speech difficulties may accompany the disease. If the functional handicap affects motor performance primarily, it is assumed that the basal ganglia of the brain are predominantly involved, and the patient is said to

---

*"Palsy" is a distorted form of *paralysis*, which comes from the Greek word meaning "to loosen," because the muscles seem to fall loose.

have *cerebral* palsy specifically. Causes are varied, from infection and malnutrition of the mother to prolonged labor, brain infection, or circulatory problems. There is no known cure.

Cerebrovascular Accident (CVA)

A *cerebrovascular accident* (CVA), commonly called a *stroke,* is a sudden withdrawal of sufficient blood supply to the brain caused by the impairment of incoming blood vessels. The resulting oxygen deficiency causes brain tissue to be damaged or even destroyed (*infarction*). Unconsciousness results if the blood and oxygen supply to the brain is insufficient for as little as 10 seconds. Irreversible brain damage can occur if the brain is deprived of oxygen for 5 minutes or more. CVA kills about half of the people it strikes, and about half of those who survive are permanently disabled. Cerebrovascular accident is the most common brain disorder, and the third most common cause of death in the United States (after cardiovascular disease and cancer).

The major causes of CVA are *thrombosis* (Gr. a clotting), the clotting of blood in a blood vessel; *embolism* (Gr. "to throw in"), a blockage in a blood vessel; and *hemorrhage,* a rupture of a cerebral artery as the result of an *aneurysm,* local dilation or ballooning of an artery. As a result, the blood supply to the brain is reduced, and blood leaks through the dura mater and puts harmful pressure on brain tissue, sometimes destroying it. The risk of thrombosis, embolism, or hemorrhage is increased by atherosclerosis (artery disease), hypertension (high blood pressure), diabetes mellitus, a high-fat diet, cigarette smoking, and lack of exercise.

Epilepsy

Epilepsy (Gr. *epilambanein,* seize upon) is a nervous disorder characterized by recurring attacks of motor, sensory, or psychological malfunction, with or without unconsciousness or convulsive movements. The term *epilepsy* refers to a group of symptoms with many causes, rather than to a specific disease. In *symptomatic epilepsy,* seizures can be traced to one of several known causes, including a brain tumor or abscess, diseases that affect central blood vessels, and poisons. Epilepsy is often caused by brain damage before, during, or shortly after birth. The more common type of *epilepsy* is *idiopathic* (Gr. a disease of unknown cause), in which brain cells act abnormally for no apparent reason. The disease is known as *grand mal* (Fr. great illness) when the motor areas of the brain are affected and severe spasms and loss of consciousness are involved. It is called *petit mal* (small illness) when the sensory areas are affected, without convulsions and prolonged unconsciousness. During an epileptic seizure, neurons in the brain fire at unpredictable times, even without a stimulus.

Headache

Headaches related to muscle tension were discussed in Chapter 11. Here we consider headaches that are connected with cranial nerves, blood vessels, and meninges.

The brain itself is not sensitive to pain, but the veins on the surface of the brain, the cerebral arteries, cranial nerves, and parts of the dura mater are. If any of these areas is disturbed, a *headache* may result. Pressure on cranial veins, arteries, or meninges is sometimes produced by tumors, hemorrhage, meningitis, or an inflamed trigeminal nerve root. However, some of the more typical causes are emotional stress, increased blood pressure, and food allergies that make blood vessels dilate or constrict, stimulating the pain-sensitive nerve endings in the vessels. The resulting headache may be accompanied by dizziness or vertigo.

Migraine headaches are severe, recurring headaches that usually affect only one side of the head. They are often preceded by fatigue, nausea, vomiting, and the vision of zigzag lines or brightness; they may be accompanied by intense pain, nausea, vomiting, and sensitivity to light and noise. Among the causes are emotional stress, hypertension, menstruation, and certain foods such as chocolate, animal fats, and alcohol. Migraine headaches frequently occur within families (suggesting a genetic basis in some cases), in compulsive, tense people, and, interestingly, on weekends and holidays, when the normal rhythm is disrupted.

Trigeminal neuralgia (*tic douloureux*) is a stabbing pain in one side of the face, along the path of one or more branches of the trigeminal nerve. It usually occurs sporadically. The cause is unknown.

Cranial arteritis is marked by intense pain and tenderness in the temples when the arteries in that region become inflamed. Untreated cranial arteritis may block the artery leading to the retina (central retinal artery) and cause blindness, especially in the elderly.

Parkinson's Disease

Parkinson's disease (also called Parkinsonism, paralysis agitans, and shaking palsy) is a progressive neurological disease characterized by stiff posture, an expressionless face, slowness in voluntary movements, tremor at "rest," and a shuffling gait. The involuntary tremor, especially of the hands, often disappears when the upper limb moves. The tremor is accompanied by "pill-rolling" actions of the thumb and fingers. Parkinson's disease results from a deficiency of the neurotransmitter *dopamine,* released by neurons with their cell bodies in the substantia nigra and their axons terminating in the striatum (caudate nucleus and putamen). Such a shortage of dopamine usually occurs when the neurons projecting from the substantia nigra to the striatum of the basal ganglion degenerate. The deficiency prevents brain cells from performing their usual inhibitory functions within the intricate circuitry of the basal ganglia in the central nervous system.

Parkinson's disease affects more men than women, typically those over 60. It usually progresses for about 10 years, at which time death generally occurs from other causes, such as pneumonia or another infection.

At present, there is no known cure. Treatment consists of drugs and physical therapy that keep the patient functional as long as possible. Short-term drug therapy usually includes *levodopa* (L-dopa), a dopamine substitute, although unpleasant side effects, including nausea and vomiting, are common. Dopamine itself cannot be administered as a drug because it does not penetrate the blood-brain barrier. Impassable drugs may be combined with fat-soluble drugs that allow them to pass through the barrier and then separate after the drug has entered the brain. New drugs that help to counter the side effects of levodopa are currently being used in conjunction with that drug.

Disorders of Cranial Nerves and Bell's Palsy

Disorders involving cranial nerves range from infections that may damage nerve fibers to serious injuries, cerebral strokes, or tumors that may produce paralysis and the loss of some sensations [TABLE 14.2].

Bell's palsy results from the dysfunction of the facial nerve (VII). The disorder may occur without any known cause. It often follows prolonged exposure to a cold breeze, and is thought to result from the swelling of the nerve within the facial canal. Symptoms occur on the same side as the lesion. They include weakness or paralysis on one side of the face, facial distortion, the inability to close the eye (the eye rolls upward when closing is attempted; called Bell's phenomenon) or wrinkle the forehead, a loss of taste on the anterior two-thirds of the tongue, and a drooping mouth (which leads to excessive drooling). Bell's palsy is most common between the ages of 40 and 65. Recovery is usually spontaneous and complete in less than two months, but partial recovery occurs in about 10 percent of the cases. Treatment with predni-

TABLE 14.2 SOME DISORDERS OF CRANIAL NERVES

Nerve	Disorder
I Olfactory	Fracture of cribriform plate of ethmoid bone or lesions along olfactory pathway may produce total inability to smell *(anosmia);* as a result, food tastes flat.
II Optic	Trauma to orbit or eyeball, or fracture involving optic foramen produces inability of pupil to constrict. Certain poisons (such as wood alcohol), or infections (such as syphilis) may damage nerve fibers. Increase in eye fluid (aqueous humor) pressure produces glaucoma, causing partial or complete loss of eyesight. Pressure on optic pathway or injury or clot in temporal, parietal, or occipital lobes of cerebrum may cause blindness in certain regions of visual fields.
III Oculomotor	Lesion or pressure on nerve produces dilated pupil, drooping upper eyelid *(ptosis)*, absence of direct pupil reflex, squinting *(strabismus)*, double vision *(diplopia)*.
V Trigeminal	Injury to terminal branches or trigeminal ganglion causes loss of pain and touch sensation, abnormal tingling, itching, numbness *(paresthesias)*, inability to contract chewing muscles, deviation of mandible to side of lesion when mouth is opened. Severe spasms in nerve branches cause nerve pain *(trigeminal neuralgia, or tic douloureux)*.
VI Abducens	Nerve lesion causes inability of eye to move laterally, with eye cocked in; diplopia; strabismus.
VII Facial	Damage causes Bell's palsy, loss of taste on anterior two-thirds of tongue on side of lesion. Cerebral stroke may produce paralysis of lower muscles of facial expression (below eye) on opposite side of lesion. Sounds are louder on side of injury because stapes muscle is paralyzed.
VIII Vestibulocochlear	Tumor or other injury to nerve produces progressive hearing loss, noises in ear *(tinnitus)*, involuntary rapid eye movement *(nystagmus)*, whirling dizziness (vertigo).
IX Glossopharyngeal	Injury to nerve produces loss of gag reflex, difficulty in swallowing *(dysphagia)*, loss of taste on posterior third of tongue (not noticed by patient unless tested), loss of sensation on affected side of soft palate, decrease in salivation.
X Vagus	Unilateral lesion of nerve produces sagging of soft palate, hoarseness due to paralysis of vocal fold, dysphagia.
XI Accessory	Laceration of neck produces inability to contract sternocleidomastoid and upper fibers of trapezius muscles, drooping shoulders, inability to rotate head *(wry neck)*.
XII Hypoglossal	Damage to nerve produces protruded tongue deviated toward affected side, moderate difficulty in speaking *(dysarthia)*, chewing, and dysphagia.

sone, a steroid drug, reduces tissue swelling near the nerve and improves blood flow and nerve conduction.

A cerebral stroke may produce paralysis of the muscles of facial expression below the eye on the side opposite the lesion.

Dyslexia

Dyslexia (L. *dys*, faulty + Gr. *lexis*, speech) is an extreme difficulty in learning to identify printed words. It is most commonly seen as a reading and writing disability, and it is usually identified in children when they are learning to read and write. Typical problems include letters that appear to be backwards, words that seem to move on the page, and transposed letters. The disorder is not related to intelligence.

Reading and writing are usually dominated by the left cerebral hemisphere, but dyslexics appear to have an overdeveloped right hemisphere that competes with the left hemisphere for the control of language skills. Apparently, the abnormal distribution of nerve cells takes place during the middle third of pregnancy, when the outer cortex of the fetal brain is formed. Prenatal disturbances such as a small stroke, maternal stress, or a viral infection may underly the abnormal fetal development. An early discovery of reading and writing problems and improved remedial teaching techniques appear to be the main factors in treating dyslexics.

Encephalitis

Encephalitis is an inflammation of the brain, usually involving a virus transmitted by a mosquito or tick. Several other viral causes are known. Brain tissue is infiltrated by an increased number of infection-fighting white blood cells called lymphocytes, cerebral edema occurs, ganglion cells in the brain degenerate, and neuron destruction may be widespread. Symptoms generally include the sudden onset of fever, progressing to headache and neck pain (indicating meningeal inflammation), drowsiness, coma, and paralysis. Extreme cases may be fatal, especially in herpes encephalitis.

CHAPTER SUMMARY

General Structure of the Brain (p. 348)

1 The major parts of the brain are the **cerebrum, diencephalon, cerebellum,** and **brainstem.** The cerebrum consists of the two *cerebral hemispheres.* The diencephalon is composed primarily of the thalamus, hypothalamus, epithalamus, and ventral thalamus (subthalamus). The brainstem is composed of the midbrain, pons, and medulla oblongata.

2 Each hemisphere of the cerebrum consists of outer gray matter, deep white matter, and deep gray matter **(basal ganglia).** The outer gray matter, called the **cerebral cortex,** has many folds and convolutions. The **corpus callosum,** a bundle of nerve fibers, connects the hemispheres and relays messages between them.

Meninges, Ventricles, and Cerebrospinal Fluid (p. 348)

1 The brain is covered by the same three **meninges** that protect the spinal cord. The outermost **dura mater** consists of an inner dura mater that is continuous with the spinal dura mater and an outer dura mater, which is actually the periosteum of the skull bones. The cranial dura mater contains venous blood vessels called *dural sinuses.* The potential space between the dura mater and the arachnoid is the subdural space.

2 The middle **arachnoid** is a thin layer between the dura mater and pia mater. Between the arachnoid and pia mater is the *subarachnoid space,* which contains cerebrospinal fluid.

3 The innermost **pia mater** directly covers the surface of the brain. In certain locations, the pia mater joins together with modified ependymal cells to form the **choroid plexuses,** networks of small blood vessels.

4 *Cerebrospinal fluid* supports and cushions the brain and helps control the chemical environment of the central nervous system.

5 Cerebrospinal fluid circulates through the *ventricular system* of the brain, which consists of a series of connected cavities called **ventricles.**

6 Cerebrospinal fluid is essentially an ultrafiltrate of blood. Most of it is formed continuously at the choroid plexuses of the lateral, third, and fourth ventricles.

7 Cerebrospinal fluid circulates from the choroid plexuses through the ventricles and then the subarachnoid space in the vicinity of the medulla oblongata, until it reaches the subarachnoid villi, through which it passes back into the venous blood of the superior sagittal sinus.

Nutrition of the Brain (p. 353)

1 A lack of oxygen or glucose will damage brain tissue faster than any other tissue. Glucose, the body's chief source of usable energy, cannot be stored in the brain.

2 The brain has built-in regulating devices that make it almost impossible to constrict the blood vessels that carry the incoming blood supply.

3 The **blood-brain barrier** is an anatomical and physiological network of selectively permeable membranes that prevent some substances from entering the brain, while allowing relatively free passage to others.

Brainstem (p. 356)

1 The **brainstem** is the stalk of the brain, relaying messages between the spinal cord and the cerebrum.

2 Within the brainstem are ascending

and descending pathways between the spinal cord and parts of the brain.

3 Contained within the core of the brainstem is the **reticular formation,** with its pathways and integrative functions.

4 Of the 12 cranial nerves, all but the olfactory and optic nerves emerge from the brainstem.

5 The lowermost portion of the brainstem is the **medulla oblongata.** It contains vital cardiac, vasomotor, and respiratory centers; the medulla oblongata also regulates vomiting, sneezing, coughing, and swallowing.

6 The **pons** forms a connecting bridge between the medulla oblongata and the midbrain. It also contains fibers that project to the hemispheres of the cerebellum and link the cerebellum with the cerebrum. Within the reticular formation of the pons are respiratory centers that are integrated with those in the medulla oblongata.

7 The **midbrain** connects the pons and cerebellum with the cerebrum. It contains all the ascending fibers projecting to the cerebrum. It also contains oculomotor nerves, which are involved with movements of the eyes, focusing, and constriction of the pupils of the eyes.

Cerebellum (p. 361)

1 The **cerebellum** is located behind the pons in the posterior cranial fossa. It is mainly a coordinating center for muscular movement, involved with balance, precision, timing, and body positions. It has no role in perception, conscious sensation, or intelligence.

2 The cerebellum is divided into the unpaired midline **vermis,** the small bilateral **flocculonodular lobes,** and the two large **lateral lobes,** or hemispheres.

3 The cerebellum is covered by a surface layer of gray matter called the **cerebellar cortex,** which is composed of a neuronal network of circuits.

4 The cerebellum is connected to the brainstem by bundles of nerve fibers called **cerebellar peduncles.**

Cerebrum (p. 363)

1 The **cerebrum** is the largest and most complex structure of the nervous system. It is composed of the cerebral *hemispheres, white matter,* and *basal ganglia.* Each hemisphere is further divided into six lobes—the frontal, parietal, occipital, temporal, limbic, and insula.

2 The surface of the cerebrum is a mantle of gray matter called the **cerebral cortex.** The cortex contains ridges *(gyri),* slitlike grooves *(sulci),* and deep *fissures,* all of which increase the surface area of the cerebrum.

3 A deep groove called the *longitudinal fissure* runs between the cerebral hemispheres. The **corpus callosum,** a bridge of nerve fibers between the hemispheres, relays nerve impulses between them.

4 Beneath the cortex is a thick layer of white matter consisting of association, commisural, and projection fibers.

5 Each cerebral hemisphere contains a large core of *gray matter* called the **basal ganglia,** which help to coordinate muscle movements by relaying neural inputs from the cerebral cortex to the thalamus and finally back to the motor cortex of the cerebrum.

6 The **frontal lobe** is involved with the motor control of voluntary movements and the control of a variety of emotional expressions and moral behavior. The **parietal lobe** is concerned with the evaluation of the general senses and of taste. The **temporal lobe** has critical roles in hearing, equilibrium, and to a certain degree, emotion and memory. The **occipital lobe** is involved with vision and its associated forms of expression. The **insula** (central lobe) appears to be associated with gastrointestinal and other visceral activities.

7 The limbic lobe is an integral part of the **limbic system,** which is involved with emotions, behavioral expressions, recent memory, and smell.

Diencephalon (p. 370)

1 The **diencephalon** is located deep to the cerebrum; it connects the midbrain with the cerebral hemispheres. It is composed of the thalamus, hypothalamus, epithalamus, and ventral thalamus (subthalamus).

2 The **thalamus** is the intermediate relay point and processing center for all sensory impulses (except smell) going to the cerebral cortex.

3 The **hypothalamus** is the highest integrating center for the autonomic nervous system and regulates many physiological and endocrine activities through its relationship with the *pituitary gland* (hypophysis). It is involved with the regulation of body temperature, water balance, sleep-wake patterns, food intake, behavioral responses associated with emotion, endocrine control, and sexual responses.

4 The **epithalamus** acts as a relay station to the midbrain. It contains the *pineal body.*

5 The **ventral thalamus,** or **subthalamus,** is functionally integrated with the basal ganglia.

Cranial Nerves (p. 373)

1 The 12 pairs of **cranial nerves** are the peripheral nerves of the brain. Their numbers indicate the sequential order in which they emerge from the brain.

2 The **motor (efferent) fibers** of the cranial nerves emerge from the brainstem. They arise from groups of neurons called **motor nuclei.** The **sensory (afferent) fibers** of cranial nerves arise from cell bodies located in **sensory ganglia** or in peripheral sense organs such as the eye. Axons of sensory neurons synapse with **sensory nuclei.** Fibers transmit input to the cerebrum, where sensations are brought to a conscious level.

3 The cranial nerves, in ascending numerical order, are the **olfactory** (I), **optic** (II), **oculomotor** (III), **trochlear** (IV), **trigeminal** (V), **abducens** (VI), **facial** (VII), **vestibulocochlear** (VIII), **glossopharyngeal** (IX), **vagus** (X), **accessory** (XI), and **hypoglossal** (XII).

Developmental Anatomy of the Brain (p. 383)

1 The CNS develops from the embryonic **neural plate,** which is soon transformed into the **neural tube.**

2 By about week 4, the neural tube develops into cavities called the **prosencephalon** (forebrain), **mesencephalon** (midbrain), and **rhombencephalon** (hindbrain). By about week 7, the prosencephalon divides into the **telencephalon** (which develops into the cerebrum) and **diencephalon,** and the rhombencephalon divides into the **metencephalon** (cerebellum and pons) and the **myelencephalon** (medulla oblongata).

3 By the end of month 3, the overall form of the brain is recognizable. *Sulci* begin to develop during month 4, and major sulci are present by month 8.

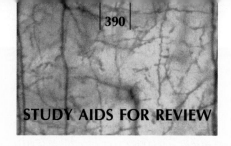

MEDICAL TERMINOLOGY

AGNOSIA (Gr. *a*, without + *gnosis*, knowledge) Inability to recognize objects.

ANALGESIA (Gr. *an*, without + *algesia*, sense of pain) Inability to feel pain.

ANESTHESIA (Gr. *an*, without + *aisthesis*, feeling) Partial or complete loss of sensation, especially tactile sensibility, induced by disease or an anesthetic.

APHASIA (Gr. *a*, without + *phasis*, speech) Partial or total loss of the ability to speak and understand words, resulting from brain damage.

APOPLEXY (Gr. *apoplessein*, to cripple by a stroke) Sudden loss of muscular control, with partial or total loss of sensation and consciousness, resulting from rupture or blocking of a blood vessel in the brain.

ATAXIA (Gr. *a*, without + *taktos*, order) Lack of coordination of voluntary muscles.

BRADYKINESIA (Gr. *bradus*, slow + *kinesis*, movement) A condition characterized by abnormally slow movements.

CEREBRAL ANEURYSM (Gr. *aneurusma*, dilation) A bubble-like sac formed by the enlargement of a cerebral blood vessel.

COMA (Gr. *koma*, deep sleep) A state of deep, prolonged unconsciousness, usually the result of injury, disease, or poison.

CONVULSION (L. *convellere*, to pull violently) Violent involuntary muscular contractions and relaxations.

CRANIECTOMY The removal of a portion of the skull bone.

CRANIOSTOMY (Gr. *tomos*, cut) An incision into the cranium to open the skull in order to have access to the brain.

DYSARTHIA (L. *dys*, faulty + Gr. *arthron*, joint) Lack of coordination of the muscles that control speech.

DYSKINESIA (L. *dys*, faulty + Gr. *kinesis*, movement) Difficulty in performing voluntary muscular movements due to brain lesion.

ENCEPHALOPATHY A general term for any brain disease.

LETHARGY (Gr. *lethargos*, forgetful) A state of sluggish indifference or unconsciousness resembling deep sleep into which a person relapses after being roused.

MENINGIOMA A meningeal tumor.

MENINGOCELE A protrusion of the meninges through an opening in the skull or spinal column.

NARCOLEPSY (Gr. *narke*, numbness + *lepsia*, to seize) A condition of uncontrollable sleepiness and sleep.

PARALYSIS (Gr. to loosen) Loss of muscle function (loss of contractility) caused by damage to the brain, spinal cord, or nerves.

PARALYSIS AGITANS (Gr. to loosen + L. *agitare*, to agitate) A technical (and contradictory) term for Parkinson's disease.

PARESIS (Gr. *parienai*, to let go) Partial or complete paralysis (less than total loss of contractility), with the muscles appearing to be relaxed.

SCISSORS GAIT A manner of walking in which each leg crosses in front of the other; occurs in cerebral palsy.

SPASTIC (Gr. *spastikos*, to pull) Pertaining to hyperactive muscular contractions, spasms, or convulsions.

STUPOR (L. *stupere*, to be stunned) A state of reduced sensibility or mental confusion; lethargy.

SUBDURAL HEMATOMA (Gr. *haimato*, blood + *oma*, tumor) A localized pool of venous blood in the subdural space.

TORPOR (L. *torpere*, to be stiff) A condition of mental or physical inactivity or insensibility; lethargy.

TREMOR (L. *tremere*, to tremble) An involuntary shaking or trembling.

UNDERSTANDING THE FACTS

1 Name the four major parts of the brain, and their main subdivisions.

2 What is the function of the corpus callosum?

3 What features of the brain provide support and protection?

4 In which specific part of the brainstem are the respiratory and cardiovascular centers located?

5 What is the blood-brain barrier?

6 Name the six cerebral lobes.

7 Give the location of the following (be specific):
  a corpus callosum
  b cerebral aqueduct
  c nuclei (of nervous system)
  d corpora quadrigemina
  e insula
  f thalamus
  g reticular formation

8 Which of the cerebral lobes is largely involved with vision?

9 What two major functions are associated with the limbic system?

10 Which part of the brain could be described as the intermediate relay point for almost all sensory impulses to the cerebral cortex?

11 Which cranial nerve(s) originate from the cerebrum?

12 Name the cranial nerves that are purely sensory in function.

13 What is the optic chiasma?

14 Which cranial nerves are involved with vision, including the movement of the eyeballs?

15 Is senility a normal aspect of aging?

16 Briefly define Alzheimer's disease, cerebral palsy, cerebrovascular accident, epilepsy, Parkinson's disease, dyslexia, encephalitis, and Bell's palsy.

UNDERSTANDING THE CONCEPTS

1 Describe the circulation of cerebrospinal fluid through the brain.

2 What is the importance of a proper oxygen and glucose supply to the brain?

3 If an abnormality occurred within the medulla oblongata, what body functions might be affected?

4 What are the functional implications of the pyramidal decussation?

5 If a cerebellar lesion occurs in the right side of the brain, on which side of the body will symptoms occur? How do you explain this fact?

6 What is the functional significance of the gyri and sulci of the cerebral cortex?

7 Compare the structure and functions of the two cerebral hemispheres.

8 Compare the proportion of the brain's gray matter devoted to manipulating the thumb to that allotted to controlling the thorax and abdomen. Explain.

9 Why is your hypothalamus so important to your well-being, and, in fact, to your survival?

SELF-QUIZ

Multiple-Choice Questions

14.1 Of the following, which is *not* associated with the medulla oblongata?
  - **a** pyramids
  - **b** vasomotor center
  - **c** decussation
  - **d** apneustic center
  - **e** olive

14.2 Which of the following is *not* a function of the cerebellum?
  - **a** initiation of body movements
  - **b** balance
  - **c** timing of body movements
  - **d** body positions
  - **e** processing input from sensory receptors in the head

14.3 Which of the following are parts of the cerebellum?
  - **a** the vermis
  - **b** flocculonodular lobes
  - **c** lateral lobes
  - **d** a and b
  - **e** a, b, and c

14.4 How many lobes comprise each cerebral hemisphere?
  - **a** one
  - **b** two
  - **c** three
  - **d** five
  - **e** six

14.5 Which of the following is *not* an important function of the hypothalamus?
  - **a** integration with the autonomic nervous system
  - **b** temperature regulation
  - **c** water and electrolyte balance
  - **d** regulation of heart rate
  - **e** food intake

14.6 Which of the following is an extraocular nerve?
  - **a** oculomotor
  - **b** trochlear
  - **c** abducens
  - **d** a and b
  - **e** a, b, and c

True-False Statements

14.7 The dural sinuses drain venous blood from the brain.

14.8 The subdural space contains cerebrospinal fluid.

14.9 Most cerebrospinal fluid is formed at the choroid plexuses.

14.10 The arachnoid villi permit a one-way bulk flow of cerebrospinal fluid into the superior sagittal sinus.

14.11 The colliculi in the midbrain are known collectively as the corpora quadrigemina.

14.12 The heavily black-pigmented nucleus in the reticular formation is called the nucleus ruber.

14.13 The glossopharyngeal nerve is cranial nerve IX.

Completion Exercises

14.14 The outer portion of the cerebrum is called the _____.

14.15 The bundle of nerve fibers that connects the two cerebral hemispheres is the _____.

14.16 A network of blood vessels called the _____ is formed where the ependyma contacts the pia mater.

14.17 The three structures of the brainstem are the _____, _____, and _____.

14.18 That part of the brain that forms the connecting bridge between the medulla oblongata and the midbrain is the _____.

14.19 The portion of the brain that is located between the diencephalon and pons is the _____.

14.20 There are _____ pairs of cranial nerves.

Matching

14.21 _____ gyri
14.22 _____ sulci
14.23 _____ fissures
14.24 _____ basal ganglia
14.25 _____ frontal lobe
14.26 _____ parietal lobe
14.27 _____ temporal lobe
14.28 _____ occipital lobe
14.29 _____ central lobe
14.30 _____ limbic lobe
14.31 _____ diencephalon
14.32 _____ thalamus

a associated with visceral activities
b gray matter
c extremely deep depressions
d houses third ventricle
e motor control over voluntary movements
f raised ridges of the cerebral cortex
g involved with hearing
h olfaction
i contains visual cortex
j evaluation of the general senses
k slitlike grooves
l attached to hypophysis
m reflex center
n an intermediate relay center
o white matter

Problem-Solving/Critical Thinking Questions

14.33 Which of the following brain parts is incorrectly matched with a function it helps control?
  a frontal lobes of cerebrum—vision
  b reticular formation—arousal
  c cerebellum—coordination of muscles
  d medulla oblongata—coordination of heart and respiratory rates
  e hypothalamus—emotions

14.34 The complex network of tiny islands of gray matter within the brain that acts as a filter for incoming sensory impulses is the:
  a dentate nucleus
  b reticular formation
  c limbic system
  d corpora quadrigemina
  e none of the above

14.35 If you were inside the limbic system, you could be in contact with all of the following *except* the:
  a putamen
  b amygdala
  c hippocampus
  d a and c
  e a and b

14.36 The thalamus
  a is a subdivision of the cerebrum
  b conducts primary motor impulses
  c has only one-way communication with the cerebrum
  d lies lateral to the third ventricle
  e a and c

14.37 The cerebellum
  a consists of a vermis and two peduncles
  b initiates motor impulses for muscle contraction
  c consists of a gray complex, underlying white matter, and deep nuclei
  d is part of the mesencephalon
  e b and c

A SECOND LOOK

1 In the drawing of the brain, label the occipital, parietal, frontal, and temporal lobes; also label the central sulcus and the lateral cerebral fissure.

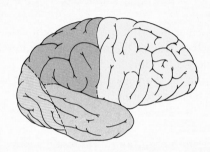

2 Label the cerebellum, midbrain, pons, medulla oblongata, spinal cord, folia cerebelli, and cerebellar cortex.

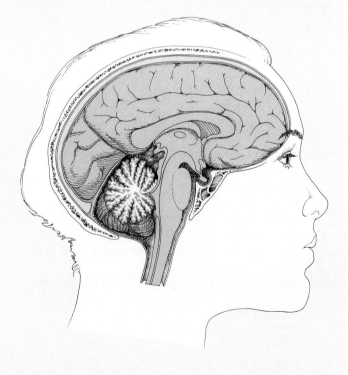

15

The Autonomic Nervous System

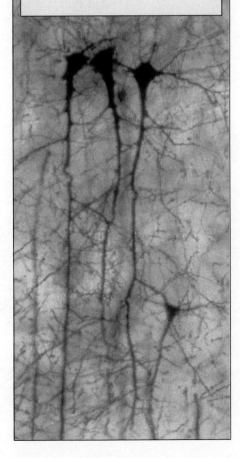

LEARNING OBJECTIVES

1 State the primary function of the autonomic nervous system.

2 Compare the structure and function of the somatic motor system with that of the autonomic nervous system.

3 Identify the two divisions of the peripheral autonomic nervous system, explain their basic functions, and describe their structural features.

4 Identify three major groups of autonomic ganglia and five important autonomic plexuses.

5 Describe the main pathways of sympathetic neurons between the central nervous system and the effectors.

6 Compare the neurotransmitters released by the sympathetic and parasympathetic divisions, and the effects of stimulation on each division.

7 Describe the pathways of parasympathetic neurons between the central nervous system and the effectors.

8 Explain how certain parts of the central nervous system exercise control over the autonomic nervous system.

9 Identify the five components of an autonomic visceral reflex arc.

When you see your favorite food, your digestive juices begin to flow. You don't have to do anything except see the food to start the secretions. This response is considered "involuntary" because you don't control it on a conscious level. Such responses occur in effectors such as (1) *cardiac muscle*, (2) the *smooth muscle* of the internal organs, and (3) in *glands*, such as salivary and sweat glands.

The ***autonomic nervous system*** (ANS) innervates these three types of effectors. It is divided into the peripheral autonomic nervous system and the central autonomic control centers. The autonomic nervous system is also known as the ***visceral efferent motor system***, because it is concerned with the internal organs, or *viscera*. Its primary function is to regulate visceral activities to maintain homeostasis.

Two important points about the autonomic nervous system must be understood at the outset:

1 The autonomic nervous system is usually defined as being exclusively the *motor* system, involved with influencing the activity of cardiac muscle, smooth muscle, and glands of the body.

2 *Sensory input* to the autonomic nervous system is derived from sensory receptors and is conveyed by *both* the somatic afferent system and the visceral afferent system.

Although the autonomic nervous system does not operate independently from the rest of the nervous system, its structure and functions are often contrasted with those of the somatic nervous system [FIGURE 15.1]. The basic function of the somatic motor (efferent) system is to regulate the skeletal-muscle activities involved with movement and the maintenance of posture. These activities are associated with adjustments to the *external* environment and are under our conscious control. In contrast, the major role of the autonomic nervous system is to regulate circulation, breathing, digestion, and other functions *not* usually subject to our conscious control, in order to maintain a relatively stable *internal* environment. The two systems often work together. For example, when you are exposed to the cold, you shiver, and the blood vessels in your skin constrict. Shivering is a reaction of the skeletal muscles that produces heat and is controlled by the somatic motor system. Constriction of blood vessels is an action of the autonomic nervous system that conserves heat.

The two systems differ significantly in the way their neurons connect to their effectors. The neurons of the somatic nervous system *directly* innervate the effectors (skeletal muscles). Only *one* set of neurons links the central nervous system with these effectors through a myoneural junction (a synapse). The neurons of the autonomic nervous system, in contrast, link the central nervous system with the effectors via *two* sets of neurons. The first neuron originates in the central nervous system and synapses with a second neuron outside the central nervous system; the second neuron reaches the effectors. The synapses occur *outside* the central nervous system in

---

FIGURE 15.1 CROSS SECTION OF THE SPINAL CORD

The drawing shows the pathways of **[A]** somatic nerves, and **[B]** visceral nerves. Afferent nerve fibers are purple, and efferent nerve fibers are in blue. Dotted lines represent efferent postganglionic fibers.

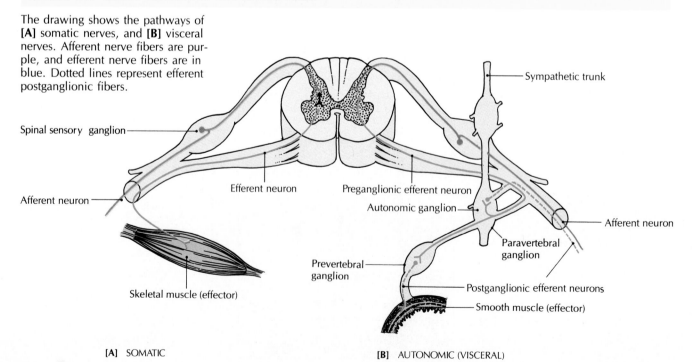

Spinal sensory ganglion

Afferent neuron

Efferent neuron

Skeletal muscle (effector)

[A] SOMATIC

Sympathetic trunk

Preganglionic efferent neuron

Autonomic ganglion

Afferent neuron

Paravertebral ganglion

Prevertebral ganglion

Postganglionic efferent neurons

Smooth muscle (effector)

[B] AUTONOMIC (VISCERAL)

clusters of cell bodies and their dendrites called *auto-nomic ganglia*. Autonomic ganglia may be thought of as relay stations between the central nervous system and the effectors.

STRUCTURE OF THE PERIPHERAL AUTONOMIC NERVOUS SYSTEM

The *peripheral autonomic nervous system* is a motor system consisting of two divisions, the *sympathetic* and *para-sympathetic*. Each division sends efferent nerve fibers to the muscle, organ, or gland it innervates. In general, but not always, the two divisions have opposite effects. In the broadest terms, we can say that the sympathetic division helps the body adjust to stressful situations, and the parasympathetic division is active when the body is operating under normal conditions. In reality, the functions of the nervous system are much too complex to set up such a clear distinction between the two divisions. They are more easily distinguished anatomically, as you will see in the following sections.

Anatomical Divisions of the Peripheral Autonomic Nervous System

The efferent nerve fibers of the autonomic nervous system that innervate the visceral organs emerge from the central nervous system at several different levels. The visceral efferent fibers that emerge through the thoracic and lumbar spinal nerves constitute the ***thoracolumbar*** (thuh-RASS-oh-LUM-bar) ***division.*** At the *cranial level*, visceral efferent fibers leave the central nervous system by way of cranial nerves III, VII, IX, and X from the brainstem. At the *sacral level*, other visceral efferent fibers leave by way of sacral spinal nerves 2, 3, and 4. These two groups make up the ***craniosacral division.*** *The craniosacral division is the parasympathetic division of the autonomic nervous system, and the thoracolumbar division is the sympathetic division* [FIGURE 15.2].

Preganglionic and Postganglionic Neurons

The linkage between the central nervous system and the effectors within the body differs in the somatic nervous system and the autonomic nervous system. As we have said, the somatic system has a *one-neuron linkage*, and the autonomic nervous system has a *two-neuron linkage*. Each somatic (lower) motor neuron has an axon that courses from its cell body in the brainstem or spinal cord through a cranial or spinal nerve, directly innervating a skeletal

FIGURE 15.2 ANATOMICAL DIVISIONS OF THE AUTONOMIC NERVOUS SYSTEM

Visceral efferent fibers that emerge from the CNS at the cranial and sacral levels make up the craniosacral, or parasympathetic, division of the ANS, while visceral efferent fibers that emerge from the thoracic and lumbar levels make up the thoracolumbar, or sympathetic, division of the ANS.

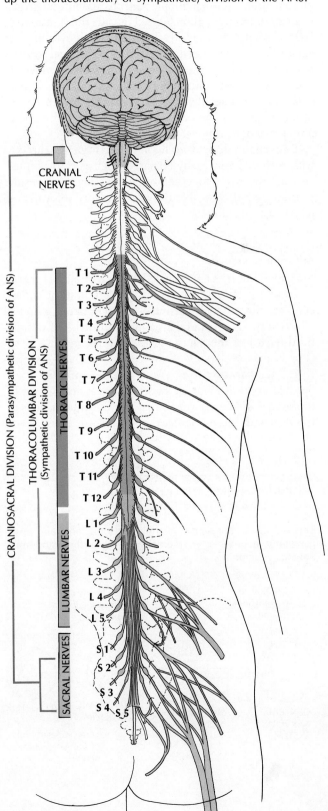

muscle. Hence, there is a one-neuron linkage between the central nervous system and an effector.

In contrast, the first neuron in the autonomic nervous system, called a ***preganglionic neuron,*** has its cell body in the brainstem or spinal cord, and a *myelinated axon* that courses through a cranial or spinal nerve and terminates by synapsing with the dendrites and cell body of one or more neurons (postganglionic neurons and interneurons) in an autonomic ganglion located outside the central nervous system [FIGURE 15.3]. (Located wholly within each ganglion are small interneurons.) The second neuron in a two-neuron linkage is called a ***postganglionic neuron.*** Its cell body is in an autonomic ganglion, and its *unmyelinated* axon courses through nerves and plexuses before terminating in a motor ending associated with cardiac muscle, smooth muscle, or a gland.

Postganglionic neurons do not always synapse directly with each effector. Many effectors, such as cardiac muscle cells, are linked together by gap junctions. As a result, a stimulated effector can convey excitatory information to an adjacent effector.

Autonomic Ganglia

Autonomic ganglia are arranged in three groups: (1) paravertebral, or lateral, ganglia, (2) prevertebral, or collateral, ganglia, and (3) terminal, or peripheral, ganglia. The first two are part of the sympathetic division, and the third is part of the parasympathetic division:

1 The ***paravertebral*** (*para,* beside) ***ganglia*** form beadlike rows of 21 or 22 swellings (3 cervical, 10 or 11 thoracic, 4 lumbar, 4 sacral) that run down both sides of the vertebral column, *outside* the spinal cord. These bead-

like ganglia are connected by intervening nerve fibers to form two vertical ganglionic chains known as the ***sympathetic trunks*** [FIGURE 15.4A]. These trunks and ganglia lie on the surface of the vertebral column and extend from the upper cervical vertebrae to the coccyx.

2 Sometimes a preganglionic fiber passes through the sympathetic trunk without forming a synapse. Instead, it joins with ***prevertebral ganglia,*** which lie in front of the vertebrae near the large thoracic, abdominal, and pelvic arteries that give them their names. The largest prevertebral ganglia are the *celiac* (solar) *ganglion* just below the diaphragm (near the celiac artery), the *superior mesenteric ganglion* in the upper abdomen (near the superior mesenteric artery), and the *inferior mesenteric ganglion* near the lower abdomen (near the inferior mesenteric artery) [FIGURE 15.4A]. Interneurons and their processes are located wholly within these ganglia.

3 ***Terminal ganglia*** are composed of small collections of ganglion cells that are very close to or within the organs they innervate. Terminal ganglia are especially common in the gastrointestinal tract and urinary bladder. They are considered to be part of the parasympathetic division.

Plexuses

Some postganglionic nerve fibers are distributed like cables to branching, interlaced networks following along the blood vessels in the thoracic, abdominal, and pelvic cavities, called ***autonomic plexuses.*** The autonomic plexuses include the following:

1 The *cardiac plexus* lies among the large blood vessels emerging from the base of the heart. It has a regulatory effect on the heart.

2 Most of the *pulmonary* (L. lung) *plexus* is located posterior to each lung. Sympathetic nerve fibers dilate the bronchi (delicate air passages leading from the trachea to the lungs), and parasympathetic fibers constrict them.

3 The *celiac* (Gr. abdomen) *plexus,* or *solar plexus,* is the largest mass of nerve cells outside the central nervous system. It lies on the aorta in the vicinity of the celiac artery, behind the stomach. A sharp blow to the solar plexus, just under the diaphragm below the sternum, may cause unconsciousness by slowing the heart rate and reducing the blood supply to the brain.

4 The *hypogastric* ("under the stomach") *plexus* connects the celiac plexus with the pelvic plexuses below. It innervates the organs and blood vessels of the pelvic region.

5 The *enteric* (Gr. intestine) *plexuses* receive sympathetic and parasympathetic fibers. Located between the longitudinal and circular muscles of the digestive system,

FIGURE 15.3 PREGANGLIONIC AND POSTGANGLIONIC NEURONS

This simplified representation shows preganglionic and postganglionic neurons in an efferent pathway of the autonomic nervous system.

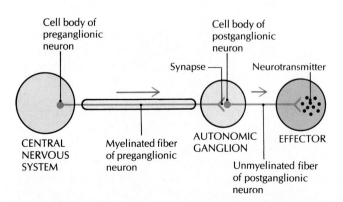

Cell body of preganglionic neuron

Cell body of postganglionic neuron

Synapse — Neurotransmitter

CENTRAL NERVOUS SYSTEM — Myelinated fiber of preganglionic neuron — AUTONOMIC GANGLION — EFFECTOR

Unmyelinated fiber of postganglionic neuron

they help regulate *peristalsis* (rhythmic contraction of the smooth muscles in the digestive system).

Certain nerves that innervate the viscera of the thorax, abdomen, and pelvis are called *splanchnic* (SPLANK-nick) (visceral) *nerves*. These nerves are composed of preganglionic or postganglionic fibers of either the sympathetic or parasympathetic nervous systems; thoracic splanchnic nerves innervate the heart and lungs, and pelvic splanchnic nerves innervate the pelvic viscera.

Sympathetic Division

The *sympathetic* (thoracolumbar, adrenergic) *division* of the autonomic nervous system arises from cell bodies in the lateral gray horn of the spinal cord. The myelinated nerve fibers emerge from the spinal cord in the ventral nerve roots of the 12 thoracic and first 2 or 3 lumbar spinal nerves. This emergence of fibers is known as the *thoracolumbar outflow*. These preganglionic fibers form small nerve bundles called *white rami communicantes* (sing. *ramus communicans*). They are white because the nerve fibers are myelinated. The fibers then pass to the paravertebral ganglia of the sympathetic trunk [FIGURE 15.1].

When a preganglionic neuron reaches the sympathetic trunk, it may take one of several pathways:*

1 It may synapse with a postganglionic neuron, which then terminates on an effector. A preganglionic neuron may synapse with postganglionic neurons within a paravertebral ganglion on the same level in the sympathetic trunk. Then the postganglionic neuron joins the spinal nerve by way of the *gray rami communicantes* before it terminates in an effector. These rami communicantes are gray because their fibers are unmyelinated.

2 In another pathway, the preganglionic neuron may course up or down to a different level of the sympathetic trunk before synapsing with postganglionic neurons within a paravertebral ganglion.

3 In still another pathway, it may pass directly through a ganglion (without synapsing) in the sympathetic trunk to synapse with postganglionic neurons within a prevertebral ganglion.

Each preganglionic fiber synapses with 20 or more postganglionic neurons in a ganglion. Some axons of the postganglionic neurons may pass directly to an effector (such as the heart or lungs), but most postganglionic axons pass through the gray rami communicantes and then travel back to the spinal nerve to be distributed for the innervation of sweat glands as well as for the smooth

muscles of blood vessels, arrector pili muscles of the skin (which make the hair stand erect), the body wall, and the limbs. Other axons from the ganglia of the chain in the neck form *perivascular plexuses* around blood vessels before terminating in the visceral structures of the head (such as the dilator muscles of the pupil and glands in the head). The postganglionic fibers from cells in the prevertebral ganglia form perivascular plexuses that innervate the viscera in the abdominal and pelvic regions. The sympathetic outflow from the spinal cord is distributed as shown in FIGURE 15.4A.

Sympathetic neurotransmitters The neurotransmitter released by the preganglionic nerve terminals is *acetylcholine* (ACh). It is inactivated to choline and acetate by the enzyme *acetylcholinesterase*. The neurotransmitter released by the postganglionic nerve terminal is *norepinephrine* (NE). However, there are a few exceptions. Sympathetic postganglionic fibers to most sweat glands and to some blood vessels release acetylcholine instead of norepinephrine.

The sympathetic division, its neurons, and its fibers comprise the *adrenergic system*, because most postganglionic neurons and fibers of this division release norepinephrine (also called *noradrenaline*). The medulla of the adrenal gland releases both norepinephrine and epinephrine into the bloodstream. Actually, cells of the adrenal medulla should be considered postganglionic neurons.

Effects of sympathetic stimulation The sympathetic division of the autonomic nervous system is anatomically and physiologically organized to affect widespread regions of the body, or even the entire body, for sustained periods of time. For example, each preganglionic neuron synapses with numerous postganglionic neurons, which in turn have long axons that terminate in neuroeffector synapses over a large area. The widespread and long-lasting sympathetic effects are directly related to the slow deactivation of norepinephrine and to the extensive distribution of norepinephrine and epinephrine released by the adrenal medulla into the bloodstream.

Parasympathetic Division

The axons of the *parasympathetic division* (Gr. *para*, beside, beyond; that is, located beside the sympathetic division) of the autonomic nervous system may also be called the craniosacral or cholinergic division. Its preganglionic neurons originate in the brainstem and sacral levels of the spinal cord. The cranial portion supplies parasympathetic innervation to the head, neck, thorax, and most of the abdominal viscera [FIGURE 15.4B]. The sacral portion supplies parasympathetic innervation to the viscera of the lower abdomen and pelvis. The body walls and limbs do not have parasympathetic innervation.

*Although we speak of several pathways, the impulse actually spreads over all of the pathways at once.

FIGURE 15.4 THE DIVISIONS OF THE AUTONOMIC NERVOUS SYSTEM

Nerve fibers actually emerge from both sides of the cord. [A] The sympathetic division. [B] The parasympathetic division.

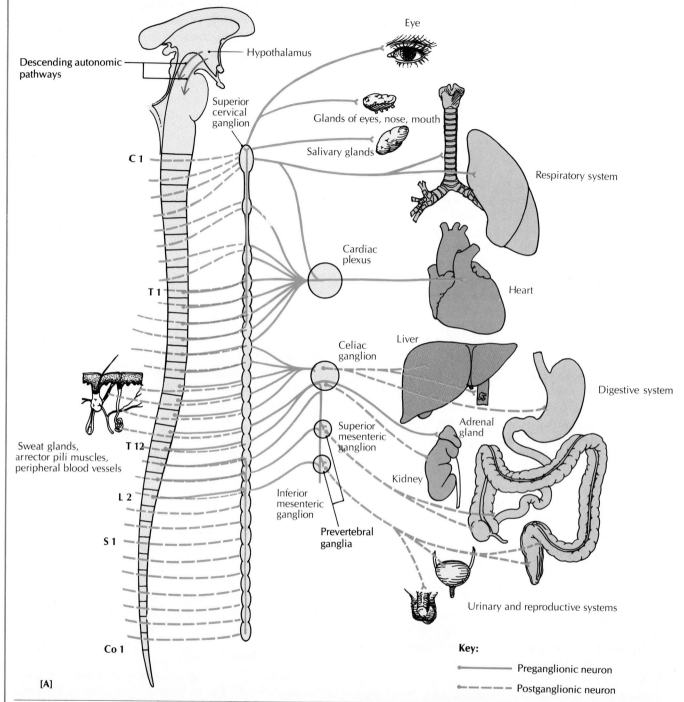

Part of body	Sympathetic effect
Eye (iris)	Dilates pupil.
Salivary glands	Increases secretion; thick, viscous saliva.
Lungs (bronchial tubes)	Causes bronchodilation.
Heart	Increases heart rate and blood pressure, dilates coronary arteries, increases blood flow to voluntary muscles.
Intestinal walls	Causes relaxation, decreases peristalsis, contracts sphincters.
Urinary bladder	Inhibits constriction of bladder wall, contracts sphincters.
Sweat glands	Increases secretion.
Peripheral blood vessels	Constricts.

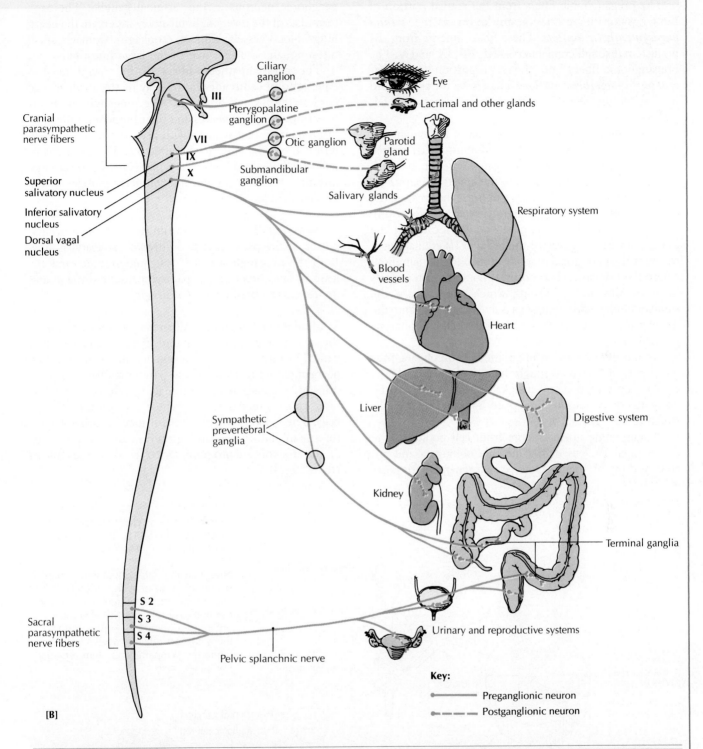

Cranial parasympathetic nerve fibers

III — Ciliary ganglion — Eye

Pterygopalatine ganglion — Lacrimal and other glands

VII — Otic ganglion — Parotid gland

IX

X — Submandibular ganglion

Salivary glands

Superior salivatory nucleus

Inferior salivatory nucleus

Dorsal vagal nucleus

Respiratory system

Blood vessels

Heart

Sympathetic prevertebral ganglia

Liver

Digestive system

Kidney

Terminal ganglia

S 2
S 3
S 4

Sacral parasympathetic nerve fibers

Urinary and reproductive systems

Pelvic splanchnic nerve

Key:

—•— Preganglionic neuron

--•-- Postganglionic neuron

[B]

Part of body	Parasympathetic effect
Eye (iris)	Constricts pupil.
Salivary glands	Increases secretion; thin, watery saliva.
Lungs (bronchial tubes)	Causes bronchoconstriction.
Heart	Decreases heart rate, constricts coronary arteries, decreases blood flow to skeletal muscles.
Intestinal walls	Causes contraction, increases peristalsis, relaxes sphincters.
Urinary bladder	Stimulates contraction of bladder wall, relaxes sphincters.
Sweat glands	No innervation.
Peripheral blood vessels	No innervation for many.

The preganglionic fibers from the cranial portion of the parasympathetic division are known as the *cranial parasympathetic outflow.* These fibers emerge from the brainstem through cranial nerves III, VII, IX, and X. The preganglionic fibers from the sacral portion are the *sacral parasympathetic outflow.* They leave the spinal cord by way of the ventral roots of spinal nerves S2, S3, and S4.

Preganglionic fibers from cell bodies located in the midbrain are conveyed by the oculomotor (III) cranial nerve to a synapse in the ciliary ganglion. The postganglionic axon terminals innervate the constrictor muscles of the pupil, as well as the ciliary muscles that change the shape of the lens to focus the eyes (accommodation).

Preganglionic fibers from cell bodies located in the lower pons leave by way of the facial (VII) cranial nerve to either the pterygopalatine or submandibular ganglion, where they synapse. The postganglionic axon terminals innervate the lacrimal glands, which secrete tears, the submandibular and sublingual salivary glands, and the other glands in the walls of the nasal, oral, and pharyngeal cavities.

From cell bodies in the upper medulla oblongata, preganglionic fibers pass through the glossopharyngeal (IX) cranial nerve to the otic ganglia. After a synapse, the postganglionic neurons innervate the parotid glands, which secrete saliva.

Preganglionic fibers emerge from cell bodies in the dorsal vagal nucleus of the medulla oblongata and are conveyed by the vagus (X) nerve to a synapse with postganglionic neurons in the terminal ganglia. The axon terminals of the postganglionic fibers innervate the heart, lungs, blood vessels, kidneys, esophagus, stomach, small intestine, and the first half of the large intestine.

The parasympathetic fibers of the sacral outflow emerge from cell bodies in the gray matter of the spinal cord at the S3 and S4 levels. The preganglionic fibers pass through spinal nerves and their branches, called *pelvic splanchnics,* and synapse with postganglionic fibers in terminal ganglia close to the effectors. The axon terminals of the postganglionic fibers innervate the lower large intestine, uterus, genitals, and sphincter muscles of the urinary bladder and urethra.

Parasympathetic neurotransmitter The neurotransmitter of the parasympathetic division is acetylcholine. Both the preganglionic and postganglionic neurons release it, and therefore the parasympathetic division and its fibers are classified as *cholinergic.*

Cholinergic receptors All preganglionic neurons of the autonomic nervous system release acetylcholine, which, in turn, excites the postganglionic neurons. The postganglionic neurons of the parasympathetic division transmit impulses to its terminals to release acetylcholine, which interacts with receptor sites on the effector cells. The response by the effector may be either excitatory or inhibitory, depending on how the receptor sites on the effectors are programmed to respond [see table in FIGURE 15.4].

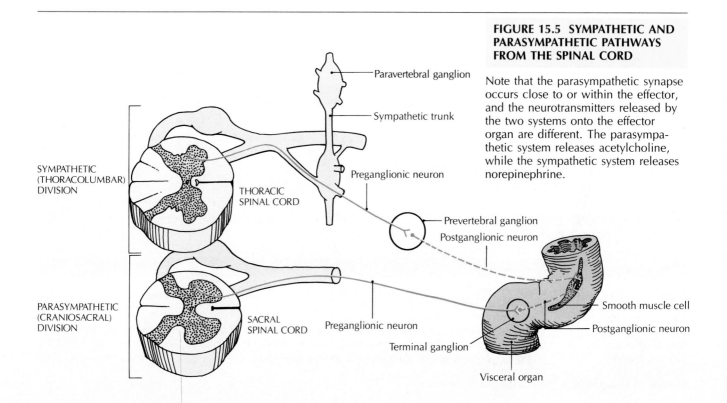

FIGURE 15.5 SYMPATHETIC AND PARASYMPATHETIC PATHWAYS FROM THE SPINAL CORD

Note that the parasympathetic synapse occurs close to or within the effector, and the neurotransmitters released by the two systems onto the effector organ are different. The parasympathetic system releases acetylcholine, while the sympathetic system releases norepinephrine.

Paravertebral ganglion

Sympathetic trunk

Preganglionic neuron

Prevertebral ganglion

Postganglionic neuron

SYMPATHETIC (THORACOLUMBAR) DIVISION

THORACIC SPINAL CORD

PARASYMPATHETIC (CRANIOSACRAL) DIVISION

SACRAL SPINAL CORD

Preganglionic neuron

Terminal ganglion

Visceral organ

Smooth muscle cell

Postganglionic neuron

Effects of parasympathetic stimulation In general, the parasympathetic division is geared to respond to a specific stimulus in a discrete region for a short time. This effect occurs because the postganglionic neurons have relatively short axons that are distributed for short distances to specific areas [FIGURE 15.5]. Also, the rapid deactivation of acetylcholine by acetylcholinesterase results in only a short-term effect by the neurotransmitter.

A S K   Y O U R S E L F

1 Why is the sympathetic division called the thoracolumbar division and the parasympathetic division called the craniosacral division?

2 What is the main function of autonomic ganglia?

3 What is the difference between a preganglionic and a postganglionic neuron?

4 What is the sympathetic trunk?

5 Where in the body is the parasympathetic division distributed?

AUTONOMIC CONTROL CENTERS IN THE CENTRAL NERVOUS SYSTEM

Neural centers in many regions of the central nervous system are involved with producing specific sympathetic and parasympathetic responses in the visceral organs. These higher centers are located in the brainstem, reticular formation, spinal cord, hypothalamus, cerebral cortex, and structures of the limbic system. The hypothalamus is considered to be the highest and main subcortical regulatory center of the autonomic nervous system.

The neural influences from the higher centers are exerted upon the activities of the autonomic nervous system through the visceral reflex arcs. Sensory inputs from the receptors in the viscera are conveyed via visceral afferent fibers to the central nervous system. There, after being processed and modulated by the higher centers, appropriate efferent influences are conveyed via the motor neurons of the sympathetic and parasympathetic outflow to the viscera.

Brainstem and Spinal Cord

Regulatory centers in the medulla oblongata of the brainstem are involved with visceral reflex arcs. For example, the cardiovascular center monitors heart rate, blood pressure, carbon dioxide concentration and the pH of blood, and blood vessel tone. After receiving input from sensory receptors, the cardiovascular center processes the neural information through the autonomic preganglionic neurons in the brainstem (expressed through the cranial nerve outflow) and the spinal cord (expressed through the thoracolumbar nerve outflow).

Respiratory centers in the brainstem are involved with reflex circuits that regulate the tone of the delicate bronchial tubes of the lungs. The neurons involved with respiration are integrated into somatic, not autonomic, reflex arcs. These arcs control somatic skeletal muscles, such as the diaphragm and intercostal muscles, which are involved in breathing.

Sensory inputs from receptors in the urinary bladder to reflex centers in the sacral spinal cord regulate the tone of the muscles of the bladder, as well as the involuntary muscle sphincters. Influences from the brain are integrated into the activity that contracts the bladder when it is full.

Hypothalamus

The hypothalamus is the highest controlling neural *regulator* and is regarded as the *coordinating center* of the autonomic nervous system. In fact, any autonomic response can be evoked by stimulating a site within the hypothalamus (see the responses noted in the table in FIGURE 15.4). Allowing for a considerable overlap, the stimulation of sites in the anterior and medial hypothalamus tends to be associated with parasympathetic responses, and the stimulation of sites in the posterior and lateral hypothalamus is usually associated with sympathetic responses. Certain somatic responses involving skeletal muscles, such as shivering, are also associated with the hypothalamus.

Within the hypothalamus are many neural circuits, called *control centers*, which control such vital autonomic activities as body temperature, heart rate, blood pressure, blood osmolarity, and the desire for food and water. The hypothalamus is also involved with behavioral expressions associated with emotion, such as blushing.

Clearly, the hypothalamus is a critical participant in maintaining homeostasis. For example, the autonomic responses that control body temperature are initiated because the hypothalamus acts as a thermostat that monitors the temperature of blood flowing through a hypothalamic control center. Neurons in the hypothalamus respond to temperature changes, activating either heat-dissipating or heat-conserving control systems to maintain the normal (37°C) body temperature. The heat-dissipating center in the anterior hypothalamus activates the responses of increased sweating and dilation of skin blood vessels, thus cooling the body. The heat-conserving center in the posterior hypothalamus causes the constriction of skin blood vessels, and shivering generates heat.

Biofeedback and the Autonomic Nervous System

The activity of the autonomic nervous system in response to an ever-changing internal environment goes on without our conscious awareness or control. Some of these automatic and involuntary functions involve heart rate, blood pressure, respiratory rate, blood-glucose concentration, and body temperature. We can have some *conscious* control over these activities through **biofeedback,** a technique that allows a person to monitor and control his or her own bodily functions.

In biofeedback, a mechanical device is used to register and display signs of physiological responses to make the person aware of them and to enable him or her to monitor changes in them. In general, any physiological function that can be recorded, amplified by electronic in-

struments, and fed back to the person through any of the five senses can be regulated to some extent by that person.

Biofeedback can work because every change in a physiological state is accompanied by a responsive change in a mental (or emotional) state. Conversely, every change in a mental state is accompanied by a change in the physiological state. The autonomic nervous system thus acts as the connecting link between the mental and physiological states.

Biofeedback is not new. Yoga and Zen masters of Eastern cultures have long demonstrated their ability to control bodily functions that were thought to be strictly involuntary. Now, Western culture has devised a technique through which, under the proper conditions,

anyone can learn to duplicate these once "impossible" physiological feats.

Biofeedback control holds considerable promise as a means of treating some psychosomatic problems (illnesses caused mainly by stress or psychological factors) such as ulcers, anxiety, and phobias, as well as many other disorders. Biofeedback can also relieve muscle-tension headaches, reduce the pain of migraine headaches, achieve relaxation during childbirth, lower blood pressure, alleviate irregular heart rhythms, and control epileptic seizures. The greatest value of these biofeedback techniques is that they demonstrate that the autonomic nervous system is not entirely autonomous and automatic; some visceral responses can be controlled.

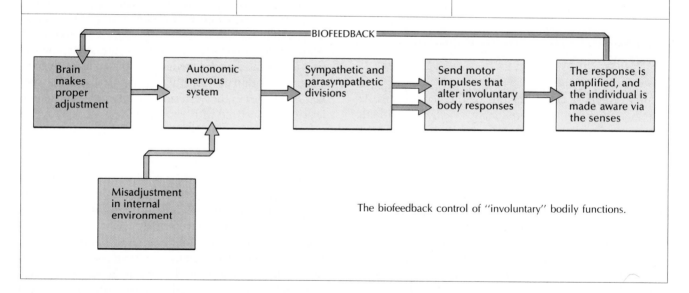

The biofeedback control of "involuntary" bodily functions.

Cerebral Cortex and Limbic System

Structures of the limbic system, such as the limbic lobe, amygdala, and hippocampus, are connected to the hypothalamus, and use the hypothalamus to express their activities. These expressions include many visceral and behavioral responses associated with self-preservation (such as feeding and fighting) and preservation of the

species (such as mating and care of the offspring). Electrical stimulation of the limbic lobe and hippocampus produces changes in the cardiovascular system, including alterations in the heart rate and tone of the blood vessels. Stimulation of the amygdala and limbic lobe may alter the secretory activity of digestive glands.

Even the cerebral cortex, which is usually considered the center of thought processes, utilizes the limbic system and hypothalamus, through its connections with the autonomic nervous system, to express some of our emotions. For example, when a person experiences anxiety, pleasure, or other emotional feelings, the cerebral cortex and limbic system become active and relay the influences to the hypothalamus. The hypothalamus responds by relaying neural influences via the descending autonomic pathways to the cardiovascular centers in the brainstem. These influences are then projected to the pools of preganglionic neurons of the cranial nerves and to the spinal cord. Depending upon which centers of the hypothalamus are stimulated, the resulting expressions can be sympathetic or parasympathetic.

Visceral Reflex Arc

A *visceral reflex* innervates cardiac muscle, smooth muscle, or glands. When stimulated, smooth muscles or cardiac muscles contract, and glands release their secretions. Such a reflex, like a somatic motor reflex, does not involve the cerebral cortex, and most visceral adjustments are made through regulatory centers, for example, in the medulla oblongata or spinal cord, without our conscious control or knowledge.

Unlike the somatic reflex, which uses only one efferent neuron, a sequence of two efferent neurons is involved in an autonomic visceral reflex arc. A *visceral reflex arc* is made up of (1) a *receptor*, (2) an *afferent neuron* that conveys sensory influences to the central nervous system, (3) *interneurons* within the gray matter of the central nervous system that connect with preganglionic neurons in the sympathetic division, (4) *two efferent neurons* that are part of the sequence composed of preganglionic neurons, an autonomic ganglion, and a postganglionic neuron, and (5) a *visceral effector*.

Some examples of autonomic visceral reflex arcs that occur in the spinal cord are the contraction of a full urinary bladder, muscular contraction of the intestines, and constriction or dilation of blood vessels. Examples of reflex arcs in the medulla oblongata include the regulation of blood pressure, heart rate, respiration, and vomiting.

ASK YOURSELF

1 What are some of the centers in the central nervous system that are involved in regulating the autonomic nervous system?

2 How does the central nervous system cooperate with the autonomic nervous system to regulate body temperature?

3 What are the components of an autonomic visceral reflex arc?

FUNCTIONS OF THE AUTONOMIC NERVOUS SYSTEM

In this section, we provide an overall picture of the autonomic nervous system as a two-part regulatory system by looking at the way the sympathetic and parasympathetic divisions balance their influences to help us react to changes and maintain our internal homeostasis. As an example, we show how the system operates during a downhill ski race.

Example of the Operation of the Autonomic Nervous System: A Ski Race

An Olympic skier on a twisting downhill slope is concentrating every part of the body to negotiate the course faster than anyone else in the world. The skier's heart, beating as much as three times faster than yours is right now, is also pumping more blood, faster, to the skeletal muscles than yours is now. The skier's pupils are dilated. The blood vessels of the skin, body organs, and salivary glands — all but those of the skeletal muscles — are constricted. The sweat glands are stimulated. Epinephrine and norepinephrine virtually pour out of the adrenal glands. Obviously ready for action, the skier shows the so-called "fright, fight, or flight" response, a state of heightened readiness.

At the same time, systems not needed in the race are practically shut down. Digestion, urination, and defecation can wait until the race is over. Blood is reserved for the skeletal muscles. The *sympathetic division* is in almost total command.

When the race is over and it is time to relax and enjoy a leisurely meal, the emphasis is on the "normal" or "maintenance" body functions and a restoration of the body's energy resources. The *parasympathetic division* is in command now. The heart rate and the force of its contractions are reduced, saliva and other digestive juices flow freely, the pupils are constricted to protect the eyes from excessive brightness, skeletal muscles are relaxed, and the body is free once again to devote time to ridding itself of wastes. Some of the glucose that was pouring out of the liver into the skeletal muscles during the race is now being diverted to other organs. So is blood. Blood vessels in the skin that were constricted to lessen the chance of serious bleeding in case of a wound are now back to normal. The extra perspiration that helped keep the skier cool during the race is not needed now, and the sweat glands relax. Blood vessels in the intestines dilate, while those in the skeletal muscles constrict to their normal diameter. The autonomic nervous system, together with the central nervous system, has kept the body in balance with its surroundings.

Coordination of the Two Divisions

The hypothalamus can increase or decrease the activity of the sympathetic and parasympathetic divisions. Small changes are constantly being made in one division or the other, with every change geared to promote homeostasis.

Many bodily activities involve either one autonomic division or the other. However, sexual activities require the coordinated, sequential involvement of both the sympathetic and parasympathetic divisions. Also, although the two divisions may be said to have opposite, or "antagonistic," effects on viscera and glands, not every structure innervated by the autonomic nervous system receives innervation from both divisions. For example, piloerector muscles in the skin may be innervated by only one (sympathetic) of the two divisions.

Many organs receive a *dual innervation*, with apparent opposite responses to stimulation from the sympathetic and parasympathetic divisions. However, such responses do not mean that the two divisions are antagonistic in the sense that they are working against each other. Rather, like antagonistic muscles, they are coordinated to achieve a single functional goal. The eye, for example, shows an interesting dual response to the degree of light intensity affecting it. The pupil dilates when the radial (dilator) smooth muscle cells of the iris are stimulated to contract by sympathetic fibers. When the sphincter (constrictor) muscle cells of the iris are stimulated to contract by parasympathetic fibers, the pupil constricts.

Responses of Specific Organs

The autonomic nervous system does not control the basic activity of the organs it innervates, but it does alter that activity. The organs innervated by the autonomic nervous system are not fully dependent upon autonomic innervation. For example, if the heart is deprived of its autonomic innervation, it will still contract, but it will not respond to the changing demands of the body and will not increase its rate when physical activity is increased. (This ability of the heart to contract without innervation is one of the reasons it can be transplanted to another person.) In contrast, a skeletal muscle deprived of its lower motor neuron innervation will not contract.

ASK YOURSELF

1 Which division of the autonomic nervous system prepares a person for intense muscular activity?

2 Do both divisions operate simultaneously? Explain.

3 What is dual innervation?

4 Does a neurotransmitter always have the same effect? Explain.

WHEN THINGS GO WRONG

Horner's Syndrome

In *Horner's syndrome*, either sympathetic preganglionics or postganglionics are interrupted. Lesions typically occur from the superior cervical ganglion to the head or the preganglionics from T1 to T4 to the superior cervical ganglion. The lesion blocks the flow of sympathetic activity to the head, resulting in a drooping eyelid (ptosis), constricted pupil, sunken eyeball, and flushed, dry skin. All of these symptoms are on the same side as the lesion.

Autonomic Dysreflexia

In *autonomic dysreflexia*, lesions occur in the spinal cord above T4 to T6 (above the sympathetic splanchnic outflow). Acute hypertension develops from the stimulation of sympathetic preganglionics by visceral or somatic afferents as they ascend the spinal cord. Blood pressure rises as high as 300 mm Hg.

Achalasia

Achalasia (Gr. failure to relax) can occur at any point along the gastrointestinal tract where the parasympathetics fail to relax the smooth muscle. In *achalasia of the esophagus*, the individual has difficulty swallowing due to a persistent contraction of the esophagus where it enters the stomach.

In *Hirschsprung's disease*, the large intestine becomes enlarged and distended because a short segment of it is continuously constricted, thereby obstructing the passage of feces. The constricted segment contains the flaw. The neurons normally present in the intestinal wall are lacking in this segment. These neurons, which are essential for normal peristalsis, are necessary for the expression of the autonomic nervous system's influences.

CHAPTER SUMMARY

The *autonomic nervous system* may also be called the *visceral efferent motor system.* Most of the functions of our internal organs (viscera) are not under our conscious control.

Structure of the Peripheral Autonomic Nervous System (p. 395)

1 The autonomic nervous system is divided into the *sympathetic* and *parasympathetic divisions.* The sympathetic division is also called the *thoracolumbar division* because its nerve fibers emerge from the thoracic and upper lumbar spinal nerves. The parasympathetic division is called the *craniosacral division* because its visceral efferent fibers leave the CNS via cranial nerves in the brainstem and spinal nerves in the sacral region of the spinal cord.

2 In the one-neuron linkage of the somatic motor system, each lower motor neuron connects directly with its effector, with no synapse outside the central nervous system. In the two-neuron linkage of the autonomic system, one neuron *(preganglionic)* synapses with a second neuron *(postganglionic)* in an *autonomic ganglion* before it stimulates its effector.

3 *Paravertebral* (lateral) and *prevertebral* (collateral) autonomic *ganglia* occur in the sympathetic division, and *terminal* (peripheral) *ganglia* in the parasympathetic division. Rows of paravertebral ganglia form the *sympathetic trunks.*

4 Some postganglionic nerve fibers are distributed to branching networks in the thoracic, abdominal, and pelvic cavities as *autonomic plexuses.*

5 The sympathetic division arises from cell bodies in the lateral gray horn of the spinal cord. The preganglionic nerve fibers emerge from the spinal cord in the ventral roots of the 12 thoracic and first 2 or 3 lumbar spinal nerves. This emergence of fibers is the *thoracolumbar outflow;* they form the *white rami communicantes.*

6 The neurotransmitter released by sympathetic preganglionic nerve terminals is *acetylcholine.* With a few exceptions, the neurotransmitter released by the sympathetic postganglionic nerve terminals is *norepinephrine* (noradrenaline) and therefore the division is called the *adrenergic division.* The effects of norepinephrine are usually widespread and relatively long lasting.

7 The preganglionic fibers from the cranial portion of the parasympathetic division are called the *cranial parasympathetic outflow.* The preganglionic fibers from the sacral portion comprise the *sacral parasympathetic outflow.*

8 The neurotransmitter of the parasympathetic division is *acetylcholine,* and the division is classified as *cholinergic.* The effects of acetylcholine are usually short range and relatively short term.

Autonomic Control Centers in the Central Nervous System (p. 401)

1 *Neural centers* in the central nervous system, including the brainstem, reticular formation, spinal cord, hypothalamus, cerebral cortex, and structures of the limbic system, are involved with producing sympathetic and parasympathetic responses in the visceral organs.

2 Regulatory centers in the medulla oblongata of the *brainstem* are involved with influencing visceral reflex arcs.

3 The *hypothalamus* is the highest and main subcortical regulatory center of the autonomic nervous system. It modulates autonomic centers in the brainstem and spinal cord.

4 Structures of the *limbic system* utilize the hypothalamus to express their activities, which include many visceral and behavioral responses associated with self-preservation and preservation of the species.

5 The *cerebral cortex* utilizes the limbic system and hypothalamus to express some emotions.

6 A *visceral reflex arc* innervates cardiac muscle, smooth muscle, and glands. Its components are a receptor, afferent neuron, interneurons, two efferent neurons, and visceral effector.

Functions of the Autonomic Nervous System (p. 403)

1 The main function of the autonomic nervous system is to promote homeostasis.

2 The sympathetic and parasympathetic divisions are coordinated into a balanced, complementary system that helps the body adjust to constantly changing environmental conditions.

3 Generally, the sympathetic division prepares the body for stressful situations, and the parasympathetic division is active when the body is at rest.

4 Many organs receive a *dual innervation* and produce opposite responses to stimulation by the sympathetic and parasympathetic divisions.

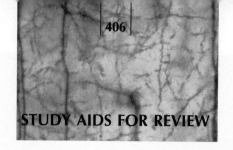

STUDY AIDS FOR REVIEW

UNDERSTANDING THE FACTS

1 What is the primary function of the autonomic nervous system?

2 What basic structural difference exists in the connection of somatic neurons to their respective effectors?

3 Name three groups of autonomic ganglia and five plexuses.

4 Name the neurotransmitters usually released by the preganglionic and postganglionic fibers in the sympathetic division.

5 Through which nerves do parasympathetic fibers emerge from the central nervous system?

6 Do autonomic visceral reflex arcs involve the cerebral cortex?

7 Does the autonomic nervous system need the cooperation of the brain for normal function? Explain.

8 Does the autonomic nervous system control the basic activity of the organs that it innervates? Explain.

UNDERSTANDING THE CONCEPTS

1 How do the general functions of the sympathetic and parasympathetic divisions differ?

2 Describe the pathways of sympathetic neurons from the spinal cord to the effectors.

3 What is the difference between the white and gray rami communicantes?

4 Make an anatomical comparison between a somatic reflex arc and an autonomic reflex arc.

5 Why are the effects of sympathetic stimulation generally widespread and long lasting, while those of parasympathetic stimulation generally local and short term?

6 When we say the sympathetic and parasympathetic systems are antagonistic, do we mean that they work against each other? Explain.

SELF-QUIZ

Multiple-Choice Questions

15.1 The autonomic nervous system consists of
  a visceral afferent fibers
  b somatic afferent fibers
  c visceral efferent fibers
  d somatic sensory fibers
  e a and b

15.2 The pathway between the central nervous system and skeletal muscles is via
  a a one-neuron linkage
  b a two-neuron linkage

  c an autonomic ganglion
  d pre- and postganglionic neurons
  e the sympathetic trunk and interneurons

15.3 In the autonomic nervous system, a preganglionic neuron
  a has its cell body in an autonomic ganglion
  b has its cell body in the brainstem or spinal cord
  c has a myelinated axon
  d has an unmyelinated axon
  e b and c

15.4 The sympathetic trunks are also known as the
  a prevertebral ganglia
  b paravertebral ganglia
  c terminal ganglia
  d collateral ganglia
  e peripheral ganglia

15.5 The sympathetic division of the autonomic nervous system is also known as the
  a thoracolumbar, or cholinergic, division
  b craniosacral, or adrenergic, division
  c splanchnic division
  d thoracolumbar, or adrenergic, division
  e craniosacral, or cholinergic, division

True-False Statements

15.6 The neurotransmitter released by postganglionic fibers of the autonomic nervous system is acetylcholine.

15.7 The effects of parasympathetic stimulation are generally localized and short-lived.

15.8 The part of the brain that exercises the greatest influence on the autonomic nervous system is the medulla oblongata.

15.9 The regulation of body temperature is a function of the hypothalamus.

15.10 Both the medulla oblongata and the hypothalamus are involved in control of blood pressure.

15.11 Unlike somatic reflex arcs, all autonomic visceral reflex arcs involve the brain.

15.12 The autonomic nervous system is not affected by emotions.

15.13 The "fright, fight, or flight" response is a function of the sympathetic nervous system.

15.14 The functions of the limbic lobe are expressed through the medulla oblongata.

15.15 The sympathetic and parasympathetic divisions of the autonomic nervous system control the basic activities of the organs they innervate.

16

The Senses

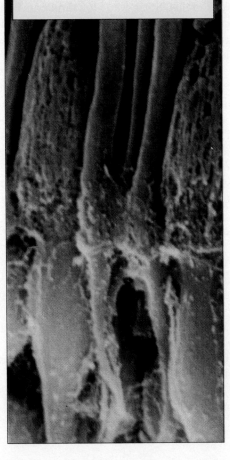

About 2000 years ago, Aristotle identified five senses — touch, taste, smell, hearing, and sight — and it is still common to refer to the "five senses" of the body. But in fact, the skin alone is involved with the sensations of fine touch, touch-pressure (deep pressure), heat, cold, and pain. Also included in a more complete list of the senses are a sense of balance, or equilibrium, and a sense of body movement. In addition, receptors in the circulatory system register changes in blood pressure and blood concentrations of carbon dioxide and hydrogen ions, and receptors in the digestive system are involved in the feelings of hunger and thirst.

Our impressions of the world are limited and defined by our senses. In fact, all knowledge and awareness depend on the reception and decoding of stimuli from the outside world and from within our bodies.

SENSORY RECEPTION

All animals have some means of sensing and responding to changes in the environment. If they didn't, they couldn't survive. We are able to cope with change because part of our nervous system is specialized to make sure we have a suitable reaction to any stimulus. Structures that are capable of perceiving and changing such stimuli are called sense organs, or **receptors.** Receptors are the body's links to the outside world and the world within us.

In terms of the nervous system, a receptor is (or is associated with) the peripheral end of the dendrites of afferent neurons. Sensory receptors are stimulated by specific stimuli. The eye (the receptor) is stimulated by visible light waves (the stimulus), and specialized receptors inside the ear are stimulated by audible sound waves. Sound waves do not stimulate the sensory receptors that are specialized to receive *light* waves, and light waves have no effect on our ears.

A favorite question is: Do sounds occur when there is no living thing to hear them? The answer is no. A "sound" is something that is received (*sensed*) by the ear and "heard" (*perceived*) by the brain. If there is no ear to receive sound waves, and no brain to translate those sound waves into what we consciously recognize as the "sound" of thunder, for example, the thunder will send out sound waves, but there will be no perceived "sound." The same is true of the other senses. The different sensations are brought about when nerve fibers connect with specialized portions of the brain. We "see" a tree, for example, not when its image enters our eyes, but when that coded image stimulates the vision centers of our brain.

All sensory receptors are structures that are capable of converting environmental information (stimuli) into nerve impulses. Thus, all receptors are *transducers*, that is, they convert one form of energy into another. Since all nerve impulses are the same, different types of receptors convert different kinds of stimuli (such as light or heat) into the same kind of impulse. The intensity of stimuli varies with the frequency of the stimulus, and the number of receptors stimulated at one time.

Basic Characteristics of Sensory Receptors

Sensory receptors may be either (1) the ending of a sensory neuron (for example, a pain receptor), or (2) a specialized receptor cell (for example, a hair cell in the cochlea of the auditory system). Although there are several types of sensory receptors, certain features are basic to all sensory receptors:

1 They *contain receptor cells* that respond to certain minimum (threshold) levels of intensity. That is, the stimulus must be strong enough to generate a receptor potential and then an action potential.

2 Their *structure is designed to receive a specific stimulus.* For example, the eye contains an elastic adjustable lens, a nonadjustable lens (cornea), and other structures suitable for directing light waves in the visible spectrum to the light-absorbing pigments in the receptor cells (rods and cones) in the back of the eye.

3 Their primary receptor cells *interact with afferent nerve fibers* that convey impulses to the central nervous system along spinal or cranial nerves. Some receptors in the skin are connected to neurons that have nerve fibers that extend for about a meter before they reach the spinal cord and form a synapse. In contrast, the primary receptor cells in the eye have short axons that synapse with other cells in the retina before projecting neural influences to the brain via the optic nerve.

After receptor cells stimulate afferent neurons, the nerve impulses are *conveyed along neural pathways* through the brainstem and diencephalon to the cerebral cortex of the brain. In the diencephalon and cerebral cortex, the original stimulus and nerve impulse are *translated* into a recognizable sensation such as sight or sound.

Classification of Sensory Receptors

Sensory receptors may be classified according to their location, type of sensation, type of stimulus, or structure (the presense or absence of a sheath).

Location of receptor Four kinds of sensory receptors are recognized on the basis of their location.

1 *Exteroceptors* (L. "received from the outside") respond to external environmental stimuli that affect the skin directly. These stimuli result in the sensations of touch-pressure, pain, and temperature.

2 *Teleceptors* (Gr. "received from a distance") are the exteroceptors located in the eyes, ears, and nose. They detect environmental changes (stimuli) that occur some distance away from the body. These stimuli are ultimately perceived as sight, sound, and smell.

3 *Interoceptors* (L. "received from the inside"), also called *visceroceptors* (L. "received from the viscera"), respond to stimuli from within the body, such as blood pressure, blood carbon dioxide, oxygen, and hydrogen-ion concentrations, and the stretching action of smooth muscle in organs and blood vessels. Interoceptors are located within organs that have motor innervation from the autonomic nervous system.

4 *Proprioceptors* (L. "received from one's own self") respond to stimuli in such deep body structures as joints, tendons, muscles, and the vestibular apparatus of the ear. They are involved with sensing where parts of the body are in relation to each other, and the position of the body in space.

Sight, hearing, equilibrium, smell, and taste are found in restricted regions of the body. Their sensory receptors are also more specialized and complex than those of the **general senses** (also called the *somatic*, or *visceral*, senses), which include touch-pressure, heat, cold, pain, body position (proprioception), and light (crude) touch.

Type of sensation Sensory receptors can detect several types of sensations that are associated with the general senses. *Thermal* sensations include cold and warmth. *Pain* sensations are the feelings initiated by harmful stimuli. Like pain, both **light touch** and **touch-pressure** sensations are also produced by mechanical stimuli that come in contact with the body. Light touch involves a finer discrimination than touch-pressure does. *Position sense* is elicited by the movement of joints and muscles. It includes both the sense of position when the body is not moving, and the sense of body movement, called *kinesthesia* (Gr. *kinesis*, motion + *esthesis*, perception).

Type of stimulus Another way to classify sensory receptors is by the stimuli to which they respond. *Thermoreceptors* (Gr. "heat receivers") respond to temperature changes. *Nociceptors* (NO-see; L. "injury receivers") respond to potentially harmful stimuli that produce pain. *Chemoreceptors* (L. "chemical receivers") respond to chemical stimuli that result in taste and smell and to changes in the concentrations of carbon dioxide, oxygen, and hydrogen ions in the blood, as well as other chemical changes. *Photoreceptors* (Gr. "light receivers") in the ret-

ina of the eye respond to the visual stimuli of visible light waves. *Mechanoreceptors* (Gr. "mechanical receivers") respond to and monitor such physical stimuli as touch-pressure, muscle tension, joint position changes, air vibrations in the cochlear system of the ear that produce hearing, and head movements detected by the vestibular system of the ear that result in sensing body equilibrium. Mechanoreceptors are the most widespread of all the sensory receptors, and are also the most varied in sensitivity and structure. *Baroreceptors* (Gr. "pressure receivers") are mechanoreceptors that respond to changes in blood pressure.

Sensory endings of receptors The neuronal terminals of spinal nerves end in sensory receptors, or *nerve endings*. Two kinds of nerve endings are usually distinguished:

1 Terminals that lack neurolemmocytes, myelin, and other cellular coverings are called **free nerve endings**. Free nerve endings are the naked telodendria in the surface epithelium of the skin, connective tissues, blood vessels, and other tissues. They are the sensors for such perceived sensations as pain, light touch, and temperature.

2 Receptors that are covered with various types of capsules are known as **encapsulated endings**. Encapsulated endings are located in the skin, muscles, tendons, joints, and body organs. Two such endings are *lamellated (Pacinian) corpuscles*, which are involved with vibratory sense and touch-pressure on the skin, and *tactile (Meissner's) corpuscles*, which are skin sensors that detect light touch [FIGURE 16.1].

Corpuscles of Ruffini and *bulbous corpuscles (of Krause)*, previously thought to be the skin receptors for temperature, are now believed to be sensors for touch-pressure, position sense of a body part, and movement. They are probably variants of other encapsulated endings, such as lamellated corpuscles. Other important encapsulated receptors are the *neuromuscular spindles* and *tendon (Golgi) organs*, which respectively monitor the stretch and tension in muscles and tendons and are involved with skeletal muscle reflexes.

Sensory Receptors and the Brain

The main purpose of the senses is to inform us about environmental conditions and changes that may be beneficial or harmful. The sensory system is an effective survival mechanism because it converts information from the environment into appropriate reactions.

A stimulus of sufficient strength is converted into action potentials that may influence neural activity (reflex arcs) that eventually stimulates an effector: a muscle contracts, a gland releases its secretion, a blood vessel constricts. The role of the brain in this process is to receive

FIGURE 16.1 NERVE ENDINGS AND SENSORY RECEPTORS IN THE SKIN

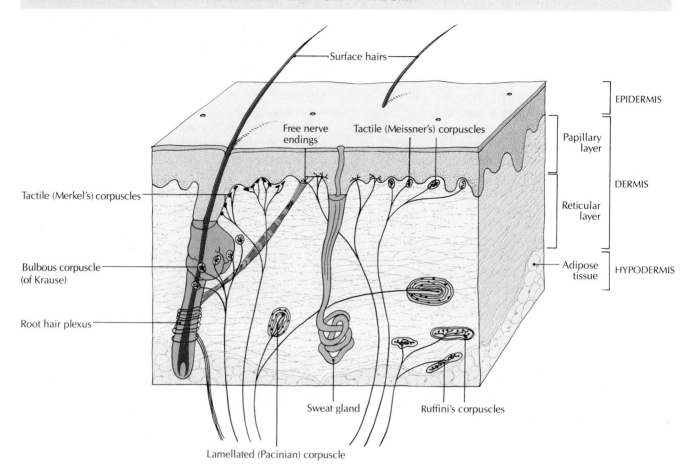

the neural influences, suppress what is irrelevant, compare it with information already stored, and coordinate the final impulses to the effector. But often the brain mediates sensory impulses and translates them into perceptible sensations. The brain's involvement between the stimulus and the response enables us to have complex behavioral patterns.

Individual nerve impulses are essentially identical, but an impulse for sight is distinguished from an impulse for taste by *the relationship between where the impulse is coming from and the places in the brain where it is going.* For example, taste receptors in the tongue receive stimuli and convert them into nerve impulses. Afferent fibers carry the impulses to a location in the brain that is specialized to receive only taste impulses. The type of sensation elicited by the brain depends upon *where* the brain is stimulated, not *how* it is stimulated.

Q: *How much outside stimuli does the nervous system screen out as irrelevant or distracting?*

A: The brain suppresses about 99 percent of the input it receives from sensory receptors.

A S K   Y O U R S E L F

1 What is a sensory receptor?

2 What is a transducer?

3 What are the basic characteristics of a sensory receptor?

4 What are the minimal requirements for the perception of a sensation?

5 What are the types of sensory receptors based on location?

6 What are the types of sensory receptors based on the types of stimuli to which they respond?

7 What is the role of the brain in the sensory system?

GENERAL SENSES

Sensory receptors in the skin detect stimuli that the brain interprets as light touch, touch-pressure, vibration, heat, cold, and pain. Several other general sensations, such as itch and tickle, will also be described.

Light Touch

Light touch is perceived when the skin is touched, but not deformed. Receptors for light touch are most numerous in the dermis, especially in the tips of the fingers and toes, the tip of the tongue, and lips.

Receptors of light touch include free nerve endings and tactile (Merkel's) corpuscles in the epidermis, and tactile (Meissner's) corpuscles just below, in the uppermost (papillary) layers of the dermis [FIGURE 16.1]. *Free nerve endings* are the most widely distributed receptors in the body, and are involved with pain and thermal stimuli as well as light touch. The next most numerous cutaneous receptors are those for touch, cold, and heat.

There are many sensory receptors called *root hair plexuses* around hair follicles. When hairs are bent, they act as levers, and the slight movement stimulates the free nerve endings surrounding the follicles, which act as detectors of touch and movement. For this reason, a tiny insect crawling along a hairy arm will be felt even if its feet never touch the skin.

Tactile (Merkel's) corpuscles are modified epidermal cells with free nerve endings attached. They are found in the deep epidermal layers of the palms of the hands and the soles of the feet. As shown in FIGURE 16.1, the "disk" portion of a tactile (Merkel's) corpuscle is formed when the unmyelinated terminal branches of myelinated afferent nerve fibers penetrate the basal layer of the epidermis. Once inside the epidermis, the fibers lose their covering of myelin and expand into a terminal disk attached to the base of a tactile (Merkel's) corpuscle.

Tactile (Meissner's) corpuscles are egg-shaped encapsulated nerve endings found in abundance on the palms of the hands, soles of the feet, lips, eyelids, external genitals, and nipples. These receptors are situated in the papillary layer of the dermis, just below the epidermis. Tactile (Meissner's) corpuscles contain flattened cells that are probably modified connective tissue, nerve endings that intertwine among them, and spiraling terminal branches that lose their myelin sheaths before they enter the corpuscle at its base. The flattened cells and accompanying fibers are enclosed within a capsule of connective tissue.

Touch-Pressure (Deep Pressure)

The difference between light touch and touch-pressure (deep pressure) on your skin can be shown by gently touching a pencil (light touch) and then sqeezing it hard (touch-pressure). Touch-pressure results from a deformation of the skin, no matter how slight. Sensations of touch-pressure last longer than sensations of light touch and are felt over a larger area. Receptors for touch-pressure are primarily *lamellated,* or *Pacinian* (pah-SIN-ee-an) *corpuscles* [FIGURE 16.1]. They are mechanoreceptors that actually measure *changes* in pressure, rather than pressure itself. Lamellated corpuscles are distributed throughout the dermis and subcutaneous layer, especially in the fingers, external genitals, and breasts, but are also found in muscles, joint capsules, the wall of the urinary bladder, and other areas that are regularly subjected to pressure.

In contrast to such sensitive areas of the skin as the fingertips, the torso (especially the back) and back of the neck are relatively insensitive to light touch. Sensitivity can be measured with a test called *two-point discrimination,* which measures the minimal distance that two stimuli must be separated to be felt as two distinct stimuli. Usually, one or two points of a compass are applied to the skin without the subject seeing how many points are being used. In areas where sensory receptors are abundant, two distinct compass points may be felt when they are separated by only 2 or 3 mm. Where there are few receptors far apart, the points may have to be separated by as much as 60 or 70 mm before they can be felt as two points. What this means is that there is virtually no spot on the fingertips that is insensitive to tactile stimuli.

Vibration

Most tactile receptors are involved to some degree in the detection of vibration. The term *vibration* refers to the continuing periodic change in a displacement with respect to a fixed reference. This change per unit time is termed the *frequency*. Different receptors detect different frequencies. For example, lamellated corpuscles can detect vibrations (frequencies) as high as 700 cycles per second (cps). Tactile (Meissner's) corpuscles and corpuscles of Ruffini, on the other hand, respond to low-frequency vibrations up to 100 cps.

Heat and Cold

Until recently it was believed that the cutaneous receptors for heat were the corpuscles of Ruffini and that the receptors for cold were the bulbous corpuscles (of Krause), but further investigation has disproved those beliefs. Currently, the cutaneous receptors for heat and cold are considered to be naked nerve endings. Cold receptors respond to temperatures below skin temperature, and heat receptors respond to temperatures above skin temperature.

So-called *cold spots* and *warm spots* are found over the surface of the entire body, with cold spots being more numerous. A *spot* refers to a small area that, when stimulated, yields a temperature sensation of warmth or cold. A spot is associated with several nerve endings. The lips have both cold and warm spots, but the tongue is only slightly sensitive to warmth. Nerve endings that innervate teeth are usually sensitive to cold, but much less sensitive to heat. The face is less sensitive to cold than other parts of the body that are usually covered by clothing.

Pain

The subjective sensation we call *pain* is a warning signal that alerts the body of a harmful or unpleasant stimulus. The sensation may be initiated by receptors that are sensitive to mechanical, thermal, electrical, or chemical stimuli. Pain receptors are **specialized free nerve endings** [FIGURE 16.1] that are present in most parts of the body (the intestines and brain tissue have no pain receptors). There are presumed to be about 3 million such *nociceptors* distributed over the surface of the body.

Types of pain include (1) fast-conducted, sharp, prickling pain, (2) slow-conducted, burning pain, and (3) deep, aching pain in joints, tendons, and viscera. Other distinctions are sometimes made: *superficial somatic pain* originates from stimulation of skin receptors; *deep somatic pain* arises from stimulation of receptors in joints, tendons, and muscles; *visceral pain* originates from stimulation of receptors in body organs.

Some tissues are more sensitive to pain than others. A needle inserted into the skin produces great pain, but the same needle probed into a muscle produces little pain. An arterial puncture is painful, but a venous puncture is almost painless. A kidney stone that distends a ureter (the tube leading from the kidney to the urinary bladder) produces excruciating pain. In contrast, the intestines are not sensitive to pain if they are cut or burned, but *are* sensitive if they are distended or markedly contracted (cramps).

We adapt to most of our senses so that they don't become a bother. If we didn't, we would be continuously aware of the touch-pressure of clothing on our skin, for example. (This phenomenon of *adaptation* refers to the decline in the response of receptors to a continuous, even stimulation.) However, it is to our benefit that we do not get completely used to pain. If pain is to be useful as a warning signal that prevents serious tissue damage, it must be felt each time it occurs, even to some extent when we are asleep. For example, who has not turned over in the middle of the night to relieve the pain caused by sleeping on a twisted arm?

Referred pain Pain that originates in a body organ or structure is usually perceived to be on the body surface, often at a site away from the visceral source. A visceral pain felt subjectively in a somatic area is known as **referred pain**. For example, the pain of coronary heart disease (*angina pectoris*) may be felt in the left shoulder, arm, and armpit; an irritation of the gallbladder may be felt under the shoulder blades.

One possible explanation for the brain's "misinterpretation" of most visceral pain is that certain neurons use a common dorsal root to innervate both the visceral and somatic locations involved in referred pain. It is thought that the visceral sensory fibers and somatic sensory fibers both discharge into a common pool of neurons in the central nervous system. In a sense, the brain interprets the source of visceral pain as a region of the skin because pain impulses originate in the skin more frequently than in the viscera.

Another explanation of referred pain is that the area of the body to which the pain is referred usually is part of the body that develops from the same embryological structure (somite) as the real source of the pain. These somites are supplied by branches of the same peripheral nerves [see FIGURE 13.16]. Using the example above, the heart and left arm are derived from the same dermatome. Thus, angina pectoris can be felt in a part of the left arm.

Phantom pain Another unusual phenomenon in the sensing of pain is the **phantom pain** that is felt in an amputated limb (*phantom limb*). Such pain may be intense, and it is actually felt. The sensations of pain, "pins and needles," and temperature change are often felt by amputees in their amputated limbs for several months. Ordinarily, the pain is felt more in the joints than in other regions of the phantom limb, and more in the distal portion of the amputated segment than in the proximal portion. Phantom pain usually persists longest in those regions that have the largest representation in the cerebral cortex: the thumb, hand, and foot.

The neural mechanism for phantom pain is not known completely. It appears that pools of neurons associated with the sensations of the missing limb persist in the brain, and are somehow activated, and result in the perception. Impulses in the pools of neurons may be triggered by the irritation of peripheral nerves in the proximal stump.

Q: *Do analgesic painkillers work by desensitizing receptors?*

A: No. Analgesics have no effect on sensory receptors. They modify the perception of pain or the emotional reaction to pain.

Proprioception

Receptors in muscles, tendons, and joints transmit impulses about our position sense up the dorsal columns of the spinal cord. These impulses help us to be aware of the position of our body and its parts without actually seeing them. This sense of position is called *proprioception* (L. *proprius*, one's self + receptor), or the *kinesthetic sense*. The receptors in or near joints that are responsible for proprioception are specialized "spray" endings. Lamellated corpuscles in the synovial membranes and ligaments may be involved.

Itch and Tickle

Itch is probably produced by the repetitive, low-key stimulation of slow-conducting nerve fibers in the skin. *Tickle* is caused by a mild stimulation of the same type of fibers, especially when the stimulus moves across the skin. Receptors for both sensations are found almost exclusively in the superficial layers of the skin. It is thought that the sensations result from the activation of several sensory endings and that the information is conveyed via a combination of pathways. Like the areas most sensitive to pain, itch usually occurs where naked endings of unmyelinated fibers are abundant. Itch occurs on the skin, in the eyes, and in certain mucous membranes (such as in the nose and rectum), but not in deep tissues or viscera.

Although itching is usually produced by a repetitive, mechanical stimulation of the skin, it is often produced by chemical stimuli such as polypeptides known as *kinins*, and by the histamine that the body releases during an allergy attack or inflammatory response.

Stereognosis

The ability to identify unseen objects by handling them is *stereognosis* (STEHR-ee-og-NO-sis; Gr. *stereos*, solid, three-dimensional; *gnosis*, knowledge). This ability depends on the sensations of touch and pressure, as well as on sensory areas in the parietal lobe of the cerebral cortex. Damage to certain areas of the cortex of the parietal lobe usually impairs stereognosis, even if the cutaneous sensations are intact.

Corpuscles of Ruffini and Bulbous Corpuscles (of Krause)

Corpuscles of Ruffini are now considered to be variants of touch-pressure receptors. They are located deep within the dermis and subcutaneous tissue, especially in the soles of the feet [FIGURE 16.1]. They are thought to be mechanoreceptors that respond to the displacement of the surrounding connective tissue within the corpuscle, and appear to be sensors for touch-pressure, position sense, and movement. A large, myelinated afferent nerve fiber enters the corpuscle, loses its myelin sheath, and forms many treelike terminal branches that intertwine with the collagen fibers within the core of the corpuscle.

Bulbous corpuscles (of Krause) are found in the dermis of the conjunctiva (the covering of the whites of the eyes and the lining of the eyelids), tongue, and external genitals. They are thought to be mechanoreceptors.

Neural Pathways for General Senses

The neural pathways involved in relaying influences from specific general sensory receptors to the cerebral cortex include the *dorsal column-medial lemniscus pathway*, the *spinothalamic tracts*, and the *trigeminothalamic tract*. The other ascending tracts and pathways for taste, smell, hearing, and vision are discussed later in this chapter.

In general, those afferent nerves that convey highly localized and discriminative sensations are larger, have more myelin, and conduct faster than those that convey less-defined sensations. They travel in the dorsal column-medial lemniscus and trigeminothalamic pathways. The afferent nerves that convey the less-defined sensations travel in the spinothalamic tracts.

Neural pathways for light touch The sensory area of the brain specialized for touch is located in the general sensory region of the parietal lobe of the cerebral cortex. Light touch is mediated by at least three neural pathways from the spinal cord to the cerebral cortex, including (1) the *dorsal column-medial lemniscal pathway,* (2) the *spinocervicothalamic pathway* of the dorsal column-medial lemniscal system, and (3) the *anterior spinothalamic tracts* of the anterolateral system.

Neural pathways for touch-pressure Touch-pressure is mediated by the *dorsal column-medial lemniscal pathway,* and probably by the *spinocervicothalamic pathway* of the posterior column-medial lemniscal system.

Neural pathways for temperature The sensory area of the brain for temperature is the same as for touch, the parietal lobe of the cerebral cortex. Crude sensations of temperature may be experienced in the thalamus. The sensation of temperature change is mediated by the *lateral spinothalamic tract* of the anterolateral system.

Q: *Why does scratching relieve the itching of insect bites?*

A: Scratching an insect bite temporarily soothes the naked endings of unmyelinated fibers near the surface of the skin. However, scratching usually makes the inflammation worse by irritating the skin and releasing the chemical stimulant that caused the itch in the first place.

Neural pathways for pain It is thought that pain impulses are conveyed by two or more pathway systems, including (1) the *lateral spinothalamic tract,* which consists of a sequence of at least three neurons with long axonal processes that relay pain impulses from the spinal cord to the thalamus, and (2) the *indirect spinoreticulothalamic pathway,* which consists of a sequence of many neurons that relay pain impulses to the reticular formation and thalamus. The pathways are in the anterolateral system. Fibers from the spinotectal tract to the midbrain may also be involved. The perception of pain occurs in the thalamus, but the discrimination (judgment) of the type of pain and its intensity occur in the parietal lobe of the cerebral cortex.

Neural pathways for proprioception Many of the proprioceptive impulses are relayed to the cerebellum, but some are conveyed to the cerebral cortex through the *medial lemniscal pathway* and from *thalamic projections.* The sensory area for conscious position sense is located in the parietal lobe of the cerebral cortex. The neural pathways are the spinocerebellar tracts (unconscious perception) and dorsal column-medial lemniscal pathway (conscious perception). Degenerative diseases of the dorsal column of the spinal cord produce *ataxia* (lack of muscular coordination associated with inadequate sensory input) because proprioceptive impulses are not conveyed to the cerebellum and to the thalamus and cerebral cortex.

Neural pathways for somesthetic sensations from the head Those afferent nerves that convey highly localized discriminative sensations from the face, mouth, nasal cavities, and associated structures, as well as those conveying cruder sensations, form the *trigeminothalamic tracts.* These tracts originate from the spinal trigeminal nucleus (pain and temperature) and principal sensory nuclei (touch-pressure), respectively. The tracts parallel the spinothalamic and the dorsal column-medial lemniscus tracts. The sensations are primarily felt on the skin and the mucosal membranes of the nasal and oral cavities. They involve the sensory region of the parietal lobe of the cerebral cortex.

A S K   Y O U R S E L F

1 What are the receptors for light touch? Touch-pressure? Pain?

2 How do light touch and touch-pressure differ?

3 What types of stimuli can initiate pain?

4 How do somatic and visceral pain differ?

5 What are the three ascending neural pathways involved in the general senses?

TASTE (GUSTATION)

The receptors for taste, or **gustation** (L. *gustus,* taste), and smell are both chemoreceptors, and the two sensations are clearly interrelated. (A person whose nasal passages are "blocked" by a cold cannot "taste" food as effectively.) Despite such similarities, taste and smell are separate, distinct senses, and are treated separately here.

Structure of Taste Receptors

The surface of the tongue is covered with many small protuberances called **papillae** (puh-PILL-ee; sing. *papilla,* puh-PILL-uh; L. diminutive of *papula,* nipple, pimple). Papillae give the tongue its bumpy appearance [FIGURE 16.2]. They are most numerous on the dorsal surface of the tongue, and are also found on the palate (roof of the mouth), throat, and posterior surface of the epiglottis.

The three main types of papillae are the following: (1) *Fungiform* (L. mushroomlike) *papillae* are scattered singly, especially near the tip of the tongue. Each fungiform papilla contains 1 to 8 taste buds. (2) From 10 to 12 *circumvallate* (L. "wall around") *papillae* form two rows parallel to the V-shaped sulcus terminalis near the posterior third of the tongue. Each circumvallate papilla may contain 90 to 250 taste buds. (3) *Filiform* (L. threadlike) *papillae* are pointed structures near the anterior two-thirds of the tongue. Filiform papillae do not necessarily contain taste buds.

Located within the crevices of most papillae are approximately 10,000 receptor organs for the sense of taste, popularly called **taste buds.** They are barrel-shaped clusters of *chemoreceptor* (taste or gustatory) *cells* and *sustentacular* (supporting) *cells,* arranged like alternating segments of an orange [FIGURE 16.2B]. Each taste bud contains about 25 receptor cells. The more numerous supporting cells act as reserve cells, replenishing the receptor cells when they die. Mature receptor cells have a life of only about 10 days, and they usually can be replaced from reserve cells in about 10 hours. Taste receptor cells are replaced with decreasing frequency as we get older. This explains, in part, why our sense of taste may diminish with age, and may also explain why babies dislike spicy foods and tend to favor relatively bland baby foods.

Q: *Why do some astronauts have trouble tasting food while traveling in space?*

A: Apparently, the loss of taste is caused by the lack of gravity. The heart, accustomed to pumping against gravity, forces more blood into the head than is necessary, and other body fluids accumulate there as well. The result is congestion, the same feeling we feel when we have a head cold.

FIGURE 16.2 TASTE-SENSITIVE AREAS AND TASTEBUDS OF THE TONGUE

[A] The specific taste-sensitive areas are shown in color. Sour sensations (green) are perceived most acutely on the sides of the tongue, stimulated by hydrogen ions in acids. Saltiness (pink) is tasted mainly at the sides and tip of the tongue. Bitter sensations (yellow) are perceived mainly at the back of the tongue. Sweetness (purple) is tasted optimally at the tip of the tongue. The center of the tongue, with only a few taste buds, is relatively insensitive to taste. **[B]** Drawing of section through taste bud. **[C]** Scanning electron micrograph of taste buds. ×350.

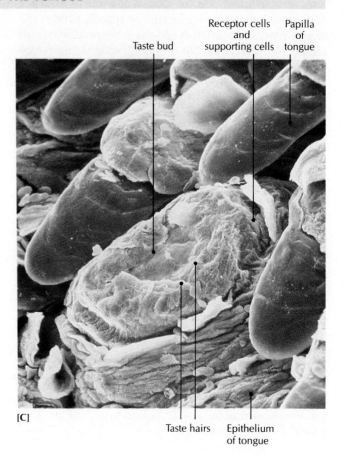

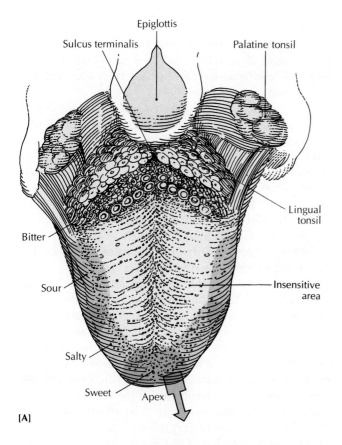

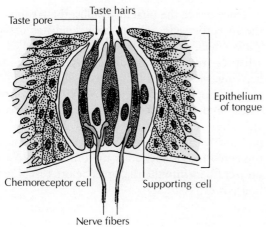

Extending from the free end of each receptor cell are short *taste hairs* (microvilli) that project through the tiny outer opening of the taste bud, called the *taste pore*, into the surface epithelium of the oral cavity. It is thought that gustatory sensations are initiated on the taste hairs, but before a substance can be tasted it must be in solution. Saliva containing ions or dissolved molecules of the substance to be tasted enters the taste pore and interacts with receptor sites on the taste hairs.

Basic Taste Sensations

Although all taste cells are structurally identical, each cell has many different types of receptor sites. Because the proportion of different types varies from cell to cell, each taste cell can respond to a variety of stimuli. The four generally recognized basic taste sensations are sweet, sour, bitter, and salty. We can taste many subtle flavors because of combinations of the four basic sensations, complemented by an overlay of odors. Taste perception is also helped by information about the texture, temperature, spiciness, and odor of food. The areas of response to the four basic tastes are located on specific parts of the tongue [FIGURE 16.2A]. Salt and sweet are perceived most acutely on the tongue, but bitter and sour are perceived more acutely on the palate.

Neural Pathways for Gustation

Taste impulses are conveyed from the anterior two-thirds of the tongue to the brain by a branch of the facial nerve (cranial nerve VII). Impulses from the posterior third of the tongue are carried to the brain by the glossopharyngeal (IX) nerve, and from the palate and pharynx by the vagus (X) nerve. The taste fibers of all three cranial nerves terminate in the *nucleus solitarius* in the medulla oblongata. From there, axons project to the thalamus, and then to the "taste center" in the parietal lobe of the cerebral cortex.

A S K   Y O U R S E L F

1 What are papillae?

2 Can you describe the structure of a taste bud?

3 What is the sensory role of taste hairs?

4 Which areas of the tongue respond to each of the four basic taste sensations?

SMELL (OLFACTION)

Our sense of smell, or **olfaction** (L. *olere*, to smell + *facere*, to make), is perhaps as much as 20,000 times more sensitive than our sense of taste. For example, we can taste quinine in a concentration of 1 part in 2 million, but we can smell mercaptans (the type of chemical released by skunks) in a concentration of 1 part in 30 *billion*. Adults can usually sense up to 10,000 different odors, and children can do even better. Unfortunately, our sense of smell is not perfect. Several poisonous gases, including carbon monoxide, are not detectable by our olfactory receptors.

Structure of Olfactory Receptors

The **olfactory receptor cells** are located high in the roof of the nasal cavity, in specialized areas of the nasal mucosa

called the *olfactory epithelium* [FIGURE 16.3]. Each nostril contains a small patch of pseudostratified, columnar olfactory epithelium about 2.5 sq cm (about the size of a thumbnail). The epithelium consists of three types of cells: (1) *receptor cells*, which actually are the olfactory neurons, (2) *sustentacular* (supporting) *cells*, and (3) a thin layer of small *basal cells*. These basal cells are capable of undergoing cell division and replacing degenerating receptor and sustentacular cells. Each receptor cell has a lifetime of about 30 days. They are replaced by basal cells, which are continually differentiating into new olfactory neurons and forming new synaptic connections in the olfactory bulb.

Because olfactory cells are the only neurons exposed to the external environment, they can be damaged or destroyed rather easily by disease or other trauma. This may account, in part, for about 1 percent of our olfactory receptor cells (neurons) that are lost every year, without replacement.

We have more than 25 million bipolar receptor cells (a hunting dog has about 220 million), each of which is surrounded by sustentacular cells. Each thin receptor cell has a short dendrite extending from its superficial end to the surface epithelium. The receptor cell ends in a bulbous *olfactory vesicle* [FIGURE 16.3C]. From this swelling, 6 to 20 long *cilia* project through the mucuslike fluid that covers the surface epithelium. The fluid is secreted by the sustentacular cells and *olfactory (Bowman's) glands*. It is important because odoriferous substances need to be dissolved before they can stimulate receptor sites.

The receptive sites of the cilia are exposed to the molecules responsible for the odors. The axons of the bipolar receptive cells (neurons) pass through the *basal lamina* and join other axons to form fascicles of the olfactory nerve (cranial nerve I). These unmyelinated olfactory nerve fibers are among the smallest and slowest-conducting fibers of the nervous system. They pass through the foramina in the cribriform plate of the ethmoid bone on their way to the olfactory bulbs. The **olfactory bulbs** are specialized structures of gray matter, stemlike extensions of the olfactory region of the brain.

Once inside the olfactory bulbs, the terminal axons of the receptor cells synapse with dendrites of *tufted cells*, *granule cells*, and *mitral cells*. These complex, ball-like synapses are called **olfactory glomeruli** (gluh-MARE-you-lie; sing. *glomerulus*; L. *glomus*, ball). Each glomerulus receives impulses from about 26,000 receptor cell axons. These impulses are conveyed along the axons of mitral and tufted cells, which form the *olfactory tract* running posteriorly to the olfactory cortex in the temporal lobe of the cerebrum. Olfaction is the only sense that does not project fibers into the thalamus before reaching the cerebral cortex.

Neural Pathways for Olfaction

Mitral cells in the olfactory bulbs project axonal branches through the olfactory tract to the primary olfactory cortex (see FIGURE 14.13B). The *primary olfactory cortex* is composed of the cortex of the uncus and adjacent areas, located in the temporal lobe of the cerebral cortex.

A S K   Y O U R S E L F

1 What are the three types of cells in the olfactory epithelium?

2 What are olfactory bulbs?

3 What happens in the olfactory glomeruli?

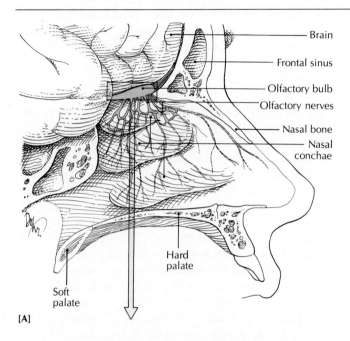

[A]

FIGURE 16.3 OLFACTORY RECEPTORS

[A] Olfactory receptive area in the roof of the nasal cavity; medial view. **[B]** Enlarged drawing of the olfactory epithelium, showing receptor cells, sustentacular cells, basal cells, and the formation of the olfactory tract inside the olfactory bulb. **[C]** Scanning electron micrograph of olfactory epithelial surface, showing olfactory vesicle, cilia extending into the nasal mucosa, and microvilli of sustentacular cells. ×9000.

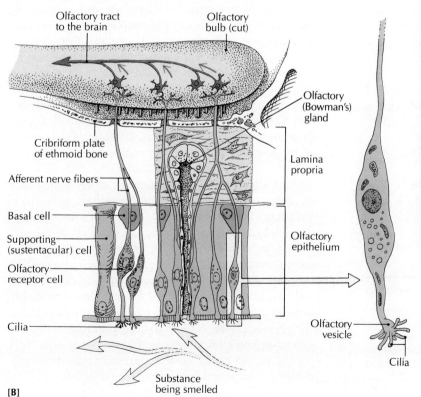

[B]

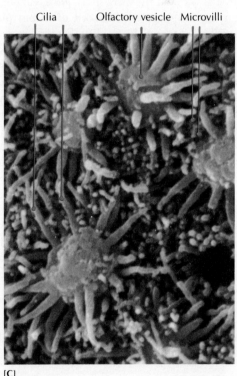

[C]

HEARING AND EQUILIBRIUM

Hearing and equilibrium are considered in the same section because both sensations are received in the same organ: the inner ear. The ear actually has two functional units: (1) the *auditory apparatus,* concerned with hearing, and (2) the *vestibular apparatus,* concerned with posture and balance. The auditory apparatus is innervated by the *cochlear nerve,* and the vestibular apparatus is innervated by the *vestibular nerve.* The two nerves are collectively known as the *vestibulocochlear nerve* (cranial nerve VIII).

Structure of the Ear

The auditory system is organized to detect several aspects of sound, including pitch, loudness, and direction. The anatomical components of this system are the external ear, the middle ear, and the inner ear [FIGURE 16.4].

External ear   The *external ear* is the part you can see. It is also called the *auricle* (L. *auris,* ear) or *pinna* (PIHN-uh; L. wing). It is composed of a thin plate of elastic cartilage covered by a close-fitting layer of skin. The funnel-like curves of the auricle are well designed to collect sound waves and direct them to the middle ear. The deepest depression, the *concha* (KONG-kuh; L. conch shell), leads directly to the external auditory canal (meatus). The area of the concha is partly covered by two small projections, the *tragus* in front (TRAY-guhss; Gr. *tragos,* goat, because hairs said to resemble a goat's beard may grow at the entrance of the external auditory canal) and the *antitragus* behind.

The *helix* is the prominent ridge that forms the rim of the uppermost portion of the auricle. The *antihelix* is a curved ridge, more or less concentric to the helix, that surrounds the concha. It is separated from the helix by a furrow called the *scapha.* The *lobule,* or earlobe, is the fatty, lowermost portion of the auricle. It is the only part of the external ear without any cartilage.

The *external auditory canal* (meatus) is a slightly curved canal, extending about 2.5 cm (1 in.) from the floor of the concha to the tympanic membrane, which separates the external ear from the middle ear. The outer third of the wall of the external auditory canal is composed of elastic cartilage, and the inner two-thirds is carved out of the temporal bone [FIGURE 16.4]. The canal and tympanic membrane are covered with skin. Fine hairs in the external ear are directed outward, and sebaceous glands and modified sweat glands (ceruminous glands) secrete *cerumen* (suh-ROO-muhn), or earwax. The hairs and wax make it difficult for tiny insects and other foreign matter to enter the canal. Cerumen also prevents the skin of the external ear from drying out. The canal also acts as a buffer against humidity and temperature changes that can alter the elasticity of the eardrum.

Middle ear   The *middle ear* is a small chamber between the tympanic membrane and the inner ear. It consists of the tympanic cavity [Gr. *tympanon,* drum) and contains the auditory ossicles (earbones).

The *tympanic membrane,* popularly called the *eardrum,* forms a partition between the external ear and middle ear. It is a thin layer of fibrous tissue continuous externally with skin and internally with the mucous membrane that lines the middle ear. Between its concave external surface and convex internal surface is a layer of circular and radial fibers that give the membrane its firm elastic tension. The tympanic membrane is attached to a ring of bone (the tympanic annulus) and vibrates in response to sound waves entering the external auditory canal. The tympanic membrane is well endowed with blood vessels and nerve endings, so a "punctured eardrum" usually produces considerable bleeding and pain.

The *tympanic cavity* (middle-ear cavity) is a narrow, irregular, air-filled space in the temporal bone. It is separated laterally from the external auditory canal by the tympanic membrane, and medially from the inner ear by the bony wall, which has two openings, the *oval window* and the *round window* [FIGURE 16.4]. An opening in the posterior wall of the cavity leads into the *tympanic antrum,* a chamber that is continuous with the small air cells of the mastoid process. When an infection of the middle ear progresses through the tympanic antrum into the mastoid cells, it can cause *mastoiditis.*

In the anterior wall of the tympanic cavity is the *auditory tube,* commonly called the *Eustachian tube* (yoo-STAY-shun). It leads downward and inward from the tympanic cavity to the nasopharynx, the space above the soft palate that is continuous with the nasal passages. The mucous membrane lining the nasopharynx is also continuous with the membrane of the tympanic cavity. As a result, an infection may spread from the nose or throat into the middle ear, producing a middle-ear infection, or *otitis media.* (It is a good idea to keep your mouth slightly open when you blow your nose. Blowing your nose hard with the mouth closed creates a pressure that may force infectious microorganisms into the middle ear.)

Q: *Why does your voice sound different on a tape recording?*

A: When you hear yourself speak, you are hearing some extra resonance produced by the conduction of sound waves through the bones of your skull. Your voice as played by a tape recorder is the way it sounds to a listener, who receives the sound waves only through air conduction.

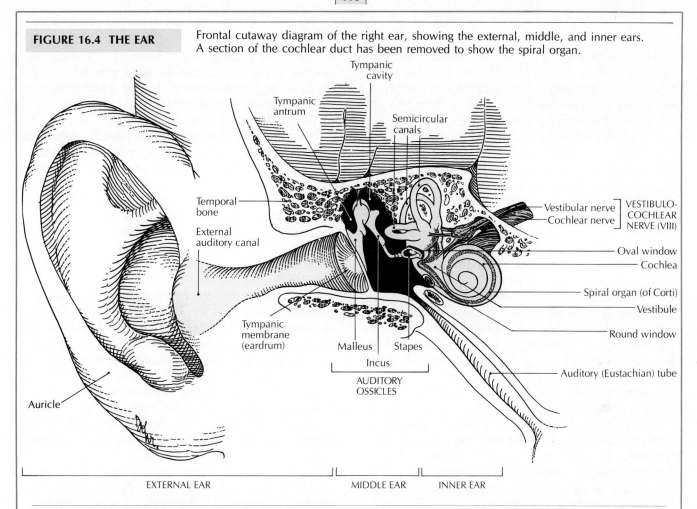

FIGURE 16.4 THE EAR

Frontal cutaway diagram of the right ear, showing the external, middle, and inner ears. A section of the cochlear duct has been removed to show the spiral organ.

Structure	Description	Structure	Description
EXTERNAL EAR		**INNER EAR (LABYRINTH)**	
Auricle (pinna)	Cartilaginous, exterior "flap" of ear; conveys sound waves to middle ear.	Vestibule	Central chamber of inner ear; includes utricle and saccule filled with endolymph and surrounded by perilymph.
External auditory canal (meatus)	Canal leading from floor of concha in outer ear to tympanic membrane.	Semicircular ducts*	Three small ducts lying at right angles to each other; posterior to the vestibule; suspended in perilymph. Each duct has an expanded end, the ampulla, which contains a receptor structure, the crista ampullaris.
MIDDLE EAR			
Tympanic membrane (eardrum)	Fibrous tissue extending across deep inner end of external auditory canal, forming partition between external and middle ears.		
Tympanic cavity	Air-filled space in temporal bone; separated from external auditory canal by tympanic membrane, and from inner ear by bony wall with round and oval windows.	Cochlea	Spiral structure containing perilymph-filled scala vestibuli and scala tympani, and endolymph-filled scala media (cochlear duct); anterior to the vestibule.
Auditory tube (Eustachian tube)	Tube leading downward and inward from tympanic cavity to nasopharynx.	Spiral organ (of Corti)	Organ of hearing resting on the basilar membrane of the cochlea; a complex of supporting cells and hair cells.
Auditory ossicles (earbones)	Malleus, incus, and stapes form chain from tympanic membrane to oval window of inner ear.		

*There are three semicircular canals surrounded by bone. Within each canal is a semicircular duct (about one-quarter the diameter of the canal), which contains endolymph. Surrounding each duct, and contained within each canal, is perilymph.

The main function of the auditory tube is to maintain equal air pressure on both sides of the tympanic membrane by permitting air to pass from the nasal cavity into the middle ear. The pharyngeal opening of the tube remains closed when the external pressure is greater, but opens during swallowing, yawning, and nose blowing, so that minor differences in pressure are adjusted without conscious effort. The tube may remain closed when the pressure change is sudden, as when an airplane takes off or lands. However, the pressure can usually be equalized, and the discomfort relieved, by swallowing or yawning. This maneuver stimulates the tensor veli palatini muscle to contract, pulling on a portion of the cartilage of the auditory tube, causing the tube to open.

The three *auditory ossicles* (earbones) of the middle ear form a chain of levers extending from the tympanic membrane to the inner ear [FIGURE 16.5]. This lever system transmits sound waves from the external ear to the inner ear. From the outside in, the tiny, movable bones are the *malleus* (hammer), *incus* (anvil), and *stapes* (STAY-peez) (stirrup). The ear bones are the smallest bones in the body, with the stapes being the smallest of all.

The auditory ossicles are held in place and attached to each other by ligaments. Two tiny muscles, the tensor tympani and the stapedius, are attached to the ear bones. The *tensor tympani* is attached to the handle of the malleus. When this muscle contracts, it pulls the malleus inward, increasing the tension on the tympanic membrane and reducing the amplitude of vibrations transmit-

ted through the chain of auditory ossicles. The *stapedius* attaches to the neck of the stapes. Its contraction pulls the footplate of the stapes, decreasing the amplitude of vibrations at the oval window.

Inner ear The inner ear is also called the *labyrinth* (Gr. maze) because of its intricate structure of interconnecting chambers and passages [FIGURE 16.6]. It consists of two main structural parts, one inside the other: (1) the *bony labyrinth* is a series of channels hollowed out of the petrous portion of the temporal bone. It is filled with a fluid called *perilymph*. (2) The bony labyrinth surrounds the inner *membranous labyrinth*, which contains a fluid called *endolymph* and all the sensory receptors for hearing and equilibrium.

The membranous labyrinth consists of three semicircular ducts, as well as the utricle, saccule, and cochlear duct, all of which are filled with endolymph and contain various sensory receptors (cristae ampullaris, maculae, and spiral organ). The semicircular ducts are located within the semicircular canals of the bony labyrinth. They are about one-quarter the diameter of the canals. Perilymph is located in the space between the ducts and the bony walls of the canals. Because the membranous labyrinth fits inside the labyrinth, these two bony channels have the same basic shape.

The bony labyrinth consists of the vestibule, three semicircular canals, and the spirally coiled cochlea. The *vestibule* (L. entrance) is the central chamber of the labyrinth. Within the vestibule are the two endolymph-filled

FIGURE 16.5 THE AUDITORY OSSICLES: MALLEUS, INCUS, AND STAPES

Right lateral view.

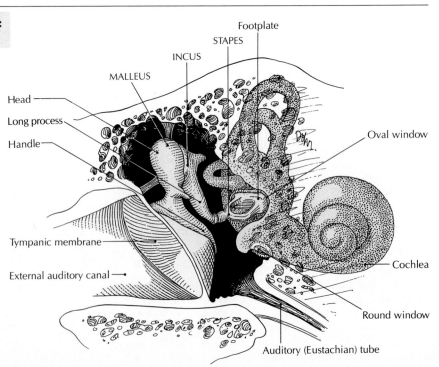

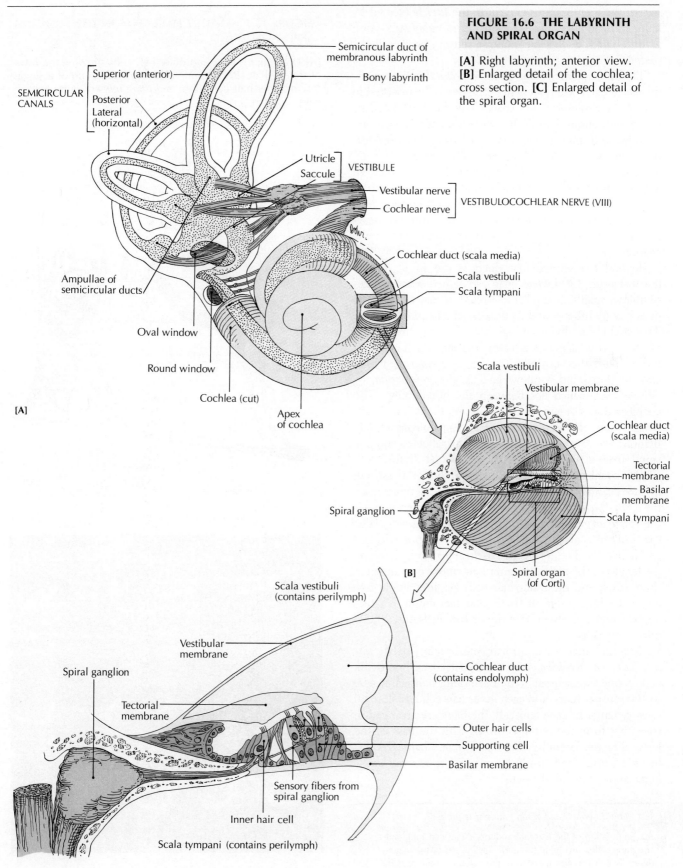

FIGURE 16.6 THE LABYRINTH AND SPIRAL ORGAN

[A] Right labyrinth; anterior view. **[B]** Enlarged detail of the cochlea; cross section. **[C]** Enlarged detail of the spiral organ.

[A]

SEMICIRCULAR CANALS
Superior (anterior)
Posterior
Lateral (horizontal)

Semicircular duct of membranous labyrinth
Bony labyrinth

Utricle
Saccule
VESTIBULE

Vestibular nerve
Cochlear nerve
VESTIBULOCOCHLEAR NERVE (VIII)

Cochlear duct (scala media)
Scala vestibuli
Scala tympani

Ampullae of semicircular ducts

Oval window

Round window

Cochlea (cut)

Apex of cochlea

[B]

Scala vestibuli
Vestibular membrane
Cochlear duct (scala media)
Tectorial membrane
Basilar membrane
Scala tympani

Spiral ganglion

Spiral organ (of Corti)

[C]

Scala vestibuli (contains perilymph)

Vestibular membrane

Spiral ganglion

Tectorial membrane

Cochlear duct (contains endolymph)

Outer hair cells
Supporting cell
Basilar membrane

Sensory fibers from spiral ganglion

Inner hair cell

Scala tympani (contains perilymph)

sacs of the membranous labyrinth, the ***utricle*** (YOO-trih-kuhl; L. little bottle) and the smaller ***saccule*** (SACK-yool; L. little sack). Each sac contains a sensory patch called a *macule*.

The three ***semicircular ducts*** and ***canals*** are perpendicular to each other, allowing each one to be oriented in one of the three planes of space. The ducts are lined by the membranous labyrinth, while the canals are surrounded and lined by bone. On the basis of their locations, the semicircular ducts are called *superior, lateral,* and *posterior.* Each duct has an expanded end called an *ampulla* (am-POOL-uh), which contains a receptor structure, the *crista ampullaris.* The utricle, saccule, and semicircular ducts are concerned with equilibrium, not hearing, and will be discussed in further detail later in the chapter.

Beyond the semicircular ducts is the spiral ***cochlea*** (KAHK-lee-uh; Gr. *kokhlos,* snail), so named because it resembles a snail's shell [FIGURE 16.6A]. It may be thought of as a bony tube wound $2\frac{3}{4}$ times in the form of a spiral. The cochlea is divided longitudinally into three spiral ducts: (1) the *scala* (L. staircase) *vestibuli,* which communicates with the vestibule; (2) the *scala tympani,* which ends at the round window; and (3) the *scala media,* or *cochlear duct,* which lies between the other ducts. The cochlear duct contains endolymph and the spiral organ, whereas the scala vestibuli and scala tympani contain perilymph. The three ducts, arranged in parallel, ascend in a spiral around the bony core, or *modiolus* (L. hub).

The cochlear duct is separated from the scala vestibuli by the *vestibular membrane,* and from the scala tympani by the *basilar membrane.* Resting on the basilar membrane is the ***spiral organ (of Corti),*** the organ of hearing. The spiral organ is an organized complex of supporting cells and hair cells. The hair cells are arranged in rows along the length of the coil. The outer hair cells are arranged in three rows, and the inner hair cells are in a single row along the inner edge of the basilar membrane [FIGURE 16.7]. There are about 3500 inner hair cells and 20,000 outer hair cells.

Both the inner and outer hair cells have bristlelike *sensory hairs,* or *stereocilia,* which are specialized microvilli, and on one side, a basal body. Each outer hair cell has 80 to 100 sensory hairs, and each inner hair cell has 40 to 60 sensory hairs. In each hair cell, the hairs are arranged in rows that form the letter W or U, with the base of the letter directed laterally. The tips of many hairs are embedded within, and firmly bound to, the *tectorial membrane* above the spiral organ.

Q: *Why do you cough when your ear is probed?*

A: Stimulation of the external auditory canal can initiate a coughing reflex involving the auricular branch of the vagus nerve, which supplies the skin of the external auditory canal.

FIGURE 16.7 SENSORY HAIR CELLS IN THE SPIRAL ORGAN

[A] Drawing of inner and outer hair cells showing the rows of stereocillia. [B] Scanning electron micrograph of a single row of inner hair cells (top) and three rows of outer hair cells. ×3400.

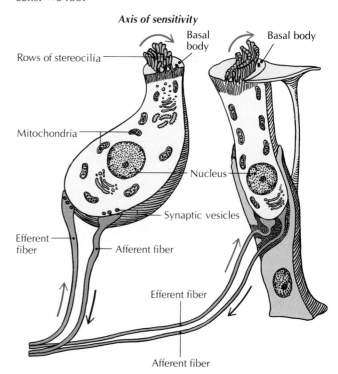

Axis of sensitivity

Rows of stereocilia — Basal body — Basal body

Mitochondria

Nucleus

Synaptic vesicles

Efferent fiber — Afferent fiber

Efferent fiber

Afferent fiber

[A] INNER HAIR CELL OUTER HAIR CELL

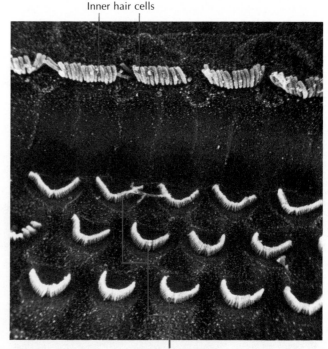

Inner hair cells

Outer hair cells

[B]

The pressure waves created by the vibrating stapes are transmitted to the perilymph, and then (1) up the scala vestibuli through an opening between the scala vestibuli and scala tympani (the helicotrema), (2) into and down the scala tympani through the $2\frac{1}{2}$ turns of the cochlea (3) to the vestibular membrane, basilar membrane, and endolymph of the cochlear duct, (4) to the spiral organ and tectorial membrane, and (5) to the round window, where the pressure waves are dampened.

Neural Pathways for Hearing

Axons of neurons with cell bodies in the spiral ganglion of the cochlear nerve extend from the spiral organ and terminate centrally in the dorsal and ventral cochlear nuclei. The cochlear nerve enters the brainstem at the junction of the pons and medulla oblongata and terminates in the cochlear nuclear complex. After entering, each fiber divides into two main branches.

Axons of neurons in the dorsal and ventral cochlear nuclei ascend as crossed and uncrossed fibers in the lateral lemnisci to the inferior colliculus. Thus, the auditory pathways ascend bilaterally. From cell bodies in the inferior colliculus, the pathway continues as the brachium of the inferior colliculus to the medial geniculate body of the thalamus. From cell bodies in the medial geniculate body, axons project via the auditory radiations to the primary auditory cortex (transverse temporal gyrus of Heschl, areas 41 and 42; see FIGURE 14.12).

A lesion of the auditory pathway on one side results in a decrease of hearing acuity in both ears. This decrease is related to the fact that the ascending auditory pathways are composed of both crossed and uncrossed projections.

Projecting from the auditory cortex and other nuclei of the auditory pathways are *descending pathways,* which accompany the ascending pathways described above. The descending fibers have a role in processing ascending auditory influences, enhancing the signals and suppressing the "noise" in the auditory pathways. Finally, neurons in the brainstem nuclei have *cochlear efferent fibers,* which are integrated into the feedback system that passes in the vestibulocochlear nerve before terminating upon the hair cells in the spiral organ. The inhibitory influences conveyed by these fibers suppress the activity of the afferent fibers of the cochlear nerve. Vestibular efferent fibers are also present.

Vestibular Apparatus and Equilibrium

The inner ear helps the body cope with changes in position, acceleration, and deceleration. The purpose of this *vestibular apparatus* [FIGURE 16.8] is to signal changes in the *motion* of the head (*dynamic equilibrium,* also called *kinetic equilibrium*), and in the *position* of the head with respect to gravity (*static equilibrium,* or *posture*). The main components of the vestibular apparatus are the utricle, saccule, and the three fluid-filled semicircular ducts of the membranous labyrinth. The receptors in the utricle and saccule regulate static equilibrium, and the receptors in the ampullae of the semicircular ducts respond to movements of the head. The equilibrium system also receives input from the eyes and from some proprioceptors in the body, especially the joints. (Try standing on your toes with your eyes closed. Without your eyes to guide your body, you invariably begin to fall forward.)

Specialized proprioceptors of the vestibular apparatus, known as *hair cells,* are arranged in clusters of *hair bundles* [FIGURE 16.9]. Hair cells are extremely sensitive receptors that convert a mechanical force applied to a hair cell into nerve impulses that are relayed to the brain. An extremely slight movement of a hair bundle can cause the hair cells to respond. Two types of sensory hairs are present in hair bundles: (1) *stereocilia,* which are actually modified microvilli, and (2) *kinocilia,* which are modified cilia. Each hair cell has about 100 stereocilia in a tuft, and 1 kinocilium at the edge of the tuft [FIGURE 16.9B]. Because of its asymmetry, a receptor hair cell is said to be polarized.

The electrical signal triggered by the movement of a hair bundle depends on the direction of the movement.

FIGURE 16.8 THE VESTIBULAR APPARATUS

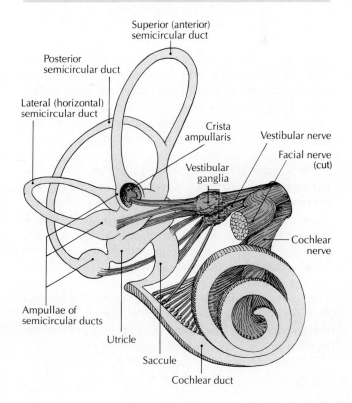

Superior (anterior) semicircular duct

Posterior semicircular duct

Lateral (horizontal) semicircular duct

Crista ampullaris

Vestibular nerve

Facial nerve (cut)

Vestibular ganglia

Cochlear nerve

Ampullae of semicircular ducts

Utricle

Saccule

Cochlear duct

FIGURE 16.9 MACULA AND HAIR CELLS

[A] Macula, receptor region of utricle and saccules. The stereocilia of hair cells extend into the adjacent gelatinous otolithic membrane, which contains embedded otoconia crystals. [B] Drawing of a hair cell in cross section. [C] Scanning electron micrograph of a hair bundle from the inner ear of a bullfrog. ×9000.

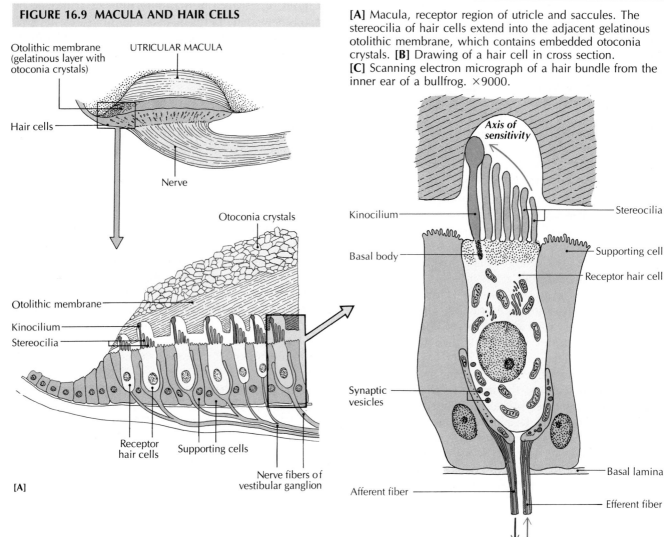

When the hairs of the hair cell are bent in the direction of the kinocilium (known as the *axis of sensitivity*), the hair cell can generate a receptor potential in the nerve ending synapsing with the hair cell. When the hairs of the hair cell are bent in the direction opposite to the axis of sensitivity, the nerve ending is not stimulated.

Static equilibrium: utricles and saccules The receptor region of the utricles and saccules, called the *macula*, contains receptor hair cells embedded in a jellylike *otolithic* (Gr. "ear stones") *membrane* [FIGURE 16.9A]. Loosely attached to the membrane, and piled on top of it, are hundreds of thousands of calcium carbonate crystals called *otoconia*, or *statoliths* ("standing stones"). The utricles and saccules are both filled with endolymph. Hair cells in the utricle respond to the motion changes that occur during back-and-forth and up-and-down movements of the head. The hair cells also monitor the position of the head in space, controlling posture.

Saccules may also function like utricles, as well as serving as auditory receptors for low frequencies.

As efficient as human utricles and saccules are, they are certainly not perfect. This lack of efficiency is demonstrated when an airplane pilot does not realize that he or she is flying through clouds upside down. Also, the slowness of the utricle in registering deceleration can be seen in the delayed stumbling response of a standee in a bus when the bus stops suddenly.

Dynamic equilibrium: semicircular ducts The utricles and saccules are *organs of gravitation*, responding to movements of the head in a straight line: forward, backward, up, or down. In contrast, the crista ampullaris of the semicircular ducts responds to changes in acceleration in the *direction of head movements*, specifically turning and rotating. These movements are called angular movements, in contrast to straight-line movements.

Because each of the three ducts is situated in a different plane, at right angles to each other, at least one duct is affected by every head movement. Each duct has a bulge, the ***ampulla,*** that contains a patch of hair cells and supporting cells embedded in the ***crista ampullaris*** [FIGURE 16.10; see also FIGURE 16.8]. The hairs of the hair cells project into a gelatinous flap called the ***cupula*** (KYOO-pyuh-luh; L. little cask or tub). The cupula acts like a swinging door, with the crista as the hinge. The free edge of the cupula brushes against the curved wall of the ampulla. When the head rotates, the endolymph in the semicircular ducts lags behind due to inertia, displacing the cupula and the hairs projecting into it in the opposite direction. (The semicircular ducts do not sense movement at slow, steady speeds, because the head and the endolymph move at the same rate.)

As a result of the slight displacement of the cupula, the hairs bend in the direction of the axis of sensitivity. This stimulates the nerve endings to generate action potentials in each nerve fiber. The brain receives the impulses and signals the appropriate muscles to contract in order to maintain the body's equilibrium.

A feeling of dizziness occurs when you spin about or move violently and then stop suddenly. Because of inertia, the endolymph keeps moving (the way you keep moving forward when your car stops quickly), and it continues to stimulate the hair cells. Although you know you have stopped moving, the signals being sent from your inner ear to your brain make you feel that motion is still occurring, but in the reverse direction. In other words, while in motion the inertia of the endolymph bends the hairs in the direction opposite to the direction of the movement. Immediately after stopping, the endolymph continues to move relative to the ducts. As a result, the hairs are now bent in the direction of the prior movement. Because the inner ears of deaf mutes are not functional, they are immune to dizziness and motion sickness, and rely on visual cues for the maintenance of normal locomotion and posture. (Without visual cues, a swimmer who has lost the use of the labyrinth may navigate down instead of up to reach the surface.)

FIGURE 16.10 DYNAMIC EQUILIBRIUM

Displacement of the hair bundle stimulates the hair cells in the crista ampullaris of the semicircular ducts.

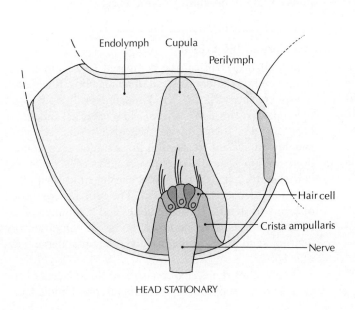

HEAD STATIONARY

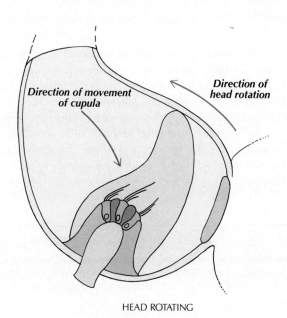

HEAD ROTATING

Neural Pathways for Equilibrium

The vestibular tracts consist of pathways to the brainstem, spinal cord, cerebellum, and cerebral cortex. The 19,000 nerve fibers of each vestibular nerve have their cell bodies in the vestibular ganglion near the membranous labyrinth. The primary vestibular fibers from the vestibular nerve pass into the upper medulla oblongata and terminate (1) in each of the four vestibular nuclei in the upper lateral medulla oblongata, and (2) in specific regions of the cerebellum.

The sensory signals from the vestibular sensors of the labyrinth are indicators of the position and movements of the head. These inputs from the vestibular receptors are critical in (1) generating compensatory movements to maintain balance and an erect posture in response to gravity, (2) producing the conjugate (coupled) movements of the eyes that compensate for changes in the position of the constantly moving head, and (3) supplying information for the conscious awareness of position, acceleration, deceleration, and rotation. The vestibular functions are supplemented by proprioceptive inputs from the muscles and joints, as well as the visual system.

ASK YOURSELF

1 What are the main parts of the middle ear?

2 What is the spiral organ?

3 What is the function of otoconia?

4 What is the importance of the hair cells of the ear?

DEVELOPMENTAL ANATOMY OF THE INNER EAR

The inner ear is the first part of the ear to develop in the embryo, usually during week 3 after fertilization. First, the ectoderm on both sides of the developing hindbrain thickens to become the *otic placodes* [FIGURE 16.11A]. By week 4, the otic placodes invaginate to form the *otic pits* [FIGURE 16.11B], and the *otocysts* (otic vesicles) are formed as the otic pits separate from the surface ectoderm and deepen further to form a pinched-off closed sac [FIGURE 16.11C]. The otocysts will later develop into the

---

Q: *Do the vestibular organs work during space travel?*

A: The utricle and saccule do not work under zero gravity conditions, but the semicircular canals do. About 40 percent of astronauts become motion sick at some time during space travel.

---

membranous labyrinth. Soon thereafter, the hollow, elongated *endolymphatic ducts* grow out from the otocysts [FIGURE 16.11D], and the saccular and utricular portions of the otocysts become recognizable. During weeks 5 through 8, the *semicircular ducts* begin to form as three disklike diverticula [FIGURE 16.11E]; during the same period the *cochlear ducts* form from the ventral saccular portion of the otocysts, soon developing into the cochlea [FIGURE 16.11F]. The *spiral organ* forms from differentiated cells in the cochlear duct. The spiral organ and other internal structures of the cochlea continue to develop into week 20.

VISION

We live primarily in a visual world, and sight is our most dominant sense. The specialized exteroceptors in our eyes comprise about 70 percent of the receptors of the entire body, and the optic nerves contain about one-third of all the afferent nerve fibers carrying information to the central nervous system.

Although we can rely on our eyes to bring us many of the sights of the external world, they are not able to reveal everything. We "see" only those objects that emit or are illuminated by light waves in our receptive range, representing only 1/70 of the entire electromagnetic spectrum. Some organisms, such as insects, are sensitive to shorter wavelengths in the range of ultraviolet, and other organisms can "see" longer wavelengths, in the infrared range of the spectrum.

Light reaches our light-sensitive "film," or *retina*, through a transparent window, the *cornea*. In addition to the basic and accessory structures of the eyeball, vision involves the brain and the optic nerve.

Structure of the Eyeball

The human eyeball can be compared to a simple, old-fashioned box camera. Instead of being a box, the eyeball is a sphere about 2.5 cm (1 in.) in diameter. In both cases, light passes through a lens. The external image is brought to a focus on the sensitive retina, which is roughly equivalent to the film in a camera. Over a hundred million specialized photoreceptor cells (rods and cones) convert light waves into electrochemical impulses, which are decoded by the brain. The retina is composed of layers of slender rods and cones and a complex of interacting, processing neurons. Some of these sensory neurons send axons to the brain via the optic nerve [FIGURE 16.12].

The wall of the eyeball consists of three layers of tissue: the outer supporting layer, the vascular middle layer,

FIGURE 16.11 DEVELOPMENTAL ANATOMY OF THE INNER EAR

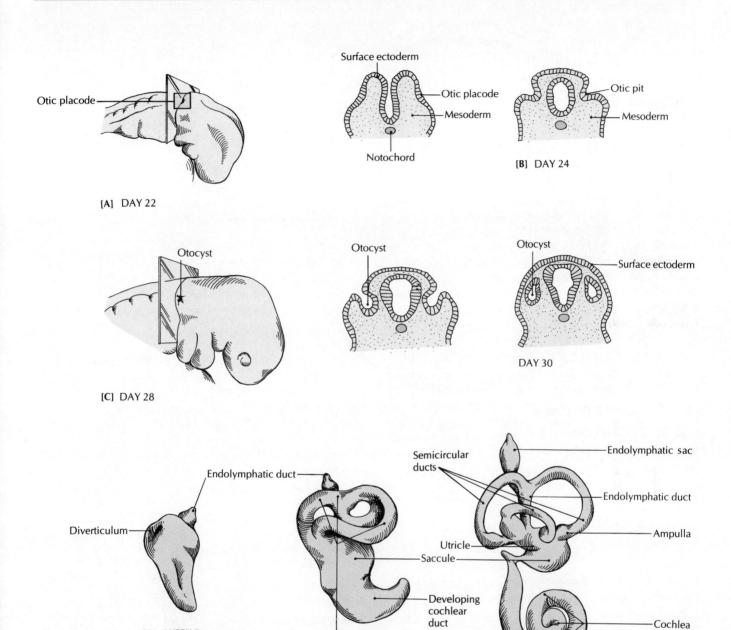

[A] DAY 22

Surface ectoderm
Otic placode
Mesoderm
Notochord

[B] DAY 24

Otic pit
Mesoderm

Otocyst

[C] DAY 28

Otocyst

Otocyst
Surface ectoderm

DAY 30

Endolymphatic duct
Diverticulum

[D] WEEK 5

Semicircular ducts
Developing semicircular ducts
Developing cochlear duct
Saccule
Utricle

[E]

Endolymphatic sac
Endolymphatic duct
Ampulla
Cochlea

[F] WEEK 8

FIGURE 16.12 THE HUMAN EYE

[A] Horizontal section through the eye. [B] Horizontal section through the anterior portion of the eye (enlarged drawing).

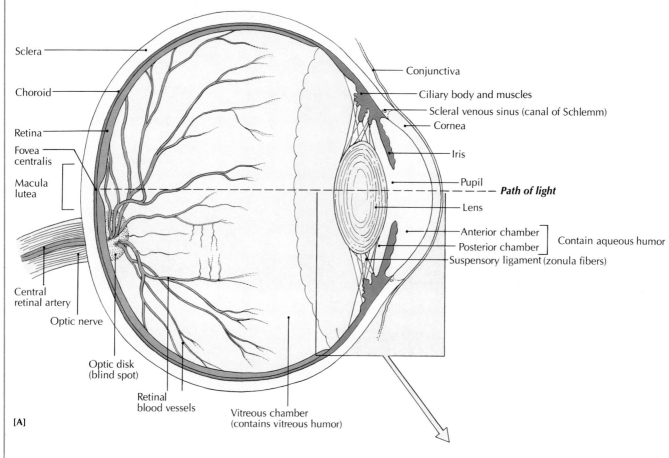

Sclera

Choroid

Retina

Fovea centralis

Macula lutea

Central retinal artery

Optic nerve

Optic disk (blind spot)

Retinal blood vessels

Vitreous chamber (contains vitreous humor)

Conjunctiva

Ciliary body and muscles

Scleral venous sinus (canal of Schlemm)

Cornea

Iris

Pupil — *Path of light*

Lens

Anterior chamber

Posterior chamber

Contain aqueous humor

Suspensory ligament (zonula fibers)

[A]

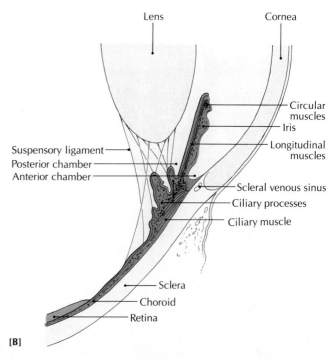

Lens

Cornea

Suspensory ligament

Posterior chamber

Anterior chamber

Circular muscles

Iris

Longitudinal muscles

Scleral venous sinus

Ciliary processes

Ciliary muscle

Sclera

Choroid

Retina

[B]

Structure	Description
SUPPORTING LAYER OF EYEBALL	
Sclera	Opaque layer of connective tissue over posterior five-sixths of outer layer of eyeball; "white" of the eye. Gives eyeball its shape, protects inner layers. Perforated to allow optic nerve fibers to exit.
Cornea	Transparent anterior portion of outer layer of eyeball. Light enters eye through cornea, a nonadjustable lens.
VASCULAR LAYER OF EYEBALL	
Choroid	Thin membrane of blood vessels and connective tissue between sclera and retina. Posterior two-thirds of vascular layer of eyeball.
Ciliary body	Thickened vascular layer in anterior portion of eyeball. Ciliary muscles help lens to focus by either increasing or decreasing tension on suspensory ligament of lens. Produces aqueous humor and some elements of vitreous humor.
Ciliary processes	Inward extensions of ciliary body. Help hold lens in place.
Lens	Elastic, colorless, transparent biconvex body of epithelial cells posterior to iris. Shape modified to focus on subjects at different distances (accommodation) through action of ciliary muscles. The lens is adjustable.
Iris	Colored part of eye. Thin, muscular layer; anterior extension of choroid. Regulates size of pupil, and thus amount of light entering eye, by controlling degree of constriction and dilation of iris.
Pupil	Adjustable circular opening in iris. Opened and closed reflexively relative to amount of light available (adaptation).
RETINAL LAYER OF EYEBALL	
Retina	Multilayered, light-sensitive membrane; innermost layer of eyeball. Connected to brain by optic nerve. Consists of neural layer and pigmented layer; which prevents reflection from back of retina. Receives focused light waves, transduces them into nerve impulses that the brain converts into visual perceptions.
Fovea (fovea centralis)	Depressed area in center of retina containing only cones. Area of most acute image formation and color vision.
CAVITIES OF EYEBALL	
Aqueous chambers	
Anterior chamber	Between cornea and iris. Contains aqueous humor.
Posterior chamber	Between iris and lens. Contains aqueous humor.
Vitreous chamber	Largest cavity, fills entire space behind lens. Contains vitreous humor.

and the inner retinal layer. The eyeball is divided into three cavities: the anterior chamber, the posterior chamber, and the vitreous chamber.

Supporting layer The outer *supporting layer* of the eyeball consists mainly of a thick membrane of tough, fibrous connective tissue. The posterior segment, which comprises five-sixths of the tough outer layer, is the opaque white *sclera* (Gr. *skleros*, hard). The sclera forms the "white" of the eye, giving the eyeball its shape and protecting the delicate inner layers. The anterior segment of the supporting layer is the transparent *cornea* (L. *corneus*, horny tissue), which comprises the modified anterior one-sixth of the outer layer. The cornea bulges slightly. If you close your eyes, place your finger lightly on your eyelid, and move your eye, you will feel the bulge. Light enters the eye through the cornea. The sclera and cornea are continuous. The cornea of this supporting layer contains no blood vessels. The supporting layer completely encloses the eyeball, except for the posterior portion, where small perforations in the sclera allow the fibers of the optic nerve to leave the eyeball on their way to the brain.

Cornea transplants from one individual to another have been very successful; about 30,000 cornea transplants were performed in the United States in 1988, with a success rate of approximately 95 percent. The typical problem of tissue rejection is usually avoided because the cornea has no blood or lymphatic vessels. As a result, antibodies and white blood cells that cause rejection cannot reach the cornea.

Vascular layer Because the middle layer of the eyeball contains many blood vessels, it is called the *vascular layer.* The dark color of the middle layer is produced by pigments that help to lightproof the wall of the eye by absorbing stray light and reducing reflection. The posterior two-thirds of the vascular layer consists of a thin membrane called the *choroid,* which is essentially a layer of blood vessels and connective tissue sandwiched between the sclera and the retina.

The vascular layer becomes thickened toward the anterior portion to form the *ciliary body.* Extending inward from the ciliary body are the fine *ciliary processes.* The smooth muscles in the ciliary body (*ciliary muscles*) contract to ease the tension on the suspensory ligament of the lens, which consists of fibrils that extend from the ciliary processes to the lens. The *lens* of the eye is a flexible, transparent, colorless, avascular body of the epithelial cells behind the iris, the colored part of the eye. The lens is held in place by the *suspensory ligament of the lens,* and by the ciliary processes. The shape of the lens can be adjusted so that objects at different distances can be brought into focus on the retina. This mechanism is called *accommodation.* The lens loses much of its elasticity with aging, making it difficult to focus efficiently without corrective eyeglasses.

The anterior extension of the choroid is a thin muscular layer called the *iris* (Gr. rainbow) because it is the colored part of the eyeball that can be seen through the cornea. In the center of the iris is an adjustable circular aperture, the *pupil* (L. doll), so called because when you look into someone else's eyes, you can see a reflected image of yourself that looks like a little doll. The pupil appears black because most of the light that enters the eye is not reflected outward. The iris, acting as a diaphragm, is able to regulate the amount of light entering the eye; it contains smooth muscles that contract or dilate in an involuntary reflex in response to the amount of light available, causing the pupil to become larger or smaller. The smaller the pupil, the less light entering the eye. This mechanism is called *adaptation.*

Retinal layer The innermost layer of the eyeball is the *retina* (REH-tin-uh; L. *rete*, net), an egg-shaped, multilayered, light-sensitive membrane containing a network of specialized nerve cells [FIGURE 16.13]. It is connected to the brain by a circuit of over a million neurons in the optic nerve. The retina has a thick and a thin layer. The thick layer is nervous tissue, called the *neuroretina,* that connects with the optic nerve. Behind it is a thin layer of pigmented epithelium that prevents reflection from the back of the retina. The pigmented layer, along with the choroid, actually absorbs stray light (light that is not used by the photoreceptor cells) and prevents reflection back to the neuroretina. Stray light in the eye can restimulate the photoreceptors. Albinos, who have no eye pigment, are abnormally sensitive to light because the stray light is not absorbed by pigment.

The function of the neuroretina is to receive focused light waves and convert them into nerve impulses that can be conveyed to the brain and converted into visual perceptions. The neuroretina does not extend into the anterior portion of the eyeball, where light could not be focused on it.

The neuroretina consists of highly specialized photoreceptor nerve cells, the *rods* and *cones* [FIGURE 16.14]. The outer segment of a rod or cone contains most of the elements necessary (including light-sensitive photopigments) to absorb light and produce a graded local potential. The inner segment contains mitochondria, Golgi apparatuses, endoplasmic reticulum, the nucleus, and other structures necessary for generating energy and renewing molecules in the outer segment. The inner segment also contains a synaptic terminal, which allows the photoreceptor cells to communicate chemically with other retinal cells.

In addition to rods and cones, the neuroretina contains several other cells, which are actually neurons.

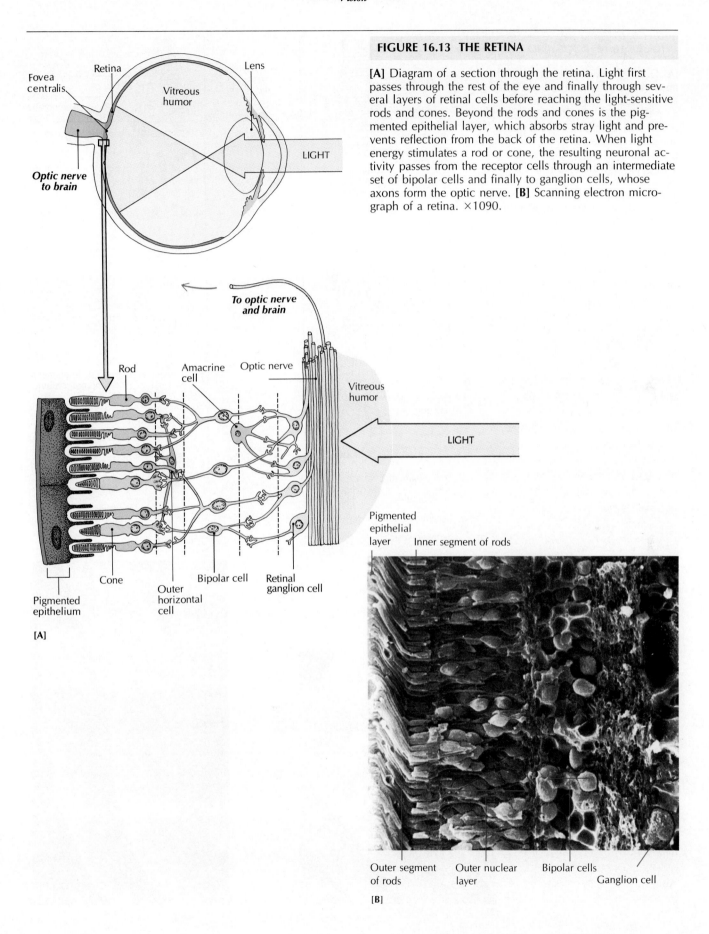

FIGURE 16.13 THE RETINA

[A] Diagram of a section through the retina. Light first passes through the rest of the eye and finally through several layers of retinal cells before reaching the light-sensitive rods and cones. Beyond the rods and cones is the pigmented epithelial layer, which absorbs stray light and prevents reflection from the back of the retina. When light energy stimulates a rod or cone, the resulting neuronal activity passes from the receptor cells through an intermediate set of bipolar cells and finally to ganglion cells, whose axons form the optic nerve. [B] Scanning electron micrograph of a retina. ×1090.

FIGURE 16.14 RODS AND CONES

[A] Detailed drawing of rod cell. The outer segment of each rod contains approximately 2000 disks stacked in an orderly pile. The disks contain most of the light-absorbing protein molecules that initiate the generation of a local potential. [B] Scanning electron micrograph of rods. [C] Detailed drawing of a cone cell. [D] Scanning electron micrograph of cones. The shapes of the rods and cones give them their names.

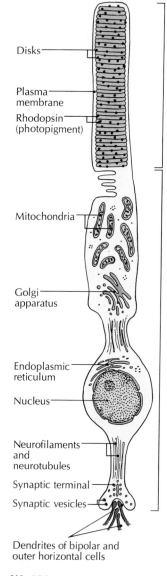

OUTER SEGMENT

Disks
Plasma membrane
Rhodopsin (photopigment)

INNER SEGMENT

Mitochondria
Golgi apparatus
Endoplasmic reticulum
Nucleus
Neurofilaments and neurotubules
Synaptic terminal
Synaptic vesicles
Dendrites of bipolar and outer horizontal cells

[A] ROD

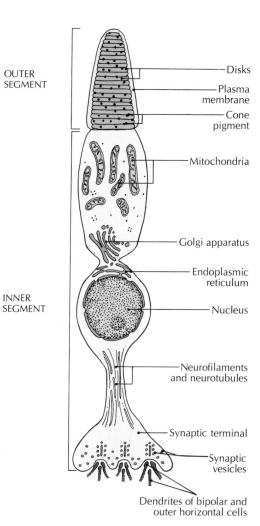

Disks
Plasma membrane
Cone pigment
Mitochondria
Golgi apparatus
Endoplasmic reticulum
Nucleus
Neurofilaments and neurotubules
Synaptic terminal
Synaptic vesicles
Dendrites of bipolar and outer horizontal cells

[C] CONE

[B]

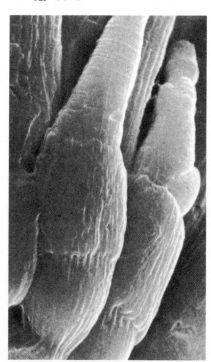

[D]

These include bipolar cells, outer horizontal cells, amacrine cells (inner horizontal cells), and ganglion neurons. These neurons form the complex neuronal circuitry for the processing of light waves within the retina [FIGURE 16.13]. The ganglion neurons contain axons that leave the eye and comprise the nerve fibers of the optic nerve, optic chiasma, and optic tracts.

Each eye has about 125 million rods and 7 million cones. Most of the cones are concentrated in the center of the retina directly behind the lens in an area called the *macula lutea* (L. "yellow spot"), especially in a small depressed rod-free area called the *fovea,* or *fovea centralis* (FOE-vee-uh; L. small pit) [FIGURE 16.12]. The rods, and some cones, are located in the remainder of the retina, called the *peripheral retina.* The rods are the sensors for the perception of black-to-white shades, and the cones are the sensors for the perception of color. Night vision is almost totally rod vision since the color-sensitive cones require 50 to 100 times more stimulation than rods do. (In order to distinguish the functions of the rods and cones, remember that the *c* in *cone* can stand for *color*.)

Vision is sharpest, and color perception is optimal, on the fovea of the macula lutea. The nonmacular retina (for peripheral vision) is sensitive to weak light intensities, and is associated with black and white vision. In normal light we can see best by looking directly at an object, so that the image falls on the cones in the fovea. However, in poor light we can see best by *not* looking directly at an object (as one does when star gazing), so that the image falls on the light-sensitive rods, located on the nonmacular portion of the retina. The portion of the retina where the optic nerve exits from the eyeball contains neither rods nor cones, and is called the optic disk, or *blind spot,* because it cannot respond to light.

Cavities of the eyeball The eyeball is divided into three cavities. The region between the cornea and iris is the *anterior chamber* [FIGURE 16.12]. The *posterior chamber* lies between the iris and lens. Both chambers are filled with *aqueous humor,* a thin, watery fluid that is essentially an ultrafiltrate of blood similar to cerebrospinal fluid. Aqueous humor is largely responsible for maintaining a constant pressure within the eyeball. It also provides such essential nourishment as oxygen, glucose, and amino acids for the lens and cornea, which do not have blood vessels to nourish them. Aqueous humor is produced by capillaries in the ciliary body. It passes through the posterior and then the anterior chamber and diffuses into a drainage vein called the *scleral venous sinus* (canal of Schlemm) at the base of the cornea.

The third and largest cavity of the eyeball is the *vitreous chamber,* which occupies about 80 percent of the eyeball. It fills the entire space behind the lens. This chamber contains *vitreous humor,* a gelatinous substance with the consistency of raw egg white. The humor is actually a modified connective tissue. Its function is to keep the eyeball from collapsing as a result of external pressure. Except for the addition of collagen and hyaluronic acid, the chemical composition is similar to that of aqueous humor. The vitreous humor also provides another source of nourishment for the lens and possibly the retina. The vitreous humor is formed by the ciliary body.

Accessory Structures of the Eye

Most of the accessory structures of the eye are either protective structures or muscles. They include the bony orbits of the skull, eyelids, eyelashes, eyebrows, conjunctiva, lacrimal (tear) apparatus, and muscles that move the eyeball and eyelid.

Orbit The eye is enclosed in an orbital cavity, or *orbit,* which protects it from external buffeting. The floor of the orbit is composed of parts of the maxilla and zygomatic and palatine bones. The roof is composed of the orbital plate of the frontal bone and the lesser wing of the sphenoid bone. Several openings in the bones of the orbit allow the passage of nerves and blood vessels. Between the orbit and the eyeball is a layer of fatty tissue that cushions the eyeball and permits its smooth rotation.

Eyelids, eyelashes, eyebrows The *eyelids* (palpebrae) are folds of skin that create an almond-shaped opening around the eyeball when the eye is open. The points of the almond, where the upper and lower eyelids meet, are called *canthi* (KAN-thigh). The medial (or inner) canthus is the one closest to the nose, and the lateral (or outer) canthus is the point closest to the ear. The eyelid may be

Q: *Why do most newborn babies have blue eyes?*

A: Eye color depends upon the number and placement of pigment cells (melanocytes) in the eye. Darker eye color is a result of a greater concentration of melanocytes. At birth, melanocytes are still being distributed in the eyes; they first appear at the back of the iris. However, the eyes appear brown only when melanocytes are deposited in the *anterior* part of the iris, in front of the muscles of the iris. This deposition does not occur until a few months after birth. Babies of dark-skinned, dark-eyed parents are usually born with dark eyes.

Q: *Why do we sometimes see "spots" in front of our eyes?*

A: Such "spots" are called *muscae volitantes,* which is Latin for "flying flies." They are actually the shadows of either red blood cells that have escaped from the capillaries in the retina or of collagenous particles. These particles and cells move slowly through the vitreous humor of the eye. Their presence is normal and harmless.

divided into four layers: (1) a skin layer contains the eyelashes, (2) a muscular layer contains the orbicularis oculi muscle, which lowers the eyelid to close the eye, (3) a fibrous connective tissue layer contains many modified sebaceous glands, whose secretions keep the eyelids from sticking together, and (4) the innermost layer is composed of a portion of the lining of the eyelid, the conjunctiva.

Eyelids protect the eyeball from dust and other harmful external objects. In addition, the periodic blinking of the eyelids sweeps glandular secretions (tears) over the eyeball, keeping the cornea moist. During sleep, the closed eyelids prevent evaporation of the secretions. The eyelids also protect the eye by closing reflexively when an external object threatens the eye, as when a piece of paper is suddenly blown toward your face.

The edges of the eyelids are lined with short, thick hairs, the *eyelashes,* which act as strainers to prevent foreign materials from entering the eye. The eyelids of each eye contain about 200 eyelashes. Each eyelash lasts three to five months before it is shed and replaced. Eyelashes are the only hairs that do not whiten with age. *Eyebrows* are thickened ridges of skin over the protruding frontal bone, covered with short, flattened hairs. They protect the eye from sweat, excessive sunlight, and foreign materials, and also help to absorb the force of blows to the eye and forehead.

Conjunctiva The *conjunctiva* (L. connective) is a thin, transparent mucous membrane that lines the eyelids and bends back over the surface of the eyeball, terminating at the transparent cornea, which is uncovered. The portion that lines the eyelid is the *palpebral conjunctiva,* and the portion that covers the white of the eye is the *bulbar conjunctiva.* Between both portions of the conjunctiva are two recesses called the *conjunctival sacs.* The looseness of the sacs makes movement of the eyeball and eyelid possible. Your ophthalmologist usually pulls back your lower eyelid to place eyedrops in the inferior conjunctival sac.

Lacrimal apparatus The *lacrimal apparatus* (LACK-ruh-mull; L. "tear") is made up of the lacrimal gland, lacrimal sac, lacrimal ducts, and nasolacrimal duct [FIGURE 16.15]. The eyeball is kept moist by the secretions of the *lacrimal gland,* or tear gland, located under the upper lateral eyelid and extending inward from the outer canthus of each eye. The lacrimal glands of infants take about four months to develop fully. As a result, the eyes of a newborn baby should be protected from dust, bright light, and other irritants.

The eye blinks every 2 to 10 seconds, with each blink lasting only 0.3 to 0.4 seconds. Blinking stimulates the lacrimal gland to secrete a sterile fluid that serves at least four functions: (1) It washes foreign particles off the eye.

FIGURE 16.15 THE LACRIMAL APPARATUS

Anterior view.

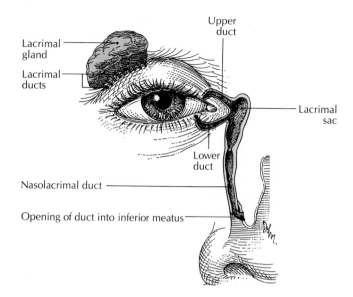

(2) It kills invading bacteria with a mild antibacterial enzyme, lysozyme. (3) It distributes water and nutrients to the cornea and lens. (4) It gives the eyeball a clear, moist, and smooth surface. Tears are composed of salts, water, and mucin (organic compounds produced by mucous membranes). In addition to the steady secretion of tears, *reflex tears* are produced in emergencies, as when the fumes from a sliced onion irritate the eyes.

Approximately 3 to 12 *lacrimal ducts* lead from each gland onto the superior conjunctival sac in the upper eyelid. From there tears flow down across the eye into small openings near the inner canthus called *lacrimal puncta.* The puncta open into the *lacrimal ducts,* which drain excess tears from the area of the inner canthus to the *lacrimal sac,* the dilated upper end of the *nasolacrimal duct* [FIGURE 16.15]. The nasolacrimal duct is a longitudinal tube that delivers excess tears into the nasal cavity.

Ordinarily, tears are carried away by the nasolacrimal duct to the nasal cavity, but when a person is crying or has conjunctivitis or hay fever, the tears form faster than they can be removed. In such cases, tears run down the cheeks, and the nasal cavity becomes overloaded. This is why you have to blow your nose when you cry. Also, a watery fluid sometimes flows out of the nose after blowing it. This fluid is tears flowing out of the unplugged nasolacrimal duct.

Q: *What causes "bloodshot eyes"?*

A: Although the bulbar conjunctiva is normally colorless, its blood vessels can become dilated and congested by infection or external irritants such as smoke. "Bloodshot eyes" is the result.

Muscles of the eye and eyelid A set of six muscles moves the eyeball in its socket. [see FIGURE 11.7]. The muscles are the four *rectus muscles* and the *superior* and *inferior oblique muscles*. They are called **extrinsic,** or **extraocular, muscles** because they are outside the eyeball (*extra* = outside). One end of each muscle is attached to a skull bone, and the other end is attached to the sclera of the eyeball. The extraocular muscles are coordinated and synchronized so that both eyes move together in order to center on a single image. These movements are called the *conjugate movements* of the eyes.

Other muscles move the eyelid. The *orbicularis oculi* lowers the eyelid to close the eye, and the *levator palpebrae superioris* raises the eyelid to open the eye. The *superior tarsal* muscle is a smooth muscle innervated by the sympathetic nervous system. It helps to raise the upper eyelid, and when it is paralyzed (as in Horner's syndrome, see Chapter 15) it causes a slight drooping (ptosis) of the upper eyelid.

Inside the eyes are three smooth **intrinsic muscles.** The *ciliary muscle* eases tension on the suspensory ligaments of the lens and allows the lens to change its shape in order for the eye to focus (accommodate) properly. The *circular muscle* (sphincter pupillae) of the iris constricts the pupil, and the *radial muscle* (dilator pupillae) dilates it.

Physiology of Vision

The visual process can be subdivided into five phases:

1 Refraction (bending) of light rays entering the eye.

2 Focusing of images on the retina by accommodation of the lens and convergence of the images.

3 Conversion of light waves by photochemical activity into neural impulses.

4 Processing of neural activity in the retina, and transmission of coded impulses through the optic nerve.

5 Processing in the brain, culminating in perception.

Neural Pathways for Vision

All visual information originates with the stimulation of the rods and cones of the retina and is conveyed to the brain by way of the axons of the ganglion cells. These axons form a visual pathway that begins in the eyes and ends in the occipital lobes of the cerebral cortex.

The field of vision is the environment viewed by the eyes. For each eye it is divided into (1) an outer, lateral (*temporal*) half and an inner, medial (*nasal*) half, and (2) an upper half and a lower half. Rays of light entering the eye move diagonally across the eyeball. Because the lens in the eye acts like the lens in a camera, the field of vision is reversed in both the vertical and horizontal planes. Thus, in the horizontal plane, the light waves from the temporal visual field fall on the nasal half of each retina, and the light waves from the nasal field fall on the temporal half of the retina [FIGURE 16.16]. In the vertical plane, the rays of light from the upper half of the visual field fall on the lower half of the retina, and rays from the lower half of the visual field fall on the upper half of the retina.

In the retina the rods or cones respond to light waves and generate neural signals that are processed by other retinal cells and stimulate the ganglion cells to produce action potentials. Then, nerve fibers of the ganglion cells from both eyes carry the impulses along two optic nerves. These two nerves meet at the **optic chiasma,** where the fibers from the nasal half of each retina cross over; the fibers from the temporal half of each retina do not cross over. Because half the nerve fibers from each eye cross over at the optic chiasma, each side of the brain receives visual messages from both eyes. After passing through the optic chiasma, the nerve fibers are called the **optic tracts.** Each optic tract contains nasal fibers from the opposite side, and temporal fibers from the same side.

Each optic tract continues posteriorly until it reaches a nucleus in the thalamus called the **lateral geniculate body.** From there, the axons of neurons in the lateral geniculate body project via the optic radiations to the primary visual cortex in the occipital lobe of the cerebral cortex (FIGURE 16.16).

A reflex pathway proceeds from the retinas directly to the superior colliculus in the midbrain. This reflex system is involved in unconscious movements of the eye, including contraction and dilation of the pupil and coordinated (conjugate) movements of both eyes.

Neural cells of the retina The retina is composed of five main classes of neurons, which are interconnected by synapses. The three types of retinal neurons that form a *direct pathway* from the retina to the brain are **photoreceptor cells** (rods and cones), **bipolar cells,** and **ganglion cells.** The two remaining classes of retinal neurons, **horizontal cells** and **amacrine cells,** synapse with the bipolar and ganglion cells. These five classes of neurons have many subtypes of cells, bringing the total number of functional neural elements in the retina to about 60.

The neural elements in the retina interact with each other and begin processing the stimuli received by the photoreceptors. In the direct pathway, this evaluation is accomplished by the convergence of signals from a number of photoreceptors onto a single ganglion cell whose axon projects via the optic nerve to the brain. The horizontal and amacrine cells, acting laterally, connect adjacent neurons and allow them to modify the signals as they are conveyed along the direct pathway to the brain.

FIGURE 16.16 THE AFFERENT VISUAL PATHWAY

[A] The output from the retina is conveyed to the lateral geniculate body by ganglion cell axons of the optic nerves. Half of the axons of each eye cross over at the optic chiasma, so that each half of the brain receives impulses from both eyes. The right half of the visual field is projected to the primary visual cortex of the left occipital lobe, and the left half of the visual field to the primary visual cortex of the right occipital lobe. [B] Photograph of the afferent visual pathway; ventral aspect of brain.

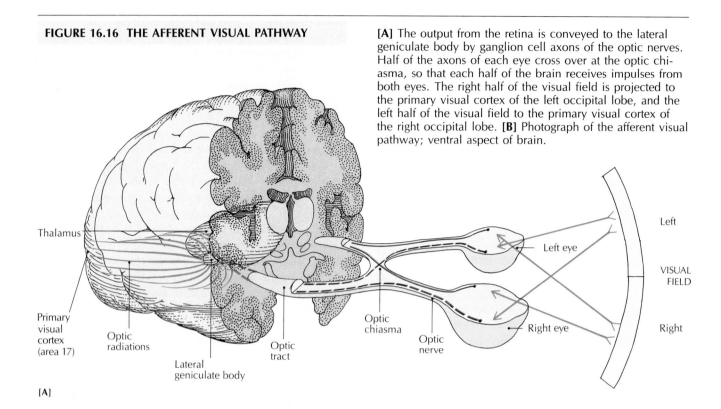

Thalamus

Primary visual cortex (area 17)

Optic radiations

Lateral geniculate body

Optic tract

Optic chiasma

Optic nerve

Left eye

Right eye

Left

VISUAL FIELD

Right

[A]

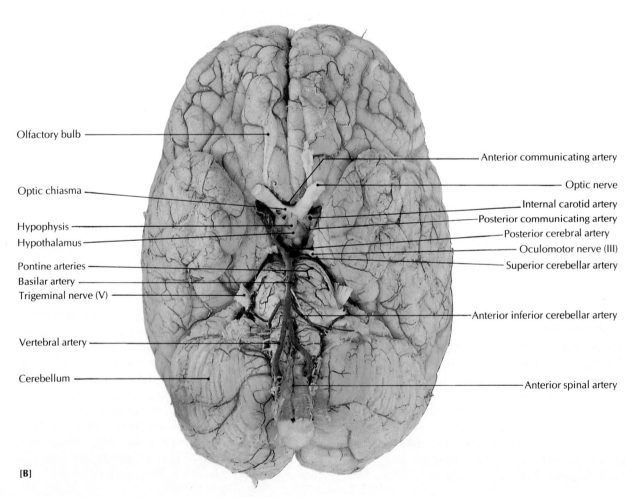

Olfactory bulb

Optic chiasma

Hypophysis

Hypothalamus

Pontine arteries

Basilar artery

Trigeminal nerve (V)

Vertebral artery

Cerebellum

Anterior communicating artery

Optic nerve

Internal carotid artery

Posterior communicating artery

Posterior cerebral artery

Oculomotor nerve (III)

Superior cerebellar artery

Anterior inferior cerebellar artery

Anterior spinal artery

[B]

ASK YOURSELF

1 How much of the entire electromagnetic spectrum do we see?

2 What are the three layers of the eyeball?

3 What is the importance of the fovea?

4 What are the differences between aqueous humor and vitreous humor?

5 Why do we blink our eyes periodically?

6 Compare the functions of rods and cones.

7 What is the function of the optic chiasma?

8 What are the five main classes of retinal neurons?

DEVELOPMENTAL ANATOMY OF THE EYE

The eyes develop from three embryonic sources: neuroectoderm, surface ectoderm, and mesoderm. The first sign of eye development appears in the 22-day embryo when a pair of shallow *optic sulci* (grooves) form from the neuroectoderm in the forebrain at the cephalic end of the embryo [FIGURE 16.17A]. The sulci develop into a pair of lateral outpockets called *optic vesicles* [FIGURE 16.17B]. The distal portion of each optic vesicle dilates, while the proximal portion remains constricted to form the *optic stalk.* Eventually, axons of ganglion cells elongate and extend into the optic stalk to become the *optic*

FIGURE 16.17 EARLY DEVELOPMENTAL ANATOMY OF THE EYE

The drawings show the developmental progression of the optic nerve, cornea, lens, neuroretina, and pigmented layer of the retina.

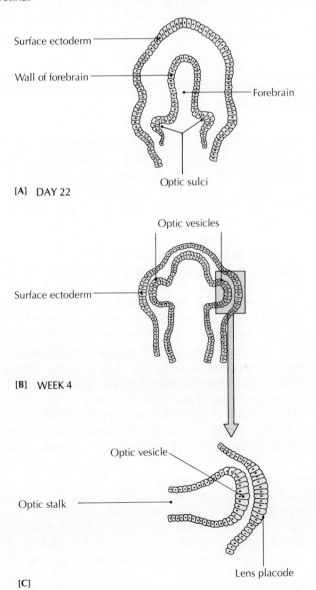

[A] DAY 22

[B] WEEK 4

[C]

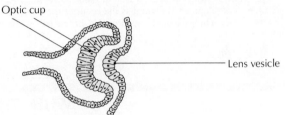

[D] WEEK 5 (early)

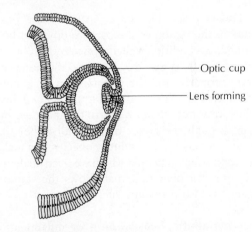

[E] WEEK 5 (late)

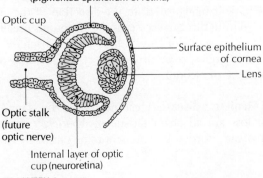

[F] WEEK 6

nerve. The optic vesicle soon reaches the surface ecto-derm and induces* it to thicken into a *lens placode* [FIG-URE 16.17C]. By the beginning of week 5, the lens placode, in turn, induces each optic vesicle to turn in on itself (invaginate), forming a double-layered *optic cup* [FIGURE 16.17D]. The internal layer of the optic cup differentiates into the complex *neural layer* (neuroretina) of the retina (which includes rods, cones, bipolar cells, horizontal cells, amacrine cells, and ganglion cells), while the external layer develops into the *pigmented epithelium* of the retina. The retina is well developed at birth except for the central foveal region, which completes its development by the fourth month after birth.

The optic cup induces the lens placode to invaginate and to form a *lens vesicle* that is partially surrounded by the optic cup [FIGURE 16.17D]. The vesicle sinks below the outer surface and pinches off like a submerged bub-

---

*The principle of *induction* occurs when one tissue is caused, or *induced,* to change its developmental pattern as the result of the influence of a different tissue. These changes are probably produced by chemical stimuli.

ble. This "bubble," now almost completely surrounded by the optic cup, forms the *lens,* while the surface ecto-derm becomes the thin surface epithelial layer of the *cornea* [FIGURE 16.17E, F]. The *iris* eventually forms from the rim of the optic cup, which partially covers the lens.

Linear grooves called *optic fissures* develop on the inferior surface of the optic cups and optic stalks. Blood vessels develop within these fissures. The hyaloid artery supplies the optic cup and lens vesicle, and the hyaloid vein returns blood. The hyaloid vessels are eventually enclosed within the optic nerve when the edges of the optic fissure fuse. The open end of the optic cup now forms a circular opening, the future *pupil.* The distal portions of the hyaloid vessels eventually degenerate, but the proximal portions become the central artery and vein of the retina.

The optic cup is surrounded by mesenchymal (meso-derm) cells that differentiate into an inner vascular layer, comprising the *choroid, ciliary body, iris,* and an outer fibrous layer comprising the *sclera* and *cornea.* The choroid is continuous with the arachnoid layer of the brain, and the sclera is continuous with the dura mater.

WHEN THINGS GO WRONG

The Ear

Otosclerosis The stapes may become fused to the bone in the region of the oval window and become immobile, a condition called *otosclerosis* (Gr. *otos,* ear + *skleros,* hard). When this happens, the bones cannot vibrate properly, and the transmission of sound waves to the inner ear may be almost impossible. A hearing aid, which transmits sound waves to the inner ear by conduction through the bones of the skull rather than by air conduction along the auditory ossicles, can make hearing almost normal in people who have been deafened by otosclerosis. An operation that allows the stapes to move again has also had some success.

Labyrinthitis *Labyrinthitis,* or inflammation of the inner ear, can cause discharge from the ear, vomiting, hearing loss, and vertigo (the feeling of spinning) and other forms of dizziness. Vertigo upsets the balance, and the patient tends to fall in the direction of the affected ear. Within three to six weeks the symptoms usually subside. The causes of labyrinthitis include bacterial infections, allergies, toxic drugs, severe fatigue, and overindulgence of alcohol.

Ménière's disease Head noises, dizziness, and hearing loss are all characteristic of *Ménière's disease,* an inner ear disorder. It is thought to be caused by an excess

of endolymph and dilation of the labyrinth. The basilar membrane is distorted, the semicircular ducts are affected by pressure, and some cochlear hair cells degenerate. A patient may have residual tinnitus (ringing or whistling in the ears) and hearing loss after repeated attacks over many years.

Motion sickness Many people know the familiar queasy feeling of *motion sickness,* which results from the sensation of motion or from repeated rhythmic movements. Symptoms of nausea, vomiting, pallor, and cold sweats are common when people travel on boats, planes, cars, and trains. However, suffering from one type of motion sickness, such as seasickness, does not necessarily mean that a person is susceptible to other forms. Motion sickness arises from the excessive stimulation of the labyrinthine receptors of the inner ear. Tension, fear, offensive odors, and visual stimuli are also important factors.

Another theory of motion sickness suggests that it results from conflicting perceptions. For example, while seated in the lounge of a ship, a person's vestibular system may detect the rocking motion of the ship, but the visual system may not. Alternatively, in a car, the vestibular system may not detect motion (since the forward motion of the car is constant), while the visual system detects motion by observing the passing scene through the side windows. According to this theory, matching the two perceptions should help to relieve the motion sickness. In other words, watch the horizon list with the ship, or look out of the front windshield of the car to perceive less motion.

Motion sickness can be prevented by drugs such as dimenhydrinate (Dramamine) that suppress the vestibular function.

Otitis media *Otitis media,* inflammation of the middle ear, causes the tympanic membrane to redden and bulge out. If untreated, the membrane may rupture. Otitis media is commonly seen in children and is most prevalent between the ages of 6 and 24 months. If the inflammation is severe enough or is prolonged, scarring, structural damage, and hearing loss may result. This disorder seems to arise from a malfunction of the auditory tube, in which bacteria from the nasopharynx enters the middle ear. Otitis media may be acute or chronic. Both forms are bacterial infections caused by organisms such as *Streptococcus pneumoniae* or *Hemophilus influenzae.*

The Eye

Detached retina Normally, no space exists between the two retinal layers, but sometimes the layers become separated. Such a condition is commonly called a *detached retina.* The portion of the neuroretina that is detached usually stops functioning because its blood supply is impaired. New surgical procedures, including photocoagulation by laser beam and cryosurgery, can successfully reattach the retinal layers in about 90 percent of the cases, and thus arrest further detachment.

Cataract One of the most common causes of blindness is *cataract formation,* in which the lens becomes opaque. As the opaque areas increase, there is a progressive loss of vision, and if the cataract is severe enough, total blindness may result. The molecular weight of the proteins within the lens increases with age, and consequently the proteins may become cloudy and then opaque. The result is a cataract. Cataracts are generally associated with people over 70, but they can occur at any age. There are various types of cataracts, including se-

nile, congenital, traumatic, and toxic. Cataracts are also frequent in people with diabetes, since the abnormal glucose metabolism of diabetes may affect the vitality of the lens. Treatment usually begins with frequent changes in eyeglasses to compensate for gradual vision loss. When new prescription eyeglasses are no longer helpful, cataract surgery is the preferred procedure. During surgery the lens is removed from its capsule, and an artificial lens is permanently implanted in its place. Soon after the operation, the patient can see and does not have to wear a contact lens. Newly prescribed eyeglasses can be used to make minor adjustments.

Conjunctivitis *Conjunctivitis,* or inflammation of the conjunctiva, is a common eye disorder. It may have several causes. Bacteria, viruses, pollen, smoke, pollutants, and excessive glare all affect the conjunctiva, causing discharge, tearing, and pain. Vision, however, generally is not affected. "Pinkeye," a form of conjunctivitis caused by pneumococci or staphylococci bacteria, is contagious. In such cases, affected people should avoid spreading the infection by not rubbing the infected eye or sharing towels and pillows. Conjunctivitis can be acute or chronic, and treatment varies with the cause.

Glaucoma *Glaucoma* (glaw-KOH-muh) is a leading cause of blindness in the United States, affecting over 1 million people. The disease strikes people of all ages, but mainly those over 40; women are more susceptible than men. Glaucoma occurs when the aqueous humor does not drain properly. Since more is formed than is drained, the fluid builds up in the eyeball and increases the intraocular pressure. If the pressure continues, it destroys the neurons of the retina, and blindness usually results. Glaucoma may be chronic (90 percent of cases) or acute. Chronic forms result in progressively reduced vision. A common symptom is a vision defect in which lights appear to be surrounded by halos. Acute forms can occur suddenly at any age, causing pain, pressure, and blurring. Chronic glaucoma may be genetically linked. Close relatives of patients with glaucoma are five times more susceptible to developing the disease than those with no glaucoma in their family history.

CHAPTER SUMMARY

Sensory Reception (p. 408)

1 Sense organs, or **receptors,** are the body's link to the outside world and to changes within the body.

2 All receptors are *transducers,* that is, they convert one form of energy into another.

3 Receptor cells that respond directly to stimuli may be either a specialized neuroepithelial cell or the ending of a sensory neuron.

4 The sensory neurons respond initially with *receptor (graded) potentials,* which are converted into action potentials. All stimulated receptors generate the same type of *action potential.* Different sensations occur when nerve fibers connect with specialized portions of the central nervous system.

5 The sensory neurons associated with specific perceived sensations, such as pain and sound, convey neural influences to specific *neural pathways* within the central nervous system. These pathways process the neural influences, which are finally perceived as the relevant sensations.

6 Sensory receptors may be classified according to their *location (exteroceptor, teleceptor, interoceptor, proprioceptor); type of sensation (thermal, pain, light touch, touch-pressure, position sense); type of stimulus (thermoreceptor, nociceptor, chemoreceptor, photoreceptor, mechanoreceptor);* or *sensory ending (free nerve ending, encapsulated ending).*

7 The **general senses** include light touch, touch-pressure (deep pressure), two-point discrimination, vibratory sense, heat, cold, pain, and body position. Taste, smell, hearing and equilibrium, and vision are called the **special senses** because they originate from sensors in restricted (special) regions of the head.

8 Individual nerve impulses are essentially identical, but an impulse for sight is distinguished from an impulse for taste by the *relationship between where the impulse is coming from and the place in the brain where it is going.*

General Senses (p. 411)

1 Sensory receptors in the skin detect stimuli that the brain interprets as light touch, touch-pressure, heat, cold, pain, proprioception, and stereognosis. Other miscellaneous cutaneous sensations include itch and tickle.

2 Two receptors for light touch are **tactile (Merkel's) corpuscles** and **tactile (Meissner's) corpuscles.** Receptors for touch-pressure are **lamellated (Pacinian) corpuscles. Corpuscles of Ruffini** and **bulbous corpuscles (of Krause)** are mechanoreceptors for touch.

3 The cutaneous receptors for heat and cold are probably free nerve endings.

4 Pain receptors are **specialized free nerve endings** present in most parts of the body.

5 **Referred pain** is a visceral pain that is felt in a somatic area. **Phantom pain** is felt in an amputated limb.

6 The sense of position of body parts in relation to each other is called **proprioception. Itch** and **tickle** are probably produced by the activation of several sensory types of nerve endings. The information resulting in the sensation is probably conveyed by a combination of pathways. **Stereognosis** is the ability to identify unseen objects merely by handling them.

Taste (Gustation) (p. 414)

1 The receptors for taste and smell are both chemoreceptors, and the two sensations are interrelated, but taste **(gustation)** and smell **(olfaction)** are separate.

2 The surface of the tongue is covered with many small protuberances called **papillae,** including three types, filiform, fungiform, and circumvallate. Located within the crevices of papillae are **taste buds,** the receptor organs for the sense of taste.

3 The four basic taste sensations are sweet, sour, bitter, and salty.

Smell (Olfaction) (p. 416)

1 The **olfactory receptor cells** are located high in the roof of the nasal cavity, in specialized areas of the nasal mucosa

called the *olfactory epithelium.* The epithelium consists of *receptor cells, sustentacular (supporting) cells,* and *basal cells.*

2 The olfactory receptor cell (neuron) ends in a bulbous *olfactory vesicle,* from which extend *cilia* that project through the mucuslike fluid that covers the surface epithelium.

3 Olfactory nerve fibers extend through foramina in the cribriform plate of the ethmoid bone and terminate in the **olfactory bulbs,** from which axons of neurons project to the olfactory cortex of the cerebrum.

Hearing and Equilibrium (p. 418)

1 The ear consists of two functional units, the *auditory (acoustic) apparatus,* concerned with hearing, and the *vestibular apparatus,* concerned with posture and balance.

2 The anatomical components of the auditory apparatus are the external ear, the middle ear, and the inner ear. The *external ear* is composed of the **auricle** and **external auditory canal;** the *middle ear* is made up of the **tympanic membrane** (eardrum), **tympanic cavity, auditory (Eustachian) tube,** and the three **auditory ossicles** (earbones); the *inner ear,* or **membranous labyrinth,** is composed of the **vestibule** (which contains the **utricle** and **saccule), semicircular canals** and **ducts,** and **cochlea** (which contains the **spiral organ of Corti).**

3 The main receptors for *equilibrium* are hair cells in the utricle, saccule, and semicircular ducts in the inner ear. The equilibrium system also receives input from the eyes and from some proprioceptors in the skin and joints.

4 The function of the **vestibular system** is to signal changes in the motion of the head **(dynamic equilibrium),** and in the position of the head with respect to gravity **(static equilibrium,** or posture).

5 Specialized receptor cells of the vestibular sense organs are **hair cells,** which are arranged in clusters called hair bundles. *Stereocilia* and a *kinocilium* are present in each hair bundle. When hairs are bent in the direction of the stereocilia, the hair cells convert a mechanical force into nerve impulses that are conveyed to the brain via the vestibular nerve.

6 When the head moves in a change of posture, calcium carbonate crystals called *otoconia* in the inner ear respond to gravity, resulting in the bending of the hairs of hair cells. The bending stimulates nerve fibers to generate action potentials, which are transmitted to the brain. The brain signals appropriate muscles to contract, and body posture is adjusted to follow the new head position.

7 The utricles and saccules are *organs of gravitation,* responding to movements of the head in a straight line: forward, backward, up, or down. In contrast, the *crista ampullaris* of the semicircular ducts responds to changes in the *direction of head movements,* including turning, rotating, and bending. Hair cells in the crista project into a gelatinous flap called the *cupula.* When the head rotates, the endolymph in the semicircular canals lags behind, displacing the cupula and the hairs projecting into it. The resulting action potentials are sent to the neural centers in the brain, which signals certain muscles to respond appropriately to maintain the body's equilibrium.

Developmental Anatomy of the Inner Ear (p. 426)

1 The inner ear begins to develop before the external or middle ear, beginning with an *otic placode* that forms from the ectoderm on the sides of the developing hindbrain.

2 The otic placodes invaginate to form *otic pits,* which then pinch off from the surface ectoderm to form *otocysts.* Otocysts eventually develop into the membranous labyrinth.

3 *Semicircular ducts* and *cochlear ducts* develop during weeks 5 through 8, with the cochlear duct developing into the *cochlea.*

4 The *spiral organ of Corti* and other internal structures of the cochlea continue to develop into week 20.

Vision (p. 426)

1 The human eye can ''see'' only those objects that emit or are illuminated by light waves representing about 1/70 of the electromagnetic spectrum.

2 The wall of the eyeball consists of three layers of tissue: the outer *supporting layer,* the *vascular layer,* and the inner *retinal layer.*

3 The posterior five-sixths of the supporting layer is the *sclera* (''white'' of the eye). It gives the eyeball its shape, and protects the delicate inner layers. The anterior segment of the supporting layer is the transparent *cornea,* through which light enters the eye.

4 The vascular layer contains many blood vessels. The posterior two-thirds is a thin membrane, the *choroid.* The *ciliary body* is the thickened anterior portion of the choroid layer. The *lens* is an elastic body that changes shape to focus on objects at different distances.

5 The anterior extension of the choroid is a muscular layer, the *iris,* which is the colored part of the eye. An adjustable opening in the center of the iris is the *pupil.* It opens and closes in response to the amount of light available.

6 The innermost portion of the eyeball is the *retina.* It contains (1) a *layer of nervous tissue (neuroretina)* that receives focused light waves, transduces them, and processes their effects into neural impulses, which are converted into visual perceptions in the brain, and (2) a *pigmented layer* behind the neural layer that absorbs light not utilized by rods and cones, and thus prevents reflection of light that could restimulate the rods and cones.

7 The retina contains highly specialized photoreceptor nerve cells, the *rods* and *cones,* as well as other types of nerve cells and fibers. Rod cells respond to the entire visual spectrum and produce black-to-white vision. They function in dim light and in peripheral vision. The color-sensitive cone cells are mainly concentrated in the center of the retina in the *fovea centralis.*

8 The hollow eyeball is divided into three cavities. The *anterior chamber* lies between the iris and cornea. The *posterior chamber* lies between the iris, lens, and vitreous chamber. Both chambers are filled with *aqueous humor.* The largest cavity of the eyeball is the *vitreous chamber* (the space behind the lens), which contains *vitreous humor.*

9 The accessory structures of the eye are mainly protective and supportive. They include the *orbits, eyelids, eyelashes, eyebrows, conjunctiva, lacrimal apparatus,* and *muscles* that act on the eyeball and eyelid.

10 The field of vision for each eye is divided into (1) an outer, lateral (*temporal*) half and an inner, medial (*nasal*) half, and (2) an upper half and a lower half. Nerve fibers of ganglion cells in the retina carry impulses from both eyes along two optic nerves. The nerves meet at the *optic chiasma,* where the fibers from the nasal half of each retina cross over; the temporal fibers do not cross over. As a result of the crossover, each side of the brain receives visual messages from both eyes.

11 After passing through the optic chiasma, the nerve fibers are called the *optic tracts.* Each optic tract continues until it reaches the *lateral geniculate body* in the thalamus. From there, axons pass to the primary visual cortex in the cerebral cortex.

12 The retina is composed of five main classes of synaptically interconnected neurons. *Photoreceptors, bipolar cells,* and *ganglion cells* form a direct pathway from the retina to the brain; *horizontal cells* and *amacrine cells* help process visual information.

Developmental Anatomy of the Eye (p. 437)

1 The eyes develop from neuroectoderm, surface ectoderm, and mesoderm, starting about day 22, when optic sulci form. The sulci develop into *optic vesicles.*

2 The optic vesicle induces the surface ectoderm to thicken into *lens placodes,* which then induce each optic vesicle to invaginate, forming *optic cups.*

3 The internal layer of the optic cups differentiate into the *neuroretina,* while the external layer develops into the *pigmented epithelium* of the retina.

4 The optic cups induce the lens placodes to invaginate, forming *lens vesicles* that eventually form the *lens* and *cornea.*

5 Blood vessels develop within optic fissures.

MEDICAL TERMINOLOGY

The Ear

EUSTACHITIS Inflammation of the auditory (Eustachian) tube.

EXTERNAL OTITIS (Gr. *ous,* ear + *itis,* inflammation) Inflammation of the outer ear.

IMPACTED CERUMEN An accumulation of cerumen, or earwax, that blocks the ear canal and prevents sound waves from reaching the tympanic membrane.

MYRINGITIS Inflammation of the tympanic membrane or eardrum. Also known as tympanitis.

OTOGLIA (Gr. *ous,* ear + *algia,* pain) Earache.

OTOPLASTY (Gr. *ous,* ear + *plastos,* molded) Surgery of the outer ear.

OTORRHEA (Gr. *ous,* ear + *rrhea,* discharge) Fluid discharge from the ear.

OTOSCOPE (Gr. *ous,* ear + *skopein,* to see) Instrument used to look into the ear.

The Eye

ACHROMATOPSIA (Gr. *a,* not + *kroma,* color + *ope,* vision) Complete color blindness.

AMBLYOPIA (Gr. *amblus,* dim + *ops,* eye) Dimness of vision from not using the eye. Also called "lazy eye."

AMETROPIA (Gr. *ametros,* without measure + *ops,* eye) Inability of the eye to focus images correctly on the retina, resulting from a refractive disorder.

ANOPIA (Gr. *a,* no + *ops,* eye) No vision, especially in one eye.

BLEPHARITIS (Gr. *blepharon,* eyelid + *itis,* inflammation) Inflammation of the eyelid.

CHALAZION (Gr. *khalaza,* hard lump) Swelling in the tarsal or Meibomian glands of the eyelid.

DIPLOPIA (Gr. *di,* double + *ope,* vision) Double vision.

ESOTROPIA Medial deviation (turning inward) of the eyeball, resulting in diplopia, caused by a muscular defect or weakness in coordination. Also called "crosseye" and convergent strabismus.

EXOPHTHALMIA (Gr. *ex,* out + *ophalmos,* eye) Abnormal protrusion of the eyeball.

EXOTROPIA (Gr. *ex,* out + *tropos,* turn) Lateral deviation (turning) of the eyeball, resulting in diplopia. Also called "walleye" and divergent strabismus.

KERATITIS Inflammation of the cornea.

KERATOPLASTY Plastic surgery of the cornea, such as corneal transplant.

MIOTIC Drug that causes the pupil to contract.

MYDRIASIS Dilation of the pupil.

MYDRIATIC Drug that causes the pupil to dilate.

NYSTAGMUS (Gr. *nustagmos,* drowsiness) Usually, a side-to-side involuntary movement of the eye, with slow movement in one direction and fast movement in the opposite direction.

OPHTHALMOSCOPE Instrument used to see the interior of the eyeball.

OPTIC NEURITIS Inflammation of the optic nerve.

PTOSIS Drooping (prolapse) of an organ or part. Specifically, drooping of the eyelid.

RETINITIS Inflammation of the retina.

SCLERITIS Inflammation of the sclera.

SCOTOMA (Gr. *skotos,* darkness) A blind spot or dark spot seen as a result of vision loss in part of the visual field.

STRABISMUS (Gr. *strabos,* squinting) "Crossed eyes." An eye muscle defect in which the eyes are not aimed in the same direction, resulting from dysfunction of the extrinsic eye muscles.

TRACHOMA Disease caused by *Chlamydia trachomatis,* a bacterium spread by insects, body contact, poor hygiene, and contaminated water. It is a leading cause of blindness in many countries.

UNDERSTANDING THE FACTS

1 Distinguish between the special senses and the general senses.
2 Define *proprioception.*
3 What are *free,* or *naked, nerve endings?*
4 With which stimuli are free nerve endings involved?
5 With what sensation are lamellated (Pacinian) corpuscles associated?
6 Which areas of the body lack pain receptors?
7 What is meant by *referred pain? Phantom pain?* Give examples.
8 Distinguish between an interoceptor and an exteroceptor, and give examples of each.
9 List the three qualities that a substance must have in order to be smelled.
10 Which nerve innervates the auditory apparatus? The vestibular apparatus?
11 What is the main function of the auditory tube?
12 The bony labyrinth is filled with _____, whereas the membranous labyrinth is filled with _____.
13 In what areas of the brain do the neural pathways for hearing and equilibrium terminate?
14 What is the function of the pupil? The retina?
15 What structures make up the lacrimal apparatus?
16 List four functions of tears.
17 What parts of the ear are affected in motion sickness? In otitis media?

UNDERSTANDING THE CONCEPTS

1 What important features do all sensory receptors have in common?

2 What is meant by the statement, "All receptors are transducers"?

3 What are some implications of two-point discrimination?

4 Describe the neural pathways for gustatory and olfactory sensations.

5 Why do we not adapt to pain to the same extent that we adapt to most of our other sensations?

6 Describe some of the protective features of the ear.

7 List the structures in the body that play a role in equilibrium.

8 Describe the origin, circulatory route, and final disposition of aqueous humor.

9 Trace the pathway of tears from their production to final disposition.

10 Name the intrinsic and extrinsic muscles of the eye and give their functions.

SELF-QUIZ
Multiple-Choice Questions

16.1 Which of the following body parts have no pain receptors?
a lungs
b brain
c stomach
d intestines
e b and d

16.2 Light touch is mediated by which of the following neural pathways?
a dorsal column-medial lemniscal pathway
b spinocervicothalamic tract
c anterior spinothalamic tract
d trigeminothalamic tract
e all but d

16.3 The neural pathway(s) for pain include the
a lateral spinothalamic tract
b indirect spinoreticulothalamic pathway
c lemniscal pathway
d a and b
e a, b, and c

16.4 Which of the following help the body deal with changes in position and acceleration?
a utricle
b saccule
c semicircular ducts
d proprioceptors
e all of the above

16.5 Which of the following is *not* a skeletal muscle of the eye?
a ciliary muscle
b medial rectus
c lateral rectus
d superior oblique
e levator palpebrae superioris

True-False Statements

16.6 The special senses are also called the somatic senses.

16.7 The type of sensation elicited by the brain depends on how it is stimulated.

16.8 The cutaneous receptors for heat and cold are naked nerve endings.

16.9 The vestibular apparatus of the ear functions in hearing.

16.10 Touch-pressure is mediated by the dorsal column-medial lemniscal pathway and probably the spinocervicothalamic pathway.

16.11 From superficial to deep, the three auditory ossicles are the incus, malleus, and stapes.

16.12 Maintenance of the position of the head with respect to gravity is called dynamic equilibrium.

16.13 The eyeball is divided into two cavities.

16.14 The layers of the wall of the eyeball are the supporting layer, the vascular layer, and the retinal layer.

Completion Exercises

16.15 Visceral pain felt subjectively in a somatic area is called _____.

16.16 The inner ear is also called the _____ because of its intricate structure.

16.17 Calcium carbonate crystals involved in static equilibrium are called _____ or _____.

16.18 The transparent membrane lining the inner eyelid is the _____.

16.19 The multilayered light-sensitive membrane of the eye is the _____.

16.20 The eyeball is kept moist by secretions of the _____.

Matching

16.21 _____ auricle
16.22 _____ retina
16.23 _____ cochlea
16.24 _____ cornea
16.25 _____ fovea
16.26 _____ scala media
16.27 _____ auditory tube
16.28 _____ iris

a in the inner ear
b prominent ridge of ear
c depressed area made up only of cones
d transparent anterior portion of eyeball
e light-sensitive membrane
f external ear
g colored part of eye
h in the middle ear
i color-sensitive cells
j contains endolymph

A SECOND LOOK
1 In the following drawing, label the external auditory canal, malleus, oval window, spiral organ, middle ear, and inner ear.

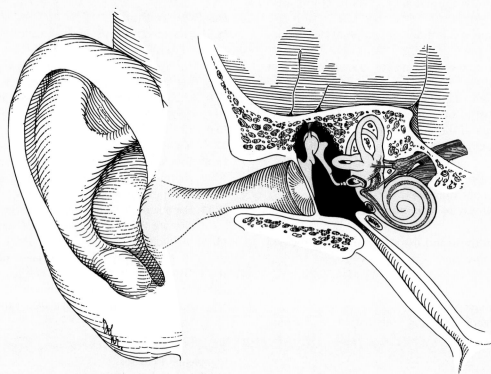

2 In the following drawing of the eye, label the iris, pupil, retina, optic nerve, cornea, and vitreous chamber.

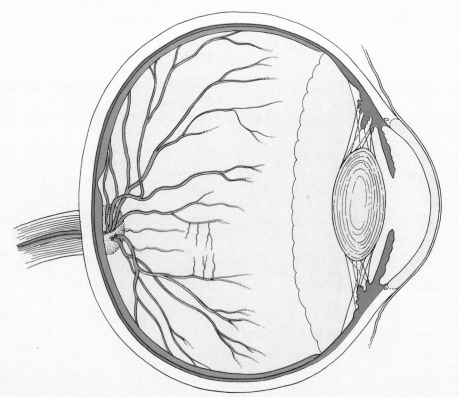

17

The Endocrine System

LEARNING OBJECTIVES

1 Describe the locations of the major endocrine glands.

2 Define a hormone, and identify its four chemical categories.

3 Explain the relationship between the pituitary gland and the hypothalamus.

4 Describe the anatomy of the thyroid gland.

5 Describe the anatomy of the adrenal cortex.

6 Describe the anatomy of the adrenal medulla.

7 Describe the anatomy of the pancreas.

8 Describe the pineal gland.

9 Explain the relationship of the thymus gland to the immune system.

10 Describe some disorders of the endocrine glands.

The *endocrine system* is a control system that is concerned mainly with three functions: (1) It helps to maintain homeostasis by regulating activities such as the concentration of chemicals in body fluids, and the metabolism of proteins, lipids, and carbohydrates. (2) Its secretions act in concert with the nervous system to help the body react to stress properly. (3) It is a major regulator of growth and development, including sexual development and reproduction.

The endocrine system is made up of tissues or organs called *endocrine glands* [FIGURE 17.1; TABLE 17.1, pages 454–455]. These glands secrete chemicals called hormones into extracellular spaces, from which they enter the bloodstream and circulate throughout the body to their target areas. In contrast, the secretions of *exocrine glands*, such as sweat and salivary glands, empty into *ducts* that transport them to specific locations.

FIGURE 17.1 MAJOR ENDOCRINE GLANDS

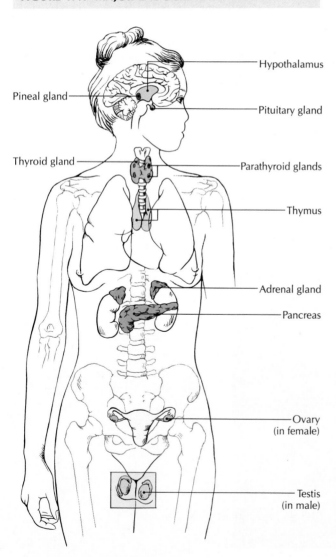

A *hormone* is a specialized chemical "messenger," produced and secreted by an endocrine cell or tissue. When liberated into the bloodstream, hormones travel to all parts of the body, but are effective only in specific *target cells* — cells with compatible receptors on the surface of their plasma membrane or within their cytoplasm. Typically, one hormone influences, depends upon, and balances another in a controlling feedback network. Hormones are usually lipid-soluble steroids, derivatives of amino acids (amines), or water-soluble proteins (peptides); a few are glycoproteins, such as those secreted by the pancreas.

PITUITARY GLAND (HYPOPHYSIS)

The *pituitary gland,** also known as the *hypophysis* (Gr. "undergrowth"), is located directly below the hypothalamus, and rests in the *sella turcica,* a depression in the sphenoid bone. This important gland is protected on three sides by bone and on top by a tough membrane. It is about 1.0 cm long, 1.0 to 1.5 cm wide, and 0.5 cm thick — about the size and shape of a plump lima bean. Because of the closeness of the pituitary to the optic chiasma, an enlarged pituitary may affect vision by impinging upon the optic pathways.

The pituitary gland has two distinct lobes: the anterior lobe, or *adenohypophysis,* and the posterior lobe, or *neurohypophysis* [FIGURE 17.2]. The adenohypophysis is the larger lobe, accounting for about 75 percent of the total weight of the gland. A more significant difference between the two lobes is the abundance of functional secretory cells in the adenohypophysis (Gr. *aden,* gland) and the presence of only supporting pituicytes in the neurohypophysis. The neurohypophysis, as its name suggests, has a greater supply of large nerve endings. Secretory cells produce and secrete hormones directly from the adenohypophysis, while the neurohypophysis contains hormones produced by neurosecretory cells in the hypothalamus. These modified nerve cells project their axons down a stalk of nerve cells and blood vessels called the *infundibular* (L. funnel) *stalk,* or *infundibulum,* into the pituitary gland. In this way, a direct link exists between the nervous system and the endocrine system [FIGURE 17.2].

Other hypothalamic neurosecretory cells secrete releasing hormones into the portal vessels of the infundibular stalk. The portal vessels carry releasing hormones to the secretory cells in the adenohypophysis, and the secretory cells respond by either stimulating or suppressing the secretion of hormones.

*The pituitary gland received its name, which means "mucus" in Latin, because it was thought to transfer mucus from the brain into the nose through the cribriform plate of the ethmoid bone.

FIGURE 17.2 PITUITARY GLAND AND HYPOTHALAMUS

The blood vessels that make up the hypothalamic-hypophyseal portal system provide the link between the hypothalamus and the adenohypophysis of the pituitary. Target areas for each hormone are shown under the relevant box. Note the detailed drawing of the hypothalamus.

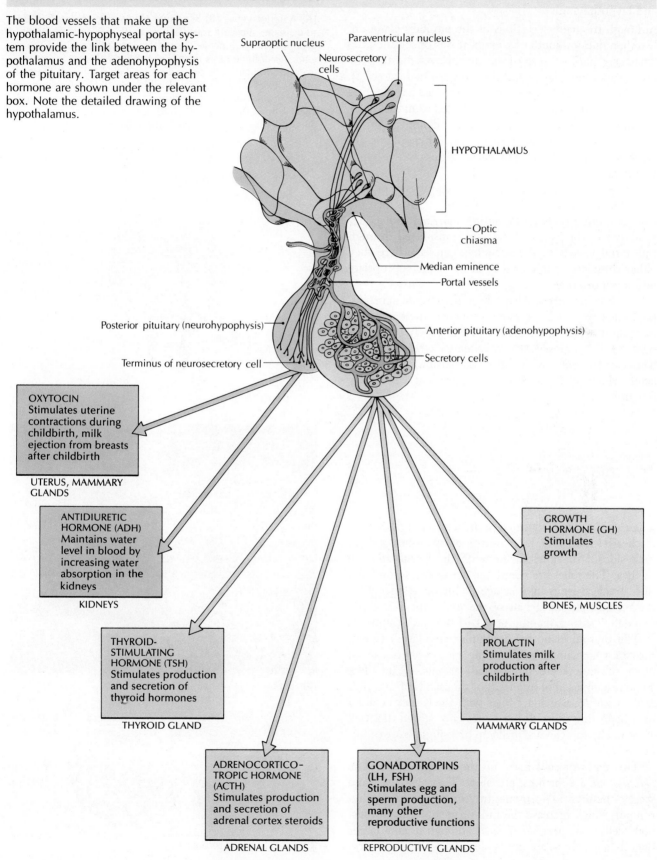

Relationship between the Pituitary and Hypothalamus

Although the daily secretions of the pituitary gland are less than one-millionth of a gram, it was once called the "master gland" because of its control over most of the other endocrine glands and body organs [FIGURE 17.2]. In truth, the *hypothalamus* might better deserve the title "master gland," since substances released from the hypothalamus control the secretions of the adenohypophysis, and hormones secreted from the neurohypophysis are synthesized and regulated by nerve centers in the hypothalamus.

The hypothalamus and the adenohypophysis are connected by an extensive system of blood vessels called the *hypothalamic-hypophyseal portal system.* Hormones produced in the hypothalamus are transported through the portal vessels to the adenohypophysis, where they either stimulate or inhibit the release of the appropriate pituitary hormones.

The link between the hypothalamus and the neurohypophysis relies on nerve impulses; hence, the name *neuro*hypophysis for the posterior pituitary. The neurohypophysis is composed of unmyelinated axons of nerves whose cell bodies are in the hypothalamus, and pituicytes, which have a supporting rather than a secretory function.

THYROID GLAND

The *thyroid gland* is located in the neck, anterior to the trachea [FIGURE 17.3A]. It consists of two lobes, one on each side of the junction between the larynx and the trachea. The lobes are connected across the second and third tracheal rings by a bridge of thyroid tissue called the *isthmus*. In about half of the cases, there is a pyramidal process extending upward from the isthmus.

The thyroid gland has a complex circulatory system through which amino acids, iodine, gland secretions, and other substances are transported [FIGURE 17.3B]. The gland is composed of hundreds of thousands of spherical sacs, or *follicles*, which are filled with a gelatinous colloid in which the thyroid hormones are stored [FIGURE 17.3C]. The follicles are made up of a single layer of follicular cells.

Two types of cells make up the thyroid gland. The *follicular cells* are the most prevalent. They synthesize and secrete the thyroid hormones (thyroxine and triiodothyronine), which increase the rate of metabolism in most body cells. The *parafollicular cells*, though usually larger than follicular cells, are not as plentiful. Clusters of para-

FIGURE 17.3 THYROID GLAND

[A] Anterior view. **[B]** Scanning electron micrograph of thyroid tissue, showing the distinct capillary plexus covering each follicle. Amino acids, iodine, and other substances enter the follicle cells from the capillary plexus. Secretions from the follicle cells enter the circulation through the capillary plexus. Veins are also visible. ×140. **[C]** Cross section of a thyroid follicle. The cuboidal epithelial cells surround a central lumen in which thyroid hormones are stored. Some parafollicular cells lie outside the follicle.

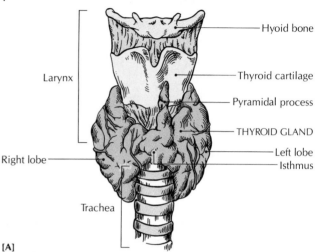

Hyoid bone
Larynx
Thyroid cartilage
Pyramidal process
THYROID GLAND
Right lobe
Left lobe
Isthmus
Trachea

[A]

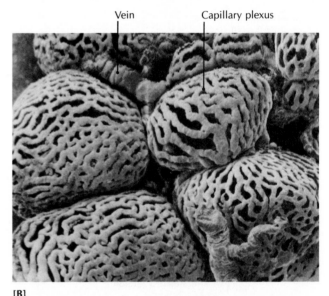

Vein   Capillary plexus

[B]

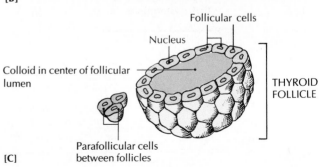

Follicular cells
Nucleus
Colloid in center of follicular lumen
THYROID FOLLICLE
Parafollicular cells between follicles

[C]

follicular cells are found between follicles. Single cells may appear within the wall of a follicle, but are not in contact with the colloid inside the lumen. Parafollicular cells synthesize and secrete the hormone calcitonin, which decreases the concentrations of calcium and phosphate in the blood.

The thyroid is the only endocrine gland that is able to store its secretions outside its principal cells, and the stored form is different from the actual hormone that is secreted into the bloodstream. The stored chemical is broken down by several enzymes before it is released into the bloodstream. The rich supply of blood vessels extending through the thyroid gland makes it relatively easy for the hormones to move through the follicle cells into the capillaries.

PARATHYROID GLANDS

The *parathyroid glands* are tiny, lentil-sized glands embedded in the posterior of the thyroid lobes, usually two glands in each lobe [FIGURE 17.4A]. The parathyroid glands are so small that they were not discovered until 1850. Before then, the parathyroids were frequently removed unknowingly during goiter surgery, and the patients died "mysteriously." The adult parathyroids are composed mainly of small *principal* (or *chief*) *cells*, which secrete most of the parathyroid hormone (parathormone, PTH), and larger *oxyphilic cells*, which contain a reserve supply of hormone [FIGURE 17.4B]. The primary effects of parathormone are (1) to increase calcium concentration in the blood and extracellular fluids, and (2) to reduce the concentration of phosphates in the blood and extracellular fluids.

ADRENAL GLANDS

The two *adrenal* (uh-DREEN-uhl; L. "upon the kidneys") *glands* rest like tilted berets on the superior tip of each kidney [FIGURE 17.5A]. Each adrenal gland is actually made up of two separate endocrine glands. The inner portion of each gland is the *medulla* (L. "marrow," meaning inside); the outer portion, which surrounds the medulla, is the *cortex* (L. "bark," as in the outer bark of a tree) [FIGURE 17.5B]. The medulla and cortex not only produce different hormones, but also have separate target organs.

FIGURE 17.4 PARATHYROID GLANDS

[A] Posterior view. [B] Section of a human parathyroid gland showing the small principal cells and the larger oxyphilic cells. ×600.

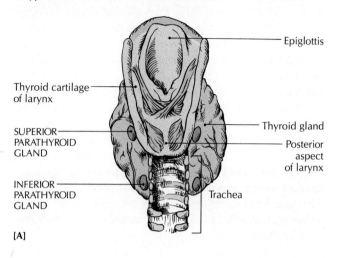

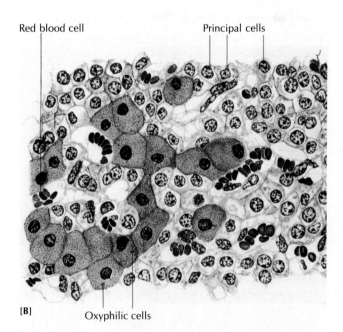

Adrenal Cortex

The ***adrenal cortex*** accounts for about 90 percent of the weight of the adrenal gland, which weighs from 5 to 7 g. Like other glands with a mesodermal embryonic origin, its hormones are steroids. The adrenal cortex secretes three classes of general steroid hormones: *glucocorticoids,* a single *mineralocorticoid* (aldosterone), and small quantities of *gonadocorticoids,* or sex hormones. The corticoids of the adrenal cortex affect the metabolism of glucose and other nutrients, as well as sodium and potassium.

The tissue of the cortex has three distinct zones lying beneath the outermost capsule [FIGURE 17.5C]. Directly beneath the capsule is the thin *zona glomerulosa,* which supplies cells for all three zones if regeneration is necessary. It also contains the enzymes necessary for the production of the mineralocorticoid hormone aldosterone. The next level down is the thick *zona fasciculata,* which makes up the bulk of the adrenal cortex. The glucocorti-coids corticosterone and cortisol, as well as a small amount of gonadocorticoids, are produced by the zona fasciculata. Cholesterol, the precursor of steroid hormones, is more concentrated in the cells of the zona fasciculata than in any other part of the body. These cells also contain considerable amounts of vitamin C. The *zona reticularis* is the deepest layer of the adrenal cortex. Although the cells here are not arranged as orderly as those in the zona fasciculata, they are similar.

Adrenal Medulla

The ***adrenal medulla,*** the inner portion of the adrenal gland, should be thought of as an entirely separate endocrine gland from the adrenal cortex. It secretes different hormones, has a different tissue structure, is derived from different embryonic tissue, and unlike the adrenal cortex, is not essential to life.

FIGURE 17.5 ADRENAL GLANDS

[A] The adrenals rest on top of the kidneys. **[B]** Cross section. **[C]** Enlarged section of the adrenal cortex of an adult man showing the layered regions from the outer capsule to the zona reticularis, which merges with the inner medulla.

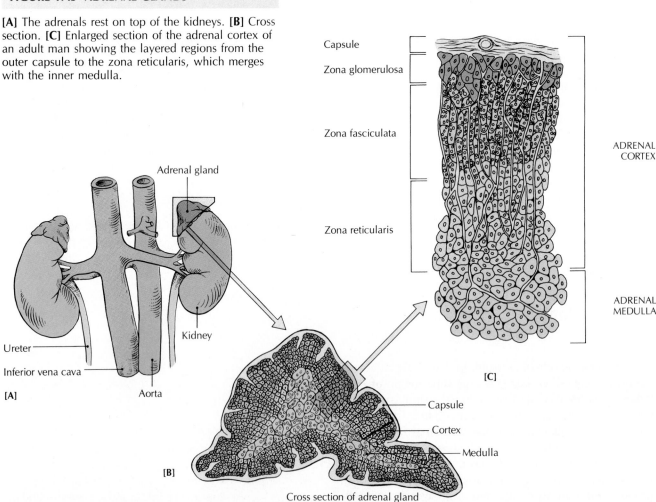

The cells of the adrenal medulla may extend somewhat into the innermost portion of the zona reticularis [FIGURE 17.5]. They are usually grouped in clumps, and are surrounded by blood vessels, especially capillaries. Granules in the cells store high concentrations of two hormones, epinephrine and norepinephrine.

The secretory cells of the adrenal medulla (sometimes called *chromaffin cells* because of their tendency to stain a dark color) are derived from the same embryonic ectodermal tissue as the ganglia of the sympathetic nervous system. Because of this common origin, it is not surprising that the two types of cells have similar functions and effects. Adrenal chromaffin cells synthesize, store, and secrete epinephrine and norepinephrine, which work in concert with glucocorticoids. They affect the heart rate and cardiac output, blood pressure, blood glucose concentration, fat metabolism, smooth muscle activity, and many other body functions.

PANCREAS

The *pancreas* (Gr. "all flesh") is an elongated (12 to 15 cm), fleshy organ consisting of a head, body, and tail [FIGURE 17.6A]. The area where the head and body join is the neck. The pancreas is located posterior to the stomach, with the head tucked into the curve of the duodenum (where the stomach meets the small intestine). The body and tail extend laterally to the left, with the tail making contact with the spleen.

The pancreas is considered a mixed gland because it functions both with ducts, as an exocrine gland, and without ducts, as an endocrine gland. As an exocrine gland, it acts as a digestive organ, secreting digestive enzymes and alkaline materials into a duct that empties into the small intestine. As an endocrine gland, it secretes its hormones into the bloodstream. The endocrine portion of the pancreas makes up only about 1 percent of the total weight of the gland. This portion synthesizes, stores, and secretes hormones from clusters of cells called the *pancreatic islets* (islets of Langerhans) [FIGURE 17.6B].

The adult pancreas contains between 200,000 and 2,000,000 pancreatic islets scattered throughout the gland. The islets contain four special groups of cells called alpha, beta, delta, and F cells [FIGURE 17.6C]. *Alpha cells* produce glucagon, and *beta cells* produce insulin. Both hormones help to regulate glucose metabolism and are usually secreted simultaneously. Beta cells are the most common type. They are generally located near the center of the islet and are surrounded by the other cell types. *Delta cells* secrete somatostatin, the hypothalamic growth-hormone inhibiting hormone that also inhibits the secretion of both glucagon and insulin. *F cells* secrete pancreatic polypeptide, which is released into the bloodstream after a meal. The endocrine function of pancreatic polypeptide is not known.

GONADS

The *gonads* (Gr. *gonos*, offspring), which are the *ovaries* in the female and the *testes* in the male, secrete hormones that help to regulate reproductive functions. These sex hormones include the male *androgens* and the female *estrogens*, *progestins*, and *relaxin*. These hormones affect the development and maintenance of sexual characteristics and all aspects of reproduction. The gonads are discussed in detail in Chapter 25.

OTHER SOURCES OF HORMONES

In addition to the major endocrine glands, other glands and organs carry on hormonal activity. In the following sections we briefly discuss the kidneys, pineal gland, thymus gland, heart, digestive system, and placenta.

Kidneys

The paired *kidneys* are primarily organs for the excretion of wastes, but they also produce several hormones, including erythropoietin (a substance that stimulates the production of red blood cells), 1,25-dihydroxyvitamin D_3, prekallikreins, and prostaglandins. In addition, the kidneys produce *renin* (*renes* is Latin for "kidneys"), an enzyme whose natural substrate is the plasma protein angiotensinogen I. Renin has an important role in the formation of angiotensin, a chemical agent that raises blood pressure.

Pineal Gland

The *pineal gland* (PIHN-ee-uhl; L. *pinea*, pine cone) is also known as the *pineal body*. It is a pea-sized body located in the roof of the diencephalon, deep within the cerebral hemispheres of the brain, at the posterior end of

FIGURE 17.6 PANCREAS

[A] Location and anatomy. [B] Section of pancreatic tissue showing a pancreatic islet. ×600. [C] A highly magnified human pancreatic islet with alpha, beta, and delta cells clearly delineated. ×5000.

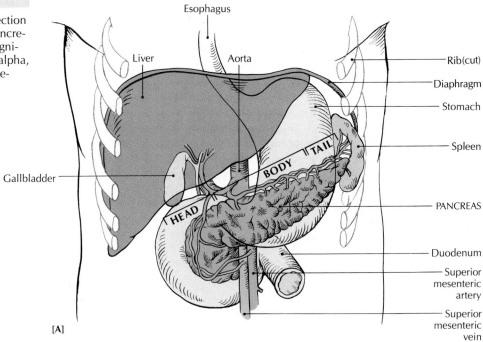

Esophagus

Liver

Aorta

Rib(cut)

Diaphragm

Stomach

Spleen

Gallbladder

TAIL

BODY

HEAD

PANCREAS

Duodenum

Superior mesenteric artery

Superior mesenteric vein

[A]

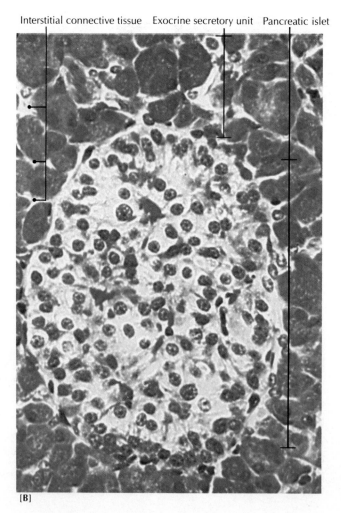

Interstitial connective tissue   Exocrine secretory unit   Pancreatic islet

[B]

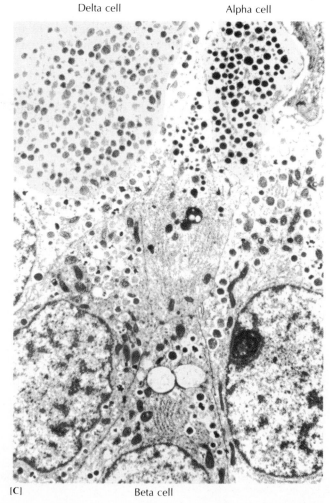

Delta cell   Alpha cell

[C]   Beta cell

the third ventricle [FIGURE 17.1]. The pineal gland has been called a "neuroendocrine transducer"—a system that converts a signal received through the nervous system (dark and light, for instance) into an endocrine signal (shifting concentrations of hormone secretion). Information about daily cycles of light and dark is detected by the eyes, and then travels from the optic nerves and tracts to the hypothalamus; sympathetic nerves convey the signals to the pineal gland.

Several chemicals that have hormonal activity have been isolated from the pineal gland, but the functions of these substances have not been fully established. One of these is *melatonin,* which is derived from serotonin. The pineal gland produces steady secretions of melatonin throughout the night; light inhibits the production of melatonin. It has been observed that melatonin causes drowsiness, and the pineal gland may affect the sleep cycle. The typical lack of light during long winters may contribute to *seasonal affective disorder* (SAD), a condition characterized by lack of energy and mood swings that border on depression. Some researchers speculate that the pineal gland is involved in SAD, but no firm evidence is available.

Thymus Gland

The *thymus gland* is a double-lobed lymphoid organ located behind the sternum in the anterior mediastinum [FIGURE 17.1]. It has an outer cortex containing many lymphocytes, and an inner medulla containing fewer lymphocytes, as well as clusters of cells called thymic (Hassall's) corpuscles, whose function is unknown. The thymus gland is well supplied with blood vessels, but has only a few nerve fibers. The thymus gland has a central role in the immune response, providing an environment for the diffentiation of T lymphocytes, which are critical to the proper functioning of the immune system.

The thymus gland is large and active only during childhood, reaching its maximum effectiveness during early adolescence. After that time, the gland begins to atrophy because of the action of sex hormones, and its glandular tissue is replaced by fatty tissue. Prolonged stress usually hastens atrophy. This happens because stress factors release adrenocortical hormones that have a destructive effect on thymus tissue. The thymus gland finally ceases activity altogether after 50 years or so, and it may therefore play an important role in the process of aging and the accompanying decrease in function of the immune system.

Heart

Recent findings have revealed that in addition to being the pump that maintains circulation, the heart also acts as an endocrine organ. Cardiac muscle cells within both atria (the upper chambers of the heart) contain secretory granules that produce, store, and secrete a peptide hormone called *atriopeptin* (formerly called *atrial natriuretic factor,* or *ANF*). It is believed that atriopeptin helps to maintain the proper balance of fluids and electrolytes and lowers excessively high blood pressure and volume.

Digestive System

Among the major hormones of the digestive system are gastrin, secretin, and cholecystokinin. *Gastrin* is a polypeptide secreted by the mucosa (lining) of the stomach. Its function is to stimulate the production of hydrochloric acid and the digestive enzyme pepsin when food enters the stomach. Thus the stomach is both the producer and the target organ of gastrin.

Secretin is a polypeptide secreted by the mucosa of the duodenum. It stimulates a bicarbonate-rich secretion from the pancreas that neutralizes stomach acid as the acid passes to the small intestine. Secretin was the first hormone to be discovered (by the British scientists William M. Bayliss and Ernest H. Starling in 1902), and the first substance to actually be called a "hormone."

Cholecystokinin (CCK; koh-lee-sis-TOE-kine-in) is secreted from the wall of the duodenum. It stimulates the contraction of the gallbladder, which releases bile when food (particularly fats) enters the duodenum. It also stimulates the secretion of enzyme-rich digestive juices from the pancreas.

Placenta

The *placenta* is a specialized organ that develops in a pregnant female as a source of nourishment for the developing fetus (see Chapter 25). It secretes estrogen, progesterone, and human chorionic gonadotropin (hCG), which help to maintain pregnancy. Target areas are the ovaries, mammary glands, and uterus.

THE EFFECTS OF AGING ON THE ENDOCRINE SYSTEM

Although most organs of the endocrine system are relatively unaffected by aging, some changes do occur. For example, the thyroid becomes smaller, and the basal metabolic rate (the amount of energy used up at rest) is lowered. The metabolism of protein and sugar is diminished as the pancreas secretes less insulin. The body burns fuel less rapidly, but efficiency is not decreased since less fuel is needed. The greatest changes occur in the activities of the reproductive organs. These changes are described in Chapter 25, which discusses the reproductive systems in detail.

TABLE 17.1 MAJOR ENDOCRINE GLANDS: HORMONES AND FUNCTIONS

Gland	Hormone	Function of hormone	Means of control
Posterior pituitary (neurohypophysis)	Antidiuretic hormone (ADH; vasopressin)	Increases water absorption from kidney tubules; raises blood pressure.	Synthesized in hypothalamus, released from neurohypophysis
	Oxytocin	Stimulates contractions of pregnant uterus, milk ejection from breasts after childbirth.	Synthesized in hypothalamus, released from neurohypophysis
Anterior pituitary (adenohypophysis)	Growth hormone (GH; somatotropic hormone, STH)	Stimulates growth of bone, muscle; promotes protein synthesis, fat mobilization; slows carbohydrate metabolism.	Hypothalamic growth-hormone releasing hormone (GHRH); growth-hormone inhibiting hormone (GHIH)
	Prolactin	Promotes breast development during pregnancy, milk production after childbirth.	Hypothalamic prolactin-inhibiting hormone (PIH); prolactin-releasing hormone (PRH)
	Thyroid-stimulating hormone (TSH)	Stimulates production and secretion of thyroid hormones.	Hypothalamic thyrotropin-releasing hormone (TRH)
	Adrenocorticotropic hormone (ACTH)	Stimulates production and secretion of adrenal cortex steroids.	Hypothalamic corticotropin-releasing hormone (CRH)
	Luteinizing hormone (LH)	*Female:* stimulates development of corpus luteum, release of oocyte, production of progesterone and estrogen. *Male:* stimulates secretion of testosterone, development of interstitial tissue of testes.	Hypothalamic gonadotropin-releasing hormone (GnRH)
	Follicle-stimulating hormone (FSH)	*Female:* stimulates growth of ovarian follicle, ovulation. *Male:* stimulates sperm production.	Hypothalamic gonadotropin-releasing hormone (GnRH)
	Melanocyte-stimulating hormone (MSH)	Apparently involved with skin color (melanocytes) in combination with ACTH; role uncertain.	Uncertain
Thyroid (follicular cells)	Thyroid hormones: thyroxine (T_4), triiodothyronine (T_3)	Increase metabolic rate, sensitivity of cardiovascular system to sympathetic nervous activity; affect maturation, homeostasis of skeletal muscle.	Thyroid-stimulating hormone (TSH) from adenohypophysis; TSH regulated by thyrotropin-releasing hormone (TRH) from brain
Thyroid (parafollicular cells)	Calcitonin	Lowers blood calcium and phosphate concentrations; acts on bone, kidney, and other cells.	Blood calcium concentration
Parathyroid	Parathormone (PTH; parathyroid hormone)	Increases blood calcium concentration, decreases blood phosphate level; acts on bone, intestine, kidney, and other cells.	Blood calcium concentration
Adrenal medulla	Epinephrine (adrenaline)	Increases heart rate; blood pressure; regulates diameter of arterioles; stimulates contraction of smooth muscle. Increases blood glucose concentration.	Sympathetic nervous system
	Norepinephrine (noradrenaline)	Constricts arterioles; increases metabolic rate.	Sympathetic nervous system

Gland	Hormone	Function of hormone	Means of control
Adrenal cortex	Glucocorticoids, mainly cortisol (hydrocortisone), corticosterone, 11-deoxycorticosterone	Affect metabolism of all nutrients; regulate blood glucose concentration; anti-inflammatories; affect growth; decrease effects of stress, ACTH secretion.	Corticotropin-releasing hormone (CRH) from hypothalamus, ACTH from adenohypophysis
	Mineralocorticoids, mainly aldosterone	Control sodium retention and potassium loss in kidney tubules.	Angiotensin II, blood potassium concentration
	Gonadocorticoids (adrenal sex hormones)	Slight effect on ovaries and testes.	ACTH
Pancreas (beta cells in pancreatic islets)	Insulin	Lowers blood glucose by facilitating glucose transport across plasma membranes and increasing glycogen storage; affects muscle, liver, adipose tissue.	Blood glucose concentration
Pancreas (alpha cells in pancreatic islets)	Glucagon	Increases blood glucose concentration.	Blood glucose concentration
Ovaries (follicle)	Estrogens	Affect development of sex organs and female characteristics, initiates development of ovarian follicle.	Follicle-stimulating hormone (FSH)
Ovaries (corpus luteum)	Progesterone, estrogens	Influence menstrual cycle; stimulates growth of uterine wall, maintains pregnancy.	Luteinizing hormone (LH)
Placenta	Estrogens, progesterone, human chorionic gonadotropin (hCG)	Maintains pregnancy.	Uncertain
Testes	Androgens, mainly testosterone	Affect development of sex organs and male characteristics, aids sperm production.	Luteinizing hormone (LH)
Thymus	Thymosin alpha, thymosin B_1 to B_5, thymopoietin I and II, thymic humoral factor (THF), thymostimulin, factor thymic serum (FTS)	Help develop T cells in thymus, maintain T cells in other lymphoid tissue; involved in development of some B cells into antibody-producing plasma cells.	Uncertain
Digestive system	Secretin	Stimulates release of pancreatic juice to neutralize stomach acid.	Acid in small intestine
	Gastrin	Produces digestive enzymes and hydrochloric acid in stomach.	Food entering stomach
	Cholecystokinin (CCK)	Stimulates release of pancreatic enzymes, gallbladder contraction.	Food in duodenum
Heart	Atriopeptin (atrial natritic factor, ANF)	Helps maintain balance of fluids, electrolytes. Decreases blood pressure and volume.	Salt concentration, blood pressure, blood volume

DEVELOPMENTAL ANATOMY OF THE PITUITARY GLAND

The pituitary gland develops from *ectoderm* of the primitive mouth cavity, which gives rise to the adenohypophysis, and from *neuroectoderm* of the diencephalon, which develops into the neurohypophysis. During week 4, the ***hypophyseal*** (Rathke's) ***pouch*** develops from ectoderm and grows toward the brain [FIGURE 17.7B, C]. By week 5, the pouch has elongated and made contact with the ***neurohypophyseal bud,*** which develops from the floor of the diencephalon. The ***infundibulum*** is the stalk connecting the pituitary gland to the hypothalamus. It develops from the neurohypophyseal bud about week 8 [FIGURE 17.7D].

The connection between the hypophyseal pouch and the oral cavity is called the ***hypophyseal stalk.*** It disappears between weeks 6 and 8 as the pouch closes up completely.

After week 8, cells of the anterior wall of the hypophyseal pouch develop into the ***pars distalis*** [FIGURE 17.7E]. Later, a small extension from the pars distalis called the ***pars tuberalis*** forms and surrounds the infundibular stalk. Cells of the posterior wall of the hypophyseal pouch develop into a thin layer called the ***pars intermedia.*** These three structures eventually make up the ***adenohypophysis.*** The ***neurohypophysis*** develops from the infundibulum, as it extends downward toward the developing adenohypophysis, and from the ***pars nervosa,*** which is in contact with the adenohypophysis via specialized nerve cells and fibers.

FIGURE 17.7 DEVELOPMENTAL ANATOMY OF THE PITUITARY GLAND

[A] Lateral view. [B–E] Midsagittal views.

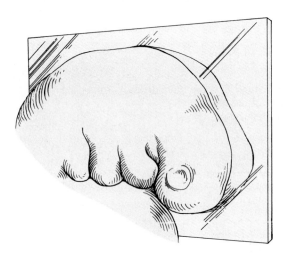

[A] WEEK 4

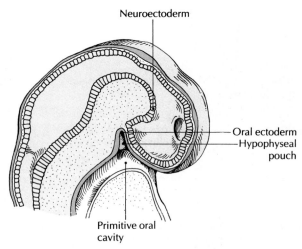

[B] WEEK 4

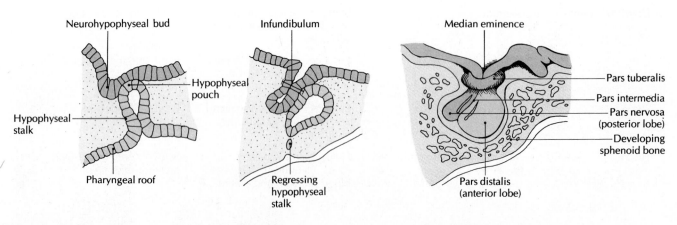

[C] WEEK 5   [D] WEEK 8   [E] FETAL STAGE—POST WEEK 8

WHEN THINGS GO WRONG

Endocrine disorders are usually caused by underfunctioning *(hypo-)* or overfunctioning *(hyper-)*, inflammation, or tumors of the endocrine glands. Certain disorders arise in the hypothalamus or pituitary; others develop in the endocrine glands that come under the influence of these higher control centers. Some of the major disorders of the endocrine system are reviewed here briefly.

Pituitary Gland: Adenohypophysis

Among the disorders of the adenohypophysis is *giantism,* caused by the oversecretion of growth hormone during the period of skeletal development (see photo). The body of a person with this disorder grows much

A normal-sized person (center) with a "giant," whose pituitary gland was overactive, and a pituitary dwarf, whose pituitary gland was underactive.

larger than normal, sometimes to more than $2\frac{1}{2}$ meters (8 ft) and 180 kilograms (400 lb). In most cases, death occurs before the age of 30.*

Acromegaly (ak-roh-MEG-uh-lee; Gr. *akros*, extremity + *megas*, big) is a form of giantism that affects adults after the skeletal system is fully developed. It is caused by a pituitary tumor. A tumor of a gland is called an *adenoma* (Gr. *aden*, gland + *oma*, tumor), and is usually benign. After maturity, oversecretion of growth hormone does not lengthen the skeleton, but does cause some cartilage and bone to thicken. The jaw, hands, feet, eyebrow ridges, and soft tissues may widen noticeably (see photos), and enlargement of the heart, liver, and other internal organs is also possible. Blood pressure may increase, and subsequent congestive heart failure is likely. Muscles grow weak, and osteoporosis and painful enlargements at the joints may occur. Acromegaly may be treated with some success by the surgical removal of the adenohypophysis, or by using radiation treatments.

Persons with undersecretion of growth hormone are known as **pituitary dwarfs.** Although their intelligence and body proportions are normal, they do not grow any taller than a normal 6-year-old child. Some pituitary dwarfs become prematurely senile, and most die before the age of 50. Usually, their sexual organs and reproductive ability are not fully developed, but some pituitary dwarfs are capable of producing normal-sized children. Some dwarfism is caused by a deficiency of the thyroid hormones rather than by a deficiency of pituitary growth hormone. A treatment of hormonal replacement may increase growth somewhat and help retain fertility in sexually active patients.

*The tallest person for whom there are authenticated records was Robert Wadlow of Illinois. He was 8 ft, 11 in. tall and still growing when he died at the age of 22.

The effects of acromegaly, such as thickening of the jaw, nose, and hands, are shown here at three different stages of an afflicted woman's life.

Pituitary Gland: Neurohypophysis

When ADH (vasopressin) is undersecreted, excessively large amounts of water are excreted in the urine *(polyuria)*. This condition, known as **diabetes insipidus,** is accompanied by dehydration and unrelenting thirst *(polydipsia)*. Diabetes insipidus can often be treated successfully by administering controlled doses of ADH.

Thyroid Gland

Overactivity of the thyroid gland, known as **hyperthyroidism** (Graves' disease), may be caused by long-acting thyroid stimulator (LATS). LATS is an antibody that acts on the thyroid gland in the same way that TSH does. Hyperthyroidism may also be caused by immunoglobulins such as human thyroid stimulator and LATS protector, or by a thyroid tumor. Among the symptoms are nervousness, irritability, increased heart rate and blood pressure, weakness, weight loss, and high oxygen use, even at rest. The bulging eyes *(exophthalmos)* typical of this condition are due partly to increased fluid behind the eyes caused by an exophthalmos-producing substance (EPS). Drug therapy that inhibits thyroxine production, and administration of radioactive iodine have successfully replaced surgery as treatments for hyperthyroidism.

Underactivity of the thyroid, or **hypothyroidism,** is often associated with *goiter* (L. "throat"), an enlarged thyroid (see photo). Such a swelling in the neck is caused when insufficient iodine in the diet forces the thyroid to expand in an attempt to produce more thyroxine. The adaptive responses are triggered, in large part, by an increased secretion of TSH, which attempts to stimulate the iodine-trapping mechanism and the subsequent steps in the metabolism of iodine. Most cases of goiter used to be found in areas away from the ocean, where iodine content in the soil and water supply is low. With the addition of minute amounts of iodine to ordinary table salt and drinking water in recent years, this type of goiter has practically disappeared as a common ailment. Symptoms of hypothyroidism include decreased heart rate, blood pressure, and body temperature; lowered basal metabolic rate; and underactivity of the nervous system.

A woman with a massive goiter. Such a large goiter would not occur with proper medical treatment.

Underactivity of the thyroid during the development of the fetus after week 12 causes **cretinism** (CREH-tinism), characterized by mental retardation and irregular development of bones and muscles. The skin is dry, eyelids are puffy, hair is brittle, and the shoulders sag. Ordinarily, cretinism cannot be cured, but early diagnosis and treatment with L-thyroxine (a drug form of thyroxine) may arrest the disease before the nervous system is damaged.

If the thyroid becomes underactive during adulthood, **myxedema** results, producing swollen facial features, dry skin, low basal metabolic rate, tiredness, possible mental retardation, and intolerance to cold in spite of increased body weight. Like cretinism, myxedema may be corrected if it is treated early.

Another form of dwarfism is the irreversible result of an underactive thyroid gland that interferes with the normal growth of cartilage. People with this condition are usually intelligent and normally active; they have stubby arms and legs, a relatively large chest and head, and flattened facial features.

Parathyroid Glands

The most common causes of underactivity of the parathyroid glands *(hypoparathyroidism)* are damage to or removal of the parathyroids during surgery and parathyroid adenoma. Hypoparathyroidism produces low concentrations of calcium in blood plasma and an overabundance of phosphorus. Such an imbalance may cause faulty transmission of nerve impulses, osteoporosis and hampered bone growth, and paralysis of muscles (tetany). Controlled doses of vitamin D and calcium salts may restore normal calcium concentrations. Low-phosphate diets and drugs that increase the excretion of phosphorus in urine may also be successful.

Overactivity of the parathyroid glands *(hyperparathyroidism)* is usually caused by an adenoma. It results in excessive amounts of calcium and lowered amounts of phosphorus. As in hypoparathyroidism, osteoporosis is evident, and many other general symptoms may occur, including loss of appetite, nausea, weight loss, personality changes, stupor, kidney stones, duodenal ulcers, kidney failure, increased blood pressure, and congestive heart failure. Treatment may include the surgical removal of excess parathyroid tissue, drugs that lower the calcium concentration, and in severe cases of kidney failure, an artificial kidney or kidney transplant.

Adrenal Cortex

Two diseases caused by overactivity of the adrenal cortex *(hyperadrenalism)* are Cushing's disease and adrenogenital syndrome. **Cushing's disease** is usually caused by a cortical tumor that overproduces glucocorticoids. The tumor, in turn, is caused by an increased secretion of ACTH by the anterior pituitary. Symptoms include fattening of the face, chest, and abdomen (the limbs remain normal), accompanied by abdominal striations and a tendency toward diabetes caused by increased blood glucose. Protein is lost, and the muscles become weak. Surgical removal of the causative tumor usually brings about a remission and reduces the secretion of ACTH.

Adrenogenital syndrome is also caused by an overactive adrenocortical tumor, which stimulates excessive production of the cortical male sex hormones known as *androgens*. These androgens cause male characteristics to appear in a female, and accelerate sexual development in a male. Hormonal disturbances during the fetal development of a female child may cause a distortion of the genitals, so that the clitoris and labia become enlarged and resemble a penis and scrotum. In a mature woman, an extreme case of adrenogenital syndrome may produce a beard. (Such pronounced male characteristics in a female may be caused by other hormonal malfunctions besides defects in the adrenal cortex.) Adrenogenital syndrome is often treated with daily doses of cortisone, which usually reinstates the normal steroid balance.

Underactivity of the adrenal cortex *(hypoadrenalism)* produces **Addison's disease,** whose symptoms include anemia (deficiency of red blood cells), weakness and fatigue, increased blood potassium, and decreased blood sodium. Skin color becomes bronzed because excess ACTH, produced by the pituitary in an effort to restore a normal cortical hormone level, also induces α-MSH secretion, and the two hormones together induce abnormal deposition of skin pigment. Until recently, Addison's disease was usually fatal, but now it can be controlled with regular doses of cortisone and aldosterone.

Pancreas

About 4 percent of the United States population will develop **diabetes mellitus** at some time in their lives. It can occur either as *Type I (insulin-dependent) diabetes*, which usually begins early in life, or *Type II (insulin-independent) diabetes*, which occurs later in life, mainly in overweight people. Heredity plays a major role in the development of both types. Type I diabetes results when beta cells do not produce enough insulin. When this happens, glucose accumulates in the blood and spills into the urine, but does not enter the cells. Excess glucose in the urine is a diuretic and causes dehydration. Because the cells are unable to use the accumulated glucose (the most readily available energy source in the body), the body actually begins to starve. Appetite may increase, but eventually the body consumes its own tissues, literally eating itself up.

In Type II diabetes there is a near-normal plasma concentration of insulin. The problem is hyporesponsiveness, or no response, to insulin, termed *insulin resistance.* Insulin's target cells do not respond normally to the circulating insulin because of alteration in the insulin receptors. An insufficient number of insulin receptors per target cell seems to be one of the causes of insulin resistance in Type II diabetics, especially those who are overweight. The majority of diabetics are Type II.

Because the removal of glucose from the kidneys requires large amounts of water, the diabetic person produces excessive sugary urine and may excrete as much as 20 L of sugary urine per day.* In response to the increased urine production, with the possibility of serious body dehydration, diabetics become extremely thirsty and drink huge amounts of liquids.

The use of fats (to replace glucose) for energy production in the diabetic causes the accumulation of acetoacetic acid, β-hydroxybutyric acid, and keto acids in the blood and body fluids. This leads to acidosis, which can lead to coma and death.

Diabetes is incurable in any form, but mild diabetes can usually be controlled by strict dietary regulation and exercise. More serious cases may require treatment with regular injections of insulin (Type I), or oral hypoglycemic agents (Type II). If the disease is untreated, almost every part of the diabetic's body will be affected, and gangrene, hardening of the arteries (arteriosclerosis), other circulatory problems, and further complications may occur.

Low blood glucose, **hypoglycemia,** may be caused by the excessive secretion of insulin. Glucose is not released from the liver, and the brain is deprived of its necessary glucose. Hypoglycemia can be controlled by regulating the diet, especially carbohydrate intake. On a short-term basis, the glucose in a glass of orange juice may restore the normal glucose balance in the blood, but a long-term treatment usually consists of a reduction of carbohydrates in the diet. Carbohydrates tend to stimulate large secretions of insulin, thereby removing glucose from the bloodstream too quickly.

---

*The word *diabetes* comes from the Greek word for "siphon" or "to pass through," referring to the seemingly instant elimination of liquids. The word was actually used by the Greeks as early as the first century. In the seventeenth century, the sweetness of diabetic urine was discovered, and the name of the disease was lengthened to diabetes mellitus. *Mellitus* comes from Greek *meli*, honey.

CHAPTER SUMMARY

The endocrine system and the nervous system together constitute the two great regulatory systems of the body. The endocrine system is made up of tissues or organs called **endocrine glands,** which secrete chemicals directly into the bloodstream. These chemical secretions are called **hormones.**

Hormones travel through the bloodstream to all parts of the body, but affect only those **target cells** that have compatible receptors. Most hormones are steroids, derivatives of amino acids, or proteins. A few are glycoproteins.

Pituitary Gland (Hypophysis) (p. 446)

1 The **pituitary gland,** or **hypophysis,** consists of an anterior lobe **(adenohypophysis)** and a posterior lobe **(neurohypophysis).** The adenohypophysis has an abundance of secretory cells, while the neurohypophysis contains many nerve endings.

2 The pituitary is connected to the hypothalamus by a stalk of nerve cells and blood vessels called the *infundibular stalk,* which provides a direct link between the nervous system and the endocrine system.

3 The **hypothalamus** releases regulating substances called **releasing** and **inhibiting hormones** that control the secretions of the adenohypophysis. Nerve centers in the hypothalamus regulate secretions from the neurohypophysis. The connection between the hypothalamus and the adenohypophysis is facilitated by a system of blood vessels called the **hypothalamic-hypophyseal portal system;** whereas the hypothalamic link with the neurohypophysis relies on nerve impulses.

4 The hypothalamus synthesizes **antidiuretic hormone** (ADH) and **oxytocin,** which are stored in and secreted from the neurohypophysis.

5 The true endocrine portion of the pituitary is the adenohypophysis, which synthesizes and secretes seven separate hormones: **growth hormone** (GH), **prolactin, thyroid-stimulating hormone** (TSH), **adrenocorticotropic hormone** (ACTH), **luteinizing hormone** (LH), **follicle-stimulating hormone** (FSH), and **melanocyte-stimulating hormone** (MSH).

Thyroid Gland (p. 448)

1 The **thyroid gland** secretes the hormones **thyroxine** and **triiodothyronine** from its follicular cells and **calcitonin** from its parafollicular cells. Thyroid hormones increase basal metabolism rate, accelerate growth, and stimulate cellular differentiation and protein synthesis.

2 Calcitonin lowers calcium and phosphate concentrations in the blood.

Parathyroid Glands (p. 449)

Parathormone, the hormone of the **parathyroid glands,** increases the concentration of calcium in the blood and decreases the concentration of phosphate.

Adrenal Glands (p. 449)

1 The **adrenal glands** are composed of an inner **medulla** and an outer **cortex.** The cortex secretes three types of steroid hormones: glucocorticoids, mineralocorticoids, and gonadocorticoids.

2 The adrenal medulla secretes **epinephrine** and **norepinephrine,** which stimulate the sympathetic nervous system.

Pancreas (p. 451)

1 The **pancreas** functions as an exocrine gland, secreting digestive enzymes into ducts, and as an endocrine organ, secreting hormones into the bloodstream. The endocrine portion synthesizes, stores, and secretes hormones from the **pancreatic islets.**

2 The pancreatic islets contain alpha cells that secrete **glucagon,** beta cells that secrete **insulin,** delta cells that secrete somatostatin, and F cells that secrete pancreatic polypeptide.

Gonads (p. 451)

1 The **gonads, ovaries** in a female and **testes** in a male, secrete hormones that control reproductive functions.

2 The major hormones are **testosterone** in males and **estrogens, progestins,** and **relaxin** in females.

Other Sources of Hormones (p. 451)

1 The **kidneys** secrete several hormones, including erythropoietin, 1,25-dihydroxyvitamin D_3, prekallikreins, and prostaglandins. In addition, the kidneys produce **renin,** an enzyme that causes blood pressure to increase.

2 The **pineal gland** contains several chemicals, whose functions have not been established. The secretion of **melatonin,** which may affect skin pigmentation, seems to be affected by light and may be involved with the sleep cycle, seasonal affective disorder, and puberty.

3 The main function of the **thymus gland** is the processing of T cells. The gland begins to atrophy with age, which may play a role in the aging process and the accompanying decline of the immune system.

4 The **heart** produces, stores, and secretes the peptide hormone **atriopeptin.** It is believed that atriopeptin helps to maintain the proper balance of fluids and electrolytes and lowers excessively high blood pressure and volume.

5 The **digestive system** secretes several digestive hormones, especially **gastrin,** which aids digestion in the stomach; **secretin,** which helps to neutralize stomach acid; and **cholecystokinin,** which stimulates the release of bile and enzymes from the gallbladder and pancreas.

6 The **placenta** produces hormones that help maintain pregnancy.

The Effects of Aging on the Endocrine System (p. 453)

1 The endocrine system is relatively unaffected by aging.

2 Some changes include the lowering of the basal metabolic rate as the thyroid becomes smaller.

Developmental Anatomy of the Pituitary Gland (p. 456)

1 The embryonic pituitary gland develops from **ectoderm** of the primitive mouth cavity and **neuroectoderm** of the diencephalon.

2 The ectoderm develops into the **adenohypophysis,** consisting of the **pars distalis, pars tuberalis,** and **pars intermedia.** The neuroectoderm develops into the **neurohypophysis,** made up of the **infundibulum** and **pars nervosa.**

STUDY AIDS FOR REVIEW

MEDICAL TERMINOLOGY

HYPERPLASIA (Gr. *hyper*, over + *plasis*, change, growth) A nontumorous growth in a tissue or organ.

HYPOPLASIA Incomplete or arrested development of a tissue or organ.

POSTPRANDIAL (L. *post*, after + *prandium*, late breakfast, meal) Usually pertaining to an examination of blood glucose content after a meal.

RADIOIMMUNOASSAY (RIA) (RAY-dee-oh-im-myu-noh-ASS-ay) A technique that measures minute quantities of a substance, such as a hormone, in the blood.

REPLACEMENT THERAPY A method of treatment where insufficient secretions of hormones are replaced with natural or synthetic chemicals.

SIMMONDS' DISEASE A disorder characterized by a total absence of pituitary hormones, usually caused by a tumor or blockage of the blood vessels supplying the pituitary gland.

STEROID THERAPY Use of steroids to treat certain endocrine disorders.

VIRILISM (L. *vir*, man) Masculinization in women.

UNDERSTANDING THE FACTS

1 Name the three chemical categories into which most hormones fit.
2 Which portion of the pituitary actually synthesizes and secretes hormones?
3 Which is the only endocrine gland able to store its secretions outside its principal cells?
4 Why is the concentration of cholesterol so great in the cells of the zona fasciculata of the adrenal cortex?
5 Locate the following structures:
  a infundibular stalk
  b pancreatic islets
  c chromaffin cells
  d follicular cells
  e pars intermedia
6 Name the endocrine structures of the pancreas and the two major hormones they secrete.
7 Name the part of the brain that is connected to part of the endocrine system; describe the structural and functional relationships of these two structures.
8 Describe the structures of the adrenal glands and the pancreatic islets.
9 What organs other than those in the endocrine system are known to produce hormones?

UNDERSTANDING THE CONCEPTS

1 Compare the basic approach to body regulation of the endocrine system with that of the nervous system. Do they complement each other, and if so, in what way?
2 In acromegaly, why do the long bones thicken but not lengthen? (You may have to think back to your study of the skeletal system.)
3 Why do some feel that the thymus gland may play an important role in the aging process?
4 What is the significance of the fact that the entrance of fatty foods into the duodenum serves as a stimulant for the secretion of cholecystokinin?
5 What is the hypothalamic-hypophyseal portal system?

SELF-QUIZ
Multiple-Choice Questions
17.1 Epinephrine and norepinephrine are secreted by the
  a stomach mucosa
  b adrenal cortex
  c pancreatic islets
  d adrenal medulla
  e anterior pituitary
17.2 The major hormone affecting metabolic rate is produced by the
  a thyroid
  b parathyroids
  c posterior pituitary
  d adrenal cortex
  e thymus
17.3 The release of hormones from the adenohypophysis is controlled by secretions of the
  a adrenal medulla
  b kidneys
  c hypothalamus
  d posterior pituitary
  e small intestine
17.4 The part of the pituitary that contains secretory cells is the
  a sella turcica
  b anterior lobe
  c infundibulum
  d posterior lobe
  e hypothalamic-hypophyseal portal system
17.5 The thyroid is made up of spherical sacs, or
  a parafollicular cells   **d** chief cells
  b follicles   **e** oxyphilic cells
  c principal cells

Completion Exercise

17.6 The anterior and posterior lobes of the pituitary gland are also known as the _____ and _____.

17.7 The hormones of the neurohypophysis are synthesized in the _____.

17.8 The only endocrine structure that is directly connected to the nervous system is the _____.

17.9 Hormones synthesized in the hypothalamus are carried to the anterior pituitary by the _____.

17.10 The parathyroid glands are embedded in the _____.

17.11 The endocrine glands located at the superior end of each kidney are the _____.

17.12 The endocrine portions of the pancreas are called the _____.

17.13 Sex hormones are secreted by the _____ and _____.

17.14 The outer portion of the adrenal gland is the _____, and the inner portion is the _____.

17.15 Digestive organs that secrete hormones are the _____ and _____.

17.16 Sweat and salivary glands are called _____ glands.

17.17 The cells that a hormone acts on are called _____ cells.

17.18 Another name for the pituitary gland is the _____.

17.19 The connection between the hypothalamus and the adenohypophysis is the _____.

17.20 Hypothalamic nerve cells are called _____ cells.

A SECOND LOOK

1 In the following drawing, label the thyroid gland, hypothalamus, pituitary gland, thymus gland, adrenal glands, pancreas, and male and female gonads.

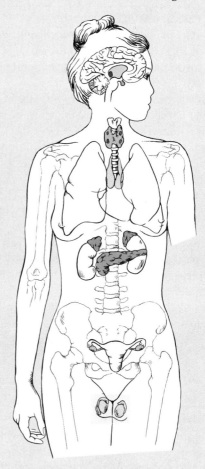

2 Label the hypothalamus, neurohypophysis, adenohypophysis, secretory cells, and neurosecretory cells.

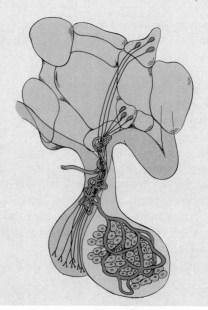

IV

TRANSPORT, MAINTENANCE, AND REPRODUCTION

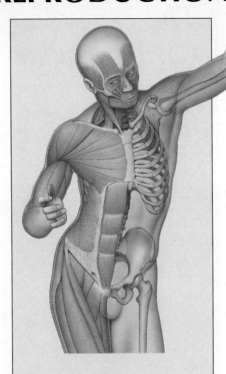

18

The Cardiovascular System: Blood

LEARNING OBJECTIVES

1 Discuss the major functions of blood, and explain how blood helps to maintain homeostasis in the body.

2 Describe some important properties of blood.

3 List the main components of blood plasma, and give their functions.

4 Explain how the structure of a red blood cell (erythrocyte) relates to its major functions.

5 Discuss the transport of oxygen and carbon dioxide by the blood.

6 Define erythropoiesis, and tell where it takes place.

7 Describe the life span, destruction, and removal of erythrocytes.

8 Name the two basic categories of leukocytes (white blood cells), and the types of leukocytes included in each.

9 Compare the five types of leukocytes as to number, structure, origin, and function.

10 Describe the structure, origin, and function of platelets.

11 Define hemostasis.

For practical reasons, it is to our advantage to discuss human anatomy in terms of separate, functional body systems. But we can see that such a classification is artificial when we attempt to classify blood. In the strictest sense, the *cardiovascular system* is made up of the heart (as a pump) and blood vessels (as a means of transport). But to exclude blood from the cardiovascular system would be like considering the respiratory system without including the air we breathe. So we will include blood, at least nominally, as part of the cardiovascular system.

Blood can also be classified as part of the *circulatory system*, which is a more general term that includes not only the blood, blood vessels, and heart, but also the lymph and lymph vessels. Blood is also part of the *hematologic (blood) system*, which includes active bone marrow, the lymph nodes, the spleen, and specialized cells such as *macrophages*.

FUNCTIONS OF BLOOD

Although higher animals have evolved various ways of transporting nutrients, gaseous waste products, and hormones throughout their bodies, the most common method uses a circulating fluid called **blood**. As blood moves throughout the body, tissues are continuously adding to it their waste products, secretions, and metabolites, and taking from it vital nutrients, oxygen, hormones, and other substances. Overall, blood performs the following functions:

1 *Transports* oxygen from the lungs to body tissues, and the waste products of cellular metabolism from body tissues to the kidneys, liver, lungs, and sweat glands for eventual removal from the body; it also transports nutrients, hormones, and enzymes throughout the body.

2 *Regulates* blood clotting to stop bleeding; body temperature, mainly by increasing or decreasing blood flow to the skin; acid-base balance (pH) through the distribution of buffers; and the amount of water and electrolytes in body fluids.

3 *Protects* against harmful microorganisms and other substances by contributing white blood cells, proteins, and antibodies to the inflammatory and immune responses.

PROPERTIES OF BLOOD

The blood volume of a healthy person fluctuates very little. Even when blood is lost, it is replaced rapidly. The blood of an average adult is about 7 to 9 percent of total body weight, or 79 mL/kg of body weight. An average man has 5 to 6 L of blood, and an average woman has 4 to 5 L.

Because blood contains red blood cells, it is thicker, denser, and more adhesive than water, and flows four to five times more slowly. This comparative resistance to flow is called *viscosity*. The more red blood cells and blood proteins, the higher the viscosity and the slower the flow. The viscosity of blood ranges between 3.5 and 5.5, compared to a constant 1.000 for water. The *specific gravity*, or density, of blood is between 1.045 and 1.065, as compared with 1.000 for water.

The red color of arterial blood is due to oxygenated *hemoglobin*, a globular protein carried by the red blood cells. When oxygen is removed, the blood appears bluish or darker. White blood cells and platelets in the blood are clear, and the plasma is yellowish.

Blood is slightly alkaline, with a pH that usually ranges between 7.35 and 7.45. Arterial blood is more alkaline than venous blood because it has less carbon dioxide. As the concentration of carbon dioxide increases, it reacts with water to form carbonic acid, which lowers the blood pH. The temperature of blood averages about 38°C (100.4°F).

ASK YOURSELF

1 What is the normal viscosity of blood? The pH? The temperature?

2 Why is blood usually red?

COMPONENTS OF BLOOD

Blood consists of a liquid part, known as *plasma* (about 55 percent), and a solid part, or the **formed elements** (about 45 percent), which are mostly blood cells suspended in the plasma. The formed elements include red blood cells (erythrocytes), white blood cells (leukocytes), and fragmented cells called platelets (thrombocytes) [FIGURE 18.1].

Plasma

Plasma is the liquid part of blood. It is about 90 percent water, and provides the solvent for dissolving and transporting nutrients. A group of proteins comprise another 7 percent of the plasma. The remaining 3 percent is composed of electrolytes, amino acids, glucose and other nutrients, various enzymes, hormones, metabolic wastes, and traces of many other organic and inorganic molecules.

FIGURE 18.1 SCANNING ELECTRON MICROGRAPH OF HUMAN BLOOD CELLS (×4000)

Thrombocytes (platelets) Erythrocytes (red blood cells)

Leukocytes (white blood cells)

Water in plasma The water in blood plasma is readily available to cells, tissues, and extracellular fluids of the body as needed to maintain the normal state of hydration. Water is also the solvent for both extracellular and intracellular chemical reactions. Thus, many of the properties of blood plasma can be correlated with the basic properties of water. The water in blood plasma also contains many solutes whose concentrations are changing constantly to meet the needs of the body.

Plasma proteins The proteins in blood plasma are referred to as *plasma proteins.* The total protein component of blood plasma can be divided into the *albumins, fibrinogen,* and the *globulins.*

The most abundant plasma proteins (about 60 percent of all plasma proteins) are the **albumins** (al-BYOO-mihnz), which are synthesized in the liver. The main function of albumins is to promote water retention in the blood, which, in turn, maintains normal blood volume and pressure. If the amount of albumin in the plasma decreases, fluid leaves the bloodstream and accumulates in the surrounding tissue, causing a swelling known as *edema* (ih-DEE-muh). Albumins also act as carrier mole-

cules by binding to other molecules, such as hormones, that are transported in plasma.

The **globulins** (GLAHB-yoo-lihnz) (about 36 percent of all plasma proteins) are divided into three classes, based on their structure and function: alpha (α), beta (β), and gamma (γ). The alpha and beta globulins are produced by the liver. Their function is to transport lipids and fat-soluble vitamins in the blood. One form of these transport molecules, *low-density lipoproteins* (LDLs), transports cholesterol from its site of synthesis in the liver to various body cells. Another transporter, *high-density lipoproteins* (HDLs), removes cholesterol and triglycerides from arteries, preventing their deposition there.

Gamma globulins are the immunoglobulins, antibodies that help to prevent diseases such as measles, tetanus, and poliomyelitis. (An *antibody* is any one of millions of distinct proteins produced by the body that is capable, along with white blood cells, of inactivating a specific bacterium, virus, protein, or cancer cell that it "recognizes" to be foreign. The antibody combines with the foreign body, known as an *antigen,* forming an *antigen-antibody complex.*)

Also produced by the liver is **fibrinogen** (fye-BRIHN-uh-jehn) (about 4 percent of all plasma proteins), a plasma protein essential for blood clotting. When fibrinogen and several other proteins involved in clotting are removed from plasma, the remaining liquid is called **serum.**

Plasma electrolytes *Electrolytes* are inorganic molecules that separate into ions when they are dissolved in water. The ions are either positively charged *(cations)* or negatively charged *(anions).* The major cation of plasma is sodium (Na^+). The principal anion is chloride (Cl^-). The other major ions in plasma are potassium (K^+), calcium (Ca^{2+}), phosphate (PO_4^{2-}), iodide (I^-), and magnesium (Mg^{2+}).

Nutrients and waste products Although *glucose* appears in plasma in low concentrations, it is the body's most readily available source of usable energy. Glucose enters the body in carbohydrate foods, and is also synthesized in the body from other kinds of nutrients, including fats and amino acids.

Like glucose, *amino acids* have a high turnover rate; that is, they are used rapidly by body cells, and usually appear in the plasma in low concentrations. Amino acids are also important because they provide the building blocks for protein synthesis.

Lipids are found in plasma in the form of phospholipids, triglycerides, free fatty acids, and cholesterol. They are important components of nerve cells and steroid hormones, and some serve the body as an excellent source of fuel. Lipids are introduced into the body in food.

Several *metabolic wastes* are transported by the plasma,

especially lactic acid, and some nitrogenous waste products of protein metabolism.

Gases and buffers Oxygen, nitrogen, and especially carbon dioxide are the principal gases dissolved in plasma. Carbon dioxide is transported by red blood cells, and in the plasma, both in the dissolved state and in the form of the bicarbonate ion (HCO_3^-). Nitrogen is transported in the dissolved state by plasma.

Red Blood Cells (Erythrocytes)

Red blood cells, or *erythrocytes* (ih-RITH-roh-sites; Gr. *erythros*, red + cells), make up about half the volume of human blood. There are about 25 trillion erythrocytes in the body, and each cubic millimeter of blood contains 4 to 6 million erythrocytes. (The average man has about 5.5 million erythrocytes per cubic millimeter of blood, and the average woman has about 4.8 million.) Erythrocytes measure about 7 micrometers in diameter, and about 2 micrometers thick.* If 5 or 6 red blood cells were placed in a row, they would reach across the period at the end of this sentence.

An erythrocyte is shaped like a disk and is slightly concave on top and bottom, like a doughnut without the hole poked completely through [FIGURE 18.2]. This shape provides a larger surface area for gas diffusion than a flat disk or a sphere. When an erythrocyte is mature, it no longer has a nucleus or many organelles, such as mitochondria. Thus, it must rely on its store of already-produced proteins, enzymes, and RNA.

Hemoglobin Almost the entire weight of an erythrocyte consists of *hemoglobin* (HEE-moh-gloh-bihn; Gr. *haima*, blood + L. *globulus*, little globe), often abbreviated as Hb, an oxygen-carrying globular protein.

*A micrometer (μm) equals $\frac{1}{1000}$ of a millimeter; it replaces the older term *micron* (μ).

Each adult hemoglobin molecule consists of 5 percent *heme*, an iron-containing pigment, and 95 percent *globin*, a polypeptide protein.* Males usually have more hemoglobin than females do. Attached to each of hemoglobin's four polypeptide chains is a heme group, which gives blood its color. Attached to each heme group is an iron atom, which is the binding site for oxygen. The iron atom plays a key role in hemoglobin's function as an oxygen-carrying substance by binding oxygen and releasing it to tissues at the appropriate times.

The function of hemoglobin depends on its ability to combine with oxygen where the oxygen concentration is high (in the lungs), and to release oxygen where the concentration is low (in the body tissues). Hemoglobin also carries waste carbon dioxide from the tissues to the lungs, where it is exhaled. By transporting oxygen and carbon dioxide, hemoglobin also helps to maintain a stable acid-base balance in the blood.

Hemoglobin has a bluish-purple color when it is deoxygenated, but it becomes red when it is loaded with oxygen. (Oxygenated hemoglobin is called *oxyhemoglobin*, HbO_2, and deoxygenated hemoglobin is called *reduced hemoglobin*.) As oxygenated blood circulates from the lungs to the rest of the body, it loses more and more oxygen to the tissues, until it appears blue in the veins. The oxyhemoglobin in the erythrocytes of subcutaneous capillaries gives cheeks and lips their pink color.

The blood of a fetus contains *fetal hemoglobin* (HbF), which has a greater affinity for oxygen than adult hemoglobin does. As a result, the vital movement of oxygen from mother to fetus is enhanced. Fetal hemoglobin is usually replaced by adult hemoglobin within a few days after birth.

Transport of oxygen and carbon dioxide The major role of erythrocytes is to transport oxygen from the lungs to the body tissues. The abundant oxygen in the alveoli of the lungs combines reversibly with the iron in hemoglobin to form oxyhemoglobin. The oxyhemoglobin is then transported in the red blood cells to the tissues. Because oxygen is constantly being used up by cellular oxidation, there is more oxygen in the blood coming from the lungs than there is in the tissues. Thus, oxygen in oxyhemoglobin tends to diffuse into the cells of the tissues.

The other major function of erythrocytes is to transport carbon dioxide from the tissues to the lungs, where it will be exhaled from the body. Carbon dioxide is more soluble in water than oxygen is, and it diffuses easily through capillary walls from the body tissues into the plasma. Once in the blood, carbon dioxide is transported

*Hemoglobin got its name in a roundabout way. Early microscopes showed the shape of a red blood cell as a sphere, not the biconcave disk we can identify today. As a result, the cells were called "globules," the proteins they contained were called "globulins" or "globins," and the substance was named "hemoglobin" or "blood protein."

FIGURE 18.2 ERYTHROCYTE STRUCTURE

Drawing of an erythrocyte cut open to show its biconcave shape.

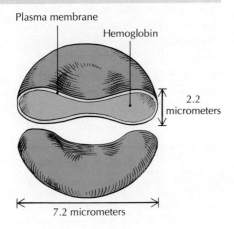

Plasma membrane

Hemoglobin

2.2 micrometers

7.2 micrometers

in three ways: (1) About 60 percent of the carbon dioxide reacts with water to form carbonic acid (H_2CO_3). (2) About 30 percent of the carbon dioxide reacts directly with hemoglobin to form *carbaminohemoglobin* ($HbCO_2$), which is carried from the tissues to the lungs. When carbaminohemoglobin arrives in the lungs, the carbon dioxide is exchanged for oxygen. (3) The remaining 10 percent of the carbon dioxide is dissolved in the plasma and red blood cells as molecular carbon dioxide (CO_2).

Erythrocyte production (erythropoiesis)   Before birth, the fetus produces blood cells progressively in the yolk sac, liver, and spleen. During the fifth fetal month, blood-cell production decreases in these sites and increases in the bone marrow cavities. After birth, erythrocytes are manufactured primarily, and continuously, in the red marrow of certain bones, especially in the vertebrae, ribs, sternum, pelvis, and the upper ends of the femur and humerus. The process is called *erythropoiesis* (ih-RITH-roh-poy-EE-sis; Gr. red + to make).

Erythrocytes are derived from large embryonic cells in the bone marrow called *committed stem cells,* or *hemocytoblasts,* which are destined to become different kinds of blood cells [FIGURE 18.3]. Some hemocytoblasts differentiate within the bone marrow into *common myeloid progenitor cells,* some of which then differentiate into *erythroblasts* and begin synthesizing hemoglobin. After several cell divisions, the erythroblast loses its nucleus and is now an immature red blood cell called a *reticulocyte* (Gr. "network cell") because it contains an intricate netlike pattern of endoplasmic reticulum. The reticulocyte leaves the bone marrow and enters the bloodstream, where it continues to synthesize hemoglobin. About 2 or 3 days later, the reticulocyte loses its endoplasmic reticulum, mitochondria, and ribosomes as it matures into an *erythrocyte.*

The production and destruction (in the spleen) of erythrocytes are maintained at an equal rate. If red blood cells are lost from the circulatory system, the rate of erythropoiesis is increased until the normal erythrocyte number is regained.

Destruction and removal of erythrocytes   The life span of an erythrocyte is only about 80 to 120 days, primarily because it has no nucleus and is unable to replace the enzymes and other proteins that it needs to function properly. Although erythrocytes are able to use some glucose as a source of energy, they cannot synthesize much protein. As the cells age, their protein goes through a normal process of degradation, which cannot be repaired by the cells. As a result, the plasma membrane begins to leak, and leaks more and more as the integrity of the protein continues to be lost.

As these aged and fragile erythrocytes pass through the narrow capillaries (sinusoids) of the spleen, their leaky membranes rupture, and the cellular remnants are engulfed by phagocytic cells called *macrophages.* The macrophages digest the hemoglobin into smaller amino acids, which are then returned to the body's amino acid pool for the future synthesis of new proteins. The iron from hemoglobin is also recovered and reused.

White Blood Cells (Leukocytes)

White blood cells, or *leukocytes* (LYOO-koh-sites; Gr. *leukos,* white + cells), serve as scavengers that destroy microorganisms at infection sites, help remove foreign molecules, and remove debris that results from dead or injured tissue cells. Leukocytes range from slightly larger to much larger than erythrocytes [FIGURE 18.3]. Unlike erythrocytes, leukocytes do have nuclei, and the cells are able to move about independently and pass through blood vessel walls into the tissues.

In adults, there are about 1000 erythrocytes for every leukocyte. The normal adult leukocyte count is between 4000 and 11,000 per cubic millimeter, which may increase as a result of infection to 25,000 per cubic millimeter, about the same number that newborn infants have.

The production of white blood cells is called *leukopoiesis.* Some leukocytes originate from the same undifferentiated hemocytoblasts of blood-forming bone marrow that erythrocytes do. However, leukopoiesis also occurs throughout the body in lymphoid tissues such as lymph nodes, spleen, and tonsils.

The term *leukocyte* is used to cover a number of different cell types that circulate in the blood. The two basic classifications of leukocytes are granulocytes (polymorphonuclear leukocytes, or polymorphs) and agranulocytes (mononuclear leukocytes).

Granulocytes   The most numerous of the white blood cells are *granulocytes,* so named because they contain large numbers of granules in the cytoplasm outside their multilobed nuclei [FIGURE 18.3]. The three types of granulocytes are *neutrophils, eosinophils,* and *basophils.* When stained with Wright's stain, the granules of neutrophils appear light pink to blue-black, eosinophil granules appear red to red-orange, and granules of basophils are blue-black to red-purple.*

---

*The names of granulocytes are derived from the type of stain with which they are most easily stained for laboratory preparations. The suffix *phil* comes from the Greek *philos,* which means "loving" or "having a preference for." Neutrophils "prefer" a *neutral* dye, eosinophils an *eosin* (acid) dye, and basophils a *basic* dye.

Q: *If the blood in our veins is bluish, why does it appear red when we cut a vein and begin to bleed?*

A: The deoxygenated, venous blood turns bright red as soon as it is exposed to oxygen in the air.

Neutrophils represent about 60 percent of the granulocytes. Neutrophils are scavenger cells that move like amoebas and send out long projections called pseudopods ("false feet") to engulf and destroy microorganisms and other foreign materials. The granules inside the cytoplasm are packets of enzymes (lysozymes) that break down and eventually destroy them. As a result of this process, which is called *phagocytosis* ("cell eating"), the neutrophils may also be destroyed as their granules are depleted. Dead microorganisms and neutrophils make up the thick, whitish fluid we call *pus*.

Eosinophils have B-shaped nuclei. Like neutrophils, eosinophils are phagocytes that have an amoeboid movement. Their granules contain lysosomal enzymes and peroxidase that, when released, destroy phagocytized material. For unknown reasons, the eosinophil count increases during allergy attacks, with certain parasitic infections, some autoimmune diseases (the production of antibodies that attack one's own tissues), and in certain types of cancer. In addition, eosinophils contain the protein plasminogen, which helps dissolve blood clots.

Basophils are granulocytes with elongated, indistinctly lobed nuclei. They are the least numerous of all the granulocytes. Their granules contain *heparin* (an anticoagulant), *histamine* (which dilates general body blood vessels and constricts blood vessels in the lungs), and *slow-reacting substance* (SRS-A) of allergies. SRS-A produces some of the allergic symptoms, such as bronchial constriction. Basophils play an important role in providing immunity against parasites.

Granulocytes develop in the bone marrow from common myeloid progenitor cells that differentiate into committed cells called *myeloblasts* [FIGURE 18.3]. The further stages of differentiation are *promyelocytes, myelocytes,* and *granulocytes.* The promyeloblasts are characterized by azurophilic granules (granules that stain with blue dyes) that differ from the granules of mature granulocytes. The myelocytes contain some characteristic granules that distinguish each of the three types of granulocytes from each other. The myelocytes continue to divide by mitosis until they differentiate into so-called *metamyelocytes,* at which stage they have lost their capacity for mitosis before differentiating into mature granulocytes, namely neutrophils, eosinophils, and basophils. During this stage such changes as indentations of the nuclei into lobes and the accumulation of more granules occur.

The bone marrow contains millions of mature granulocytes that can be released into the blood when necessary. The bone marrow of an average person produces about 100 billion neutrophils a day. Most of the neutrophils in the bloodstream are not actually circulating, but instead are clinging to the inner walls of blood vessels, ready to move toward an infection or injury site. The number of neutrophils in the bloodstream may actually quintuple in the first few hours of a serious infection.

Most granulocytes have a life span of about 5 to 10 days, and survive only a few hours after they enter the bloodstream. The level of granulocytes is probably regulated hormonally, but the precise mechanism is not known.

Agranulocytes Despite their name, *agranulocytes* (or *nongranular leukocytes*) usually have a few nonspecific lysosome granules in their cytoplasm. The two types of agranulocytes are monocytes and lymphocytes, which comprise about 5 and 30 percent of the leukocytes, respectively.

Monocytes, which are mobile phagocytes, are the largest blood cells. They have large, folded nuclei and often have fine granules in the cytoplasm [FIGURE 18.3].

Monocytes develop in bone marrow from *monoblasts,* enter the bloodstream for about 30 to 70 hours, and then leave. Once in the tissue spaces, monocytes enlarge to 5 to 10 times their normal size and become phagocytic macrophages. These macrophages form a key portion of the *reticuloendothelial system,* which lines the vascular portions of the liver, lungs, lymph nodes, thymus gland, and bone marrow. (The reticuloendothelial system is the system of macrophages throughout the body.) In the connective tissue of these regions, macrophages phagocytize microorganisms and cellular debris. Macrophages also play a role in the immune system by processing specific antigens.

Lymphocytes are small, mononuclear, agranular leukocytes with a large, round nucleus that occupies most of the cell [FIGURE 18.3]. They get their name from lymph, the fluid that transports them. Lymphocytes move sluggishly, and do not travel the same routes through the bloodstream as other leukocytes. They originate from the hemocytoblasts of bone marrow and then invade lymphoid tissues, where they establish colonies. These colonies then produce additional lymphocytes *without* involving the bone marrow. Most lymphocytes are found in the body's tissues, especially in lymph nodes, the spleen, thymus gland, tonsils, adenoids, and the lymphoid tissue of the gastrointestinal tract. They also differ from other leukocytes in that they can leave as well as re-enter the circulatory system, and can leave the blood more easily than other cells to enter lymphoid tissue. Some lymphocytes live for years, recirculating between blood and lymphoid organs. The biggest difference between lymphocytes and other white blood cells is that lymphocytes are not phagocytes.

Two distinct types of lymphocytes are recognized: B cells and T cells [FIGURE 18.3]. *B cells* originate in the bone marrow and colonize lymphoid tissue. In contrast, *T cells* are associated with, and influenced by, the thymus gland before they colonize lymphoid tissue. Both B cells and T cells regulate the cellular immune response that protects the body from its own defense system, and they

FIGURE 18.3 DEVELOPMENT OF BLOOD CELLS

Dashed lines indicate the omission of some intermediate stages. The mature cells are shown enlarged.

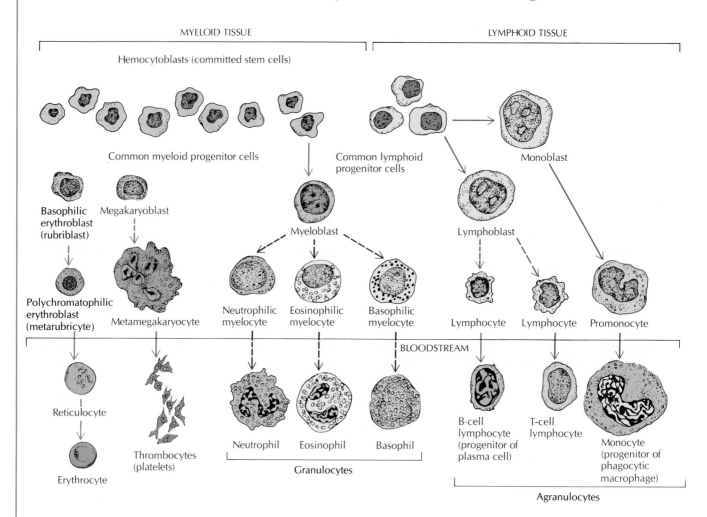

Type of cell	Description
ERYTHROCYTE (RED BLOOD CELL)	
	Biconcave disk; no nucleus, principal component hemoglobin; diameter about 7 μm. About 50% of total blood volume (4–6 million/mm^3).
LEUKOCYTE (WHITE BLOOD CELL)	
	Several forms of cells; capable of independent amoeboid movement. Less than 1% of total blood volume (4000–11,000/mm^3).
Granulocytes (polymorphonuclear)	Larger than erythrocytes; contain many granules in cytoplasm. About 52–75% of total leukocytes.
Neutrophil	Multilobed nucleus; granules appear pink to blue-black in a neutral stain; diameter about 10 μm. About 50–70% of total leukocytes.
Eosinophil	B-shaped nucleus; granules appear red to red-orange in an acid stain; diameter about 10 μm. About 1–4% of total leukocytes.
Basophil	Lobed nucleus; granules appear blue-black to red-purple in basic stain; diameter about 10 μm. Less than 1% of total leukocytes.
Agranulocyte (mononuclear)	Contains only occasional granules in cytoplasm.
Monocyte	Largest blood cells, approximately two to three times larger than erythrocytes; diameter about 20 μm. About 2–8% of total leukocytes.
Lymphocyte	Large, round nucleus that nearly fills the cell; diameter about 10 μm. About 20–40% of total leukocytes.
THROMBOCYTE (PLATELET)	
	Fragments of megakaryocytes; no nucleus; diameter about 2–4 μm. Less than 1% of total blood volume (130,000–360,000/mm^3).

secrete chemicals that destroy harmful microscopic intruders such as bacteria, poisons, viruses, and tissue and chemical debris. When B cells are activated, they enlarge and become *plasma cells.* Plasma cells have much more cytoplasm than B cells do, which enables them to accommodate the biochemical machinery necessary for the production of antibodies.

Platelets (Thrombocytes)

Platelets (so named because of their platelike flatness), or *thrombocytes* (Gr. *thrombus,* clot + cells), are disk-shaped, and about one-quarter the size of erythrocytes. Their main function is to start the intricate process of blood clotting. Platelets are much more numerous than leukocytes, averaging about 350,000 per cubic millimeter. (An average adult has about a trillion platelets.) Platelets lack nuclei, and are incapable of cell division, but they have a complex metabolism and internal structure [FIGURE 18.4]. Once in the bloodstream, platelets have a life span of 7 to 8 days.

About 200 billion platelets are produced every day. Like other blood cells, they originate from committed hemocytoblasts in the bone marrow [FIGURE 18.3]. The hemocytoblasts involved in platelet formation develop into myeloid progenitor cells from which large cells called **megakaryoblasts** arise. The megakaryoblasts differentiate into *megakaryocytes.* Platelets break off from the pseudopods of these cells in the bone marrow, and then enter the bloodstream. Thus, the platelets that appear in the circulating blood are not actually blood cells, but are cellular fragments of megakaryocytes. After entering the bloodstream, platelets begin to pick up and store chemical substances that can be released later to help seal vessel breaks.

Platelets adhere to each other, and to the collagen in connective tissue, but not to red or white blood cells, a property that is directly related to blood clotting and the overall process of *hemostasis,* the prevention and control of bleeding. When a blood vessel is injured (capillaries rupture many times a day), platelets immediately move to the site and begin to clump together, attaching themselves to the damaged area. The platelets release granules that contain *serotonin,* which constricts broken or injured vessels and retards bleeding, and *adenosine diphosphate* (ADP), which attracts more platelets to the damaged area. If the break is small enough, it is repaired by the platelet plug [FIGURE 18.4B]. However, if there are not enough platelets to make the repair, they begin the complex process of blood clotting, or **coagulation** (L. *coagulare,* to curdle).

FIGURE 18.4 PLATELETS

[A] Flash-contact x-ray micrograph of a human platelet. This new technique shows *live* cells. Note the forming pseudopod, which will become part of the meshwork essential for blood clotting. ×30,000. **[B]** Electron micrograph of cross section of an injured capillary. Note the platelet plugging a tiny break in the capillary wall. Larger breaks in blood vessels attract many platelets to the injured site. Sections of four erythrocytes appear above the platelet. ×8000.

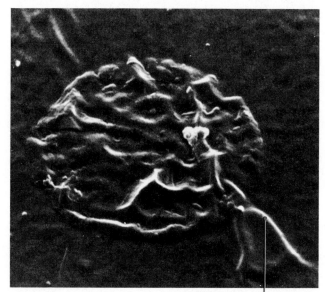

[A]   Pseudopod

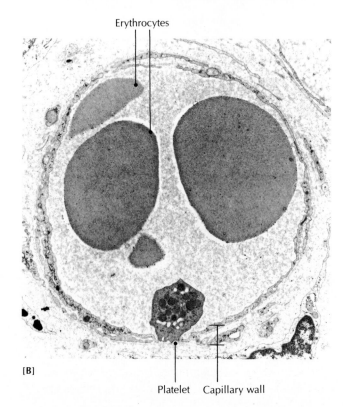

Erythrocytes

[B]

Platelet   Capillary wall

HEMOSTASIS: PREVENTION OF BLOOD LOSS

One drawback of a circulatory system such as ours, where the liquid blood is under high pressure, is that serious bleeding can take place even after a slight injury. To prevent the possibility of uncontrolled bleeding, we have a three-part hemostatic mechanism consisting of (1) the constriction of blood vessels, (2) the clumping together (aggregation) of platelets, and (3) blood clotting. Overall, *hemostasis* is a specific type of homeostasis that prevents blood loss.

Vasoconstrictive Phase

Normally, when a tissue is damaged and blood escapes from a blood vessel, the vessel wall constricts in order to narrow the opening of the vessel and slow the flow of blood. This vasoconstriction is due to contraction of the smooth muscle of the vessel wall as a direct result of the injury and the release of vasoconstrictor chemicals from platelets. Proper vasoconstriction is also enhanced by pain reflexes, producing constriction in proportion to the extent of the injury. Constriction of capillaries, which do not have muscular layers, is due to the vascular compression caused by the pressure of lost blood that accumulates in surrounding tissues. Injured blood vessels may continue to constrict for 20 minutes or more.

Platelet Phase

The next event in hemostasis is the escape from blood vessels of platelets, which swell and adhere to the collagen in adjacent connective tissues. This attachment stimulates vasoconstriction. By now, the platelets have become very sticky, so that as more and more of them move into the injured area, they stick together. In about a minute they can clog a small opening in the vessel with a platelet plug. The process is called *platelet aggregation*. It is important partly because it successfully stops hundreds of small hemorrhages every day and partly because it triggers the blood-clotting mechanism.

Hemostasis and the Nervous System

The sympathetic nervous system helps to control massive blood loss when a blood clot is inadequate to stop the flow. When the body loses more than 10 percent of its blood, there is a sudden drop in blood pressure, and the body goes into shock. The decreased blood pressure triggers reflexes in the sympathetic nervous system that constrict veins and the small terminal branches of arteries (arterioles) in an attempt to limit the decrease in blood pressure. Also, the heart rate may rise from a normal 72 to 84 beats a minute to as many as 200, increasing the blood flow to counteract the reduced blood pressure. Blood flow to the brain and heart is especially increased. Without the perfect functioning of the sympathetic reflexes, a person would probably die after losing 15 to 20 percent of the total blood volume. When the feedback system is operating properly, a person may still be alive after losing as much as 40 percent of the total blood volume. If the lost blood is not replaced quickly, however, death will follow.

WHEN THINGS GO WRONG

Anemias
Anemia is a condition in which the number of red blood cells or the concentration of hemoglobin in the red blood cells is below normal. In general, anemia decreases the blood's oxygen-carrying capacity. When anemia is due to the excessive destruction of erythrocytes, there is usually an abnormally high amount of bilirubin in the plasma, producing a characteristic jaundice of the skin and a darkening of the feces. Anemia is usually a symptom of some underlying disease. Descriptions of several types of anemia follow.

Hemorrhagic anemia results from heavy blood loss and is sometimes seen in cases of heavy menstrual bleed-ing, severe wounds, or the sort of internal bleeding that accompanies a serious stomach ulcer.

Iron-deficiency anemia is the most common type of anemia. It can be caused by a long-term blood loss, low intake of iron, or faulty iron absorption. If iron is not available to make erythrocytes, a *hypochromic* (low concentration in erythrocytes) *microcytic* (small erythrocytes) anemia develops.

Aplastic anemia is characterized by failure of the bone marrow to function normally. It is usually caused by poisons such as lead, benzene, or arsenic, that hamper the production of red blood cells, or by radiation, such as x-rays or atomic radiation, that damages the bone marrow. With the marrow not functioning properly, the total count of erythrocytes and leukocytes decreases drastically.

Hemolytic anemia is produced when an infecting organism such as the malarial parasite enters red blood cells and reproduces until the cell actually bursts. Such a rupture is called *hemolysis* ("blood destruction") and causes the anemia. (Other causes may be hereditary, or due to sickle-cell anemia or an adverse reaction to a drug.) *Thalassemia* is the name of a group of hereditary hemolytic anemias characterized by impaired hemoglobin production. As a result, the synthesis of erythrocytes is also impaired. It is most common among people of Mediterranean descent. Sickle-cell anemia (discussed below) is another hereditary hemolytic anemic condition.

Pernicious anemia is usually caused by an improper absorption of dietary vitamin B_{12}, which is required for the complete maturation of red blood cells. The majority of cases are associated with antibodies that destroy *intrinsic factor* in the stomach mucosa. Intrinsic factor is a protein that combines with vitamin B_{12} and facilitates its absorption by the small intestine. Without intrinsic factor, immature red blood cells accumulate in the bone marrow. The number of mature red blood cells circulating in the blood may drop to as low as 20 percent of normal, and immature cells may enter the bloodstream. If diagnosed early, pernicious anemia is usually treated successfully with intramuscular injections of vitamin B_{12} and an improved diet. If left untreated, the disease may affect the nervous system, causing decreased mobility, general weakness, and damage to the brain and spinal cord.

Sickle-cell anemia is a genetic hemolytic anemia that occurs most often in blacks; it affects 1 in 600 American blacks. It results from the inheritance of the hemoglobin S-producing gene, which causes the substitution of the amino acid valine for glutamic acid in two of the four chains that compose the hemoglobin molecule. Sickle-cell anemia is so named because the abnormal red blood cells become rigid, rough, and crescent-shaped like a

sickle (see photo). Sickled cells do not carry or release oxygen as well as normal erythrocytes. Such cells make the blood more viscous by clogging up capillaries and other small blood vessels, reducing blood supply to some tissues and producing swelling, pain, and tissue destruction. No effective permanent treatment has yet been discovered. Chronic fatigue and pain, increased susceptibility to infection, decreased bone marrow activity, and even death are common results of sickle-cell anemia.

Hemophilia

Hemophilia is a hereditary disease in which blood fails to clot or clots very slowly. The gene that causes the condition is carried by females, who are not themselves affected, and is expressed mostly in males. Females can be born with the disease only when the mother is a carrier and the father already has hemophilia. The untreated hemophiliac bleeds very easily, especially into joints. Until recently, hemophilia created a severe disability, but excellent transfusion therapy to reduce fatal bleeding is now available. However, there is no known permanent cure.

Leukemia

Leukemia is a malignant (cancerous) disease characterized by uncontrolled leukocyte production. Leukocytes increase in number as much as 50- or 60-fold, and millions of abnormal, immature leukocytes are released into the bloodstream and lymphatic system. The white blood cells tend to use the oxygen and nutrients that normally go to other body cells, causing the death of otherwise healthy cells. Infiltration of the bone marrow by abnormal leukocytes prevents the normal production of white blood cells and red blood cells, causing anemia. Because almost all the new white blood cells are immature and incapable of normal function, victims of leukemia have little resistance to infection. In most cases, the combination of starving cells and a deficiency in the immune system is enough to cause death. The cause of leukemia is unknown, although a viral infection is considered likely.

The two major forms of leukemia are lymphogenous and myelogenous. *Acute lymphogenous leukemia* appears suddenly and progresses rapidly. It occurs most often in children and is much more serious than *chronic lymphogenous leukemia*, which occurs later in life. Both of these leukemias are caused by the cancerous production of lymphoid cells in lymph nodes, which spread to other parts of the body. In *myelogenous leukemia*, the cancer begins in myelogenous cells, such as early neutrophils and monocytes in the bone marrow, and then spreads throughout the body, where white blood cells can be formed in extraneous organs. Acute myelogenous leukemia can occur at any age, whereas chronic myelogenous leukemia occurs during middle age.

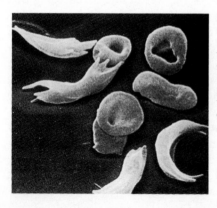

Scanning electron micrograph of erythrocytes from a patient suffering from sickle-cell anemia. Instead of being normal biconcave disks, these cells are variously distorted. ×2000.

Bone Marrow Transplantation

A possible method of combating disorders of blood-forming tissues such as aplastic anemia, leukemia, and lymphoma, is **bone marrow transplantation.** The purpose of bone marrow transplantation is to introduce normal blood cells into the bloodstream of the patient in an attempt to counteract the effects of the disease. To ensure a high degree of compatibility between donor and recipient so that the recipient will not reject the donated marrow, physicians try to match six tissue factors called human leukocyte antigens (HLA). In bone marrow transplantation, about 500 to 700 mL of marrow from a compatible donor (preferably a twin or sibling) are aspirated (removed by suction) from the donor's pelvic bones (usually the iliac crest) with a large hypodermic needle attached to a syringe. Only a small incision is necessary. The removed marrow is mixed with an anticoagulant (heparin) and then filtered to separate the donor's T cells, specialized cells of the immune system that would otherwise encourage rejection of the marrow by the recipient. The recipient receives the donor's marrow through an intravenous infusion. Two units of the donor's blood, which were removed during two sessions about a week before the transplantation, are transfused back into the donor to replace the blood aspirated along with the marrow.

CHAPTER SUMMARY

Functions of Blood (p. 465)

1 **Blood** is the circulating fluid that transports nutrients, oxygen, carbon dioxide, gaseous waste products, and regulatory substances throughout the body.

2 Blood also defends against harmful microorganisms, is involved in inflammation, coagulation, and the immune response, and helps regulate the pH of body fluids.

Properties of Blood (p. 465)

1 The average person has about 5 L of blood.

2 *Blood viscosity* and *specific gravity* are greater than those of water.

3 Blood *pH* ranges from 7.35 to 7.45.

4 Blood *temperature* averages about 38°C (100.4°F).

Components of Blood (p. 465)

1 Blood consists basically of two parts, liquid **plasma** and solid **formed elements** (red blood cells, white blood cells, platelets) suspended in the plasma.

2 **Plasma** provides the solvent for dissolved nutrients. It is about 90 percent water, 7 percent dissolved plasma proteins (albumin, globulins, fibrinogen), and 3 percent electrolytes, amino acids, glucose and other nutrients, enzymes, antibodies, hormones, metabolic wastes, and traces of other materials.

3 Red blood cells, or **erythrocytes,** make up about half the volume of blood. Their biconcave shape provides a large surface area for gas diffusion. They contain the globular protein **hemoglobin,** which transports oxygen from the lungs to all body cells, and helps remove waste carbon dioxide. Hemoglobin contains a small amount of *heme,* an iron-containing globular protein that binds to oxygen and releases it at appropriate times.

4 The production of erythrocytes in bone marrow is called **erythropoiesis.**

5 White blood cells, or **leukocytes,** destroy microorganisms at infection sites and remove foreign substances and body debris. Leukocytes use amoeboid movement to creep along the inner walls of blood vessels, and they can pass through blood vessel walls into tissue spaces.

6 The two basic classes of leukocytes are **granulocytes** and **agranulocytes.** The three types of granulocytes are neutrophils, eosinophils, and basophils. **Neutrophils** destroy harmful microorganisms and other foreign particles, **eosinophils** help destroy parasites and antibody-antigen complexes, and **basophils** are involved in allergic reactions and inflammation, although their specific function is unknown.

7 The two types of agranulocytes are monocytes and lymphocytes. **Monocytes** can become macrophages that ingest and destroy harmful substances. **Lymphocytes** are involved in the immune response and the synthesis of antibodies.

8 Platelets, or **thrombocytes,** initiate the blood-clotting process when a blood vessel is injured; they are important in the overall process of **hemostasis,** the prevention and control of bleeding. Platelets are fragmented from large cells called **megakaryocytes.**

Hemostasis: Prevention of Blood Loss (p. 473)

1 The body has a three-phase hemostatic mechanism consisting of constriction of blood vessels, clumping **(aggregation)** of platelets, and blood clotting **(coagulation).**

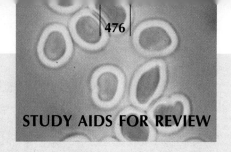
STUDY AIDS FOR REVIEW

MEDICAL TERMINOLOGY

ANOXIA (*an-*, without + oxygen) An oxygen deficiency in the blood.

APHERESIS (uh-FUR-ee-sis; Gr. *aphairein*, to take away from) A medical technique for cleansing the blood, in which a portion (such as plasma) suspected of containing harmful substances is removed and replaced with fresh ingredients.

DIRECT (IMMEDIATE) TRANSFUSION Transfer of blood directly from one person to another, without exposure of the blood to air.

DRIED PLASMA Normal plasma that has been vacuum dried to prevent microorganisms from growing in it.

ERYTHROPENIA (Gr. red + poverty, lack) Decreased red blood cell count due to disease or hemorrhage.

EXCHANGE TRANSFUSION Direct transfer of blood from donor to replace blood as it is removed from recipient. This technique is used in poisonings and other conditions.

FRACTIONED BLOOD Blood separated into its components.

HEMORRHAGE (Gr. blood + bursting forth) A discharge of blood from a broken vessel.

INDIRECT (MEDIATE) TRANSFUSION Transfer of blood in which whole or fractioned donor blood is stored for later delivery to a recipient. The blood can be separated into components, and patients receive only the needed portions.

LEUKOPENIA (Gr. white + poverty, lack) Decreased white blood cell count due to disease or hemorrhage.

NORMAL PLASMA Plasma that has had blood cells removed, but retains the normal concentrations of solutes; employed to bring blood volume back to normal levels.

PACKED RED CELLS The concentrated solution of erythrocytes that remains when plasma is removed from whole blood.

PLATELET CONCENTRATE Platelets separated from fresh whole blood; used for platelet-deficiency disorders.

POLYCYTHEMIA (pahl-ee-sigh-THEE-mee-uh; "many cells in the blood") A condition in which the number of red blood cells is above normal.

PURPURA (Gr. *porphura*, shellfish yielding a purple dye) Purple spots on the skin resulting from escaped erythrocytes from capillaries or larger hemorrhagic areas; also appear when platelets are drastically reduced in number.

SEPTICEMIA (sehp-tih-SEE-mee-uh; Gr. *septos*, rotten + blood) The presence of harmful substances, such as bacteria or toxins, in the blood. Also called *blood poisoning*.

THROMBOCYTOPENIA Fewer platelets, resulting from impaired production or increased destruction.

THROMBOCYTOSIS An increased number of platelets, usually due to increased production accompanying diseases such as leukemia and polycythemia, or following the removal of the spleen. Also called *thrombocythemia*.

VENISECTION Opening a vein to withdraw blood. Also called *phlebotomy*.

WHOLE BLOOD Blood that has all its components (formed elements, plasma, and plasma solutes) in natural concentration.

UNDERSTANDING THE FACTS

1 About how much blood does the average adult have?
2 Which of the formed elements in blood is the largest? The most abundant?
3 What is the most abundant plasma protein?
4 A plasma protein that is essential for blood clotting is

_____ .

5 Name the primitive cells of the red bone marrow from which erythrocytes are derived.
6 Where and how are old red blood cells destroyed?
7 Which is the most abundant leukocyte?
8 Name the two basic categories of leukocytes and the types of leukocytes in each category.
9 What is the origin of platelets?
10 Define hemostasis.

UNDERSTANDING THE CONCEPTS

1 Discuss the major functions of blood and how these functions contribute to homeostasis.
2 How is oxygen transported from the lungs to body tissues? How is carbon dioxide transported from the tissues to the lungs?
3 List the five types of leukocytes, and compare their functions.
4 How do platelets contribute to hemostasis?
5 Describe the types of anemia studied, and give the basic causes of each.
6 Why do leukemia victims have difficulty fighting infections when they have a greatly increased number of white blood cells?

SELF-QUIZ
Multiple-Choice Questions

18.1 The proteins of blood plasma are
   a. albumins   d a and b
   b globulins   e a, b, and c
   c fibrinogen

18.2 Lipids are found in the plasma in the form of
   a phospholipids   d cholesterol
   b triglycerides   e all of the above
   c free fatty acids

18.3 Carbon dioxide is transported in the blood in the form of
  a carbonic acid **d** dissolved CO_2
  b bicarbonate ions **e** all of the above
  c $HbCO_2$

18.4 Which of the following is *not* a type of granulocyte?
  a polymorph **d** eosinophil
  b neutrophil **e** basophil
  c monocyte

18.5 Granulocytes develop in bone marrow from undifferentiated cells called
  a myeloblasts **d** eosinophilic myelocytes
  b promyelocytes **e** none of the above
  c metamyelocytes

18.6 The largest blood cells are
  a basophils **d** neutrophils
  b monocytes **e** erythrocytes
  c eosinophils

True-False Statements

18.7 The most abundant plasma proteins are the albumins.

18.8 When the proteins involved in clotting are removed from the plasma, the remaining liquid is called serum.

18.9 Erythrocytes are derived from bone marrow stem cells called hemocytoblasts.

18.10 The short life span of erythrocytes is due to their inability to utilize glucose as an energy source.

18.11 When T cells are activated, they enlarge and become plasma cells.

18.12 The main function of blood platelets is to initiate blood clotting.

18.13 The sympathetic nervous system helps to control massive blood loss when a blood clot is inadequate to stop the flow.

18.14 The most numerous blood cells are thrombocytes.

18.15 Blood clotting is also called phagocytosis.

18.16 Platelets are formed by the fragmentation of megakaryocytes.

18.17 Erythrocytes can leave the bloodstream and move through body tissues.

Completion Exercises

18.18 The comparative resistance to flow of a fluid, such as blood, is called _____.

18.19 The liquid portion of blood is called _____.

18.20 The proteins in blood plasma are referred to as _____.

18.21 A protein essential for blood clotting is _____.

18.22 The production of red blood cells is called _____.

18.23 Erythrocyte destruction occurs in the _____.

18.24 The production of white blood cells is termed _____.

18.25 The most numerous of the white blood cells are the _____.

18.26 The most numerous blood cells are _____.

18.27 Monocytes are white blood cells that can develop into _____.

18.28 Plasma cells produce _____.

18.29 B cells and T cells are types of _____.

A SECOND LOOK

In the following drawing, identify a monoblast, myeloblast, reticulocyte, lymphocyte, and monocyte; also identify myeloid and lymphoid cells.

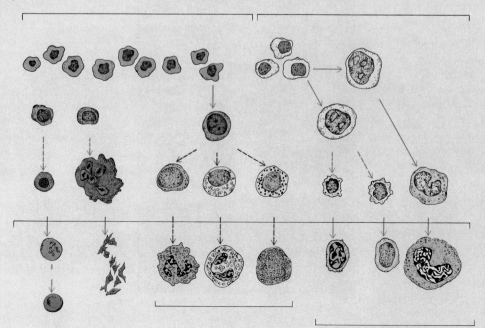

19

The Cardiovascular System: The Heart

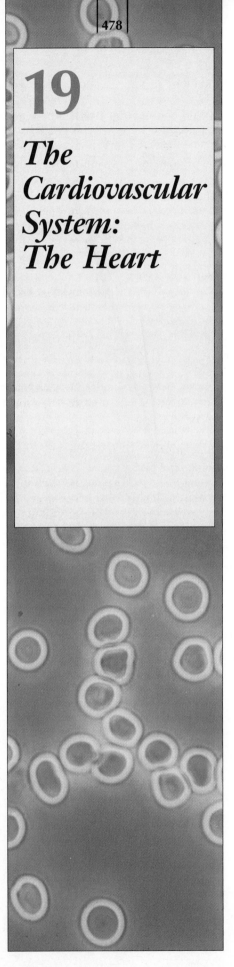

LEARNING OBJECTIVES

1 Explain the basic functions of the heart, and give its size, shape, position, and location.

2 Describe the covering of the heart, the heart wall, and the cardiac skeleton.

3 Relate the structure of the heart chambers to their functions.

4 Describe the structure and functions of the different heart valves.

5 Name the veins and arteries that supply the heart muscle and the vessels that emerge from the heart.

6 Describe how cardiac muscle differs from skeletal muscles.

7 Describe the impulse-conducting system of the heart.

8 Define cardiac cycle, and describe the path of the blood through the heart during the cycle.

9 Discuss the cause and significance of the different heart sounds.

10 Explain the role of the nervous and endocrine systems in the control of heart rate.

11 Describe the causes and symptoms of two degenerative heart disorders.

12 Describe some forms of valvular heart disease.

As you saw in the last chapter, blood is the connecting link between the outside environment and the cells of your body. In order to maintain homeostasis throughout the body, blood must flow continuously. Your heart is the pump that provides the energy to keep blood circulating, and your blood vessels are the pipelines through which blood is channeled to all parts of your body. This chapter describes the structure of the heart, and the next one discusses the blood vessels.

Human blood flows in a *closed system* of vessels, remaining essentially within the vessels that carry it. Blood travels in a circle, pumped from the muscular heart out through elastic arteries, through tiny capillaries, and then through veins back again to the heart. Each side of the heart contains an elastic upper chamber called an *atrium*, where blood enters the heart, and a lower pumping chamber called a *ventricle*, where blood leaves the heart. Thus, the heart is actually a *double pump*. The oxygen-poor blood returning to the right atrium from the body tissues is pumped by the right ventricle (the first pump) into the lungs, where carbon dioxide is exchanged for oxygen. The oxygenated blood moves from the lungs into the left atrium and then into the left ventricle (the second pump), which pumps blood to all parts of the body except the lungs. The circulation between the heart and the lungs is called the **pulmonary circulation** (L. *pulmo*, lung), and the circulation between the heart and the other parts of the body is called the **systemic circulation** [FIGURE 19.1].

STRUCTURE OF THE HEART

The **heart** (Gr. *kardia*, as in *cardiac*, L. *cor*)* is shaped like a blunt cone. It is about the size of the clenched fist of its owner. It averages about 12 cm long and about 9 cm wide (5.0 in. by 3.5 in.). The heart of an adult male weighs about 250 to 390 g (8.8 to 13.8 oz), and the heart of an adult female usually weighs between 200 and 275 g (7.0 to 9.7 oz).

The heart, together with its vessels, takes shape and begins to function long before any other major organ. It begins beating in a human embryo during the fourth week of development, and continues throughout the life

*The word *coronary* is often used in reference to structures and events involving the heart. (A "coronary thrombosis" is a blockage of a coronary artery by a blood clot. The resultant heart attack is informally called a "coronary.") However, the term actually comes from the Latin word *corona*, which means "crown." If you look at FIGURE 19.5A, you will see that the two coronary ateries sit, slightly askew, on top of the heart like a crown.

of a person at a rate of about 70 times a minute, 100,000 times a day, or about 2.5 billion times during a 70-year lifetime. The specialized cardiac muscle tissue that performs this extraordinary feat is found only in the wall of the heart.

The human body contains 4 to 6 L (about 1 to 1.5 gal) of blood, but the heart takes only a little more than a minute to pump a complete cycle of blood throughout the body. In times of strenuous exercise, the heart can quintuple this output. Yet, the heart is a relatively small organ. On the average, the adult heart pumps about 7500 L (2000 gal) of blood throughout the body every day.

Location of the Heart

The heart is located in the center of the chest. It is slanted diagonally, with about two-thirds of its bulk to the left of the body's midline [FIGURE 19.2]. The heart is turned on its longitudinal axis so that the right ventricle is partially in front of the left, directly behind the sternum. The left ventricle faces the left side and the back of the thorax. The heart lies closer to the front of the thorax than the back.

The surfaces of the heart are called the *anterior* (sternocostal, next to the sternum and ribs), *inferior* (diaphragmatic, against the diaphragm), *left side* (pulmonary, next to the left lung), and *posterior* (base) surfaces.

The pointed end of the blunt cone is called the **apex.** It extends forward, downward, and to the left. Normally the apex is located between the fifth and sixth ribs (the fifth intercostal space) on the midclavicular line (a perpendicular line from the middle of the clavicle to the diaphragm). The uppermost part of the heart, the **base,** extends upward, backward, and to the right. Anteriorly, it lies just below the second rib.

The base is in a relatively fixed position because of its attachments to the great vessels, but the apex is able to move. When the ventricles contract, they change shape just enough so that the apex moves forward and strikes the left chest wall near the fifth intercostal space. This thrust of the apex is what we normally feel from the outside as a heartbeat.

Covering of the Heart: Pericardium

The heart does not hang freely in the chest. It hangs by the great blood vessels inside a protective sac called the **pericardium** ("around the heart") [FIGURE 19.3]. This sac is composed of an outer, fibrous layer of connective tissue, the *fibrous pericardium*, and an inner layer of serous

FIGURE 19.1 SYSTEMIC AND PULMONARY CIRCULATIONS

The systemic circulation starts in the heart, flows through the muscles, organs, and tissues of the body, and then returns to the heart. The pulmonary circulation flows only from the heart to the lungs and back to the heart. Oxygenated blood is shown in red, deoxygenated blood in blue.

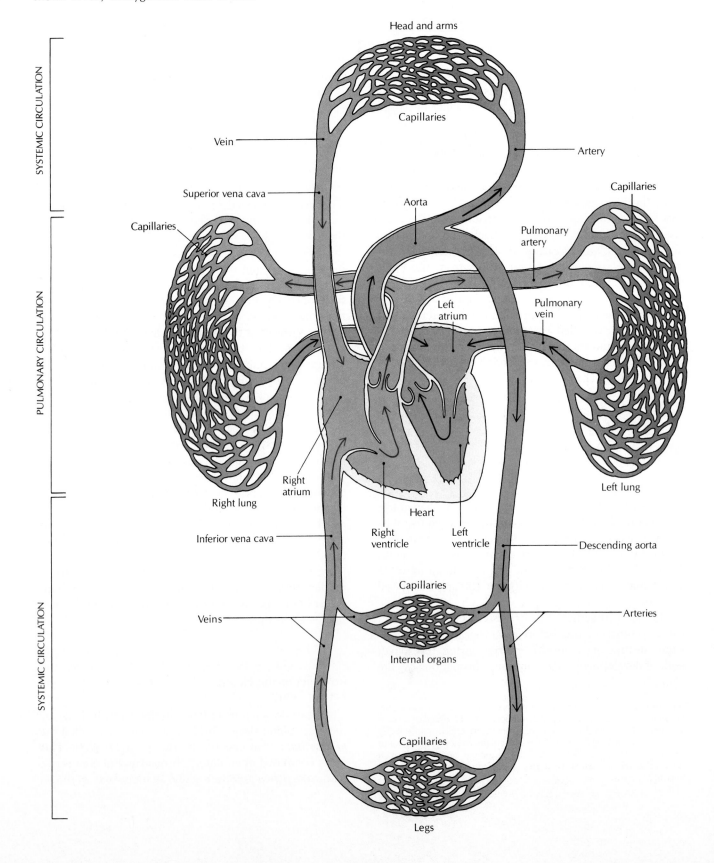

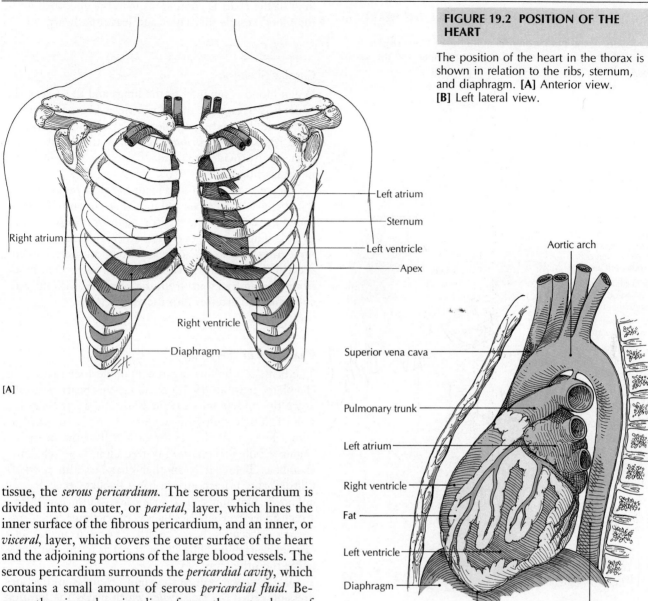

FIGURE 19.2 POSITION OF THE HEART

The position of the heart in the thorax is shown in relation to the ribs, sternum, and diaphragm. [A] Anterior view. [B] Left lateral view.

tissue, the *serous pericardium*. The serous pericardium is divided into an outer, or *parietal*, layer, which lines the inner surface of the fibrous pericardium, and an inner, or *visceral*, layer, which covers the outer surface of the heart and the adjoining portions of the large blood vessels. The serous pericardium surrounds the *pericardial cavity*, which contains a small amount of serous *pericardial fluid*. Because the visceral pericardium forms the outer layer of the heart wall, it is usually called the *epicardium* ("upon the heart").

The heart is held securely in place by connective tissue that binds the pericardium to the sternum, spinal column, and other parts of the chest cavity. The pericardium is tough and inelastic, yet loose-fitting enough to allow the heart to move in a limited way. The serous pericardial fluid moistens the sac and minimizes friction between the membranes as the heart moves during its contraction-relaxation phases.

Q: *During a physical examination, why does a physician tap the chest wall while listening with a stethoscope?*

A: A physician can estimate the size of the heart by tapping the chest wall progressively and listening for sound changes.

Wall of the Heart

The wall of the heart is made up of three layers: (1) the outer *epicardium* (*epi*, upon), (2) the middle *myocardium*, or muscular layer (*myo*, muscle), and (3) the inner *endocardium* (*endo*, inside) [FIGURE 19.3].

If you were to cut away the parietal pericardium, you would see that the surface of the heart itself is reddish and shiny. This shiny membrane is the **epicardium**. Inside the epicardium, and often surrounded with fat, are the main coronary blood vessels that supply and drain blood from the heart.

FIGURE 19.3 COVERING AND WALL OF THE HEART

[A] Layers of the pericardium and the heart wall. [B] Enlarged view of the structure of the pericardium and the ventricular heart wall.

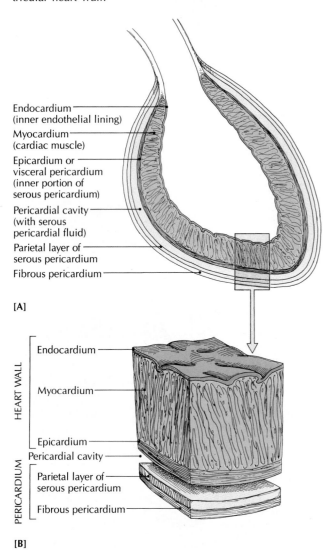

Endocardium
(inner endothelial lining)

Myocardium
(cardiac muscle)

Epicardium or
visceral pericardium
(inner portion of
serous pericardium)

Pericardial cavity
(with serous
pericardial fluid)

Parietal layer of
serous pericardium

Fibrous pericardium

[A]

HEART WALL
Endocardium
Myocardium
Epicardium

PERICARDIUM
Pericardial cavity
Parietal layer of
serous pericardium
Fibrous pericardium

[B]

Directly beneath the epicardium is the middle layer, the ***myocardium*** ("heart muscle"), which is a thick layer of cardiac muscle that gives the heart its special pumping ability. The myocardium has three spiral layers of cardiac muscle, which are attached to a fibrous ring (fibrous trigone) that forms the cardiac skeleton. The spiral is the most effective arrangement for squeezing blood out of the heart's chambers.

The inside cavities of the heart and all of the associated valves and muscles are covered with the ***endocardium*** ("inside the heart"). The endocardium is a thin, fibrous layer lined with simple squamous epithelial tissue (endothelium) and some connective tissue. The endothe-

lium of the heart is continuous with the endothelium of the blood vessels that enter and leave the heart.

Cardiac Skeleton

When blood is pumped to the lungs and the rest of the body, it is wrung out of the ventricles like water from a wet cloth. This wringing motion is possible because the heart has a fibrous ***cardiac skeleton*** of tough connective tissue that provides attachment sites for the valves and muscular fibers.

The cardiac skeleton is made up of a *fibrous trigone* (Gr. *trigonos*, triangular) and four rings, or cuffs, one surrounding each of the heart's four openings [FIGURE 19.6c]. These heart openings are regulated by valves. Four rings of the cardiac skeleton support the valves and prevent them from stretching. The valves prevent blood from flowing backward from the ventricles into the atria and from the arteries into the ventricles.

Chambers of the Heart

The heart is a hollow organ containing four cavities, or chambers [FIGURE 19.4]. Dividing the heart vertically down the middle into a ***right heart*** and a ***left heart*** is a wall of muscle called the ***septum*** (L. *sepire*, to separate with a hedge). At the top of each half of the heart is a chamber called an ***atrium*** (AY-tree-uhm; L. porch, antechamber). Below it is another chamber, the ***ventricle*** ("little belly"). Each atrium has a flaplike appendage of unknown function called an *auricle* ("little ear"). The atria lead to the ventricles by way of openings called *atrioventricular orifices* (L. *or*, mouth + *facere*, to make). In a normal heart, blood flows from the atria to the ventricles, never the other way.

On the septal wall of the right atrium is an oval depression called the *fossa ovalis*, which marks the location of the fetal foramen ovale, which closes at birth. The lining of both atria is smooth except for some ridges called *musculi pectinati* (pectinate muscles), which are formed by parallel muscle bundles that look much like a comb. The musculi pectinati are located on both right and left auricles and on the anterior wall of the right atrium.

The right and left ventricles have walls that contain ridges called *trabeculae carneae* (L. "little beams of flesh"), which are formed by coarse bundles of cardiac muscle fibers. Both right and left ventricles contain papillary muscles and supportive cords called *chordae tendineae*, which are attached to the atrioventricular valves. The upper part of the right ventricular wall, called the *conus*, or *infundibulum*, is smooth and funnel-shaped. This area leads to the pulmonary artery. The left ventricle does not have this conus.

FIGURE 19.4 CHAMBERS OF THE HEART

[A] Schematic drawing of interior of the heart, with blood vessels removed to show the four chambers; anterior view. **[B]** Photograph of heart chambers, internal anterior view.

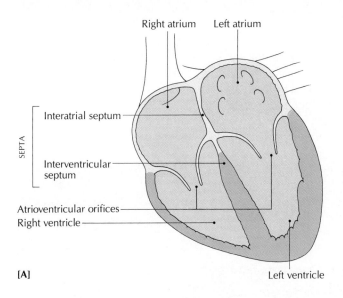

Right atrium   Left atrium

Interatrial septum

SEPTA

Interventricular septum

Atrioventricular orifices

Right ventricle

Left ventricle

[A]

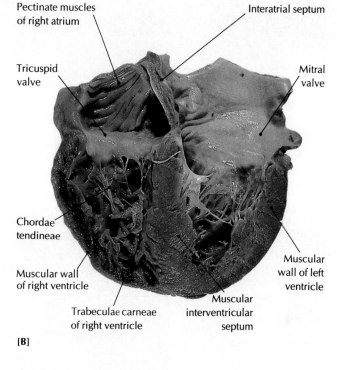

Pectinate muscles of right atrium

Tricuspid valve

Interatrial septum

Mitral valve

Chordae tendineae

Muscular wall of right ventricle

Trabeculae carneae of right ventricle

Muscular interventricular septum

Muscular wall of left ventricle

[B]

Because the heart is a closed system of chambers, when the left and right ventricles pump almost simultaneously, equal amounts of blood must enter and leave the heart. The wall of the left ventricle is thicker, however, because it must be strong enough to supply blood to the systemic circulation, whereas the right ventricle supplies blood only to the lungs. The walls of the atria are thinner than the ventricular walls because less pressure is required to pump blood against less resistance a short distance into the ventricles.

On the surface of the heart are some *sulci* (depressions) that are helpful in locating certain features of the ventricles and atria, as well as coronary vessels [FIGURE 19.5A]. The **coronary sulcus** encircling the heart indicates the border between the atria and ventricles. Embedded in fat within the coronary sulcus are the *right* and *left coronary arteries.*

On the anterior surface of the heart is the **anterior interventricular** ("between the ventricles") **sulcus,** which marks the location of the *interventricular septum,* the anterior part of the septum separating the ventricles. Embedded within the anterior interventricular sulcus are the *anterior interventricular artery* and the *great cardiac vein.* On the posterior surface of the heart [FIGURE 19.5B] is the **posterior interventricular sulcus,** which marks the location of the posterior interventricular septum. Embedded within the posterior interventricular sulcus are the *posterior interventricular descending artery* and the *middle cardiac vein.*

Valves of the Heart

The four heart valves allow blood to flow through the heart in the proper direction. The two *atrioventricular valves* direct the flow of blood from the atria to the ventricles, and the two *semilunar valves* direct the flow of blood from the ventricles to the pulmonary artery (on the way to the lungs) and to the aorta (on the way to the rest of the body) [FIGURE 19.6].

Atrioventricular valves The two **atrioventricular (AV) valves** differ in structure, but they operate in the same way. The right atrioventricular is known as the **tricuspid valve** of the right heart because it consists of three cusps, or flaps (L. *tri,* three + *cuspis,* point) [FIGURE 19.6B]. The left atrioventricular, or **bicuspid valve** of the left heart, has two cusps (*bi* = two). Because the cusps resemble a bishop's miter (a tall, pointed hat), the valve is also called the **mitral** (MY-truhl) **valve.**

Each cusp is a thin, strong, fibrous flap covered by endocardium. Its broad base is anchored into a ring of the cardiac skeleton [FIGURE 19.6C]. Attached to its free end are strong, yet delicate, tendinous cords called *chordae tendineae* (KOR-dee ten-DIHN-ee-ee) [FIGURE 19.6E, F]. The chordae tendineae, which resemble the cords of a parachute, are continuous with the nipplelike papillary muscles (L. *papilla,* nipple) in the wall of the ventricle [FIGURE 19.6E, F].

FIGURE 19.5 EXTERNAL HEART

[A] Anterior external view of the heart, showing surface features and great vessels. [B] Photograph of anterior external heart, actual size.

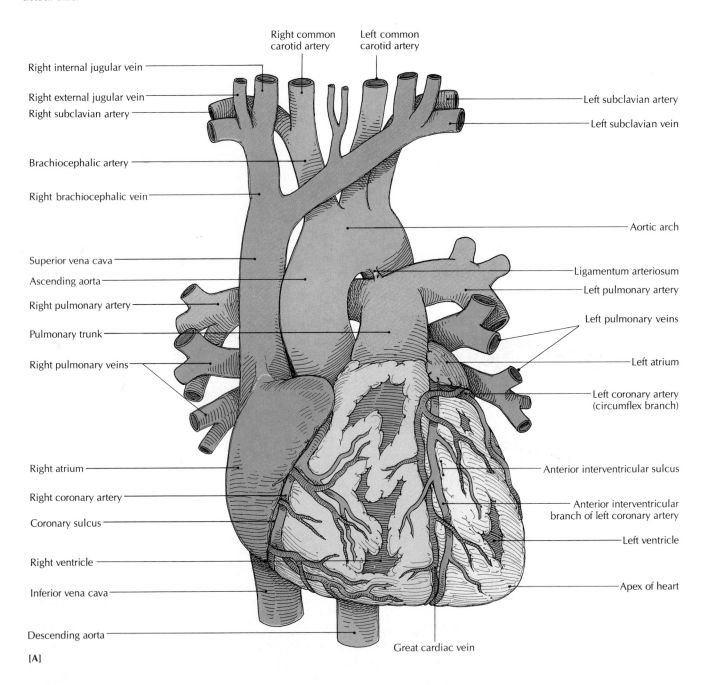

Right common carotid artery

Left common carotid artery

Right internal jugular vein

Right external jugular vein

Right subclavian artery

Left subclavian artery

Left subclavian vein

Brachiocephalic artery

Right brachiocephalic vein

Aortic arch

Superior vena cava

Ligamentum arteriosum

Ascending aorta

Left pulmonary artery

Right pulmonary artery

Left pulmonary veins

Pulmonary trunk

Right pulmonary veins

Left atrium

Left coronary artery (circumflex branch)

Right atrium

Anterior interventricular sulcus

Right coronary artery

Anterior interventricular branch of left coronary artery

Coronary sulcus

Left ventricle

Right ventricle

Inferior vena cava

Apex of heart

Descending aorta

Great cardiac vein

[A]

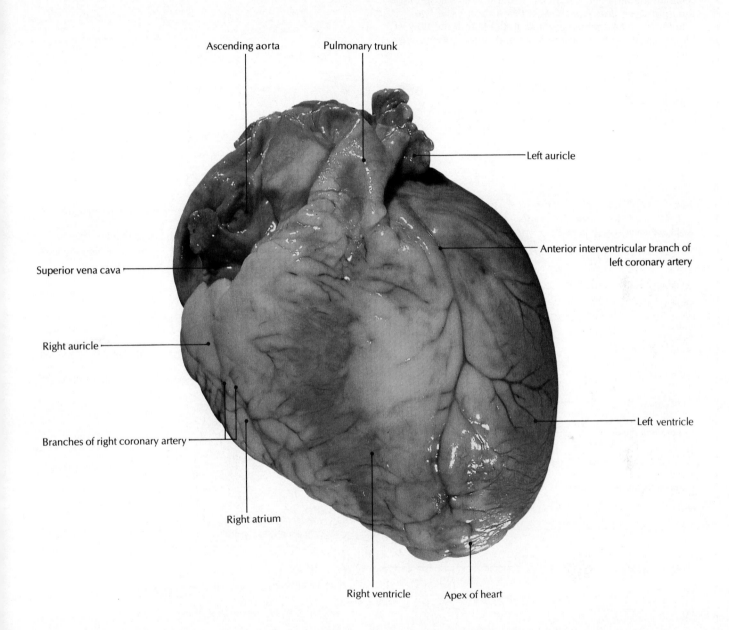

Ascending aorta

Pulmonary trunk

Left auricle

Anterior interventricular branch of
left coronary artery

Superior vena cava

Right auricle

Left ventricle

Branches of right coronary artery

Right atrium

Right ventricle

Apex of heart

FIGURE 19.6 VALVES OF THE HEART

[A] The two atrioventricular valves (tricuspid and bicuspid) separate the atria and ventricles. (Note that one cusp of the tricuspid valve does not show in this section.) The two semilunar valves (pulmonary and aortic) permit the flow of blood out of the heart to the lungs and body; anterior view.

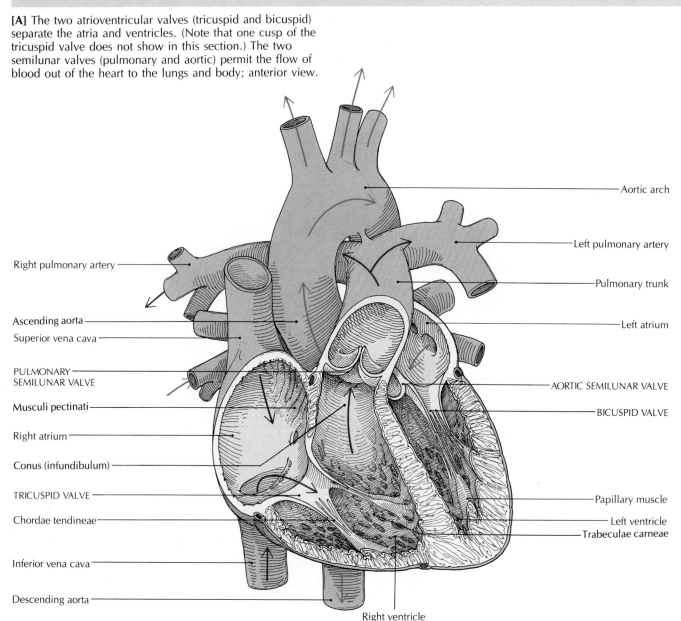

Aortic arch

Left pulmonary artery

Pulmonary trunk

Left atrium

AORTIC SEMILUNAR VALVE

BICUSPID VALVE

Papillary muscle

Left ventricle

Trabeculae carneae

Right pulmonary artery

Ascending aorta

Superior vena cava

PULMONARY SEMILUNAR VALVE

Musculi pectinati

Right atrium

Conus (infundibulum)

TRICUSPID VALVE

Chordae tendineae

Inferior vena cava

Descending aorta

Right ventricle

[A]

When blood is flowing from the atria to the ventricles, the cusps of the AV valves lie open against the ventricular walls. When the ventricles contract, the cusps are brought together by the increasing ventricular blood pressure, and the atrioventricular openings are closed [FIGURE 19.6D]. At the same time, the papillary muscles contract, putting tension on the chordae tendineae. The chordae tendineae pull on the cusps, preventing them from being forced upward into the atria. Otherwise, blood would flow backward from the ventricles into the atria.

Semilunar valves The *semilunar* ("half-moon") *valves* prevent blood in the pulmonary trunk and aorta from flowing back into the ventricles. The left semilunar valve is larger and stronger than the right.

The right semilunar valve is in the opening between the right ventricle and the pulmonary trunk, and is called the *pulmonary semilunar valve.* It allows oxygen-poor blood to enter the pulmonary trunk on its way to the lungs from the right ventricle. The left semilunar valve is the *aortic semilunar valve,* which allows freshly oxygenated blood to enter the aorta from the left ventricle.

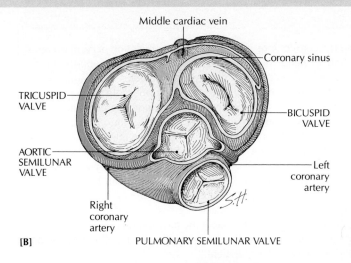

Middle cardiac vein

Coronary sinus

TRICUSPID VALVE

BICUSPID VALVE

AORTIC SEMILUNAR VALVE

Left coronary artery

Right coronary artery

PULMONARY SEMILUNAR VALVE

S.H.

[B]

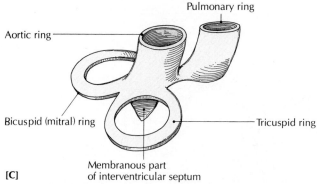

Pulmonary ring

Aortic ring

Bicuspid (mitral) ring

Tricuspid ring

Membranous part of interventricular septum

[C]

[B] The closed valves of the heart, viewed from above; the atria have been removed. **[C]** The fibrous skeleton of the heart. **[D]** The seemingly fragile pulmonary valve opens (left) and closes (right) about once a second. **[E]** Cusps of the tricuspid (right atrioventricular) valve remain open, allowing blood to flow from the right atrium into the right ventricle, when the chordae tendineae and papillary muscles are relaxed. The valve closes, blocking blood flow, when the chordae tendineae are taut and the papillary muscles are contracted. **[F]** A photo inside the right ventricle showing the branching chordae tendineae rising from strong papillary muscles.

[D]   Valve open   Valve closed

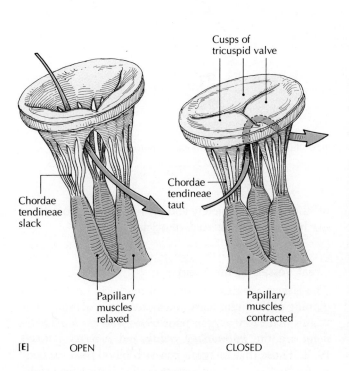

Cusps of tricuspid valve

Chordae tendineae taut

Chordae tendineae slack

Papillary muscles relaxed

Papillary muscles contracted

[E]   OPEN   CLOSED

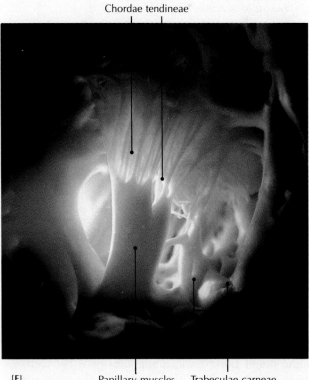

Chordae tendineae

[F]   Papillary muscles   Trabeculae carneae

Cardiac Catheterization

A precise way of looking *inside* the heart is via **cardiac catheterization** (KATH-uh-tuh-rihz-AY-shun). A *catheter* (Gr. something inserted) is a hollow, flexible tube. It is inserted into a vein, usually in the arm, thigh, or neck, and then moved along slowly with the aid of a fluoroscope into the chambers of the heart (see drawings), and, if necessary, into the lung.

With cardiac catheterization, it is possible to measure the blood flow through the heart and lungs, the volume and pressure of blood in the heart, the pressure of blood passing through a heart valve, and the oxygen content of blood in the heart and its major blood vessels. It is also possible to inject harmless radiopaque dyes through the catheter; the dye can be viewed on an x-ray to show septal openings and other congenital disorders in the aorta and pulmonary artery. This technique is called *angiocardiography*. Currently a motion-picture x-ray machine is commonly used with dye injected through cardiac catheterization. This technique, called *cineangiocardiography*, can indicate a heart abnormality by tracing the path of dye through the heart and its blood vessels.

A recent application of cardiac catheterization is *balloon angioplasty*. In this procedure, a catheter with a small, uninflated balloon on its tip is inserted into an obstructed artery and inflated. The material causing the obstruction is pushed aside, the catheter is withdrawn, and blood flow is improved. Also, clot-dissolving enzymes (such as streptokinase and urokinase) can be injected directly into coronary arteries after a heart attack. Cardiac catheterization does not affect the functioning of the heart, and is usually not uncomfortable for the patient.

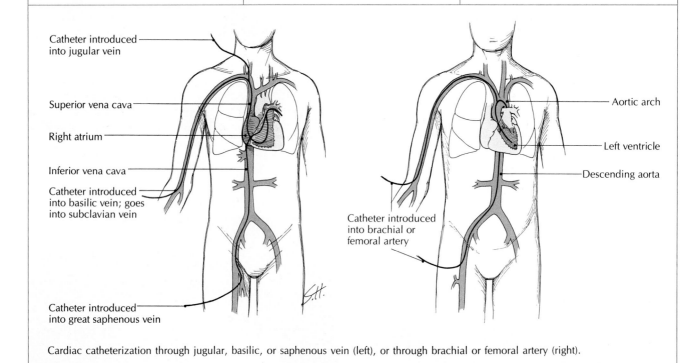

Cardiac catheterization through jugular, basilic, or saphenous vein (left), or through brachial or femoral artery (right).

Each valve contains three cusps. The base of each cusp is anchored to the fibrous ring of the cardiac skeleton. The free borders of the cusps curve outward and extend into the respective artery. These cusps do not have chordae tendineae attached to their free margins, as the atrioventricular valves do. During a ventricular contraction, the blood rushes out of the ventricles and pushes the cusps up and against the wall of each artery, allowing blood to flow into the artery. When the ventricles relax, some blood flows back, filling the space between the cusps and the walls of the aorta and pulmonary trunk, forcing the cusps down and together. This motion closes off the artery completely and prevents blood from flowing back into the relaxed ventricles.

Great Vessels of the Heart

The largest arteries and veins of the heart form the beginning of the pulmonary and systemic circulations. The vessels that carry oxygen-poor blood *from the heart to the lungs* are the **pulmonary trunk** and **arteries** [FIGURE 19.7]. Those that carry oxygen-rich blood *from the lungs to the heart* are the **pulmonary veins.** Draining venous blood from the upper and lower parts of the body to the heart are the *superior vena cava* and the *inferior vena cava,* respectively.* The artery carrying oxygenated blood away from the heart is the *aorta.*

Blood Supply to the Heart

The heart muscle needs more oxygen than any organ except the brain. To obtain this oxygen, the heart must have a generous supply of blood. Like other organs, the heart receives its blood supply from arterial branches that arise from the aorta. The flow of blood that supplies the heart tissue itself is the *coronary circulation.* The heart needs its own separate blood supply because blood in the pulmonary and systemic circulations cannot seep through the lining of the heart (endocardium) from the heart chambers to nourish cardiac tissue. The heart pumps about 380 L (100 gal) to its own muscle tissue every day, or about 5 percent of all the blood pumped by the heart.

Coronary arteries Blood is supplied to the heart by the right and left *coronary arteries,* which are the first branches off the aorta. The aortic semilunar valve partially covers the openings of these arteries while blood is being pumped into the aorta. Blood can pass into the

*The superior and inferior veins are called *cava* (L. hollow, cavern) because of their great size.

coronary arteries only when the left ventricle is relaxed and the cusps do not cover the openings. The branching of the coronary arteries varies from person to person, but the following arrangement is the common one [FIGURE 19.7]:

1 The *right coronary artery* arises from the aorta near the right cusp of the aortic semilunar valve and runs between the right atrium and ventricle. Its major branches include (1) the nodal branch, which supplies the sinoatrial node; (2) the right marginal branch, which feeds the dorsal and ventral surfaces of the right ventricle; (3) a nodal branch to the atrioventricular node; (4) small branches to the right branch of the AV bundle; and (5) branches on the posterior aspect of the heart that anastomose (join) with terminals of the anterior descending branch and the circumflex artery of the left coronary artery.

2 The *left coronary artery* arises from the aorta near the left cusp of the aortic semilunar valve; it shortly divides into two branches. (1) The anterior interventricular artery runs along the anterior interventricular sulcus and anastomoses with the terminals of the right coronary artery in the vicinity of the apex. Small branches supply the left branches of the AV bundle. (2) The circumflex artery courses in the coronary sulcus to the lateral and posterior aspects of the heart. Its terminal branches anastomose with the right coronary artery.

FIGURE 19.7 CORONARY VEINS AND ARTERIES

Anterior view.

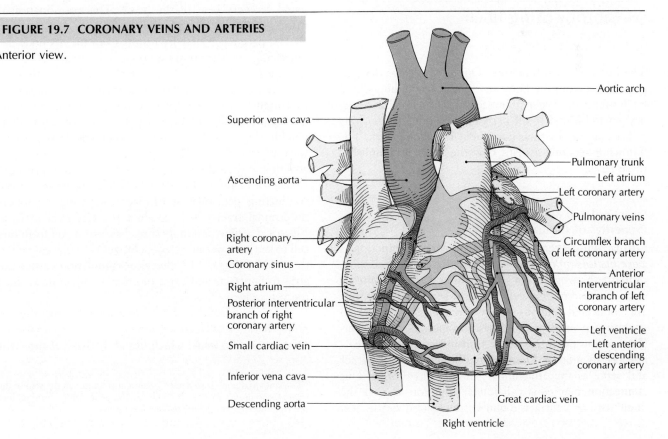

Superior vena cava

Ascending aorta

Right coronary artery

Coronary sinus

Right atrium

Posterior interventricular branch of right coronary artery

Small cardiac vein

Inferior vena cava

Descending aorta

Aortic arch

Pulmonary trunk

Left atrium

Left coronary artery

Pulmonary veins

Circumflex branch of left coronary artery

Anterior interventricular branch of left coronary artery

Left ventricle

Left anterior descending coronary artery

Great cardiac vein

Right ventricle

Coronary sinus and veins Most of the cardiac veins drain into the **coronary sinus,** a large vein located in the coronary sulcus on the posterior surface of the heart [FIGURE 19.7]. The coronary sinus empties into the right atrium. The coronary sinus receives blood from the *great cardiac vein,* which drains the anterior portion of the heart, the *middle cardiac vein* and *oblique vein,* which drain the posterior aspect of the heart, and the *small cardiac vein,* which drains the right side of the heart. Among the other veins is the *anterior cardiac vein,* which drains directly into the right atrium. Tiny veins called *thebesian veins* open directly into each of the four heart chambers.

PHYSIOLOGY OF THE HEART

The heart has two functions. One is to receive oxygen-poor blood from the body and send it to the lungs for a fresh supply of oxygen. The second is to pump the newly oxygenated blood to all parts of the body, where body cells can use it for their day-to-day metabolic work. The following sections explain how the heart accomplishes these tasks.

Structural and Metabolic Properties of Cardiac Muscle

Cardiac muscle cells in the atria or ventricles function as a *coordinated unit* in response to physiological stimulation, rather than as a group of separate units as skeletal muscle does. Cardiac muscle cells act in this way because they are connected end-to-end by *intercalated disks,* which contain gap junctions and desmosomes. Gap junctions allow action potentials to be transmitted from one cardiac cell to another. Desmosomes hold the cells together and serve as the attachment sites for myofibrils. This connection maintains cell-to-cell cohesion so that the "pull" of one contractile unit is transmitted to the next

one. Over all, this series of interconnected cells forms a latticework called a *syncytium* (sihn-SIE-shum). The importance of a syncytial muscle mass is that when either the atrial or ventricular muscle mass is stimulated, the action potential spreads over the entire atrial or ventricular syncytium and causes the muscle cells in the entire chamber muscle mass to contract in near unison.

Cardiac muscle contains a large number of mitochondria, which provide a constant source of ATP energy for the hardworking heart muscle. It also has an abundant blood supply and a high concentration of *myoglobin,* a muscle protein that stores oxygen.

Conduction System of the Heart

You saw in Chapter 10 that skeletal muscles of the body cannot contract unless they receive an electrochemical stimulation from the nervous system. Although the central nervous system does exert some control over the heart, cardiac muscle has its own built-in electrochemical activator, called a **pacemaker,** and can initiate a beat independently of the central nervous system.

The electrical stimulation that starts the heartbeat and controls its rhythm originates in the superior wall of the right atrium, near the entry point of the superior vena cava; it originates in a mass of specialized heart muscle tissue called the **sinoatrial, or SA, node.*** Although the muscles of the atria are not continuous with the muscles of the ventricles, the atria and ventricles must be coordinated at every beat of the heart. This coordination is made possible by the SA node.

The pacemaker activity results from the SA node depolarizing spontaneously at regular intervals, 70 to 80 times a minute. The SA node makes contact with adjacent atrial muscle cells and causes them to be depolarized by conduction of ions through the gap junctions of the intercalated disks. These atrial cells, in turn, cause their neighboring cells to start action potentials. In this way, a wave of electrical activity spreads throughout the right atrium and then the left atrium, much like the spreading ripples on a pond. In addition, some specialized atrial conducting pathways of cardiac conducting myofibers (internodal tracts) help conduct the electrical activity. Electrical activity is simply spread as ions travel from one cell to the next, as one cell depolarizes the adjacent cell to threshold voltage. The electrical stimulation causes the atria to contract, and extra blood is pumped down into the ventricles.

A few hundredths of a second after leaving the SA node, the wave of electrical activity reaches the **atrioventricular, or AV, node,** which lies at the base of the right

---

*The sinoatrial node is so named because it is located in the wall of the *sinus venosus* during early embryonic development, and is absorbed into the right atrium along with the sinus venosus.

FIGURE 19.8 CONDUCTION SYSTEM OF THE HEART

The impulse-conducting system of the heart is composed of specialized cardiac muscle cells within the myocardium. This system, through its sensitivity and autorhythmicity, synchronizes the events of the cardiac cycle. The system consists of impulses conveyed in an orderly sequence from (1) the sinoatrial (SA) node that sweep across the cardiac muscle of the atria after the atria contract, (2) the atrioventricular (AV) node, and then (3) descends via the left and right AV bundles (bundles of His) and spreads throughout the myocardium of the ventricles. Ventricular contraction results.

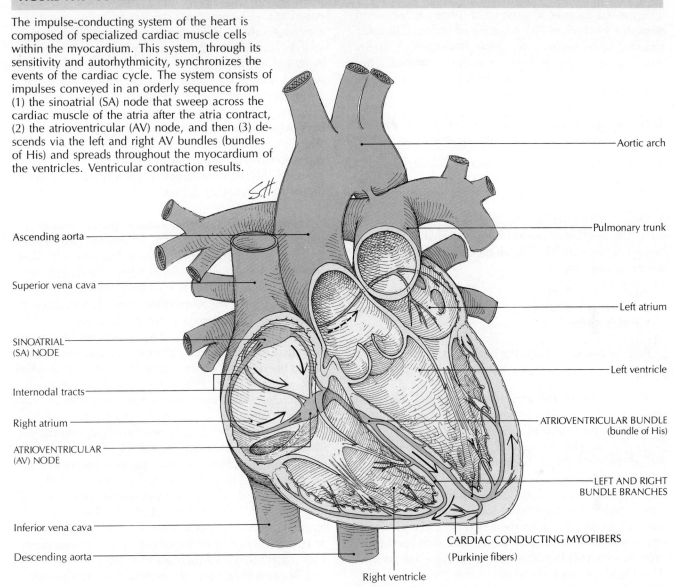

atrium, between the atrium and ventricle [FIGURE 19.8]. The AV node delays the electrical activity another few hundredths of a second before allowing it to pass into the ventricles. This delay allows time for the atria to force blood into the ventricles.

From the AV node, a group of conducting fibers in the interventricular septum called the *atrioventricular bundle* (*bundle of His*) runs for a short distance, and then divides into two branches that spread along the septum, one branch for each ventricle. Because a sheet of connective tissue separates the atria from the ventricles, the atrioventricular bundle is the only electrical link between the atria and ventricles. When the branches reach the apex of the ventricles, they divide into hundreds of tiny special-

ized cardiac muscle fibers called *cardiac conducting myofibers,* or *Purkinje fibers,* that follow along the muscular walls of the ventricles.* Such an arrangement directs the electrical impulse in a definite pathway so that it can make contact with all areas of the ventricular muscle. Therefore, an impulse traveling along the cardiac conducting myofibers is conducted rapidly and directly into the cardiac muscle, and each syncytium contracts in unison with the others, producing a coordinated pumping effort.

*Cardiac conducting myofibers are actually modified cardiac muscle cells that conduct action potentials at higher velocities than normal cardiac muscle cells do. They are continuous with the myocardial muscle cells, but have only feeble contractile properties.

Cardiac Cycle

The **cardiac cycle** is a carefully regulated sequence of steps that we think of as the beating of a heart. The cycle includes the *contraction*, or **systole** (SISS-toe-lee), of the atria and ventricles, and the *relaxation*, or **diastole** (die-ASS-toe-lee), of the atria and ventricles. The cardiac cycle proceeds in four stages:

1 During *atrial systole* (which lasts 0.1 sec), both atria contract, forcing extra blood into the ventricles.

2 During *ventricular systole* (0.3 sec), both ventricles contract, forcing blood out through the pulmonary artery (to the lungs) and aorta (to the rest of the body).

3 During *atrial diastole* (0.7 sec), or relaxation of the atria, the ventricles remain contracted, and the atria begin refilling with blood from the large veins leading to the heart from the body.

4 *Ventricular diastole* (0.5 sec), or relaxation of the ventricles, begins before atrial systole, allowing the ventricles to fill with blood from the atria.

Path of blood through the heart   The path of blood flow through the heart may be described as follows:

1 ***Blood enters the atria*** [FIGURE 19.9A]. Oxygen-poor blood from the body flows into the right atrium at about the same time as newly oxygenated blood from the lungs flows into the left atrium: (a) The *superior vena cava* returns blood from all body structures above the diaphragm (except the heart and lungs). (b) The *inferior vena cava* returns almost all blood to the right atrium from all regions below the diaphragm. (c) The *coronary sinus* returns about 85 percent of the blood from the coronary circulation to the right atrium. (d) The *pulmonary veins* carry oxygenated blood from the lungs to the left atrium. The blood entering the right atrium (blue in FIGURE 19.9A) is low in oxygen and high in carbon dioxide because it has just returned from supplying oxygen to the body tissues. The blood entering the left atrium (red in FIGURE 19.9A) is rich in oxygen because it has just passed through the lungs, where it has picked up a new supply of oxygen. (This is the only place in adults where *venous* blood is highly *oxygenated* because it is coming to the heart directly from the lungs.)

2 ***Blood is pumped into the ventricles*** [FIGURE 19.9B]. The heart's natural pacemaker (the SA node) fires an electrical impulse that coordinates the contractions of both atria (atrial systole). Blood is forced through the one-way AV valves into the relaxed ventricles.

3 ***The ventricles, filled with blood, hesitate for an instant*** [FIGURE 19.9C].

4 ***The ventricles contract, sending blood to the body (systemic circulation) and lungs (pulmonary circulation)***

[FIGURE 19.9D]. The ventricular contraction creates a pressure wave that closes the atrioventricular valves between the atria and ventricles, while opening the two semilunar valves leading out of the ventricles. The right ventricle forces blood low in oxygen out through the right and left pulmonary arteries to the lungs. The left ventricle pumps the newly oxygenated blood through the aortic semilunar valve into the aorta. The aorta branches into the ascending and descending arteries that carry oxygenated blood to all parts of the body [FIGURE 19.9D]. The left and right ventricles pump almost simultaneously, so that equal amounts of blood enter and leave the heart. By this time, the atria have already started to refill, preparing for another cardiac cycle.

Heart sounds   Detectable **heart (valve) sounds** are produced with each heartbeat. Physicians first studied the sounds by putting their ears to the patient's chest, later by listening with a solid wooden stick shaped like a trumpet, and then by listening with a stethoscope. These sounds represent the *auscultatory events* (L. *auscultare*, to listen to) of the cardiac cycle, and can be heard best in the areas indicated in FIGURE 19.10. Heart sounds can be amplified and recorded by placing an electronically amplified microphone on the chest. The recording is called a *phonocardiogram*. It shows heart sounds as waves.

There are four heart sounds associated with the cardiac cycle, although only the first and second sounds (traditionally referred to as *lubb* and *dupp*, respectively) can be heard easily with a stethoscope. The ***first heart sound*** is more complex, lower in pitch, and lasts longer than the second sound. The first sound occurs when the ventricles have been filled, and the atrioventricular valves of both atria close, the aortic and pulmonic valves open, and blood begins to be ejected into the aorta and pulmonary arteries. The first heart sound can be heard most clearly with a stethoscope in the area of the apex.

The ***second heart sound*** is high in pitch. It is produced by the slamming of the semilunar valves after the ventricles have pumped their blood to the lungs and body, and have begun to contract. Ordinarily, the aortic valve closes a split second before the pulmonic valve, but in a healthy heart the two sounds are usually perceived as one. The second heart sound is heard best over the second intercostal space, where the aorta is closest to the surface. Immediately after the second sound there is a short interval of silence.

A low-pitched ***third heart sound*** is heard occasionally. It is caused by the vibration of ventricular walls after the atrioventricular valves open and blood gushes into the ventricles. The sound is heard best in the tricuspid area.

A ***fourth heart sound*** is usually not heard with an unamplified stethoscope in normal hearts because of its frequency. It is caused by blood rushing into the ventricles. It is best heard in the mitral area.

FIGURE 19.9 THE CARDIAC CYCLE AND THE PATH OF BLOOD THROUGH THE HEART

[A] Blood enters the atria. [B] Blood is pumped into the ventricles. [C] The ventricles relax. [D] The ventricles contract, pumping blood through the pulmonary trunk and aorta to the lungs and to the rest of the body.

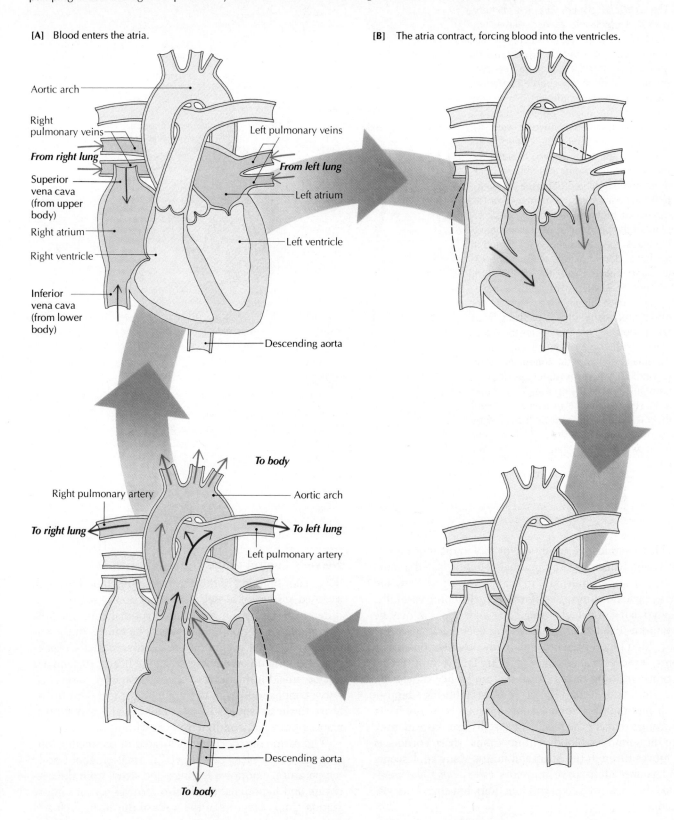

[A] Blood enters the atria.

[B] The atria contract, forcing blood into the ventricles.

Aortic arch

Right pulmonary veins

From right lung

Superior vena cava (from upper body)

Right atrium

Right ventricle

Inferior vena cava (from lower body)

Left pulmonary veins

From left lung

Left atrium

Left ventricle

Descending aorta

To body

Right pulmonary artery

Aortic arch

To right lung

To left lung

Left pulmonary artery

Descending aorta

To body

[D] The ventricles contract, pumping blood into the pulmonary trunk and aorta.

[C] The filled ventricles relax.

Artificial Pacemaker

The **artificial pacemaker** is a battery-operated electronic device that is implanted in the chest with electrical leads to the heart of a person whose natural pacemaker (the SA node) has become erratic. In a relatively simple operation, electrode leads (catheters) from the pacemaker are passed beneath the skin, through the external jugular vein (or other neck vein), into the superior vena cava, into the right atrium, through the tricuspid valve, and into the myocardium of the right ventricle (see drawing). If the patient's veins are damaged or too narrow to receive the typical chest implant of the pacemaker with its connecting wires, the pacemaker is implanted in the left abdominal area, with a connecting lead inserted into the epicardium.

Three basic types of artificial pacemakers are available. The first type delivers impulses when the patient's heart rate is slower than that set for the pacemaker, and shuts off when the natural pacemaker is working adequately. The second is a fixed-rate model that delivers constant electrical impulses at a preset rate. The third is a transistorized model that picks up impulses from the patient's SA node and operates at 72 beats per minute when the natural pacemaker fails.

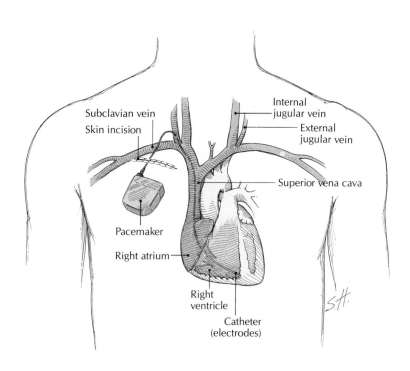

An implanted artificial pacemaker.

Heart sounds are an important tool in diagnosing valvular abnormalities. Any unusual sound is called a *murmur*, but not all murmurs indicate a valve problem, and many have no clinical significance. By listening carefully, a physician can detect *resting heart murmurs* that may be symptoms of valvular malfunctions, congenital heart disease, high blood pressure, and many other serious problems. If the atrioventricular valves are faulty, for instance, a gentle blowing or hissing sound can be heard between the first and second heart sounds: lubb-hiss-dupp — lubb-hiss-dupp — . Some whooshing sounds are not heart murmurs that are the result of heart disease. Instead, they are the sounds of blood swirling around sharp corners as it moves through the heart and lungs. Many adolescents and young adults have murmurs called *functional murmurs* (because the valves still function), but soon outgrow them.

Nervous Control of the Heart

The major function of the heat is to pump blood through a closed system of vessels. This function is regulated at different levels by the cerebrum, hypothalamus, medulla oblongata, and autonomic nerves. The effects of the autonomic nervous system on the heart are strictly *regulatory*, speeding up or slowing down the heart rate, and are not essential for the heart to beat. (If you were to sever all nerve connections from the brain and spinal cord to the heart, the heart would still beat. This explains why a transplanted heart can continue to beat without innervation.)

The main control center is located in the medulla oblongata, which receives sensory information about body temperature, emotions, feelings, and stress from the cerebrum and hypothalamus. It also receives sensory information from receptors in the walls of the aortic arch and

FIGURE 19.10 LOCATION OF HEART SOUNDS

The chest areas where adult heart sounds can be heard most clearly. The ribs are numbered for clarity.

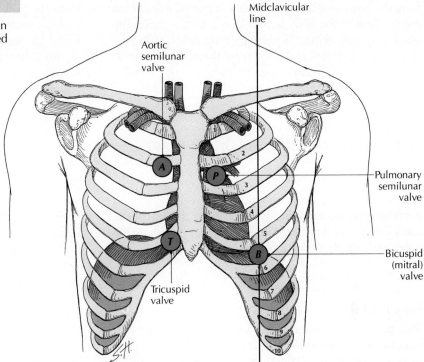

carotid artery sinuses. Information about the changing chemical composition of blood is transmitted from chemically sensitive receptors called *chemoreceptors*, and information about how arteries are stretched by changes in blood pressure is received from pressure-sensitive clusters of cells called *baroreceptors*. Both types of receptors help the body to make the adjustments necessary to restore normal conditions.

The upper part of the medulla oblongata contains an area called the *cardioacceleratory center (CAC)*, or *pressor center*; the lower part contains the *cardioinhibitory center (CIC)*, or *depressor center*. Because the neurons of both centers interact to regulate the heart rate, they are collectively called the **cardioregulatory center.**

Endocrine Control of the Heart

Chemical transmitters are used by the nervous system to regulate the activity of the heart. Any generalized sympathetic activity within the autonomic nervous system affects the medullary region of the adrenal glands. When cells of the adrenal medulla are stimulated, they produce norepinephrine, some of which enters the blood. The major portion of norepinephrine, however, is converted in the adrenal medulla into epinephrine, which then enters the bloodstream, and along with norepinephrine, increases the heart rate and strength of contraction.

A S K   Y O U R S E L F

1 What are some of the properties of cardiac muscle?

2 How does an electrical impulse flow through the heart?

3 What events occur during the cardiac cycle?

4 Why is the SA node considered to be the heart's natural pacemaker?

5 What role does the autonomic nervous system play in regulating cardiac output?

THE EFFECTS OF AGING ON THE HEART

Although the heart usually does not decrease in size with age, its pumping efficiency is reduced because some muscle and valve tissues are replaced by fibrous tissue. Blood pressure is usually raised, and the heart rate does not compensate as well in response to stress as it used to. If hardening of the arteries occurs, the arterial walls become less elastic, and the heart must work harder to pump the same amount of blood. As a result, the heart becomes enlarged, and if the heart reaches its maximum limits, heart failure occurs. Congestive *heart failure* occurs when a weakening or scarring of cardiac muscle affects the heart to the point where it is unable to pump as much blood as the body needs. At age 45, the maximum heart output is about 94 percent of optimum (100 percent at age 25), and it decreases to about 87 percent at age 65, and 81 percent at age 85.

DEVELOPMENTAL ANATOMY OF THE HEART

The embryonic circulatory system forms and becomes functional before any other system, and by the end of the third week after fertilization the system is fulfilling the nutritional needs of the embryo.

Early Development of the Heart

The heart begins as a network of cells, from which two longitudinal **heart cords** (also called *cardiogenic cords*) are formed about 18 or 19 days after fertilization. On about day 20, the heart cords develop canals, forming **heart tubes** (also called *endocardial heart tubes*). On days 21 and 22 the heart tubes fuse, forming a single median **endocardial heart tube** [FIGURE 19.11A, B]. Immediately following, the endocardial heart tube elongates; dilations and constrictions develop, forming first the **bulbus cordis, ventricle,** and **atrium,** and then the **truncus arteriosus** and **sinus venosus** [FIGURE 19.11C]. At this time (day 22), the sinus venosus develops *left and right horns*, the venous end of the heart is established as the **pacemaker,** heart contractions resembling peristaltic waves begin in the sinus venosus, and on about day 28 coordinated contractions of the heart begin to pump blood in one direction only.

Each horn of the sinus venosus receives an **umbilical vein** from the chorion (primitive placenta), a **vitelline vein** from the yolk sac, and a **common cardinal vein** from the embryo itself [FIGURE 19.11D]. Eventually, the left horn forms the coronary sinus, and the right horn becomes incorporated into the wall of the right atrium.

Growth is rapid, and because the heart tube is anchored at both ends within the pericardium, by day 25 the tube bends back upon itself, forming first a U-shape and then an S-shape. The tube develops alternate dilations and constrictions along with further bending, until it begins to resemble the fully developed heart [FIGURE 19.11C, D]. During the first weeks of embryonic growth, the heart is about nine times as large in proportion to the whole body as it is in the adult. Also, its position is higher in the thorax than the permanent position it will assume later.

Partitioning of the Heart

On about day 25, partitioning of the atrioventricular orifice, atrium, and ventricle begins, and is completed about 10 to 20 days later [FIGURE 19.11E]. Most of the wall of the left atrium develops from the pulmonary vein, and as mentioned above, the right horn of the sinus venosus becomes incorporated into the wall of the right atrium. By about day 28, the heart wall has formed its three layers (endocardium, pericardium, epicardium). At about day 32, the **interatrial septum** forms, dividing the single atrium into left and right atria. An opening in the septum, the **foramen ovale,** closes at birth. The partitioning of the single ventricle into left and right ventricles begins with the formation of an upward fold in the floor of the ventricle, called the **interventricular septum** [FIGURE 19.11E], and is completed about day 48. The development of the atrioventricular valves, chordae tendineae, and papillary muscles proceeds from about the fifth week until the fifth month. The external form of the heart continues to develop from about day 28 to day 60.

WHEN THINGS GO WRONG

Heart disease is a commonly used term for any disease that affects the heart. A more appropriate term is *cardiovascular disease*, which includes both heart and blood vessel disorders. About 40 million Americans have some form of cardiovascular disease, and it is responsible for more deaths than all other causes of death *combined*.* Although the American death rate from cardiovascular dis-

ease is decreasing, it is still one of the highest in the world. This year more than 1.5 million Americans will suffer heart attacks, and of the 550,000 who survive, 100,000 will have another, fatal, attack within a year.

This section concentrates on cardiovascular diseases that involve the heart primarily. The next chapter discusses diseases that originate in the blood vessels.

Degenerative Heart Disorders

Degenerative heart disorders result from the deterioration of tissues or organs of the cardiovascular system.

---

*The cardiovascular diseases that cause the most deaths each year are heart attack (56.3 percent), stroke (17.4 percent), hypertensive disease (3.1 percent), rheumatic fever and rheumatic heart disease (0.8 percent), all other cardiovascular diseases (22.4 percent).

FIGURE 19.11 EARLY EMBRYONIC DEVELOPMENT OF THE HEART

[A, B] The primitive heart tubes fuse together and form a single endocardial heart tube during days 21 and 22 after fertilization; ventral views. [C] On about day 22 or 23, the first major structures form, and the heart tube begins to bend and twist; ventral view. [D] On about day 25, the tube has formed an S-shape. [E] At about 32 days, the inte-

rior partitions can be seen clearly in this frontal section. [F] A frontal view at about 5 weeks shows the remaining three pairs of aortic arches. [G] A frontal section at 6 weeks shows the aorta and pulmonary trunk after the bulbus cordis has been incorporated into the ventricles to become the infundibulum. Black arrows indicate the movement of the heart, and red arrows indicate the flow of blood.

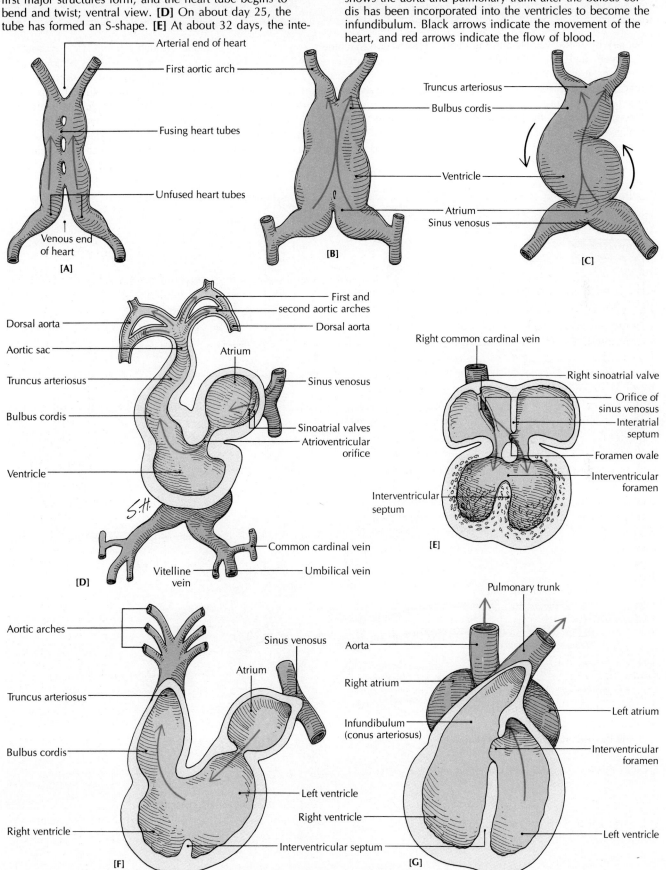

Heart Surgery

Surgery on blood vessels and the heart has been done since the early 1930s. However, it was not until the heart-lung machine was introduced in 1953 that true open-heart surgery was possible. In December of 1982, the first operation in which an artificial heart was placed into a person took place.

Heart-Lung Machine

Open-heart surgery is possible only if the heart is quiet and empty of blood. The *heart-lung machine* takes over the job of pumping *and* oxygenating blood while open-heart surgery is in progress (see drawing A). Tubes are inserted into the inferior and superior vena cavae to lead blood through a pump and oxygenator, where carbon dioxide is removed and oxygen is added, just as occurs in the lungs. A pump then returns the oxygenated blood into the arterial circulation by way of the aorta or one of its branches.

While surgery is taking place and blood is being circulated through the

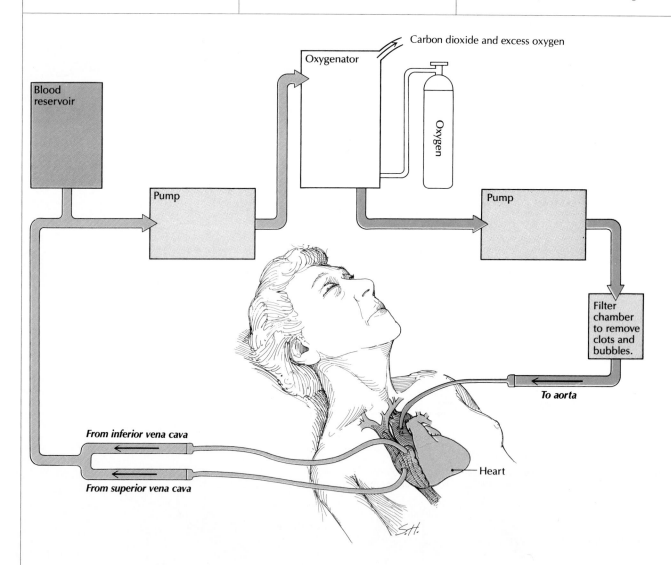

[A] The heart-lung machine (simplified).

heart-lung machine, blood clotting is prevented by introducing the anticoagulant heparin into the circulation. When the natural circulation through the heart is restored, the effect of heparin is reversed by the introduction of protamine sulfate.

An important technique used with the heart-lung machine is *hypothermia,* in which the patient's body is cooled enough to induce ventricular fibrillation, thus providing a quiet state of the heart. Recent techniques allow cooling of only the heart. The newer techniques elimi-nate all heart activity during the operation, and reduce metabolism so that absolutely no strain is placed on the myocardium during surgery.

Coronary Bypass Surgery

The most common serious heart disease is obstruction or narrowing of the coronary arteries by atherosclerosis. When coronary arteries become blocked, the flow of blood to the heart is reduced or cut off completely, and angina pectoris or myocardial infarction may result. In **coronary bypass surgery,** the surgeon removes the diseased portion of the coronary artery and replaces it with a segment of the internal thoracic artery or saphenous vein from the patient's own body (see drawing B). (Because the replacement vessel is taken from the patient's body, there is virtually no danger of tissue rejection.) One end of the vein is sutured to the aorta, and the other end is sutured to a coronary artery beyond the point of obstruction. Thus, the diseased artery is "bypassed," and normal blood flow is re-established.

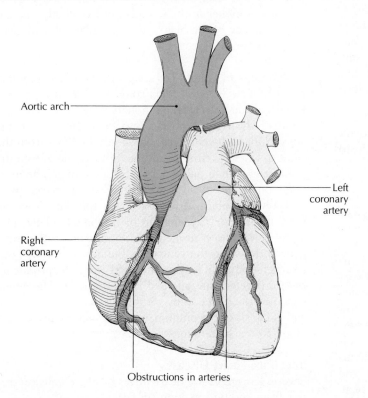
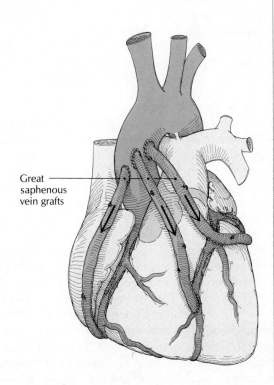

[B] Coronary arteries before (left) and after (right) coronary bypass surgery.

Myocardial infarction (heart attack) When the blood flow through a coronary artery is reduced for any reason (usually because of a clot or plaque* build-up), the myocardium is deprived of oxygen, and begins to die. The result is a ***myocardial infarction*** (L. *infercire*, to stuff), or *heart attack*. An *infarct* is an area of tissue that has died because of an inadequate blood supply. Although a heart attack is always serious, it is not always diagnosed easily. The usual symptoms include pain in the midchest (*angina pectoris*, described on page 501), which travels up the neck or out through the shoulders and arms, especially the left. Sometimes there may be only a shortness of breath, or no symptoms at all.

Congestive heart failure *Congestive heart failure* (CHF) occurs when either ventricle fails to pump blood out of the heart as quickly as it enters the atria, or when the ventricles do not pump equal amounts of blood. For example, left heart failure occurs if the left ventricle is weakened and pumps less blood than normal; the right ventricle will then pump more blood *into* the lungs than can be pumped by the left ventricle. As a result, the lungs become engorged with blood in a condition called *pulmonary edema*, which also results in kidney failure. The old-fashioned term *dropsy* refers to the accumulation of fluid in the abdomen and legs that accompanies heart failure.

Valvular Heart Diseases
As you saw earlier, the two major heart sounds provide information about the heart's valves. Abnormal heart sounds, or *murmurs*, may be indicative of a **valvular heart disease,** in which one or more cardiac valves operate improperly. Certain types of valvular malfunction may produce *regurgitation* (backflow of blood through an incompletely closed valve) or *stenosis* (an incompletely opened valve). Either type of malfunction can lead to congestive heart failure. Some forms of valvular heart disease are (1) *mitral insufficiency* (blood from the left ventricle flows back into the left atrium during systole), (2) *mitral stenosis* (narrowing of the valve reduces flow from the left atrium to the left ventricle), (3) *aortic insufficiency* (blood flows back into the left ventricle during diastole), (4) *aortic stenosis* (pressure in the left ventricle increases in response to a narrowed valve opening, and ischemia—localized tissue anemia—results because of an insufficient oxygen supply), (5) *pulmonary insufficiency* (blood flows back from the pulmonary artery into the right ventricle during diastole), (6) *pulmonary stenosis* (blood is prevented from leaving the right ventricle, causing ventricular hypertrophy and eventual failure), (7) *tricuspid insufficiency* (blood flows back into the right atrium during systole), and (8) *tricuspid stenosis* (blood is obstructed from flowing from the right atrium to the right ventricle). Valvular heart diseases may sometimes be corrected with the surgical substitution of an artificial valve.

*Plaque in an artery is a build-up of cholesterol and other fatty acids.

Congenital Heart Diseases
A congenital disease is one that is present at birth, but is not necessarily hereditary.

Ventricular septal defect *Ventricular septal defect* is the most common congenital heart disease. It is an opening that is present at birth in the ventricular septum that allows blood to move back and forth between the ventricles. Small openings usually either close naturally, before permanent damage occurs, or are repaired surgically. Large openings are not always reparable, and patients often die in their first year from biventricular congestive heart failure or secondary complications.

Interatrial septal defect In an ***interatrial septal defect,*** the foramen ovale between the two atria fails to close at birth, and the child is born with an opening between the right and left atria that allows blood to flow through the foramen. After birth, blood usually flows from the left atrium to the right atrium, because atrial pressure is higher in the left atrium. Eventually, the entire right side of the heart becomes enlarged, and heart failure occurs.

Tetralogy of Fallot ***Tetralogy of Fallot*** is a combination of four (Gr. *tetra*, four) congenital defects: (1) ventricular septal defect, (2) pulmonary stenosis, (3) enlargement of the right ventricle, and (4) emergence of the aorta from both ventricles. Blood usually flows from the right to the left ventricle, although it may flow from the left to right when pulmonary stenosis is mild. The most common symptom of tetralogy of Fallot is *cyanosis* (Gr. dark blue), a bluish discoloration of the skin, resulting from inadequate oxygenation of the blood. Ventricular septal defect allows oxygenated blood to mix with deoxygenated blood in the ventricles. Pulmonary stenosis obstructs blood from leaving the right ventricle, producing the third defect, an enlarged right ventricle. The emergence of the aorta from within the ventricles, coupled with stenosis of the pulmonary artery, allows insufficient blood to reach the lungs.

Infectious Heart Diseases
Severe damage to the heart valves or heart walls can result from certain infectious diseases.

Rheumatic fever and rheumatic heart disease *Rheumatic fever* is a severe infectious disease occurring mostly in children. It is characterized by fever and painful inflammation of the joints, and frequently results in permanent damage to the heart valves. Rheumatic fever is a hypersensitive reaction to a specific streptococcal bacterial infection in which antibodies cause inflammatory reactions in the joints and heart. ***Rheumatic heart disease*** refers to the secondary complications that affect the heart. They include *pancarditis* (*pan* = all, because it encompasses pericarditis, myocarditis, and endocarditis) in the early acute phase, and valvular heart disease later.

Pericarditis, myocarditis, endocarditis Inflammation of the three layers of the heart wall is called, from the outside in, pericarditis, myocarditis, and endocarditis.
Pericarditis is an inflammation of the parietal and/or visceral pericardium. When it is caused by a bacterial, fungal, or viral infection, it is known as *infectious pericarditis*. It may also be caused by uremia, high-dose radiation, rheumatic fever, cancer, drugs, and trauma such as myocardial infarction. Pain is typical, especially when the heart presses against its covering membranes.
Myocarditis is an inflammation of the myocardium. It may result from viral, bacterial, parasitic, or other infections; hypersensitive immune reactions; large doses of radiation; and poisons such as alcohol. Myocarditis usually proceeds without complications, but occasionally it leads to heart failure or arrhythmias.
Endocarditis is caused by a bacterial or fungal infection of the endocardium, heart valves, and endothelium of adjacent blood vessels. Untreated, endocarditis is usually fatal.

Cardiac Complications
Among the complications that may affect the heart are circulatory shock, rapid rise in pressure, and abnormal rhythms.

Circulatory shock *Circulatory shock* refers to a generalized inadequacy of blood flow throughout the body. It may be caused by *reduced blood volume*, as a result of hemorrhage, severe diarrhea or vomiting, lack of water, or severe burns. Another possible cause is *increased capacity of veins*, as a result of bacterial toxins. A third cause is *damage to the myocardium*, usually as a result of a heart attack.
When blood volume drops below about four-fifths of normal, one of the results is **hypovolemic shock,** which impairs circulation and the flow of liquids to the tissues. Such a condition is usually caused by hemorrhage or severe burns. Hypovolemic shock is usually accompanied by low blood pressure, rapid heart rate, rapid and shallow breathing, and cold, clammy skin.
In contrast to hypovolemic shock, **cardiogenic shock** occurs when the cardiac output is decreased, resulting in too little fluid reaching the tissues. It is caused by severe failure of the left ventricle, usually as a complication of myocardial infarction. Most patients die within 24 hours.

Cardiac tamponade In **cardiac (pericardial) tamponade** (Fr. *tampon*, plug), a rapid rise in pressure occurs inside the pericardial sac, usually because of the accumulation of blood or other fluid as a result of an infection or severe hemorrhage in the chest area.

Cardiac arrhythmias Abnormal electrical conduction or changes in heart rate and rhythm are called *cardiac arrhythmias.* In *paroxysmal ventricular tachycardia* (Gr. *paroxunein*, to stimulate; *takhus*, swift), the heart suddenly races at a steady rhythm of 200 to 300 beats per minute. In medical terms, *paroxysm* refers to something that starts and stops suddenly, and *tachycardia* is a heart rate faster than 100 beats per minute. The AV node may fire abnormally for minutes, hours, or days.
Atrial fibrillation (L. *fibrilla*, fibril) occurs when the atria beat with fast, but feeble, twitching movements. The ventricles contract normally for a short time, but their beat soon becomes irregular also. Because of the irregular beat, the ventricles may contract when they are not full, reducing cardiac output. Blood may stagnate and clot in the atria. If clots get into the circulation, they can clog arteries and cause heart attacks and stroke. The leading cause of atrial fibrillation is mitral stenosis.
Atrial flutter occurs when the atria beat regularly, but at a very rapid, "flapping" pace of 200 to 300 beats per minute. Every third or fourth atrial beat stimulates a ventricular beat, but the relationship can vary from two to one to as much as six to one. The causes of atrial flutter are usually similar to those of atrial fibrillation.
Atrioventricular block (or *AV block*) occurs when diseased cells of the atrioventricular node cannot transmit adequate electrical impulses. The condition may progress from a pattern where the impulse rhythm is almost imperceptibly slowed to an irregular, slow atrial beat and a very slow ventricular beat that may be accompanied by dizziness. If the cells of the AV node totally lose their ability to conduct, the beat of the heart is taken over by one of the emergency pacemakers in the atria or ventricles. At this stage, called *complete heart block*, the atria and ventricles beat independently. If the ventricles stop beating for a few seconds, fainting and convulsions may occur.
Ventricular fibrillation is caused by a continuous recycling of electrical waves through the ventricular myocardium. As a result, abnormal contraction patterns with varied rates are set up. During ventricular fibrillation, the strong, steady contractions of the ventricles are replaced by a feeble twitching that pumps little or no blood. These effects can be produced by damage to the myocardium, usually from an inadequate blood supply. Without treatment, the victim usually dies in a matter of minutes.

Angina Pectoris
Angina pectoris (L. strangling + chest) is an example of referred pain. It occurs when not enough blood gets to the heart muscle due to a damaged or blocked artery. Sometimes exercising or stress will cause angina (pain). Angina does not always lead to a heart attack. Sometimes collateral circulation develops, more blood reaches the heart muscle, and the pain decreases. Angina may disappear if the heart muscle is receiving enough blood.

CHAPTER SUMMARY

Human blood flows in a *closed system,* remaining essentially within the vessels that carry it. The heart is a *double pump,* pumping oxygen-poor blood to the lungs, and pumping oxygen-rich blood to the rest of the body. The circulation of the blood between the heart and the lungs is the **pulmonary circulation** and between the heart and the rest of the body is the **systemic circulation.**

Structure of the Heart (p. 479)
1 The cone-shaped heart is about the size of a fist. It is located in the center of the chest. It is oriented obliquely, with about two-thirds of its bulk to the left of the body's midline.

2 The heart lies within a protective sac called the **pericardium.** It is composed of an outer *fibrous pericardium* and an inner *serous pericardium.* Pericardial fluid between the sac and the heart helps to minimize friction when the heart moves. The serous pericardium is divided into an outer *parietal layer* and an inner *visceral layer,* separated by the *pericardial cavity.* The visceral layer forms part of the heart wall.

3 The wall of the heart is composed of the **epicardium** or outer layer (the visceral pericardium), the **myocardium** or middle muscular layer, and the **endocardium** or inner layer.

4 The **cardiac skeleton** is a structure of tough connective tissue inside the heart. It provides attachment sites and support for the valves and muscular fibers that allow the heart to wring blood out of the ventricles.

5 The heart is made up of two separate, parallel pumps, often called the **right heart** and **left heart.** Each of the two pumps has a receiving chamber on top called an **atrium** and a discharge pumping chamber below called a **ventricle.** Separating the left and right hearts is a thick wall of muscle called the **septum.**

6 Visible features on the surface of the heart include some sulci and coronary veins and arteries that carry blood to and from the heart.

7 The two **atrioventricular (AV) valves** permit blood to flow from the atria to the ventricles, and the two **semilunar valves** permit the flow from the ventricles to the pulmonary artery and aorta. The atrioventricular valves are the **tricuspid valve** of the right heart, and the **bicuspid,** or **mitral, valve** of the left heart. The right semilunar valve is the **pulmonary semilunar valve,** and the left semilunar valve is the **aortic semilunar valve.**

8 The great vessels of the heart are the **superior vena cava, inferior vena cava, pulmonary artery, pulmonary veins,** and **aorta.**

9 Circulation to and from the tissues of the heart is the **coronary circulation.** Blood is supplied to the heart by the right and left **coronary arteries.** Most of the cardiac veins drain into the **coronary sinus.**

Physiology of the Heart (p. 490)
1 Cardiac muscle cells function as a single unit in response to physiological stimulation because they are connected by *intercalated disks.* The interconnected cells form a latticework called a *syncytium.* Action potentials spread over a syncytium, causing all the cardiac muscle cells to contract in unison.

2 The impulse-conducting system of the heart consists of the **sinoatrial (SA) node** (*pacemaker*) in the right atrium, the **atrioventricular (AV) node** between the atrium and ventricle, a tract of conducting fibers called the **atrioventricular bundle** that divides into a branch for each ventricle, and modified nerve fibrils called **cardiac conducting myofibers,** or **Purkinje fibers,** in the walls of the ventricles.

3 The **cardiac cycle** is the carefully regulated sequence of steps that comprises a heartbeat. A complete cardiac cycle consists of an atrial contraction, or **systole,** a ventricular systole, an atrial relaxation, or **diastole,** and a ventricular diastole.

4 The path of blood through the heart proceeds as follows: (1) Oxygen-poor blood from the body enters the right atrium, and oxygen-rich blood from the lungs enter the left atrium. (2) Blood from the atria is forced into the ventricles. (3) After hesitating for an instant, the ventricles contract. The right ventricle pumps oxygen-poor blood to the lungs, and the left ventricle pumps oxygen-rich blood (which just entered the heart from the lungs) through the aorta to the body. By this time, the atria have started to refill, and another cardiac cycle is about to begin.

5 *Heart sounds* are caused by the closing of the heart valves and vibrations in the heart wall. These sounds can be used to diagnose cardiovascular abnormalities. An unusual sound is called a *murmur.*

6 The central mechanism regulating the heartbeat, rate, and volume is the **cardioregulatory center** in the medulla oblongata. The autonomic system speeds or slows the heartbeat, but does not initiate it. Endocrine control is involved also.

7 Nervous control operates through a negative feedback system involving **baroreceptors** and **chemoreceptors** in the carotid sinuses and aorta.

The Effects of Aging on the Heart (p. 495)
1 The pumping efficiency of the heart is reduced with age, and congestive heart failure may occur.

2 The heart usually becomes enlarged with age, and it does not respond to stress as well as it used to.

3 Hardening of the arteries may cause the heart to work harder to achieve the same results as before.

Developmental Anatomy of the Heart (p. 496)
1 The embryonic circulatory system forms and becomes functional before any other system, fulfilling the nutritional needs of the embryo by the end of the third week after fertilization.

2 The external form of the heart continues to develop until about day 60.

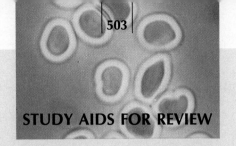

STUDY AIDS FOR REVIEW

MEDICAL TERMINOLOGY

ANEURYSM A bulging of a portion of the heart, aorta, or other artery.

ASYSTOLE Failure of the myocardium to contract.

AUSCULTATION Listening for heart sounds.

BRADYCARDIA A slower-than-normal heartbeat.

CARDIAC ARREST Cessation of normal, effective heart action, usually caused by asystole or ventricular fibrillation.

CARDIAC MASSAGE Manual stimulation of the heart when asystole occurs.

CARDIOMEGALY Enlargement of the heart.

CARDIOTONIC DRUG Drug that strengthens the heart.

CHRONOTROPIC DRUG Drug that changes the timing of the heart rhythm. A positive chronotropic drug increases the heart rate, and a negative drug decreases the heart rate.

COMMISSUROTOMY An operation to widen a heart valve that has been thickened by scar tissue.

CONSTRICTIVE PERICARDITIS A condition in which the heart muscle cannot expand and contract properly because the pericardium has shrunk and thickened.

CORONARY OCCLUSION Blockage in the circulation of the heart.

COR PULMONALE Heart disease resulting from disease of the lungs in which the right ventricle hypertrophies and there is pulmonary hypertension.

DEFIBRILLATOR Instrument that corrects abnormal cardiac rhythms by applying electric shock to the heart.

EMBOLISM (Gr. *emballaein,* to throw in) Obstruction or occlusion of a blood vessel by an air bubble, blood clot, mass of bacteria, or other foreign material.

PALPITATION Skipping, pounding, or racing heartbeats.

SINUS RHYTHM Normal cardiac rhythm regulated by the SA node.

STETHOSCOPE Instrument used to listen to sounds in the chest.

STOKES-ADAMS SYNDROME Sudden seizures of unconsciousness which may accompany heart attacks.

TACHYCARDIA A faster-than-normal heartbeat.

THROMBUS (Gr. a clotting) A blood clot obstructing a blood vessel or heart cavity; the condition is *thrombosis.*

UNDERSTANDING THE FACTS

1 What is meant by a closed system of blood vessels?
2 Give the location of the following (be specific):
   a myocardium
   b chordae tendineae

   c fossa ovalis
   d atrioventricular bundle
   e mitral valve
3 Describe the shape, size, position, and location of the heart.
4 Why are the walls of the ventricles thicker than those of the atria?
5 What are the functions of pericardial fluid?
6 Is there an advantage to having many collateral branches of the coronary arteries? Explain.
7 Distinguish between systole and diastole.
8 Specifically when does your heart rest?
9 What is the actual cause of the heart sounds?
10 Name the great vessels of the heart
11 Baroreceptors are sensitive to changes in _____.
12 Define myocardial infarction.
13 Define stenosis.
14 What is the relationship of rheumatic fever to heart disease?

UNDERSTANDING CONCEPTS

1 Describe the basic functions of the heart.
2 Trace the pathway of blood through the heart, starting with the right atrium. Be sure to include all the valves in the pathway.
3 Describe the cardiac skeleton and its functions.
4 Give the location and describe the function of the papillary muscles.
5 Explain the impulse-conducting system of the heart.
6 Does the human heart need nervous stimulation in order to beat? Explain.
7 How do the properties of cardiac muscle affect heart function?

SELF-QUIZ
Multiple-Choice Questions

19.1 The part of the serous pericardium that lines the outer portion of the pericardial sac is called the
   a visceral pericardium
   b parietal pericardium
   c epicardium
   d endocardium
   e myocardium

19.2 The outermost layer of the heart is called the
   a myocardium
   b endocardium
   c epicardium
   d myocardial cortex
   e visceral pericardium

19.3 The oval depression on the septal wall of the right atrium is the
a musculi pectinati
b trabeculae carneae
c conus
d fossa ovalis
e infundibulum

19.4 Which of the following valves allows deoxygenated blood to enter the pulmonary artery?
a atrioventricular valve
b aortic semilunar valve
c pulmonary semilunar valve
d bicuspid valve
e tricuspid valve

19.5 Of the following, which is *not* one of the great vessels of the heart?
a pulmonary arteries
b coronary arteries
c pulmonary veins
d superior vena cava
e inferior vena cava

19.6 The tiny veins that open directly into each of the four heart chambers are the
a great cardiac veins
b small cardiac veins
c thebesian veins
d oblique veins
e anterior cardiac veins

19.7 Which of the following is *not* part of the heart's conducting system?
a SA node
b AV node
c atrioventricular bundle
d cardiac conducting myofibers
e nervous tissue

19.8 Which heart sound usually requires amplification in order to be heard?
a first
b second
c third
d fourth
e fifth

19.9 The main control center for the heart is located in the
a cerebrum
b hypothalamus
c medulla oblongata
d cerebellum
e pons

19.10 The skeleton of the heart consists of
a bone within the myocardium
b bone within the interatrial septum
c fibrous connective tissue in the pulmonary trunk
d fibrous connective tissue encircling the atrioventricular orifices
e mostly cartilage

19.11 Which endocrine gland produces a hormone that affects the heart rate?
a thymus gland
b pineal gland
c thyroid gland
d pancreas
e adrenal gland

True-False Statements

19.12 The circulation between the heart and the lungs is called the systemic circulation.
19.13 The apex is the top portion of the heart.
19.14 The heart muscle needs more oxygen than any organ except the brain.
19.15 Most of the cardiac veins drain into the coronary sinus.
19.16 The superior vena cava returns blood from the heart and lungs.

Completion Exercises

19.17 The inside cavities of the heart are covered with _____.
19.18 The tough connective tissue that provides attachment sites for the heart valves is the _____.
19.19 The specialized tissue that initiates the heartbeat is called the _____.
19.20 The heart functions as a syncytium because of _____ between cells.

Matching

19.21 _____ chordae tendineae
19.22 _____ trabeculae carneae
19.23 _____ sulci
19.24 _____ musculi pectinati
19.25 _____ atrial systole
19.26 _____ ventricular systole
19.27 _____ atrial diastole
19.28 _____ ventricular diastole

a ridges
b muscle ridges
c blood vessels
d depressions
e supportive cords
f begins before atrial systole
g blood is forced into the ventricles
h the atria begin refilling with blood
i blood is forced into the systemic circulation
j lasts about 3 seconds

Problem-Solving/Critical Thinking Questions

19.29 Which of the following statements about the heart valves is/are correct?

a The closing of the AV valves is responsible for the second heart sound.

b Small muscles attached to the AV valves are responsible for opening and closing the valves with each cycle of contraction.

c The aortic valve opens when the pressure in the left ventricle becomes slightly greater than the pressure in the aorta.

d If the aortic valve became damaged and was unable to close completely, one would find a decreased diastolic pressure in the aorta.

e c and d

19.30 Cardiac contraction is characterized by which of the following?

a During contraction, the internal volume of the heart is reduced.

b Contractions of the individual muscle cells must be coordinated so that they contract almost simultaneously.

c The initial stimulus for contraction normally arises within a special area of the myocardium.

d a and b

e a, b, and c

A SECOND LOOK

1 In the following drawing, label the endocardium, myocardium, epicardium, and pericardial cavity.

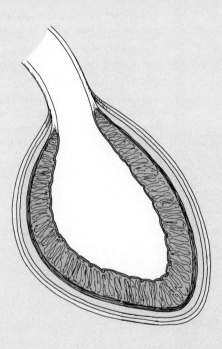

2 In the following drawing, label the common carotid arteries, ascending and descending aorta, aortic arch, and inferior vena cava.

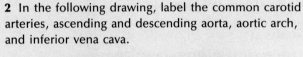
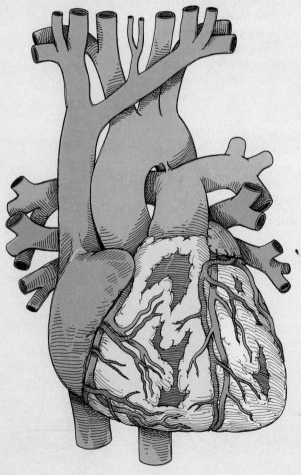

20

The Cardiovascular System: Blood Vessels

LEARNING OBJECTIVES

1 Describe the structure of arteries, arterioles, capillaries, venules, and veins.

2 Describe three different types of capillaries and relate their structures to their functions.

3 Compare the structure of the walls in the different types of blood vessels.

4 Explain microcirculation.

5 Describe the structure of pulmonary circulation, and name the major vessels in this circuit.

6 Discuss the structure and major components of each division of the systemic circulation.

7 Describe the portal systems, and name their major components.

8 Explain the main differences between fetal and adult circulation.

9 Identify the major arteries and veins, and describe the region supplied or drained by each.

10 Discuss the causes or effects of stroke, aneurysms, atherosclerosis, coronary artery disease, hypertension, thrombophlebitis, and varicose veins.

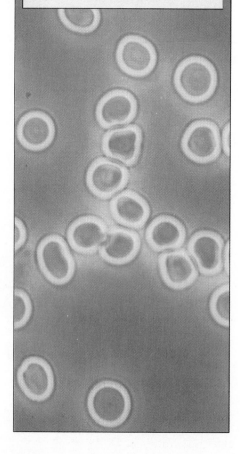

Although the center of the cardiovascular system is the heart, it is the blood vessels that carry blood throughout the body. As blood flows from the heart through these vessels, it supplies tissues with substances needed for metabolism and homeostatic regulation. Waste products that accumulate in the cells pass into the blood, which carries them off through other vessels for removal by the kidneys, skin, and lungs.

TYPES OF BLOOD VESSELS

Arteries are blood vessels that carry blood *away from* the heart to the organs and tissues of the body. Because different arteries contain varying amounts of elastic and smooth muscle tissue, some are called *elastic*, and others are called *muscular*. The walls of elastic arteries expand slightly with each heartbeat. The muscular arteries branch into smaller *arterioles*. These vessels play important roles in determining the amount of blood going to any organ or tissue, and in regulating blood pressure. The muscular arteries and arterioles can be either dilated or constricted by nervous control. Arterioles branch into smaller vessels called *metarterioles*, which carry blood either into venules (tiny veins) or into the smallest vessels in the body, the capillaries.

Capillaries are microscopic vessels with walls mostly one cell thick. The thin capillary wall is porous, and allows the passage of water and small particles of dissolved materials. Capillaries are distributed throughout the body, except in the dead outer layers of skin and in such special places as the lenses of the eyes. Capillaries converge into larger vessels called *venules*, which merge to form larger vessels called veins.

Veins carry blood *toward* the heart. They are generally more flexible than arteries, and they collapse if blood pressure is not maintained.

How Blood Vessels Are Named

The majority of blood vessels are named according to the major organ or anatomical site supplied (by arteries) or drained (by veins), or according to their location in the body. For example, the *renal artery* supplies the kidneys, and the *renal vein* drains the kidneys. The depth of the vessel is also used. For example, the *internal jugular vein* lies deeper in the neck than the *external jugular vein* does. In some instances, the name of a blood vessel changes as it passes into a different part of the body. For example, the *subclavian artery* runs underneath the clavicle in the neck region. When it enters the region of the armpit (axilla), its name changes to the *axillary artery*, and it becomes the *brachial artery* when it enters the arm (brachium). In general, veins run parallel to most arteries

and are given the same names, although there are exceptions.

Arteries

Most *arteries** are efferent vessels that carry blood *away from* the heart to the capillary beds throughout the body. In the adult, all arteries except the pulmonary trunk and pulmonary arteries carry oxygenated blood. The left and right pulmonary arteries carry *deoxygenated* blood from the heart to the lungs. Thus, the most reliable way to classify blood vessels is by the *direction* in which they carry blood, either toward the heart or away from it.

The great arteries that emerge from the heart are often called *trunks*. The major arterial trunks are the **aorta** (ascending aorta) from the left ventricle and the **pulmonary trunk** from the right ventricle.

The central canal of all blood vessels, including arteries, is called the *lumen*. Surrounding the lumen of an artery is a thick wall composed of three layers, or *tunicae* (TEW-nih-see; pl. of L. *tunica*, covering) [FIGURE 20.1].

The **tunica intima** ("innermost covering") has a lining of endothelial cells (simple squamous epithelium), a thin subendothelial layer of fine areolar tissue, and an internal elastic layer called the internal elastic lamina.

*The word *artery* comes from the Greek word for "windpipe." When the ancient Greeks dissected corpses, they found blood in the veins, but none in the arteries. As a result, they thought the arteries carried air.

FIGURE 20.1 TYPICAL ELASTIC ARTERY

Small arteries do not contain vasa vasorum, but do have the three layers shown in this cutaway drawing.

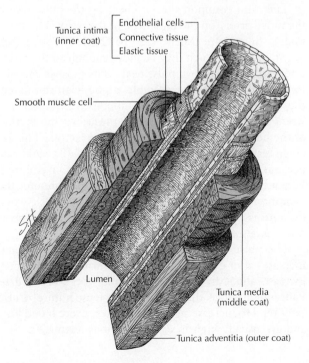

Tunica intima (inner coat)
Endothelial cells
Connective tissue
Elastic tissue
Smooth muscle cell
Lumen
Tunica media (middle coat)
Tunica adventitia (outer coat)

The *tunica media* ("middle covering") is the thickest layer of arterial wall in the large arteries. It is composed mainly of connective tissue, smooth muscle cells, and elastic fibers. The walls of the largest arteries (elastic arteries) have elastic tissue rather than smooth muscle. In the smaller arteries (muscular arteries), the elastic fibers in the tunica media are replaced by smooth muscle cells.

The *tunica adventitia* ("outermost covering") is composed mainly of collagen fibers and elastic fibers. Occasional smooth muscle fibers run longitudinally next to the outer border of the tunica media. Nerves and lymphatic vessels are also found within this layer. The walls of the large arteries (greater than 20 mm) are nourished by small blood vessels called the *vasa vasorum* ("vessels of the vessels"), which form capillary networks within the tunica adventitia and outer part of the tunica media.

The strong elastic walls of the largest arteries allow these vessels to adjust to the great pressure created by the contraction of the ventricles during systole. Blood is ejected into the aorta at a speed of about 30 to 40 cm/sec (almost a mile an hour).

Pulse When an artery lies close to the surface of the skin, a *pulse* can be felt that corresponds to the beating of the heart and the alternating expansion and elastic recoil of the arterial wall. The pulse is produced when the left ventricle forces blood against the wall of the aorta, and the impact creates a continuing pressure wave along the branches of the aorta and the rest of the elastic arterial walls. A venous pulse occurs only in the largest veins. It is produced by the changes in pressure that accompany atrial contractions. All arteries have a pulse, but it can be felt most easily at the points shown in FIGURE 20.2.

The most common site for checking the pulse rate is the radial artery on the underside of the wrist. The three middle fingers are usually used to check the pulse (but not the thumb, which may be close enough to the radial artery to reflect its own pulse beat). The pulse is checked for several reasons. For example, a physician can detect the number of heartbeats per minute (heart rate), the strength of the beat, the tension of the artery, the rhythm of the beat, and several other diagnostic factors. The average resting pulse rate can range between 70 to 90 beats per minute in adults and from 80 to 140 in children. When the pulse rate exceeds 100 beats per minute, the condition is called *tachycardia* (Gr. *takhus*, swift); when the rate is below 60 beats per minute, the condition is called *bradycardia* (Gr. *bradus*, slow).

The pulse rate normally decreases during sleep and increases after eating or exercising. During a fever it may increase about five beats per minute for every degree Fahrenheit above the normal body temperature. Pulse rates tend to increase significantly after severe blood loss, and are usually high in cases of serious anemia.

FIGURE 20.2 THE ARTERIAL PULSE

The pulse can be felt most easily at the points shown.

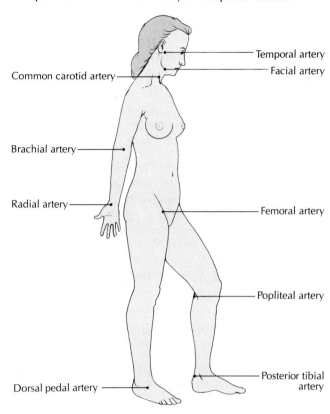

Temporal artery
Facial artery
Common carotid artery
Brachial artery
Radial artery
Femoral artery
Popliteal artery
Posterior tibial artery
Dorsal pedal artery

Arterioles

The arteries nearest the heart are the largest. As their distance from the heart increases, the arteries branch into smaller and smaller arteries, then into *arterioles* shortly before reaching the capillary networks. Arterioles are covered by the three tunicae. Because their walls contain smooth muscle cells, arterioles can dilate or constrict, thus controlling the flow of blood from arteries into capillaries and later into the organs as they require blood. If necessary, an arteriole can dilate to increase blood flow to capillaries by as much as 400 percent.

Terminal arterioles (those closest to a capillary) are muscular and well supplied with nerves, but they do not have an internal elastic layer or their own blood vessels (vasa vasorum), as larger arteries do. The tunica media is particularly well supplied with sympathetic nerves, which cause the smooth muscle cells to contract and the lumen to constrict.

Q: *Since blood is already within large arteries and veins, why do these vessels need their own blood supply?*

A: The walls of large arteries and veins are so thick that nutrients cannot diffuse far enough to reach all of their cells.

Capillaries

Terminal arterioles branch out to form *capillaries* (L. *capillus*, hair), which connect the arterial and venous systems [FIGURE 20.3]. Capillaries are generally composed of only a single layer (tunica intima) of endothelial cells on a thin basement membrane of glycoprotein. Capillaries are the smallest and most numerous blood vessels in the body. If all the capillaries in an adult body were connected, they would stretch about 96,000 km (60,000 mi). This abundance of capillaries makes an enormous surface area available for the exchange of gases, fluids, nutrients, and wastes between the blood and nearby cells.

Capillaries are about 600 times narrower than a medium-sized vein, and about 500 times narrower than a medium-sized artery. The diameter of capillaries varies with the function of the tissue.

Types of capillaries At least three different types of capillaries are recognized, and each one performs a specific function [FIGURE 20.4].

Continuous capillaries are found in skeletal muscle tissue. Their walls are made up of one continuous endothelial cell, with the ends overlapping in a tight endothelial cell junction [FIGURE 20.4A]. Apparently, intracellular

FIGURE 20.4 TYPES OF CAPILLARIES

[A] Continuous capillary in muscle, showing the overlapping ends of the single endothelial cell. **[B]** Fenestrated capillary of an endocrine gland, showing two endothelial cells connected by thin stretches of membrane called fenestrations. **[C]** Discontinuous capillaries (sinusoids) in the liver, showing open, irregular structure and a phagocytic stellate reticuloendothelial (Kupffer) cell.

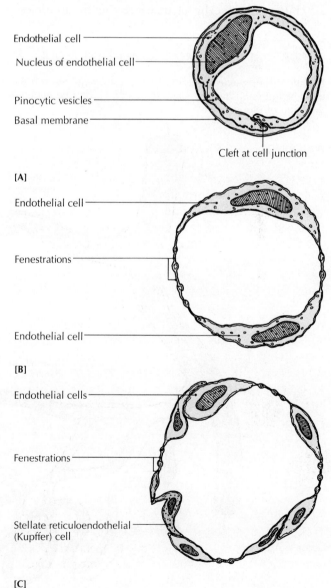

Endothelial cell
Nucleus of endothelial cell
Pinocytic vesicles
Basal membrane
Cleft at cell junction
[A]

Endothelial cell
Fenestrations
Endothelial cell
[B]

Endothelial cells
Fenestrations
Stellate reticuloendothelial (Kupffer) cell
[C]

FIGURE 20.3 CAPILLARY

Electron micrograph of a cross section of a capillary. The capillary wall is composed of two endothelial cells, and the large nucleus of one of the cells is prominent. Note the clefts at the junctions between the endothelial cells. ×41,000.

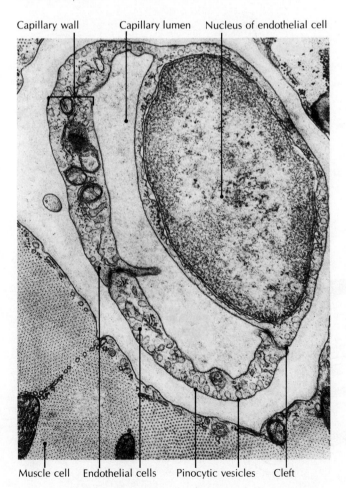

Capillary wall   Capillary lumen   Nucleus of endothelial cell

Muscle cell   Endothelial cells   Pinocytic vesicles   Cleft (pinocytic) vesicles help to move liquids across the membrane of the cell by exocytosis and endocytosis. The walls of **fenestrated capillaries** (L. *fenestra,* window) consist of two or more adjacent endothelial cells connected by thin endothelial membranes called fenestrations, or pores [FIGURE 20.4B]. Fenestrated capillaries are usually found in the kidneys, endocrine glands, and intestines. **Discontinuous capillaries,** also known as **sinusoids** or **vascular sinuses,** have fenestrations and a much wider lumen than the other types [FIGURE 20.4C]. Such an open, irregular structure is highly permeable. This type of capillary system is found in the liver and spleen. Liver sinusoids also contain active phagocytes called *stellate reticuloendothelial cells (Kupffer cells;* KOOP-fur), which are part of the reticuloendothelial system.

Capillary blood flow Blood leaves the heart traveling about 30 to 40 cm/sec (12 in./sec), but it is slowed to only 2.5 cm/sec (1 in./sec) by the time it reaches the arterioles,

and less than 1 mm/sec (0.04 in./sec) in the capillaries. Blood remains in the capillaries for only a second or two, but given the short length of capillaries (about 1 mm), that is long enough for the crucial exchanges of nutrients and wastes.

Because capillary walls are usually only one cell thick, certain materials pass through rather easily. Small molecules, including gases such as oxygen and carbon dioxide, certain waste products, salts, sugars, and amino acids, pass through freely, but the large molecules of plasma proteins pass through only with difficulty, if at all. Red blood cells cannot pass through the capillary walls.

Microcirculation of the blood The capillaries and their associated structures (including terminal arterioles, metarterioles, and venules) constitute the **microcirculation** of the blood [FIGURE 20.5]. This name reflects the extremely narrow diameters of the vessels.

A *metarteriole* is a vessel that emerges from an arteri-

FIGURE 20.5 MICROCIRCULATION OF THE BLOOD

Metarterioles provide the path of least resistance between the arterioles and venules. Note the absence of smooth muscle in the true capillaries. Arrows indicate direction of blood flow.

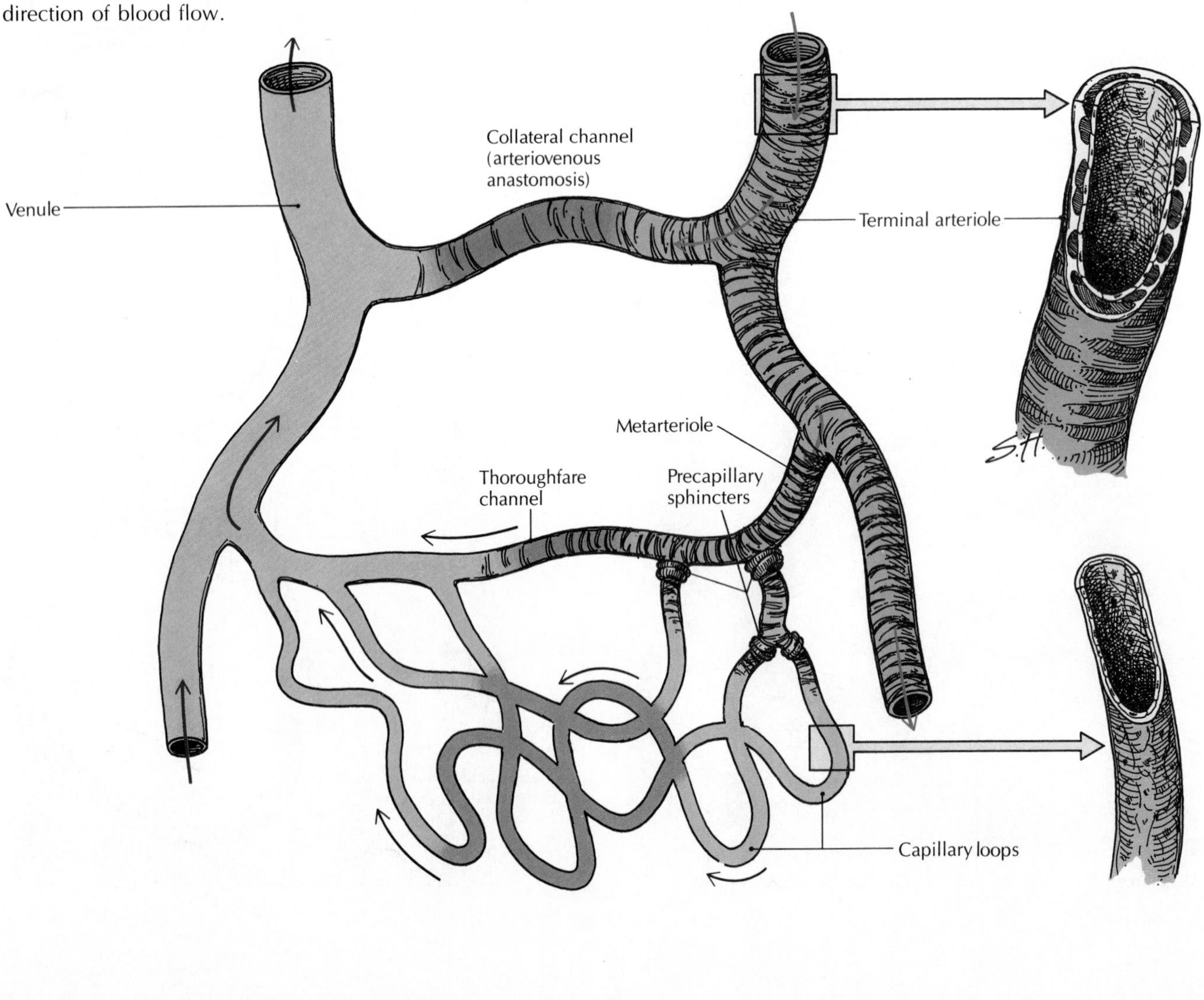

ole, traverses the capillary network, and empties into a venule. At the junction of the metarteriole and capillary is the *precapillary sphincter*, a ringlike smooth muscle cell that regulates the flow of blood into the capillaries.

Some metarterioles connect directly to venules by way of *thoroughfare channels*. Similar to thoroughfare channels are *collateral channels (capillary shunts* or *arteriovenous anastomoses)*, which bypass the capillary beds and act as direct links between arterioles and venules. Collateral channels are numerous in the fingers, palms, and earlobes, where they control heat loss. Because these collateral channels have thick walls, they are not exchange vessels in the way that capillaries are. They are heavily innervated, and are muscular at the arteriole end and elastic at the venule end.

Q: *Why do we sometimes have dark circles under our eyes?*

A: In the skin below your eyes are hundreds of tiny blood vessels, which help drain blood from the head. When you are ill or tired, blood circulation may slow down, concentrating blood in the vessels and causing them to swell. Because the skin under the eyes is very thin, the engorged vessels become visible. The dark circles are actually pools of blood.

Venules

Blood drains from the capillaries into **venules** (VEHN-yoolz), tiny veins that unite to form larger venules and veins. The transition from capillaries to venules occurs gradually. The immediate postcapillary venules consist mainly of endothelium and a thin tunica adventitia. Postcapillary venules play an important role in the exchange between blood and interstitial fluid. Unlike capillaries, these venules are easily affected by inflammation and allergic reactions. They respond to these conditions by opening their pores, allowing water, solutes, and white blood cells to move out into the extracellular space.

Veins

Venules join together to form **veins.** Superficial veins are found in areas where blood is collected near the surface of the body, and are especially abundant in the limbs. Veins become larger and less branched as they move away from the capillaries and toward the heart.

Most veins are relatively large [FIGURE 20.6]. They carry *deoxygenated* blood from the body tissues to the

---

FIGURE 20.6 VEINS

[A] Cutaway drawing of a medium-sized vein, showing a valve and the three-layered wall. Compare the thickness of the layers with those of the artery shown in FIGURE 20.1. **[B]** Scanning electron micrograph of a medium-sized vein and artery, showing the larger lumen and thinner wall of the vein. Connective tissue surrounds the vessels. ×305.

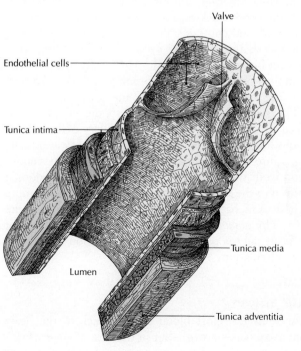

[A]

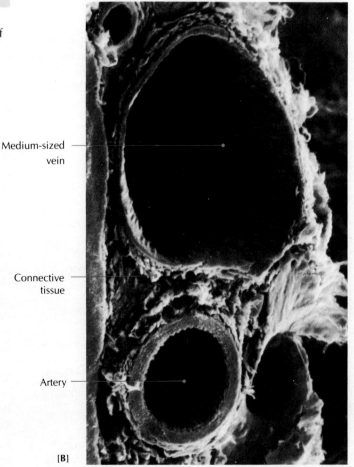

[B]

heart. However, there are three exceptions:

1 The four *pulmonary veins* carry *oxygenated* blood from the lungs to the left atrium of the heart.

2 The *hepatic portal system* of veins carries blood from the capillaries of the intestines to the capillaries and sinusoids of the liver.

3 In the *hypothalamic-hypophyseal portal system*, capillaries of the hypothalamus unite to form veins that divide into a second set of sinusoidal capillaries in the anterior pituitary gland.

The walls of veins contain the same three layers (tunicae) as arterial walls, but the tunica media is much thinner [FIGURE 20.6]. Also, venous walls contain less elastic tissue, collagenous tissue, and smooth muscle. As a result, veins are very distensible and compressible. The smooth muscle cells that *are* found in veins are arranged in either a circular or longitudinal pattern. Like arteries, veins are nourished by small vasa vasorum.

Veins usually contain paired semilunar bicuspid valves that permit blood to flow in only one direction, restricting any backflow [FIGURE 20.6A]. Blood pressure in the veins is low, and the venous blood is helped along by the skeletal muscle pump (the compression of the venous walls by the contraction of surrounding skeletal muscles). The venous valves, which are derived from folds of the tunica intima, are especially abundant in the lower limbs, where gravity opposes the return of blood to the heart. There are no valves in veins narrower than 1 mm or in regions of great muscular pressure, such as the thoracic and abdominal cavities.

A S K   Y O U R S E L F

1 What are the basic functions of arteries? Veins?

2 What are the three layers of arterial walls?

3 What are arterioles? Venules?

4 Why are capillaries sometimes called exchange vessels?

5 What are the three types of capillaries?

6 What is meant by the microcirculation of the blood?

7 How do venous valves function?

8 Can you describe some of the ways that blood vessels are named?

CIRCULATION OF THE BLOOD

Blood circulates throughout the body in two main circuits: the pulmonary and systemic circuits [see FIGURE 19.1]. In the following sections we will describe these two circuits, as well as some subdivisions or special areas of circulation. (Coronary circulation was described in Chapter 19.)

Pulmonary Circulation

The ***pulmonary circulation*** supplies blood only to the lungs. It carries deoxygenated blood from the heart to the lungs, where carbon dioxide is removed and oxygen is added. It then returns the oxygenated blood to the heart for distribution to the rest of the body [FIGURE 20.7]. Pulmonary circulation takes 4 to 8 seconds.

The major blood vessels of the pulmonary circulation are the *pulmonary trunk* and two *pulmonary arteries*, which carry deoxygenated blood from the right ventricle of the heart to the lungs; *pulmonary capillaries*, which are the site of the exchange of oxygen and carbon dioxide in the lungs; and the four *pulmonary veins* (two from each lung), which carry oxygenated blood from the lungs to the left atrium of the heart. Each of the pulmonary veins enters the left atrium through a separate opening.

Systemic Circulation

The ***systemic circulation*** supplies all the cells, tissues, and organs of the body with oxygenated blood, and it returns deoxygenated blood from the body tissues to the heart [FIGURE 20.8]. The systemic circuit from the heart and back again takes about 25 to 30 seconds. The systemic circulation is divided into the arterial and venous divisions. The main vessel of the ***arterial division*** is the *aorta*, which emerges from the left ventricle as the *ascending aorta*, curves backward over the top of the heart as the *aortic arch*, and continues down through the thorax and abdomen as the *descending aorta*. The descending aorta gives off branches to internal organs in the thoracic and abdominal cavities, and then terminates in the two *common iliac arteries*, which supply the lower extremities.

The ***venous division*** of the systemic circulation is linked to the arterial system by capillary beds. All the venous blood from the upper part of the body eventually drains into the large *superior vena cava*, and the venous blood from the lower extremities, pelvis, and abdomen drains into the *inferior vena cava*. Both venae cavae empty their deoxygenated blood into the right atrium of the heart. The *coronary sinus* drains blood from the walls of the heart, and is part of the venous division of the systemic circulation.

Portal Systems

Most veins transport blood directly back to the heart from a capillary network, but in the case of a ***portal system,*** the blood passes through *two* sets of capillaries on its

FIGURE 20.7 PULMONARY CIRCULATION

The freshest, most highly oxygenated blood in the body enters the left heart from the lungs through the pulmonary

veins, not the pulmonary arteries. Arrows indicate direction of blood flow.

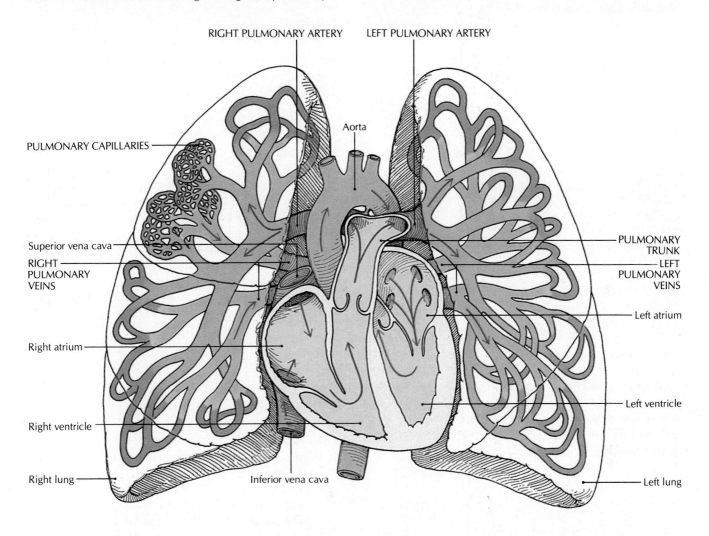

way to the venous system. The human body has two portal systems. The **hypothalamic-hypophyseal portal system** moves blood from the capillary bed of the hypothalamus directly, by way of veins, to the sinusoidal bed of the pituitary gland (hypophysis; see FIGURE 17.2). The **hepatic portal system** (Gr. *hepatikos,* liver) moves blood from the capillary beds of the intestines to the sinusoidal beds of the liver [FIGURE 20.9].

The hepatic portal system varies considerably from one person to the next, but it follows a basic pattern. The **hepatic portal vein** is formed behind the neck of the pancreas by the union of the **splenic** and **superior mesenteric veins.** The splenic mesenteric vein returns blood from the spleen, and the superior mesenteric vein returns blood from the small intestine. The hepatic portal vein also receives coronary, cystic, and pyloric branches.

The splenic vein receives the left *gastroepiploic vein* (or *short gastric vein*) from the great curvature of the stomach, and the *inferior mesenteric vein,* which receives blood from the veins of the large intestine.

The *cystic vein* brings blood from the gallbladder to the hepatic portal vein. The *right gastric (pyloric) vein* arises from the stomach and then enters the hepatic portal vein. The *left gastric (coronary) vein* also empties into the hepatic portal vein.

Because the veins of the hepatic portal system lack valves, the veins are vulnerable to the excessive pressures of venous blood. This situation can produce *hemorrhoids* (dilated veins in the anal region) in the internal hemorrhoidal plexus [FIGURE 20.9].

Much of the venous blood returning to the heart from the capillaries of the spleen, stomach, pancreas, gallblad-

FIGURE 20.8 SYSTEMIC CIRCULATION

The drawing shows the major arteries and veins of the systemic circulation. The pulmonary circulation (between the heart and lungs) is not shown.

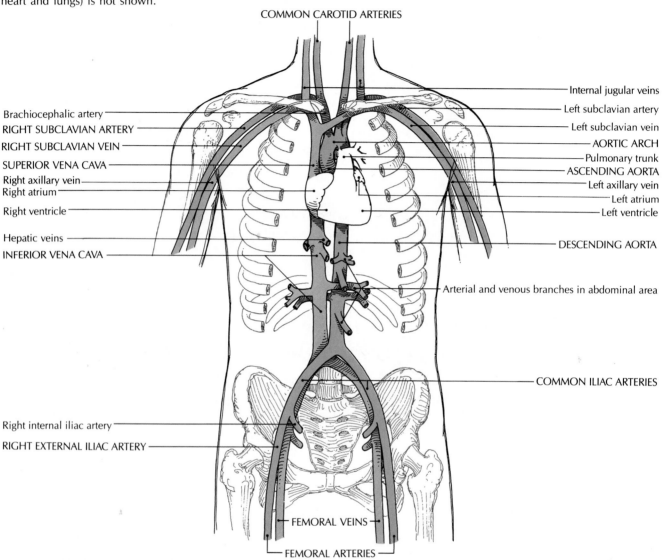

COMMON CAROTID ARTERIES

Internal jugular veins
Left subclavian artery
Left subclavian vein
AORTIC ARCH
Pulmonary trunk
ASCENDING AORTA
Left axillary vein
Left atrium
Left ventricle

DESCENDING AORTA

Arterial and venous branches in abdominal area

COMMON ILIAC ARTERIES

Brachiocephalic artery
RIGHT SUBCLAVIAN ARTERY
RIGHT SUBCLAVIAN VEIN
SUPERIOR VENA CAVA
Right axillary vein
Right atrium
Right ventricle

Hepatic veins
INFERIOR VENA CAVA

Right internal iliac artery
RIGHT EXTERNAL ILIAC ARTERY

FEMORAL VEINS
FEMORAL ARTERIES

der, and intestines contains products of digestion. This nutrient-rich blood is carried by the *hepatic portal vein* to the *sinusoids* of the liver. After leaving the sinusoids of the liver, the blood is collected in the **hepatic veins** and drained into the **inferior vena cava,** which returns the blood to the right atrium of the heart.

Cerebral Circulation

The brain is supplied with blood by four major arteries: two **vertebral arteries** and two **internal carotid arteries** [FIGURE 20.13A]. Blood from the vertebral arteries supplies the cerebellum, brainstem, and posterior (occipital)

part of the cerebrum. The internal carotid arteries supply the rest of the cerebrum and both eyes.

The vertebral arteries merge on the ventral surface of the brain to form the *basilar artery*, which terminates by forming the left and right *posterior cerebral arteries*. The internal carotid arteries enter the cranial cavity and then branch into the *ophthalmic arteries*, which supply the eyes, and the *anterior* and *middle cerebral arteries*, which supply the medial and lateral parts of the cerebral hemispheres, respectively.

All the blood entering the cerebrum must first pass through the **cerebral arterial circle** (*circle of Willis*) [FIGURE 20.13B]. This circular *anastomosis*, or shunt, at the base of the brain consists of the proximal portion of the posterior cerebral arteries, the posterior communicating arteries, the internal carotid arteries, the anterior cerebral arteries, and the anterior communicating artery. A blockage of blood within a vessel of the cerebral arterial circle may be partially overcome by a reversal of the blood flow.

Q: *Why do your nose and cheeks turn red on a cold day?*

A: When the skin is very cold, oxygen in the capillaries is not needed for metabolism by the relatively inactive skin. Thus, large numbers of red blood cells containing red hemoglobin accumulate in the capillaries, showing through the skin.

Blood flows from the brain capillaries into large *venous sinuses* located in the folds of the dura mater. The sinuses then empty into the *internal jugular veins* on each side of the neck. Blood returns to the heart by way of the *brachiocephalic veins*. The junction of the left and right brachiocephalic veins forms the *superior vena cava*, which conveys blood to the right atrium of the heart.

Cutaneous Circulation

The arrangement of blood vessels in the skin allows for the increase or decrease of heat radiation from the integumentary system. When the body temperature increases, more blood flows to the superficial layers, from which

---

FIGURE 20.9 HEPATIC PORTAL CIRCULATION

The hepatic portal system transports venous blood containing absorbed nutrients from the gastrointestinal tract to the liver. The liver also receives oxygenated blood from the hepatic artery. Arrows indicate the direction of blood flow.

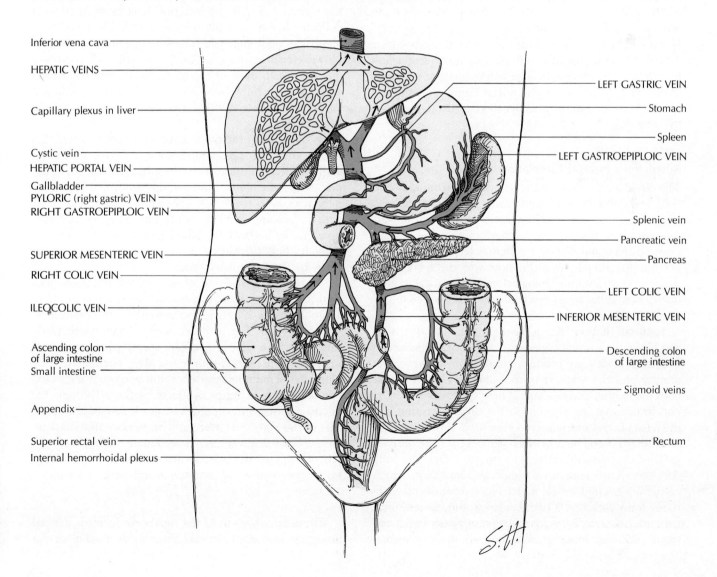

heat radiates from the body. In contrast, when the body needs to conserve heat, blood is shunted away from the surface of the skin through deep arteriovenous anastomoses (vessels that bypass capillary beds; FIGURE 20.5). In addition to these shunts, the skin has an extensive system of venous plexuses that can hold a great deal of blood. From these plexuses, heat is radiated to the surface of the skin. The amount of blood flowing through the plexuses can be controlled by either the constriction or dilation of the appropriate vessels. The diameter of the vessels is regulated by the hypothalamic temperature control center by way of sympathetic nerves to the vessels.

Skeletal Muscle Circulation

Blood nourishes skeletal muscles and also removes wastes during both physical activity and rest. Because the total body mass of skeletal muscle is so large, the blood vessels play an important role in homeostasis, especially during physical activity. Neural regulation of blood flow in skeletal muscles is controlled largely by complex interactions of neurotransmitters of the sympathetic nervous system. Norepinephrine or large doses of epinephrine produce vasoconstriction; low doses of epinephrine or acetylcholine produce vasodilation. The autoregulatory response of precapillary sphincters also dilates blood vessels in response to a decrease in the oxygen concentration in active muscles.

Adaptations in Fetal Circulation

The circulatory system of the fetus differs from that of a child and adult for two main reasons: (1) the fetus gets oxygen and nutrients and eliminates carbon dioxide and waste products through the mother's blood. (2) The fetal lungs, kidneys and digestive system (except for the liver) are not functional. At birth, however, the baby must make several rapid physiological and anatomical adjustments as it shifts to an essentially adult circulation.

Fetal circulation The *placenta,* a thick bed of tissues and blood vessels embedded in the wall of the mother's uterus, provides an indirect connection between the mother and the fetus. It contains arteries and veins of both the mother and the fetus. These vessels intermingle, but do not join. As a result of this close intermingling, the fetus obtains nutrients and eliminates wastes through its mother's blood instead of its own organs. The *umbilical cord* connects the placenta with the fetus. The fully developed cord contains a single *umbilical vein,* which carries oxygenated, nutrient-rich blood from the placenta to the fetus [FIGURE 20.10A]. Coiled within the umbilical cord are two *umbilical arteries,* which carry deoxygenated blood and waste material from the fetus to the placenta.

The umbilical vein, carrying fully oxygenated blood, enters the abdomen of the fetus, where it branches. One branch joins the fetal portal vein, which goes to the liver, while the other branch, called the *ductus venosus,* joins the inferior vena cava and bypasses the liver [FIGURE 20.10A]. Before the oxygenated blood that enters the vena cava actually enters the heart, it becomes mixed with deoxygenated blood being returned from the lower extremities of the fetus.

The fetal inferior vena cava empties into the right atrium. A large opening called the *foramen ovale* ("oval window") connects the two atria. Most of the blood passes through the foramen ovale into the left atrium. Flaps of tissue in the left atrium act as a valve, and prevent the backflow of blood into the right atrium. From the left atrium, the blood flows into the left ventricle, and from there it is pumped into the aorta for distribution primarily to the brain and upper extremities, with some going to the rest of the body.

Blood from the fetal head and upper extremities enters the right atrium via the superior vena cava, where it is deflected mainly into the right ventricle. The blood then leaves the right ventricle through the pulmonary trunk. However, since the fetal lungs are still collapsed and nonfunctional, most of the blood passes into the *ductus arteriosus,* a shunt between the pulmonary trunk and aortic arch, where it mixes with blood coming from the left ventricle.

By connecting the pulmonary artery to the aorta, the ductus arteriosus allows fetal blood to bypass the lungs, and to flow from the aorta to the umbilical arteries to the placenta. Some blood, however, does go from the pulmonary arteries to the lungs, not only to supply lung cells with oxygen and other nutrients, but to ensure a circulation that will accept the blood flow when the newborn takes its first breaths. The blood is drained from the lungs by the pulmonary veins.

Because of the mixing of oxygenated and deoxygenated blood, fetal blood has a lower oxygen content than adult blood does. To help compensate for this deficiency, *fetal hemoglobin* can combine with oxygen more easily than adult hemoglobin can. This special property of fetal hemoglobin is lost within a few days after birth. The blood in the pulmonary artery is low in oxygen and nutrients and high in waste products. It flows through the ductus arteriosus into the aorta, down toward the lower part of the body, and eventually to the two umbilical arteries. The umbilical arteries terminate in the blood vessels of the placenta, where carbon dioxide is exchanged for oxygen, and waste products are exchanged for nutrients.

Circulatory system of the newborn At birth, several important changes must take place in the cardiovascular

FIGURE 20.10 FETAL CIRCULATION

[A] Before birth the lungs are essentially bypassed because the foramen ovale permits the passage of deoxygenated blood from the right atrium to the left atrium in the fetal heart, and the ductus arteriosus carries blood from the pulmonary artery (from the right ventricle) directly into the aorta. Note that much of the blood bypasses the liver through the ductus venosus.

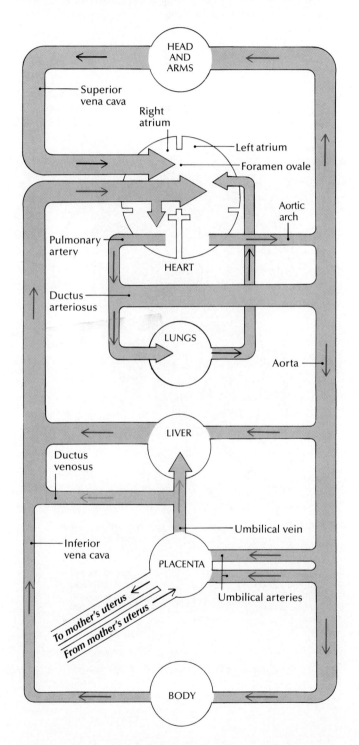

[A]

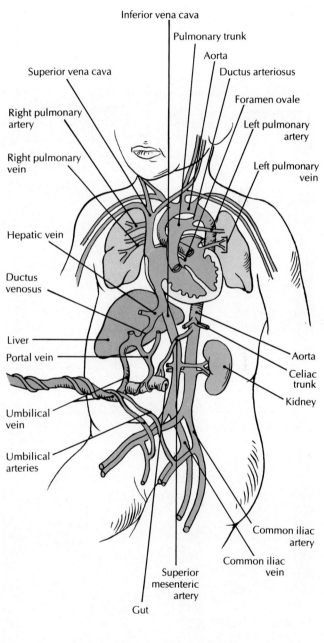

(Figure 20.10 continues on the following page)

FIGURE 20.10 FETAL CIRCULATION *(Continued)*

[B] After birth, the foramen ovale closes, and the ductus arteriosus is converted into the ligamentum arteriosum, resulting in an adult-type circulation. Also, the ductus veno- sus becomes the ligamentum venosum, the umbilical vein becomes the round ligament, and the umbilical arteries become the lateral umbilical ligaments.

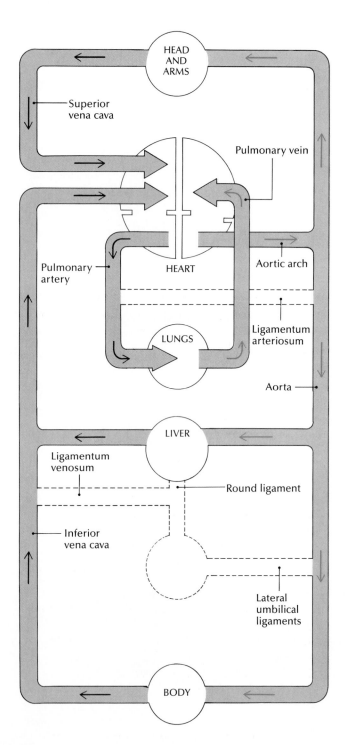
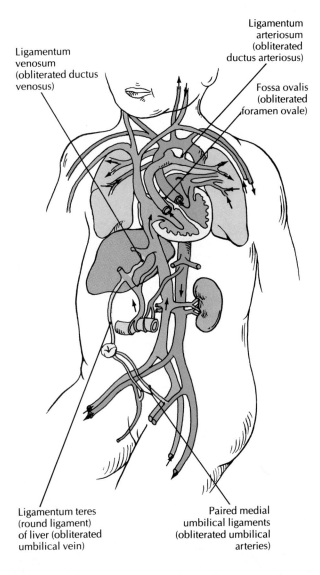

[B]

system of the newborn infant [FIGURE 20.10B]. When an infant takes its first breath, the collapsed lungs inflate, and the pulmonary vessels become functional. As a result of the increased pulmonary flow, the pressure in the left atrium is raised above that of the right atrium. The increased pressure closes the foramen ovale by pressing on the overlapping valves of the interatrial septum. The heart then begins to function like a double pump. It takes several months for the foramen ovale to fuse completely. Eventually, only a depression called the *fossa ovalis* remains at the former site of the foramen ovale. At the same time that the foramen ovale is closing, the ductus arteriosus is gradually constricting. In about six weeks it becomes the *ligamentum arteriosum.*

During the first few days after birth, the stump of the umbilical cord dries up and drops off. Inside the body, the umbilical vessels atrophy. The umbilical vein becomes the *round ligament* of the liver, the ductus venosus forms the *ligamentum venosum* (a fibrous cord embedded in the wall of the liver), and the umbilical arteries become the *lateral umbilical ligaments.*

If either the foramen ovale or ductus arteriosus does not close off, the oxygen content of the blood will be low, and the baby's skin will appear slightly blue. A newborn baby with this condition is therefore called a "blue baby." Most often, a defective ductus arteriosus must be corrected surgically, usually a few days after birth.

ASK YOURSELF

1 How does the systemic circulation differ from the pulmonary circulation?

2 What is the hepatic portal system? The hypothalamic-hypophyseal portal system?

3 What are some special areas of circulation?

4 What are the two main differences between fetal circulation and a young child's circulation?

5 What is the ductus arteriosus?

6 What causes the condition that produces a "blue baby"?

MAJOR ARTERIES AND VEINS

Two arterial trunks emerge from the heart. The *pulmonary trunk,* arising from the right ventricle, supplies blood to the respiratory portions of the lungs. The *ascending aorta,* emerging from the left ventricle, supplies blood to the rest of the body. FIGURE 20.11A gives an overall view of the major arteries of the body. The main segments of the aorta are illustrated in FIGURES 20.11A and 20.12, and are summarized in the table accompanying FIGURE 20.12. The major arteries of specific parts of the body are presented in FIGURES 20.13 through 20.15.

Four large veins drain blood from the body into the heart. The *pulmonary veins* drain blood from the lungs into the left atrium. The other three veins drain into the right atrium. The *superior vena cava* drains the head, neck, upper limbs, and thorax; the *inferior vena cava* drains the abdomen, pelvis, and lower limbs; and the *coronary sinus* receives blood from the veins that drain the heart. FIGURE 20.11B is an overall view of the major veins of the body. More detail about specific veins is given in FIGURES 20.16 through 20.19.

ASK YOURSELF

1 What are the major segments of the aorta?

2 Which arteries supply blood to the head and neck?

3 What region is supplied by the axillary artery?

4 What are the major arteries of the lower limb?

5 What are the major veins of the head and neck?

6 Which veins of the upper limb are deep veins? Superficial veins?

7 What region is drained by the azygos vein and its tributaries?

THE EFFECTS OF AGING ON BLOOD VESSELS

Cerebral arteriosclerosis is a disease in which the arteries that supply blood to the brain harden and narrow. If blood vessels leading to the brain become blocked or leak so that the brain is denied blood for even a short amount of time, serious brain damage and even death may occur.

Arteriosclerosis and *hypertension* (high blood pressure) are the chief causes of *cerebral hemorrhage* (stroke). In the United States, about 98 percent of the 100,000 people who die of a stroke each year are over 50.

Transient ischemic attacks (TIAs, or "little strokes") occur when momentary blood clots form in arteries leading to the brain; *multiple infarct dementia* occurs when recurrent small clots in the brain destroy brain tissue, impairing thinking and memory. Hypertension usually accompanies *atherosclerosis,* a condition in which plaque builds up in the arteries, narrowing the lumen and reducing the elasticity of the vessels.

FIGURE 20.11 MAJOR ARTERIES AND VEINS

Anterior view of **[A]** the aorta and its principal branches;
[B] principal veins.

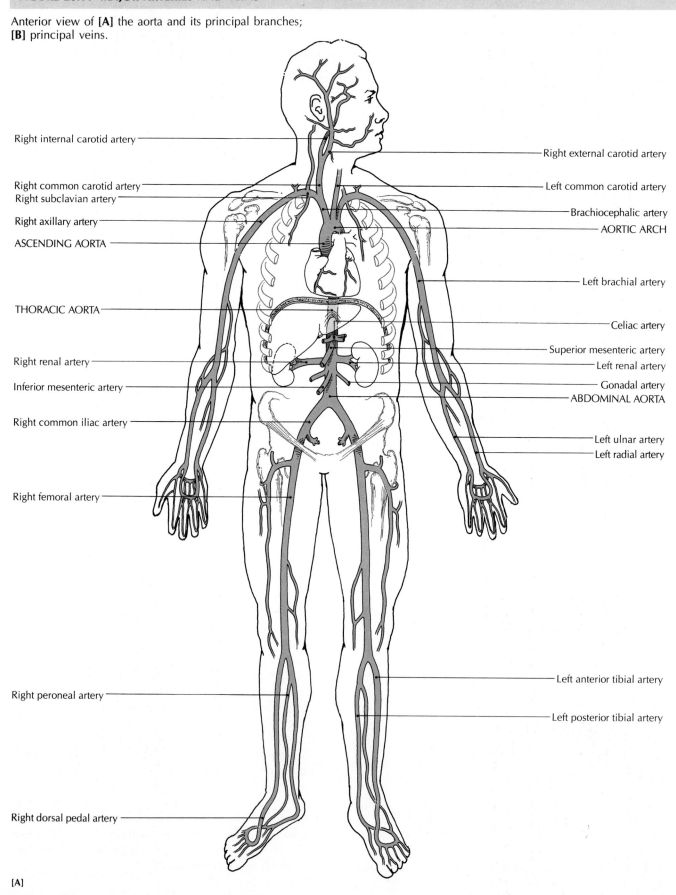

Right internal carotid artery

Right common carotid artery
Right subclavian artery

Right axillary artery

ASCENDING AORTA

THORACIC AORTA

Right renal artery

Inferior mesenteric artery

Right common iliac artery

Right femoral artery

Right peroneal artery

Right dorsal pedal artery

Right external carotid artery

Left common carotid artery

Brachiocephalic artery
AORTIC ARCH

Left brachial artery

Celiac artery

Superior mesenteric artery
Left renal artery

Gonadal artery
ABDOMINAL AORTA

Left ulnar artery
Left radial artery

Left anterior tibial artery

Left posterior tibial artery

[A]

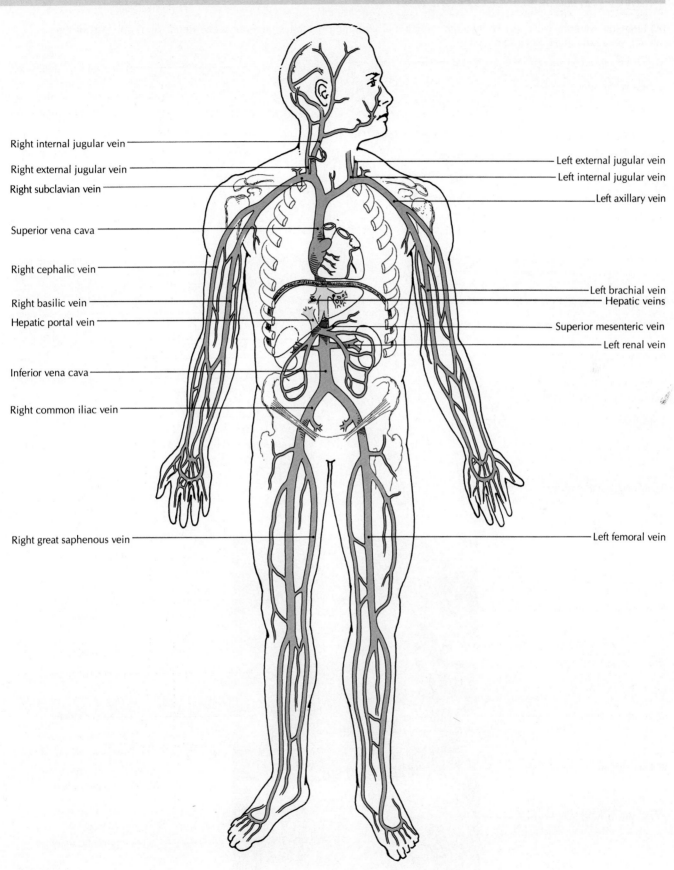

Right internal jugular vein

Right external jugular vein

Right subclavian vein

Superior vena cava

Right cephalic vein

Right basilic vein

Hepatic portal vein

Inferior vena cava

Right common iliac vein

Right great saphenous vein

Left external jugular vein

Left internal jugular vein

Left axillary vein

Left brachial vein

Hepatic veins

Superior mesenteric vein

Left renal vein

Left femoral vein

[B]

FIGURE 20.12 MAJOR BRANCHES OF THE AORTA

[A] Drawing, anterior view. [B] Photograph, abdominal aorta and branches, anterior view. Renal veins and part of the inferior vena cava have been removed.

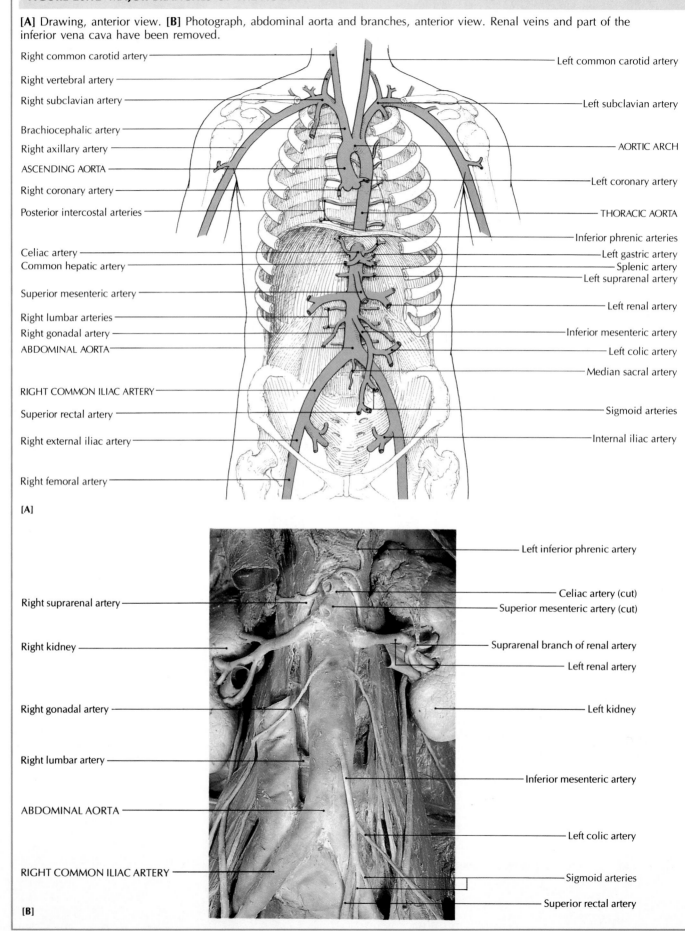

[A]

[B]

Artery	Major branches and region supplied
ASCENDING AORTA	
Right coronary artery	Marginal and posterior interventricular branches; nutrient arteries to heart wall
Left coronary artery	Circumflex and anterior interventricular branches; nutrient arteries to heart wall
AORTIC ARCH	
Brachiocephalic trunk (innominate artery)	Right common carotid artery*; head, neck Right subclavian artery; right upper limb
Left common carotid artery*	Head, neck
Left subclavian artery	Left upper limb
THORACIC AORTA (thoracic portion of descending aorta)	
Bronchial arteries (unpaired)	Bronchi, bronchioles
Esophageal arteries (unpaired)	Esophagus
Intercostal arteries (9 pairs)	Dorsal branch; spinal cord, muscles and skin of back
Posterior intercostal arteries (1 pair below rib 12)	Muscular branch; chest muscles Cutaneous branch; skin of thorax Mammary branch; breasts
Superior phrenic arteries	Diaphragm
ABDOMINAL AORTA (abdominal portion of descending aorta)	
Inferior phrenic arteries	Diaphragm, lower esophagus
Celiac artery (trunk) (unpaired)	Left gastric artery; stomach, esophagus Common hepatic artery; liver, stomach, pancreas, duodenum, gallbladder, bile duct Splenic artery; stomach, pancreas, spleen, omentum
Superior mesenteric artery (unpaired)	Inferior pancreaticoduodenal artery; head of pancreas, duodenum Jejuneal and ileal arteries; small intestine Ileocolic artery; lower ileum, appendix, ascending and part of transverse colon Right colic and middle colic arteries; transverse and ascending colon
Inferior mesenteric artery (unpaired)	Left colic artery; transverse and descending colon Sigmoid artery; descending and sigmoid colon Superior rectal artery; rectum, anal region
Renal arteries	Inferior suprarenal abdominal artery; kidneys, ureters, adrenal glands
Gonadal (testicular or ovarian) arteries	Testes or ovaries
Lumbar arteries (4 or 5 pairs)	Skin, muscle, and vertebrae of lumbar region of back, spinal cord and meninges of lower back, caudal equina
Median sacral (middle) artery	Sacrum, coccyx
Common iliac arteries	External iliac arteries; lower limb Internal iliac arteries; viscera, walls of pelvis, perineum, rectum, gluteal region

*Right and left common carotid arteries ascend into the neck and then bifurcate at the upper part of the larynx into right and left external and internal carotid arteries.

524

FIGURE 20.13 MAJOR ARTERIES OF THE HEAD AND NECK

[A] Arteries of the head and neck, right lateral view.
[B] Ventral surface of the brain showing the cerebral arterial circle (circle of Willis).

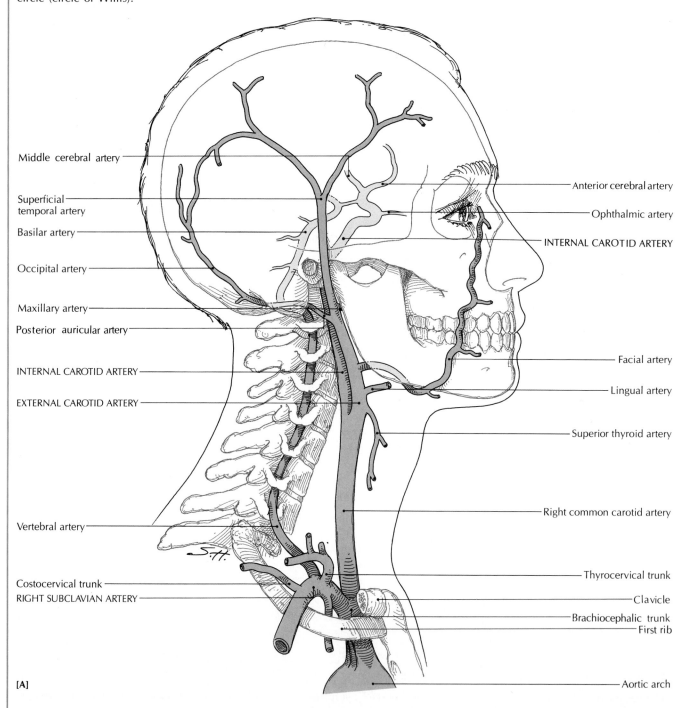

[A]

Artery	Major branches and region supplied	Comments
Internal carotid artery*	Anterior cerebral artery; brain Middle cerebral artery; brain	Internal carotid has no neck branches; passes through carotid canal into cranial cavity, gives off ophthalmic artery, terminates by dividing into anterior and middle cerebral arteries.

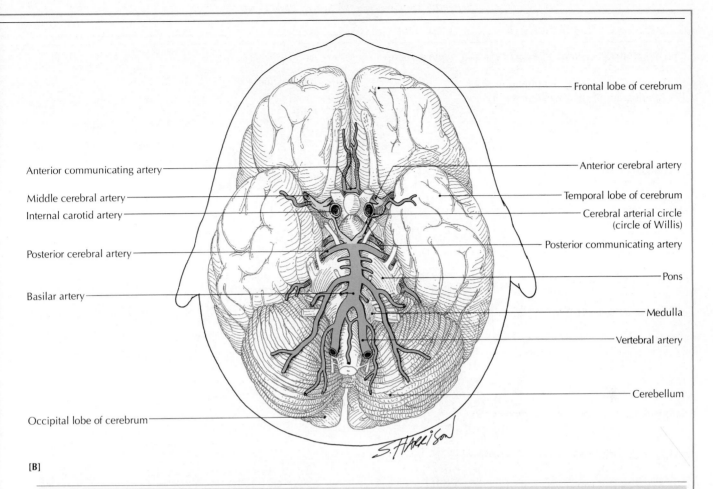

Anterior communicating artery

Middle cerebral artery

Internal carotid artery

Posterior cerebral artery

Basilar artery

Occipital lobe of cerebrum

Frontal lobe of cerebrum

Anterior cerebral artery

Temporal lobe of cerebrum

Cerebral arterial circle (circle of Willis)

Posterior communicating artery

Pons

Medulla

Vertebral artery

Cerebellum

[B]

Artery	Major branches and region supplied	Comments
Subclavian artery	Vertebral artery; brain, spinal cord (cervical region), vertebrae, other deep neck structures Thyrocervical trunk; trachea, esophagus, thyroid gland, neck muscles Costocervical trunk; muscles of back of neck, neck vertebrae	Right and left vertebral arteries unite to form one basilar artery, which bifurcates and terminates as the paired posterior cerebral arteries.
Internal thoracic artery	Anterior thoracic wall, muscles, skin, breasts	
External carotid artery*	Superior thyroid artery; thyroid gland, neighboring muscles, larynx Lingual artery; floor of mouth, tongue, neighboring muscles, salivary glands Facial artery; face below eyes, including soft palate, tonsils, and submandibular gland Occipital artery; skin, muscles, and associated tissues of back of head (scalp), brain meninges Posterior auricular artery; skin, muscles, and associated tissues of and near ear and posterior scalp Superficial temporal artery; outer ear, parotid gland, forehead, scalp and muscles in temporal region Maxillary artery; face, including upper and lower jaws, teeth, palate, nose, muscles of mastication, and brain dura mater	External carotid terminates in region of parotid gland by bifurcating into superficial temporal and maxillary arteries.

*The internal and external carotid arteries are branches of the common carotid artery.

FIGURE 20.14 MAJOR ARTERIES OF THE UPPER LIMB

[A] Drawing of arteries, right anterior view.
[B] Arteriogram of right wrist, anterior view.

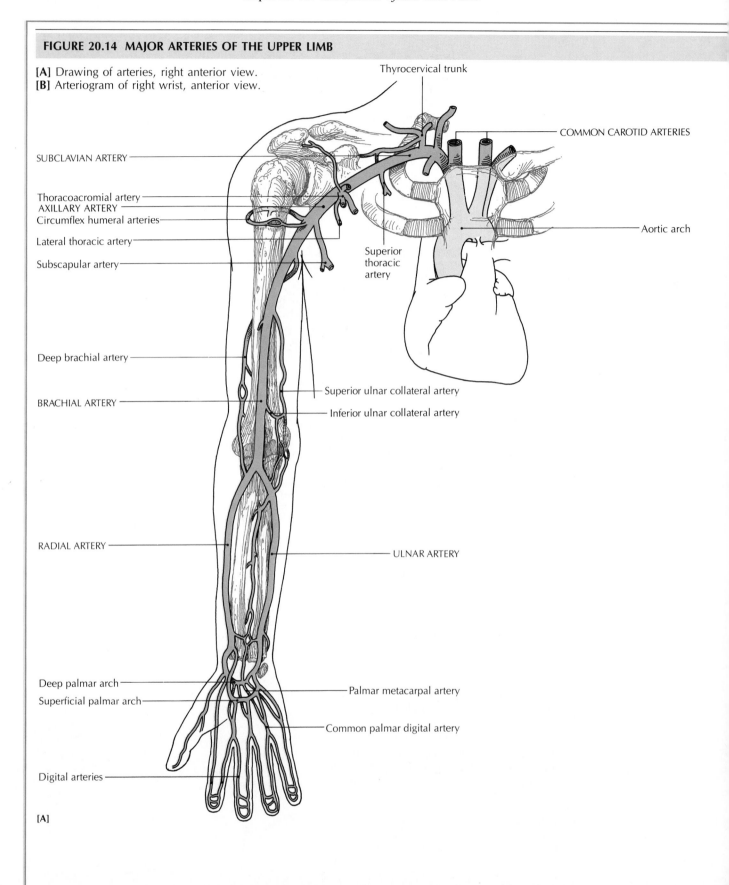

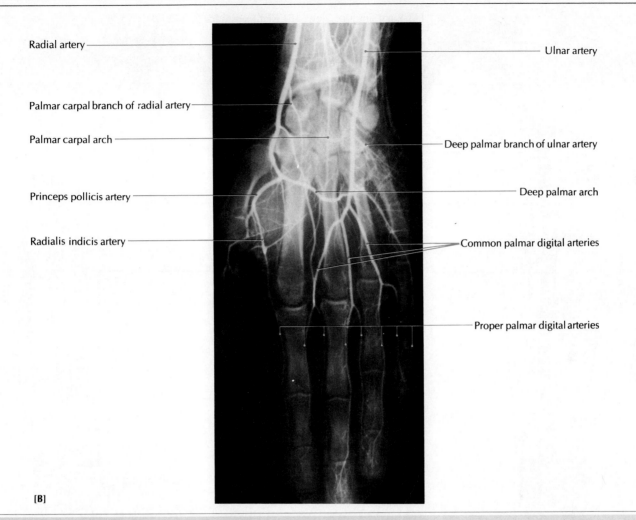

Radial artery

Palmar carpal branch of radial artery

Palmar carpal arch

Princeps pollicis artery

Radialis indicis artery

Ulnar artery

Deep palmar branch of ulnar artery

Deep palmar arch

Common palmar digital arteries

Proper palmar digital arteries

[B]

Artery	Major branches and region supplied	Comments
Axillary artery	Superior thoracic artery, thoracoacromial artery, lateral thoracic artery, subscapular artery, circumflex humeral artery; chest wall, muscles of chest wall, shoulder girdle, shoulder joint	
Brachial artery	Deep (profunda) brachial artery, ulnar collateral arteries; muscles of arm, upper forearm, humerus, and elbow joint	Brachial artery in anterior elbow is site used for measurement of blood pressure.
Radial artery	Branches in forearm; elbow joint, radius, muscles of radial (thumb) side of forearm Continues as deep palmar arch and palmar branches; carpal and metacarpal bones and muscles in palm Arch (along with superior palmar arch) provides palmar membranes, which continue as digital arteries; fingers	Pulse of radial artery can be felt at base of thumb and on back of wrist between tendons of thumb muscles. Anastomoses among arteries of lower forearm and those of branches of arches of palm are substantial.
Ulnar artery	Branches in forearm; elbow joint, ulna, muscles of ulnar (small-finger) side of forearm Continues as superficial palmar arch and palmar branches; carpal and metacarpal bones and muscles in palm Arch (along with deep palmar arch) provides palmar and digital branches; structures of palm and fingers	Pulse of ulnar artery can be felt on the ulnar side of wrist.

FIGURE 20.15 MAJOR ARTERIES OF THE PELVIS AND LOWER LIMB

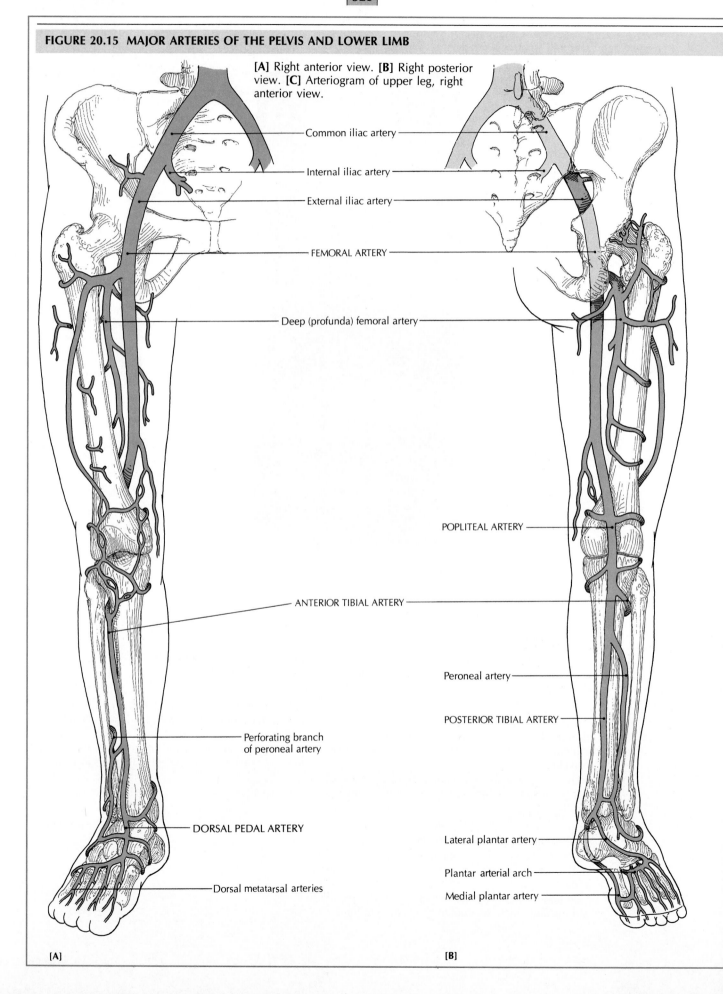

[A] Right anterior view. [B] Right posterior view. [C] Arteriogram of upper leg, right anterior view.

Common iliac artery

Internal iliac artery

External iliac artery

FEMORAL ARTERY

Deep (profunda) femoral artery

POPLITEAL ARTERY

ANTERIOR TIBIAL ARTERY

Peroneal artery

POSTERIOR TIBIAL ARTERY

Perforating branch of peroneal artery

DORSAL PEDAL ARTERY

Lateral plantar artery

Plantar arterial arch

Medial plantar artery

Dorsal metatarsal arteries

[A]

[B]

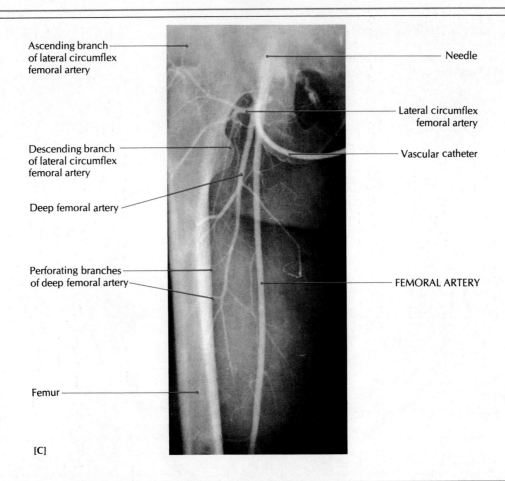

Ascending branch of lateral circumflex femoral artery

Descending branch of lateral circumflex femoral artery

Deep femoral artery

Perforating branches of deep femoral artery

Femur

Needle

Lateral circumflex femoral artery

Vascular catheter

FEMORAL ARTERY

[C]

Artery	Major branches and region supplied	Comments
Common iliac artery	Internal iliac artery; pelvic viscera, including urinary bladder, prostate gland, ductus deferens, uterus, vagina, rectum. Peroneal region, including anal canal, external genitalia; walls of pelvic cavity, including muscles. Gluteal region (buttocks), including muscles, hip joint. External iliac artery; lower limb.	External iliac artery enters thigh behind inguinal ligament and becomes femoral artery.
Femoral artery	Muscular branches, including deep (profunda) femoral artery; skin and muscles of thigh, including anterior flexors, medial adductors, extensors (hamstrings).	Femoral artery, located on anterior thigh, passes backward through adductor magnus to become popliteal artery on back of knee (popliteal fossa).
Popliteal artery	Muscular branches; knee joint, gastrocnemius and soleus muscles.	Popliteal artery terminates just beyond knee joint by dividing into anterior tibial artery and posterior tibial artery.
Anterior tibial artery	Muscular branches; knee joint, muscles, and skin in front of leg, ankle joint.	Anterior tibial artery continues over ankle to become dorsal pedal artery.
Dorsal pedal artery	Muscles, skin, and joints (including ankle) on dorsal side of foot.	
Posterior tibial artery	Muscular branches, including peroneal artery on lateral leg; knee joint, muscles and skin on back of leg, ankle joint. Peroneal, medial malleolar, calcaneal, and plantar arteries; fibula, muscles of fibula, ankle joint, heel, toes.	

FIGURE 20.16 MAJOR VEINS OF THE HEAD AND NECK

[A] Drawing, right lateral view. [B] Cast of blood vessels of the head and neck, anterior view.

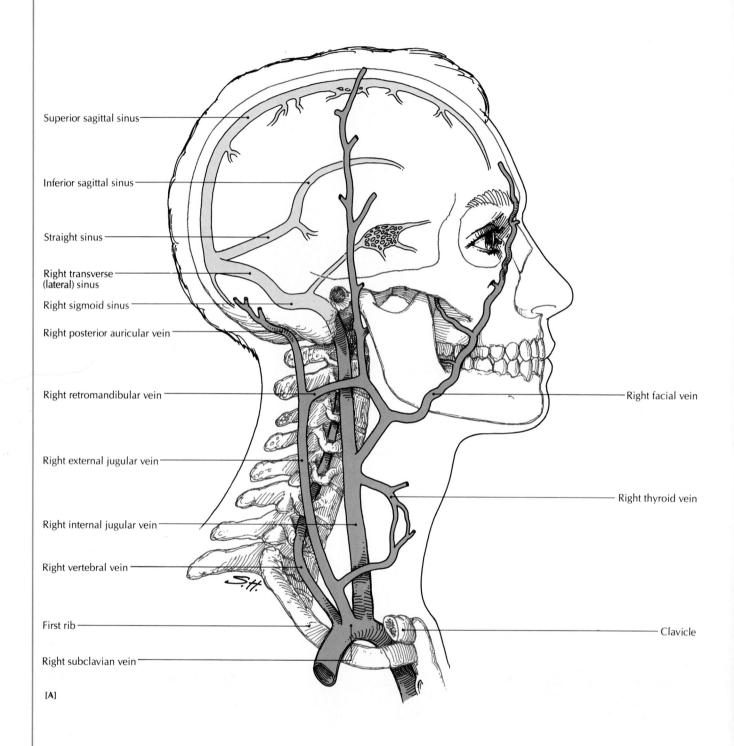

Superior sagittal sinus

Inferior sagittal sinus

Straight sinus

Right transverse (lateral) sinus

Right sigmoid sinus

Right posterior auricular vein

Right retromandibular vein

Right external jugular vein

Right internal jugular vein

Right vertebral vein

First rib

Right subclavian vein

Right facial vein

Right thyroid vein

Clavicle

[A]

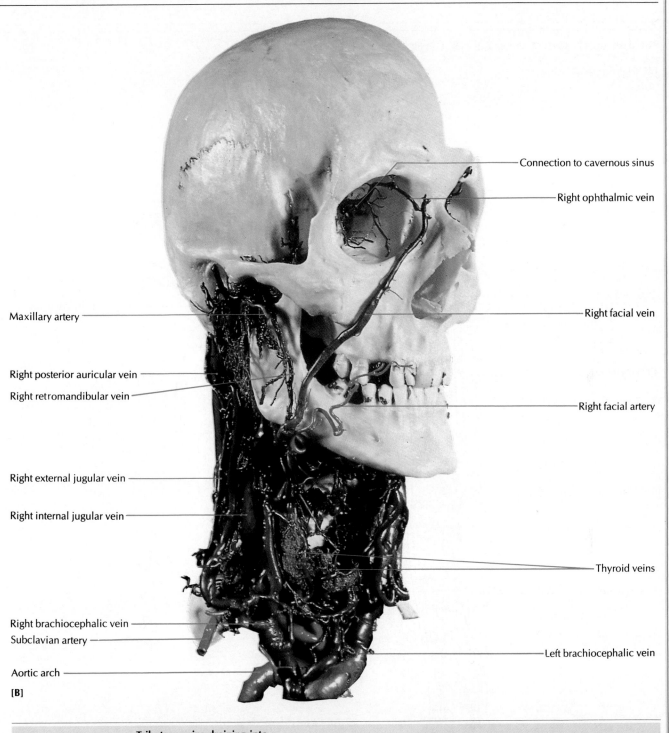

Connection to cavernous sinus

Right ophthalmic vein

Maxillary artery

Right facial vein

Right posterior auricular vein

Right retromandibular vein

Right facial artery

Right external jugular vein

Right internal jugular vein

Thyroid veins

Right brachiocephalic vein

Subclavian artery

Aortic arch

Left brachiocephalic vein

[B]

Major vein	Tributary veins draining into major veins and region drained	Comments
Internal jugular vein	Dural sinuses, including superior sagittal sinus, inferior sagittal sinus, straight sinus, transverse sinuses, sigmoid sinuses; brain, essentially all venous blood within cranial cavity. Facial and thyroid veins; face, neck.	Sigmoid sinuses at base of skull continue through jugular foramina, become paired internal jugular veins; venous dural sinuses are between layers of dura mater. Internal jugular veins join subclavian veins to form brachiocephalic veins, which drain into superior vena cava.
External jugular vein	Retromandibular vein; posterior face. Posterior auricular vein; scalp behind ear. Other veins; parotid glands, scalp, neck muscles.	External jugular formed by retromandibular and posterior auricular veins and drains into subclavian vein.

FIGURE 20.17 MAJOR VEINS OF THE UPPER LIMB

[A] Right anterior view.

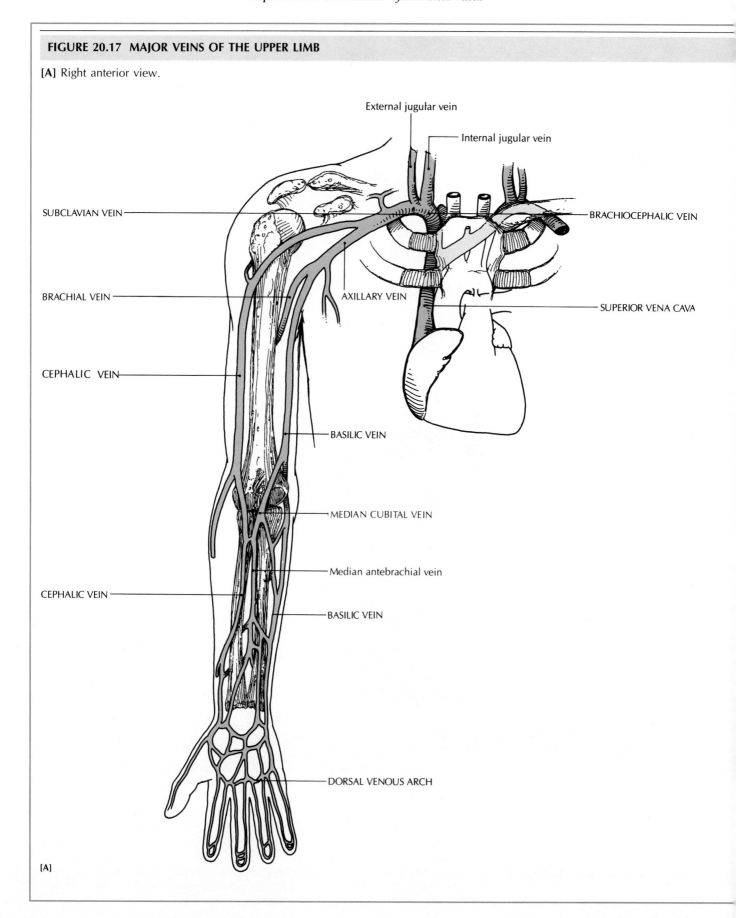

External jugular vein

Internal jugular vein

SUBCLAVIAN VEIN

BRACHIOCEPHALIC VEIN

BRACHIAL VEIN

AXILLARY VEIN

SUPERIOR VENA CAVA

CEPHALIC VEIN

BASILIC VEIN

MEDIAN CUBITAL VEIN

Median antebrachial vein

CEPHALIC VEIN

BASILIC VEIN

DORSAL VENOUS ARCH

[A]

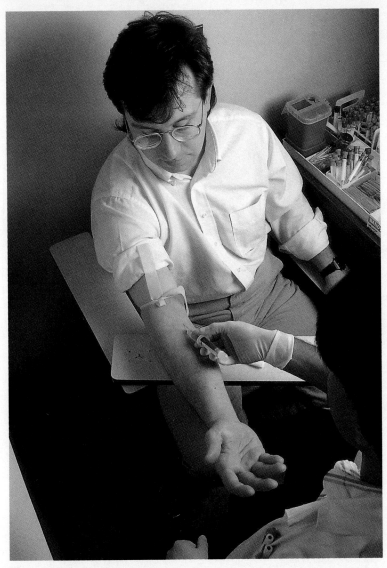

[B] The photograph shows a medical technician drawing blood from the patient's median cubital vein into a hypodermic needle. The tourniquet above the elbow should be snug, but not tight, to close off (occlude) the superficial venous return, distending the vein. If the tourniquet is too tight, the brachial artery will be occluded, obstructing the flow of arterial blood to the forearm and hand.

[B]

Major veins	Region drained	Comments
Deep veins Brachial vein Axillary vein Subclavian vein	Upper limb	Deep veins may accompany an artery and have transverse anastomoses (*venae comitantes*) with arteries. Axillary and subclavian veins accompany corresponding artery; right and left subclavian veins join with internal jugular veins to form brachiocephalic veins.
Superficial veins Dorsal venous arch	Back of hand	Superficial veins are cutaneous and have anastomoses with deep veins.
Basilic vein	Medial forearm and arm	Basilic vein originates from dorsal venous arch, extends along medial side of limb, and terminates by joining brachial vein to form axillary vein.
Cephalic vein	Lateral forearm and arm	Cephalic vein originates from dorsal venous arch, extends along lateral side of limb, and terminates by emptying into axillary vein near lateral end of clavicle.
Median cubital vein	Anterior elbow	Median cubital vein communicates between cephalic and basilic veins, which form variable patterns; cubital vein is often used when vein must be punctured for injection, transfusion, or withdrawal of blood.

FIGURE 20.18 MAJOR VEINS OF THE LOWER LIMB

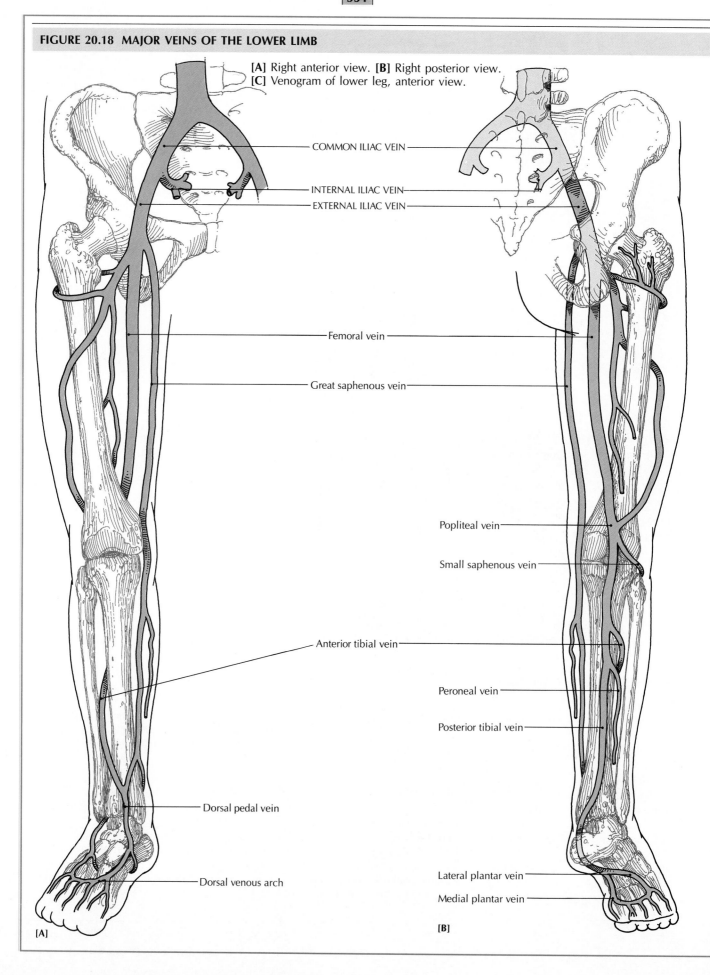

[A] Right anterior view. [B] Right posterior view.
[C] Venogram of lower leg, anterior view.

COMMON ILIAC VEIN

INTERNAL ILIAC VEIN
EXTERNAL ILIAC VEIN

Femoral vein

Great saphenous vein

Popliteal vein

Small saphenous vein

Anterior tibial vein

Peroneal vein

Posterior tibial vein

Dorsal pedal vein

Dorsal venous arch

Lateral plantar vein

Medial plantar vein

[A]

[B]

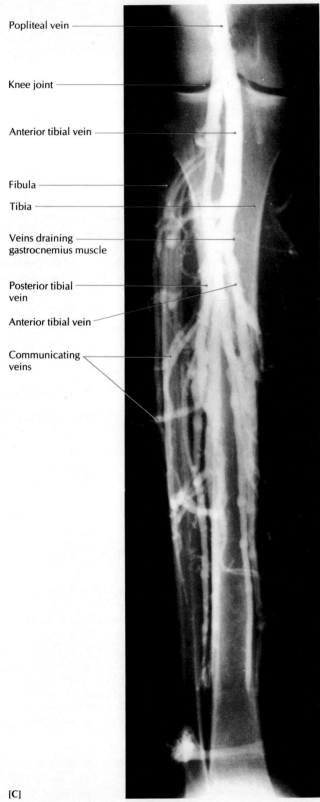

Popliteal vein

Knee joint

Anterior tibial vein

Fibula

Tibia

Veins draining
gastrocnemius muscle

Posterior tibial
vein

Anterior tibial vein

Communicating
veins

[C]

Major veins	Region drained	Comments
Deep veins		
Femoral vein	Lower limb	Deep veins are organized as *venae comitantes*. Femoral vein passes under inguinal ligament to become external iliac vein of pelvis.
Popliteal vein		
Anterior tibial vein		
Posterior tibial vein		
Peroneal vein		
Dorsal pedal vein		
Medial plantar vein		
Lateral plantar vein		
Superficial veins		
Small saphenous vein	Lateral side of leg	Superficial veins are cutaneous and have anastomoses with deep veins. The small saphenous vein arises at lateral side of dorsal venous arch of foot, ascends on posterolateral side of calf, and joins deep popliteal vein.
Great saphenous vein	Medial side of lower limb	Great saphenous vein arises from medial side of dorsal venous arch of the foot, ascends upward on medial aspect of leg and thigh, and joins deep femoral vein in groin.

FIGURE 20.19 MAJOR VEINS OF THE THORAX, ABDOMEN, AND PELVIS

[A] Anterior view. [B] Photograph of interior vena cava, anterior view.

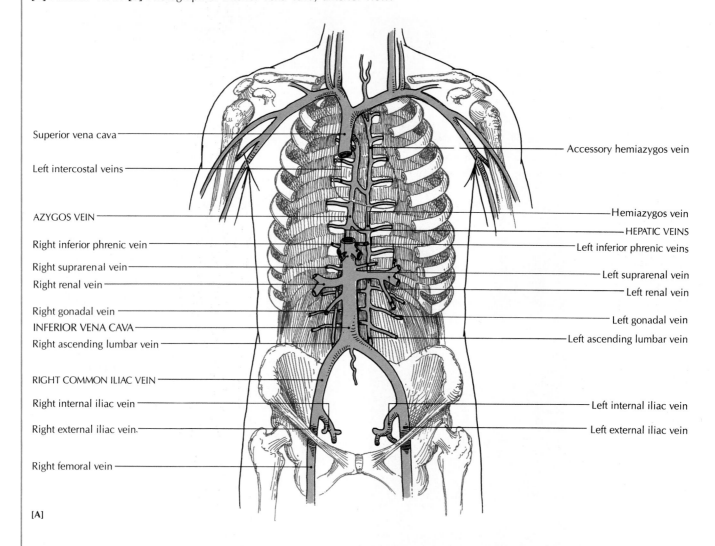

Superior vena cava

Left intercostal veins

AZYGOS VEIN

Right inferior phrenic vein

Right suprarenal vein

Right renal vein

Right gonadal vein

INFERIOR VENA CAVA

Right ascending lumbar vein

RIGHT COMMON ILIAC VEIN

Right internal iliac vein

Right external iliac vein

Right femoral vein

Accessory hemiazygos vein

Hemiazygos vein

HEPATIC VEINS

Left inferior phrenic veins

Left suprarenal vein

Left renal vein

Left gonadal vein

Left ascending lumbar vein

Left internal iliac vein

Left external iliac vein

[A]

Major vein	Tributary veins draining into major veins and region drained	Comments
Common iliac vein	External iliac vein; lower limb	Common iliac vein is continuation of femoral vein.
	Internal iliac vein; pelvis, buttocks, pelvic viscera, including gluteal muscles, rectum, urinary bladder, prostate, uterus, vagina, external genitalia	Common iliac veins unite to form inferior vena cava.
Inferior vena cava (abdominal area)	Ascending lumbar veins; body wall Renal veins; kidneys	Inferior vena cava is largest blood vessel in body (approximately 3.5 cm in diameter). It receives tributaries from lumbar, visceral, and hepatic areas, but not directly from digestive tract, pancreas, or spleen.
	Gonadal veins (ovarian and testicular veins).	Left gonadal vein drains directly into renal vein; right gonadal vein drains directly into inferior vena cava.
	Suprarenal veins.	Left suprarenal vein drains into renal vein.
	Inferior phrenic veins; diaphragm	
	Hepatic veins; liver	Hepatic vein drains blood from liver to inferior vena cava.

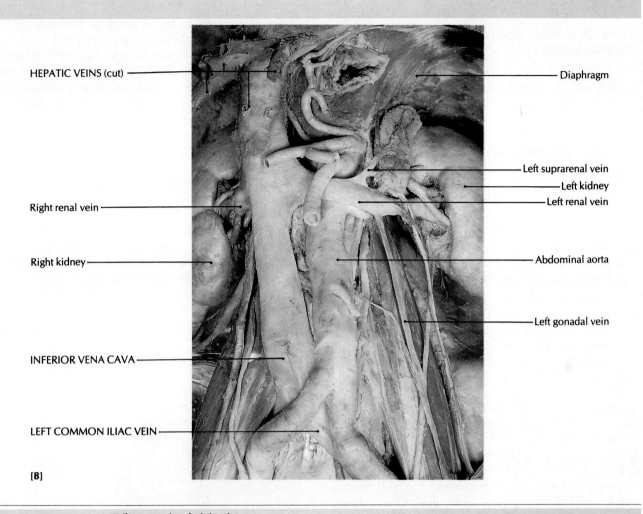

HEPATIC VEINS (cut)

Diaphragm

Left suprarenal vein

Left kidney

Right renal vein

Left renal vein

Right kidney

Abdominal aorta

Left gonadal vein

INFERIOR VENA CAVA

LEFT COMMON ILIAC VEIN

[B]

Major vein	Tributary veins draining into major veins and region drained	Comments
Portal system (hepatic portal vein)	Splenic vein; spleen Superior mesenteric vein; small intestine, part of large intestine Inferior mesenteric vein; large intestine	Portal system drains abdominal and pelvic viscera, which are supplied by celiac, superior mesenteric, and inferior mesenteric arteries. Venous blood of portal system drains into liver and its sinusoids and then into hepatic vein.

Azygos vein	Azygos vein
	  Intercostal veins; chest wall
	  Right ascending lumbar vein; abdominal body wall
	  Right bronchial vein; bronchi
		Azygos vein drains into superior vena cava.

	Hemiazygos vein
	  Intercostal veins (lower 4 or 5); chest wall
	  Left ascending lumbar vein; abdominal body wall
	  Left bronchial vein; bronchi
		Hemiazygos vein drains into azygos vein at level of ninth thoracic vein.

	Accessory hemiazygos vein
	  Intercostal veins (upper 3 or 4); chest wall
	  Right bronchial vein; bronchi
		Accessory hemiazygos vein joins azygos vein at level of eighth thoracic vein.
		Azygos system drains blood from thoracic wall and posterior wall of abdomen. It has venous connections with veins draining inferior vena cava, including those of esophagus, mediastinum, pericardium, and bronchi.
		When inferior vena cava of hepatic portal vein is obstructed, venous blood can be diverted to azygos system and superior vena cava.

DEVELOPMENTAL ANATOMY OF MAJOR BLOOD VESSELS

Blood leaves the embryonic heart through the arterial end (truncus arteriosus) and returns through the horns of the sinus venosus at the venous end, passing into the single atrium through the sinoatrial orifice. The orifice is equipped with two flaps of endocardium that function as primitive valves to prevent the backflow of blood. Blood from the atrium flows into the ventricle through the atrioventricular orifice. When the ventricle contracts, blood is pumped into the bulbus cordis. The bulbus cordis narrows to become the truncus arteriosus, which pierces the roof of the pericardium to open into the *aortic sac.* From the somewhat dilated aortic sac emerge the *aortic arches* [see FIGURE 19.11F].

Six pairs of aortic arches develop, but only the left arch of the fourth pair forms part of the arch of the mature aorta. The first, second, and fifth pairs disappear during embryonic life. The third pair forms the common carotid arteries and the internal carotid arteries. The right arch of the fourth pair becomes the right subclavian artery. The proximal part of the left sixth aortic arch develops into the proximal part of the left pulmonary artery, and the distal part forms the ductus arteriosus; the proximal part of the right sixth aortic arch develops into the proximal part of the right pulmonary artery, while the distal part of the arch degenerates.

At about day 35 to 42, the truncus arteriosus is divided into the "great arteries," the *pulmonary trunk* and *aorta* [see FIGURE 19.11G]. The pulmonary trunk carries blood to the lungs, and the aorta carries blood to the blood vessels that supply the rest of the body.

WHEN THINGS GO WRONG

Aneurysm

An *aneurysm* (Gr. "to dilate") is a balloonlike, blood-filled dilation of a blood vessel. It occurs most often in the larger arteries. Cerebral aneurysms are also common. Aneurysms happen when the muscle cells of the tunica media become elongated, and the wall weakens. The wall may also be weakened by degenerative changes related to such diseases as *arteriosclerosis* (hardening of the arteries). If an aneurysm is not treated, the vessel will eventually burst, and if a cerebral vessel is affected, a stroke (cerebrovascular accident, CVA) may result.

Atherosclerosis

Atherosclerosis (Gr. "porridgelike hardening") is the leading cause of coronary artery disease. It is characterized by deposits of fat, fibrin, cellular debris, and calcium on the inside of arterial walls (see photo). These built-up materials, called *atheromatous plaques,* narrow the lumen and reduce the elasticity of the vessel.

Although the basic cause of atherosclerosis is not known, the earliest stage in its development is believed to be damage to the endothelial cells and tunica intima of the vessel wall. Once the damage occurs, the endothelial cells proliferate and attract lipid substances. Several factors increase its progress. Among them are cigarette smoking and animal fat and cholesterol in the diet. Other factors that may have an effect are hypertension, diabetes, age, and heredity.

There are three ways in which atherosclerosis can cause a heart attack: (1) It can completely clog a coronary artery. (2) It can provide a rough surface where a blood clot *(thrombus)* can form and grow so it closes off the artery and causes a *coronary thrombosis.* (3) It can partially

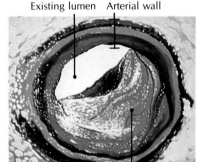

Existing lumen   Arterial wall

Atherosclerotic plaque can narrow the lumen of an artery substantially.

Plaque

block blood to the myocardium and regions of the impulse-conducting system of the heart; this may result in the heart not beating rhythmically.

Coronary Artery Disease

The typical effect of *coronary artery disease* (CAD) is a reduced supply of oxygen and other nutrients to the myocardium. The condition is usually brought on by atherosclerosis. As a result, less blood reaches the heart, and *myocardial ischemia* (ih-SKEE-mee-uh; Gr. *iskhein,* to keep back + *haima,* blood) or local anemia in cardiac muscle, occurs.

Hypertension (High Blood Pressure)

Hypertension, commonly called *high blood pressure,* is characterized by systolic or diastolic pressure that is above normal all the time (not just as a result of specific activities or conditions). About 35 million Americans have hypertension, and more than a million die annually of related diseases. The two types of hypertension are the more common *primary* or *essential (idiopathic) hypertension,* an above-normal blood pressure that cannot be attributed to any particular cause, and *secondary hypertension,* which results from a disorder such as kidney disease or arteriosclerosis.

Hypertensive arterial walls are hard and thick, and their elasticity is reduced, forcing the heart to work harder to pump enough blood. If hypertension persists, the heart may become enlarged, a condition called *hypertensive heart disease.* High blood pressure can also cause a stroke if the extra force of the blood ruptures an artery in the brain and a cerebral hemorrhage occurs.

Although hypertension may be accompanied by headache, dizziness, or other symptoms, one of the problems with treating hypertension is that often there are no external signs that blood pressure is abnormally high. Prognosis is optimistic if the disease is diagnosed and treated early, before complications develop. If untreated, hypertension is accompanied by a high mortality rate.

Hypertension may be an inherited problem, but it is also related to stress, obesity, a diet that is high in sodium and saturated fats, aging, lack of physical activity, race (it is most common in blacks), and the use of tobacco and oral contraceptives (especially if used together).

Stroke

A *stroke* occurs when the blood supply to a portion of the brain is cut off. There are several ways a stroke may occur: (1) Atherosclerosis in the arteries of the brain or neck may block the flow of blood. (2) A thrombus may form in the atherosclerotic vessel, closing off the artery and causing a *cerebral thrombosis.* (3) A traveling blood clot *(embolus)* can become wedged in a small artery of the brain or neck; this kind of stroke is an *embolism.* (4) A weak spot in a blood vessel may break; this is a *cerebral hemorrhage.* (5) In rare cases, a brain tumor may press on a blood vessel and shut off the blood supply.

Thrombophlebitis

Thrombophlebitis (Gr. *thrombos,* clot + *phleps,* blood vessel + *itis,* inflammation) is an acute condition characterized by clot formation and inflammation of deep or superficial veins. Thrombophlebitis is usually the result of an alteration in the endothelial lining of the vein. Platelets begin to gather on the roughened surface, and the consequent formation of fibrin traps red blood cells, white blood cells, and more platelets. The result is a thrombus. If untreated, the thrombus may become detached from the vein and begin to move through the circulatory system. Such a mobile blood clot is called an *embolus.* The result may be pulmonary embolism, a blockage of the pulmonary veins that is a potentially fatal condition. Some causes of thrombophlebitis include surgery, trauma, childbirth, oral contraceptives, prolonged bed rest, infection, and intravenous drug abuse.

Varicose Veins

Varicose veins (L. *varix,* swollen veins) are abnormally dilated and twisted veins. The saphenous veins and their branches in the legs are often affected. They become permanently dilated and stretched when the one-way valves in the legs weaken. As a result, some blood flows backward and pools in the veins. Varicose veins can result from a hereditary weakness of the valves, but can also be caused or aggravated by vein inflammation *(phlebitis),* blood-clot formation *(thrombophlebitis),* pregnancy, lack of exercise, loss of elasticity in old age, smoking, low-fiber diet, and occupations that require long periods of standing. Untreated varicose veins may lead to edema in the ankles and lower legs, leg ulcers, dizziness, pain, fatigue, and nocturnal cramps.

Portal Cirrhosis of the Liver

Portal cirrhosis of the liver results from chronic alcoholism, the ingestion of poison, viral diseases (such as infectious hepatitis), or other infections in the bile ducts. It is a condition in which large amounts of fibrous tissue develop within the liver. This fibrous tissue destroys many of the functioning liver cells, and eventually grows around the blood vessels, constricting them and greatly impeding the flow of portal blood through the liver. This impediment blocks the return of blood from the intestines and spleen. As a result, blood pressure increases so much that fluid moves out through the capillary fluid shift into the peritoneal cavity, leading to *ascites,* the accumulation of free fluid in the peritoneal cavity.

Patent Ductus Arteriosus

Patent ductus arteriosus occurs when the ductus arteriosus of the fetus, which carries blood from the pulmonary artery to the descending aorta, does not shut down at birth (see page 516). The fetal ductus arteriosus bypasses the nonfunctioning lungs. But when the lungs begin to function at birth (and the baby is no longer relying on the placenta for oxygen and the release of carbon dioxide), the blood normally passes through the pulmonary artery to the lungs. Patent ductus arteriosus creates a left-to-right movement of some arterial blood from the aorta to the pulmonary artery, which recirculates arterial blood through the lungs.

Transposition of the Great Vessels

In *transposition of the great vessels,* the two arterial trunks emerging from the heart are reversed. The aorta emerges from the right ventricle, and the pulmonary artery emerges from the left ventricle. As a result, oxygenated blood returning to the left side of the heart is carried back to the lungs by the pulmonary artery, and unoxygenated blood traveling to the right side of the heart is carried into the systemic circulation by the transposed aorta.

CHAPTER SUMMARY

Types of Blood Vessels (p. 507)

1 *Arteries* carry blood away from the heart to capillary beds throughout the body. Arterial blood is oxygenated, with the exception of the blood in the *pulmonary arteries,* which carry deoxygenated blood from the heart to the lungs.

2 A *pulse* is a beat felt on the surface of the skin over a nearby artery. It corresponds to the beat of the heart and the alternating expansion and recoil of the arterial wall.

3 The major arterial trunks are the **aorta** from the left ventricle and the **pulmonary trunk** from the right ventricle.

4 Arterial walls are composed of three layers: the inner **tunica intima,** the middle **tunica media,** and the outer **tunica adventitia.**

5 Arteries branch into smaller arteries and then into smaller **arterioles** shortly before reaching the capillary networks. Terminal arterioles control the flow of blood from arteries into capillaries.

6 Arterioles enter the body tissues and branch out further to form **capillaries,** the bridge between the arterial and venous systems.

7 The **microcirculation** of the blood consists of the capillaries, terminal arterioles, metarterioles, and venules.

8 The three types of capillaries are **continuous capillaries, fenestrated capillaries,** and **sinusoids,** each differentiated to perform a specific function.

9 Blood drains from capillaries into **venules,** tiny veins that unite to form larger venules and veins. *Veins* carry deoxygenated blood from the body tissues to the heart, with the exception of the **pulmonary veins,** the *hepatic portal system* that carries blood from the capillaries of the intestines to the capillaries of the liver, and the *hypothalamic-hypophy-* *seal portal system* in which veins formed from the capillaries of the hypothalamus divide into the capillaries of the anterior pituitary gland.

10 Veins usually contain paired semilunar bicuspid valves that permit blood to flow only toward the heart. Their walls contain the same layers as arterial walls.

11 Venous blood pressure is low, and blood is assisted toward the heart by the skeletal muscle pump.

Circulation of the Blood (p. 512)

1 The **pulmonary circulation** carries deoxygenated blood from the heart to the lungs, where carbon dioxide is removed and oxygen is added. It then carries oxygenated blood back to the heart.

2 The *systemic circulation* supplies the tissues of the body with blood rich in oxygen and also removes blood high in carbon dioxide. The main vessels of the **arterial division** are the aorta, ascending aorta, aortic arch, descending aorta, and common iliac arteries. The main vessels of the **venous division** are the superior vena cava, inferior vena cava, and coronary sinus.

3 Veins ordinarily transport blood directly back to the heart from a capillary network, but the two **portal systems** of the body (hepatic and hypothalamic-hypophyseal) transport the blood to a second set of capillaries on its way to the venous system.

4 The brain is supplied with blood by two **vertebral arteries** and two **internal carotid arteries.** All the blood entering the cerebrum must first pass through the **cerebral arterial circle.**

5 The arrangement of blood vessels in the skin allows for the increase or decrease of heat radiation from the integumentary system.

6 Blood nourishes skeletal muscles and also removes wastes. The main controllers of blood flow are the sympathetic vasodilator fibers that cause the blood vessels to dilate.

7 The *circulatory system of a fetus* differs from that of a child or adult in that the fetus gets nutrients and removes its wastes through the placenta, and its lungs, kidneys, and digestive system (except for the liver) do not function.

8 The fetus has an opening in the septum between the atria called the **foramen ovale,** and a vessel called the **ductus arteriosus** that bypasses the lungs by carrying blood from the pulmonary artery to the aorta. In a normal child, both close at birth.

Major Arteries and Veins (p. 519)

The major arteries and veins of the body are illustrated in FIGURES 20.12 through 20.19, and are summarized in the accompanying tables.

The Effects of Aging on Blood Vessels (p. 519)

1 The effects of aging generally cause blood vessels to become narrowed and less elastic.

2 Some major disorders associated with aging and blood vessels are cerebral arteriosclerosis, cerebral hemorrhage, hypertension, thrombophlebitis, and varicose veins.

Developmental Anatomy of Major Blood Vessels (p. 538)

1 Six pairs of *aortic arches* develop from the aortic sac, but only the left arch of the fourth pair forms part of the mature aortic arch.

2 At about day 35 to 42, the truncus arteriosus is divided into the *pulmonary trunk* and *aorta.*

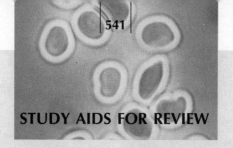

STUDY AIDS FOR REVIEW

MEDICAL TERMINOLOGY

ANGIOGRAM An x-ray of blood vessels, taken after a radiopaque dye is injected into the vessels.

ARTERIOGRAPHY Technique of x-raying arteries after a radiopaque dye is injected.

AVASCULAR NECROSIS Condition in which tissue dies from the lack of blood (from the blood vessels).

CLAUDICATION Improper circulation of blood in vessels of a limb, which causes pain and lameness.

CYANOSIS A bluish discoloration of tissue due to an oxygen deficiency in the systemic blood.

EDEMA Swelling due to abnormal accumulation of fluid in intercellular tissue spaces.

ENDARTERECTOMY Removal of an obstructing region of the inner wall of an artery.

HEMATOMA Tumor or swelling in tissue due to an accumulation of blood from a break in the wall of a blood vessel.

OCCLUSION Clot or closure in the lumen of a blood vessel or other structure.

PHLEBOSCLEROSIS Thickening or hardening of walls of veins.

SHUNT Connection between two blood vessels or between two sides of the heart.

THROMBECTOMY Removal of a blood clot from a blood vessel.

VALVOTOMY Cutting into a valve.

VENIPUNCTURE Inserting a catheter or needle into a vein.

UNDERSTANDING THE FACTS

1 Arteries branch into smaller vessels called _____, which then branch into smaller _____.
2 What is the pulse?
3 Define vasa vasorum.
4 What is the function of stellate reticuloendothelial cells (Kupffer cells)?
5 What functions do precapillary sphincters perform?
6 The _____ drains blood from the walls of the heart.
7 List the major blood vessels of the pulmonary circulation.
8 Much of the blood in the fetus bypasses the liver by flowing through the _____.
9 List the major arteries that branch from the aorta.
10 How do the walls of veins differ from the walls of arteries?
11 What is microcirculation?
12 How does exercise affect venous circulation?
13 Define aneurysm.
14 List some factors that can increase the progress of atherosclerosis.
15 What factors are often related to hypertension?
16 Give the location of the following (be specific):
   a external carotid artery
   b vasa vasorum
   c capillary channels
   d common iliac vein
   e foramen ovale
   f great saphenous vein

UNDERSTANDING THE CONCEPTS

1 Why is it better to classify blood vessels by direction of blood flow rather than by the type of blood they carry?
2 Describe the structure of the arterial wall and how it relates to the function of arteries.
3 How do the arterioles help the body to cope with environmental heat?
4 Why must the muscular system be nourished by the blood on a priority basis?
5 The arterial system holds about 20 percent of the blood, while the venous system holds about 75 percent. How do you account for this difference?
6 Discuss the modifications of the circulatory system that occur in the fetus.
7 What are some of the arterial conditions that can bring about an increase in blood pressure?
8 How does dehydration of the body affect circulation?

SELF-QUIZ
Multiple-Choice Questions

20.1 The great arteries that emerge from the heart are often called
   a major arteries
   b aortas
   c trunks
   d arterioles
   e vessels
20.2 Which of the following tunics contains connective tissue and smooth muscle cells?
   a tunica intima
   b tunica media
   c tunica adventitia
   d a and b
   e a, b, and c

20.3 Of the following, which is *not* a function of arterioles?

 a help regulate body temperature
 b direct the flow of blood
 c respond to emotional stimuli
 d allow gas exchange to take place
 e act as shunts

20.4 Which type of capillary has walls made up of one continuous endothelial cell?

 a sinusoids
 b continuous
 c fenestrated
 d arteriole
 e venule

20.5 Veins that do not carry deoxygenated blood are the

 a pulmonary veins
 b hepatic veins
 c hypophyseal veins
 d a and b
 e a, b, and c

20.6 Which of the following is associated with the microcirculation?

 a terminal arterioles
 b metarterioles
 c venules
 d a and b
 e a, b, and c

20.7 In a portal system, the blood passes through how many sets of capillaries?

 a one
 b two
 c three
 d four
 e five

20.8 All the blood entering the cerebrum must first pass through the

 a vertebral arteries
 b carotid arteries
 c basilar artery
 d cerebral arterial circle
 e cerebral arteries

20.9 The umbilical cord contains

 a one umbilical vein
 b two umbilical veins
 c one umbilical artery
 d two umbilical arteries
 e a and d

20.10 Which of the following vessels has the greatest overall cross-sectional area?

 a arteries
 b arterioles
 c capillaries
 d venules
 e veins

True-False Statements

20.11 Terminal arterioles are those closest to a capillary.

20.12 Capillary walls are selectively permeable.

20.13 The tunica media in veins is much thicker than in arteries.

20.14 Like arteries, veins are nourished by small vasa vasorum.

20.15 Venous valves are especially abundant in the abdominal viscera.

20.16 Collateral channel arteries are exchange vessels.

20.17 Blood flow in large arteries does not show a pulse.

Completion Exercises

20.18 Some arteries are elastic and some are _____.

20.19 Arterioles branch into smaller vessels called _____.

20.20 The walls of the large arteries are nourished by small blood vessels called the _____.

20.21 The immediate postcapillary venules are called _____ venules.

20.22 _____ venules generally accompany muscular arteries.

20.23 Blood is drained from the walls of the heart by the _____.

Matching

20.24 _____ ductus venosus
20.25 _____ foramen ovale
20.26 _____ ductus arteriosus
20.27 _____ fossa ovalis
20.28 _____ celiac artery
20.29 _____ iliac artery
20.30 _____ subclavian artery
20.31 _____ brachial artery
20.32 _____ hemiazygos vein
20.33 _____ external iliac vein
20.34 _____ inferior phrenic vein
20.35 _____ inferior mesenteric vein

 a connects the two artria
 b aortic arch shunt
 c bypasses the liver
 d shallow depression
 e found in the left ventricle
 f supplies the neck
 g supplies the arm
 h supplies the stomach
 i supplies the muscles of the abdomen
 j supplies the lower limbs
 k continuation of the femoral vein
 l drains diaphragm
 m drains large intestine
 n drains kidneys
 o drains into the azygous vein

A SECOND LOOK

1 In the following drawing, label the internal and external carotid arteries and the right subclavian artery.

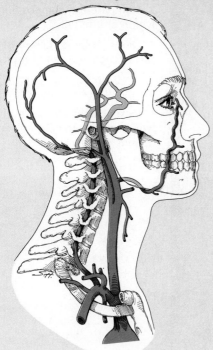

2 In the following drawing, label the brachial, ulnar, and axillary arteries and the common carotid arteries.

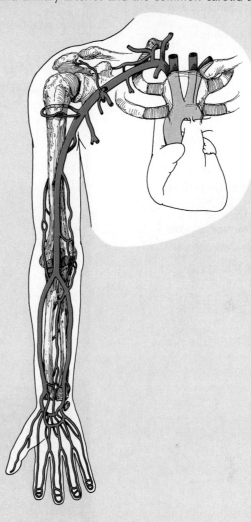

3 In the following drawing, label the common iliac vein, external iliac vein, internal iliac vein, femoral vein, and the great saphenous vein.

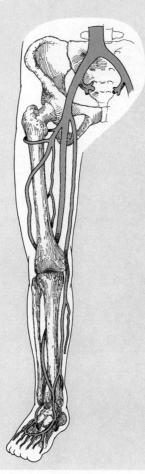

21

The Lymphatic System

LEARNING OBJECTIVES

1 Give the four major functions of the lymphatic system.

2 Compare the compositions of lymph and blood.

3 Describe the structure of the lymphatic capillaries and other lymph vessels.

4 Trace the circulation of lymph.

5 Describe the structure of the lymph nodes, tonsils, spleen, and thymus gland.

Both the cardiovascular system and the lymphatic system move fluid throughout the body, but there are distinct differences between them. The lymphatic system is not a closed, circular system, and it does not have a central pump. It is made up of a network of thin-walled vessels that carry a clear fluid [FIGURE 21.1]. The lymphatic system performs four major functions:

1 It collects excess water and proteins from the interstitial fluids that bathe cells throughout the body and returns them as lymph to the bloodstream.

2 It transports fats from the tissue surrounding the small intestine to the bloodstream.

3 It destroys microorganisms and other foreign substances.

4 It provides long-term protection against microorganisms and other foreign substances.

THE LYMPHATIC SYSTEM

The *lymphatic system* begins with very small vessels called *lymphatic capillaries*, which are in direct contact with interstitial fluid and surrounding tissues [FIGURE 21.1B, C]. The system collects and drains most of the fluid that is forced out of the bloodstream and accumulates in the spaces between cells. The small lymphatic capillaries merge to form larger lymphatic vessels called *lymphatics*, which pass through specialized structures called *lymph nodes*. These larger vessels converge to form two main drainage ducts that return the excess fluid to the blood circulation through the subclavian veins above the heart. All the tissues of the body except those of the central nervous system and the cornea are drained by the lymphatic system.

In addition to lymphatic capillaries, lymphatics, and lymph nodes, the lymphatic system consists of lymphoid organs: the *spleen* and *thymus gland*. The three pairs of *tonsils* are actually lymphatic tissue of the throat region. Lymphatic tissue is a special type of tissue, set apart from the fundamental tissue types already described. It is a variety of reticular connective tissue containing varying amounts of specialized leukocytes called lymphocytes. All lymphatic organs contain relatively large numbers of lymphocytes within a framework of reticular cells and fibers. Finally, *aggregated lymph nodules*, which are scattered throughout the small intestine, help initiate the secretion of antibodies.

Lymph

The blood hydrostatic pressure inside blood vessels, which is generated by the ventricular contraction of the heart, causes water, small proteins (albumin), and other materials to be forced out of capillaries into the spaces between cells. This *interstitial fluid* bathes and nourishes surrounding body tissues. However, if there were not some means of draining it from the tissue, excess interstitial fluid would cause the tissues to swell, producing *edema*. Normally, however, excess fluid moves from the body tissues into the far-reaching small lymphatic capillaries and through the lymphatic system until it returns to the bloodstream. Once inside the lymphatic capillaries, the fluid is called *lymph* (L. "clear water").

Composition and cells of lymph The body contains about 1 to 2 L of lymph, or about 1 to 3 percent of body weight. Lymph is similar to blood, but it does not contain erythrocytes and most of the proteins found in blood. The cells of the lymph fluid are *leukocytes*, and they perform much the same protective functions as leukocytes do in the blood (see Chapter 18). Most of the leukocytes essential to the function of lymph are produced in the bone marrow; leukocytes are found in the blood as well as in the organs and lymph nodes of the lymphatic system.

Monocytes are a class of leukocytes capable of developing into phagocytic *macrophages*. Macrophages are collectively referred to as the *reticuloendothelial system* ("network of endothelial cells"). These cells are found in lymph, in lymph nodes, and adhering to the walls of blood vessels and lymph vessels. Macrophages also migrate into and become attached to many other tissues, and are then referred to as *tissue macrophages* or *histocytes* ("tissue cells"). Histocytes are common in the walls of the lung (alveolar macrophages), the sinusoids of the liver (stellate reticuloendothelial, or Kupffer, cells), the kidney (mesangial phagocytes), the spleen (the littoral cells), and the bone marrow. These cells group into large clusters that surround and isolate foreign particles that are too large to phagocytize. This "walling-off" process occurs in certain chronic infections, such as tuberculosis, and limits the spread of the infectious microorganism.

Macrophages in the lymph nodes are active phagocytes and play a major role in resistance against *all* invading microorganisms. They also are important in resistance against a specific foreign microorganism or its toxin. However, specific defenses are accomplished by another class of leukocyte, the *lymphocyte*.

The two fundamental types of lymphocytes are *B cells* and *T cells* (also called **B lymphocytes** and **T lymphocytes**). The B cells produce specific antibodies, and the T

FIGURE 21.1 THE LYMPHATIC SYSTEM

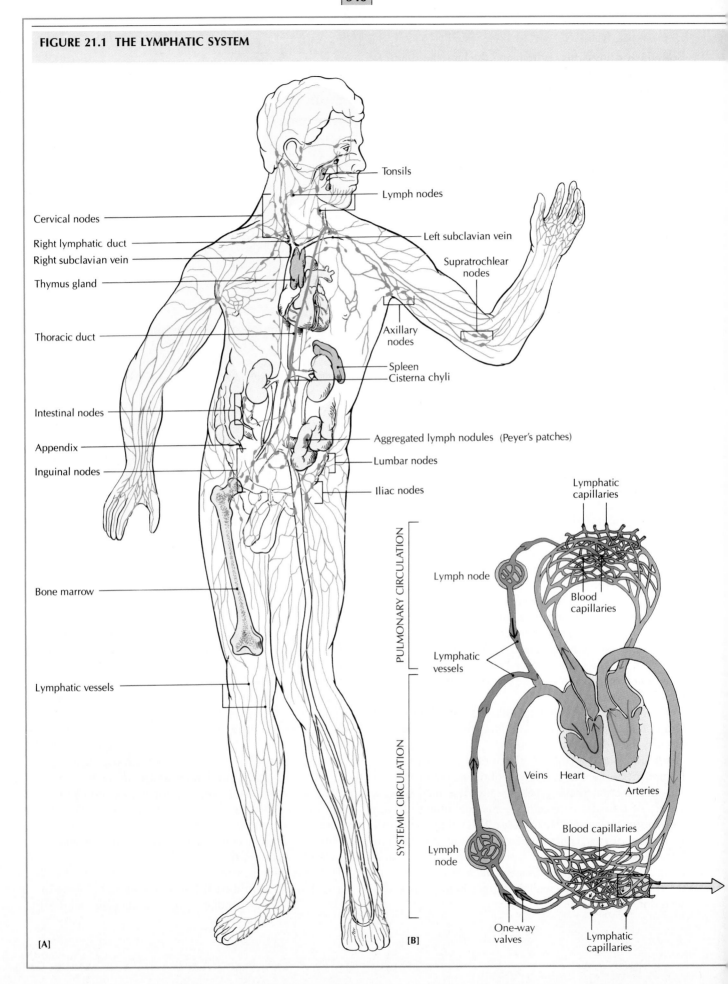

Tonsils

Lymph nodes

Cervical nodes

Right lymphatic duct

Right subclavian vein

Thymus gland

Thoracic duct

Intestinal nodes

Appendix

Inguinal nodes

Bone marrow

Lymphatic vessels

Left subclavian vein

Supratrochlear nodes

Axillary nodes

Spleen
Cisterna chyli

Aggregated lymph nodules (Peyer's patches)

Lumbar nodes

Iliac nodes

Lymphatic capillaries

Lymph node

Blood capillaries

Lymphatic vessels

PULMONARY CIRCULATION

SYSTEMIC CIRCULATION

Veins   Heart

Arteries

Lymph node

Blood capillaries

One-way valves

Lymphatic capillaries

[A]

[B]

[A] Major components of the lymphatic system. The right lymphatic duct drains the upper right quadrant of the body (yellow), and the thoracic duct drains the rest of the body. [B] Relationship of the blood vessels and lymphatic vessels (green). Excess fluid and proteins in the tissue spaces enter lymphatic capillaries and return to the venous system by lymphatic vessels. Arrows indicate direction of fluid flow. [C] Enlargement showing the relationship of lymphatic and blood capillaries. [D] Basic structure of lymphatic tissue.

Structure	Major functions
Lymphatic capillaries	Collect excess interstitial fluid from tissues.
Lymphatics (collecting vessels)	Carry lymph from lymphatic capillaries to veins in neck.
Lymph nodes	Situated along collecting lymphatic vessels; filter foreign material from lymph.
Axillary nodes	Drain arms, most of thoracic wall, breasts, upper abdominal wall.
Supratrochlear nodes	Drain hands, forearms.
Cervical nodes	Drain scalp, face, nasal cavity, pharynx.
Intestinal nodes	Drain abdominal viscera.
Inguinal nodes	Drain legs, external genitalia, lower abdominal wall.
Iliac nodes	Drain pelvic viscera.
Lumbar nodes	Drain pelvic viscera.
Tonsils	Destroy foreign substances at upper entrances of respiratory and digestive systems.
Spleen	Filters foreign substances from blood, manufactures phagocytic lymphocytes, stores red blood cells, releases blood to body in case of extreme blood loss.
Thymus gland	Forms antibodies in newborn; involved in initial development of immune system; site of differentiation of lymphocytes into T cells; produces thymosin.
Aggregated lymph nodules (Peyer's patches)	Respond to antigens in intestine by generating plasma cells that secrete antibodies.

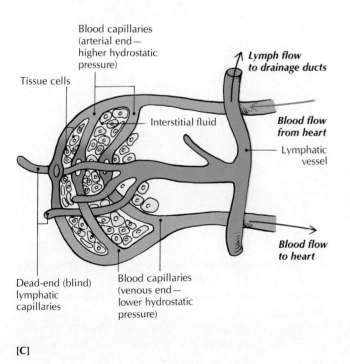

[C]

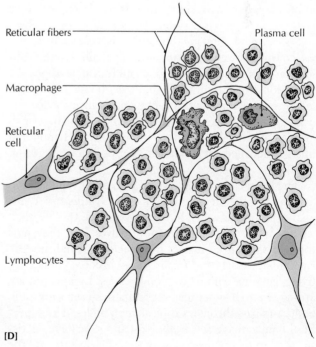

[D]

cells attack specific foreign cells. The human body contains about 2 trillion lymphocytes. They are the backbone of the immune system, and are the basis of the immune response.

Lymphatic Capillaries and Other Vessels

Lymphatic capillaries and blood capillaries are somewhat similar in structure. Both consist of a single layer of endothelial tissue that permits fluid absorption and *diapedesis*, the passage of white blood cells through vessel walls. The major structural difference is an important one: lymphatic capillaries are one-way vessels with a dead-end (blind) terminal end, not part of a circuit of vessels as the blood capillaries are.

Lymphatic capillaries are slightly wider than blood capillaries. The endothelial cells that make up the walls of lymphatic capillaries regulate the passage of materials into and out of the lymph. The ends of adjacent endothelial cells overlap to form *flap valves* that open to permit fluid to enter the capillary, and close when the capillary contracts [FIGURE 21.2].

Lymphatic capillaries are most abundant near the innermost and outermost surfaces of the body, for example, the dermis of the skin, and the mucosal and submucosal layers of the respiratory and digestive systems. Lymphatic capillaries are also numerous beneath the mucous membrane that lines the body cavities and covers the surface of organs. Very few lymphatic capillaries are found in muscles, bones, or connective tissue. There are none in the central nervous system or cornea of the eyeball.

Specialized lymphatic capillaries called *lacteals* (L. *lacteus*, of milk) extend into the intestinal villi. Lacteals absorb fat from the small intestine and transport it into the blood for distribution throughout the body. The lymph in the lacteals takes on a milky appearance (hence the name *lacteal*) because of the presence of many small droplets of fat. At that point, the mixture of lymph and finely emulsified fat is known as *chyle* (KILE; L. *chylus*, juice).

The lymphatic capillaries join with other capillaries to become larger collecting vessels called *lymphatics.* Lymphatics resemble veins, but their walls are thinner than venous walls, they contain more valves, and they pass through specialized masses of tissue (the lymph nodes). Lymphatic vessels are usually found in loose connective tissue, running parallel to blood vessels. Lymphatics are arranged into a superficial set and a deep set within the body, and pass through various lymph nodes. The superficial lymphatic vessels in the skin and subcutaneous tissue tend to follow the course of superficial veins, and the deeper vessels follow the deep veins and arteries.

FIGURE 21.2 LYMPHATIC CAPILLARIES

[A] Flap valves between adjacent endothelial cells in lymphatic capillaries open to permit tissue fluid to enter and close to prevent leakage. The red arrow indicates the direction of fluid flow. [B] Electron micrograph of "loose junction" between the overlapping ends of two endothelial cells in the wall of a lymphatic capillary. ×64,000.

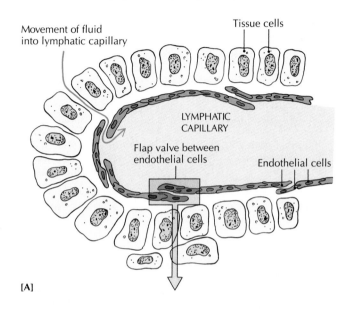

[A]

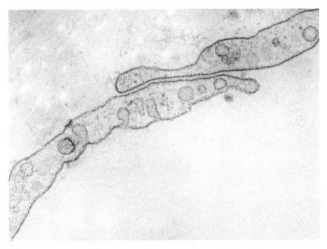

[B]

Lymphatics merge, eventually forming two large ducts, the ***right lymphatic duct*** and the ***thoracic duct,*** that empty their contents into the subclavian veins above the heart.

Lymph travels in only one direction because of valves within the lymphatics that do not allow fluid to flow back. The lymphatic valves operate in the same way as the one-way valves in veins.

Circulation of Lymph

Since there is no central pump to circulate the lymph, the actual movement is accomplished primarily by three other forces: (1) The action of circular and some longitudinal smooth muscles in the lymphatic vessels other than capillaries. (2) The squeezing action of skeletal muscles during normal body movement helps to move lymph through the vessels. Lymph flow may increase 5- to 15-fold during vigorous exercise. (3) The lymphatic system runs parallel with the venous system in the thorax, where a subatmospheric pressure exists. This pressure gradient creates a "pull factor," called the auxiliary respiratory pump, which aids lymph flow.

All lymph vessels are directed toward the thoracic cavity. The upper right quadrant of the body contains lymphatics that drain their contents into the right subclavian vein through the *right lymphatic duct* [FIGURE 21.3]. This drainage includes the right side of the head, right upper extremity, right thorax and lung, right side of the heart, and the upper portion of the liver. The remainder of the lymphatic system returns its fluid to the left subclavian vein through the *thoracic* (left lymphatic) *duct,* the largest of the lymphatics. Following is a summary of the circulation of lymph through the lymphatic system.

FIGURE 21.3 LYMPHATIC CIRCULATION

[A] Drawing showing how lymph drains from the lymphatic system (green) into the bloodstream; anterior view. Arrows indicate the direction of lymph flow into the subclavian veins. [B] Photograph of the thoracic duct and cisterna chyli; anterior view.

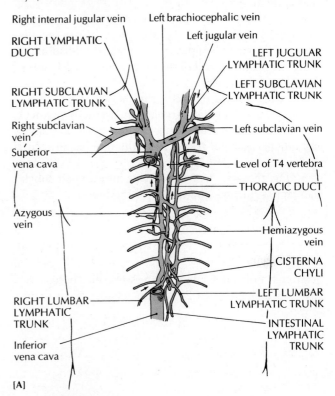

[A]

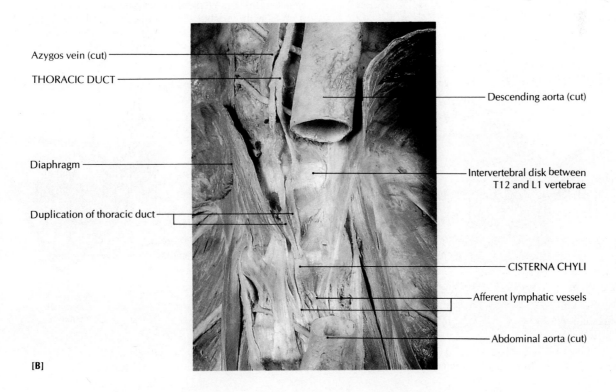

[B]

1 The upper right quadrant of the body drains into the right lymphatic duct.

2 The right lymphatic duct empties into the right subclavian vein.

3 The three-quarters of the body not drained by the right subclavian vein is drained by the thoracic (left lymphatic) duct.

4 The thoracic duct begins as a dilated portion called the *cisterna chyli* within the abdomen, below the diaphragm. It extends upward through the diaphragm, along the posterior wall of the thorax, into the left side of the base of the neck.

5 At the base of the neck, the thoracic duct receives the left jugular lymphatic trunk from the head and neck, the left subclavian trunk from the left upper extremity, and other lymphatic vessels from the thorax and related parts.

6 The thoracic duct then opens into the left subclavian vein, returning the lymph to the blood.

Lymph Nodes

Scattered along the lymphatic vessels, like beads on a string, are small (1 to 25 mm in length), bean-shaped masses of tissue called ***lymph nodes*** [FIGURE 21.4]. The nodes are found in the largest concentrations at the neck, axilla, thorax, abdomen, and groin. Lesser concentrations are found behind the elbow and knee. The *superficial lymph nodes* are located near the body surface in the neck, axilla, and groin. The *deep lymph nodes* are located deep within the groin area, near the lumbar vertebrae, at the base of the lungs, attached to the tissue surrounding the small intestines, and in the liver. Most lymph passes through at least one lymph node on its way back to the bloodstream.

Lymph nodes, which are sometimes incorrectly called glands,* filter out harmful microorganisms and other foreign substances from the lymph, trapping them in a mesh of reticular fibers. Lymph nodes are also the initiating sites for the specific defenses of the immune response. These functions are intimately related to the structure of the lymph node and the cells it contains.

*Lymph nodes were originally called *lymph glands* (L. *glans*, acorn) because they seemed to resemble acorns. Soon, other small bits of tissue came to be called glands, even if they did not look anything like acorns. When it was discovered that some of these bits of tissue secreted various fluids, it was decided that a gland was a structure that formed secretions. Ironically, the original "glands," the lymph glands, do not secrete fluids, and were misnamed in the first place. The word *node*, derived from the Latin word for "knob," seems to describe the structure more appropriately.

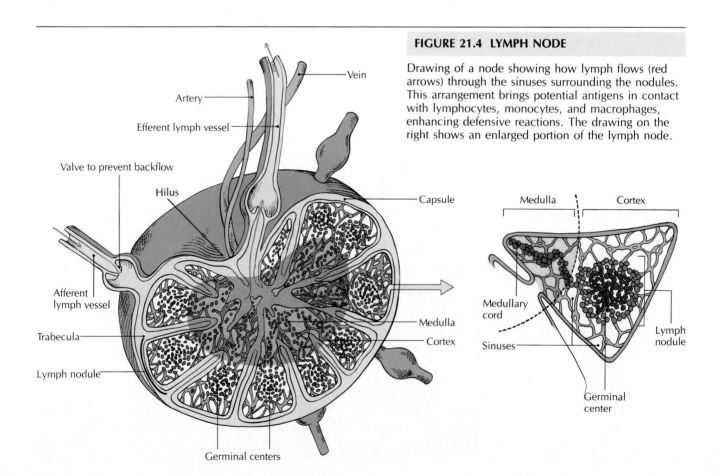

FIGURE 21.4 LYMPH NODE

Drawing of a node showing how lymph flows (red arrows) through the sinuses surrounding the nodules. This arrangement brings potential antigens in contact with lymphocytes, monocytes, and macrophages, enhancing defensive reactions. The drawing on the right shows an enlarged portion of the lymph node.

Lymph nodes are covered by a *capsule* of fibrous connective tissue. Projections of connective tissue, *trabeculae*, extend inward from the capsule toward the center of the lymph node, dividing it into compartments. The outer portion of each compartment is the *cortex* of the node. It contains lymphocytes in dense clusters called *lymph nodules*. In the middle of each nodule is the *germinal center*, where lymphocytes are produced by cell division. The lymph nodes produce about 10 billion lymphocytes every day. The inner part of a lymph node is the *medulla*. It contains strands of lymphocytes extending from the nodule. These strands are called the *medullary cords*.

Running from the cortex to the medulla and surrounding the nodules and medullary cords are *medullary sinuses*, through which the lymph flows before it leaves the node. The lymph node effectively funnels foreign materials in the lymph through the sinuses so that the lymph comes in contact with lymphocytes and macrophages. Afferent (*to* the node) lymphatic vessels with lymph enter the node at various points along the outer capsule. Efferent (*from* the node) lymphatic vessels leave the node from a small depressed area called the *hilus*. Blood vessels enter, as well as leave, through the hilus.

Tonsils

The *tonsils* are aggregates of lymphatic nodules enclosed in a capsule of connective tissue. There are three types of tonsils: (1) the *pharyngeal tonsil* in the upper posterior wall of the pharynx behind the nose, (2) the *palatine tonsils* on each side of the soft palate, and (3) the *lingual tonsils* at the base of the tongue.

The tonsils lack afferent lymphatic vessels, a feature that distinguishes them from lymph nodes. The efferent lymphatics of the tonsils contribute many lymphocytes to the lymph. These cells are capable of leaving the tonsils and destroying invading microorganisms in other parts of the body. Together the tonsils form a band of lymphoid tissue that is strategically placed at the upper entrances to the digestive and respiratory systems, where foreign substances may enter easily. Most infectious microorganisms are either killed by lymphocytes at the surface of the pharynx, or are killed later, after the initial defenses are set in motion by the tonsils. The presence of plasma cells within the tonsils indicates the formation of antibodies.

Although tonsils usually function to prevent infection, they may become infected repeatedly themselves. In such cases, some physicians recommend their removal. The palatine tonsils are the ones most frequently removed (the familiar operation is called a *tonsillectomy*). The lingual tonsils are rarely removed. Pharyngeal tonsils, popularly called *adenoids*, may also become enlarged. If so, obstruction of the nasal pharynx caused by swelling can interfere with breathing, necessitating the removal of the adenoids (*adenoidectomy*).

Spleen

The *spleen* is the largest lymphoid organ in the body. It is about the size and shape of a clenched fist, measuring about 12 cm in length, and is purplish in color. Located inferior to the diaphragm on the left side of the body, it rests on portions of the stomach, left kidney, and large intestine [FIGURE 21.1A].

The main functions of the spleen are filtering blood and manufacturing phagocytic lymphocytes and monocytes. It also contributes to the functioning of the cardiovascular and lymphatic systems as follows:

1 Macrophages, abundant in the spleen, help remove damaged or dead erythrocytes and platelets, microorganisms, and other debris from the blood as it circulates through the spleen. Macrophages also remove iron from the hemoglobin of old red blood cells and return it to the circulation for use by the bone marrow in producing new red blood cells. The breakdown of hemoglobin also allows the production of the pigment bilirubin, which can be circulated to the liver, where it becomes a constituent of bile.

2 Antigens in the blood of the spleen activate lymphocytes that develop into cells that either produce antibodies or are otherwise involved in the immune reaction.

3 The spleen produces red blood cells during fetal life. In later life, it stores newly formed red blood cells and platelets and releases them into the bloodstream as they are needed.

4 Because the spleen contains a large volume of blood, it serves as a blood reservoir. If the body loses blood suddenly, the spleen contracts and adds blood to the general circulation. It also relieves the venous pressure on the heart by releasing stored blood into the circulation during bursts of physical activity. (The spleen is capable of releasing approximately 200 mL of blood into the general circulation in one minute.)

The tissue structure of the spleen is similar to that of lymph nodes [FIGURE 21.5]. Surrounded by a *capsule* of connective tissue, the spleen is divided by *trabeculae* into compartments called *lobules*. The functional part of the medulla consists of *splenic pulp*, which contains small islands of white pulp scattered throughout red pulp. The *white pulp* is made up of compact masses of lymphocytes surrounding small branches of the splenic artery. These masses, which occur at intervals, are called *splenic nodules (Malpighian corpuscles)*. Within the *red pulp* are *venous sinusoids* filled with blood and lined with monocytes and macrophages. (The pulp is red because of the many erythrocytes in the blood.) This arrangement brings the blood and any foreign materials it may contain to the lymphocytes, monocytes, and macrophages for cleansing. Since the spleen does not receive lymphatics and lymph,

FIGURE 21.5 SPLEEN

[A] Drawing of the spleen showing blood flow (colored arrows) through the white pulp to the venous sinusoids in the red pulp. Note that the white pulp consists of nodules and lymphocytes, while the red pulp is an open mesh with venous sinuses running through it. [B] Scanning electron micrograph of a section through the spleen. ×1100.

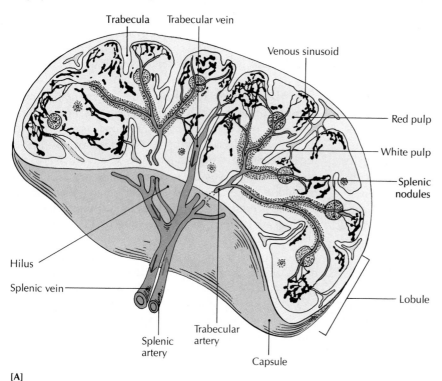

[A]

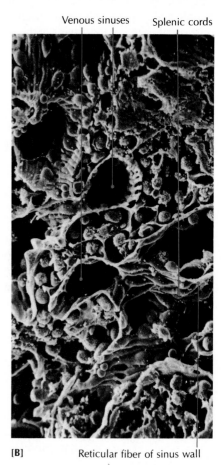

[B]

it cannot be considered a filter in the same way that a lymph node is a filter. Efferent lymphatics, which leave the spleen, contribute lymphocytes, monocytes, and macrophages to the lymph.

Thymus Gland

The ***thymus gland*** is a ductless, pinkish-gray mass of flattened lymphoid tissue located in the thorax. It overlies the heart and its major blood vessels. The thymus gland consists of two *lobes* joined by connective tissue and surrounded by a *capsule* of connective tissue. Each lobe is divided into *lobules* by coarse *trabeculae*, with each lobule having an outer *cortex* and an inner *medulla* [FIGURE 21.6]. The cortex contains many lymphocytes, most of which degenerate before they ever leave the thymus gland. The only blood vessels within the thymus gland are capillaries. The medulla contains fewer lymphocytes than the cortex does, making the epithelial reticular cells in the medulla more obvious.

The fetal thymus gland is already involved with the production of lymphocytes, and assists in the formation of antibodies elsewhere. Undifferentiated lymphocytes migrate to the thymus gland to become specialized, or immune-competent T cells (T for "thymus-dependent") with defined roles in specific defenses against foreign cells. The thymus gland is relatively large at birth (about 12 to 15 g), forms antibodies in the newborn, and plays a major role in the early development of the body's immune system. The thymus gland increases in size from birth to puberty, and it is most active in childhood and early adolescence. The thymus gland also secretes a group of hormones collectively called *thymosin*, which may be necessary for the differentiation of T cells from stem cells.

The thymus gland shrinks progressively during adulthood, and degenerates into adipose and connective tissue in old age. Like the tonsils and spleen, the thymus gland has efferent lymph vessels, but no afferent ones. As a result, no lymph drains into it.

FIGURE 21.6 INTERNAL ANATOMY OF THYMUS GLAND

The drawing of a lobule of the thymus gland shows mature lymphocytes in the medulla and many more immature lymphocytes in the cortex.

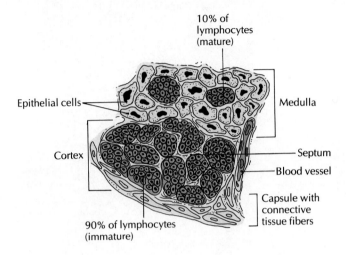

FIGURE 21.7 AGGREGATED LYMPH NODULES (PEYER'S PATCHES)

Light micrograph of tissue from an aggregated lymph nodule. ×60.

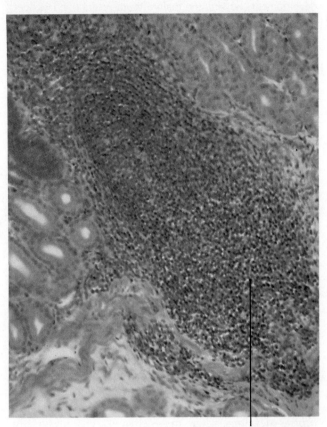

Aggregated lymph nodules

Aggregated Lymph Nodules (Peyer's Patches)

Aggregated lymph nodules (Peyer's patches) are clusters of unencapsulated lymphoid tissue found in the intestine and appendix. Because of their location, they are also called *gut-associated lymphoid tissue* (GALT). Similar clusters of tissue are found along the respiratory tract. Aggregated lymph nodules generate plasma cells that secrete antibodies in large quantities in response to antigens from the intestine [FIGURE 21.7].

DEVELOPMENTAL ANATOMY OF THE LYMPHATIC SYSTEM

The development of the lymphatic system begins at the end of the fifth week after fertilization. Lymphatic vessels emerge from six primary lymph sacs [FIGURE 21.8A]; they branch from the paired *jugular lymph sacs* to the head, neck, and arms; from the paired *iliac lymph sacs* to the lower trunk and lower limbs; and from the single *retroperitoneal lymph sac* and the *cisterna chyli* to the digestive system. The jugular lymph sacs are linked to the cisterna chyli by the left and right thoracic ducts; at about week 9, these ducts become connected by an anastomosis [FIGURE 21.8B].

The mature *thoracic duct* develops from the caudal portion of the right thoracic duct, the anastomosis, and the cranial portion of the left thoracic duct. The *right lymphatic duct* develops from the cranial portion of the

FIGURE 21.8 DEVELOPMENTAL ANATOMY OF THE LYMPHATIC SYSTEM

[A] Drawing of a 7-week embryo, lateral view; note the five primary lymph sacs. [B] The paired thoracic ducts are developed at about week 9; ventral view. [C] Later stage of development showing the mature thoracic duct and right lymphatic duct; ventral view.

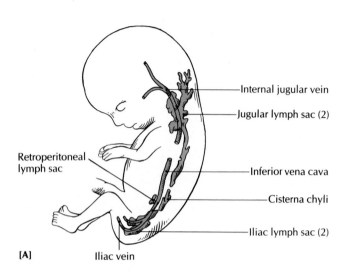

[A]   Iliac vein

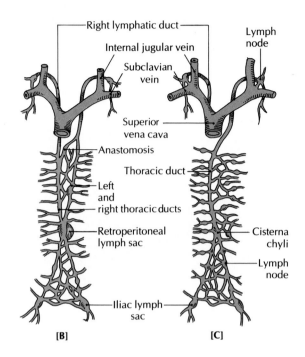

[B]   [C]

right thoracic duct [FIGURE 21.8C]. The superior portion of the cisterna chyli remains after birth. Except for this portion of the cisterna chyli, the embryonic lymph sacs develop into recognizable groups of *lymph nodes* during the early fetal period. The network of *lymphatic sinuses* develops when mesenchymal cells invade the lymph sacs and break down the sac cavities. Immature

fetal lymphocytes are produced in the bone marrow, and later migrate to the thymus gland, where they mature into lymphocytes (T cells).

The *spleen* develops from mesenchymal cells in the stomach mesentery (greater omentum), and the *tonsils* develop from various lymph nodules near the mouth and throat.

WHEN THINGS GO WRONG

Acquired Immune Deficiency Syndrome (AIDS)

As its name suggests, *acquired immune deficiency syndrome,* or *AIDS,** is a disease that cripples the body's immune system. It is caused by a virus (*human immunodeficiency virus,* or *HIV*) that kills specific helper T cells called *CD4 cells*† and some other cells of the body's immune system.

*The acronym AIDS has come to stand for a slightly shortened phrase: *acquired immunodeficiency syndrome.* The original name was purposely descriptive. The condition was *acquired* rather than inherited, produced a *deficient immune system,* and was a collection of several diseases (a *syndrome*) rather than a single, easily classified disease.

†CD4 cells are so named because their plasma membranes have CD4 receptor molecules to which HIV binds before entering and infecting the host cells. "CD4" stands for "Class Designation 4"; 8 class designations of T cells are known.

Without a sufficient number of CD4 cells, the immune system becomes ineffective, and a person infected with HIV becomes vulnerable to infections that rarely affect healthy people. (Such infections are referred to as *opportunistic infections.*) Also, relatively uncommon forms of cancer of the blood and lymphatic system seem to be unusually prevalent among AIDS victims. In fact, the first sign that a new disease had emerged was the sudden appearance in 1981 among young, middle-class white males of *Pneumocystis carinii* pneumonia (PCP) and Kaposi's sarcoma (KS; a formerly rare cancer of the lining of blood vessels that usually shows externally as purplish spots on the skin). Until then, KS appeared mainly among older Italian and Jewish men, and in Africa. The young males turned out to be predominantly homosexual, and the diseases were also spreading among drug users who used contaminated hypodermic needles, and people who received frequent blood transfusions.

As recently as 1989 the great majority (70 percent) of AIDS victims in the United States were active male homosexuals or bisexuals, and only 4 percent were heterosexuals. But the trend has changed since then, with the percentage of male homosexual and bisexual victims decreasing, and the percentage of heterosexual victims increasing. The incidence of AIDS in women and children appears to be rising also, and intravenous drug users remain at high risk.

Of all activities, HIV is transmitted most often via unprotected receptive anal intercourse. Anal intercourse is an especially effective method of transmission because rectal capillaries are so close to the skin that infection may occur through microscopic breaks in the skin. However, researchers stress that AIDS is not exclusively a disease of homosexuals, bisexuals, or drug addicts. In Africa, for example, it is a disease of heterosexuals.

The AIDS virus is an unusual kind of virus called a **retrovirus.** Like ordinary viruses, retroviruses need host cells in order to reproduce. However, the genetic material of a retrovirus is RNA, not DNA. The production of viral DNA from viral RNA requires the presence of an enzyme called *reverse transcriptase* (see drawing).

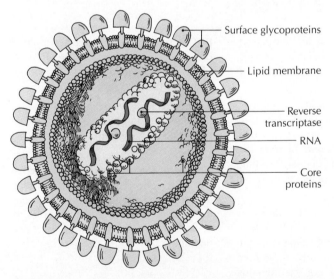

Both the RNA and reverse transcriptase from an AIDS virus are injected into a host cell, which is usually a CD4 cell (see photo). Within the cell, the viral RNA and enzyme produce viral DNA. The viral DNA enters the nucleus of the CD4 cell, where it inserts itself into the cell DNA. The viral DNA proceeds to make viral protein that is used to assemble new AIDS viruses. After producing new AIDS viruses, the CD4 cell eventually dies (probably because the budding of viruses from the cell produces tiny holes in its plasma membrane), and the newly created viruses spread to infect other CD4 cells.

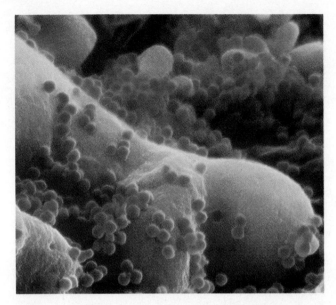

AIDS viruses (blue spheres) attack a helper T cell. ×80,000.

This sequence is repeated over and over again until the supply of CD4 cells is seriously depleted.

Because the body does not produce an unlimited supply of CD4 cells, they soon reach a low point, at which time the immune system is no longer effective. During this period, diseases that are normally repelled by a healthy immune system become established.

The first symptoms of HIV infection often resemble the symptoms of the flu or mononucleosis; they include persistent or recurrent fever, severe dry cough and sore throat, blurred vision, enlarged lymph nodes, night sweats, chronic fatigue, decreased appetite and rapid weight loss, blood or mucus in the feces, and white blemishes inside the mouth. An upper-body rash may also be present. These initial symptoms usually last from a few days to a month or more. About 85 percent of all AIDS patients eventually show signs of damage to the nervous system.

It is likely that HIV did not exist before the 1960s, but the virus may have been infecting humans long before the recognized outbreak in 1981. HIV has been traced to south central Africa before 1970, but it may have originated elsewhere. It arrived in New York City around 1977 (probably from the Caribbean), and by 1978 it had reached Los Angeles, San Francisco, and Miami. By 1990, HIV was found in all 50 states, and throughout the world. As many as 2 million Americans may be infected, and the situation is much worse in Africa and the Caribbean.

Current treatment concentrates on preventing PCP with the use of an inhalant spray called pentamidine, interrupting the reverse transcriptase as it helps assemble the viral DNA, and using immunomodulators (drugs that help restore the immune system to a healthy state). Initial use of AZT (zidovudine, azidothymidine; trade name

Retrovir), an unsuccessful anticancer drug, has been promising but expensive, and the search for an effective vaccine is underway. Prevention may be the most effective approach to conquering AIDS, and prevention has been helped considerably by the development of tests that detect the presence of antibodies to the AIDS virus in blood. Testing of blood to be used in transfusions has effectively stopped the spread of AIDS by that route.

Hodgkin's Disease

Hodgkin's disease is a form of cancer typified by the presence of large, multinucleate cells (Reed-Sternberg cells) in the affected lymphoid tissue. The first sign of Hodgkin's disease is usually a painless swelling of the lymph nodes, most commonly in the cervical area. As lymphocytes, eosinophils, and other cells proliferate, there is a progressive enlargement of the lymph nodes, the spleen, and other lymphoid tissues. Late symptoms of the disease include edema of the face and neck, anemia, jaundice, and increased susceptibility to infection.

Hodgkin's disease occurs most often in young adults between the ages of 15 to 38, and in people over 50. Interestingly, it occurs in Japan only in people over 50. Its cause is unknown but the disease is potentially curable. It is usually treated with considerable success with a combination of radiation therapy and chemotherapy. Untreated, Hodgkin's disease is fatal.

Infectious Mononucleosis

Infectious mononucleosis is a viral disease caused by the Epstein-Barr virus, a member of the herpes group. It appears most often in 15- to 25-year-olds. It is usually accompanied by fever, increased lymphocyte production, and enlarged lymph nodes in the neck. Secondary symptoms include dysfunction of the liver and increased numbers of monocytes. Infectious mononucleosis is not contagious in adults. In adolescents, however, the Epstein-Barr virus may be spread during close contact or via the exchange of body fluids, as when drinking glasses are shared or through kissing. The virus may also be spread from infant to infant in saliva, from the mother's breast, or food.

CHAPTER SUMMARY

The **lymphatic system** returns to the blood excess fluid and proteins from the spaces around cells, plays a major role in the transport of fats from the tissue surrounding the small intestine to the blood, filters and destroys microorganisms and other foreign substances, and aids in providing long-term protection for the body.

The Lymphatic System (p. 545)

1 The lymphatic system consists of vessels, lymph nodes, lymph, leukocytes, lymphatic organs (spleen and thymus gland), and specialized lymphoid tissues.

2 Excess interstitial fluid is drained from tissues by the lymphatic system. Once inside the lymphatic capillaries, the fluid is called **lymph.**

3 The composition of lymph is similar to that of blood, except that lymph lacks red blood cells and most of the blood proteins.

4 Leukocytes in lymph include monocytes, which develop into macrophages, and two types of **lymphocytes, B cells** and **T cells.**

5 **Lymphatic capillaries** are one-way vessels with a closed terminal end. They join together to form **lymphatics,** which drain into the **right lymphatic duct** and the **thoracic duct.** The ducts return fluid to the bloodstream through the right and left subclavian veins.

6 **Lymph nodes** are small masses of lymphoid tissue scattered along the lymphatic vessels. Lymphocytes in the nodes filter out harmful substances from the lymph, and they are the initiating sites for the specific defenses of the immune system.

7 The three **tonsils** (pharyngeal, palatine, and lingual) form a band of lymphoid tissue that prevents foreign substances from entering the body through the throat.

8 The **spleen** is the largest lymphoid organ. Its major functions are filtering blood and manufacturing phagocytic lymphocytes.

9 The **thymus gland** forms antibodies in the newborn and plays a major role in the early development of the body's immune system.

10 **Aggregated lymph nodules** are clusters of lymphoid tissue in the intestine and appendix. They respond to antigens from the intestine by generating plasma cells that secrete antibodies in large quantities.

Developmental Anatomy of the Lymphatic System (p. 553)

1 The development of the lymphatic system begins at the end of week 5.

2 **Lymphatic vessels** develop from **lymph sacs** and the **cisterna chyli.** The lymph sacs develop into **lymph nodes.**

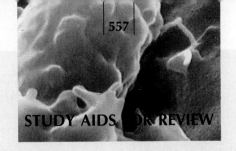

MEDICAL TERMINOLOGY

ANAPHYLAXIS (an-uh-fuh-LACK-sis; L. *ana*, intensification + Gr. *phulassein*, to guard) Hypersensitivity to a foreign substance.

ELEPHANTIASIS Obstruction caused by a parasitic filarial worm of the return of lymph to the lymphatic ducts that causes enlargement of a limb, usually a lower limb, or the genital area.

HYPERSPLENISM A condition in which the spleen is abnormally active and blood cell destruction is increased.

LYMPHADENECTOMY Removal of lymph nodes.

LYMPHADENOPATHY A disease of lymph nodes, resulting in enlarged glands.

LYMPHANGIOMA Benign tumor of lymph tissue.

LYMPHANGITIS Inflammation of lymphatic vessels.

LYMPHATIC METASTASIS A condition in which a disease travels around the body via the lymphatic system.

LYMPHOMA Tumor of the lymph nodes.

LYMPHOSARCOMA Malignant tumor of lymph tissue.

SPLENECTOMY Total removal of the spleen.

SPLENOMEGALY Enlargement of the spleen following infectious diseases such as scarlet fever, typhus, and syphilis.

VACCINE General term for the immunization preparation used against specific diseases.

UNDERSTANDING THE FACTS

1 List the four major functions of the lymphatic system.
2 Define lymph.
3 Describe the structure of a lymph node.
4 What is the main structural difference between lymphatic capillaries and blood capillaries?
5 Name and describe the vessels of the lymphatic system.
6 What is the major function of the lacteals?
7 Into which large veins does lymph drain?
8 What portion of the body is drained by the thoracic duct?
9 Which lymphatic organ serves as a blood reservoir?

UNDERSTANDING THE CONCEPTS

1 Since the lymphatic system lacks a "pump," how do you explain the movement of lymph?
2 Although lymph contains all the coagulation factors found in blood and is capable of clotting to a small degree, why does lymph lack clotting ability?
3 Why is the reticuloendothelial system important to the body?
4 Why are valves so important to the lymphatic system?

SELF-QUIZ

Multiple-Choice Questions

21.1 Excess interstitial fluid drains from body tissues into
  a the right lymphatic duct
  b the subclavian veins
  c the thoracic duct
  d lymphatic capillaries
  e aggregated lymph nodules

21.2 Which of the following cell types is *not* found in lymph?
  a monocytes
  b erythrocytes
  c leukocytes
  d macrophages
  e B cells

21.3 Flap valves are found in
  a lymphatics
  b the thoracic duct
  c lymphatic capillaries
  d endothelial cells
  e a and c

21.4 Lacteals contain fluid drained from the
  a liver
  b stomach
  c small intestine
  d large intestine
  e spleen

21.5 A lymphatic structure that serves as a blood reservoir is the
  a spleen
  b tonsil
  c thymus gland
  d aggregated lymph nodule
  e adenoid

True-False Statements

21.6 Macrophages make up the reticuloendothelial system.
21.7 Monocytes can develop into plasma cells.
21.8 The activity of the thymus gland increases with age.
21.9 Foreign matter in the intestine can stimulate the production of plasma cells by the aggregated lymph nodules.
21.10 The right lymphatic duct is formed by the merger of several lymphatic capillaries.
21.11 Lymphatics are formed by the merger of lymphatic capillaries.

21.12 The largest lymphatic vessel is the right lymphatic duct.

21.13 The thoracic duct returns fluid to the superior vena cava.

21.14 Lymph from the lower limbs and trunk is emptied into the right subclavian vein.

21.15 Lymph is moved through the vessels of the lymphatic system by the beating of the heart.

Completion Exercises

21.16 The presence of excess interstitial fluid in the body tissues produces a condition called _____.

21.17 Lymphatics pass through bean-shaped masses of tissue called _____.

21.18 Blood vessels enter and leave lymph nodes through an area called the _____.

21.19 The three pairs of tonsils are the _____, _____, and _____ tonsils.

21.20 "Adenoids" is the common name for the _____.

21.21 A lymphatic structure that filters blood as well as lymph is the _____.

21.22 The fluid forced out of the capillaries by the hydrostatic pressure of the blood is _____.

21.23 The products of fat digestion in the intestine are absorbed into _____.

21.24 Tissue macrophages are also called _____.

21.25 Within lymphatic vessels, lymph can flow in only one direction because of the presence of _____.

21.26 The two basic types of lymphocytes are _____ and _____.

21.27 The lymphoid organ that produces red blood cells in the fetus is the _____.

21.28 Disease-causing microorganisms in the throat are attacked by lymphocytes produced by the _____.

A SECOND LOOK

In the following drawing, label the thymus gland, tonsils, iliac nodes, axillary nodes, lymphatic vessels, and spleen.

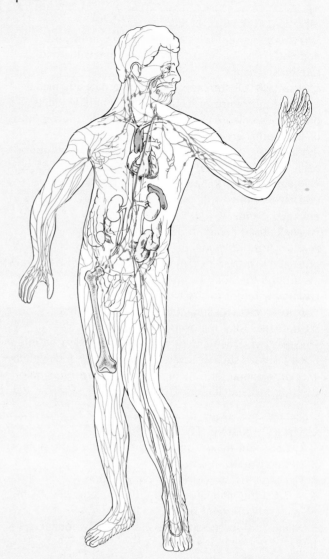

22

The Respiratory System

LEARNING OBJECTIVES

1 List the respiratory structures through which air passes before it reaches the bloodstream.

2 Describe the structure of the nose within the respiratory system.

3 Name the parts of the pharynx, and describe their locations and functions.

4 Explain the structures of the epiglottis and larynx.

5 Describe the trachea and its relationship to the respiratory tract.

6 Identify the parts of the "respiratory tree" from the largest to the smallest branches.

7 Relate the structure of the alveoli and surrounding capillary network to their functions.

8 Describe the structure of the lungs and their pleurae.

9 Summarize the processes of expiration and inspiration, naming the muscles and other structures involved.

A human body can survive without food for as much as several weeks, and without water for several days, but if breathing stops for three to six minutes, death is likely. Body tissues, especially the heart and brain, require a constant supply of oxygen. The *respiratory system* delivers air containing oxygen to the blood and removes carbon dioxide, the gaseous waste product of metabolism. The respiratory system includes the lungs, the several passageways leading from outside to the lungs, and the muscles that move air into and out of the lungs.

The term *respiration* has several meanings: (1) *Cellular respiration* is the sum of biochemical processes by which the chemical energy of foods is released to provide energy for life's processes. (2) *External respiration,* which occurs in the lungs, is the exchange of gases between the blood and the lungs; oxygen from the lungs moves into the blood, and carbon dioxide and water move from the blood to the lungs. (3) *Internal respiration* is the exchange of gases in the body tissues; carbon dioxide from the cells is exchanged for oxygen from the blood. In contrast to these physiological processes, *ventilation,* or *breathing,* is the mechanical process that moves air into and out of the lungs. It includes two phases, *inspiration* and *expiration.*

It is helpful to think of the action of the respiratory system as a series of steps, even though in a living body they all take place at the same time and continuously. First, inhaled air, rich in oxygen and poor in carbon dioxide, travels through the respiratory tract deep into the terminal portions of the lungs; this is *inspiration.* There, oxygen diffuses from the lungs into the blood. From the lungs, oxygenated blood is carried to the heart, and then via the systemic circulatory system to all parts of the body. In the body tissues, oxygen moves from the blood into the cells, and carbon dioxide and other wastes are released from the cells into the blood. Finally, deoxygenated venous blood, carrying its load of wastes, is forced back to the lungs, where carbon dioxide is exhaled during *expiration.*

RESPIRATORY TRACT

Air flows through respiratory passages because of differences in pressure produced by chest and trunk muscles during breathing. Except for the beating of cilia in the respiratory lining, the passageways are simply a series of openings through which air is forced. These passageways and the lungs make up the *respiratory tract* [FIGURE 22.1].

Nose

Air normally enters the respiratory tract through the nose. The external nose is supported at the bridge by *nasal bones.* The *septal cartilage* separates the two external openings of the *nostrils,* or *external nares* (NAIR-eez; L. nostrils). Along with the vomer and perpendicular plate of the ethmoid bone, the septal cartilage forms the *nasal septum,* which divides the nasal cavity into bilateral halves.

The *nasal cavity* fills the space between the base of the skull and the roof of the mouth. It is supported above by the *ethmoid bones* and on the sides by the *ethmoid, maxillary,* and *inferior conchae bones* [FIGURE 22.2]. The superior, middle, and inferior conchae, also known as the turbinate bones, bear longitudinal ridges that are covered with highly vascular respiratory mucosa. Between the ridges of the conchae are folds called the *superior, middle,* and *inferior meatuses* (sing., *meatus*). These meatuses serve as air passageways. The blood vessels in this area, called venous sinuses, bring a large quantity of blood to the mucous membrane that adheres to the underlying periosteum. The blood is a constant source of heat that warms the air inhaled through the nose. Since warm air holds more water than cold air, the air is also moistened as it moves through the nasal cavity.

The *hard palate,* strengthened by the palatine and parts of the maxillary bones, forms the floor of the nasal cavity. If these bones do not grow together normally by the third month of prenatal development, the nasal cavity and mouth are not separated adequately. The condition is known as a *cleft palate.* The immediate problem facing a newborn infant with a cleft palate is that not enough suction can, be created to nurse properly. A cleft palate can usually be corrected surgically.

The *soft palate* [FIGURE 22.2] is a flexible muscular sheet that separates the oropharynx from the nasal cavity. It is continuous with the posterior border of the hard palate, and consists of several skeletal muscles covered by mucous membrane. Depression of this area during breathing enlarges the air passage into the pharynx. When the soft palate is elevated during swallowing, it prevents food from entering the nasal pharynx.

The *frontal, sphenoidal, maxillary,* and *ethmoid sinuses* are blind sacs that open into the nasal cavity. They are not present in the newborn infant. Together they are called the *paranasal sinuses.* They give the voice a full, rich tone. A *nasolacrimal* (L. "nose" and "tear") *duct* leads from each eye to the nasal cavity, through which excessive tears from the surface of the eye are drained into the nose; that is why weeping may be accompanied by a watery flow from the nose (runny nose). At the back of the nasal cavity, the two *internal nares* open into the pharynx.

FIGURE 22.1 THE RESPIRATORY SYSTEM

The interior of the right lung (not in scale) is revealed.

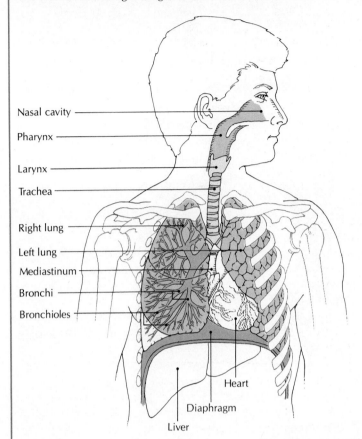

Structure	Major functions
Nasal cavity	Filters, warms, moistens incoming air; passageway to pharynx.
Paranasal sinuses	Produce mucus; act as resonators for sound; lighten skull.
Pharynx	Passageway for air between nose and larynx, and for food from mouth to esophagus.
Larynx (voice box)	Passageway for air between pharynx and rest of respiratory tract; produces sound; protects trachea from foreign objects.
Trachea (windpipe)	Passageway for air to and from thoracic cavity; traps and expels foreign matter.
Diaphragm	Enlarges thoracic cavity to allow for inspiration, returns to original position for expiration.
Bronchi	Passageways for air to and from lungs; filter air.
Bronchioles	Passageways for air to and from alveoli.
Alveoli	Site of gas exchange; functional units of lungs.
Lungs	Major respiratory organs.
Pleurae	Protect, compartmentalize, and lubricate outer surfaces of lungs; enclose pleural cavities.

"Smoker's Cough"

The human body has several defenses against dirty air. Hairs in the nose filter out large particles. The mucus that is secreted continuously by the goblet cells in the upper respiratory tract washes out or dissolves smaller irritants, and traps particles that enter the respiratory tract. Also, the many cilia in the upper tract sweep away foreign particles and excessive mucus. But air pollutants (especially cigarette smoke, ozone, sulfur dioxide, and nitrogen dioxide) can stiffen, slow, or even destroy the cilia, causing the mucus to clog, a reaction frequently resulting in "smoker's cough." When cilia in the nose are impaired by pollutants or are partially paralyzed by cold air, there is an overproduction of mucus, which is not swept back into the throat as usual. When this happens, the mucus drips forward and causes a runny nose.

As a result of inefficient cilia, microorganisms and other foreign particles can penetrate the lung tissue and cause respiratory infections, and possibly even lung cancer. As lung tissue becomes more irritated by air pollutants, mucus flows more freely to help remove the inhaled pollutants. A normal coughing mechanism expels the contaminated air and mucus. Smoking or air pollution can trigger excessive mucus flow that eventually blocks the air passages, which in turn causes more coughing—a type of positive feedback. Over time, the muscles surrounding the bronchial tree can weaken from prolonged, strong coughing, and breathing becomes progressively more difficult. If this cycle continues, *chronic bronchitis* (a chronic inflammation of the mucous membranes of the respiratory tree) can occur.

FIGURE 22.2 UPPER RESPIRATORY SYSTEM

[A] Right midsagittal section through the head showing the upper portion of the respiratory system: the nasal cavity, pharynx, larynx, and part of the trachea. **[B]** A photograph of a sagittal section through the head.

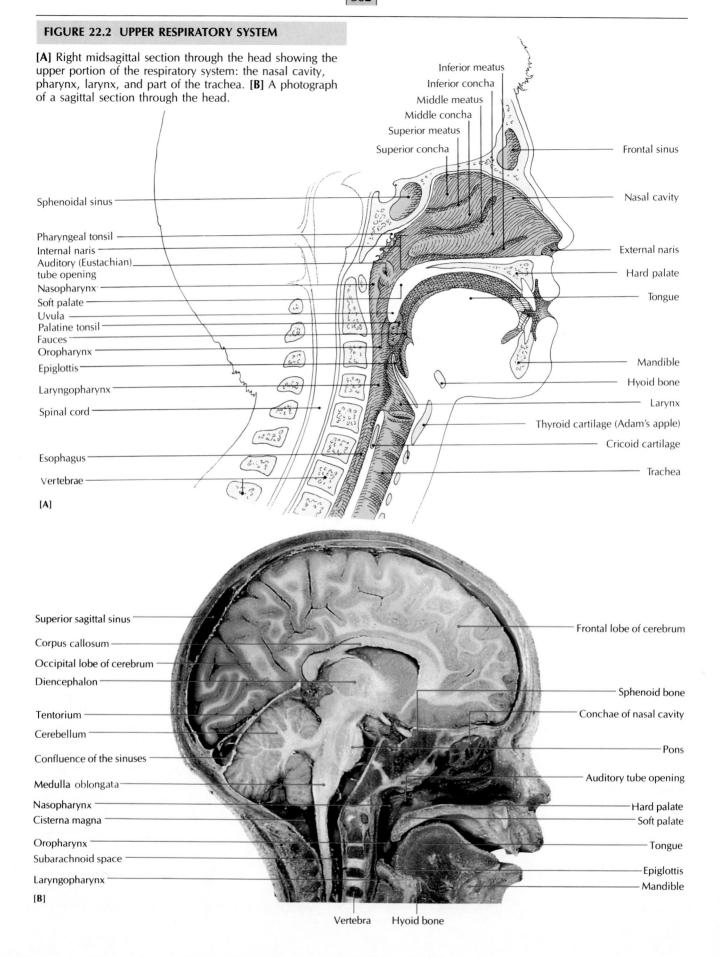

The surface of the nasal cavity is lined with two types of epithelium: (1) *Nasal epithelium* warms and moistens inhaled air. It is supplied with mucus-secreting *goblet cells*, which keep the surfaces wet. The mucus catches many of the small dust particles and microorganisms carried in air that get past the coarse hairs in the nostrils. The nasal epithelium is a pseudostratified columnar tissue bearing millions of cilia, which also help capture minute particles and microorganisms [FIGURE 22.3]. (2) *Olfactory epithelium* in the upper part of the nasal cavity contains sensory endings of the neurons whose axons form the olfactory nerve. Aromatic substances dissolved in the moist coating of the epithelial surface stimulate the nerve endings and cause the sensation of smell.

Pharynx

The *pharynx* (throat; FAR-ingks) leads to the respiratory passage and esophagus. It is connected to the nasal cavity through the internal openings of the nares, and also to the mouth, or *buccal (oral) cavity* [FIGURE 22.2A]. The pharynx is divided anatomically into three portions, from superior to inferior: (1) nasopharynx, (2) oropharynx, and (3) laryngopharynx. The three parts of the pharynx are continuous with each other.

Nasopharynx The *nasopharynx* is the portion of the pharynx superior to the soft palate. Like the nasal cavity, the nasopharynx is lined with ciliated epithelium that aids in cleaning inspired air. The auditory tubes open into the nasopharynx just posterior and lateral to the internal nares. These tubes are air passages that equalize the air pressure on both sides of the eardrums. If these tubes are not open, changes in atmospheric pressure, such as those felt when one goes quickly up or down in an airplane, can cause the ears to "pop." This temporary imbalance can usually be alleviated by chewing gum or swallowing.

Oropharynx The *oropharynx,* at the posterior part of the mouth, extends from the soft palate to the epiglottis, where the respiratory and digestive tracts separate into the trachea and the esophagus. The oropharynx has muscular walls, and along with the laryngopharynx initiates the act of swallowing. It is separated from the mouth by a pair of narrow passageways called *fauces* (FAW-seez; sing. *faux*) that spread open during swallowing or panting, when most of the air entering the respiratory tract is drawn in through the mouth rather than the nostrils. Like most tubes that are subject to abrasion, the oropharynx is lined with stratified squamous epithelium, which is constantly being shed and renewed.

Laryngopharynx The lowermost portion of the pharynx is the *laryngopharynx* (luh-RING-go-fair-inks). It extends from the level of the hyoid bone posteriorly to the larynx. It is at that point that the respiratory and digestive systems become distinct, with air moving anteriorly into the larynx and food moving posteriorly into the esophagus past the flexible glottis.

FIGURE 22.3 NASAL EPITHELIUM

[A] Light micrograph of pseudostratified columnar epithelium in the nasal cavity. ×160. [B] Scanning electron micrograph of the ciliated epithelium that lines the nasal cavity. ×3500.

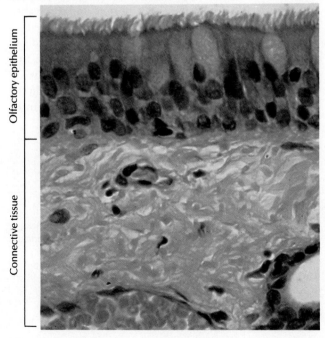

[A]

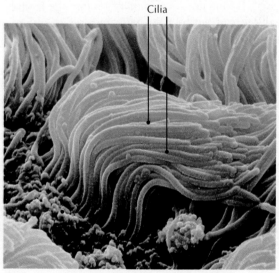

Cilia

[B]

Tonsils *Tonsils* are structures that are part of both the pharynx and the lymphatic system. *Lingual tonsils* are at the base of the tongue; *palatine tonsils* are lateral to the oropharynx; and *pharyngeal tonsils* (adenoids) hang from the roof of the nasopharynx. If the adenoids become inflammed and swollen, they can interfere with breathing through the nose, may block the opening of the auditory tube, and even make speech difficult.

Larynx

The *larynx* (voice box; LAR-ingks) is an air passage from the pharynx to the rest of the respiratory tract, and produces most of the sound used in speaking and singing [FIGURE 22.4].

Structure The larynx is supported by cartilages, of which the most prominent is the *thyroid cartilage* (Adam's apple), which consists of two laminae that fuse to form an acute angle at the midline. Because the thyroid cartilage is larger in males than in females, it is visible in the front of the neck in men, but is less conspicuous in women. The larynx is supported above the thyroid cartilage by ligaments connected to the hyoid bone. Below the thyroid cartilage is the *cricoid cartilage*, a ring of cartilage that connects the thyroid cartilage with the trachea below.

The *epiglottis* is a flap of cartilage that folds down over the opening into the larynx, called the **glottis,** during swallowing and swings back up when the act of swallowing ceases. Since air must pass from the pharynx (which is *posterior* to the mouth) to the larynx (which is *anterior* to the esophagus), there is a crossover between the respiratory and digestive tracts The epiglottis works well most of the time, mainly because people involuntarily inhale before swallowing, getting air into the lungs, and then exhale after swallowing, thus clearing the air passage. If food does accidentally get into the larynx, it is usually forced out by a strong cough reflex, although it may have to be dislodged by using the Heimlich maneuver (see "The Heimlich Maneuver" on page 575).

Humans produce sound mainly by vibrations of the **vocal cords** [FIGURE 22.4B]. These paired strips of stratified squamous epithelium at the base of the larynx have a front-to-back slit between them, the *glottis.* Above and beside the vocal cords is a pair of *vestibular folds*, usually called *false vocal cords*, that protrude into the vestibule of the larynx, hence *vestibular* folds. The true vocal cords are held in place and are regulated by a pair of *arytenoid cartilages*, which in turn are held by a pair of *cuneiform cartilages.* Still another pair of cartilages, the *corniculate cartilages*, lies between the arytenoids and the epiglottis.

Sound production Sound production is the result of a complex coordination of muscles. The immediate source of most human sound is the vibration of the vocal cords. The pitch is regulated mainly by the tension put on the cords, with greater tension producing higher pitch. The tension on the vocal cords is controlled by the forward motion of the thyroid cartilage, relative to the fixed location of the arytenoid cartilages. But the actual size of the cords also contributes to pitch. The longer, thicker cords of most men give lower pitches than the shorter, thinner cords of most women. The fundamental tone from the cords is only a part of the final quality of the sound, as overtones are added by changes in the positions of lips, tongue, and soft palate.

The variable shapes of the nasal cavity and sinuses give voices their individual qualities. In fact, each person has such a distinct set of vocal overtones that every human voice is as unique as a set of fingerprints. The intensity, volume, or "loudness" of vocal sounds is regulated by the amount of air passing over the vocal cords, and that in turn is regulated by the pressure applied to the lungs, mainly by the abdominal muscles.

Trachea

The *trachea* (TRAY-kee-uh), or windpipe, is an open tube about 2.5 cm (1 in.) in diameter and 10 to 12 cm (4 to 5 in.) long. It extends from the base of the larynx to the top of the lungs, where it forks into two branches, the right and left **bronchi** (BRONG-kee; sing., *bronchus*) [FIGURE 22.4D]. The trachea is kept open by 16 to 20 C-shaped cartilaginous rings that are open on the dorsal side next to the esophagus. The rings are connected by fibroelastic connective tissue and longitudinal smooth muscle, making the trachea both flexible and extensible.

The inside surface of the trachea is lined with pseudostratified ciliated columnar epithelium, which produces moist mucus [FIGURE 22.5]. It contains upward-beating cilia. Those dust particles and microorganisms that are not caught in the nose and pharynx may be trapped in the trachea and carried up to the pharynx by the cilia to be swallowed or spat out *(expectorated).*

Q: *Why are nosebleeds so common?*

A: Bleeding from the nose is common because the nose sticks out far enough to be struck easily, the blood supply is abundant in the nasal membranes, and the mucosal layer is delicate. Bleeding can usually be controlled by packing the external or internal nares, or both. Nosebleeds are more common in cold weather because the lower absolute humidity of the inspired air promotes drying, cracking, and bleeding of the nasal epithelium.

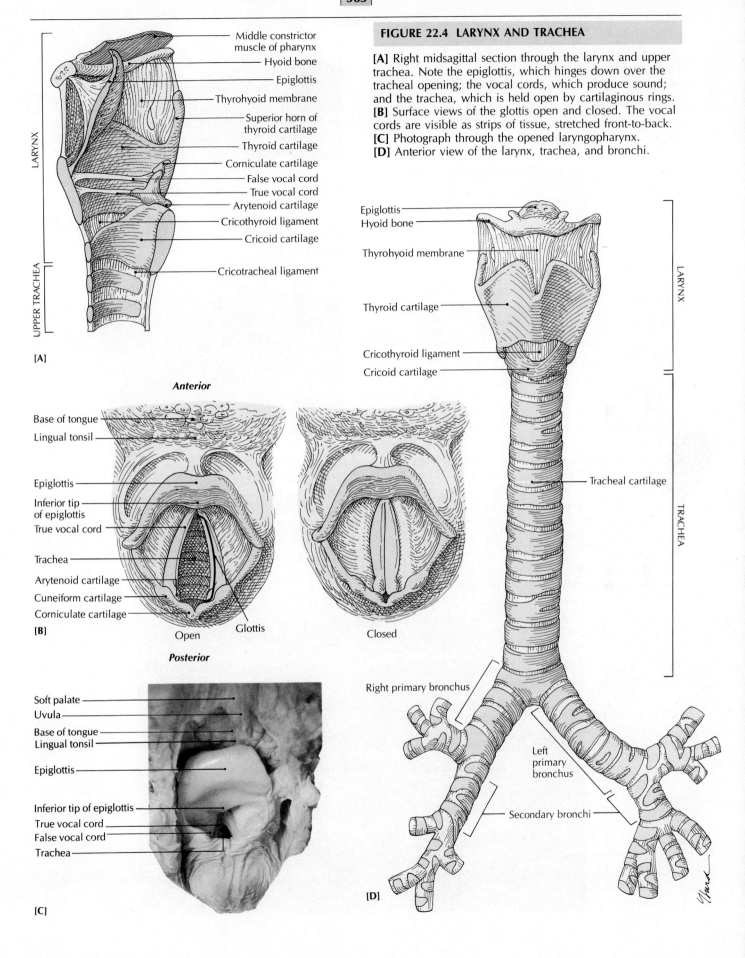

FIGURE 22.4 LARYNX AND TRACHEA

[A] Right midsagittal section through the larynx and upper trachea. Note the epiglottis, which hinges down over the tracheal opening; the vocal cords, which produce sound; and the trachea, which is held open by cartilaginous rings.
[B] Surface views of the glottis open and closed. The vocal cords are visible as strips of tissue, stretched front-to-back.
[C] Photograph through the opened laryngopharynx.
[D] Anterior view of the larynx, trachea, and bronchi.

[A]

LARYNX

UPPER TRACHEA

- Middle constrictor muscle of pharynx
- Hyoid bone
- Epiglottis
- Thyrohyoid membrane
- Superior horn of thyroid cartilage
- Thyroid cartilage
- Corniculate cartilage
- False vocal cord
- True vocal cord
- Arytenoid cartilage
- Cricothyroid ligament
- Cricoid cartilage
- Cricotracheal ligament

[B]

Anterior

- Base of tongue
- Lingual tonsil
- Epiglottis
- Inferior tip of epiglottis
- True vocal cord
- Trachea
- Arytenoid cartilage
- Cuneiform cartilage
- Corniculate cartilage
- Glottis

Open

Closed

Posterior

[C]

- Soft palate
- Uvula
- Base of tongue
- Lingual tonsil
- Epiglottis
- Inferior tip of epiglottis
- True vocal cord
- False vocal cord
- Trachea

[D]

- Epiglottis
- Hyoid bone
- Thyrohyoid membrane
- Thyroid cartilage
- Cricothyroid ligament
- Cricoid cartilage

LARYNX

- Tracheal cartilage

TRACHEA

- Right primary bronchus
- Left primary bronchus
- Secondary bronchi

FIGURE 22.5 TRACHEA

[A] Cross section of the trachea. [B] Light micrograph of pseudostratified epithelium in the trachea. ×100. [C] Scanning electron micrograph of a section through the wall of the trachea. ×300.

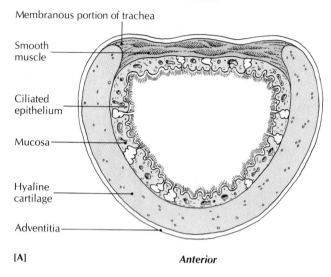

Posterior

Membranous portion of trachea

Smooth muscle

Ciliated epithelium

Mucosa

Hyaline cartilage

Adventitia

[A]

Anterior

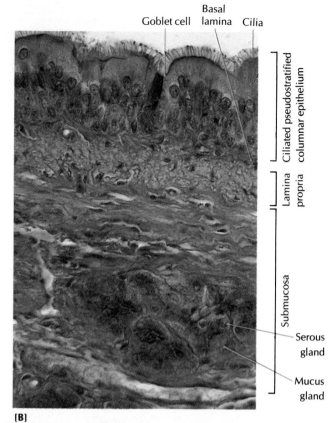

Goblet cell — Basal lamina — Cilia

Ciliated pseudostratified columnar epithelium

Lamina propria

Submucosa

Serous gland

Mucus gland

[B]

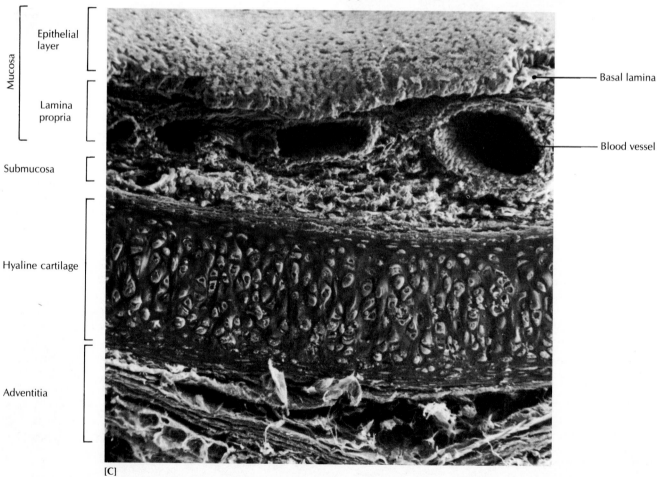

Mucosa

Epithelial layer

Lamina propria

Basal lamina

Blood vessel

Submucosa

Hyaline cartilage

Adventitia

[C]

Lungs: The Respiratory Tree

The trachea branches into the right and left bronchi, the *primary bronchi* that enter the right and left lungs. Each primary bronchus divides into smaller *secondary bronchi*, which in turn divide into *tertiary (segmental) bronchi*. These bronchi continue to branch into smaller and smaller tubes called **bronchioles** and then **terminal bronchioles** [FIGURE 22.6A, B]. The whole system inside the lung looks so much like an upside-down tree that it is commonly called the "respiratory tree."

Bronchi have cartilage in their walls, but bronchioles do not. Bronchioles contain more smooth muscle than the bronchi, however. Both bronchioles and bronchi are lined with ciliated columnar epithelium.

The bronchioles continue to branch, finally ending in *respiratory bronchioles* [FIGURE 22.6B]. They, in turn, branch into many *alveolar ducts*, which lead into microscopic air sacs called *alveoli*, where gas exchange takes place. Alveolar ducts and alveoli are lined with simple squamous epithelium.

In summary, it can be said that the *conducting portion* of the lung is composed of the bronchi and bronchioles. Beyond the terminal bronchioles (the most distal branches of the bronchial tree), the thinner-walled respiratory bronchioles begin the *respiratory portion* of the lung, where the exchange of respiratory gases occurs. The respiratory portion consists of the alveolar ducts, alveolar sacs, and alveoli, which lead from the respiratory bronchioles.

Alveoli The functional units of the lungs are the **alveoli** (al-VEE-oh-lye; sing. *alveolus*). Each lung contains over 350 million alveoli, each surrounded by many capillaries. Alveoli are clustered in bunches like grapes [FIGURE 22.6C, D], and provide enough surface area to allow ample gas exchange. The total interior area of adult lungs provides about 60 to 70 sq m, 20 times greater than the surface area of the skin (FIGURE 22.7B). Every time you inhale, you expose to fresh air an area of lung roughly equal to that of a tennis court.

A single alveolus looks like a bubble, which is supported by a *basement membrane*. A group of several alveoli with a common opening into an alveolar duct is called an *alveolar sac*. The lining epithelium of alveoli consists mainly of a single layer of squamous cells (squamous pul-

Q: *Why does yelling make us hoarse?*

A: Yelling vibrates the vocal cords and slams them together with such strong force that the cords may swell or even bleed. Swollen cords do not close properly, and hoarseness results as air leaks between them.

monary epithelial cells), also called *type I cells*. It also has *septal cells*, also called *type II cells*, which are smaller, scattered, cuboidal secretory cells [FIGURE 22.8]. Type II cells secrete a detergentlike chemical called *lung surfactant*, which helps keep alveoli inflated by reducing surface tension. Alveoli also contain phagocytic *alveolar macrophages* that adhere to the alveolar wall or circulate freely in the lumen of alveoli. These macrophages ingest and destroy microorganisms and other foreign substances that enter the alveoli. The foreign material either is moved upward by ciliary action to be expelled by coughing, or enters lymphatics to be carried to the lymph nodes at the hilum of the lung.

Since the alveoli are the sites of gas exchange with the blood by diffusion, the membranes of both alveolar walls and the capillary walls that line them must be thin enough to give maximum permeability. Alveoli are about 25 μm in diameter, and their walls are about 4 μm thick, which is much thinner than a sheet of paper. The capillaries surrounding each alveolus are also thin-walled. Gas exchange is facilitated by the small size and the large number of capillaries. Capillaries are so small that red blood cells flow through them in single file, giving each cell maximum exposure to the alveolar walls. There are so many capillaries that at any instant almost a liter of blood is being processed in the lungs.

Alveolar surface tension At the surface of water, especially where air and water meet, the water molecules are more attracted to one another than they are to the air. The result is a thin layer of greater density on the surface than beneath the surface of the water. This film of strongly attracted molecules produces a **surface tension** that tends to make the surface contract to its smallest possible area. (That is why small raindrops are spherical.) An alveolus is so wet that its watery lining would tend to make the alveolus collapse if there were not some way to reduce the surface tension. That tendency is counteracted by lung surfactant, which reduces the surface tension of the water on the inner surface of each alveolus. As a result, the alveolar walls can be thin without collapsing.

Because babies have no need to breathe before they are born, they do not need lung surfactant, but they do need it just before or just after birth so that the alveoli do not collapse. Without it they may suffer from insufficient opening of the alveoli. If the alveoli fail to open, or if they do open and collapse, the condition is known as *atelectasis* (imperfect expansion). Because the lungs develop toward the end of gestation, premature babies are at a high risk to develop hyaline membrane disease (see When Things Go Wrong). Atelectasis can also result from a number of pathological conditions, such as pneumonia, tuberculosis, or cancer.

FIGURE 22.6 LUNGS

[A] Lower portion of the respiratory tract—the respiratory tree. Note that the left lung, with no horizontal fissure, has only two lobes, the superior and inferior; the right lung has three lobes, including a middle lobe. **[B]** Lung lobule showing alveolar sacs. **[C]** Detail of alveolar sac. **[D]** Clusters of alveoli are surrounded by a capillary network.

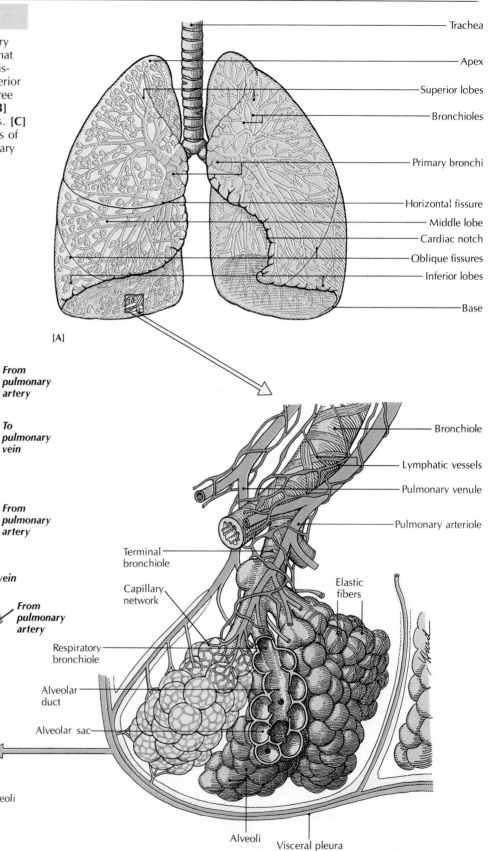

Trachea

Apex

Superior lobes

Bronchioles

Primary bronchi

Horizontal fissure

Middle lobe

Cardiac notch

Oblique fissures

Inferior lobes

Base

[A]

Alveolar cavity

From pulmonary artery

To pulmonary vein

From pulmonary artery

[D]

To pulmonary vein

From pulmonary artery

Alveoli

[C]

Bronchiole

Lymphatic vessels

Pulmonary venule

Pulmonary arteriole

Terminal bronchiole

Capillary network

Elastic fibers

Respiratory bronchiole

Alveolar duct

Alveolar sac

Alveoli

Visceral pleura

[B]

FIGURE 22.7 LUNG TISSUE

[A] Light micrograph of lung tissue. ×40. [B] Scanning electron micrograph of the inside of a lung. The darker holes are aveoli and alveolar ducts, which lead into deeper alveoli. The micrograph gives an idea of the convolutions that add to the surface area available for gas exchange. ×180.

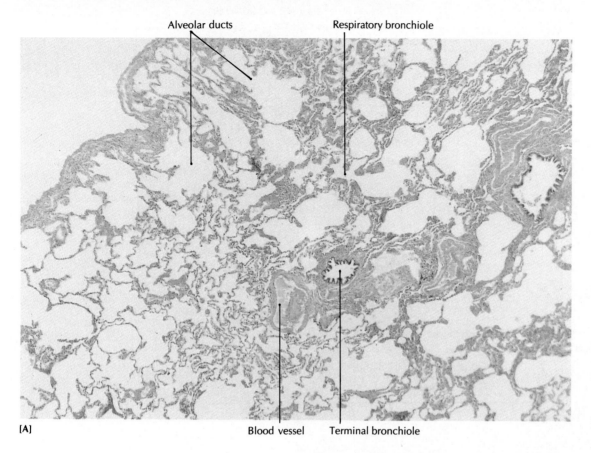

[A]

Alveolar ducts
Respiratory bronchiole
Blood vessel
Terminal bronchiole

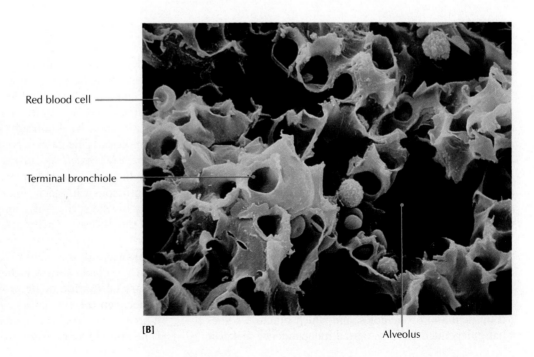

[B]

Red blood cell
Terminal bronchiole
Alveolus

FIGURE 22.8 CELLS OF LUNG TISSUE

This schematic drawing of a section of lung tissue shows an alveolus and surrounding capillaries, as well as various types of cells.

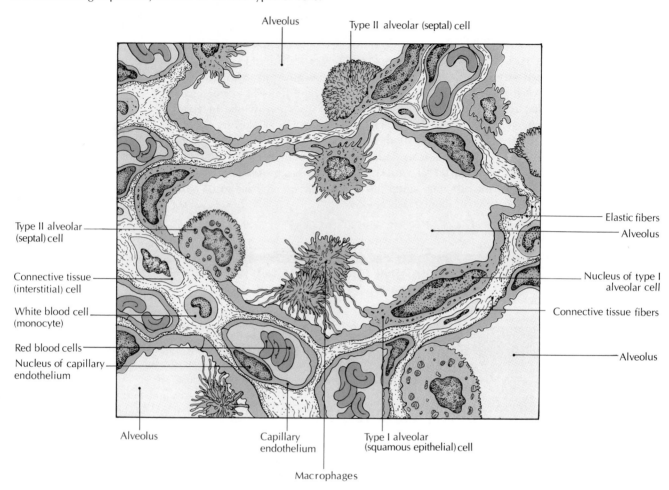

Alveolus
Type II alveolar (septal) cell
Type II alveolar (septal) cell
Connective tissue (interstitial) cell
White blood cell (monocyte)
Red blood cells
Nucleus of capillary endothelium
Elastic fibers
Alveolus
Nucleus of type I alveolar cell
Connective tissue fibers
Alveolus
Alveolus
Capillary endothelium
Type I alveolar (squamous epithelial) cell
Macrophages

Lungs: Lobes and Pleurae

The **lungs** fill the thoracic cavity except for a midventral region, called the *mediastinum*, where the heart and major blood vessels lie. Each lung is somewhat pointed at the *apex* (the top) and concave at the *base*, where it lies against the diaphragm [FIGURES 22.6A and 22.9]. Because the heart takes more space on the left side than on the right, and because the liver bulges up somewhat on the right side, the two lungs are not symmetrical. The left one is thinner and longer than the right.

The right lung has three main lobes, the *superior, middle*, and *inferior*. The left lung has no middle lobe, but has a concavity, the *cardiac notch*, into which the heart fits. Each lobe is further subdivided into 10 *bronchopulmonary segments* [FIGURE 22.9C, D], which are served by individual bronchi and bronchioles, and which function somewhat independently. This partial independence makes it

possible to remove one diseased segment without incapacitating the rest of the lung. Bronchopulmonary segments contain smaller subdivisions, called *lobules*, which are surrounded by elastic connective tissue [FIGURE 22.6B]. A lobule has hundreds of alveoli. It is served by a bronchiole carrying gases, a lymphatic vessel, a small vein (a *pulmonary venule*), and an artery (a *pulmonary arteriole*).

The lungs are covered by two pleural membranes that are continuous with each other: the inner *visceral pleura* directly on the surface of a lung, and the *parietal pleura* lining the thoracic cavity [FIGURE 22.10]. The small moisture-filled potential space between the visceral and parietal pleura is the **pleural cavity**. With each breath, the visceral pleura slides on the parietal pleura, with the necessary lubrication being provided by the thin film of fluid between the two layers.

Three functions are associated with the pleurae: (1) The thin film of moisture from the membranes within

FIGURE 22.9 EXTERNAL STRUCTURE OF THE LUNGS

[A] Right lung, anterior view. [B] Left lung, anterior view. [C] Bronchopulmonary segments of the right lung and [D] the left lung. The medial basal segments are hidden in [C] and [D].

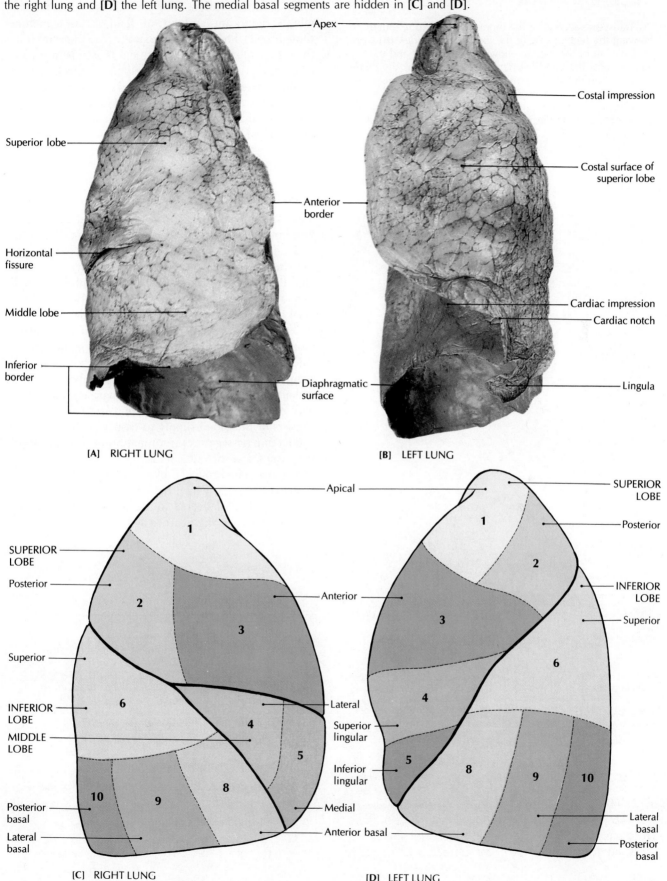

Apex

Superior lobe

Horizontal fissure

Middle lobe

Inferior border

Anterior border

Costal impression

Costal surface of superior lobe

Cardiac impression

Cardiac notch

Diaphragmatic surface

Lingula

[A] RIGHT LUNG

[B] LEFT LUNG

Apical

SUPERIOR LOBE

Posterior

SUPERIOR LOBE

Posterior

Superior

INFERIOR LOBE

MIDDLE LOBE

Posterior basal

Lateral basal

Anterior

Lateral

Superior lingular

Inferior lingular

Medial

Anterior basal

SUPERIOR LOBE

Posterior

INFERIOR LOBE

Superior

Lateral basal

Posterior basal

[C] RIGHT LUNG

[D] LEFT LUNG

FIGURE 22.10 PLEURAE

[A] Transverse section of the thorax as seen from above, showing the relationship of the lungs and pleurae. [B] Coronal section of the thorax, showing the pleurae and the separation of right and left pleural membranes by the mediastinum. Portions of the parietal pleura are named according to the structures to which the portions fuse: the mediastinal pleura next to the mediastinum, the costal pleura next to the ribs, and the diaphragmatic pleura next to the diaphragm.

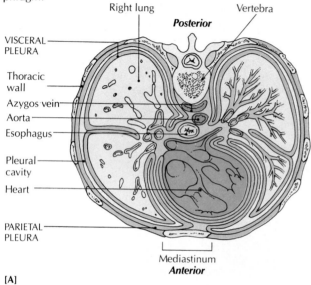

Right lung   Vertebra
Posterior
VISCERAL PLEURA
Thoracic wall
Azygos vein
Aorta
Esophagus
Pleural cavity
Heart
PARIETAL PLEURA
Mediastinum
Anterior

[A]

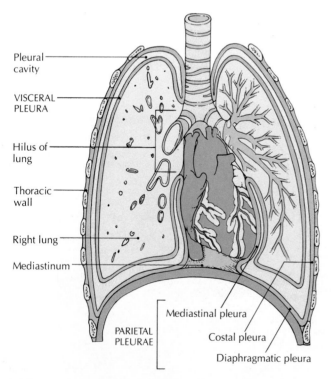

Pleural cavity
VISCERAL PLEURA
Hilus of lung
Thoracic wall
Right lung
Mediastinum
Mediastinal pleura
PARIETAL PLEURAE
Costal pleura
Diaphragmatic pleura

[B]

the pleural cavity acts as a lubricant for the lungs, which are in constant motion. (The pleural cavity does not contain a significant amount of fluid, as is sometimes thought.) (2) The air pressure in the pleural cavity is less than the atmospheric pressure, and thus aids in the mechanics of breathing. (3) The pleurae effectively separate the lungs from the medially located mediastinum, which contains the other thoracic organs. These organs include the heart, the esophagus, the thoracic duct, nerves, and major blood vessels.

Nerve and Blood Supply

The smooth muscle of the tracheobronchial tree is innervated by the autonomic nervous system. Branches of the vagus nerve run alongside the pulmonary blood vessels, carrying parasympathetic fibers that constrict the bronchi. Branches from the thoracic sympathetic ganglia carry fibers that dilate the bronchi. In addition, the bronchial walls contain parasympathetic ganglia.

The respiratory tree is supplied with blood by the pulmonary and bronchial circulations. The pulmonary circuit begins in the right ventricle and ends in the left atrium. Overall, this circuit includes the pulmonary trunk and arteries, the capillaries of the lungs, and the pulmonary veins. More specifically, the *bronchial circulation* consists of the blood supply to tissues of the lung's air-conducting passageways (terminal bronchioles, supporting tissues, and the outer walls of the pulmonary arteries and veins). The bronchial blood supply comes from the bronchial arteries, which branch off the thoracic aorta. Since the blood delivered to the alveoli comes from the right ventricle of the heart, it is low in oxygen.

A S K   Y O U R S E L F

1 What are the basic components of the respiratory system?

2 Why is the lining of the nasal cavity important in respiration?

3 What is the major function of the larynx?

4 What is the function of the epiglottis?

5 What are bronchi and bronchioles?

6 Why are alveoli called the functional units of the lungs?

7 What is the function of lung surfactant?

MECHANICS OF BREATHING

The layer of air covering the surface of the earth exerts a pressure of about 760 mm Hg, or approximately 1 atmosphere (atm) at sea level. Most of the air that we breathe is inert (chemically unreactive) nitrogen, which has no importance in normal respiration. The atmospheric gases of interest in respiration are oxygen and carbon dioxide.

Changes in the size of the thoracic cavity, and thus of the lungs, allow us to inhale and exhale air, in a process called *pulmonary ventilation*, or more commonly, *breathing*. When the thoracic cavity expands and the air pressure inside is lowered, the greater pressure outside causes a flow of air into the lungs. When the thoracic cavity shrinks, the increased pressure inside causes some contained air to flow out.

Muscular Control of Breathing

Breathing requires continual work by the muscles that increase the volume of the thoracic cavity and expand the lungs. During *inspiration* (inhalation), air flows into the lungs to equalize a reduced air pressure caused by the enlargement of the thoracic cavity. The mechanism operates in the following way [FIGURE 22.11]:

1 Several sets of muscles contract, the main ones being the *diaphragm* and the *external intercostal muscles.* The intercostal muscles extend from rib to rib, and when they contract, they pull the ribs upward and outward, enlarging the thoracic cavity.

2 The thoracic cavity is enlarged further when the muscles of the diaphragm contract, lowering the diaphragm. Although the intercostal and abdominal muscles

FIGURE 22.11 INSPIRATION AND EXPIRATION

The thorax at full inspiration **[A]** and at the end of expiration **[B].** Note the differences in the size of the thoracic cavity, the position of the sternum and ribs, and the shape of the diaphragm and abdominal wall.

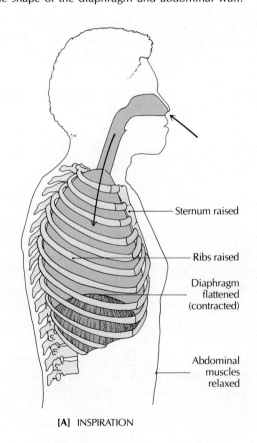

- Sternum raised
- Ribs raised
- Diaphragm flattened (contracted)
- Abdominal muscles relaxed

[A] INSPIRATION

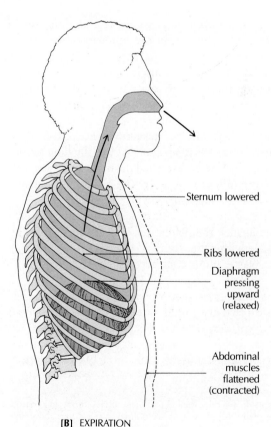

- Sternum lowered
- Ribs lowered
- Diaphragm pressing upward (relaxed)
- Abdominal muscles flattened (contracted)

[B] EXPIRATION

FIGURE 22.12 THE EFFECT OF POSTURE ON INSPIRATION

[A] When a patient is lying on his back, gravity forces the abdominal organs against the diaphragm, making it difficult for the diaphragm to contract (lower). [B] When the patient is propped up, the abdominal organs actually assist in lowering the diaphragm, thus assisting inspiration.

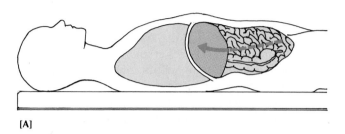

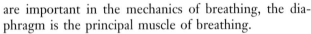

[A]

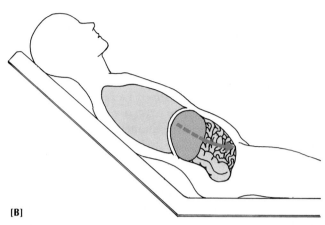

[B]

are important in the mechanics of breathing, the diaphragm is the principal muscle of breathing.

3 To compensate for the compression of abdominal organs by the lowered diaphragm, the abdominal muscles relax.*

4 The increased size of the thoracic cavity causes the pressure in the cavity to drop below the atmospheric pressure, and air rushes through the respiratory passages into the lungs, equalizing the pressure.

Ordinary *expiration* (exhalation), or the expulsion of air from the lungs, occurs in the following way:

1 External intercostal muscles and the diaphragm relax, allowing the thoracic cavity to return to its original, smaller size by *extrinsic elastic recoil* provided by the costal cartilages, thereby increasing the pressure within the thoracic cavity. The decrease in volume of the cavity is due partly to the *intrinsic elastic recoil* of the lung tissues themselves, which were stretched during inspiration, and partly to the upward push of the diaphragm, which bulges passively under pressure of the viscera.

2 Abdominal muscles contract, pushing the abdominal viscera against the diaphragm and further increasing the pressure within the thoracic cavity.

3 The elastic lungs contract as air is expelled.

During *forced expiration*, which occurs during strenuous physical activity, other muscles contract. **Internal intercostal muscles,** which run at right angles to the external ones, reduce the lung space when they contract. Also, the abdominal muscles can contract forcibly, press-

ing the viscera against the passive diaphragm and further reducing lung space. Although the expiratory muscles work during forced expiration, the main ones are the abdominal muscles, which force additional air out of the lungs [FIGURE 22.12].

The visceral pleura covering the lungs and the parietal pleura lining the thoracic cavity are separated only by the moisture from these membranes. Nevertheless, the pleurae are free to slide slightly past one another, but as no air is present between them, they do not separate when the thorax expands. If the thorax is punctured, and air enters the pleural cavity, the wet connection between the pleurae is broken, and the lung collapses. This condition is called a *pneumothorax*.

ASK YOURSELF

Q: Why do you sometimes get a "stitch" in your side when you run?

A: If the tissues of the diaphragm do not receive enough oxygen (as can happen when you are running and breathing heavily) the muscles in the diaphragm may tire and cause you to feel a sharp pain in your side.

OTHER ACTIVITIES OF THE RESPIRATORY SYSTEM

Several other activities involve the respiratory system. A **coughing reflex,** involving short, forceful exhalations through the mouth, occurs when sensitive parts of the air passages to the lungs are irritated, either by solids or liquids. The reflex persists even during deep coma.

A **sneeze** is an involuntary explosive expiration, mainly through nasal passages, and is usually caused by irritation

---

*Many people do not relax their abdominal muscles during inspiration, contracting them instead. Singers are trained to relax their abdominal muscles during inspiration, gradually contracting them during expiration.

The Heimlich Maneuver

Choking on food is such a common accident, and so often causes death from asphyxiation, that many restaurants post notices on how to perform the **Heimlich maneuver.** If a person gets a particle stuck in the glottis or larynx and cannot either exhale or inhale, a second person can help dislodge the particle by using the Heimlich maneuver. The aim is to compress the remaining air in the lungs violently enough to blow the particle free.

If the victim can stand, the rescuer stands behind the victim, places a fist against the abdomen between the ribs and navel, covers the fist with the other hand, and gives a strong, sudden squeeze inward and upward (see drawings). If the victim is unable to stand, the rescuer kneels astride the victim and applies pressure with the heel of one hand between the ribs and navel, press-

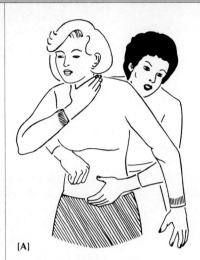 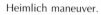 

[A]

[B]

Heimlich maneuver.

ing down on that lower hand with the other.

Some danger accompanies the Heim-

lich maneuver because too forceful or badly placed pressure can injure ribs, the diaphragm, or the liver.

of the mucous membranes in the nose. Particles ejected by a robust sneeze travel about 165 km/hr (103 mph).

A *hiccup* is a sudden, involuntary contraction of the diaphragm that usually results from a disruption in the normal pattern of breathing. Hiccups may be caused by the stimulation of nerves in the digestive system, but the respiratory system is involved. When the diaphragm is suddenly contracted, air is "sucked in" so abruptly that the epiglottis snaps shut, producing the sound of a hiccup. Hiccupping usually lasts 2 to 5 minutes, although some highly unusual cases persist for months or even years. It serves no known function. As a last resort to relieve chronic hiccups, one phrenic nerve in the neck (anterior surface of the anterior scalene muscle) is sometimes crushed. The crushed nerve can regenerate and reinnervate half of the paralyzed diaphragm to restore activity.

Q: *Why can a sneeze be stopped by pressing between the upper lip and the base of the nose?*

A: The nerve receptor for the stimulation of sneezing is located there. Firm pressure deadens the nerve and suppresses the sneeze.

Snoring can occur when muscles in the throat relax during sleep, and the loose tissue of the soft palate and uvula (the flap of tissue hanging down from the soft palate) partially obstruct the upper airway.

Sighing, sobbing, crying, yawning, and *laughing* are variants of ordinary breathing and are so closely associated with subjective emotional conditions that they defy exact definition.

THE EFFECTS OF AGING ON THE RESPIRATORY SYSTEM

As people grow older, some alveoli in the lungs are replaced by fibrous tissue, and the exchange of oxygen and carbon dioxide is reduced. Breathing capacity and usable lung capacity are also reduced when muscles in the rib cage are weakened and the lungs lose their full elasticity. (By age 45, lung capacity is usually reduced to about 82 percent of optimum; at age 65 it is about 62 percent, and it drops all the way to about 50 percent at age 85.) Aging also causes a decrease in gas exchange across pulmonary membranes.

Emphysema is a fairly common disease among the elderly, especially those who are heavy smokers or who live in air-polluted cities. *Lung cancer, chronic bronchitis,* and *pneumonia* also become more prevalent with age, and men's voices usually become higher as their vocal cords stiffen and vibrate faster.

DEVELOPMENTAL ANATOMY OF THE RESPIRATORY TREE

About four weeks after fertilization, at about the same time that the laryngotracheal tube and esophagus are separating into distinct structures, an endodermal *bronchial bud* forms at the caudal end of the developing laryngotracheal tube [FIGURE 22.13A]. The single bronchial bud soon divides into two buds, the right one being larger [FIGURE 22.13B]. These bronchial buds develop into the left and right primary bronchi of the future lungs during the fifth week [FIGURE 22.13C]. The right bud divides to form three secondary bronchi, and the left bud forms two. The right bronchus continues to be larger and more branched than the left during embryonic de-

velopment, and by the eighth week the bronchi show their basic mature form, with three secondary bronchi and three lobes in the right lung, but only two secondary bronchi and two lobes in the left lung [FIGURE 22.13D]. The bronchi subsequently branch repeatedly to form more bronchi and bronchioles.

By week 24, the epithelium of the bronchioles thins markedly, and highly vascularized primitive *alveoli* (terminal air sacs) have formed. A baby born prematurely during week 26 has a fairly good chance to survive without the help of artificial respiratory devices because its lungs have developed an alveolar mechanism for gas exchange. The development of pulmonary blood vessels is as important to the survival of premature infants as the thinness of the alveolar epithelium, which allows an adequate gas exchange. A newborn infant has only about 15 percent of the adult number of alveoli. The number and size of alveoli continue to increase until a child is about eight years old. Also at about 24 weeks, type II alveolar epithelial cells begin to secrete *lung surfactant*, a liquid that prevents alveoli from collapsing by reducing their surface tension. After two or three weeks of secretion, the alveoli are strong enough to retain air and remain open when breathing finally begins at birth.

FIGURE 22.13 EMBRYONIC DEVELOPMENT OF THE BRONCHI

[A] 4 weeks (early), **[B]** 4 weeks (late), **[C]** 5 weeks, and **[D]** 8 weeks. Ventral views.

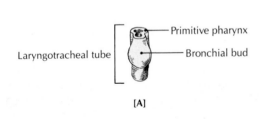

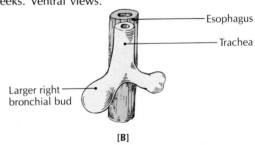

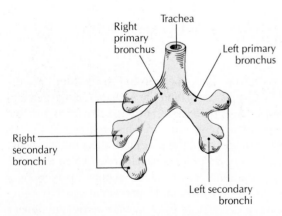

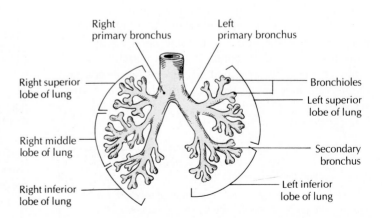

WHEN THINGS GO WRONG

A constant supply of oxygen and the ready removal of waste carbon dioxide are so necessary to continued life that some sort of respiratory failure is the immediate cause of most deaths. Any interference with breathing is serious and requires quick attention.

Rhinitis

Any inflammation of the nose can be called *rhinitis* (Gr. *rhin-*, nose). The most common nasal trouble is what is known as the "common cold." This virus-induced malady (or several maladies) causes excessive mucus secretion and swelling of the membranes of the nose and pharynx, and can spread into the nasolacrimal ducts, sinuses, auditory tubes, and larynx. It can cause fever, general malaise, and breathing difficulty, but is seldom grave except in people weakened by some other illness.

Laryngitis

A local infection of the larynx is *laryngitis.* It can totally incapacitate the vocal cords, making normal speech impossible, but is usually not painful. It may be *idiopathic*, that is, not associated with any other disease. Irritants such as tobacco smoke can cause swelling of the vocal cords to such an extent that chronic hoarseness results.

Asthma

Asthma (Gr. no wind) is a general term for difficulty in breathing. *Bronchial asthma* is a result of the constriction of smooth muscles in the bronchial and bronchiolar walls, accompanied by excess mucus secretion and insufficient contraction of the alveoli. Air is held in the lungs, and the victim cannot exhale normally. The condition is usually caused by an allergic reaction, but may also be brought on by an emotional upset. Asthma is not always dangerous, though over a long period of time it may result in permanently damaged alveoli. An acute attack, if not treated immediately, can cause death by asphyxiation.

Hay Fever

Hay fever is characterized by abundant tears and a runny nose. Like asthma, it is caused by an allergic reaction. So many people are affected that pollen counts of the atmosphere are regularly reported in the public media when pollen is abundant. The worst plant offenders are wind-pollinated trees, grasses, and ragweed.

Emphysema

Emphysema (Gr. blown up) is an abnormal, permanent enlargement of the bronchi, accompanied by the destruc-

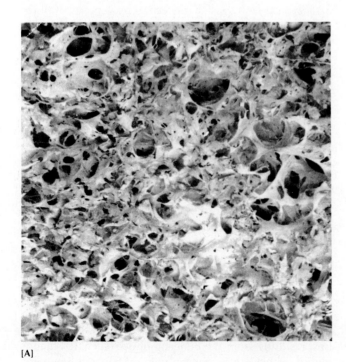

[A]

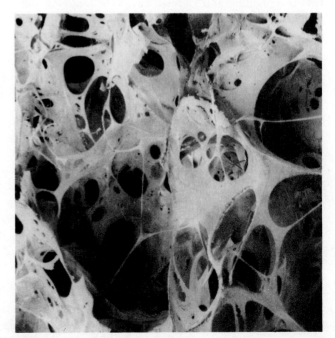

[B]

[A] Normal lung tissue. ×100. [B] Emphysematous lung tissue. ×100.

tion of the alveolar walls. Breathing difficulties, especially during expiration, are caused by the destruction of alveolar septae, the partial collapse of the airways, and a loss of elasticity in the lungs. A person suffering from emphysema cannot exhale normal amounts of air even when trying forcibly to do so. As a result, undesirable levels of carbon dioxide may build up. In long-standing cases, the alveoli deteriorate, their walls harden, and alveolar capillaries are damaged (see photos on page 577). Fortunately, because a normal adult has about eight times as many alveoli as needed for routine activities, acute cases of emphysema are not as prevalent as they might be otherwise. As the disease progresses, the respiratory center in the brain loses its effectiveness, the right ventricle of the heart becomes overworked in its efforts to force blood through the constricted lung capillaries, and the victim runs short of breath even after slight exercise.

The most common cause of emphysema is air pollution, including tobacco smoke, industrial solvents, and agricultural dust. It can develop in old people as a result of hardening of alveolar walls, without any apparent environmental stimulus.

Cardiopulmonary Resuscitation (CPR)
In a respiratory emergency, breathing stops or is reduced enough so that the victim's body tissues do not get sufficient oxygen to support life. If breathing is not restored, the heart can stop beating (cardiac arrest). *Cardiopulmonary resuscitation (CPR)* is a technique that combines rescue breathing with external cardiac compression to restore breathing and circulation. *Rescue breathing* supplies oxygen to the victim's lungs, and *cardiac compression* circulates blood by compressing the heart between the sternum and vertebrae.

Pneumonia
Pneumonia is a general term for any condition that results in the filling of the alveoli with fluid. It can be caused by a number of factors: chemicals, bacteria, viruses, protozoa, or fungi, but the usual infective agent is a *Streptococcus* bacterium. Pneumonia most readily attacks people who are already weakened by illness or whose lungs are damaged. Blood infections, chronic alcoholism, inhalation of fluids into the lungs, or even prolonged immobility in bed can predispose a person to microbial infection of the lungs.

Tuberculosis
Tuberculosis, a disease caused by *Myobacterium tuberculosis,* can occur in almost any part of the body, but because it most commonly affects the lungs, it is usually regarded as a respiratory ailment. The infectious bacterium is usu-

ally inhaled, and if it becomes established, the lungs can be damaged in a variety of ways.

Lung Cancer
Lung cancer, or *pulmonary carcinoma,* usually seems to be caused by environmental factors, cigarette smoking being the one most commonly mentioned. Any inhaled irritant apparently can stimulate some cells to begin abnormal growth. As has been pointed out frequently, early diagnosis is critical. Practically all sufferers from lung cancer cough, most breathe with difficulty, and many have chest pains and spit blood. Any such combination of symptoms calls for chest x-rays, bronchoscopy, and biopsy. Damaged lung tissue shows up in x-rays, sometimes as a discrete region and sometimes in such scattered spots as to suggest a "snowstorm." Lung cancer can spread to other organs, or other organs can send malignant cells to the lungs. In either case, the outlook is grave. Since it is known that the average survival time for a victim of lung cancer after diagnosis is about nine months, and that cigarette smokers run about 20 times the risk of nonsmokers, cigarette smoking is clearly a risky habit.

Pleurisy
Pleurisy is an infection of the pleurae. It is incapacitating, but is not necessarily dangerous in itself. There are several stages and types of the disease. *Fibrinous,* or *dry, pleurisy* causes intense pain in the parietal pleurae (the visceral pleurae are insensitive), and results in audible crackling or grating when the patient breathes. *Serofibrinous pleurisy* is characterized by deposits of fibrin and by an accumulation of watery fluid (up to 5 L in extreme cases) in the pleural cavity. It can be detected by percussion, but x-rays are most useful. In *purulent pleuritis,* or *empyema,* a pus-laden secretion accumulates in the pleural cavity. All types of pleurisy are caused by microorganisms.

Drowning
One of the most common accidental causes of respiratory failure is *drowning.* A drowning person suffers a laryngeal spasm trying to inhale under water. Little water enters the lungs, but much may be swallowed. Death occurs within minutes unless oxygen is delivered to the tissues promptly.

Cyanosis
Whenever breathing is stopped and the pulse is weak or absent, *cyanosis* (Gr. dark blue) occurs. Cyanosis is the development of a bluish color of the skin, especially the lips, resulting from the build-up of deoxygenated hemoglobin, which is less crimson than oxygenated hemoglobin and appears bluish through the skin.

Hyaline Membrane Disease

Hyaline membrane disease, or "glassy-lung" disease, is a failure of newborn infants to produce enough surfactant to allow alveoli to fill with air properly. At birth, lungs contain no air, but they must quickly become inflated and stay inflated. In cases of too little pulmonary surfactant, the lungs may be lined with a hyaline (transparent) coating, hence the name of the disease.

Sudden Infant Death Syndrome (SIDS)

Another malady of infants is "crib death," or *sudden infant death syndrome* (SIDS). It claims about 10,000 victims under 1 year of age annually in the United States, but its cause remains unknown. It is included here among respiratory ailments because one likely cause is respiratory failure, either from spasmodic closure of the air passages, or some malfunction of the respiratory center in the brain. The fact that crib death is most common in the autumn seems to indicate that an infectious agent, such as a virus, is responsible.

CHAPTER SUMMARY

The *respiratory system* delivers oxygen to the body tissues, and removes gaseous wastes, mainly carbon dioxide.

Respiratory Tract (p. 560)

1 The *respiratory tract* includes the *nose, nasal cavities, pharynx, larynx, trachea,* and *bronchi,* which lead by way of *bronchioles* to the *lungs.*

2 Air usually enters the body through the nose. The respiratory mucosa of the **nasal cavity** is specialized to moisten and warm air and to capture particles like dust. The olfactory mucosa is specialized to sense odors.

3 The **pharynx** connects the nasal cavity and mouth with the rest of the respiratory tract and the esophagus. It is divided into the **nasopharynx, oropharynx,** and **laryngopharynx.**

4 The *larynx* contains the **vocal cords,** which are largely responsible for producing sound. At the opening of the larynx, the *glottis* closes during swallowing and opens to allow air to pass.

5 The *trachea,* or windpipe, carries air from the larynx to two *bronchi.* Its mucosal and ciliated lining traps and removes dust particles and microorganisms before they can enter the lungs.

6 These two bronchi divide into smaller and smaller bronchi, and then into even smaller **bronchioles,** forming the "respiratory tree." Around the tiniest branches, called respiratory bronchioles, are minute air sacs known as alveoli.

7 *Alveoli* are the functional units of the lungs. Exchange of oxygen and carbon dioxide takes place through the walls of the alveoli and the walls of the pulmonary capillaries.

8 In the thoracic cavity, the two **lungs** are separated by the *mediastinum.* The right lung has three *lobes,* and the left lung has two. Each lobe is further divided into *bronchopulmonary segments* and then into *lobules.*

9 The lungs are covered by the *visceral pleura,* and the thoracic cavity by the *parietal pleura.* Between these membranes is the **pleural cavity.**

10 The smooth muscle of the tracheobronchial tree is innervated by the autonomic nervous system. The respiratory tree is supplied with blood by the pulmonary and bronchial circulations.

Mechanics of Breathing (p. 573)

1 *Ventilation* is the mechanical process that moves air into and out of the lungs.

2 During **inspiration,** air is pushed into the lungs by the pressure of the outside air when the size of the thoracic cavity is increased by contraction of the **external intercostal muscles** and the **diaphragm** and the raising of the ribs.

3 During ordinary **expiration,** air is expelled from the lungs when the *respiratory muscles* are relaxed and the volume of the thoracic cavity is reduced.

Other Activities of the Respiratory System (p. 574)

Other activities of the respiratory system include coughing, sneezing, crying, laughing, yawning, hiccuping, and snoring.

The Effects of Aging on the Respiratory System (p. 575)

1 Aging is usually accompanied by a decrease in gas exchange, and a reduction in breathing and lung capacities.

2 Emphysema, lung cancer, chronic bronchitis, and pneumonia may become more prevalent with age.

Developmental Anatomy of the Respiratory Tree (p. 576)

1 Bronchial buds that emerge about week 4 develop into the left and right primary bronchi of the future lungs. The bronchi show their basic mature form about week 8, and they branch repeatedly to form more bronchi and bronchioles.

2 Alveoli form by week 24, and surfactant begins to be secreted. After two or three weeks of secretion, the alveoli have expanded enough to be functional.

MEDICAL TERMINOLOGY

ANOXIA Absence of oxygen.

ANTITUSSIVE A drug that controls coughing.

BRONCHITIS Inflammation of the bronchi.

BRONCHODILATOR A drug that dilates the bronchi.

BRONCHOSCOPY Examination of the interior of the bronchi.

EMPYEMA A pleural space filled with pus.

EXPECTORANT A drug that encourages the expulsion of mucus from the respiratory tree.

HEMOTHORAX A pleural space filled with blood.

HYPERVENTILATION Excessive movement of air into and out of the lungs.

HYPOXIA Insufficient oxygen in organs and tissues.

LOBECTOMY Removal of one of the lobes of the lungs.

NASAL POLYPS Tumors caused by abnormal growth of the nasal mucous membranes.

PNEUMONECTOMY Removal of one of the lungs.

PULMONARY EDEMA A condition in which there is excessive fluid in the lungs.

SPUTUM (SPYOO-tuhm) Liquid substance containing mucus and saliva.

TRACHEOSCOPY Examination of the interior of the trachea.

TRACHEOTOMY A surgical procedure to make a hole and insert a breathing tube in the trachea.

UNDERSTANDING THE FACTS

1 What do the conchae contribute functionally to the respiratory system?

2 What is meant by a cleft palate?

3 What factors often contribute to "smoker's cough"?

4 Name the three parts of the pharynx, and give their locations and functions.

5 What is the function of the trachea?

6 What are the two tubes into which the trachea divides?

7 What roles do the macrophages in the alveoli play?

8 Distinguish between visceral and parietal pleurae.

9 What muscles are used during inspiration? During expiration?

10 Define pneumonia.

11 What are some of the causes of emphysema?

12 What is the usual cause of asthma?

13 Describe the Heimlich maneuver.

14 Give the location of the following (be specific):

  a buccal cavity

  b thyroid cartilage

  c mediastinum

  d intercostal muscles

UNDERSTANDING THE CONCEPTS

1 What are the functions of the respiratory system?

2 How do the structures of the larynx contribute to sound production?

3 What roles do the pleurae of the lungs play in respiration?

4 What is lung surfactant, and why is it important in the lungs?

5 List in order the structures through which a molecule of oxygen in the atmosphere would pass in order to reach the bloodstream.

6 Compare the processes of internal and external respiration.

SELF-QUIZ

Multiple-Choice Questions

22.1 Which of the following bones is *not* associated with the nasal cavity?

  a ethmoid

  b maxillary

  c inferior conchae

  d mandible

  e turbinate

22.2 Of the following, which is *not* a paranasal sinus?

  a frontal

  b mandible

  c sphenoidal

  d maxillary

  e ethmoid

22.3 Surfactant is secreted by

  a type I cells

  b type II cells

  c goblet cells

  d basement membranes

  e none of the above

22.4 Which of the following muscles is/are *not* involved with normal inspiration?

  a internal intercostals

  b external intercostals

  c diaphragm

  d a and b

  e b and c

22.5 The membrane that covers the lungs is the

  a alveolar membrane

  b pleura

  c peritoneum

  d viscera

  e mediastinum

True-False Statements

22.6 The olfactory epithelium contains sensory nerve endings that are connected to the olfactory nerve.

22.7 The most prominent cartilage of the larynx is the cricoid cartilage.

22.8 The lowermost portion of the pharynx is the laryngopharynx.

22.9 The exchange of respiratory gases between the blood and the body cells is called cellular respiration.

22.10 The nasal cavity is divided into bilateral halves by the nasal bones.

22.11 The soft palate separates the oropharynx from the nasal cavity.

22.12 The trachea branches into two terminal bronchioles.

22.13 Each lung has a superior, middle, and inferior lobe.

22.14 The lungs occupy the mediastinum.

22.15 The main muscles involved in expiration are the diaphragm and the external intercostals.

Completion Exercises

22.16 When the palatine and maxillary bones do not grow together by the third month of prenatal development, a _____ might result.

22.17 Another name for the oral cavity is the _____.

22.18 The body has _____ pair(s) of tonsils.

22.19 The _____ is a flap of cartilage that folds down over the opening into the larynx during swallowing.

22.20 The correct anatomical term for the windpipe is the _____.

22.21 In the lungs, gas exchange occurs within the _____.

22.22 The common term for pulmonary ventilation is _____.

22.23 The exchange of gases between the blood and alveoli is termed _____ respiration.

22.24 Air is warmed and moistened in the _____.

22.25 The internal intercostals contract in _____.

A SECOND LOOK

1 In the following drawing, label the epiglottis, thyroid cartilage, larynx, trachea, and primary and secondary bronchi.

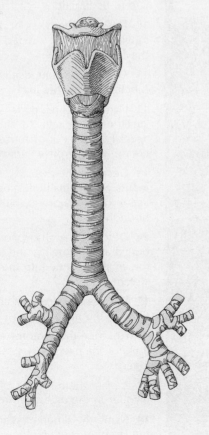

2 In the following drawing, label the capillary network, terminal bronchiole, alveolar duct, and alveolar sac.

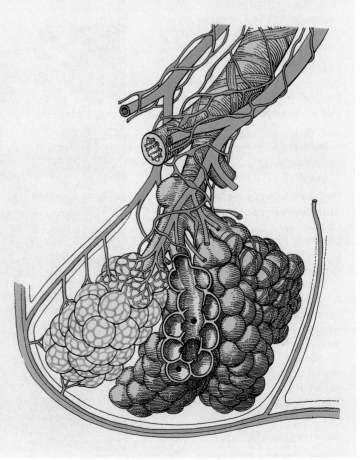

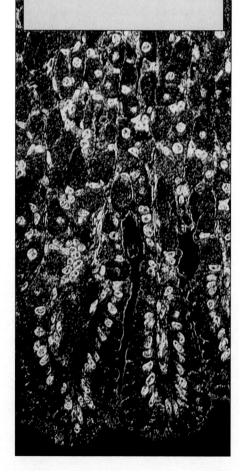

23

The Digestive System

LEARNING OBJECTIVES

1 Differentiate between digestion and absorption.

2 List the major organs of the alimentary canal and the associated structures of digestion, and give the basic functions of each.

3 Describe the four layers of tissue in the wall of the digestive tract.

4 Discuss the digestive activities that occur in the mouth.

5 Describe the three main parts of a tooth, the composition of teeth, and the types of permanent teeth.

6 Name the three pairs of salivary glands, and explain the functions of saliva.

7 Describe the stages of swallowing, along with the structures involved in the process.

8 Discuss the structure of the abdominopelvic cavity and the peritoneum.

9 Identify the four major regions of the stomach and its two openings, and describe its tissues as they relate to its functions.

10 Describe the types of digestive movements of the stomach and how they are regulated.

11 Discuss the composition and functions of gastric juice.

12 Name the three parts of the small intestine, and explain how the special features of its mucosa increase its absorptive surface.

13 Describe the digestive movements of the small intestine and the action of its major enzymes and hormones.

14 Describe the absorption of carbohydrates, proteins, and fats from the small intestine into the bloodstream.

15 Discuss the structure and functions of the pancreas.

16 Describe the anatomy and functions of the liver as they relate to digestion.

17 Explain the functions of the gallbladder in the digestive process.

18 Name the different segments of the large intestine and describe their functions.

The digestive tract, which includes the mouth, pharynx, esophagus, stomach, and intestines, is like a tube within the body from the mouth to the anus [FIGURE 23.1]. Nutrients that are within this tube are not inside the body proper until they are absorbed from the intestines. After food is ingested, it undergoes *digestion,* the mechanical and chemical processes that break down large food molecules into smaller ones. But the small molecules are useless unless they can get into the individual cells of the body. This is accomplished when the small food molecules pass through the cells of the small intestine into the bloodstream and lymphatic system. Water, minerals (salts), and some vitamins enter the bloodstream via the large intestine. This second process involving the intestines is called *absorption.*

After digestion and absorption have taken place, the small molecules can be used by the cells of the body. Some of the molecules are used as a source of energy. Others, such as amino acids, are used by cells to rebuild, repair, and reproduce themselves. Materials that are not digested and absorbed are finally eliminated from the body. The processes of digestion and absorption include:

1 *Ingestion,* or eating.

2 *Peristalsis,* or the involuntary, sequential smooth muscle contractions that move ingested nutrients along the digestive tract.

3 *Digestion,* or the conversion of large nutrient molecules into small molecules.

4 *Absorption,* or the passage of usable nutrient molecules from the intestines into the bloodstream and lymphatic system.

5 *Defecation,* or the elimination from the body of undigested and unabsorbed material as solid waste.

INTRODUCTION TO THE DIGESTIVE SYSTEM

In simple terms, digestion is the process of breaking down large molecules of food that cannot be used by the body into small, soluble molecules that can be absorbed and used by cells. *Mechanical digestion* involves physical means, such as chewing, peristalsis, and the churning movements of the stomach and small intestine to mix the food with enzymes and digestive juices. *Chemical digestion* breaks down food particles through a series of metabolic reactions involving enzymes. The digestive process takes place in the **digestive tract,** or **alimentary canal.** The part of the digestive tract inferior to the diaphragm is called the *gastrointestinal (GI) tract.*

The alimentary canal consists of the mouth, pharynx, esophagus, stomach, small intestine, large intestine, rectum, anal canal, and anus. From mouth to anus this canal is about 9 m (30 ft) long. The *associated structures* of the digestive system include the salivary glands, pancreas, liver, and gallbladder.

Basic Functions

The digestive system is divided into compartments, each adapted to perform a specific function [FIGURE 23.1]. In the *mouth,* the breakdown of food begins with chewing and enzymatic action. The *pharynx* (throat) performs the act of swallowing, and food passes through the *esophagus* to the stomach. The *stomach* stores food and breaks it down with acid and some enzymes. The *small intestine* is where most of the large molecules are chemically broken down into smaller ones, and where most of the nutrients are absorbed into the bloodstream and lymphatic system. The *large intestine* absorbs minerals and some vitamins, carries undigested food, removes additional water from it, and releases solid waste products through the *anus.* The *associated structures* assist in the breakdown and conversion of food particles in a variety of ways, both mechanical and chemical.

Tissue Structure

Despite the compartmentalization of the digestive system, the walls of the various portions of the tract have the same basic organization. The wall of the tube (from the inside out) consists of four main layers of tissue: **mucosa, submucosa, muscularis externa,** and **serosa.** They are shown and described in FIGURE 23.2.

MOUTH

By following a mouthful of food through the digestive tract, we can observe the digestive process in detail. Food enters the digestive system through the **mouth,** also called the **oral cavity.** The mouth has two parts: (1) The small, outer **vestibule,** or **buccal** (BUCK-uhl) **cavity,** [FIGURE 23.3] is bounded by the lips and cheeks on the outside and the teeth and gums on the inside. (2) The **oral cavity proper** extends from behind the teeth and gums to the *fauces* (FAW-seez; L. throat), the opening that leads to the *pharynx.* The fauces has sensory nerve endings that trigger the involuntary phase of swallowing.

In the mouth, food is *masticated* (chewed) by the ripping and grinding action of the teeth. As much pressure as 500 kg/sq cm (7000 lb/sq in.) can be exerted by the

FIGURE 23.1 DIGESTIVE SYSTEM

[A] The drawing shows structures of the digestive tract and associated organs. [B] Photograph of the digestive system, anterior view.

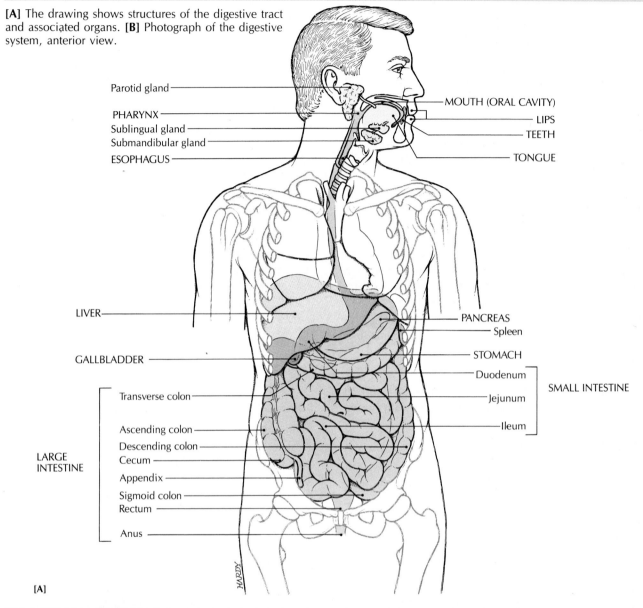

Parotid gland
PHARYNX
Sublingual gland
Submandibular gland
ESOPHAGUS

MOUTH (ORAL CAVITY)
LIPS
TEETH
TONGUE

LIVER
GALLBLADDER

PANCREAS
Spleen
STOMACH
Duodenum
Jejunum
Ileum
SMALL INTESTINE

Transverse colon
Ascending colon
Descending colon
Cecum
Appendix
Sigmoid colon
Rectum
Anus
LARGE INTESTINE

HARDY

[A]

Structure	Major digestive secretions	Major digestive functions
Lips and cheeks	None	Hold food in position to be chewed. Help identify food textures. Buccal muscles contribute to chewing process.
Teeth	None	Break food into pieces, exposing surface for digestive enzymes.
Tongue	None	Assists chewing action of teeth. Helps shape food into bolus, pushes bolus toward pharynx to be swallowed. Contains taste buds.
Palate	None	Hard palate helps crush and soften food. Soft palate closes off nasal opening during swallowing.
Salivary glands	Saliva—salivary amylase, mucus, water, various salts	Salivary amylase in saliva begins breakdown of cooked starches into soluble sugars maltose and dextrin. Helps form bolus and lubricates it prior to swallowing. Dissolves food for tasting. Moistens mouth. Helps prevent tooth decay.
Pharynx	None	Continues swallowing activity when bolus enters from mouth, propels bolus from mouth into esophagus.

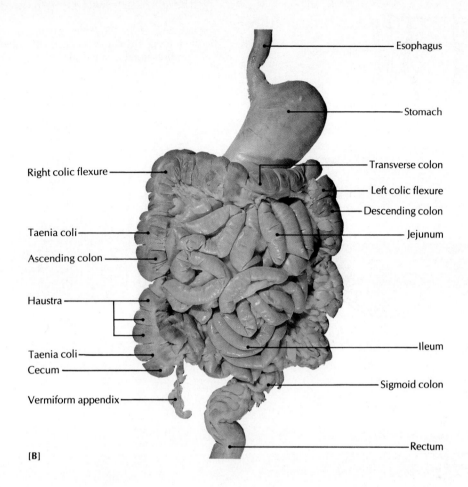

Esophagus

Stomach

Right colic flexure

Transverse colon

Left colic flexure

Descending colon

Taenia coli

Jejunum

Ascending colon

Haustra

Taenia coli

Ileum

Cecum

Vermiform appendix

Sigmoid colon

Rectum

[B]

Structure	Major digestive secretions	Major digestive functions
Esophagus	Mucus	Propels food from pharynx into stomach.
Stomach	Hydrochloric acid, pepsin, mucus, some lipase, gastrin, intrinsic factor	Stores, mixes, digests food, especially protein. Regulates even flow of food into small intestine. Acid kills bacteria.
Small intestine	Enzymes: enterokinase, peptidases, maltase, lactase, sucrase, intestinal amylase, intestinal lipase; salts, water, mucus; hormones: cholecystokinin, gastric inhibitory peptide, secretin	Site of most chemical digestion and absorption of most water and nutrients into bloodstream.
Pancreas	Enzymes: trypsin, chymotrypsin, car-boxypeptidase, pancreatic amylase, pancreatic lipase, nuclease; bicarbonate	Secretes many digestive enzymes. Neutralizes stomach acid with alkaline bicarbonate secretion.
Liver	Bile, bicarbonate	Secretes bile. Detoxifies harmful substances. Converts nutrients into usable forms and stores them for later use. Neutralizes stomach acid with alkaline bicarbonate secretion.
Gallbladder	Mucin	Stores and concentrates bile from liver and releases it when needed.
Large intestine	Mucus	Removes salts and water from undigested food. Releases feces through anus. Aids synthesis of vitamins B_{12} and K.
Rectum	None	Removes solid wastes by process of defecation.

FIGURE 23.2 THE WALL OF THE DIGESTIVE TRACT

The drawings show the basic organization of the wall of the digestive tract. **[A]** Cross section. **[B]** Enlargement of the inner portion (tunica mucosa).

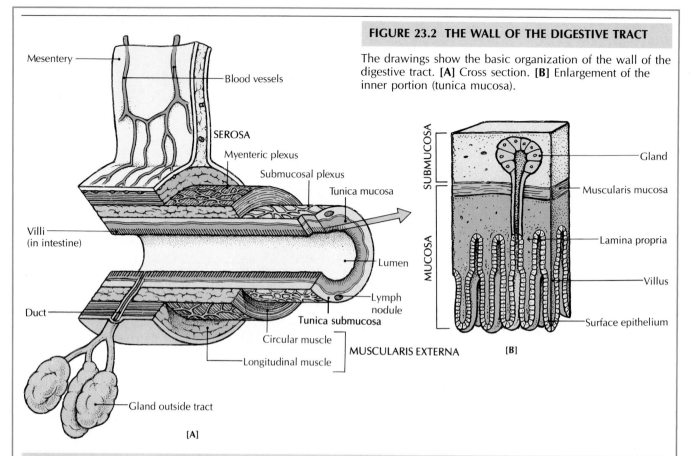

Layer	Description
MUCOSA (tunica mucosa, mucous membrane)	
Epithelium Lamina propria Muscularis mucosa	Innermost layer of digestive tract. *Epithelium* is site of interaction between ingested food and body. *Lamina propria* is layer of connective tissue supporting epithelium. *Muscularis mucosa* consists of two thin muscular layers and contains a nerve plexus. Surface area of epithelial layer is increased by large folds (rugae in the stomach and plicae circulares and villi in the small intestine), indentations (crypts), and glands. Acts as lubricating, secreting, absorbing layer. Lubricates solid contents and facilitates their passage through digestive tract. Contains absorptive cells and secretory cells that produce mucus, enzymes, various ions.
SUBMUCOSA (tunica submucosa)	
	Highly vascular layer of connective tissue between mucosa and muscle layers; contains many nerves and in certain areas, lymphatic nodules. Submucosal glands are present in esophagus and duodenum.
MUSCULARIS EXTERNA (tunica muscularis)	
Circular muscle Longitudinal muscle Oblique muscle	Main muscle layer, consisting in most regions of inner *circular* layer and outer *longitudinal* layer of mostly smooth muscle. Stomach contains additional *oblique* layer internal to other layers. Upper esophagus and sphincters of anus consist of skeletal muscle fibers. Moves food through lumen of digestive tract by waves of muscular contraction called *peristalsis*.
SEROSA (tunica serosa)	
Visceral peritoneum Adventitia	Outermost lamina, consisting of thin connective tissue, and in many places epithelium covering digestive tube and digestive organs. Where epithelium is lacking (as in esophagus), serosa is called *adventitia*. Portion covering viscera is *visceral peritoneum*. Double-layered *mesentery* is portion of serous membrane that connects intestines to dorsal abdominal wall. Contains blood vessels, nerves, and lymphatics.

FIGURE 23.3 MOUTH

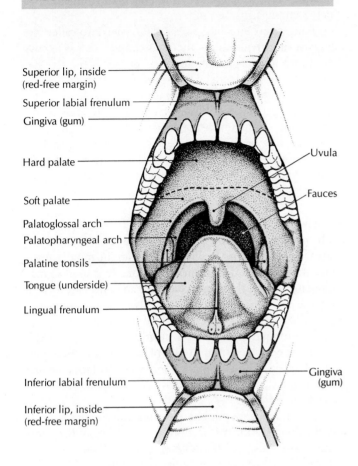

Superior lip, inside (red-free margin)

Superior labial frenulum

Gingiva (gum)

Hard palate

Soft palate

Palatoglossal arch

Palatopharyngeal arch

Palatine tonsils

Tongue (underside)

Lingual frenulum

Inferior labial frenulum

Inferior lip, inside (red-free margin)

Uvula

Fauces

Gingiva (gum)

Lips, like the anus at the other end of the digestive system, form a transition between the external skin and the mucous membrane of the moist epithelial lining of the internal passageways. The skin of the outer lip surface (beneath the nose and above the chin) contains the usual sweat glands, sebaceous glands, and hair follicles. The *red free margin*, the portion we normally think of as "lips," is covered by a thin, translucent epidermis that allows the capillaries underneath to show through, giving the lips a reddish color. The skin of the red free margin does not contain sweat glands, sebaceous glands, or hair follicles.

The lips help to place food in the mouth and to keep it in the proper position for chewing. They also contain sensory receptors that make it relatively easy for us to identify specific textures of foods.

The moist mucous membrane of the inner surface of the lips leads directly to the mucous membrane of the inner surface of the cheeks. The *cheeks* are the fleshy part of either side of the face, below the eye and between the nose and ear. The mucous membrane lining the lips and cheeks is a thick stratified squamous epithelium (nonkeratinized). Such an epithelium is typical of moist epithelial surfaces that are subjected to a great deal of abrasive force. As soon as the surface cells are worn away, they are replaced by the rapidly dividing cells underneath.

The cheeks, like the lips, help to hold food in a position where the teeth can chew it conveniently. Also, the muscles of the cheeks contribute to the chewing process.

molars. At the same time, the food is moistened by saliva, which intensifies its taste and eases its passage down the delicate tissue of the esophagus. Saliva, secreted by the salivary glands, contains an enzyme that begins the breakdown of large carbohydrate molecules into small sugar molecules.

Lips and Cheeks

The *lips* are the two fleshy, muscular folds that surround the opening of the mouth. They consist mainly of fibroelastic connective tissue and skeletal muscle fibers. The orbicularis oris muscle makes the lips capable of versatile movement. The lips are also extremely sensitive, and are abundantly supplied with blood vessels, lymphatic vessels, and sensory nerve endings from the trigeminal nerve.

Each lip is connected at its midline to the gum by a fold of mucous membrane called a *labial frenulum* (L. *frenum*, bridle) [FIGURE 23.3]. The frenulum under the tongue is the *lingual frenulum*. It limits the backward movement of the tongue.

Teeth and Gums

Six months or so after birth, the first ***deciduous**** teeth (baby teeth, milk teeth) erupt through the gums [FIGURE 23.4A]. A normal child will eventually have 20 "baby" teeth, each jaw holding 10 teeth: 4 *incisors* (for cutting), 2 *canines* (for tearing), and 4 *premolars* (for grinding). The deciduous teeth are lost when the permanent teeth are ready to emerge. Both sets of teeth are usually present in the gums at birth, or shortly afterward, with the permanent teeth lying under the deciduous teeth [FIGURE 23.4B]. By the time a permanent tooth is ready to erupt, the root of the deciduous tooth above it has been completely resorbed by osteoclasts. The six permanent molars in each jaw have no deciduous predecessors. The shedding of deciduous teeth and the appearance of permanent teeth follow a fairly consistent pattern.

The 32 permanent teeth (16 in each jaw) are arranged in two arches, one in the upper jaw (maxilla), and the other in the lower jaw (mandible). Each jaw holds 4 *incisors* (cutting teeth), 2 *canines* (cuspid, with one point or cusp in its crown), 4 *premolars* (bicuspids, each with two

---

*Deciduous (dih-SIHDJ-oo-us) means "to fall off," and is typically used to describe trees that shed their leaves in the autumn.

cusps), and *6 molars* (millstone teeth). Third molars, commonly called *wisdom teeth*, usually erupt between the ages of 17 and 21. But they often remain within the alveolar bone and do not erupt. In such cases the teeth are said to be *impacted*, and may have to be removed surgically.

Because the upper incisors are wider than the lower ones, the lower grinding teeth are usually aligned slightly in front of the upper grinders. This arrangement enhances the grinding motion between the upper and lower teeth.

The teeth are held in their sockets by bundles of connective tissue called *periodontal ligaments* [FIGURE 23.4C]. The collagenous fibers of each ligament extend from the alveolar bone into the cement of the tooth, and they allow for some normal movement of the teeth during chewing. Nerve endings in the ligaments monitor the pressures of chewing and relay the information to the brain centers involved with chewing movements.

Parts of a tooth All teeth consist of the same three parts: (1) a *root,* embedded in a socket (alveolus) in the alveolar process of a jaw bone, (2) a *crown,* projecting upward from the gum, and (3) a narrowed *neck* between the root and crown, which is surrounded by the gum [FIGURE 23.4C]. The incisors, canines, and premolars have a single root, although the first upper premolar may initially have a double root. The lower molars have two flattened roots, and the upper molars have three conical roots. At the apex of each root is the *apical foramen,* which leads successively into the *root canal* and the *root cavity* (pulp cavity).

Composition of teeth and gums Each tooth is composed of dentine, enamel, cement, and pulp. The *dentine* is the extremely sensitive yellowish portion surrounding the pulp cavity. It forms the bulk of the tooth. The *enamel* is the insensitive white covering of the crown. It is the hardest substance in the body. The *cement* is the bonelike covering of the neck and root. The *pulp* is the soft core of connective tissue that contains the nerves and blood vessels of the tooth. Lining the innermost surface of the dentine are odontoblasts of the pulp. Their processes extend through the dentine to the border with the

enamel. The teeth's sensitivity to pain is attributed to the odontoblasts, which convey the effects of stimulation to nerve endings in the pulp.

The *gum* (Old Eng. *goma,* palate, jaw), also called the *gingiva* (jihn-JYE-vuh), is the firm connective tissue covered with mucous membrane that surrounds the alveolar processes of the teeth. The stratified squamous epithelium of the gums is slightly keratinized to withstand friction during chewing. The gums are usually attached to the enamel of the tooth somewhere along the crown, but the gum line gradually recedes as we get older. In fact, the gum line may recede so far in elderly people that the gum is attached to the cement instead of the enamel.*

Tongue

The *tongue* functions in mechanical digestion, mainly in chewing and helping to move food from the mouth down into the pharynx. The front of the tongue is used to manipulate the food during chewing, and the base of the tongue aids in swallowing. It is also a sensitive tactile organ and plays an important role in speech.

The tongue is composed mostly of skeletal muscle, and is covered by a smooth film of mucous membrane on the underside. The irregular dorsal (top) surface contains papillae, taste buds, and other structures associated with the sense of taste.

The mucous membrane covering the tongue is ordinarily divided into two sections, the *oral part* of the anterior two-thirds, and the *pharyngeal part* of the posterior third. The separate sections are delineated by the V-shaped sulcus terminalis [see FIGURE 16.2]. The oral part, corresponding to the *body* of the tongue, contains the three types of papillae. The pharyngeal part, representing the *root* of the tongue, contains the lymphatic nodules of the lingual tonsil.

The interlacing muscles of the tongue are so arranged that it can be moved in any direction, and even slightly shortened or lengthened. The tongue contains three bilateral pairs of extrinsic muscles (muscles with attachments outside the tongue), and three pairs of intrinsic muscles (muscles wholly within the tongue). The extrinsic muscles move food within the mouth to form it into a round mass, or *bolus,* and the intrinsic muscles assist in swallowing. These muscles are innervated by the hypoglossal cranial nerve (XII).

Palate

The *palate* (PAL-iht), or "roof of the mouth," is one of the many examples of structures that are perfectly suited

---

Q: *Why are canine teeth called "eyeteeth"?*

A: Early anatomical schemes often named body parts according to their relation to other structures or functions. Eyeteeth were so named probably because they lie directly under the eyes. (As another example, many people in the Western world wear wedding rings on the fourth finger of the left hand because it was believed that this finger was connected directly to the heart.)

---

*The expression "long in the tooth," relating to elderly people or animals, refers to this phenomenon of the recession of the gums, which exposes more and more of the full length of the tooth.

FIGURE 23.4 TEETH

[A] Deciduous and permanent teeth in the upper and lower jaws. [B] X-ray showing the permanent second premolar (arrow) in place beneath the deciduous tooth of a 10-year-old child. [C] Sagittal section of a tooth.

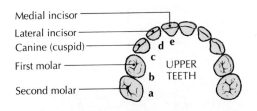

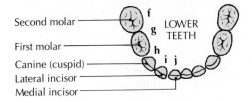

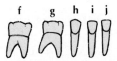

[A] DECIDUOUS "BABY" TEETH

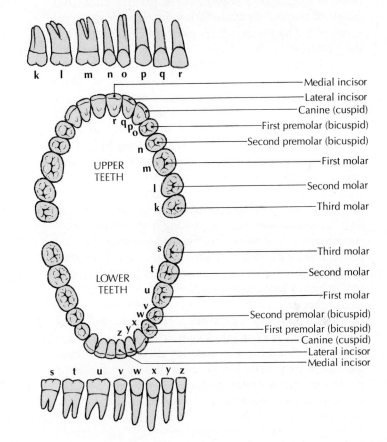

PERMANENT TEETH

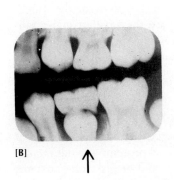

[B]

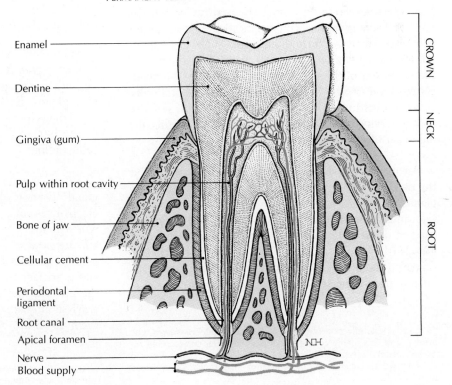

[C]

to their functions. It has two sections [FIGURE 23.3]: (1) The anterior **hard palate,** bordered by the upper teeth, is formed by a portion of the palatine bones and maxillae. Its upper surface forms the floor of the nasal cavity. (2) The posterior **soft palate** is continuous with the posterior border of the hard palate. It extends between the oral and nasal portions of the pharynx; a small fleshy cone called the *uvula* (YOO-vyoo-luh; L. small grapes) hangs down from the center of its lowermost border [FIGURE 23.3]. The uvula helps to keep food from entering the nasal passages during swallowing. Extending laterally and downward from each side of the base of the soft palate are two curved folds of mucous membranes called the *palatoglossal* and *palatopharyngeal arches.* The palatine tonsils lie between the two arches.

When food is being chewed and moistened with saliva, the tongue is constantly pushing it against the tough surface of the *hard palate,* crushing and softening the food before it is swallowed. The hard palate is covered with a firmly anchored mucous membrane and the same tough epithelium of keratinized stratified squamous epithelium as the cheeks.

The *soft palate* has a very different structure and function. It is composed of interlacing skeletal muscle that is involved with raising and lowering the soft palate over the nasopharynx when it is raised. In this way it functions to close off the nasopharynx during swallowing, preventing food from being forced into the nasal cavity.

Salivary Glands

Many glands secrete saliva into the oral cavity, but the term **salivary glands** usually refers to the three largest pairs: the parotid, submandibular, and sublingual glands [FIGURE 23.5]. **Saliva** is the watery, tasteless mixture of the secretions of the salivary glands and the oral mucous glands. It lubricates chewed food, moistens the oral walls, contains salts to buffer chemicals in the mouth, and also contains salivary amylase, the enzyme that begins the digestion of carbohydrates.

Parotid glands The **parotid glands** (Gr. *parotis,* "near the ear") are the largest of the three main pairs of salivary glands.* They lie in front of the ears, covering the masseter muscle posteriorly. The long ducts from the glands (*parotid,* or *Stensen's, ducts*) pass forward over the masseter muscle (they can be felt as a ridge by moving the tip of a finger up and down over the muscle) and end in the vestibule alongside the second upper molar tooth.

---

*A specific viral infection of the parotid glands produces *mumps.* The incidence of this highly contagious disease has been considerably reduced since the advent of the combined measles, mumps, and German measles vaccine, which is routinely given to children when they are 15 months old.

FIGURE 23.5 PAROTID, SUBMANDIBULAR, AND SUBLINGUAL SALIVARY GLANDS

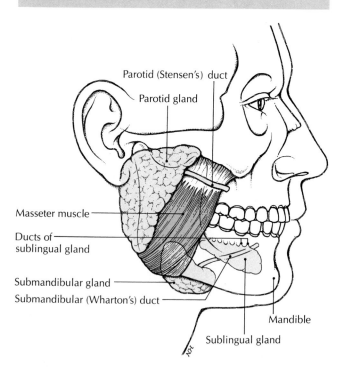

Submandibular glands The **submandibular** ("under the mandible") **glands** are located on the medial side of the mandible [FIGURE 23.5]. They are about the size of a walnut, roughly half the size of the parotid glands. Their ducts, called *submandibular (Wharton's) ducts,* open into a papilla on the floor of the mouth beside the lingual frenulum behind the lower incisors.

Sublingual glands The **sublingual** ("under the tongue") **glands** are located in the floor of the mouth beneath the tongue.† They are the smallest major salivary glands. Each gland drains through a dozen or so small ducts located on the summit of a fold on either side of the lingual frenulum, just behind the orifice of the submandibular duct.

Saliva and its functions The secretions of salivary glands contain about 99 percent water and 1 percent electrolytes and proteins. The proteins include *mucin,* which helps to form mucus, and at least one enzyme, *salivary amylase.* The interaction of mucin with the water in saliva produces the highly viscous mucus that moistens and lubricates the food particles so they can slide down the pharynx and esophagus during swallowing. Salivary amylase immediately acts on carbohydrates, breaking them down into maltose and dextrin. However, it is in-

---

†Parts of the submandibular and sublingual glands merge to form a submandibular-sublingual complex.

activated by the stomach acids only minutes after it is secreted.

Saliva also cleanses the mouth and teeth of cellular and food debris, helps keep the soft parts of the mouth supple, and buffers the acidity of the oral cavity. The high bicarbonate concentration of saliva helps to reduce *dental caries* (cavities) by neutralizing the acidity of food in the mouth. Remember also that taste buds cannot be stimulated until the food molecules are dissolved. Saliva provides the necessary solvent.

Control of salivary secretion Because the mouth and throat must be kept moist at all times, there is a continuous, spontaneous secretion of saliva, which is stimulated by the parasympathetic nerve endings that terminate in the salivary glands. In addition to this continuous low-level secretion, the presence of food in the mouth and stomach and the act of chewing (which presses on chemoreceptors and pressoreceptors in the mouth), and even the smell, taste, sight, or thought of food, usually stimulate the secretion of saliva.

Salivary secretion is entirely under the control of the autonomic nervous system, with no hormonal stimulation. (All other digestive secretions are under both nervous and hormonal control.) The major stimulation of all salivary glands comes from the parasympathetic system, and to a lesser degree from the sympathetic system. Unpleasant stimuli related to food (such as the smell of rotten fish) inhibit the parasympathetic system and cause the mouth and pharynx to become dry. As a result, the mouth feels dry during stressful situations.

Different types of stimuli produce different amounts or compositions of saliva. Parasympathetic stimulation produces a thick, mucus-rich, viscous secretion. Sympathetic stimulation acts to produce a watery secretion.

ASK YOURSELF

1 What is the difference between digestion and absorption?

2 What are the parts of the alimentary canal?

3 What are the parts of a tooth, and what are they composed of?

4 How do the functions of the hard and soft palates differ?

5 What are the main functions of saliva?

Q: *What causes the dryness of "morning mouth"?*

A: Ordinarily, a constant flow of saliva flushes out the mouth and keeps the papillae on the tongue short. But salivation is reduced during sleep, and the papillae grow longer, trapping food and bacteria and producing "morning mouth."

PHARYNX AND ESOPHAGUS

The act of swallowing has a voluntary phase, which occurs in the mouth, and an involuntary phase, which involves the pharynx and esophagus.

Pharynx

Food moves from the mouth into the pharynx. The *pharynx* serves as an air passage during breathing and a food passage during swallowing. It extends from the base of the skull to the larynx, where it becomes continuous inferiorly with the esophagus and anteriorly with the nasal cavity [FIGURES 23.6, 23.1]. Thus, the pharynx can be divided into three parts: (1) the *nasopharynx*, superior to the soft palate, (2) the *oropharynx*, from the soft palate to the epiglottis, and (3) the *laryngopharynx*, posterior to the epiglottis, which joins the esophagus.

The muscles of the pharynx, called the *pharyngeal constrictors*, are skeletal muscles, and their contraction during swallowing is controlled by the somatic motor system. However, the musculature of the pharynx and the upper half of the esophagus, though skeletal in structure, is not under voluntary control. For this reason, swallowing is considered to be automatic, not autonomic.

Esophagus

The *esophagus* (ih-SOFF-uh-guss; Gr. gullet) is a muscular, membranous tube, about 25 cm (10 in.) long, through which food passes from the pharynx into the stomach [FIGURE 23.1].

The inner *mucosa* of the esophagus is lined with nonkeratinized stratified squamous epithelium arranged in longitudinal folds. Several mucous glands in the mucosa and *submucosa* provide a film of lubricating mucus that facilitates the passage of food to the stomach. The submucosa also contains blood vessels. The middle *muscularis externa* consists wholly of skeletal muscle in the upper third of the esophagus, a combination of smooth and skeletal muscle in the middle third, and wholly smooth muscle in the lower third. The outer fibrous layer is called the *adventitia* (instead of serosa) because it lacks an epithelial layer.

The esophagus is located just in front of the vertebral column and behind the trachea. It passes through the lower neck and thorax before penetrating the diaphragm and joining the stomach.

Each end of the esophagus is closed by a sphincter muscle* when the tube is at rest and collapsed. The upper

---

*A *sphincter*, which comes from the Greek term for "to bind tight," is usually in a state of contraction, like the tightened drawstrings of a purse or a bag of marbles.

FIGURE 23.6 PHARYNX

[A] The pharynx extends from behind the nasal cavities to the larynx. It carries both air and food. **[B]** Muscles of the pharynx. Right lateral views.

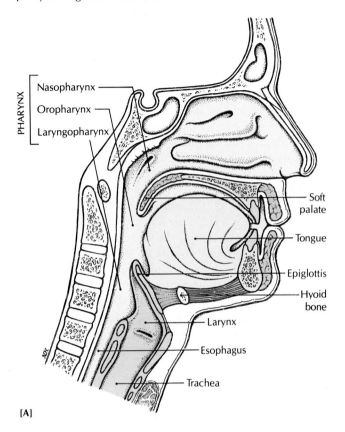

PHARYNX
- Nasopharynx
- Oropharynx
- Laryngopharynx

Soft palate

Tongue

Epiglottis

Hyoid bone

Larynx

Esophagus

Trachea

[A]

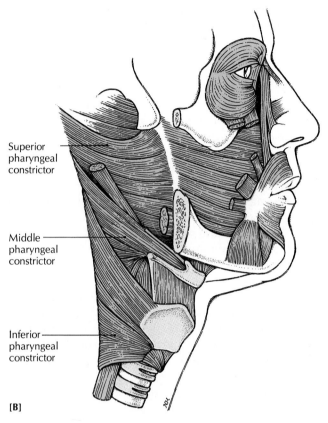

Superior pharyngeal constrictor

Middle pharyngeal constrictor

Inferior pharyngeal constrictor

[B]

sphincter is the *superior esophageal sphincter.* Closing of this sphincter is caused not by active muscular contraction but rather by the passive elastic tension in the wall of the esophagus when the esophageal muscles are relaxed. The *lower esophageal sphincter* is a band of smooth muscle that includes the last 4 cm of the esophagus just before it connects to the stomach. The lower sphincter relaxes only long enough to allow food and liquids to pass into the stomach. The rest of the time it is contracted to prevent food and hydrochloric acid from being forced back into the esophagus when pressure increases in the abdomen. Such pressure typically increases when the abdominal muscles contract during breathing, during the late stages of pregnancy, and during the normal stomach contractions during digestion. If the lower esophageal sphincter does not close, the hydrochloric acid in the stomach can be forced up into the lower esophagus. The resultant irritation of the lining of the esophagus is commonly known as *heartburn,* so called because the painful sensation is referred, and appears to be located near the heart.

Swallowing (Deglutition)

After food in the mouth is chewed, and moistened and softened by saliva, it is formed into a rounded mass known as a **bolus,** and is then ready to be swallowed. The first stage of swallowing, or **deglutition** (L. *deglutire,* to swallow down), is voluntary. In fact, it is the final voluntary digestive movement until feces are expelled during defecation. The lips, cheeks, and tongue all help to form the food into a bolus, which is pushed by the tongue against the hard palate and into the pharynx. The voluntary phase of swallowing ends when the bolus touches the entrance to the pharynx.

The involuntary stage of swallowing consists of the following three *simultaneous* movements:

1 *Voluntary oral phase.* Contractions of the mylohyoid and digastric muscles raise the hyoid bone and the tongue toward the hard palate of the mouth. In turn, the intrinsic tongue muscles elevate the tip of the tongue against the upper incisor teeth and maxilla and then squeeze the bolus toward the pharynx (like toothpaste from a tube).

2 *Pharyngeal phase.* The pharyngeal phase is triggered in the region of the fauces by the stimulation of the glossopharyngeal nerve. The soft palate is raised to prevent the bolus from entering the nasopharynx and nasal cavity. The base of the tongue is then thrust backward (retracted), propelling the bolus into the oropharynx. The sequential contractions initiate a muscular wave called *peristalsis* that propels the bolus along recesses on either side of the larynx on its way to the esophagus. The contractions of the stylohyoid and digastric muscles raise the hyoid bone and draw the larynx under the tongue so that the flap of cartilage called the *epiglottis* is pushed from a vertical to a horizontal position, and the larynx rises to close its opening against the epiglottis. The paired arytenoid cartilages "pivot" so that each vocal fold approaches its mate to close the larynx. This action can prevent the bolus from entering the trachea.

3 *Involuntary esophageal phase.* In this phase the inferior pharyngeal muscles contract, initiating a wave of peristalsis that propels the bolus through the esophagus and into the stomach. The relaxation of the palatine, tongue, and pharyngeal muscles results in the opening of the passages between the nasal cavities and the pharynx, and between the oral and pharyngeal cavities. The larynx is drawn down by the contraction of the infrahyoid muscles, and the contraction of the hyoglossus and genioglossus muscles returns the tongue to the floor of the mouth. The epiglottis returns to its vertical position as the hyoid bone is lowered, and the laryngeal passage is opened as the vocal folds separate.

> **ASK YOURSELF**
>
> **1** What are the functions of the pharynx and esophagus?
>
> **2** What are the different regions of the pharynx?
>
> **3** What is heartburn?
>
> **4** What are the main stages of swallowing?

ABDOMINAL CAVITY AND PERITONEUM

The *abdominal cavity* is the portion of the trunk cavities that lies inferior to the diaphragm [see FIGURE 1.20]. If the pelvic cavity is included, it is referred to as the *abdominopelvic cavity,* with an imaginary plane separating the abdominal and pelvic cavities. The abdominal viscera include the liver, gallbladder, stomach, pancreas, spleen, kidneys, and small and large intestines. The pelvic viscera include the rectum, urinary bladder, and internal reproductive organs.

Serous membranes line the closed abdominal cavity and the viscera contained within the cavity. A serous membrane consists of a smooth sheet of simple squamous epithelium (called *mesothelium*) and an adhering layer of loose connective tissue containing capillaries. The serous membrane of the abdominal cavity is the **peritoneum** (per-uh-tuh-NEE-uhm). The *parietal peritoneum* lines the abdominal cavity, and the *visceral peritoneum* covers most of the organs in the cavity [FIGURE 23.7].

FIGURE 23.7 PERITONEUM

The drawing shows the female peritoneum (blue) and its relationship to some major structure in the abdominopelvic cavity; sagittal section.

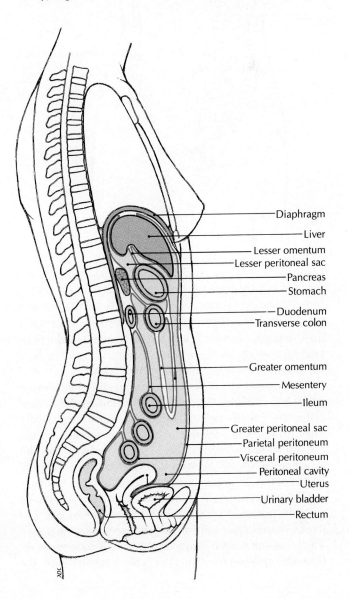

- Diaphragm
- Liver
- Lesser omentum
- Lesser peritoneal sac
- Pancreas
- Stomach
- Duodenum
- Transverse colon
- Greater omentum
- Mesentery
- Ileum
- Greater peritoneal sac
- Parietal peritoneum
- Visceral peritoneum
- Peritoneal cavity
- Uterus
- Urinary bladder
- Rectum

FIGURE 23.8 MESENTERIES AND GREATER AND LESSER OMENTA

[A] Greater omentum, lifted with an instrument to show the intestines underneath.

[B] Mesentery, with greater omentum deleted for clarity. The intestines are lifted to reveal the mesentery.

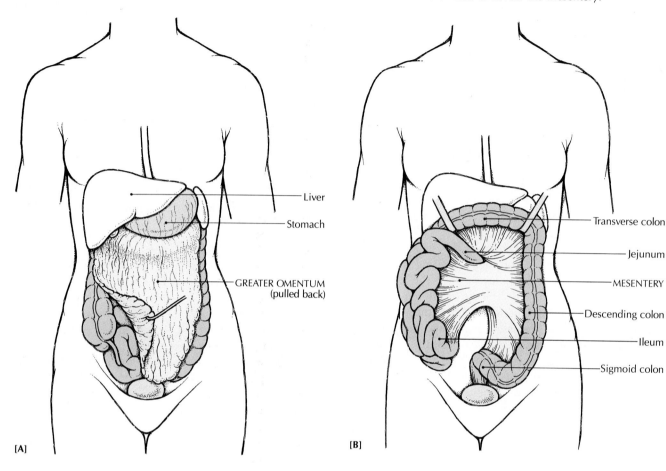

[A]

— Liver

— Stomach

— GREATER OMENTUM (pulled back)

[B]

— Transverse colon

— Jejunum

— MESENTERY

— Descending colon

— Ileum

— Sigmoid colon

Between the parietal and visceral membranes is a space called the *peritoneal cavity.* It usually contains a small amount of *serous fluid,* which is secreted by the peritoneum and allows the nearly frictionless movement of the abdominal organs and their membranes.

Abdominal organs that lie posterior to the peritoneal cavity and are covered, but not surrounded, by peritoneum are called *retroperitoneal.* Retroperitoneal structures include the pancreas, most of the duodenum, the abdominal aorta, inferior vena cava, ascending and descending colons of the large intestine, and the kidneys. Retroperitoneal organs are more or less fixed in place.

Other organs in the abdominal cavity are suspended from the posterior abdominal wall by two fused layers of serous membrane called **mesenteries** [FIGURE 23.8]. These are the *intraperitoneal organs,* and include the liver, stomach, spleen, most of the small intestine, and the

transverse colon of the large intestine. Intraperitoneal organs are not fixed firmly in position. In addition to providing a point of attachment for organs, the mesenteries carry the major arteries, veins, and nerves of the digestive tract, liver, pancreas, and spleen.

The *coelom* can be considered a *peritoneal sac,* subdivided into greater and lesser sacs by the stomach and two special mesenteries, the *greater* and *lesser omenta* (L. "fat skin"; sing. omentum). The greater and lesser sacs contain some fluid and a few cells within their potential spaces, but no organs. The **greater omentum** is an extensive folded membrane, extending from the greater curvature of the stomach to the back wall and down to the pelvic cavity. It contains large quantities of fat. Excess fat in the skin and greater omentum, plus a loss of muscle tone in the abdominal wall, produce the characteristic "pot belly."

[C] Lesser omentum, with greater omentum deleted and liver and gallbladder lifted.

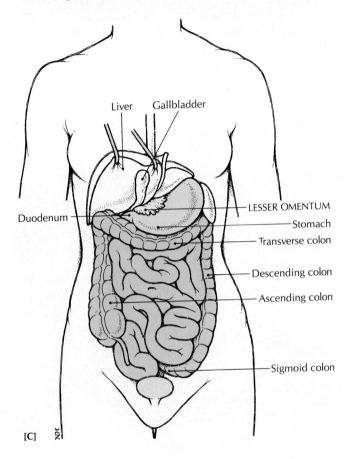

Liver Gallbladder

Duodenum

LESSER OMENTUM
Stomach
Transverse colon
Descending colon
Ascending colon
Sigmoid colon

[C]

The greater omentum hangs down like a "fatty" apron over the abdominal organs, protecting and insulating them [FIGURE 23.8]. It may extend down to the pelvic cavity. The *lesser omentum* extends from the liver to the lesser curvature of the stomach; a small portion extends from the duodenum to the liver.

STOMACH

The bolus is propelled from the esophagus into the *stomach*,* the saclike portion of the alimentary canal. The stomach is the most expandable part of the digestive tract. It stores, mixes, and digests ingested nutrients.

Anatomy of the Stomach

The stomach is usually described as a J-shaped sac with a maximum length of 25 cm (10 in.) and a width of 15 cm (6 in.). The average adult stomach has a capacity of about 1.5 L, but there are considerable variations in size and shape. As shown in FIGURE 23.9, the stomach has a *greater curvature* toward the left side of the body and a *lesser curvature* on the right.

The stomach is located much higher than some people think. Rather than lying behind the navel, it is directly under the dome of the diaphragm and is protected by the rib cage [FIGURE 23.1].

The bolus of food from the esophagus enters the stomach through an opening called the *lower esophageal* (cardiac) *orifice,* so called because it is located near the heart. Partially digested nutrients leave the stomach and enter the small intestine through an opening called the *pyloric orifice* (or *pylorus*).† The cardiac and pyloric orifices both contain sphincters. The pyloric sphincter is more powerful than the lower esophageal sphincter, with a substantial amount of circular smooth muscle. The pyloric orifice is usually opened slightly to permit fluids, but not solids, to pass into the duodenum.

The stomach is subdivided into four major regions [FIGURE 23.9A]: (1) The small *cardiac region* is near the lower esophageal orifice. (2) The *fundus* is a small, rounded area above the level of the lower esophageal orifice; it usually contains some swallowed air. (3) The *body* is the large central portion. (4) The *pyloric region* (or *antrum*), which includes the *pyloric canal,* is a narrow portion leading to the pyloric orifice.

The *muscularis externa* of the stomach consists of three layers of smooth muscle fibers [FIGURE 23.9B]: (1) The outermost *longitudinal* layer is continuous with the muscles of the esophagus and is most prominent along the curvatures of the stomach. (2) The middle *circular* layer is wrapped around the body of the stomach and becomes thickened at the pylorus to form the pyloric

*The word *stomach* comes from the Greek word for "throat." The Greek *gaster* for "stomach" gives us the stem *gastero-,* as in gastric.

†The ancient anatomists were not totally without a sense of humor when it came to naming the parts of the body. *Pylorus* means "gatekeeper" in Greek.

FIGURE 23.9 STOMACH

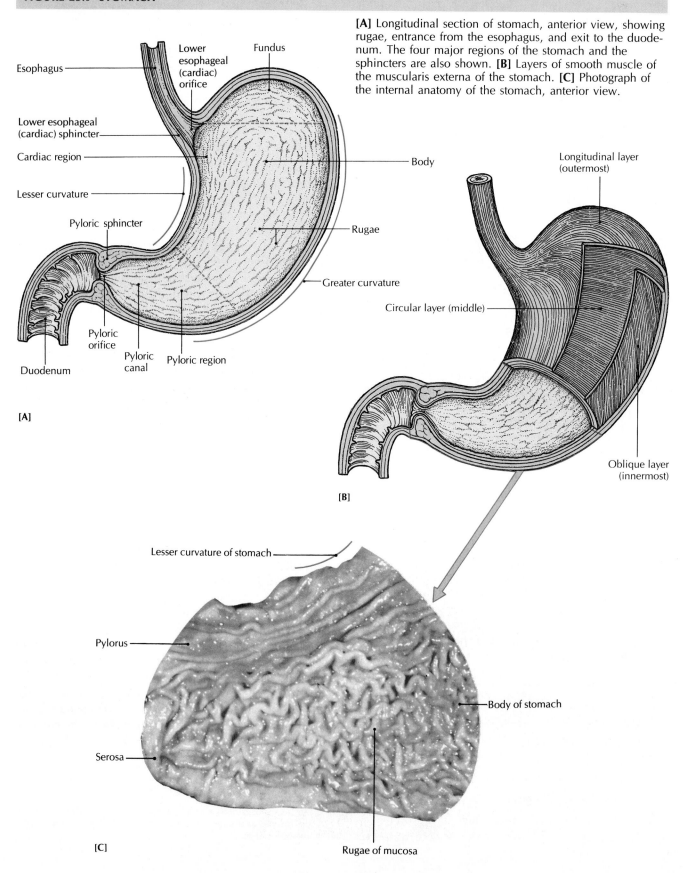

[A] Longitudinal section of stomach, anterior view, showing rugae, entrance from the esophagus, and exit to the duodenum. The four major regions of the stomach and the sphincters are also shown. **[B]** Layers of smooth muscle of the muscularis externa of the stomach. **[C]** Photograph of the internal anatomy of the stomach, anterior view.

Esophagus

Lower esophageal (cardiac) orifice

Fundus

Lower esophageal (cardiac) sphincter

Cardiac region

Lesser curvature

Body

Pyloric sphincter

Rugae

Greater curvature

Pyloric orifice

Pyloric canal

Pyloric region

Duodenum

[A]

Longitudinal layer (outermost)

Circular layer (middle)

Oblique layer (innermost)

[B]

Lesser curvature of stomach

Pylorus

Body of stomach

Serosa

Rugae of mucosa

[C]

sphincter. (3) The innermost *oblique* layer covers the fundus; it runs parallel with the lesser curvature of the stomach along the anterior and posterior walls.

When the stomach is empty, its inner mucous membrane contains branching wrinkles called ***rugae*** (ROO-jee; L. folds), which gradually flatten as the stomach becomes filled [FIGURES 23.9C and 23.10C]. The stomach is lined with a layer of simple columnar epithelium, which is in-

dented by millions of *gastric pits* [FIGURE 23.10B]. Extending from each of these pits are three to eight tubular *gastric glands.*

There are three types of regional gastric glands: the *cardiac* glands near the lower esophageal orifice, the *pyloric* glands in the pyloric canal, and the *fundic* glands in both the fundus and body of the stomach. Of these three types, the fundic glands are the most numerous.

FIGURE 23.10 STOMACH WALL

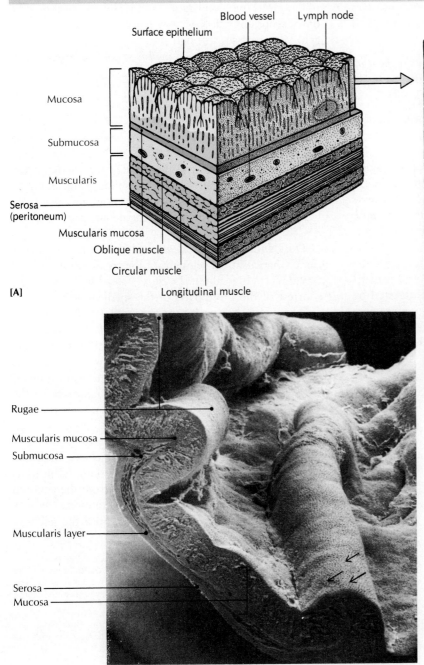

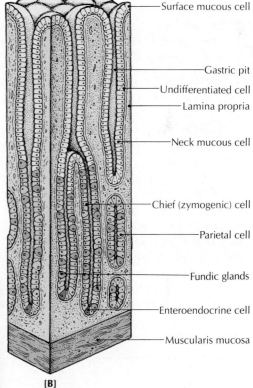

[A] Transverse section through the stomach wall at the fundus. [B] Stomach cells near the body of the stomach at a higher magnification. [C] Scanning electron micrograph of the stomach wall. The arrows point to the openings into gastric pits, which are mostly in the mucosa. ×55.

The six functionally active cell types lining the surface, pits, and glands of the stomach are shown in FIGURE 23.10.

Functions of the Stomach

The stomach has three main functions:

1 The stomach *stores ingested nutrients until they are released into the small intestine* with steady, regulated spurts (at a rate that is physiologically appropriate for the relatively slow processes of digestion and absorption). The stomach is well suited for storage because its muscles have little tone, and it can expand to contain up to 4 L if necessary.

2 The stomach *churns ingested nutrients, breaks them up into small particles, and mixes them with gastric juices* to form a soupy liquid mixture called *chyme* (KIME).

3 The stomach *secretes hydrochloric acid and enzymes that initiate the digestion of proteins.*

Secretion of gastric juices *Gastric juice* is a clear, colorless fluid secreted by the stomach mucosa in response to food. Typically, more than 1.5 L of gastric juice are secreted daily. It is composed of *hydrochloric acid, mucus,* and several enzymes, especially *pepsinogen,* a precursor of the active enzyme *pepsin.*

The major function of gastric juice is to begin the digestion of protein. Although the stomach wall is composed mainly of protein, it is normally not digested by gastric juices because it is covered with a protective coat of mucus secreted by the surface epithelial cells and neck mucous cells. A break in the mucus lining often results in a stomach sore, or *gastric ulcer* (see When Things Go Wrong at the end of the chapter).

Q: *What are "hunger pangs"?*

A: When the stomach is empty, peristaltic waves cease, but after about 10 hours of fasting, new waves may occur in the pyloric region of the stomach. These waves can cause "hunger pangs" (or hunger pains) as sensory vagal fibers carry impulses to the brain.

ASK YOURSELF

1 What are the main parts of the stomach?

2 What are the six types of stomach cells?

3 What are the functions of the stomach?

4 How do the muscular layers of the stomach contribute to peristalsis?

5 What are the components of gastric juice?

SMALL INTESTINE

After one to three hours in the stomach, the liquefied food mass, now called *chyme,* moves into the **small intestine,** where further contractions continue to mix it. It takes 1 to 6 hours for chyme to move through the 6-m (20-ft) long small intestine [FIGURE 23.11]. It is here that digestion of carbohydrates and proteins is completed, and where the digestion of most fats occurs. The small intestine absorbs practically all of the digested molecules of food into the bloodstream and lymphatic system. The only exception is alcohol, which is absorbed by the stomach. Water passes through the stomach almost immediately, but remains longer in the small intestine, where much of it is absorbed.

Anatomy of the Small Intestine

The small intestine, like the large intestine, lies within the abdominopelvic cavity.* It is subdivided into three parts, the duodenum, jejunum, and ileum. Most of the digestive processes take place in the duodenum, and most of the absorption of nutrients into the bloodstream and lymphatic system occur in the duodenum and jejunum.

Duodenum The *duodenum* (doo-oh-DEE-nuhm)† is the C-shaped initial segment of the small intestine [FIGURE 23.11]. It is about 25 cm (10 in.) long, and is the shortest of the three parts of the small intestine.

The duodenum itself has four parts. The first, or *superior,* part is about 5 cm (2 in.) long. It is the most common site of duodenal (peptic) ulcers. The second, or *descending,* part is between 7 and 8 cm (3 in.) long. The third, or *horizontal,* part is about 10 cm (4 in.) long. It crosses the third lumbar vertebra and connects to the *fourth* part, about 2.5 cm (1 in.) long, as it joins the jejunum.

Jejunum and ileum The *jejunum* (jeh-JOO-nuhm),‡ between the duodenum and ileum, is about 2.5 m (8 ft) long. The *ileum* (ILL-ee-uhm) extends from the jejunum to the cecum, the first part of the large intestine [FIGURE 23.11]. It is about 3.5 m (12 ft) long. Both the jejunum

---

*The small intestine and large intestine are named for their relative diameters (about 4 cm and 6 cm, respectively), not their lengths. The small intestine is about 6 m (20 ft) long, and the large intestine is about 1.5 m (5 ft) long.

†*Duodenum* comes from the Greek word meaning "twelve fingers wide," referring to its length, and from the Latin *duodecim,* meaning "twelve."

‡*Jejunum* comes from a Latin word meaning "fasting intestine," so named because it was always found empty when a corpse was dissected.

FIGURE 23.11 SMALL AND LARGE INTESTINES

[A] The drawing shows that the small intestine begins at the pyloric orifice, where it joins the stomach, and ends at the ileocecal valve, where it joins the large intestine. [B] Radiograph of the small intestine three hours after a barium meal. Anterior views.

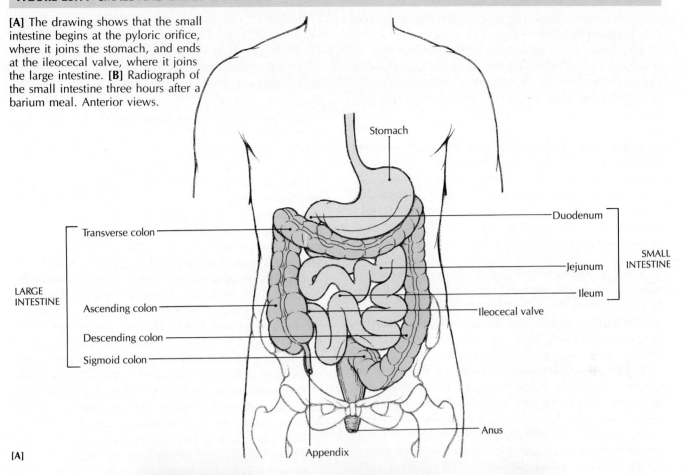

Stomach

Duodenum

Transverse colon

SMALL INTESTINE

Jejunum

Ileum

LARGE INTESTINE

Ascending colon

Ileocecal valve

Descending colon

Sigmoid colon

Anus

Appendix

[A]

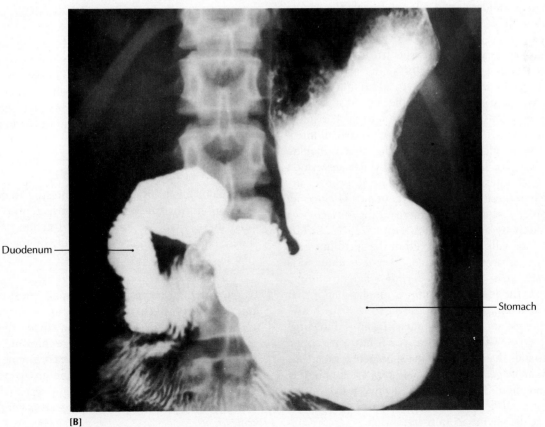

Duodenum

Stomach

[B]

and ileum are suspended from the posterior abdominal wall by the mesentery.

The ileum joins the cecum at the *ileocecal valve*, a sphincter that ordinarily remains constricted, regulating the entrance of chyme into the large intestine and preventing the contents of the cecum from flowing back into the ileum.

Adaptations of the mucosa of the small intestine

The wall of the small intestine is composed of the same four layers as the rest of the digestive tract [FIGURE 23.2]. However, the mucosa has three distinctive features that enhance the digestion and absorption processes that take place in the small intestine — plicae circulares, villi, and glands that secrete intestinal juice:

1 *Plicae circulares* (PLY-see sir-cue-LAR-eez) are circular folds that increase the surface area available for absorption. Unlike the rugae of the stomach, the plicae circulares are permanent, and do not disappear when the mucosa is distended. They are most abundant near the junction of the duodenum and jejunum.

2 The absorptive surface area of the mucosa is increased by millions of fingerlike protrusions called *villi* (VILL-eye; L. "shaggy hairs"), which look like the velvety pile of a rug [FIGURE 23.12A, B]. The surface area is further increased by the infolding of the epithelium between the bases of the villi, forming tubular *intestinal glands (crypts of Lieberkühn)* [FIGURE 23.12B, C]. Each villus contains blood capillaries and a lymph vessel called a *lacteal* [FIGURE 23.12C]. Lacteals are important in the absorption of fats.

The villi are very effective structures. Only about 5 percent of fats and 10 percent of amino acids are not absorbed from the small intestine. In addition to increasing the surface area of the mucosa, the villi aid absorption by adding their constant waving motion to the peristaltic movement already occurring throughout the small intestine. The thousands of columnar epithelial cells making up each villus are the units through which absorption from the small intestine takes place. The absorptive surface of each of these cells contains thousands of *microvilli* (brush border), which further increase the surface area for absorption by about 30-fold [FIGURE 23.12C]. The folded mucosa, villi, and microvilli of the small intestine increase the absorptive surface so that it is about 600 times greater than it would be with a smooth lining.

3 Within the mucosa are simple, tubular glands that secrete several enzymes that aid the digestion of carbohydrates, proteins, and fats. The mucosal glands (intestinal glands, or crypts of Lieberkühn) reach into the lamina propria. Glands that reach into the submucosa, called *duodenal submucosal glands (Brunner's glands)*, are found only in the duodenum. Their viscous, alkaline mucus secretion acts as a lubricating barrier to protect the mucosa from the acidic chyme from the stomach.

The lamina propria of the intestinal mucosa in the ileum has regions where lymphoid tissue is concentrated. Large lymphatic nodules form round or oval patches called *aggregated lymph nodules*, which help to destroy microorganisms absorbed from the small intestine.

Cell types in the small intestine The epithelium of the intestinal mucosa is simple columnar epithelium. It contains several types of cells, including (1) columnar absorptive cells, (2) undifferentiated columnar cells, (3) mucous goblet cells, (4) Paneth cells, and (5) enteroendocrine cells.

Columnar absorptive cells are involved in the absorption of sugars, amino acids, and fats from the lumen of the small intestine. They also produce enzymes for the terminal digestion of carbohydrates and proteins.

Undifferentiated cells are found in the depths of the crypts. The entire epithelial surface of the small intestine is replaced about every five days. Undifferentiated cells at the base of the villi divide, become differentiated, and migrate upward to replace the other cell types as needed.

Mucous goblet cells appear in the depths of the crypts and migrate upward, secreting and accumulating mucus until they swell into the bulbous shape of a goblet. After these cells release their mucus into the lumen, they die. Mucous goblet cells are especially abundant in the duodenum, where mucus protects the tissue from the high acidity of the chyme entering from the stomach.

Paneth cells lie deep within the intestinal crypts. They secrete enzymes (peptidases) for the digestion of proteins.

Enteroendocrine cells (formerly called argentaffin, argyrophylic, or enterochromaffin cells) are found not only in the small intestine, but also in the stomach, large intestine, appendix, ducts of the pancreas and liver, and even respiratory passages. They are responsible for the synthesis of more than 20 gastrointestinal hormones.

Functions of the Small Intestine

The major functions of the small intestine, digestion and absorption, are made possible by movements of the intestinal muscles and by the chemical action of enzymes. These enzymes are secreted not only by the small intestine itself, but also by the pancreas.

Digestive movements of the small intestine Rhythmic movements of circular and longitudinal smooth muscle in the small intestine cause the chyme to be mixed, and ultimately pushed into the large intestine. The mixing movements reduce large particles of chyme to smaller ones, exposing as much of the chyme surface to digestive enzymes as possible. The two main types of muscular activity in the small intestine are *segmenting* and *peristaltic contractions*.

FIGURE 23.12 WALL AND LINING OF THE SMALL INTESTINE

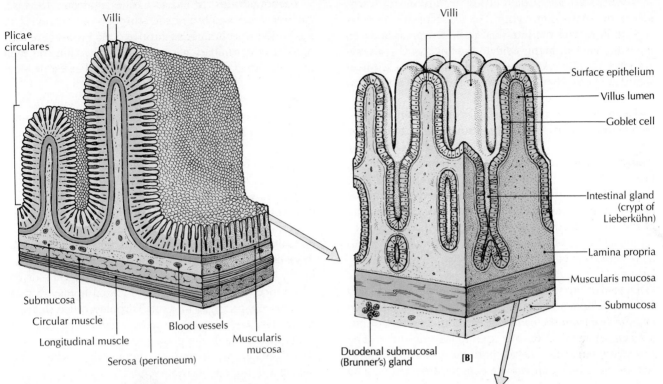

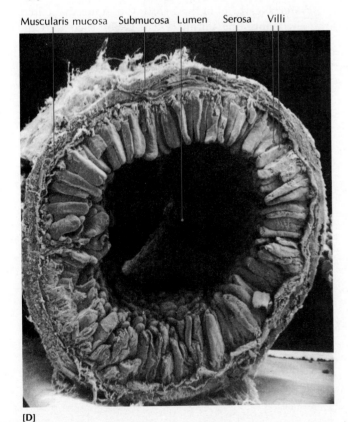

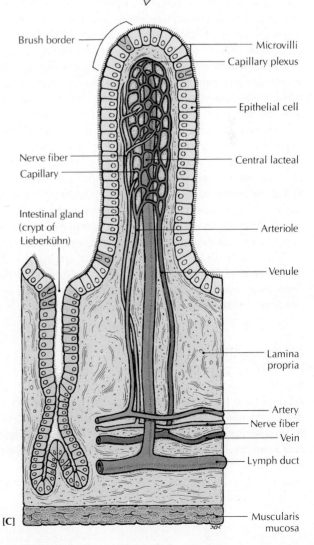

[A] Lining of the small intestine. [B] Enlarged portion of transverse section of the wall of the duodenum. [C] Internal structure of villus. Microvilli extending from the villus form a brush border. [D] Scanning electron micrograph of a section of the small intestine. ×30.

Segmenting contractions divide the intestine into segments by sharp contractions of areas of the circular smooth muscle in the intestinal wall. The areas between the contracted segments contain the bolus of chyme, and are relaxed. The duodenum and ileum start to segment simultaneously when chyme first enters the small intestine.

Peristaltic contractions (propulsive contractions) propel chyme through the small intestine and into the large intestine with weak, repetitive waves that start at the stomach and move short distances down the intestine before dying out. Each wave takes 100 to 150 minutes to migrate from its initiation site to the end of the small intestine. After the peristaltic contractions reach the end of the small intestine, new waves replace them at the initiation site. It is believed that the hormone *motilin* regulates peristaltic contractions.

Digestive enzymes in the small intestine Digestion within the lumen of the small intestine is accomplished by enzymes from the pancreas, with the digestion of lipids being enhanced by bile secreted by the liver. The pancreatic enzymes and bile both enter the small intestine through the common bile duct [FIGURE 23.13]. The various salts in bile break down lipid droplets into particles small enough to be attacked by enzymes. The physical breakdown of lipid particles is called *emulsification* (the same process by which soap breaks down grease).

The bile and pancreatic juices help to change the chemical environment in the small intestine from acidic to alkaline. The new alkaline environment stops the action of pepsin, and allows the intestinal enzymes to function properly. These enzymes continue the breakdown of carbohydrates, proteins, and lipids. Carbohydrates are broken down into disaccharides and monosaccharides, proteins are broken down into small peptide fragments and some amino acids, and lipids are completely broken down into their absorbable units (glycerol and free fatty acids). The digestion of lipids is now completed, but carbohydrates and proteins need further digestive action.

The microvilli of the small intestinal epithelial cells have special actin-stiffened hairlike projections called the *brush border*. Within the brush border are enzymes for the final digestion of carbohydrates and proteins.

Absorption from the Small Intestine

All the products of carbohydrate, lipid, and protein digestion, as well as most of the ingested electrolytes, vitamins, and water, are normally absorbed by the small intestine, with most absorption occurring in the duodenum and jejunum. The ileum is the primary site for the absorption of bile salts and vitamin B_{12}.

Carbohydrates Monosaccharides (glucose, fructose, galactose) are readily absorbed through the microvilli on the border of columnar absorptive cells. From there they pass into the capillary network within the villi, and then into the blood vessels in the underlying lamina propria.

Proteins After proteins have been broken down, their constituent amino acids are absorbed through columnar absorptive cells into capillary networks within the villi.

Lipids The absorption of lipids into the central lacteal of the intestinal villus begins with their emulsification by bile salts, which facilitates their digestion by enzymes (lipases). The products of digestion (fatty acids, monoglycerides) then combine under the further influence of bile salts to form submicroscopic particles called *micelles* (my-SEHLZ; L. *mica*, grain). Micelles, unlike undigested lipids, can be absorbed into the absorptive cells of the villi. Once inside the cells, the digested lipids are reassembled and packaged into small droplets called *chylomicrons* (Gr. *chylos*, juice + *micros*, small). The chylomicrons pass through the plasma membrane into a central lacteal, and eventually into the thoracic duct, which empties into the left subclavian vein.

Water Nearly all of the 5 to 10 L of water that enters the small intestine during a day is absorbed into the bloodstream, most of it from the duodenum. The remaining half liter or so enters the large intestine.

Other substances Several other substances, including vitamins, nucleic acids, ions, and trace elements, are absorbed from the small intestine. The fat-soluble vitamins (A, D, E, K) are dissolved by the fat droplets that enter the small intestine from the stomach. They can be absorbed with micelles into the absorptive cells of the villi. The water-soluble vitamins are readily absorbed. Only vitamin B_{12} requires a specialized protein carrier, known as *gastric intrinsic factor*, for absorption in the terminal ileum by the process of endocytosis.

A S K   Y O U R S E L F

1 What are the main parts of the small intestine?

2 In addition to digestion, what is the main function of the small intestine?

3 What are the main types of cells in the small intestine?

4 How do segmenting contractions differ from peristaltic contractions?

FIGURE 23.13 PANCREAS

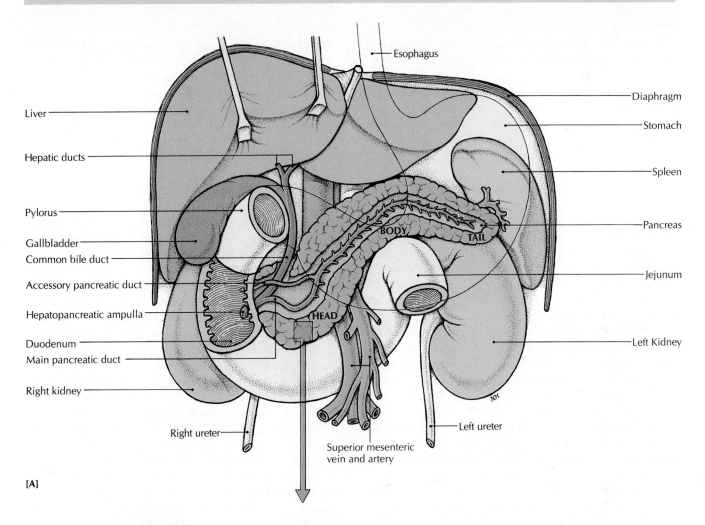

Liver

Hepatic ducts

Pylorus

Gallbladder
Common bile duct

Accessory pancreatic duct

Hepatopancreatic ampulla

Duodenum
Main pancreatic duct

Right kidney

Right ureter

Esophagus

Diaphragm
Stomach
Spleen

Pancreas

Jejunum

Left Kidney

Left ureter

Superior mesenteric
vein and artery

BODY TAIL
HEAD

[A]

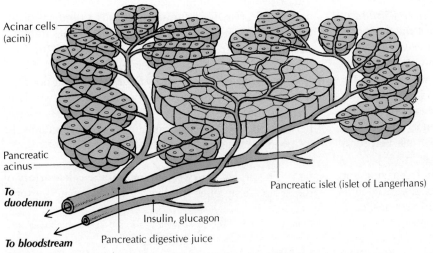

Acinar cells
(acini)

Pancreatic
acinus

**To
duodenum**

To bloodstream

Pancreatic islet (islet of Langerhans)

Insulin, glucagon

Pancreatic digestive juice

[B]

[A] Position of the pancreas, with its head tucked into the curve of the duodenum and its tail extending to the spleen. Note the duct system of the pancreas. The liver, gallbladder, and pancreas all secrete their products into ducts that merge as they enter the duodenum. [B] Cells of a pancreatic islet surrounded by acini. Acini secrete digestive fluids that pass into the small intestine.

PANCREAS AS A DIGESTIVE ORGAN

When the chyme is emptied into the small intestine, it is mixed with secretions of the pancreas and liver, in addition to the juices secreted by the intestinal cells. In the following sections we will discuss the important digestive roles of these accessory organs, along with the gallbladder, before continuing with a description of the large intestine.

Gross anatomy The *pancreas* is a soft, pinkish-gray gland about 12 to 15 cm (5 to 6 in.) long. It lies transversely across the posterior abdominal wall, behind the stomach [FIGURE 23.13A]. It is completely retroperitoneal. The *endocrine* function of the pancreas is described in Chapter 17, but FIGURE 23.13B shows how little clumps of endocrine cells are clearly separated from the exocrine cells.

The pancreas consists of three parts: (1) The broadest part, or *head*, lies on the right side of the abdominal cavity and fits into the C-shaped concavity of the duodenum. (2) The main part, or *body*, lies between the head and tail, behind the stomach and in front of the second and third lumbar vertebrae. (3) The narrow *tail* extends to the left all the way to the spleen.

Microscopic anatomy The *exocrine* cells of the pancreas form groups of cells called *acini* (ASS-ih-nye; L. grapes, so named because they resemble bunches of grapes), which secrete fluid containing digestive enzymes into the small intestine. The exocrine cells are clustered around tiny ducts that pass through the pancreas from the tail to the head. Each acinus has a central lumen that connects to the *main pancreatic duct (duct of Wirsung)*, which carries the digestive enzymes toward the descending part of the duodenum. The *accessory pancreatic duct (duct of Santorini)* empties a small amount of pancreatic enzymes into the duodenum, about an inch above the main duct. The main pancreatic duct does not empty its contents directly into the duodenum. Instead, it joins the common bile duct, and finally, a duct called the *hepatopancreatic ampulla (ampulla of Vater)*, just before it enters the wall of the duodenum [FIGURE 23.13A].

Pancreatic fluid contains approximately 15 enzymes or precursors of enzymes. These enzymes are produced and secreted by acinar cells for the digestion of proteins, carbohydrates, and lipids in the small intestine.

LIVER AS A DIGESTIVE ORGAN

The liver* is a large, compound, tubular gland, weighing about 1.5 kg (3 lb) in the average adult.† Though it lies outside the digestive tract, it has many functions that are relevant to digestion and absorption. It is involved with modifying nutrients to make them usable by body tissues.

Anatomy of the Liver

The liver is reddish, wedge-shaped, and covered by a network of connective tissue called *Glisson's capsule*. It is located under the diaphragm in the upper region of the abdominal cavity, mostly on the right side. The undersurface faces the stomach, the first part of the duodenum, and the right side of the large intestine [FIGURE 23.14A].

The liver is divided into two main lobes by the *falciform ligament*, a mesentery attached to the anterior mid-abdominal wall [FIGURE 23.14B, C]. The **right lobe,** which is about six times larger than the left lobe, is situated over the right kidney and the right colic (hepatic) flexure of the large intestine. The **left lobe** lies over the stomach. In the free border of the falciform ligament, extending from the liver to the umbilicus, is the *ligamentum teres* (round ligament), a fibrous cord that is a remnant of the left fetal umbilical vein [see FIGURE 20.10].

The right lobe is further subdivided into a small *quadrate lobe* and a small *caudate lobe* on its ventral (visceral) surface [FIGURE 23.14C]. The quadrate lobe is flanked by the gallbladder on the right and the ligamentum teres on the left. The quadrate lobe partially envelops and cushions the gallbladder. The caudate lobe is flanked by the inferior vena cava on the right and the *ligamentum venosus* on the left. The ligamentum venosus is a remnant of the fetal ductus venosus, a venous shunt from the umbilical vein to the inferior vena cava [see FIGURE 20.10].

Vessels of the liver The *porta hepatis* ("liver door") is the area through which the blood vessels, nerves, lymphatics, and ducts enter and leave the liver. It is located between the quadrate and caudate lobes, and contains the following vessels and ducts [FIGURE 23.15]:

1 The *hepatic artery* is a branch of the celiac artery of the aorta. It supplies the liver with oxygenated arterial

---

*The Greek for "liver" is *hepar*, a word used in the combining form *hepato-* in such words as *hepatic* and *hepatitis*.

†An infant usually has a pudgy abdomen because of the disproportionately large size of its liver. In most children, the liver occupies about 40 percent of the abdominal cavity, and is responsible for approximately 4 percent of the total body weight. In an adult, the liver represents about 2.5 percent of the total body weight.

blood, which represents about 20 percent of the blood flow to the liver.

2 The *hepatic portal vein* drains venous blood into the liver from the entire gastrointestinal tract. It supplies the remaining 80 percent of the liver's blood. This blood contains the nutrients absorbed by the small intestine.

3 The ***hepatic vein*** drains blood from the liver into the inferior vena cava.

4 Bile ducts, called ***bile canaliculi,*** are formed by the ***bile capillaries*** that unite after collecting bile from the liver cells [FIGURE 23.16]. Bile is secreted by the liver into the bile canaliculi, which drain into the ***right*** and ***left hepatic ducts.*** The ducts, in turn, converge with the ***cystic duct*** from the gallbladder to form the ***common bile duct.***

The common bile duct joins the main pancreatic duct, enlarges into the hepatopancreatic ampulla, and then joins the duodenal papilla, which opens into the second part of the duodenum [FIGURE 23.17].

Figure 23.15A shows the four main vessels and ducts of the liver, with the hepatic artery and hepatic portal vein entering, and the hepatic vein and common bile duct leaving.

Microscopic anatomy The functional units of the liver are five- or six-sided ***lobules*** that contain a branch of the *hepatic vein* (the *central vein*) running longitudinally

FIGURE 23.14 LIVER

[A] Position of the liver in the abdominal cavity. [B] Right and left lobes, anterior view. [C] Caudate and quadrate lobes, posteroinferior view.

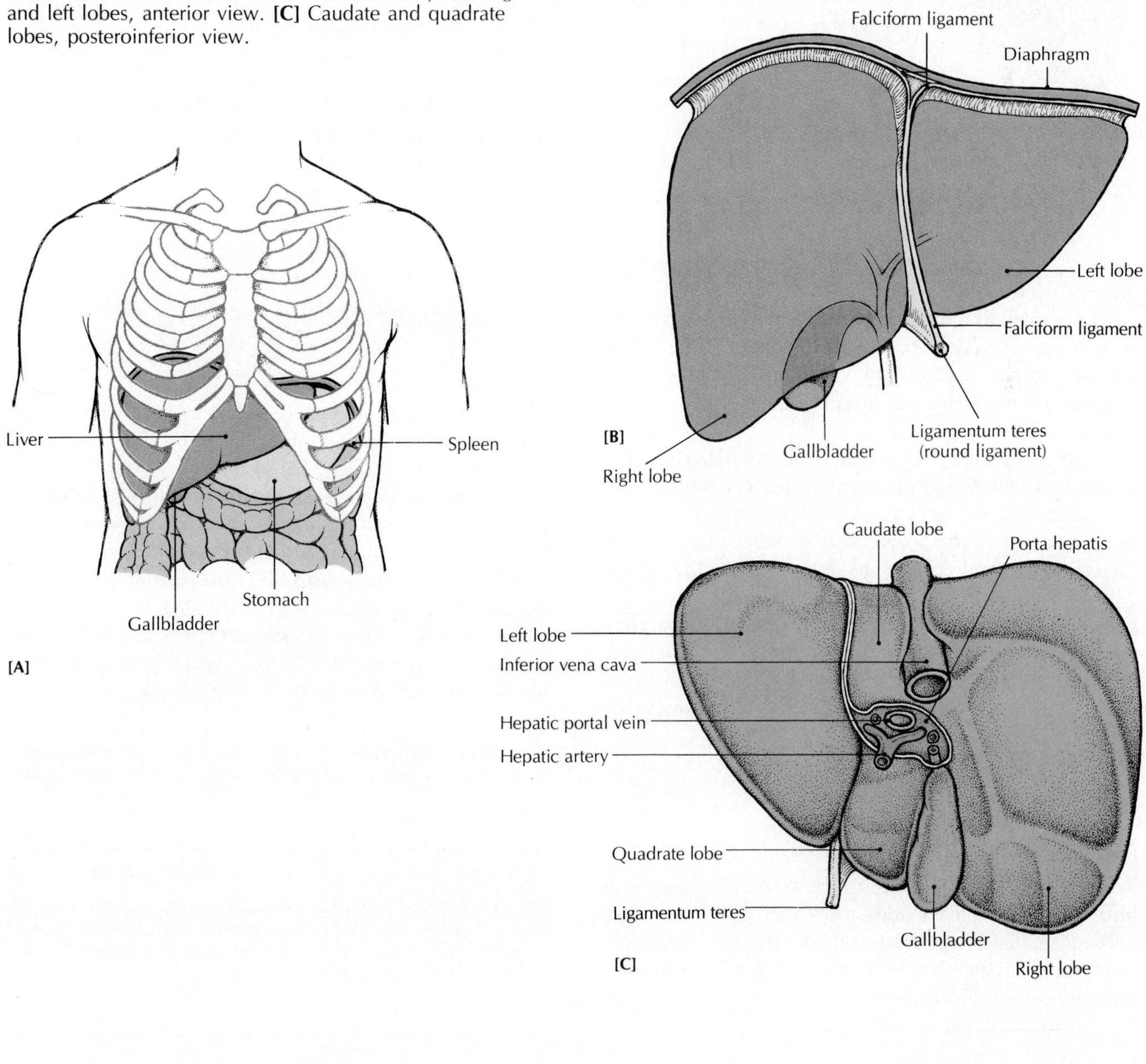

FIGURE 23.15 VESSELS AND DUCTS OF THE LIVER

Arrows indicate the direction of blood flow; anterior view.

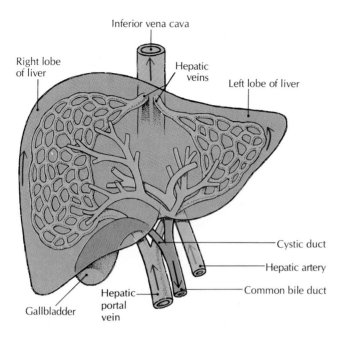

Inferior vena cava

Right lobe of liver

Hepatic veins

Left lobe of liver

Cystic duct

Hepatic artery

Common bile duct

Hepatic portal vein

Gallbladder

Metabolic functions of the liver The liver performs so many metabolic functions that only the major ones can be presented here:

1 Formation of urea from amino acids; urea is toxic and is excreted by the kidneys.

2 Manufacture of blood (plasma) proteins.

3 Formation of erythrocytes during fetal life.

4 Synthesis of the blood-clotting agent fibrinogen.

5 Detoxification of many poisons.

6 Synthesis of bile and bile salts for emulsification of fats in the small intestine.

ASK YOURSELF

1 Where is the liver located?

2 What are the vessels of the liver?

3 How does a liver lobule resemble a wheel?

4 What are the components of a portal area?

5 What are some of the major metabolic functions of the liver?

---

through it [FIGURE 23.16]. Most lobules are about 1 mm (0.04 in.) in diameter. Liver cells, known as ***hepatic cells,*** or hepatocytes, within the lobules are arranged in one-cell-thick platelike layers that radiate from the central vein to the edge of the lobule. Each corner of the lobule usually contains a *portal area*, a complex composed of branches of the portal vein, hepatic artery, bile duct, and nerve.

Between the radiating rows of cells are delicate blood channels called ***sinusoids,*** which transport blood from the portal vein and hepatic artery. Blood flows from the artery and vein in the portal areas into the sinusoids, and then to the central vein, which drains it from the lobule. The walls of the sinusoids are lined with endothelial cells. Attached to these lining cells are phagocytic stellate reticuloendothelial cells that engulf and digest worn-out red blood cells and white blood cells, microorganisms, and other foreign particles passing through the liver.

GALLBLADDER AND BILIARY SYSTEM

The gallbladder* is a small, pear-shaped, saclike organ situated in a depression under the right lobe of the liver [FIGURES 23.17 and 22.13A].

The gallbladder has an important storage function. Bile is secreted continuously by the liver. However, more bile is produced than is ordinarily required, and the excess is stored in the gallbladder until needed in the duodenum. The capacity of the gallbladder is between 30 and 60 mL. When the gallbladder is empty, its mucosa is thrown into folds (rugae), which permit the gallbladder to expand to hold the bile.

* *Gallbladder* is derived from the Latin word *galbinus*, meaning greenish yellow, which is the usual color of *bile*. *Gall* is the archaic term for bile.

Functions of the Liver

The liver chemically converts the nutrients from food into usable substances and stores them until they are needed. In addition to the metabolic, storage, and secretory functions described here, the liver also is involved in excretion, the subject of Chapter 24.

Q: *Is the gallbladder essential for the digestive process?*

A: It is not, and a diseased gallbladder can be removed in a surgical procedure called a cholecystectomy. The hepatic ducts and common bile duct can dilate sufficiently to take over the storage function of the removed gallbladder.

FIGURE 23.16 LIVER LOBULES

[A] Cross section showing typical six-sided shape of lobules. [B] Cutaway view of lobule showing channels of flow, including bile canaliculi and sinusoids, which convey their contents in opposite directions. [C] Light micrograph of liver tissue. ×25. [D] Scanning electron micrograph of a liver lobule, showing hepatocytes radiating toward the central vein. ×400.

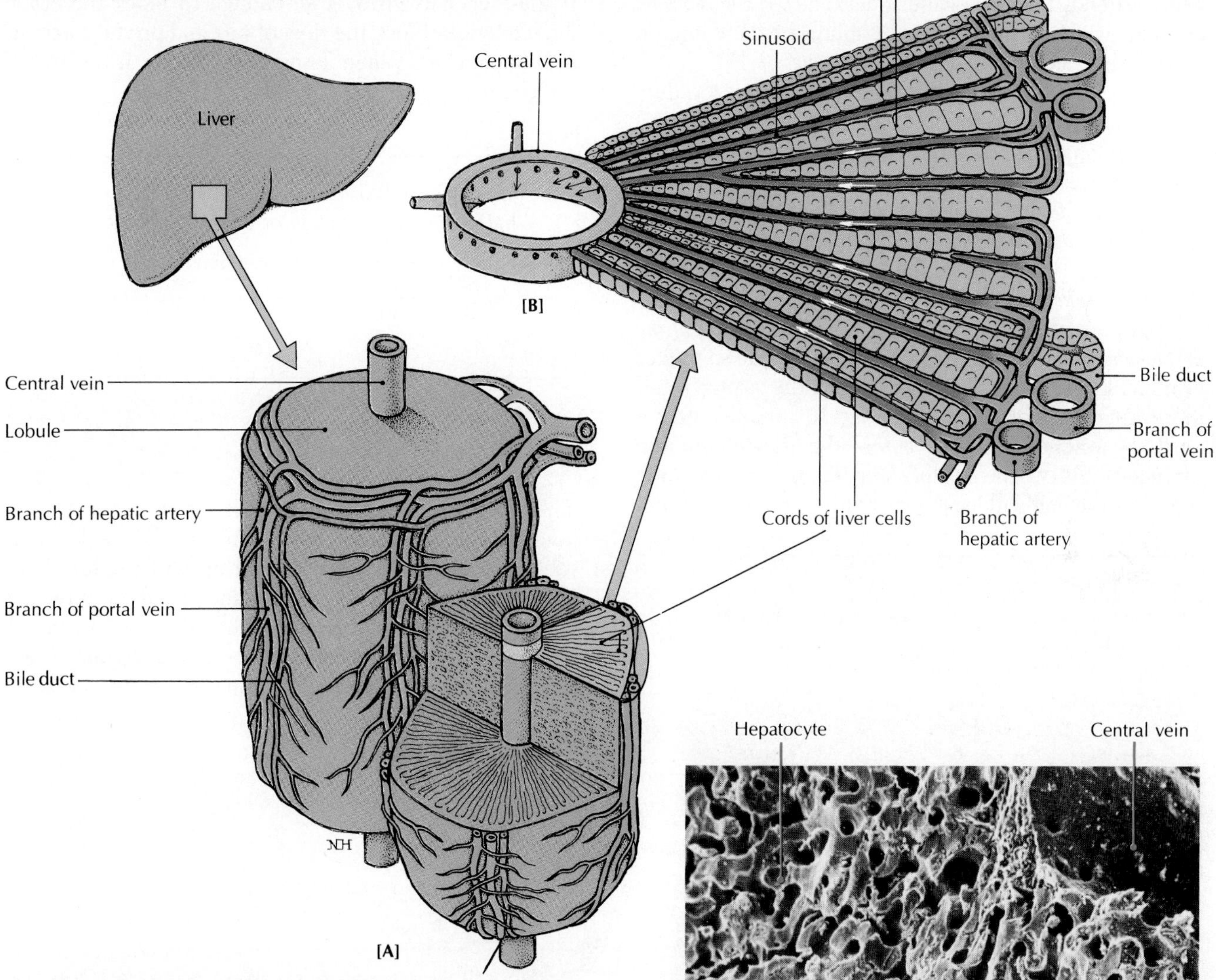

Liver

Central vein

Bile canaliculi

Sinusoid

[B]

Bile duct

Branch of portal vein

Cords of liver cells

Branch of hepatic artery

Central vein

Lobule

Branch of hepatic artery

Branch of portal vein

Bile duct

[A]

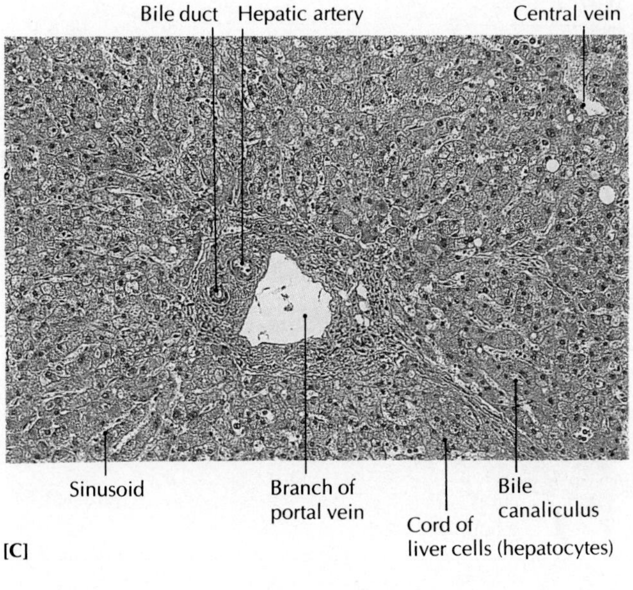

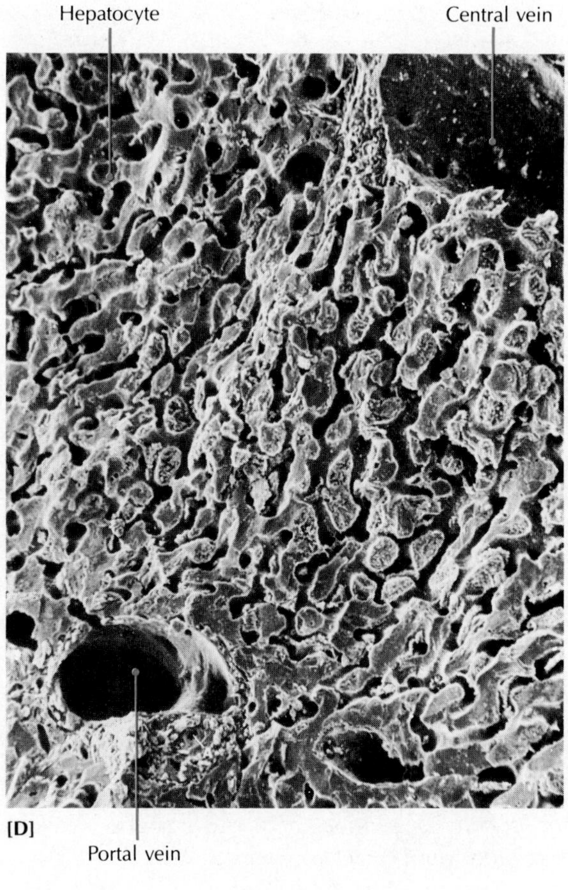

Hepatocyte

Central vein

Bile duct   Hepatic artery

Central vein

Sinusoid

Branch of portal vein

Cord of liver cells (hepatocytes)

Bile canaliculus

[C]

[D]

Portal vein

The so-called **biliary system** consists of (1) the *gallbladder*, (2) the *left* and *right hepatic ducts*, which come together as the *common hepatic duct*, (3) the *cystic duct*, which extends from the gallbladder, and (4) the *common bile duct*, which is formed by the union of the common hepatic duct and the cystic duct [FIGURE 23.17].

The common bile duct and the main pancreatic duct join at an entrance to the duodenum about 10 cm (4 in.) from the pyloric orifice. They fuse to form the *hepatopancreatic ampulla*. The ampulla travels obliquely through the duodenal wall and opens into the duodenum through the *duodenal papilla*. A sphincter located at the outlet of the common bile duct is called the *sphincter of the common bile duct (sphincter of Boyden)*, and the muscle below it, near the duodenal papilla, is the *sphincter of the hepatopancreatic ampulla (sphincter of Oddi)*. The sphincter of the common bile duct appears to be the stronger and more important of the two. About 30 minutes after a meal, or whenever chyme enters the duodenum, the sphincters relax, the gallbladder contracts, and bile stored in the gallbladder is squirted into the duodenum.

Q: *What is usually meant by "bowels"?*

A: The bowels (L. *botulus*, sausage) are usually considered to be the large intestine and rectum, but they may also refer to the digestive tract below the stomach.

Bile is rich in cholesterol, a rather insoluble fatty substance, and concentrated bile in the gallbladder may form crystals of cholesterol that are commonly called *gallstones*. If these crystals grow large enough to block the cystic duct, they can block the flow of bile and produce a great deal of pain (see When Things Go Wrong at the end of the chapter).

ASK YOURSELF

1 What is the main function of the gallbladder?

2 What components make up the biliary system?

LARGE INTESTINE

The chyme remains in the small intestine for 1 to 6 hours. It then passes in liquid form through the ileocecal valve into the *cecum*, the first part of the large intestine. By now, digestion is complete, and the large intestine functions to remove more water and salts from the liquid matter. Removal of water, along with the action of microorganisms, converts liquid wastes into *feces* (FEE-seez;

FIGURE 23.17 GALLBLADDER AND ITS CONNECTING DUCTS

Bile secreted from the liver reaches the gallbladder via the cystic duct.

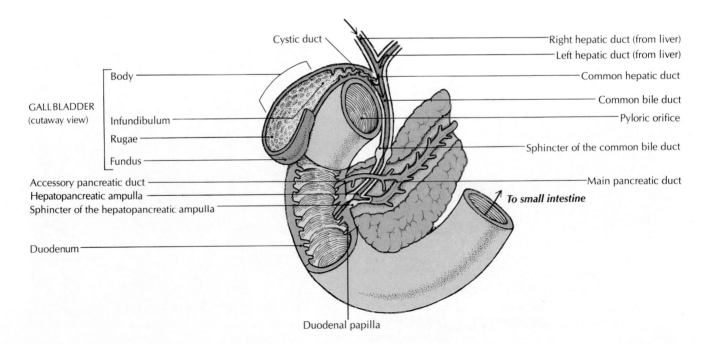

L. *faex*, dregs), a semisolid mixture that is stored in the large intestine until ready to be eliminated through the anus during defecation. The undigestible products of digestion remain in the large intestine from 12 to 36 hours.

Anatomy of the Large Intestine

The *large intestine* is the part of the digestive system between the ileocecal orifice and the anus. It consists of the cecum; vermiform appendix; ascending, transverse, descending, and sigmoid colons; and the rectum [FIGURE 23.18A]. The entire large intestine (sometimes called the *colon*) forms a rectangle that frames the tightly packed small intestine.

Just on the other side of the ileocecal orifice is the *cecum* (SEE-kuhm; L. blind), a cul-de-sac pouch about 6 cm (2.5 in.) long. Opening into the cecum, about 2 cm below the *ileocecal valve*, is the *vermiform appendix* (L. *vermis*, worm, hence "wormlike"), popularly called the *appendix*. It is the narrowest part of the intestines, and can range in length from 5 to 15 cm (2 to 6 in.). Because of these physical characteristics, bacteria and indigestible material may become trapped in the appendix, leading to inflammation *(appendicitis)* and one of the most common of all surgical procedures, an *appendectomy*.

The *ascending colon* extends upward from the cecum. Under the liver it makes a right-angle bend known as the *right colic* (or *hepatic*) *flexure* [FIGURE 23.18A]. It continues as the *transverse colon,* which extends across the abdominal cavity from right to left, where it makes a right-angle downward turn at the spleen, known as the *left colic* (or *splenic*) *flexure.*

The *descending colon* extends from the left colic flexure down to the rim of the pelvis, where it becomes the sigmoid colon. The *sigmoid colon* (Gr. *sigma*, the letter S) is usually S-shaped. It travels transversely across the pelvis to the right to the middle of the sacrum, where it continues to the rectum.

External anatomy of the large intestine Besides the differences in diameter and length between the large and small intestines, the large intestine has three distinctive structural features:

1 In the small intestine, an external longitudinal smooth muscle layer completely surrounds the intestine; but in the large intestine, an incomplete layer of longitudinal muscle forms three separate bands of muscle called *taeniae coli* (TEE-nee-ee KOHL-eye; L. ribbons + Gr. intestine) along the full length of the intestine.

2 Because the taeniae coli are not as long as the large intestine itself, the wall of the intestine becomes puckered with bulges called *haustra* (sing. *haustrum*).

3 Fat-filled pouches called *epiploic appendages* are formed at the points where the visceral peritoneum is attached to the taeniae coli in the serous layer.

Microscopic anatomy of the large intestine The microscopic anatomy of the large intestine reflects its primary functions: the reabsorption of any remaining water and some salts, and the accumulation and movement (excretion) of undigested substances as feces. Elimination is aided by the secretion of mucus from the numerous goblet cells in the mucosal layer. Mucus acts as a lubricant and protects the mucosa from the semisolid dehydrated contents.

Because the mucosa of the large intestine does not contain villi or plicae circulares, its smooth absorptive surface is only about 3 percent of the absorptive surface of the small intestine. The lamina propria and submucosa contain lymphatic tissue in the form of many lymphoid nodules. As part of the body's immune system, aggregated lymph nodules *(gut-associated lymphoid tissue)* along with epithelial cells, helps defend against the everchanging mixture of antigens, microorganisms, and potentially harmful substances that enter the digestive tract.

Rectum, Anal Canal, and Anus

The terminal segments of the large intestine are the rectum, anal canal, and anus [FIGURE 23.19]. The *rectum* (L. *rectus*, straight) extends about 15 cm (6 in.) from the sigmoid colon to the anus. Despite its name, the rectum is not straight. Three lateral curvatures occur because the longitudinal muscle coat of the rectum is shorter than the other layers of the wall. The rectum is a retroperitoneal structure, and has no mesentery, appendixes, epiploic appendages, haustra, or taeniae coli. The rectum continues as the *anal canal,* pierces the muscular pelvic diaphragm, and turns sharply downward and backward. The anal canal is compressed by the anal muscles as it progresses for about 4 cm (1.5 in.) and finally opens to the outside as a slit, the *anus* (L. ring).

The anal canal and anus are open only during defecation. At all other times they are held closed by an involuntary *internal anal sphincter* of circular smooth muscle and a complex *external anal sphincter* of skeletal muscle [FIGURE 23.19]. The external sphincter is under voluntary control.

The mucosa and circular muscle of the muscularis of the rectum form shelves within the tract called *plicae transversales.* The plicae must be avoided when diagnostic and therapeutic tubes and instruments are inserted into the rectum. The upper part of the anal canal contains 5 to 10 permanent longitudinal columns of mucous membrane known as *anal* (or *rectal*) *columns.* They are united by folds called *anal valves.* The mucosa above the

FIGURE 23.18 LARGE INTESTINE

[A] Anterior view with part of the cecum and ascending colon removed to show the ileocecal valve.
[B] Transverse section of large intestine.

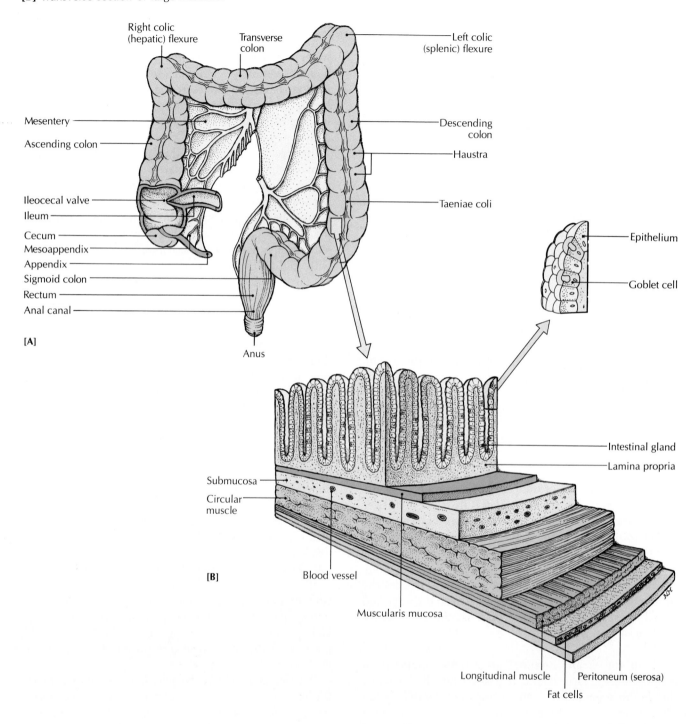

Right colic (hepatic) flexure

Transverse colon

Left colic (splenic) flexure

Mesentery

Ascending colon

Descending colon

Haustra

Ileocecal valve

Ileum

Cecum

Mesoappendix

Appendix

Sigmoid colon

Rectum

Anal canal

Taeniae coli

Epithelium

Goblet cell

[A]

Anus

Intestinal gland

Lamina propria

Submucosa

Circular muscle

[B]

Blood vessel

Muscularis mucosa

Longitudinal muscle

Fat cells

Peritoneum (serosa)

FIGURE 23.19 RECTUM, ANAL CANAL, ANUS

Longitudinal section of rectum, anal canal, and anus; anterior view.

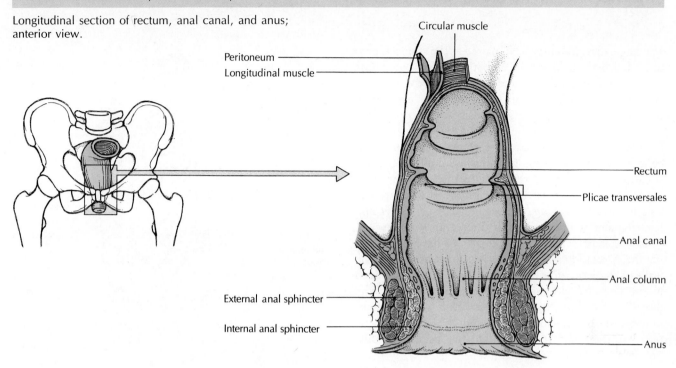

valves is lined by columnar epithelium, with a scattering of goblet cells and crypts. The mucosa and submucosa of the rectum contain a rich network of veins called the *hemorrhoidal plexus*. When these veins become enlarged, twisted, and blood-filled, the condition is called *hemorrhoids*.

Functions of the Large Intestine

By the time chyme reaches the large intestine, digestion is complete, and only some water, salts, and vitamins remain to be absorbed. The amount of chyme that enters the large intestine daily varies from less than 100 mL to 500 mL, and only about one-third of it is excreted as feces. The rest (mostly water) is absorbed back into the body in the ascending or transverse colons, and the unabsorbed water becomes part of the feces. Reabsorption of water in the large intestine helps to avoid dehydration.

Formation of feces Approximately 150 g of feces (about 100 g of water and 50 g of solids) are normally eliminated from the body daily. The normal brown color of feces is caused by products of the breakdown of red blood cells into bile pigments (bilirubin). Excessive fat in the diet causes feces to be a pale color, and blood and other foods containing large amounts of iron will darken feces.

Movements of the large intestine and defecation Most of the digestive movements of the large intestine are slow and nonpropulsive. The primary motility comes from *haustral contractions* that depend upon the autonomous rhythmicity of the smooth muscle cells. These movements are similar to the segmentations of the small intestine, but usually take 20 to 30 minutes between contractions. This slow movement allows for the final absorption of the remaining water and electrolytes.

Three to four times a day, generally following meals, motility increases markedly, as large segments of the ascending and tranverse colons contract simultaneously, driving the feces one-half to three-quarters the length of the colon in a few seconds. These "sweeping" peristaltic waves are called *mass movements*. They drive the feces into the descending colon, where waste material is stored until defecation occurs.

ASK YOURSELF

1 What are the main parts of the large intestine?

2 What are taeniae coli?

3 What are the main functions of the large intestine?

THE EFFECTS OF AGING ON THE DIGESTIVE SYSTEM

With age, digestion may be impaired as the stomach produces less hydrochloric acid, and other factors contribute to a general breakdown of the normal digestive process. If the protective mucous barrier breaks down, excess acid may cause stomach ulcers. Peristalsis slows down, tooth decay declines as the enamel hardens, but diseased gums (peridontal disease) and loss of teeth make chewing difficult, and even swallowing becomes more difficult. Constipation is a frequent result. The intestines produce fewer digestive enzymes, and the intestinal walls are less able to absorb nutrients. Defecation may be slower and less frequent as the muscles of the rectum weaken.

About 25 percent of all people over 65 have *diverticulosis*, the formation of small pockets (diverticuli) in the walls of the large intestine. When food becomes lodged in the pockets, and the pockets become inflamed, the disease is called *diverticulitis. Hemorrhoids* (varicose veins of the lower rectum and anus) frequently occur in people over 50, especially if constipation has been a chronic problem.

DEVELOPMENTAL ANATOMY OF THE DIGESTIVE SYSTEM

The embryonic digestive system begins as a hollow tube. The outside of the tube is covered with ectoderm, and the inside is lined with endoderm. The internal cavity develops into the mature digestive system. The mouth forms at about day 22 as a depression in the surface ectoderm called the *stomodeum*. The **primitive gut** forms during week 4 as the internal cavity of the embryonic yolk sac is enclosed by the lateral walls of the embryo. It is derived from endoderm and develops into three specific regions of the digestive system: (1) the anterior extension into the head region forms the *foregut*, (2) the central region forms the *midgut*, which opens into a pouch called the yolk sac, and (3) the portion extending posteriorly forms the *hindgut* [FIGURE 23.20]. At about day 24, the membrane separating the stomodeum from the foregut ruptures, allowing amniotic fluid to enter the embryo's mouth. At about week 5, the yolk sac constricts at the point where it is attached to the midgut, and seals off the midgut.

FIGURE 23.20 EARLY DEVELOPMENT OF THE DIGESTIVE SYSTEM

About day 28; sagittal section.

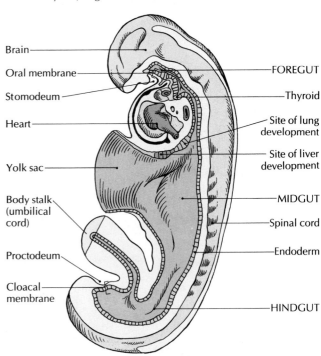

The *foregut* develops into the stomach, duodenum up to the entrance of the common bile duct, liver, gallbladder, and pancreas. The *midgut* develops into the small intestine, cecum and appendix, ascending colon, and most of the proximal portion of the transverse colon. The *hindgut* develops into the remaining distal portion of the transverse colon, descending colon, sigmoid colon, rectum, and superior portion of the anal canal. The primitive anus develops as an indentation at the caudal end of the digestive tube called the *proctodeum*. The membrane separating the proctodeum from the hindgut ruptures at about week 7, forming a complete tube from the mouth to the anus.

Later, hollow endodermal buds in the foregut move into the mesoderm and develop into the salivary glands, liver, gallbladder, and pancreas. Each of these accessory organs become connected to the central digestive tract by way of ducts.

WHEN THINGS GO WRONG

Anorexia Nervosa and Bulimia

Anorexia nervosa and bulimia are examples of eating disorders with a psychological basis. ***Anorexia nervosa*** (Gr. *an*, without + *orexis*, a longing) is characterized by self-imposed starvation and subsequent nutritional deficiency disorders. Victims of this disorder, most often adolescent females from upwardly mobile families, usually have a normal weight to begin with, but are convinced that they are grossly overweight. Excessive dieting is often accompanied by compulsive physical activity, resulting in a weight loss of 20 percent or more. The exact cause of anorexia nervosa is not known, but it is often brought on by anxiety, fear, or peer or family pressure that encourages thinness. Generally accompanying weight loss and chronic undernourishment are atrophy of skeletal muscle tissue, constipation, hypotension, dental caries, irregular or absent menstrual periods, and susceptibility to infection, among other disorders. If the victim does eat, it may be followed by self-induced vomiting or an excessive dose of laxatives or diuretics. Anorexia nervosa is fatal in 5 to 15 percent of the cases, and feelings of despondency lead to a higher-than-normal suicide rate.

Bulimia (Gr. *bous*, ox; *limos*, hunger) also occurs mainly in upwardly mobile young females of normal weight. It is also called *gorge-purge syndrome* and *bulimia nervosa*. It is characterized by an insatiable appetite and gorging on food four or five times a week (ingesting as many as 40,000 calories), followed by self-induced vomiting or an overuse of laxatives or diuretics that may lead to dehydration and metabolic imbalances as in anorexia nervosa. It may also impair kidney and liver function and cause dry skin, frequent infections, muscle spasms, and other disorders. Apparently, bulimarectics suffer from a decreased secretion of cholecystokinin, a hormone that induces a sense of fullness after a meal.

Cholelithiasis (Gallstones)

When the proportions of cholesterol, lecithin, and bile salts are altered, ***gallstones*** (biliary calculi) may form in the gallbladder, a condition known as ***cholelithiasis*** (KOH-lee-lih-thigh-uh-sihss; bile, gall + stone). Gallstones are composed of cholesterol and bilirubin, and are usually formed when there are insufficient bile salts to dissolve the cholesterol in micelles. Cholesterol, being insoluble in water, crystallizes and becomes hardened into "gallstones" (photo). Gallstones often pass out of the gallbladder and lodge in the hepatic and common bile ducts, obstructing the flow of bile into the duodenum

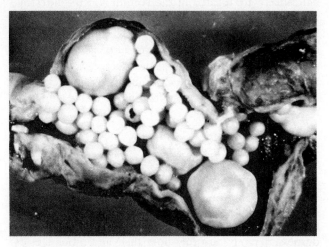

Section of gallbladder showing many gallstones.

and interfering with fat absorption. Jaundice may occur if a stone blocks the common bile duct because bile pigments cannot be excreted, and surgery may be required if stones become impacted in the cystic duct, where pain is usually maximal because the contractions of the gallbladder press on the stones. Other common sites of gallstones are the pancreatic ducts and the hepatopancreatic ampulla.

The production of gallstones is sometimes initiated by pregnancy, diabetes mellitus, celiac disease, cirrhosis of the liver, pancreatitis, or the use of oral contraceptives.

A recent nonsurgical technique for removing gallstones employs a device called a *lithotripter* (Gr. "stone crusher"), which uses high-pressure shock waves to pulverize the gallstones after they are located with ultrasound. Endoscopic removal is also becoming common.

Cirrhosis of the Liver

Cirrhosis (sih-ROE-siss; L. "orange-colored disease," from the color of the diseased liver) is a chronic liver disease characterized by the destruction of hepatic cells and their replacement by abnormal fibrous connective tissues. These changes disrupt normal blood and lymph flow, and eventually result in blockage of portal circulation, portal hypertension, liver failure, and death. Cirrhosis is most prevalent among chronic alcoholics over 50, especially if their diets are poor. The mortality rate is high. Among the several types of cirrhosis, the most common is *Laennec's type* (also called portal, nutritional, or alcoholic cirrhosis), which accounts for up to half of all persons suffering from the disease. Of these people, about 90 percent are heavy consumers of alcohol.

Colon-Rectum Cancer

More than half of all cancers of the large intestine (colon) are found in the rectum, and about 75 percent of intestinal cancers are located in the rectum and the intestinal region (sigmoid colon) just above it. Although the number of *colon-rectum cancer* cases is second only to lung cancer, early diagnosis and treatment make it possible to save 4 out of 10 patients. The exact cause of colon-rectum cancer is unknown, but increasing evidence points to a low-fiber diet containing excessive animal fat (especially beef).

Constipation and Diarrhea

Constipation (L. "to press together") is a condition in which feces move through the large intestine too slowly. As a result, too much water is reabsorbed, the feces become hard, and defecation is difficult. Constipation may be caused by spasms that reduce intestinal motility, nervousness, a temporary low-fiber diet, and other factors. Constipation may produce abdominal discomfort and distension, loss of appetite, and headaches, but it does *not* increase the amount of toxic matter in the body.

Diarrhea (Gr. "a flowing through") is a condition in which watery feces move through the large intestine rapidly and uncontrollably in a reaction sometimes called a *peristaltic rush*. Among the causes are viral, bacterial, or protozoan infections, extreme nervousness, ulcerative colitis, and excessive use of cathartics, all of which increase fluid secretion and intestinal motility. Severe diarrhea is dangerous, especially in infants, because it increases the loss of body fluids and ions.

Crohn's Disease

Crohn's disease (named after Dr. Burrill Crohn, who first diagnosed the disease in the early 1900s) is a chronic intestinal inflammation that most often affects the ileum. It extends through all layers of the intestinal wall, and is generally accompanied by abdominal cramping, fever, and diarrhea. The most common victims are 20- to 40-year-old adults, especially Jews.

Diverticulosis and Diverticulitis

Diverticula are bulging pouches in the gastrointestinal wall that push the mucosal lining through the muscle in the wall. Diverticula most commonly form in the sigmoid colon, but other likely sites are the duodenum near the hepatopancreatic ampulla and the jejunum. Diverticula appear most often in men over 40, especially if their diet is low in roughage.

In *diverticulosis,* pouches are present but do not present symptoms. In *diverticulitis,* the diverticula are inflamed and may become seriously infected if undigested food and bacteria become lodged in the pouches. The subsequent blockage of blood can lead to infection, peritonitis, and possible hemorrhage.

Food Poisoning

The term *food poisoning* is commonly used to describe gastrointestinal diseases caused by eating food contaminated with either infectious microorganisms or toxic substances produced by the microorganisms. Most food poisonings are associated with the production of toxins.

Hepatitis

Hepatitis, an infection of the liver, may have either a viral or a nonviral cause. Viral hepatitis has three forms: type A, type B, and type C. *Type A* is infectious, occurs most commonly during the fall and winter in children and young adults, and is transmitted through contaminated food (especially seafood), water, milk, semen, tears, and feces. Onset is sudden, and the incubation period is relatively short (15 to 45 days), but the overall prognosis for a complete cure is good. A person who contracts type A viral hepatitis does not become a carrier, but a patient afflicted with type B does carry the disease for an indefinite period. *Type B* viral hepatitis can affect people of any age, at any time of the year, and usually has a slow prolonged onset (40 to 100 days). It is transmitted through serum, blood and blood products, semen, and feces. Unlike type A, type B becomes worse with age, and permanent immunity does not result. *Type C* hepatitis usually results from contaminated commercial blood donations. Because liver cells are capable of regeneration, liver cell destruction resulting from hepatitis is normally overcome.

Nonviral hepatitis is usually caused by exposure to drugs or chemicals, such as carbon tetrachloride, poisonous mushrooms, and vinyl chloride. Symptoms usually resemble those of viral hepatitis.

Hernia

A *hiatal hernia* (L. gap + rupture) is a defect in the diaphragm that permits a portion of the stomach to protrude into the thoracic cavity. The common cause of a hernia is connected with a typical weakening of muscle tissue that often comes with aging. An *inguinal hernia* (L. *inguen,* groin) is the protrusion of the small or large intestine, omentum, or urinary bladder through the inguinal canal in the area of the groin.

Jaundice

Ordinarily, yellow bile pigment (bilirubin) is excreted by the liver in the bile, and not enough of it circulates in the blood to affect the color of the skin. Occasionally, however, pigment levels rise sufficiently to produce a yellowish tint in the skin, mucous membranes, and whites of the eyes. The condition is known as *jaundice* (Old Fr. *jaune,* yellow), or in common usage, *yellow jaundice.*

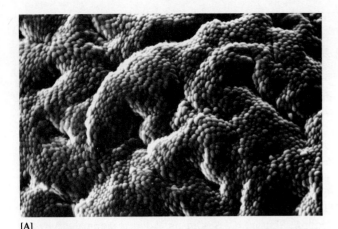

[A]

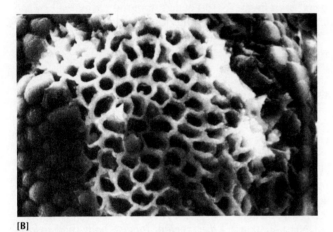

[B]

Scanning electron micrographs of **[A]** normal stomach lining, with a continuous coating of mucus (×350), and **[B]** damaged stomach tissue that results in ulcers (×1150).

Peptic Ulcers

Peptic ulcers (Gr. *peptein*, to digest) are lesions in the mucosa of the digestive tract (see photos). They are caused by the digestive action of gastric juice, especially hydrochloric acid. Most peptic ulcers occur in the duodenum, before gastric hydrochloric acid is neutralized by the alkaline secretions of the small intestine.

People who are most susceptible to *duodenal ulcers* secrete high amounts of gastric juices between meals, when there is little or no food in the stomach to buffer the acidity. Men between the ages of 20 and 50 are most vulnerable. *Gastric ulcers*, in contrast, seem to be most common in men over 50 and among people whose stomach mucosa has a reduced resistance to digestion rather than an overabundance of gastric secretions. Large amounts of aspirin and alcohol decrease the resistance of the mucosa, and thus may lead to gastric ulcers.

Peritonitis

Peritonitis is an inflammation of the peritoneum, usually resulting from a perforation in the gastrointestinal tract that allows bacteria to enter the normally sterile peritoneum. Such a perforation often results from a ruptured appendix, but can also be caused by diverticulitis, a peptic ulcer, or any other disease or physical trauma that breaks through the wall of the digestive tract.

Tooth Decay (Dental Caries)

Bacteria that cause *dental caries,* or tooth decay, produce a gluelike enzyme called *glucosyl transferase* (GTF) that converts ordinary sugar into *dextran*, a sticky substance that clings to the tooth's enamel. GTF also helps the bacteria adhere to the enamel. Dextran is involved in the formation of *plaque*, a destructive film that builds up on teeth. Once plaque is formed, the bacteria on the teeth produce another enzyme that promotes erosion of the enamel.*

Vomiting

Vomiting (emesis) is the forceful expulsion of part or all of the contents of the stomach and duodenum through the mouth, usually in a series of involuntary spasms. During vomiting, the pyloric region of the stomach goes into spasm, and the usual downward motion of peristalsis is reversed. Meanwhile, the body of the stomach and lower esophageal sphincter are relaxed. The duodenum also goes into spasm, forcing its contents into the stomach.

The final thrust of vomit comes when the abdominal muscles contract, lifting the diaphragm and increasing the pressure in the abdomen. At the same time, slow, deep breathing with the glottis partially closed reduces the pressure in the thorax. As a result, the stomach is squeezed between the diaphragm and compressed abdominal cavity, forcing the contents of the stomach and duodenum past the lower esophageal sphincter into the esophagus, and out through the mouth. When this series of events is repeated several times, without vomiting, it is called *retching*.

During actual vomiting, the breathing rate decreases, the glottis is closed, and the soft palate is raised, all ensuring that the vomit does not enter the breathing passages and cause suffocation.

Although it seems clear that vomiting may be the body's way of ridding itself of a harmful food substance, it is not fully understood why motion sickness or pain should induce vomiting. Also, it is not certain why some women become nauseous and vomit during early pregnancy ("morning sickness"). Excessive vomiting, like prolonged diarrhea, can lead to a dangerous depletion of fluids and salts.

*Our prehistoric ancestors had cavities too, but a study of skeletons that have been dated to before the Iron Age indicates that humans had about 2 to 4 percent decay then, compared to 40 to 70 percent today. However, public health officials predict the virtual end of tooth decay among children and young adults by the end of this century due to improved dental technology and the widespread use of fluoridation in city water supplies.

CHAPTER SUMMARY

Food is ingested into the digestive tract and undergoes **digestion,** the process of breaking down large food molecules into smaller ones. However, small nutritive molecules must leave the digestive system and enter the body proper before they can be used. This is accomplished by a second process, called **absorption,** when the small food molecules pass through the plasma membranes of the small intestine into the bloodstream.

Introduction to the Digestive System (p. 583)

1 The digestive process takes place in the **digestive tract,** or **alimentary canal,** which extends from the lips to the anus.

2 The digestive system is compartmentalized, with each part adapted to a specific function. The alimentary canal consists of the mouth, pharynx, esophagus, stomach, small intestine, large intestine, rectum, anal canal, and anus. The associated structures include the salivary glands, pancreas, liver, and gallbladder.

3 Parts of the digestive tract have specialized functions, but all are composed of the same basic layers of tissue. The wall of the tube (from the inside out) is composed of the **mucosa, submucosa, muscularis externa,** and **serosa.**

Mouth (p. 583)

1 The **mouth** (**oral,** or **buccal, cavity**) consists of two parts, the small, outer **vestibule,** and the *oral cavity* proper.

2 The **lips** and **cheeks** aid the digestive process by holding the food in position to be chewed.

3 There are 20 **deciduous** (''baby'') teeth and 32 **permanent** teeth, with specialized shapes for cutting, tearing, and grinding food. Each tooth consists of a **root, crown,** and **neck,** and is composed of **dentine, enamel, cement,** and **pulp.** The **gum** is the connective tissue that surrounds the alveolar processes of the teeth.

4 The **tongue** aids in digestion by helping to form the moistened, chewed food

into a **bolus,** and pushing it toward the pharynx to be swallowed. It also contains taste buds.

5 The tongue also pushes the food against the **hard palate,** where it is crushed and softened. The movable **soft palate** prevents food from entering the nasal passages by closing over the nasopharynx.

6 The three largest pairs of **salivary glands** are the **parotid, submandibular,** and **sublingual glands.** They secrete saliva, which contains water, salts, proteins, and at least one enzyme, *salivary amylase,* which begins the digestion of carbohydrates. Saliva also moistens and lubricates food so it can be swallowed easily, and allows us to taste food by dissolving food molecules.

Pharynx and Esophagus (p. 591)

1 The **pharynx** leads from the mouth to the esophagus. It is the common pathway for the passage of food (swallowing) and air. The **esophagus** is the tube that carries food from the pharynx to the stomach.

2 Swallowing action, or **deglutition,** in the pharynx initiates a muscular wave called **peristalsis,** which pushes the food along the esophagus and into the stomach. Peristalsis continues to assist the passage of food through the rest of the digestive system.

Abdominal Cavity and Peritoneum (p. 593)

1 The **abdominal cavity** is the portion of the trunk cavities that lies inferior to the diaphragm. When considered as including the pelvic cavity, it is called the **abdominopelvic cavity.**

2 The serous membrane of the abdominal cavity is the **peritoneum.** The *parietal peritoneum* lines the abdominal cavity, and the *visceral peritoneum* covers most of the organs within the cavity.

3 The space between the parietal and visceral membranes is the *peritoneal cavity.* It usually contains a small amount of *peritoneal fluid,* which re-

duces the friction as abdominal organs move.

4 Abdominal organs that lie posterior to the peritoneal cavity and are covered, but not surrounded, by the peritoneum are called *retroperitoneal.* Those surrounded by the peritoneum are called *intraperitoneal.*

5 Organs in the abdominal cavity are suspended from the cavity wall by the **mesenteries,** or *visceral ligaments.* The mesentery connected to the stomach is the **omentum.**

Stomach (p. 595)

1 The **stomach** is the most expandable portion of the alimentary tract. It stores, mixes, and digests food.

2 Food enters the stomach from the esophagus through the **lower esophageal orifice** and empties into the small intestine through the **pyloric orifice.**

3 The stomach is divided into the **cardiac region, fundus, body,** and **pyloric region.**

4 The stomach stores large quantities of food, uses a churning action to mix food into a soupy mixture called **chyme,** releases the chyme in regular spurts into the small intestine, and secretes gastric juices, which initiate the digestion of protein.

5 The **muscularis externa** of the stomach has three layers of smooth muscle: the outermost *longitudinal* layer, the middle *circular* layer, and the innermost *oblique* layer.

6 **Gastric juice** is composed of *hydrochloric acid, mucus,* and several enzymes, including *pepsinogen* (a precursor of the active enzyme **pepsin**). The major function of gastric juice is to begin the digestion of protein.

Small Intestine (p. 598)

1 The **small intestine** is subdivided into the **duodenum,** where most of the remaining digestion takes place, and the **jejunum** and **ileum,** where most of the absorption of nutrients and water into the bloodstream and lymphatic system occurs.

2 The absorptive surface of the small intestine is increased substantially by protrusions called *villi,* which contain additional protrusions called *microvilli,* and circular folds called *plicae circulares.*

3 *Mucosal* and *submucosal glands* secrete *intestinal juice* and several enzymes that aid the digestion of carbohydrates, proteins, and lipids. Large *aggregated lymph nodules* combat microorganisms.

4 The two main types of muscular activity in the small intestine are *segmenting contractions* and *peristaltic contractions.*

5 The *intestinal juice* contains water, salts, mucus, and enzymes that complete the breakdown of proteins, carbohydrates, and lipids.

6 The products of carbohydrate, protein, and lipid digestion, as well as most ingested electrolytes, vitamins, and water, are absorbed by the small intestine. After carbohydrates and proteins have been broken down by digestion, they are readily absorbed into columnar absorptive cells in the small intestine and then into blood capillaries.

7 The absorption of lipids is more complex than that of carbohydrates and proteins. It involves the breakdown of water-insoluble triglyceride droplets into water-soluble particles called *micelles,* which are absorbed by cells. Once inside cells, the breakdown products of lipids are resynthesized into triglycerides and packaged into tiny droplets called *chylomicrons,* which move from the cells into lymphatic vessels and then into the blood for distribution throughout the body.

Pancreas as a Digestive Organ (p. 604)

1 The *pancreas* consists of a head, body, and tail. It is composed of both exocrine and endocrine secretory cells. The exocrine cells form groups of cells called *acini,* which secrete digestive juices into the small intestine.

2 Three main types of pancreatic digestive enzymes in the digestive juices are transported through the pancreatic duct to the duodenum, where they act on proteins, carbohydrates, and lipids during the final steps of digestion.

Liver as a Digestive Organ (p. 604)

1 The liver, the largest glandular organ in the body, is divided into two main lobes by the *falciform ligament,* with the *right lobe* being six times larger than the *left lobe.* The right lobe is further subdivided into a *quadrate lobe* and a *caudate lobe.*

2 The *porta hepatis* is the area through which blood vessels, nerves, and ducts enter and leave the liver. The *hepatic artery* and *hepatic portal vein* enter, and the *hepatic vein* and *bile duct* exit.

3 The functional units of the liver are *lobules* that contain *hepatic cells* arranged in plates that radiate from a *central vein.* Between the rows of cells are blood channels called *sinusoids;* in the corners of the five- or six-sided lobules are *portal areas* that contain branches of the portal vein, hepatic artery, bile duct, and nerve.

4 The many functions of the liver include the secretion of *bile,* storage of excess glucose as glycogen, removal of amino acids from organic compounds, conversion of excess amino acids into urea, synthesis of certain amino acids, and conversion of carbohydrates and proteins into fat.

Gallbladder and Biliary System (p. 606)

1 The *gallbladder* concentrates and stores bile from the liver until it is needed for digestion.

2 The *biliary system* consists of the gallbladder, hepatic ducts, cystic duct, and common bile duct.

Large Intestine (p. 608)

1 When chyme leaves the small intestine, digestion is complete, and the *large intestine* functions to remove water and salts from the liquid chyme. Removal of water converts liquid wastes into *feces.*

2 The large intestine consists of the *cecum* (which contains the *vermiform appendix*), *ascending colon, transverse colon, descending colon, sigmoid colon,* and *rectum.*

3 The large intestine has an incomplete layer of longitudinal smooth muscle that forms three separate bands of muscle called *taeniae coli.* The intestinal wall contains bulges called *haustra* and fat-filled pouches called *epiploic appendages.*

4 The terminal segments of the large intestine are the *rectum, anal canal,* and *anus.* The elimination of feces from the anus is called *defecation.*

The Effects of Aging on the Digestive System (p. 612)

Digestion may become less efficient with age, as less hydrochloric acid and fewer enzymes are produced, peristalsis and other muscular actions slow down, and peridontal disease occurs.

Developmental Anatomy of the Digestive System (p. 612)

1 The digestive system begins as a hollow tube lined with endoderm. The *primitive gut* forms during week 4, and develops into the anterior *foregut,* the central *midgut,* and the posterior *hindgut.*

2 The mouth develops from a surface depression called the *stomodeum,* and the anus develops from an indentation at the caudal end of the digestive tract called the *proctodeum.*

3 Accessory organs, such as the pancreas, are connected to the central digestive tract via ducts.

MEDICAL TERMINOLOGY

ACHOLIA (*a*, without + Gr. *khole*, bile) Absence of bile secretion.

ANTIEMETIC (Gr. *anti*, against + *emetos*, vomiting) A substance used to control vomiting.

ANTIFLATULENT (Gr. *anti*, against + L. *flatus*, a breaking wind) A drug that prevents the retention of air in the digestive tract.

ASCITES (uh-SEE-teez; Gr. "bag") An accumulation of serous fluid in the abdominal cavity.

BARIUM ENEMA EXAMINATION Fluoroscopic and radiographic examination of the large intestine after administering a barium sulfate mixture via the rectum. Commonly called *lower GI series*.

CHOLECYSTITIS (Gr. *khole*, bile + *cystis*, bladder + *-itis*, inflammation) Inflammation of the gallbladder.

COLECTOMY (L. *colon*, intestine + *-ectomy*, removal) Surgical removal of the colon. A *hemicolectomy* removes half the colon.

COLITIS Inflammation of the colon.

COLOSTOMY (L. *colon*, intestine + Gr. *stoma*, opening) Surgical creation of an opening in the large intestine, through which feces can be eliminated. Also, the opening thus created.

CREPITUS (L. *crepitare*, to crackle) Discharge of flatus from the intestine.

DYSPEPSIA (Gr. *dus*, faulty + *pepsia*, digestion) A condition in which digestion is difficult. Commonly called *indigestion*.

DYSPHAGIA (Gr. *dus*, faulty + *phagein*, to eat) Difficulty in swallowing.

EMETIC (Gr. *emetos*, vomiting) A substance used to induce vomiting.

ENDODONTICS The branch of dentistry that deals with diseases of the dental pulp and associated processes.

ENTERITIS (Gr. *enteron*, intestine + inflammation) Inflammation of the intestines, particularly the small intestine.

GASTRECTOMY (Gr. *gaster*, belly + *-ectomy*, removal) Surgical removal of part or all of the stomach.

GASTROENTEROLOGIST A physician specializing in treating stomach and intestinal (gastrointestinal) disorders.

GASTROSCOPY (Gr. *gaster*, belly + *skopein*, to examine) A procedure in which a viewing device is inserted into the stomach for exploratory purposes.

GASTROSTOMY (Gr. *gaster*, belly + *stoma*, opening) Surgical creation of an opening in the stomach, usually for the purpose of inserting a feeding tube.

GINGIVITIS (L. *gingiva*, gum + inflammation) Inflammation of the gums.

HALITOSIS (L. *halitus*, breath) Bad breath.

HYPEROREXIA (Gr. *hyper*, over + *orexis*, a longing) Abnormal appetite.

HYPOCHLORHYDRIA A deficiency of hydrochloric acid in the stomach.

IRRITABLE BOWEL SYNDROME A generalized gastrointestinal disorder, characterized by a spastic colon, alternating diarrhea and constipation, and excessive mucus in the feces.

LAPAROTOMY (Gr. *lapara*, flank + *tome*, incision) Cutting into the abdomen in order to explore the abdominal cavity.

LAXATIVE (L. *laxus*, loose) A substance used to loosen the contents of the large intestine so that defecation can occur.

MALOCCLUSION (L. *mal*, badly + *occludere*, to close) A condition in which the upper and lower teeth do not fit properly when the jaws are closed.

ORTHODONTIA (Gr. *orthos*, straight + *odous*, tooth) The branch of dentistry that specializes in aligning teeth properly.

PROCTOCELE (Gr. *proktos*, anus + *koilos*, hollow) Hernia of the rectum.

PROCTOSCOPY (Gr. *proktos*, anus + *skopein*, to examine) A procedure in which a viewing device is inserted into the anus and rectum for exploratory purposes.

PROLAPSE OF THE RECTUM A condition where the weakened walls of the rectum fall outward or downward.

PRURITIS ANI (L. *prurire*, to itch + anus) Chronic itching of the anus.

PYORRHEA (L. *pyo*, pus + Gr. *rrhoos*, flowing) A condition in which pus oozes from infected gums.

ROOT CANAL THERAPY The dental procedure that removes infected or damaged pulp, and then cleans, sterilizes, fills, and seals the root cavity.

SIALITIS Inflammation of a salivary duct or gland.

SMALL INTESTINE SERIES Serial radiograms of the small intestine during the passage of ingested barium. Commonly called *upper GI series*.

STOMATITIS (Gr. *stoma*, mouth + inflammation) Inflammation of the mouth.

TRACHEOSTOMY (trachea + opening) Surgical creation of a permanent opening in the trachea.

TRENCH MOUTH (VINCENT'S INFECTION) A painful inflammation of the gums, usually including fever, ulcerations, and bleeding.

UNDERSTANDING THE FACTS

1 Which organs form the digestive tract? What is the basic function of each?

2 What are the two main processes involved in converting food into nutrients that can be used by body cells?

3 List the major accessory organs and associated structures of the digestive system. How are they related to digestion?

4 Differentiate between chemical and mechanical digestion.

5 Define mucosa.

6 Which teeth have no deciduous predecessors? (Supply number and type.)

7 List the three largest pairs of glands that secrete saliva.

8 Define bolus. How is it formed?

9 Name the structure that regulates the flow of material from the stomach to the duodenum.

10 Does the gallbladder produce bile? Explain.

11 Distinguish between diverticulosis and diverticulitis.

UNDERSTANDING THE CONCEPTS

1 If the small intestine were to be cut into, through what tissue layers would the scalpel pass?

2 What digestive activities occur in the mouth? What structures and secretions are involved?

3 Describe the typical series of events that lead to tooth decay.

4 How important is gravity to the process of swallowing? Explain.

5 How does the arrangement of the muscle fibers in the walls of the stomach facilitate its functions?

6 Why do we normally not digest our stomach wall?

7 Other than its great length, what structural factors contribute to the great absorptive surface of the small intestine?

8 Sketch, label, and discuss the gross anatomy of the small intestine.

9 How does the liver contribute to the homeostasis of the body?

10 Why may severe diarrhea be a serious problem in infants?

11 What may be the relationship of aspirin and alcohol to gastric ulcers?

SELF-QUIZ

Multiple-Choice Questions

23.1 The walls of the digestive system have the following layers from the inside out:

   a serosa, muscularis externa, submucosa, mucosa

   b mucosa, submucosa, muscularis externa, serosa

   c submucosa, mucosa, muscularis externa, serosa

23.2 Normally, a child has how many baby teeth?

   a 10   **d** 18

   b 12   **e** 20

   c 15

23.3 An adult has all of the following *except*

   a four incisors   **d** two cuspids

   b two canines   **e** six molars

   c four premolars

23.4 The largest of the three pairs of salivary glands are the

   a parotid   **d** sublingual

   b submandibular   **e** subparotid

   c mandibular

23.5 How many sphincters are associated with the esophagus?

   a one   **d** four

   b two   **e** five

   c three

23.6 Which of the following is *not* one of the major regions of the stomach?

   a cardiac   **d** esophageal

   b fundus   **e** pyloric

   c body

23.7 The stomach contains how many layers of smooth muscle fibers?

   a one   **d** four

   b two   **e** five

   c three

True-False Statements

23.8 The tongue is composed mainly of smooth muscle.

23.9 The uvula is associated with the hard palate.

23.10 The serous membrane of the abdominal cavity is the peritoneum.

23.11 Serous fluid is found within the coelom.

23.12 Duodenal submucosal glands are found in the stomach.

23.13 Segmental contractions of the small intestine are controlled by pacemaker cells in the smooth muscle layer.

23.14 The ileum is the primary site for the absorption of bile salts and vitamin B_{12}.

23.15 Absorption of carbohydrates occurs mostly in the duodenum and jejunum.

23.16 The products of protein and lipid breakdown pass from the intestine into lacteals.

Completion Exercises

23.17 The part of the digestive tract below the diaphragm is called the _____ tract.

23.18 Teeth are held in their sockets by bundles of connective tissue called _____.

23.19 _____ is the watery, tasteless mixture of salivary gland and oral mucous gland secretions.

23.20 The portion of the digestive system that serves as both an air passage and a food passage is the _____.

23.21 The serous membrane of the abdominal cavity is the _____.

23.22 The extensive folded membrane of the abdominal cavity is the _____.

23.23 The absorptive surface area of the intestinal mucosa is increased by millions of fingerlike projections called _____.

23.24 The liver is divided into two main lobes by the _____ ligament.

23.25 The functional units of the liver are the _____.

23.26 In the large intestine, removal of water converts liquid wastes into _____.

23.27 The large intestine contains bulges called _____.

Matching

23.28 _____ dentine

23.29 _____ enamel

23.30 _____ cement

23.31 _____ pulp

a bonelike covering of the neck and root

b surrounds the pulp cavity

c soft core of connective tissue

d the white covering of the crown

e embedded in skull bones

Problem-Solving/Critical Thinking Questions

23.32 Before reaching the mucosa of the stomach, a surgeon's knife would pass through the (in order)
a peritoneum, abdominal cavity, serosa, smooth muscle, submucosa
b abdominal cavity, serosa, smooth muscle, submucosa, peritoneum
c peritoneum, abdominal cavity, smooth muscle, serosa, submucosa
d serosa, peritoneum, smooth muscle, submucosa, abdominal cavity

23.33 Which of the following types of teeth are present in the jaw of an adult but not a child?
a incisors
b canines
c premolars
d molars
e a and b

23.34 The small intestine
a terminates at the iliocecal junction
b is mostly intraperitoneal
c begins at the pylorus
d is supplied by the superior and inferior mesenteric arteries
e all but d

A SECOND LOOK

1 In the following photograph, label the stomach, transverse colon, descending colon, ileum, jejunum, sigmoid colon, and rectum.

2 In the following drawing of the small intestine, label the brush border, capillary plexus, central lacteal, intestinal gland, lamina propria, and muscularis mucosa.

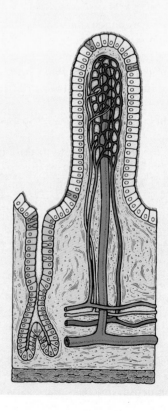

24

The Urinary System

LEARNING OBJECTIVES

1 Name the major components of the urinary system, and describe their functions.

2 Explain how the kidney (or nephron) functions to maintain homeostasis in the body.

3 Describe the location and external anatomy of the kidney.

4 Discuss the internal anatomy of the kidney.

5 Describe the blood, lymphatic, and nerve supply of the kidney.

6 Relate the components of the nephron to their specific functions.

7 Describe the components and function of the juxtaglomerular apparatus.

8 Relate the structure of the ureters, urinary bladder, and urethra to their functions.

9 Explain how kidney function changes from birth to old age.

Like other land-dwelling organisms, human beings must deal with the problem of obtaining and conserving water. We must also have a way of removing excess salts and metabolic wastes, especially the nitrogen-containing end products of protein metabolism, such as urea, uric acid, ammonia, and creatinine. Our urinary system accomplishes both of these crucial tasks: it removes waste products through the production of urine, and it carefully regulates the liquid content of the body.

What we excrete has much more influence on homeostasis than what we eat. *Excretion* (L. *excernere*, to sift out) is the elimination of the waste products of metabolism, and the removal of surplus substances from body tissues. The urinary system eliminates the wastes of protein metabolism in the urine, but it does much more than that. It also regulates the amount of water in the body and, in turn, the amount of salts in the blood. These regulatory functions are accomplished by the kidneys, which form and eliminate varying amounts of urine containing varying amounts of salts and water. In addition, many foreign substances, including drugs, are excreted by the kidneys.

Excretion is performed in several ways by several organs. The lungs excrete carbon dioxide and water when we exhale. The skin releases some salts and organic substances such as urea and ammonia during periods of heavy sweating. The large intestine excretes a small amount of water, salts, and some microorganisms and undigested food. However, the prime regulator of the proper balance between water and other substances is the urinary system, with the kidneys as its main component.

THE URINARY SYSTEM: COMPONENTS AND FUNCTIONS

The **urinary system,** also called the *renal system* (L. *renes,* kidneys), consists of (1) two *kidneys*, which remove dissolved waste and excess substances from the blood and form urine, (2) two *ureters*, which transport urine from the kidneys to (3) the *urinary bladder*, which acts as a reservoir for urine, and (4) the *urethra*, the duct through which urine from the bladder flows to the outside of the body during urination [FIGURE 24.1].

Every day the kidneys filter about 1700 L of blood. Each kidney contains over a million *nephrons*, the functional units. Fortunately, we have many more nephrons than we actually need for these purposes, and if necessary, we could lead a normal life with only one kidney.

How do the nephrons accomplish such a controlled regulation? First, they **filter** water and soluble components from the blood. Second, they selectively **reabsorb** some of the components back into the blood to maintain

the normal blood concentration. Finally, they selectively **secrete** wastes into the urine that were not filtered from the blood efficiently. The water and other substances not reabsorbed are the true wastes, which constitute the urine.

ANATOMY OF THE KIDNEYS

The average adult takes in about 2.7 L of water every day, most of it by ingesting foods and liquids. Under normal conditions, the same amount of water is given off daily, about a third of it evaporated from the skin and exhaled from the lungs, and more than half excreted in urine. The remarkable organs that maintain this constant water balance are the **kidneys,** a pair of reddish-brown, bean-shaped organs, each one about the size of a large bar of soap. They are usually about 11.25 cm (4.4 in.) long, 5 cm (2 in.) wide, and 2.5 cm (1 in.) thick.

Location and External Anatomy

The kidneys are located in the posterior wall of the abdominal cavity, one on each side of the vertebral column. They usually span the distance from vertebra T12 to vertebra L3. The paired kidneys are protected, at least partially, by the last pair of ribs and are capped by the *adrenal* ("upon the kidney") *glands.*

Just as the stomach is higher than most people think, so are the kidneys. Rather than being located in the small of the back, the upper portion of each kidney is in contact with the diaphragm, and the left kidney also touches the spleen. The right kidney is in extensive contact with the liver, and because of the liver's large right lobe, the right kidney is slightly lower than the left. The kidneys are retroperitoneal [FIGURE 24.2] and are well supplied with blood vessels, lymphatics, and nerves.

The bean shape of the kidney is medially concave and laterally convex. On the medial concave border is the **hilus** (HYE-luhss), a small, indented opening where arteries, veins, nerves, and the ureter enter and leave the kidney [FIGURE 24.3].

Covering and supporting each kidney are three layers of tissue: (1) The innermost layer is a tough, fibrous material called the *renal capsule*. It is continuous with the surface layer of the ureters. (2) The middle layer is the *adipose capsule*, composed of perirenal ("around the kidney") fat, which gives the kidney a protective cushion against impacts and jolts. (3) The outer layer is subserous fascia called the *renal fascia*. Surrounding the renal fascia is another layer of fat, the *pararenal fat*. The renal fascia is composed of connective tissue that surrounds the kidney

FIGURE 24.1 THE URINARY SYSTEM

Anterior view. Differences in the lower
portions of the male and female systems are shown in FIGURE 24.9.

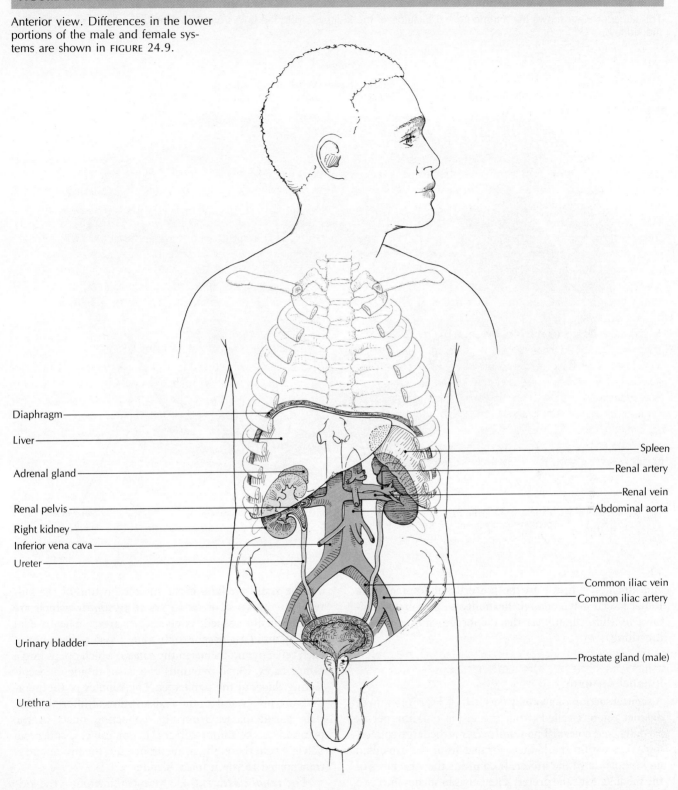

Diaphragm

Liver

Adrenal gland

Renal pelvis

Right kidney

Inferior vena cava

Ureter

Urinary bladder

Urethra

Spleen

Renal artery

Renal vein

Abdominal aorta

Common iliac vein

Common iliac artery

Prostate gland (male)

FIGURE 24.2 TRANSVERSE SECTION THROUGH UPPER ABDOMEN

This superior view shows the retroperitoneal location of the kidneys. Note the layers of adipose tissue and fasciae surrounding the kidneys.

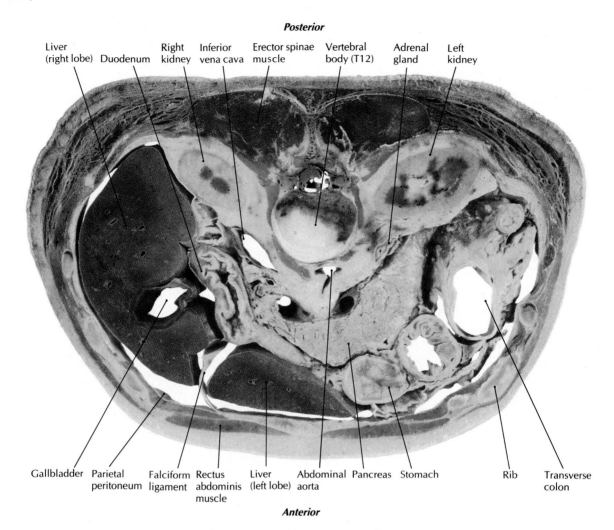

Posterior

Liver (right lobe) Duodenum Right kidney Inferior vena cava Erector spinae muscle Vertebral body (T12) Adrenal gland Left kidney

Gallbladder Parietal peritoneum Falciform ligament Rectus abdominis muscle Liver (left lobe) Abdominal aorta Pancreas Stomach Rib Transverse colon

Anterior

and attaches it firmly to the posterior abdominal wall. Renal fasciae have enough flexibility to permit the kidneys to shift slightly as the diaphragm moves during breathing.

Internal Anatomy

A sagittal section of a kidney [FIGURE 24.3A] reveals three distinct regions called (from the inside out) the pelvis, medulla, and cortex. The *renal pelvis* is the large collecting space within the kidney, formed from the expanded upper portion of the ureter. It connects the structures of the medulla with the ureter. The pelvis branches into two smaller cavities, the *major calyces* and *minor calyces* [KAY-luh-seez; sing., *calyx*, KAY-licks; Gr. cup). There are usually 2 to 3 major calyces and 8 to 18 minor calyces in each kidney.

The *renal medulla* is the middle portion of the kidney. It consists of 8 to 18 *renal pyramids,* which are longitudinally striped, cone-shaped areas. The base of each pyramid is adjacent to the outer cortex. The apex of each renal pyramid ends in the *papilla,* which opens into a minor calyx. Renal pyramids consist of tubules and collecting ducts of the nephrons. The tubules of the pyramids are involved with the reabsorption of filtered materials. Urine passes from the collecting ducts in the pyramid to the minor calyces, major calyces, and renal pelvis. From there, the urine drains into the ureter and is transported to the urinary bladder.

The *renal cortex* is the outermost portion of the kidney. It is divided into two regions, the outer *cortical region* and the inner *juxtamedullary* ("next to the medulla") *region.* The cortex has a granular, textured appearance; it extends from the outermost *capsule* to the base of the

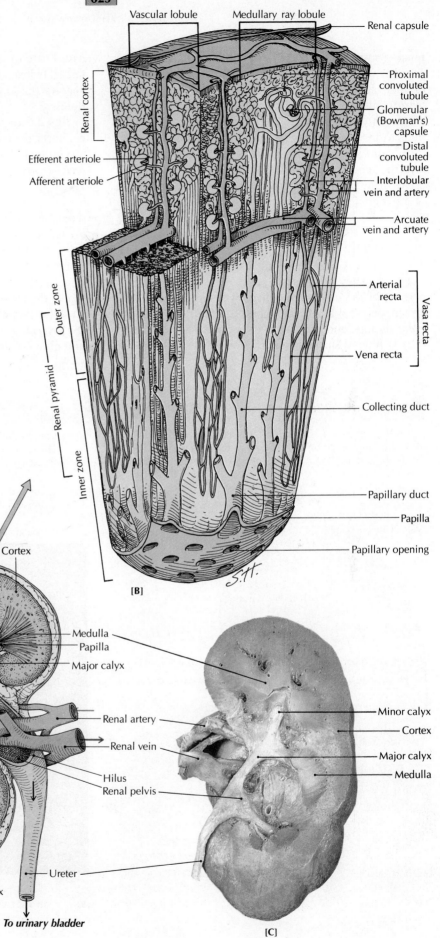

FIGURE 24.3 INTERNAL STRUCTURE OF THE KIDNEY

[A] Sagittal section of a kidney showing major structures and blood supply. Veins and arteries pass through the renal pelvis and medulla, branching extensively to form capillary beds in the renal cortex. Arrows indicate the direction of fluid flow. [B] Enlarged drawing, showing the details of a renal pyramid and cortex. [C] Dissection of kidney.

Vascular lobule • Medullary ray lobule • Renal capsule • Proximal convoluted tubule • Glomerular (Bowman's) capsule • Distal convoluted tubule • Interlobular vein and artery • Arcuate vein and artery • Renal cortex • Efferent arteriole • Afferent arteriole • Outer zone • Renal pyramid • Inner zone • Arterial recta • Vena recta • Vasa recta • Collecting duct • Papillary duct • Papilla • Papillary opening

[B]

Renal fascia • Adipose tissue • Cortex • Renal capsule • Arcuate artery • Interlobular artery • Interlobular vein • Interlobar artery • Arcuate vein • Interlobar vein • Nephron • Renal column • Minor calyx • Renal pyramid • Medulla • Papilla • Major calyx • Renal artery • Renal vein • Hilus • Renal pelvis • Ureter • To urinary bladder

[A]

Minor calyx • Cortex • Major calyx • Medulla

[C]

renal pyramids [FIGURE 24.3]. Within the cortex, the granular appearance is caused by spherical bundles of capillaries and associated structures of the nephron that help to filter blood. The cortical tissue that penetrates the depth of the renal medulla between the renal pyramids forms **renal columns.** The columns are composed mainly of **kidney tubules** that drain and empty the urine into the lumen of a minor calyx. Each minor calyx receives urine from a *lobe* of the kidney. Each lobe consists of a pyramid and the associated cortical tissue above it.

Blood Supply

The kidneys receive more blood in proportion to their weight than any other organ in the body. About 20 to 25 percent of the cardiac output goes to the kidneys. Approximately 1.2 L of blood pass through the kidneys every minute, and the body's entire blood volume (4 to 6 L) is filtered through the kidneys about 340 times a day. Little of this blood supplies the kidneys' nutritive needs. Instead, this large blood flow through the kidneys enables the kidneys to maintain normal levels of body fluids.

Blood comes to the kidneys directly from the abdominal aorta through the **renal artery** [FIGURE 24.4]. Once inside the kidney, the renal artery branches into the **interlobar arteries,** which pass through the renal columns. When the arteries reach the juncture of the cortex and medulla, they turn and run parallel to the bases of the renal pyramids. At the turning point, the arteries are the **arcuate arteries** (L. *arcuare,* to bend like a bow), which make small arcs around the boundary between the cortex and medulla. The arcuate arteries branch further into the **interlobular arteries,** which ascend into the cortex to supply the renal corpuscle (the renal capsule and glomerulus). Further branching produces numerous small **afferent arterioles,** which carry blood *to* the site of filtration (the glomerulus).

FIGURE 24.4 CIRCULATORY PATHWAY THROUGH KIDNEY

[A] In this schematic drawing, veins and arteries beyond the incoming and outgoing renal artery and vein are actually inside the kidney; the actual locations of vessels are shown in FIGURE 24.3A. Arrows indicate the direction of blood flow. **[B]** Cast of the right renal artery, calyces, renal pelvis, and ureter; anterior view.

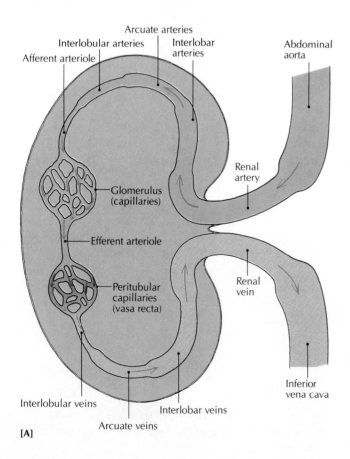

Each afferent arteriole branches extensively to form a tightly coiled ball (tuft) of capillaries called a *glomerulus* (glow-MARE-yoo-luhss; L. ball). This capillary bed is where the blood is filtered. Glomerular capillary loops eventually join together to form a single *efferent arteriole,* which carries blood *away from* the glomerulus. The efferent arteriole is narrower than the afferent arteriole, and this difference tends to increase blood pressure in the glomerulus. The increased blood pressure produces efficient filtration from the glomerulus.

The efferent arteriole eventually branches to form a second capillary bed (the glomerulus was the first one), made up of the *peritubular* ("around the tubules") *capillaries.* These capillaries reabsorb some of the water and solutes that were filtered from the blood in the glomerulus. The peritubular capillaries unite to form the *interlobular veins,* which carry blood out of the cortex to the *arcuate veins.* The small arcuate veins join to form the larger *interlobar veins* in the renal columns, and the interlobar veins eventually come together to form the single *renal vein* that carries cleansed blood from each kidney to the inferior vena cava and into the circulatory system.

The *arterial rectae* ("straight arteries") are extensions of the efferent arterioles that surround a portion of the nephron called the loop of the nephron (loop of Henle) [FIGURE 24.3B]. They follow the loop of the nephron deep into the medulla and drain into the *venae rectae,* which carry blood to the interlobular veins. Only about 1 or 2 percent of the total renal blood flows through these vessels, collectively called the *vasa recta.*

Like all arterioles, the afferent and efferent arterioles that join the glomerular and peritubular capillary beds contain smooth muscle in their walls (tunica media). This muscle permits the arterioles to constrict or dilate in response to nervous or hormonal stimulation. Constriction or dilation not only allows for the control of blood pressure in the glomerulus, it also helps to maintain filtration.

Lymphatics

Lymphatic vessels accompany the larger renal blood vessels, and are more prominent around the arteries than the veins. The lymphatic vessels converge to form several large vessels that leave the kidney at the renal hilus.

Nerve Supply

Nerves from the *renal plexus* of the autonomic nervous system enter the kidney at the hilus and follow the arteries to innervate the smooth muscle of the afferent and efferent arterioles. This vasomotor supply functions almost entirely as a vasoconstrictor. Generally, the afferent arteriole constricts more than the efferent arteriole.

Changes in posture (gravity), physical activity, and stress also increase the activity of these sympathetic nerves. Overall, vasomotor nerves help to control kidney function by regulating blood pressure in the glomerulus, as well as blood distribution throughout the kidney.

In a male, the nerves from the kidneys also communicate with nerves from the testes. This explains why kidney disorders often produce pain in the testes.

The Nephron

Each *nephron* (Gr. *nephros,* kidney) is an independent urine-making unit, and each kidney contains approximately 1 million nephrons. As the functional unit of the kidney, the nephron accomplishes the initial filtration of blood, the selective reabsorption back into the blood of filtered substances that are useful to the body, and the secretion of unwanted substances.

Two types of nephrons are recognized, *cortical* and *juxtamedullary* [FIGURE 24.5B]. The tubular structures of the cortical nephron extend only into the base of the renal pyramid of the medulla, while the longer loop of the nephron of the juxtamedullary nephron projects deeper into the renal pyramid. Cortical nephrons are about seven times more numerous than juxtamedullary nephrons.

A nephron consists of (1) a tubular component, and (2) a vascular component. The *tubular component* starts with the glomerular capsule and includes the excretory tubules, which are the proximal convoluted tubule, loop of the nephron, and distal convoluted tubule [FIGURE 24.5C]. The excretory tubules of a nephron are coiled and winding. All the tubules from all the nephrons in the body have a combined length of about 80 km (50 mi). Each of the excretory tubules leads into a large collecting duct (which is not part of the nephron) that transports the resulting renal filtrate. The entire tubular portion of the nephron is composed of a single layer of epithelial cells. All along the tubule, cell structure changes as the functions of the tubule change.

The *vascular component* of a nephron is made up of blood vessels. These include the glomerulus and the peritubular capillaries, which surround the excretory tubules. The reabsorption of substances from the excretory tubules into the blood takes place in the peritubular capillaries and vasa recta.

Glomerular (Bowman's) capsule The *glomerular (Bowman's) capsule* (originally named for Sir William Bowman, the nineteenth-century English anatomist who first described this structure) is the portion of the nephron that encloses the glomerulus like a hand wrapped around a ball [FIGURE 24.6]. Together, the capsule and glomerulus form the *renal corpuscle.* The glomerular

FIGURE 24.5 NEPHRON

[A] Entire kidney. [B] Enlarged section showing locations of juxtamedullary and cortical nephrons. [C] Structure of a cortical nephron, including blood vessels. Arrows indicate the direction of fluid flow.

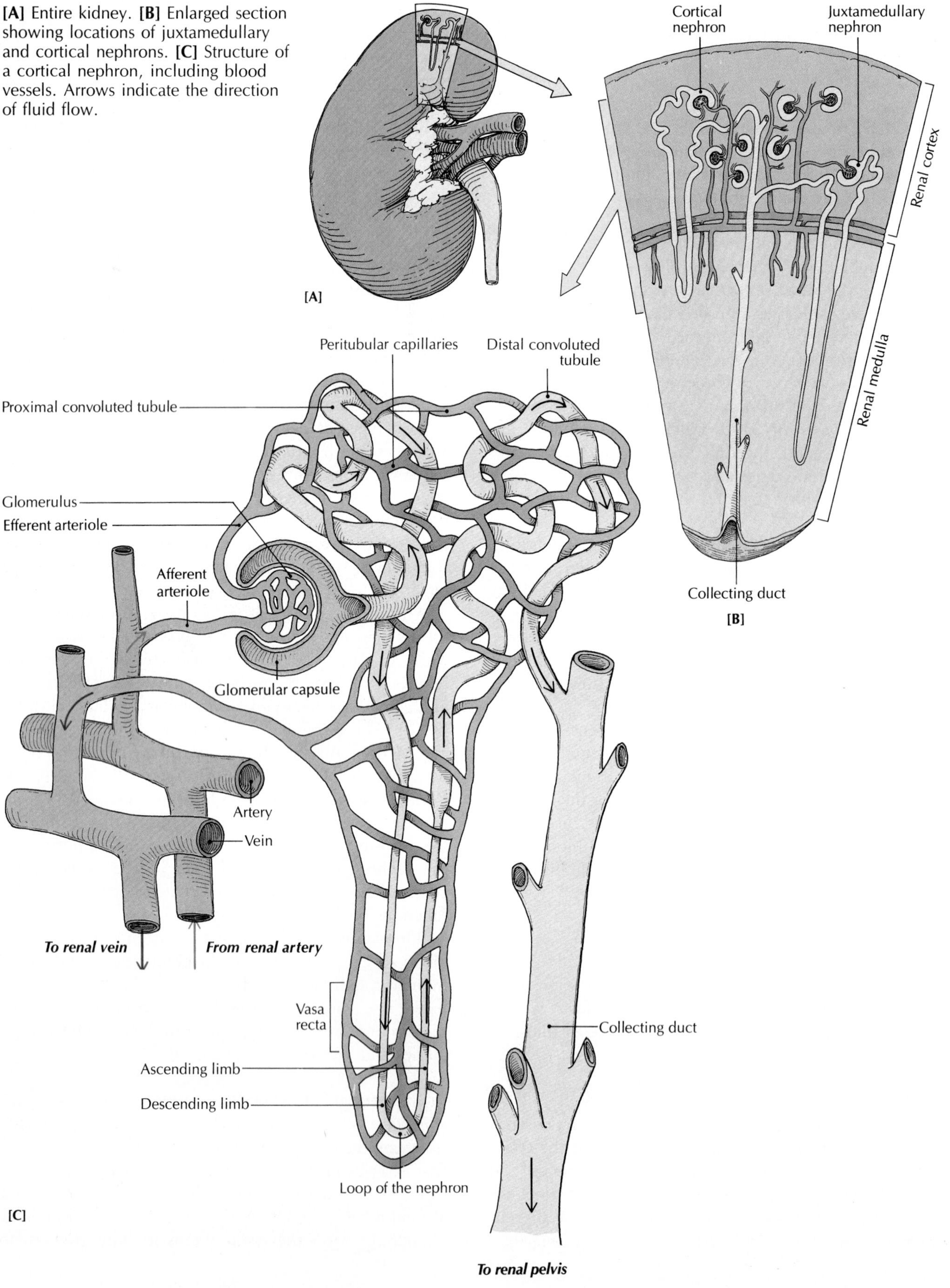

[A]

Cortical nephron

Juxtamedullary nephron

Renal cortex

Renal medulla

Collecting duct

[B]

Peritubular capillaries

Distal convoluted tubule

Proximal convoluted tubule

Glomerulus

Efferent arteriole

Afferent arteriole

Glomerular capsule

Artery

Vein

To renal vein

From renal artery

Vasa recta

Ascending limb

Descending limb

Loop of the nephron

Collecting duct

[C]

To renal pelvis

FIGURE 24.6 GLOMERULAR (BOWMAN'S) CAPSULE

[A] Drawing of the renal corpuscle. Arrows indicate direction of blood flow. [B] Enlarged drawing of part of a glomerular capillary. [C] Scanning electron micrograph showing podocyte cell processes on glomerular capillaries.

[D] Scanning electron micrograph showing a transected glomerular capillary loop surrounded by podocytes. ×17,200. [E] Electron micrograph of a portion of the wall of a glomerular capillary. ×34,000.

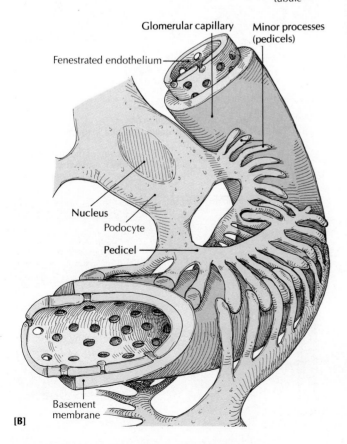

Juxtaglomerular cells
Endothelium
Afferent arteriole
Smooth muscle
Podocyte
Efferent arteriole
Capsular space
Glomerulus
Glomerular capillaries
Basement membrane
Glomerular capsule
Proximal convoluted tubule

[A]

Glomerular capillary
Minor processes (pedicels)
Fenestrated endothelium
Nucleus
Podocyte
Pedicel
Basement membrane

[B]

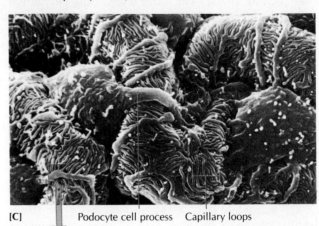

Podocyte cell process   Capillary loops

[C]

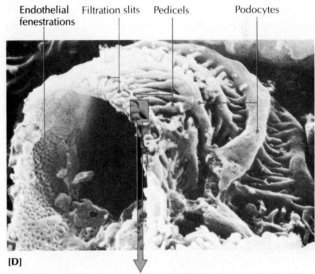

Endothelial fenestrations   Filtration slits   Pedicels   Podocytes

[D]

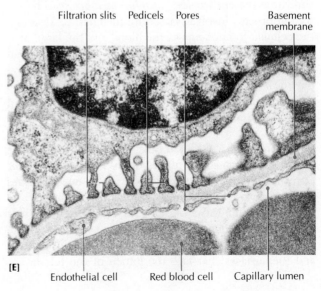

Filtration slits   Pedicels   Pores   Basement membrane

Endothelial cell   Red blood cell   Capillary lumen

[E]

capsule is always located in the cortex of the kidney and is the first part of the tubular component of a nephron. The inner and outer walls form a cavity called the *capsular space* [FIGURE 24.6A]. The outer layer of the glomerular capsule is the *parietal layer*, which is composed of simple squamous epithelial cells that have a basement membrane. The inner *visceral layer* is composed of specialized epithelial cells called *podocytes* (POH-doh-sites; "footlike cells"), which surround the glomerular capillaries [FIGURE 24.6B].

Filtration of the blood takes place in the renal corpuscle across three layers:

1 The first layer is the *endothelium* of the glomerular capillaries, which contains tiny pores called *fenestrations* ("windows").

2 The middle layer is the *basement membrane* (basal lamina) of the glomerulus. Like all capillary basement membranes, it is composed of fibrous proteins.

3 The third layer is the *visceral layer* of the glomerular capsule and the podocytes. The podocytes are relatively large, nucleated cells, with several cell processes that spread from the cell body. These processes eventually branch to form many smaller, fingerlike processes called *foot processes*, or *pedicels* (PEHD-ih-sehlz; L., "little feet") [FIGURE 24.6D]. The pedicels are the portions of the podocytes that are in contact with the glomerular capillaries. The small region between pedicels exposes the underlying basement membrane; it is called a *filtration slit*, or *slit pore* [FIGURE 24.6E]. A thin *slit membrane* extends between the foot processes of adjacent cells, forming a barrier (the *filtration barrier*) that restricts the passage of certain molecules through the filtration slit, but allows the passage of others. (The filtration barrier is actually made up of fenestrated endothelium, basement membrane, and filtration slits.)

Taken together, the three layers of the renal corpuscle constitute the *endothelial capsular membrane.* The filtration process involves the entire membrane. Although the cellular components of the blood and the large proteins do not normally pass through, water and dissolved solutes (electrolytes, sugars, urea, amino acids, and polypeptides) have no trouble passing from the blood into the capsular space. The fluid filtered from the blood is called the *glomerular filtrate.*

Proximal convoluted tubule From the glomerular capsule, the glomerular filtrate drains into the *proximal convoluted tubule* [FIGURE 24.5C]. The name describes its location (it is the portion of the excretory tubule *proximal* to the glomerular capsule) and its appearance (it is coiled in an irregular, *convoluted* way).

The epithelial cells making up the tubule are cuboidal, and the surfaces that face the lumen of the tubule are lined with microvilli, forming a brush border [FIGURE 24.7]. The microvilli enormously increase the epithelial surface area over which transport can take place. The proximal convoluted tubule is the site for the reabsorption of many substances filtered from the blood, such as water, salts, glucose, and some amino acids and polypeptides.

Loop of the nephron (loop of Henle) After passing through the proximal convoluted tubule, the glomerular filtrate enters a straightened portion of the excretory tubule called the *loop of the nephron (loop of Henle)* [FIGURE 24.5C], originally named for the nineteenth-century German anatomist Friedrich Henle. The loop of the nephron is composed of descending and ascending limbs of cuboidal epithelium connected by a thin limb of simple squamous epithelium. The *descending limb* extends

FIGURE 24.7 EXCRETORY TUBULE

Types of epithelial cells in different portions of the nephron.

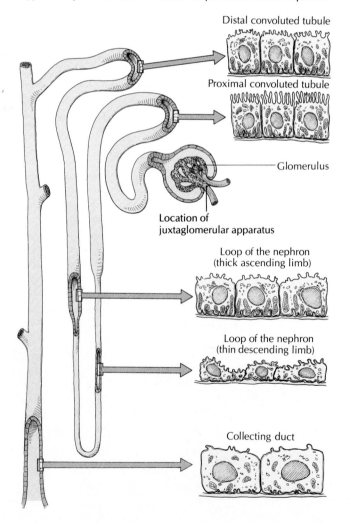

into the medulla of the kidney and becomes very narrow. At that point it makes an abrupt upward U-turn and widens again. This *ascending limb* passes through the medulla into the cortex.

Distal convoluted tubule After the glomerular filtrate passes through the entire length of the loop of the nephron, it moves into the ***distal convoluted tubule*** in the cortex [FIGURE 24.5C]. As the name suggests, it is an irregularly shaped tubule located far from the origin of the excretory tubules at the glomerular capsule. The cuboidal epithelial cells of the distal tubule are similar in size to those of the proximal tubule, but they have very few microvilli. The cells are abundantly supplied with mitochondria near their basal surfaces, where potassium and hydrogen ions are actively transported into the glomerular filtrate.

Collecting duct The glomerular filtrate passes from the distal convoluted tubule into the ***collecting duct*** [FIGURE 24.5]. Actually, the distal tubules of many nephrons

may empty into a single collecting duct, which travels through the medulla, roughly parallel to the limbs of the loop of the nephron. The collecting ducts join to form larger and larger tubes until they reach the minor calyx. From there, the final filtrate (now called urine) drains into the renal pelvis, which acts as a funnel that directs the urine into the ureter, and subsequently to the urinary bladder.

Juxtaglomerular apparatus In the cortex, the distal convoluted tubule makes intimate contact with the afferent and efferent arterioles of the glomerulus [FIGURE 24.8]. Here, the smooth muscle cells of the tunica media of the arterioles have cytoplasm that contains more granules than myofilaments. These specialized smooth muscle cells, known as ***juxtaglomerular cells,*** are in contact with a group of epithelial cells of the distal tubule called the ***macula densa*** (L. "dense spot"). The cells of the macula densa are longer and narrower than the epithelial cells of the distal tubule, and their nuclei are closer together.

FIGURE 24.8 JUXTAGLOMERULAR APPARATUS

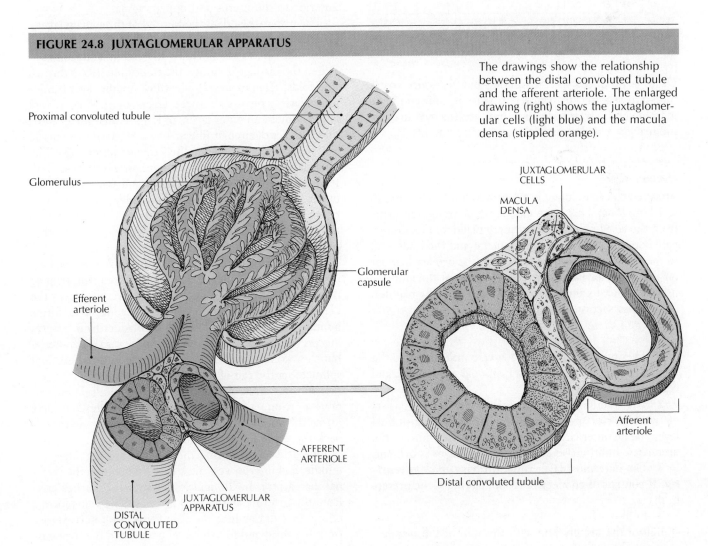

The drawings show the relationship between the distal convoluted tubule and the afferent arteriole. The enlarged drawing (right) shows the juxtaglomerular cells (light blue) and the macula densa (stippled orange).

Together, the juxtaglomerular cells, macula densa, and afferent and efferent arterioles make up the *juxtaglomerular apparatus.* This structure is the source of the enzyme *renin* that helps to regulate blood pressure.

ASK YOURSELF

1 What is the relationship between the renal pelvis and the calyces?

2 Where are the renal pyramids located?

3 What are the two regions of the renal cortex?

4 What is the glomerulus? The vasa recta?

5 What are the main functions of a nephron?

6 What is the glomerular capsule, and how is it related to the proximal convoluted tubule, loop of the nephron, distal convoluted tubule, and collecting duct?

7 What is the juxtaglomerular apparatus?

ACCESSORY EXCRETORY STRUCTURES

Urine is produced in the kidneys, but accessory structures are required to transfer it, store it, and eliminate it from the body. These structures are the two ureters, the urinary bladder, and the urethra [FIGURE 24.1].

Ureters

Attached to each kidney is a tube about 25 to 30 cm (10 to 12 in.) long. This is the *ureter,* which transports urine from the renal pelvis to the urinary bladder. The ureters pass between the parietal peritoneum and the body wall to the pelvic cavity, where they enter the urinary bladder on the posterior lateral surfaces. Narrow at the kidneys, ureters widen to about 1.7 cm (0.5 in.) near the bladder.

In cross section, the lumen of the ureter has a star shape, and three layers of tissue:

1 The innermost layer, the *tunica mucosa,* faces the lumen, and is made up of transitional epithelium and connective tissue.

2 The middle layer, the *tunica muscularis,* consists of two layers of smooth muscle—an inner longitudinal layer and an outer circular layer. In the lower third of the ureter, an additional outer longitudinal layer is present.

3 The outermost layer is the fibrous *tunica adventitia.* It consists of connective tissue that holds the ureters in place.

Before the ureters join with the wall of the bladder, they run obliquely for a short distance within the bladder wall. This oblique course functions as a check valve, preventing the backflow of urine from the bladder into the ureters; however, it does permit the flow of urine from the ureters into the bladder.

The ureters are supplied by branches from the renal, testicular, internal iliac, and inferior vesical arteries. There are no specific veins. The nerves are derived from the inferior mesenteric, testicular, and pelvic plexuses.

Urinary Bladder

The *urinary bladder* is the hollow, muscular sac that collects urine from the ureters and stores it until it is excreted from the body through the urethra. It usually accumulates 300 to 400 mL of urine before emptying, but it can expand enough to hold 600 to 800 mL. The bladder is located on the floor of the pelvic cavity, and like the kidneys and ureters, it is retroperitoneal. In males, it is anterior to the rectum and above the prostate gland [FIGURE 24.9]. In females, it is located much lower, anterior to the uterus and upper vagina.

The urinary bladder is composed of three main layers: (1) It is lined with *transitional epithelium* (the tunica mucosa) that allows the bladder to stretch and contract. When the bladder is empty, the tunica mucosa is thrown into folds called *rugae.* (2) The thick middle layer (tunica muscularis) consists of three layers of meshed *smooth muscle:* inner and outer layers of longitudinal fibers and a middle layer of circular fibers. The three layers of muscle are collectively known as the *detrusor muscle.* (3) The outer layer of the bladder is the *adventitia* (tunica serosa). It is derived from the peritoneum and covers only the upper and lateral surfaces of the bladder. It is a moist tissue that lubricates the upper surface, eliminating friction when the distended bladder presses against other organs.

The openings of the ureters and urethra into the cavity of the bladder outline a triangular area, the *trigone* [FIGURE 24.10]. At the site where the urethra leaves the bladder, the smooth muscle in the bladder wall forms bundles that contract to prevent the bladder from emptying prematurely. These bundles *function* as a sphincter, but they do not form a true anatomical internal urethral sphincter at the exit of the urethra from the urinary bladder. Farther along the urethra, in the middle membranous portion, a circular sphincter of skeletal muscle forms the voluntary *external urethral sphincter* that holds back the urine until urination is convenient.

The arteries supplying the bladder are the superior, middle, and inferior vesical. All are branches of the internal iliac artery. In the female, additional branches arise from the urterine and vaginal arteries. The veins form a plexus that drains into the iliac veins. The nerves arise from the third and fourth sacral nerves and the hypogastric plexus.

FIGURE 24.9 BLADDER AND URETHRA

[A] Sagittal section of the female pelvis showing the bladder and urethra. As the uterus enlarges during pregnancy, it pushes down on the bladder. **[B]** Sagittal section of the male pelvis. The longer male urethra also carries semen from the testes and accessory reproductive glands through the full length of the penis.

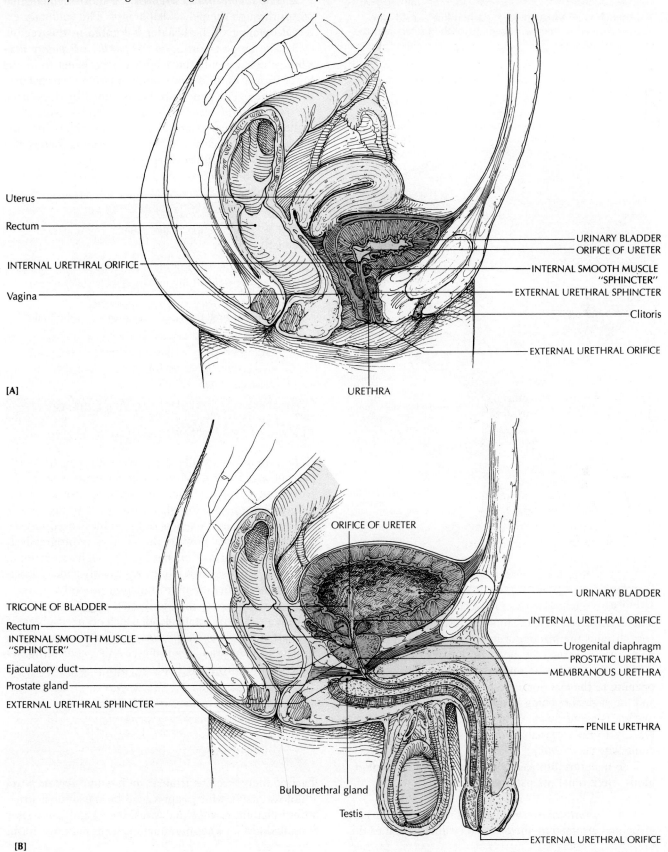

[A]

Uterus

Rectum

INTERNAL URETHRAL ORIFICE

Vagina

URINARY BLADDER
ORIFICE OF URETER

INTERNAL SMOOTH MUSCLE "SPHINCTER"

EXTERNAL URETHRAL SPHINCTER

Clitoris

EXTERNAL URETHRAL ORIFICE

URETHRA

[B]

ORIFICE OF URETER

TRIGONE OF BLADDER

Rectum

INTERNAL SMOOTH MUSCLE "SPHINCTER"

Ejaculatory duct

Prostate gland

EXTERNAL URETHRAL SPHINCTER

Bulbourethral gland

Testis

URINARY BLADDER

INTERNAL URETHRAL ORIFICE

Urogenital diaphragm
PROSTATIC URETHRA
MEMBRANOUS URETHRA

PENILE URETHRA

EXTERNAL URETHRAL ORIFICE

FIGURE 24.10 TRIGONE AND URETHRAL SPHINCTERS

The trigone is a triangular area (dashed lines) inside the bladder formed by the two openings of the ureters and the urethral orifice.

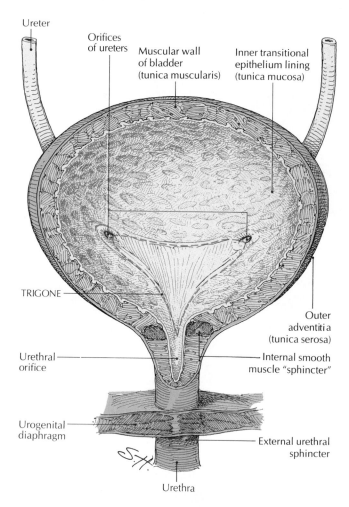

juncture, the male urethra serves both reproductive and excretory functions.

2 The *membranous urethra* is a short segment that passes through the pelvic diaphragm. The voluntary external opening of the bladder is located in this region.

3 The longest portion is the *penile*, or *spongy urethra*, which extends the length of the penis from the lower surface of the pelvic diaphragm to the external urethral orifice. The penile urethra is joined by ducts from the *bulbourethral glands* (Cowper's glands), which, together with the prostate gland, secrete fluids into the semen during ejaculation. (Mucus-secreting glands also empty into the urethra along its full length.)

ASK YOURSELF

1 What is the function of the ureters? The urinary bladder?

2 What is the function of the internal bundles of smooth muscle in the walls of the urinary bladder near the urethra?

3 What is the urethra? How does it differ in the male and female?

THE EFFECTS OF AGING ON THE URINARY SYSTEM

Kidney function usually decreases with age, from 100 percent at 25 years old to about 88 percent at 45, 78 percent at 65, and 69 percent at 85, as arteries to the kidneys become narrowed. *Kidney stones* (renal calculi) can reduce the effectiveness of the kidneys even further. The renal plasma flow to the kidneys is progressively reduced to about half by age 70. Although some kidney tissue may be lost, each kidney needs only about 25 percent of its original tissue to function properly.

Frequent urination in elderly men may be due to enlargement of the prostate gland, which occurs in about 75 percent of men over 55. *Incontinence* (lack of control over urination) may result when muscles that control the release of urine from the bladder become weakened.

DEVELOPMENTAL ANATOMY OF THE KIDNEYS

Each kidney develops from an embryonic *metanephros* ("hindkidney"), which is preceded by transitional structures called the *pronephros* ("forekidney") and *mesonephros* ("midkidney"). The nonfunctional pronephros forms

Urethra

The *urethra* is a tube of smooth muscle lined with mucosa. It joins the bladder at its lowest surface and transports urine outside the body during urination. In the female, the urethra is only about 4 cm (1.5 in.) long, opening to the exterior at an orifice between the clitoris and the vaginal opening [FIGURE 24.9A]. In the male, the urethra is about 20 cm (8 in.) long, extending from the bladder to the external urethral orifice at the tip of the penis [FIGURE 24.9B].

The male urethra passes through three different regions, after which its parts are named:

1 The *prostatic urethra* passes through the prostate gland and is joined by the ejaculatory duct. Distal to this

early in the fourth week after fertilization, and degenerates late in the fourth week [FIGURE 24.11A]. Most of the pronephric duct is used by the newly formed mesonephros, which appears caudally to the degenerated pronephros. The mesonephros possibly functions for about a month, although this is not certain.

Mesonephric ducts and tubules form late in the fourth week, and the metanephros begins to form early in the fifth week.* The metanephros develops from the *metanephric diverticulum* (ureteric bud) and the *metanephric mesodermal mass* (metanephrogenic blastema) [FIGURE 24.11A].

The metanephric diverticulum induces the formation of many important structures of the kidney. As the metanephric diverticulum extends dorsocranially, the meta-

---

*Although the mature kidneys are not yet developed, urine is formed throughout the fetal period. The urine mixes with the amniotic fluid surrounding the fetus. The fetus ingests amniotic fluid, and the urine is absorbed by the intestines. It is eventually deposited in the mother's bloodstream and eliminated in her urine.

nephric mesodermal mass forms a cap over it [FIGURE 24.11B]. This metanephric cap becomes larger, and internal differentiation occurs rapidly. At the same time, the stalk of the metanephric diverticulum and its extended cranial end form the *ureter* and *renal pelvis*, respectively [FIGURE 24.11C]. The pelvis then divides into the *major* and *minor calyces* [FIGURE 24.11D], with *collecting tubules* soon emerging from the minor calyces [FIGURE 24.11E]. The tubules branch repeatedly as they move closer to the outer edges of the metanephros.

At about 8 weeks, the metanephros begins to function as mesodermal cells become arranged in small vesicular masses at the blind end of the collecting tubules [FIGURE 24.11F]. These vesicular masses differentiate into a urine-carrying tubule that drains into its nearest collecting tubule. By week 13, the proximal portion of the S-shaped vesicular mass develops into the *distal* and *proximal convoluted tubules* and the *loop of the nephron;* the distal end of the vesicular mass becomes the *glomerulus* and *glomerular capsule* [FIGURE 24.11G]. The glomeruli are fully developed by week 36, about a month prior to birth.

FIGURE 24.11 EMBRYONIC DEVELOPMENT OF THE KIDNEY

The kidney develops from a primitive metanephros. The drawings show a progression of development from early in week 4 [A] to week 13 [G]; lateral views.

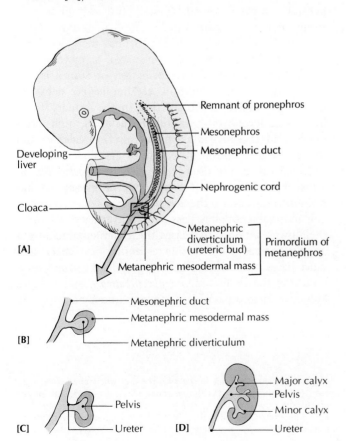
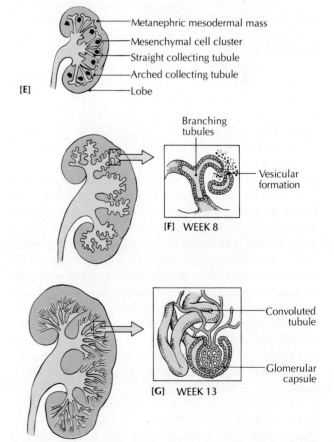

WHEN THINGS GO WRONG

Acute and Chronic Renal Failure

Acute renal failure is the total or near-total stoppage of kidney function. Little or no urine is produced, and substances that are normally eliminated from the body are retained. It is often caused by a diminished blood supply to the kidneys, which may be brought on by a serious blood loss due to an injury or hemorrhage, a heart attack, or a thrombosis. Another common cause of acute renal failure is a high level of toxic materials, such as mercury, arsenic, carbon tetrachloride, and insecticides, which build up in the kidneys. Other causes include obstruction (kidney stones, for example) and damage to the kidneys themselves. Recovery usually takes from 7 to 10 days and can be assisted by dialysis. Although complete recovery is not uncommon, residual kidney damage may lead to chronic renal failure.

In contrast to acute renal failure, *chronic renal failure* develops slowly and progresses over many years. Its most common causes are bacterial inflammation of the interstitial area and renal pelvis *(pyelonephritis)*, kidney inflammation involving the structures around the renal pelvis or glomeruli *(glomerulonephritis)*, and renal damage due to high blood pressure or obstructions in the lower urinary tract. The condition is characterized by progressive destruction of nephrons, which may lead to reduced amounts of urine, dilute urine, thirstiness, severe high blood pressure, poor appetite and vomiting, frequent urination, depletion of bone calcium, coma, and convulsions. A low-protein diet is usually prescribed, and dialysis may be necessary in some cases.

Glomerulonephritis

Acute glomerulonephritis (also known as AGN or Bright's disease) is an inflammation of the glomeruli that often follows a streptococcal infection. The body's normal response to this infection produces antibodies that cause an autoimmune inflammation in the glomerulus and damage the endothelial capsular membrane. Symptoms include presence of blood cells and plasma proteins in the urine (hematuria and proteinuria), and retention of water and salts. Most patients recover fully (especially children), but chronic renal failure is not an uncommon side effect.

Chronic glomerulonephritis (CGN) is a progressive disease that usually leads to renal failure. It is usually irreversible by the time it produces symptoms. The kidneys atrophy as the disease progresses, and recovery is unlikely.

Pyelonephritis

Acute *pyelonephritis* is one of the most common kidney diseases. It is caused by a bacterial infection, usually from the intestinal bacterium *Escherichia coli*, spreading from the bladder to the ureters, and then to the kidneys. The infection begins as an inflammation of the renal tissue between the nephrons and progresses to the glomeruli, tubules, and blood vessels. The disease occurs most often in females, probably because their urethras are short and closer to the rectal and vaginal openings. Symptoms include a fever of 44°C (101°F) or higher, back pain, increased leukocytes in the blood, painful urination *(dysuria)*, cloudy urine with an ammonia smell, and presence of bacteria in the urine. The disease is further complicated for diabetics, whose glucose in the urine *(glycosuria)* provides a ready energy source for bacterial growth.

Renal Calculi

Renal calculi are commonly called *kidney stones.* They may appear anywhere in the urinary tract, but are most common in the renal pelvis or calyx. The most usual type of calculus, which accounts for about 80 percent of all cases, forms from the precipitated salts of calcium (calcium oxalate and calcium phosphate). Kidney stones formed from calcium oxalate and trapped in the ureter are usually the ones that cause an intense, stabbing pain because of their jagged shapes.* Calcium phosphate stones, in contrast, grow quickly and may occupy a large part of the renal pelvis. Kidney stones may also form from the salts of magnesium, uric acid, or cysteine.

The exact cause of the formation of stones is not known, but several conditions are implicated: dehydration, renal infection, obstruction in the urinary tract, hyperparathyroidism, renal tubular acidosis, high levels of uric acid, excessive intake of vitamin D or calcium, and ineffective metabolism of oxalate or cysteine.

Renal calculi are usually revealed in a *pyelogram,* an x-ray of the kidney and ureters after an opaque dye has been introduced into the urinary system. Most renal calculi are small enough to pass out of the urinary system on their own, but in other cases the treatment includes a greatly increased water intake, antibiotics, analgesics, diuretics, a low-calcium diet, surgery, and *extracorporeal* ("outside the body") *shock-wave lithotripsy* that reduces kidney stones to passable particles without harming the

---

*Kidney stones may lodge in one of three sites where the ureters narrow: (1) at the junction with the renal pelvis, (2) where the ureter crosses the pelvic rim, and (3) where the ureters enter the urinary bladder. The ureter distends proximal to the lodged stone, resulting in intense pain.

body. Lithotripsy is usually effective in the removal of renal calculi less than 2 cm in diameter, which includes most stones. The characteristic extreme pain caused by the passage of stones is called *renal colic*, because it resembles intestinal (colic) pain.

Infection of the Urinary Tract

Two forms of a lower urinary tract infection are ***cystitis,*** inflammation of the bladder, and ***urethritis,*** inflammation of the urethra. Both are much more prevalent in women than in men; older men are usually affected as they begin to encounter prostate problems. Such inflammations may be trivial, but some urinary infections persist, causing permanent discomfort. The most common cause is an intestinal bacterium such as *Escherichia coli* or *Proteus mirabilis*, but some infections are caused by several different bacteria. Symptoms include frequent urges to urinate, spasms of the bladder, discharge from the penis in males, pain during urination, and excessive urination during the night *(nocturia)*.

A form of upper urinary tract infection is *pyelitis*, an inflammation of the renal pelvis and calyces.

Urinary Incontinence

The inability to retain urine in the urinary bladder and control urination is called ***urinary incontinence.*** Temporary incontinence may be caused by emotional stress. Permanent incontinence usually involves an injury to the nervous system, bladder infections, or tissue damage to either the bladder or urethra.

Kidney and Bladder Cancer

Kidney cancer occurs most often between the ages of 50 and 60 and is twice as prevalent in men as in women. Early symptoms are blood in the urine *(hematuria)*, pain in one side or the other, and a firm, painless growth of tissue. Kidney cancer often spreads, or *metastasizes*, to other parts of the body. Children under 6 or 7 may develop a variation of adult kidney cancer called *Wilms' tumor*, which causes the abdomen to swell noticeably. It is thought that Wilms' tumor originates in the embryo and then remains dormant for several years.

Bladder cancer appears most often in industrial cities, where such environmental carcinogens as benzidine, nitrates, tobacco smoke, and other chemical inhalants are common. Its symptoms resemble those of kidney cancer, and like kidney cancer, it develops most often in people over 50, especially in men.

Nephroptosis (Floating Kidney)

When a kidney is no longer held in place by the peritoneum, it usually begins to move to the abdominal area above. This condition is called ***nephroptosis,*** or floating kidney. The moving kidney sometimes twists its ureter, which may lead to blockage of urine. Nephroptosis occurs frequently among truck drivers, horseback riders, and motorcyclists.

Congenital Abnormalities of the Urinary System

Abnormalities of the urinary system occur in approximately 12 percent of all newborns. Some common abnormalities include the absence of a kidney *(renal agenesis)*, location of a kidney in the abdominal region *(renal ectopia)*, and fusion of the kidneys across the midline *(horseshoe kidney)*. When a kidney contains many cysts it is called a *polycystic kidney*. The most common abnormality of the urethra occurs when the male urethra opens on the ventral surface of the penis instead of the glans *(hypospadias)*. When the urethra fails to close on the dorsal surface of the penis, the condition is called *epispadias*.

Duplication of the ureters (two ureters from each kidney) occurs in about 1 in 200 births.

CHAPTER SUMMARY

Excretion, the elimination of metabolic waste products, is accomplished in part by the lungs, skin, and large intestine, but the prime regulator of water balance and waste elimination is the urinary system.

The Urinary System: Components and Functions (p. 622)

1 The *urinary system* consists of two kidneys, two ureters, the urinary bladder, and the urethra. Urine is formed in the kidneys, carried by the ureters, stored in the urinary bladder, and expelled through the urethra.

2 Each kidney contains over a million **nephrons,** the functional units that filter water and soluble substances from the blood, selectively reabsorb some of them back into the blood to maintain a proper balance, and selectively secrete wastes into the urine.

Anatomy of the Kidneys (p. 622)

1 The paired *kidneys* are retroperitoneal, located in the posterior part of the abdomen, lateral to the vertebral column.

2 The medial concave border of each kidney contains a *hilus,* an indented opening where blood vessels, nerves, and ureter join the kidney.

3 The innermost layer of the kidney is the *renal capsule;* the middle layer is the *adipose capsule;* the outer layer is the *renal fascia,* which attaches the kidney to the abdominal wall.

4 The kidney contains three regions: the innermost **renal pelvis,** which branches into the **major** and **minor calyces;** the middle **renal medulla,** consisting of several **renal pyramids,** which open into the calyces; and the outermost **renal cortex.**

5 Blood enters the kidney through the **renal artery,** which branches into **interlobar arteries, arcuate arteries, interlobular arteries,** and then **afferent arterioles,** which carry blood to the filtration site.

6 Each afferent arteriole branches extensively to form a ball of capillaries called a **glomerulus,** where filtration

starts. Glomerular capillaries join to form an **efferent arteriole,** which carries blood away from the glomerulus.

7 The efferent arteriole branches to form the **peritubular capillaries,** which unite to form the **interlobular veins, arcuate veins, interlobar veins,** and the **renal vein,** which carries waste-free blood from the kidney to the inferior vena cava.

8 *Vasa recta* are extensions of the efferent arterioles that provide the kidney with an emergency system to maintain blood pressure and urine concentration.

9 Each **nephron** is an independent urine-making unit. It consists of a *vascular component* (the glomerulus) and a *tubular component,* including a glomerular capsule, proximal convoluted tubule, loop of the nephron, distal convoluted tubule, and collecting duct. The **renal corpuscle** consists of the glomerulus and the **glomerular capsule.**

10 Filtration of blood takes place through the three layers of the renal corpuscle (constituting the **endothelial capsular membrane**) from the capillaries of the glomerulus into the glomerular capsule.

11 From the glomerular capsule, the fluid filtered from the blood **(glomerular filtrate)** moves into the **proximal convoluted tubule** where glucose, proteins, and certain other solutes filtered from the blood are absorbed.

12 The filtrate passes from the proximal convoluted tubule to the **loop of the nephron,** which is responsible for the reabsorption of water and the concentration of urine.

13 From the loop of the nephron, the filtrate moves into the **distal convoluted tubule.**

14 The filtrate moves from the distal tubule into the **collecting duct,** where the dilute filtrate is concentrated and passed on to the minor calyx.

15 The **juxtaglomerular apparatus** is made up of **juxtaglomerular cells** and the **macula densa.** The juxtaglomerular apparatus secretes the enzyme renin, which helps regular blood pressure.

Accessory Excretory Structures (p. 632)

1 The paired **ureters** carry urine from the renal pelvis of the kidney to the urinary bladder. Their tissue layers are the innermost **tunica mucosa,** the middle **tunica muscularis,** and the outermost **tunica adventitia.**

2 The muscular **urinary bladder** is an expandable sac that collects and stores urine until it is excreted. Its tissue layers resemble those of the ureter. At the site where the urethra leaves the bladder, involuntary smooth muscle in the bladder wall contracts to prevent the bladder from emptying prematurely, and a voluntary **external urethral sphincter** keeps the urine from leaving the urethra until it is time to urinate.

3 The **urethra** is the tube that transports urine from the bladder to the outside during urination. It is much longer in males than in females. The portions of the male urethra are the **prostatic, membranous,** and **penile** (spongy) **urethra.**

The Effects of Aging on the Urinary System (p. 634)

1 Kidney function usually decreases with age, as arteries to the kidneys become narrowed.

2 Elderly men may urinate more frequently as the prostate gland enlarges, and incontinence may occur when muscles that control urination are weakened.

Developmental Anatomy of the Kidneys (p. 634)

1 Each kidney develops from an embryonic **metanephros,** which begins to function at about week 8.

2 Urine is formed throughout the fetal period, even before the mature kidneys are developed. It is eliminated via the amniotic fluid.

3 The **proximal convoluted tubules** and the **loop of the nephron** develop by about week 13, and the **glomeruli** are fully developed by week 36.

MEDICAL TERMINOLOGY

ANURIA The absence of urine, such as when the ureters are obstructed.

CYSTECTOMY Surgical removal of the urinary bladder.

CYSTO- A prefix meaning "bladder."

CYSTOMETRY An examination of the bladder to evaluate its efficiency.

CYSTOSCOPY An examination of the interior of the bladder with a fiberoptic scope.

CYSTOURETHROGRAPHY X-ray examination, after introducing a contrast dye, to determine the size and shape of the bladder and urethra.

NEPHRECTOMY Surgical removal of a kidney.

NOCTURNAL ENURESIS Involuntary urination during sleep. Commonly called "bed wetting."

POLYURIA Frequent urination.

RENAL ANGIOGRAPHY Examination of renal blood vessels, using contrast dye injected into a catheter in the femoral artery or vein.

RENAL HYPERTENSION High blood pressure of the kidney.

RENAL INFARCTION Formation of a clotted area of dead kidney tissue as a result of blockage of renal blood vessels.

RENAL VEIN THROMBOSIS Blood clotting in the renal vein.

RENOVASCULAR HYPERTENSION A rise in systemic blood pressure as a result of blockage of renal arteries.

UREMIA A condition in which waste products that are normally excreted in the urine are found in the blood.

VESICOURETERAL REFLUX A backflow of urine from the bladder into the ureters and renal pelvis.

UNDERSTANDING THE FACTS

1 What makes up the renal pyramids?
2 What are the renal columns, and what is their composition?
3 Within the kidney, where does the actual filtration of blood occur?
4 What makes up the renal corpuscle?
5 What structural modifications in the proximal convoluted tubule fit it for its function of reabsorption?
6 Of the water filtered, what percentage is reabsorbed?
7 What portion of the nephron is the most important in the reabsorption of nutritionally important substances?
8 List the accessory organs of the excretory system.
9 What prevents the backflow of urine from the urinary bladder into the ureters?

10 Give the location of the following (be specific):
   a hilus   e ureters
   b renal plexus   f trigone
   c vasa recta   g juxtaglomerular apparatus
   d podocytes
11 What are some common causes of acute renal failure?
12 What is acute glomerulonephritis?
13 What is a "floating kidney"?

UNDERSTANDING THE CONCEPTS

1 Describe the location of the kidneys.
2 The fact that the efferent arteriole leading from the glomerulus is smaller than the afferent arteriole has considerable physiological significance. Explain.
3 What is the significance of the fact that the efferent arteriole leading from the glomerulus contains smooth muscle in its walls?
4 What role does the autonomic nervous system play in regulating kidney function?
5 Discuss the structure of the nephron.
6 Explain how the structure of the endothelial capsular membrane is related to its main function.

SELF-QUIZ
Multiple-Choice Questions

24.1 The apex of each renal pyramid ends in the
   a cortical region   d capsule
   b papilla   e tubule
   c juxtamedullary region

24.2 Approximately what percentage of the cardiac output goes to the kidneys?
   a 5 percent   d 30 percent
   b 10 percent   e 50 percent
   c 20 percent

24.3 Specialized epithelial cells in the visceral layer of the glomerular capsule are called
   a fenestrations   d endothelia
   b capsules   e none of the above
   c podoytes

24.4 The juxtaglomerular apparatus consists of
   a juxtaglomerular cells   d a and b only
   b the macula densa   e a, b, and c
   c arterioles

24.5 The innermost layer of the ureter is the
   a mucosa   d longitudinal layer
   b muscularis   e circular layer
   c adventitia

24.6 Which is true of kidney structure?

   a The nephron consists of both a vascular and a tubular component.

   b The glomerulus is involved in filtration, secretion, and reabsorption.

   c The renal pelvis is continuous with the collecting ducts and empties into the ureter.

   d The composition of urine in the ureter is the same as that in the urethra.

   e all but **b**

True-False Statements

24.7 The left kidney is slightly lower than the right one.

24.8 In the kidney, the lymphatic vessels accompany the large renal vessels and are more prominent around arteries than the veins.

24.9 The nerve supply to the kidney is almost entirely vasoconstrictor in function.

24.10 The kidney cannot function adequately without its nerve supply.

24.11 The ascending loop of the nephron is permeable to water.

Completion Exercises

24.12 The kidney's outer layer is subserous fascia called the _____.

24.13 The renal _____ is the large collecting space within the kidney.

24.14 The renal _____ is the outermost portion of the kidney.

24.15 Once inside the kidney, the renal artery branches into the _____ arteries.

24.16 The arterial rectae and venae rectae are collectively called the _____.

24.17 The three layers of the renal corpuscle together constitute the _____ membrane.

24.18 From the glomerular capsule, the glomerular filtrate moves into the _____ tubule.

24.19 The openings of the urethra and ureters into the cavity of the urinary bladder outline a triangular area called the _____.

24.20 The _____ "sphincter" is the more important one for the containment of urine in the urinary bladder.

A SECOND LOOK

1 In the following photograph, label the renal artery and veins, minor and major calyces, cortex, medulla, renal pelvis, and ureter.

2 In the following drawing, label the glomerulus, proximal convoluted tubule, vasa recta, collecting tube, and distal convoluted tubule.

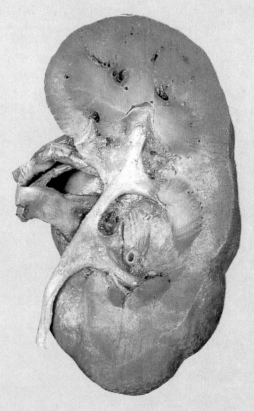

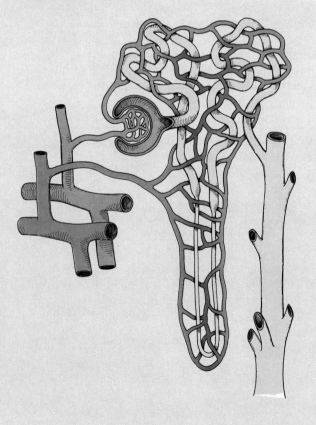

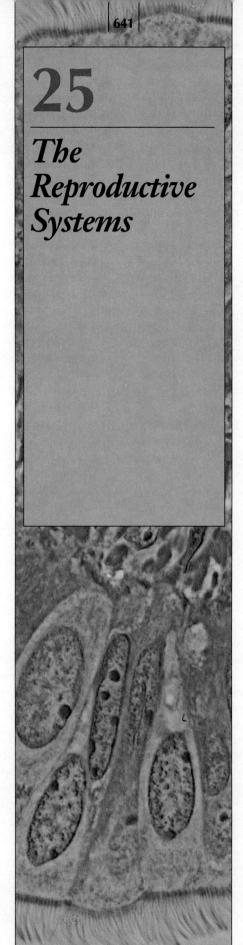

25

The Reproductive Systems

LEARNING OBJECTIVES

1 Name the principal male and female reproductive organs, and describe their structure and functions.

2 Describe the system of ducts and specialized cells within the testes, emphasizing their functional roles.

3 Describe the relationship of the male accessory ducts to one another, relating structure and function.

4 Explain the contribution of the three male accessory glands to the composition of semen.

5 Relate the structure of the penis to its functions.

6 Explain how specific structures in the ovaries are related to the production of ova and the secretion of female hormones.

7 Describe the gross and microscopic structure of the uterus, and explain its functions.

8 Describe the structure and functions of female accessory structures, including the vulva and mammary glands.

9 Outline the major steps in spermatogenesis and oogenesis, and compare them.

10 Explain the processes of fertilization and sex determination.

11 Discuss several of the most common sexually transmitted diseases.

Reproduction produces new human beings and allows hereditary traits to be passed from both parents to their children. Sexual reproduction always involves the union of two sex cells, an egg (ovum) from the mother and a sperm from the father. This allows the hereditary material (DNA) from both parents to combine, forming a new individual with a unique combination of genes.

In this chapter we discuss male and female reproductive anatomy. We also consider the formation of sperm and egg cells, and conception.

MALE REPRODUCTIVE ANATOMY

The reproductive role of the male is to produce sperm and deliver them to the vagina of the female. These functions require four different types of structures [FIGURE 25.1]:

1 The *testes* produce sperm and the primary male sex hormone, testosterone.

2 *Accessory glands* furnish a fluid for carrying the sperm to the penis. This fluid plus sperm is called *semen*.

3 *Accessory ducts* store and carry secretions from the testes and accessory glands to the penis.

4 The *penis* deposits semen into the vagina during sexual intercourse.

Testes

The paired **testes** (TESS-teez; sing., *testis*) are the male reproductive organs (*gonads*), which produce *sperm* [FIGURE 25.2]. Testes are often called *testicles* (from a Latin diminutive form of *testes*).

During fetal development, the testes are formed just below the kidneys inside the abdominopelvic cavity. By the third fetal month each testis has descended from its original site in the abdomen to the *inguinal canal*. During the seventh fetal month the testes pass through the inguinal canal [FIGURE 25.3]. The inguinal canal is a passageway leading to the **scrotum** (SCROH-tuhm; L. *scrautum*, a leather pouch for arrows), an external sac of skin that hangs between the thighs. The testes complete their descent into the scrotum shortly before or after birth. The inguinal canal is usually sealed off after the testes pass through. If the canal fails to close properly, or if the area is strained or torn, an *inguinal hernia* or *rupture* may result.

One testis (usually the left) hangs slightly lower than the other, so the testes do not collide during normal activities. Because the testes hang outside the body, their

temperature is about 3°F cooler than body temperature. The lower temperature is necessary for active sperm production and survival. The sperm remain nonviable if they are retained inside the body cavity. For the same reason, a fever can kill hundreds of thousands of sperm. In warm temperatures, the skin of the scrotum hangs loosely, and the testes are held in a low position. Sweat glands also help to cool the testes. In cold temperatures, the *cremaster* muscles under the skin of the scrotum contract, and pull the testes closer to the warm body. In this way, the temperature of the testes remains somewhat constant.

The interior of the scrotum is divided into two separate compartments by a fibrous *median septum*. One testis lies in each compartment. The line of the median septum is visible on the outside as a ridge of skin, the *perineal raphe* (RAY-fee; Gr. seam), which continues forward to the underside of the penis, and backward to the *perineum*, the diamond-shaped area between the legs and back to the anus.

Each testis is oval-shaped, weighs about 10 to 14 g, and measures about 4 to 5 cm (2 in.) long and 2.5 cm (1 in.) wide in an adult. It is enclosed in a fibrous sac called the **tunica albuginea** (al-byoo-JIHN-ee-uh; L. *albus*, white). This sac extends into the testis as *septae*, which divide it into compartments called *lobules* [FIGURE 25.2]. The tunica albuginea is lined by the *tunica vasculosa* and covered by the *tunica vaginalis*. The tunica vasculosa contains a network of blood vessels, and the tunica vaginalis is a continuation of the membrane that lines the abdominopelvic cavity.

Each testis contains over 800 tightly coiled **seminiferous tubules,** which produce thousands of sperm each second in a healthy young man. The combined length of the seminiferous tubules in both testes is about 225 m (750 ft).

The walls of the seminiferous tubules are lined with *germinal tissue*, which contains two types of cells, spermatogenic cells and sustentacular (Sertoli) cells. The **spermatogenic cells,** including spermatogonia, spermatocytes, and spermatids, eventually develop into mature sperm [FIGURE 25.2D]. The development of sperm, or **spermatogenesis,** is discussed on page 656. The **sustentacular cells** provide nourishment for the germinal sperm as they mature. Sustentacular cells also secrete a fluid into the tubules that provides a liquid medium for the developing sperm. Recent evidence suggests that as the

Q: *During a routine physical examination for a male, why does the physician insert a finger in the superficial inguinal ring under the scrotum and ask the patient to cough?*

A: Coughing exerts pressure on the internal abdominal organs, forcing them against the inguinal canal. A loose inguinal ring, felt by the physician, may indicate a developing hernia.

FIGURE 25.1 MALE REPRODUCTIVE ANATOMY

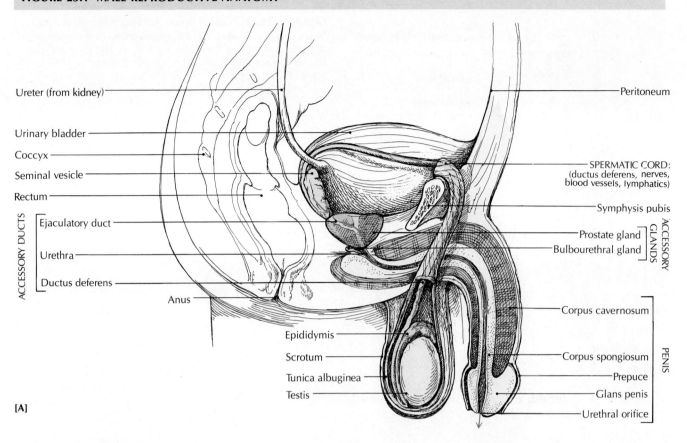

Ureter (from kidney)

Urinary bladder

Coccyx

Seminal vesicle

Rectum

ACCESSORY DUCTS
Ejaculatory duct

Urethra

Ductus deferens

Anus

Peritoneum

SPERMATIC CORD:
(ductus deferens, nerves,
blood vessels, lymphatics)

Symphysis pubis

ACCESSORY GLANDS
Prostate gland
Bulbourethral gland

PENIS
Corpus cavernosum

Corpus spongiosum

Epididymis

Scrotum

Tunica albuginea

Testis

Prepuce

Glans penis

Urethral orifice

[A]

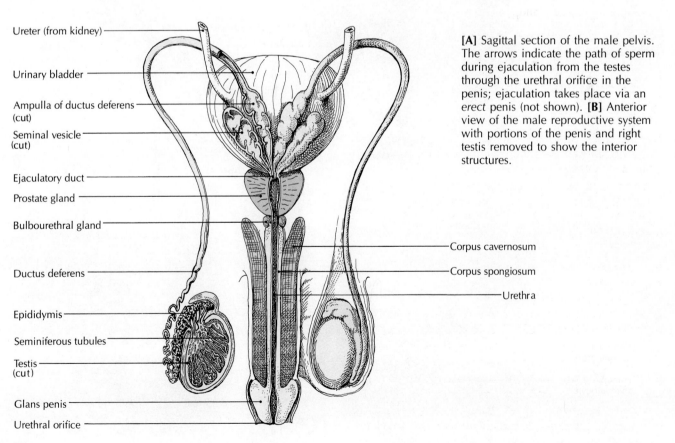

Ureter (from kidney)

Urinary bladder

Ampulla of ductus deferens
(cut)

Seminal vesicle
(cut)

Ejaculatory duct

Prostate gland

Bulbourethral gland

Ductus deferens

Epididymis

Seminiferous tubules

Testis
(cut)

Glans penis

Urethral orifice

Corpus cavernosum

Corpus spongiosum

Urethra

[A] Sagittal section of the male pelvis.
The arrows indicate the path of sperm
during ejaculation from the testes
through the urethral orifice in the
penis; ejaculation takes place via an
erect penis (not shown). [B] Anterior
view of the male reproductive system
with portions of the penis and right
testis removed to show the interior
structures.

[B]

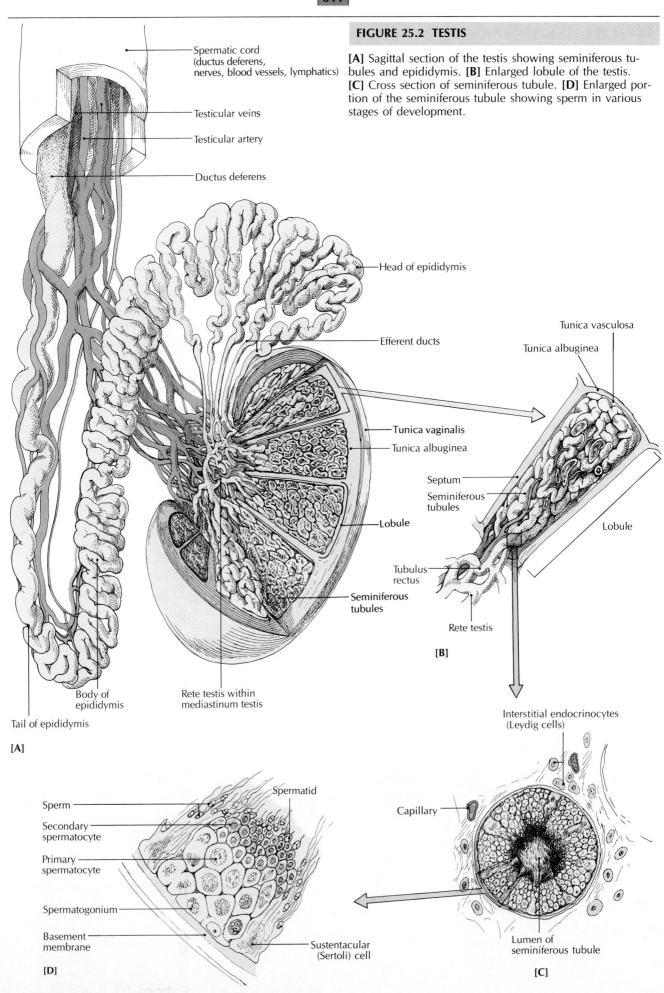

FIGURE 25.2 TESTIS

[A] Sagittal section of the testis showing seminiferous tubules and epididymis. [B] Enlarged lobule of the testis. [C] Cross section of seminiferous tubule. [D] Enlarged portion of the seminiferous tubule showing sperm in various stages of development.

Spermatic cord (ductus deferens, nerves, blood vessels, lymphatics)

Testicular veins

Testicular artery

Ductus deferens

Head of epididymis

Efferent ducts

Tunica vaginalis

Tunica albuginea

Lobule

Seminiferous tubules

Rete testis within mediastinum testis

Body of epididymis

Tail of epididymis

[A]

Tunica vasculosa

Tunica albuginea

Septum

Seminiferous tubules

Lobule

Tubulus rectus

Rete testis

[B]

Spermatid

Sperm

Secondary spermatocyte

Primary spermatocyte

Spermatogonium

Basement membrane

Sustentacular (Sertoli) cell

[D]

Interstitial endocrinocytes (Leydig cells)

Capillary

Lumen of seminiferous tubule

[C]

FIGURE 25.3 DESCENT OF THE TESTES

The testes descend from the abdominal cavity into the scrotum during fetal development. **[A]** Six-month fetus. **[B]** Seven-month fetus. **[C]** At birth.

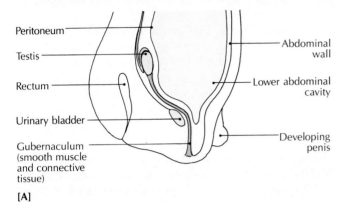

Peritoneum
Testis
Rectum
Urinary bladder
Gubernaculum (smooth muscle and connective tissue)
Abdominal wall
Lower abdominal cavity
Developing penis

[A]

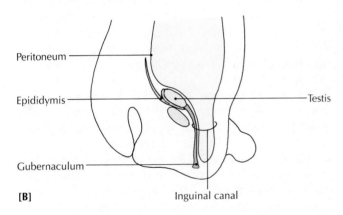

Peritoneum
Epididymis
Gubernaculum
Testis
Inguinal canal

[B]

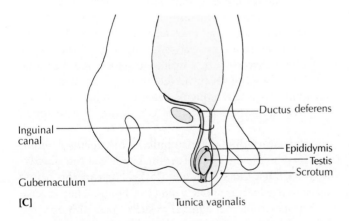

Inguinal canal
Gubernaculum
Ductus deferens
Epididymis
Testis
Scrotum
Tunica vaginalis

[C]

number of sustentacular cells decreases (usually after the age of about 60), the infertility rate increases proportionately.

Between the seminiferous tubules are clusters of endocrine cells called **interstitial endocrinocytes,** or *Leydig cells.* They secrete the male sex hormones, called **androgens,** of which **testosterone** is the most important.

Sperm (Spermatozoa)

Mature **sperm,** or **spermatozoa** (sing., spermatozoan; Gr. *sperma,* seed + *zoon,* animal), have a *head,* a *middle piece,* and a *tail* [FIGURE 25.4]. At the tip of the head is an *acrosome* (formed from the Golgi body), containing several enzymes that help the sperm penetrate an egg. In the center of the head is a compact *nucleus,* which contains the chromosomes (and therefore all the genetic material). The middle piece consists mainly of spiraling mitochondria, which supply ATP to provide the energy for movement. The beating movement of the undulating tail (flagellum) drives the swimming sperm forward.

A sperm is one of the smallest cells in the body. Its length from the tip of the head to the tip of the tail is about 0.05 mm. Although it appears to be structurally simple — basically a nucleus with a tail — each sperm requires more than two months for its complete development.

Normally, 300 to 500 million sperm are released during an ejaculation. A male who releases less than about 20 to 30 million normal sperm tend to be infertile. Men who produce a subnormal sperm count may still father children, as compared with men who are *sterile,* who do not produce viable sperm. Sterility may result from sexually transmitted diseases, which may interfere with sperm production, or by such diseases as mumps, which destroys the lining of the seminiferous tubules.

Sperm are constantly being produced in the seminiferous tubules, which always contain sperm in various stages of development. The final maturation of a sperm

FIGURE 25.4 SPERM

[A] Schematic drawing of a sperm showing its internal structure and parts, including the head, middle piece, and tail. **[B]** Cross section of tail.

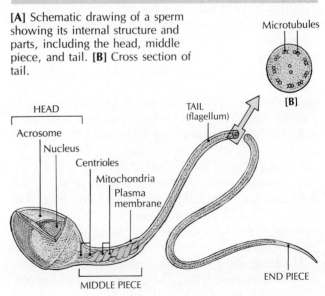

Microtubules

[B]

HEAD
Acrosome
Nucleus
Centrioles
Mitochondria
Plasma membrane
MIDDLE PIECE
TAIL (flagellum)
END PIECE

[A]

takes place in the epididymis lying on the surface of each testis. The epididymis and other accessory ducts and glands are described in the following sections.

Accessory Ducts

The sperm produced in the testes are carried to the point of ejaculation from the penis by a system of ducts. These ducts lead from the testes into the abdominopelvic cavity, where they eventually join the prostatic portion of the urethra just prior to reaching the penis.

Epididymis The seminiferous tubules merge in the central posterior portion of the testis [FIGURE 25.2A]. This area is called the *mediastinum testis* (L. "being in the middle of the testis"). The seminiferous tubules straighten to become the *tubuli recti* ("straight tubules"), which open into a network of tiny tubules called the *rete testis* (L. *rete*, net). The rete testis, at its upper end, drains into 15 to 20 tubules called the *efferent ducts (ductuli efferentes)*.

The efferent ducts penetrate the tunica albuginea at the upper posterior part of the testis; they extend upward into the convoluted mass of tubules that forms a crescent shape as it passes over the top of the testis and down along its side. This coiled tube is the *epididymis* (pl. *epididymides;* Gr. "upon the twin," the "twin" being the testis). The tightly coiled epididymis is bunched up along a length of only about 4 cm (1.5 in.), but if it were straightened out, it would extend about 6 m (20 ft). The epididymis has three main functions: (1) It stores sperm until they are mature and ready to be ejaculated. (2) It serves as a duct system for the passage of sperm from the testis to the ejaculatory duct. (3) It contains circular smooth muscle that helps propel mature sperm toward the penis by peristaltic contractions.

The efferent ducts are lined with alternating groups of ciliated tall pseudostratified columnar epithelium and nonciliated short columnar epithelium. The movement of the cilia helps to propel sperm toward the epididymis. (Sperm do not become motile until they enter the vagina.) The epididymis itself has a thinner muscular coat than the efferent duct and is lined with nonmotile cilia.

Each epididymis has a head, body, and tail. The *head,* which fits over the top of the testis, consists mostly of convoluted efferent ducts. The *body,* extending down the posterolateral border of the testis, consists mainly of the *ductus epididymis.* The *tail* extends almost to the bottom of the testis, where it becomes less and less convoluted until it finally dilates and turns upward as the *ductus deferens* [FIGURE 25.2A].

Maturing sperm leave the seminiferous tubules and move into the epididymis. The journey through the epididymis may take as little as 10 days or as long as 4 to 5 weeks, until the sperm are mature. While being stored,

the sperm are nourished by the lining of the epididymis. When the sperm are mature, they enter the ductus deferens on their way to the ejaculatory duct. Sperm can remain fertile in the epididymis and ductus deferens for about a month. If they are not ejaculated during that time, they degenerate and are resorbed by the body.

Ductus deferens The *ductus deferens* (pl., *ductus deferentia;* L. *deferre,* to carry away), formerly called the *vas deferens* or *sperm duct,* is the dilated continuation of the epididymis. It is about 45 cm (18 in.) long. The paired ductus deferentia extend between the epididymis of each testis and the ejaculatory duct [FIGURE 25.1]. As each one passes from the tail of the epididymis, it is covered by the *spermatic cord,* which contains the testicular artery, veins, autonomic nerves, lymphatics, and connective tissue from the anterior abdominal wall [FIGURE 25.2A]. Continuing upward after leaving the scrotum, the ductus deferens enters the lower part of the abdominal wall by way of the inguinal canal [FIGURE 25.3]. There, the ductus deferens becomes free of the spermatic cord and passes behind the urinary bladder, where it travels alongside an accessory gland called the *seminal vesicle* and becomes the *ejaculatory duct.*

Just before reaching the seminal vesicle, the ductus deferens widens into an enlarged portion called the *ampulla* [FIGURES 25.1B and 25.5]. Sperm are probably stored in the ampulla prior to ejaculation.

The ductus deferens is the main carrier of sperm. It is lined with pseudostratified columnar epithelium, and contains three thick layers of smooth muscle. Some of the sympathetic nerves from the pelvic plexus terminate on this smooth muscle. Stimulation of these nerves produces peristaltic contractions that move sperm forward, toward the ejaculatory duct near the base of the penis.

The thick coating of smooth muscle gives the ductus deferens a characteristic cordlike structure, and the initial portion of the duct can be felt through the skin of the scrotum. Because the ductus deferens can be located easily here, it is the site that is cut during a vasectomy.

Ejaculatory duct The ampulla of the ductus deferens is joined by the duct of the seminal vesicle at the *ejaculatory duct* [FIGURES 25.1 and 25.5]. Each of the paired ejaculatory ducts is about 2 cm (1 in.) long. They receive secretions from the seminal vesicles, pass through the prostate gland on its posterior surface, where they receive additional secretions, and finally join the single urethra.

Urethra The male *urethra* is the final section of the reproductive duct system. It leads from the urinary bladder, through the prostate gland, and into the penis [FIGURES 25.1 and 25.7]. Its reproductive function is to transport semen outside the body during ejaculation. As you

FIGURE 25.5 SEMINAL VESICLE

[A] Drawing of seminal vesicle, ductus deferens, and ejaculatory duct. **[B]** Scanning electron micrograph of a seminal vesicle, showing large outpockets and numerous folds of the mucosa. ×50.

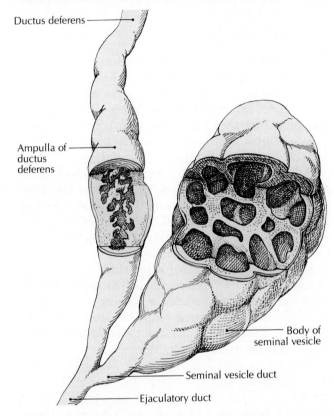

Ductus deferens

Ampulla of ductus deferens

Body of seminal vesicle

Seminal vesicle duct

Ejaculatory duct

[A]

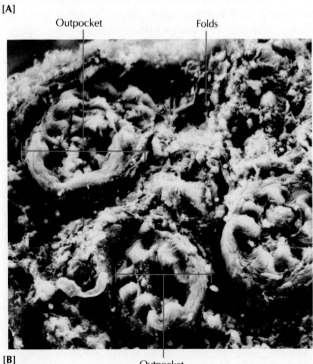

Outpocket   Folds

[B]   Outpocket

may remember, the male urethra also carries urine from the urinary bladder during urination. However, it is physically impossible for a man to urinate and ejaculate at the same time because just prior to ejaculation the internal urethral "sphincter" closes off the opening of the bladder. The muscle does not relax until the ejaculation is completed. The closing of this sphincter-like muscle also prevents the backflow of semen into the urinary bladder.

The male urethra consists of prostatic, membranous, and penile portions. The **prostatic urethra** (about 2.5 cm) starts at the base of the urinary bladder and proceeds through the prostate gland. Here it receives secretions from the small ducts of the prostate gland and the two ejaculatory ducts. Upon its exit from the prostate gland, the urethra passes through the urogenital diaphragm and is consequently called the **membranous urethra** (about 0.5 cm). The external urethral sphincter muscle is located here. The **penile,** or **spongy, urethra** (about 15 cm) extends the full length of the spongy portion of the penis to the external urethral orifice on the glans penis, where either urine or semen leaves the penis. The wall of the urethra has a lining of mucous membrane and a thick, outer layer of smooth muscle. Within the wall are *urethral glands*, which secrete mucus into the urethral canal.

Accessory Glands

After the ductus deferens passes around the urinary bladder, several accessory glands add their secretions to the sperm as they are propelled through the remaining ducts. These accessory glands are the seminal vesicles, prostate gland, and bulbourethral glands. The fluid that results from the combination of sperm and glandular secretions is *semen*. FIGURE 25.1 shows the major complement of accessory ducts and glands.

Seminal vesicles The paired **seminal vesicles** are secretory sacs that lie next to the ampullae of the ductus deferentia [FIGURES 25.1 and 25.5]. Their alkaline secretions, which provide the bulk of the seminal fluid, contain mostly water, fructose, prostaglandins, and vitamin C. The secretions are produced by the mucous membrane lining of the glands. The seminal vesicles are innervated by sympathetic nerves from the pelvic plexus. Stimulation during sexual excitement and ejaculation causes the seminal fluid to be emptied into the ejaculatory ducts by muscular contractions of the smooth muscle layers. The seminal fluid provides an energy source for the sperm and helps to neutralize the natural acidity of the vagina.

The wall of each seminal vesicle has an outer layer of connective tissue, a middle layer of smooth muscle, and a lining of mucosa. The mucosa contains foldings and outpockets that extend from a single folded and coiled tube. The folds provide a large area of cuboidal secretory epithelium.

Prostate gland The ***prostate gland*** lies inferior to the urinary bladder and surrounds the first 3 cm (1.2 in.) of the urethra [FIGURE 25.7]. It is a rounded mass about the size of a chestnut—about 4 cm (1.6 in.) across, 3 cm (1.2 in.) high, and 2 cm (0.8 in.) deep. The smooth muscles of the prostate can contract like a sponge to squeeze the prostatic secretions through tiny openings into the urethra. These secretions help make sperm motile and also help to neutralize vaginal acidity.

As the urethra leaves the bladder, it passes through the prostate, where it receives prostatic secretions continually. Some of these secretions are passed off with the urine, but most of them are released with the semen during ejaculation. The secretions are released when smooth muscle fibers in the wall of the prostate are stimulated by sympathetic nerves from the pelvic plexus. The ejaculatory ducts also pass through the prostate gland and receive its secretions during ejaculation.

The prostate is surrounded by a thin but firm capsule of fibrous connective tissue and smooth muscle. Inside, the prostate is made up of many individual glands, which release their secretions into the prostatic urethra through separate ducts [FIGURE 25.6]. There are three types of glands inside the prostate.

1 The inner *mucosal glands* secrete mucus. These small glands sometimes become inflamed and enlarged *(prostatitis)* in older men and make urination difficult by pressing on the urethra.

2 The middle *submucosal glands.*

3 The *main* (external) *prostatic glands* supply the major portion of the prostatic secretions. The secretion contains mainly water, acid phosphatase (an enzyme of unknown function), cholesterol, buffering salts, and phospholipids. Prostate cancer, one of the more common types of cancer, usually occurs only in the main glands.

Bulbourethral glands The paired ***bulbourethral glands*** (also called *Cowper's glands*) are about the shape and size of a pea. They lie directly below the prostate gland, one on each side of the undersurface of the urethra [FIGURES 25.1 and 25.7]. Each gland has a duct that opens into the penile part of the urethra.

With the onset of sexual excitement, bulbourethral glands secrete clear alkaline fluids into the urethra to neutralize the acidity of any remaining urine. These fluids also act as a lubricant within the urethra to facilitate the ejaculation of semen and to lubricate the tip of the penis prior to sexual intercourse.

FIGURE 25.6 PROSTATE GLAND

This cross section of the prostate gland shows the three types of inner glands and their relationships to the urethra.

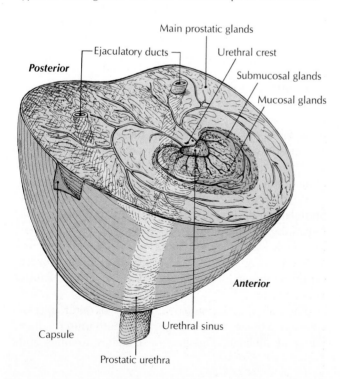

Semen

Secretions from the epididymis, seminal vesicles, prostate gland, and bulbourethral glands, together with the sperm, make up the ***semen*** (SEE-muhn; L. seed). Sperm make up only about 1 percent of the semen. The rest is fluid from the accessory glands, which provides fructose to nourish the sperm, an alkaline medium to help neutralize urethral and vaginal acidity that could otherwise inactivate the sperm, and buffering salts and phospholipids that make the sperm motile.

Semen is over 90 percent water, but contains many substances, most notably energy-rich fructose. The known vitamins it contains are vitamin C and inositol, and trace elements include calcium, zinc, magnesium, copper, and sulfur. The odor of semen is caused by amines (derivatives of ammonia), which are produced in the testes. The consistency of semen varies from thick and viscous to almost watery. The thinner consistency is

Q: *Is it possible for a woman to become pregnant if the man does not ejaculate during sexual intercourse?*

A: Yes, because secretions from the bulbourethral glands and epididymis (released after sexual excitement, but *before* ejaculation) may contain sperm.

usually the result of frequent ejaculations, but variations occur among men. The average ejaculation produces about 3 or 4 mL of semen (about a teaspoonful) and contains 300 to 500 million sperm.

Penis

The *penis* (PEE-nihss; L. tail) has two functions: (1) it carries urine through the urethra to the outside during urination, and (2) it transports semen through the urethra during ejaculation. In addition to the urethra, the penis contains three cylindrical strands of *erectile tissue*: two *corpora cavernosa,* which run parallel on the dorsal part, and the *corpus spongiosum,* which contains the urethra. The corpora cavernosa are surrounded by a dense, relatively inelastic, connective tissue called the *tunica albuginea.* The corpus spongiosum does not have such a surrounding layer. The corpora cavernosa contain numerous vascular cavities called *venous sinusoids* [FIGURE 25.7].

The corpus spongiosum extends distally beyond the corpus cavernosa, and becomes expanded into the tip of the penis, called the **glans penis.** Because the glans penis is a sensitive area containing many nerve endings, it is an important source of sexual arousal. The penile nerve endings are especially rich in the *corona,* the ridged proximal edge of the glans.

The loosely fitting skin of the penis is folded forward over the glans to form the **prepuce,** or **foreskin.** *Circumcision* is the removal, for religious or health reasons, of the prepuce. Today many circumcisions are performed in the belief that the operation may decrease the occurrence of cancer of the penis, although a controversy currently surrounds the practicality of circumcision.

Just below the corona, on each side of a ridge of tissue called the *frenum,* are the paired *Tyson's glands,* which are modified sebaceous glands. Their secretions, together with old cells shed from the glans and corona, form a cheeselike substance called *smegma.* An accumulation of smegma, which is more likely to occur in an uncircumcised male, may be a source of infection.

Ordinarily, the penis is soft and hangs limply. During tactile or mental stimulation, a parasympathetic reflex causes marked vasodilation of the arteries that contain the blood that flows into the sinusoids of the corpora cavernosa, which become engorged with blood under high pressure. The distended sinusoids and the increased pressure compress the veins that usually drain blood away, because of the resistance of the enveloping tunica albuginea. This resistance impedes the drainage of blood. In contrast, semen can pass through the urethra during ejaculation because the corpus spongiosum is not surrounded by the inelastic tunica albuginea. This dual action prevents the blood from escaping, and the penis becomes enlarged and firm in an **erection** [FIGURE 25.7B and C].

FIGURE 25.7 PENIS

[A] Coronal section of penis. [B] Cross section of flaccid penis. [C] Cross section of erect penis.

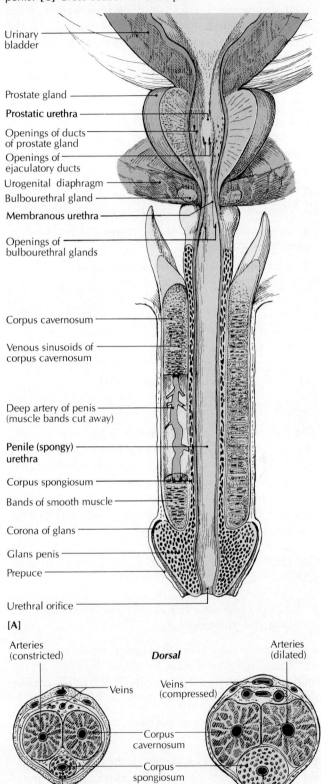

Urinary bladder

Prostate gland

Prostatic urethra

Openings of ducts of prostate gland

Openings of ejaculatory ducts

Urogenital diaphragm

Bulbourethral gland

Membranous urethra

Openings of bulbourethral glands

Corpus cavernosum

Venous sinusoids of corpus cavernosum

Deep artery of penis (muscle bands cut away)

Penile (spongy) urethra

Corpus spongiosum

Bands of smooth muscle

Corona of glans

Glans penis

Prepuce

Urethral orifice

[A]

Arteries (constricted)

Dorsal

Arteries (dilated)

Veins

Veins (compressed)

Corpus cavernosum

Corpus spongiosum

Urethra

[B]

Ventral

[C]

When sexual stimulation ceases, the penile arteries constrict, pressure on the veins is relieved as blood is drained away, and the penis returns to its nonerect, *flaccid* (FLACK-sihd) state. This return to the flaccid state is called *detumescence*. The flaccid dimension of a penis bears little relation to the erect dimension, and the smaller flaccid penis erects to a proportionally larger size than a large flaccid penis. The greatest variation in penis size occurs in the *girth* of the erect penis, not the length.

ASK YOURSELF

1 What are the basic reproductive functions of the male?

2 What are the main functions of the seminiferous tubules?

3 What are the three main functions of the epididymis?

4 How do the secretions of the seminal vesicles and prostate gland aid sperm?

5 What are the major constituents of semen?

6 What is the mechanism that produces an erect penis?

FEMALE REPRODUCTIVE ANATOMY

The reproductive role of the female is far more complex than that of the male. Not only do females produce egg cells (ova), but after fertilization they also nourish, carry, and protect the developing embryo. Then they usually nurse the infant for a time after it is born. Another difference between the sexes is the monthly rhythmicity of the female reproductive system.

The female reproductive system consists of a wide variety of structures with specialized functions [FIGURE 25.8]:

1 Two *ovaries* produce ova and the female sex hormones (estrogen and progesterone).

2 Two *uterine tubes*, also called *Fallopian tubes* or *oviducts*, one from each ovary, carry ova from the ovary to the uterus. Fertilization usually occurs in one of the uterine tubes.

3 The *uterus* houses and nourishes the developing embryo.

4 The *vagina* receives semen from the penis during sexual intercourse, is the exit point for menstrual flow, and is the canal through which the baby passes from the uterus during childbirth.

5 The *external genitalia*, collectively called the *vulva*, have protective functions and play a role in sexual arousal.

6 The *mammary glands*, contained in the paired *breasts*, produce milk for the newborn baby.

Ovaries

The female gonads are the paired *ovaries* (L. *ovum*, egg), which produce ova and female hormones [FIGURE 25.8]. These elongated, somewhat flattened bodies are about 2.5 to 5.0 cm (1.0 to 2.0 in.) long, 1.5 to 3.0 cm (0.6 to 1.2 in.) wide, and 0.6 to 1.5 cm (0.24 to 0.6 in.) thick—about the size and shape of an unshelled almond.

The ovaries are located in the pelvic part of the abdomen, one on each side of the uterus. Hanging almost free, the ovaries are attached by a mesentery called the *mesovarium* to the back side of the *broad ligament,* a double fold of peritoneum attached to the uterus. A thickening in the border of the mesovarium called the *ovarian ligament* extends from the ovary to the uterus. The mesovarium contains veins, arteries, lymphatics, and nerves to and from the *hilum* (opening) of the ovary. The ovary is suspended from the pelvic wall by the *suspensory ligament.*

The ovaries are covered by the *germinal layer,* a layer of specialized epithelial cells. Beneath it is the *stroma,* a mass of connective tissue that contains ova in various stages of maturity.

Follicles and corpus luteum A cross section of an ovary reveals a *cortex* and a vascular *medulla.* The cortex contains round epithelial vesicles called *follicles,* the actual centers of ovum production, or *oogenesis.* Each follicle contains an immature ovum called a *primary oocyte,* and follicles are always present in several stages of development [FIGURE 25.9]. The outer part of the cortex, directly beneath the epithelial coating, forms a zone of thin connective tissue called the *tunica albuginea.*

Follicles are classified as either primordial or vesicular (Graafian). *Primordial follicles* are not yet growing, while *vesicular follicles* are almost ready to release a secondary oocyte (commonly called an ovum) in the process called *ovulation.* Follicles are usually located directly beneath the cortex of the ovary, but once they begin to mature they migrate toward the inner medulla. The medulla consists of layers of soft stromal tissue, which contains a rich supply of blood vessels, nerves, and lymph vessels. Follicles secrete estrogen. After a secondary oocyte is discharged, the lining of the follicle grows inward, forming the *corpus luteum* ("yellow body"), temporary endocrine tissue that secretes female sex hormones.

The corpus luteum secretes *estrogen* and *progesterone,* which stop further ovulation and stimulate the thicken-

FIGURE 25.8 FEMALE REPRODUCTIVE ANATOMY

[A] Midsagittal section through the female pelvis. The arrows trace the path of an ovum from the ovary to the uterus and out of the vagina. **[B]** Anterior view of female reproductive system.

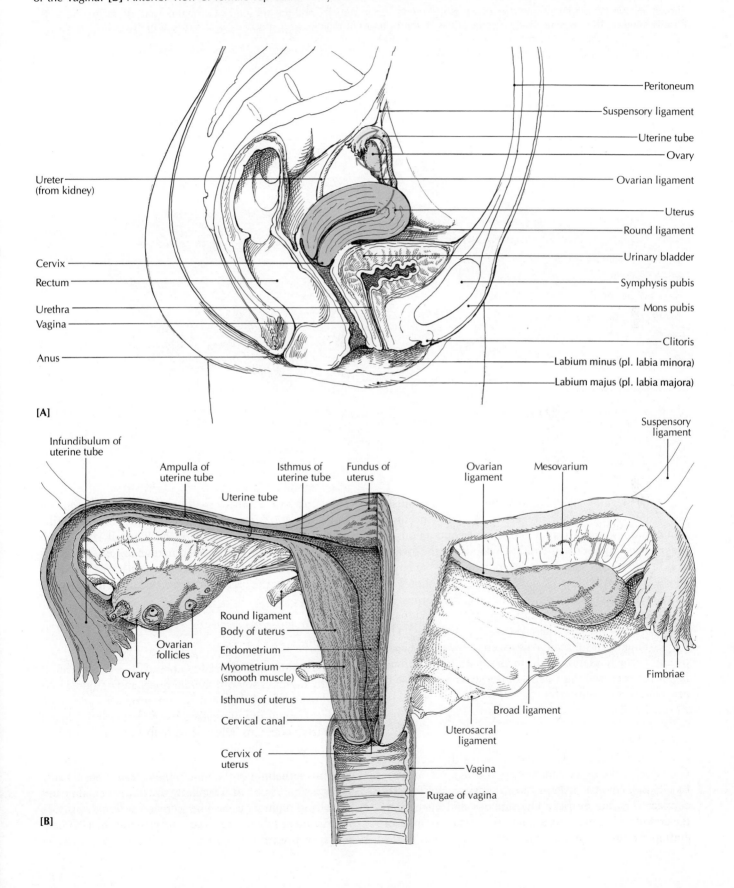

[A]

Ureter (from kidney)

Cervix

Rectum

Urethra

Vagina

Anus

Peritoneum

Suspensory ligament

Uterine tube

Ovary

Ovarian ligament

Uterus

Round ligament

Urinary bladder

Symphysis pubis

Mons pubis

Clitoris

Labium minus (pl. labia minora)

Labium majus (pl. labia majora)

[B]

Infundibulum of uterine tube

Ampulla of uterine tube

Isthmus of uterine tube

Uterine tube

Fundus of uterus

Ovarian ligament

Mesovarium

Suspensory ligament

Ovarian follicles

Ovary

Round ligament

Body of uterus

Endometrium

Myometrium (smooth muscle)

Isthmus of uterus

Cervical canal

Cervix of uterus

Uterosacral ligament

Broad ligament

Vagina

Rugae of vagina

Fimbriae

FIGURE 25.9 SCHEMATIC DIAGRAM OF AN OVARY

Shown are the development and eventual rupture of an ovarian follicle, and the subsequent formation and disintegration of a corpus luteum. The sequence of events begins with the primordial follicles and proceeds in the direction of the arrows.

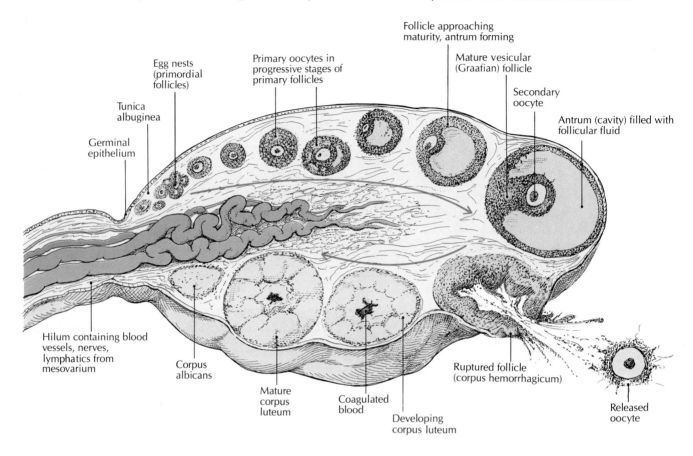

ing of the uterine wall and development of the mammary glands in preparation for pregnancy. A high concentration of progesterone also inhibits uterine contractions. If pregnancy does not occur within 14 days after the formation of the corpus luteum, it degenerates into the *corpus albicans*, a form of scar tissue. Menstruation follows almost immediately. If pregnancy occurs, the corpus luteum persists for 2 to 3 months, and then it eventually degenerates as the placenta takes over its activities. The corpus luteum and placenta also secrete *relaxin*. Relaxin promotes the relaxation of the birth canal and the softening of the cervix and ligaments of the symphysis pubis in preparation for childbirth.

Uterine Tubes

One of the paired ***uterine tubes*** receives the secondary oocyte from the ovary and conveys it to the uterus [FIGURE 25.8B]. The 10-cm-long (4-in.) tubes are not directly connected to the ovaries. The superior end opens into the abdominal cavity, very close to the ovary; the inferior end opens into the uterus. Each uterine tube lies in the

upper part of a transverse double fold of peritoneum called the *broad ligament*. Three distinct portions of the tube are the funnel-shaped ***infundibulum*** (L. funnel) near the ovary, the thin-walled middle ***ampulla,*** and the ***isthmus,*** which opens into the uterus [FIGURE 25.8].

The wall of the uterine tube is made up of three layers:

1 The outer *serous membrane* is part of the visceral peritoneum.

2 The middle *muscularis* is composed of an inner layer of spirally oriented smooth muscle fibers and an outer longitudinal layer. Hormonal action produces peristaltic contractions in the muscularis close to the time when a secondary oocyte is released to help move it down the uterine tube to the uterus.

3 The epithelium of the inner *mucous membrane* is made up of a single layer of irregularly distributed ciliated and secretory columnar cells. The secretory cells may provide nourishment for the secondary oocyte, and the cilia help propel it toward the uterus.

FIGURE 25.10 UTERINE TUBE

Cross section.

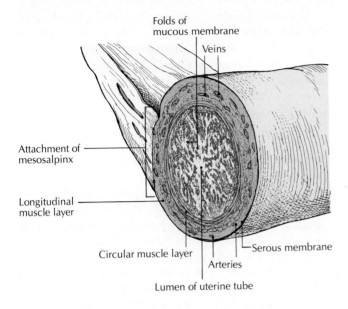

- Folds of mucous membrane
- Veins
- Attachment of mesosalpinx
- Longitudinal muscle layer
- Circular muscle layer
- Arteries
- Serous membrane
- Lumen of uterine tube

The infundibulum is fringed with feathery *fimbriae* (FIHM-bree-ee; L. threads), which may actually overlap the ovary. Each month as a secondary oocyte is released, it is effectively swept across a tiny gap between the tube and the ovary into the infundibulum by the motion of cilia in the fimbriae. Few oocytes are ever lost in the abdominal cavity. Apparently there is no particular pattern that selects one ovary over the other each month.

Unlike sperm, the secondary oocyte is unable to move on its own. Instead, it is carried along the uterine tube toward the uterus by the peristaltic contractions of the tube and the waving movements of the cilia in the mucous membrane [FIGURE 25.10]. Fertilization of the secondary oocyte usually occurs in the ampulla. An unfertilized oocyte will degenerate in the uterine tube. A fertilized oocyte (zygote) continues its journey toward the uterus, where it will become implanted. The journey takes four to seven days.

Occasionally, a fertilized ovum adheres to the wall of the uterine tube. This type of implantation is called an *ectopic pregnancy* (Gr. *ektopos*, out of place). Such a pregnancy is doomed because of a lack of nourishment and space to develop, and the embryo is discharged into the abdominal cavity through the opening of the ampulla or through a rupture in the wall of the uterine tube. This *miscarriage* is usually accompanied by hemorrhage. In some cases, surgery may be required to remove the implanted embryo, or to repair the ruptured uterine tube.

Uterus

The uterine tubes terminate in the *uterus* (L. womb), a hollow, muscular organ located in front of the rectum and behind the urinary bladder [FIGURE 25.8]. It is shaped like an inverted pear when viewed anteriorly, and it is pear-sized as well [about 7.5 cm (3 in.) long and 5 cm (2 in.) wide]. However, it increases three to six times in size during the 9 calendar months of pregnancy.

Just below the entrance of the uterine tubes are the *round ligaments* of the uterus. They help to keep the uterus tilted forward over the bladder. The uterus is attached to the lateral wall of the pelvis by two *broad ligaments,* which extend from the uterus to the floor and lateral walls of the pelvic cavity. Two *uterosacral ligaments* extend from the upper part of the cervix to the sacrum. The *posterior ligament* attaches the uterus to the rectum, and the *anterior ligament* attaches the uterus to the urinary bladder.

The wide upper portion of the uterus is called the *fundus.* The uterine tubes enter the uterus below the fundus [FIGURE 25.8B]. The tapering middle portion of the uterus is the *body,* which terminates in the narrow *cervix* (L. neck), the juncture between the uterus and the vagina. The constricted region between the body and the cervix is the *isthmus.* The *cervical canal,* the interior of the cervix, opens into the vagina. The uterus of a woman who is not pregnant is somewhat flattened, so that the interior of the uterus, or *uterine cavity,* is just a slit [FIGURE 25.11A]. In its usual position, the flattened uterus tilts forward (anteriorly) over the urinary bladder, at almost a right angle to the vagina.

The uterus is made up of three layers of tissue [FIGURE 25.11B]:

1 The outer *serosal layer* extends to form the two broad ligaments that stretch from the uterus to the lateral walls of the pelvis.

2 The middle, muscular layer, called the *myometrium* (Gr. *myo,* muscle + *metra,* womb), makes up the bulk of the uterine wall. It is composed of three layers of smooth muscle fibers. From the outside in, they are arranged longitudinally, randomly in all directions, and both longitudinally and spirally. The interweaving muscles contract downward during labor with more force than any other muscle. These muscles are capable of stretching during pregnancy to accommodate one or more growing fetuses; they also contract during a woman's orgasm. The myometrium is almost a centimeter thick, but becomes even thicker during pregnancy in preparation for childbirth.

3 The innermost layer of the uterus is composed of a specialized mucous membrane called the *endometrium,*

FIGURE 25.11 UTERUS

[A] Sagittal view. [B] Scanning electron micrograph of cross section of uterus. ×50.

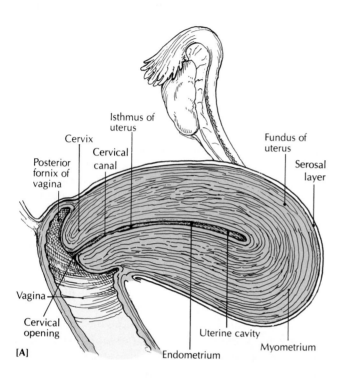

[A]

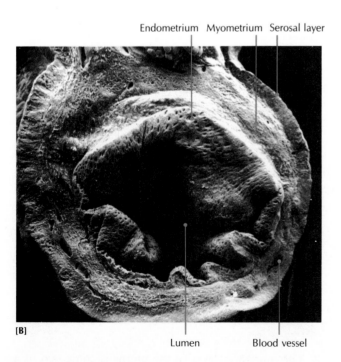

[B]

which is deep and velvety in texture. It contains an abundant supply of blood vessels and is pitted with simple tubular glands.

The endometrium is composed of two layers, the *stratum functionalis* and the *stratum basalis*. Every month, in response to estrogen secretion, the endometrium is built up in preparation for the possible implantation of a fertilized ovum (the beginning of *pregnancy*). Secretions of progesterone help the endometrium to develop into an active gland, rich in nutrients and ready to receive a fertilized ovum. If implantation does not occur, the *stratum functionalis* layer of the endometrium is shed together with blood and glandular secretions through the cervical canal and vagina. This breakdown of the endometrium makes up the **menstrual flow** (L. *mensis*, monthly), and the process is **menstruation.** The stratum basalis layer is permanent, and from it a new stratum functionalis regenerates after the 3- to 5-day menstrual period.

If fertilization and implantation do occur, the uterus houses, nourishes, and protects the developing fetus within its muscular walls. As the pregnancy continues, estrogen secretions develop the smooth muscle in the uterine walls in preparation for the expulsive action of childbirth.

Vagina

The uterus leads downward to the **vagina** (L. sheath), a muscle-lined tube about 8 to 10 cm (3 to 4 in.) long [FIGURE 25.8]. The vagina is the site where semen from the penis is deposited during sexual intercourse, the channel for the removal of menstrual flow, and the birth canal for the baby during childbirth. It lies behind the urinary bladder and urethra, and in front of the rectum and anus; it angles upward and backward.

The wall of the vagina is composed mainly of smooth muscle and fibroelastic connective tissue. It is lined with mucous membrane containing many folds called **rugae.** Stratified squamous nonkeratinized epithelium covers the vagina and also the cervix of the uterus. Ordinarily the wall of the vagina is collapsed, but the vagina can enlarge to accommodate an erect penis during sexual intercourse or the passage of a baby during childbirth.

The mucus that lubricates the vagina comes from glands in the cervix of the uterus. The acidic environment of the vagina, which helps prevent infection, is derived from the fermentation action of bacteria from the vaginal epithelium. The acidic environment is hostile to sperm, but alkaline fluids from male accessory sex glands help to neutralize vaginal acidity.

A fold of vaginal mucosa called the *hymen** (Gr. mem-

*Hymen was the mythical Greek god of marriage.

brane) partially blocks the vaginal entrance. The hymen is usually ruptured during the female's first sexual intercourse, but it may be broken earlier during other physical activities, or by the insertion of a tampon.

External Genital Organs

The *external genitalia* include the mons pubis, labia majora, labia minora, vestibular glands, clitoris, and vestibule of the vagina [FIGURE 25.12]. As a group, these organs are called the *vulva* (L. womb, covering).

The *mons pubis* (L. mountain + pubic), also called *mons veneris*, is a mound of fatty tissue that covers the symphysis pubis. At puberty, the mons pubis becomes covered with pubic hair. Unlike the pubic hair of a male, which may extend upward in a thin line as far as the navel, the upper limit of female pubic hair lies horizontally across the lower abdomen.

Just below the mons pubis are two longitudinal folds of skin, the *labia majora* ("major lips"; sing. *labium majus*), which form the outer borders of the vulva. They contain fat, smooth muscle, areolar tissue, sebaceous glands, and many sensory receptors. After puberty, their outer surface contains hairs.

The *labia minora* ("minor lips"; sing. *labium minus*) are two smaller folds of skin that lie between the labia majora. Together with the labia majora, they surround and protect the vaginal and urethral openings. The labia minora contain sebaceous glands and many blood vessels, but no hair or fat. They also contain many nerve endings and are sensitive to the touch. The labia merge at the top to form the *foreskin,* or *prepuce,* of the clitoris.

The *clitoris* (KLIHT-uh-rihss; Gr. small hill) is a small (less than 2.5 cm long), erectile organ at the upper end of the vulva, below the mons pubis, where the labia minora meet. Like the penis, the clitoris contains many nerve endings, and has two corpora cavernosa that can fill with blood during sexual stimulation, causing the clitoris to become enlarged. It is one of the major sources of sexual arousal. The clitoris is capped by a sensitive *glans,* but does not contain a urethra as the penis does.

The *vestibule* of the vagina is the space between the labia minora. Its floor contains the *greater vestibular glands* (Bartholin's glands) and the openings for the urethra and vagina. (Note that in the female, unlike the male, the urinary and reproductive systems are entirely separate.) Small *lesser vestibular glands* (Skene's glands) open by way of ducts into the anterior part of the vestibule between the urethral and vaginal orifices. During sexual excitement, the greater and lesser vestibular glands secrete an alkaline mucus solution that provides some lubrication and also offsets some of the natural acidity of the vagina. Most of the lubrication, however, comes from secretions of glands in the cervix.

The diamond-shaped *perineum* (per-uh-NEE-uhm) is the region bound anteriorly by the symphysis pubis, posteriorly by the inferior tip of the coccyx, and laterally by the ischial tuberosities. The anterior portion of the perineal diamond is called the *urogenital region (triangle)*, and the posterior portion is the *anal region (triangle)* [FIGURE 25.12]. The perineum exists in males also.

FIGURE 25.12 VULVA

Shown are the borders of the diamond-shaped perineum and the urogenital and anal triangles.

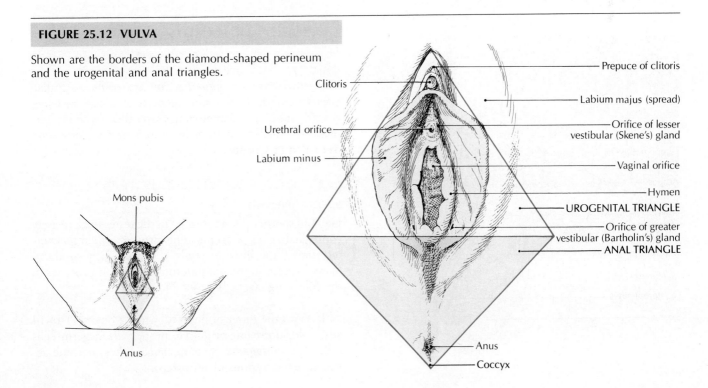

Mammary Glands

The *mammary glands* (L. *mammae,* breasts) within the paired breasts of a woman are modified sweat glands that produce and secrete milk. The breasts rest upon deep fascia covering the pectoralis major and minor muscles, each breast extending from the lateral border of the sternum to the middle of the axilla [FIGURE 25.13]. The breasts are held in place by *suspensory ligaments of the breast (Cooper's ligaments).* They contain varying amounts of adipose tissue. The amount of adipose tissue determines the size of the breasts, but the amount of mammary tissue does not vary widely from one woman to another.

Breasts are present in an underdeveloped form in children and men, but during puberty in the adolescent female the breasts begin their development as potential milk-producing organs. During pregnancy, secretions of estrogen and progesterone cause the mammary glands to develop further. The actual milk-producing hormone is *prolactin.*

Each mammary gland is composed of 15 to 20 lobes of compound areolar glands that radiate out from the nipple. These clusters of glands look like bunches of grapes. *Lactiferous* ("milk-carrying") *ducts* carry milk from the glands. The ducts from many glands converge into larger ducts. Just before a lactiferous duct reaches the nipple, it dilates into a *lactiferous sinus,* and then constricts as it enters the nipple. Milk is stored in the lactiferous sinuses.

The pigmented area around the nipple is the *areola.* It enlarges and darkens during pregnancy, and retains the darkened color permanently. The nipple consists of dense connective tissue and smooth muscle fibers, as well as many blood vessels and sensitive nerve endings that make it an important area of sexual stimulation.

The breasts contain an extensive drainage system made up of many lymph vessels. Because cells of malignant breast tumors may spread via the lymphatics, frequent self-examination of the breasts for lumps should be a routine practice in order to locate tumors before they spread.

ASK YOURSELF

1 What is an ovarian follicle?

2 What are the dual functions of ovaries?

3 What is the function of a corpus luteum?

4 In what part of the female reproductive system does fertilization usually take place?

5 What are the functions of the uterus? The vagina?

6 What is the vulva?

7 What is the function of the clitoris?

FIGURE 25.13 BREAST AND MAMMARY GLANDS

The mammary glands and areola are shown; right anterior view, partial midsagittal section.

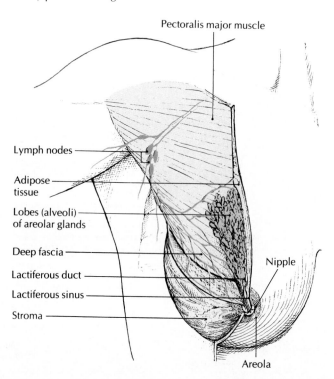

Pectoralis major muscle

Lymph nodes

Adipose tissue

Lobes (alveoli) of areolar glands

Deep fascia

Lactiferous duct

Lactiferous sinus

Stroma

Nipple

Areola

FORMATION OF SEX CELLS (GAMETOGENESIS)

The formation of *gametes,* or sex cells, is called *gametogenesis.* Male gametogenesis is called *spermatogenesis,* and female gametogenesis is called *oogenesis.* The phases of nuclear division in mitosis and meiosis were described in Chapter 3.

Spermatogenesis

Haploid sperm cells are formed in the testes in a precisely controlled series of events. This process, called *spermatogenesis* ("the birth of seeds"), continuously produces mature sperm in the seminiferous tubules. It proceeds in the following stages [FIGURE 25.14]:

1 With the onset of puberty, when a boy is 11 to 14 years old, dormant, primitive, unspecialized germ cells called *spermatogonia* (sing. *spermatogonium*) become activated by secretions of testosterone.

2 Each spermatogonium divides by *mitosis* to produce two daughter cells, each containing the full complement of 46 chromosomes. (The term "daughter cells" has nothing to do with gender. These cells could just as easily be called "offspring cells.")

3 One of the two daughter cells is a spermatogonium, which continues to produce daughter cells. The other daughter cell is a ***primary spermatocyte***, a large cell that moves toward the lumen of the seminiferous tubule.

4 The primary spermatocyte undergoes *meiosis* to produce two smaller ***secondary spermatocytes***, each with 23 chromosomes: 22 body cell chromosomes and one X or Y sex chromosome.

5 Both secondary spermatocytes undergo the second meiotic division to form four primitive germinal cells, the ***spermatids***, which still have only 23 chromosomes.

6 The spermatids develop into mature ***sperm*** without undergoing any further cell division. Each sperm has 23 chromosomes. The entire process of spermatogenesis takes about 64 days.

Oogenesis

Oogenesis (OH-oh-jehn-ih-sihss), the maturation of ova in the ovary, differs from the maturation of sperm in several ways. Usually only one ovum at a time matures each month, but billions of sperm may mature during that time. The process of cellular division differs too. Whereas one primary spermatocyte yields four sperm after meiosis, a primary oocyte yields only one ovum. Oogenesis proceeds through various stages [FIGURE 25.15].

FIGURE 25.14 SPERMATOGENESIS

Spermatogenesis in the seminiferous tubules.

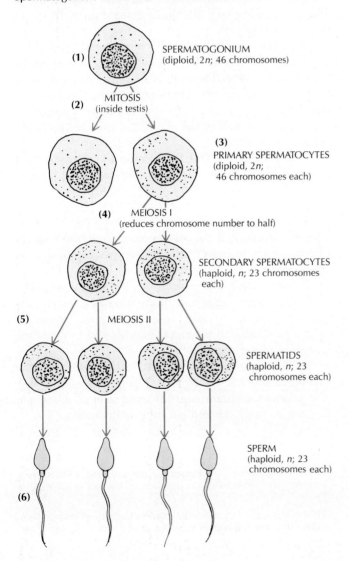

FIGURE 25.15 OOGENESIS

Oogenesis in the ovary and uterine tube.

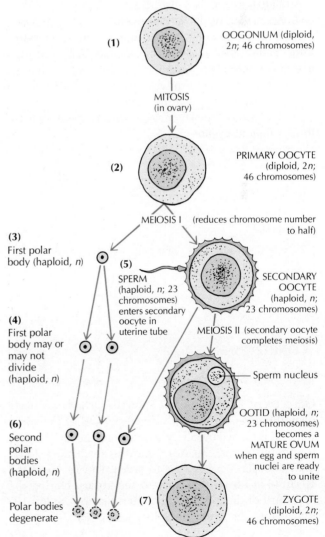

1 The *oogonium* (OH-oh-go-nee-uhm), the diploid precursor cell of the ovum, is enclosed in a follicle within the ovary.

2 The oogonium develops into a *primary oocyte* (OH-oh-site), which contains 46 chromosomes. The primary oocyte undergoes meiosis, which produces two daughter cells of unequal size.

3 The larger of the daughter cells is the haploid *secondary oocyte*. It is perhaps a thousand times as large as the other cell and contains most of the primary oocyte's cytoplasm, which provides nourishment for the developing ovum.

4 The smaller of the two daughter cells is the *first polar body*. It may divide again, but eventually it degenerates.

5 The large secondary oocyte leaves the ovarian follicle during ovulation and enters the uterine tube. If the secondary oocyte is fertilized, it begins to go through a second meiotic division, and a *second polar body* is "pinched off." It, too, is destined to die. If fertilization does not occur, menstruation follows shortly, and the cycle begins again.

6 As a result of the second meiotic division, the secondary oocyte has the haploid number of 23 chromosomes and is called an *ootid* (OH-oh-tihd). When the haploid sperm and ovum nuclei are finally ready to merge, the ootid is considered to have reached its final stage of nuclear maturity as a *mature ovum*.

7 The haploid nuclei of the ovum and sperm unite to form a diploid *zygote*.

A S K   Y O U R S E L F

1 What are the main steps in spermatogenesis? In oogenesis?

2 How does oogenesis differ from spermatogenesis?

CONCEPTION

Once every 28 days or so, a single secondary oocyte (ovum) emerges from a weakened portion of the ovary wall and is carried by ciliary action along the uterine tube toward the uterus. The ovum, moving much more slowly than the sperm, takes four to seven days to travel the 10 cm (4 in.) from the ovary to the uterus. During the first third of the journey, where fertilization is most advantageous, the ovum slows its pace.

Fertilization

The process of fertilization and the subsequent establishment of pregnancy are collectively referred to as *conception.* Fertilization must occur no more than 24 hours after ovulation, since the ovum will remain viable for only that period of time.* Sperm may remain viable in the female tract for up to 72 hours.†

During ejaculation, hundreds of millions of sperm enter the female's vaginal canal. If sexual intercourse takes place at about the same time as ovulation, some of these sperm travel toward the opposite-moving ovum, but only one sperm may eventually enter and fertilize the ovum. If only one sperm may enter, why are millions discharged? Most likely, more than enough sperm are discharged to increase the chances that one will successfully complete its mission. The sperm must travel from the vagina all the way to the uterine tubes, a trip of 15 to 20 cm that may take as long as a few hours. They must resist the spermicidal acidity of the vagina and overcome opposing fluid currents in the uterus and uterine tubes. So it is not surprising that the mortality rate of the sperm is enormously high.

It is also assumed that the quantity of sperm discharged is so large so that several hundred will reach the ovum and act together to provide sufficient amounts of *acrosin* [FIGURE 25.16]. This enzyme, which is found in the head of the sperm, helps to break down a portion of the outermost wall of the ovum, the *zona pellucida* (L. "transparent zone").

The penetration and fertilization of an ovum by a sperm take place in the following sequence [FIGURE 25.17]:

1 A sperm approaches the ovum and binds with it, as specific proteins on the plasma membrane of the sperm form a complementary "fit" with sperm-receptor sites on the surface of the zona pellucida.

2 Immediately after a sperm binds to the zona pellucida, the plasma membrane of the sperm's acrosome begins to break up, releasing vesicles of acrosin, which digest a path through the zona pellucida and facilitate the entrance of the sperm.

3 The exposed acrosome eventually fuses with the plasma membrane of the ovum. The head passes through the plasma membrane into the cytoplasm of the ovum, leaving its midpiece and tail outside the ovum.

---

*It is ironic that after lying dormant for anywhere from 10 to 50 years, an unfertilized ovum released from an ovary has only about a day to live.

† If semen is stored in a laboratory or sperm bank at very cold temperatures, the sperm will remain viable for years.

FIGURE 25.16 MEETING OF SPERM AND OVUM

Although the ovum is enormous compared with the sperm, they both contribute exactly the same amount of genetic material to the zygote. ×1500.

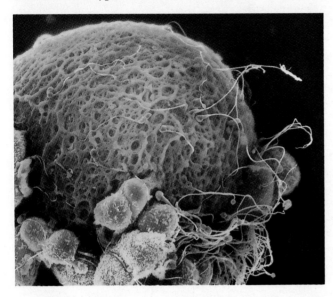

4 Cortical granules within the ovum release enzymes that thicken the zona pellucida and make it impenetrable, preventing other sperm from entering the ovum. At the same time, the sperm-binding sites on the zona pellucida become inactivated, providing further assurance that only one sperm will enter.

A few hours after the sperm penetrates the ovum, the ovum undergoes a cell division that produces two cells, each with 23 chromosomes. The smaller of the cells disintegrates, and the other cell, now called a haploid *pronucleus*, fuses with the nucleus of the sperm (now also called a diploid pronucleus). Fertilization is complete as a diploid zygote with 46 chromosomes is formed. Now embryonic development begins.

If more than one sperm somehow enters the ovum (a condition called *polyspermy*), the development of the zygote is quickly stopped. This cessation occurs because the extra mitotic spindles cause the abnormal segregation of chromosomes during cleavage.

Q: *Can women be allergic to sperm?*

A: Yes, some women are allergic to their partner's sperm, and in fact, scientists do not completely understand why the immune systems of *all* women do not reject sperm as foreign bodies. Some men are infertile because they produce antibodies against their own sperm, especially following vasectomy.

Human Sex Determination

A new individual begins its development with a full complement of hereditary material. Each parent supplies 23 chromosomes, giving the zygote 23 pairs. In the male and female sets of chromosomes, 22 always match in size and shape, and determine the same traits. The 22 matching pairs are called **autosomes**. The other pair are the **sex chromosomes,** and they determine the sex of the new individual. Because the pair of sex chromosomes look alike in females, they are designated **XX**. Because the sex chromosomes do *not* look alike in males, they are designated **XY**.

After meiosis in males, half of the sperm contains 22 autosomes and an X chromosome, and the other half contains 22 autosomes and a Y chromosome [FIGURE 25.18]. (After meiosis in the female, *all* ova contain 22 autosomes and an X chromosome.) If an ovum is fertilized by a sperm bearing an X chromosome, an XX zygote results, and the zygote will develop into a female. Fertilization of an ovum by a Y-bearing sperm produces an XY zygote, which will develop into a male. So the father actually determines the sex of the child, since only his sperm cells contain the variable, the Y chromosome. Also, it should be noted that the sex of the child is determined only at conception, never before or after.

ASK YOURSELF

1 Why is the mortality rate of sperm high after they enter the female system?

2 What is the zona pellucida?

3 What is acrosin?

4 How does the genetic material of the mother and father combine to make a new individual?

5 How is the sex of a new individual determined?

THE EFFECTS OF AGING ON THE REPRODUCTIVE SYSTEMS

As women grow older, the ovaries decrease in weight and begin to atrophy. Menstrual flow declines in frequency and quantity, and menopause usually occurs in the late 40s or early 50s. The vagina also decreases in length and width, its lining usually becomes less moist, and infections and vaginal discharges may result. Fibrous tissue becomes abundant, ovarian cysts are relatively common, and blood vessels harden. The uterus weighs about half

FIGURE 25.17 PENETRATION OF AN OVUM BY A SPERM

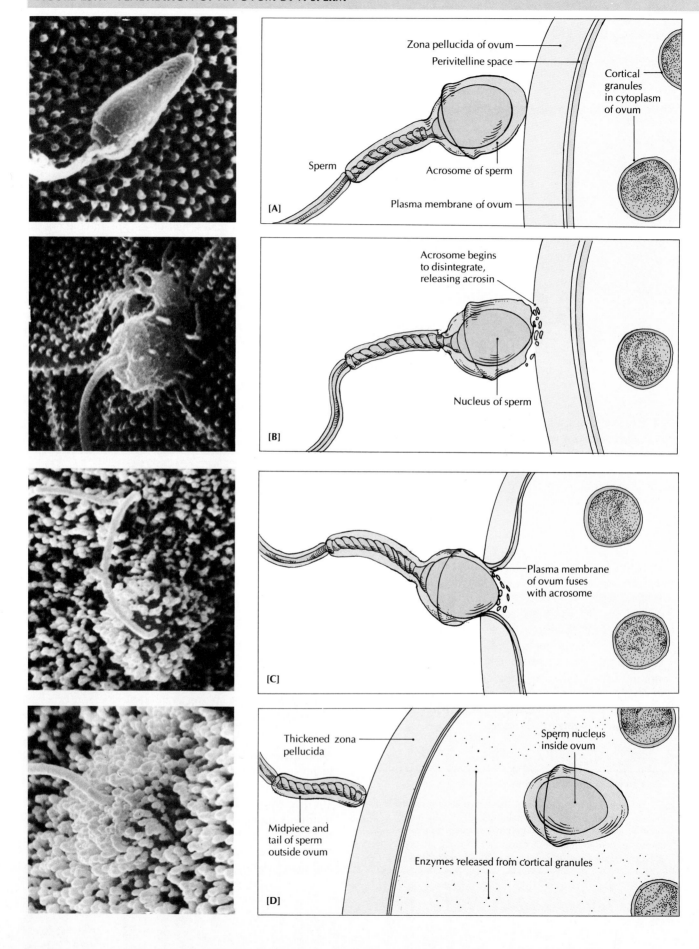

[A]
Zona pellucida of ovum
Perivitelline space
Cortical granules in cytoplasm of ovum
Sperm
Acrosome of sperm
Plasma membrane of ovum

[B]
Acrosome begins to disintegrate, releasing acrosin
Nucleus of sperm

[C]
Plasma membrane of ovum fuses with acrosome

[D]
Thickened zona pellucida
Sperm nucleus inside ovum
Midpiece and tail of sperm outside ovum
Enzymes released from cortical granules

Scanning electron micrographs and comparative drawings. [A] Acrosome of sperm touches the microvilli on the surface of the zona pellucida of an ovum, and binds with a sperm-receptor site. [B] The acrosome of the sperm begins to disintegrate, releasing acrosin, which allows the acrosome to move through the zona pellucida. [C] The acrosome fuses with the plasma membrane of the ovum, and the sperm head is engulfed by the plasma membrane. [D] The head is pulled into the cytoplasm of the ovum, leaving the midpiece and tail outside. Cortical granules within the ovum release enzymes that make the zona pellucida impenetrable to other sperm. Eventually, the sperm and ovum will fuse to form a zygote.

FIGURE 25.18 SEX DETERMINATION

The sex of a child is determined by the type of sperm entering the ovum. An X-bearing sperm fertilizing an ovum (all of which ova have an X chromosome) produces an XX zygote, which develops into a female. A Y-bearing sperm fertilizing an ovum produces an XY zygote, which develops into a male.

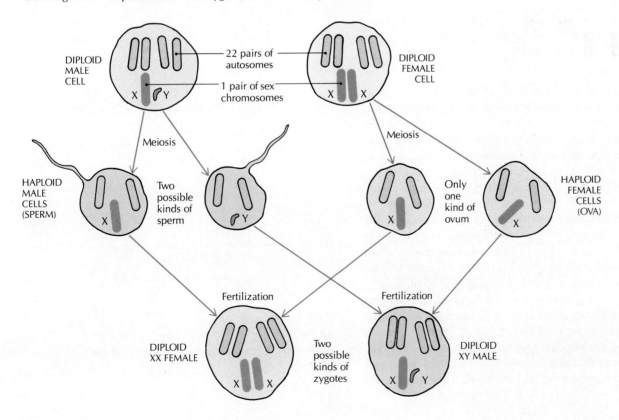

as much at age 50 as it did at 30, and its elasticity is lost as elastic tissues are replaced by clumps of fibrous tissue. Ligaments that hold the uterus, urinary bladder, and rectum in place may weaken in older women and allow these organs to drop down. A *cystocele* is a protrusion of the urinary bladder into the vagina, *prolapse* of the uterus is its protrusion into the vagina, and a *rectocele* is a protrusion of the rectum that presses on the vagina.

Hormonal levels of estrogen drop at menopause. The breasts sag and become flattened as ligaments supporting the breasts become lax, fibrous cells replace milk glands and fat tissue is lost. The amount of pubic hair gradually decreases, as does the layer of fat under the skin in the pubic region.

Reproductive changes due to age are less evident in men than in women. Although the testes do not necessarily decrease in size and weight, the specialized cells that produce and nourish sperm cells gradually become fewer and less active. Sperm count gradually decreases to about 50 percent, but fertility is frequently retained past the age of 80. Accessory glands and organs such as the seminal vesicles and prostate gland begin to atrophy, and the secretion of testosterone decreases. Sexual desire in men and women is usually not affected by old age.

Certain cancers become more prevalent in elderly men and women. Women over 60 are more likely to get cancer of the breast or uterus than younger women, and men over 60 are particularly prone to cancer of the prostate gland.

DEVELOPMENTAL ANATOMY OF THE REPRODUCTIVE SYSTEMS

Until about week 10 after fertilization, the external genitalia of the male fetus do not differ greatly from those of the female [FIGURE 25.19A]. By week 12 in the female, the bud that derived from the genital tubercle is beginning to develop into the clitoris, the genital fold begins to develop into the labia, and the urogenital opening begins to divide into the separate openings for the urethra and vagina. By week 12 in the male, the bud is clearly developing into the penis, and the genital fold fuses over the urogenital opening to become the scrotum. The testes begin to descend into the scrotum when the fetus is about 7 months old [FIGURE 25.3B]. At week 34, male and female genitals look very much as they will at birth.

The male and female internal reproductive systems are still undifferentiated at week 8, but have undergone distinct changes by week 10 [FIGURE 25.19B]. The undifferentiated gonads develop into the testes and ovaries, the mesonephric ducts become the male epididymis and the female uterine tube, and male seminal vesicles begin to form.

FIGURE 25.19 DEVELOPMENT OF REPRODUCTIVE ORGANS

[A] Development of the external genitalia. **[B]** Development of the internal reproductive systems.

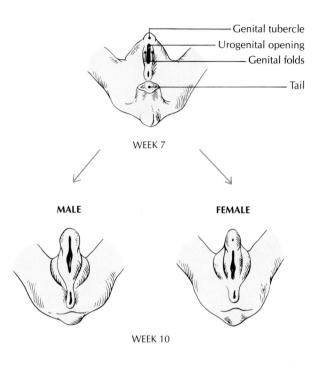

WEEK 7

MALE   FEMALE

WEEK 10

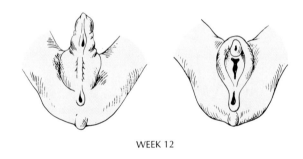

WEEK 12

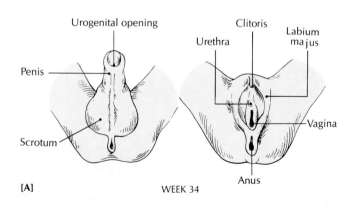

[A]   WEEK 34

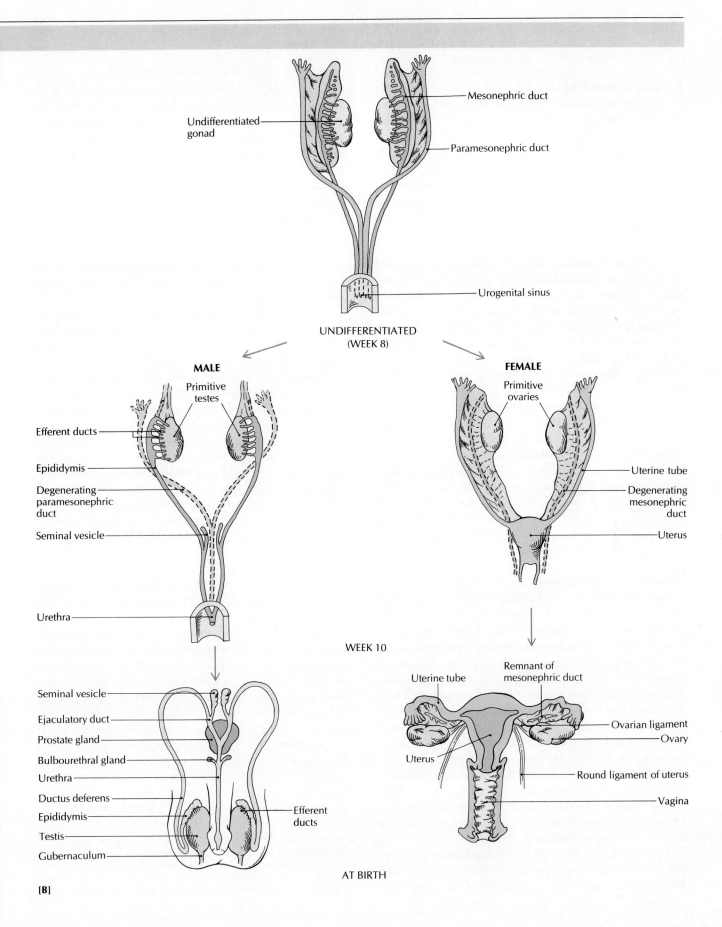

Mesonephric duct

Undifferentiated gonad

Paramesonephric duct

Urogenital sinus

UNDIFFERENTIATED
(WEEK 8)

MALE

Primitive testes

Efferent ducts

Epididymis

Degenerating paramesonephric duct

Seminal vesicle

Urethra

FEMALE

Primitive ovaries

Uterine tube

Degenerating mesonephric duct

Uterus

WEEK 10

Seminal vesicle

Ejaculatory duct

Prostate gland

Bulbourethral gland

Urethra

Ductus deferens

Epididymis

Testis

Gubernaculum

Efferent ducts

Uterine tube

Remnant of mesonephric duct

Ovarian ligament

Ovary

Uterus

Round ligament of uterus

Vagina

AT BIRTH

[B]

WHEN THINGS GO WRONG

Sexually Transmitted Diseases

Some microorganisms and agents are transferred from person to person mainly by sexual contacts. The diseases they cause are known as *sexually transmitted diseases* (STD), or *venereal* ("of Venus") *diseases* (VD). Contrary to some beliefs, these diseases are rarely contracted by casual, dry contact with persons or objects. The most common sexually transmitted diseases are *nongonococcal urethritis* (NGU), caused by a bacterium; *type II herpes simplex*, caused by a virus; and *gonorrhea* and *syphilis*, both caused by bacteria.

The most common venereal disease in the United States is *nongonococcal urethritis (NGU),* or *chlamydial disease,* named for the bacterium that causes it, *Chlamydia trachomatis.* Although it afflicts between 3 and 10 million Americans every year, NGU is not well known because its symptoms are difficult to diagnose, and the disease is easily confused with gonorrhea.

NGU in men produces infection in the urethra, prostate gland, and epididymis. If left untreated, it can cause sterility. Women are afflicted with a cervical infection and inflammation of the uterine tubes, which, if untreated, can result in scarring, sterility, or ectopic pregnancy. Women who acquire the disease during pregnancy may transmit it to their babies as they pass through the birth canal. Affected infants may have conjunctivitis or pneumonia, and there is some evidence that the risk of premature or stillborn infants is increased. Both men and women may be affected by conjunctivitis, pneumonia, and enlarged lymph nodes in the groin. The preferred treatment is an extended dose of erythromycin until the infection is gone, usually in about a week.

Type II herpes (Gr. "a creeping") *simplex* (also called herpes genitalis) ordinarily appears as blisterlike sores on or near the external genitalia about a week after intercourse with an infected partner. Fever, muscle aches, and swollen lymph nodes may also be apparent. When the blisters rupture, they produce painful ulcers and release millions of infectious viruses. The blisters usually heal after a week or two, but the viruses retreat to nerves near the lower spinal cord, where they remain dormant until the next attack. Infected people may harbor the infectious viruses for months, years, or a lifetime.

Perhaps the most serious complication of type II herpes simplex is the infection of a baby as it passes through the birth canal. For this reason, if a pregnant woman is known to have type II herpes simplex, the baby is usually delivered by cesarean section.

Gonorrhea (Gr. "flow of seed") is primarily an infection of the urinary and reproductive tracts (especially the urethra and cervix), but any moist part of the body, espe-

cially the eyes, may suffer. The causative bacterium, *Neisseria gonorrhoeae,* can be recognized by microscopic examination. Symptoms include urethritis (inflammation of the urethra) and urethral discharge in males, and a greenish-yellow cervical discharge in females. Itching, burning, redness, and dysuria (painful urination) are also common symptoms, but some infected people, especially women, are without symptoms. A complication of gonorrhea in men is epididymitis (inflammation of the epididymis), with an accompanying swelling of the testes. The leading cause of arthritis in young adults of both sexes is gonorrhea. If left untreated, gonorrhea becomes difficult to cure, but it can usually be cured by antibiotics, especially if it is diagnosed early.

Before routine treatment of the eyes of newborn infants with 1 percent silver nitrate was established, thousands of babies were blinded by gonorrheal infection acquired during passage through the vaginal canal.

*Syphilis** is a more dangerous disease than gonorrhea. It is caused by a motile, corkscrew-shaped bacterium, *Treponema pallidum.* It begins in the mucous membranes and spreads quickly to lymph nodes and the bloodstream. The early symptom of syphilis is a sore, the hard chancre, at the place where infection occurred. Other symptoms may include fever, general body pain, and skin lesions. Sometimes these symptoms disappear, even without treatment, leaving the victim with the false impression that the disease is gone. But later, circulatory or nervous tissue may degenerate so that paralysis, insanity, and death follow.

One of the most unfortunate aspects of sexually transmitted infection is the transfer of the microorganisms from a mother to her baby. The syphilis bacterium is able to cross the placenta during pregnancy, whereas the gonorrhea bacterium seems unable to do so. Thus, the developing fetus can contract syphilis early in its development and exhibit some of the serious symptoms of the disease at birth. If the baby contracts syphilis during the actual birth process, it is not likely to exhibit any symptoms at the time of birth. However, a baby infected with syphilis will grow poorly, be mentally retarded, and die early.

Like gonorrhea, syphilis can usually be cured by antibiotics if treatment is started early. However, it is well to remember that the occurrence of sexually transmitted diseases in this country has reached epidemic proportions (more than 5 million cases of NGU and gonorrhea are

---

*Syphilis is the title character in a poem by Girolamo Fracastoro, a Veronese physician and poet. In the poem, published in 1530, Syphilis is stricken with a disease that was known at the time as the "great pox." Ever since 1530, however, the disease has been called syphilis.

reported annually, and the reported cases of syphilis are increasing), especially among teenagers and young adults. False feelings of security about not contracting the diseases, quick antibiotic cures, and lack of knowledge about the seriousness of the consequences of the diseases should not be encouraged. It is true that sexually transmitted diseases can be cured with greater ease than ever before, especially if they are reported and treated early, but the most effective treatment is still intelligent prevention.

Trichomoniasis is a sexually transmitted disease caused by the *Trichomonas vaginalis* protozoan. The infection affects the lower genitourinary tract, especially the vagina and urethra of females, and the lower urethra of males. Because the infecting organism is most productive in an alkaline environment, it is aided by excessive douching that disrupts the normal acidity of the vagina, pregnancy, and the use of oral contraceptives. Symptoms include itching, swelling, and frothy vaginal discharge in females, and urethritis and dysuria in males.

Pelvic Inflammatory Disease

Pelvic inflammatory disease (PID) is a general term referring to inflammation of the uterus and uterine tubes (*salpingitis*) with or without inflammation of the ovary (*oophoritis*), localized pelvic peritonitis, and abscess formation (*tuboovarian abscess*). It is the most severe complication of the sexually transmitted diseases caused by *N. gonorrhoeae* and *C. trachomatis*. The spread of the infection can be controlled with antibiotics, if treatment is begun early. Severe cases of PID can lead to peritonitis.

Some Common Cancers of the Reproductive Systems

Breast cancer is a common cancer in women and is one of the leading causes of death by cancer among females. It is especially prevalent between the ages of 55 and 74, and usually does not occur before 35. Breast cancer kills three times as many women as uterine or ovarian cancer, spreading through lymphatics and blood vessels to other parts of the body, including the lungs, liver, bones, adrenal glands, and brain. The most common route is through the lymphatics that lead to the axillary lymph nodes.

The warning signs of breast cancer include a hard lump in the breast, a change in the shape or size of one breast, a change in skin texture, discharge from the nipple, itching or other changes in the nipple, an increase in the skin temperature of the breast, and breast pain. Self-examination of the breasts and regular physical examinations are highly recommended by physicians as a way of early detection of these warning signs.

The type of surgical treatment for breast cancer depends upon the stage at which the cancer is diagnosed. If the diagnosis is early, the cancerous cells can often be removed successfully by a *lumpectomy*, in which the tumor

and axillary lymph nodes are removed. The most drastic surgical treatment is a *radical mastectomy*, in which the entire breast and underlying fascia, the pectoral muscles, and all the axillary lymph nodes are removed. This is performed only as a last resort. Less drastic procedures include a *modified radical mastectomy*, in which the breast and axillary lymph nodes are removed, and a *simple mastectomy*, in which only the breast is removed. Chemotherapy and radiation therapy are frequently used in conjunction with surgery. The current favored treatment for early breast cancer is a lumpectomy and postoperative radiation.

Cervical cancer is one of the most common cancers among females. When cancer cells invade the basement membrane and spread to adjacent pelvic areas or to distant sites through lymphatic channels, it is classified as *invasive*. When only the epithelium is affected, it is *preinvasive*. If detected early, preinvasive cancer is curable 75 to 90 percent of the time. While preinvasive cancer produces no apparent symptoms, invasive cancer is characterized by unusual vaginal bleeding or discharge and postcoital pain or bleeding. The most effective method of detection is the Pap test (Papanicolau stain slide test), a microscope examination of cells taken from the cervix. Advanced cases of invasive cervical cancer may call for a *hysterectomy*, the surgical removal of the uterus.

Prostatic cancer is one of the most common cancers among males. Most prostatic cancers originate in the posterior portion of the gland. When the cancer spreads beyond the prostate itself, it usually travels along the ejaculatory ducts in the spaces between the seminal vesicles. Because prostatic carcinomas rarely produce symptoms until they are well advanced, prostatic cancer is often fatal. Annual or semiannual rectal examinations may detect a small, hard nodule while it is still localized, and in such cases the recovery rate is high. Regular examinations are especially important in men over 40.

Prostate Disorders

The prostate gland may be affected by inflammation (acute or chronic infections), enlargement, and benign growths. Benign enlargement of the prostate gland is known as *benign prostatic hypertrophy (BPH)*, or benign prostatic hyperplasia. It is the most common type of benign tumor, affecting about 75 percent of all men over 50. An enlarged prostate may be caused by inflammation, metabolic and nutritional imbalances, or other factors. When the prostate becomes enlarged, it compresses the urethra and obstructs normal urinary flow. Surgery that removes part or all of the prostate usually relieves the obstruction. A relatively new surgical technique does not sever the nerves near the prostate that influence erections. As a result, sexual potency is retained. Hormonal treatment is under investigation.

Ovarian Cysts

Ovarian cysts generally occur either in the follicles or within the corpus luteum. Although most ovarian cysts are not dangerous, they must be examined thoroughly as possible cancer sites.

Follicular cysts are usually small, distended bubbles of tissue that are filled with fluid. Ordinarily, small follicular cysts do not produce symptoms unless they rupture, but large or multiple cysts may cause pelvic pain, abnormal uterine bleeding, and irregular ovulation. If follicular cysts are present at menopause, they secrete excessive amounts of estrogen in response to the increased menopausal secretions of other hormones.

Granulosa-lutein cysts are produced when an excessive amount of blood accumulates during menstruation. If they appear in early pregnancy, they may cause pain on one side of the pelvis, and if they rupture, there will be massive hemorrhaging within the peritoneum. Granulosa-lutein cysts in nonpregnant women may cause irregular menstrual periods and abnormal bleeding.

Because most ovarian cysts disappear of their own accord, typical treatment consists of observation to detect early malignancies.

Endometriosis

Endometriosis is the abnormal location of endometrial tissue in sites such as the ovaries, pelvic peritoneum, and the outer surface of the small intestine. Most cases probably develop as a result of retrograde passage of bits of menstrual endometrium through the opening of the uterine tube into the peritoneal cavity. This condition is usually associated with dysmenorrhea (painful menstruation), pelvic pain, infertility, and dyspareunia (painful coitus). Treatment ranges from symptomatic relief of pain to surgical removal of the endometrial implants, including the use of a laparoscope and laser beam.

CHAPTER SUMMARY

Male Reproductive Anatomy (p. 642)

1 The reproductive function of the male is to produce sperm and deliver them to the vagina of the female.

2 The paired **testes** are held outside the body in the **scrotum**, a sac of skin between the thighs. The **seminiferous tubules** in the testes produce sperm, and **interstitial endocrinocytes** secrete testosterone, the primary male sex hormone.

3 A mature **sperm** consists of a *head* that has an *acrosome* that contains enzymes to aid the sperm in penetrating an ovum, and a *nucleus* that contains DNA; a *middle piece* that contains the mitochondria that provide ATP energy; and a *tail* that drives the sperm forward.

4 The seminiferous tubules merge to form the **tubuli recti**, the **rete testis**, and then the **efferent ducts**, which pass into the tightly coiled epididymis in each testis. The **epididymis** stores sperm as they mature, serves as a duct for the passage of sperm from the testis to the ejaculatory duct, and propels sperm toward the penis with peristaltic contractions.

5 The paired **ductus deferentia** are the dilated continuations of the epididymis. They extend to the ejaculatory ducts and are the main carriers of sperm.

6 The paired **ejaculatory ducts** receive secretions from the seminal vesicles and prostate gland. They carry semen to the urethra during ejaculation.

7 The **urethra** is a single tube through the penis that includes *prostatic, membranous,* and *penile portions.* It transports sperm outside the body during ejaculation.

8 Secretions of the paired **seminal vesicles** provide nourishment for the sperm and help to neutralize vaginal acidity. Secretions of the **prostate gland** make the sperm motile and help to neutralize vaginal acidity. Secretions of the paired **bulbourethral glands** neutralize any urine in the urethra prior to ejaculation and lubricate the urethra to facilitate ejaculation.

9 Secretions from the epididymis, seminal vesicles, prostate gland, and bulbourethral glands, together with the sperm, make up the **semen.**

10 The **penis** carries urine to the outside during urination and transports semen during ejaculation. It contains the urethra and three strands of erectile tissue: two **corpora cavernosa** and the **corpus spongiosum.** The tip of the penis is the **glans penis.** During sexual stimulation, the penis becomes enlarged and firm in an **erection.**

Female Reproductive Anatomy (p. 650)

1 The reproductive functions of the female are to produce ova; to nourish, carry, and protect the developing fetus; and to nurse the newborn baby.

2 The female gonads are the paired **ovaries,** which produce ova and the female sex hormones. The centers of ovum production in the ovaries are the **follicles,** which are always present in various stages of development. *Primordial follicles* are not yet growing. *Vesicular follicles* are almost ready to release an ovum in the monthly process of **ovulation.** After a mature ovum is discharged from a ruptured follicle, the follicle becomes the **corpus luteum,** a temporary endocrine gland that secretes *estrogen, progesterone,* and *relaxin.*

3 The paired **uterine tubes** receive mature ova from the ovary and convey them to the uterus. Ova are usually fertilized while they are still in the uterine tube, and the fertilized ovum *(zygote)* is then transported to the uterus.

4 The uterine tubes terminate in the *uterus,* a hollow, muscular organ that is the site of implantation of a fertilized egg. If pregnancy occurs, the uterus houses, nourishes, and protects the developing fetus. The inner lining of the uterus is the **endometrium,** which is built up every month in preparation for a possible pregnancy. If pregnancy does not occur, the stratum functionalis of the endometrium is shed in the monthly process of **menstruation.**

5 The uterus leads downward to the *vagina,* a muscle-lined tube that is the site where sperm from the penis are deposited during sexual intercourse, the exit point for menstrual flow, and the birth canal for the baby during childbirth.

6 The *external genitalia,* collectively called the *vulva,* are the mons pubis, labia majora, labia minora, vestibular glands, clitoris, and vestibule of the vagina.

7 *Mammary glands* are modified sweat glands contained in the breasts. They produce and secrete milk for the newborn baby.

Formation of Sex Cells (Gametogenesis) (p. 656)

1 The process in which haploid sperm cells are formed in the testes is called **spermatogenesis.** A **spermatogonium** undergoes mitotic division to produce a primary spermatocyte and another spermatogonium. The **primary spermatocyte** undergoes meiosis to produce two **secondary spermatocytes,** each with 22 autosomes and one X or Y sex chromosome. The second meiotic division forms four haploid **spermatids,** the final primitive sex cells, which develop into mature **sperm.**

2 The maturation of ova in the ovaries is **oogenesis.** It begins with a basic cell, the **oogonium,** which develops into a diploid **primary oocyte** with 46 chromosomes. After meiosis, a haploid **secondary oocyte** and the first **polar body** are produced. The polar body degenerates, and the secondary oocyte leaves the ovary and enters the uterine tube. If the secondary oocyte is penetrated by a sperm, it undergoes a second meiotic division and produces a second polar body and a haploid **ootid.** When the haploid sperm and ovum nuclei are ready to merge, the ootid is considered to be a **mature ovum.**

Conception (p. 658)

1 Hundreds of millions of sperm are ejaculated into the vagina during coitus. Although only one sperm will penetrate the ovum, the release of *acrosin* by many are required to dissolve the outer shell of the ovum to allow penetration.

2 Once one sperm enters the ovum, an impenetrable membrane forms around the ovum, and no other sperm can enter.

3 About an hour after fertilization, the haploid (23 chromosomes) nuclei of the ovum and sperm fuse to form a single diploid (46 chromosomes) cell, the **zygote.**

4 From the parents, the child receives 22 pairs of matching chromosomes called **autosomes,** and another pair called the **sex chromosomes.** The pair of sex chromosomes look alike in females and are designated **XX.** The sex chromosomes do not look alike in males and are designated **XY.**

5 After meiosis in males, half of the sperm contain an X chromosome, and half contain a Y chromosome. An ovum fertilized by an X-bearing sperm produces an XX zygote, which develops into a female. Fertilization of an ovum by a Y-bearing sperm produces an XY zygote, which develops into a male.

The Effects of Aging on the Reproductive Systems (p. 659)

1 Major changes in the female reproductive system due to aging include ovarian atrophy, menopause, the prevalence of ovarian cysts and breast or uterine cancer, and the weakening of the ligaments that support the uterus and breasts.

2 Male changes due to aging include a decreased sperm count and testosterone secretion, atrophy of accessory reproductive glands and organs, and an increased susceptibility to cancer of the prostate.

Developmental Anatomy of the Reproductive Systems (p. 662)

1 Male and female external genitalia do not differ greatly until about week 10. The genitalia are almost fully formed at week 34.

2 The internal reproductive systems are still undifferentiated at week 8, but have undergone significant anatomical changes by week 10.

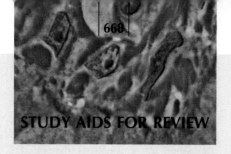

STUDY AIDS FOR REVIEW

MEDICAL TERMINOLOGY

AMENORRHEA The absence of menstrual flow.

ANOVULATION A condition in which ovulation does not occur.

CRYPTORCHIDISM (krip-TOR-kye-dizm; Gr. *kryptos*, hidden + *orchis*, testis, because of the shape of the orchid root) A condition in which the testes do not descend from their fetal position inside the abdominal cavity into the external scrotum.

DILATION AND CURETTAGE A procedure that dilates the cervix in order to scrape the lining of the uterus. Also called *D and C.*

DYSMENORRHEA Painful or difficult menstruation.

FIBROADENOMA A fibroid breast tumor.

HERMAPHRODITISM A condition in which male and female sex organs are contained in one person.

HYSTERECTOMY The surgical removal of the uterus. A *total* hysterectomy usually removes all the female reproductive organs (including the ovaries and uterine tubes) except the vagina.

IMPOTENCE The inability of a man to have an erection sufficient for sexual intercourse.

LEUKORRHEA A whitish, viscous discharge from the vagina or uterus.

OLIGOSPERMIA A condition in which only a small amount of sperm is produced.

OOPHORECTOMY The surgical removal of an ovary.

OOPHORITIS An inflammation of an ovary.

ORCHITIS An inflammation of the testes, produced by a bacillus or staphylococcus infection.

PREMENSTRUAL SYNDROME (PMS) A condition characterized by nervousness, irritability, and abdominal bloating 1 to 2 weeks before menstruation.

PROSTATECTOMY The surgical removal of all or part of the prostate gland.

SALPINGECTOMY The surgical removal of a uterine tube.

SPONTANEOUS ABORTION The loss of a fetus through natural causes. Usually called a *miscarriage.*

STERILITY The inability to reproduce.

UTERINE LEIOMYOMA A smooth muscle uterine tumor, including myomas, fibromyomas, and fibroids.

VAGINITIS An inflammation of the vagina.

UNDERSTANDING THE FACTS

1 What are the two basic functions of the testes?

2 List the accessory reproductive organs of the male.

3 Give the location of the following (be specific):

   a corpora cavernosa   d epididymis
   b inguinal hernia   e bulbourethral glands
   c acrosome

4 Describe the function of the fimbriae of the uterine tube.

5 Define vulva.

6 Give the location of the following (be specific):

   a corpus luteum   e mons pubis
   b infundibulum   f areola
   c round ligaments   g zygote
   d cervix

7 Why is a cesarean section often performed on a pregnant woman who has type II herpes simplex?

UNDERSTANDING THE CONCEPTS

1 Why is it important that the sperm contain mitochondria?

2 Trace the pathway of a sperm cell from the site of production until it leaves the body of the male.

3 Why do we say that the reproductive function in the female is more complex than in the male?

SELF-QUIZ

True-False Statements

25.1 The outer layer of the uterine tube is a typical serous membrane that is part of the visceral peritoneum.

25.2 If the secondary oocyte is not fertilized, it passes out of the female's body.

25.3 Mammary glands are modified sweat glands.

25.4 The pigmented area around the nipple of the breast is the lactiferous duct.

25.5 Secondary spermatocytes undergo meiosis and produce primary spermatocytes.

25.6 In oogenesis, one primary oocyte yields four ova.

25.7 An acrosome is found in the middle piece of a sperm.

25.8 The final maturation of sperm cells occurs in the epididymis.

25.9 The corpus luteum develops from the ovarian follicle.

25.10 The part of the uterine tube nearest the ovary is called the isthmus.

Completion Exercises

25.11 If the inguinal area fails to close properly after descent of the testes, a(n) _____ may result.

25.12 The diamond-shaped area between the legs and back to the anus is called the _____.

25.13 Sperm cells are produced in the _____.

25.14 The _____ was formerly called the vas deferens.

25.15 Circumcision is the removal of the _____.

25.16 Bulbourethral glands produce _____.

25.17 The female gonads are the _____.

25.18 If a fertilized egg adheres to the wall of the uterine tube, a(n) _____ results.

25.19 The uterus is attached to the lateral wall of the pelvis by two _____.

25.20 The innermost layer of the uterus is a specialized mucous membrane called the _____.

25.21 The greater vestibular glands are also called _____ glands.

25.22 The broad ligaments attach the _____ to the pelvis.

25.23 The narrow portion of the uterus at the juncture with the vagina is called the _____.

25.24 The cell formed as a result of fertilization is the _____.

25.25 The 22 pairs of chromosomes that are not involved in sex determination are called _____.

A SECOND LOOK

1 In the following drawing, label the ductus deferens, epididymis, prostate gland, bulbourethral gland, spermatic cord, and corpus cavernosum.

2 In the following drawing, label the uterine cervix, vagina, uterine tube, ovary, mons pubis, labium minus, and clitoris.

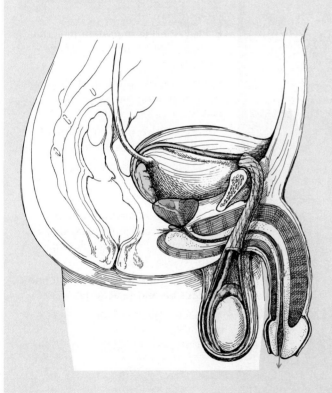

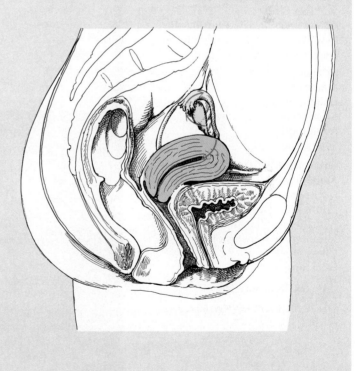

26

Developmental Anatomy: A Life Span Approach

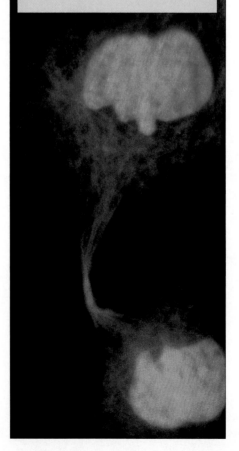

LEARNING OBJECTIVES

1 Outline in sequence the events that take place from fertilization to birth.

2 Describe the stages of development during cleavage, with the related structural changes in the zygote.

3 Describe the process of implantation in the uterine wall.

4 Explain the structure and functions of the extraembryonic membranes.

5 Discuss the roles of the placenta and the umbilical cord in prenatal development.

6 Describe the major body structures that begin to form during the embryonic period.

7 Trace the development of major body structures during the fetal period.

8 Describe the important maternal changes that occur during pregnancy.

9 Describe the process of childbirth, including the three stages of labor.

10 Explain two major adjustments that the newborn must make.

11 Discuss the changes that occur during each of the major stages of postnatal life from birth to old age.

12 Describe several important methods for examining and diagnosing problems in the developing embryo and fetus.

13 Discuss the uses of intrauterine transfusion and fetal surgery.

The development of a human being may be divided into *prenatal* ("before birth") and *postnatal* ("after birth") *periods*. During the prenatal period, the developing individual is called a *zygote* at fertilization, an *embryo* for the next eight weeks, and a *fetus* from nine weeks until birth. It becomes an *infant* at birth.

The time of prenatal activity is called the **gestation period** (L. *gestare*, to carry). For as long as the new individual lives, it will never again experience such a dramatic burst of growth and development. During the 9-month gestation period, the weight increases from the fertilized ovum to the newborn baby about six *billion* times. In contrast, the body weight increases only about 20 times from birth to age 20. During the first month of prenatal development, the embryo increases in weight about a million percent. From that time on, the rate of growth continues to decrease, though there is a growth spurt during the third month.

Prenatal development is usually divided into two periods: the **embryonic period,** from fertilization to the end of the eighth week, and the **fetal period,** from the beginning of the ninth week until birth.

EMBRYONIC DEVELOPMENT

When a sperm penetrates a secondary oocyte, it sheds its tail and midpiece and works its head into the oocyte's cytoplasm. Meiosis in the oocyte, which had stopped at metaphase of the second division, is resumed after entry of the sperm. When meiosis is complete, the second polar body is pinched off, and the remaining haploid structure is the *ovum*. The ovum and sperm nuclei then fuse to form the diploid nucleus of the zygote. A **zygote** is the single cell resulting from the fertilization of an ovum by a sperm. The zygote nucleus contains all the genetic material, DNA, which subsequent mitotic divisions will copy and distribute to all cells of the embryo.

Cleavage

When fertilization is complete, the haploid nuclei of the sperm and ovum (each called a *pronucleus*) fuse to form a diploid zygote with a complete set of 23 pairs of chromosomes. Then a series of mitotic cell divisions called **cleavage** begins [FIGURE 26.1]. From the time cleavage begins until the end of the eighth week, the developing individual is referred to as an embryo.

The first cleavage is complete about 36 hours after fertilization, and subsequent cleavages take place about twice a day. The two-celled embryo that results from the first cleavage is approximately 0.1 mm in diameter, still only about the size of the period at the end of this sen-

tence. The daughter cells are called **blastomeres** (Gr. *blastos*, bud + *meros*, a part). After about 50 hours there are four cells, and mitotic divisions continue. Because the cells do not receive additional nutrients during these early cleavages, they do not grow as they divide. As each blastomere divides, it forms two cells that are each half the size of the original cell. As a result, the overall size of the organism stays approximately the same as the original zygote.

The first six to eight cleavages occur while the embryo is still enclosed within the zona pellucida. About day 3 after fertilization, a solid ball of about 8 to 50 blastomeres is formed into a mulberry-shaped **morula** (MORE-uh-luh; L. *morum*, mulberry tree). By this time, the embryo is completing its 10-cm (4-in.) journey along the uterine tube and is approaching the entrance to the uterus.

About day 4 to 5 after fertilization, the morula develops into a fluid-filled hollow sphere called a **blastocyst** (Gr. *blasto*, germ + *kystis*, bladder) with an inner cavity called the **blastocoel** (BLASS-toh-seel; Gr. *koilos*, hollow). The distinguishing feature of the blastocyst is that it has differentiated at one pole into an **inner cell mass** from which the embryo will form, and a surrounding epithelial layer called the **trophectoderm,** which is composed of *trophoblast cells*. The trophectoderm will later develop into the *chorion* (a fetal membrane that develops from trophoblast cells) and become part of the membrane system that transports nutrients to the embryo and removes wastes from it.

The blastocyst floats freely in the uterine cavity for one or two days while it continues to divide and grow, nourished by fluids secreted by the glands of the endometrium. Now the blastocyst is ready to shed its zona pellucida, allowing it to become attached to the maternal uterus. It will soon become embedded in the uterine wall in the process of *implantation.*

With the hormonal changes that accompany ovulation and the formation of the corpus luteum, the development of the endometrium continues. This building up of the uterine wall prepares the uterus to accept the blastocyst. An unfertilized ovum is not able to attach and implant itself into the endometrium.

Implantation in the Uterus

About day 6 to 7 after fertilization, implantation begins as the blastocyst attaches to the endometrium [FIGURE 26.2A]. The inner cell mass faces toward the epithelium, while the trophectoderm actually attaches to the epithelium, usually to the upper, posterior wall of the uterus. At this point, the blastocyst is not yet completely implanted. Within 24 hours after the blastocyst becomes attached to the uterine wall, the trophoblast cells differentiate into *cytotrophoblasts*, which surround the inner cell mass, and

FIGURE 26.1 CLEAVAGE

[A] The single-celled zygote still has two polar bodies attached to it. Successive divisions result in two **[B]**, four **[C]**, eight **[D]** cells, and so on. Continued divisions and rearrangement form a morula **[E]** and then a blastocyst **[F, G]**. The cells present at the end of cleavage are much smaller than the zygote, which is a large cell. SEMs ×200.

[A] ZYGOTE BEFORE CLEAVAGE

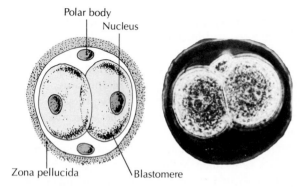

Polar body
Nucleus
Zona pellucida
Blastomere

[B] 36 HOURS AFTER FERTILIZATION
FIRST DIVISION, 2 CELLS

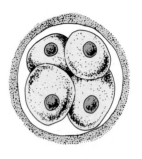

[C] SECOND DIVISION, 4 CELLS

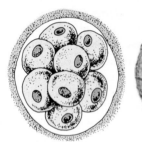

[D] THIRD DIVISION, 8 CELLS

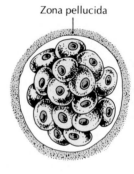
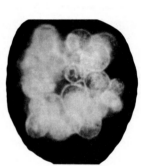

Zona pellucida

[E] 3 TO 5 DAYS AFTER FERTILIZATION
MORULA

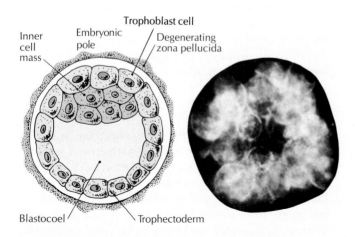

Inner cell mass
Embryonic pole
Trophoblast cell
Degenerating zona pellucida
Blastocoel
Trophectoderm

[F] 5 TO 6 DAYS AFTER FERTILIZATION
BLASTOCYST (EARLY)

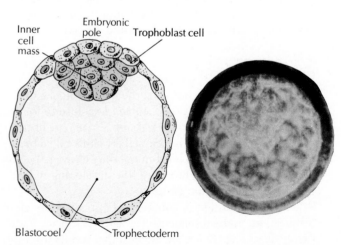

Inner cell mass
Embryonic pole
Trophoblast cell
Blastocoel
Trophectoderm

[G] BLASTOCYST (LATE)

FIGURE 26.2 IMPLANTATION

[A] About day 6 after fertilization, trophoblast cells of the developing blastocyst attach to endometrial cells of the uterine wall. **[B]** About day 7, implantation proper begins as the syncytiotrophoblast burrows into the endometrial epithelium as the blastocyst begins to move into the uterine wall. Hypoblasts of the inner cell mass will eventually differentiate into fetal tissues. **[C]** Implantation is almost complete about day 9 as the endometrial epithelium grows over the implanted blastocyst. The amnion, amniotic cavity, primitive yolk sac, bilaminar embryonic disk, and extraembryonic mesoderm are already present.

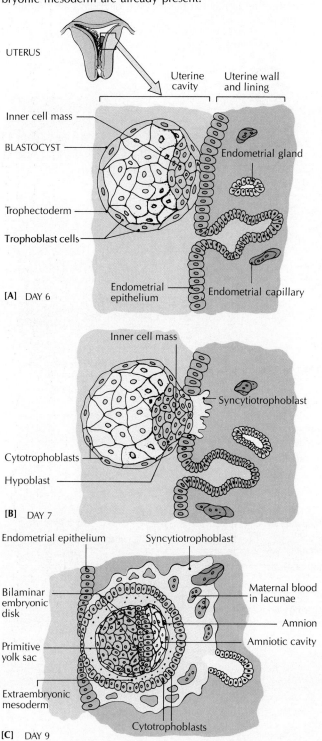

[A] DAY 6

[B] DAY 7

[C] DAY 9

the *syncytiotrophoblast*, a large multinucleated structure. The syncytiotrophoblast implants the blastocyst in the uterine wall by invading the nutritious *inner* portion of the endometrium [FIGURE 26.2B]. On about day 9, the blastocyst is completely enclosed by endometrial cells [FIGURE 26.2C], and it can continue to grow and develop. This complete implantation is the point at which *pregnancy* is considered to begin. FIGURE 26.3 shows an overall view of the processes of fertilization, cleavage, and implantation.

Maintenance of the endometrium Before implantation can begin, the trophoblast cells must secrete the hormone **human chorionic gonadotropin (hCG),** which maintains the corpus luteum at a time when it would otherwise begin to degenerate. The corpus luteum continues its secretion of estrogen and progesterone, preventing menstruation and the breakdown of the endometrium and allowing implantation to take place.

Eventually, the cytotrophoblasts and syncytiotrophoblast will form the *placenta*, the structure that transfers nutrients from the mother to the fetus. In the meantime, the embryo receives nutrients from the many capillaries in the uterine wall.

As implantation continues during the second week, the inner cell mass changes shape, and the blastocyst assumes a flattened disk shape to form the **bilaminar** (two-layered) **embryonic disk.** These two layers are composed of endoderm and ectoderm. As the bilaminar disk is forming, several supporting membranes and other structures also develop. These are the amniotic cavity, yolk sac, body stalk, and chorion, which will be described in the next section.

Cell division stops briefly after implantation in the uterus. When cell division resumes, *cell differentiation* begins, rearranging the inner cell mass of the blastocyst into the three primary germ layers: endoderm, ectoderm, and mesoderm. See FIGURE 3.25 and TABLE 3.3 for a review of the formation and development of the primary germ layers.

Extraembryonic Membranes

During the first two weeks, the embryo does not have a functional circulatory system. During the implantation process, enzymes released by the trophoblast cells destroy some maternal capillaries in the endometrium. Blood from these capillaries comes into direct contact with the trophoblast cells, thus providing a temporary source of nutrition. In the third week, four membranes begin to form from tissues that were once the trophoblast. Because these membranes are not part of the embryo itself, they are called **extraembryonic,** or **fetal, membranes.** They include the yolk sac, amnion, allantois, and chorion.

FIGURE 26.3 FERTILIZATION, CLEAVAGE, AND IMPLANTATION

[A] Development from ovulation to implantation: (1) A secondary oocyte is released (ovulation) from the ovary when a mature follicle ruptures. The follicle then becomes a corpus luteum. (2) The secondary oocyte is swept into the uterine tube. (3) Meiosis produces the haploid chromosome number, and the first polar body is formed. The corona radiata consists of follicle cells surrounding the mature oocyte. (4) A sperm penetrates the oocyte. The second meiotic division occurs, and the second polar body is formed. (5) The male and female nuclei fuse, forming the zygote. (6) The zygote begins cleavage. (7, 8) Cleavage results in the formation of a morula. (9) An early blastocyst is formed. (10) About four days after fertilization, a late blastocyst is formed. (11–13) About a week after fertilization, the blastocyst begins the process of implantation into the uterine wall. (14) Once implantation has been accomplished, the uterine wall starts contributing the outer portion of the placenta; the embryo itself contributes the inner part. **[B]** Photograph of an implanted blastocyst 6 or 8 days after fertilization. ×20.

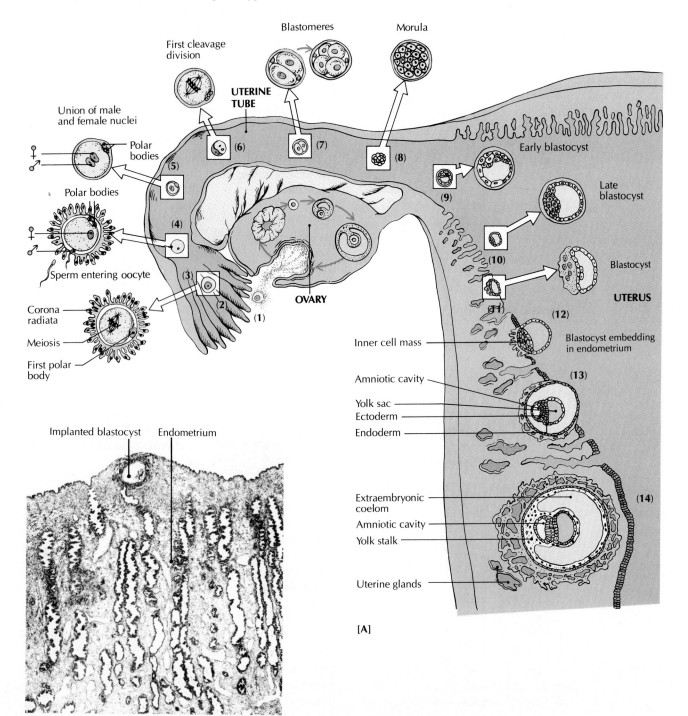

As the trophoblast grows, it branches and extends into the tissue of the uterus. The two kinds of tissue, embryonic and maternal, grow until there is sufficient surface contact to ensure the adequate passage of nourishment and oxygen from the mother, as well as the removal of metabolic wastes, including carbon dioxide, from the embryo. After the embryo is implanted into the endometrium, small *chorionic villi* grow outward into the maternal tissue from the *chorion,* the protective sac around the embryo [FIGURE 26.4]. The chorion makes up most of the placenta. It contains many villi, which allow the exchange of nutrients, gases, and metabolic wastes between the maternal blood and that of the embryo. Eventually, the fully differentiated villi of the embryo and fetus remain to make contact with the maternal endometrium, and later become the mature placenta.

The *yolk sac* is a primitive respiratory and digestive structure that is reduced by the sixth week of development to the thin *yolk stalk,* which then becomes incorporated into the umbilical cord via the body stalk. The yolk sac appears to be involved in transporting nutrients to the embryo during the second and third weeks, before the placental transfer is fully developed. It also is the initial site for the formation of blood cells until the liver assumes that responsibility during the fifth week. Finally, *primordial germ cells* in the yolk sac eventually migrate to the developing gonads, where they become spermatogonia or oogonia.

The *amnion* (Gr. sac) is a tough, thin, transparent membrane that envelops the embryo. Its interior space, the *amniotic cavity,* becomes filled with a watery *amniotic fluid,* which is made up of shed fetal epithelial cells, protein, carbohydrates, fats, enzymes, hormones, and pigments. Later, it also contains fetal excretions. The amniotic fluid suspends the embryo in a relatively shock-free environment, prevents the embryo from adhering to the amnion and producing malformations, permits the developing fetus to move freely (which aids the proper development of the muscular and skeletal systems), and provides a relatively stable temperature for the embryo.

The *allantois* (uh-LAN-toh-ihz; Gr. "sausage") usually appears early in the third week as a small fingerlike outpocket, or diverticulum, of the caudal wall of the yolk sac. It is involved with the formation of blood cells and the development of the urinary bladder, and its blood vessels become the vein and arteries of the umbilical cord. The allantois is eventually transformed into the median umbilical ligament, which extends from the urinary bladder to the navel.

Placenta and Umbilical Cord

The essential exchange mechanism between mother and embryo is in place by the beginning of the fourth week. At the site of implantation, the chorion joins intimately and intricately with the endometrium to develop into the

FIGURE 26.4 DEVELOPMENT OF EXTRAEMBRYONIC MEMBRANES

The extraembryonic membranes are the chorion, yolk sac, and amnion. **[A]** Week 3. **[B]** Week 4. **[C]** Week 10. **[D]** Week 20.

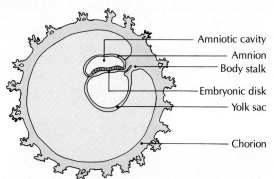

[A] WEEK 3

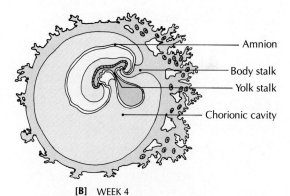

[B] WEEK 4

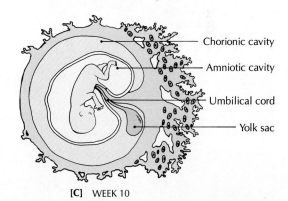

[C] WEEK 10

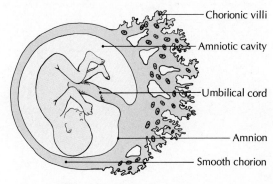

[D] WEEK 20

FIGURE 26.5 DEVELOPMENT OF THE PLACENTA

[A] The blastocyst is implanted in the wall of the uterus. [B] Enlarged view showing the trophoblast cells and the penetration of chorionic villi into the endometrium. Arrows indicate developing blood vessels. [C] The trophectoderm differentiates into two layers, and chorionic villi reach the maternal blood supply. [D] The yolk stalk of the developing embryo will eventually become the umbilical cord. [E] At about 5 weeks, the umbilical blood vessels reach the embryo. [F] At about 8 weeks, the chorion and amnion form, and the placenta and umbilical cord are highly developed.

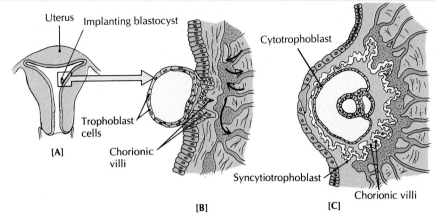

Uterus Implanting blastocyst

Trophoblast cells

Chorionic villi

[A]

Cytotrophoblast

Syncytiotrophoblast

Chorionic villi

[B]

[C]

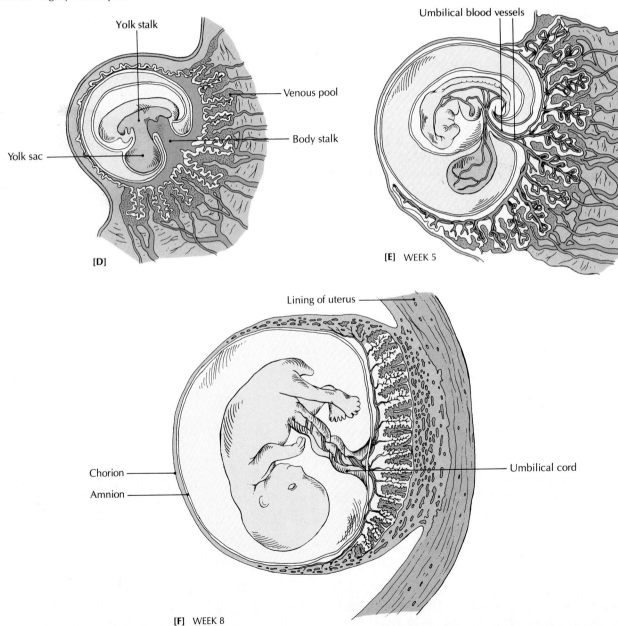

Yolk stalk

Venous pool

Body stalk

Yolk sac

[D]

Umbilical blood vessels

[E] WEEK 5

Lining of uterus

Chorion

Amnion

Umbilical cord

[F] WEEK 8

placenta (L. flat cake, because of its disklike shape) [FIGURE 26.5]. The placenta grows rapidly until the fifth month, when it is almost fully developed. A full-term placenta is about 2.5 cm (1 in.) thick and 20.5 cm (8 in.) in diameter. It weighs about 0.45 kg (1 lb).

The side of the placenta facing the fetus is relatively smooth, with the umbilical cord usually attached somewhere near the center. The side of the placenta that faces the mother has grooves and protuberances. The irregular surface increases the area for the interchange between fetal and maternal circulation. The surface area of the rough side is about 13 sq m (140 sq ft), more than three times greater than the smooth side. The connection between the fetal part of the placenta and the maternal part is close, with thousands of chorionic villi on the fetal part embedded in the maternal part, adding to the contact surface enormously. The fetal capillaries come close to the maternal capillaries to effect the exchange of substances. There is no actual blood flow between the fetal and maternal circulations, and there is no nervous connection.

The placenta has three main functions:

1 It transports materials between the mother and fetus. The transported materials include *gases* (such as oxygen, carbon dioxide), *nutrients* (such as water, vitamins, glucose), *hormones* (especially steroids such as testosterone), *antibodies* (which bestow passive immunity), *wastes* (such as carbon dioxide, urea, uric acid, bilirubin), *drugs* (most drugs pass easily, especially alcohol), and *infectious agents* (such as rubella, measles, encephalitis, poliomyelitis, and AIDS viruses).

2 It synthesizes glycogen and fatty acids and probably contributes nutrients and energy to the embryo and fetus, especially during the early stages of pregnancy.

3 It secretes hormones, especially the *protein hormones* human chorionic gonadotropin (hCG) and human chorionic somatomammotropin (hCS) and, with the cooperation of the fetus, the *steroid hormones* progesterone and estrogen.

The inner lining of the placenta is made up of the extensive blood vessels and connective tissue of the chorion. These blood vessels are formed from the embryo and are connected to the embryo by way of the **umbilical cord** [FIGURE 26.6]. The umbilical cord (L. *umbilicus*, navel) is formed from the body stalk, yolk stalk, and other extraembryonic membranes during the fifth week. The cord contains two arteries, which carry carbon dioxide and nitrogen wastes from the embryo and fetus to the placenta, and a vein, which carries oxygen and nutrients from the placenta to the embryo and fetus. A gelatinous cushion of embryonic connective tissue surrounds the vessels of the umbilical cord. This resilient pad, together with the pressure of blood and other liquids pulsating through the cord, prevents the cord from twisting shut when the fetus turns around in the uterus. An umbilical cord is usually 1 to 2 cm (0.4 to 0.8 in.) in diameter and 50 to 55 cm (19 to 22 in.) in length.

There is normally no direct connection between the embryonic and the maternal tissue, at least no actual blood flow and no nerve connection. Nutrients, water, oxygen, and hormones can cross the placental barrier, as can infectious agents, toxic substances (such as lead and

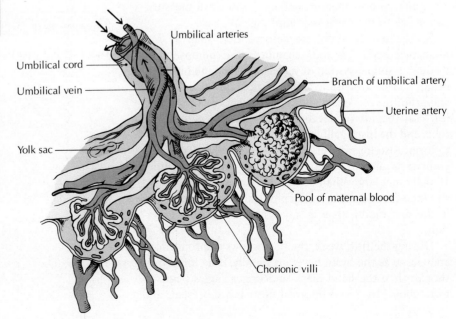

FIGURE 26.6 UMBILICAL CORD AND PLACENTAL CIRCULATION

Section of uterine wall with connected umbilical cord. The left and center chorionic villi are shown in a cutaway view to reveal the internal structure of the blood vessels. In addition to the exchange of respiratory gases, the placenta makes possible the elimination of metabolic wastes from the fetus through the umbilical arteries (downward arrows) and allows the entry into the fetus of foods, vitamins, and electrolytes through the umbilical vein (upward arrows).

insecticides), and drugs. Because these substances can pass into the fetal blood, the fetus can be infected, poisoned, or become addicted to drugs such as cocaine. (In fact, a newborn baby can show drug withdrawal symptoms if its mother used heroin during pregnancy.) Nevertheless, the growing embryo is well insulated from most of the possibly harmful influences to which the mother is exposed. The placenta is eventually shed after the baby is born as part of the *afterbirth*.

Weeks 1 through 8: Formation of Body Systems

As already described, the first week of prenatal development is devoted to cleavage, while during the second week of prenatal life, the cells start to differentiate structurally and functionally, as the primary germ layers begin to take form. Within the developing chorionic villi, the first blood vessels are growing.

Rapid development occurs during the third week. Differentiation of the endoderm, mesoderm, and ectoderm is complete, and the body, head, and tail can be distinguished. The primitive streak, a thickened dorsal longitudinal strip of ectoderm and mesoderm in the embryo, appears about day 15. The notochord also becomes apparent at this time. It is the dorsal, rodlike structure that runs the length of the embryo and serves as its internal skeleton. The neural tube will eventually form dorsal to the notochord and develop into the brain, spinal cord, and the rest of the nervous system [FIGURE 26.7]. The primitive body cavities and the cardiovascular system also take form. Villi continue to develop for an improved exchange between the embryo and mother. The primordial heart begins to pulsate, circulating oxygenated blood and other nutrients to the developing embryonic structures. The embryo now measures about 2.5 mm (all measurements are from crown to rump).

By the fourth week, the embryo is C-shaped. The U-shaped heart has four chambers and pumps blood through a simple system of vessels. Upper limb buds and primitive ears are visible about day 26. Lower limb buds and primitive eye lenses appear about day 28. The umbilical cord begins to develop. The intestine is a simple tube, and the liver, gallbladder, pancreas, and lungs begin to form. Also forming are the eyes, nose, and brain. The embryo is about 5 mm from crown to rump, about 10,000 times larger than the fertilized ovum. The relative size increase and the extent of physical change are greater in the first month than at any other time during gestation.

During the fifth week, the head grows disproportionately large as the brain develops rapidly. The forearm is shorter than the hand plate, and finger ridges begin to form about day 33. Primordial nostrils are present, and

FIGURE 26.7 EMBRYO AFTER 3 WEEKS

Three-layered embryo; posterior view. ×47. The drawing (left) shows the actual size of the embryo.

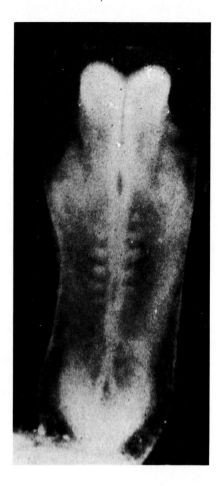

FIGURE 26.8 EMBRYO AFTER 5 WEEKS (5–8 mm)

The inset drawing shows the embryo at actual size.

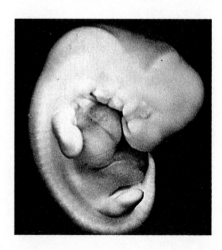

FIGURE 26.9 EMBRYO AFTER 6 WEEKS (10–14 mm)

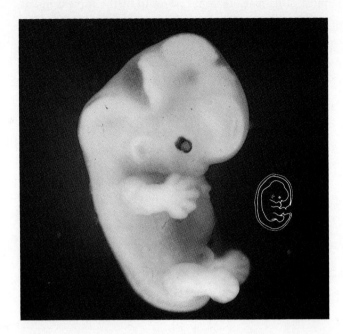

FIGURE 26.10 EMBRYO DURING THE SEVENTH WEEK (17–22 mm)

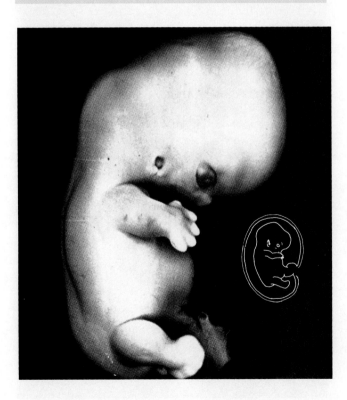

the tail is prominent [FIGURE 26.8]. Primordial kidneys, the upper jaw, and the stomach begin to form. The nose continues to develop. The intestine elongates into a loop, and the genital ridge bulges. Primordial blood vessels extend into the head and limbs, and the spleen is visible. Spinal nerves are formed, and cranial nerves are developing. Premuscle masses appear in the head, trunk, and limbs. The epidermis is gaining a second layer. The cerebral hemispheres are bulging. At this stage, drugs taken by the mother and diseases, such as German measles, may be transmitted to the embryo, affecting its development. The embryo measures about 8 mm.

By the sixth week, the components of the upper jaw are prominent but separate, and the lower jaw halves are fused. The limb buds differentiate noticeably, and the development of the upper limbs is more rapid than that of the lower limbs. The head is larger than the trunk, which begins to straighten [FIGURE 26.9]. The external ears appear, and the eyes continue to develop and become accentuated as the retinal pigment is added. Simple nerve reflexes are established. The heart and lungs acquire their definitive shapes. The embryo is about 12 mm (0.5 in.).

During the seventh and eighth weeks, the yolk sac is reduced to the yolk stalk. The face and neck begin to form, and the fingers and toes are differentiated. The back straightens, and the upper limbs extend over the chest [FIGURE 26.10]. The tail is regressing, and the three segments of the upper limbs are evident. The jaws are formed and begin to ossify. The stomach is attaining its final shape, and the brain is becoming large. The muscles are differentiating rapidly throughout the body, assuming their final shapes and relations. The eyelids are beginning to form. At the end of the embryonic period, the embryo is about 17 mm (0.75 in.).

ASK YOURSELF

1 What is cleavage, and when does it occur?

2 What is a morula? A blastocyst?

3 When does implantation take place?

4 What are extraembryonic membranes?

5 What is the difference between the chorion and the amnion?

6 What are the main functions of the placenta?

7 What is the vascular structure of the umbilical cord?

8 Describe the major anatomical changes during the first eight weeks of prenatal life.

FETAL DEVELOPMENT

Typically, the embryonic heart begins to contract by the twenty-second day. By the end of the eighth week, the major external features (ears, eyes, mouth, upper and lower limbs, fingers, toes) are formed, and the major organ systems have been established. Once this stage is reached, the developing individual is referred to as a *fetus* (FEE-tuhss). The fetus, although only 3 cm (1.25 in.) long, looks recognizably human after two months [FIGURE 26.11]. After three months in the uterus, the fetus is about 5 to 6 cm (2.0 to 2.4 in.) long and contains all the organ systems characteristic of an adult.

The last six months of pregnancy are devoted to the increase in size and maturation of the organs formed during the first three months. By the time the fetus is 10 cm (4 in.) long, it can move and be felt by the mother. It is thin, wrinkled, hairy, and moist. As it ages, the fetus loses most of its hair, its bones begin to ossify, it picks up fat, and it becomes mature enough to be born. It is said to have come to *term*.

Third Lunar Month*

By the ninth week, the nose of the fetus is flat, and the eyes are far apart. The tongue muscles are well differentiated, and the earliest taste buds form. The ear canals are distinguishable. The fingers and toes are well formed, and the head is elevating [FIGURE 26.12]. The growth of the intestines makes the body evenly round. The small intestine is coiling within the umbilical cord, and intestinal villi are developing. The liver is relatively large, and the testes or ovaries are distinguishable as such. The main blood vessels assume their final organization, and

*Because the reproductive cycles of women are closer to the lunar (moon) month of 28 days than to the calendar month, fetal development is conventionally described in terms of lunar rather than calendar months.

FIGURE 26.11 FETUS AFTER 8 WEEKS (32–50 mm)

A fetus in its amniotic sac after being removed from the chorionic sac; actual size.

Yolk sac   Chorionic villi

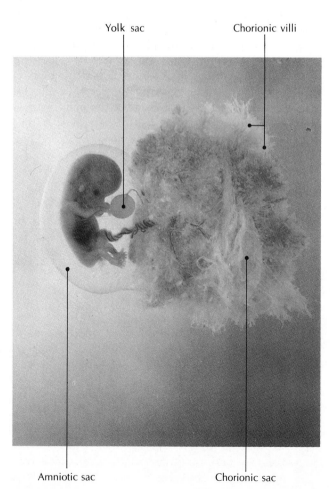

Amniotic sac   Chorionic sac

FIGURE 26.12 FETUS AFTER 9 WEEKS (25–40 mm)

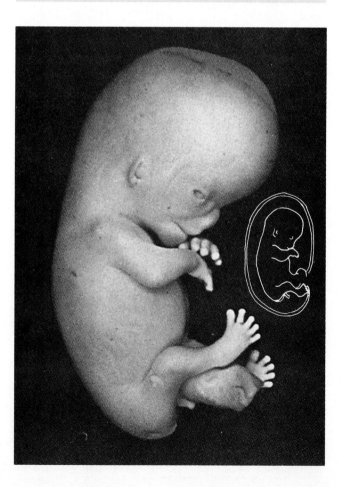

there is some bone formation. The muscles of the trunk, limbs, and head are well represented, and the fetus is capable of some movement. All five major subdivisions of the brain are formed, but the brain lacks the convolutions that are characteristic of later stages. The external, middle, and internal ears are assuming their final forms. The fetus is about 23 mm (1 in.).

All embryos appear female during early development because the genitals are not yet differentiated. Distinct male characteristics begin to differentiate through the action of male sex hormones (androgens) at about the sixth week of embryonic development. The differentiation of sexes is completed by the third month of fetal life.

During the tenth to eleventh weeks, the head is held erect. The limbs are nicely molded [FIGURE 26.13], and the fingernail and toenail folds are indicated. The eyelids are fused, the lips have separated from the jaws, and the nasal passages are partitioned. The intestines withdraw from the umbilical cord, and the anal canal is formed. The kidneys begin to function, the urinary bladder expands as a sac, and urination is now possible. The vaginal sac and rudimentary sex ducts are forming. Early lymph glands appear, and red blood cells predominate in the blood. The earliest hair follicles begin to develop on the face, and tear glands are budding. During this period the spinal cord attains its definitive internal structure. A pulse is now detectable. The placenta is producing progesterone, which was formerly produced by the corpus luteum. The fetus is about 40 mm (1.5 in.) from crown to rump.

FIGURE 26.13 FETUS AT ABOUT 10 WEEKS (43–61 mm)

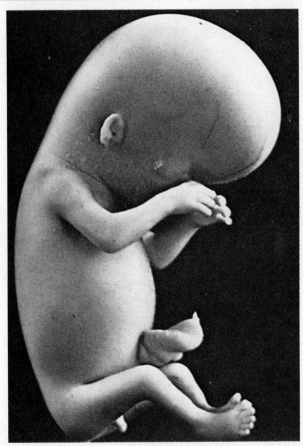

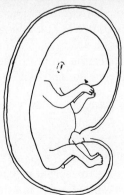

Fourth Lunar Month

The head is still dominant. The nose gains its bridge, tooth buds and bones form, the cheeks are represented, and the nasal glands form. The external genitalia attain their distinctive features. The lungs acquire their final shape, and in the male the prostate gland and seminal vesicles appear. Blood formation begins in the marrow, and some bones are well outlined. The epidermis is triple-layered, and the characteristic organization of the eyes is attained. The brain develops its general structural features. By the sixteenth week, all the vital organs are formed. The enlargement of the uterus can be felt by the mother, and the fetus measures about 56 mm (2.25 in.).

Fifth Lunar Month

At the beginning of the fifth lunar month the face looks "human," and hair appears on the head. Muscles become active spontaneously, and the body grows faster than the head. The hard and soft palates are differentiating, and the gastric and intestinal glands begin to develop. The kidneys attain their typical shape and plan. In the male, the testes are in position for their later descent into the scrotum. In the female, the uterus and vagina are recognizable as such. Blood formation is active in the spleen, and most bones are distinctly indicated throughout the body. Stretching movements by the fetus are now felt by the mother. The epidermis begins adding other layers to form the skin. Body hair starts developing, and sweat glands appear. The general sense organs begin to differentiate. The crown-to-rump measurement is about 112 mm (4.5 in.).

At the end of the fifth lunar month, the body is covered with downy hair called *lanugo* (luh-NEW-go; L. *lana*, wool). The nasal bones begin to harden. In the female, the vaginal passageway begins to develop. Until birth, blood formation continues to increase in the bone marrow. Fetal heart sounds can be heard with a stethoscope, and the heart beats at twice the adult rate. The lungs are formed, but do not function. The gripping reflex of the hand begins to develop, and kicking movements and hiccupping may be felt by the mother. The fetus is about 160 mm (6.5 in.).

Sixth Lunar Month

The body is now lean and better proportioned than before. The internal organs occupy their normal positions, and the large intestine becomes recognizable. The nostrils open. The cerebral cortex now gains its typical layers. Thumb sucking may begin. FIGURE 26.14 shows the development of the hand from the fifth week to the eighth week.

FIGURE 26.14 DEVELOPMENT OF THE HAND

[A] At 5 weeks, the forearm is shorter than the hand, and finger ridges begin to form. [B] At 6 weeks, clearly delineated fingers have begun to develop. [C] At 7 to 8 weeks, the fingers, thumb, and fingerprints are well formed.

[A] 5 WEEKS

[B] 6 WEEKS

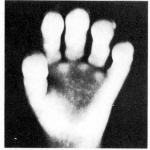

[C] 7–8 WEEKS

FIGURE 26.15 FETUS AT ABOUT 7 MONTHS

The fetus is shown smaller than life size.

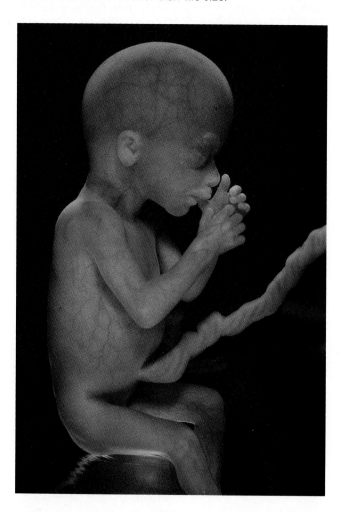

Seventh Lunar Month

The fetus is lean, wrinkled, and red [FIGURE 26.15]. The eyelids open, and eyebrows and eyelashes form. Lanugo is prominent. The scrotal sac develops, and the testes begin to descend, concluding their descent in the ninth month. The nervous system is developed enough so that the fetus practices controlled breathing and swallowing movements. The brain enlarges, and cerebral fissues and convolutions are appearing rapidly. The retinal layers of the eyes are completed, so light can be perceived. If delivered at this stage, the baby would have at least a 10-percent chance of survival.

Eighth Lunar Month

The testes settle into the scrotum. Fat is collecting so that wrinkles are smoothing out, and the body is round-ing. A sense of taste is present. The weight increase slows down. If delivered at this stage, the baby has about a 70-percent change of survival.

Ninth Lunar Month

The nails reach the tips of the fingers and toes. Additional fat accumulates.

Tenth Lunar Month

Pulmonary branching is only two-thirds complete, and the lungs do not function until birth. Some fetal blood passages are discontinued. The lanugo is shed. The digestive tract is still immature. The lack of space in the mother's uterus causes a decrease in fetal activity. Usually, the fetus turns to a head-down position [FIGURE 26.16]. The maternal blood supplies antibodies. The placenta regresses, and placental blood vessels degenerate. The baby is now at *full term,* with a crown-to-rump measurement of about 350 mm (14 in.), and is ready to be born. (Measurements in the final months can vary greatly.) The newborn baby's eyes react to light, but they will not assume their final coloration until about a month after birth. In developed countries about 99 percent of full-term infants survive.

ASK YOURSELF

1 At what point is the developing individual considered to be a fetus?

2 When can the sex of the fetus first be distinguished?

3 At what point is the development of the hand nearly completed so that thumb sucking often begins?

MATERNAL EVENTS OF PREGNANCY

The first sign of pregnancy is usually a missed menstrual period about three weeks after the coitus that resulted in conception. Because irregular menstruation may not be a drastic change for some women, it is not a foolproof diagnostic tool. A test of the woman's urine may detect pregnancy about 12 days after the first missed period.

During the fifth or sixth week of pregnancy, some superficial veins may become prominent, the breasts may begin to enlarge and feel tender, and the areola may darken. The temporary nausea called "morning sickness" may begin about the seventh week, and "stretch marks" on the breasts may appear as the breasts continue to en-

large. At eight weeks, a physician can diagnose pregnancy through a physical examination by detecting an enlarged uterus and a softened cervix (Hagar's sign).

At the beginning of the ninth week, *colostrum* (a thin, milky fluid secreted by the mammary glands before milk is released) may be squeezed from the breasts, and during the tenth week the physician may be able to feel weak uterine contractions. The nausea that started during the seventh week usually stops during the twelfth week.

During the sixteenth week, the abdomen usually begins to protrude as the uterus and fetus grow larger. The genitals may darken and turn bluish, and the forehead and cheeks may also darken in a pigmentation change called the "mask of pregnancy."

Some time between the seventeenth and twentieth weeks, the mother will probably feel the fetus moving in the uterus. By about the twenty-sixth week, the movements of the fetus may be noticeable from the outside. By the thirtieth week the abdomen protrudes considerably, the breasts are considerably enlarged, and an overall weight gain of almost 9 kg (20 lb) is not unusual. The cardiac output is now increased, and the heartbeat is accelerated. Many other changes are obvious, too. The pregnant woman tires more easily. Her hands feel clammy because of an increase of blood flow to them. A general feeling of hotness, increased thirst, and the need to urinate more frequently are common. Varicose veins or hemorrhoids may develop as venous pressure increases. The woman may experience shortness of breath, and her back may ache as the back muscles strain to support the extra abdominal weight.

As the final month of pregnancy approaches, the ribs spread out to accommodate the lungs as they are displaced by the still-expanding uterus. As the room for internal organs becomes scarce, pressure may be put on some nerves, causing pain. At this point, or even earlier, the attending physician will usually urge the pregnant woman to limit her travel plans, especially air travel, because of reduced oxygen at high altitudes. In the thirty-sixth week, four weeks before delivery, the uterus has risen to the rib level, and the woman has to lean back to keep her balance. Uterine contractions become more frequent.

About 266 days after fertilization, uterine contractions become regular and strong, the uterine membranes rupture and the amniotic fluid is discharged, the cervix dilates, and after a variable period of labor, the baby is born. The placenta is expelled about 30 minutes later, and pregnancy is over.

Pregnancy Testing

The external signs of pregnancy are significant, but not always dependable, and a woman usually does not conclude that she is pregnant until laboratory tests confirm

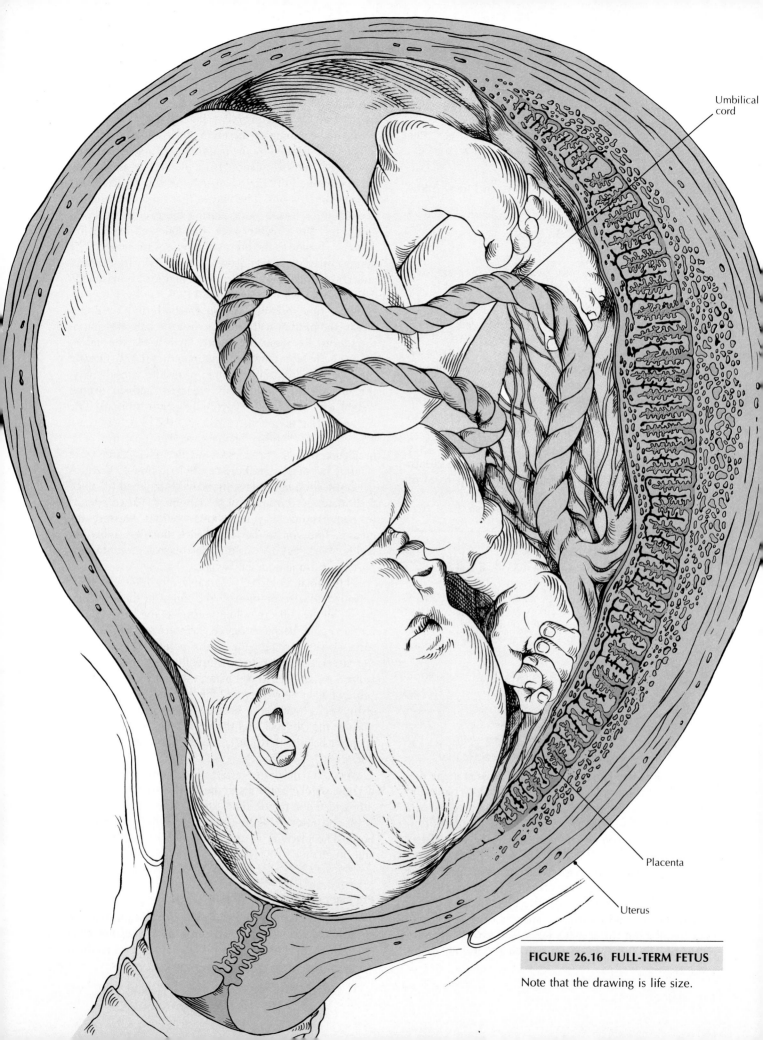

Umbilical cord

Placenta

Uterus

FIGURE 26.16 FULL-TERM FETUS

Note that the drawing is life size.

her suspicions. Pregnancy may be detected by some tests as early as 12 days after the first missed menstrual period, but tests are most accurate after the fortieth day of pregnancy. Most of these tests depend on the fact that *human chorionic gonadotropin* (hCG), a hormone produced by the placenta, is abundant in the blood and urine of a pregnant woman. The level of hCG is highest during the first three months of pregnancy and then decreases.

In testing, a drop of a chemical that neutralizes hCG (anti-hCG) is combined with a drop of the woman's urine on a glass slide, and a minute later, latex rubber particles are added. If the woman is not pregnant, the anti-hCG will attach itself to the latex particles, and curdlike lumps will form. If the woman *is* pregnant, the hCG will be neutralized by the anti-hCG before the latex particles are added, and no hCG will be left to form lumps.

ASK YOURSELF

1 What is usually the first sign of pregnancy?

2 How is hCG related to pregnancy tests?

BIRTH

Childbirth, or *parturition* (L. *parturire*, to be in labor), usually occurs about 266 days after fertilization, or about 280 days from the first day of the menstrual period preceding fertilization. During the last month of pregnancy, the myometrium of the uterus becomes increasingly sensitive to the hormone oxytocin, and a pregnant woman usually begins to experience irregular uterine contractions at that time.

Although the sensitivity to oxytocin may be caused by either an increased level of uterine prostaglandins or a decreased level of progesterone and estrogens, the exact cause is not known. The signal for the initiation of labor may come from the fetus itself, but no firm evidence supports this idea. One theory for the onset and maintenance of labor is that the fetus finally reaches a point in its development where its head begins to press down on the cervix. This pressure causes the cervix to dilate, which produces uterine contractions and the secretion of oxytocin. Oxytocin stimulates further uterine contractions until the baby is finally pushed down past the cervix and out through the vagina.

Process of Childbirth

The initial contractions stimulate the liberation of oxytocin from the pituitary gland, which further stimulates

even more powerful uterine contractions. Waves of muscular contractions spread down the walls of the uterus, forcing the fetus toward the cervix [FIGURE 26.17]. By now, the cervix is dilated close to its maximum diameter of about 10 cm (4 in.).

The amniotic sac may burst at any time during labor, or it may have to be ruptured by the attending physician. Either way, the ruptured sac releases the amniotic fluid. If this loss of fluid occurs very early, it may signal the onset of labor. Another possible indication that labor has begun is the release of the cervical plug of mucus from the vagina. Ordinarily, either the bursting of the amniotic sac or the release of the cervical plug (termed "show") will occur early enough to provide ample time to prepare for childbirth, but great variations exist among pregnant women.

During pregnancy, the muscle cells of the uterus grow to as much as 40 times their former size, transforming the uterus into an enormously powerful muscle. The sturdy walls of the uterus harbor and nourish the fetus during pregnancy, and finally, through the muscular contractions during labor, the uterus forcefully expels the fetus outward through the vaginal canal.

Babies are born head first, in the so-called *cephalic position*, about 95 percent of the time. Other possible birth positions almost always produce complications that require the intercession of the attending physician. Remember that the skull bones of a newborn are not yet fused, and the skull is still pliable. If it were not, the baby's head would not be small enough to pass through the vaginal canal.

After the baby is born, it is still attached to the placenta by the umbilical cord. Immediately after the baby is expelled from the uterus, the umbilical cord is clamped and cut below the clamp. Ordinarily, the umbilical cord is cut immediately after birth, but some physicians believe that severing the cord should take place only after the afterbirth is expelled and all of the placental blood drains into the baby's circulation. Such a procedure gives the baby an additional 80 to 90 mL of blood, about 25 to 30 percent more than it would receive otherwise.

Human fetuses are born long before they are able to care for themselves — in a sense, they are always born prematurely. But if gestation continued for longer than 9 or 10 months, the baby's relatively large head (and brain) could not pass through the vaginal canal.

Three stages of *labor* are usually described:

1 The *first stage of labor* starts with regular uterine contractions. Contractions occur every 20 or 30 minutes at first and become more and more frequent until they occur every 2 or 3 minutes. The first stage usually lasts about 14 hours for the birth of a first child, and gets shorter for subsequent births. Its main function is to dilate the cervix to its maximum.

2 The *second stage of labor* is the actual birth of the child. It lasts anywhere from several hours to several minutes, with the average time being two hours. If the fetal membranes have not ruptured during the first stage, they do so now.

3 The *third stage of labor*, which takes about 20 minutes, is the delivery of the placenta, fetal membranes, and any remaining uterine fluid, collectively called the **afterbirth.** It usually follows within 30 minutes of the expulsion of the baby and is accomplished by further uterine contractions. These final contractions also help to close off the blood vessels that were ruptured when the placenta was torn away from the uterine wall. About half a liter (1 pt) of blood is usually lost at this time.

The period after the placenta has been delivered is called the **puerperal period** (L. *puerperus,* bearing young; *puer,* child + *parere,* to give birth to). It is the time when the mother's body reverts back to its nonpregnant state, and it usually lasts at least a month. For example, the uterus and vagina revert back to their normal sizes about six weeks after childbirth. If a woman is not breastfeeding her baby, it will take anywhere from 6 to 24 weeks for the first menstruation.

Premature and Late Birth

On the average, pregnancy lasts for about 266 days after fertilization, but many variations occur, and babies have survived after being in the uterus for only 180 to 200 days. These *premature* babies usually grow to be healthy, normal children, but they may require about three years to catch up developmentally to their full-term contemporaries. Babies born after 240 days are not considered to be premature as long as they weigh at least 2.4 kg (5.5 lb).

Premature babies are usually poorly proportioned compared with healthy full-term babies. They usually have a weak cry, poor temperature regulation, wrinkled and dull red skin, closed eyes, and short fingernails and toenails. Because their muscular system is not fully developed, their movements are labored, and breathing may be difficult. Yet, a baby born a month early has at least a 70-percent chance of survival. It is interesting that a 1.8-kg (4-lb) baby born after 34 or 35 weeks, and nursed by its mother, has a better chance of being normal than a full-term baby of the same weight. This is so because when a baby is born prematurely its mother's milk contains more protein, antibodies, sodium, and chloride than it would have at full term.

When a fetus remains in the uterus for longer than the normal full term, the attending physician may decide to induce labor with intravenous injections of pitocin (synthetic oxytocin) or other methods, or to perform a cesarean section. Many physicians prefer to let the baby be born whenever it is ready, as long as there are no complications. Babies born late who weigh more than 4 or 4.5 kg (9 or 10 lb) usually are not as healthy as normal-sized full-term babies, probably because the mother's body cannot provide adequate nourishment in such circumstances.

Q: *Why are some babies born prematurely?*

A: Babies can be born prematurely for many reasons, including fetal malformations and genetic defects, but the most common cause is the mother's poor health. Maternal factors include malnutrition, alcohol or drug abuse, smoking, and hypertension.

FIGURE 26.17 BIRTH OF A BABY

The internal events are shown in a series of six models, **[A]–[F]**. (Photographs of live births show only external events.)

[A]

[D]

Multiple Births

Multiple births are always an exception in human beings. Twins are born once in 86 births, triplets once in 7400 births, and quadruplets once in 635,000 births. The chances of quintuplets being born are about 1 in 55 million. In most cases, one or more of the children born as quintuplets or sextuplets do not survive beyond infancy.

Women who have taken "fertility drugs" (which stimulate the ovaries), women who are older than 35, and women who have had children previously may have a higher-than-average rate of multiple births. This increase in multiple births may be caused by irregular ovu-

lation patterns in older women and the resultant release of more than one ovum at a time. Apparently, the tendency to release more than one ovum at a time is an inherited trait. However, the cell separation that produces identical twins is not due to a hereditary factor, and its cause is not known.

Twins may be identical or fraternal. *Fraternal (dizygotic) twins* are formed when more than one ovum is released from the ovary or ovaries at the same time and two ova are fertilized [FIGURE 26.18A]. Each fraternal twin usually has its own placenta, umbilical cord, chorion, and amnion. Fraternal twins are the most common multiple births. About 70 percent of all twins are frater-

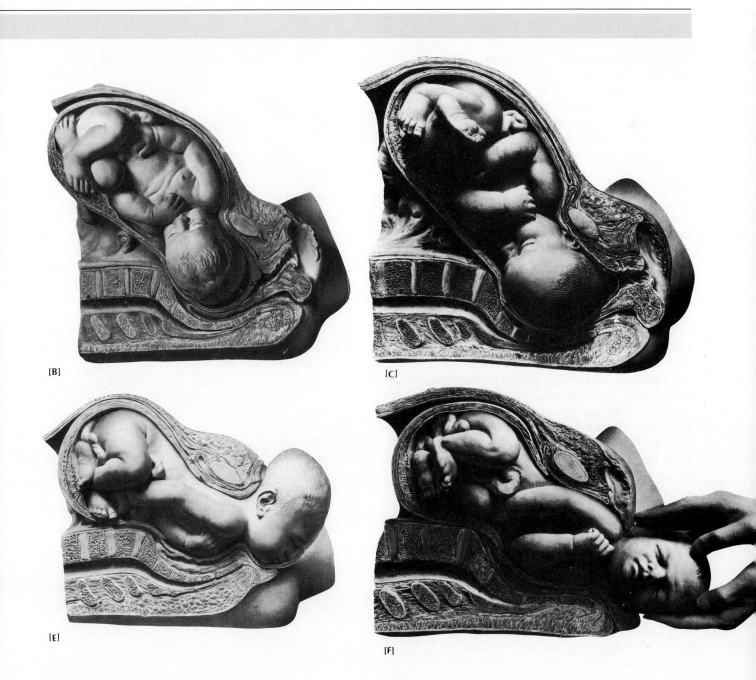

[B]

[C]

[E]

[F]

FIGURE 26.18 MULTIPLE BIRTHS

[A] Fraternal twins result when two ova are fertilized. Each usually has its own placenta and chorion. **[B]** Identical twins result when the inner cell mass of one fertilized ovum divides in half. Identical twins share the same placenta and chorion.

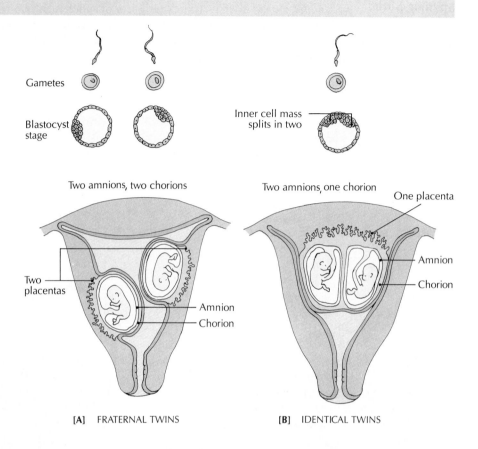

[A]   FRATERNAL TWINS   [B]   IDENTICAL TWINS

nal, and 30 percent are identical. Aside from being the same age, fraternal twins resemble one another no more than any other brothers or sisters, and may be the same sex or different sexes.

FIGURE 26.18B shows the formation of ***identical (monozygotic) twins.*** Because identical twins result from the same fertilized ovum, they are always the same sex and are genetically identical. Apparently, identical twins are formed when a fertilized ovum divides into two identical cells before implantation (about day 6 or 7), but there is also some evidence that identical twins may also form after implantation. Because human identical twins develop from a single fertilized ovum, they are contained within a common chorionic sac and have a common placenta. However, the umbilical cords are separate and the amnions are also individual, except in rare instances.

Triplets may be either identical or fraternal, and they may be formed in several ways. For example, three separate ova may be fertilized by three sperm. Two ova may be fertilized, and one may split, producing identical twins. The other possibility is that one ovum is fertilized and then splits into three parts, producing identical triplets.

Conjoined twins, commonly called *Siamese twins,** are identical twins whose embryonic disks do not separate completely. Sometimes the twins share only skin or tissue, but frequently they share an organ, such as a liver or an anus. Conjoined twins usually die before they reach old age. They are more often females than males and are more common than quintuplets.

Adjustments to Life Outside the Uterus

Most of the body systems of the late fetus are developed and ready to function, but they are not used until the baby is born. Although the sudden shift from total dependency in the uterus to near-independence upon birth is drastic, the newborn baby usually adapts to its new

---

*In 1811, Eng and Chang were born as conjoined twins in Bangkok, the capital of Siam (now Thailand). They shared a liver and were joined together at the lower end of the sternum. When they were 18, they were taken to the United States by P. T. Barnum, who made the twins famous in his circus. Eng and Chang (which mean "left" and "right" in Siamese) were not actually Siamese, since their parents were Chinese. The Siamese called them the "Chinese twins."

environment smoothly and without apparent trauma. The major changes affect breathing and circulation.

At birth, the baby must adapt quickly to life in air by making several rapid physiological and anatomical adjustments. It must shift from a fetal circulation, which depends on the placenta, to an essentially adult circulation, which depends on the lungs for gas exchange. Before birth, the fetal heart has an opening, the *foramen ovale,* that conveys blood from the right to the left atrium. Also, the *ductus arteriosus* conveys blood from the pulmonary artery to the thoracic aorta, thus bypassing the lungs [see FIGURE 20.12].

Both the atrial opening and the pulmonary bypass close at birth, allowing the blood to pass through the pulmonary artery to the newly functioning lungs. If either the foramen ovale or ductus arteriosus does not close off completely, the oxygen content of the blood will be low, and the baby's skin will appear slightly blue. A neonate with this condition is therefore called a "blue baby." Usually a defective ductus arteriosus must be corrected surgically after birth.

The start of actual breathing can be helped if it does not begin immediately at birth. Fluid can be cleaned from the nose and mouth of the neonate, the baby can be held upside down to help drain the breathing passage, and if necessary, it can be given a smack on its bottom to shock it into crying, its first gasp of air. Today it is more common to stimulate breathing with a small whiff of carbon dioxide.

ASK YOURSELF

1 What are the three stages of labor?

2 What are some of the problems that premature babies often face?

3 How do fraternal and identical twins differ?

4 What are some of the physiological adjustments the neonate has to make?

POSTNATAL LIFE CYCLE

From the time of fertilization, human development is a highly organized combination of three processes: growth, morphogenesis, and cellular differentiation. *Growth* is an increase in overall size, usually from an increase in the number of cells as the result of cell division. These cells must be arranged into specific structures that ultimately form the tissues and organs of the body. The cellular movements that bring about tissue and organ development are called *morphogenesis* (Gr. *morphe,* shape + *gennan,* to produce). After cells have become organized into specific forms, they must become specialized in order to perform the specific functions of tissues. This specialization process is called *cellular differentiation. Development* refers to the changes produced by these processes in successive phases of the life cycle, which begins with fertilization and ends with death. In the earlier sections of this chapter, we focused on prenatal development and birth. Here we will outline the stages of *postnatal* ("after birth") development, from the first weeks of life to old age.

Neonatal Stage

The newborn baby is called a *neonate* during its first four weeks of life. This period is one of many adjustments. In addition to the adjustments in the respiratory and circulatory systems already described, a neonate may face some special functional problems:

1 Because its bones are still being transformed from a skeleton of cartilage, a neonate requires an adequate supply of calcium and also of vitamin D, which increases the absorption of calcium. It needs iron to ensure that its liver will continue to synthesize red blood cells, and it needs vitamin C (which is not stored) to aid the proper formation of cartilage, bone, and intercellular materials.

2 The neonatal rates of respiration, cardiac output, and body metabolism are about twice as high as adult rates, in proportion to body weight, and as a result, these body functions may be unstable.

3 Because the temperature-regulating system of a neonate is unreliable, body temperature can fluctuate greatly.

4 Fluid intake and excretion are about seven times greater in a neonate than an adult, making diarrhea and vomiting major problems.

5 The liver may not function adequately during the first postnatal week, causing a lowered concentration of plasma proteins, a subnormal excretion of bilirubin (producing jaundice), an insufficient synthesis of glucose, and poor blood coagulation.

6 Secretions of pancreatic amylase are too low for the proper digestion of starches, and fat absorption is poor.

7 Although the neonate has acquired antibodies from its mother during intrauterine development, its own production of antibodies is not fully operating for at least a year. When the infant does begin to produce antibodies, severe allergies may develop, resulting in serious eczema, gastrointestinal abnormalities, or even anaphylaxis (immediate hypersensitivity that can be serious).

Infancy

The neonatal period is followed by *infancy,* which continues until the infant can sit erectly and walk, usually between 10 and 14 months after birth. The body weight generally triples during the first year, from an average birth weight of 5.4 kg (7 lb 8 oz) for both boys and girls, to an average year-old weight of 9.5 kg (21 lb) for boys and 9.1 kg (20 lb) for girls. The average height for boys and girls is 45.7 cm (18 in.) at birth and 73.7 cm (29 in.) at 1 year.

Gradual physiological changes occur during infancy, especially in the cardiovascular and respiratory systems. Myelinization of the nervous system begins during this period, and motor activities become more and more coordinated. Vision and other functions of the cerebral cortex are refined during the first few postnatal months. The size of the brain increases from one-quarter to one-half of adult size, and the incisor teeth erupt. Because of the immaturity of the liver and kidneys, the infant may be unable to excrete toxic substances, and in general, the infant is susceptible to viral and bacterial diseases before its immune system is functioning at its total capability.

Childhood

Childhood is the period from the end of infancy to the beginning of adolescence (12 to 13 years). In early childhood, ossification is rapid. Bone growth slows down in late childhood and then increases again during the prepubertal period. Deciduous teeth erupt and are replaced by permanent teeth during late childhood, motor coordination becomes more fully developed, and language and other intellectual skills are refined.

Adolescence and Puberty

The hormone-secreting gonads of both sexes are inactive until the final maturation of the reproductive system. This period of maturation, development, and growth is called *adolescence* (L. *adolescere,* to grow up). It spans the years from the end of childhood to the beginning of adulthood. Adolescence and puberty are not the same. *Puberty* (L. *puber,* adult) is the period when the individual becomes physiologically capable of reproduction. It usually takes place during the early years of adolescence. The average age of puberty is 12 for girls and 14 for boys.

Before puberty, between the ages of 7 and 10, there is usually a slow increase in the secretion of estrogens and androgens. The concentrations of these sex hormones remain inadequate, however, to promote the development of *secondary sex characteristics,* such as body hair

or enlarged genital organs. At the start of puberty, the hypothalamus begins to release luteinizing hormone-releasing hormone (LHRH), which stimulates the secretion of FSH and LH from the pituitary. It is these pituitary gonadotropic hormones that stimulate the testes and ovaries to secrete androgens and estrogens at full capacity.

New evidence suggests that the pineal gland is an important factor in the hormonal changes that initiate puberty. The pineal gland secretes the hormone melatonin, which inhibits the production of sex hormones during early childhood. At the onset of puberty, however, there is an abrupt drop in the secretion of melatonin, allowing estrogens and androgens to promote sexual changes.

Sequence of body changes The events of puberty [TABLE 26.1] usually occur in a definite sequence. The typical pattern for a boy begins about the age of 11, when he may begin to get spontaneous erections, with no apparent cause. He may also accumulate deposits of fat prior to the actual changes of puberty. The events of puberty may take as long as four years. The average 18-year-old American boy is 1.76 m (5 ft 9.5 in.) tall and weighs 68 kg (150 lb). He will probably stop growing when he is about 22, having added about 1.27 cm (0.5 in.) to his height.

The changes that accompany puberty in a girl start about two years earlier than those for boys, and they also happen more rapidly and closer together. The whole sequence of changes, starting with the breast buds increasing in size and the nipples beginning to protrude, usually takes two and a half to three years. When an average American girl reaches physical maturity at about 18, she is 1.64 m (5 ft 4.5 in.) tall and weighs 55.8 kg (123 lb).

Menarche The first menstrual period, the *menarche* (meh-NAR-kee; Gr. "beginning the monthly"), comes during the latter stages of puberty, though ovulation may not take place until a year or so later. The menarche seems to occur when a girl reaches a weight of about 48 kg (106 lb), rather than at a certain age or height. At the start of adolescence, a female has five times as much lean tissue as fat tissue, but the amount of fat more than doubles by the time of menarche. Should there be a shortage of food, the extra fat is enough to nourish a fetus through 9 calendar months of pregnancy, and to provide milk for the newborn baby for about a month.

The average age of menarche is 12.8 years, but it usually occurs later (13.1 years) in areas of high altitude. Contrary to some beliefs, the menarche is not hastened by a warm climate, but good nutrition seems to encourage an early menarche, and is probably the most important cause of the increasingly lower ages for menarche.

TABLE 26.1 MAJOR BODY CHANGES AT PUBERTY: SECONDARY SEX CHARACTERISTICS

Area of body	Description of changes (male)	Description of changes (female)
General bodily changes	Shoulders broaden, muscles thicken, height increases. Body odor from armpits, genitals. Skeletal growth ceases by about age 21.	Pelvis widens. Fat distribution increases in hips, buttocks, breasts. Skeletal growth ceases by about age 18.
External genital organs	Penis increases in size, becomes more pigmented. Scrotum becomes enlarged, more pigmented, more wrinkled.	Breasts enlarge. Vagina enlarges and vaginal walls thicken.
Internal genital organs	Testes enlarge. Sperm production increases in testes. Seminal vesicles, prostate gland, bulbourethral glands enlarge, begin to secrete.	Uterus enlarges. Ovaries secrete estrogens. Ova in ovaries begin to mature. Menstruation begins.
Skin	Secretions of sebaceous glands thicken and increase, often causing acne. Skin thickens.	Estrogen secretions keep sebaceous secretions fluid, inhibit development of acne and blackheads.
Hair growth	Hair appears on face, pubic area, armpits, chest, around anus. General body hair increases. Hairline recedes in the lateral frontal regions.	Hair appears on pubic area, armpits. Scalp hair increases with childhood hairline retained.
Voice	Voice becomes deeper as larynx enlarges and vocal cords become longer and thicker.	Voice remains relatively high pitched as larynx grows only slightly.

Adulthood

Adulthood begins between about 18 and 25 and spans the years until old age. The peak of sexual capacity for males and the maximum secretion of testosterone are reached between the ages of 18 and 20, and muscular strength reaches its peak at about 25.

Typical age changes Although individuals vary, certain changes are common at various ages. Scalp hair usually remains thick throughout the 20s, but height may begin decreasing by 25. The level of thymic hormones decreases steadily.

Between the ages of 30 and 40, the functional capacity of the body declines by about 0.8 percent per year. The heart muscle begins to thicken, and the skin loses some elasticity, producing wrinkles. Vertebral disks begin to degenerate, causing the vertebrae to move closer together. Hearing, which was most efficient at 10, begins to decline.

Between 40 and 50, the back may begin to hunch over. A 40-year-old man is probably 4.5 to 9 kg (10 to 20 lb) heavier and ⅛ in. shorter than at 20. Hair begins to turn gray and becomes thinner. Most people become farsighted in their late 40s. Lymphocytes are less effective against cancer cells, and the immune system declines in effectiveness.

Between 50 and 55, the skin continues to loosen and wrinkle, and the sense of taste declines. Nearsighted people may have temporarily normal eyesight as their eyes become less effective at close range.

Rapid changes begin between the ages of 55 and 60. Muscles and other tissues begin to deteriorate, and body weight may decrease. As the metabolism slows, fat may accumulate to balance the weight loss. Although billions of neurons in the brain have been lost by now, memory loss is minimal in a healthy person. The volume of semen decreases, and the male voice usually sounds higher as the vocal cords stiffen and vibrate at a higher frequency.

Menopause Women between 45 and 55 usually stop producing and releasing ova, and the monthly menstrual cycle stops. This cessation of menstrual periods is called the *menopause* ("ceasing the monthly"), and it signals the end of reproductive ability. The average age for menopause is currently 52 and is increasing. Menopause is not an abrupt change; it usually takes about two years of irregular menstrual periods before menstruation and ova production stop permanently.

The decrease of estrogen that accompanies menopause is thought to be the cause of several physical problems, including "hot flashes" of the skin, caused by changes in the vasomotor system that dilate blood vessels and increase blood flow. Estrogen decrease may also cause osteoporosis and other irregularities in bone metabolism, as well as dizziness, fatigue, headaches, chest and neck pains, and insomnia. Before menopause, diseases such as hardening of the arteries are quite rare in women, probably because estrogen lowers cholesterol in the blood. But as estrogen secretions continue to diminish after menopause, the incidence of cardiovascular disease becomes almost equal in men and women. Psychological disturbances, including depression, may accompany the physical problems. Some of the symptoms of menopause may be relieved by small doses of estrogen, administered under the strict supervision of a physician.

Although males usually experience a gradual decrease in testosterone secretion after they reach 40 or 50, they do not experience as drastic hormonal changes as women do at that age. The most likely cause of psychological problems during this period is not hormonal, but rather the fear of impotency and old age. Despite the decrease in testosterone, normal males may retain sexual potency in old age.

Senescence

The indeterminate period when an individual is said to grow old is called *senescence* (L. *senescere*, to grow old). By the age of 70, height is usually a full inch less than it was in the 20s or 30s. Between 70 and 80, body strength decreases to half of what it was at 25, lung capacity decreases to half, and about 65 percent of a person's taste buds become inactive. The nose, ears, and earlobes are up to a half inch longer. Life expectancy is currently 71.4 for males and 78.7 for females. Current evidence suggests that the maximum human life span is about 110 years.

ASK YOURSELF

1 What are the three main processes of human development?

2 Define neonate, infant, child, and adult.

3 What is the difference between adolescence and puberty?

4 What changes occur during puberty in boys and girls?

WHEN THINGS GO WRONG

During the 9-month gestation period, *anything* can go wrong, but fortunately, everything of importance usually goes right. Several developmental problems have already been described in this chapter, so we will concentrate on how embryonic and fetal problems are detected and even corrected.

Amniocentesis

Amniocentesis (am-nee-oh-sehn-TEE-sihss; Gr. sac + puncture) is the technique of obtaining cells from a fetus. First, the fetus is located by bouncing high-frequency sound waves off it and recording the echoes. This is a relatively safe procedure, unlike the use of potentially dangerous x-rays. Then, a hypodermic needle is inserted into the amnion, usually directly through the abdominal wall of the mother (see drawing). The amnion surrounding the fetus is filled with fluid in which a number of loose cells float. Because all of these cells are derived from the original zygote, they are genetically alike. Some of the amniotic fluid may be grown as a tissue culture. The cells in the fluid may be studied directly to save time, but cultured cells can be made to yield more information.

By measuring the amount of protein that leaks from the neural tube into the amniotic fluid, such diseases as spina bifida and anencephaly (absence of the forebrain)

may be detected. Hemolytic diseases (including hemolytic disease of the newborn) may be detected. Inborn errors of metabolism may be identified by studying cell cultures derived from the amniotic fluid.

Since chromosomes are examined, the sex of the fetus can be determined, but medically it is more important to discover any chromosomal abnormalities that might indicate possible disorders. The specific chromosome characteristics, or *karyotype*, can be read with fair accuracy. For example, if three chromosomes of chromosome number 13 are present instead of the usual pair, the fetus, if allowed to come to term, will probably not survive infancy. If there is a history of abnormality in the family, or the parents are known to be at risk for some disorder, amniocentesis may reassure the parents by showing that the chromosomes are normal.

The procedure of amniocentesis is reasonably safe for both fetus and mother. Amniocentesis is advisable if a mother is known to carry a chromosomal aberration or if she is over 40, since women over 40 have a disproportionately high number of babies with chromosomal defects.

A promising new technique provides more information than amniocentesis by allowing physicians to locate the fetus by ultrasound and place a needle into the blood

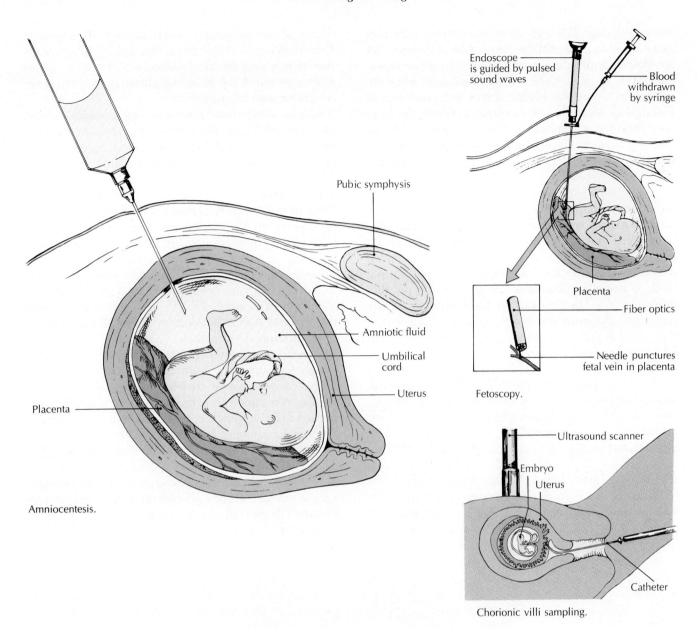

Pubic symphysis

Amniotic fluid

Umbilical cord

Uterus

Placenta

Amniocentesis.

Endoscope is guided by pulsed sound waves

Blood withdrawn by syringe

Placenta

Fiber optics

Needle punctures fetal vein in placenta

Fetoscopy.

Ultrasound scanner

Embryo

Uterus

Catheter

Chorionic villi sampling.

vessels of the umbilical cord to take samples of fetal blood, to inject drugs, or to perform transfusions. This technique can be used as early as the eighteenth week of pregnancy.

Fetoscopy

Fetoscopy ("seeing the fetus") allows the physician to view the fetus directly. The fetus is located with ultrasound waves, and then an *endoscope* (an instrument for examining the interior of hollow organs) is inserted into the uterus through a small abdominal incision (see drawing). Physicians can also insert tiny forceps to take a fetal skin sample from the fetal blood vessels that lie on the surface of the placenta. Like any other intrauterine examination, fetoscopy may damage fetal or maternal tissue if done improperly.

Chorionic Villi Sampling

Chorionic villi sampling is a technique that can be performed in a physician's office as early as the fifth week of pregnancy. A catheter is inserted through the mother's cervix into the uterus, where a sample of chorionic villi tissue is collected (see drawing). The tissue is identical to fetal tissue, and its DNA can be examined for such diseases as Down's syndrome and sickle-cell anemia. The technique appears to have a low risk factor and can be performed 6 weeks earlier than amniocentesis.

Alpha-fetoprotein Test

Alpha-fetoprotein is a substance that is excreted by the fetus into the amniotic fluid and enters the maternal bloodstream. It can be measured by the *alpha-fetoprotein test* on a sample of blood from the pregnant woman.

Abnormally high levels of alpha-fetoprotein in the maternal blood indicate that the neural tube of the fetus has not closed completely, allowing large quantities of alpha-fetoprotein to "leak." The test is administered when the fetus is 16 weeks old. It may detect such neurological disorders as *anencephaly*, a condition in which the newborn child has a primitive brain or no brain at all, and *spina bifida*, a condition that exposes a portion of the spine. Neural-tube defects occur in about 1 out of every 1000 births.

Intrauterine Transfusion

An *intrauterine transfusion* is an injection of a concentrate of red blood cells into the peritoneal cavity of the fetus. The concentrate is prepared from Rh-negative whole blood to help the fetus combat hemolytic anemia (such as hemolytic disease of the newborn), a possibly fatal condition. The cells pass into the fetal circulation by way of the diaphragmatic lymphatics.

Ultrasonography

Ultrasonography is a technique that uses high-frequency sound waves to locate and examine the fetus. Ultrasound images are projected onto a viewing monitor. Ultrasonography allows the physician to see the embryonic sac during the embryonic period, the size of the fetus and its placenta, multiple fetuses, and any abnormal fetal positions. Serious abnormalities of the central nervous system, such as hydrocephaly, can also be detected.

Recent refinements in ultrasonography permit physicians to see *inside* the fetal heart. The new technique can detect severe congenital problems, such as abnormal heart chamber and valve formation, and any other condition that would usually produce a "blue baby." This technique allows the attending physician to prepare for treatment soon after birth.

Critics of electronic fetal monitoring, which includes ultrasonography, suggest that ultrasound waves may produce cell damage, and they recommend that sound waves be used only when complications are suspected.

Fetal Surgery

Fetal surgery, surgery performed on the fetus *before* it is born, is still in its experimental stages, but the early results have been promising. In 1982 the first successful operation on a fetal kidney was performed at the University of Connecticut Health Center. The kidney was threatened by an accumulation of excess urine because of a blockage of the ureter. The kidney was drained through the pregnant woman's abdominal wall with the guidance of an ultrasound video scan.

About 3 weeks later, another sonogram showed that fluid was accumulating again, and doctors decided to perform a cesarean section to deliver the 8-month-old fetus so that conventional corrective surgery could be performed. The surgery was successful, and the infant left the hospital a month later with two healthy kidneys.

Fetal surgery has corrected hydrocephalus and other diseases. In an extreme case, a 5-month fetus was removed from the uterus, operated on for a urinary blockage, and then returned to the uterus. It was born healthy 4 months later.

CHAPTER SUMMARY

Embryonic Development (p. 671)

1 The fusion of ovum and sperm nuclei produces the diploid *zygote.* The daughter cells, produced by a series of mitotic cell divisions called **cleavage,** are *blastomeres.* After cleavage begins, the developing organism is called an *embryo* until the end of the eighth week. During the third day, about the time the embryo is entering the uterus, a solid ball of about 16 blastomeres, which is called the *morula,* is formed.

2 After 2 or 3 days, the morula develops into a fluid-filled hollow sphere called a *blastocyst.* The blastocyst is composed of about 100 cells. Its outer covering of cells *(trophoblasts)* is the **trophectoderm;** the blastocyst cavity is the *blastocoel;* a grouping of cells at one pole is the *inner cell mass,* from which the embryo will grow.

3 About a week after fertilization, the blastocyst becomes implanted in the uterine lining. *Implantation* is the actual start of pregnancy.

4 During week 2, the blastocyst develops into the *bilaminar embryonic disk,* composed of the endoderm and ectoderm.

5 The *extraembryonic membranes* include the yolk sac, amnion, allantois, and chorion.

6 The *placenta* synthesizes glycogen and other nutrients, transports materials between the mother and embryo, and secretes protein and steroid hormones. The blood vessels in the *umbilical cord* carry nutrients from the maternal placenta to the fetus and wastes from the fetus to the placenta.

7 The embryonic heart is beating by the fourth week, and by the end of 8 weeks, the major organ systems have become established. The embryo is now called a *fetus.*

Fetal Development (p. 680)

The fetal period begins with the ninth week. Growth is rapid at first but slows down between the seventeenth and twentieth weeks, as development continues and systems become mature enough to sustain life outside the uterus.

Maternal Events of Pregnancy (p. 683)

1 The first maternal sign of pregnancy is usually a missed menstrual period. Pregnancy tests, which depend on the abundance of hCG in the blood and urine of pregnant women, confirm or deny pregnancy.

2 The pregnant woman's body goes through a series of physical and hormonal changes to accommodate the developing and enlarging fetus. Weak uterine contractions may begin about week 10, becoming stronger as the pregnancy progresses, and frequent during week 36.

3 About day 266, uterine contractions become regular and strong, the uterine membranes rupture and the amniotic fluid is discharged, the cervix dilates, and after a variable period of labor, the baby is born. Pregnancy officially ends about 30 minutes later, with the expulsion of the placenta.

Birth (p. 685)

1 *Parturition* is the birth of the child. The initial uterine contractions stimulate the pituitary gland to secrete oxytocin, which increases the force of the contractions until the baby is born.

2 The three stages of *labor* are usually described as the start and continuation of regular uterine contractions, the birth of the child, and the delivery of the afterbirth. The period from the delivery of the afterbirth until the mother's body reverts to its nonpregnant state is the *puerperal period.*

3 A baby is considered *premature* if it is born before 240 days and weighs less than 2.4 kg.

4 *Multiple births* are always an exception in human beings. *Fraternal* (dizygotic) *twins* are formed when two ova are released and are fertilized. *Identical* (monozygotic) *twins* result from the same fertilized ovum and are genetically identical and the same sex. *Conjoined twins* are identical twins whose embryonic disks do not separate completely.

5 Among the adjustments the newborn makes to extrauterine life, the major ones are breathing on its own and the shift from a fetal circulation to an adult circulation, which depends on the lungs for gas exchange.

Postnatal Life Cycle (p. 689)

1 Human development consists of *growth,* an increase in size and number of cells as a result of mitosis; *morphogenesis,* the cellular movements that bring about tissue and organ development; and *cellular differentiation,* the specialization of cells that allows them to perform the specific functions of tissues.

2 The first four weeks after birth are the *neonatal period.* It is followed by *infancy,* which continues until the infant can sit erectly and walk, usually at 10 to 14 months. *Childhood* is the period from the end of infancy to the beginning of adolescence at 12 or 13 years.

3 *Adolescence* is the period of maturation, development, and growth from the end of childhood to the beginning of adulthood. *Puberty* is the period during early adolescence when an individual becomes physiologically capable of reproduction. It occurs about two years earlier in females than in males.

4 At the start of puberty, the release of GnRH from the hypothalamus stimulates the secretion of FSH and LH from the pituitary. These pituitary hormones stimulate the testes and ovaries to secrete androgens and estrogens at full capacity.

5 The events of puberty usually occur in a definite sequence. The major body changes involve the development of *secondary sex characteristics.* The first menstrual period, or *menarche,* usually occurs during the latter stages of puberty.

6 *Adulthood* begins between about 18 and 25 and spans the years until old age. The cessation of menstrual periods, *menopause,* usually occurs between the ages of 45 and 55.

7 The indeterminate period when an individual is said to grow old is *senescence.*

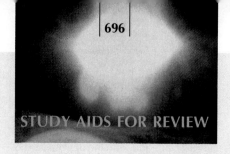

MEDICAL TERMINOLOGY

ABORTION (L. *aboriri*, to disappear) The fatally premature expulsion of an embryo or fetus from the uterus.

ABRUPTIO PLACENTAE The premature separation of the placenta from the uterus.

CESAREAN SECTION An incision through the abdominal wall and the uterine wall *(hysterotomy)* to remove a fetus.

CONGENITAL DEFECT A condition existing at birth, but not hereditary.

DYSTOCIA Abnormal or difficult childbirth.

EPISIOTOMY The cutting of the vulva at the lower end of the vaginal orifice at the time of delivery in order to enlarge the birth canal.

FETAL MONITORING The continuous recording of the fetal heart rate, using a transcervical catheter.

FETOLOGY The study of the fetus.

GALACTORRHEA A persistence of lactation.

GYNECOMASTIA Breast development in a male.

LAPAROSCOPE (Gr. *lapara*, flank) A long, thin, light-bearing instrument used during a *laparotomy*, which is the surgical incision into any part of the abdominal wall.

LOCHIA (LOE-kee-uh; Gr. *lokhios*, of childbirth) The normal discharge of blood, tissue, and mucus from the vagina after childbirth.

PERINATOLOGY (Gr. *peri* around + L. *natus*, born) The study and care of the fetus and newborn infant during the perinatal period, usually considered to be from about the fourth month of pregnancy to about a month after childbirth.

PLACENTA PREVIA A condition in which the placenta is closer to the cervix than the fetus is.

PROLAPSE OF THE UMBILICAL CORD Expulsion of the umbilical cord before the fetus is delivered, reducing or cutting off the fetal blood supply.

TOXEMIA An abnormal condition caused by toxic substances in the mother's blood that may be harmful to the fetus.

UNDERSTANDING THE FACTS

1 Into what two main periods may the development of a human be divided?

2 During what period in human development do we use the term zygote? Embryo? Fetus?

3 About how long after fertilization does the first cell division occur?

4 Define trophoblast cells, and explain what they will develop into.

5 When does pregnancy actually begin?

6 Into what structure does the blastocyst develop just after implantation?

7 What is the significance of the chorionic villi?

8 In which structure are blood cells initially developed?

9 Briefly describe the role of the placenta.

10 To what is the last six months of pregnancy devoted?

11 Give two or three indications that labor is imminent or has just begun.

12 Why do we usually state development periods in lunar instead of calendar months?

13 Give the location of the following (be specific):

   a morula   d ductus arteriosis

   b blastocoel   e lanugo

   c foramen ovale

UNDERSTANDING THE CONCEPTS

1 What is the significance of the fact that the overall size of the organism when it consists of several blastomeres is approximately the same as that of the original zygote?

2 What is meant by differentiation?

3 What are several benefits of having the developing individual suspended in the amniotic fluid?

4 The vein within the umbilical cord transports oxygen and nutrients; the arteries transport carbon dioxide and nitrogen wastes. Can you reconcile this fact with the usual function of arteries and veins?

5 What is the significance of the foramen ovale and the ductus arteriosus?

6 Why may hemorrhoids and varicose veins often develop during late pregnancy?

7 List some of the information that can be obtained by amniocentesis.

8 What are some uses of intrauterine transfusions? Fetal surgery?

SELF-QUIZ

Multiple-Choice Questions

26.1 During the third through fifth day of embryonic development, a solid ball of cells forms called a

   a zygote   d gastrula

   b blastomere   e blastocyst

   c morula

26.2 The mass of cells that will develop into the chorion and placenta is termed the

   a trophectoderm   d zona pellucida

   b embryoblast   e inner cell mass

   c blastocyst

26.3 Which of the following is *not* one of the extra-embryonic membranes?

    a yolk sac   **d** allantois

    b coelom   **e** chorion

    c amnion

26.4 Which of the following is *not* a function of amniotic fluid?

    a suspends the embryo

    b prevents the embryo from adhering to the amnion

    c permits the developing fetus to change position

    d provides nutrients to the fetus

    e provides a stable temperature

26.5 The placenta functions in

    a producing and contributing nutrients and energy to the embryo

    b transporting material between the mother and embryo

    c secreting hormones

    d a and b

    e a, b, and c

26.6 The diagnostic technique in which cells obtained from an unborn fetus are examined is called

    a fetoscopy   **d** ultrasonography

    b fetal surgery   **e** amniocentesis

    c endoscopy

26.7 An intrauterine transfusion could be used to treat

    a Down's syndrome

    b muscular dystrophy

    c spina bifida

    d hemolytic disease of the newborn

    e PKU

26.8 The cessation of menstrual periods is called

    a amenorrhea   **d** a and b

    b menopause   **e** a, b, and c

    c menarche

True-False Statements

26.9 The yolk stalk becomes incorporated into the umbilical cord via the body stalk.

26.10 The site where the chorion joins intimately with the endometrium develops into the placenta.

26.11 Gases pass through the placenta from the maternal blood to the fetus by passive diffusion and specific carrier molecules.

26.12 The last six months of prenatal development are devoted to the increase in size and maturation of the organs.

26.13 At the beginning of the third lunar month, the face of a fetus looks recognizably human.

26.14 At the end of the sixth lunar month the body is covered with downy hair called lanugo.

Completion Exercises

26.15 Four or five days after fertilization, the embryo is in the form of a hollow sphere called a _____.

26.16 The process in which the embryo becomes embedded in the uterine wall is _____.

26.17 The embryonic membrane that develops as an outpocketing of the yolk sac is the _____.

26.18 The daughter cells that result from division of the zygote are called _____.

26.19 The process by which cells develop into specialized tissues and organs is termed _____.

26.20 The protective sac around the embryo is termed the _____.

26.21 Blood vessels that carry materials between the embryo and the mother are found in the _____.

26.22 Birth occurs about _____ days from conception.

26.23 A baby is at full term by the _____ lunar month.

A SECOND LOOK

1 In the following figure, label the amnion, body stalk, yolk stalk, and chorionic cavity.

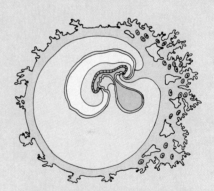

2 Label the chorionic villi, umbilical cord, amnion, and amniotic cavity.

APPENDIXES / REFERENCES / CREDITS
GLOSSARY / INDEX

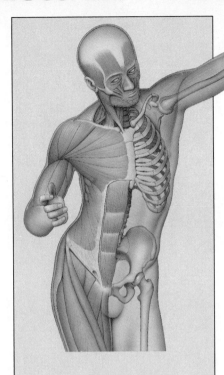

APPENDIXES

A

An Overview of Regional
Human Anatomy

B

Prefixes, Combining Word Roots,
and Suffixes

C

Reclassification of
Eponymous Terms

D

Some Useful Abbreviations

E

Answers to Self-Quizzes

SELECTED ANATOMICAL REFERENCES

CREDITS

GLOSSARY

INDEX

APPENDIX A

An Overview of Regional Human Anatomy

The following photographs of dissected, preserved human cadavers convey a sense of the actual appearance and relations of grossly observable anatomical structures. The photographs are in contrast to the drawings in textbooks and atlases that, no matter how accurately illustrated, are merely representations of the actual body. The photographs in this appendix provide the student with an opportunity to view an actual dissection of the human body. The photographs begin with external views, proceed to views of the superficial and deep muscle layers, and then show the thoracic viscera, the posterior abdominal and thoracic wall, the abdominal cavity with viscera exposed, and finally, male and female pelvic cavities with reproductive organs exposed.

These photographs are reminders of the profound role played by the dissection of the human body in the advancement of science. Leonardo da Vinci (1452–1519) produced his legendary (but not always accurate) anatomical sketches during the Renaissance, and in 1543 a scientific breakthrough occurred with the publication of *De corporis humani fabrica* ("On the Structure of the Human Body") by Andreas Vesalius (1514–1564). The anatomically accurate drawings in this monumental work were based on dissections of the human body, and the illustrations of the muscles in particular were so accurate that they are still used by students of anatomy and medical illustration. Coincidentally, in the same year, the Polish astronomer Nicolas Copernicus (1473–1543) published his great book *De revolutionibus orbium coelestium libri VI* ("Six Books On the Revolutions of the Celestial Spheres") that proposed that the earth and other planets revolve around the sun in relatively fixed orbital paths. Together, the two books provided the foundation for the birth of the Scientific Revolution.

FIGURE A.1 EXTERNAL ANTERIOR VIEW OF THE BODY

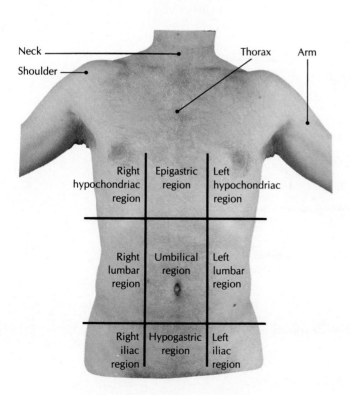

FIGURE A.2 EXTERNAL POSTERIOR VIEW OF THE BODY

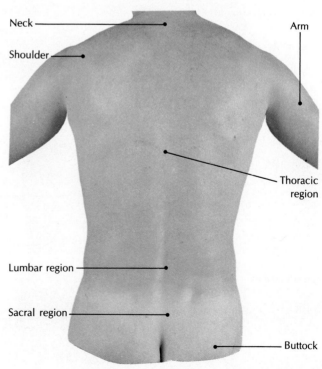

FIGURE A.3 SUPERFICIAL MUSCLE LAYER EXPOSED

Anterior view.

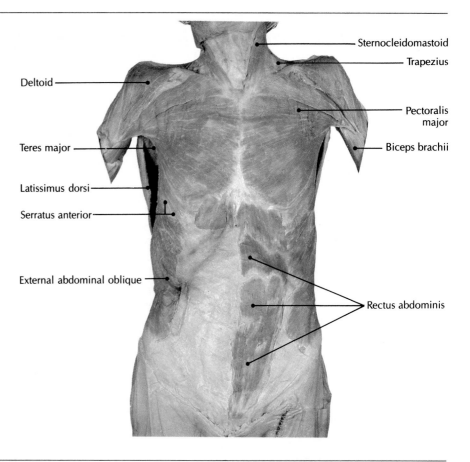

Sternocleidomastoid

Trapezius

Deltoid

Pectoralis major

Biceps brachii

Teres major

Latissimus dorsi

Serratus anterior

External abdominal oblique

Rectus abdominis

Infraspinatus

FIGURE A.4 SUPERFICIAL MUSCLE LAYER EXPOSED

Posterior view.

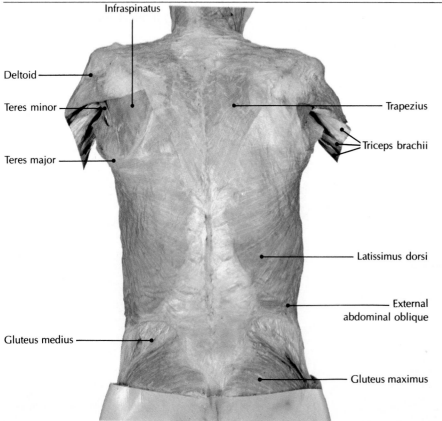

Deltoid

Teres minor

Teres major

Trapezius

Triceps brachii

Latissimus dorsi

External abdominal oblique

Gluteus medius

Gluteus maximus

FIGURE A.5 DEEP MUSCLE LAYER EXPOSED

Anterior view.

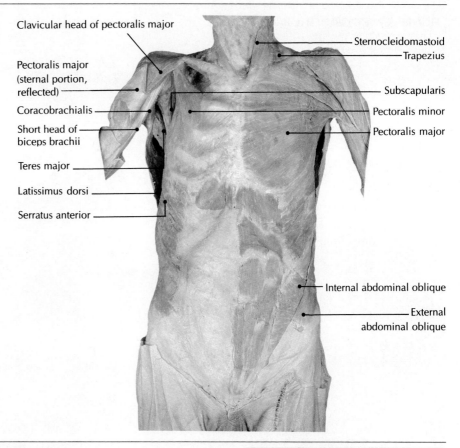

Clavicular head of pectoralis major

Sternocleidomastoid

Trapezius

Pectoralis major (sternal portion, reflected)

Subscapularis

Coracobrachialis

Pectoralis minor

Short head of biceps brachii

Pectoralis major

Teres major

Latissimus dorsi

Serratus anterior

Internal abdominal oblique

External abdominal oblique

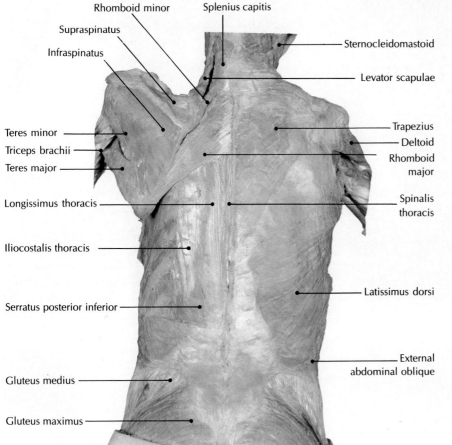

Rhomboid minor

Splenius capitis

Supraspinatus

Infraspinatus

Sternocleidomastoid

Levator scapulae

Teres minor

Triceps brachii

Teres major

Trapezius

Deltoid

Rhomboid major

Longissimus thoracis

Spinalis thoracis

Iliocostalis thoracis

Serratus posterior inferior

Latissimus dorsi

External abdominal oblique

Gluteus medius

Gluteus maximus

FIGURE A.6 DEEP MUSCLE LAYER EXPOSED

Posterior view.

FIGURE A.7 EXPOSED THORACIC VISCERA

Anterior view.

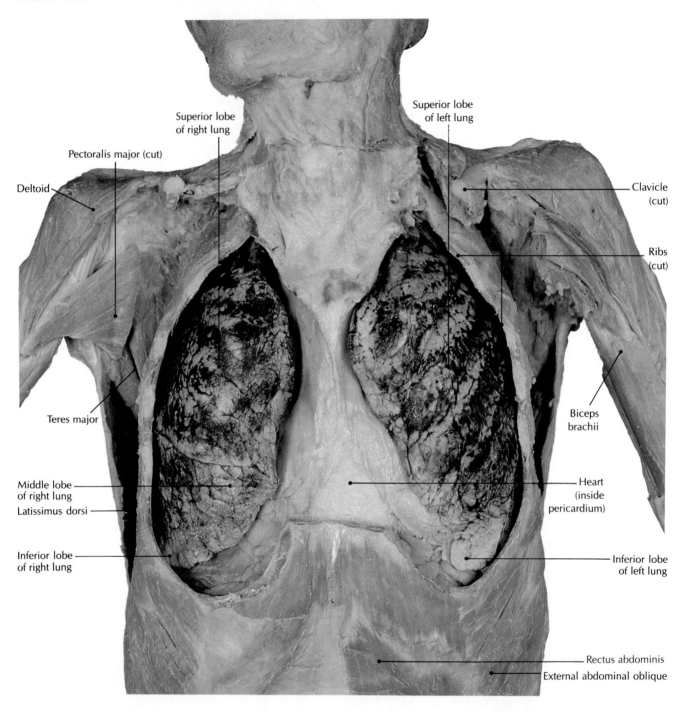

Superior lobe of right lung

Superior lobe of left lung

Pectoralis major (cut)

Deltoid

Clavicle (cut)

Ribs (cut)

Teres major

Biceps brachii

Middle lobe of right lung

Latissimus dorsi

Heart (inside pericardium)

Inferior lobe of right lung

Inferior lobe of left lung

Rectus abdominis

External abdominal oblique

FIGURE A.8 EXPOSED POSTERIOR WALL OF THORAX AND ABDOMEN

Anterior view.

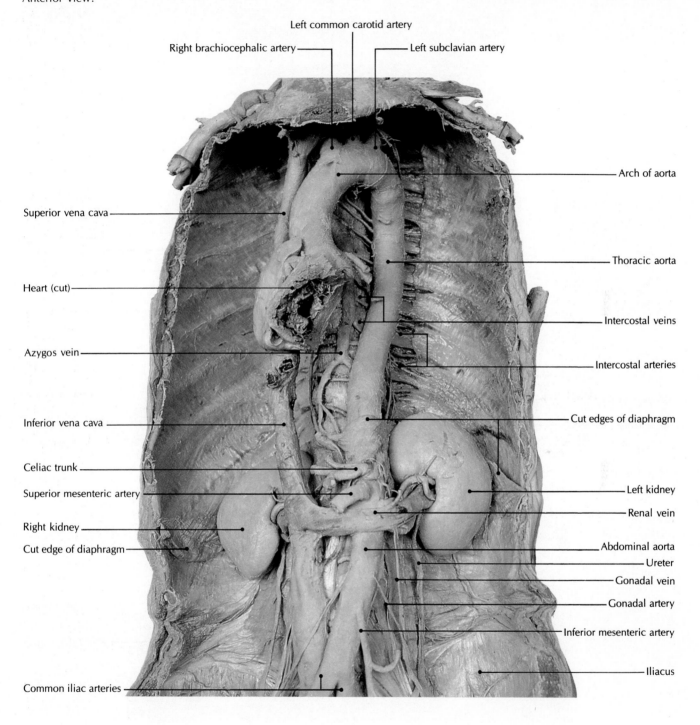

Left common carotid artery

Right brachiocephalic artery

Left subclavian artery

Arch of aorta

Superior vena cava

Thoracic aorta

Heart (cut)

Intercostal veins

Azygos vein

Intercostal arteries

Inferior vena cava

Cut edges of diaphragm

Celiac trunk

Superior mesenteric artery

Left kidney

Renal vein

Right kidney

Abdominal aorta

Cut edge of diaphragm

Ureter

Gonadal vein

Gonadal artery

Inferior mesenteric artery

Iliacus

Common iliac arteries

FIGURE A.9 EXPOSED FEMALE PELVIC CAVITY AND REPRODUCTIVE ORGANS

Sagittal view.

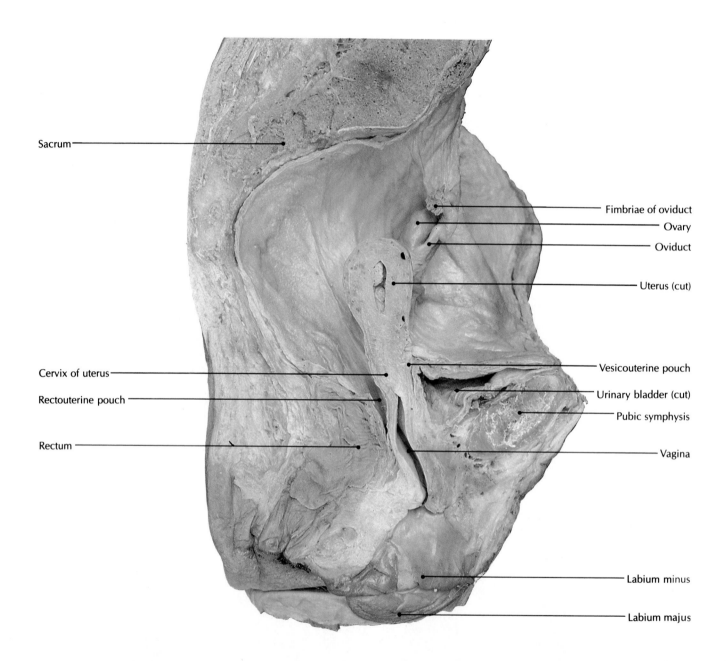

Sacrum

Cervix of uterus

Rectouterine pouch

Rectum

Fimbriae of oviduct

Ovary

Oviduct

Uterus (cut)

Vesicouterine pouch

Urinary bladder (cut)

Pubic symphysis

Vagina

Labium minus

Labium majus

FIGURE A.10 EXPOSED MALE PELVIC CAVITY AND REPRODUCTIVE ORGANS

Sagittal view.

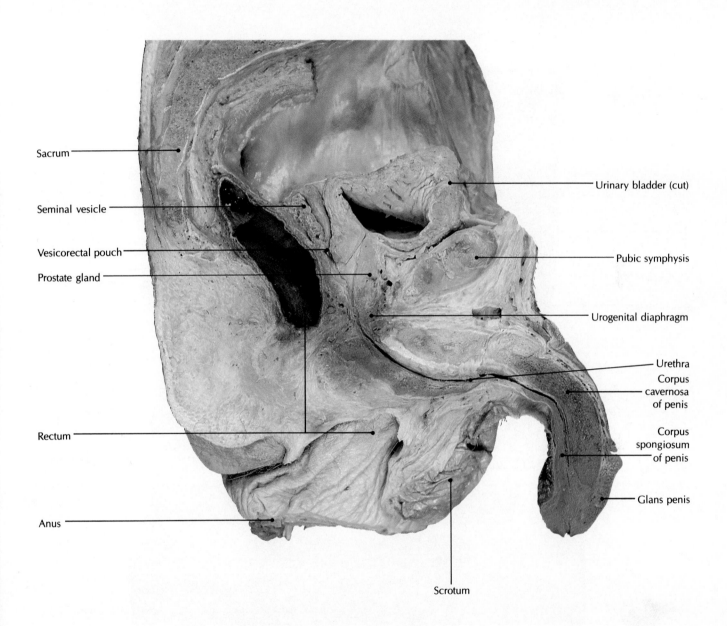

Sacrum

Seminal vesicle

Vesicorectal pouch

Prostate gland

Rectum

Anus

Urinary bladder (cut)

Pubic symphysis

Urogenital diaphragm

Urethra
Corpus cavernosa of penis

Corpus spongiosum of penis

Glans penis

Scrotum

FIGURE A.11 ABDOMINAL CAVITY WITH VISCERA EXPOSED

Anterior view.

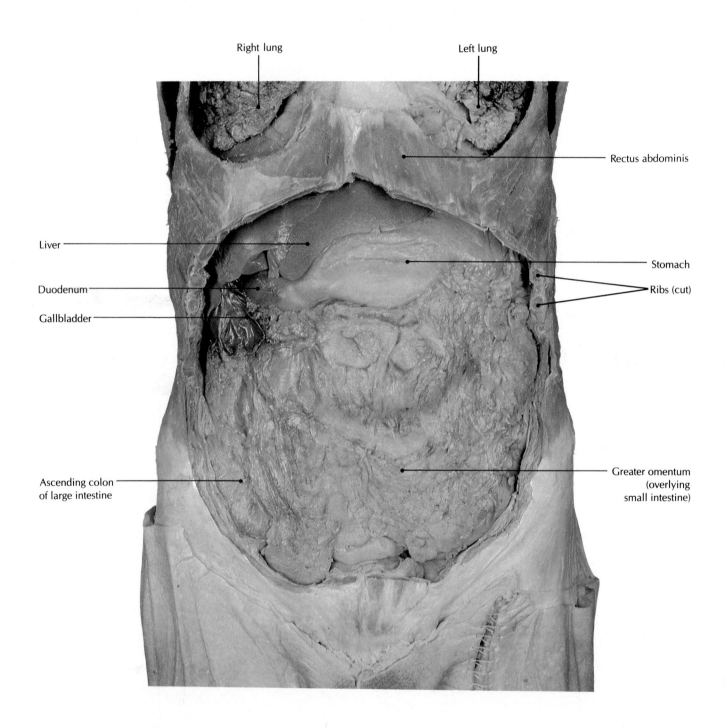

Right lung

Left lung

Rectus abdominis

Liver

Stomach

Duodenum

Ribs (cut)

Gallbladder

Ascending colon
of large intestine

Greater omentum
(overlying
small intestine)

APPENDIX B

Prefixes, Combining Word Roots, and Suffixes

PREFIXES AND COMBINING WORD ROOTS

a-, an- without, absent (*apnea:* temporary absence of respiration).

ab- away from (*abduct:* move away from).

abdomin(o)- abdomen (*abdominal:* pertaining to the abdomen).

acou- hearing (*acoustics:* study of sound).

acr(o)- extremity, tip (*acrocyanosis:* bluish color of hands and feet).

ad- to, toward, near (*adduct:* move toward).

aden- gland (*adenoma:* glandular tumor).

af- to, toward (*afferent:* toward a given point).

alba- white (*albino:* lacking color, appearing white).

alg- pain (*neuralgia:* nerve pain).

alve- channel (*alveolus:* air channel in lung).

ambi-, amphi- both, on both sides (*ambidextrous:* using both hands; *amphiarthrosis:* an articulation permitting only slight motion).

ambly- dull (*amblyopia:* diminished vision).

ana- up, back again (*anabolism:* building up).

andr(o)- male (*androgen:* male sex hormone).

angi(o)- vessel, duct, usually a blood vessel (*angiology:* study of blood and lymph vessels).

ankyl(o)- crooked, bent, fused, stiff (*ankylosed:* fused, as a joint).

ante- before, in front of (*antepartum:* before delivery).

anti- against, counteracting (*antidote:* treatment to counteract effects of a poison).

arthr(o)- joint (*arthritis:* inflammation of a joint).

atel- imperfect, incomplete (*atelectases:* imperfect expansion of lungs).

auto- self (*autoregulatory:* self-regulating).

bi- two, twice (*binocular:* pertaining to both eyes).

bio- life (*biochemistry:* chemistry of living matter).

blasto- growth (*blastocyst:* hollow ball of embryonic cells).

brachi(o)- arm (*brachialis:* muscle for flexing forearm).

brachy- short (*brachydactyly:* abnormal shortness of fingers or toes).

brady- slow (*bradycardia:* slow heart rate).

bronch- windpipe (*bronchitis:* inflammation of the bronchus).

cac(o)- bad, ill (*cachexia:* generally poor condition of the body).

capit- head (*capitulum:* small eminence, or little head, on a bone by which it articulates with another bone).

carcin(o)- cancer (*carcinogenic:* cancer causing).

cata-, kata- down, through (*catabolism:* breaking down).

caud- tail (*caudal:* toward the posterior part of the body).

centi- 1/100, 100 (*centimeter:* 1/100 meter).

cephal(o)- head (*encephalitis:* inflammation of the brain).

cerebr(o)- brain (*cerebral:* pertaining to the brain).

cervic- neck (*cervix:* neck-shaped outer end of the uterus).

chondr(o)-, chondri-, chondrio- cartilage (*perichondrium:* membrane surrounding cartilage).

chromo-, chromato- color (*chromocyte:* colored cell).

circum- around (*circumduction:* circular motion around a joint).

cirrh- yellow (*cirrhosis:* disease causing yellowing of the liver).

co-, com-, con- with, together (*congenital:* born with).

coel- cavity within a body or body organ (*coelom:* cavity formed by splitting the mesoderm into two layers).

contra- opposite, against (*contralateral:* opposite side).

crani(o)- skull (*craniotomy:* surgical opening in the skull).

cryo- cold (*cryosurgery:* surgery performed with application of extreme cold to tissues).

crypt(o)- hidden (*cryptorchidism:* undescended testicle).

cyan(o)- blue (*cyanosis:* bluish discoloration of the skin).

cyst- sac, bladder (*cystitis:* inflammation of the bladder).

cyt(o)- cell (*cytology:* study of cells).

de- away, not, down (*dehydrate;* take water away from).

deca- 10 (*decaliter:* 10 liters).

deci- 1/10 (*deciliter:* 1/10 liter).

dent(i)- tooth (*dentiform:* tooth-shaped).

derm- skin (*dermatitis:* inflammation of the skin).

dextro- right (*dextrogastria:* displacement of the stomach to the right side of body).

di- two, twice, doubly (*diphasic:* occurring in two stages).

dia- apart, across, through, between, completely (*diastasis:* separation of normal joined parts).

diplo- double (*diplopia:* double vision).

dis- apart, away from (*disarticulate:* come apart at the joints).

dorso-, dorsi- back (*dorsiflexion:* bending to the back of hand, foot, or spine).

duct- lead, conduct (*ductus deferens:* tube that conducts sperm from testes or ejaculatory duct).

dys- diseased, difficult (*dysmenorrhea:* difficult menses).

ecto- outside (*ectoderm:* outer germ layer).

edem- swelling (*edema:* swelling).

em-, en- in (*encapsulated:* enclosed in a capsule).

endo- within (*endoderm:* inner germ layer).

enter(o)- intestine (*gastroenteritis:* inflammation of the digestive system).

epi- on, over, upon (*epidermis:* outer layer of skin).

erythr(o)- red (*erythropoiesis:* formation of red blood cells).

eu- good, normal (*eupnea:* easy, normal breathing).

ex- out of (exhale: breathe out).

exo- outside (*exocytosis:* process of expelling large particles from a cell).

extra- outside of, in addition to, beyond (*extracellular:* outside a cell).

fasci- band (*fascia:* sheet of fibrous tissue).

fore- before, in front (*forebrain:* front part of brain).

gastr(o)- stomach (*gastritis:* inflammation of the stomach).

glyco-, gluco- sugar, sweet (*glycolysis:* breakdown of glucose).

gyn(o), gyne(co)- female, woman (*gynecology:* study of the female reproductive system).

haplo- single, simple (*haploid:* having a single set of chromosomes).

hema-, hemato-, hemo- blood (*hematocyst:* a cyst containing blood).

hemi- one-half (*hemiplegia:* paralysis of one side of the body).

hepat(o)- liver (*hepatitis:* inflammation of the liver).

hetero- other, different (*heterosexual:* other sex).

hist(o)- tissue (*histology:* study of tissues).

homeo- unchanged (*homeostasis:* state of inner stability of the body).

homo- same (*homolateral:* on the same side).

hydr(o)- water (*hydrocele:* accumulation of serous fluid in a body cavity).

hyper- above, excessive (*hyperactivity:* excessive activity).

hypo- beneath, under, deficient (*hypogastric:* below the gastric region).

idio- peculiar to the individual, distinct (*idiopathic:* disease with no apparent cause).

in-, im- in, within, into; not (*indigestion:* difficulty in digesting food; *immerse:* dip into).

infra- below (*infracostal:* below the ribs).

inter- between, among (*intercellular:* between cells).

intra- within (*intracellular:* within a cell).

ir- against, into, toward (*irradiate:* emit rays into).

is(o)- equals, same (*isopia:* equal vision in both eyes).

labi- lip (*labia major:* external folds or lips of the vulva).

lacri- tears (*nasolacrimal apparatus:* tear-producing and tear-carrying apparatus).

lact(o)- milk (*lactose:* milk sugar).

leuk(o)- white (*leukocyte:* white blood cell).

lip(o)- fat, fatty (*lipoma:* benign tumor of fatty cells).

macr(o)- large (*macrophage:* large phagocytic cell).

mal- bad (*malabsorption:* impaired absorption).

medi- middle (*medial:* near the midline of the body).

mega- large (*megakaryocyte:* large cellular forerunner of platelet).

melan(o)- black (*melanin:* dark skin pigment).

meso- middle (*mesoderm:* middle layer of skin).

meta- beyond, next to (*metacarpal:* bone of the hand that is next to the wrist).

micr(o)- small (*photomicrograph:* photograph of a very small object as seen through a microscope).

milli- 1/1000 (*millimeter:* 1/1000 meter).

mon(o)- one (*monocular:* pertaining to one eye).

morph(o)- form, shape (*morphology:* study of the form and structure of organisms and parts of organisms).

multi- many (*multinuclear:* having many nuclei).

my(o)- muscle (*myocardium:* heart muscle).

myel(o)- marrow, spinal cord (*myeloma:* malignant tumor of bone marrow).

neo- new, strange (*neonatal:* newborn).

nephr(o)- kidney (*nephrectomy:* surgical removal of a kidney).

neur(o)- nerve (*neuron:* nerve cell).

non- not (*noninfectious:* not able to spread).

ob- against (*obstruction:* blocking of a structure).

ocul(o)- eye (*oculomotor:* pertaining to eye movement).

olig(o)- little, few, scanty (*oliguria:* secretion of a small amount of urine).

oo- egg (*oocyte:* egg cell).

ophthalmo- eye (*ophthalmology:* study of the eye).

orth(o)- straight, normal, correct (*orthodontic:* pertaining to the straightening of teeth).

oste(o)- bone (*osteitis:* inflammation of the bone).

ot(o)- ear (*otology:* study of the ear).

ovi-, ovo- egg (*ovum:* egg cell).

para- beside, beyond (*paraspinal:* beside the spine).

path(o)- disease, suffering (*pathogenic:* capable of producing disease).

ped(ia)- child (*pediatrician:* physician who specializes in childhood diseases).

pend- hang down (*appendicular skeleton:* portion of the skeleton that hangs from the pectoral and pelvic girdles).

per- through (*permeate:* to pass through).

peri- around (*pericardium:* membrane surrounding the heart).

phag(o)- eat, consume (*phagocyte:* cell that ingests other cells or particles).

phleb(o)- vein (*phlebitis:* inflammation of a vein).

pleuro-, pleura- rib, side (*pleurisy:* inflammation of pleura, the membrane covering the lungs and lining the thoracic sac).

plur(i)- many, more (*pluriglandular:* several glands).

pneumato-, pneuma- air (*pneumatometer:* instrument for measuring pressure of inspiration and expiration of lungs).

pneumo-, pneumono- lung, respiratory organs (*pneumonia:* chronic lung infection).

pod- foot (*podiatrist:* a specialist in care and treatment of the foot).

poly- many (*polycystic:* having many cysts).

post- after (*postpartum:* after delivery).

pre-, pro- before (*prenatal:* before birth).

pro- for, in front of, before (*prophylaxis:* preventive treatment, to keep guard before).

proct- rectum, anus (*proctoscope:* instrument for examining the rectum).

pseudo- false (*pseudocyst:* false cyst).

pulmo- lung (*pulmonary:* pertaining to the lungs).
py(o)- pus (*pyocyst:* pus-filled cyst).

quadr(i)- four (*quadriplegia:* paralysis of all four limbs).

re- back, again (*reflex:* bend back).
ren- kidney (*adrenal:* relating to the position of the adrenal gland atop the kidney).
retro- behind, backward (*retrograde:* going backward).
rhin(o)- nose (*rhinitis:* inflammation of nasal mucous membranes).

sclero- hard (*sclerosis:* hardening of tissue).
semi- one-half, partly (*semilunar valve:* half-moon-shaped valve).
steno- contracted, narrow (*stenosis:* condition of being narrowed).
sub- beneath (*subcutaneous:* beneath the skin).
super- above (*superior:* toward the head).
supra- above, over (*suprapubic:* above the pubic bone).
sym-, syn- together, union (*synapse:* meeting between two or more nerve cells).

tachy- fast (*tachycardia:* abnormally fast heart rate).
telo- end (*telephase:* final phase of mitosis).
tetr(a)- four (*tetrad:* group of four chromosomes formed during meiosis).
therap- treatment (*therapeutic:* having a healing effect).
therm(o)- heat, warmth (*thermometer:* device to measure heat).
thorac(o)- chest (*thoracic cavity:* chest cavity).
thromb(o)- lump, clot (*thrombosis:* formation of a blood clot in a vessel or heart cavity).
tox- poison (*toxemia:* blood poisoning).
trans- across, through (*transepidermal:* occurring through or across the skin).
tri- three (*triceps:* three-headed muscle of the posterior of arm).

ultra- excessive, extreme (*ultrasonic:* sound waves beyond the audio-frequency range).
un(i)- one (*unilateral:* one side).
uria-, uro- urine, urinary tract (*polyuria:* production of excess urine).

vas(o)- vessel (*cardiovascular:* vessel related to the heart).
viscera-, viscero- organ (*visceral:* pertaining to a body organ).

SUFFIXES

-ac pertaining to (*celiac:* pertaining to the abdominal region).
-ad toward, in the direction of (*cephalad:* toward the head).
-agra severe pain, seizure (*podagra:* severe pain in the foot).
-al pertaining to (*digital:* pertaining to the finger or toe).

-an, -ian pertaining to, characteristic of, belonging to (*ovarian:* pertaining to the ovary).
-ar of or relating to, being, resembling (*valvular:* relating to a valve).
-ary of or relating to, connected with (*biliary:* relating to bile).
-ase enzyme (*lactase:* enzyme that catalyzes conversion of lactose into glucose and galactose).
-ate that which is acted upon, marked by having, to act on (*substrate:* substance acted on by enzyme; *lobate:* having lobes; *separate:* to keep apart).
-atresia abnormal closure (*proctatresia:* closed anus).

-blast sprout, growth (*osteoblast:* cell from which bone develops).

-cele swelling, tumor (*cystocele:* hernia of urinary bladder).
-centesis puncture of a cavity (*paracentesis:* puncture of a space around an organ or within a cavity to remove fluid).
-cide kill (*germicide:* killer of germs).
-cis cut (*excise:* cut out).
-clasis, -clasia breaking up (*bacterioclasis:* breaking up of bacteria).
-cleisis closure (*colpocelisis:* operation for closure of vagina).

-desis binding, fusion (*tenodesis:* fixation of a loose tendon).
-dynia pain (*pleurodynia:* pain on the side of the chest).

-ectasia, -ectasis dilation, expansion (*telangiectasis:* dilation of capillaries).
-ectomy excision, cutting out (*laryngectomy:* removal of larynx).
-emia condition of the blood (*leukemia:* blood condition characterized by excess leukocytes).

-facient making or causing to become (*febrifacient:* causing a fever).
-ferent carrying (*efferent:* carrying away from).
-form structure, shape (*ossiform:* resembing the form of bone).
-fugal driving or traveling away from (*centrifugal:* moving away from the center).
-ful full of, characterized by (*stressful:* causing stress).

-gen, -gene producer (*mutagen:* substance that increases the frequency of mutation).
-genesis production of, origin of (*glycogenesis:* production of glycogen).
-gram record (*electrocardiogram:* record of electrical activity in the heart).
-graph instrument for making records (*electroencephalograph:* apparatus for recording brain waves).

-ia condition (*anuria:* condition of lack of urine).
-iasis a diseased condition (*cholelithiasis:* condition of bile stones).
-ic pertaining to (*colonic:* pertaining to the colon).
-ician, -ist specialist, practitioner (*pediatrician:* specialist in childhood diseases).

-ile having the qualities of or capability for (*febrile:* feverish).
-ion act of, state of, result of (*incision:* act of cutting or result of cutting).
-ism condition, act of, process of (*dwarfism:* condition of being a dwarf).
-itis inflammation (*phlebitis:* inflammation of a vein).
-ity quality of, state of (*obesity:* state of being obese).
-ive that which performs or tends toward (*tardive disease:* disease with late-appearing symptoms).

-logy study of (*cardiology:* study of the heart).
-lysis breaking down (*glycolysis:* breakdown of glucose).

-malacia softening (*osteomalacia:* softening of bone).
-meter instrument or means of measuring (*thermometer:* instrument to measure heat).

-ness quality of, state of (*illness:* state of being ill).

-oid having the appearance of (*cuboid:* resembling a cube).
-oma tumor (*carcinoma:* malignant tumor).
-opia eye disorder (*myopia:* nearsightedness).
-or that which, one who (*receptor:* that which receives).
-ory process of, pertaining to, function of (*sensory:* pertaining to the senses).
-ose full of, having the qualities of (*comatose:* having the qualities of a coma).
-osis action, state, process, condition (*halitosis:* condition of having bad breath).
-ostomy creation of an opening (*colostomy:* creation of a new opening between the bowel and abdominal wall).
-otomy cutting into (*tracheotomy:* cut into the trachea).
-ous, -ious having the qualities of, capable of, full of (*infectious:* capable of being transmitted).

-pathy disease (*cardiopathy:* heart disease).
-penia deficiency, lack (*leukopenia:* lack of white blood cells).
-pexy fixation (*sigmoidopexy:* fixation of large intestine).
-philia love of, tendency toward (*hemophilia:* "love of blood," a blood disease).

-phobia abnormal fear (*claustrophobia:* fear of confinement).
-plasty molding or shaping (*osteoplasty:* plastic surgery on bone).
-plegia paralysis (*paraplegia:* paralysis of lower half of body).
-poiesis production, formation of (*hematopoiesis:* production of blood cells).
-ptosis dropping, downward displacement (*carpoptosis:* wrist drop).

-rrhagia bursting forth, excessive discharge (*hemorrhage:* escape of blood from vessels).
-rrhaphy closure of by suturing, repair (*cystorrhaphy:* suture of bladder).
-rrhea flow, discharge (*galactorrhea:* excessive or spontaneous flow of milk).

-scope looking at, examining (*bronchoscope:* instrument used to view inside breathing tubes).
-sect cut (*dissect:* cut into parts).
-sis process, action (*dialysis:* separation of particles through a semipermeable membrane).
-stasis state of being at a standstill (*hemostasis:* stopping of blood flow).
-sthenia strength (*asthenia:* loss of strength).
-stomy surginal opening (*colostomy:* surgical opening in the colon).

-tion act of, result of, process of (*elongation:* process of making longer).
-tomy to cut (*lobotomy:* incision into a lobe).
-tonia stretching, putting under tension (*hypertonia:* excessive tension).
-tripsy rubbing, crushing (*lithotripsy:* surgical crushing of kidney stones).
-trophic related to nutrition, growth, development (*dystrophic:* faulty nutrition).
-tropic turning toward, changing (*hydrotropic:* turning toward water).

-y having the nature or quality of (*gouty:* goutlike).

APPENDIX C

Reclassification of Eponymous Terms

An *eponym* is a person for whom something is named, and an *eponymous term* is the term that used an eponym — *Achilles tendon*, for example. Because eponymous terms are not anatomically or physiologically descriptive, most of them have been replaced in this book with currently accepted terms that help the student understand the meaning of the term. Eponymous terms are shown in parentheses the first time the current term is used. In a few cases, because the eponym is used so extensively, it has been retained, with the new term in parentheses — *Broca's speech area (anterior speech area)*, for example. Two eponymous terms have been retained because of the lack of an appropriate substitution: *Tyson's glands* and *corpuscles of Ruffini*. (For the record, the champion eponym is the French chemist Georges Deniges, with 78 eponymous terms named after him.)

Eponymous term	Preferred term
Achilles reflex	plantar reflex
Achilles tendon	calcaneal tendon
Adam's apple	thyroid cartilage
ampulla of Vater	hepatopancreatic ampulla
aqueduct of Sylvius	cerebral aqueduct
axis of Henke	subtalar axis
Bartholin's glands	greater vestibular glands
Bowman's capsule	glomerular capsule
Bowman's glands	olfactory glands
Broca's area	Broca's speech area (anterior speech area)
Brodmann area 4	primary motor cortex
Brunner's glands	duodenal submucosal glands
bundle of His	atrioventricular bundle
canal of Schlemm	scleral venous sinus
circle of Willis	cerebral arterial circle
Cooper's ligaments	suspensory ligaments of the breast
Cowper's glands	bulbourethral glands
crypts of Lieberkühn	intestinal glands
duct of Santorini	accessory pancreatic duct
duct of Wirsung	pancreatic duct
Eustachian tube	auditory tube
Fallopian tube	uterine tube
fissure of Rolando	central sulcus
fissure of Sylvius	lateral cerebral sulcus
foramen of Monro	interventricular foramen
Golgi tendon organ	tendon organ

Eponymous term	Preferred term
Graafian follicle	vesicular ovarian follicle
Graves' disease	hyperthyroidism
gyrus of Heschl	transverse temporal gyrus
Hassal's corpuscles	thymic corpuscles
Haversian canal	central canal
Haversian system	osteon
islet of Langerhans	pancreatic islet
Krause's end bulbs	bulbous corpuscles (of Krause)
Krebs cycle	citric-acid cycle
Kupffer cells	stellate reticuloendothelial cells
Langerhans' cell	nonpigmented granular dendrocyte
Langer's lines	cleavage lines
Leydig cells	interstitial endocrinocytes
loop of Henle	loop of the nephron
Malpighian corpuscles	splenic nodules
Meissner's corpuscles	tactile (Meissner's) corpuscles
Merkel's disks	tactile (Merkel's) corpuscles
Müllerian duct	paramesonephric duct
Müller's tubercle	sino-utricular node
Nissl bodies	chromatophilic substance
nodes of Ranvier	neurofibral nodes
organ of Corti	spiral organ
Pacinian corpuscle	lamellated (Pacinian) corpuscle
Peyer's patches	aggregated lymph nodules
Purkinje fibers	cardiac conducting myofibers (Purkinje fibers)
Rathke's pouch	Hypophyseal pouch
Ruffini's corpuscles	corpuscles of Ruffini
Schwann cell	neurolemmocyte
Schwann sheath	neurilemma
Sertoli cells	sustentacular cells
Sharpey's fibers	periosteal perforating fibers
Skene's glands	lesser vestibular glands
sphincter of Boyden	sphincter of the common bile duct
sphincter of Oddi	sphincter of the hepatopancreatic ampulla
Stensen's ducts	parotid ducts
Volkmann's canal	perforating canal, nutrient canal
Wernicke's area	Wernicke's area (posterior speech area)
Wharton's duct	submandibular duct
Wharton's jelly	mucous connective tissue
Wormian bone	sutural bone

APPENDIX D

Some Useful Abbreviations

A&P	auscultation and percussion		**Cysto**	cystoscopic examination
ABO	main blood group system		**D&C**	dilation and curettage
ac	before meals (*ante cibum*)		**DERM**	dermatology
ACCOM	accommodation		**Diff**	differential (white blood cell) count
ACh	acetylcholine		**DIP joint**	distal interphalangeal joint
ACL	anterior cruciate ligament		**DMD**	Duchenne muscular dystrophy
ACTH	adrenocorticotropic hormone		**DNA**	deoxyribonucleic acid
AD	Alzheimer's disease		**DNR**	do not resuscitate
ADH	antidiuretic hormone		**DPT**	diphtheria, pertussis, and tetanus vaccine
ADP	adenosine diphosphate		**DRG**	dorsal root ganglia
AIDS	acquired immune deficiency syndrome		**DSA**	digital subtraction angiography
ALS	amyotrophic lateral sclerosis		**DSR**	dynamic spatial reconstructor
AMP	adenosine monophosphate		**DUB**	dysfunctional uterine bleeding
ANS	autonomic nervous system		**ECF**	extracellular fluid
ARDS	adult respiratory distress syndrome (caused by illness or injury)		**EEF**	electroencephalogram
			EM	electron micrograph; electron microscope
ARF	acute renal failure		**EMG**	electromyogram
ASTIG	astigmatism		**ENT**	ear, nose, throat
ATP	adenosine triphosphate		**Eosins**	eosinophils
AV	atrioventricular		**ER**	endoplasmic reticulum
Ba	barium		**ESR**	erythrocyte sedimentation rate
BaE	barium enema		**ESWL**	extracorporeal shock-wave lithotripsy
Basos	basophils		**FHT**	fetal heart tones
BBB	blood-brain barrier		**FRAs**	flexor reflex afferents
BPH	benign prostatic hypertrophy		**FSH**	follicle-stimulating hormone
Broncho	bronchoscopy		**GALT**	gut-associated lymphoid tissue (aggregated lymph nodules or Peyer's patches)
BX	biopsy			
C	cervical		**GH**	growth hormone
CA	cancer		**GI**	gastrointestinal (as in GI tract)
CAC	cardioacceleratory center		**GU**	genitourinary
CAD	coronary artery disease		**Hb**	hemoglobin
cAMP	cyclic AMP		**HbO$_2$**	oxyhemoglobin
CAT scan	computer-assisted tomography		**hCG**	human chorionic gonadotropin
CBC	complete blood count		**HCT**	hematocrit
CCK	cholecystokinin		**HD**	hemodialysis
CCU	coronary care unit		**HSG**	heterosalpingography
CEA	carcinoembryonic antigen		**HSV**	herpes simplex virus
CIC	cardioinhibitory center		**IBS**	irritable bowel syndrome
CMG	cystometrogram (study of the urinary bladder in contraction)		**ICSH**	interstitial cell-stimulating hormone
			ID	intradermal
CNS	central nervous system		**IM**	intramuscular
COAD	chronic obstructive airways disease		**IP joint**	interphalangeal joint
COPD	chronic obstructive pulmonary disease		**IU**	international units
CPR	cardiopulmonary resuscitation		**IUD**	intrauterine device
CRF	chronic renal failure		**IV**	intravenous
CS,C section	cesarean section		**IVP**	intravenous pyelogram
CSF	cerebrospinal fluid		**KUB**	kidney, ureter, bladder
CVA	cerebrovascular accident		**L**	lumbar
CVS	chorionic villi sampling		**LA**	left atrium
CXR	chest x-ray		**LDL**	low-density lipoprotein

LH	luteinizing hormone	**RE**	reticuloendothelial system
LLL	left lower lobe of lung	**REM**	rapid eye movement
LLQ	left lower quadrant	**Rh**	blood factor important in transfusions (from *Rh*esus monkey in which it was first found)
LM	light micrograph; light microscope		
LMP	last menstrual period	**RLL**	right lower lobe of lung
LP	lumbar puncture	**RLQ**	right lower quadrant
LRI	lower respiratory infection	**RML**	right middle lobe of lung
LTH	lactogenic hormone; luteotropic hormone	**RNA**	ribonucleic acid
LUL	left upper lobe of lung	**RP**	retrograde pyelogram
LUQ	left upper quadrant	**rRNA**	ribosomal ribonucleic acid
LV	left ventricle	**RUL**	right upper lobe of lung
Lymphs	lymphocytes	**RUQ**	right upper quadrant
MG	myasthenia gravis	**RV**	right ventricle
MH	marital history	**S**	sacral
MI	myocardial infarction	**SA**	sinoatrial node
MP joint	metacarpophalangeal joint	**SEM**	scanning electron micrograph; scanning electron microscope
MRI	magnetic resonance imaging		
mRNA	messenger ribonucleic acid	**SI joint**	sacroiliac joint
MS	multiple sclerosis	**SIDS**	sudden infant death syndrome
MSH	melanocyte-stimulating hormone	**SOB**	shortness of breath
MYOP	myopia	**SPF**	sun protection factor
N	normality	**STD**	sexually transmitted disease
NB	newborn	**STH**	somatotropic hormone
NE	norepinephrine	**SUBCU**	subcutaneous
NG tube	nasogastric tube	**T**	thoracic
NGU	nongonococcal urethritis	**T tubule**	transverse tubule
NMR	nuclear magnetic resonance	**TB**	tuberculosis
NPO	nothing by mouth (*nil per os*)	**TDF**	testis-determining factor
OB	obstetrics	**TEM**	transmission electron microscope; transmission electron micrograph
OPHTH	ophthalmology		
ORTHO	orthopedics	**TIA**	temporary ischemic attack
OT	oxytocin	**TM joint**	temporomandibular joint
OTO	otology	**tRNA**	transfer ribonucleic acid
pc	after meals (*post cibum*)	**TSH**	thyroid-stimulating hormone
PCT	positron-computed tomography	**TURP**	transurethral resection of the prostate
PCV	packed cell volume	**UA**	urinalysis
PET	positron-emission tomography	**UC**	uterine contractions
PID	pelvic inflammatory disease	**UGI**	upper gastrointestinal
PIP joint	proximal interphalangeal joint	**UNG**	ointment
PMN	polymorphonuclear leukocyte	**URI**	upper respiratory infection
PNS	peripheral nervous system	**UTI**	urinary tract infection
po	given orally (*per os*)	**UV**	ultraviolet
PP	after meals (postprandial)	**UVB**	ultraviolet B
PRL	prolactin	**VA**	visual acuity
PT	physical therapy	**VD**	venereal disease
PTH	parathyroid hormone	**VF**	visual field
PU	pregnancy urine	**VLDL**	very low-density lipoprotein
RA	rheumatoid arthritis	**WBC**	white blood cell
RAS	reticular activating system	**WBC**	white blood cell count
RBC	red blood cell; red blood cell count	**XX**	female sex chromosome
RDS	respiratory distress syndrome	**XY**	male sex chromosome

APPENDIX E

Answers to Self-Quizzes

CHAPTER 1

1.1 e, 1.2 c, 1.3 e, 1.4 b, 1.5 b,
1.6 T, 1.7 F, 1.8 T, 1.9 F, 1.10 T,
1.11 connective, 1.12 integumentary,
1.13 peripheral, 1.14 lymph,
1.15 plantar, 1.16 transverse,
1.17 proximal, 1.18 ventral,
1.19 lungs, 1.20 heart, 1.21 g,
1.22 e, 1.23 c, 1.24 i, 1.25 f,
1.26 h, 1.27 d, 1.28 j, 1.29 a,
1.30 b

CHAPTER 2

2.1 c, 2.2 d, 2.3 c, 2.4 c, 2.5 a

CHAPTER 3

3.1 c, 3.2 e, 3.3 e, 3.4 e, 3.5 c,
3.6 a, 3.7 F, 3.8 T, 3.9 F, 3.10 F,
3.11 T, 3.12 T, 3.13 T,
3.14 glycocalyx, 3.15 selective
permeability, 3.16 endoplasmic
reticulum (ER), 3.17 lysosomes,
3.18 metastasis, 3.19 carcinogen,
3.20 mutation, 3.21 d, 3.22 b,
3.23 f, 3.24 j, 3.25 h, 3.26 c,
3.27 i, 3.28 g, 3.29 e, 3.30 a,
3.31 k, 3.32 l

CHAPTER 4

4.1 e, 4.2 c, 4.3 d, 4.4 e, 4.5 e,
4.6 e, 4.7 d, 4.8 e, 4.9 e, 4.10 a,
4.11 F, 4.12 T, 4.13 T, 4.14 T,
4.15 F, 4.16 F, 4.17 T, 4.18 T,
4.19 F, 4.20 T, 4.21 mesothelium,
4.22 secretion, 4.23 pseudostratified,
4.24 elastic, 4.25 fibroblasts,
4.26 fibers, 4.27 mesenchyme,
4.28 chondrocytes, 4.29 hyaline,
4.30 collagen, 4.31 b, 4.32 d,
4.33 e, 4.34 a, 4.35 g, 4.36 h,
4.37 i, 4.38 f, 4.39 a

CHAPTER 5

5.1 d, 5.2 c, 5.3 e, 5.4 e, 5.5 e,
5.6 c, 5.7 a, 5.8 T, 5.9 F, 5.10 T,
5.11 F, 5.12 F, 5.13 F,
5.14 epidermis, 5.15 stratum
germinativum, 5.16 stratum corneum,
5.17 papillae, 5.18 cleavage lines
(Langer's lines), 5.19 hypodermis,
5.20 melanocytes, 5.21 d

CHAPTER 6

6.1 T, 6.2 F, 6.3 F, 6.4 F, 6.5 F,
6.6 T, 6.7 T, 6.8 T, 6.9 F, 6.10 T,
6.11 cartilage; bone, 6.12 cancellous,
6.13 osteon, 6.14 sutural (Wormian),
6.15 tables, 6.16 red blood cells,
6.17 fat, 6.18 periosteal perforating
(Sharpey's fibers), 6.19 central
(Haversian), 6.20 lacunae,
6.21 osteoclasts, 6.22 ossification,
6.23 intramembranous ossification,
6.24 c, 6.25 d, 6.26 e, 6.27 b,
6.28 g, 6.29 i, 6.30 h, 6.31 f,
6.32 o, 6.33 k, 6.34 n

CHAPTER 7

7.1 a, 7.2 b, 7.3 b, 7.4 d, 7.5 a,
7.6 b, 7.7 b, 7.8 e, 7.9 c, 7.10 c,
7.11 T, 7.12 F, 7.13 F, 7.14 F,
7.15 T, 7.16 T, 7.17 sutural,
7.18 paranasal sinuses, 7.19 vomer,
7.20 maxillae, 7.21 mandible,
7.22 malleus; incus; stapes, 7.23 atlas,
7.24 coccyx, 7.25 12, 7.26 parietal;
coccyx, 7.27 b, 7.28 a

CHAPTER 8

8.1 d, 8.2 b, 8.3 d, 8.4 e, 8.5 c,
8.6 e, 8.7 a, 8.8 d, 8.9 T, 8.10 F,
8.11 T, 8.12 T, 8.13 F, 8.14 T,
8.15 T, 8.16 humerus, 8.17 tibia,
8.18 tibia, 8.19 carpal, 8.20 patella,
8.21 interosseous membrane,
8.22 lesser pelvis, 8.23 ilium; ischium;
pubis, 8.24 sacrum; coccyx; hip bones,
8.25 femur

CHAPTER 9

9.1 c, 9.2 d, 9.3 e, 9.4 e, 9.5 e,
9.6 a, 9.7 d, 9.8 d, 9.9 e, 9.10 e,
9.11 e, 9.12 e, 9.13 T, 9.14 F,
9.15 T, 9.16 T, 9.17 T, 9.18 F,
9.19 F, 9.20 T, 9.21 T, 9.22 T,
9.23 T, 9.24 T, 9.25 a, 9.26 i,
9.27 f, 9.28 d, 9.29 c, 9.30 g,
9.31 b, 9.32 j

CHAPTER 10

10.1 F, 10.2 T, 10.3 T, 10.4 T,
10.5 T, 10.6 T, 10.7 F, 10.8 F,
10.9 F, 10.10 T, 10.11 aponeurosis,
10.12 motor unit, 10.13 synaptic cleft,
10.14 visceral, 10.15 cardiac,
10.16 mesoderm, 10.17 smooth,
10.18 muscles; muscles,
10.19 epimysium, 10.20 nerve

CHAPTER 11

11.1 a, 11.2 e, 11.3 b, 11.4 d,
11.5 e, 11.6 b, 11.7 e, 11.8 e,
11.9 e, 11.10 d, 11.11 d, 11.12 e,
11.13 c, 11.14 e, 11.15 F, 11.16 F,
11.17 T, 11.18 T, 11.19 F, 11.20 T,
11.21 belly, 11.22 tendon,
11.23 aponeuroses, 11.24 origin,
11.25 insertion, 11.26 fascicles,
11.27 flexion, 11.28 pronator,
11.29 invertor, 11.30 first, 11.31 j,
11.32 e, 11.33 l, 11.34 f, 11.35 k,
11.36 d, 11.37 g, 11.38 a, 11.39 h,
11.40 c

CHAPTER 12

12.1 d, 12.2 e, 12.3 e, 12.4 e,
12.5 b, 12.6 c, 12.7 T, 12.8 F,
12.9 F, 12.10 T, 12.11 T, 12.12 F,
12.13 neurotransmitters, 12.14 brain;
spinal cord, 12.15 somatic; visceral,
12.16 sympathetic; parasympathetic,
12.17 neurofibral nodes (nodes of
Ranvier), 12.18 ganglion,
12.19 threshold stimulus,
12.20 all-or-none principle,
12.21 saltatory conduction,
12.22 synapse, 12.23 i, 12.24 b,
12.25 j, 12.26 g, 12.27 c, 12.28 d,
12.29 h, 12.30 f

CHAPTER 13

13.1 d, 13.2 e, 13.3 e, 13.4 e,
13.5 e, 13.6 e, 13.7 e, 13.8 b,
13.9 e, 13.10 F, 13.11 F, 13.12 T,
13.13 T, 13.14 F, 13.15 T, 13.16 F,
13.17 conus terminalis or medullaris,
13.18 filum terminale, 13.19 dorsal
root ganglion or spinal ganglion,
13.20 funiculi, 13.21 decussation,
13.22 reflex, 13.23 below,
13.24 plexuses, 13.25 dermatome,
13.26 intercostal

CHAPTER 14

14.1 d, 14.2 a, 14.3 e, 14.4 e,
14.5 d, 14.6 e, 14.7 T, 14.8 F,

14.9 T, **14.10** T, **14.11** T, **14.12** F,
14.13 T, **14.14** cerebral cortex,
14.15 corpus callosum, **14.16** choroid
plexus, **14.17** medulla oblongata; pons;
midbrain, **14.18** pons,
14.19 midbrain, **14.20** 12, **14.21** f,
14.22 k, **14.23** c, **14.24** b, **14.25** e,
14.26 j, **14.27** g, **14.28** i, **14.29** a,
14.30 h, **14.31** d, **14.32** n, **14.33** e,
14.34 b, **14.35** a, **14.36** d, **14.37** c

CHAPTER 15

15.1 c, **15.2** b, **15.3** b, **15.4** b,
15.5 d, **15.6** T, **15.7** T, **15.8** T,
15.9 T, **15.10** T, **15.11** F, **15.12** F,
15.13 T, **15.14** F, **15.15** T

CHAPTER 16

16.1 e, **16.2** e, **16.3** d, **16.4** e,
16.5 a, **16.6** F, **16.7** F, **16.8** T,
16.9 F, **16.10** T, **16.11** F, **16.12** F,
16.13 T, **16.14** T, **16.15** referred
pain, **16.16** labyrinth,
16.17 octoconia; statoliths,
16.18 conjunctiva, **16.19** retina,
16.20 lacrimal gland, **16.21** f,
16.22 e, **16.23** a, **16.24** d, **16.25** c,
16.26 j, **16.27** h, **16.28** g

CHAPTER 17

17.1 d, **17.2** a, **17.3** c, **17.4** b,
17.5 b, **17.6** adenohypophysis;
neurohypophysis, **17.7** hypothalamus,
17.8 hypothalamus,
17.9 hypothalamic-hypophyseal portal
system, **17.10** thyroid, **17.11** adrenal
glands, **17.12** pancreatic islets (islets of
Langerhans), **17.13** ovaries; testes,
17.14 cortex; medulla oblongata,
17.15 stomach; small intestine,
17.16 exocrine, **17.17** target,
17.18 hypophysis,
17.19 infundibulum,
17.20 neurosecretory

CHAPTER 18

18.1 e, **18.2** e, **18.3** e, **18.4** c,
18.5 a, **18.6** b, **18.7** T, **18.8** T,
18.9 T, **18.10** F, **18.11** F, **18.12** T,
18.13 T, **18.14** F, **18.15** F, **18.16** T,
18.17 F, **18.18** viscosity,
18.19 plasma, **18.20** plasma proteins,
18.21 fibrinogen, **18.22** erythropoiesis,
18.23 spleen, **18.24** leukopoiesis,
18.25 neutrophils, **18.26** erythrocytes,
18.27 plasma cells, **18.28** antibodies,
18.29 lymphocytes

CHAPTER 19

19.1 b, **19.2** c, **19.3** d, **19.4** c,
19.5 b, **19.6** c, **19.7** e, **19.8** d,
19.9 c, **19.10** d, **19.11** e, **19.12** F,
19.13 F, **19.14** T, **19.15** T, **19.16** F,
19.17 endocardium, **19.18** cardiac
skeleton, **19.19** SA node, **19.20** gap
junctions, **19.21** e, **19.22** a, **19.23** d,
19.24 b, **19.25** g, **19.26** h, **19.27** i,
19.28 f, **19.29** e, **19.30** e

CHAPTER 20

20.1 c, **20.2** b, **20.3** d, **20.4** b,
20.5 e, **20.6** e, **20.7** b, **20.8** d,
20.9 e, **20.10** c, **20.11** T, **20.12** T,
20.13 F, **20.14** T, **20.15** F, **20.16** F,
20.17 F, **20.18** muscular,
20.19 metarterioles, **20.20** vasa vasora,
20.21 pericytic, **20.22** muscular,
20.23 coronary sinus, **20.24** c,
20.25 a, **20.26** b, **20.27** d, **20.28** h,
20.29 j, **20.30** f, **20.31** g, **20.32** o,
20.33 k, **20.34** l, **20.35** m

CHAPTER 21

21.1 d, **21.2** b, **21.3** a, **21.4** c,
21.5 a, **21.6** T, **21.7** T, **21.8** F,
21.9 T, **21.10** F, **21.11** T, **21.12** T,
21.13 F, **21.14** T, **21.15** F,
21.16 edema, **21.17** lymph nodes,
21.18 hilus, **21.19** pharyngeal;
palatine; lingual, **21.20** pharyngeal
tonsils, **21.21** spleen, **21.22** plasma,
21.23 lacteals, **21.24** histocytes,
21.25 valves, **21.26** B cells; T cells,
21.27 spleen, **21.28** tonsils

CHAPTER 22

22.1 d, **22.2** b, **22.3** b, **22.4** a,
22.5 b, **22.6** T, **22.7** F, **22.8** T,
22.9 F, **22.10** T, **22.11** T, **22.12** F,
22.13 F, **22.14** F, **22.15** T,
22.16 cleft palate, **22.17** buccal cavity,
22.18 three, **22.19** epiglottis,
22.20 trachea, **22.21** alveoli,
22.22 breathing, **22.23** external,
22.24 nasal passages, **22.25** inspiration

CHAPTER 23

23.1 b, **23.2** e, **23.3** d, **23.4** a,
23.5 b, **23.6** d, **23.7** c, **23.8** F,
23.9 F, **23.10** T, **23.11** T, **23.12** F,
23.13 T, **23.14** T, **23.15** T, **23.16** F,
23.17 GI, **23.18** periodontal
ligaments, **23.19** saliva,
23.20 pharynx, **23.21** peritoneum,
23.22 greater omentum,
23.24 falciform, **23.25** lobules,
23.26 feces, **23.27** haustra, **23.28** b,
23.29 d, **23.30** a, **23.31** c, **23.32** a,
23.33 d, **23.34** e

CHAPTER 24

24.1 b, **24.2** c, **24.3** c, **24.4** e,
24.5 a, **24.6** e, **24.7** F, **24.8** T,
24.9 T, **24.10** F, **24.11** F,
24.12 renal fascia, **24.13** pelvis,
24.14 cortex, **24.15** interlobar,
24.16 vasa recta, **24.17** endothelial
capsular, **24.18** proximal convoluted,
24.19 trigone, **24.20** external
voluntary

CHAPTER 25

25.1 T, **25.2** F, **25.3** T, **25.4** F,
25.5 F, **25.6** F, **25.7** F, **25.8** T,
25.9 T, **25.10** F, **25.11** hernia
(inguinal), **25.12** perineum,
25.13 seminiferous tubules,
25.14 ductus deferens, **25.15** prepuce
(foreskin), **25.16** smegma,
25.17 ovaries, **25.18** ectopic
pregnancy, **25.19** broad ligaments,
25.20 endometrium, **25.21** greater
vestibular (Bartholin's) glands,
25.22 uterus, **25.23** isthmus,
25.24 zygote, **25.25** somatic

CHAPTER 26

26.1 c, **26.2** a, **26.3** b, **26.4** d,
26.5 e, **26.6** e, **26.7** d, **26.8** d,
26.9 T, **26.10** T, **26.11** T, **26.12** T,
26.13 T, **26.14** F, **26.15** morula,
26.16 implantation, **26.17** amnion,
26.18 blastomeres,
26.19 differentiation, **26.20** chorion,
26.21 umbilical cord, **26.22** 266,
26.23 tenth

SELECTED ANATOMICAL REFERENCES

GENERAL BOOKS

Angevine, Jay B., Jr., and Carl W. Cotman. *Principles of Neuroanatomy.* New York: Oxford, 1981.

Basmajian, John V., and Charles E. Slonecker. *Grant's Method of Anatomy,* 11th ed. Baltimore: Williams & Wilkins, 1989.

Brodal, A. *Neurological Anatomy in Relation to Clinical Medicine.* New York: Oxford, 1981.

Carpenter, M. B., and J. Sutin. *Human Neuroanatomy,* 8th ed. Baltimore: Williams & Wilkins, 1983.

Chewning, E. B. *Anatomy Illustrated.* New York: Simon and Schuster, 1979.

Clemente, Carmine D. *Anatomy: A Regional Atlas of the Human Body,* 3d ed. Baltimore: Urban and Schwarzenberg, 1987.

Clemente, Carmine D. *Gray's Anatomy of the Human Body,* 30th ed. Philadelphia: Lea & Febiger, 1984.

Ferner, H., and J. Staubesand, eds. *Sobotta Atlas of Human Anatomy,* 10th ed. Baltimore: Urban and Schwarzenberg, 1983.

Fischer, Harry W. *Radiographic Anatomy: A Working Atlas.* New York: McGraw-Hill, 1988 (paperback).

Gosling, J. A., P. F. Harris, J. R. Humpherson, I. Whitmore, and P. L. T. Willan. *Atlas of Human Anatomy with Integrated Text.* Philadelphia: Lippincott, 1985.

Knight, Bernard. *Discovering the Human Body.* New York: Lippincott & Crowell, 1980.

Leeson, C. Roland, and Thomas S. Leeson. *Human Structure.* Toronto: Decker, 1989.

Lyons, A. S., and R. J. Petrucelli. *Medicine, an Illustrated History.* New York: Abrams, 1978.

McMinn, R. M. H., and R. T. Hutchings. *Color Atlas of Human Anatomy.* Chicago: Year Book Medical Publishers, 1977.

Montgomery, Royce L. *Basic Anatomy for the Allied Health Professions.* Baltimore: Urban and Schwarzenberg, 1981.

Moore, Keith L. *Clinically Oriented Anatomy.* Baltimore: Williams & Wilkins, 1980.

Netter, Frank H. *Atlas of Human Anatomy.* Summit, NJ: Ciba-Geigy, 1989.

Noback, C. R., N. L. Strominger, and R. J. Demarest. *The Human Nervous System: Introduction and Review,* 4th ed. Philadelphia: Lea and Febiger, 1991.

Persaud, T. V. N. *Early History of Human Anatomy from Antiquity to the Beginning of the Modern Era.* Springfield, IL: Charles C Thomas, 1984.

Romanes, G. J., ed. *Cunningham's Textbook of Anatomy,* 12th ed. New York: Oxford, 1981.

Shipman, Pat, Alan Walker, and David Bishell. *The Human Skeleton.* Cambridge, MA: Harvard University Press, 1985.

Singer, C. *A Short History of Anatomy and Physiology from the Greeks to Harvey,* 2nd ed. New York: Oxford, 1957.

Snell, Richard S. *Clinical Anatomy for Medical Students,* 3rd ed. Boston: Little, Brown, 1986.

Squire, J. *The Structural Basis of Muscle Contraction.* New York: Plenum, 1981.

Thompson, C. W. *Manual of Structural Kinesiology,* 9th ed. St. Louis: Mosby, 1981.

Williams, P. I., et al., eds. *Gray's Anatomy,* 37th ed. New York: Churchill, 1989.

Wilson, F. C. *The Musculoskeletal System: Basic Processes and Disorders,* 2nd ed. Philadelphia: Lippincott, 1983.

Woodburne, Russell T., and William E. Burkel. *Essentials of Human Anatomy,* 8th ed. New York: Oxford, 1988.

HISTOLOGY BOOKS

Cormack, David H. *Ham's Histology,* 9th ed. Philadelphia: Lippincott, 1987.

DiFiore, M. S. H. *An Atlas of Histology,* 5th ed. Philadelphia: Lea and Febiger, 1989.

Fawcett, Don W. *Bloom and Fawcett: A Textbook of Histology,* 11th ed. Philadelphia: Saunders, 1986.

Kelly, Douglas E., and Allen C. Enders. *Bailey's Textbook of Microscopic Anatomy,* 18th ed. Baltimore: Williams & Wilkins, 1984.

Snell, Richard S. *Clinical and Functional Histology for Medical Students.* Boston: Little, Brown, 1984.

Weiss, L., ed. *Cell and Tissue Biology: A Textbook of Histology,* 6th ed. New York: Urban and Schwarzenberg, 1988.

Wheater, P. R., H. G. Burkitt, and V. G. Daniels. *Functional Histology: A Text and Colour Atlas,* 2nd ed. London: Churchill, Livingstone, 1987.

ARTICLES

Borg, E., and S. A. Counter. "The Middle-ear Muscles." *Scientific American,* August 1989.

Caplan, Arnold I. "Cartilage." *Scientific American,* October 1981.

Goss, C. M. "On the Anatomy of Muscles for Beginners by Galen of Pergamon." *The Anatomical Record* 145 (1963):477–502.

Hudspeth, A. J. "The Hair Cells of the Inner Ear." *Scientific American,* January 1983.

Lake, James A. "The Ribosome." *Scientific American,* August 1981.

McIntosh, J. Richard, and Kent L. McDonald. "The Mitotic Spindle." *Scientific American*, October 1989.

Moog, Florence. "The Lining of the Small Intestine." *Scientific American*, November 1981.

Morell, Pierre, and William T. Norton. "Myelin." *Scientific American*, May 1980.

Parker, Donald E. "The Vestibular Apparatus." *Scientific American*, November 1980.

Rothman, J. E. "The Compartmental Organization of the Golgi Apparatus." *Scientific American*, September 1985.

Scientific American, September 1979. (Available in book form.) The entire issue is devoted to the brain.

Unwin, N., and R. Henderson. "The Structure of Proteins in Cell Membranes." *Scientific American*, February 1984.

CREDITS

CHAPTER 16 FIGURE **16.2:** [C] Bruce Coleman. FIGURE **16.3:** [C] Gene Shih and Richard Kessel, *Living Images: Biological Microstructures Revealed by SEM*, Boston, Science Books International, 1982. FIGURE **16.7:** [B] CNRI/Science Photo Library/Photo Researchers. FIGURE **16.9:** [C] Courtesy of Robert S. Kimura, Ph.D., Massachusetts Eye and Ear Infirmary, Boston. FIGURE **16.13:** [B] Gene Shih and Richard Kessel, *Living Images: Biological Microstructures Revealed by SEM*, Boston, Science Books International, 1982. FIGURE **16.14:** [B, D] E. R. Lewis, Y. Y. Zeevi, and F. S. Werblin, *Brain Research* 15(1969):559–562. FIGURE **16.16:** [B] Lester V. Bergman & Associates.

CHAPTER 17 FIGURE **17.3:** [B] Fujita, Tanaka, and Tokunaga, *SEM Atlas of Cells and Tissues*, Tokyo, Igaku-Shoin, 1981. FIGURE **17.4:** [B] W. Bloom and D. W. Fawcett, *A Textbook of Histology*, 11th ed., Philadelphia, Saunders, 1986. FIGURE **17.6:** [B] Biophoto/Photo Researchers; [C] Courtesy of K. Kovacs and E. Horvath. PAGE **457:** (left) Michael Serino/The Picture Cube; (bottom) Reprinted with permission from "Clinicopathologic Conference," *The American Journal of Medicine*, Vol. 20, No. 1, January 1956. PAGE **458:** John Paul Kay/Peter Arnold.

CHAPTER 18 FIGURE **18.1:** Bruce Wetzel and Harry Schaefer, National Cancer Institute. FIGURE **18.3:** (top to bottom) Manfred Kage/Peter Arnold; Lester V. Bergman & Associates; Cabisco/Visuals Unlimited; Lester V. Bergman & Associates; John D. Cunningham/Visuals Unlimited; Biophoto Associates/Science Source/Photo Researchers; Lester V. Bergman & Associates. FIGURE **18.4:** [A] Ralph Feder, IBM, Thomas J. Watson Research Center; [B] Hans R. Baumgartner, Hoffman-LaRoche & Company, Basel. PAGE **474:** N. Calder/Science Source/Photo Researchers.

CHAPTER 19 FIGURE **19.4:** [B] Martin M. Rotker/Phototake. FIGURE **19.5:** [B] Martin M. Rotker/Phototake. FIGURE **19.6:** [D] Dr. Wallace McAlpine; [F] Lennart Nilsson, *Behold Man*, Boston, Little, Brown, 1974.

CHAPTER 20 FIGURE **20.3:** D. W. Fawcett/Photo Researchers. FIGURE **20.6:** [B] Richard G. Kessel and Randy H. Kardon, *Tissues and Organs: A Text-Atlas of Scanning Electron Microscopy*, San Francisco, Freeman, 1979. FIGURE **20.12:** [B] Courtesy of Gower Medical Publishing, a division of Lippincott Company, London. From J. A. Gosling et al., *Atlas of Human Anatomy with Integrated Text*, 2d ed., Philadelphia, Lippincott, 1991. FIGURE **20.14:** [B] From H. W. Fischer, *Radiographic Anatomy: A Working Atlas*, New York, McGraw-Hill, 1988. FIGURE **20.15:** [C] Stanley M. Rogoff, M.D., Department of Radiology, Strong Memorial Hospital, Rochester, NY. FIGURE **20.16:** [B] John Watney/Photo Researchers. FIGURE **20.17:** [B] Albert Paglialunga, Bergenfield, NJ. FIGURE **20.18:** [C] From H. W. Fischer, *Radiographic Anatomy: A Working Atlas*, New York, McGraw-Hill, 1988. FIGURE **20.19:** [B] Courtesy of Gower Medical Publishing, a division of Lippincott Company, London. From J. A. Gosling et al., *Atlas of Human Anatomy with Integrated Text*, 2d ed., Philadelphia, Lippincott, 1991. PAGE **538:** Courtesy of NIH.

CHAPTER 21 FIGURE **21.2:** [B] Johannes A. G. Rhodin, M.D., Ph.D., Department of Anatomy, College of Medicine, Tampa, Florida. FIGURE **21.3:** [B] Courtesy of Gower Medical Publishing, a division of Lippincott Company, London. From J. A. Gosling et al., *Atlas of Human Anatomy with Integrated Text*, 2d ed., Philadelphia, Lippincott, 1991. FIGURE **21.5:** [B] Fujita, Tanaka, and Tokunaga, *SEM Atlas of Cells and Tissues*, Tokyo, Igaku-Shoin, 1981. FIGURE **21.7:** Lester V. Bergman & Associates. PAGE **555:** Courtesy of Lennart Nilsson, © Boehringer-Ingelheim International, GmbH.

CHAPTER 22 FIGURE **22.2:** [B] CNRI/Phototake. FIGURE **22.3:** [A] Biophoto Associates/Photo Researchers; [B] Lennart Nilsson, *Behold Man*, Boston, Little, Brown, 1974. FIGURE **22.4:** [C] Courtesy of Gower Medical Publishing, a division of Lippincott Company, London. From J. A. Gosling et al., *Atlas of Human Anatomy with Integrated Text*, 2d ed., Philadelphia, Lippincott, 1991. FIGURE **22.5:** [B] R. Calentine/Visuals Unlimited; [C] Richard Kessel and Randy Kardon, *Tissues and Organs: A Text-Atlas of Scanning Electron Microscopy*, San Francisco, Freeman, 1979. FIGURE **22.7:** [A] Bruce Iverson/Visuals Unlimited; [B] Manfred Kage/Peter Arnold. FIGURE **22.9:** [A, B] Courtesy of Gower Medical Publishing, a division of Lippincott Company, London. From J. A. Gosling et al., *Atlas of Human Anatomy with Integrated Text*, 2d ed., Philadelphia, Lippincott, 1991. PAGE **577:** [A, B] Kenneth A. Siegesmund, Ph.D., Department of Anatomy and Cellular Biology, Medical College of Wisconsin.

CHAPTER 23 FIGURE **23.1:** [B] Thomas Bednarek/Custom Medical Stock Photo. FIGURE **23.4:** [B] Dr. Leonard B. Zaslow, courtesy of Matthew Carola. FIGURE **23.9:** [C] Relam/Custom Medical Stock Photo. FIGURE **23.10:** [C] Gene Shih and Richard Kessel, *Living Images: Biological Microstructures Revealed by SEM*, Boston, Science Books International, 1982. FIGURE **23.11:** [B] Lester V. Bergman & Associates. FIGURE **23.12:** [D] Gene Shih and Richard Kessel, *Living Images: Biological Microstructures Revealed by SEM*, Boston, Science Books International, 1982. FIGURE **23.16:** [C] Lester V. Bergman & Associates; [D] Courtesy of Dr. M. Muto, Department of Anatomy, Niigata University Medical School. PAGE **613:** L. J. Schoenfield/Visuals Unlimited. PAGE **615:** [A, B] Dr. Gene M. Riddle, Director, Rheumatology Research Lab, Henry Ford Hospital, Detroit, Michigan.

CHAPTER 24 FIGURE **24.2:** Courtesy of Stephen A. Kieffer, M.D., Department of Radiology, SUNY Health Science Center, Syracuse. FIGURE **24.3:** [C] Courtesy of Gower Medical Publishing, a division of Lippincott Company, London. From J. A. Gosling et al., *Atlas of Human Anatomy with Integrated Text*, 2d ed., Philadelphia, Lippincott, 1991. FIGURE **24.4:** [B] John Watney/Photo Researchers. FIGURE **24.6:** [C] David M. Phillips/Visuals Unlimited; [D] Reprinted with permission from Leon Weiss, ed., *Histology: Cell and Tissue Biology*, 5th ed., New York, Elsevier, 1973; [E] Daniel S. Friend, M.D.

CHAPTER 25 FIGURE **25.5:** [B] Custom Medical Stock Photos. FIGURE **25.11:** [B] Gene Shih and Richard Kessel, *Living Images: Biological Microstructures Revealed by SEM*, Boston, Science Books International, 1982. FIGURE **25.16:** © Lennart Nilsson, *A Child Is Born*, Bonnier Fakta, Stockholm. FIGURE **25.17:** [A, C, D] Mia Tegner, Scripps Institution of Oceanography; [B] Courtesy of Gerald Schatten.

CHAPTER 26 FIGURE **26.1:** [A, G] Biophoto Associates/Photo Researchers; [B, C, E, F] Landrum B. Shettles, M.D.; [D] Petit Format/Nestle/Photo Researchers. FIGURE **26.3:** [B] Photo by A. T. Hertig and J. Rock, reprinted with permission from Leon Weiss, ed., *Histology: Cell and Tissue Biology*, 5th ed., New York, Elsevier, 1973. FIGURES **26.7; 26.9; 26.11; 26.15:** Landrum B. Shettles, M.D. FIGURES **26.8; 26.10; 26.12; 26.13:** Reprinted with permission from Donald D. Ritchie and Robert Carola, *Biology*, 2d ed., Reading, Mass., Addison-Wesley, 1983, from Roberts Rugh, Landrum B. Shettles with Richard Einhorn, *From Conception to Birth: The Drama of Life's Beginnings*, New York, Harper & Row, 1971. FIGURE **26.14:** Courtesy of Carnegie Institute of Washington. FIGURE **26.17:** Maternity Center Association, *The Birth Atlas*.

APPENDIXES FIGURE **F.1–F.11:** McGraw-Hill photographs by Custom Medical Stock Photo.

GLOSSARY

A

ABDOMINAL CAVITY Superior portion of the abdominopelvic cavity; contains the liver, gallbladder, stomach, pancreas, spleen, and intestines.

ABDOMINOPELVIC CAVITY Portion of the ventral cavity below the diaphragm; divided by an imaginary line into *abdominal* and *pelvic cavities*.

ABDUCTION (L. *ab*, away from) Movement of a limb away from the midline of the body or of fingers or toes from the medial longitudinal axis of the hand or foot.

ABSORPTION Passage of usable nutrient molecules from the small intestine into the bloodstream and lymphatic system.

ACCOMMODATION Reflex that adjusts the lens of the eye to permit focusing of images from different distances.

ACETABULUM [ass-eh-TAB-yoo-luhm; L. vinegar cup] Socket of the ball-and-socket joint of the femur.

ACETYLCHOLINE [us-SEET-uhl-KOH-leen] Chemical neurotransmitter released by motor neurons at a neuromuscular junction; may be excitatory or inhibitory; plays a major role in muscle contraction.

ACETYLCHOLINESTERASE Enzyme that breaks down acetylcholine into acetic acid (acetate) and choline to halt continuous muscle contractions.

ACINAR GLAND (L. *acinus*, grape, berry) See *alveolar gland*.

ACINI [ASS-ih-nye; sing. **ACINUS**; L. grapes] Clustered exocrine cells in the pancreas that secrete digestive juices.

ACROSOME (Gr. *akros*, topmost, extremity + *soma*, body) A region at the tip of the head of a sperm, containing several enzymes that help the sperm penetrate an ovum.

ACTIN Protein that makes up part of the thin myofilament of muscle fibers; forms light I bands of skeletal muscle.

ACTION POTENTIAL Spread of an impulse along the axon following *depolarization* of the plasma membrane; also called *nerve impulse*.

ACTIVE PROCESSES Processes that move substances through a selectively permeable membrane from an area of low concentration to an area of higher concentration; require cellular energy.

ACTIVE TRANSPORT Movement of molecules through a membrane against the *concentration gradient*; requires cellular energy.

ADDUCTION (L. *ad*, toward) Movement of a limb toward or beyond the midline of the body or of fingers or toes toward the midline of a body part.

ADENOHYPOPHYSIS The anterior lobe of the pituitary (hypophysis), containing many secretory cells.

ADENOSINE TRIPHOSPHATE (ATP) [uh-DEN-uh-seen try-FOSS-fate] An organic compound that stores energy in its chemical bonds and serves as an energy source for chemical reactions in cells.

ADIPOSE CELL (L. *adeps*, fat) A fixed connective tissue cell that synthesizes and stores lipids; also called *fat cell*.

ADIPOSE TISSUE Tissue composed almost entirely of clustered adipose cells supported by strands of collagenous and reticular fibers; provides a reserve food supply, cushions organs, and helps prevent loss of body heat.

ADOLESCENCE (L. *adolescere*, to grow up) The period from the end of childhood to the beginning of adulthood.

ADRENAL CORTEX The outer larger portion of the adrenal gland. It secretes *glucocorticoids*, a *mineralocorticoid*, and *gonadocorticoids*.

ADRENAL GLAND [uh-DREEN-uhl; L. "upon the kidneys"] One of a pair of endocrine glands, one resting upon each kidney.

ADRENAL MEDULLA The inner, smaller portion of the adrenal gland. Its main secretion is *epinephrine*.

ADRENERGIC Pertaining to neurons and fibers of the sympathetic division of the autonomic nervous system; so called because the postganglionic neurons and fibers release norepinephrine (nor*adrenaline*).

AFFERENT ARTERIOLES Branchings of the *interlobular arteries* in the kidney; they carry blood to the glomerulus.

AFFERENT NEURON (L. *ad*, toward, + *ferre*, to bring) A nerve cell that conveys impulses from sensory receptors to the central nervous system; also called *sensory neuron*.

AFTERBIRTH Collective name for the placenta, fetal membranes, and any remaining uterine fluid after their postnatal delivery.

AGGREGATED LYMPH NODULES Clusters of unencapsulated lymphoid tissue in the intestine and appendix; also called *Peyer's patches*, *gut-associated lymphoid tissue (GALT)*.

AGONIST (Gr. *agonia*, contest, struggle) A muscle that is primarily responsible for a movement; also called *prime mover*.

ALIMENTARY CANAL The digestive tract, from mouth to anus.

ALL-OR-NONE PRINCIPLE The tendency of muscle fibers to contract fully or not at all, and of neurons to fire at full power or not at all.

ALLANTOIS [uh-LAN-toh-ihz; Gr. "sausage"] Small fingerlike outpocket of the caudal wall of the embryonic yolk sac; involved with blood cell formation and the development of the urinary bladder.

ALVEOLAR [al-VEE-uh-lur; L. *alveolus*, hollow, cavity] Pertaining to a rounded portion.

ALVEOLAR GLAND An exocrine gland whose secretory portion is rounded; also called *acinar gland*.

ALVEOLI [al-VEE-oh-lie; sing. **ALVEOLUS**; L. a cavity] Functional units of the lungs; sites of gas exchange.

AMINO ACID [uh-MEE-noh] A chemical compound containing an amino group (—NH₂) and a carboxyl group (—COOH), plus a variable portion; the structural subunit of proteins.

AMNIOCENTESIS [am-nee-oh-sehn-TEE-sihss; Gr. sac + puncture] Technique of obtaining cells of a fetus from its amniotic fluid.

AMNION (Gr. sac) A tough, thin, transparent membrane enveloping the embryo.

AMPHIARTHROSIS (Gr. *amphi*, on both sides, + *arthrosis*, articulation) A slightly movable joint.

AMPULLA A small dilation in a canal or duct.

ANAL CANAL The continuation of the rectum that pierces the pelvic diaphragm and turning sharply downward and backward, opening into the anus.

ANAL COLUMNS About 5 to 10 permanent longitudinal columns of mucous membrane in the upper part of the anal canal; also called *rectal columns*.

ANAPHASE The third and shortest stage of mitosis; chromosome pairs separate and move toward opposite poles.

ANATOMICAL POSITION The universally accepted position from which locations of body parts can be described. The body is standing erect and facing forward, feet together, arms hanging at sides with palms facing forward.

ANATOMY (Gr. *ana* + *temnein*, to cut)

The study of the body and its parts.

ANTAGONIST ("against the agonist") A muscle that opposes the movement of a prime mover; see *agonist*.

ANTERIOR (L. *ante*, before) A relative directional term: toward the front of the body.

ANTIBODY A protein produced by lymphocytes in response to an antigen.

ANTIGEN A substance against which an antibody is produced.

ANUS (L. ring) Opening from the rectum to the outside; the site of feces excretion.

AORTA The major arterial trunk emerging from the left ventricle of the heart.

AORTIC BODY Chemoreceptor in the sinuses of the aorta that responds to a lowered blood pH by increasing the respiratory rate.

APNEUSTIC AREA A breathing-control area of the pons which, when stimulated, causes strong inhalations and weak exhalations; see *respiratory center*.

APONEUROSIS A broad, flat sheet of dense connective tissue that attaches to two or more muscles that work together or to the convering of a bone.

APPENDICULAR (L. *appendere*, to hang from) Pertaining to the *appendicular part* of the body, which includes the *upper* and *lower extremities*.

APPENDICULAR SKELETON The bones of the upper and lower extremities.

APPENDIX See *vermiform appendix*.

AQUEOUS HUMOR Thin, watery fluid in the posterior and anterior chambers of the eye.

ARACHNOID (Gr., cobweblike) The delicate middle layer of the meninges.

ARBOR VITAE [VYE-tee; L. tree of life] A branched arrangement of white matter within the vermis of the cerebellum.

AREOLA (L. *dim*. of *area*, open place) A small space in a tissue; a small, dark-colored area around a center portion; the pigmented area around a nipple.

AREOLAR TISSUE [uh-REE-uh-lur; L. *areola*, open place] Most common connective tissue; contains tiny extracellular spaces usually filled with ground substance and tissue fluids.

ARRECTOR PILI Muscle that contracts to pull a follicle and its hair to an erect position, elevating the skin above and producing a "goose bump."

ARTERIOLE Small artery that branches before reaching capillary networks.

ARTERY A blood vessel that carries blood away from the heart.

ARTHROSCOPY A diagnostic and surgical technique in which a small fiberoptic scope is used to look into a joint.

ARTICULAR CAPSULE A fibrous capsule that lines the synovial cavity in the noncartilaginous parts of the joint, permitting considerable movement.

ARTICULAR CARTILAGE The smooth cartilage that caps the bones facing the synovial cavity; also called *hyaline cartilage*.

ARTICULAR DISK A fibrocartilage disk that (1) acts as a shock absorber for a joint, (2) adjusts the uneven articulating surfaces, and (3) allows two kinds of movements to occur simultaneously.

ARTICULATION (L. *articulus*, joint) A joint, the place where bones meet, or where cartilages or bones and cartilages meet.

ASCENDING COLON The portion of the large intestine extending upward from the cecum.

ASSOCIATION FIBERS Nerve fibers that link one area of the cerebral cortex to another area of the cortex of the same cerebral hemisphere.

ASTROCYTE The largest, most numerous glial cell; sustains neurons.

ATLAS The first cervical vertebra (spinal bone), which supports the head.

ATOM (Gr. *atomos*, indivisible) The basic unit of all matter, consisting of protons, neutrons, and electrons.

ATP See *adenosine triphosphate*.

ATRIOVENTRICULAR BUNDLE A group of conducting fibers in the interventricular septum that branch into each ventricle, finally branching into hundreds of tiny specialized muscle fibers called *cardiac conducting myofibers* (Purkinje fibers); formerly called *bundle of His*.

ATRIOVENTRICULAR (AV) NODE A mass of specialized heart tissue that delays the electrical activity of the heart a few hundredths of a second before allowing it to pass into the ventricles.

ATRIOVENTRICULAR (AV) VALVES Heart valves that allow blood to flow in one direction, from atria to ventricles.

ATRIUM [AY-tree-uhm; L. porch, antechamber] One of two upper heart chambers.

AUDITORY CORTEX Portion of temporal lobe of cerebrum involved with basic sounds and the feeling of dizziness.

AUDITORY OSSICLES The three small bones (malleus, incus, stapes) of the middle ear that transmit sound from the external to the inner ear.

AUDITORY TUBE Tube in anterior wall of *tympanic cavity*; permits air to pass from nasal cavity into middle ear, maintaining equal air pressure on both sides of the tympanic membrane; commonly called the *Eustachian tube*.

AURICLE (L. *auris*, ear) 1. external part of the ear; also called *pinna*. 2. Flaplike appendage of atrium of heart.

AUTONOMIC GANGLIA Clusters of cell bodies and dendrites that occur outside the central nervous system; relay stations between the CNS and effectors.

AUTONOMIC NERVOUS SYSTEM (ANS) The efferent motor division of the visceral nervous system; consists of *sympathetic* and *parasympathetic nervous systems*.

AUTONOMIC PLEXUSES (L. braids) Branched, interlaced networks of nerves in the thoracic, abdominal, and pelvic cavities.

AUTOSOMES The 22 matching pairs of chromosomes that determine genetic traits, but not gender.

AXIAL (L. *axis*, hub, axis) Pertaining to the axis, or trunk, of the body. The *axial part* is composed of the head, neck, thorax, abdomen, and pelvis.

AXIAL SKELETON The portion of the skeleton forming the longitudinal axis of the body; includes the skull, vertebral column, sternum, and ribs.

AXIS The second cervical vertebra, on which the *atlas* rests, allowing the skull and axis to move together in a "no" motion.

AXON A long, specialized process of a neuron that carries nerve impulses away from the cell body to the next neuron, muscle cell, or gland cell.

B

B CELL A type of lymphocyte that produces specific antibodies; also called *B lymphocyte*.

B LYMPHOCYTE See *B cell*.

BALL-AND-SOCKET JOINT A multiaxial joint in which the globular head of one bone fits into a cuplike cavity of another bone, such as in the hip.

BARORECEPTORS (Gr. "pressure receivers") Clusters of cells in the walls of the aortic arch and carotid artery sinuses; sensory receptors that respond to changes in blood pressure; also called *pressoreceptors*.

BARTHOLIN'S GLANDS See *greater vestibular glands*.

BASAL Pertaining to, located at, or forming a base.

BASAL GANGLIA Deep, large cores of gray matter beneath the white matter of each cerebral hemisphere; cell bodies of neuron clusters that help coordinate muscle movements. Basal ganglia are actually *nuclei*.

BASAL LAMINA A homogeneous layer of polysaccharides, with no fibers; that is part of a basement membrane.

BASEMENT MEMBRANE A thin layer composed of tiny fibers nonliving polysaccharide material (the basal lamina) and a deeper layer of reticular fibers; produced by epithelial cells; anchors epithelial tissue to underlying connective tissue.

BASOPHIL A type of white blood cell involved in allergic reactions and inflammation.

BELLY The bulging part of a muscle between its two ends.

BICUSPID VALVE The left atrioventricular heart valve; also called the *mitral valve*.

BILAMINAR EMBRYONIC DISK A bilayered structure formed when the blastocyst assumes a flattened disk shape during the second embryonic week; composed of endoderm and ectoderm.

BILATERAL Pertaining to both sides of the body.

BILE An alkaline liquid secreted by the liver; aids in emulsification of lipids.

BIOPSY (Gr. *bios*, life + *-opsy*, examination) The microscopic examination of living tissue removed from the body.

BLASTOCOEL [BLASS-toh-seel; Gr. bud + *koilos*, hollow] The fluid-filled inner cavity of a *blastocyst*.

BLASTOCYST (Gr. *blasto*, germ + *kystis*, "bladder") A fluid-filled hollow sphere of cells formed about 4 to 5 days after fertilization.

BLASTOMERE (Gr. *blastos*, bud + *meros*, a part) One of the daughter cells resulting from the cleavage of a zygote.

BLOOD PRESSURE The force (energy) with which blood is pushed against the walls of blood vessels when the heart contracts.

BLOOD-BRAIN BARRIER System of tight junctions in the endothelial cells of brain capillaries that forms a semipermeable membrane, allowing only certain substances to enter the brain; also called *hematoencephalic barrier*.

BONE See *osseous tissue*.

BONE MARROW Tissue filling the porous *medullary cavity* of the *diaphysis* of bones; also called *myeloid tissue*.

BOWMAN'S CAPSULE See *glomerular capsule*.

BRACHIAL PLEXUS The ventral rami of the lower four cervical and first thoracic nerves in the lower neck and axilla.

BRAINSTEM The stalk of the brain; consists of the medulla oblongata, pons, and midbrain.

BREATHING The mechanical process that moves air into and out of the lungs; includes *inspiration* and *expiration*; also called *ventilation* or *pulmonary ventilation*.

BROAD LIGAMENTS Paired double folds of peritoneum that attach the uterus to the lateral wall of the pelvis.

BROCA'S (MOTOR SPEECH) AREA (area 44) Motor area in the frontal lobe of the cerebrum; involved in formulating spoken words; also called *anterior speech area*.

BRONCHI [BRONG-kee; sing. **BRONCHUS**; Gr. throat, windpipe] Branches of the respiratory tree that emerge from the trachea into the lungs.

BRONCHIOLES [BRONG-kee-ohlz] Small tubes emerging from branching bronchi in the lungs; they continue to branch until they end as *terminal bronchioles*.

BUCCAL CAVITY The space between the cheeks and lips, and the teeth and gums; also called vestibule; see *oral cavity*.

BULBOURETHRAL GLANDS Paired male reproductive glands that open into the urethra, secreting alkaline fluids that neutralize the acidity of urine, and act as a lubricant at the tip of the penis prior to sexual intercourse; also called *Cowper's glands*.

BULBOUS CORPUSCLES (OR KRAUSE) Sensory receptors in the skin believed to be sensors for touch-pressure, position sense of a body part, and movement; formerly called *Krause's end bulbs*.

BUNDLE OF HIS See *atrioventricular bundle*.

BURSAE [BURR-see; sing. **BURSA**, BURR-sah; Gr. purse] Flattened sacs filled with synovial fluid; eliminate friction in areas where a muscle or tendon rubs against another muscle, tendon, or bone.

C

CALVARIA (L. skull) The roof (vault) of the cranium, composed of the brow portion of the frontal bone, the parietal bones, and the occipital bone.

CALYCES [KAY-luh-seez; sing. **CALYX**; KAY-licks; Gr. cups] Two small cavities (major and minor calyces) in the kidney formed from the branching of the *renal pelvis*.

CANALICULI [KAN-uh-lick-yuh-lie; L. dim. *canalis*, channel] Small channels radiating from *lacunae* in bone; transport materials by diffusion.

CAPILLARY (L. *capillus*, hair) Tiny blood vessels that connect the arterial and venous systems.

CAPUT (L. head) See *head*.

CARBOHYDRATE A molecule composed of carbon, hydrogen, and oxygen in a ratio of 1:2:1; the main source of body energy.

CARDIAC (Gr. *kardia*, heart) Pertaining to the heart.

CARDIAC CHEMORECEPTORS Chemically sensitive receptors in the carotid si-

nuses and aortic arch that help regulate the heart rate by responding to changes in the blood levels of oxygen, carbon dioxide, or hydrogen ions.

CARDIAC CONDUCTING MYOFIBERS Modified nerve fibrils in the walls of the ventricles of the heart; help produce a coordinated pumping effort; formerly called *Purkinje fibers*.

CARDIAC CYCLE The carefully regulated sequence of steps commonly referred to as the beating of a heart; it includes contraction (systole) and relaxation (diastole).

CARDIAC MUSCLE TISSUE (Gr. *kardia*, heart) Specialized muscle tissue found only in the heart.

CARDIAC SKELETON Tough connective-tissue in the heart that provides attachment sites for the valves and corresponding muscle fibers.

CARDIOREGULATORY CENTER Sympathetic nerve centers in the medulla oblongata that regulate the heart rate.

CAROTID BODY A chemoreceptor in the sinuses of the carotid arteries that responds to a lowered blood pH by increasing the respiratory rate.

CARPUS (Gr. *karpos*, wrist) The eight short bones connected by ligaments in each wrist.

CARTILAGE Specialized type of connective tissue that provides support and aids movement at joints.

CARTILAGINOUS JOINT Joint in which bones are joined by hyaline cartilage or a fibrocartilaginous disk; allows little or no movement; includes *synchondroses* and *symphyses*.

CAT SCAN See *computer-assisted tomography*.

CATARACT Condition in which the lens of the eye becomes opaque, possibly producing blindness.

CAUDA EQUINA [KAW-duh ee-KWY-nuh; L. horse's tail] The collection of spinal nerve roots passing caudally below the conus terminalis of the spinal cord.

CECUM (L. blind) A cul-de-sac pouch on the distal side of the ileocecal orifice.

CELL The smallest independent unit of life; the component of tissues.

CELL CYCLE The period from the beginning of one cell division to the beginning of the next; the life span of a cell.

CELLULAR RESPIRATION The sum of biochemical events by which the chemical energy of foods is released to provide energy for life's processes.

CEMENT The bonelike covering of the neck and root of a tooth.

CENTER OF OSSIFICATION The site at which a ring of cells forms around a blood vessel in the development of bone tissue; also called *ossification center.*

CENTRAL CANAL A longitudinal channel in the osteon (Haversian system); contains nerves and lymphatic and blood vessels; also called *Haversian canal.*

CENTRAL LOBE See *insula.*

CENTRAL NERVOUS SYSTEM (CNS) The brain and spinal cord.

CENTRIOLES Two small organelles in the *centrosome,* a specialized region of the cytoplasm of a cell; involved with the movement of chromosomes during cell division; form basal body of *cilia* and *flagella.*

CEPHALIC [suh-FAL-ihk; Gr. *kephale,* head] Pertaining to the head.

CEREBELLAR CORTEX A surface layer of gray matter covering the cerebellar lobes; composed of a network of billions of neurons.

CEREBELLAR PEDUNCLES [peh-DUNG-kuhlz; L. little feet] Three nerve bundles that attach the cerebellum to the brainstem; composed of inferior, middle, and superior cerebellar peduncles.

CEREBELLUM (L. little brain) The second-largest part of the brain, composed of the vermis, two small flocculonodular lobes, and two large lateral lobes; located behind the pons in the posterior cranial fossa; refines and coordinates muscular movements.

CEREBRAL CORTEX A surface mantle of gray matter over the cerebrum; it is thin and convoluted, containing about 50 billion neurons and 250 billion glial cells.

CEREBRAL PEDUNCLES Two nerve fiber bundles composed of the pyramidal tract and corticobulbar and corticopontine fibers.

CEREBROSPINAL FLUID (CSF) A clear, watery ultrafiltrate solution formed from blood in the capillaries; bathes the ventricles of the brain and cavity of the spinal cord; cushions the brain and spinal cord; helps control chemical environment of central nervous system.

CEREBRUM [suh-REE-bruhm; L. brain] Largest and most complex structure of the nervous system; consists of two cerebral hemispheres and the diencephalon; each hemisphere is composed of a cortex (gray matter), white matter, and basal ganglia. Conscious living depends on the cerebrum.

CERVICAL PLEXUS The ventral rami of C1 to C4 nerves in the neck region.

CERVICAL VERTEBRAE The seven small neck bones between the skull and thoracic vertebrae; support the head and allow it to move.

CERVIX (L. neck) Any neck-shaped anatomical structure; the narrow juncture between the uterus and vagina.

CESAREAN SECTION An incision through the abdominal and uterine walls to remove a fetus.

CHEMICAL SYNAPSE A junction by which two cells communicate by way of a chemical neurotransmitter.

CHEMORECEPTOR (L. "chemical receiver") A sensory receptor that responds to chemical stimuli.

CHEMOTHERAPY Therapy that uses drugs in treating cancer.

CHILDHOOD The period from the end of infancy to the beginning of adolescence.

CHOLINERGIC Pertaining to neurons of the ANS and motor end plates of motor neurons that release *acetylcholine.*

CHOLINESTERASE Enzyme that breaks down acetylcholine into choline and acetic acid.

CHONDROCYTE (Gr. *khondros,* cartilage) A cartilage cell embedded in small cavities (lacunae) within the matrix of cartilage connective tissue.

CHORION The protective sac around the embryo.

CHORIONIC VILLI Fingerlike projections growing outward from the embryonic chorion into maternal tissue; allow for the exchange of materials between mother and embryo.

CHORIONIC VILLI SAMPLING Technique in which a sample of chorionic villi tissue is collected and examined for genetic diseases.

CHOROID The posterior two-thirds of the vascular layer of the eye, composed of blood vessels and connective tissue between the sclera and retina.

CHROMATOPHILIC SUBSTANCE Rough endoplasmic reticulum and free ribosomes in the cell bodies of neurons; involved in protein synthesis; formerly called *Nissl bodies.*

CHROMOSOME (Gr. "colored body") A threadlike nucleoprotein structure within the nucleus of a cell that contains the DNA.

CHYLOMICRON A tiny, protein-coated droplet containing triglycerides, phospholipids, cholesterol, and free fatty acids.

CHYME [KIME] A soupy liquid mixture formed in the stomach from gastric juices and broken-down food particles.

CILIA [SIHL-ee-uh; sing. CILIUM; L. eyelid, eyelash] Short processes extending from the surface of some cells; often capable of producing a rhythmic paddling motion.

CILIARY BODY The thickened part of the vascular layer of the eye, connecting the choroid with the iris.

CIRCUMDUCTION A movement in which the distal end of a bone moves

in a circular path while the proximal end remains stable, acting as a pivot.

CLEAVAGE A series of mitotic cell divisions occurring in a zygote immediately after fertilization.

CLITORIS [KLIHT-uh-rihss; Gr. small hill] A small erectile organ at the upper end of the vulva, below the mons pubis, where the labia minora meet; a major source of sexual arousal.

COAGULATION (L. *coagulare,* to curdle) The process of blood clotting.

COCCYGEAL PLEXUS Plexus formed by the coccygeal nerve plus sacral nerves S4 and S5.

COCCYX The three to five fused vertebrae at the end of the vertebral column; commonly called *tailbone.*

COCHLEA [KAHK-lee-uh; Gr. *kokhlos,* snail] A spiral bony chamber beyond the semicircular ducts of the inner ear; divided into the scala vestibuli, scala tympani, and scala media.

COITUS [KOH-uh-tuhss; L. *coire,* to come together] The act of sexual intercourse; also called *copulation.*

COLLAGEN [KAHL-uh-juhn; Gr. *kolla,* glue + *genes,* born, produced] Protein found in the fibers of bone, cartilage, and connective tissue proper.

COLLATERAL BRANCHES Side branches of an axon.

COLLECTING DUCT Portion of renal excretory tubule that receives glomerular filtrate after it has passed through the *distal convoluted tubule.*

COLOSTRUM High-protein fluid present in the mother's breasts about 3 days before milk production starts.

COMMISSURAL FIBERS Axons that project from a cortical area of one cerebral hemisphere to the corresponding cortical area of the opposite hemisphere.

COMPACT BONE TISSUE The very hard and dense portion of bone.

COMPUTER-ASSISTED TOMOGRAPHY (CAT) A scanning procedure that combines x-rays with computer technology to show cross-sectional views of internal body structures; the device is called a *CAT scanner.*

CONCENTRATION GRADIENT The difference in solute concentration on either side of the plasma membrane.

CONCEPTION The process of fertilization and the subsequent establishment of pregnancy.

CONDYLE (L. knuckle) A rounded, knuckle-shaped projection on a bone; may be concave or convex.

CONE Color-sensitive photoreceptor cell in the retina.

CONJOINED TWINS Identical (monozygotic) twins whose embryonic disks do not separate completely; commonly called *Siamese twins.*

CONJUNCTIVA (pl. CONJUNCTIVAE; L. connective) Transparent mucous membrane lining the inner surface of the eyelid and exposed surface of the eyeball.

CONNECTIVE TISSUE Supportive and protective tissue consisting of fibers, ground substance, cells, and some extracellular fluid; most abundant type of tissue.

CONTRACTILITY The ability of muscle tissue to contract; the basic physiological property of muscle tissue.

CONTRALATERAL Pertaining to opposite sides of the body.

CONUS TERMINALIS The cone-shaped lowermost end of the spinal cord; also called *conus medullaris.*

CONVERGENCE (L. *convergere,* to merge) 1. A condition in which the receptive segment of a postsynaptic neuron is excited or inhibited by the axon terminals of many presynaptic neurons. 2. The coordinated inward turning of the eyes to focus on a nearby object, so that both images fall on the corresponding points of both retinas.

CONVOLUTIONS See *gyri.*

CORNEA (L. *corneus,* horny tissue) The transparent anterior part of the outer layer of the eye; light enters the eye through the cornea.

CORONAL PLANE See *frontal plane.*

CORONARY CIRCULATION The flow of blood that supplies the heart itself.

CORONARY SINUS A large vein that receives blood from the veins that drain the heart, emptying it into the right atrium.

CORPUS CALLOSUM (L. hard body) The larger of two cerebral commissures that connect the two cerebral hemispheres.

CORPUS LUTEUM (L. "yellow body") A temporary ovarian endocrine tissue that secretes female sex hormones; formed as the lining of a follicle grows inward after the discharge of an ovum.

CORPUS STRIATUM (L. furrowed body) The largest mass of gray matter in the cerebral basal ganglia; composed of the *caudate* and *lentiform nuclei.*

CORPUSCLES OF RUFFINI Sensory receptors in skin for touch-pressure, position sense of a body part, and movement.

CORTEX (L. bark, shell) The outer layer of an organ or part.

CORTICOSPINAL TRACTS Descending motor tracts from the cerebral motor cortex to the spinal cord; also called *pyramidal tracts.*

CRANIAL NERVES Twelve pairs of peripheral nerves that carry sensory impulses to the brain, and/or motor impulses from the brain to various places in the body.

CRANIAL PARASYMPATHETIC OUTFLOW Preganglionic fibers from the cranial portion of the parasympathetic nervous system that leave the brainstem via cranial nerves III, VII, IX, and X.

CRANIOSACRAL DIVISION Visceral efferent nerve fibers that leave the central nervous system via cranial nerves III, VII, IX, and X from the brainstem, or spinal nerves S3 and S4; part of *autonomic nervous system.*

CRYPTS OF LIEBERKÜHN See *intestinal glands.*

CYTOKINESIS (Gr. *cyto,* cell + movement) The separation of the cytoplasm into two parts following mitosis; two genetically identical daughter cells are formed.

CYTOLOGY [sigh-TAHL-uh-jee; Gr. *kytos,* hollow vessel; *cyto,* cell] The microscopic study of cells.

CYTOPLASM The portion of a cell outside the nucleus, where metabolic reactions take place; the fluid portion, the *cytosol,* contains subcellular *organelles.*

CYTOPLASMIC INCLUSIONS Solid particles temporarily in cells; usually either basic food material or the stored products of the cell's metabolic activities.

CYTOSKELETON The flexible cellular framework involved with support and cellular movement; site for the binding of specific enzymes; consists of interconnecting *microfilaments, intermediate filaments,* and *microtubules;* also called *microtrabecular lattice.*

D

DECUSSATION (L. *decussis,* the number ten, X, indicating a crossing over) The crossing over of the axons of sensory and motor pathways from one side of the spinal cord or brainstem to the other.

DEEP A relative directional term: farther from the surface of the body.

DEFECATION The discharge of solid wastes in the form of feces from the rectum through the anus.

DEGLUTITION (L. *deglutire,* to swallow down) The act of swallowing.

DENDRITES (Gr. *dendron,* tree) Threadlike extensions of the cell body of a neuron; they conduct impulses toward the cell body.

DENTINE Sensitive yellowish portion of tooth surrounding the pulp cavity; forms bulk of a tooth.

DEOXYRIBONUCLEIC ACID (DNA) [dee-AHK-see rye-boh-noo-KLAY-ihk] A double-stranded nucleic acid that is a constituent of chromosomes; contains hereditary information coded in specific sequences of *nucleotides.*

DEPOLARIZATION A reversal of electrical charges on the plasma membrane of a neuron, giving the inner side of the membrane a positive charge relative to the outer side.

DEPRESSION A movement that lowers a body part, such as opening the mouth.

DERMATOLOGIST A physician specializing in the treatment of skin disorders.

DERMATOME (Gr. *derma,* skin + *tomos,* a cutting) 1. An instrument used in cutting thin slices, as in skin grafting. 2. A segment of skin with sensory fibers from a single spinal nerve; used in locating injuries to dorsal roots of the spinal cord.

DERMIS (Gr. *derma,* skin) Strong, flexible connective tissue meshwork of collagenous, reticular, and elastic fibers that makes up most of the skin.

DESCENDING COLON Portion of large intestine extending from the left colic flexure of the transverse colon down to the rim of the pelvis, where it becomes the sigmoid colon.

DESMOSOME (Gr. *desmos,* binding) A junction with no direct contact between adjacent plasma membranes; common in skin; also called *spot desmosome.*

DIAPHYSIS [die-AHF-uh-siss; Gr. *dia,* between + growth] A tubular shaft of compact bone in most adult long bones; shaft center is filled with marrow, and there is a spongy *epiphysis* at each end of the shaft.

DIARTHROSIS (Gr. *dia,* between + *arthrosis,* articulation) A freely movable joint.

DIASTOLE [die-ASS-toe-lee] The relaxation of the atria and ventricles during the *cardiac cycle.*

DIASTOLIC PRESSURE The blood pressure during the interval between heartbeats; it is the second number in a blood pressure reading.

DIENCEPHALON (L. "between brain") Part of the brain connecting the midbrain with the cerebral hemispheres; composed of the thalamus, hypothalamus, epithalamus, and ventral thalamus (subthalamus); houses the third ventricle.

DIFFERENTIATION The process by which cells develop into specialized tissues and organs.

DIFFUSION See *simple diffusi n.*

DIGESTION The chemical and mechanical conversion of large nutrient particles into small, absorbable molecules.

DIGITAL SUBTRACTION ANGIOGRAPHY (DSA) (Gr. *angeion,* vessel) A noninvasive exploratory technique that uses a digital computer to produce three-dimensional pictures of blood vessels.

DISTAL (from *distant*) A relative directional term: away from the trunk of the body (away from the attached end of a limb).

DISTAL CONVOLUTED TUBULE Portion of renal excretory tubule that receives the *glomerular filtrate* after it has passed through the entire length of the *loop of the nephron*.

DIURETIC (Gr. "to urinate through") A therapeutic drug that increases the volume of urine by decreasing the water reabsorbed from the renal collecting duct.

DIVERGENCE (L. *divergere*, to blend) A condition in which the transmissive segment of a presynaptic neuron branches to have many synaptic connections with the receptive segments of many other neurons.

DIZYGOTIC TWINS Twins formed when more than one ovum is released at the same time and two ova are fertilized; also called *fraternal twins*.

DNA See *deoxyribonucleic acid*.

DORSAL (L. *dorsalis*, back) 1. A relative directional term: toward the back of the body; opposite of *ventral*. 2. The smaller of two main body cavities; contains the cranial and spinal cavities. 3. The upper surface of the hand or foot.

DORSIFLEXION Flexion of the foot at the ankle joint.

DUCTUS ARTERIOSUS A shunt in the aortic arch where fetal blood enters and mixes with blood from the left ventricle; allows fetal blood to bypass the nonfunctional lungs.

DUCTUS DEFERENS (pl. **DUCTUS DEFERENTIA**; L. *deferre*, to bring to) The continuation of the ductus epididymis extending from the epididymis of each testis to the ejaculatory duct; formerly called *vas deferens* or *sperm duct*.

DUCTUS VENOSUS A branch of the umbilical vein (which carries oxygenated blood) that bypasses the fetal liver.

DUODENUM [doo-oh-DEE-nuhm; Gr. "twelve fingers wide"] The C-shaped initial segment of the small intestine.

DURA MATER [DYOOR-uh MAY-ter; L. hard mother] The tough, fibrous outermost layer of the meninges.

DYNAMIC EQUILIBRIUM See *vestibular apparatus*.

DYNAMIC SPATIAL RECONSTRUCTOR (DSR) A scanning device that produces three-dimensional computer-generated pictures of the brain; can be used to view blood flow through the brain.

E

ECTODERM (Gr. *ektos*, outside + *derma*, skin) The outermost of the three primary germ layers of an embryo, developing into mature epithelial and nervous tissue.

EFFECTOR A muscle or gland; receives impulses from the central nervous system that result in physical activity.

EFFERENT ARTERIOLE (L. *ex*, away from + *ferre*, to bring) Blood vessel in the kidney formed from joining of glomerular capillary loops; carries blood away from *glomerulus*.

EFFERENT NEURON Nerve cell that conveys impulses away from the central nervous system to effectors; also called *motor neuron*.

EJACULATION (L. *ejaculari*, to throw) The expulsion of semen through the penis via the urethra; the expulsion accompanies orgasm in males.

EJACULATORY DUCT A duct continuing from the ductus deferens that transports sperm and secretions from the seminal vesicles and prostate gland to the urethra.

ELECTRICAL SYNAPSE A gap junction by which two cells are electrically coupled by tiny intercellular channels.

ELECTROCARDIOGRAM (EKG or ECG) The visual recording of the electrical waves of the heart, as registered on an *electrocardiograph*.

ELECTROENCEPHALOGRAM (EEG) ("electric writing in the head") The tracing, in the form of waves, produced by an electroencephalograph; shows the electrical activity in the brain.

ELECTROLYTE Any substance whose solution conducts electricity.

ELECTRON A negatively charged subatomic particle that moves around the nucleus of an atom.

ELEVATION A movement that raises a body part.

EMBRYONIC PERIOD The prenatal period from fertilization to the end of the eighth week.

ENAMEL The insensitive white covering of the crown of a tooth.

ENDOCARDIUM ("inside the heart") The fibrous layer covering the inside cavities of the heart and all associated valves and muscles.

ENDOCHONDRAL OSSIFICATION (Gr. *endon*, within + *khondros*, cartilage) The process by which bone tissue develops by replacing *hyaline cartilage*.

ENDOCRINE GLAND (Gr. *endon*, within + *krinein*, separate or secrete) Organ with specialized secretory cells but no ducts; the cells release their secretions directly into the bloodstream.

ENDOCYTOSIS (Gr. *endon*, within + *cyto*, cell) The active process that moves large molecules or particles through a plasma membrane when the membrane forms a pocket around the material, enclosing it and drawing it into the cytoplasm. Includes *pinocytosis*, *receptor-mediated endocytosis*, and *phagocytosis*.

ENDODERM (Gr. *endon*, within + Gr. *derma*, skin) Innermost of the three primary germ layers of an embryo, developing into mature epithelial tissue.

ENDOLYMPH Thin, watery fluid in the membranous labyrinth of inner ear.

ENDOMETRIUM (inside + Gr. *metra*, womb) The innermost tissue layer of the uterus, composed of specialized mucous membrane.

ENDOMYSIUM (Gr. *endon*, inside, within + muscle) A connective tissue sheath surrounding each muscle fiber.

ENDONEURIUM ("within the nerve") Interstitial connective tissue separating individual nerve fibers.

ENDOPLASMIC RETICULUM (ER) A complex labyrinth of flattened sheets, sacs, tubules, and double membranes that branch and spread throughout the cytoplasm, creating a series of channels for intracellular transport.

ENDOSTEUM [end-AHSS-tee-uhm; Gr. *endon*, inside + *osteon*, bone] The membrane lining the internal cavities of bones.

ENZYME (Gr. *enzumos*, leavened) A protein catalyst that increases the rate of a chemical reaction.

EOSINOPHIL White blood cell that modulates allergic inflammatory reactions and destroys antibody-antigen complexes.

EPENDYMAL CELL Glial cell that forms part of the inner membranes of the neural tube during embryonic growth; secretes *cerebrospinal fluid*.

EPIDERMIS (Gr. *epi*, over + *derma*, skin) The outermost layer of the skin.

EPIDIDYMIS (Gr. "upon the twin") A coiled tube that stores immature sperm until they mature; serves as a duct for the passage of sperm from the testis to the ejaculatory duct.

EPIDURAL SPACE Space between the dura mater of the spinal cord and the periosteum of the vertebrae; contains blood vessels and fat.

EPIGLOTTIS Flap of cartilage that folds down over the opening into the larynx during swallowing and swings back up when swallowing ceases.

EPIMYSIUM (*epi*, over, upon + muscle) The connective tissue sheath below the deep fascia surrounding a muscle.

EPINEURIUM ("upon the nerve") Connective tissue sheath containing blood and lymphatic vessels that surrounds a bundle of nerve fascicles.

EPIPHYSEAL PLATE Thick plate of hyaline cartilage providing the framework for formation of spongy bone within the *metaphysis*; also called *growth plate*.

EPIPHYSIS [ih-PIHF-uh-siss; Gr. "to grow upon"] The roughly spherical end of a long bone; composed of spongy bone.

EPIPLOIC APPENDAGES Fat-filled pouches in the large intestine formed where the visceral peritoneum is attached to the taeniae coli in the serous layer.

EPISIOTOMY Cutting of the vulva at the lower end of the vaginal orifice at the time of delivery to enlarge the birth canal.

EPITHELIAL TISSUE Groups of cells that cover or line something; secretes substances that lubricate or take part in chemical reactions; also called *epithelium*.

EPITHELIUM See *epithelial tissue*.

EPONYCHIUM [epp-oh-NICK-ee-uhm; Gr. *epi*, upon + *onyx*, nail] Thin layer of epidermis covering the developing nail; commonly called the *cuticle* in mature nail.

ERECTION The state of the clitoris or penis when its spongy tissue becomes engorged with blood, usually during erotic stimulation.

ERYTHEMA Diffuse redness of the skin.

ERYTHROCYTES [ih-RITH-roh-sites; Gr. *eruthros*, red + cell] Red blood cells.

ERYTHROPOIESIS [ih-RITH-roh-poy-EE-sis; Gr. red + to make] The production of erythrocytes in bone marrow.

ESOPHAGUS [ih-SOFF-uh-guss; Gr. gullet] Muscular, membranous tube through which food passes from the pharynx to the stomach.

EUSTACHIAN TUBE [yoo-STAY-shun] See *auditory tube*.

EVERSION Movement of foot in which great toe is turned downward and sole of foot faces laterally.

EXCITABILITY Capacity of nerve or muscle cell to receive and respond to stimuli.

EXCRETION (L. *excernere*, to sift out) The elimination of the waste products of metabolism.

EXHALATION See *expiration*.

EXOCRINE GLANDS Organs with specialized cells that produce secretions, and with ducts that carry the secretions.

EXOCYTOSIS (Gr. *exo*, outside + *cyto*, cell) Active process in which endosome containing undigested particles fuses with the plasma membrane and expels unwanted particles from a cell; opposite of *endocytosis*.

EXPIRATION The process in which oxygen-poor, carbon dioxide-rich air is forced out of the alveoli and through the respiratory tract back into the atmosphere; also called *exhalation*.

EXTENSION A straightening motion that increases the angle of a joint.

EXTERNAL A relative directional term: on the outside of the body.

EXTERNAL AUDITORY CANAL A slightly curved canal separating the external ear from the middle ear; also called *external auditory meatus*.

EXTERNAL AUDITORY MEATUS See *external auditory canal*.

EXTERNAL EAR The visible part of the ear; also called *auricle* or *pinna*.

EXTERNAL NARES [NAIR-eez; L. nostrils] The two external openings of the nostrils.

EXTERNAL RESPIRATION Gas exchange in which oxygen from the lungs moves into the blood, and carbon dioxide and water move from the blood to the lungs.

EXTEROCEPTOR (L. "received from the outside") A sensory receptor that responds to external stimuli that affect the skin directly.

EXTRACELLULAR FLUID Fluid that surrounds and bathes the body's cells.

EXTRAEMBRYONIC MEMBRANES Four membranes that form outside the embryo during the third week of embryonic development, including the yolk sac, amnion, allantois, and chorion; also called *fetal membranes*.

EXTREMITIES The extremities, or appendages, of the body are the *upper extremities* (shoulders, upper arms, forearms, wrists, hands) and *lower extremities* (thighs, legs, ankles, feet).

F

FACET (Fr. little face) A small flat surface on a bone.

FACILITATED DIFFUSION The passive-transport process in which large molecules need carrier proteins to pass through the protein channels of a plasma membrane.

FALLOPIAN TUBE See *uterine tube*.

FASCIA [FASH-ee-uh; pl. FASCIAE; L. band] A sheath of fibrous tissue enclosing skeletal muscles and holding them together; may be superficial, deep, or subserous (visceral).

FASCICLE A bundle of nerve or muscle fibers.

FASCICULI (fah-SICK-yoo-lie; L. little bundles] Bundles of fibers divided into tracts within each *funiculus* in the white matter of the spinal cord.

FAT An energy-rich molecule that is a source of reserve food or long-term fuel; stored body fat provides insulation and cushioning.

FECES [FEE-seez; L. *faex*, dregs] An indigestible semisolid mixture stored in the large intestine until ready to be eliminated through the anus during defecation.

FETAL MEMBRANES See *extraembryonic membranes*.

FETAL PERIOD The prenatal period from the beginning of the ninth week until birth.

FETOSCOPY ("seeing the fetus") A procedure that allows the physician to view the fetus directly by inserting an endoscope into the mother's uterus.

FETUS [FEE-tuhss] The embryo after 8 weeks of development.

FIBRIN The insoluble, stringy plasma protein whose threads entangle blood cells to form a clot during blood clotting.

FIBRINOGEN A soluble plasma protein converted into insoluble fibrin during blood clotting.

FIBRINOLYSIS ("fibrin breaking") Blood-clot destruction.

FIBROBLASTS (L. *fibro*, fiber + Gr. *blastos*, growth) Most common connective tissue cells; the only cell type in tendons; synthesize matrix materials and are considered to be secretory; assist in wound-healing.

FIBROUS JOINT A joint that lacks a joint cavity and has its bones united by fibrous connective tissue; includes *sutures*, *syndesmoses*, and *gomphoses*.

FILTRATION A passive transport process that forces small molecules through selectively permeable membranes with the aid of hydrostatic pressure or some other externally applied force.

FILUM TERMINALE A nonneural fibrous filament extending caudally from the conus terminalis of the spinal cord; attaches to the coccyx.

FIMBRIAE [FIHM-bree-ee; L. threads] Ciliated fringes that help sweep an ovum released by an ovary into the *infundibulum* of the uterine tube.

FISSURE (L. *fissio*, split) A groove or cleft, as in bones and the brain.

FIXATOR MUSCLE A muscle that provides a stable base for the action of the prime mover; also called *postural muscle*.

FLAGELLA (sing. FLAGELLUM; L. "whip") Threadlike appendages of certain cells, usually no more than one or two per cell; used to propel the cell through a fluid environment.

FLEXION A bending motion that decreases an angle at a joint.

FLEXOR REFLEX ARC A withdrawal reflex involving sensory receptors, afferent neurons, interneurons, alpha motor neurons, and skeletal muscles.

FOLLICLE (of hair) (L. *folliculus*, little bag) The tubular structure enclosing the hair *root* and *bulb*.

FOLLICLES Centers of oogenesis in cortex of ovaries.

FONTANEL A large membranous area between incompletely ossified bones.

FORAMEN [fuh-RAY-muhn; pl. FORAMINA; L. opening] A natural opening into or through a bone.

FORAMEN OVALE ("oval window") An opening in the fetal interatrial septum that closes at birth.

FORESKIN See *prepuce.*

FORMED ELEMENTS Cellular part of blood; includes erythrocytes and leukocytes.

FOSSA (L. trench) A shallow depressed area, as in bones.

FOVEA [FOE-vee-uh; L. small pit] A depressed area in the *macula lutea* near the center of the retina; contains only cones; image formation and color vision are most acute here.

FRACTURE (L. *fractura,* broken) A broken bone.

FRATERNAL TWINS See *dizygotic twins.*

FRONTAL LOBE Cerebral lobe involved with the control of voluntary movements and a variety of emotional expressions and moral and ethical behavior; also called the *motor lobe.*

FRONTAL PLANE A plane dividing the body into anterior and posterior sections formed by making a lengthwise cut at right angles to the midsagittal plane. Also called *coronal plane.*

FULCRUM (L. *fulcire,* to support) The point or support on which a lever turns.

FUNDUS (L. bottom) The inner basal surface of an organ farthest away from the opening; the wide upper portion of the uterus.

FUNICULI [fyoo-NICK-yoo-lie; sing. FUNICULUS; L. little ropes] Three pairs of columns of myelinated fibers that run the length of the white matter of the spinal cord.

G

GAMETE (L. *gamos,* marriage) A sex cell; the female gamete is an egg, the male gamete is a sperm.

GAMETOGENESIS The formation of gametes (sex cells).

GANGLIA (sing. GANGLION) Groups of cell bodies located outside the central nervous system.

GAP JUNCTION A junction formed from several links of channel protein connecting two plasma membranes; found in interstitial epithelia.

GASTRIC (Gr. *gaster,* belly, womb) Pertaining to the stomach.

GASTROINTESTINAL (GI) TRACT The part of digestive tract below diaphragm.

GENE [JEEN; Gr. *genes,* born, to produce] A segment of DNA that controls a specific cellular function, either by determining which proteins will be synthesized or by regulating the action of other genes; a hereditary unit that carries hereditary traits.

GENERAL SENSES The senses of touch-pressure, heat, cold, pain, and body position; also called *somatic senses.*

GENOTYPE [JEEN-oh-tipe] The genetic make-up, sometimes hidden, of a person.

GERIATRICS (Gr. *geras,* old age + *iatrikos,* physician) The medical study of the physiology and ailments of old age.

GERMINAL EPITHELIUM A layer of specialized epithelial cells covering the ovaries and lining the seminiferous tubules of the testes; also called *germinal layer.*

GESTATION (L. *gestare,* to carry) The period of carrying developing offspring in the uterus during pregnancy.

GINGIVA [jihn-JYE-vuh] See *gum.*

GLANS CLITORIS (L. acorn) The small mass of sensitive tissue at the tip of the clitoris.

GLANS PENIS (L. acorn) The sensitive tip of the penis.

GLIAL CELLS See *neuroglia.*

GLIDING JOINT A small biaxial joint that usually has only one axis of rotation, permitting side-to-side and back-and-forth movements.

GLOMERULAR CAPSULE The portion of the nephron enclosing the glomerulus; also called *Bowman's capsule.*

GLOMERULAR FILTRATE The fluid filtered from the blood in the kidney.

GLOMERULAR FILTRATION Renal process that forces fluid from the glomerulus into the glomerular capsule.

GLOMERULUS [glow-MARE-yoo-luhss; pl. GLOMERULI, glow-MARE-you-lie; L. ball] Coiled ball of capillaries in the kidney formed by branching of an afferent arteriole; site of blood filtration.

GLOTTIS The space that serves as an air passage between the true vocal cords in the larynx.

GOLGI APPARATUS Flattened stacks of disklike membranes found in cytoplasm of most cells; packages glycoproteins for secretion; also called *Golgi complex, Golgi body.*

GOLGI TENDON ORGAN Encapsulated sense receptor that monitors tension and stretch in muscles and tendons; involved with skeletal muscle reflexes.

GOMPHOSIS (Gr. *gomphos,* bolt) A fibrous joint in which a peg fits into a socket.

GONADS (Gr. *gonos,* offspring) Sex organs: *ovaries* in females, and *testes* in males.

GRAAFIAN FOLLICLE See *vesicular ovarian follicle.*

GRAY COMMISSURE (L. "joining together") The pair of anterior horns that forms the "cross bar" of the H-shaped gray matter in the spinal cord.

GRAY MATTER Central part of spinal cord; consists of cell bodies and dendrites of association and efferent neurons, unmyelinated axons of spinal neurons, sensory and motor neurons, and axon terminals.

GRAY RAMI COMMUNICANTES [RAY-mee] Unmyelinated nerve fibers containing postganglionic sympathetic fibers.

GREATER OMENTUM (L. "fat skin") A folded, fatty membrane extending from greater curvature of stomach to back wall and down to pelvic cavity; protects and insulates abdominal organs.

GREATER VESTIBULAR GLANDS Paired glands located in the floor of the vaginal vestibule; during sexual arousal they secrete an alkaline mucus solution that provides some lubrication and offsets some vaginal acidity; also called *Bartholin's glands.*

GROSS ANATOMY Any branch of anatomy that can be studied without a microscope.

GROUND SUBSTANCE A homogeneous, extracellular material of tissues that provides a suitable medium for the passage of nutrients and wastes between cells and the bloodstream.

GROWTH PLATE See *epiphyseal plate.*

GUM (Old Eng. *goma,* palate, jaw) Firm connective tissue covered with mucous membrane that surrounds alveolar processes of teeth; also called *gingiva.*

GYNECOLOGY (Gr. *gune,* woman + study of) The medical science of female reproductive physiology, endocrinology, and disease.

GYRI [JRY-rye; sing. GYRUS; L. circles] Raised ridges of the cerebral cortex; also called *convolutions.*

H

HAIR A specialization of the skin that develops from the epidermis; composed of cornified threads of cells, covering almost the entire body.

HAIR CELLS Specialized proprioceptor cells of the vestibular apparatus; convert mechanical force into impulses that are relayed to the brain.

HAUSTRA (sing. HAUSTRUM) Puckered bulges in the wall of the large intestine caused by the uneven pull of the *taeniae coli.*

HAVERSIAN CANAL See *central canal.*

HAVERSIAN SYSTEM See *osteon.*

HEAD Expanded, rounded surface at proximal end of a bone; often joined to shaft by narrow neck; also called *caput.*

HEART (Gr. *kardia,* L. *cor*) The four-chambered muscular organ that pumps blood through the circulatory system.

HEMATOENCEPHALIC BARRIER See *blood-brain barrier.*

HEMOGLOBIN [HEE-moh-gloh-bihn; Gr. *haima,* blood + L. *globulus,* little globe] Iron-containing protein found in erythrocytes; transports oxygen from the lungs and carbon dioxide from tissues.

HEMOPOIETIC TISSUE [hee-muh-poy-ET-ihk; Gr. *haima,* blood + *poiein,* to make] Tissues including red bone marrow and lymphoid tissues that produce red blood cells, white blood cells, and platelets.

HEMOSTASIS The stoppage of bleeding.

HEPARIN An acidic mucopolysaccharide that inhibits blood clotting.

HEPATIC (Gr. *hepatikos,* liver) Pertaining to the liver.

HEPATIC PORTAL SYSTEM (Gr. *hepatikos,* liver) System of vessels that carries blood from capillary beds of the intestines to sinusoidal beds of the liver.

HILUS [HYE-luhss; L. *hilum,* trifle] A small, indented opening on the medial concave border of the kidney where arteries, veins, nerves, and the ureter enter and leave the kidney; also called the *hilum.*

HINGE JOINT Joint in which the convex surface of one bone fits into the concave surface of another bone, permitting only a uniaxial movement around a single axis, such as at the knee joint.

HISTOLOGY [hiss-TAHL-uh-jee; Gr. *histos,* web] The microscopic study of tissues.

HOLOCRINE GLAND (Gr. *holos,* whole) Exocrine gland that releases secretions by the detaching and dying of whole cells, which become the secretion; sebaceous glands are probably the only holocrine glands in the body.

HOMEOSTASIS [ho-mee-oh-STAY-siss; Gr. *homois,* same + *stasis,* standing still] A state of balance and stability in the body, which remains relatively constant despite external environmental changes.

HORMONE A chemical "messenger" produced and secreted by endocrine cells or tissues; circulates via the bloodstream to "target" cells or tissues, affecting their metabolic activity.

HYALINE CARTILAGE [HYE-uh-lihn; Gr. *hyalos,* glassy] The most prevalent type of cartilage, containing collagenous fibers scattered in a network filled in with ground substance.

HYDROSTATIC PRESSURE The force exerted by a fluid against the surface of the compartment containing the fluid.

HYMEN (Gr. membrane; Hymen was Greek god of marriage) Fold of skin partially blocking vaginal entrance.

HYPEREXTENSION Excessive extension beyond the straight (anatomical) position.

HYPODERMIS (Gr. *hypo,* under + *derma,* skin) Layer of loose, fibrous connective tissue lying below the dermis; also called *subcutaneous layer.*

HYPOPHYSIS (Gr. undergrowth) See *pituitary gland.*

HYPOTHALAMIC-HYPOPHYSEAL PORTAL SYSTEM An extensive system of blood vessels connecting the hypothalamus and the adenohypophysis.

HYPOTHALAMUS ("under the thalamus") The part of the brain located under the thalamus, forming the floor of the third ventricle; regulates body temperature, some metabolic processes, and other autonomic activities.

I

IDENTICAL TWINS See *monozygotic twins.*

ILEUM [ILL-ee-uhm] The portion of the small intestine extending from the *jejunum* to the *cecum,* the first part of the large intestine.

IMMUNE SYSTEM The overall defensive system of lymphocytes and lymphatic tissues and organs.

IMMUNITY (L. *immunis,* exempt) An overall protective mechanism that forms antibodies to help protect the body against foreign substances.

IMPLANTATION The process by which the blastocyst becomes embedded in the uterine wall.

INFANCY The period between the first 4 weeks of life and the time the child can sit erectly and walk.

INFERIOR (L. low) A relative directional term: toward the feet; below.

INFERIOR VENA CAVA Large vein that drains blood from the abdomen, pelvis, and lower limbs, emptying it into the right atrium.

INFLAMMATORY RESPONSE Part of the healing process, including redness, pain, swelling, scavenging by *neutrophils* and *monocytes,* and tissue repair by *fibroblasts.*

INFUNDIBULAR STALK (L. funnel) A stalk of nerve cells and blood vessels that connects the pituitary and the hypothalamus; also called *infundibulum.*

INFUNDIBULUM (L. funnel) The funnel-shaped portion of the uterine tube near the ovary; also see *infundibular stalk.*

INGESTION The taking in of nutrients by eating or drinking.

INHALATION See *inspiration.*

INNER EAR (internal ear) Portion of ear that includes the vestibule, semicircular ducts and canals, and cochlea; also called *labyrinth.*

INORGANIC COMPOUND A chemical compound composed of relatively small molecules and not containing both carbon and hydrogen.

INSERTION The point of attachment of a muscle to the bone it moves.

INSPIRATION The process in which oxygen-rich, carbon dioxide-poor air travels from the atmosphere through the respiratory tract to the terminal portion of the lungs; also called *inhalation.*

INSULA (L. island) The cerebral lobe beneath the parietal, frontal, and temporal lobes; appears to be associated with gastrointestinal and other visceral activities; also called *central lobe.*

INTERCALATED DISKS (L. *intercalatus,* to insert between) Thickenings of sarcolemma separating adjacent cardiac muscle fibers.

INTERNAL A relative directional term: inside the body.

INTERNAL RESPIRATION The process in which body cells exchange carbon dioxide for oxygen from the blood.

INTEROCEPTOR (L. "received from inside") Sensory receptor responding to stimuli originating in internal organs; also called *visceroceptor.*

INTERPHASE The period between cell divisions during which growth, cellular respiration, RNA and protein synthesis, and DNA replication take place.

INTERSTITIAL CELLS OF LEYDIG See *interstitial endocrinocytes.*

INTERSTITIAL ENDOCRINOCYTES Clusters of endocrine cells among the seminiferous tubules that secrete the male sex hormones; also called *Leydig cells.*

INTERSTITIAL FLUID The fluid between cells that bathes and nourishes surrounding body tissues.

INTESTINAL GLANDS Tubular glands in the small intestine; formerly called *crypts of Lieberkühn.*

INTRACELLULAR FLUIDS (ICF) Fluids inside a cell.

INTRAFUSAL MUSCLE FIBERS Tiny muscles within a neuromuscular spindle.

INTRAMEMBRANOUS OSSIFICATION Process by which bone develops directly from connective tissue.

INVERSION A movement of the foot in which the great toe is turned upward and the sole of the foot faces medially.

INVOLUNTARY MUSCLE See *smooth muscle tissue.*

IPSILATERAL Pertaining to the same side of the body.

IPSILATERAL REFLEX ("same side") A reflex occurring on the same side of the body and spinal cord as the stimulus.

IRIS (Gr. rainbow) The colored portion of the eye that surrounds the pupil.

ISLETS OF LANGERHANS See *pancreatic islets.*

ISTHMUS [ISS-muhss] A narrow passage connecting two larger cavities.

J

JEJUNUM [jeh-JOO-nuhm; L. "fasting intestine"] The 2.5-m-long portion of small intestine between the duodenum and the ileum.

JUNCTIONAL COMPLEXES Specialized parts that hold cells together, enabling groups of cells to function as a unit; include *desmosomes, gap junctions,* and *tight junctions.*

JUXTAGLOMERULAR APPARATUS The portion of the kidney composed of juxtaglomerular cells (specialized smooth muscle cells), the macula densa (a group of epithelial cells of the distal tubule), afferent and efferent arterioles, and a pad of cells called the *extraglomerular mesangium.*

K

KERATIN (Gr. *keras,* horn) Tough protein forming outer layer of hair and nails; soft in hair, and hard in nails.

KIDNEYS Pair of organs located in posterior wall of abdominal cavity, on either side of the vertebral column; remove wastes of metabolism from blood and maintain water and salt balance.

KINESIOLOGY The study of motion.

KINESTHESIA (Gr. *kinema,* motion + sensory ability) The sense of body movement.

KRAUSE'S END BULBS See *bulbous corpuscles (of Krause).*

L

LABIA MAJORA ("major lips"; sing. LABIUM MAJUS) Two longitudinal folds of skin just below the mons pubis, forming the outer borders of the *vulva.*

LABIA MINORA ("minor lips"; sing. LABIUM MINUS) Two relatively small folds

of skin lying between the larger *labia majora.*

LABYRINTH (Gr. maze) The intricate interconnecting chambers and passages that make up the inner ear.

LACRIMAL APPARATUS [LACK-ruh-mull; L. tear] Structure consisting of lacrimal gland and sac and nasolacrimal duct.

LACRIMAL GLAND The tear gland of the eye.

LACTATION (L. *lactare,* to suckle) Process including production of milk by the mammary glands and release of milk from the breasts.

LACTEALS (L. *lacteus,* of milk) Specialized lymphatic capillaries that extend into the intestinal villi.

LACTIC ACID Toxic waste product that builds up in muscles as they become fatigued during vigorous physical activity.

LACUNAE [luh-KYOO-nee; sing. LACUNA; L. cavities, pods] Small cavities within the connective tissue matrix containing *chondrocytes* in cartilage, and *osteocytes* in bone.

LAMELLAE (L. "thin plates") Concentric layers of bone that make up cylinders of calcified bone called *osteons.*

LAMELLATED (PACINIAN) CORPUSCLE A sensory receptor in skin, muscles, tendons, joints, and body organs; involved with vibratory sense and touch-pressure.

LANUGO (L. *lana,* fine wool) Fine, downy hair covering the fetus by the fifth month; shed before birth, except on the eyebrows and scalp, where it becomes thicker.

LARGE INTESTINE The part of the digestive system between the ileocecal orifice of the small intestine and the anus.

LARYNGOPHARYNX The lowest part of the pharynx, extending downward into the larynx.

LARYNX An air passage at the beginning of the respiratory tract; contains vocal cords; commonly called *voice box.*

LATERAL (L. *lateralis,* side) A relative directional term: away from the midline of the body; toward the *side* of the body.

LATERAL ROTATION A twisting movement in which the anterior surface of a limb or bone moves away from the body's medial plane.

LENS An elastic, colorless, transparent body of epithelial cells behind the iris; its shape is adjustable to focus on objects at different distances.

LESSER OMENTUM (L. "fat skin") A peritoneal membrane extending from the liver to the lesser curvature of the stomach.

LESSER VESTIBULAR GLANDS Paired

glands with ducts opening into the vaginal vestibule; during sexual arousal alkaline secretion of glands provides some lubrication and offsets some vaginal acidity; also called *Skene's glands.*

LEUKOCYTE [LOO-koh-site; Gr. *leukos,* clear, white] A white blood cell, usually a scavenger, that ingests foreign materials in blood and tissues.

LEYDIG CELLS See *interstitial endocrinocytes.*

LIGAMENT A fairly inelastic fibrous thickening of an articular capsule that joins a bone to its articulating mate, allowing movement at the joint.

LIGHT TOUCH The sense that is perceived where the skin is touched, but not deformed.

LIMBIC SYSTEM Group of structures in cerebrum, diencephalon, and midbrain involved in memory and emotions and associated visceral and behavioral responses.

LIPID [LIHP-ihd; Gr. *lipos,* fat] Organic compound (usually fat, oil, or wax) that is insoluble in water but can be dissolved in organic solvents.

LOOP OF HENLE See *loop of the nephron.*

LOOP OF THE NEPHRON Straightened portion of the kidney excretory tubule; receives *glomerular filtrate* from the proximal convoluted tubule; also called *loop of Henle.*

LOWER EXTREMITIES See *extremities.*

LUMBAR PLEXUS The ventral rami of L1 to L4 nerves in the interior of the posterior abdominal wall.

LUNG One of the pair of organs of respiration; on either side of the heart in the thoracic cavity.

LYMPH (L. "clear water") *Interstitial fluid* inside *lymphatic capillaries.*

LYMPH NODES Bodies of lymphoid tissue situated along lymphatic vessels; filter foreign material from lymph.

LYMPHATIC CAPILLARIES Capillaries of lymphatic system that collect excess interstitial fluid in tissues.

LYMPHATIC SYSTEM The body system that collects and drains fluid lost from the bloodstream and accumulates in the spaces between cells; returns fluid to bloodstream.

LYMPHATICS Collecting vessels that carry lymph from lymphatic capillaries to veins in the neck, where it is returned to the bloodstream.

LYMPHOCYTE A wandering connective tissue cell found under moist epithelial linings of respiratory and intestinal tracts; main producer of antibodies; also called *plasma cell.*

LYSOSOME (Gr. "dissolving body") A small, membrane-bound organelle containing digestive enzymes, which break down dead or damaged cells.

M

MACROPHAGE CELL [MACK-roh-fahj; Gr. *makros*, large; *phagein*, to eat] A connective tissue cell that is an active *phagocyte*; can be fixed or wandering.

MACULA The receptor region of utricles and saccules in the ear, containing *hair cells*.

MACULA DENSA (L. "dense spot") See *juxtaglomerular apparatus*.

MACULA LUTEA (L. "yellow spot") Area in center of the retina that contains only cones.

MAGNETIC RESONANCE IMAGING (MRI) A noninvasive diagnostic technique that uses a strong magnetic field to detect unhealthy tissues; also called *nuclear magnetic resonance (NMR)*.

MAMMARY GLANDS (L. *mammae*, breasts) Paired female modified sweat glands that produce and secrete milk for a newborn child.

MARROW See *bone marrow*.

MAST CELL A wandering connective tissue cell often found near blood vessels; contains secretory granules containing *heparin* and *histamine*.

MATRIX [MAY-triks; L. womb, mother] 1. The extracellular fibers and gound substance in connective tissues. 2. The thick layer of skin beneath the root of a nail, where new cells are generated.

MEATUS [mee-AY-tuhss; L. passage] A large, tubular channel or opening, not necessarily through a bone.

MECHANORECEPTOR ("mechanical receiver") Sensory receptor that responds to physical stimuli such as touch-pressure, muscle tension, air vibrations, and head movements.

MEDIAL [MEE-dee-uhl; L. *medius*, middle] A relative directional term: toward the midline of the body.

MEDIAL ROTATION A twisting movement in which the anterior surface of a limb or bone moves toward the medial plane of the body.

MEDIASTINUM [mee-dee-as-TIE-nuhm; L. *medius*, middle] The mass of tissues and organs between the lungs. It contains all the contents of the thoracic cavity except the lungs.

MEDULLA [meh-DULL-uh; L. marrow] The inner core of a structure.

MEDULLA OBLONGATA (L. elongated marrow) The lowermost portion of the brainstem, continuous with the spinal cord.

MEDULLARY CAVITY [MED-uh-lehr-ee; L. *medulla*, marrow] The marrow cavity inside the shaft of a long bone.

MEIOSIS [mye-OH-sihss; Gr. *meioun*, to diminish] A process that reduces the number of chromosomes in gametes to half.

MEISSNER'S CORPUSCLES See *tactile (Meissner's) corpuscles*.

MELANIN [MEHL-un-nihn; Gr. *melas*, black] A dark pigment produced by specialized cells called *melanocytes*; contributes to skin color.

MEMBRANES Thin, pliable layers of epithelial and/or connective tissue that line body cavities and cover or separate regions, structures, and organs.

MENARCHE [meh-NAR-kee; Gr. "beginning the monthly"] The first menstrual period, usually occurring during the latter stages of puberty.

MENINGES [muh-NIHN-jeez; Gr. pl. of *meninx*, membrane] Three layers of protective membranes (dura mater, arachnoid, pia mater) surrounding the brain and spinal cord.

MENOPAUSE (Gr. "ceasing the monthly") The cessation of menstrual periods.

MENSTRUAL CYCLE Monthly series of events that prepares the endometrium of uterus for pregnancy and then discharges sloughed-off endometrium, mucus, and blood in the menstrual flow if pregnancy does not occur.

MENSTRUATION (L. *mensis*, monthly) The monthly breakdown of the endometrium of a nonpregnant female.

MERKEL'S DISKS See *tactile (Merkel's) corpuscles*.

MEROCRINE GLAND (Gr. *meros*, divide) An exocrine gland that releases its secretions via *exocytosis*.

MESENCHYME [MEHZ-uhn-kime; Gr. *mesos*, middle + L. *enchyma*, cellular tissue] Embryonic *mesoderm* that develops into connective tissue.

MESENTERY [MEZZ-uhn-ter-ee; Gr. *mes*, middle + *enteron*, intestines] Fused layers of visceral peritoneum that attach abdominopelvic organs to the cavity wall; also called *visceral ligament*.

MESODERM (Gr. *mesos*, middle) The embryonic germ layer between the *ectoderm* and *endoderm*; develops into epithelial, connective, and muscle tissue.

MESOVARIUM Mesentery that attaches ovaries to broad ligament.

METABOLISM [muh-TAB-uh-lihz-uhm; Gr. *metabole*, change] All the chemical activities of the body.

METACARPAL BONES (L. behind the wrist) The five miniature long bones constituting the palm of each hand; also called *metacarpus*.

METACARPUS See *metacarpal bones*.

METAPHASE The second stage of mitosis; centromeres double, one going to each chromatid, each of which is now a single-stranded chromosome.

METAPHYSIS [muh-TAHF-uh-siss; Gr. "to grow beyond"] Area in a bone where longitudinal growth continues after

birth, between the *epiphyseal plate* and *diaphysis*.

METATARSAL BONES The five miniature long bones in each foot between the tarsal bones and those of the toes.

MICELLE [my-SELL; L. *mica*, grain] A submicroscopic water-soluble particle composed of fatty acids, phospholipids, and glycerides.

MICROFILAMENT A solid, rodlike subcellular structure containing the protein *actin*; provides cellular support and aids movement.

MICROGLIA The smallest glial cell; a macrophage that removes disintegrating products of neurons.

MICROTUBULE A slender subcellular structure that helps support the cell; involved with organelle movement, cellular shape changes, and intracellular transport.

MICROVILLI [my-krow-VILL-eye; Gr. *mikros*, small; L. *villus*, shaggy hair] Fingerlike projections from plasma membranes of some cells.

MICTURITION (L. *micturire*, to want to urinate) The process of emptying the urinary bladder; also called *urination*.

MIDBRAIN Portion of brainstem between the pons and diencephalon; connects pons and cerebellum with the cerebrum; also called *mesencephalon*.

MIDDLE EAR A small chamber between the tympanic membrane and inner ear; composed of the tympanic cavity, and contains auditory ossicles.

MIDSAGITTAL PLANE The plane that divides the left and right sides of the body lengthwise along the midline into externally symmetrical sections.

MITOCHONDRION (pl. MITOCHONDRIA; Gr. *mitos*, a thread) A double-membraned, saclike organelle; produces most of the ATP for cellular metabolism.

MITOSIS (Gr. *mitos*, a thread) The process of nuclear division; it arranges cellular material for equal distribution to daughter cells and divides the nuclear DNA equally to each new cell.

MITRAL VALVE [MY-truhl] See *bicuspid valve*.

MONOCYTE White blood cell that becomes a phagocytic macrophage that ingests and destroys foreign particles.

MONOZYGOTIC TWINS Twins that result from one fertilized ovum that divides into two identical cells before implantation; also called *identical twins*.

MONS PUBIS (L. mountain + pubic) A mound of fatty tissue covering the female symphysis pubis; also called *mons veneris*.

MONS VENERIS (L. mountain + Venus, the Roman goddess of love and beauty) See *mons pubis*.

MORPHOGENESIS (Gr. *morphe*, shape + *gennan*, to produce) The cellular changes that result in tissue and organ development.

MORULA [MORE-uh-luh; L. *morum*, mulberry tree] A solid ball of about 8 to 50 cells produced by cell divisions of a fertilized ovum.

MOTOR END PLATE Junction between a motor neuron and a muscle fiber.

MOTOR NEURON See *efferent neuron*.

MOTOR UNIT A motor neuron and the muscle fibers it innervates.

MUCOUS MEMBRANE The membrane that lines body passageways that open to the outside of the body.

MUCUS [MYOO-kuhss] The thick, protective liquid secreted by glands in the mucous membranes.

MULTIPOLAR NEURON A nerve cell with many dendrites radiating from the cell body, but with only one axon.

MUSCLE (L. *musculus*, "little mouse") A collection of muscle fibers that can contract and relax to move body parts.

MUSCLE FIBER A collection of specialized, individual muscle cells that make up skeletal muscle tissue; muscle fibers have a long, cylindrical shape and several nuclei.

MUTATION (L. *mutare*, to change) A change in genetic material.

MYELIN [MY-ih-linn; Gr. *myelos*, marrow] A lipid that forms a laminated sheath covering an axon.

MYELIN SHEATH A thick pad of insulating myelin surrounding an axon; also called *medullary sheath*.

MYELINATED FIBER A nerve fiber covered with a myelin sheath.

MYOCARDIUM ("heart muscle") The middle layer of muscle in the heart wall.

MYOFIBRIL Small units or fibers within individual threadlike muscle fibers; suspended in the sarcoplasm along with mitochondria and other multicellular material.

MYOFILAMENT A muscle filament composed of thick or thin threads that make up a *myofibril*.

MYOGLOBIN A form of hemoglobin found in muscle fibers.

MYOMETRIUM (muscle + Gr. *metra*, womb) The middle, muscular tissue layer of the uterus.

MYONEURAL JUNCTION See *neuromuscular junction*.

MYOSIN A fairly large protein that makes up the thick myofilaments of muscle fibers.

N

NAIL A modification of the epidermis, composed of hard keratin overlying the tips of fingers and toes.

NASAL CAVITY The cavity that fills the space between the base of the skull and the roof of the mouth.

NASAL SEPTUM A vertical wall dividing the nasal cavity.

NASOLACRIMAL DUCT (L. "nose" + "tear") Duct leading from each eye to the nasal cavity; drains excessive secretions of tears into the nasal cavity.

NASOPHARYNX The part of the pharynx above the soft palate.

NECROBIOSIS (Gr. *nekros*, corpse, death + *biosis*, way of life) Natural degeneration and death of cells and tissues.

NECROSIS (Gr. *nekros*, corpse, death) The death of cells or tissues due to disease or injury.

NEGATIVE FEEDBACK SYSTEM A regulatory system that produces a response that opposes or reduces the effects of the initial stimulus.

NEONATE A newborn child during the first 4 weeks after birth.

NEPHRON (Gr. *nephros*, kidney) The functional unit of a kidney; operate as independent urine-making units.

NERVE A bundle of peripheral nerve fibers enclosed in a sheath.

NERVE IMPULSE See *action potential*.

NEURILEMMA The outer layer, or sheath, of Schwann cells surrounding a nerve fiber.

NEUROFIBRAL NODES Regular gaps in a myelin sheath around a nerve fiber; also called *nodes of Ranvier*.

NEUROFILAMENT A semirigid tubular structure in cell bodies and processes of neurons; also called *microfilament*.

NEUROGLIA (Gr. nerve + glue) Nonconducting cells of the central nervous system that protect, nurture, and support the nervous system; also called *glial cells*.

NEUROHYPOPHYSIS The posterior lobe of the pituitary.

NEUROLEMMOCYTE A peripheral nerve cell that forms the myelin sheath; also called *Schwann cell*.

NEUROMUSCULAR JUNCTION The junction of a motor nerve ending and a muscle fiber; also called *myoneural junction*.

NEURON (Gr. nerve) A cell of the nervous system specialized to transmit impulses.

NEUROTRANSMITTER A chemical substance synthesized by neurons; may produce an excitatory or inhibitory response in a receptor.

NEUROTUBULE A threadlike protein structure in cell bodies and processes of neurons; involved in intracellular transport of proteins and other substances.

NEUTROPHIL A type of phagocytic white blood cell.

NISSL BODIES See *chromatophilic substance*.

NOCICEPTOR [NO-see; L. "injury receiver"] A sensory receptor responding to potentially harmful stimuli that produce pain.

NODES OF RANVIER See *neurofibral nodes*.

NORADRENERGIC Pertaining to neurons or fibers that release the neurotransmitter *norepinephrine*.

NUCLEIC ACID [noo-KLAY-ihk] Any of two groups of complex compounds (DNA and RNA) composed of bonded units called *nucleotides*; DNA is the carrier of hereditary information.

NUCLEOLUS [new-KLEE-oh-luhss; pl. **NUCLEOLI**; "little nucleus"] A spherical mass in the nucleus; contains genetic material in the form of DNA and RNA; a preassembly point for ribosomes.

NUCLEOPLASM The material within the nucleus of a cell.

NUCLEOTIDE [NOO-klay-uh-tide] The structural unit of nucleic acids, composed of a phosphate group, a sugar, and a nitrogenous base.

NUCLEUS (L. nut, kernel) 1. The central portion of an atom, containing positively charged protons and uncharged neutrons. 2. The central portion of a cell, containing chromosomes. 3. A collection of nerve cells inside the central nervous system that processes afferent impulses.

NUTRIENTS The chemical components of foods that supply the energy and physical materials a body needs.

O

OCCIPITAL LOBE The posterior cerebral lobe; composed of several areas organized for vision and its associated forms of expression.

OIL GLANDS See *sebaceous glands*.

OLFACTION (L. *olfacere*, to smell) The sense of smell.

OLFACTORY Pertaining to the sense of smell.

OLFACTORY BULB A stemlike extension of the olfactory region of the brain; receives impulses from nerve fibers that have been stimulated by an odiferous substance.

OLFACTORY TRACT Axons of cells that carry impulses from the olfactory bulb posteriorly to the olfactory cortex in the brain.

OLIGODENDROCYTE Glial cell similar to a *neurolemmocyte*; produces and nurtures myelin sheath segments of nerve fibers, provides a supportive framework, and supplies nutrition for neurons.

ONCOLOGY (Gr. *onkos*, tumor + *logy*, the study of) The study of neoplasms; the study of cancer.

OOGENESIS The monthly maturation of an ovum in the ovary.

OPPOSITION The angular movement of the thumb pad touching and opposing a finger pad; occurs only at the carpometacarpal joint of the thumb.

OPTIC (Gr. *optikos*, visible) Pertaining to the eye, vision, and related topics.

OPTIC CHIASMA [kye-AZ-muh; after the X-shaped Greek letter *chi*, KYE] A point in the cranial cavity where half the fibers of each optic nerve of each eye cross over to the other side.

OPTIC TRACT Nerve fibers after they have passed through the *optic chiasma*.

ORAL CAVITY Mouth cavity; consists of a smaller outer part, the *vestibule* (buccal cavity), and a larger inner part, the oral cavity proper.

ORGAN An integrated collection of two or more kinds of tissue that combine to perform a specific function.

ORGAN OF CORTI See *spiral organ (of Corti)*.

ORGANELLE ("little organ") Various subcellular structures with specific structures and functions.

ORGANIC COMPOUND Chemical compound containing carbon and hydrogen.

ORGANISM The product of all the body systems specialized within themselves and coordinated with each other; the body; an individual living thing.

ORGASM (Gr. *orgasmos*, to swell with excitement) The climax of sexual excitement.

ORIGIN The end of a muscle attached to the bone that does not move.

OROPHARYNX The part of the pharynx at the back of the mouth, extending from the soft palate to the epiglottis, where the respiratory and alimentary tracts separate.

OSMOSIS (Gr. "pushing") The passive transport process occurring when water (or another solvent) passes through a selectively permeable membrane from an area of high concentration to an area of lower concentration.

OSSEOUS TISSUE (L. *os*, bone) A tissue composed of cells embedded in a matrix of ground substance, inorganic salts, and collagenous fibers; also called *bone tissue*.

OSSICLES The three bones (malleus, incus, stapes) of the ear.

OSSIFICATION CENTER See *center of ossification*.

OSTEOBLAST (Gr. *osteon*, bone + *blastos*, bud, growth) A bone cell capable of synthesizing and secreting new bone

matrix as needed; usually found in growing portions of bones.

OSTEOCLAST (Gr. *osteon*, bone + *klastes*, breaker) A multinuclear bone-destroying cell; usually found where bone is resorbed during growth.

OSTEOCYTE (Gr. *osteon*, bone + cell) A main cell of mature bone tissue; regulates the concentration of calcium in body fluids by helping to release calcium from bone tissue into the blood.

OSTEOGENIC CELL (Gr. *osteon*, bone + *genes*, born) Bone cell that can be transformed into an *osteoblast* or *osteoclast*.

OSTEON (Gr. bone) Concentric cylinders of calcified bone that make up compact bone; also called *Haversian system*.

OTIC (OH-tick; Gr. *otikos*, ear) Pertaining to the ear.

OTOLITHS ("ear stones") Calcium carbonate crystals piled on top of the otolithic membrane; assist in maintaining *static equilibrium*.

OVARY (L. *ovum*, egg) One of a pair of female gonads that produce ova and female hormones.

OVIDUCT See *uterine tube*.

OVULATION The monthly process in which a mature ovum is ejected from the ovary into the serous fluid of the peritoneal cavity near the uterine tube.

OXYHEMOGLOBIN The compound formed when oxygen is bound to hemoglobin.

P

PACINIAN CORPUSCLE [pah-SIHN-ee-an] See *lamellated (Pacinian) corpuscle*.

PALATE [PAL-iht] The roof of the mouth, divided into the anterior hard palate and the posterior soft palate.

PALMAR A relative directional term: surface of the palm of the hand; also called *volar*.

PANCREAS (Gr. "all flesh") Organ located posterior to the stomach; secretes the hormones *insulin* and *glucagon*; also functions as an exocrine gland.

PANCREATIC ISLETS Clusters of endocrine cells in the pancreas that produce insulin and glucagon; also called *islets of Langerhans*.

PAPANICOLAOU STAIN SLIDE TEST A diagnostic procedure that tests cells scraped from the female genital epithelium to detect malignant and premalignant conditions; also called *Pap test*, *Pap smear*.

PAPILLAE [puh-PILL-ee, sing. PAPILLA; L. nipple; dim. of *papula*, pimple] Projections on the tongue surface, pal-

ate, throat, and posterior surface of the epiglottis that contain *taste buds*.

PARANASAL SINUS An air cavity within a bone in direct communication with the nasal cavity.

PARASYMPATHETIC NERVOUS SYSTEM The portion of the *autonomic nervous system* that directs activities associated with the conservation and restoration of body resources.

PARATHYROID GLANDS Small endocrine glands embedded in the posterior thyroid; their secretions affect blood concentration of calcium and phosphate.

PARIETAL [puh-RYE-uh-tuhl; L. *paries*, wall of a room] Pertaining to the outer wall of a body cavity.

PARIETAL LOBE The cerebral lobe that lies between the frontal and occipital lobes; concerned with evaluation of the general senses and of taste.

PARIETAL PLEURA See *pleural cavity*.

PAROTID GLANDS (Gr. *parotis*, "near the ear") The largest of the three main pairs of salivary glands; located below the ears.

PARTURITION (L. *parturire*, to be in labor) Childbirth.

PASSIVE TRANSPORT The movement of molecules across cell membranes from areas of high concentration to areas of lower concentration, without the use of cellular energy.

PATELLA REFLEX (KNEE JERK) A diagnostic reflex in which the tapping of the patellar tendon produces the contraction of the quadriceps femoris muscle, causing the lower leg to jerk upward.

PATHOLOGICAL ANATOMY (Gr. *pathos*, suffering + study) The study of changes in diseased cells and tissues; also called *pathology*.

PECTORAL GIRDLE The upper limb girdle consisting of the clavicle and scapula; also called *shoulder girdle*.

PEDICEL (L. little foot) A footlike process in contact with glomerular capillaries.

PELVIC CAVITY The inferior portion of the abominopelvic cavity; contains the urinary bladder, rectum, anus, and internal reproductive organs.

PELVIC GIRDLE The paired hipbones (*ossa coxae*), formed by the fusion of the *ilium*, *ischium*, and *pubis*; also called *lower limb girdle*.

PELVIS (L. basin) The bowl-shaped bony structure formed by the sacrum and coccyx posteriorly, and the two hipbones anteriorly and laterally.

PENIS [L. tail] The male copulatory organ, which transports semen through the urethra during ejaculation; also carries urine through the urethra to the outside during urination.

PENNATE MUSCLE (L. *penna*, feather) A

muscle with many short fascicles set at an angle to a long tendon.

PERFORATING CANAL In bone, a branch running at a right angle to the *central canal*, extending the system of nerves and vessels outward to the *periosteum* and inward to the *endosteum*; also called *nutrient canal, Volkmann's canal.*

PERICARDIAL (Gr. *peri*, around + heart) Pertaining to the membranes enclosing the heart and lining the pericardial cavity.

PERICARDIUM ("around the heart") A protective sac around the heart; also called *pericardial sac.*

PERICHONDRIUM (Gr. *peri*, around + *khondros*, cartilage) A fibrous covering enclosing hyaline and elastic cartilage.

PERILYMPH The thin, watery fluid in the bony labyrinth of the inner ear.

PERIMYSIUM (Gr. *peri*, around + muscle) Connective tissue layer extending inward from the *epimysium*; encloses bundles of muscle fibers.

PERINEURIUM ("around the nerve") Connective tissue sheath surrounding a primary bundle of nerve fibers; found where *epineurium* is absent.

PERIOSTEUM [pehr-ee-AHSS-tee-uhm; Gr. *peri*, around + *osteon*, bone] Fibrous membrane covering outer surfaces of bones (except joints); contains bone-forming cells, nerves, and blood vessels.

PERIPHERAL A relative directional term used to describe structures other than internal organs that are located or directed away from the central axis of the body.

PERIPHERAL AUTONOMIC NERVOUS SYSTEM System including only peripheral efferent (motor) nerves; consisting of the *sympathetic* and *parasympathetic divisions*; each division sends efferent fibers to the muscle, gland, or organ it innervates.

PERIPHERAL NERVOUS SYSTEM (PNS) Includes the cranial and spinal nerves; may be subdivided on a functional basis into the *somatic* and *visceral nervous systems.*

PERISTALSIS The involuntary, sequential muscular contractions that move food along the digestive tract.

PERITONEUM [per-uh-tuh-NEE-uhm] The serous membrane that lines the abdominal cavity and covers most of the abdominal organs.

PEROXISOME A membrane-bound organelle containing oxidative enzymes that carry out metabolic reactions and destroy toxic hydrogen peroxide.

PET SCAN See *positron-emission tomography.*

PHAGOCYTOSIS (Gr. *phagein*, to eat + cyto, cell) Active processes by which large molecules and particles are taken into the cytoplasm through the plasma membrane; a form of *endocytosis.*

PHALANGES [fuh-LAN-jeez; sing. **PHALANX**; FAY-langks, Gr. line of soldiers] The 14 finger bones in each hand; also the 14 toe bones in each foot.

PHARYNX The tube leading from the internal nares (nostrils) and mouth to the larynx and esophagus; serves as an air passage during breathing, and a food passage during swallowing; commonly called the *throat.*

PHENOTYPE [FEE-noh-tipe; Gr. "that which shows"] The expression of a genetic trait (how a person looks or functions).

PHOTORECEPTOR (Gr. "light receiver") A light-sensitive sensory receptor in the retina; see *rod, cone.*

PHYSIOLOGY The study of how the body and its parts function.

PIA MATER (L. tender mother) The thin, highly vascular innermost layer of the meninges.

PINEAL GLAND [PIHN-ee-uhl; L. *pinea*, pine cone] A small gland in the midbrain that converts a signal received through the nervous system into an endocrine signal; it produces *melatonin*, but its exact function is uncertain; also called *pineal body, epiphysis cerebri.*

PINOCYTOSIS (Gr. *pinein*, to drink + cyto, cell) The nonspecific uptake by a cell of small droplets of extracellular fluid; a form of *endocytosis.*

PITUITARY GLAND An endocrine gland consisting of two lobes, the anterior *adenohypophysis*, containing many secretory cells, and the posterior *neurohypophysis*, containing many nerve endings; also called *hypophysis.*

PIVOT JOINT A uniaxial joint that rotates only around a central axis.

PLACENTA A thick bed of tissues and blood vessels embedded in the wall of a pregnant woman's uterus; it contains arteries and veins of mother and fetus, and provides an indirect connection between mother and fetus.

PLANTAR A relative directional term: surface of the sole of the foot; see *dorsal (3).*

PLANTAR FLEXION The downward bending of the foot at the ankle.

PLASMA The clear, yellowish liquid part of blood.

PLASMA CELL See *lymphocyte.*

PLASMA MEMBRANE The bilayered semipermeable outermost boundary of a cell.

PLASMA PROTEINS Proteins found dissolved in blood plasma; include *albumins, fibrinogen*, and *globulins.*

PLATELET A small disk or platelike structure important in blood clotting; also called *thrombocyte.*

PLEURA [PLOOR-uh; Gr. side, rib] The serous membrane lining the *pleural cavity* and covering the lungs; also lines the thoracic wall.

PLEURAL CAVITY The moisture-filled potential space between the *visceral pleura* on the lung surface and the *parietal pleura* lining the chest cavity.

PLEXUS (L. braid) A complex network of interlaced nerves.

PLICAE CIRCULARES [PLY-see sir-cue-LAR-eez] Circular folds in the small intestine that increase the surface area for absorption.

PNEUMOTAXIC AREA A breathing-control area in the pons that, when stimulated, reduces inhalations and increases exhalations.

PODOCYTE ("footlike cell") A specialized epithelial cell surrounding glomerular capillaries.

POLAR BODY The smaller of two cells resulting from the uneven distribution of cytoplasm during oogenesis.

POLARIZATION A condition of the plasma membrane of a neuron, when the intracellular fluid is negatively charged relative to the positively charged extracellular fluid.

POLYPEPTIDE A protein formed by the chemical bonding of many amino acids.

PORTA HEPATIS ("liver door") The area through which the blood vessels, nerves, lymphatics, and ducts enter and leave the liver.

POSITIVE FEEDBACK SYSTEM A reaction that reinforces a stimulus; can disrupt homeostasis, so it is rare.

POSITRON-EMISSION TOMOGRAPHY (PET) A scanning procedure that produces pictures that reveal the metabolic state of the organ being viewed; the device is called a *PET scanner.*

POSTERIOR (L. *post*, behind, after) A relative directional term: toward the back of the body.

POSTGANGLIONIC NEURON The second neuron in a two-neuron sequence in the autonomic nervous system; its cell body is in an autonomic ganglion, and its unmyelinated axon terminates in a motor ending associated with smooth or cardiac muscle or a gland.

POSTSYNAPTIC NEURON ("after the synapse") A nerve cell that carries impulses away from a synapse.

POTENTIAL DIFFERENCE The difference in electrical charge on either side of the plasma membrane.

PREGANGLIONIC NEURON ("before the ganglion") The first neuron in a two-neuron sequence in the autonomic nervous system; its cell body is in the brainstem or spinal cord, and its myelinated axon terminates in an autonomic ganglion outside the central nervous system.

PREGNANCY A series of events initiated by the fertilization of an ovum and its implantation in the uterine wall, continuing through fetal intrauterine development and ending with childbirth.

PREPUCE [PREE-pyoos] The loose-fitting skin folded over the glans of the clitoris and an uncircumcized penis; also called *foreskin.*

PRESSORECEPTORS See *baroreceptors.*

PRESYNAPTIC NEURON ("before the synapse") A nerve cell carrying impulses toward a synapse.

PREVERTEBRAL GANGLIA Autonomic ganglia lying in front of the vertebrae; also called *collateral ganglia.*

PRIMARY CENTER OF OSSIFICATION The site near the middle of what will become the diaphysis, where bone cell development occurs by the second or third prenatal month.

PRIMARY GERM LAYERS Three layers of embryonic tissue, endoderm, mesoderm, and ectoderm, which form the organs and tissues of the body.

PRIMARY MOTOR CORTEX A motor area in the frontal lobe of the cerebrum controlling specific voluntary muscles or muscle groups; also called *Brodmann area 4.*

PRIMARY SOMESTHETIC AREA The portion of the parietal lobe of the cerebrum that receives information about the general senses from receptors in the skin, joints, muscles, and body organs; also called *general sensory area.*

PRIMARY VISUAL CORTEX (area 17) An area in the occipital lobe of the cerebrum that receives information about visual images from the retina and conveys it to cerebral areas 18 and 19 for further processing and evaluation.

PRIME MOVER See *agonist.*

PRIMORDIAL FOLLICLE An ovarian follicle that is not yet growing.

PRONATION A pivoting movement of the forearm that turns the palm downward or backward, crossing the radius diagonally over the ulna.

PROPHASE The first stage of mitosis; centriole pairs move to opposite poles of the nucleoplasm, microtubules form a spindle from pole to pole, and chromatid pairs move to the center of the spindle.

PROPRIOCEPTOR (L. "received from one's self") A sensory receptor responding to stimuli within the body, such as those from muscles and joints.

PROSTATE GLAND (Gr. *prostates,* standing in front of) A male secretory gland whose secretions pass into the semen to make sperm motile and help neutralize vaginal acidity.

PROTEIN (Gr. *protos,* first) Any of a group of complex organic compounds that always contain carbon, hydrogen, oxygen, and nitrogen; their basic structural units are *amino acids.*

PROTRACTION A forward pushing movement.

PROXIMAL (L. *proximus,* nearest) A relative directional term: nearer the trunk of the body (toward the attached end of a limb).

PROXIMAL CONVOLUTED TUBULE The portion of the coiled excretory tubule proximal to the glomerular capsule; it receives *glomerular filtrate* from the glomerular capsule.

PUBERTY (L. *puber,* adult) The developmental period when the person becomes physiologically capable of reproduction.

PULMONARY (L. *pulmo,* lung) Pertaining to the lungs.

PULMONARY ARTERIES Blood vessels carrying oxygen-poor blood from the heart to the lungs.

PULMONARY CIRCULATION The system of blood vessels that carries blood between the heart and the lungs.

PULMONARY TRUNK Arterial trunk emerging from the right ventricle; carries blood to the lungs.

PULMONARY VEINS Large veins that drain blood from the lungs into the left atrium.

PULP The soft core of connective tissue that contains the nerves and blood vessels of a tooth.

PULSE The alternating expansion and elastic recoil of the arterial wall that can be felt where an artery lies close to the skin.

PUPIL (L. doll) The opening in the iris that is opened and closed reflexively.

PURKINJE FIBERS See *cardiac conducting myofibers.*

PYRAMID A bilateral elevated ridge in the ventral surface of the medulla oblongata; composed of fibers of motor tracts from the cerebral motor cortex to the spinal cord.

PYRAMIDAL DECUSSATION (L. *decussare,* from *dec,* ten; the Latin symbol for 10 is X, representing the crossing over) The crossing over of the *pyramidal tracts* in the lower part of the medulla oblongata to the opposite side of the spinal cord.

PYRAMIDAL SYSTEM Tracts from the cerebral motor cortex that terminate in the brainstem (corticobulbar) fibers and in the spinal cord (pyramidal tracts).

PYRAMIDAL TRACTS See *corticospinal tracts.*

R

RADIOGRAPHIC ANATOMY The study of the structures of the body using x-rays.

RAMI COMMUNICANTES [RAY-mee; sing. **RAMUS** [RAY-muhss; pl. **RAMI,** RAY-mee] A branch of a spinal nerve.

RAMUS COMMUNICANS Myelinated or unmyelinated branches of a spinal nerve; composed of sensory and motor nerve fibers associated with the autonomic nervous system.

RECEPTOR (L. *recipere,* to receive) The peripheral end of the dendrites of afferent sensory neurons specialized to receive stimuli and convert them into nerve impulses.

RECTUM (L. *rectus,* straight) The 15-cm duct in the digestive tract between the sigmoid colon and the anus; it removes solid wastes by the process of defecation.

RED BLOOD CELLS See *erythrocytes.*

REFLEX (L. to bend back) A predictable involuntary response to a stimulus.

REFLEX ARC A sequence of events leading to a *reflex action.*

REFRACTORY PERIOD The brief period after the firing of a nerve impulse when the plasma membrane of a neuron cannot generate another impulse.

REGIONAL ANATOMY The anatomical study of specific regions of the body.

RENAL (L. *renes,* kidneys) Pertaining to the kidney.

RENAL CAPSULE The outer layer of the kidney.

RENAL COLUMN A column of renal cortex tissue that penetrates the depth of the *medulla* between *pyramids;* composed mainly of tubules that drain and empty urine.

RENAL CORPUSCLE The portion of a kidney consisting of the glomerulus and glomerular capsule.

RENAL CORTEX The outermost portion of the kidney, divided into the outer cortical region and the inner juxtamedullary region.

RENAL MEDULLA The middle portion of the kidney, consisting of 8 to 18 *renal pyramids.*

RENAL PELVIS The large collecting space within the kidney formed from the expanded upper portion of the ureter; connects structures of the *renal medulla* with the ureter.

RENAL PYRAMID A longitudinally striped, cone-shaped area within the renal medulla; consists of tubules and collecting ducts; involved with the reabsorption of filtered materials; 8 to 18 pyramids make up one renal medulla.

REPOLARIZATION The restoration of the resting charge across the plasma membrane of a neuron.

RESPIRATION The overall exchange of oxygen and carbon dioxide between the atmosphere, blood, lungs, and body cells.

RESPIRATORY CENTER The complete breathing-control center, including the inspiratory and expiratory areas in the medulla oblongata and the apneustic and pneumotaxic areas in the pons.

RESTING MEMBRANE POTENTIAL The electrical potential difference across a nerve cell membrane that is not conducting impulses.

RETICULAR (L. *rete*, net) Resembling a network.

RETICULAR ACTIVATING SYSTEM (RAS) A network of branched nerve cells in the brainstem; involved with the adjustment of many behavioral activities, including the sleep-wake cycle, awareness, levels of sensory perception, emotions, and motivation; also called the *arousal system.*

RETICULAR CELL A flat, star-shaped cell that forms the framework of bone marrow, lymph nodes, and other lymphoid tissues.

RETICULAR FIBERS Fibers that form delicately branched networks in some connective tissues; similar to collagenous fibers, but not as elastic.

RETICULAR FORMATION A network of nerve cells and fibers throughout the brainstem; consists of ascending and descending pathways and cranial nerves; regulates respiratory and cardiovascular centers, as well as the brain's awareness level.

RETINA [REH-tin-uh; L. *rete*, net] The innermost layer of the eye, containing a layer of photoreceptor cells and other nervous tissue called the *neuroretina* and a thin layer of pigmented epithelium.

RETRACTION A backward movement of a body part.

RETROPERITONEAL (L. *retro*, behind + peritoneum) Located behind the abdominopelvic cavity and the peritoneum.

RIBONUCLEIC ACID (RNA) [rye-boh-noo-KLAY-ihk] A single-stranded nucleic acid containing the sugar ribose; found in both the nucleus and the cytoplasm of a cell.

RIBOSOME A subcellular structure containing RNA and protein; the site of protein synthesis.

RIGHT LYMPHATIC DUCT A large duct that drains lymph from the right side of the head; right upper extremity, tho-rax, and lung; right side of the heart; and the upper portion of the liver into the right subclavian vein.

RNA See *ribonucleic acid.*

ROD A specialized photoreceptor cell in the retina; not sensitive to color, but very sensitive to light.

ROTATION A pivoting movement that twists a body part on its long axis.

ROUND LIGAMENTS Paired bands of fibrous connective tissue just below the entrance of the uterine tubes into the uterus; help to keep the uterus tilted forward over the urinary bladder.

RUFFINI CORPUSCLES See *corpuscles of Ruffini.*

RUGAE [ROO-jee; L. folds] Folds or creases of tissue, as in the stomach and vagina.

S

SACRAL PLEXUS The ventral rami of L4, L5, and S1 to S3 nerves in the posterior pelvic wall.

SADDLE JOINT A multiaxial joint in which opposing articular surfaces of both bones are shaped like a saddle.

SAGITTAL (L. *sagitta*, arrow) An off-center longitudinal plane dividing the body into asymmetrical left and right sections.

SALIVA The secretion of salivary glands; contains the enzyme, salivary amylase.

SALIVARY GLANDS The three largest pairs of glands that secrete saliva into the oral cavity: *parotid, submandibular,* and *sublingual glands.*

SALTATORY CONDUCTION (L. *saltare*, to jump) Conduction along a myelinated nerve fiber in which the *action potential* appears to jump from one neurofibral node to the next.

SARCOLEMMA (Gr. *sarkos*, flesh + *lemma*, husk) A thin membrane enclosing each skeletal muscle fiber.

SARCOMERE (Gr. *sarkos*, flesh + *meros*, part) The fundamental unit of muscle contraction, composed of a section of muscle fiber extending from one Z line to the next.

SARCOPLASM (Gr. *sarkos*, flesh) Cytoplasm of skeletal muscle fibers.

SARCOPLASMIC RETICULUM A specialized type of endoplasmic reticulum containing a network of tubes and sacs containing calcium ions; in muscle tissue.

SCHWANN CELL See *neurolemmocyte.*

SCLERA (Gr. *skleros*, hard) The posterior segment of the outer supporting layer of the eyeball; the "white" of the eye.

SCROTUM [SCROH-tuhm; L. *scrautum*, a leather pouch for arrows] An external sac of skin that hangs between the male thighs; it contains the testes.

SEBACEOUS GLANDS [sih-BAY-shuhss; L. *sebum*, tallow, fat] Simple, branched alveolar glands in the dermis that secrete *sebum;* their main functions are lubrication and protection; also called *oil glands.*

SEBUM (L. tallow, fat) The oily secretion found at the base of a hair follicle.

SECONDARY SEX CHARACTERISTICS Sexually distinct characteristics such as body hair and enlarged genitals that develop during puberty.

SELECTIVE PERMEABILITY (L. *permeare*, to pass through) The quality of cellular membranes that allows some substances into the cell while keeping others out.

SELLA TURCICA [SEH-luh TUR-sihk-uh; L. *sella,* saddle + Turkish] A deep depression within the body of the sphenoid bone, which houses and protects the pituitary gland.

SEMEN [SEE-muhn; L. seed] Male ejaculatory fluid containing sperm and secretions from the epididymis, seminal vesicles, prostate gland, and bulbourethral glands; also called *seminal fluid.*

SEMILUNAR VALVES Heart valves that prevent blood in the pulmonary artery and aorta from flowing back into the ventricles.

SEMINAL FLUID See *semen.*

SEMINAL VESICLES Paired secretory sacs whose secretions provide an energy source for motile sperm and help to neutralize the acidity of the vagina.

SEMINIFEROUS TUBULES Tightly coiled tubules in the testes that produce sperm.

SENESCENCE (L. *senescere*, to grow old) The indeterminate period when an individual is said to grow old; also called *primary aging, biological aging.*

SENSORY NEURON See *afferent neuron.*

SEPTUM (L. *sepire*, to separate with a hedge) A wall between two cavities.

SEROUS (L. *serosus*, serum) Pertaining to the secretion of a serumlike fluid.

SEROUS MEMBRANE A double layer of loose connective tissue covered by a layer of simple squamous epithelium that lines some of the walls of the thoracic and abdominopelvic cavities and covers organs lying within these cavities; includes *peritoneum, pericardium,* and *pleura.*

SERTOLI CELL See *sustentacular cell.*

SEX CHROMOSOMES A pair of chromosomes that determine the sex of the new individual; an XX pair produces a female, and an XY pair a male.

SICKLE-CELL ANEMIA A hereditary form of anemia characterized by crescent-shaped red blood cells that do not carry or release sufficient oxygen.

SIGMOID COLON (Gr. *sigma,* the letter "*S*") The S-shaped portion of the large intestine immediately following the descending colon; transverses the pelvis to the right to the middle of the sacrum, where it continues to the rectum.

SIMPLE DIFFUSION (L. *diffundere,* to spread) A passive transport process in which molecules move randomly from areas of high concentration to areas of lower concentration until they are evenly distributed.

SINOATRIAL (SA) NODE A mass of specialized heart muscle that initiates and controls the heartbeat.

SKELETAL MUSCLE Muscle tissue that can be contracted voluntarily; it is attached to the skeleton; also known as *striated* and *voluntary muscle.*

SKENE'S GLANDS See *lesser vestibular glands.*

SKULL Bones of the head, with or without the mandible.

SMALL INTESTINE The 6-m-long portion of the digestive tract between the stomach and the large intestine.

SMOOTH MUSCLE TISSUE Nonstriated muscle tissue, controlled by the autonomic nervous system; forms sheets in the walls of large, hollow organs; also called *involuntary muscle.*

SOMATIC NERVOUS SYSTEM The portion of the *peripheral nervous system* composed of a sensory division that receives and processes input from the sense organs and a motor division that excites skeletal muscles.

SONOGRAPHY See *ultrasound.*

SPECIAL SENSES The senses of sight, hearing, equilibrium, smell, and taste.

SPERM (Gr. *sperma,* seed) Mature male sex cells; also called *spermatozoa* (Gr. *zoon,* animal).

SPERMATIC CORD A cord covering the ductus deferens as it passes from the tail of the epididymis; contains the testicular artery, veins, autonomic nerves, lymphatics, and connective tissue.

SPERMATOGENESIS ("the birth of seeds") The continuous process that forms sperm in the testes.

SPHINCTER (Gr. "that which binds tight") A circular muscle that can close off an opening or tubular structure.

SPINAL CORD The part of the central nervous system extending inferiorly from the foramen magnum; has 31 pairs of nerves and is the connecting link between the brain and most of the body.

SPINOUS PROCESS A sharp, elongated process of a bone.

SPIRAL ORGAN (OF CORTI) The organ of hearing; formerly called *organ of Corti.*

SPLEEN The largest lymphoid organ, located below the diaphragm on the left side; it filters blood and produces phagocytic lymphocytes and monocytes.

STATIC EQUILIBRIUM See *vestibular apparatus.*

STEREOGNOSIS [STEHR-ee-oh-NO-siss; Gr. *stereos,* solid, three-dimensional + *gnosis,* knowledge] The ability to identify unseen objects by handling them.

STIMULUS (L. a goad) A change in the external environment or within the body itself that initiates impulses in a receptor.

STOMACH The distensible sac that churns ingested nutrients into small particles, stores them until they can be released into the small intestine, and secretes hydrochloric acid and enzymes that initiate protein digestion.

STRATUM BASALE (L. *basis,* base) Layer of epidermis resting on basement membrane next to the dermis.

SUBARACHNOID SPACE The space between the arachnoid and pia mater layers of the meninges; contains cerebrospinal fluid and blood vessels.

SUBCUTANEOUS LAYER (L. *sub,* under + *cutis,* skin) See *hypodermis.*

SUBDURAL SPACE The potential space between the dura mater and arachnoid meninges; contains no cerebrospinal fluid.

SUBLINGUAL GLANDS ("under the tongue") Paired salivary glands located in the floor of the mouth beneath the tongue.

SUBMANDIBULAR GLANDS ("under the mandible") Paired salivary glands located on medial side of the mandible.

SUDORIFEROUS GLANDS (L. *sudor,* sweat) Commonly called sweat glands.

SULCUS [pl. **SULCI,** SUHL-kye; L. groove] A deep furrow on the surface of a structure; also called *groove.*

SUPERFICIAL A relative directional term: nearer the surface of the body.

SUPERIOR (L. *superus,* situated above) A relative directional term: toward the head; above.

SUPERIOR VENA CAVA The large vein that drains blood from the head, neck, upper limbs, and thorax, emptying it into the right atrium.

SUPINATION A pivoting movement of the forearm that turns the palm forward or upward, making the radius parallel with the ulna.

SURFACTANT A phospholipid that reduces surface tension in the alveoli; produced by the lungs.

SUSTENTACULAR CELL A type of cell in the seminiferous tubules that provides nourishment for the sperm as they mature; also called *Sertoli cell.*

SUSTENTACULUM A supportive process on a bone.

SUTURAL BONES Separate small bones in the sutures of the calvaria of the skull; also called *Wormian bones.*

SUTURE (L. *sutura,* seam) A seamlike joint that connects skull bones, making them immovable.

SWALLOWING See *deglutition.*

SWEAT GLANDS See *sudoriferous glands.*

SYMPATHETIC NERVOUS SYSTEM The division of the autonomic nervous system that mobilizes the body during emergency and stress situations; also called *thoracolumbar division* or *adrenergic division.*

SYMPHYSIS (Gr. "growing together") A cartilaginous joint in which two bony surfaces are covered by thin layers of hyaline cartilage and are cushioned by fibrocartilaginous disks; also called *secondary synchondrosis.*

SYNAPSE [SIN-apps; Gr. a connection] The junction between neurons.

SYNAPTIC BOUTON A tiny swelling on the terminal ends of *telodendria* at the distal end of an axon.

SYNAPTIC CLEFT The narrow space between an axon terminal and the receptor site of the postsynaptic cell. At a neuromuscular junction, the space between the axon terminal and sarcolemma.

SYNAPTIC DELAY The period during which a neurotransmitter bridges a synaptic cleft.

SYNAPTIC GUTTER The invaginated area of the sarcolemma under and around the axon terminal; also called *synaptic trough.*

SYNARTHROSIS (Gr. *syn,* together + *arthrosis,* articulation) An immovable joint.

SYNCHONDROSIS (Gr. "together with cartilage") A cartilaginous joint important for growth, not movement; a temporary joint of cartilage that joins the epiphysis and diaphysis of a growing bone; it is eventually replaced by bone; also called *primary cartilaginous joint.*

SYNDESMOSIS (Gr. "to bond together"; *syn* + *desmos,* bond) A fibrous joint in which bones are held close together, but not touching, by collagenous fibers or interosseous ligaments.

SYNERGISTIC MUSCLE [SIHN-uhr-jist-ihk; Gr. *syn,* together + *ergon,* work] A muscle that complements the action of a prime mover (agonist).

SYNOVIAL [sin-OH-vee-uhl; Gr. *syn,* with + L. *ovum,* egg] Pertaining to the thick, lubricating fluid secreted by membranes in joint cavities.

SYNOVIAL CAVITY The space between two articulating bones; also called *joint cavity.*

SYNOVIAL FLUID A viscous fluid that lubricates synovial joints.

SYNOVIAL JOINT An articulation in which bones move easily on each other; most permanent joints of the body are synovial.

SYNOVIAL MEMBRANE A membrane lining the cavities of joints and similar areas where friction needs to be reduced; composed of loose connective and adipose tissues covered by fibrous connective tissue.

SYSTEM A group of organs that work together to perform a major body function.

SYSTEMIC ANATOMY The anatomical study of the systems of the body.

SYSTEMIC CIRCULATION The system of blood vessels that carries blood between the heart and all parts of the body except the lungs.

SYSTOLE [SISS-toe-lee] The contraction of the atria and ventricles during the *cardiac cycle.*

SYSTOLIC PRESSURE Part of blood pressure measurement that represents the highest pressure reached during ventricular ejection; the first number in a blood pressure reading.

T

T CELL A type of lymphocyte that attacks specific foreign cells; able to differentiate into helper T cell, suppressor T cell, or killer T cell; also called *T lymphocyte.*

T LYMPHOCYTE See *T cell.*

TACTILE (MEISSNER'S) CORPUSCLES Sensory receptors in the skin that detect light pressure; formerly called *Meissner's corpuscles.*

TACTILE (MERKEL'S) CORPUSCLES Sensory receptors of light touch, located in the deep epidermal layers of the palms and soles; formerly called *Merkel's disks.*

TAENIAE COLI [TEE-nee-ee KOHL-eye; L. ribbons + Gr. intestine] Three separate bands of longitudinal muscle along the full length of the large intestine.

TARSUS The seven proximally located short bones of each foot.

TELERECEPTOR (Gr. "received from a distance") Sensory receptor that detects relatively distant environmental stimuli; in eyes, ear, and nose.

TELOPHASE The final stage of mitosis; chromosomes arrive at the poles and are surrounded by a new nuclear envelope, and nucleoli are formed.

TEMPORAL LOBE The cerebral lobe closest to the ears; functions in hearing, equilibrium, and to a certain degree, emotion and memory.

TENDON A strong cord of fibrous connective tissue that attaches muscle to the periosteum of bone.

TESTES [TESS-teez; sing. **TESTIS**, L. witness] The paired male reproductive organs, which produce sperm.

TESTIS-DETERMINING FACTOR (TDF) A specific gene on the Y chromosome that apparently determines the sex of offspring.

THALAMUS (Gr. inner chamber) Two masses of gray matter covered by a thin layer of white matter; located directly beneath the cerebrum and above the hypothalamus; forms the walls of the third ventricle; the intermediate relay point and processing center for all sensory impulses (except smell) ascending to the cerebral cortex from the spinal cord, brainstem, cerebellum, basal ganglia, and other sources.

THERMORECEPTOR (Gr. "heat receiver") A sensory receptor that responds to temperature changes.

THORACIC (Gr. *thorax*, breastplate) Pertaining to the chest.

THORACIC DUCT The largest lymphatic vessel; drains lymph not drained by the *right lymphatic duct* into the left subclavian vein; also called the *left lymphatic duct.*

THORACOLUMBAR DIVISION [thuh-RASS-oh-LUM-bar] The portion of the *autonomic nervous system* with visceral efferent nerve fibers that leave the central nervous system through thoracic and lumbar spinal nerves.

THORACOLUMBAR OUTFLOW Myelinated nerve fibers emerging from the spinal cord in the ventral nerve roots of the 12 thoracic and first two or three lumbar spinal nerves.

THORAX (Gr. breastplate) The chest portion of the *axial skeleton,* including 12 thoracic vertebrae, 12 pairs of ribs, 12 costal cartilages, and sternum.

THRESHOLD STIMULUS A stimulus strong enough to initiate an impulse in a neuron.

THROMBOCYTE See *platelet.*

THYMUS GLAND A ductless mass of flattened lymphoid tissue situated behind the top of the sternum; forms antibodies in the newborn and is involved in the development of the immune system.

THYROID GLAND An endocrine gland involved with the metabolic functions of the body.

TISSUE An aggregation of many similar cells that perform a specific function.

TOMOGRAPHY (Gr. *tomos,* a cut or section + *graphein,* to write or draw) A technique for making pictures of a section of a body part, as in *computer-assisted tomography.* The picture produced is a *tomogram.*

TONSILS Aggregates of lymphatic nodules enclosed in a capsule of connective tissue; they help destroy microorganisms that enter the digestive and respiratory systems; include *pharyngeal, palatine,* and *lingual tonsils.*

TRABECULAE (truh-BECK-yuh-lee; L. dim. *trabs,* beam] Tiny spikes of bone tissue surrounded by calcified bone matrix; prominent in the interior structure of spongy bone tissue.

TRACHEA [TRAY-kee-uh] An open tube extending from the base of the larynx to the top of the lungs, where it forks into two *bronchi;* commonly called *windpipe.*

TRACT A bundle of nerve fibers and their sheaths within the central nervous system.

TRANSVERSE COLON The portion of the large intestine immediately after the ascending colon, extending across the abdominal cavity from right to left, making a right angle downward turn at the spleen.

TRANSVERSE PLANE A plane that divides the body horizontally into superior and inferior parts; made at right angles to midsagittal, sagittal, and frontal planes.

TRANSVERSE TUBULES A series of tubes linked by extensions of the plasma membrane; they cross the sarcoplasmic reticulum at right angles within a muscle fiber; also called *T tubules.*

TRIAD The combination of a transverse tubule and a terminal cisterna of sarcoplasmic reticulum.

TRICUSPID VALVE The right *atrioventricular valve.*

TRIGONE A triangular area in the urinary bladder outlined by the openings of the ureters and urethra into the cavity of the bladder.

TROCHANTER Either of the two large, rounded processes found below the neck of the femur.

TROPHECTODERM The surrounding epithelial layer of a blastocyst, composed of cells called *trophoblasts;* later develops into a fetal membrane system.

TROPHOBLAST See *trophectoderm.*

TUBERCLE (L. small lump) A small, roughly rounded process of a bone.

TUBEROSITY (L. lump) A medium-sized, roughly rounded elevated process of a bone.

TUBULAR REABSORPTION The renal process that returns useful substances, such as water, some salts, and glucose, to the blood by active transport.

TUBULAR SECRETION The renal process

that forces waste products, such as potassium and hydrogen ions, and certain drugs into the *glomerular filtrate* before it leaves the kidney.

TUNICA ADVENTITIA ("outermost covering") The outermost covering of an artery; composed mainly of collagen fibers and elastic tissue.

TUNICA ALBUGINEA [al-byoo-JIHN-ee-uh; L. *albus,* white] The thick, bluish-white fibrous membrane that encloses a testis in the male and forms a thin connective tissue over the outer portion of an ovary in the female.

TUNICA INTIMA ("innermost covering") The innermost lining of an artery.

TUNICA MEDIA ("middle covering") The middle, and thickest, layer of the arterial wall in large arteries.

TYMPANIC CAVITY (Gr. *tympanon,* drum) A narrow, irregular, air-filled space in the temporal bone; separated from the external auditory canal by the tympanic membrane and from the inner ear by the posterior bony wall; also called *middle ear cavity.*

TYMPANIC MEMBRANE A deflectable membrane between the external and middle ear; vibrates in response to sound waves entering the ear; popularly called the *eardrum.*

U

ULTRASONOGRAPHY A technique using high-frequency sound waves to locate and examine a fetus or other internal structures.

ULTRASOUND A noninvasive exploratory technique that sends pulses of ultrahigh-frequency sound waves into designated body cavities; images are formed from echoes of the sound waves; also called *sonography.*

UMBILICAL CORD (L. *umbilicus,* navel) The structure connecting the embryo and placenta; contains blood vessels.

UNIPOLAR NEURON A nerve cell with one process dividing into two branches; one branch extends into the brain or spinal cord, while the other extends to a peripheral sensory receptor.

UPPER EXTREMITIES See *extremities.*

URETERS A pair of tubes that carries urine from the pelvis of each kidney to the urinary bladder.

URETHRA A tube that transports urine from the bladder to the outside during urination in both males and females. In males, it is also the final section of the reproductive duct system, leading from the urinary bladder, through the prostate gland, and into the penis, carrying semen outside the body during ejaculation.

URINARY BLADDER A hollow, muscular sac that collects urine from the ureters and stores it until it is excreted from the body through the urethra.

URINARY SYSTEM The body system that eliminates the wastes of protein metabolism in the urine and regulates the amount of water and salts in the blood; composed of two kidneys and ureters, the urinary bladder, and the urethra.

URINE (Gr. *ourein,* to urinate) The excretory fluid produced by the kidneys, composed mainly of water, urea, chloride, potassium, creatinine, phosphates, sulfates, and uric acid.

UTERINE TUBES A pair of tubes that receive the mature ovum from the ovary and convey it to the uterus; also called *Fallopian tubes, oviducts.*

UTEROSACRAL LIGAMENTS Along with other ligaments, provide support for the uterus by attaching to the sacrum.

UTERUS (L. womb) A hollow, muscular organ behind the urinary bladder that is the site of menstruation, and during pregnancy houses, nourishes, and protects the developing fetus; also called *womb.*

V

VAGINA (L. sheath) A muscle-lined tube from the uterus to the outside; the site of semen deposition during sexual intercourse, the channel for menstrual flow, and the birth canal for the baby during childbirth.

VAS DEFERENS See *ductus deferens.*

VASA VASORUM ("vessels of the vessels") Small blood vessels that supply nutrients to the walls of arteries and veins.

VASCULAR (L. *vasculum,* dim. of *vas,* vessel) Pertaining to blood vessels.

VASOCONSTRICTION The process of constricting blood vessels.

VASODILATION The process of dilating blood vessels.

VASOMOTION The collective term for *vasoconstriction* and *vasodilation.*

VASOMOTOR CENTER A regulatory center in the lower parts of the pons and medulla oblongata; regulates the diameter of blood vessels, especially arterioles.

VEIN A blood vessel that carries blood from the body toward the heart.

VENTILATION See *breathing.*

VENTRAL (L. *venter,* belly) A relative directional term: toward the front of the body.

VENTRAL CAVITY The larger of two main body cavities; separated into the thoracic and abdominopelvic cavities by the diaphragm.

VENTRICLE (L. "little belly") 1. A cavity in the brain filled with cerebrospinal fluid. 2. The left or right inferior heart chamber.

VENULE [VEHN-yool] A tiny vein into which blood drains from capillaries.

VERMIFORM APPENDIX (L. *vermis,* worm) The short, narrow, wormlike region of the digestive tract opening into the cecum; may be involved with the immune system; commonly called the *appendix.*

VERMIS (L. worm) The midline portion of the cerebellum, separating the lateral lobes (hemispheres); with the flocculonodular lobes, involved in maintenance of muscle tone, equilibrium, and posture.

VERTEBRAE (VER-tuh-bree; L. "something to turn on") The 26 individual bones in the spinal column; they surround and protect the spinal cord.

VERTEBRAL COLUMN Commonly called the spinal column.

VESICULAR FOLLICLE Ovarian follicle that is almost ready to release a mature ovum *(ovulation).*

VESTIBULAR APPARATUS (L. *vestibulum,* entrance) Parts of the inner ear that signal changes in the motion of the head *(dynamic equilibrium)* and the position of the head with respect to gravity *(static equilibrium,* or posture).

VESTIBULAR GLANDS See *greater* and *lesser vestibular glands.*

VESTIBULE 1. Any cavity, chamber, or channel serving as an approach or entrance to another cavity. 2. The space between the labia minora. 3. The central chamber of the labyrinth in the middle ear.

VILLI [VILL-eye; sing. **VILLUS;** L. "shaggy hairs"] Tiny fingerlike protrusions in the mucosa of the small intestine that increase the absorptive surface area.

VISCERA [VISS-ser-uh; L. body organ] The internal organs of the body.

VISCERAL Pertaining to an internal organ or a body cavity.

VISCERAL NERVOUS SYSTEM The portion of the *peripheral nervous system* composed of a motor division (autonomic nervous system) that may inhibit or excite smooth muscle, cardiac muscle, or glands, and a sensory division that receives afferent input from internal organs.

VISCERAL PLEURA See *pleural cavity.*

VITREOUS HUMOR A gelatinous substance within the large vitreous chamber of the eyeball; keeps the eyeball from collapsing.

VOCAL CORDS Paired strips of stratified squamous epithelium at base of larynx; produce sound when vibrated.

VOICE BOX See *larynx*.
VOLAR See *palmar*.
VOLKMANN'S CANAL See *perforating canal*.
VOLUNTARY MUSCLE See *skeletal muscle*.
VULVA (L. womb) The collective name for the external female genitalia, including the mons pubis, labia majora and minora, vestibular glands, clitoris, and vestibule of the vagina.

W

WANDERING CELLS Connective tissue cells usually involved with short-term activities such as protection and repair.

WHITE BLOOD CELL See *leukocyte*.
WHITE MATTER The portion of the spinal cord consisting mainly of whitish myelinated nerve fibers.
WHITE RAMI COMMUNICANTES Myelinated preganglionic fibers of the *thoracolumbar outflow* of the spinal cord that form small nerve bundles.
WINDPIPE See *trachea*.
WOMB See *uterus*.
WORMIAN BONES See *sutural bones*.

Y

YOLK SAC An extraembryonic membrane that is a primitive respiratory and digestive system before the development of the placenta; it becomes nonfunctional and is incorporated into the umbilical cord by the sixth or seventh week.

Z

ZONA PELLUCIDA (L. *perlucere*, to shine through, thus "transparent zone") The outer wall of an ovum.
ZYGOTE (Gr. *zugotos*, joined, yolked) The cell formed by the union of male and female gametes.

INDEX

Page references in **boldface** introduce or define the term; page references in *italic* indicate illustrations; page references in ***boldface italic*** indicate an illustration and term; *n* indicates a footnote.

Lateral rotators, 247, 277
Lateral spinothalamic tract, 325–326, **328, 413–414**
Lateral umbilical ligaments, *518*, 519
Latissimus dorsi, *11, 40–41, 44,* 248–249, *266–268,* **269,** *284, 286, 338,* A.4, A.5, A.6
LATS (*see* Long-acting thyroid stimulator)
Laxative, **618**
LDL (*see* Low-density lipoproteins)
Left brain, 370–371
Left colic flexure, *585,* 609–610
Left-handedness and brain dominance, 371
Legs, *50–51, 191*
  bones of, 149, *191, 193–195, 196*
  joints of, *196, 219*
  muscles of, *276–281*
Lens, *428–430, 431, 437–***438**
Lens placode, *437–***438**
Lens vesicle, *437–438*
Lentiform nucleus, **365–366**
Lesser omentum, 594–**595**
Lesser pelvis, **190,** *192*
Lesser vestibular (Skene's) glands, **655**
Lethargy, **390**
Leukemia, **78,** 474
Leukocytes (white blood cells), *95,* **96,** 468–472
  count, 468
  in leukemia, 474
  in lymph, 545
  production of, 468
Leukopenia, **476**
Leukopoiesis, **468**
Leukorrhea, **668**
Levator, 247
Levator ani, *264–265,* 286
Levator labii superioris, *252–253*
Levator palpebrae superioris, *253–254,* 377, 435
Levator scapulae, *258–259, 266–267,* A.5
Lever systems and muscle action, 250–251
Leverage, **251**
Levodopa, 387
Leydig cells, *644–645*
LH (*see* Luteinizing hormone)
Ligament, **207–208**
Ligamentum arteriosum, *484, 518,* 519
Ligamentum teres, *518,* 604–**605**
  (*See also* Round ligament)
Ligamentum venosum, *518,* 519
Ligamentum venosus, 604
Ligands, *60*
Light touch, **409, 411,** 413
Limb, **184**
Limbic lobe, *357,* **367–369**
Limbic system, *368–369,* 402–403
Linea alba, *42–43,* **262**
Linea aspera, 146, **193**
Linea of bone, 147
Linea semilunaris, *43*
Lingual frenulum, *587*
Lingual nerve, *378*
Lingual tonsils, *415,* 551, *564–565*
Lip ridge of bone, 147
Lipid, **57,** 466
  absorption of, 602
Lips, *584,* **587**
Lithotripsy, 613
Liver, *39, 452, 584–585, 603, 605–607, 623–624,* A.10
  capillary plexus in, *515*

as digestive organ, 604–607
and hepatic portal circulation, *515*
innervation of, *381, 398–399*
metabolic functions of, 606
structure of, 604–607
vagus innervation of, *381*
Liver spots, 121, **126**
Lobule, *33, 418, 551–552,* **605,** *607*
Lobectomy, **580**
Local circuit neurons, 297
Lochia, **696**
Lockjaw, 239
Locus coeruleus, *357–358*
Long-acting thyroid stimulator (LATS), 458
Long bones, **129,** *130*
Longissimus capitis, *258*
Longissimus cervicis, *259*
Longissimus lumborum, *259*
Longissimus thoracis, *259, 266,* A.5
Longitudinal fissure, *363–366*
Longitudinal section, *21*
Loop of Henle (*see* Loop of the nephron)
Loop of the nephron, *628,* **630–631,** 635
Loose connective tissue, *97–98*
Lordosis, *179*
Low-density lipoproteins (LDL), 466
Lower esophageal orifice, **595–596**
Lower esophageal sphincter, *592–596*
Lower extremity, *48,* 190–198
LTH (*see* Prolactin)
Lumbar lymphatic trunk, *549*
Lumbar nerves, *323*
Lumbar nodes, *546–547*
Lumbar plexus, *319,* **335,** *337*
Lumbar puncture, **170,** 320
Lumbar region, *19,* A.3
Lumbar vertebrae, *148, 166, 167,* **170,** *171, 323*
Lumbosacral angle, **171**
Lumbosacral enlargement, **318–319**
Lumbosacral joint, *171, 172*
Lumbosacral plexus, *308, 337*
Lumbrical muscles:
  of foot, *282, 283*
  of hand, *244, 270, 272,* **274,** *275*
Lumpectomy, 666
Lunate bone, *188, 189*
Lund-Browder method, *122,* **123**
Lungs, *13, 561, 567–572,* A.6, A.10
  cancer, **578**
  development of, *576*
  innervation, *381, 398–399*
  as respiratory tree, *567–569*
  structure of, *567–572*
  surface area of, *567*
  vagus innervation of, *381*
Lunula, **120,** *120n., 121*
Luteinizing hormone (LH), *447,* 454
Luteotropic hormone (*see* Prolactin)
Lutropin (*see* Luteinizing hormone)
Luxation of a joint, **223**
Lymph, 13, **545–548**
Lymph nodes, *545–547,* **550**–551, *554*
  of neck, *36*
Lymph nodules, *550–551*
  aggregated, *545–547,* **553**
Lymphadenectomy, **557**
Lymphadenopathy, **557**
Lymphangioma, **557**
Lymphangitis, **557**
Lymphatic capillaries, *545–***548**
Lymphatic circulation, *549–550*

Lymphatic duct, right, *546,* **548–550, 553–** *554*
Lymphatic metastasis, **557**
Lymphatics, 545, *547–***548**
Lymphatic sinuses, **554**
Lymphatic system, 9, *13,* **545–553**
  developmental anatomy of, *553–554*
  functions of, 545
Lymphatic trunks, *549*
Lymphatic vessels, *13, 546–547, 549–550*
Lymphoblasts, *470*
Lymphocytes, **469–472,** 545–548
  B and T, **545–548**
  of thymus gland, *553*
Lymphogenous leukemia, 474
Lymphoid progenitor cells, *470*
Lymphoma, **557**
Lymphosarcoma, **557**
Lysosomes, *55–56, 63*

<hr>

M

M line of muscle fiber, 230–*231*
Macrophages, *95,* **96–97,** 111, 468, **545,** *547*
Macula, **424**
Macula densa, *631*
Macula lutea, *428*
Macules, 124, 422
Main pancreatic duct, *603–604, 608*
Malar bones (*see* Zygomatic bone)
Male reproductive system, 642–650
  development of, *662–663*
Male sex characteristics, 691
Malignant carcinoma, **79**
Malignant melanoma, 124–**125**
Malleolus of bone, *50–51,* 147, **195–***196*
Malleus, 153, 157, **164,** *419–***420**
Malocclusion, **618**
Malpighian corpuscles (*see* Splenic nodules)
Mammary glands, *15, 39, 94,* **656**
Mamillary body, *366, 368, 372*
Mamillary process, of vertebrae, *168*
Mandible, *34–35, 37,* 151, 153, 156, 158, *163–164, 213, 380, 562, 590*
  angle of, **163,** *164*
  body of, **163,** *164*
  structure of, **163–164**
Mandibular foramen, **163,** *164*
Mandibular fossa, *155, 164*
Mandibular nerve, *377–378*
Manubriosternal joint, *38,* **173,** 209
Manubrium, *172–174, 186*
Marfan's syndrome, **104**
Marrow (*see* Bone marrow)
Marrow cavity, *139*
Mask of pregnancy, 683
Mass movements of large intestine, 611
Masseter muscle, *34–35,* **255,** *378, 590*
Mast cells, *95,* **97**
Mastectomy, 665
Mastication, muscles of, *255–257, 378*
Mastoid air cells, 157
Mastoid fontanel, *151*
Mastoid foramen, *155, 159*
Mastoid process, *34–35, 37, 154–155,* **157–** *158*
Mastoid region, **157**
Mastoid sinuses, 157
Mastoiditis, 157, 418
Matrix, **64**
  of bone, 132
  of connective tissues, 96
  of epithelial cells, 85

Sarcoma, **78**
  Kaposi's, 554
  osteogenic, **143**
Sarcomere, *230–231*
Sarcoplasm, **229**, *233, 237*
Sarcoplasmic reticulum, **229**, *231*
Sartorius, *11, 48–49*, 229, *248, 276, 278, 279, 280, 285, 286*
Satellite cells, **305–306**
Scab, formation of, 116
Scabies, **126**
Scala media, *421–422*
Scala tympani, *421–422*
Scala vestibuli, *421–422*
Scalenes, 260–**261**
Scalp, muscles of, *252*
Scapha, *33*, 418
Scaphoid bone, 188, *189*
Scapula, *39, 40–41, 45*, **184**, *185–186, 212, 217*
  angles of, **184**, *186*
  borders of, **184**, *186*
Scar tissue, 116
Schwann cell (*see* Neurolemmocytes)
Schwann sheath, 305–*306*
Sciatic nerve, *308, 319, 335, 336*
Sciatic notch, **190**, *193*
Sciatica, 335, **343**
Scissors gait, **390**
Sclera, *428–***430**, *438*
Scleral venous sinus, *428*
Scleritis, **442**
Sclerocorneal junction, *33*
Sclerotome, *237–238*
Scoliosis, **119**
Scotoma, **442**
Scrotum, *15*, **642**–*643, 645, 662*, A.9
Scurvy, **104**
Seasonal affective disorder (SAD), 453
Sebaceous cyst, 118
Sebaceous glands, 93, 94, *112*, **118**
Seborrheic keratosis, 121
Sebum, **118**
Second-class levers, 250–*251*
Second-order neurons, 327, *328*
Secondary sex characteristics, 690–*691*
Secretin, **453**, 455
Secretory vesicles, *63*
Segmenting contractions, **602**
Selective permeability, **57**
Sella turcica, *159*, **160**, 446
Semen, **648**–649
Semicircular canals, *379, 419*, 421–**422**
Semicircular ducts, *419, 421–423, 425*, **426**, *427*
Semilunar notch, **187**, *188*
Semilunar valves, **486–488**
Semimembranosus, *249*, 276, **277**, 278, **279**, *280, 285, 286*
Seminal vesicle glands, 94
Seminal vesicles, *15, 643*, **647–648**, A.9
  development of, *663*
Seminiferous tubules, *642–644*
Semispinalis capitis, 258–**259**, 266
Semispinalis cervicis, **259**
Semispinalis thoracis, **259**
Semitendinosus, *249*, 276, **277**, 278, **279**, *280, 285, 286*
Senescence, **692**
Senile dementia, *25*, **385**
Senility, **385**
Sensation, *369*

Senses, 407–438
  general, **409**, 411–414
Sensory fibers, **373**
Sensory ganglia, **373**
Sensory hairs of spiral organ, *422*
Sensory nerve fibers, 221
Sensory neurons, **307–**309
Sensory nuclei, **373**
Sensory pathways, 327–329, 356
Sensory reception, 408–410
Sensory receptors, 296
  and brain, 409–410
  characteristics of, 408
  classification of, 408–409
  for cold, 411–412
  for heat, 411–412
  for itch, 413
  for light touch, 411
  of skin, *410*
  for tickle, 413
  for touch-pressure, 411
  for vibration, **411**
Sensory tracts of spinal cord, 325
Septal cells of alveoli, 567, *570*
Septicemia, **476**
Septum of heart, **482–**483, 497
Serofibrinous pleurisy, 578
Serosa, **583**, *586, 596–597, 601*
Serosal layer of uterus, 653–654
Serotonin, in blood clotting, 472
Serous fluid, 103, 594
Serous glands, **93**
Serous membranes, 23, 88, **101–103**, 566
Serous pericardium, 481–*482*
Seratus anterior muscle, *39, 43, 248, 262, 266, 267, 338*, A.4, A.5
Serratus posterior interior muscle, 266, A.5
Sertoli cells (*see* Sustentacular cells)
Serum, blood, **466**
Sesamoid bone, **129**, *130*
Sex, determination of, 659, 661
Sex cells, 75, 656–658
Sex chromosomes, **659**
Sexual response and hypothalamus, 373
Sexually transmitted diseases (STD), **664**–665
Sharpey's fibers (*see* Periosteal perforating fibers)
Shin splints, **199**
Shinbone (*see* Tibia)
Shingles, **343**
Short bones, **129**, *130*
Shoulder blade (*see* Scapula)
Shoulder girdle, **44**, 184, *185*
  muscles of, 266–267
Shoulder joint, *185, 208, 214*, **216**, 217
  arteries of, *526, 527*
  bones of, **184**–*186*
  dislocation of, 223
  muscles of, 268–269
  separation of, **225**
  veins of, *532, 533*
Show, 685
Shunt, 541
Sialitis, **618**
Siamese twins, 688
Sickle-cell anemia, **474**, 693
SIDS (*see* Sudden infant death syndrome)
Sigmoid colon, *39, 584, 585, 594–595*, **609**–*610*
Sigmoid sinus, *530*
Simmonds' disease, **461**
Simple fracture, *176*

Simple gland, **93**
Sinoatrial (SA) node, 490*n.*, **490**, *491*
Sinoatrial valve, *497*
Sinuses of bone, 132
Sinus rhythm, **502**
Sinus venosus, **496**, *497*
Sinusitis, 161
Sinusoids of liver lobules, 514, **606–607**
SITS muscles, 216, *269*
Skeletal muscle, 7, **101**, **229–234**
  circulation and, 516
  development of, 237–238
  fasciae of, **231**
  structure of, 229–231
Skeletal system, *10*
  functions of, 9, 146
Skeleton:
  appendicular, *148*, **149**
  axial, 146–175, **149**
  development of, *137*
  divisions of, 149
  of embryo, 136
  major bones of, *148*
  and osteomalacia, 142
  and rickets, 142
  thoracic, 172
Skene's glands, **655**
Skin, **111**, *112–118*
  cancer of, 124–125
  color of, 115
  developmental anatomy of, 116–*117*
  disorders of, 123–125
  functions of, 114–115
  glands of, 117–118
  lesions, types of, 124
  sense receptors of, *410*–411
  structure of, 111–114
  tanning of, 115
  wound healing of, 115–*116*
  wrinkling of, 113
Skull, *33*, 148, 149, *156*
  cranial, 150, *151*
  facial, 161–163
  fractures of, *177*
  growth of, 136–137, *151*
  structure of, **150–165**
Sleep-wake cycle:
  and hypothalamus, 372
  and thalamus, 372
Slit membrane, 630
Slow-reacting substance (SRS–A), 469
Small intestine, *14, 584*, **598–602**
  absorption in 602
  digestion in, 600–602
  enzymes of, 602
  functions of, *584–585, 600–602*
  structure of, 598–600
  villi of, *601*
Small intestine series, **618**
Smegma, 649
Smoker's cough, 561
Smooth endoplasmic reticulum, 55, 61
Smooth muscle, **101**, 229, **234–236**
  of blood vessels, 234
  contraction of, 235–236
  development of, 238
  of intestine, 234
  multi-unit, **236**
  organization of, 234
  and peristalsis, 235
  properties of, 235
  rhythmicity of, 235
  single-unit, **235**